Symbol Table

$\overline{A}$	complement of event A	H	Kruskal-Wallis test statistic
H_0	null hypothesis	R	sum of the ranks for a sample; used in the Wilcoxon rank-sum test
H_1	alternative hypothesis		
α	alpha; probability of a type I error or the area of the critical region	μ_R	expected mean rank; used in the Wilcoxon rank-sum test
β	beta; probability of a type II error	σ_R	expected standard deviation of ranks; used in the Wilcoxon rank-sum test
r	sample linear correlation coefficient	G	number of runs in runs test for randomness
ρ	rho; population linear correlation coefficient	μ_G	expected mean number of runs; used in runs test for randomness
r^2	coefficient of determination		
R^2	multiple coefficient of determination	σ_G	expected standard deviation for the number of runs; used in runs test for randomness
r_s	Spearman's rank correlation coefficient		
b_1	point estimate of the slope of the regression line	$\mu_{\bar{x}}$	mean of the population of all possible sample means $\bar{x}$
b_0	point estimate of the y-intercept of the regression line	$\sigma_{\bar{x}}$	standard deviation of the population of all possible sample means $\bar{x}$
$\hat{y}$	predicted value of y	E	margin of error of the estimate of a population parameter, or expected value
d	difference between two matched values		
$\overline{d}$	mean of the differences d found from matched sample data	Q_1, Q_2, Q_3	quartiles
s_d	standard deviation of the differences d found from matched sample data	$D_1, D_2, \ldots, D_9$	deciles
		$P_1, P_2, \ldots, P_{99}$	percentiles
s_e	standard error of estimate	x	data value
T	rank sum; used in the Wilcoxon signed-ranks test		

Symbol Table

f	frequency with which a value occurs	$z_{\alpha/2}$	critical value of z
Σ	capital sigma; summation	t	t distribution
Σx	sum of the values	$t_{\alpha/2}$	critical value of t
Σx^2	sum of the squares of the values	df	number of degrees of freedom
$(\Sigma x)^2$	square of the sum of all values	F	F distribution
Σxy	sum of the products of each x value multiplied by the corresponding y value	χ^2	chi-square distribution
		χ_R^2	right-tailed critical value of chi-square
n	number of values in a sample	χ_L^2	left-tailed critical value of chi-square
$n!$	n factorial	p	probability of an event or the population proportion
N	number of values in a finite population; also used as the size of all samples combined		
		q	probability or proportion equal to $1 - p$
k	number of samples or populations or categories	$\hat{p}$	sample proportion
		$\hat{q}$	sample proportion equal to $1 - \hat{p}$
$\overline{x}$	mean of the values in a sample	$\overline{p}$	proportion obtained by pooling two samples
$\overline{R}$	mean of the sample ranges	$\overline{q}$	proportion or probability equal to $1 - \overline{p}$
μ	mu; mean of all values in a population	$P(A)$	probability of event A
s	standard deviation of a set of sample values	$P(A|B)$	probability of event A, assuming event B has occurred
σ	lowercase sigma; standard deviation of all values in a population		
		$_nP_r$	number of permutations of n items selected r at a time
s^2	variance of a set of sample values		
σ^2	variance of all values in a population	$_nC_r$	number of combinations of n items selected r at a time
z	standard score		

ELEMENTARY STATISTICS USING **THE TI-83/84 PLUS CALCULATOR**

3RD EDITION

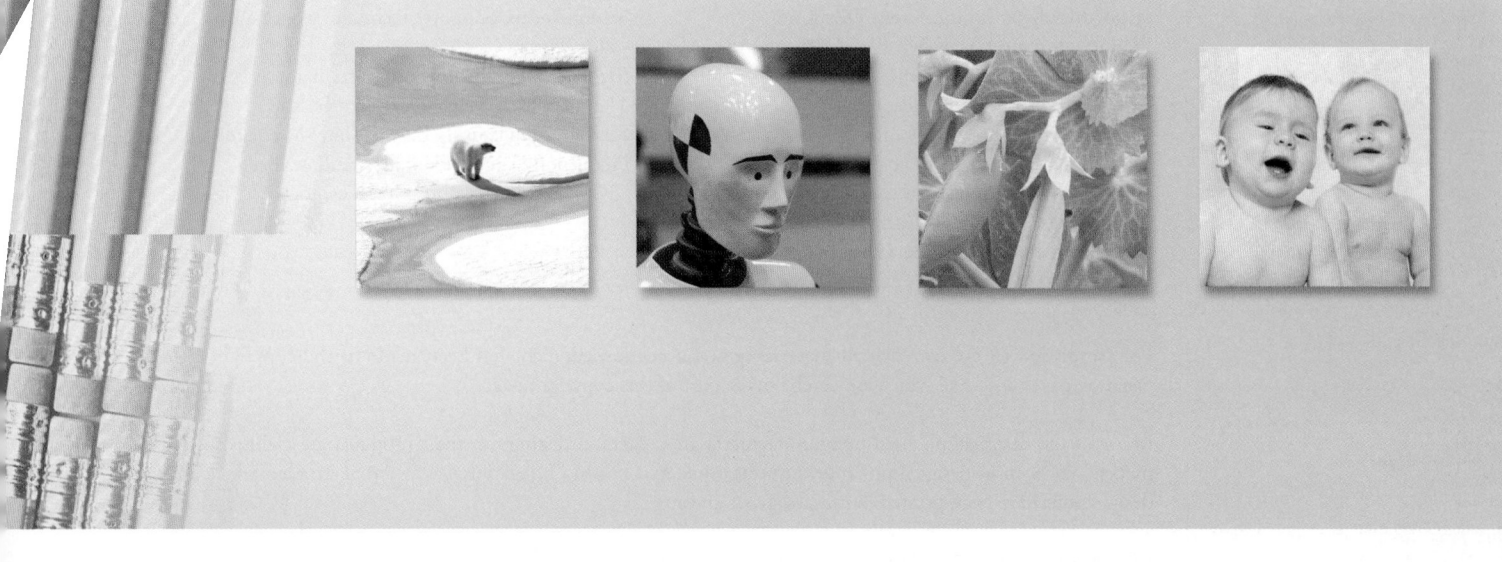

ELEMENTARY **STATISTICS**
USING **THE TI-83/84**
PLUS CALCULATOR

MARIO F. TRIOLA

3RD EDITION

Addison-Wesley

Boston   Columbus   Indianapolis   New York   San Francisco   Upper Saddle River
Amsterdam   Cape Town   Dubai   London   Madrid   Milan   Munich   Paris   Montréal   Toronto
Delhi   Mexico City   São Paulo   Sydney   Hong Kong   Seoul   Singapore   Taipei   Tokyo

Editor in Chief: Deirdre Lynch
Acquisitions Editor: Christopher Cummings
Associate Editor: Christina Lepre
Editorial Assistant: Dana Jones
Senior Managing Editor: Karen Wernholm
Production Project Manager: Beth Houston
Interior Design: Leslie Haimes
Cover Art Direction: Beth Paquin
Cover Design: Lisa Kuhn, Curio Press, LLC
Marketing Manager: Alex Gay
Marketing Assistant: Kathleen DeChavez
Cover Images: Windmills, Art Life Images; Canada,
    Nunavut Territory, Arctic, Getty Images; Crash Test
    Dummy, Pea Plant; and Pencil, Shutterstock

Photo Researcher: Beth Anderson
Media Producer: Christine Stavrou
MyStatLab Project Supervisor: Edward Chappell
QA Manager, Assessment Content: Marty Wright
Senior Author Support/Technology Specialist:
    Joe Vetere
Rights and Permissions Advisor: Shannon Barbe
Manufacturing Manager: Evelyn Beaton
Senior Manufacturing Buyer: Ginny Michaud
Production Coordination, Illustrations, and
    Composition: Nesbitt Graphics, Inc.

For permission to use copyrighted material, grateful acknowledgment has been made to the copyright holders listed on pages 849–850 which are hereby made part of this copyright page.

Many of the designations used by manufacturers and sellers to distinguish their products are claimed as trademarks. Where those designations appear in this book, and Pearson Education was aware of a trademark claim, the designations have been printed in initial caps or all caps.

Library of Congress Cataloging-in-Publication Data

Triola, Mario F.
    Elementary statistics using the TI-83/84 Plus calculator / Mario F. Triola. -- 3rd ed.
        p. cm.
    Includes bibliographical references and index.
    ISBN 0-321-64148-5
    1. Mathematical statistics--Textbooks. 2. Statistics--Graphic methods--Textbooks. 3. TI-83 (Calculator)
4. TI-84 (Calculator) I. Title.
    QA276.12.T76 2011
    519.5--dc22
                            2009013575

2 3 4 5 6 7 8 9 10—WC—14 13 12 11 10

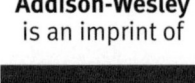

Addison-Wesley
is an imprint of

PEARSON

www.pearsonhighered.com

ISBN-13: 978-0-321-64148-9
ISBN-10: 0-321-64148-5

To
Ginny
Marc, Dushana, and Marisa
Scott, Anna, Siena, and Kaia

About the Author

Mario F. Triola is a Professor Emeritus of Mathematics at Dutchess Community College, where he has taught statistics for over 30 years. Marty is the author of *Essentials of Statistics,* 4th edition; *Elementary Statistics Using Excel,* 4th edition; *Elementary Statistics,* 11th edition; and he is a coauthor of *Biostatistics for the Biological and Health Sciences; Statistical Reasoning for Everyday Life,* 4th edition; *Business Statistics;* and *Introduction to Technical Mathematics,* 5th edition. Marty designed the original STATDISK statistical software, and he has written several manuals and workbooks for technology supporting statistics education. He has been a speaker at many conferences and colleges. Marty's consulting work includes the design of casino slot machines and fishing rods, and he has worked with attorneys in determining probabilities in paternity lawsuits, identifying salary inequities based on gender, and analyzing disputed election results. He has also used statistical methods in analyzing medical data, medical school surveys, and survey results for New York City Transit Authority. Marty has testified as an expert witness in New York State Supreme Court. The Text and Academic Authors Association has awarded Marty a "Texty" for Excellence for his work on *Elementary Statistics.*

Contents

Chapter 1 **Introduction to Statistics** 2

    1-1   Review and Preview   4

    1-2   Statistical Thinking   4

    1-3   Types of Data   11

    1-4   Critical Thinking   17

    1-5   Collecting Sample Data   26

    1-6   Introduction to the TI-83/84 Plus Calculator   37

Chapter 2 **Summarizing and Graphing Data** 50

    2-1   Review and Preview   52

    2-2   Frequency Distributions   52

    2-3   Histograms   62

    2-4   Statistical Graphics   65

    2-5   Critical Thinking: Bad Graphs   76

Chapter 3 **Statistics for Describing, Exploring, and Comparing Data** 88

    3-1   Review and Preview   90

    3-2   Measures of Center   90

    3-3   Measures of Variation   105

    3-4   Measures of Relative Standing and Boxplots   120

Chapter 4 **Probability** 142

    4-1   Review and Preview   144

    4-2   Basic Concepts of Probability   144

    4-3   Addition Rule   158

    4-4   Multiplication Rule: Basics   165

    4-5   Multiplication Rule: Complements and Conditional Probability   177

    4-6   Probabilities Through Simulations   184

    4-7   Counting   189

    4-8   Bayes' Theorem (on CD-ROM)   199

Chapter 5 **Discrete Probability Distributions** 208

    5-1   Review and Preview   210

    5-2   Random Variables   211

    5-3   Binomial Probability Distributions   225

    5-4   Mean, Variance, and Standard Deviation for the Binomial Distribution   235

    5-5   Poisson Distributions   240

Chapter 6   **Normal Probability Distributions**   254

    6-1   Review and Preview   256

    6-2   The Standard Normal Distribution   257

    6-3   Applications of Normal Distributions   270

    6-4   Sampling Distributions and Estimators   282

    6-5   The Central Limit Theorem   293

    6-6   Normal as Approximation to Binomial   305

    6-7   Assessing Normality   315

Chapter 7   **Estimates and Sample Sizes**   332

    7-1   Review and Preview   334

    7-2   Estimating a Population Proportion   334

    7-3   Estimating a Population Mean: σ Known   351

    7-4   Estimating a Population Mean: σ Not Known   361

    7-5   Estimating a Population Variance   376

Chapter 8   **Hypothesis Testing**   396

    8-1   Review and Preview   398

    8-2   Basics of Hypothesis Testing   399

    8-3   Testing a Claim About a Proportion   418

    8-4   Testing a Claim About a Mean: σ Known   431

    8-5   Testing a Claim About a Mean: σ Not Known   438

    8-6   Testing a Claim About a Standard Deviation or Variation   447

Chapter 9   **Inferences from Two Samples**   464

    9-1   Review and Preview   466

    9-2   Inferences About Two Proportions   466

    9-3   Inferences About Two Means: Independent Samples   477

    9-4   Inferences from Dependent Samples   491

    9-5   Comparing Variation in Two Samples   501

Chapter 10   **Correlation and Regression**   520

    10-1   Review and Preview   522

    10-2   Correlation   522

    10-3   Regression   540

    10-4   Variation and Prediction Intervals   555

    10-5   Multiple Regression   564

    10-6   Modeling   574

Chapter 11   **Goodness-of-Fit and Contingency Tables**   588

    11-1   Review and Preview   590

    11-2   Goodness-of-Fit   590

    11-3   Contingency Tables   602

    11-4   McNemar's Test for Matched Pairs   615

Chapter 12   **Analysis of Variance**   630

    12-1   Review and Preview   632

    12-2   One-Way ANOVA   632

    12-3   Two-Way ANOVA   646

Chapter 13   **Nonparametric Statistics**   664

    13-1   Review and Preview   666

    13-2   Sign Test   667

    13-3   Wilcoxon Signed Ranks Test for Matched Pairs   676

    13-4   Wilcoxon Rank-Sum Test for Two Independent Samples   682

    13-5   Kruskal-Wallis Test   688

    13-6   Rank Correlation   693

    13-7   Runs Test for Randomness   701

Chapter 14   **Statistical Process Control**   716

    14-1   Review and Preview   718

    14-2   Control Charts for Variation and Mean   718

    14-3   Control Charts for Attributes   732

Chapter 15   **Projects, Procedures, Perspectives**   746

    15-1   Projects   746

    15-2   Procedures   748

    15-3   Perspectives   749

Appendices   751

Appendix A: Tables   752

Appendix B: Data Sets   769

Appendix C: Bibliography of Books and Web Sites   798

Appendix D: Answers to odd-numbered section exercises, plus answers to all end-of-chapter Statistical Literacy and Critical Thinking exercises, chapter Quick Quizzes, Review Exercises, and Cumulative Review Exercises   799

Credits   849

Index   851

Preface

Statistics is used everywhere—from opinion polls to clinical trials in medicine, statistics influences and shapes the world around us. *Elementary Statistics Using the TI-83/84 Plus Calculator* illustrates the relationship between statistics and our world with a variety of real applications bringing life to abstract theory.

Which Calculators Can Be Used with This Book?

This book includes many references to the TI-83/84 Plus calculator. The instructions, screen displays, programs, and the APP (application) for the TI-83/84 Plus calculator also apply to the TI-83 Plus Silver Edition, the TI-84 Plus, and the TI-84 Plus Silver Edition calculators. Any of these calculators can be used with this book:

- TI-83 Plus

- TI-83 Plus Silver Edition

- TI-84 Plus

- TI-84 Plus Silver Edition

- The TI-Nspire with the TI-84 Plus keypad installed.

The TI-Nspire calculator comes with an interchangeable TI-84 Plus keypad that slides in, so the calculator becomes compatible with the TI-83 Plus and TI-84 Plus calculators.

Throughout this book, references to the "TI-83/84 Plus" calculator apply to any of the above calculators.

Other Texas Instruments Calculators

The TI-89, TI-89 Titanium, TI-92 Plus, and Voyager 2000 have the ability to download the free "Statistics with List Editor" application, which is available from the Texas Instruments Web site. After downloading that application, these calculators have at least as much statistics functionality as the TI-83 Plus and TI-84 Plus family of calculators. However, procedures for using the "Statistics with List Editor" application are generally different from the calculator procedures described in this book. Consequently, the TI-89, TI-89 Titanium, TI-92 Plus, and Voyager 2000 calculators could be used for many procedures in this book, but instructions, screen displays, the data application, and programs are not compatible with this book.

Which Calculators Are Approved for Use on AP, SAT, ACT, and PSAT/NMSQT Exams?

The TI-83/84 Plus calculators are on the list of calculators approved for use on AP, SAT, ACT, and PSAT/NMSQT exams, including the AP statistics exam. Installed programs are also allowed, but memory cannot be used to store notes. Other calculators approved for these exams are the TI-89, TI-92 Plus, and the TI-Nspire. For a complete list of approved calculators, see the Web site for the individual exam.

Why Use a TI-83/84 Plus Calculator?

The TI-83/84 Plus calculators have dramatically affected the teaching of introductory statistics courses. This family of calculators includes almost all of the important functions required in an introductory statistics course; they are relatively inexpensive, and easy to use in class and on tests. The TI-83/84 Plus calculators simplify otherwise tedious calculations, so professors and students can better focus on important concepts of statistics instead of time-consuming arithmetic. These calculators include features that make it possible to conduct classroom experiments that illustrate simulations and other important concepts. Instructors and students enjoy immediate results, instead of waiting to gain access to a laboratory or home computer.

Here are just three of many features that have contributed to the success of the TI-83/84 Plus calculator:

- In many cases, the calculator is designed to accept either lists of original data, or summary statistics. If the summary statistics are known, there is no need to waste time, energy, and (most of all) patience by manually entering a list of original sample values.

- The calculator has the ability to store APPS (applications). The CD-ROM included with this book contains a TI-83/84 Plus APP that contains all of the data sets found in Appendix B of this textbook. Once this APP has been downloaded to the calculator, instructors and students have easy and ready access to a variety of data sets that can be opened without manual entry of the sample values.

- Students can bring their TI-83/84 Plus calculators to each class, so that everyone has immediate and ready access to the same technology. This makes it easy for instructors to conduct experiments and projects. Also, test problems can be made more realistic with larger sets of real data.

Using the TI-83/84 Plus Calculator

Elementary Statistics Using the TI-83/84 Plus Calculator is designed to be an introduction to basic statistics. Instead of being a manual of calculator instructions, this book places strong emphasis on understanding concepts of statistics, with the TI-83/84 Plus calculator included throughout as the key technology supplement. Topics are presented with illustrative examples, identification of required assumptions, and underlying theory. TI-83/84 Plus calculator instructions are provided along with typical displays of results.

This book has key features that are critical to effective use of the calculator as a statistical tool:

- **Introduction to the TI-83/84 Plus Calculator** Section 1-6 introduces important and basic features of a TI-83/84 Plus calculator.

- **TI-83/84 Plus Instructions** Detailed TI-83/84 Plus calculator instructions are included throughout the book.

- **TI-83/84 Plus Displays** 144 samples of TI-83/84 Plus calculator displays are included throughout the book. Also, displays from Minitab®, Excel®, and STATDISK® are occasionally included so students can see different formats for displaying results.

- **TI-83/84 Plus Data APP** All data sets in Appendix B are available as part of the TI-83/84 Plus APP named TRIOLA11. That APP is on the CD-ROM packaged with this book. Some exercises use those data sets, so students can open TI-83/84 Plus calculator files instead of manually entering long lists of data.

- **TI-83/84 Plus Exercises** Exercises that are suitable for use with the TI-83/84 Plus calculator are identified with a special **icon**.

- **TI-83/84 Plus Answers** In some cases, exercise answers based on table values may be slightly different from answers based on the use of the TI-83/84 Plus calculator. For those exercises, Appendix D generally gives both answers.

- **TI-83/84 Plus Calculator Projects** are included among the variety of the end-of-chapter features.

- **TI-83/84 Plus Programs** The CD-ROM included with this book has separate TI-83/84 Plus programs for several procedures that are not included with the calculator. The programs include procedures for hypothesis tests with variances, confidence intervals with variances, multiple regression, two-way analysis of variance, Wilcoxon rank-sum test, Wilcoxon signed-ranks test, and the Kruskal-Wallis test.

- **STATDISK** For those who choose to further supplement the TI-83/84 Plus calculator, the CD-ROM includes additional **software** that many students find helpful. STATDISK is a separate computer program designed specifically for the content of this book.

Goals

In addition to the inclusion of the TI-83/84 Plus calculator, this Third Edition was written with several goals:

- Provide new and interesting data sets, examples, and exercises.

- Foster personal growth of students through critical thinking, use of technology, collaborative work, and development of communication skills.

- Incorporate the latest and best methods used by professional statisticians.

- Include information personally helpful to students, such as the best job search methods and the importance of avoiding mistakes on résumés.

- Provide the largest and best set of supplements to enhance teaching and learning.

This book reflects recommendations from the American Statistical Association and its *Guidelines for Assessment and Instruction in Statistics Education* (GAISE). Those guidelines suggest the following objectives and strategies.

1. **Emphasize statistical literacy and develop statistical thinking:** Each exercise set begins with *Statistical Literacy and Critical Thinking* exercises. Many of the book's exercises are designed to encourage statistical thinking rather than the blind use of mechanical procedures.

2. **Use real data:** 93% of the examples and 82% of the exercises use real data.

3. **Stress conceptual understanding rather than mere knowledge of procedures:** Exercises and examples involve conceptual understanding, and each chapter also includes a *Data to Decision* project.

4. **Foster active learning in the classroom:** Each chapter ends with several *Cooperative Group Activities*.

5. **Use technology for developing conceptual understanding and analyzing data:** TI-83/84 Plus calculator displays are included throughout the book. Special

subsections include instruction for using the TI-83/84 Plus calculator. Each chapter includes a *TI-83/84 Plus Calculator Project, Internet Project,* and *Applet Project.* The CD-ROM included with the book includes free text-specific software (STAT-DISK) and the Appendix B data sets formatted for several different technologies.

6. **Use assessments to improve and evaluate student learning:** Assessment tools include an abundance of section exercises, Chapter Review Exercises, Cumulative Review Exercises, Chapter Quick Quizzes, activity projects, and technology projects.

Audience/Prerequisites

This book is written for students majoring in any subject. Algebra is used minimally, but students should have completed at least a high school or college elementary algebra course. In many cases, underlying theory behind topics is included, but this book does not require the mathematical rigor more suitable for mathematics majors.

Changes in This Edition

- **Exercises** This Third Edition includes 2011 exercises (13% more than the Second Edition), and 87% of them are new. 82% of the exercises use real data (compared to 53% in the Second Edition). Each chapter now includes a 10-question Chapter Quick Quiz.

- **Examples** Of this edition's 257 examples, 86% are new, and 93% involve real data. Examples are now numbered consecutively within each section.

- **Chapter Problems** All Chapter Problems are new.

- **Organization**
    New Sections 1-2: Statistical Thinking; 2-5: Critical Thinking: Bad Graphs

    Combined Section 3-4: Measures of Relative Standing and Boxplots

    New topics added to Section 2-4: Bar graphs and multiple bar graphs

    Glossary (Appendix C in the Second Edition) has been moved to the CD-ROM and is available in MyStatLab.

- **Margin Essays** Of 122 margin essays, 15% are new; many others have been updated. New topics include *iPod Random Shuffle, Mendel's Data Falsified,* and *Speeding Out-of-Towners Ticketed More.*

- **New Features**
    Chapter Quick Quiz with 10 exercises is now included near the end of each chapter.

    ### CAUTION
    ....
    "Cautions" draw attention to potentially serious errors throughout the book.

    An **Applet Project** is now included near the end of each chapter.

Exercises

Many exercises require the *interpretation* of results. Great care has been taken to ensure their usefulness, relevance, and accuracy. Exercises are arranged in order of increasing difficulty by dividing them into two groups: (1) Basic Skills and Concepts and (2) Beyond the Basics. Beyond the Basics exercises address more difficult concepts or require a stronger mathematical background. In a few cases, these exercises introduce a new concept.

Real data: Hundreds of hours have been devoted to finding data that are real, meaningful, and interesting to students. In addition, some exercises refer to the 24 large data sets listed in Appendix B. Those exercises are located toward the end of each exercise set, where they are clearly identified.

Technology Supplements

- **On the CD-ROM:**

    STATDISK statistical software. New features include *Normality Assessment, modified boxplots,* and the ability to handle more than nine columns of data.

    Appendix B data sets formatted for Minitab, Excel, SPSS, SAS®, and JMP®, and also available as text files. Additionally, the CD-ROM contains these data sets as an APP for the TI-83/84 Plus calculator, and includes supplemental programs for the TI-83/84 Plus calculator.

    Extra data sets, applets, and Data Desk XL (DDXL, an Excel add-in).

- Separate manuals/workbooks are available for STATDISK, Minitab, Excel, SPSS (PASW), SAS, and the TI-83/84 Plus and TI-Nspire calculators.

- Study Cards are available for various technologies.

- **PowerPoint® Lecture Slides, Active Learning Questions,** and the **TestGen®** computerized test generator are available for instructors on the Instructor Resource Center.

Flexible Syllabus

This book's organization reflects the preferences of most statistics instructors, but there are two common variations:

- **Early coverage of correlation & regression:** Some instructors prefer to cover the basics of correlation and regression early in the course. *Sections 10-2 (Correlation) and 10-3 (Regression) can be covered early.* Simply limit coverage to Part 1 (Basic Concepts) in each of those two sections.

- **Minimum probability:** Some instructors prefer extensive coverage of probability, while others prefer to include only basic concepts. Instructors preferring minimum coverage can include Section 4-2 while skipping the remaining sections of Chapter 4, as they are not essential for the chapters that follow. Many instructors prefer to cover the fundamentals of probability along with the basics of the addition rule and multiplication rule, and those topics can be covered with Sections 4-1 through 4-4. Section 4-5 includes conditional probability, and the subsequent sections cover simulation methods and counting (including permutations and combinations).

Hallmark Features

Great care has been taken to ensure that each chapter of this book will help students understand the concepts presented. The following features are designed to help meet that objective:

Chapter-opening features:

- A list of chapter sections previews the chapter for the student.

- A chapter-opening problem, using real data, motivates the chapter material.

- The first section is a brief review of relevant earlier concepts, and previews the chapter's objectives.

End-of-chapter features:

A **Chapter Review** summarizes the key concepts and topics of the chapter.

Statistical Literacy and Critical Thinking exercises address chapter concepts.

A **Chapter Quick Quiz** provides ten review questions that require brief answers.

Review Exercises offer practice on the chapter concepts and procedures.

Cumulative Review Exercises reinforce earlier material.

A **TI-83/84 Plus Calculator Project** provides an activity for the TI-83/84 Plus calculator.

An **Internet Project** provides an activity for use of the Internet.

An **Applet Project** provides an activity for use of the applet included on the CD-ROM.

From Data to Decision is a capstone problem that requires critical thinking and writing.

Cooperative Group Activities encourage active learning in groups.

An **Interview** is included with a professional who uses statistics in day-to-day work

Real Data Sets Appendix B contains printed versions of 24 large data sets referenced throughout the book, including 8 that are new and 2 others that have been updated. These data sets are also available on the companion Web site and the CD-ROM bound in the back of new copies of the book.

Margin Essays The text includes 122 margin essays (15% new), which illustrate uses and abuses of statistics in real, practical, and interesting applications.

Flowcharts The text includes 20 flowcharts that appear throughout the text to simplify and clarify more complex concepts and procedures. Animated versions of the text's flowcharts are available within MyStatLab and MathXL.

Top 20 Topics The most important topics in any introductory statistics course are identified in the text with the icon. Students using MyStatLab have access to additional resources for learning these topics with definitions, animations, and video lessons.

Quick-Reference Endpapers Tables A-2 and A-3 (the normal and t distributions) are reproduced on inside cover pages. A symbol table is included at the front of the book for quick and easy reference to key symbols.

Detachable Formula and Table Card This insert, organized by chapter, gives students a quick reference for studying, or for use when taking tests (if allowed by the instructor). It also includes the most commonly used tables.

CD-ROM: The CD-ROM was prepared by Mario F. Triola and is bound into the back of every new copy of the book. It contains the data sets from Appendix B (available as txt files), Minitab worksheets, SPSS files, SAS files, JMP files, Excel workbooks, and a TI-83/84 Plus application. The CD also includes a section on Bayes' Theorem, a glossary, programs for the TI-83/84 Plus calculator, STATDISK Statistical Software (Version 11), and the Excel add-in DDXL, which is designed to enhance the capabilities of Excel's statistics programs.

Supplements

For the Student

Student's Solutions Manual, by Milton Loyer (Penn State University), provides detailed, worked-out solutions to all odd-numbered text exercises. (ISBN-13: 978-0-32164157-1; ISBN-10: 0-321-64157-4)

The following technology manuals include instructions, examples from the main text, and interpretations to complement those given in the text.

Graphing Calculator Manual for the TI-83 Plus, TI-84 Plus, TI-89 and TI-Nspire, by Patricia Humphrey (Georgia Southern University). (ISBN-13: 978-0-321-57061-1; ISBN 10: 0-321-57061-8)

Excel Student Laboratory Manual and Workbook, by Johanna Halsey and Ellena Reda (Dutchess Community College). (ISBN-13: 978-0-321-57073-4; ISBN-10: 0-321-57073-1)

MINITAB Student Laboratory Manual and Workbook, by Mario F. Triola. (ISBN-13: 978-0-321-57081-9; ISBN-10: 0-321-57081-2)

SAS Student Laboratory Manual and Workbook, by Joseph Morgan. (ISBN-13: 978-0-321-57071-0; ISBN-10: 0-321-57071-5)

SPSS Student Laboratory Manual and Workbook, by James J. Ball (Indiana State University). (ISBN-13: 978-0-321-57070-3; ISBN-10: 0-321-57070-7)

STATDISK Student Laboratory Manual and Workbook, by Mario F. Triola. (ISBN-13: 978-0-321-57069-7; ISBN-10: 0-321-57069-3)

For the Instructor

Instructor's Edition, by Mario F. Triola, contains answers to all exercises (ISBN-13: 978-0-321-64160-1; ISBN-10: 0-321-64160-4)

Instructor's Solutions Manual, by Milton Loyer (Penn State University), contains solutions to all the exercises and sample course syllabi. (ISBN-13: 978-0-321-64159-5; ISBN-10: 0-321-64159-0)

Insider's Guide to Teaching with the Triola Statistics Series, by Mario F. Triola, contains sample syllabi and tips for incorporating projects, as well as lesson overviews, extra examples, minimum outcome objectives, and recommended assignments for each chapter. (ISBN-13: 978-0-321-57078-9; ISBN-10: 0-321-57078-2)

Testing System: Not only is there an online test bank, there is also a computerized test generator, TestGen®. TestGen enables instructors to build, edit, print, and administer tests using a computerized bank of questions developed to cover all the objectives of the text. TestGen is algorithmically based, allowing instructors to create multiple but equivalent versions of the same question or test with the click of a button. Instructors can also modify test bank questions or add new questions. Tests can be printed or administered online. The software and online test bank are available for download from Pearson Education's online catalog. (Test bank ISBN-13: 978-0-321-57087-1; ISBN-10: 0-321-57087-1)

PowerPoint® Lecture Slides: Free to qualified adopters, this classroom lecture presentation software is geared specifically to the sequence and philosophy of this book. Key graphics from the book are included to help bring the statistical concepts alive in the classroom. The PowerPoint Lecture Slides are available for download within MyStatLab and from the Pearson Education online catalog.

Active Learning Questions: Prepared in PowerPoint®, these questions are intended for use with classroom response systems. Several multiple-choice questions are available for each section of the book, allowing instructors to quickly assess mastery of material in class. The Active Learning Questions are available for download from within MyStatLab® and from Pearson Education's online catalog at www.pearsonhighered.com/irc.

Technology Resources

- **On the CD-ROM**

  - Appendix B data sets formatted for Minitab, SPSS, SAS, Excel, JMP, and as text files. Additionally, the CD-ROM contains these data sets as an APP for the TI-83/84 Plus calculators, and includes supplemental programs for the TI-83/84 Plus calculator.

  - **STATDISK** statistical software. New features include *Normality Assessment, modified boxplots,* and the ability to handle more than nine columns of data.

  - Extra data sets, applets, and Data Desk XL (DDXL, an Excel add-in).

- **Videos on DVD** have been expanded and now supplement most sections in the book, with many topics presented by the author. The videos feature technologies found in the book and the worked-out Chapter Review exercises. This is an excellent resource for students who have missed class or wish to review a topic. It is also an excellent resource for instructors involved with distance learning, individual study, or self-paced learning programs. These DVDs also contain optional English and Spanish captioning. (Videos on DVD ISBN-13: 978-0-321-57079-6; ISBN-10: 0-321-57079-0).

- **Triola *Elementary Statistics* Web site:** This Web site may be accessed at http://www.pearsonhighered.com/triola, and provides Internet projects keyed to every chapter of the text, plus the book's data sets.

- **MyStatLab™** MyStatLab (part of the MyMathLab® and MathXL® product family) is a text-specific, easily customizable online course that integrates interactive multimedia instruction with textbook content. Powered by CourseCompass™ (Pearson Education's online teaching and learning environment) and MathXL (our online homework, tutorial, and assessment system), MyStatLab gives you the tools you need to deliver all or a portion of your course online, whether your students are in a lab setting or working from home. MyStatLab provides a rich and flexible set of course materials, featuring free-response tutorial exercises for unlimited practice and mastery. Students can also use online tools, such as video lectures, animations, and a multimedia textbook, to independently improve their understanding and performance. Instructors can use MyStatLab's homework and test managers to select and assign online exercises correlated directly to the textbook, and they can also create and assign their own online exercises and import

TestGen tests for added flexibility. MyStatLab's online gradebook—designed specifically for mathematics and statistics—automatically tracks students' homework and test results and gives the instructor control over how to calculate final grades. Instructors can also add offline (paper-and-pencil) grades to the gradebook. MyStatLab also includes access to Pearson Tutor Services, which provides students with tutoring via toll-free phone, fax, email, and interactive Web sessions. MyStatLab is available to qualified adopters. For more information, visit our Web site at **www.mystatlab.com** or contact your sales representative.

- **MathXL® for Statistics**

  MathXL® for Statistics is a powerful online homework, tutorial, and assessment system that accompanies Pearson textbooks in statistics. With MathXL for Statistics, instructors can create, edit, and assign online homework and tests using algorithmically generated exercises correlated at the objective level to the textbook. They can also create and assign their own online exercises and import TestGen tests for added flexibility. All student work is tracked in MathXL's online gradebook. Students can take chapter tests in MathXL and receive personalized study plans based on their test results. The study plan diagnoses weaknesses and links students directly to tutorial exercises for the objectives they need to study and retest. Students can also access supplemental animations and video clips directly from selected exercises. MathXL for Statistics is available to qualified adopters. For more information, visit **www.mathxl.com,** or contact your sales representative.

- **StatCrunch™**

  StatCrunch is a powerful online tool that provides an interactive environment for doing statistics. You can use StatCrunch for both numerical and graphical data analysis, taking advantage of interactive graphics to help you see the connection between objects selected in a graph and the underlying data. In MyStatLab, the data sets from your textbook are pre-loaded into StatCrunch. StatCrunch is also available as a tool from the online homework and practice exercises in MyStatLab and in MathXL for Statistics.

- **ActivStats®**, developed by Paul Velleman and Data Description, Inc., is an award-winning multimedia introduction to statistics and a comprehensive learning tool that works in conjunction with the book.

It complements this text with interactive features such as videos of real-world stories, teaching applets, and animated expositions of major statistics topics. It also contains tutorials for learning a variety of statistics software, including Data Desk®, Excel, JMP, Minitab, and SPSS. Homework problems and data sets from the Triola text are included (ActivStats for Windows and Macintosh ISBN-13: 978-0-321-50014-4; ISBN-10: 0-321-50014-8). Contact your Pearson Arts & Sciences sales representative for details or visit www.pearsonhighered.com/activstats.

- **The Student Edition of MINITAB** is a condensed version of the Professional release of MINITAB statistical software. It offers the full range of statistical methods and graphical capabilities, along with worksheets that can include up to 10,000 data points. Individual copies of the software can be bundled with the text (ISBN-13: 978-0-321-11313-9; ISBN-10: 0-321-11313-6) (CD only).

- **JMP Student Edition** is an easy-to-use, streamlined version of JMP desktop statistical discovery software from SAS Institute, Inc., and is available for bundling with the text. (ISBN-13: 978-0-321-67212-4; ISBN-10: 0-321-67212-7)

- **SPSS (PASW)**, a statistical and data management software package, is also available for bundling with the text. (ISBN-13: 978-0-13-215171-9; ISBN-10: 0-13-215171-5)

Acknowledgments

I would like to thank the thousands of statistics professors and students who have contributed to the success of this book. I would like to extend special thanks to Mitchel Levy of Broward College, who made extensive suggestions.

This book is truly a team effort, and I consider myself fortunate to work with the dedication and commitment of the Pearson Arts & Sciences team. I thank Deirdre Lynch, Elizabeth Bernardi, Chris Cummings, Beth Houston, Sara Oliver Gordus, Christina Lepre, Alex Gay, Kathleen DeChavez, Dana Jones, Joe Vetere, and Beth Anderson. I also thank Laura Wheel for her work as developmental editor, and I extend special thanks to Marc Triola, M.D., for his outstanding work on the STATDISK software.

I thank the following individuals for their help:

Text Accuracy Reviewers

David Lund
Kimberley Polly

For providing special help and suggestions, I thank Pierre Fabinski of Pace University and Michael Steinberg of Verizon.

For providing help and suggestions in special areas, I would like to thank the following individuals:

Vincent DiMaso	David Straayer, Sierra College
Rod Elsdon, Chaffey College	Glen Weber, Christopher Newport University

For help in testing and improving STATDISK, I thank the following individuals:

Justine Baker	Sr. Eileen Murphy	Victor Strano
Henry Feldman, M.D.	John Reeder	Gary Turner
Robert Jackson	Carolyn Renier	
Caren McClure	Cheryl Slayden	

I extend my sincere thanks for the suggestions made by the following reviewers and users of previous editions of the book:

Dan Abbey, Broward Community College
Mary Abkemeier, Fontbonne College
William A. Ahroon, Plattsburgh State
Scott Albert, College of Du Page
Jules Albertini, Ulster County Community College
Tim Allen, Delta College
Raid W. Amin, University of West Florida
Stu Anderson, College of Du Page
Jeff Andrews, TSG Associates, Inc.
Mary Anne Anthony, Rancho Santiago Community College
William Applebaugh, University of Wisconsin—Eau Claire
James Baker, Jefferson Community College
Justine Baker, Peirce College, Philadelphia, PA
David Balueuer, University of Findlay
Anna Bampton, Christopher Newport University
Donald Barrs, Pellissippi State Technical Community College
James Beatty, Burlington County College

Philip M. Beckman, Black Hawk College
Marian Bedee, BGSU, Firelands College
Marla Bell, Kennesaw State University
Don Benbow, Marshalltown Community College
Michelle Benedict, Augusta College
Kathryn Benjamin, Suffolk County Community College
Ronald Bensema, Joliet Junior College
David Bernklau, Long Island University
Maria Betkowski, Middlesex Community College
Shirley Blatchley, Brookdale Community College
Randy Boan, Aims Community College
John Bray, Broward Community College—Central
Denise Brown, Collin County Community College
Patricia Buchanan, Pennsylvania State University
John Buchl, John Wood Community College

Michael Butler, Mt. San Antonio College
Jerome J. Cardell, Brevard Community College
Keith Carroll, Benedictine University
Don Chambless, Auburn University
Rodney Chase, Oakland Community College
Monte Cheney, Central Oregon Community College
Bob Chow, Grossmont College
Philip S. Clarke, Los Angeles Valley College
Darrell Clevidence, Carl Sandburg College
Paul Cox, Ricks College
Susan Cribelli, Aims Community College
Imad Dakka, Oakland Community College
Arthur Daniel, Macomb Community College
Gregory Davis, University of Wisconsin, Green Bay
Tom E. Davis III, Daytona Beach Community College
Charles Deeter, Texas Christian University
Joseph DeMaio, Kennesaw State University
Joe Dennin, Fairfield University

Nirmal Devi, Embry Riddle Aeronautical University

Richard Dilling, Grace College

Rose Dios, New Jersey Institute of Technology

Christopher Donnelly, Macomb Community College

Dennis Doverspike, University of Akron

Paul Duchow, Pasadena City College

Bill Dunn, Las Positas College

Marie Dupuis, Milwaukee Area Technical College

Theresa DuRapau, Our Lady of Holy Cross

Evelyn Dwyer, Walters State Community College

Jane Early, Manatee Community College

Billy Edwards, University of Tennessee—Chattanooga

Wayne Ehler, Anne Arundel Community College

Sharon Emerson-Stonnell, Longwood College

Marcos Enriquez, Moorpark College

Angela Everett, Chattanooga State Technical Community College

P. Teresa Farnum, Franklin Pierce College

Ruth Feigenbaum, Bergen Community College

Vince Ferlini, Keene State College

Maggie Flint, Northeast State Technical Community College

Bob France, Edmonds Community College

Christine Franklin, University of Georgia

Joe Franko, Mount San Antonio College

Richard Fritz, Moraine Valley Community College

Maureen Gallagher, Hartwick College

Joe Gallegos, Salt Lake Community College

Sanford Geraci, Broward Community College

Mahmood Ghamsary, Long Beach City College

Tena Golding, Southeastern Louisiana University

Elizabeth Gray, Southeastern Louisiana University

Jim Graziose, Palm Beach Community College

David Gurney, Southeastern Louisiana University

Francis Hannick, Mankato State University

Sr. Joan Harnett, Molloy College

Kristin Hartford, Long Beach City College

Laura Heath, Palm Beach Community College

Leonard Heath, Pikes Peak Community College

Peter Herron, Suffolk County Community College

Mary Hill, College of Du Page

Laura Hillerbrand, Broward Community College

Larry Howe, Rowan College of New Jersey

Lloyd Jaisingh, Morehead State University

Lauren Johnson, Inver Hills Community College

Martin Johnson, Gavilan College

Roger Johnson, Carleton College

Herb Jolliff, Oregon Institute of Technology

Francis Jones, Huntington College

Toni Kasper, Borough of Manhattan Community College

Alvin Kaumeyer, Pueblo Community College

William Keane, Boston College

Robert Keever, SUNY, Plattsburgh

Alice J. Kelly, Santa Clara University

Dave Kender, Wright State University

Michael Kern, Bismarck State College

Gary King, Ozarks Technical Community College

John Klages, County College of Morris

Marlene Kovaly, Florida Community College at Jacksonville

John Kozarski, Community College of Baltimore County—Catonsville

Tomas Kozubowski, University of Tennessee

Shantra Krishnamachari, Borough of Manhattan Community College

Richard Kulp, David Lipscomb University

Linda Kurz, SUNY College of Technology

Christopher Jay Lacke, Rowan University

Tommy Leavelle, Mississippi College

Tzong-Yow Lee, University of Maryland

R. E. Lentz, Mankato State University

Timothy Lesnick, Grand Valley State University

Mickey Levendusky, Pima County Community College

Dawn Lindquist, College of St. Francis

George Litman, National-Louis University

Benny Lo, Ohlone College

Sergio Loch, Grand View College

Debra Loeffler, Community College of Baltimore County—Catonsville

Tristan Londre, Blue River Community College

Vincent Long, Gaston College

Alma Lopez, South Plains College

Barbara Loughead, National-Louis University

Rhonda Magel, North Dakota State University—Fargo

Gene Majors, Fullerton College

Hossein Mansouri, Texas State Technical College

Virgil Marco, Eastern New Mexico University

Joseph Mazonec, Delta College

Caren McClure, Santa Ana College

Phillip McGill, Illinois Central College

Marjorie McLean, University of Tennessee

Austen Meek, Canada College

Robert Mignone, College of Charleston

Glen Miller, Borough of Manhattan Community College

Kermit Miller, Florida Community College at Jacksonville

Kathleen Mittag, University of Texas—San Antonio

Mitra Moassessi, Santa Monica College

Charlene Moeckel, Polk Community College

Carla Monticelli, Camden County Community College

Theodore Moore, Mohawk Valley Community College

Rick Moscatello, Southeastern Louisiana University

Gerald Mueller, Columbus State Community College

Sandra Murrell, Shelby State Community College

Faye Muse, Asheville-Buncombe Technical Community College

Gale Nash, Western State College

Felix D. Nieves, Antillean Adventist University

Lyn Noble, Florida Community College at Jacksonville—South

Julia Norton, California State University Hayward

DeWayne Nymann, University of Tennessee

Patricia Oakley, Seattle Pacific University

Keith Oberlander, Pasadena City College

Patricia Odell, Bryant College

James O'Donnell, Bergen Community College

Alan Olinksy, Bryant College

Nasser Ordoukhani, Barry University

Michael Oriolo, Herkimer Community College

Jeanne Osborne, Middlesex Community College

Ron Pacheco, Harding University

Lindsay Packer, College of Charleston

Kwadwo Paku, Los Medanos College

Deborah Paschal, Sacramento City College

S. A. Patil, Tennessee Technological University

Robin Pepper, Tri-County Technical College

David C. Perkins, Texas A&M University—Corpus Christi

Anthony Piccolino, Montclair State University

Richard J. Pulskamp, Xavier University

Diann Reischman, Grand Valley State University

Vance Revennaugh, Northwestern College

C. Richard, Southeastern Michigan College

Don Robinson, Illinois State University

Sylvester Roebuck, Jr., Olive Harvey College

Ira Rosenthal, Palm Beach Community College—Eissey Campus

Kenneth Ross, Broward Community College

Charles M. Roy, Camden County College

Kara Ryan, College of Notre Dame

Ali Saadat, University of California—Riverside

Radha Sankaran, Passaic County Community College

Fabio Santos, LaGuardia Community College

Richard Schoenecker, University of Wisconsin, Stevens Point

Nancy Schoeps, University of North Carolina, Charlotte

Jean Schrader, Jamestown Community College

A. L. Schroeder, Long Beach City College

Phyllis Schumacher, Bryant College

Pradipta Seal, Boston University

Sankar Sethuraman, Augusta College

Rosa Seyfried, Harrisburg Area Community College

Calvin Shad, Barstow College

Carole Shapero, Oakton Community College

Adele Shapiro, Palm Beach Community College

Lewis Shoemaker, Millersville University

Joan Sholars, Mt. San Antonio College

Galen Shorack, University of Washington

Teresa Siak, Davidson County Community College

Cheryl Slayden, Pellissippi State Technical Community College

Arthur Smith, Rhode Island College

Marty Smith, East Texas Baptist University

Laura Snook, Blackhawk Community College

Aileen Solomon, Trident Technical College

Sandra Spain, Thomas Nelson Community College

Maria Spinacia, Pasco-Hernandez Community College

Paulette St. Ours, University of New England

W. A. Stanback, Norfolk State University

Carol Stanton, Contra Costra College

Richard Stephens, Western Carolina College

W. E. Stephens, McNeese State University

Terry Stephenson, Spartanburg Methodist College

Consuelo Stewart, Howard Community College

David Stewart, Community College of Baltimore County—Dundalk

Ellen Stutes, Louisiana State University at Eunice

Sr. Loretta Sullivan, University of Detroit Mercy

Tom Sutton, Mohawk College

Sharon Testone, Onondaga Community College

Andrew Thomas, Triton College

Evan Thweatt, American River College

Judith A. Tully, Bunker Hill Community College

Gary Van Velsir, Anne Arundel Community College

Randy Villa, Napa Valley College

Hugh Walker, Chattanooga State Technical Community College

Charles Wall, Trident Technical College

Dave Wallach, University of Findlay

Cheng Wang, Nova Southeastern University

Glen Weber, Christopher Newport College

David Weiner, Beaver College

Sue Welsch, Sierra Nevada College

Roger Willig, Montgomery County Community College

Gail Wiltse, St. John River Community College

Odell Witherspoon, Western Piedmont Community College

Claire Wladis, Borough of Manhattan Community College

Jean Woody, Tulsa Junior College

Carol Yin, LeGrange College

Thomas Zachariah, Loyola Marymount University

Yong Zeng, University of Missouri at Kansas City

Jim Zimmer, Chattanooga State Technical Community College

Elyse Zois, Kean College of New Jersey

Cathleen Zucco-Teveloff, Trinity College

Mark Z. Zuiker, Minnesota State University, Mankato

M.F.T.
LaGrange, New York
September, 2009

Index Of Applications

In the center column,
CP = Chapter Problem
IE = In-Text Example
M = Margin Example
E = Exercise
BB = Beyond the Basics
R = Review Exercise
CR = Cumulative Review Exercise
DD = Data to Decision
CGA = Cooperative Group Activity
TP = Technology Project
SW = Statistics at Work

Agriculture

Fertilizer (CR), 138; (IE), 496
Hens Laying Eggs (IE), 13, 212
Milk From Cows (IE), 13, 212
Phenotypes of Peas (E), 101, 117, 215; (IE), 217, 218, 222
Straw Seed (R), 512; (E), 681
Weights of Poplar Trees (E), 654

Biology

Archeological Research (SW), 519
Bear Data (BB), 573; (E), 430, 437, 580; (R), 581
Capture-Recapture Method (CGA), 206
Cricket Chirps and Temperature (IE), 70; (E), 74, 538, 554, 700
DNA Nucleotides (E), 196
E. Coli Bacteria (E), 183
Ecology, Animal Behavior and Ecotoxicology (SW), 715
Fruit Flies (BB), 158; (E), 182, 293
Genetic Disorder (E), 221
Genetics Experiment (IE), 211, 226, 228, 236; (E), 233, 416, 600; (R), 325
Genetics: Eye Color, Age and Gender (R), 44, 201; (E), 156, 222, 231, 232, 572; (CR), 203
Genotypes (IE), 148, 154; (BB), 235
Hybridization Experiment (E), 187, 313; (CP), 209; (BB), 234
Manatee Deaths (E), 578
Mendelian Genetics (E), 11, 155, 239, 347, 430; (M), 594
Plants Being Grown in Homes (CR), 138
Skull Breadths (E), 643, 691, 692
Sociality and Population of Sperm Whales (SW), 253
Weights of Seals (E), 536, 553, 562, 700
Wildlife Population Sizes (M), 353

Business and Economics

Acceptance Sampling (E), 175, 183, 234, 313
Advertising (E), 25; (CGA), 394
Bar Codes (M), 190; (R), 203
Brand Recognition (E), 187, 232; (IE), 228; (R), 248
Casino Size and Revenue (E), 537, 553
CEO Profile (BB), 80–81
Coca-Cola Directors (E), 197
Commercials (M), 432
Consumer Price Index (E), 536, 552, 562
Consumer Products (E), 538, 554, 700
Customer Waiting Times (E), 103, 118, 386; (CGA), 516, 662

Daily Oil Consumption (E), 80
Defective Items (E), 176, 181, 234, 736; (IE), 178, 734; (BB), 184, 225, 240; (CR), 738; (R), 740; (TP), 741
Dow Jones Industrial Average (IE), 71; (CGA), 743
Failure Rate of Manufactured Products (M), 213
Forecasting and Analysis of Walt Disney World (SW), 587
Google Software Engineer (SW), 395
High Cost of Low Quality (M), 736
Home Sales and Prices (R), 44; (M), 480; (E), 486, 508, 572
Manufacturing Memory Chips (E), 580
Marketing Strategy (SW), 745
Media and Advertising (E), 25
Moore's Law (E), 75; (BB), 578
Paper-Coating Machine (M), 721
Pizza and Subway Costs (CP), 521; (IE), 525, 528, 529, 531, 532, 542, 545, 547, 549, 557, 559, 560; (E), 577
Predicting Condo Prices (M), 548
Publishing Company (SW), 87
Quality Control (E), 35, 36, 181, 293; (CR), 46; (IE), 169, 170, 449, 451; (M), 732
Six Sigma in Industry (M), 734
Sony Manufacturing Compact Discs (M), 567
Statistics and Quality Management (SW), 463
Stock Market (E), 578, 708
Stockholders of IBM (R), 45
Tax Audits (E), 35
Toxicologist (SW), 663
Values of New Cars and Clothes (E), 699
Vending Machines (E), 304
Wedding Ring Prices (E), 580

Education

Absences (CGA), 744
Age of Faculty Members (E), 119
Back-to-School Spending (E), 371
Better Results with Smaller Class Sizes (M), 483
Calculators (E), 9, 182; (BB), 315; (R), 457
Class Attendance and Grades (M), 669
Class Seating Arrangement (CGA), 713
Class Size Paradox (M), 93
College Applications Online (E), 426, 473
College Graduates Live Longer (E), 23
College Tuition (E), 81, 360
College Undergraduate Enrollments (IE), 58; (E), 74, 81, 736
Course Grades (IE), 14; (E), 104
Curriculum Planning (E), 35

Curving Test Scores (BB), 281
Education and Smoking (IE), 58, 429
Genders of Students (E), 76
Grade and Seat Location (E), 598
Grade Point Average (IE), 97; (E), 104, 360; (CR), 390
Guessing on a Test (IE), 151, 190; (E), 154, 175, 181, 232, 238
IQ Scores (IE), 56, 105, 111, 316, 356; (E), 59, 277, 278, 302, 312, 360, 436, 538, 554; (M), 237, 719; (BB), 281, 447; (R), 324; (TP), 514
IQ Scores of Statistics Professors (E), 444
Length of a Classroom (CGA), 393, 461, 661-662
Major and Gender (CGA), 662, 714
Multiple Choice Quiz (E), 312
Number of Classes (BB), 62
Odd and Even Digits in PI (E), 708
Oldest College Graduate (E), 136
Predictors of Success (M), 564
Prices of College Textbooks (IE), 15
Ranking Colleges (IE), 14, 696; (CP), 665; (R), 710
Sampling Students (E), 36
SAT and ACT Tests (BB), 282; (CR), 390
SAT Scores (E), 302, 359, 384
Statistics Students (E), 238, 291, 359, 619; (BB), 240
Statistics Students Present in a Class (IE), 212
Students Suspended (IE), 20
Systems of Measurement (CGA), 516
Teacher Evaluations Correlate With Grades (M), 527
Test Scores (E), 134, 453; (R), 137
Time to Earn Bachelor's Degree (E), 102, 117, 445, 457
Working Students (CR), 514

Engineering

Axial Load of an Aluminum Can (BB), 62, 65; (R), 457, 738–739; (E), 507; (BB), 601

xxv

Designing Aircraft Seating (E), 280; (DD), 330
Designing Caskets (R), 325
Designing Doorways (IE), 271, 274
Designing Manhole Covers (CGA), 329
Designing Motorcycle Helmets (E), 303; (CR), 741
Designing Seating (E), 385
Doorway Height (E), 279, 303; (BB), 304
Electricity (E), 196
Elevator Capacity (BB), 305
Energy Consumption (CR), 250; (E), 386, 731
Home Power Supply (IE), 257
Integrated Circuits (BB), 323; (E), 359
Mars Climate Orbiter (IE), 13; (M), 720
Redesign of Ejection Seats (E), 304
Smoke Alarms (E), 200
Voltages (E), 16, 61, 65, 103, 118, 119, 135, 267, 322, 438, 445, 446, 486, 490, 500, 644, 657, 675, 732; (BB), 105; (TP), 138; (IE), 258, 380, 383

Entertainment

Ages of Oscar-Winning Actors and Actresses (DD), 86, 140, 518; (E), 132, 133, 374, 498, 538, 554, 674, 707
Deal or No Deal Television Game Show (E), 224; (R), 248
iPod Random Shuffle (M), 702; (CGA), 714
Movie Budgets and Gross (E), 16, 103, 118, 122, 123, 360, 490, 538, 554, 573, 644, 687; (IE), 126, 128; (BB), 135
Movie Data (E), 17, 37, 324, 350, 374, 385, 657; (CR), 46; (IE), 318, 319
Movie Ratings (E), 17, 430; (R), 44, 45; (CR), 660
Napster Website (IE), 18
Nielsen Television Rating (BB), 36; (SW), 49; (E), 222, 233, 312, 428; (M), 338
Number of Possible Melodies (CR), 204
Playing Times of Popular Songs (E), 23, 101, 117, 454
Routes to National Parks (IE), 191
Salaries of TV Personalities (E), 101, 116, 374
Substance Abuse in Video Games (IE), 368; (E), 374, 385; (TP), 390
Tobacco and Alcohol in Children's Movies (IE), 319; (E), 499, 674
Writing a Hit Song (E), 436, 444

Environment

Air Pollution (IE), 22
Atmospheric Carbon Dioxide (E), 578, 731
Car Emissions (E), 102, 372, 374, 445, 645, 692; (BB), 646
Carbon Monoxide (E), 416
Earthquakes (IE), 241
Earth's Temperatures (IE), 718, 719, 723, 725, 726
Global Warming (R), 45; (CP), 333, 717; (IE), 335, 340, 705; (E), 348, 429, 475, 534, 537, 554, 563, 578, 613, 708

Lake Mead Elevations (E), 729
Monitoring Lead in Air (E), 374, 385; (IE), 442
Old Faithful Geyser (R), 83; (E), 133, 321, 497, 573
Precipitation in Boston (E), 349, 358, 444, 453, 456; (BB), 701
Radioactive Decay (E), 244
Temperature and Carbon Dioxide (CR), 740
Temperatures (E), 17, 75, 708; (BB), 17; (R), 45, 711
Weather Forecasts (E), 154, 498, 681; (M), 540; (R), 711
Weights of Garbage Discarded by Households (IE), 369, 529, 567; (R), 389; (E), 60, 61, 65, 74, 103, 372, 437, 500, 535, 551; (TP), 459; (CR), 660

Finance

ATM Machine (E), 197
Auditing Checks (CR), 249
Author's Checks (E), 244, 601
Average Annual Incomes (E), 24, 302
Bankruptcies (E), 102, 117
Change for a Dollar (BB), 199
Checks and Charges (E), 487, 601
Choosing Personal Security Codes (M), 189
Credit Cards (E), 35
Credit Rating (E), 61, 64, 101, 116, 134, 182, 233, 291, 358, 360, 386, 437, 444, 447; (IE), 111, 112, 113
Dollar Bills (E), 79
Income and Education (E), 16; (IE), 77–78; (R), 513
Income Data (M), 71
Junk Bonds (BB), 225
More Stocks, Less Risks (M), 108
Mortgage Payments (E), 488
Personal Income (IE), 95; (E), 116, 324
Reporting Income (E), 163, 427; (R), 388

Food/Drink

Caffeine Nation (E), 221
Chocolate Health Food (E), 24
Coke Versus Pepsi (CGA), 393, 461; (E), 511
Coke Volume (E), 16, 135, 437, 675, 729
Filling Cans of Soda (IE), 298; (E), 304, 729
Herb Consumption (R), 457
Hershey's Kiss Candies (CGA), 206
Italian Food (E), 416
M&M's (E), 238, 303, 314, 321, 349, 384, 385, 430, 436, 600, 644; (BB), 375, 430, 510; (CR), 458
Pancake Experiment (E), 656
Protein Energy Bars (R), 45
Scale for Rating Food (BB), 17
Sugar in Oranges (M), 368
Weights of Coke and Diet Coke (E), 61, 65, 134, 322, 490; (CR), 203
Weights of Steak (R), 136; (CR), 203
Wine Tasting (E), 35

Games

Card Playing (E), 195
Casino Dice (E), 154; (IE), 219
Counting Cards (M), 146
Drawing Cards (E), 154; (BB), 177
Florida Lottery (IE), 194
Fundamental Principle of Gambling (M), 172
Gambling Strategy (BB), 314
How Many Shuffles? (M), 192
Illinois Pick 3 Game (E), 224; (IE), 243
Jumble Puzzle (E), 197
Kentucky Pick 4 Lottery (R), 248
Labeling Dice (BB), 225
Loaded Die (E), 221, 437, 599
Magazine Sweepstakes (R), 248
Monty Hall Problem (BB), 189; (CGA), 206
Multiple Lottery Winners (M), 270
New Jersey's Pick 4 Game (E), 224
Picking Lottery Numbers (E), 200; (M), 216; (BB), 235
Roller Coaster (BB), 184
Rolling Dice (E), 9, 187, 244; (IE), 283, 285, 286
Roulette (E), 154, 224; (BB), 157; (IE), 153, 218
Schemes to Beat the Lottery (M), 307
Slot Machine (E), 9, 233, 598
Solitaire (BB), 157
Tossing Coins (BB), 184, 350; (IE), 186; (E), 187; (TP), 204; (CGA), 394
Winning the Lottery (E), 154, 195, 196, 198; (R), 202

General Interest

Age of Books (CGA), 392, 460
Alarm Clock Redundancy (E), 183
Analysis of Pennies (E), 426, 445, 454, 598, 599
Anchoring Numbers (CGA), 139, 516
Area Codes (E), 198
Areas of States and Countries (E), 16, 79
Authors Identified (M), 54
Bed Length (R), 325
Birthdays (E), 155, 187; (BB), 177, 184, 189; (IE), 171, 185
Coincidences (M), 178
Combination Lock (E), 195
Comparing Ages (E), 511
Comparing Readability (R), 512; (E), 643; (CR), 711–712
Cost of Laughing Index (M), 121
Deaths from Horse Kicks (E), 244
Definition of a Second (M), 105
Dropping Thumbtacks (CGA), 206
Effect of Blinding (R), 512
Elbow and Wrist Breadths of Women (IE), 6, 22; (E), 323, 436
Evaluating a Statistical Study (M), 5
Fabric Flammability Tests (E), 692
Foot Breadths of Men (E), 453
Friday the 13th (E), 499, 681
Grip Reach (R), 137

Handshakes and Round Tables (BB), 198

Head Circumference and Forearm Length (CGA), 584

Height and Arm Span (CGA), 584, 714

Height and Navel Height (CGA), 584, 714

Heights of Martians (BB), 376

Judges of Marching Bands (E), 699, 700

Lefties Die Sooner? (M), 439

Leg Length of Adults (E), 358

Length of Men's Hands (E), 452

Length of Screws (E), 61, 65, 103, 118, 438, 446, 675; (IE), 299

Length of Straws (CGA), 48

Lightning Deaths (R), 624

Meta-Analysis (M), 440

Misleading Statistics in Journalism (M), 18

Monkey Typists (M), 186

Name Recognition (E), 349

National Statistics Day (R), 202

Number of English Words (E), 100, 116, 134, 446

Origin of Statistics (M), 12

Penny Weights (IE), 57, 70; (E), 59, 436, 445, 453, 507; (R), 389

Periodic Table (E), 16

Points on a Stick (BB), 158

Probabilities that Challenge Intuition (M), 145

Random Telephone Numbers (E), 223, 278, 322, 360, 384, 597; (BB), 430

Reliability of Systems (M), 166; (IE), 172; (E), 176; (BB), 176

Safe Combination (E), 197

Scheduling Assignments (E), 196

Scheduling Routes (E), 195, 196; (BB), 198–199

Shakespeare's Vocabulary (M), 159

Sitting Distance (E), 280

Sitting Heights (R), 137; (CR), 137, 459; (E), 437; (CGA), 461

Six Degrees of Separation (DD), 48

Stuck in an Elevator (IE), 148

Struck by Lightening (E), 155

Thanksgiving Day (E), 9; (IE), 150

The Random Secretary (M), 193

Twins in Twinsburg (M), 494

UFO Sightings (E), 599

Upper Leg Lengths (BB), 135

Wedding Months (E), 598

Weights of One-Dollar Coins (R), 327; (E), 372

Weights of Quarters (E), 61, 65, 134, 675; (IE), 503; (BB), 510

Win $1,000,000 for ESP (M), 422

Word *Ginormous* added to Dictionary (E), 511

Words Per Sentence (E), 644

Wristwatch Time (CGA), 461

Years (IE), 15

Zip Codes (E), 135–136

Health

Adverse Effect of Viagra (E), 155, 475

Amalgam Tooth Fillings (E), 612

Aspirin Preventing Heart Attacks (M), 399

Atkins Weight Loss Program (IE), 7; (E), 360, 373

Bayer Aspirin (E), 415, 735, 736

Bednets to Reduce Malaria (E), 473

Birth Genders (IE), 145; (E), 154, 181, 182; (M), 158

Birth Rate (E), 737

Births (E), 292, 599

Birth Weights (E), 244, 279, 359, 372, 384, 453; (IE), 272, 275

Blood Cell Counts (E), 133, 359, 360, 384, 436, 445, 489, 509

Blood Groups and Types (E), 61, 75, 175, 232, 313; (BB), 235

Blood Pressure (E), 16, 116, 281, 303, 323, 360, 486, 499, 536, 553; (CR), 203; (BB), 555; (IE), 696

Blood Testing (R), 202; (E), 233

BMI and Gender (CGA), 85; (IE), 370

Body Mass Index (E), 102, 118, 446, 454, 489, 498, 509, 687

Body Temperatures (IE), 15, 672, 679; (R), 44, 580, 710; (CR), 40, 660; (E), 103, 118, 133, 280, 304, 373, 436, 446, 499

Carbon Monoxide in Cigarettes (R), 659

Cardiovascular Effects (BB), 37

Carpal Tunnel Syndrome: Splinting or Surgery (E), 11, 613; (R), 511

Cell Phones and Cancer (E), 239, 313, 348, 428

Children's Respiratory Systems (R), 624

Cholesterol Levels (E), 323, 372, 507, 551, 656; (R), 326; (IE), 364; (CR), 711

Cholesterol Reducing Drug (E), 314

Cigarette Tar and Nicotine (E), 539, 555, 572, 701

Cleanliness (CR), 625

Clinical Trials (M), 20, 575; (IE), 193, 414; (E), 427, 472, 473, 600, 612, 613; (R), 624

Colorblindness (E), 163

Cotinine in Smokers (CR), 84

Crash Hospital Costs (E), 372

Disease Clusters (E), 245

Drug to Lower Blood Pressure (R), 512

Effectiveness of Acupuncture for Migraines (E), 373

Effectiveness of an HIV Training Program (CR), 249

Effectiveness of Crest in Reducing Cavities (M), 491

Effectiveness of Dozenol (E), 497

Effectiveness of Echinacea (E), 35, 373, 474; (BB), 36; (IE), 602, 605

Effectiveness of Flu Vaccine (E), 613

Effectiveness of Hip Protectors (E), 617, 620

Effectiveness of Humidity in Treating Croup (IE), 478; (E), 486

Effectiveness of the Salk Vaccine (IE), 26, 31–32, 152; (E), 610

Effects of Alcohol (BB), 491; (E), 508

Effects of Cocaine on Children (E), 512

Effects of Marijuana Use on College Students (E), 488, 508

Eight-Year False Positive (M), 602

Expensive Diet Pill (M), 481

Freshman Weight Gain (IE), 5, 6, 7, 477, 492-494, 495, 496, 670-671, 678, 704; (R), 44; (E), 101, 117, 349, 430, 509; (CP), 465

Gender Gap in Drug Testing (M), 670

Growth Charts Updated (M), 56

Hawthorne and Experimenter Effects (M), 28

Health Plans (E), 35

Heart Attacks and Statistics (M), 449

Heart Pacemakers (E), 156, 473

Heart Patient Survival (E), 181

Heartbeats (CGA), 461

Height and Pulse Rate (E), 535, 551; (CGA), 714

High-Dose Nicotine Patch Therapy (IE), 344

HIV Infections (E), 183, 197

Hormone Therapy (M), 27

Internist Specializing in Infectious Diseases (SW), 141

Length of Pregnancy (E), 133, 280, 303

Lipitor (M), 67; (E), 231, 473, 486, 614, 737, 738; (DD), 662

Magnet Treatment of Pain (E), 373, 488, 489, 508

Medical Malpractice (E), 347

Nasonex Treatment (BB), 157; (E), 188, 197; (CR), 327

Nicotine in Cigarettes (E), 10, 16, 61, 64, 102, 118, 375, 435, 487, 488, 573, 645, 687, 693; (CR), 46; (TP), 47

Norovirus on Cruise Ships (E), 474, 613

PET/CT Compared to MRI (E), 621

Placebo Effect (M), 257

Polio Experiment (M), 470

Predicting Measles Immunity (E), 620

Pregnancy Test Results (DD), 205

Prescription Pills (E), 36

Process of Drug Approval (M), 423

Pulse Rates (R), 44, 82; (IE), 53, 54, 56, 57, 66, 67, 68, 129, 131; (E), 64, 73, 74, 81, 103, 119, 361, 375, 385, 453, 454, 482, 497; (BB), 65, 76, 555; (CGA), 84, 516, 585, 743; (TP), 84

Radiation Effects (E), 475

Radiation in Baby Teeth (E), 60, 64, 74, 102, 118, 134, 489, 508, 686

Reye's Syndrome (BB), 305

SIDS (E), 25

Smoking and Cancer (E), 79, 695

Smoking, Body Temperature, and Gender (R), 513, 659

Smoking Treatments (E), 60, 239, 619–620; (BB), 417; (R), 624

Tar in Cigarettes (E), 59, 75, 103, 119, 359, 445, 453, 487, 645, 687, 693

Testing a Treatment (E), 621

Testing for Adverse Reactions (R), 623

Testing for Syphilis (M), 179

Treating Athlete's Foot (E), 620, 621

Treating Syphilis (E), 35

Weight (E), 79, 599; (IE), 592–593
Weight Lost on Different Diets (E), 9, 10, 436, 445, 453, 508, 644; (R), 658
Weight Watchers Weight Loss Program (E), 534
X-Linked Disorder (E), 157, 181
Zinc Treatment (E), 507

Labor

Drug Testing of Applicants and Employees (CR), 249, 390; (E), 387
Earnings Between Men and Women (IE), 68; (E), 79
Employee Perceptions (E), 429, 475, 476
Finding a Job Through Networking (IE), 69
Hiring Job Applicants (E), 313, 429
Interviewing and Job Application Mistakes (IE), 6, 22, 234; (E), 16, 74, 75, 348, 429
Job Interviews (E), 223
Job Longevity (E), 234, 239
Occupational Injuries (E), 599
Reasons for Being Fired (R), 248
Salary of Teachers (E), 79
Unemployment (IE), 28
Wise Action for Job Applicants (CR), 625

Law

Biased Test (E), 23
Bribery in Jai Alai (M), 723
Burglaries Cleared by Arrests
Convicted by Probability (M), 169
Crime and Strangers (E), 611
Death Penalty (E), 156; (CGA), 461; (M), 471
Detecting Fraud (E), 223, 314, 601; (R), 248; (CGA), 252
Firearm Rejections (CR), 514
Identifying Thieves (M), 477
Identity Theft (IE), 189; (E), 196
Is the Nurse a Serial Killer? (CP), 589; (IE), 606
Jury Selection (DD), 251; (E), 429
Lie Detectors (M), 404, 407; (E), 428, 611
Murders Cleared by Arrest (IE), 162
Percentage of Arrests (E), 427, 429
Polygraph Test (CP), 143; (E), 154–155, 163, 174, 182, 314; (IE), 159, 160, 167, 179
Prisoners (E), 35; (M), 451
Prosecutor's Fallacy (M), 180
Ranking DWI Judges (E), 700
Sentence Independent of Plea (E), 612
Sobriety Checkpoint (E), 35, 163
Solved Robberies (E), 182
Speed Limits (IE), 125
Speeding Tickets (E), 9, 103; (M), 523
Speeds of Drivers Ticketed on an Interstate (E), 384
Supreme Court (E), 9
Testifying in Supreme Court (M), 469
Violent Crimes (E), 600, 736
Voice Identification of a Criminal (E), 175

People and Psychology

Adoptions from China (E), 80
Ages of New York City Police Officers (E), 321
Ages of U.S. Presidents (E), 64, 292, 375; (CGA), 392, 461
Ages of Winners of the Miss America Pageant (E), 119
Census Data (E), 24
Children's Defense Fund (M), 317
Extrasensory Perception (ESP), (M), 177; (CGA), 251, 461
Florence Nightingale (M), 68
Gender in a Family (M), 282
Gender of Children (IE), 149, 154, 177, 289–290; (E), 154, 156; (BB), 293; (CGA), 329
Gender Selection (IE), 8, 184, 192, 399, 400, 403, 405, 407, 410, 411, 413, 419, 670, 671; (E), 16, 155, 162, 175, 188, 197, 231, 238, 239, 312, 313, 347, 415, 428, 674, 675; (BB), 165, 417, 418; (CGA), 392, 460; (CP), 397
Height (E), 59, 456; (M), 120, 294
Height Requirement for Soldiers (E), 133, 279
Heights of Girls (R), 137
Heights of Men and Women (IE), 115, 120; (E), 119, 133, 321, 323, 358, 499, 506, 509, 561, 654; (BB), 281, 323; (CGA), 516; (R), 581; (CR), 582, 625
Heights of Presidents (E), 132, 500, 536, 553
Heights of Rockettes (R), 325
Heights of Statistics Students (E), 60
Heights of Supermodels (E), 446, 453, 487, 535, 551
Household Size (IE), 288; (E), 292; (BB), 305
Left-Handedness (E), 34, 188; (CR), 327
Life Insurance Policy (E), 224, 245; (SW), 207
Life Spans (CGA), 661
Longevity (BB), 104, 491; (E), 489, 686; (R), 659; (CR), 659–660
Measuring Disobedience (M), 13
Mortality Study (R), 202
Most Average Person in the United States (M), 91
Number of Children (CGA), 252
Number of Girls (E), 221, 223
Pain Intensity (DD), 586
Palm Reading (M), 526
Parent/Child Heights (E), 535, 551; (IE), 565, 569
Postponing Death (R), 247; (E), 347, 428; (M), 546
Predicting Sex of Baby (IE), 422–423, 424; (E), 428, 429, 675
Prospective National Children's Study (M), 30
Racial Profiling (E), 23
Reaction Time (BB), 501; (CGA), 516, 585

Tall Clubs International (E), 279
Touch Therapy (E), 34, 240, 348, 426
Twins (E), 183, 680
Victims of Terrorism (BB), 36
Weights of Men and Women (IE), 115, 120, 132, 353; (E), 132, 359, 561, 654; (R), 581
Weights of Supermodels (CR), 390; (E), 535, 551
Word Counts of Men and Women (CP), 89; (IE), 91, 108-109, 128, 477, 479, 481; (E), 490, 539, 554, 701; (CR), 513, 514; (TP), 661

Politics

Captured Tank Serial Numbers (M), 354
Draft Lottery (E), 708; (DD), 713; (CGA), 714
Interpreting Political Polling (BB), 17
Keeping the United Nations in the United States (E), 6, 24
Line Item Veto (IE), 21
Political Contributions (E), 601
Political Party Choice (CGA), 627
Presidential Election (E), 36, 154, 239, 427, 708; (R), 45; (IE), 148; (M), 185
Senators in Congress (IE), 12, 27; (E), 16, 155
Tax Returns and Campaign Funds (E), 475
Voter's Opinion of a Candidate (E), 24
Voting Behavior (IE), 20; (E), 737
World War II Bombs (R), 249; (E), 600

Social Issues

Accepting a Date (E), 181
Affirmative Action Program (E), 233
Age Discrimination (E), 488, 509, 686
American Habits (R), 44
Cell Phones (CR), 46; (E), 75, 156, 349, 707; (IE), 214
Changing Populations (M), 52, 92
Crime and Abortions (CGA), 48
Crowd Size (M), 367
Deaths From Motor Vehicles and Murders (R), 202; (E), 244, 578; (IE), 705
Drug Testing (E), 164, 238, 427
Drug Use in College (E), 474
Ergonomics (E), 35; (R), 136; (CR), 625
Firearm Injuries (CR), 660
Gender Discrimination (E), 223, 234; (R), 326; (CR), 514
Guns and Murder Rate (E), 23
Homicide Deaths (E), 245; (R), 740
Households in the United States (IE), 22
Marriage and Divorce Rates (E), 75
Marriage Rate (CGA), 744
Money Spent on Welfare (IE), 20
Morality and Marriage (E), 474
Personal Calls at Work (E), 156
Population Control (BB), 189; (CGA), 206
Population in 2020 (IE), 575
Population in 2050 (BB), 579
Population Size (E), 577

Queues (M), 241
Rate of Smoking (R), 45, 456
Rebuilding the World Trade Center Towers (CR), 138
Self-Esteem Levels (E), 655
Television Households (E), 222

Sports

Baseball (E), 437, 538, 554; (TP), 582–583
Baseball Player's Hits (M), 106; (BB), 315
Basketball Foul Shots (CGA), 743
Golf Scores (CR), 625
Gondola Safety (E), 302
Heights of NBA Basketball Players (E), 358
Helmet Use and Head Injuries (E), 200; (R), 201
Home Field Advantage (M), 424; (E), 610
Horse Racing (E), 195
Icing the Kicker (M), 569
Injury from Sports (E), 153
Kentucky Derby (BB), 157; (IE), 192; (E), 599
NBA Salaries and Performances (M), 568
NCAA Basketball Tournament (E), 198
NCAA Football Coach Salaries (E), 101, 117, 359, 437; (CR), 327
Olympic Winners (CR), 457
Parachuting (M), 319
Shaquille O'Neal (CR), 46; (E), 187
Shirt Numbers of Basketball Players (IE), 12
Sports Columnist (SW), 629
Sports Hot Streaks (M), 704
Steroid Testing (E), 181
Super Bowls (E), 9, 100, 134, 534, 577, 708; (M), 482, 556
Tennis Instant Replay (E), 164, 427, 475, 611
Triathlon Times (E), 645
World Record (M), 94
World Series Games (E), 222, 470, 611, 707, 708; (IE), 594; (R), 710

Surveys and Opinion Polls

America Online Survey (E), 17, 24, 25, 324, 348, 388, 426, 429; (CGA), 48; (IE), 149; (CR), 203, 741
Bad Question (E), 24
Cloning Survey (E), 17; (R), 389
Consumer Survey (E), 17; (CGA), 48
Curbstoning (M), 336
Detecting Phony Data (M), 19
"Do Not Call" Registry (DD), 394
Ethics in Reporting (R), 202; (M), 682
Exit Polls (E), 35, 36; (CR), 138
Falsifying Data (BB), 26
Gallup Poll (IE), 19; (E), 25, 35, 43, 174, 231, 237, 238; (R), 44
Good Housekeeping Survey (E), 24
Health Survey (E), 17
Influence of Gender (E), 36; (CGA), 516; (IE), 608–609; (BB), 614
Internet Survey (IE), 309–312; (E), 313, 415

Literary Digest Poll (CP), 3; (IE), 4, 12, 19, 26; (E), 43
Mail Survey (E), 23, 312; (IE), 307–309
Merrill Lynch Client Survey (E), 24
Poll Accuracy (BB), 350
Poll Confidence Level (E), 175
Poll Resistance (M), 647
Poll Results (E), 16, 73, 346; (DD), 462
Polls and Psychologists (M), 688
Pre-Election Poll (R), 46; (M), 468
Public Polling (SW), 331
Push Polling (M), 377
Questionnaires to Women's Groups (M), 410
Repeated Callbacks (M), 226
Smoking Survey (E), 11, 35, 472
Stem Cell Survey (E), 156, 428, 675
Student Survey (BB), 36; (E), 231
Sudoku Poll (R), 202
Survey Medium Can Affect Results (M), 635
Survey of Car Owners (R), 45, 389; (E), 349; (CR), 514
Survey of Executives (CR), 83, 84
Survey of Married Couples (CGA), 516
Survey of Politicians (E), 231
Survey of Voters (E), 313, 347
Survey Refusals and Age Bracket (E), 164
Survey Responses (IE), 14, 22, 23, 24
Survey Results (CP), 51; (IE), 77
Telephone Polls and Surveys (E), 17, 371
What's Wrong With This Picture? (BB), 26

Technology

Computer Design (E), 196
Computer Intelligence (BB), 199
Computer Variable Names (BB), 198
Internet Use (IE), 342–343; (E), 348, 349, 429
Satellites (E), 102, 117, 323; (IE), 106; (BB), 119
Scientific Thermometers (IE), 261–265; (E), 268, 269
Space Shuttle Flights (E), 82, 101, 117, 132, 281, 322
Unauthorized Downloading (E), 59

Transportation

Age of Cars Driven by Students (R), 137
Ages of Faculty and Student Cars (E), 507
Aircraft Altimeter Errors (E), 454
Aircraft Safety Standards (R), 326
Airline Passengers with Carry-on Baggage (IE), 403
ATV Accidents (BB), 11
Average Speed (BB), 104
Braking Distances (E), 80, 323, 358, 487, 508, 684
Bumped from a Flight (E), 156
Bumper Stickers (E), 223
Car Acceleration Times (E), 83
Car Crashes (E), 100, 116, 154, 182
Car Crash Costs (E), 446

Car Crash Tests (CP), 631; (IE), 634, 647, 650, 651, 689-690; (R), 389, 658–659; (E), 657
Car Cylinders (E), 16
Car Ignition Systems (E), 176
Car Mileage (CR), 327; (E), 498, 537, 553
Car Reliability Data (IE), 72
Car Repair Costs (E), 499, 537, 553
Car Weight (E), 16; (R), 457, 658
Car Weight and Braking Distance (E), 83, 539, 554
Car Weight and Fuel Consumption (E), 10, 561, 562
Cell Phones and Crashes (E), 36; (CR), 712
Cheating Gas Pumps (E), 428, 675
Colors of Cars (R), 201; (CGA), 206
Commuters and Parking Spaces (E), 537, 553
Cost of Flying (E), 102, 118, 498, 537, 553, 674
Distances Traveled by Cars (IE), 15; (E), 64
Do Air Bags Save Lives? (IE), 468, 470
Driving and Cell Phones (E), 427
Driving and Texting (E), 426
Femur Injury in a Car Crash (E), 645, 692; (R), 711
Flat Tire and Missed Class (E), 598
Ford and Mazda Producing Similar Transmissions (M), 503
Fuel Consumption Rate (BB), 11; (E), 101, 117, 498, 500, 674, 692
Head Injury in a Car Crash (R), 389; (E), 645, 655, 692
Highway Speeds (E), 444, 445
Jet Engines (M), 167
Lost Baggage (IE), 20
Motor Vehicles Produced in the U.S. (E), 577
Motorcycle Fatalities (CGA), 85
Motorcycle Helmets and Injuries (E), 24, 614; (CR), 46
Navigation Equipment Used in Aircraft (M), 362
Operational Life of an Airplane (M), 147
Overbooking Flights (E), 221, 233, 314; (TP), 250; (BB), 315
Pedestrian Walk Buttons (E), 16
Probability of a Car Crash (IE), 147
Reaction Time (E), 371, 384
Safe Loads in Aircraft and Boats (CP), 255; (E), 314; (IE), 432, 433, 434, 440
Safest Airplane Seats (M), 595
Safest Car Seats (M), 593
Seat Belt Use (E), 416, 474, 614
Tests of Child Booster Seats (E), 100, 116, 446
Titanic Survivors (E), 16, 60; (CR), 250; (DD), 628
Traffic Lights (R), 137
Train Derailments (E), 61, 75
Travel Time to Work (E), 43
Water Taxi Safety (IE), 296; (E), 280, 302

ELEMENTARY **STATISTICS** USING **THE TI-83/84** **PLUS CALCULATOR**

3RD EDITION

1-1 Review and Preview

1-2 Statistical Thinking

1-3 Types of Data

1-4 Critical Thinking

1-5 Collecting Sample Data

1-6 Introduction to the TI-83/84 Plus Calculator

1

Introduction to Statistics

Why was the *Literary Digest* poll so wrong?

Founded in 1890, the *Literary Digest* magazine was famous for its success in conducting polls to predict winners in presidential elections. The magazine correctly predicted the winners in the presidential elections of 1916, 1920, 1924, 1928, and 1932. In the 1936 presidential contest between Alf Landon and Franklin D. Roosevelt, the magazine sent out 10 million ballots and received 1,293,669 ballots for Landon and 972,897 ballots for Roosevelt, so it appeared that Landon would capture 57% of the vote. The size of this poll is extremely large when compared to the sizes of other typical polls, so it appeared that the poll would correctly predict the winner once again. James A. Farley, Chairman of the Democratic National Committee at the time, praised the poll by saying this: "Any sane person cannot escape the implication of such a gigantic sampling of popular opinion as is embraced in *The Literary Digest* straw vote. I consider this conclusive evidence as to the desire of the people of this country for a change in the National Government. *The Literary Digest* poll is an achievement of no little magnitude. It is a poll fairly and correctly conducted." Well, Landon received 16,679,583 votes to the 27,751,597 votes cast for Roosevelt. Instead of getting 57% of the vote as suggested by the *Literary Digest* poll, Landon received only 37% of the vote. The results for Roosevelt are shown in Figure 1-1. The *Literary Digest* magazine suffered a humiliating defeat and soon went out of business.

In that same 1936 presidential election, George Gallup used a much smaller poll of 50,000 subjects, and he correctly predicted that Roosevelt would win. How could it happen that the larger *Literary Digest* poll could be wrong by such a large margin? What went wrong? As you learn about the basics of statistics in this chapter, we will return to the *Literary Digest* poll and explain why it was so wrong in predicting the winner of the 1936 presidential contest.

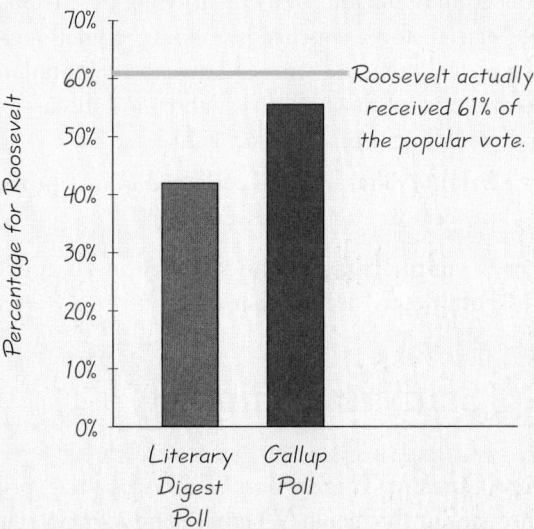

Figure 1-1 Poll Results for the Roosevelt–Landon Election

1-1 Review and Preview

The first section of each of the Chapters 1 through 14 begins with a brief review of what preceded the chapter, and a preview of what the chapter includes. This first chapter isn't preceded by much of anything except the Preface, and we won't review that (most people don't even read it in the first place). However, we can review and formally define some statistical terms that are commonly used. The Chapter Problem discussed the *Literary Digest* poll and George Gallup's poll, and both polls used sample data. Polls collect data from a small part of a larger group so that we can learn something about the larger group. This is a common and important goal of statistics: Learn about a large group by examining data from some of its members. In this context, the terms *sample* and *population* have special meanings. Formal definitions for these and other basic terms are given here.

DEFINITION

Data are collections of observations (such as measurements, genders, survey responses).

Statistics is the science of planning studies and experiments, obtaining data, and then organizing, summarizing, presenting, analyzing, interpreting, and drawing conclusions based on the data.

A **population** is the complete collection of all individuals (scores, people, measurements, and so on) to be studied. The collection is complete in the sense that it includes *all* of the individuals to be studied.

A **census** is the collection of data from *every* member of the population.

A **sample** is a *subcollection* of members selected from a population.

For example, the *Literary Digest* poll resulted in a sample of 2.3 million respondents. Those respondents constitute a *sample*, whereas the *population* consists of the entire collection of all adults eligible to vote. In this book we demonstrate how to use sample data to form conclusions about populations. It is *extremely* important to obtain sample data that are representative of the population from which the data are drawn. As we proceed through this chapter and discuss types of data and sampling methods, we should focus on these key concepts:

- **Sample data must be collected in an appropriate way, such as through a process of *random* selection.**

- **If sample data are not collected in an appropriate way, the data may be so completely useless that no amount of statistical torturing can salvage them.**

 ## 1-2 Statistical Thinking

Key Concept This section introduces basic principles of statistical thinking used throughout this book. Whether conducting a statistical analysis of data that we have collected, or analyzing a statistical analysis done by someone else, we should not rely on blind acceptance of mathematical calculations. We should consider these factors:

- Context of the data

- Source of the data

- Sampling method

• Conclusions

• Practical implications

In learning how to think statistically, common sense and practical considerations are typically much more important than implementation of cookbook formulas and calculations.

Statistics involves the analysis of data, so let's begin by considering the data in Table 1-1.

Table 1-1   Data Used for Analysis

x	56	67	57	60	64
y	53	66	58	61	68

After completing an introductory statistics course, we are armed with many statistical tools. In some cases, we are "armed and dangerous" if we jump in and start calculations without considering some critically important "big picture" issues. In order to properly analyze the data in Table 1-1, we must have some additional information. Here are some key questions that we might pose to get this information: What is the context of the data? What is the source of the data? How were the data obtained? What can we conclude from the data? Based on statistical conclusions, what practical implications result from our analysis?

Context As presented in Table 1-1, the data have no context. There is no description of what the values represent, where they came from, and why they were collected. Such a context is given in Example 1.

EXAMPLE 1

Context for Table 1-1 The data in Table 1-1 are taken from Data Set 3 in Appendix B. The entries in Table 1-1 are weights (in kilograms) of Rutgers students. The x values are weights measured in September of their freshman year, and the y values are their corresponding weights measured in April of the following spring semester. For example, the first student had a September weight of 56 kg and an April weight of 53 kg. These weights are included in a study described in "Changes in Body Weight and Fat Mass of Men and Women in the First Year of College: A Study of the 'Freshman 15,'" by Hoffman, Policastro, Quick, and Lee, *Journal of American College Health,* Vol. 55, No. 1. The title of the article tells us the goal of the study: Determine whether college students actually gain 15 pounds during their freshman year, as is commonly believed according to the "Freshman 15" legend.

The described context of the data in Table 1-1 shows that they consist of matched pairs. That is, each x-y pair of values has a "before" weight and an "after" weight for one particular student included in the study. An understanding of this context will directly affect the statistical procedures we use. Here, the key issue is whether the changes in weight appear to support or contradict the common belief that college students typically gain 15 lb during their freshman year. We can address this issue by using methods presented later in this book. (See Section 9-4 for dealing with matched pairs.)

If the values in Table 1-1 were numbers printed on the jerseys of Rutgers basketball players, where the x-values are from the men's team and the y-values are from the women's team, then this context would suggest that there is no meaningful statistical

Should You Believe a Statistical Study?

In *Statistical Reasoning for Everyday Life*, 3rd edition, authors Jeff Bennett, William Briggs, and Mario Triola list the following eight guidelines for critically evaluating a statistical study. (1) Identify the goal of the study, the population considered, and the type of study. (2) Consider the source, particularly with regard to a possibility of bias. (3) Analyze the sampling method. (4) Look for problems in defining or measuring the variables of interest. (5) Watch out for confounding variables that could invalidate conclusions. (6) Consider the setting and wording of any survey. (7) Check that graphs represent data fairly, and conclusions are justified. (8) Consider whether the conclusions achieve the goals of the study, whether they make sense, and whether they have practical significance.

Ethics in Statistics

Misuses of statistics often involve ethical issues. It was clearly unethical and morally and criminally wrong when researchers in Tuskegee, Alabama, withheld the effective penicillin treatment

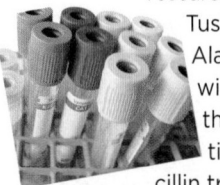

to syphilis victims so that the disease could be studied. That experiment continued for a period of 27 years.

Fabricating results is clearly unethical, but a more subtle ethical issue arises when authors of journal articles sometimes omit important information about the sampling method, or results from other data sets that do not support their conclusions. John Bailar was a statistical consultant to the *New England Journal of Medicine* when, after reviewing thousands of medical articles, he observed that statistical reviews often omitted critical information. The effect was that the authors' conclusions appear to be stronger than they should have been.

Some basic principles of ethics are: (1) all subjects in a study must give their informed consent; (2) all results from individuals must remain confidential; (3) the well-being of study subjects must always take precedence over the benefits to society.

procedure that could be used with the data (because the numbers don't measure or count anything). *Always consider the context of the data, because that context affects the statistical analysis that should be used.*

Source of Data   Consider the source of the data, and consider whether that source is likely to be objective or there is some incentive to be biased.

> **EXAMPLE 2**   **Source of the Data in Table 1-1** Reputable researchers from the Department of Nutritional Sciences at Rutgers University compiled the measurements in Table 1-1. The researchers have no incentive to distort or spin results to support some self-serving position. They have nothing to gain or lose by distorting results. They were not paid by a company that could profit from favorable results. We can be confident that these researchers are unbiased and they did not distort results.

Not all studies have such unbiased sources. For example, Kiwi Brands, a maker of shoe polish, commissioned a study that led to the conclusion that wearing scuffed shoes was the most common reason for a male job applicant to fail to make a good first impression. Physicians who receive funding from drug companies conduct some clinical experiments of drugs, so they have an incentive to obtain favorable results. Some professional journals, such as *Journal of the American Medical Association,* now require that physicians report such funding in journal articles. We should be vigilant and skeptical of studies from sources that may be biased.

Sampling Method If we are collecting sample data for a study, the sampling method that we choose can greatly influence the validity of our conclusions. Sections 1-4 and 1-5 will discuss sampling methods in more detail, but for now note that voluntary response (or self-selected) samples often have a bias, because those with a special interest in the subject are more likely to participate in the study. In a *voluntary response sample,* the respondents themselves decide whether to be included. For example, the ABC television show *Nightline* asked viewers to call with their opinion about whether the United Nations headquarters should remain in the United States. Viewers then decided themselves whether to call with their opinions, and those with strong feelings about the topic were more likely to call. We can use sound statistical methods to analyze voluntary response samples, but the results are not necessarily valid. There are other sampling methods, such as random sampling, that are more likely to produce good results. See the discussion of sampling strategies in Section 1-5.

> **EXAMPLE 3**   **Sampling Used for Table 1-1** The weights in Table 1-1 are from the larger sample of weights listed in Data Set 3 of Appendix B. Researchers obtained those data from subjects who were volunteers in a health assessment conducted in September of their freshman year. All of the 217 students who participated in the September assessment were invited for a follow-up in the spring, and 67 of those students responded and were measured again in the last two weeks of April. This sample is a voluntary response sample. The researchers wrote that "the sample obtained was not random and may have introduced self-selection bias." They elaborated on the potential for bias by specifically listing particular potential sources of bias, such as the response of "only those students who felt comfortable enough with their weight to be measured both times."

Not all studies and articles are so clear about the potential for bias. It is very common to encounter surveys that use self-selected subjects, yet the reports and conclusions fail to identify the limitations of such potentially biased samples.

Conclusions When forming a conclusion based on a statistical analysis, we should make statements that are clear to those without any understanding of statistics and its terminology. We should carefully avoid making statements not justified by the statistical analysis. For example, Section 10-2 introduces the concept of a *correlation,* or association between two variables, such as smoking and pulse rate. A statistical analysis might justify the statement that there is a correlation between the number of cigarettes smoked and pulse rate, but it would not justify a statement that the number of cigarettes smoked *causes* a person's pulse rate to change. Correlation does not imply causality.

EXAMPLE 4 **Conclusions from Data in Table 1-1** Table 1-1 lists before and after weights of five subjects taken from Data Set 3 in Appendix B. Those weights were analyzed with conclusions included in "Changes in Body Weight and Fat Mass of Men and Women in the First Year of College: A Study of the 'Freshman 15,'" by Hoffman, Policastro, Quick, and Lee, *Journal of American College Health,* Vol. 55, No. 1. In analyzing the data in Table 1-1, the investigators concluded that the freshman year of college is a time during which weight gain occurs. But the investigators went on to state that in the small nonrandom group studied, the weight gain was less than 15 pounds, and this amount was not universal. They concluded that the "Freshman 15" weight gain is a myth.

Practical Implications In addition to clearly stating conclusions of the statistical analysis, we should also identify any practical implications of the results.

EXAMPLE 5 **Practical Implications from Data in Table 1-1** In their analysis of the data collected in the "Freshman 15" study, the researchers point out some practical implications of their results. They wrote that "it is perhaps most important for students to recognize that seemingly minor and perhaps even harmless changes in eating or exercise behavior may result in large changes in weight and body fat mass over an extended period of time." Beginning freshman college students should recognize that there could be serious health consequences resulting from radically different diet and exercise routines.

The *statistical significance* of a study can differ from its *practical significance.* It is possible that, based on the available sample data, methods of statistics can be used to reach a conclusion that some treatment or finding is effective, but common sense might suggest that the treatment or finding does not make enough of a difference to justify its use or to be practical.

EXAMPLE 6 **Statistical Significance versus Practical Significance** In a test of the Atkins weight loss program, 40 subjects using that program had a mean weight loss of 2.1 lb after one year (based on data from "Comparison of the Atkins, Ornish,

continued

Weight Watchers, and Zone Diets for Weight Loss and Heart Disease Risk Reduction," by Dansinger et al., *Journal of the American Medical Association,* Vol. 293, No. 1). Using formal methods of statistical analysis, we can conclude that the mean weight loss of 2.1 is statistically significant. That is, based on statistical criteria, the diet appears to be effective. However, using common sense, it does not seem worthwhile to pursue a weight loss program resulting in such relatively insignificant results. Someone starting a weight loss program would likely want to lose considerably more than 2.1 lb. Although the mean weight loss of 2.1 lb is statistically significant, it does not have practical significance. The statistical analysis suggests that the weight loss program is effective, but practical considerations suggest that the program is basically ineffective.

Statistical Significance *Statistical significance* is a concept we will consider at length throughout this book. To prepare for those discussions, Examples 7 and 8 illustrate the concept in a simple setting.

> **EXAMPLE 7** **Statistical Significance** The Genetics and IVF Institute in Fairfax, Virginia developed a technique called MicroSort, which supposedly increases the chances of a couple having a baby girl. In a preliminary test, researchers located 14 couples who wanted baby girls. After using the MicroSort technique, 13 of them had girls and one couple had a boy. After obtaining these results, we have two possible conclusions:
>
> 1. The MicroSort technique is not effective and the result of 13 girls in 14 births occurred by chance.
>
> 2. The MicroSort technique is effective, and couples who use the technique are more likely to have baby girls, as claimed by the Genetics and IVF Institute.
>
> When choosing between the two possible explanations for the results, statisticians consider the *likelihood* of getting the results by chance. They are able to determine that if the MicroSort technique has no effect, then there is about 1 chance in 1000 of getting results like those obtained here. Because that likelihood is so small, statisticians conclude that the results are statistically significant, so it appears that the MicroSort technique is effective.

> **EXAMPLE 8** **Statistical Significance** Instead of the result in Example 7, suppose the couples had 8 baby girls in 14 births. We can see that 8 baby girls is more than the 7 girls that we would expect with an ineffective treatment. However, statisticians can determine that if the MicroSort technique has no effect, then there are roughly two chances in five of getting 8 girls in 14 births. Unlike the one chance in 1000 from the preceding example, two chances in five indicates that the results could *easily occur by chance.* This would indicate that the result of 8 girls in 14 births is *not statistically significant.* With 8 girls in 14 births, we would not conclude that the technique is effective, because it is so easy (two chances in five) to get the results with an ineffective treatment or no treatment.

What Is Statistical Thinking? Statisticians universally agree that statistical thinking is good, but there are different views of what actually constitutes statistical thinking. In this section we have described statistical thinking in terms of the ability to see the big picture and to consider such relevant factors as context, source of data, and sampling method, and to form conclusions and identify practical implications. Statistical thinking involves critical thinking and the ability to make sense of results. Statistical thinking might involve determining whether results are statistically significant, as in Examples 7 and 8. Statistical thinking is so much more than the mere ability to execute complicated calculations. Through numerous examples, exercises, and discussions, this book will develop the statistical thinking skills that are so important in today's world.

1-2 Basic Skills and Concepts

Statistical Literacy and Critical Thinking

1. Voluntary Response Sample What is a voluntary response sample?

2. Voluntary Response Sample Why is a voluntary response sample generally not suitable for a statistical study?

3. Statistical Significance versus Practical Significance What is the difference between statistical significance and practical significance?

4. Context of Data You have collected a large sample of values. Why is it important to understand the *context* of the data?

5. Statistical Significance versus Practical Significance In a study of the Weight Watchers weight loss program, 40 subjects lost a mean of 3.0 lb after 12 months (based on data from "Comparison of the Atkins, Ornish, Weight Watchers, and Zone Diets for Weight Loss and Heart Disease Risk Reduction," by Dansinger et al., *Journal of the American Medical Association,* Vol. 293, No. 1). Methods of statistics can be used to verify that the diet is effective. Does the Weight Watchers weight loss program have statistical significance? Does it have practical significance? Why or why not?

6. Sampling Method In the study of the Weight Watchers weight loss program from Exercise 5, subjects were found using the method described as follows: "We recruited study candidates from the Greater Boston area using newspaper advertisements and television publicity." Is the sample a voluntary response sample? Why or why not?

In Exercises 7–14, use common sense to determine whether the given event is (a) impossible; (b) possible, but very unlikely; (c) possible and likely.

7. Super Bowl The New York Giants beat the Denver Broncos in the Super Bowl by a score of 120 to 98.

8. Speeding Ticket While driving to his home in Connecticut, David Letterman was ticketed for driving 205 mi/h on a highway with a speed limit of 55 mi/h.

9. Traffic Lights While driving through a city, Mario Andretti arrived at three consecutive traffic lights and they were all green.

10. Thanksgiving Thanksgiving day will fall on a Monday next year.

11. Supreme Court All of the justices on the United States Supreme Court have the same birthday.

12. Calculators When each of 25 statistics students turns on his or her TI-84 Plus calculator, all 25 calculators operate successfully.

13. Lucky Dice Steve Wynn rolled a pair of dice and got a total of 14.

14. Slot Machine Wayne Newton hit the jackpot on a slot machine each time in ten consecutive attempts.

In Exercises 15–18, refer to the data in the table below. The x-values are nicotine amounts (in mg) in different 100 mm filtered, non-"light" menthol cigarettes; the y-values are nicotine amounts (in mg) in different king-size nonfiltered, nonmenthol, and non-"light" cigarettes. (The values are from Data Set 4 in Appendix B.)

Nicotine Amounts from Menthol and King-Size Cigarettes

x	1.1	0.8	1.0	0.9	0.8
y	1.1	1.7	1.7	1.1	1.1

15. Context of the Data Refer to the table of nicotine amounts. Is each x value matched with a corresponding y value, as in Table 1-1 on page 5? That is, is each x value associated with the corresponding y value in some meaningful way? If the x and y values are not matched, does it make sense to use the difference between each x value and the y value that is in the same column?

16. Source of the Data The Federal Trade Commission obtained the measured amounts of nicotine in the table. Is the source of the data likely to be unbiased?

17. Conclusion Note that the table lists measured nicotine amounts from two different types of cigarette. Given these data, what issue can be addressed by conducting a statistical analysis of the values?

18. Conclusion If we use suitable methods of statistics, we conclude that the average (mean) nicotine amount of the 100 mm filtered non-"light" menthol cigarettes is less than the average (mean) nicotine amount of the king-size nonfiltered, nonmenthol, non-"light" cigarettes. Can we conclude that the first type of cigarette is safe? Why or why not?

In Exercises 19–22, refer to the data in the table below. The x-values are weights (in pounds) of cars; the y-values are the corresponding highway fuel consumption amounts (in mi/gal). (The values are from Data Set 16 in Appendix B.)

Car Weights and Highway Fuel Consumption Amounts

Weight (lb)	4035	3315	4115	3650	3565
Highway Fuel Consumption (mi/gal)	26	31	29	29	30

19. Context of the Data Refer to the given table of car measurements. Are the x values matched with the corresponding y values, as in Table 1-1 on page 5? That is, is each x value somehow associated with the corresponding y value in some meaningful way? If the x and y values are matched, does it make sense to use the difference between each x value and the y value that is in the same column? Why or why not?

20. Conclusion Given the context of the car measurement data, what issue can be addressed by conducting a statistical analysis of the values?

21. Source of the Data Comment on the source of the data if you are told that car manufacturers supplied the values. Is there an incentive for car manufacturers to report values that are not accurate?

22. Conclusion If we use statistical methods to conclude that there is a correlation (or relationship or association) between the weights of cars and the amounts of fuel consumption, can we conclude that adding weight to a car causes it to consume more fuel?

In Exercises 23–26, form a conclusion about statistical significance. Do not make any formal calculations. Either use results provided or make subjective judgments about the results.

23. Statistical Significance In a study of the Ornish weight loss program, 40 subjects lost a mean of 3.3 lb after 12 months (based on data from "Comparison of the Atkins, Ornish, Weight Watchers, and Zone Diets for Weight Loss and Heart Disease Risk Reduction," by

Dansinger et al., *Journal of the American Medical Association,* Vol. 293, No. 1). Methods of statistics can be used to show that if this diet had no effect, the likelihood of getting these results is roughly 3 chances in 1000. Does the Ornish weight loss program have statistical significance? Does it have practical significance? Why or why not?

24. Mendel's Genetics Experiments One of Gregor Mendel's famous hybridization experiments with peas yielded 580 offspring with 152 of those peas (or 26%) having yellow pods. According to Mendel's theory, 25% of the offspring peas should have yellow pods. Do the results of the experiment differ from Mendel's claimed rate of 25% by an amount that is statistically significant?

25. Secondhand Smoke Survey In a Gallup poll of 1038 randomly selected adults, 85% said that secondhand smoke is somewhat harmful or very harmful, but a representative of the tobacco industry claims that only 50% of adults believe that secondhand smoke is somewhat harmful or very harmful. Is there statistically significant evidence against the representative's claim? Why or why not?

26. Surgery versus Splints A study compared surgery and splinting for subjects suffering from carpal tunnel syndrome. It was found that among 73 patients treated with surgery, there was a 92% success rate. Among 83 patients treated with splints, there was a 72% success rate. Calculations using those results showed that if there really is no difference in success rates between surgery and splints, then there is about 1 chance in 1000 of getting success rates like the ones obtained in this study.

a. Should we conclude that surgery is better than splints for the treatment of carpal tunnel syndrome?

b. Does the result have statistical significance? Why or why not?

c. Does the result have practical significance?

d. Should surgery be the recommended treatment for carpal tunnel syndrome?

1-2   Beyond the Basics

27. Conclusions Refer to the city and highway fuel consumption amounts of different cars listed in Data Set 16 of Appendix B. Compare the city fuel consumption amounts and the highway fuel consumption amounts, then answer the following questions without doing any calculations.

a. Does the conclusion that the highway amounts are greater than the city amounts appear to be supported with statistical significance?

b. Does the conclusion that the highway amounts are greater than the city amounts appear to have practical significance?

c. What is a practical implication of a substantial difference between city fuel consumption amounts and highway fuel consumption amounts?

28. ATV Accidents The Associated Press provided an article with the headline, "ATV accidents killed 704 people in '04." The article noted that this is a new record high, and compares it to 617 ATV deaths the preceding year. Other data about the frequencies of injuries were included. What important value was not included? Why is it important?

1-3   Types of Data

Key Concept A goal of statistics is to make inferences, or generalizations, about a population. In addition to the terms *population* and *sample,* which we defined at the start of this chapter, we need to know the meanings of the terms *parameter* and *statistic.* These new terms are used to distinguish between cases in which we have data for an entire population, and cases in which we have data for a sample only.

Origin of "Statistics"

The word *statistics* is derived from the Latin word *status* (meaning "state"). Early uses of statistics involved compilations of data and graphs describing various aspects of a state or country. In 1662, John Graunt published statistical information about births and deaths. Graunt's work was followed by studies of mortality and disease rates, population sizes, incomes, and unemployment rates. Households, governments, and businesses rely heavily on statistical data for guidance. For example, unemployment rates, inflation rates, consumer indexes, and birth and death rates are carefully compiled on a regular basis, and the resulting data are used by business leaders to make decisions affecting future hiring, production levels, and expansion into new markets.

We also need to know the difference between *quantitative data* and *categorical data,* which distinguish between different types of numbers. Some numbers, such as those on the shirts of basketball players, are not quantities because they don't measure or count anything, and it would not make sense to perform calculations with such numbers. In this section we describe different types of data; the type of data determines the statistical methods we use in our analysis.

In Section 1-1 we defined the terms *population* and *sample*. The following two terms are used to distinguish between cases in which we have data for an entire population, and cases in which we have data for a sample only.

> **DEFINITION**
>
> A **parameter** is a numerical measurement describing some characteristic of a *population*.
>
> A **statistic** is a numerical measurement describing some characteristic of a *sample*.

> **EXAMPLE 1**
>
> 1. **Parameter:** There are exactly 100 Senators in the 109th Congress of the United States, and 55% of them are Republicans. The figure of 55% is a *parameter* because it is based on the entire population of all 100 Senators.
>
> 2. **Statistic:** In 1936, *Literary Digest* polled 2.3 million adults in the United States, and 57% said that they would vote for Alf Landon for the presidency. That figure of 57% is a *statistic* because it is based on a sample, not the entire population of all adults in the United States.

Some data sets consist of numbers representing counts or measurements (such as heights of 60 inches and 72 inches), whereas others are nonnumerical (such as eye colors of green and brown). The terms *quantitative data* and *categorical data* distinguish between these types.

> **DEFINITION**
>
> **Quantitative** (or **numerical**) **data** consist of *numbers* representing counts or measurements.
>
> **Categorical** (or **qualitative** or **attribute**) **data** consist of names or labels that are not numbers representing counts or measurements.

> **EXAMPLE 2**
>
> 1. **Quantitative Data:** The ages (in years) of survey respondents
>
> 2. **Categorical Data:** The political party affiliations (Democrat, Republican, Independent, other) of survey respondents
>
> 3. **Categorical Data:** The numbers 24, 28, 17, 54, and 31 are sewn on the shirts of the LA Lakers starting basketball team. These numbers are substitutes for names. They don't count or measure anything, so they are categorical data.

When we organize and report quantitative data, it is important to use the appropriate units of measurement, such as dollars, hours, feet, or meters. When we examine statistical data that others report, we must observe the information given about the units of measurement used, such as "all amounts are in *thousands of dollars*," "all times are in *hundredths of a second*," or "all units are in *kilograms*," to interpret the data correctly. To ignore such units of measurement could lead to very wrong conclusions. NASA lost its $125 million Mars Climate Orbiter when it crashed because the controlling software had acceleration data in *English* units, but they were incorrectly assumed to be in *metric* units.

Quantitative data can be further described by distinguishing between *discrete* and *continuous* types.

> **DEFINITION**
>
> **Discrete data** result when the number of possible values is either a finite number or a "countable" number. (That is, the number of possible values is 0 or 1 or 2, and so on.)
>
> **Continuous (numerical) data** result from infinitely many possible values that correspond to some continuous scale that covers a range of values without gaps, interruptions, or jumps.

EXAMPLE 3

1. **Discrete Data:** The numbers of eggs that hens lay are *discrete* data because they represent counts.

2. **Continuous Data:** The amounts of milk from cows are *continuous* data because they are measurements that can assume any value over a continuous span. During a year, a cow might yield an amount of milk that can be any value between 0 and 7000 liters. It would be possible to get 5678.1234 liters because the cow is not restricted to the discrete amounts of 0, 1, 2, . . . , 7000 liters.

When describing smaller amounts, correct grammar dictates that we use "fewer" for discrete amounts, and "less" for continuous amounts. It is correct to say that we drank *fewer* cans of cola and, in the process, we drank *less* cola. The numbers of cans of cola are discrete data, whereas the volume amounts of cola are continuous data.

Another common way of classifying data is to use four levels of measurement: nominal, ordinal, interval, and ratio. In applying statistics to real problems, the level of measurement of the data helps us decide which procedure to use. There will be some references to these levels of measurement in this book, but the important point here is based on common sense: Don't do computations and don't use statistical methods that are not appropriate for the data. For example, it would not make sense to compute an average of Social Security numbers, because those numbers are data that are used for identification, and they don't represent measurements or counts of anything.

> **DEFINITION**
>
> The **nominal level of measurement** is characterized by data that consist of names, labels, or categories only. The data cannot be arranged in an ordering scheme (such as low to high).

Measuring Disobedience

How are data collected about something that doesn't seem to be measurable, such as people's level of disobedience? Psychologist Stanley Milgram devised the following experiment: A researcher instructed a volunteer subject to operate a control board that gave increasingly painful "electrical shocks" to a third person. Actually, no real shocks were given, and the third person was an actor. The volunteer began with 15 volts and was instructed to increase the shocks by increments of 15 volts. The disobedience level was the point at which the subject refused to increase the voltage. Surprisingly, two-thirds of the subjects obeyed orders even though the actor screamed and faked a heart attack.

EXAMPLE 4   Here are examples of sample data at the nominal level of measurement.

1. **Yes/no/undecided:** Survey responses of *yes, no,* and *undecided* (as in the Chapter Problem)

2. **Political Party:** The political party affiliations of survey respondents (Democrat, Republican, Independent, other)

Because nominal data lack any ordering or numerical significance, they should not be used for calculations. Numbers such as 1, 2, 3, and 4 are sometimes assigned to the different categories (especially when data are coded for computers), but these numbers have no real computational significance and any average calculated from them is meaningless.

DEFINITION

Data are at the **ordinal level of measurement** if they can be arranged in some order, but differences (obtained by subtraction) between data values either cannot be determined or are meaningless.

 EXAMPLE 5   Here are examples of sample data at the ordinal level of measurement.

1. **Course Grades:** A college professor assigns grades of A, B, C, D, or F. These grades can be arranged in order, but we can't determine differences between the grades. For example, we know that A is higher than B (so there is an ordering), but we cannot subtract B from A (so the difference cannot be found).

2. **Ranks:** *U.S. News and World Report* ranks colleges. Those ranks (first, second, third, and so on) determine an ordering. However, the differences between ranks are meaningless. For example, a difference of "second minus first" might suggest $2 - 1 = 1$, but this difference of 1 is meaningless because it is not an exact quantity that can be compared to other such differences. The *difference* between Harvard and Brown cannot be quantitatively compared to the *difference* between Yale and Johns Hopkins.

Ordinal data provide information about relative comparisons, but not the magnitudes of the differences. Usually, ordinal data should not be used for calculations such as an average, but this guideline is sometimes violated (such as when we use letter grades to calculate a grade-point average).

DEFINITION

The **interval level of measurement** is like the ordinal level, with the additional property that the difference between any two data values is meaningful. However, data at this level do not have a *natural* zero starting point (where *none* of the quantity is present).

EXAMPLE 6 These examples illustrate the interval level of measurement.

1. **Temperatures:** Body temperatures of 98.2°F and 98.6°F are examples of data at this interval level of measurement. Those values are ordered, and we can determine their difference of 0.4°F. However, there is no natural starting point. The value of 0°F might seem like a starting point, but it is arbitrary and does not represent the total absence of heat.

2. **Years:** The years 1492 and 1776. (Time did not begin in the year 0, so the year 0 is arbitrary instead of being a natural zero starting point representing "no time.")

DEFINITION

The **ratio level of measurement** is the interval level with the additional property that there is also a natural zero starting point (where zero indicates that *none* of the quantity is present). For values at this level, differences and ratios are both meaningful.

EXAMPLE 7 The following are examples of data at the ratio level of measurement. Note the presence of the natural zero value, and also note the use of meaningful ratios of "twice" and "three times."

1. **Distances:** Distances (in km) traveled by cars (0 km represents no distance traveled, and 400 km is twice as far as 200 km.)

2. **Prices:** Prices of college textbooks ($0 does represent no cost, and a $100 book does cost *twice* as much as a $50 book.)

Hint: This level of measurement is called the ratio level because the zero starting point makes ratios meaningful, so here is an easy test to determine whether values are at the ratio level: Consider two quantities where one number is twice the other, and ask whether "twice" can be used to correctly describe the quantities. Because a 400-km distance is *twice* as far as a 200-km distance, the distances are at the ratio level. In contrast, 50°F is *not twice* as hot as 25°F, so Fahrenheit temperatures are *not* at the ratio level. For a concise comparison and review, see Table 1-2.

Table 1-2 Levels of Measurement

Ratio:	There is a natural zero starting point and ratios are meaningful.	*Example:* Distances ~~Time spent in prison~~
Interval:	Differences are meaningful, but there is no natural zero starting point and ratios are meaningless.	*Example:* Body temperatures in degrees Fahrenheit or Celsius
Ordinal:	Categories are ordered, but differences can't be found or are meaningless.	*Example:* Ranks of colleges in *U.S. News and World Report*
Nominal:	Categories only. Data cannot be arranged in an ordering scheme.	*Example:* Eye colors

Hint: Consider the quantities where one is twice the other, and ask whether "twice" can be used to correctly describe the quantities. If yes, then the ratio level applies.

1-3   Basic Skills and Concepts

Statistical Literacy and Critical Thinking

1. Parameter and Statistic How do a parameter and a statistic differ?

2. Quantitative/Categorical Data How do quantitative data and categorical data differ?

3. Discrete/Continuous Data How do discrete data and continuous data differ?

4. Identifying the Population Researchers studied a sample of 877 surveyed executives and found that 45% of them would not hire someone with a typographic error on a job application. Is the 45% value a statistic or a parameter? What is the population? What is a practical implication of the result of this survey?

In Exercises 5–12, determine whether the given value is a statistic or a parameter.

5. Income and Education In a large sample of households, the median annual income per household for high school graduates is $19,856 (based on data from the U.S. Census Bureau).

6. Politics Among the Senators in the current Congress, 44% are Democrats.

7. Titanic A study of all 2223 passengers aboard the *Titanic* found that 706 survived when it sank.

8. Pedestrian Walk Buttons In New York City, there are 3250 walk buttons that pedestrians can press at traffic intersections. It was found that 77% of those buttons do not work (based on data from the article "For Exercise in New York Futility, Push Button," by Michael Luo, *New York Times*).

9. Areas of States If the areas of the 50 states are added and the sum is divided by 50, the result is 196,533 square kilometers.

10. Periodic Table The average (mean) atomic weight of all elements in the periodic table is 134.355 unified atomic mass units.

11. Voltage The author measured the voltage supplied to his home on 40 different days, and the average (mean) value is 123.7 volts.

12. Movie Gross The author randomly selected 35 movies and found the amount of money that they grossed from ticket sales. The average (mean) is $123.7 million.

In Exercises 13–20, determine whether the given values are from a discrete or continuous data set.

13. Pedestrian Buttons In New York City, there are 3250 walk buttons that pedestrians can press at traffic intersections, and 2500 of them do not work (based on data from the article "For Exercise in New York Futility, Push Button," by Michael Luo, *New York Times*).

14. Poll Results In the *Literary Digest* poll, Landon received 16,679,583 votes.

15. Cigarette Nicotine The amount of nicotine in a Marlboro cigarette is 1.2 mg.

16. Coke Volume The volume of cola in a can of regular Coke is 12.3 oz.

17. Gender Selection In a test of a method of gender selection developed by the Genetics & IVF Institute, 726 couples used the XSORT method and 668 of them had baby girls.

18. Blood Pressure When a woman is randomly selected and measured for blood pressure, the systolic blood pressure is found to be 61 mm Hg.

19. Car Weight When a Cadillac STS is randomly selected and weighed, it is found to weigh 1827.9 kg.

20. Car Cylinders A car is randomly selected at a traffic safety checkpoint, and the car has 6 cylinders.

In Exercises 21–28, determine which of the four levels of measurement (nominal, ordinal, interval, ratio) is most appropriate.

21. Voltage measurements from the author's home (listed in Data Set 13 in Appendix B)

22. Types of movies (drama, comedy, adventure, documentary, etc.)

23. Critic ratings of movies on a scale from 0 star to 4 stars

24. Actual temperatures (in degrees Fahrenheit) as listed in Data Set 11 in Appendix B

25. Companies (Disney, MGM, Warner Brothers, Universal, 20th Century Fox) that produced the movies listed in Data Set 7 in Appendix B

26. Measured amounts of greenhouse gases (in tons per year) emitted by cars listed in Data Set 16 in Appendix B

27. Years in which movies were released, as listed in Data Set 9 in Appendix B

28. Ranks of cars evaluated by Consumer's Union

In Exercises 29–32, identify the (a) **sample** *and (b)* **population**. *Also, determine whether the sample is likely to be representative of the population.*

29. USA Today Survey The newspaper *USA Today* published a health survey, and some readers completed the survey and returned it.

30. Cloning Survey A Gallup poll of 1012 randomly surveyed adults found that 9% of them said cloning of humans should be allowed.

31. Some people responded to this request: "Dial 1-900-PRO-LIFE to participate in a telephone poll on abortion. ($1.95 per minute. Average call: 2 minutes. You must be 18 years old.)"

32. AOL Survey America Online asked subscribers to respond to this question: "Which slogan do you hate the most?" Responders were given several slogans used to promote car sales, and Volkswagon's slogan received 55% of the 33,160 responses. The Volkswagon slogan was "Relieves gas pains."

1-3 Beyond the Basics

33. Interpreting Temperature Increase In the *Born Loser* cartoon strip by Art Sansom, Brutus expresses joy over an increase in temperature from $1°$ to $2°$. When asked what is so good about $2°$, he answers that "it's twice as warm as this morning." Explain why Brutus is wrong yet again.

34. Interpreting Poll Results For the poll described in the Chapter Problem, assume that the respondents had been asked for their political party affiliation, and the responses were coded as 0 (for Democrat), 1 (for Republican), 2 (for Independent), or 3 (for any other response). If we calculate the average (mean) of the numbers and get 0.95, how can that value be interpreted?

35. Scale for Rating Food A group of students develops a scale for rating the quality of cafeteria food, with 0 representing "neutral: not good and not bad." Bad meals are given negative numbers and good meals are given positive numbers, with the magnitude of the number corresponding to the severity of badness or goodness. The first three meals are rated as 2, 4, and -5. What is the level of measurement for such ratings? Explain your choice.

1-4 Critical Thinking

Key Concept This section is the first of many throughout the book in which we focus on the meaning of information obtained by studying data. The aim of this section is to improve our skills in interpreting information based on data. It's easy to enter data into a computer and get results; unless the data have been chosen carefully, however, the result may be "GIGO"—garbage in, garbage out. Instead of blindly using formulas and procedures, we must *think carefully* about the context of the data, the source of the data, the method used in data collection, the conclusions reached,

Misleading Statistics in Journalism

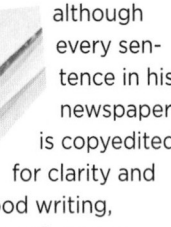

New York Times reporter Daniel Okrant wrote that although every sentence in his newspaper is copyedited for clarity and good writing, "numbers, so alien to so many, don't get nearly this respect. The paper requires no specific training to enhance numeracy, and no specialists whose sole job is to foster it." He cites an example of the *New York Times* reporting about an estimate of more than $23 billion that New Yorkers spend for counterfeit goods each year. Okrant writes that "quick arithmetic would have demonstrated that $23 billion would work out to roughly $8000 per city household, a number ludicrous on its face."

and the practical implications. This section shows how to use common sense to think critically about data and statistics.

Although this section focuses on misuse of statistics, this is not a book about the misuse of statistics. The remainder of this book will investigate the very meaningful uses of valid statistical methods. We will learn general methods for using sample data to make inferences about populations; we will learn about polls and sample sizes; and we will learn about important measures of key characteristics of data.

Quotes like the following are often used to describe the misuse of statistics.

- "There are three kinds of lies: lies, damned lies, and statistics."—Benjamin Disraeli

- "Figures don't lie; liars figure."—Attributed to Mark Twain

- "Some people use statistics as a drunken man uses lampposts—for support rather than illumination."—Historian Andrew Lang

- "Statistics can be used to support anything—especially statisticians." —Franklin P. Jones

- Definition of a statistician: "A specialist who assembles figures and then leads them astray."—*Esar's Comic Dictionary*

- "There are two kinds of statistics, the kind you look up, and the kind you make up."—Rex Stout

- "58.6% of all statistics are made up on the spot."—Unknown

There are typically two ways in which the science of statistics is used for deception: (1) evil intent on the part of dishonest persons; (2) unintentional errors on the part of people who don't know any better. As responsible citizens and as more valuable professional employees, we should learn to distinguish between statistical conclusions that are likely to be valid and those that are seriously flawed, regardless of the source.

Graphs/Misuse of Graphs   Statistical data are often presented in visual form— that is, in graphs. Data represented graphically must be interpreted carefully, and we will discuss graphing in Section 2-5. In addition to learning how to organize your own data in graphs, we will examine misleading graphs.

Bad Samples   Some samples are bad in the sense that the method used to collect the data dooms the sample, so that it is likely to be somehow *biased*. That is, it is not representative of the population from which it has been obtained. The following definition refers to one of the most common and most serious misuses of statistics.

DEFINITION

A **voluntary response sample** (or **self-selected sample**) is one in which the respondents themselves decide whether to be included.

CAUTION

Do not use voluntary response sample data for making conclusions about a population.

 EXAMPLE 1   **Voluntary Response Sample** *Newsweek* magazine ran a survey about the Napster Web site, which had been providing free access to downloading copies of music CDs. Readers were asked this question: "Will you still use Napster if you have to pay a fee?" Readers could register their responses on the Web site

newsweek.msnbc.com. Among the 1873 responses received, 19% said yes, it is still cheaper than buying CDs. Another 5% said yes, they felt more comfortable using it with a charge. When *Newsweek* or anyone else runs a poll on the Internet, individuals decide themselves whether to participate, so they constitute a voluntary response sample. But people with strong opinions are more likely to participate, so it is very possible that the responses are not representative of the whole population.

These are common examples of voluntary response samples which, by their very nature, are seriously flawed because we should not make conclusions about a population based on such a biased sample:

- Polls conducted through the Internet, in which subjects can decide whether to respond

- Mail-in polls, in which subjects can decide whether to reply

- Telephone call-in polls, in which newspaper, radio, or television announcements ask that you voluntarily call a special number to register your opinion

With such voluntary response samples, we can only make valid conclusions about the specific group of people who chose to participate, but a common practice is to incorrectly state or imply conclusions about a larger population. From a statistical viewpoint, such a sample is fundamentally flawed and should not be used for making general statements about a larger population.

EXAMPLE 2 **What went wrong in the *Literary Digest* poll?** *Literary Digest* magazine conducted its poll by sending out 10 million ballots. The magazine received 2.3 million responses. The poll results suggested incorrectly that Alf Landon would win the presidency. In his much smaller poll of 50,000 people, George Gallup correctly predicted that Franklin D. Roosevelt would win. The lesson here is that it is not necessarily the *size* of the sample that makes it effective, but it is the *sampling method*. The *Literary Digest* ballots were sent to magazine subscribers as well as to registered car owners and those who used telephones. On the heels of the Great Depression, this group included disproportionately more wealthy people, who were Republicans. But the real flaw in the *Literary Digest* poll is that it resulted in a voluntary response sample. Gallup used an approach in which he obtained a representative sample based on demographic factors. (Gallup modified his methods when he made a wrong prediction in the famous 1948 Dewey/Truman election. Gallup stopped polling too soon, and he failed to detect a late surge in support for Truman.) The *Literary Digest* poll is a classic illustration of the flaws inherent in basing conclusions on a voluntary response sample.

Correlation and Causality Another way to misinterpret statistical data is to find a statistical association between two variables and to conclude that one of the variables *causes* (or directly affects) the other variable. Recall that earlier we mentioned that it may seem as if two variables, such as smoking and pulse rate, are linked. This relationship is called a *correlation*. But even if we found that the number of cigarettes was linked to pulse rate, we could not conclude that one variable caused the other. Specifically, *correlation does not imply causality*.

Detecting Phony Data

A class is given the homework assignment of recording the results when a coin is tossed 500 times. One dishonest student decides to save time by just making up the results instead of actually flipping a coin. Because people generally cannot make up results that are really random, we can often identify such phony data. With 500 tosses of an actual coin, it is extremely likely that you will get a run of six heads or six tails, but people almost never include such a run when they make up results.

Another way to detect fabricated data is to establish that the results violate Benford's law: For many collections of data, the leading digits are not uniformly distributed. Instead, the leading digits of 1, 2, . . . , 9 occur with rates of 30%, 18%, 12%, 10%, 8%, 7%, 6%, 5%, and 5%, respectively. (See "The Difficulty of Faking Data," by Theodore Hill, *Chance*, Vol. 12, No. 3.)

Publication Bias

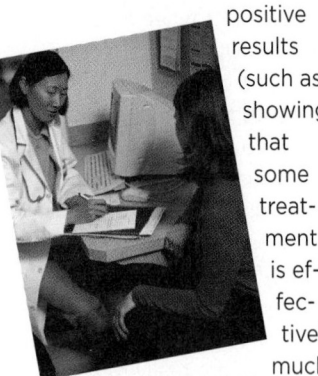

There is a "publication bias" in professional journals. It is the tendency to publish positive results (such as showing that some treatment is effective) much more often than negative results (such as showing that some treatment has no effect). In the article "Registering Clinical Trials" (*Journal of the American Medical Association*, Vol. 290, No. 4), authors Kay Dickersin and Drummond Rennie state that "the result of not knowing who has performed what (clinical trial) is loss and distortion of the evidence, waste and duplication of trials, inability of funding agencies to plan, and a chaotic system from which only certain sponsors might benefit, and is invariably against the interest of those who offered to participate in trials and of patients in general." They support a process in which *all* clinical trials are registered in one central system.

CAUTION

Do not use a correlation between two variables as a justification for concluding that one of the variables is the cause of the other.

The media frequently report a newfound correlation with wording that directly indicates or implies that one of the variables is the cause of the other, but such media reports are wrong.

Reported Results When collecting data from people, it is better to take measurements yourself instead of asking subjects to report results. Ask people what they weigh and you are likely to get their *desired* weights, not their actual weights. If you really want accurate weight data, use a scale and weigh the people.

EXAMPLE 3 **Voting Behavior** When 1002 eligible voters were surveyed, 70% of them said that they had voted in a recent presidential election (based on data from ICR Research Group). However, voting records show that only 61% of eligible voters actually did vote.

Small Samples Conclusions should not be based on samples that are far too small.

EXAMPLE 4 **Small Sample** The Children's Defense Fund published *Children Out of School in America,* in which it was reported that among secondary school students suspended in one region, 67% were suspended at least three times. But that figure is based on a sample of only *three* students! Media reports failed to mention that this sample size was so small. (In Chapters 7 and 8 you will see that we can *sometimes* make some inferences from small samples, but we should be careful to verify that the necessary requirements are satisfied.)

Sometimes a sample might seem relatively large (as in a survey of "2000 randomly selected adult Americans"), but if conclusions are made about subgroups, such as the 21-year-old male Republicans from Pocatello, such conclusions might be based on samples that are too small. Although it is important to have a sample that is sufficiently large, it is just as important to have sample data that have been collected in an appropriate way. Even large samples can be bad samples.

Percentages Some studies will cite misleading or unclear percentages. Keep in mind that 100% of some quantity is *all* of it, but if there are references made to percentages that exceed 100%, such references are often not justified.

EXAMPLE 5 **Misused Percentage** In referring to lost baggage, Continental Airlines ran ads claiming that this was "an area where we've already improved 100% in the last six months." In an editorial criticizing this statistic, the *New York Times* correctly interpreted the 100% improvement to mean that no baggage is now being lost—an accomplishment not yet enjoyed by Continental Airlines.

The following list identifies some key principles to use when dealing with percentages. These principles all use the basic notion that % or "percent" really means "divided by 100." The first principle is used often in this book.

- **Percentage of:** To find a *percentage of* an amount, drop the % symbol and divide the percentage value by 100, then multiply. This example shows that 6% of 1200 is 72:

$$6\% \text{ of } 1200 \text{ responses} = \frac{6}{100} \times 1200 = 72$$

- **Fraction → Percentage:** To *convert from a fraction to a percentage,* divide the denominator into the numerator to get an equivalent decimal number, then multiply by 100 and affix the % symbol. This example shows that the fraction 3/4 is equivalent to 75%:

$$\frac{3}{4} = 0.75 \rightarrow 0.75 \times 100\% = 75\%$$

- **Decimal → Percentage:** To *convert from a decimal to a percentage,* multiply by 100%. This example shows that 0.250 is equivalent to 25.0%:

$$0.250 \rightarrow 0.250 \times 100\% = 25\%$$

- **Percentage → Decimal:** To *convert from a percentage to a decimal number,* delete the % symbol and divide by 100. This example shows that 85% is equivalent to 0.85:

$$85\% = \frac{85}{100} = 0.85$$

Loaded Questions If survey questions are not worded carefully, the results of a study can be misleading. Survey questions can be "loaded" or intentionally worded to elicit a desired response.

EXAMPLE 6 **Effect of the Wording of a Question** See the following actual "yes" response rates for the different wordings of a question:

97% yes: "Should the President have the line item veto to eliminate waste?"

57% yes: "Should the President have the line item veto, or not?"

In *The Superpollsters,* David W. Moore describes an experiment in which different subjects were asked if they agree with the following statements:

- Too little money is being spent on welfare.
- Too little money is being spent on assistance to the poor.

Even though it is the poor who receive welfare, only 19% agreed when the word "welfare" was used, but 63% agreed with "assistance to the poor."

Order of Questions Sometimes survey questions are unintentionally loaded by such factors as the order of the items being considered.

EXAMPLE 7   **Effect of the Order of Questions** These questions are from a poll conducted in Germany:

- Would you say that traffic contributes more or less to air pollution than industry?

- Would you say that industry contributes more or less to air pollution than traffic?

When traffic was presented first, 45% blamed traffic and 27% blamed industry; when industry was presented first, 24% blamed traffic and 57% blamed industry.

Nonresponse   A *nonresponse* occurs when someone either refuses to respond to a survey question or is unavailable. When people are asked survey questions, some firmly refuse to answer. The refusal rate has been growing in recent years, partly because many persistent telemarketers try to sell goods or services by beginning with a sales pitch that initially sounds like it is part of an opinion poll. (This "selling under the guise" of a poll is now called *sugging.*) In *Lies, Damn Lies, and Statistics,* author Michael Wheeler makes this very important observation:

> People who refuse to talk to pollsters are likely to be different from those who do not. Some may be fearful of strangers and others jealous of their privacy, but their refusal to talk demonstrates that their view of the world around them is markedly different from that of those people who will let poll-takers into their homes.

Missing Data   Results can sometimes be dramatically affected by missing data. Sometimes sample data values are missing because of random factors (such as subjects dropping out of a study for reasons unrelated to the study), but some data are missing because of special factors, such as the tendency of people with low incomes to be less likely to report their incomes. It is well known that the U.S. Census suffers from missing people, and the missing people are often from the homeless or low income groups. In years past, surveys conducted by telephone were often misleading because they suffered from missing people who were not wealthy enough to own telephones.

Self-Interest Study   Some parties with interests to promote will sponsor studies. For example, Kiwi Brands, a maker of shoe polish, commissioned a study that resulted in this statement printed in some newspapers: "According to a nationwide survey of 250 hiring professionals, scuffed shoes was the most common reason for a male job seeker's failure to make a good first impression." We should be very wary of such a survey in which the sponsor can enjoy monetary gains from the results. Of growing concern in recent years is the practice of pharmaceutical companies paying doctors who conduct clinical experiments and report their results in prestigious journals, such as the *Journal of the American Medical Association.*

CAUTION

When assessing the validity of a study, always consider whether the sponsor might influence the results.

Precise Numbers   "There are now 103,215,027 households in the United States." Because that figure is very precise, many people incorrectly assume that it is also *accurate.* In this case, that number is an estimate, and it would be better to state that the number of households is about 103 million.

Deliberate Distortions   In the book *Tainted Truth,* Cynthia Crossen cites an example in which the magazine *Corporate Travel* published results showing that among car rental companies, Avis was the winner in a survey of people who rent cars. When Hertz requested detailed information about the survey, the actual survey responses disappeared and the magazine's survey coordinator resigned. Hertz sued Avis (for false advertising based on the survey) and the magazine; a settlement was reached.

In addition to the cases cited above, there are many other examples of the misuse of statistics. Books such as Darrell Huff's classic *How to Lie with Statistics,* Robert Reichard's *The Figure Finaglers,* and Cynthia Crossen's *Tainted Truth* describe some of those other cases. Understanding these practices will be extremely helpful in evaluating the statistical data found in everyday situations.

1-4   Basic Skills and Concepts

Statistical Literacy and Critical Thinking

1. Voluntary Response Sample What is a voluntary response sample, and why is it generally unsuitable for methods of statistics?

2. Voluntary Response Sample Are all voluntary response samples bad samples? Are all bad samples voluntary response samples?

3. Correlation and Causality Using data collected from the FBI and the Bureau of Alcohol, Tobacco, and Firearms, methods of statistics showed that for the different states, there is a correlation (or association) between the number of registered automatic weapons and the murder rate. Can we conclude that an increase in the number of registered automatic weapons causes an increase in the murder rate? Can we reduce the murder rate by reducing the number of registered automatic weapons?

4. Large Number of Responses Typical surveys involve about 500 people to 2000 people. When author Shere Hite wrote *Woman and Love: A Cultural Revolution in Progress,* she based conclusions on a relatively large sample of 4500 replies that she received after mailing 100,000 questionnaires to various women's groups. Are her conclusions likely to be valid in the sense that they can be applied to the general population of all women? Why or why not?

In Exercises 5–8, use critical thinking to develop an alternative or correct conclusion. For example, consider a media report that BMW cars cause people to be healthier. Here is an alternative conclusion: Owners of BMW cars tend to be wealthier than others, and greater wealth is associated with better health.

5. College Graduates Live Longer Based on a study showing that college graduates tend to live longer than those who do not graduate from college, a researcher concludes that studying causes people to live longer.

6. Selling Songs Data published in *USA Today* were used to show that there is a correlation between the number of times songs are played on radio stations and the numbers of times the songs are purchased. Conclusion: Increasing the times that songs are played on radio stations causes sales to increase.

7. Racial Profiling? A study showed that in Orange County, more speeding tickets were issued to minorities than to whites. Conclusion: In Orange County, minorities speed more than whites.

8. Biased Test In the judicial case *United States v. City of Chicago,* a minority group failed the Fire Captain Examination at a much higher rate than the majority group. Conclusion: The exam is biased and causes members of the minority group to fail at a much higher rate.

In Exercises 9–20, use critical thinking to address the key issue.

9. Discrepancy Between Reported and Observed Results When Harris Interactive *surveyed* 1013 adults, 91% of them said that they washed their hands after using a public restroom. But when 6336 adults were *observed,* it was found that 82% actually did wash their hands. How can we explain the discrepancy? Which percentage is more likely to accurately indicate the true rate at which people wash their hands in a public restroom?

10. O Christmas Tree, O Christmas Tree The Internet service provider America Online (AOL) ran a survey of its users and asked if they preferred a real Christmas tree or a fake one. AOL received 7073 responses, and 4650 of them preferred a real tree. Given that 4650 is 66% of the 7073 responses, can we conclude that about 66% of people who observe Christmas prefer a real tree? Why or why not?

11. Chocolate Health Food The *New York Times* published an article that included these statements: "At long last, chocolate moves toward its rightful place in the food pyramid, somewhere in the high-tone neighborhood of red wine, fruits and vegetables, and green tea. Several studies, reported in the *Journal of Nutrition,* showed that after eating chocolate, test subjects had increased levels of antioxidants in their blood. Chocolate contains flavonoids, antioxidants that have been associated with decreased risk of heart disease and stroke. Mars Inc., the candy company, and the Chocolate Manufacturers Association financed much of the research." What is wrong with this study?

12. Census Data After the last national census was conducted, the *Poughkeepsie Journal* ran this front-page headline: "281,421,906 in America." What is wrong with this headline?

13. "900" Numbers In an ABC *Nightline* poll, 186,000 viewers each paid 50 cents to call a "900" telephone number with their opinion about keeping the United Nations in the United States. The results showed that 67% of those who called were in favor of moving the United Nations out of the United States. Interpret the results by identifying what we can conclude about the way the general population feels about keeping the United Nations in the United States.

14. Loaded Questions? The author received a telephone call in which the caller claimed to be conducting a national opinion research poll. The author was asked if his opinion about Congressional candidate John Sweeney would change if he knew that in 2001, Sweeney had a car crash while driving under the influence of alcohol. Does this appear to be an objective question or one designed to influence voters' opinions in favor of Sweeney's opponent, Kirstin Gillibrand?

15. Motorcycle Helmets The Hawaii State Senate held hearings while considering a law requiring that motorcyclists wear helmets. Some motorcyclists testified that they had been in crashes in which helmets would not have been helpful. Which important group was not able to testify? (See "A Selection of Selection Anomalies," by Wainer, Palmer, and Bradlow in *Chance,* Vol. 11, No. 2.)

16. Merrill Lynch Client Survey The author received a survey from the investment firm of Merrill Lynch. It was designed to gauge his satisfaction as a client, and it had specific questions for rating the author's personal Financial Consultant. The cover letter included this statement: "Your responses are extremely valuable to your Financial Consultant, Russell R. Smith, and to Merrill Lynch. . . . We will share your name and response with your Financial Consultant." What is wrong with this survey?

17. Average of Averages The *Statistical Abstract of the United States* includes the average per capita income for each of the 50 states. When those 50 values are added, then divided by 50, the result is $29,672.52. Is $29,672.52 the average per capita income for all individuals in the United States? Why or why not?

18. Bad Question The author surveyed students with this request: "Enter your height in inches." Identify two major problems with this request.

19. Magazine Survey *Good Housekeeping* magazine invited women to visit its Web site to complete a survey, and 1500 responses were recorded. When asked whether they would rather have more money or more sleep, 88% chose more money and 11% chose more sleep. Based on these results, what can we conclude about the population of all women?

20. SIDS In a letter to the editor in the *New York Times,* Moorestown, New Jersey, resident Jean Mercer criticized the statement that "putting infants in the supine position has decreased deaths from SIDS." (SIDS refers to sudden infant death syndrome, and the *supine* position is lying on the back with the face upward.) She suggested that this statement is better: "Pediatricians advised the supine position during a time when the SIDS rate fell." What is wrong with saying that the supine position *decreased* deaths from SIDS?

Percentages. *In Exercises 21–28, answer the given questions that relate to percentages.*

21. Percentages

a. Convert the fraction $5/8$ to an equivalent percentage.

b. Convert 23.4% to an equivalent decimal.

c. What is 37% of 500?

d. Convert 0.127 to an equivalent percentage.

22. Percentages

a. What is 5% of 5020?

b. Convert 83% to an equivalent decimal.

c. Convert 0.045 to an equivalent percentage.

d. Convert the fraction $227/773$ to an equivalent percentage. Express the answer to the nearest tenth of a percent.

23. Percentages in a Gallup Poll

a. In a Gallup poll, 49% of 734 surveyed Internet users said that they shop on the Internet frequently or occasionally. What is the actual number of Internet users who said that they shop on the Internet frequently or occasionally?

b. Among 734 Internet users surveyed in a Gallup poll, 323 said that they make travel plans on the Internet frequently or occasionally. What is the percentage of responders who said that they make travel plans on the Internet frequently or occasionally?

24. Percentages in a Gallup Poll

a. In a Gallup poll of 976 adults, 68 said that they have a drink every day. What is the percentage of respondents who said that they have a drink every day?

b. Among the 976 adults surveyed, 32% said that they never drink. What is the actual number of surveyed adults who said that they never drink?

25. Percentages in AOL Poll America Online posted this question on its Web site: "How much stock do you put in long-range weather forecasts?" Among its Web site users, 38,410 chose to respond.

a. Among the responses received, 5% answered with "a lot." What is the actual number of responses consisting of "a lot?"

b. Among the responses received, 18,053 consisted of "very little or none." What percentage of responses consisted of "very little or none?"

c. Because the sample size of 38,410 is so large, can we conclude that about 5% of the general population puts "a lot" of stock in long-range weather forecasts? Why or why not?

26. Percentages in Advertising A *New York Times* editorial criticized a chart caption that described a dental rinse as one that "reduces plaque on teeth by over 300%." What is wrong with that statement?

27. Percentages in the Media In the *New York Times Magazine,* a report about the decline of Western investment in Kenya included this: "After years of daily flights, Lufthansa and Air France had halted passenger service. Foreign investment fell 500 percent during the 1990s." What is wrong with this statement?

28. Percentages in Advertising In an ad for the Club, a device used to discourage car thefts, it was stated that "The Club reduces your odds of car theft by 400%." What is wrong with this statement?

1-4   Beyond the Basics

29. Falsifying Data A researcher at the Sloan-Kettering Cancer Research Center was once criticized for falsifying data. Among his data were figures obtained from 6 groups of mice, with 20 individual mice in each group. These values were given for the percentage of successes in each group: 53%, 58%, 63%, 46%, 48%, 67%. What's wrong with those values?

30. What's Wrong with This Picture? The *Newport Chronicle* ran a survey by asking readers to call in their response to this question: "Do you support the development of atomic weapons that could kill millions of innocent people?" It was reported that 20 readers responded and 87% said "no" while 13% said "yes." Identify four major flaws in this survey.

1-5   Collecting Sample Data

Key Concept The methods we discuss in this section are important because the method used to collect sample data influences the quality of our statistical analysis. Of particular importance is the *simple random sample.* We use this sampling measure in this section and throughout the book. As you read this section, keep this concept in mind:

> **If sample data are not collected in an appropriate way, the data may be so completely useless that no amount of statistical torturing can salvage them.**

The first part of this section introduces the basics of data collection, and the second part of the section refines our understanding of two types of studies—observational studies and experiments.

Part 1: Basics of Collecting Data

Statistical methods are driven by the data that we collect. We typically obtain data from two distinct sources: *observational studies* and *experiments.*

> **DEFINITION**
>
> In an **observational study,** we observe and measure specific characteristics, but we don't attempt to *modify* the subjects being studied.
>
> In an **experiment,** we apply some *treatment* and then proceed to observe its effects on the subjects. (Subjects in experiments are called **experimental units.**)

> **EXAMPLE 1**   **Observational Study and Experiment**
>
> **Observational Study:** A good example of an observational study is a poll in which subjects are surveyed, but they are not given any treatment. The *Literary Digest* poll in which respondents were asked who they would vote for in the presidential election is an observational study. The subjects were asked for their choices, but they were not given any type of treatment.
>
> **Experiment:** In the largest public health experiment ever conducted, 200,745 children were given a treatment consisting of the Salk vaccine, while 201,229 other children were given a placebo. The Salk vaccine injections constitute a treatment that modified the subjects, so this is an example of an experiment.

Whether conducting an observational study or an experiment, it is important to select the sample of subjects in such a way that the sample is likely to be representative of the larger population. In Section 1-3 we saw that a voluntary response sample is one in which the subjects decide themselves whether to respond. Although voluntary response samples are very common, their results are generally useless for making valid inferences about larger populations.

STATISTICS IN THE NEWS

Clinical Trials vs. Observational Studies

In a *New York Times* article about hormone therapy for women, reporter Denise Grady wrote about a report of treatments tested in randomized controlled trials. She stated that "Such trials, in which patients are assigned at random to either a treatment

or a placebo, are considered the gold standard in medical research. By contrast, the observational studies, in which patients themselves decide whether to take a drug, are considered less reliable. . . . Researchers say the observational studies may have painted a falsely rosy picture of hormone replacement because women who opt for the treatments are healthier and have better habits to begin with than women who do not."

> ### DEFINITION
>
> A **simple random sample** of *n* subjects is selected in such a way that every possible *sample of the same size n* has the same chance of being chosen.

Throughout this book, we will use various statistical procedures, and we often have a requirement that we have collected a *simple random sample*, as defined above.

The following definitions describe two other types of samples.

> ### DEFINITION
>
> In a **random sample** members from the population are selected in such a way that each *individual member* in the population has an equal chance of being selected.
>
> A **probability sample** involves selecting members from a population in such a way that each member of the population has a known (but not necessarily the same) chance of being selected.

Note the difference between a random sample and a simple random sample. Exercises 21 to 26 will give you practice in distinguishing between a random sample and a simple random sample.

With random sampling we expect all components of the population to be (approximately) proportionately represented. Random samples are selected by many different methods, including the use of computers to generate random numbers. Unlike careless or haphazard sampling, random sampling usually requires very careful planning and execution.

EXAMPLE 2

Sampling Senators Each of the 50 states sends two senators to Congress, so there are exactly 100 senators. Suppose that we write the name of each *state* on a separate index card, then mix the 50 cards in a bowl, and then select one card. If we consider the two senators from the selected state to be a sample, is this result a random sample? Simple random sample? Probability sample?

SOLUTION

The sample is a random sample because each individual senator has the same chance (one chance in 50) of being selected. The sample is *not* a simple random sample because not all samples of size 2 have the same chance of being chosen. (For example, this sampling design makes it impossible to select 2 senators from different states.) The sample is a probability sample because each senator has a known chance (one chance in 50) of being selected.

Hawthorne and Experimenter Effects

The well-known placebo effect occurs when an untreated subject incorrectly believes that he or she is receiving a real treatment and reports an improvement in symptoms. The Hawthorne effect occurs when treated subjects somehow respond differently, simply because they are part of an experiment. (This phenomenon was called the "Hawthorne effect" because it was first observed in a study of factory workers at Western Electric's Hawthorne plant.) An experimenter effect (sometimes called a Rosenthall effect) occurs when the researcher or experimenter unintentionally influences subjects through such factors as facial expression, tone of voice, or attitude.

Other Sampling Methods   In addition to random samples and simple random samples, there are other sampling techniques. We describe the common ones here. Figure 1-2 compares the different sampling approaches.

 DEFINITION

> In **systematic sampling,** we select some starting point and then select every kth (such as every 50th) element in the population.
>
> With **convenience sampling,** we simply use results that are very easy to get.
>
> With **stratified sampling,** we subdivide the population into at least two different subgroups (or strata) so that subjects within the same subgroup share the same characteristics (such as gender or age bracket), then we draw a sample from each subgroup (or stratum).
>
> In **cluster sampling,** we first divide the population area into sections (or clusters), then randomly select some of those clusters, and then choose *all* the members from those selected clusters.

It is easy to confuse stratified sampling and cluster sampling, because they both use subgroups. But cluster sampling uses *all* members from a *sample* of clusters, whereas stratified sampling uses a *sample* of members from *all* strata. An example of cluster sampling is a preelection poll, in which pollsters randomly select 30 election precincts from a large number of precincts and then survey all the people from each of those precincts. This is much faster and much less expensive than selecting one person from each of the many precincts in the population area. Pollsters can adjust or weight the results of stratified or cluster sampling to correct for any disproportionate representations of groups.

For a fixed sample size, if you randomly select subjects from different strata, you are likely to get more consistent (and less variable) results than by simply selecting a random sample from the general population. For that reason, pollsters often use stratified sampling to reduce the variation in the results. Many of the methods discussed later in this book require that sample data be a *simple random sample,* and neither stratified sampling nor cluster sampling satisfies that requirement.

Multistage Sampling Professional pollsters and government researchers often collect data by using some combination of the basic sampling methods. In a **multistage sample design,** pollsters select a sample in different stages, and each stage might use different methods of sampling.

> **EXAMPLE 3**   **Multistage Sample Design** The U.S. government's unemployment statistics are based on surveyed households. It is impractical to personally visit each member of a simple random sample, because individual households would be spread all over the country. Instead, the U.S. Census Bureau and the Bureau of Labor Statistics combine to conduct a survey called the Current Population Survey. This survey obtains data describing such factors as unemployment rates, college enrollments, and weekly earnings amounts. The survey incorporates a multistage sample design, roughly following these steps:
>
> 1. The surveyors partition the entire United States into 2007 different regions called *primary sampling units* (PSU). The primary sampling units are metropolitan areas, large counties, or groups of smaller counties.   *continued*

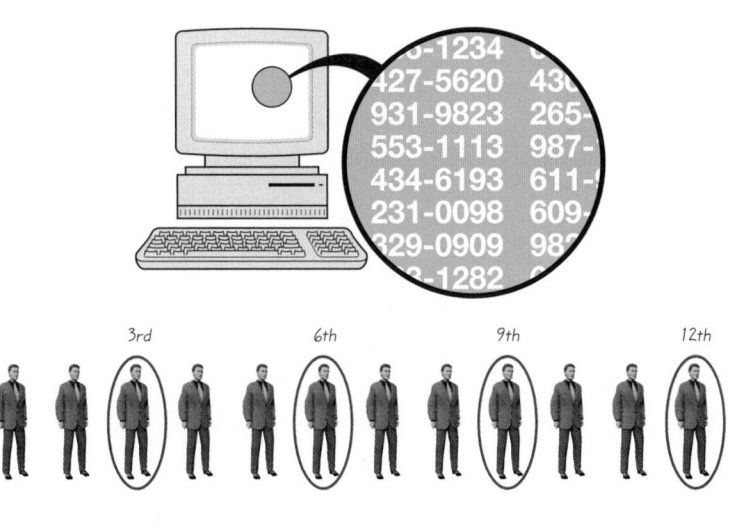

Random Sampling:
Each member of the population has an equal chance of being selected. Computers are often used to generate random telephone numbers.

Simple Random Sampling:
A sample of n subjects is selected in such a way that every possible sample of the same size n has the same chance of being chosen.

Systematic Sampling:
Select some starting point, then select every kth (such as every 50th) element in the population.

Convenience Sampling:
Use results that are easy to get.

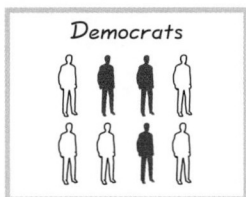

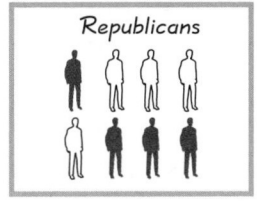

Stratified Sampling:
Subdivide the population into at least two different subgroups (or strata) so that subjects within the same subgroup share the same characteristics (such as gender or age bracket), then draw a sample from each subgroup.

Cluster Sampling:
Divide the population into sections (or clusters), then randomly select some of those clusters, and then choose all members from those selected clusters.

Figure 1-2 Common Sampling Methods

Prospective National Children's Study

A good example of a prospective study is the National Children's Study begun in 2005. It is tracking 100,000 children from birth to age 21. The children

are from 96 different geographic regions. The objective is to improve the health of children by identifying the effects of environmental factors, such as diet, chemical exposure, vaccinations, movies, and television. The study will address questions such as these: How do genes and the environment interact to promote or prevent violent behavior in teenagers? Are lack of exercise and poor diet the only reasons why many children are overweight? Do infections impact developmental progress, asthma, obesity, and heart disease? How do city and neighborhood planning and construction encourage or discourage injuries?

2. The surveyors select a sample of primary sampling units in each of the 50 states. For the Current Population Survey, 792 of the primary sampling units are used. (All of the 432 primary sampling units with the largest populations are used, and 360 primary sampling units are randomly selected from the other 1575.)

3. The surveyors partition each of the 792 selected primary sampling units into blocks, and they then use stratified sampling to select a sample of blocks.

4. In each selected block, surveyors identify clusters of households that are close to each other. They randomly select clusters, and they interview all households in the selected clusters.

This multistage sample design includes random, stratified, and cluster sampling at different stages. The end result is a complicated sampling design, but it is much more practical and less expensive than using a simpler design, such as using a simple random sample.

Part 2: Beyond the Basics of Collecting Data

In this part, we refine what we've learned about observational studies and experiments by discussing different types of observational studies and experiment design.

There are various types of observational studies in which investigators observe and measure characteristics of subjects. The definitions below, which are summarized in Figure 1-3, identify the standard terminology used in professional journals for different types of observational studies.

> **DEFINITION**
>
> In a **cross-sectional study,** data are observed, measured, and collected at one point in time.
>
> In a **retrospective** (or **case-control**) **study,** data are collected from the past by going back in time (through examination of records, interviews, and so on).
>
> In a **prospective** (or **longitudinal** or **cohort**) **study,** data are collected in the future from groups sharing common factors (called *cohorts*).

The sampling done in retrospective studies differs from that in prospective studies. In retrospective studies we go back in time to collect data about the characteristic that is of interest, such as a group of drivers who died in car crashes and another group of drivers who did not die in car crashes. In prospective studies we go forward in time by following groups with a potentially causative factor and those without it, such as a group of drivers who use cell phones and a group of drivers who do not use cell phones.

Design of Experiments

We now consider experiment design, starting with an example of an experiment having a good design. We use the experiment first mentioned in Example 1, in which researchers tested the Salk vaccine. After describing the experiment in more detail, we identify the characteristics of that experiment that typify a good design.

Figure 1-3
Types of Observational Studies

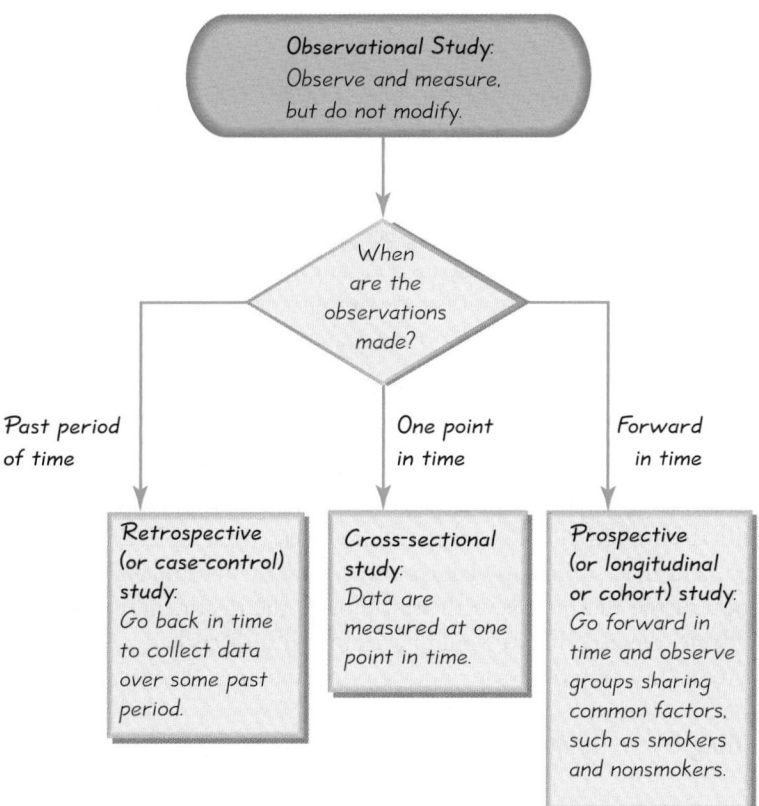

EXAMPLE 4 **The Salk Vaccine Experiment** In 1954, a large-scale experiment was designed to test the effectiveness of the Salk vaccine in preventing polio, which had killed or paralyzed thousands of children. In that experiment, 200,745 children were given a treatment consisting of Salk vaccine injections, while a second group of 201,229 children were injected with a placebo that contained no drug. The children being injected did not know whether they were getting the Salk vaccine or the placebo. Children were assigned to the treatment or placebo group through a process of random selection, equivalent to flipping a coin. Among the children given the Salk vaccine, 33 later developed paralytic polio, but among the children given a placebo, 115 later developed paralytic polio.

Randomization is used when subjects are assigned to different groups through a process of random selection. The 401,974 children in the Salk vaccine experiment were assigned to the Salk vaccine treatment group or the placebo group through a process of random selection, equivalent to flipping a coin. In this experiment, it would be extremely difficult to directly assign children to two groups having similar characteristics of age, health, sex, weight, height, diet, and so on. There could easily be important variables that we might not realize. The logic behind randomization is to use chance as a way to create two groups that are similar. Although it might seem that we should not leave anything to chance in experiments, randomization has been found to be an extremely effective method for assigning subjects to groups.

Replication is the repetition of an experiment on more than one subject. Samples should be large enough so that the erratic behavior that is characteristic of very small samples will not disguise the true effects of different treatments. Replication is used effectively when we have enough subjects to recognize differences from different treatments. (In another context, *replication* refers to the repetition or duplication of an experiment so that results can be confirmed or verified.) With replication, the large sample sizes increase the chance of recognizing different treatment effects. However, a large sample is not necessarily a good sample. Although it is important to have a sample that is sufficiently large, it is more important to have a sample in which subjects have been chosen in some appropriate way, such as random selection.

> **Use a sample size that is large enough to let us see the true nature of any effects, and obtain the sample using an appropriate method, such as one based on *randomness*.**

In the experiment designed to test the Salk vaccine, 200,745 children were given the actual Salk vaccine and 201,229 other children were given a placebo. Because the actual experiment used sufficiently large sample sizes, the researchers could observe the effectiveness of the vaccine. Nevertheless, though the treatment and placebo groups were very large, the experiment would have failed if subjects had not been assigned to the two groups in a way that made both groups similar in the ways that were important to the experiment.

Blinding is a technique in which the subject doesn't know whether he or she is receiving a treatment or a placebo. Blinding allows us to determine whether the treatment effect is significantly different from a **placebo effect,** which occurs when an untreated subject reports an improvement in symptoms. (The reported improvement in the placebo group may be real or imagined.) Blinding minimizes the placebo effect or allows investigators to account for it. The polio experiment was **double-blind,** meaning that blinding occurred at two levels: (1) The children being injected didn't know whether they were getting the Salk vaccine or a placebo, and (2) the doctors who gave the injections and evaluated the results did not know either.

Controlling Effects of Variables Results of experiments are sometimes ruined because of *confounding*.

 DEFINITION

Confounding occurs in an experiment when you are not able to distinguish among the effects of different factors.

Try to plan the experiment so that confounding does not occur.

See Figure 1-4(a), where confounding can occur when the treatment group of women shows strong positive results. Because the treatment group consists of women and the placebo group consists of men, confounding has occurred because we can't determine whether the treatment or the sex of the subjects causes the positive results. It is important to design experiments to control and understand the effects of the variables (such as treatments). The Salk vaccine experiment in Example 4 illustrates one method for controlling the effect of the treatment variable: Use a *completely randomized experimental design*, whereby randomness is used to assign subjects to the treatment group and the placebo group. The objective of this experimental design is to control the effect of the treatment, so that we are able to clearly recognize the difference between the effect of the Salk vaccine and the effect of the placebo. Completely randomized experimental design is one of the following four methods used to control effects of variables.

Completely Randomized Experimental Design: Assign subjects to different treatment groups through a process of *random selection*. See Figure 1-4(b).

Randomized Block Design: A **block** is a group of subjects that are similar, but blocks differ in ways that might affect the outcome of the experiment. (In designing an experiment to test the effectiveness of aspirin treatments on heart disease, we might form a block of men and a block of women, because it is known that hearts of men and women can behave differently.) If testing one or more different treatments with different blocks, use this experimental design (see Figure 1-4(c)):

1. Form blocks (or groups) of subjects with similar characteristics.

2. Randomly assign treatments to the subjects within each block.

Rigorously Controlled Design: Carefully assign subjects to different treatment groups, so that those given each treatment are similar in the ways that are important to the experiment. In an experiment testing the effectiveness of aspirin on heart disease, if the placebo group includes a 27-year-old male smoker who drinks heavily and consumes an abundance of salt and fat, the treatment group should also include a person with similar characteristics (which, in this case, would be easy to find). This approach can be extremely difficult to implement, and we might not be sure that we have considered all of the relevant factors.

Matched Pairs Design: Compare exactly two treatment groups (such as treatment and placebo) by using subjects matched in pairs that are somehow related or have similar characteristics. A test of Crest toothpaste used matched pairs of twins, where one twin used Crest and the other used another toothpaste. The matched pairs might also consist of measurements from the same subject before and after some treatment.

Bad experimental design:
Treat all women subjects, and don't treat men. (Problem: We don't know if effects are due to sex or to treatment.)

Completely randomized experimental design:
Use randomness to determine who gets the treatment.

Randomized block design:
1. Form a block of women and a block of men.
2. Within each block, randomly select subjects to be treated.

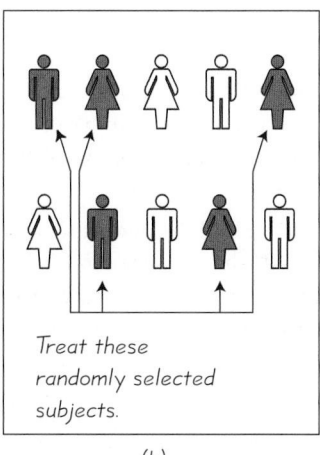

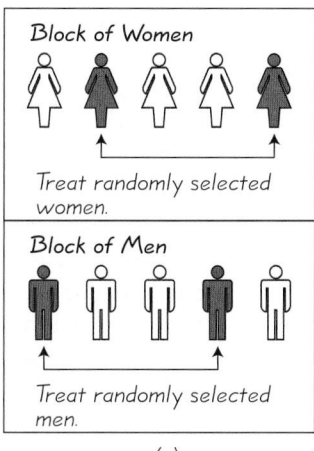

Figure 1-4 **Controlling Effects of a Treatment Variable**

Summary   Three very important considerations in the design of experiments are the following:

1. Use *randomization* to assign subjects to different groups.

2. Use *replication* by repeating the experiment on enough subjects so that effects of treatments or other factors can be clearly seen.

3. *Control the effects of variables* by using such techniques as blinding and a completely randomized experimental design.

Sampling Errors   No matter how well you plan and execute the sample collection process, there is likely to be some error in the results. For example, randomly select 1000 adults, ask them if they graduated from high school, and record the sample percentage of "yes" responses. If you randomly select another sample of 1000 adults, it is likely that you will obtain a *different* sample percentage.

● DEFINITION

A **sampling error** is the difference between a sample result and the true population result; such an error results from chance sample fluctuations.

A **nonsampling error** occurs when the sample data are incorrectly collected, recorded, or analyzed (such as by selecting a biased sample, using a defective measurement instrument, or copying the data incorrectly).

If we carefully collect a sample so that it is representative of the population, we can use methods in this book to analyze the sampling error, but we must exercise extreme care to minimize nonsampling error.

Experimental design requires much more thought and care than we can describe in one relatively brief section. Taking a complete course in the design of experiments is a good way to learn much more about this important topic.

1-5   Basic Skills and Concepts

Statistical Literacy and Critical Thinking

1. Random Sample and Simple Random Sample What is the difference between a random sample and a simple random sample?

2. Observational Study and Experiment What is the difference between an observational study and an experiment?

3. Simple Random Convenience Sample A student of the author listed his adult friends, then he surveyed a simple random sample of them. Although this is a simple random sample, are the results likely to be representative of the general population of adults in the United States? Why or why not?

4. Convenience Sample The author conducted a survey of the students in his classes. He asked the students to indicate whether they are left-handed or right-handed. Is this convenience sample likely to provide results that are typical of the population? Are the results likely to be good or bad? Does the quality of the results in this survey reflect the quality of convenience samples in general?

In Exercises 5–8, determine whether the given description corresponds to an observational study or an experiment.

5. Touch Therapy Nine-year-old Emily Rosa was an author of an article in the *Journal of the American Medical Association* after she tested professional touch therapists. Using a cardboard partition, she held her hand above the therapist's hand, and the therapist was asked to identify the hand that Emily chose.

6. Smoking Survey A Gallup poll surveyed 1018 adults by telephone, and 22% of them reported that they smoked cigarettes within the past week.

7. Treating Syphilis In a morally and criminally wrong study, 399 black men with syphilis were *not* given a treatment that could have cured them. The intent was to learn about the effects of syphilis on black men. The subjects were initially treated with small amounts of bismuth, neoarsphenamine, and mercury, but those treatments were replaced with aspirin.

8. Testing Echinacea A study of the effectiveness of echinacea involved 707 cases of upper respiratory tract infections. Children with 337 of the infections were given echinacea, and children with 370 of the infections were given placebos (based on data from "Efficacy and Safety of Echinacea in Treating Upper Respiratory Tract Infections in Children," by Taylor et al., *Journal of the American Medical Association,* Vol. 290, No. 21).

In Exercises 9–20, identify which of these types of sampling is used: **random, systematic, convenience, stratified,** *or* **cluster.**

9. Ergonomics A student of the author collected measurements of arm lengths from her family members.

10. Testing Echinacea A study of the effectiveness of echinacea involved upper respiratory tract infections. One group of infections was treated with echinacea and another group was treated with placebos. The echinacea and placebo groups were determined through a process of random assignment (based on data from "Efficacy and Safety of Echinacea in Treating Upper Respiratory Tract Infections in Children" by Taylor et al., *Journal of the American Medical Association,* Vol. 290, No. 21).

11. Exit Polls On the day of the last presidential election, ABC News organized an exit poll in which specific polling stations were randomly selected and all voters were surveyed as they left the premises.

12. Sobriety Checkpoint The author was an observer at a Town of Poughkeepsie Police sobriety checkpoint at which every fifth driver was stopped and interviewed. (He witnessed the arrest of a former student.)

13. Wine Tasting The author once observed professional wine tasters working at the Consumer's Union testing facility in Yonkers, New York. Assume that a taste test involves three different wines randomly selected from each of five different wineries.

14. Recidivism The U.S. Department of Corrections collects data about returning prisoners by randomly selecting five federal prisons and surveying all of the prisoners in each of the prisons.

15. Quality Control in Manufacturing The Federal-Mogul Company manufactures Champion brand spark plugs. The procedure for quality control is to test every 100th spark plug from the assembly line.

16. Credit Card Data The author surveyed all of his students to obtain sample data consisting of the number of credit cards students possess.

17. Tax Audits The author once experienced a tax audit by a representative from the New York State Department of Taxation and Finance, which claimed that the author was randomly selected as part of a "statistical" audit. (Isn't that ironic?) The representative was a very nice person and a credit to humankind.

18. Curriculum Planning In a study of college programs, 820 students are randomly selected from those majoring in communications, 1463 students are randomly selected from those majoring in business, and 760 students are randomly selected from those majoring in history.

19. Study of Health Plans Six different health plans were randomly selected, and all of their members were surveyed about their satisfaction (based on a project sponsored by RAND and the Center for Health Care Policy and Evaluation).

20. Gallup Poll In a Gallup poll, 1003 adults were called after their telephone numbers were randomly generated by a computer, and 20% of them said that they get news on the Internet every day.

Random Samples and Simple Random Samples. *Exercises 21–26 relate to random samples and simple random samples.*

21. Sampling Prescription Pills Pharmacists typically fill prescriptions by scooping a sample of pills from a larger batch that is in stock. A pharmacist thoroughly mixes a large batch of Lipitor pills, then selects 30 of them. Does this sampling plan result in a random sample? Simple random sample? Explain.

22. Systematic Sample A quality control engineer selects every 10,000th M&M plain candy that is produced. Does this sampling plan result in a random sample? Simple random sample? Explain.

23. Cluster Sample ABC News conducts an election day poll by randomly selecting voting precincts in New York, then interviewing all voters as they leave those precincts. Does this sampling plan result in a random sample? Simple random sample? Explain.

24. Stratified Sample In order to test for a gender gap in the way that citizens view the current President, the Tomkins Company polls exactly 500 men and 500 women randomly selected from adults in the United States. Assume that the numbers of adult men and women are the same. Does this sampling plan result in a random sample? Simple random sample? Explain.

25. Convenience Sample NBC News polled reactions to the last presidential election by surveying adults who were approached by a reporter at a location in New York City. Does this sampling plan result in a random sample? Simple random sample? Explain.

26. Sampling Students A classroom consists of 36 students seated in six different rows, with six students in each row. The instructor rolls a die to determine a row, then rolls the die again to select a particular student in the row. This process is repeated until a sample of 6 students is obtained. Does this sampling plan result in a random sample? Simple random sample? Explain.

1-5 Beyond the Basics

In Exercises 27–30, identify the type of observational study (**cross-sectional, retrospective, prospective**).

27. Victims of Terrorism Physicians at the Mount Sinai Medical Center studied New York City residents with and without respiratory problems. They went back in time to determine how those residents were involved in the terrorist attacks in New York City on September 11, 2001.

28. Victims of Terrorism Physicians at the Mount Sinai Medical Center plan to study emergency personnel who worked at the site of the terrorist attacks in New York City on September 11, 2001. They plan to study these workers from now until several years into the future.

29. TV Ratings The Nielsen Media Research Company uses people meters to record the viewing habits of about 5000 households, and today those meters will be used to determine the proportion of households tuned to *CBS Evening News*.

30. Cell Phone Research University of Toronto researchers studied 699 traffic crashes involving drivers with cell phones (based on data from "Association Between Cellular-Telephone Calls and Motor Vehicle Collisions," by Redelmeier and Tibshirani, *New England Journal of Medicine*, Vol. 336, No. 7). They found that cell phone use quadruples the risk of a collision.

31. Blinding A study funded by the National Center for Complementary and Alternative Medicine found that echinacea was not an effective treatment for colds in children. The experiment involved echinacea treatments and placebos, and blinding was used. What is blinding, and why was it important in this experiment?

32. Sampling Design You have been commissioned to conduct a job survey of graduates from your college. Describe procedures for obtaining a sample of each type: random, systematic, convenience, stratified, cluster.

33. Confounding Give an example (different from the one in the text) illustrating how confounding occurs.

34. Sample Design In "Cardiovascular Effects of Intravenous Triiodothyronine in Patients Underdoing Coronary Artery Bypass Graft Surgery" (*Journal of the American Medical Association,* Vol. 275, No. 9), the authors explain that patients were assigned to one of three groups: (1) a group treated with triidothyronine, (2) a group treated with normal saline bolus and dopamine, and (3) a placebo group given normal saline. The authors summarize the sample design as a "prospective, randomized, double-blind, placebo-controlled trial." Describe the meaning of each of those terms in the context of this study.

Introduction to the TI-83/84 Plus Calculator

This book is designed for TI-83 Plus, TI-83 Plus Silver Edition, TI-84 Plus, and TI-84 Plus Silver Edition calculators. Instead of making repeated reference to each of these calculators throughout this book, we will use "TI-83/84 Plus" as the general reference to the TI-83 Plus, TI-83 Plus Silver Edition, TI-84 Plus, and TI-84 Plus Silver Edition calculators.

Throughout this book, "TI-83/84 Plus" refers to any of the following calculators:

- TI-83 Plus
- TI-83 Plus Silver Edition
- TI-84 Plus
- TI-84 Plus Silver Edition
- TI-Nspire with the TI-84 Plus keypad installed

Instructions, screen displays, programs, and the APP (application) for the "TI-83/84 Plus" calculator also apply to the TI-83 Plus Silver Edition and the TI-84 Plus Silver Edition calculators.

TI-83 Plus versus TI-84 Plus Instructions for TI-83 Plus calculators and TI-84 Plus calculators are essentially identical, but TI-84 Plus calculators have a few functions that are not available on the TI-83 Plus calculators.

TI-83 Calculator The older TI-83 calculator (not the TI-83 Plus version) lacks the ability to use APPS (applications), so users of TI-83 calculators will not be able to use the APP included on the CD-ROM accompanying new copies of this textbook. That APP includes the data sets in Appendix B of this textbook, so TI-83 users will sometimes need to manually enter data instead of using downloaded data. However, TI-83 users can download the programs included on the CD-ROM.

TI Operating System To check your calculator's current operating system, press **2ND** (+) (to select the memory menu), then select the first item labeled **About.** As of this writing, the latest operating system for the TI-84 Plus is 2.43 and the latest operating system for the TI-83 Plus is 1.19. If you wish to update your calculator's operating system, Google either "TI-84 Plus operating system" or "TI-83 Plus operating system" and click on the appropriate link to get the free download along with instructions for installing it on your calculator.

TI-89 and TI-Nspire Calculators The TI-89 calculator has a few statistical functions, but statistics students should download the extensive and free "Statistics with List Editor" application from this Texas Instruments Web site:

http://education.ti.com/us/product/tech/89/apps/appslist.html.

(The TI-89 Titanium edition has the statistics application already installed.) The TI-Nspire calculator essentially becomes a TI-84 Plus calculator when it has the TI-84 Plus keypad installed. For more detailed instructions for the TI-89 and TI-Nspire calculators, see this printed supplement for the Triola Statistics Series: *Graphing Calculator Manual for the TI-83 Plus, TI-84 Plus, TI-89, and TI-Nspire* by Patricia Humphrey.

The "TI-83/84" Plus calculator is an excellent tool that is extremely helpful in an introductory statistics course. It is easy to carry to class, and it has an abundance of statistical tools that can be used throughout the entire course. Using the TI-83/84 Plus calculator makes the statistics course easier by allowing students to focus on the selection of procedures and the interpretation of results, instead of being crushed by arithmetic calculations.

Setting the Calculator Mode

Before starting, it would be helpful to verify that your TI-83/84 Plus calculator has the mode settings suitable for the introductory statistics course. Press the **MODE** key and compare your display to the one shown in the margin. If your mode settings are not as shown here, change them by using the ⌄ and ⌃ keys to scroll to the desired row, use the ⟩ and ⟨ keys to highlight the desired row entry, and then press **ENTER**. When finished, quit by pressing **2ND** **MODE** (which activates the QUIT option, which is depicted above the **MODE** key). This mode setting can be selected once, and it will remain in effect unless it is changed later.

Screen Darkness and Batteries

If the displays on the screen appear to be too light or too dark, adjust the contrast as follows:

Increase darkness of display: Press and release the **2ND** key and then press and hold the ⌃ key.

Decrease darkness of display: Press and release the **2ND** key and then press and hold the ⌄ key.

The screen display will gradually fade as the batteries become tired, and the darkness of the display will need to be increased. Eventually, a low battery–warning message will be displayed, so replace batteries when necessary. A low-battery message usually indicates that the four AAA batteries need replacement. The lithium battery should last a few years, and it should not be replaced at the same time that the AAA batteries are being replaced.

Special Keys: **2ND**, **ALPHA**, and **CLEAR**

The typical key on the TI-83/84 Plus can be used for two or three different purposes. For example, the ④ key has 4 on its face, but it also has L4 above it in yellow or blue, as well as the letter T in green. Pressing ④ causes the calculator to use the number 4, pressing **2ND** ④ causes it to refer to the list label L4, and pressing **ALPHA** ④ causes the letter T to be used. The use of the key is easy to follow with the color coding of the yellow or blue **2ND** key and the green **ALPHA** key. The color of the **2ND** key corresponds to the entry above the key being used, and the green **ALPHA** key refers to the green alphabetic character (or other green special symbol) placed above the key being used.

Press **CLEAR** to clear (what else?) the screen display.

Basic Calculations

Many inexpensive calculators use a ⟨ = ⟩ key to indicate the end of calculation instructions, but the TI-83/84 Plus uses **ENTER** instead. For example, the addition of

2 + 7 is accomplished with this key sequence: ②　＋　⑦　**ENTER**. The result should be 9.

Like any good scientific calculator, the TI-83/84 Plus is designed to execute arithmetic operations with a hierarchy that causes multiplications and divisions to be computed before additions and subtractions. This feature can be overcome through the use of parentheses. See the examples below:

2 ＋ 3 × 4 **ENTER**: *Result:*　14 (because the multiplication is done first)

(2 ＋ 3 **)** × 4 **ENTER**: *Result:*　20 (the parentheses cause the addition to be done first)

When unsure about the calculator's order of operations, use parentheses to ensure that the calculation is executed exactly as you desire.

Difference Between ⊖ and ⊝

It is very important to understand the difference between the ⊖ and ⊝ keys. They are described below:

⊖ **key:**　Arithmetic operation of *subtraction,* as in 8 − 5

⊝ **key:**　*Negative sign,* as in −6

For example, the operation of 8 − 5 is done by pressing 8 ⊖ 5 **ENTER**, and the result should be 3. If you incorrectly enter 8 ⊝ 5 **ENTER**, you will get an error message, because the negative sign makes no sense in this context. Be careful to use ⊖ for the operation of subtraction so that the ⊖ key is placed between two numbers, whereas the ⊝ key is a negative sign that applies to the number that follows. It is easy to remember that ⊖ represents the operation of subtraction, because it is located with the other operations of ⊘, ⊗, and ⊕.

Scientific Notation

Some results in later chapters are displayed by the TI-83/84 Plus calculator as values expressed in *scientific notation.* For example, 3.518437209**E**13 is a value displayed in scientific notation. This value can be expressed as $3.518437209 \times 10^{13}$ or as 35,184,372,090,000. This might seem complicated, but note that we can express 3.518437209**E**13 in ordinary notation simply by moving the decimal place to the right 13 places. See the following two examples:

With a *positive* number following **E**, move the decimal place to the *right,* as in 2.345**E**15, which can be expressed as 2,345,000,000,000,000.

With a *negative* number following **E**, move the decimal place to the *left,* as in 2.345**E**⁻15, which can be expressed as 0.000000000000002345.

Clearing a List

Instead of using single numerical values, the typical statistics course uses collections of data. The TI-83/84 Plus calculator is designed so that we can use *lists* of data. The TI-83/84 Plus calculator already has lists named L1, L2, L3, L4, L5, and L6. Also, you can assign a list of data to a name of your choice. If entering data in L1, or L2, or L3, etc., it is a good practice to first *clear* the list of any data that might be there from some prior procedure. Clear a list using the following sequence of steps:

1. Press **STAT**.
2. Use ⊽ to scroll down to the fourth menu item identified as 4: ClrList.
3. Press **ENTER**.

4. The screen should now display ClrList, so press 2ND 1 to get the display of ClrList L1, which causes list L1 to be cleared. (Note that 2ND 1 activates the yellow or blue item *above* the 1 key, which is L1 for List 1.)

5. Press ENTER.

Entering a List of Data

Let's assume that we want to work with the following lengths (in seconds) of scenes showing tobacco use in animated movies from Universal Studios.

0   223   0   176   0   548

To enter such a list of values, first clear the list as described above. The following procedure describes the entry of a list labeled as L1, but we could also use L2, L3, L4, L5, or L6, as well as using a name that we can enter. The following procedure applies to the list labeled as L1:

1. Press STAT.

2. We want the first menu item of Edit, and that is the default, so simply press ENTER to select that option.

3. If the cursor is not positioned in column L1, use ◁ and ▷ to get to the column list L1.

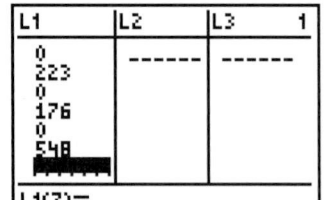

4. Enter a data value and then press ENTER. Enter the second data value and press ENTER again. Continue to enter individual data values followed by ENTER. (If you make a mistake in a data value, simply use △ and ▽ to highlight the wrong entry and then enter the correct value and press ENTER.) The screen should appear as shown in the accompanying screen display (although lists L2 and L3 might have other data displayed).

5. Press 2ND ENTER to QUIT the list editor.

Even though Step 5 caused us to quit, the list will remain as entered until it is cleared or changed. List L1 will remain in the calculator's memory even after the calculator is turned off.

Naming a List

Sometimes we want to manually enter a long list of data that will be used in future applications. However, the lists L1, L2, L3, L4, L5, and L6 may be used often, and their labels do not suggest the nature of the data. There is a procedure for entering a list and saving it with a name other than L1, L2, L3, L4, L5, or L6. Here is one procedure for saving a list with a desired name:

1. Enter the data enclosed within braces and separated by commas, as in {0, 223, 0, 176, 0, 548}.

2. Press STO▸.

3. Use ALPHA to enter the alphabetic characters that constitute the desired name. The name TIME can be entered with ALPHA T ALPHA I ALPHA M ALPHA E. (*Shortcut hint:* Instead of pressing ALPHA before each letter, you could press 2ND ALPHA to lock the alpha key so that it remains in effect until you unlock it by entering ALPHA a second time.)

4. Press ENTER.

The preceding three steps result in this sequence of keystrokes:

{0, 223, 0, 176, 0, 548} STO▸ ALPHA T ALPHA I ALPHA M ALPHA E ENTER and the effect of this sequence of keystrokes is to have the list of values assigned to the name of TIME.

Another way to name a list of data is to first enter the data in list L1 and then enter the expression L1→**L**TIME, where the symbol **L** indicates that the name represents a *list* of values (instead of a single value). The symbol **L** can be obtained by pressing **2ND** **STAT** and selecting OPS and then the last item of **L**. If you now enter **L** TIME, you will see the list of values.

APPS

TRIOLA11 One of the great features of the TI-83/84 Plus is its ability to handle APPs, which are software applications. (APPs cannot be used with the older TI-83 calculators.) One really helpful APP is the collection of data sets found in Appendix B. That data set is an APP on the CD-ROM that comes with this book, and it is also available on the Triola Web site at www.aw.com/triola. It is necessary to install the TRIOLA11 APP just once. That APP will remain in the calculator until it is removed.

To download the APP from another TI-83/84 Plus calculator to your calculator, you need the cable that came with your calculator. Follow these steps:

1. Use the cable to connect the two calculators.

2. On the receiving calculator, enter **2ND** **X,T,Θ,n**, which calls upon the LINK function. Press **▷** to highlight RECEIVE and then press **ENTER** to get a message of Waiting.

3. On the sending calculator, enter **2ND** **X,T,Θ,n**, which calls upon the LINK function. Use the **▽** key to select the menu item of C:Apps and press **ENTER**. Now use **▽** to select the item of TRIOLA11 and press **ENTER**. Finally, press **▷** to select TRANSMIT and press **ENTER**.

To download the APP from a computer to your calculator, first obtain download information from the Texas Instruments Web Site at www.ti.com. The procedure, software, and connection devices will depend on whether you have a TI-83 Plus, TI-83 Plus Silver Edition, TI-84 Plus, or TI-84 Plus Silver Edition calculator. Whereas the TI-83 Plus calculator might require a special TI-GRAPHLINK™ cable, the newer TI-84 Plus and TI-84 Plus Silver Edition have a USB port that simplifies connection to a computer by using a standard cable.

CtlgHelp In addition to the TRIOLA11 application, it is also recommended that you download the CtlgHelp (catalog help) application. This application comes with the TI-83 Plus Silver Edition and the TI-84 Plus Silver Edition. It can be downloaded to the TI-83 Plus calculator and the TI-84 Plus calculator. One nice feature of this catalog help application is that it can be used to identify the format required for many important functions to be described later in this book.

Memory Error It is possible that in the attempt to download an APP, you may get the error message of ERR:MEMORY. If this happens, it is necessary to clear some memory by erasing data stored in lists or by removing other APPs. Because the newer calculators have more memory, they are less likely to create memory errors. (The TI-84 Plus calculator has three times as much memory as the TI-83 Plus calculator. The TI-83 Plus Silver Edition and the TI-84 Plus Silver Edition have about nine times as much memory as the TI-83 Plus calculator.)

Important: This section contains much information that might seem to be overwhelming. However, it is the author's experience that students quickly and easily learn to make effective use of their TI-83/84 Plus calculators. They find that their statistics course becomes much easier by allowing them to focus on the selection of procedures and the interpretation of results, without being tortured with mind-numbing tedious calculations.

1-6   Basic Skills and Concepts

Statistical Literacy and Critical Thinking

1. APP In the terminology of Texas Instruments calculators, what is an APP?

2. TI The TI-83 Plus, TI-84 Plus, TI-89, and TI-Nspire calculators all have model identifications that begin with the letters TI. What do the letters TI represent?

3. Negative and Subtraction On a TI-83/84 calculator, what is the difference between the keys ⬚ and ⬚?

4. ALPHA Key What does "ALPHA" on the **ALPHA** key represent?

In Exercises 5–12, use a TI-83/84 Plus calculator to find the results.

5. $12 - (-4)$

6. $(-5) - (-8)$

7. $\sqrt{1024}$

8. $\sqrt{0.015129}$

9. $\sqrt{5^2 + 6^2}$

10. $\dfrac{(12 - 3)^2}{3}$

11. $\sqrt{\dfrac{(8 - 2)^2 + (9 - 2)^2}{2}}$

12. 0.5^8

In Exercises 13–16, find the TI-83/84 Plus calculator result and explain how the result occurred. For example, the result of $4 + 5 \times 6$ would be 34, and the calculator used a hierarchy in which the multiplication was executed first so that the result is $4 + 30$, or 34.

13. **2ND** **x²** 9 **ENTER**

14. 25 **x²** **ENTER**

15. 5 **⌃** 2 **ENTER**

16. **2ND** **⌃** **ENTER**

*In Exercises 17–20, enter the values of 0, 223, 0, 176, 0, and 548 in list L1, then press **2ND** **MODE** to quit the LIST editor, and then describe the results of the given expressions. Also, describe how the result relates to the original list.*

17. **2ND** 1 **ENTER**

18. 5 **+** **2ND** 1 **ENTER**

19. **2ND** 1 **+** **2ND** 1 **ENTER**

20. **2ND** 1 **x²** **ENTER**

Scientific Notation In Exercises 21–24, use a TI-83/84 Plus calculator to find the result and then express the result in ordinary notation (not scientific notation using E).

21. 2^{45} (which can be entered as 2 **⌃** 45 **ENTER**)

22. 0.2^{12} (which can be entered as 0.2 **⌃** 12 **ENTER**)

23. 0.5^{10}

24. 5^{16}

1-6   Beyond the Basics

25. Using an APP Using the CD-ROM included with this book, using the Triola Web site, or using another TI-83/84 Plus calculator, download the TRIOLA11 application (APP).

a. What does the screen show when the **APPS** key is pressed?

b. Press the **APPS** key, use ⬚ to scroll to the entry of TRIOLA11, and then press **ENTER**. Describe the resulting screen display.

c. After doing parts (a) and (b), use ⬚ to scroll to the entry of CIGARET, and press **ENTER**. Describe the screen display.

d. After completing parts (a)–(c), use ⬚ to highlight All and press **ENTER**. Now use ⬚ to highlight Load, use ⬚ to select 2:Load, and press **ENTER**. Quit and then select List by pressing **2ND** **STAT**. Describe the result.

e. After completing parts (a)–(d), clear the screen and select List by pressing **2ND** **STAT**, use ⌄ to highlight: KGTAR, and press **ENTER** twice. Describe the result. (*Hint:* See Data Set 4 in Appendix B.)

TI **26. Creating a Named List** Assign the list of values below to the name ABC and describe the result of entering the following:

a. ABC

b. **L** ABC (where the symbol **L** can be obtained by pressing **2ND** **STAT**, selecting OPS, and selecting the last item of **L**).

c. **L** ABC followed by pressing the .

0   −2   5   8   1

Review

Instead of presenting formal statistics procedures, this chapter emphasizes a general understanding of some important issues related to uses of statistics. Definitions of the following terms were presented in this chapter, and they should be known and clearly understood: *sample, population, statistic, parameter, quantitative data, categorical data, voluntary response sample, observational study, experiment,* and *simple random sample.* Section 1-2 introduced statistical thinking, and addressed issues involving the context of data, source of data, sampling method, conclusions, and practical implications. Section 1-3 discussed different types of data, and the distinction between categorical data and quantitative data should be well understood. Section 1-4 dealt with the use of critical thinking in analyzing and evaluating statistical results. In particular, we should know that for statistical purposes, some samples (such as voluntary response samples) are very poor. Section 1-5 introduced important items to consider when collecting sample data. On completing this chapter, you should be able to do the following:

• Distinguish between a population and a sample and distinguish between a parameter and a statistic

• Recognize the importance of good sampling methods in general, and recognize the importance of a *simple random sample* in particular. Understand that if sample data are not collected in an appropriate way, the data may be so completely useless that no amount of statistical torturing can salvage them.

Statistical Literacy and Critical Thinking

1. Election Survey *Literary Digest* magazine mailed 10 million sample ballots to potential voters, and 2.3 million responses were received. Given that the sample is so large, was it reasonable to expect that the sample would be representative of the population of all voters? Why or why not?

2. Movie Data Data Set 9 in Appendix B includes a sample of movie titles and their lengths (in minutes).

a. Are the lengths categorical or quantitative data?

b. Are the lengths discrete or continuous?

c. Are the data from an observational study or an experiment?

d. What is the level of measurement of the titles (nominal, ordinal, interval, ratio)?

e. What is the level of measurement of the lengths (nominal, ordinal, interval, ratio)?

3. Gallup Poll The typical Gallup poll involves interviews with about 1000 subjects. How must the survey subjects be selected so that the resulting sample is a simple random sample?

4. Sampling The U.S. Census Bureau provided the average (mean) travel time to work (in minutes) for each state and the District of Columbia for a recent year. If we find the average (mean) of those 51 values, we get a result of 22.4 minutes. Is this result the average (mean) travel time to work for the United States? Why or why not?

Chapter Quick Quiz

1. True or false: The collection of all cars registered in the United States is an example of a population.

2. Are weights of motorcycles discrete data or continuous data?

3. True or false: Selecting every fifth name on a list results in a simple random sample.

4. True or false: The average (mean) age of people who respond to a particular survey is an example of a parameter.

5. For a study in which subjects are treated with a new drug and then observed, is the study observational or is it an experiment?

6. True or false: Eye colors are an example of ordinal data.

7. Fill in the blank: A parameter is a numerical measurement describing some characteristic of a _____ .

8. Are movie ratings of G, PG-13, and R quantitative data or categorical data?

9. What is the level of measurement of data consisting of the book categories of science, literature, mathematics, and history (nominal, ordinal, interval, ratio)?

10. A pollster calls 500 randomly selected people, and all 500 respond to her first question. Because the subjects agreed to respond, is the sample a voluntary response sample?

Review Exercises

1. Sampling Seventy-two percent of Americans squeeze their toothpaste tube from the top. This and other not-so-serious findings are included in *The First Really Important Survey of American Habits*. Those results are based on 7000 responses from the 25,000 questionnaires that were mailed.

a. What is wrong with this survey?

b. As stated, the value of 72% refers to all Americans, so is that 72% a statistic or a parameter? Explain.

c. Does the survey constitute an observational study or an experiment?

2. Gallup Polls When Gallup and other polling organizations conduct polls, they typically contact subjects by telephone. In recent years, many subjects refuse to cooperate with the poll. Are the poll results likely to be valid if they are based on only those subjects who agree to respond? What should polling organizations do when they encounter a subject who refuses to respond?

3. Identify the level of measurement (nominal, ordinal, interval, ratio) used in each of the following.

a. The pulse rates of women listed in Data Set 1 of Appendix B

b. The genders of the subjects included in the Freshman 15 Study Data (Data Set 3 in Appendix B)

c. The body temperatures (in degrees Fahrenheit) of the subjects listed in Data Set 2 of Appendix B

d. A movie critic's ratings of "must see, recommended, not recommended, don't even think about going"

4. Identify the level of measurement (nominal, ordinal, interval, ratio) used in each of the following.

a. The eye colors of all fellow students in your statistics class

b. The ages (in years) of homes sold, as listed in Data Set 23 of Appendix B

c. The age brackets (under 30, 30–49, 50–64, over 64) recorded as part of a Pew Research Center poll about global warming

d. The actual temperatures (in degrees Fahrenheit) recorded and listed in Data Set 11 of Appendix B

5. IBM Survey The computer giant IBM has 329,373 employees and 637,133 stockholders. A vice president plans to conduct a survey to study the numbers of shares held by individual stockholders.

a. Are the numbers of shares held by stockholders discrete or continuous?

b. Identify the level of measurement (nominal, ordinal, interval, ratio) for the numbers of shares held by stockholders.

c. If the survey is conducted by telephoning 20 randomly selected stockholders in each of the 50 United States, what type of sampling (random, systematic, convenience, stratified, cluster) is being used?

d. If a sample of 1000 stockholders is obtained, and the average (mean) number of shares is calculated for this sample, is the result a statistic or a parameter?

e. What is wrong with gauging stockholder views about employee benefits by mailing a questionnaire that IBM stockholders could complete and mail back?

6. IBM Survey Identify the type of sampling (random, systematic, convenience, stratified, cluster) used when a sample of the 637,133 stockholders is obtained as described. Then determine whether the sampling scheme is likely to result in a sample that is representative of the population of all 637,133 stockholders.

a. A complete list of all stockholders is compiled and every 500th name is selected.

b. At the annual stockholders' meeting, a survey is conducted of all who attend.

c. Fifty different stockbrokers are randomly selected, and a survey is made of all their clients who own shares of IBM.

d. A computer file of all IBM stockholders is compiled so that they are all numbered consecutively, then random numbers generated by computer are used to select the sample of stockholders.

e. All of the stockholder zip codes are collected, and 5 stockholders are randomly selected from each zip code.

7. Percentages

a. Data Set 9 in Appendix B includes a sample of 35 movies, and 12 of them have ratings of R. What percentage of these 35 movies have R ratings?

b. In a study of 4544 students in grades 5 through 8, it was found that 18% had tried smoking (based on data from "Relation between Parental Restrictions on Movies and Adolescent Use of Tobacco and Alcohol," by Dalton et al., *Effective Clinical Practice,* Vol. 5, No. 1). How many of the 4544 students tried smoking?

8. JFK

a. When John F. Kennedy was elected to the presidency, he received 49.72% of the 68,838,000 votes cast. The collection of all of those votes is the population being considered. Is 49.72% a parameter or a statistic?

b. Part (a) gives the total votes cast in the 1960 presidential election. Consider the total numbers of votes cast in all presidential elections. Are those values discrete or continuous?

c. What is the number of votes that Kennedy received when he was elected to the presidency?

9. Percentages

a. The labels on U-Turn protein energy bars include the statement that these bars contain "125% less fat than the leading chocolate candy brands" (based on data from *Consumer Reports* magazine). What is wrong with that claim?

b. In a Pew Research Center poll on driving, 58% of the 1182 respondents said that they like to drive. What is the actual number of respondents who said that they like to drive?

c. In a Pew Research Center poll on driving, 331 of the 1182 respondents said that driving is a chore. What percentage of respondents said that driving is a chore?

10. Why the Discrepancy? A Gallup poll was taken two years before a presidential election, and it showed that Hillary Clinton was preferred by about 50% more voters than Barack Obama. The subjects in the Gallup poll were randomly selected and surveyed by telephone. An America Online (AOL) poll was conducted at the same time as the Gallup poll, and it showed that Barack Obama was preferred by about twice as many respondents as Hillary Clinton. In the AOL poll, Internet users responded to voting choices that were posted on the AOL site. How can the large discrepancy between the two polls be explained? Which poll is more likely to reflect the true opinions of American voters?

Cumulative Review Exercises

For Chapters 2–15, the Cumulative Review Exercises include topics from preceding chapters. For this chapter, we present *calculator warm-up exercises,* with expressions similar to those found throughout this book. Use your TI-83/84 Plus calculator to find the indicated values.

TI 1. Cigarette Nicotine Refer to the nicotine amounts (in milligrams) of the 25 king-size cigarettes listed in Data Set 4 in Appendix B. What value is obtained when those 25 amounts are added, and the total is then divided by 25? (This result, called the *mean,* is discussed in Chapter 3.)

TI 2. Movie Lengths Refer to the lengths (in minutes) of the 35 movies listed in Data Set 9 in Appendix B. What value is obtained when those 35 amounts are added, and the total is then divided by 35? (This result, called the *mean,* is discussed in Chapter 3.) Round the result to one decimal place.

TI 3. Height of Shaquille O'Neal Standardized The given expression is used to convert the height of basketball star Shaquille O'Neal to a standardized score. Round the result to two decimal places.

$$\frac{85 - 80}{3.3}$$

TI 4. Quality Control for Cola The given expression is used for determining whether a sample of cans of Coca Cola are being filled with amounts having an average (mean) that is less than 12 oz. Round the result to two decimal places.

$$\frac{12.13 - 12.00}{\frac{0.12}{\sqrt{24}}}$$

TI 5. Determining Sample Size The given expression is used to determine the size of the sample necessary to estimate the proportion of adults who have cell phones.

$$\left[\frac{1.96 \cdot 0.25}{0.01}\right]^2$$

TI 6. Motorcycle Helmets and Injuries The given expression is part of a calculation used to study the relationship between the colors of motorcycle helmets and injuries. Round the result to four decimal places.

$$\frac{(491 - 513.174)^2}{513.174}$$

TI 7. Variation in Body Temperatures The given expression is used to compute a measure of variation (variance) of three body temperatures.

$$\frac{(98.0 - 98.4)^2 + (98.6 - 98.4)^2 + (98.6 - 98.4)^2}{3 - 1}$$

TI 8. Standard Deviation The given expression is used to compute the standard deviation of three body temperatures. (The standard deviation is introduced in Section 3-3.) Round the result to three decimal places.

$$\sqrt{\frac{(98.0 - 98.4)^2 + (98.6 - 98.4)^2 + (98.6 - 98.4)^2}{3 - 1}}$$

Scientific Notation *In Exercises 9–12, the given expressions are designed to yield results expressed in a form of scientific notation. For example, the calculator displayed result of 1.23E5 can be expressed as 123,000, and the result of 4.56E-4 can be expressed as 0.000456. Perform the indicated operation and express the result as an ordinary number that is not in scientific notation.*

 **9.** 0.4^{12}　　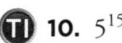 **10.** 5^{15}　　**TI 11.** 9^{11}　　**TI 12.** 0.25^{6}

TI-83/84 Plus Calculator Project

a. Begin by downloading the APP named TRIOLA11.

b. Data Set 8 in Appendix B lists counts of words spoken by samples of males and females. The APP name for Data Set 8 is WORDS. Using the TRIOLA11 APP, load the twelve individual files in the APP named WORDS. (The individual files are named M1, F1, M2, F2, and so on.) *Hint:* When you see the menu items of Sel, All, and Load at the top, select **All** and press **ENTER**, then select **Load**, select the second menu item of **Load**, and press **ENTER**.

c. Quit by pressing **2ND** **MODE**.

d. Press **STAT**, then select **CALC** and press **ENTER** to select the default menu item of **1-Var Stats**. Access the list of variable names by pressing **2ND** **STAT**, then continue to press ⌄ until reaching the name of M1. Press **ENTER** **ENTER**.

e. The bottom line of the display should show $n =$ followed by a number. What is that number? What does it represent?

f. Identify the values of $\bar{x}$ and Sx. The value of $\bar{x}$ is the *mean* of the values listed in the first column of Data Set 8. (The *mean* is defined in Section 3-2.) The value of Sx is the *standard deviation* of the values listed in the first column of Data Set 8. (The *standard deviation* is defined in Section 3-3.)

INTERNET PROJECT

Web Site

Go to: **www.aw.com/triola**

In this section of each chapter, you will be instructed to visit the home page on the Web site for this textbook. From there you can reach the pages for all the Internet Projects accompanying this book. Go to this Web site now and familiarize yourself with all of the available features for the book.

Each Internet Project includes activities, such as exploring data sets, performing simulations, and researching true-to-life examples found at various Web sites. These activities will help you explore and understand the rich nature of statistics and its importance in our world. Visit the book site now and enjoy the explorations!

APPLET PROJECT

The CD included with this book contains applets designed to help visualize various concepts. Open the Applets folder on the CD and click on **Start.** Select the menu item of **Sample from a population.** Use the default distribution of **Uniform,** but change the sample size to $n = 1000$. Proceed to click on the button labeled **Sample** several times and comment on how much the results change. (Ignore the values of the mean, median, and standard deviation, and consider only the shape of the distribution of the data.) Are the changes more dramatic with a sample size of $n = 10$? What does this suggest about samples in general?

Critical Thinking

The concept of "six degrees of separation" grew from a 1967 study conducted by psychologist Stanley Milgram. His original finding was that two random residents in the United States are connected by an average of six intermediaries. In his first experiment, he sent 60 letters to subjects in Wichita, Kansas, and they were asked to forward the letters to a specific woman in Cambridge, Massachusetts. The subjects were instructed to hand deliver the letters to acquaintances who they believed could reach the target person either directly or through other acquaintances.

Of the 60 subjects, 50 participated, and three of the letters reached the target. Two subsequent experiments had low completion rates, but Milgram eventually reached a 35% completion rate and he found that for completed chains, the mean number of intermediaries was around six. Consequently, Milgram's original data led to the concept referred to as "six degrees of separation."

Analyzing the Results

1. Did Stanley Milgram's original experiment have a good design, or was it flawed? Explain.

2. Do Milgram's original data justify the concept of "six degrees of separation?"

3. Describe a sound experiment for determining whether the concept of six degrees of separation is valid.

Cooperative Group Activities

1. In-class activity From the cafeteria, obtain 18 straws. Cut 6 of them in half, cut 6 of them into quarters, and leave the other 6 as they are. There should now be 42 straws of 3 different lengths. Put them in a bag, mix them up, then select one straw, find its length, then replace it. Repeat this until 20 straws have been selected. (*Important:* Select the straws without looking into the bag, and select the first straw that is touched.) Find the average (mean) of the lengths of the sample of 20 straws. Now remove all of the straws and find the mean of the lengths of the population. Did the sample provide an average that was close to the true population average? Why or why not?

2. In-class activity In mid-December of a recent year, the Internet service provider America Online (AOL) ran a survey of its users. This question was asked about Christmas trees: "Which do you prefer?" The response could be "a real tree" or "a fake tree." Among the 7073 responses received by the Internet users, 4650 indicated a real tree, and 2423 indicated a fake tree. We have already noted that because the sample is a voluntary response sample, no conclusions can be made about a population larger than the 7073 people who responded. Identify other problems with this survey question.

3. In-class activity Identify the problems with the following:

• A recent televised report on *CNN Headline News* included a comment that crime in the United States fell in the 1980s because of the growth of abortions in the 1970s, which resulted in fewer unwanted children.

• *Consumer Reports* magazine mailed an Annual Questionnaire about cars and other consumer products. Also included were a request for a voluntary contribution of money and a ballot for the Board of Directors. Responses were to be mailed back in envelopes that required postage stamps.

4. Find a professional journal with an article that uses a statistical analysis of an experiment. Describe and comment on the design of the experiment. Identify one particular issue and determine whether the result was found to be statistically significant. Determine whether that same result has practical significance.

NAME:	Peter Katsingris
JOB:	Vice President
COMPANY:	National Audience Insights, The Nielsen Company

Peter Katsingris, Vice President of National Audience Insights at the Nielsen Company, provides insight on audience estimates and researches issues related to these estimates. Depending on the situation, different statistical approaches are used to dig deep into the data to determine what is occurring.

Q: Is your use of probability and statistics increasing, decreasing, or remaining stable?

A: With all the changes occurring within the media industry, more questions arise about audience measurements, and it is very likely we will increase the amount of statistical analyses we conduct.

Q: How critical do you find your knowledge of statistics for performing your responsibilities?

A: Understanding statistical implications of data is extremely important in our research. Knowing why things happen and then explaining the results is crucial to us and our clients. I consider what I do to be similar to the scientific crime scene investigators we see on television. We use the data and statistical methods to determine the causes for the issues we face.

Q: Cite an example of how your data are used.

A: Recently, ratings for a particular channel were increasing, without any apparent reason. A client from another channel came to us with concerns that this increase might be coming at the expense of their own programming. After lengthy research, we found that this was not the case. We discovered a correlation between the number of households with a particular viewing device and the ratings for that particular channel. As homes with this device increased, the ratings also increased. As the use of this device stabilizes, the ratings will tend to be more consistent.

Q: In terms of statistics, what would you recommend for prospective employees?

A: In my line of work, I would recommend an introductory course in statistics. Any additional training can be learned later on the job or through additional courses.

Q: Do you recommend statistics for today's college students?

A: Absolutely. A basic understanding of statistics is necessary in any job you might have. The use of statistics can help provide answers to many questions you might come across in life, whether it be in starting and running your own business, buying a home, or purchasing life insurance. I hope that students today keep a basic understanding of statistics with them. It will prove helpful, even though they might not know that now.

Q: Which other skills are important for today's college students?

A: Being a math and statistics major, I feel I was lacking in the areas of writing and making presentations. I have certainly developed these over time, but having some experience while in college would definitely have been beneficial.

2-1 Review and Preview

2-2 Frequency Distributions

2-3 Histograms

2-4 Statistical Graphics

2-5 Critical Thinking: Bad Graphs

2

Summarizing and Graphing Data

Are the survey results presented in a way that is fair and objective?

At age 26, Terri Schiavo was married and was seeking to have a child when she collapsed from respiratory and cardiac arrest. Attempts to revive her were unsuccessful and she went into a coma. She was declared to be in a persistent vegetative state in which she appeared to be awake but unaware. She remained in that state for 15 years, unable to communicate or care for herself in any way. She was kept alive through the insertion of a feeding tube. There were intense debates about her situation, with some arguing that she should be allowed to die without the feeding tube, while others argued that her life should be preserved with the feeding tube and any other necessary means. After many legal battles, her feeding tube was removed, and Terri Schiavo died 13 days later at the age of 41. Although there were very different and strong opinions about Terri Schiavo's medical treatment, there was universal sympathy for her.

In the midst of the many debates about the removal of Terri Schiavo's feeding tube, there was a CNN/*USA Today*/Gallup poll in which respondents were asked this question: "Based on what you have heard or read about the case, do you agree with the court's decision to have the feeding tube removed?" The survey was conducted by telephone and there were 909 responses from adults in the United States. Respondents were also asked about their political party affiliations, and a bar graph similar to Figure 2-1 was placed on the CNN Web site. Figure 2-1 shows the poll results broken down by political party. Based on Figure 2-1, it appears that responses by Democrats were substantially different from responses by Republicans and Independents.

We will not address the human issues related to the removal of the feeding tube, although it raises important questions that everyone should carefully consider. Instead, we will focus on the graph in Figure 2-1. Our understanding of graphs and the information they convey will help us answer this question: Does Figure 2-1 fairly represent the survey results?

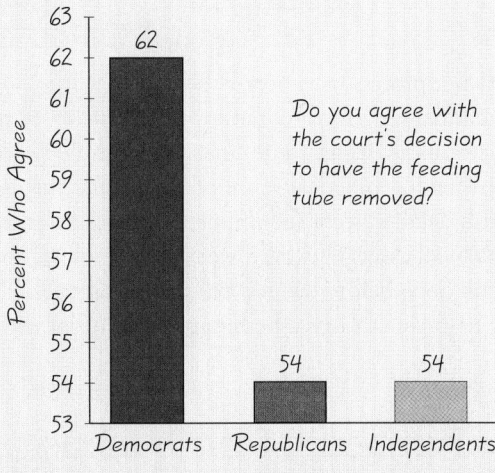

Figure 2-1 **Survey Results by Party**

No Phones or Bathtubs

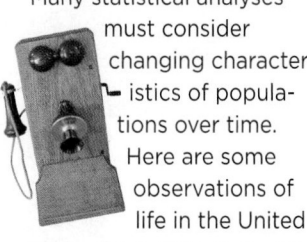

Many statistical analyses must consider changing character- istics of popula- tions over time. Here are some observations of life in the United States from 100 years ago:

- 8% of homes had a tele- phone.
- 14% of homes had a bathtub.
- The mean life expectancy was 47 years
- The mean hourly wage was 22 cents.
- There were approximately 230 murders in the entire United States.

Although these observations from 100 years ago are in stark contrast to the United States of today, statistical analyses should always con- sider changing population characteristics which might have more subtle effects.

Review and Preview

Chapter 1 discussed statistical thinking and methods for collecting data and identifying types of data. Chapter 1 also discussed consideration of the context of the data, the source of the data, and the sampling method. Samples of data are often large; to analyze such large data sets, we must organize, summarize, and represent the data in a conve- nient and meaningful form. Often we organize and summarize data numerically in ta- bles or visually in graphs, as described in this chapter. The representation we choose de- pends on the type of data we collect. However, our ultimate goal is not only to obtain a table or graph, but also to analyze the data and understand what it tells us. In this chap- ter we are mainly concerned with the *distribution* of the data set, but that is not the only characteristic of data that we will study. The general characteristics of data are listed here. (Note that we will address the other characteristics of data in later chapters.)

Characteristics of Data

1. **Center:** A representative or average value that indicates where the middle of the data set is located.

2. **Variation:** A measure of the amount that the data values vary.

3. **Distribution:** The nature or shape of the spread of the data over the range of values (such as bell-shaped, uniform, or skewed).

4. **Outliers:** Sample values that lie very far away from the vast majority of the other sample values.

5. **Time:** Changing characteristics of the data over time.

Study Hint: Blind memorization is not effective in remembering information. To re- member the above characteristics of data, it may be helpful to use a memory device called a mnemonic for the first five letters **CVDOT.** One such mnemonic is "**C**omputer **V**iruses **D**estroy **O**r **T**erminate." Memory devices are effective in recalling key words related to key concepts.

Critical Thinking and Interpretation: Going Beyond Formulas and Manual Calculations

Statistics professors generally believe that it is not so important to memorize formulas or manually perform complex arithmetic calculations. Instead, they focus on obtain- ing results by using some form of technology (calculator or computer software), and then making practical sense of the results through critical thinking. This chapter in- cludes detailed steps for important procedures, but it is not necessary to master those steps in all cases. However, we recommend that in each case you perform a few manual calculations before using a technological tool. This will enhance your understanding and help you acquire a better appreciation of the results obtained from the technology.

Frequency Distributions

Key Concept When working with large data sets, it is often helpful to organize and summarize the data by constructing a table called a *frequency distribution,* defined be- low. Because computer software and calculators can automatically generate frequency distributions, the details of constructing them are not as essential as what they tell us about data sets. In particular, a frequency distribution helps us understand the nature of the *distribution* of a data set.

DEFINITION

A **frequency distribution** (or **frequency table**) shows how a data set is partitioned among all of several categories (or classes) by listing all of the categories along with the number of data values in each of the categories.

Consider pulse rate measurements (in beats per minute) obtained from a simple random sample of 40 males and another simple random sample of 40 females, with the results listed in Table 2-1 (from Data Set 1 in Appendix B). Our pulse is extremely important, because it's difficult to function without it! Physicians use pulse rates to assess the health of patients. A pulse rate that is abnormally high or low suggests that there might be some medical issue; for example, a pulse rate that is too high might indicate that the patient has an infection or is dehydrated.

Table 2-1 Pulse Rates (beats per minute) of Females and Males

Females																			
76	72	88	60	72	68	80	64	68	68	80	76	68	72	96	72	68	72	64	80
64	80	76	76	76	80	104	88	60	76	72	72	88	80	60	72	88	88	124	64

Males																			
68	64	88	72	64	72	60	88	76	60	96	72	56	64	60	64	84	76	84	88
72	56	68	64	60	68	60	60	56	84	72	84	88	56	64	56	56	60	64	72

Table 2-2 is a frequency distribution summarizing the pulse rates of females listed in Table 2-1. The **frequency** for a particular class is the number of original values that fall into that class. For example, the first class in Table 2-2 has a frequency of 12, indicating that 12 of the original pulse rates are between 60 and 69 beats per minute.

Some standard terms used in discussing and constructing frequency distributions are defined here.

Table 2-2 Pulse Rates of Females

Pulse Rate	Frequency
60–69	12
70–79	14
80–89	11
90–99	1
100–109	1
110–119	0
120–129	1

DEFINITION

Lower class limits are the smallest numbers that can belong to the different classes. (Table 2-2 has lower class limits of 60, 70, 80, 90, 100, 110, and 120.)

Upper class limits are the largest numbers that can belong to the different classes. (Table 2-2 has upper class limits of 69, 79, 89, 99, 109, 119, 129.)

Class boundaries are the numbers used to separate the classes, but without the gaps created by class limits. Figure 2-2 shows the gaps created by the class limits from Table 2-2. In Figure 2-2 we see that the values of 69.5, 79.5, . . . , 119.5 are in the centers of those gaps. These are the class boundaries. Following the pattern established, we see that the lowest class boundary is 59.5, and the highest class boundary is 129.5. So, the complete list of class boundaries is 59.5, 69.5, 79.5, . . . , 119.5, 129.5.

Class midpoints are the values in the middle of the classes. (Table 2-2 has class midpoints of 64.5, 74.5, 84.5, 94.5, 104.5, 114.5, and 124.5.) Each class midpoint is found by adding the lower class limit to the upper class limit and dividing the sum by 2.

Class width is the difference between two consecutive lower class limits or two consecutive lower class boundaries in a frequency distribution. (Table 2-2 uses a class width of 10.)

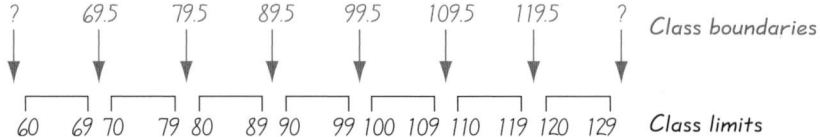

Figure 2-2   Finding Class Boundaries

CAUTION

The definitions of class width and class boundaries are a bit tricky. Be careful to avoid the easy mistake of making the class width the difference between the lower class limit and the upper class limit. See Table 2-2 and note that the class width is 10, not 9. You can simplify the process of finding class boundaries by understanding that they basically split the difference between the end of one class and the beginning of the next class, as depicted in Figure 2-2.

Procedure for Constructing a Frequency Distribution

We construct frequency distributions so that (1) large data sets can be summarized, (2) we can analyze the nature of data, and (3) we have a basis for constructing graphs (such as *histograms,* introduced in the next section). Although technology allows us to automatically generate frequency distributions, the steps for manually constructing them are as follows:

1. Determine the number of classes. The number of classes should be between 5 and 20, and the number you select might be affected by the convenience of using round numbers.

2. Calculate the class width.

$$\text{Class width} \approx \frac{(\text{maximum data value}) - (\text{minimum data value})}{\text{number of classes}}$$

   Round this result to get a convenient number. (We usually round *up.*) If necessary, change the number of classes so that they use convenient values.

3. Choose either the minimum data value or a convenient value below the minimum data value as the first lower class limit.

4. Using the first lower class limit and the class width, list the other lower class limits. (Add the class width to the first lower class limit to get the second lower class limit. Add the class width to the second lower class limit to get the third lower class limit, and so on.)

5. List the lower class limits in a vertical column and then enter the upper class limits.

6. Take each individual data value and put a tally mark in the appropriate class. Add the tally marks to find the total frequency for each class.

When constructing a frequency distribution, be sure the classes do not overlap. Each of the original values must belong to exactly one class. Include all classes, even those with a frequency of zero. Try to use the same width for all classes, although it is sometimes impossible to avoid open-ended intervals, such as "65 years or older."

EXAMPLE 1   **Pulse Rates of Females** Using the pulse rates of females in Table 2-1, follow the above procedure to construct the frequency distribution shown in Table 2-2. Use 7 classes.

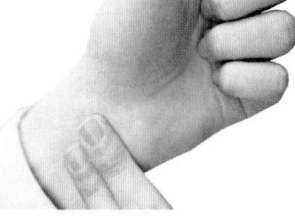

SOLUTION

Step 1: Select 7 as the number of desired classes.

Step 2: Calculate the class width. Note that we round 9.1428571 up to 10, which is a much more convenient number.

$$\text{Class width} \approx \frac{(\text{maximum data value}) - (\text{minimum data value})}{\text{number of classes}}$$
$$= \frac{124 - 60}{7} = 9.1428571 \approx 10$$

Step 3: Choose 60, which is the minimum data value and is also a convenient number, as the first lower class limit.

Step 4: Add the class width of 10 to 60 to get the second lower class limit of 70. Continue to add the class width of 10 to get the remaining lower class limits of 80, 90, 100, 110, and 120.

Step 5: List the lower class limits vertically as shown in the margin. From this list, we identify the corresponding upper class limits as 69, 79, 89, 99, 109, 119, and 129.

Step 6: Enter a tally mark for each data value in the appropriate class. Then add the tally marks to find the frequencies shown in Table 2-2.

Relative Frequency Distribution

A variation of the basic frequency distribution is a **relative frequency distribution.** In a relative frequency distribution, the frequency of a class is replaced with a relative frequency (a proportion) or a percentage frequency (a percent). Note that when percentage frequencies are used, the relative frequency distribution is sometimes called a *percentage frequency distribution.* In this book we use the term "relative frequency distribution" whether we use a relative frequency or a percentage frequency. Relative frequencies and percentage frequencies are calculated as follows.

$$\text{relative frequency} = \frac{\text{class frequency}}{\text{sum of all frequencies}}$$
$$\text{percentage frequency} = \frac{\text{class frequency}}{\text{sum of all frequencies}} \times 100\%$$

In Table 2-3 the corresponding relative frequencies expressed as percents replace the actual frequency counts from Table 2-2. With 12 of the 40 data values falling in the first class, that first class has a relative frequency of $12/40 = 0.3$ or 30%. The second class has a relative frequency of $14/40 = 0.35$ or 35%, and so on. If constructed correctly, the sum of the relative frequencies should total 1 (or 100%), with some small discrepancies allowed for rounding errors. (A sum of 99% or 101% is acceptable.)

The sum of the relative frequencies in a relative frequency distribution must be close to 1 (or 100%).

Cumulative Frequency Distribution

The **cumulative frequency** for a class is the sum of the frequencies for that class and all previous classes. The *cumulative frequency distribution* based on the frequency distribution of Table 2-2 is shown in Table 2-4. Using the original frequencies of 12, 14, 11, 1, 1, 0, and 1, we add 12 + 14 to get the second cumulative frequency of 26, then

Table 2-3 Relative Frequency Distribution of Pulse Rates of Females

Pulse Rate	Relative Frequency
60–69	30%
70–79	35%
80–89	27.5%
90–99	2.5%
100–109	2.5%
110–119	0
120–129	2.5%

Table 2-4 Cumulative Frequency Distribution of Pulse Rates of Females

Pulse Rate	Cumulative Frequency
Less than 70	12
Less than 80	26
Less than 90	37
Less than 100	38
Less than 110	39
Less than 120	39
Less than 130	40

Growth Charts Updated

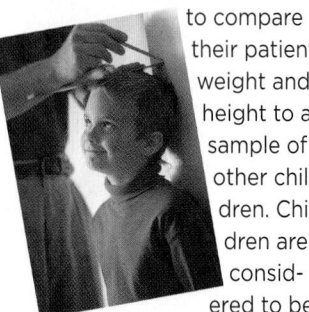

Pediatricians typically use standardized growth charts to compare their patient's weight and height to a sample of other children. Children are considered to be in the normal range if their weight and height fall between the 5th and 95th percentiles. If they fall outside of that range, they are often given tests to ensure that there are no serious medical problems. Pediatricians became increasingly aware of a major problem with the charts: Because they were based on children living between 1929 and 1975, the growth charts were found to be inaccurate. To rectify this problem, the charts were updated in 2000 to reflect the current measurements of millions of children. The weights and heights of children are good examples of populations that change over time. This is the reason for including changing characteristics of data over time as an important consideration for a population.

we add $12 + 14 + 11$ to get the third, and so on. See Table 2-4 and note that in addition to using cumulative frequencies, the class limits are replaced by "less than" expressions that describe the new ranges of values.

Critical Thinking: Interpreting Frequency Distributions

In statistics we are interested in the distribution of the data and, in particular, whether the data have a *normal distribution*. (We discuss normal distributions in detail in Chapter 6.) A frequency distribution is often one of the first tools we use in analyzing data, and it often reveals some important characteristics of the data. Here we use a frequency distribution to determine whether the data have approximately a normal distribution. Data that have an approximately normal distribution are characterized by a frequency distribution with the following features:

Normal Distribution

1. The frequencies start low, then increase to one or two high frequencies, then decrease to a low frequency.

2. The distribution is approximately symmetric, with frequencies preceding the maximum being roughly a mirror image of those that follow the maximum.

EXAMPLE 2   **Normal Distribution** IQ scores from 1000 adults were randomly selected. The results are summarized in the frequency distribution of Table 2-5. The frequencies start low, then increase to a maximum frequency of 490, then decrease to low frequencies. Also, the frequencies are roughly symmetric about the maximum frequency of 490. It appears that the distribution is approximately a normal distribution.

Table 2-5   IQ Scores of 1000 Adults

Normal distribution: **The frequencies start low, reach a maximum, then become low again. Also, the frequencies are roughly symmetric about the maximum frequency.**

IQ Score	Frequency	Normal Distribution:
50–69	24	← Frequencies start low, . . .
70–89	228	
90–109	490	← increase to a maximum, . . .
110–129	232	
130–149	26	← decrease to become low again.

Table 2-5 illustrates data with a normal distribution. The following examples illustrate how frequency distributions are used to describe, explore, and compare data sets.

EXAMPLE 3   **Describing Data: How Were the Pulse Rates Measured?** The frequency distribution in Table 2-6 summarizes the *last digits* of the pulse rates of females from Table 2-1 on page 53. If the pulse rates are measured by counting the number of heartbeats in 1 minute, we expect that the last digits should occur with frequencies that are roughly the same. But note that the frequency distribution shows that the last digits are all *even* numbers; there are *no* odd numbers present! This suggests that the pulse rates were not counted for 1 minute. Upon further examination of the *original* pulse rates, we can see that every original value is a multiple of four, suggesting that the number of heartbeats was counted for 15 seconds, then that count was multiplied by 4. It's fascinating and interesting that we are able to deduce something about the measurement procedure through an investigation of characteristics of the data.

Table 2-6 Last Digits of Female Pulse Rates

Last Digit	Frequency
0	9
1	0
2	8
3	0
4	6
5	0
6	7
7	0
8	10
9	0

Table 2-7 Randomly Selected Pennies

Weights (grams) of Pennies	Frequency
2.40–2.49	18
2.50–2.59	19
2.60–2.69	0
2.70–2.79	0
2.80–2.89	0
2.90–2.99	2
3.00–3.09	25
3.10–3.19	8

Table 2-8 Pulse Rates of Women and Men

Pulse Rate	Women	Men
50–59	0%	15%
60–69	30%	42.5%
70–79	35%	20%
80–89	27.5%	20%
90–99	2.5%	2.5%
100–109	2.5%	0%
110–119	0%	0%
120–129	2.5%	0%

EXAMPLE 4   **Exploring Data: What Does a Gap Tell Us?** Table 2-7 is a frequency distribution of the weights (grams) of randomly selected pennies. Examination of the frequencies reveals a large *gap* between the lightest pennies and the heaviest pennies. This suggests that we have two different populations. Upon further investigation, it is found that pennies made before 1983 are 97% copper and 3% zinc, whereas pennies made after 1983 are 3% copper and 97% zinc, which explains the large gap between the lightest pennies and the heaviest pennies.

Gaps Example 4 illustrates this principle: *The presence of gaps can show that we have data from two or more different populations.* However, the converse is not true, because data from different populations do not necessarily result in gaps such as that in the example.

EXAMPLE 5   **Comparing Pulse Rates of Women and Men** Table 2-1 on page 53 lists pulse rates of simple random samples of 40 females and 40 males. Table 2-8 shows the relative frequency distributions for those pulse rates. By comparing those relative frequencies, we see that pulse rates of males tend to be lower than those of females. For example, the majority (57.5%) of the males have pulse rates below 70, compared to only 30% of the females.

So far we have discussed frequency distributions using only quantitative data sets, but frequency distributions can also be used to summarize qualitative data, as illustrated in Example 6.

Table 2-9 Colleges of Undergraduates

College	Relative Frequency
Public 2-Year	36.8%
Public 4-Year	40.0%
Private 2-Year	1.6%
Private 4-Year	21.9%

EXAMPLE 6 **College Undergraduate Enrollments** Table 2-9 shows the distribution of undergraduate college student enrollments among the four categories of colleges (based on data from the U.S. National Center for Education Statistics). The sum of the relative frequencies is 100.3%, which is slightly different from 100% because of rounding errors.

EXAMPLE 7 **Education and Smoking: Frequency Distribution?** Table 2-10 is a type of table commonly depicted in media reports, but it is *not* a relative frequency distribution. (Table 2-10 is based on data from the Centers for Disease Control and Prevention.) The definition of a frequency distribution given earlier requires that the table shows how a data set is distributed among all of several categories, but Table 2-10 does not show how the population of smokers is distributed among the different education categories. Instead, Table 2-10 shows the percentage of smokers in each of the different categories. Also, the sum of the frequencies in Table 2-10 is 157%, which is clearly different from 100%, even after accounting for any rounding errors. Table 2-10 has value for conveying important information, but it is not a frequency distribution.

Table 2-10 Education and Smoking

Education	Percentage Who Smoke
0–12 (no diploma)	26%
GED diploma	43%
High school graduate	25%
Some college	23%
Associate degree	21%
Bachelor's degree	12%
Graduate degree	7%

Using the TI-83/84 Plus Calculator to Construct a Frequency Distribution

You can use a TI-83/84 Plus calculator to construct a frequency distribution from a list of sample data. The basic approach is to generate a histogram (a type of graph described in the following section), then use the (TRACE) and (WINDOW) keys to identify the class frequencies and boundaries. Use the following procedure.

TI-83/84 Plus

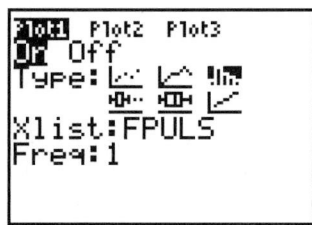

1. Enter or open a list of data. For example, from the TRIOLA11 APP, open the data set FHEALTH (Data Set 1 in Appendix B), and select FPULS, which contains the same female pulse rates listed in Table 2-1.

2. Select the STAT PLOT function by pressing **2ND** (Y=).

3. Press **ENTER** and use the arrow keys ⏷ ⧀ ⧁ ⏶ to turn Plot1 to the On state, and also highlight the graph with the bars. The screen display should be as shown in the margin.

4. *Using default settings:* If you want to let the calculator automatically determine the class width and class boundaries, press (ZOOM) (9).

    Use your own class width and boundaries: If you want to set your own class width and class boundaries, press (WINDOW) and proceed to enter the values that you desire. The Xscl value will be the class width. The entries in the displayed screen in the margin include 60 for the minimum *x* value, 135 for the upper boundary, and a class width of 10 (as in Table 2-2). You might have to estimate (guess) a value for Ymax, which is the highest class frequency, but if the choice doesn't work well, it can be easily changed.

5. Press (GRAPH), then press (TRACE). If using the settings shown in the display from Step 4, the display in the margin will appear. The blinking cursor is positioned at the top of the first bar, and for this first class we can see from the display that the class frequency is $n = 12$, and the first class includes values greater than or equal to 60 and less than 70 (as indicated by "min = 60" and "max < 70"). Because the data consist of whole numbers only, the first class has limits of 60–69, and it has a frequency of 12.

6. Now press (▷) to move the cursor to the next cell. Using the settings described above, the next cell will display a frequency of $n = 14$, and the next cell will include values greater than or equal to 70 and less than 80. The second cell therefore corresponds to a class with limits of 70–79 and a frequency of 14.

7. Continue to press (▷) to find that the remaining classes have the same class limits and frequencies given in Table 2-2.

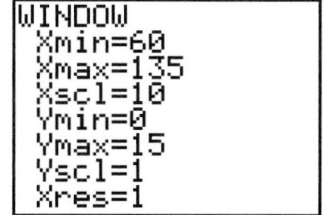

TI-83/84 Plus

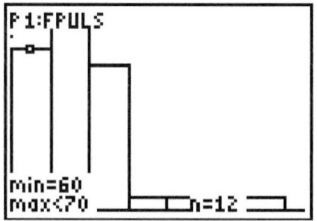

TI-83/84 Plus

2-2   Basic Skills and Concepts

Statistical Literacy and Critical Thinking

1. Frequency Distribution Table 2-7 on page 57 is a frequency distribution summarizing the weights of 72 different pennies. Is it possible to identify the original list of the 72 individual weights from Table 2-7? Why or why not?

2. Relative Frequency Distribution After constructing a relative frequency distribution summarizing IQ scores of college students, what should be the sum of the relative frequencies?

3. Unauthorized Downloading A Harris Interactive survey involved 1644 people between the ages of 8 years and 18 years. The accompanying table summarizes the results. Does this table describe a relative frequency distribution? Why or why not?

4. Analyzing a Frequency Distribution The accompanying frequency distribution summarizes the heights of a sample of people at Vassar Road Elementary School. What can you conclude about the people included in the sample?

In Exercises 5–8, identify the **class width, class midpoints,** *and* **class boundaries** *for the given frequency distribution. The frequency distributions are based on data from Appendix B.*

Table for Exercise 3

Downloaded Material	Percent
Music	32%
Games	25%
Software	14%
Movies	10%

Table for Exercise 4

Height (in.)	Frequency
35–39	6
40–44	31
45–49	67
50–54	21
55–59	0
60–64	0
65–69	6
70–74	10

5.

Tar (mg) in Nonfiltered Cigarettes	Frequency
10–13	1
14–17	0
18–21	15
22–25	7
26–29	2

6.

Tar (mg) in Filtered Cigarettes	Frequency
2–5	2
6–9	2
10–13	6
14–17	15

7. Weights (lb) of Discarded Metal	Frequency		**8.** Weights (lb) of Discarded Plastic	Frequency
0.00–0.99	5		0.00–0.99	14
1.00–1.99	26		1.00–1.99	20
2.00–2.99	15		2.00–2.99	21
3.00–3.99	12		3.00–3.99	4
4.00–4.99	4		4.00–4.99	2
			5.00–5.99	1

Critical Thinking. *In Exercises 9–12, answer the given questions that relate to Exercises 5–8.*

9. Identifying the Distribution Using a strict interpretation of the relevant criteria on page 56, does the frequency distribution given in Exercise 5 appear to have a normal distribution? Does the distribution appear to be normal if the criteria are interpreted very loosely?

10. Identifying the Distribution Using a strict interpretation of the relevant criteria on page 56, does the frequency distribution given in Exercise 6 appear to have a normal distribution? Does the distribution appear to be normal if the criteria are interpreted very loosely?

11. Comparing Relative Frequencies Construct one table (similar to Table 2-8 on page 57) that includes relative frequencies based on the frequency distributions from Exercises 5 and 6, then compare the amounts of tar in nonfiltered and filtered cigarettes. Do the cigarette filters appear to be effective?

12. Comparing Relative Frequencies Construct one table (similar to Table 2-8 on page 57) that includes relative frequencies based on the frequency distributions from Exercises 7 and 8, then compare the weights of discarded metal and plastic. Do those weights appear to be about the same or are they substantially different?

In Exercises 13 and 14, construct the **cumulative frequency distribution** *that corresponds to the frequency distribution in the exercise indicated.*

13. Exercise 5   **14.** Exercise 6

In Exercises 15 and 16, use the given qualitative data to construct the relative frequency distribution.

15. Titanic Survivors The 2223 people aboard the *Titanic* include 361 male survivors, 1395 males who died, 345 female survivors, and 122 females who died.

16. Smoking Treatments In a study, researchers treated 570 people who smoke with either nicotine gum or a nicotine patch. Among those treated with nicotine gum, 191 continued to smoke and the other 59 stopped smoking. Among those treated with a nicotine patch, 263 continued to smoke and the other 57 stopped smoking (based on data from the Centers for Disease Control and Prevention).

TI 17. Analysis of Last Digits Heights of statistics students were obtained by the author as part of a study conducted for class. The last digits of those heights are listed below. Construct a frequency distribution with 10 classes. Based on the distribution, do the heights appear to be reported or actually measured? What do you know about the accuracy of the results?

0 0 0 0 0 0 0 0 0 1 1 2 3 3 3 4 5 5 5 5 5 5 5 5 5 5 5 5 5 5 5 6 6 8 8 8 9

TI 18. Radiation in Baby Teeth Listed below are amounts of strontium-90 (in millibecquerels) in a simple random sample of baby teeth obtained from Pennsylvania residents born after 1979 (based on data from "An Unexpected Rise in Strontium-90 in U.S. Deciduous Teeth in the 1990s," by Mangano, et. al., *Science of the Total Environment*). Construct a frequency distribution with eight classes. Begin with a lower class limit of 110, and use a class width of 10. Cite a reason why such data are important.

155 142 149 130 151 163 151 142 156 133 138 161 128 144 172 137 151 166 147 163

145 116 136 158 114 165 169 145 150 150 150 158 151 145 152 140 170 129 188 156

19. Nicotine in Nonfiltered Cigarettes Refer to Data Set 4 in Appendix B and use the 25 nicotine amounts (in mg) listed for the nonfiltered king-size cigarettes. Construct a frequency distribution. Begin with a lower class limit of 1.0 mg, and use a class width of 0.20 mg.

20. Nicotine in Filtered Cigarettes Refer to Data Set 4 in Appendix B and use the 25 nicotine amounts (in mg) listed for the filtered and nonmenthol cigarettes. Construct a frequency distribution. Begin with a lower class limit of 0.2 mg, and use a class width of 0.20 mg. Compare the frequency distribution to the result from Exercise 19.

21. Home Voltage Measurements Refer to Data Set 13 in Appendix B and use the 40 home voltage measurements. Construct a frequency distribution with five classes. Begin with a lower class limit of 123.3 volts, and use a class width of 0.20 volt. Does the result appear to have a normal distribution? Why or why not?

22. Generator Voltage Measurements Refer to Data Set 13 in Appendix B and use the 40 voltage measurements from the generator. Construct a frequency distribution with seven classes. Begin with a lower class limit of 123.9 volts, and use a class width of 0.20 volt. Using a very loose interpretation of the relevant criteria, does the result appear to have a normal distribution? Compare the frequency distribution to the result from Exercise 21.

23. How Long Is a 3/4 in. Screw? Refer to Data Set 19 in Appendix B and use the 50 screw lengths to construct a frequency distribution. Begin with a lower class limit of 0.720 in., and use a class width of 0.010 in. The screws were labeled as having a length of 3/4 in. Does the frequency distribution appear to be consistent with the label? Why or why not?

24. Weights of Discarded Paper As part of the Garbage Project at the University of Arizona, the discarded garbage for 62 households was analyzed. Refer to the 62 weights of discarded paper from Data Set 22 in Appendix B and construct a frequency distribution. Begin with a lower class limit of 1.00 lb, and use a class width of 4.00 lb. Do the weights of discarded paper appear to have a normal distribution? Compare the weights of discarded paper to the weights of discarded metal by referring to the frequency distribution given in Exercise 7.

25. FICO Scores Refer to Data Set 24 in Appendix B for the FICO credit rating scores. Construct a frequency distribution beginning with a lower class limit of 400, and use a class width of 50. Does the result appear to have a normal distribution? Why or why not?

26. Regular Coke and Diet Coke Refer to Data Set 17 in Appendix B. Construct a relative frequency distribution for the weights of regular Coke. Start with a lower class limit of 0.7900 lb, and use a class width of 0.0050 lb. Then construct another relative frequency distribution for the weights of Diet Coke by starting with a lower class limit of 0.7750 lb, and use a class width of 0.0050 lb. Then compare the results to determine whether there appears to be a significant difference. If so, provide a possible explanation for the difference.

27. Weights of Quarters Refer to Data Set 20 in Appendix B and use the weights (grams) of the pre-1964 quarters. Construct a frequency distribution. Begin with a lower class limit of 6.0000 g, and use a class width of 0.0500 g.

28. Weights of Quarters Refer to Data Set 20 in Appendix B and use the weights (grams) of the post-1964 quarters. Construct a frequency distribution. Begin with a lower class limit of 5.5000 g, and use a class width of 0.0500 g. Compare the frequency distribution to the result from Exercise 27.

29. Blood Groups Listed below are blood groups of O, A, B, and AB of randomly selected blood donors (based on data from the Greater New York Blood Program). Construct a table summarizing the frequency distribution of these blood groups.

O A B O O O O O AB O O O O B O B O A A A O A A B AB

A B A A A A O A O O A A O O A O O O O A A A A A AB

30. Train Derailments An analysis of 50 train derailment incidents identified the main causes listed below, where T denotes bad track, E denotes faulty equipment, H denotes human error, and O denotes other causes (based on data from the Federal Railroad Administration). Construct a table summarizing the frequency distribution of these causes of train derailments.

T T T E E H H H H H O O H H H E E T T T E T H O T

T T T T T H T T H E E T T E E T T T H T T O O O

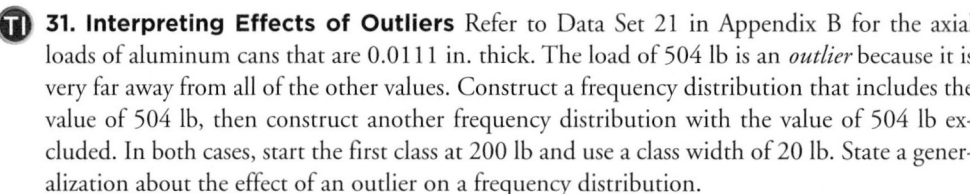

2-2   Beyond the Basics

 **31. Interpreting Effects of Outliers** Refer to Data Set 21 in Appendix B for the axial loads of aluminum cans that are 0.0111 in. thick. The load of 504 lb is an *outlier* because it is very far away from all of the other values. Construct a frequency distribution that includes the value of 504 lb, then construct another frequency distribution with the value of 504 lb excluded. In both cases, start the first class at 200 lb and use a class width of 20 lb. State a generalization about the effect of an outlier on a frequency distribution.

32. Number of Classes According to Sturges's guideline, the ideal number of classes for a frequency distribution can be approximated by $1 + (\log n)/(\log 2)$, where n is the number of data values. Use this guideline to complete the table in the margin.

Table for Exercise 32

Number of Data Values	Ideal Number of Classes
16–22	5
23–45	6
?	7
?	8
?	9
?	10
?	11
?	12

2-3   Histograms

Key Concept In Section 2-2 we introduced the frequency distribution as a tool for summarizing a large data set and determining the distribution of the data. In this section we discuss a visual tool called a *histogram,* and its significance in representing and analyzing data. Because many statistics computer programs and calculators can automatically generate histograms, it is not so important to master the mechanical procedures for constructing them. Instead we focus on the information we can obtain from a histogram. Namely, we use a histogram to analyze the shape of the distribution of the data.

> **DEFINITION**
>
> A **histogram** is a graph consisting of bars of equal width drawn adjacent to each other (without gaps). The horizontal scale represents classes of quantitative data values and the vertical scale represents frequencies. The heights of the bars correspond to the frequency values.

A histogram is basically a graphic version of a frequency distribution. For example, Figure 2-3 shows the histogram corresponding to the frequency distribution in Table 2-2 on page 53.

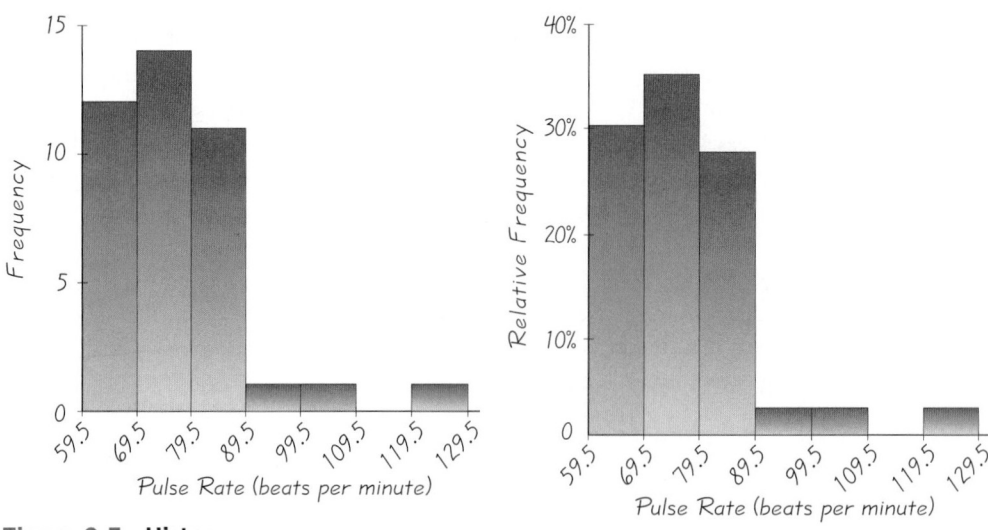

Figure 2-3   Histogram

Figure 2-4   Relative Frequency Histogram

The bars on the horizontal scale are labeled with one of the following: (1) class boundaries (as shown in Figure 2-3); (2) class midpoints; or (3) lower class limits.

The first and second options are technically correct, while the third option introduces a small error. Both axes should be clearly labeled.

> **Horizontal Scale for Histogram:** Use class boundaries or class midpoints.
>
> **Vertical Scale for Histogram:** Use the class frequencies.

Relative Frequency Histogram

A **relative frequency histogram** has the same shape and horizontal scale as a histogram, but the vertical scale is marked with relative frequencies (as percentages or proportions) instead of actual frequencies, as in Figure 2-4.

Critical Thinking: Interpreting Histograms

Remember that the objective is not simply to construct a histogram, but rather to *understand* something about the data. Analyze the histogram to see what can be learned about CVDOT: the center of the data, the variation (which will be discussed at length in Section 3-3), the distribution, and whether there are any outliers (values far away from the other values). Examining Figure 2-3, we see that the histogram is centered roughly around 80, the values vary from around 60 to 130, and the shape of the distribution is heavier on the left. The bar at the extreme right appears to represent a questionable pulse rate of about 125 beats per minute, which is exceptionally high.

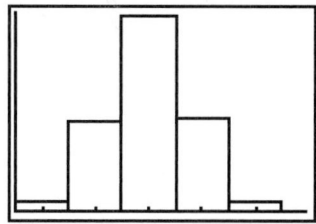

Normal Distribution When graphed, a normal distribution has a "bell" shape. Characteristics of the bell shape are (1) the frequencies increase to a maximum, and then decrease, and (2) symmetry, with the left half of the graph roughly a mirror image of the right half. The TI-83/84 Plus histogram in the margin corresponds to the frequency distribution of Table 2-5 on page 56, which was obtained from a simple random sample of 1000 IQ scores of adults in the United States. Many statistical methods require that sample data come from a population having a distribution that is approximately a normal distribution, and we can often use a histogram to determine whether this requirement is satisfied.

Using the TI-83/84 Plus Calculator to Construct Histograms

A TI-83/84 Plus calculator can be used to generate a histogram, such as the one shown in Figure 2-3. Use the following procedure.

1. Enter or open a list of data. For example, use the TRIOLA11 APP to open the FHEALTH data set (Data Set 1 in Appendix B), then select FPULS, which consists of the female pulse rates listed in Table 2-1. We illustrate the construction of a histogram by using this list of female pulse rates.

2. Select the "STAT PLOT" function by pressing (2ND) (Y=).

3. Press (ENTER) and use the arrow keys (↓)(◁)(▷)(△) to turn Plot1 to the On state, and also highlight the graph (Type) with the bars. The screen display should be as shown in the margin.

4. *Using default settings:* If you want to let the calculator determine the class width and class boundaries, press (ZOOM) (9).

   Use your own class width and boundaries: If you want to set your own class width and class boundaries, press (WINDOW) and proceed to enter the values that you desire. The Xscl value will be the class width. The entries in the displayed WINDOW screen in the margin include 60 for the minimum x value, 135 for the upper boundary, and a class width of 10 (as in Table 2-2). You might have to estimate (guess) a value for Ymax, which is the highest class frequency, but if the choice doesn't work well, it can be easily changed.

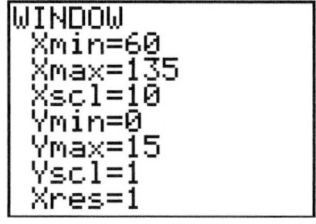

TI-83/84 Plus

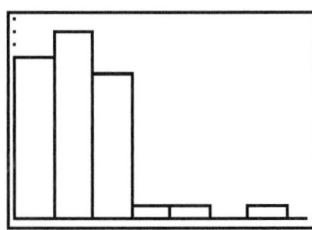

5. Press GRAPH. If using the settings shown in the display from Step 4, the result will be the top histogram in the margin. You can press TRACE to identify the heights of the bars as well as their boundaries. After pressing TRACE, you can move the cursor by pressing ▷. For example, the second histogram in the margin indicates that the third class has a frequency of 11, and it includes values greater than or equal to 80 (indicated by "min = 80") and less than 90 (indicated by "max < 90").

TI-83/84 Plus

2-3   Basic Skills and Concepts

Statistical Literacy and Critical Thinking

1. Histogram Table 2-2 is a frequency distribution summarizing the pulse rates of females listed in Table 2-1, and Figure 2-3 is a histogram depicting that same data set. When trying to better understand the pulse rate data, what is the advantage of examining the histogram instead of the frequency distribution?

2. Voluntary Response Sample The histogram in Figure 2-3 on page 62 is constructed from a *simple random sample* of women. If you construct a histogram with data collected from a *voluntary response sample*, will the distribution depicted in the histogram reflect the true distribution of the population? Why or why not?

3. Small Data The population of ages at inauguration of all U. S. Presidents who had professions in the military is 62, 46, 68, 64, 57. Why does it not make sense to construct a histogram for this data set?

4. Normal Distribution When referring to a normal distribution, does the term "normal" have the same meaning as in ordinary language? What criterion can be used to determine whether the data depicted in a histogram have a distribution that is approximately a normal distribution? Is this criterion totally objective, or does it involve subjective judgment?

TI-83/84 Plus

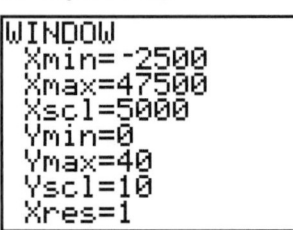

In Exercises 5–8, answer the questions by referring to the accompanying TI-83/84 Plus displays. The window settings in the first display are used for the histogram below it. (Hints: The dots on the vertical scale are 10 units apart, and the first bar in the histogram begins at a value of 2500 on the horizontal scale.) The histogram represents the numbers of miles driven by automobiles in New York City in one year.

TI-83/84 Plus

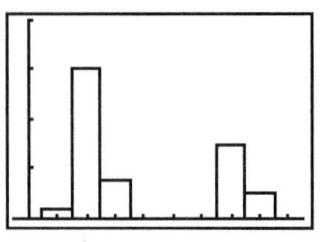

Ⓣ **5. Sample Size** How many automobiles are included in the histogram? How many of the automobiles traveled more than 20,000 miles?

Ⓣ **6. Class Width and Class Limits** What is the class width? What are the approximate lower and upper class limits of the first class?

Ⓣ **7. Variation** What is the minimum possible number of miles traveled by an automobile included in the histogram? What is the maximum possible number of miles traveled?

Ⓣ **8. Gap** What is a reasonable explanation for the large gap in the histogram?

Ⓣ **9. Analysis of Last Digits** Use the frequency distribution from Exercise 17 in Section 2-2 to construct a histogram. What can you conclude from the distribution of the digits? Specifically, do the heights appear to be reported or actually measured?

Ⓣ **10. Radiation in Baby Teeth** Use the frequency distribution from Exercise 18 in Section 2-2 to construct a histogram.

Ⓣ **11. Nicotine in Nonfiltered Cigarettes** Use the frequency distribution from Exercise 19 in Section 2-2 to construct a histogram.

Ⓣ **12. Nicotine in Filtered Cigarettes** Use the frequency distribution from Exercise 20 in Section 2-2 to construct a histogram. Compare this histogram to the histogram from Exercise 11.

Ⓣ **13. Home Voltage Measurements** Use the frequency distribution from Exercise 21 in Section 2-2 to construct a histogram. Does the result appear to be a normal distribution? Why or why not?

TI 14. Generator Voltage Measurements Use the frequency distribution from Exercise 22 in Section 2-2 to construct a histogram. Using a very loose interpretation of the relevant criteria, does the result appear to be a normal distribution? Compare this histogram to the histogram from Exercise 13.

TI 15. How Long Is a 3/4 in. Screw? Use the frequency distribution from Exercise 23 in Section 2-2 to construct a histogram. What does the histogram suggest about the length of 3/4 in., as printed on the labels of the packages containing the screws?

TI 16. Weights of Discarded Paper Use the frequency distribution from Exercise 24 in Section 2-2 to construct a histogram. Do the weights of discarded paper appear to have a normal distribution?

TI 17. FICO Scores Use the frequency distribution from Exercise 25 in Section 2-2 to construct a histogram. Does the result appear to be a normal distribution? Why or why not?

TI 18. Regular Coke and Diet Coke Use the relative frequency distributions from Exercise 26 in Section 2-2 to construct a histogram for the weights of regular Coke and another histogram for the weights of diet Coke. Compare the results and determine whether there appears to be a significant difference.

TI 19. Weights of Quarters Use the frequency distribution from Exercise 27 in Section 2-2 to construct a histogram.

TI 20. Weights of Quarters Use the frequency distribution from Exercise 28 in Section 2-2 to construct a histogram. Compare this histogram to the histogram from Exercise 19.

2-3 Beyond the Basics

21. Back-to-Back Relative Frequency Histograms When using histograms to compare two data sets, it is sometimes difficult to make comparisons by looking back and forth between the two histograms. A *back-to-back relative frequency histogram* uses a format that makes the comparison much easier. Instead of frequencies, we should use relative frequencies (percentages or proportions) so that the comparisons are not distorted by different sample sizes. Complete the back-to-back relative frequency histograms shown here by using the data from Table 2-8 on page 57. Then use the result to compare the two data sets.

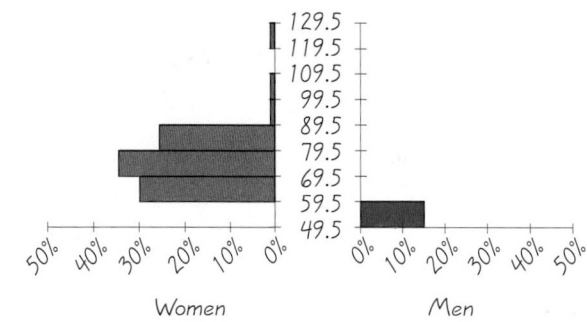

TI 22. Interpreting Effects of Outliers Refer to Data Set 21 in Appendix B for the axial loads of aluminum cans that are 0.0111 in. thick. The load of 504 lb is an *outlier* because it is very far away from all of the other values. Construct a histogram that includes the value of 504 lb, then construct another histogram with the value of 504 lb excluded. In both cases, start the first class at 200 lb and use a class width of 20 lb. State a generalization about the effect an outlier might have on a histogram.

2-4 Statistical Graphics

Key Concept In Section 2-3 we discussed histograms. In this section we discuss other types of statistical graphs. Our objective is to identify a suitable graph for representing a data set. The graph should be effective in revealing the important characteristics of the data. Although most of the graphs presented here are standard statistical graphs, statisticians are developing new types of graphs for depicting data. We examine one such graph later in the section.

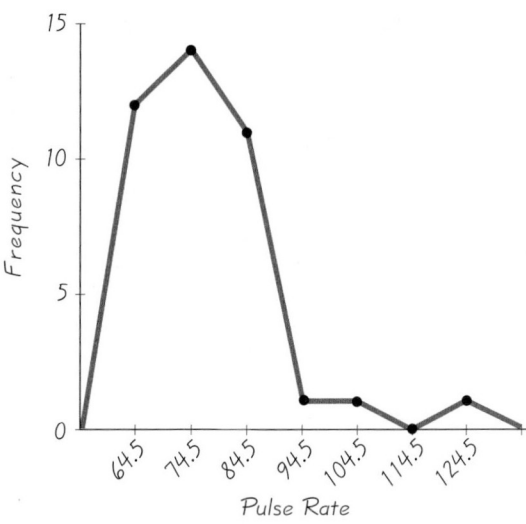

Figure 2-5   Frequency Polygon: Pulse Rates of Women

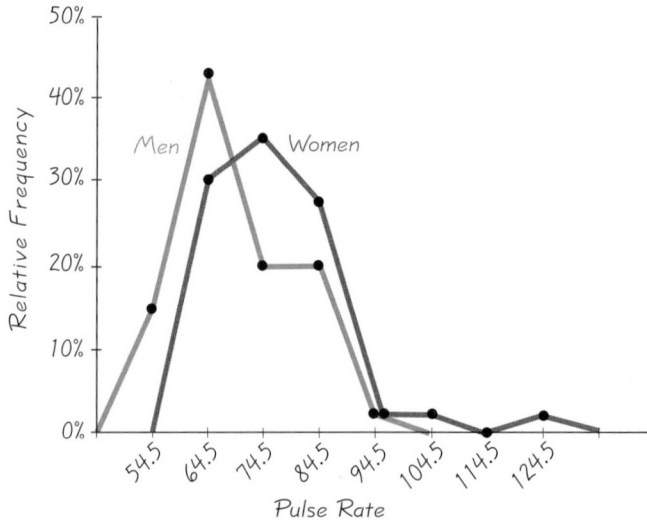

Figure 2-6   Relative Frequency Polygons: Pulse Rates of Women and Men

Frequency Polygon

One type of statistical graph involves the class midpoints. A **frequency polygon** uses line segments connected to points located directly above class midpoint values. We construct a frequency polygon from a frequency distribution as shown in Example 1. Instructions for using a TI-83/84 Plus calculator to construct a frequency polygon are given near the end of this section.

> **EXAMPLE 1**   **Frequency Polygon: Pulse Rates of Women** See Figure 2-5 for the frequency polygon corresponding to the pulse rates of women summarized in the frequency distribution of Table 2-2 on page 53. The heights of the points correspond to the class frequencies, and the line segments are extended to the right and left so that the graph begins and ends on the horizontal axis. Just as it is easy to construct a histogram from a frequency distribution table, it is also easy to construct a frequency polygon from a frequency distribution table.

A variation of the basic frequency polygon is the **relative frequency polygon,** which uses relative frequencies (proportions or percentages) for the vertical scale. When trying to compare two data sets, it is often very helpful to graph two relative frequency polygons on the same axes.

> **EXAMPLE 2**   **Relative Frequency Polygon: Pulse Rates** See Figure 2-6, which shows the relative frequency polygons for the pulse rates of women and men as listed in Table 2-1 on page 53. Figure 2-6 makes it clear that the pulse rates of men are less than the pulse rates of women (because the line representing men is farther to the left than the line representing women). Figure 2-6 accomplishes something that is truly wonderful: It enables an understanding of data that is not possible with visual examination of the lists of data in Table 2-1. (It's like a good poetry teacher revealing the true meaning of a poem.)

Ogive

Another type of statistical graph called an *ogive* (pronounced "oh-jive") involves cumulative frequencies. Ogives are useful for determining the number of values below some particular value, as illustrated in Example 3. An **ogive** is a line graph that depicts *cumulative* frequencies. An ogive uses class boundaries along the horizontal scale, and cumulative frequencies along the vertical scale.

> **EXAMPLE 3** **Ogive: Pulse Rate of Females** Figure 2-7 shows an ogive corresponding to Table 2-4 on page 55. From Figure 2-7, we see that 26 of the pulse rates are less than 79.5.

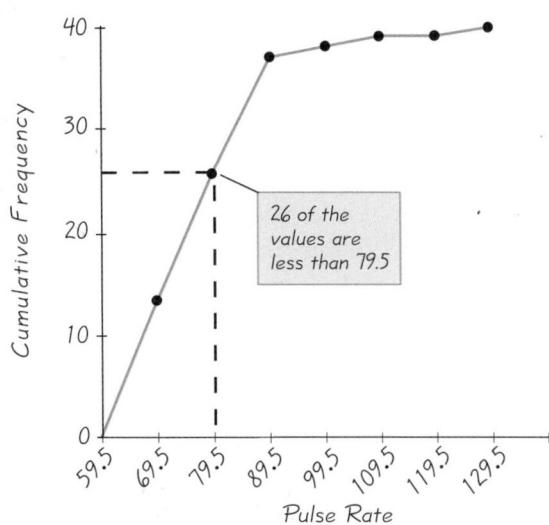

Figure 2-7 Ogive

Dotplots

A **dotplot** consists of a graph in which each data value is plotted as a point (or dot) along a scale of values. Dots representing equal values are stacked.

> **EXAMPLE 4** **Dotplot: Pulse Rate of Females** A Minitab-generated dotplot of the pulse rates of females from Table 2-1 on page 53 appears below. The three stacked dots at the left represent the pulse rates of 60, 60, and 60. The next four dots are stacked above 64, indicating that there are four pulse rates of 64 beats per minute. This dotplot reveals the distribution of the pulse rates. It is possible to recreate the original list of data values, since each data value is represented by a single point.

MINITAB

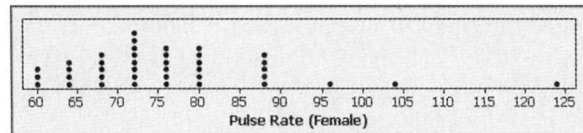

Stemplots

A **stemplot** (or **stem-and-leaf plot**) represents quantitative data by separating each value into two parts: the stem (such as the leftmost digit) and the leaf (such as the rightmost digit).

Florence Nightingale

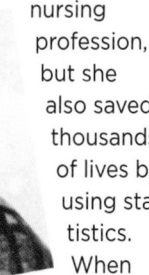

Florence Nightingale (1820–1910) is known to many as the founder of the nursing profession, but she also saved thousands of lives by using statistics. When she encountered an unsanitary and under-supplied hospital, she improved those conditions and then used statistics to convince others of the need for more widespread medical reform. She developed original graphs to illustrate that, during the Crimean War, more soldiers died as a result of unsanitary conditions than were killed in combat. Florence Nightingale pioneered the use of social statistics as well as graphics techniques.

> **EXAMPLE 5**
>
> **Stemplot: Pulse Rate of Females** The following stemplot depicts the pulse rates of females listed in Table 2-1 on page 53. The pulse rates are arranged in increasing order as 60, 60, 60, 64, . . . , 124. The first value of 60 is separated into its stem of 6 and leaf of 0, and each of the remaining values is separated in a similar way. Note that the stems and leaves are arranged in increasing order, not the order in which they occur in the original list.

Stemplot Stem (tens)	Leaves (units)	
6	000444488888	← Data values are 60, 60, 60, 64, . . . , 68.
7	22222222666666	
8	00000088888	
9	6	← Data value is 96.
10	4	← Data value is 104.
11		
12	4	

By turning the stemplot on its side, we can see a distribution of these data. One advantage of the stemplot is that we can see the distribution of data and yet retain all the information in the original list. If necessary, we could reconstruct the original list of values. Another advantage is that construction of a stemplot is a quick way to *sort* data (arrange them in order), which is required for some statistical procedures (such as finding a median, or finding percentiles).

The rows of digits in a stemplot are similar in nature to the bars in a histogram. One of the guidelines for constructing frequency distributions is that the number of classes should be between 5 and 20, and the same guideline applies to histograms and stemplots for the same reasons. Better stemplots are often obtained by first rounding the original data values. Also, stemplots can be *expanded* to include more rows and can be *condensed* to include fewer rows. See Exercise 28.

Bar Graphs

A **bar graph** uses bars of equal width to show frequencies of categories of qualitative data. The vertical scale represents frequencies or relative frequencies. The horizontal scale identifies the different categories of qualitative data. The bars may or may not be separated by small gaps. For example, Figure 2-1 included with the Chapter Problem is a bar graph. A **multiple bar graph** has two or more sets of bars, and is used to compare two or more data sets.

> **EXAMPLE 6**
>
> **Multiple Bar Graph of Gender and Income** See the following Minitab-generated multiple bar graph of the median incomes of males and females in different years (based on data from the U.S. Census Bureau). From this graph we see that males consistently have much higher median incomes than females, and that both males and females have steadily increasing incomes over time. Comparing the heights of the bars from left to right, the ratios of incomes of males to incomes of females are ratios that appear to be decreasing, which indicates that the gap between male and female median incomes is becoming smaller.

MINITAB MULTIPLE BAR GRAPH

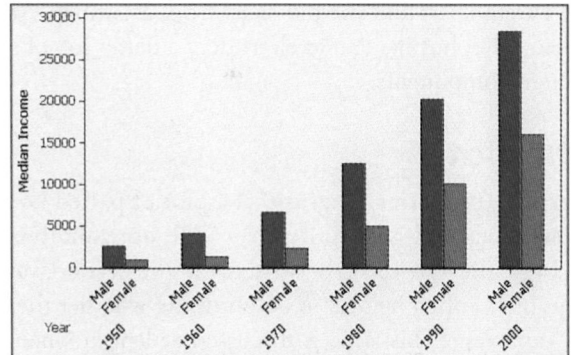

Pareto Charts

When we want to draw attention to the more important categories, we can use a *Pareto chart*. A **Pareto chart** is a bar graph for qualitative data, with the added stipulation that the bars are arranged in descending order according to frequencies. The vertical scale in a Pareto chart represents frequencies or relative frequencies. The horizontal scale identifies the different categories of qualitative data. The bars decrease in height from left to right.

EXAMPLE 7   **Pareto Chart: How to Find a Job** The following Minitab-generated Pareto chart shows how workers found their jobs (based on data from The Bernard Haldane Associates). We see that networking was the most successful way workers found their jobs. This Pareto chart suggests that instead of relying solely on such resources as school job placement personnel or newspaper ads, job applicants should actively pursue networking as a means for getting a job.

MINITAB PARETO CHART   **MINITAB PIE CHART**

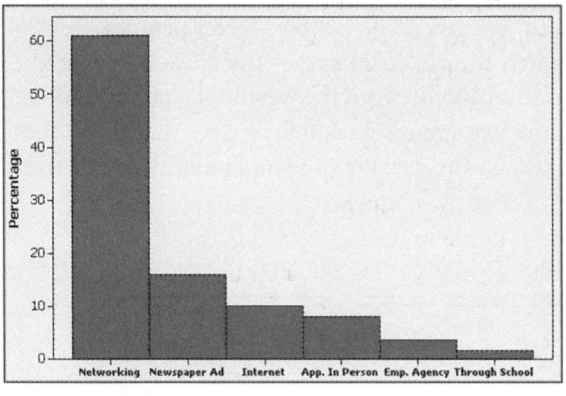   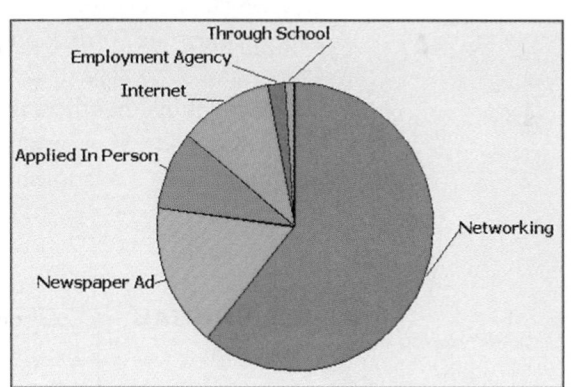

Pie Charts

A **pie chart** is a graph that depicts qualitative data as slices of a circle, in which the size of each slice is proportional to the frequency count for the category.

EXAMPLE 8   **Pie Chart: How to Find a Job** The above Minitab-generated pie chart is based on the same data used for the Pareto chart in Example 7. Construction of a pie chart involves slicing up the circle into the proper proportions that represent relative frequencies. For example, the category of networking represents 61% of the total, so the slice representing networking should be 61% of the total (with a central angle of $0.61 \times 360° = 220°$).

The Pareto chart and the pie chart from Examples 7 and 8 depict the same data in different ways, but the Pareto chart does a better job of showing the relative sizes of the different components.

Scatterplots

A **scatterplot** (or **scatter diagram**) is a plot of paired (*x, y*) quantitative data with a horizontal *x*-axis and a vertical *y*-axis. The horizontal axis is used for the first (*x*) variable, and the vertical axis is used for the second variable. The pattern of the plotted points is often helpful in determining whether there is a relationship between the two variables. (This issue is discussed at length when the topic of correlation is considered in Section 10-2.) Instructions for using a TI-83/84 Plus calculator to generate a scatterplot are given near the end of this section.

TI-83/84 Plus

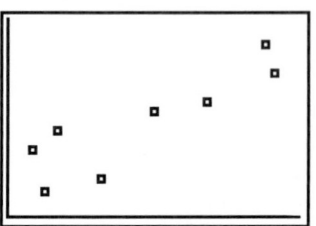

> **EXAMPLE 9**   **Scatterplot: Crickets and Temperature** One classic use of a scatterplot involves numbers of cricket chirps per minute paired with temperatures (°F). Using data from *The Song of Insects* by George W. Pierce (Harvard University Press), the TI-83/84 Plus scatterplot is shown here. There does appear to be a relationship between chirps and temperature, with increasing numbers of chirps corresponding to higher temperatures. Crickets can therefore be used as thermometers.

> **EXAMPLE 10**   **Clusters and a Gap** Consider the Minitab-generated scatterplot of paired data consisting of the weight (grams) and year of manufacture for each of 72 pennies. This scatterplot shows two very distinct clusters separated by a gap, which can be explained by the inclusion of two different populations: pre-1983 pennies are 97% copper and 3% zinc, whereas post-1983 pennies are 3% copper and 97% zinc. If we ignored the characteristic of the clusters, we might incorrectly think that there is a relationship between the weight of a penny and the year it was made. If we examine the two groups separately, we see that there does *not* appear to be a relationship between the weights of pennies and the years they were made.

MINITAB

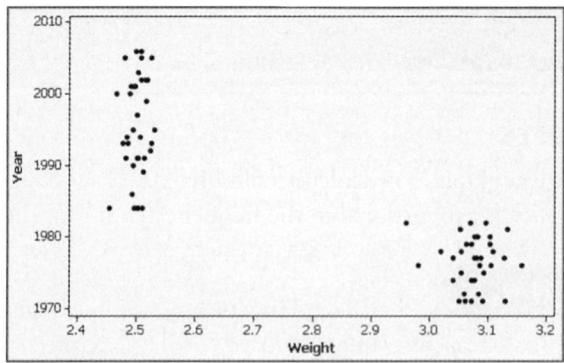

Time-Series Graph

A **time-series graph** is a graph of *time-series data,* which are quantitative data that have been collected at different points in time.

EXAMPLE 11 **Time-Series Graph: Dow Jones Industrial Average** The accompanying SPSS-generated time-series graph shows the yearly high values of the Dow Jones Industrial Average (DJIA) for the New York Stock Exchange. This graph shows a steady increase between the years 1980 and 2007, but the DJIA high values have not been so consistent in more recent years.

SPSS TIME-SERIES GRAPH

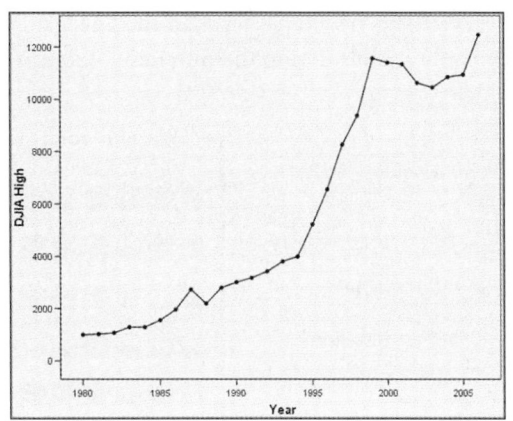

Help Wanted: Statistical Graphics Designer

In addition to the graphs we have discussed, there are many other useful graphs—some of which have not yet been created. Our society desperately needs more people who can create original graphs that give us insight into the nature of data. Currently, graphs found in newspapers, magazines, and television are too often created by reporters with a background in journalism or communications, but with little or no background in working with data.

For some really helpful information about graphs, see *The Visual Display of Quantitative Information,* second edition, by Edward Tufte (Graphics Press, P.O. Box 430, Cheshire, CT 06410). Here are a few of the important principles suggested by Tufte:

- For small data sets of 20 values or fewer, use a table instead of a graph.

- A graph of data should make the viewer focus on the true nature of the data, not on other elements, such as eye-catching but distracting design features.

- Do not distort the data; construct a graph to reveal the true nature of the data.

- Almost all of the ink in a graph should be used for the data, not for other design elements.

- Don't use screening consisting of features such as slanted lines, dots, or cross-hatching, because they create the uncomfortable illusion of movement.

- Don't use areas or volumes for data that are actually one-dimensional in nature. (For example, don't use drawings of dollar bills to represent budget amounts for different years.)

- Never publish pie charts, because they waste ink on nondata components, and they lack an appropriate scale.

Missing Data

Samples are commonly missing some data. Missing data fall into two general categories: (1) Missing values that result from random causes unrelated to the data values, and (2) missing values resulting from causes that are not random. Random causes include factors such as the incorrect entry of sample values or lost survey results. Such missing values can often be ignored because they do not systematically hide some characteristic that might significantly affect results. It's trickier to deal with values missing because of factors that are not random. For example, results of an income analysis might be seriously flawed if people with very high incomes refuse to provide those values because they fear income tax audits. Those missing high incomes should not be ignored, and further research would be needed to identify them.

EXAMPLE 12   **Car Reliability Data** Figure 2-8 exemplifies excellence in originality, creativity, and effectiveness in helping the viewer easily see complicated data in a simple format. It shows a comparison of two different cars and is based on graphs used by *Consumer's Report* magazine. See the key at the bottom of the figure showing that red is used for bad results and green is used for good results, so the color scheme corresponds to the "go" and "stop" used for traffic signals that are so familiar to drivers. (The *Consumer's Report* graphs use red for good results and black for bad results.) We see that over the past several years, the Firebrand car appears to be generally better than the Speedster car. Such information is valuable for consumers considering the purchase of a new or used car.

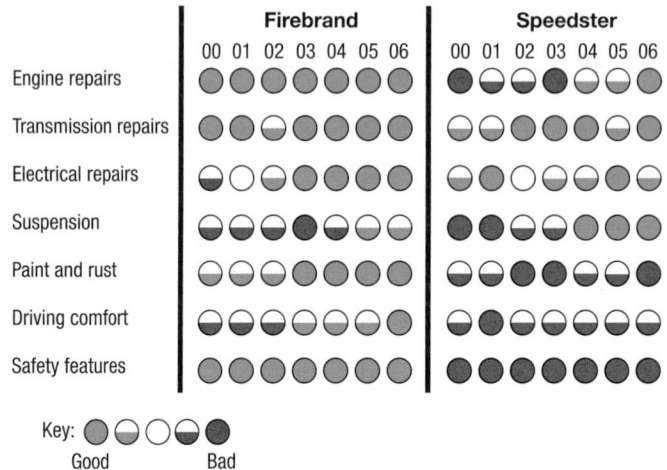

Figure 2-8   Car Reliability Data

Conclusion

In this section we saw that graphs are excellent tools for describing, exploring, and comparing data.

Describing data: In a histogram, for example, consider the distribution, center, variation, and outliers (values that are very far away from almost all of the other data values). (Remember the mnemonic of CVDOT, but the last element of time doesn't apply to a histogram, because changing patterns of data over time cannot be seen in a histogram). What is the approximate value of the center of the distribution, and what is the approximate range of values? Consider the overall shape of the distribution. Are the values evenly distributed? Is the distribution skewed (lopsided) to the right or left? Does the distribution peak in the middle? Is there a large gap, suggesting that the data might come from different populations? Identify any extreme values and any other notable characteristics.

Exploring data: We look for features of the graph that reveal some useful and/or interesting characteristics of the data set. For example, the scatterplot included with Example 9 shows that there appears to be a relationship between temperature and how often crickets chirp.

Comparing data: Construct similar graphs to compare data sets. For example, Figure 2-6 shows a frequency polygon for the pulse rates of females and another frequency polygon for pulse rates of males, and both polygons are shown on the same set of axes. Figure 2-6 makes the comparison easy.

Using the TI-83/84 Plus Calculator to Construct a Frequency Polygon

A TI-83/84 Plus calculator can be used to generate a frequency polygon, such as the one shown in Figure 2-5. Use the following procedure.

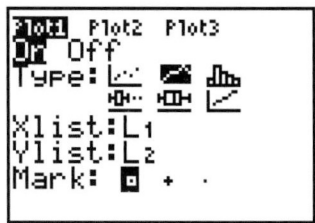

1. After constructing a frequency distribution (such as Table 2-2), enter the class midpoint values in list L1 and enter the corresponding frequencies in list L2.

2. Select the "STAT PLOT" function by pressing **2ND** (Y=).

3. Press **ENTER** and use the arrow keys ⌄ ⟨ ⟩ ⌃ to turn Plot1 to the On state, and also highlight the graph type as shown in the accompanying display. Also enter L1 for the Xlist entry and enter L2 for the Ylist entry, as shown in the first display in the margin.

TI-83/84 Plus

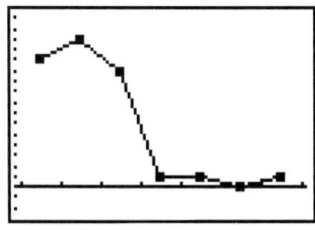

4. Press (ZOOM) (9) to obtain the second display. (Pressing (ZOOM) (9) causes the generation of a graph with settings that are automatically determined to fit the screen.) You can use the (TRACE) and arrow keys to identify frequencies and positions of the points. Note that unlike Figure 2-5, this display does not show a frequency polygon that starts and ends on the horizontal axis. To fix this feature, modify lists L1 and L2 so that they have an additional row at the top and an additional row at the bottom. For these two additional rows, use class midpoint values that follow the same pattern as the other class midpoint values, and use frequencies of 0. The result is the third display in the margin.

TI-83/84 Plus

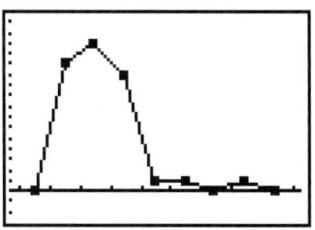

Using the TI-83/84 Plus Calculator to Construct a Scatterplot

You can use a TI-83/84 Plus calculator to construct a scatterplot from a list of paired sample data. Use the following procedure.

1. Enter or open a two lists of data with the corresponding entries paired. For example, the TRIOLA11 software application (APP) includes the data set MHEALTH with lists named MWAST and MWT, which are lists of waist sizes and weights for the 40 males in Data Set 1 in Appendix B. We will illustrate the construction of a scatterplot by using these two lists.

2. Select the STAT PLOT function by pressing **2ND** (Y=).

TI-83/84 Plus

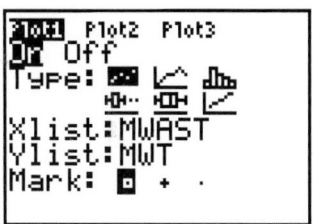

3. Press **ENTER** and use the arrow keys ⌄ ⟨ ⟩ ⌃ to turn Plot1 to the On state, and also highlight the graph type as shown in the third line of the display in the margin. Also enter the list names, as shown in the display.

4. Press (ZOOM) (9) to obtain the scatterplot, such as the one shown here. (Pressing (ZOOM) (9) causes the generation of a graph with settings that are automatically determined to fit the screen.) You can press (WINDOW) to change the settings as you desire.

TI-83/84 Plus

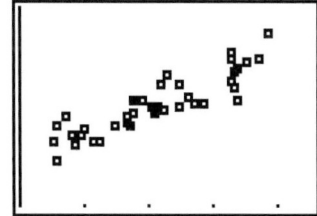

2-4   Basic Skills and Concepts

Statistical Literacy and Critical Thinking

1. Frequency Polygon Versus Dotplot Example 1 includes a frequency polygon depicting pulse rates of women, and Example 4 includes a dotplot of the same data set. What are some advantages of the dotplot over a frequency polygon?

2. Scatterplot Example 9 includes a scatterplot of temperature/chirps data. In general, what type of data is required for the construction of a scatterplot, and what does the scatterplot reveal about the data?

3. Relative Frequency Polygon Figure 2-6 includes relative frequency polygons for the pulse rates of females and males. When comparing two such data sets, why is it generally better to use relative frequency polygons instead of frequency polygons?

4. Pie Chart Versus Pareto Chart Examples 7 and 8 show a Pareto chart and pie chart for job procurement data. For such data, why is it generally better to use a Pareto chart instead of a pie chart?

In Exercises 5–8, use the listed amounts of Strontium-90 (in millibecquerels) in a simple random sample of baby teeth obtained from Pennsylvania residents born after 1979 (based on data from "An Unexpected Rise in Strontium-90 in U.S. Deciduous Teeth in the 1990s," by Mangano, et. al., Science of the Total Environment).

155 142 149 130 151 163 151 142 156 133 138 161 128 144 172 137 151 166 147 163

145 116 136 158 114 165 169 145 150 150 150 158 151 145 152 140 170 129 188 156

5. Dotplot Construct a dotplot of the amounts of Strontium-90. What does the dotplot suggest about the distribution of those amounts?

6. Stemplot Construct a stemplot of the amounts of Strontium-90. What does the stemplot suggest about the distribution of those amounts?

TI **7. Frequency Polygon** Construct a frequency polygon of the amounts of Strontium-90. For the horizontal axis, use the midpoints of the class intervals in the frequency distribution in Exercise 18 from Section 2-2: 110–119, 120–129, 130–139, 140–149, 150–159, 160–169, 170–179, 180–189.

TI **8. Ogive** Construct an ogive of the amounts of Strontium-90. For the horizontal axis, use the class boundaries corresponding to the class limits given in Exercise 7. How many of the amounts are below 150 millibecquerels?

In Exercises 9–12, use the 62 weights of discarded plastic listed in Data Set 22 of Appendix B.

9. Stemplot Use the weights to construct a stemplot. What does the stemplot suggest about the distribution of the weights?

10. Dotplot Construct a dotplot of the weights of discarded plastic. What does the dotplot suggest about the distribution of the weights?

11. Ogive Use the weights to construct an ogive. For the horizontal axis, use these class boundaries: –0.005, 0.995, 1.995, 2.995, 3.995, 4.995, 5.995. (*Hint:* See Exercise 8 in Section 2-2.) How many of the weights are below 4 lb?

TI **12. Frequency Polygon** Use the weights of discarded plastic to construct a frequency polygon. For the horizontal axis, use the midpoints of these class intervals: 0.00–0.99, 1.00–1.99, 2.00–2.99, 3.00–3.99, 4.00–4.99, 5.00–5.99.

13. Pareto Chart for Undergraduate Enrollments Table 2-9 (based on data from the U.S. National Center for Education Statistics) shows the distribution of undergraduate college student enrollments. Construct a Pareto chart for the data in Table 2-9.

14. Pie Chart for Undergraduate Enrollments Construct a pie chart for the data in Table 2-9. Compare the pie chart to the Pareto chart in Exercise 13. Which graph is more effective in showing the information in Table 2-9?

15. Pie Chart of Job Application Mistakes Chief financial officers of U.S. companies were surveyed about areas in which job applicants make mistakes. Here are the areas and the frequency of responses: interview (452); résumé (297); cover letter (141); reference checks (143); interview follow-up (113); screening call (85). These results are based on data from Robert Half Finance and Accounting. Construct a pie chart representing the given data.

Table 2-9

College	Relative Frequency
Public 2-Year	36.8%
Public 4-Year	40.0%
Private 2-Year	1.6%
Private 4-Year	21.9%

16. Pareto Chart of Job Application Mistakes Construct a Pareto chart of the data given in Exercise 15. Compare the Pareto chart to the pie chart. Which graph is more effective in showing the relative importance of the mistakes made by job applicants?

17. Pie Chart of Blood Groups Construct a pie chart depicting the distribution of blood groups from Exercise 29 in Section 2-2.

18. Pareto Chart of Blood Groups Construct a Pareto chart depicting the distribution of blood groups from Exercise 29 in Section 2-2.

19. Pareto Chart of Train Derailments Construct a Pareto chart depicting the distribution of train derailments from Exercise 30 in Section 2-2.

20. Pie Chart of Train Derailments Construct a pie chart depicting the distribution of train derailments from Exercise 30 in Section 2-2.

In Exercises 21 and 22, use the given paired data from Appendix B to construct a scatterplot.

21. Cigarette Tar/CO In Data Set 4, use tar in king-size cigarettes for the horizontal scale and use carbon monoxide (CO) in the same king-size cigarettes for the vertical scale. Determine whether there appears to be a relationship between cigarette tar and CO in king-size cigarettes. If so, describe the relationship.

22. Energy Consumption and Temperature In Data Set 12, use the 22 average daily temperatures and use the corresponding 22 amounts of energy consumption (kWh). (Use the temperatures for the horizontal scale.) Based on the result, is there a relationship between the average daily temperatures and the amounts of energy consumed? Try to identify at least one reason why there is (or is not) a relationship.

23. Time Series Graph for Moore's Law In 1965, Intel cofounder Gordon Moore proposed what has since become known as *Moore's law:* the number of transistors per square inch on integrated circuits will double approximately every 18 months. The table below lists the number of transistors per square inch (in thousands) for several different years. Construct a time-series graph of the data.

Year	1971	1974	1978	1982	1985	1989	1993	1997	1999	2000	2002	2003
Transistors	2.3	5	29	120	275	1180	3100	7500	24,000	42,000	220,000	410,000

24. Time-Series Graph for Cell Phone Subscriptions The following table shows the numbers of cell phone subscriptions (in thousands) in the United States for various years. Construct a time-series graph of the data. "Linear" growth would result in a graph that is approximately a straight line. Does the time-series graph appear to show linear growth?

Year	1985	1987	1989	1991	1993	1995	1997	1999	2001	2003	2005
Number	340	1231	3509	7557	16,009	33,786	55,312	86,047	128,375	158,722	207,900

25. Marriage and Divorce Rates The following table lists the marriage and divorce rates per 1000 people in the United States for selected years since 1900 (based on data from the Department of Health and Human Services). Construct a multiple bar graph of the data. Why do these data consist of marriage and divorce *rates* rather than total numbers of marriages and divorces? Comment on any trends that you observe in these rates, and give explanations for these trends.

Year	1900	1910	1920	1930	1940	1950	1960	1970	1980	1990	2000
Marriage	9.3	10.3	12.0	9.2	12.1	11.1	8.5	10.6	10.6	9.8	8.3
Divorce	0.7	0.9	1.6	1.6	2.0	2.6	2.2	3.5	5.2	4.7	4.2

26. Genders of Students The following table lists (in thousands) the numbers of male and female higher education students for different years. (Projections are from the U.S. National Center for Education Statistics.) Construct a multiple bar graph of the data, then describe any trends.

Year	2004	2005	2006	2007	2008	2009	2010
Males	7268	7356	7461	7568	7695	7802	7872
Females	9826	9995	10,203	10,407	10,655	10,838	10,944

2-4   Beyond the Basics

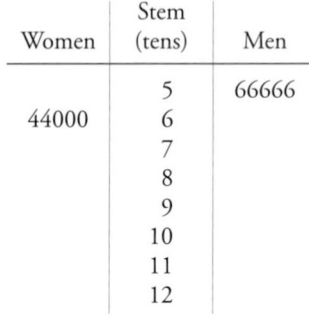

Women	Stem (tens)	Men
	5	66666
44000	6	
	7	
	8	
	9	
	10	
	11	
	12	

27. Back-to-Back Stemplots A format for *back-to-back stemplots* representing the pulse rates of females and males from Table 2-1 (on page 53) is shown in the margin. Complete the back-to-back stemplot, then compare the results.

28. Expanded and Condensed Stemplots Refer to the stemplot in Example 5 to complete the following.

a. The stemplot can be *expanded* by subdividing rows into those with leaves having digits of 0 through 4 and those with digits 5 through 9. The first two rows of the expanded stemplot are shown. Identify the next two rows.

Stem	Leaves
6	0004444 ← For leaves of 0 through 4.
6	88888 ← For leaves of 5 through 9.

b. The stemplot can be condensed by combining adjacent rows. The first row of the condensed stemplot is shown below. Note that we insert an asterisk to separate digits in the leaves associated with the numbers in each stem. Every row in the condensed plot must include exactly one asterisk so that the shape of the reduced stemplot is not distorted. Complete the condensed stemplot by inserting the remaining entries.

Stem	Leaves
6–7	000444488888*22222222666666

Critical Thinking: Bad Graphs

Key Concept   Some graphs are bad in the sense that they contain errors, and some are bad because they are technically correct, but misleading. It is important to develop the ability to recognize bad graphs and to identify exactly how they are misleading. In this section we present two of the most common types of bad graphs.

Nonzero Axis   Some graphs are misleading because one or both of the axes begin at some value other than zero, so that differences are exaggerated, as illustrated in Example 1.

EXAMPLE 1   **Misleading Bar Graph** Figure 2-1 (reproduced here) is a bar graph depicting the results of a CNN poll regarding the case of Terri Schiavo. Figure 2-9 depicts the *same survey results.* Because Figure 2-1 uses a vertical scale that does not start at zero, differences among the three response rates are exaggerated. This graph creates the incorrect impression that significantly more Democrats agreed with the court's decision than Republicans or Independents. Since Figure 2-9 depicts the data objectively, it creates the more correct impression that the differences are not very substantial. A graph like Figure 2-1 was posted on the CNN Web site, but many Internet users complained that it was deceptive, so CNN posted a modified graph similar to Figure 2-9.

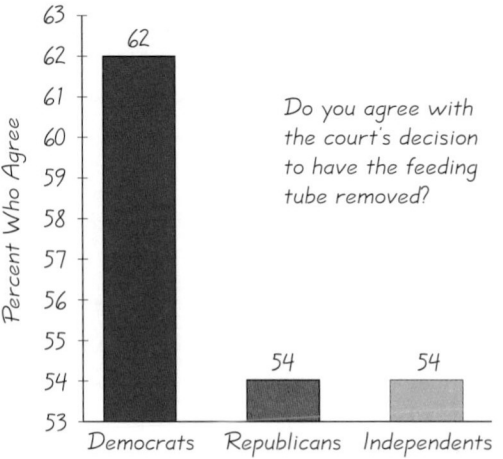

Figure 2-1   Survey Results by Party

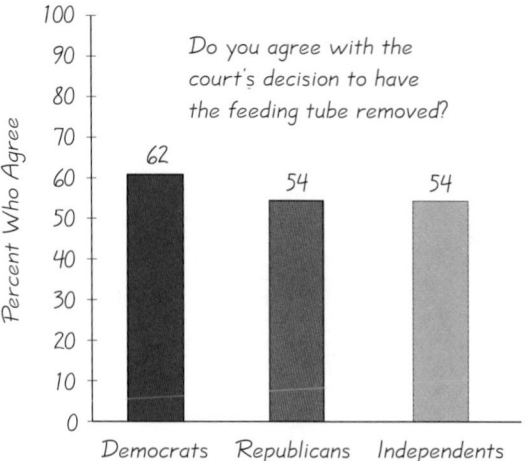

Figure 2-9   Survey Results by Party

Pictographs Drawings of objects, called *pictographs,* are often misleading. Three-dimensional objects—such as moneybags, stacks of coins, army tanks (for military expenditures), people (for population sizes), barrels (for oil production), and houses (for home construction)—are commonly used to depict data. When drawing such objects, artists can create false impressions that distort differences. (If you double each side of a square, the area doesn't merely double; it increases by a factor of four; if you double each side of a cube, the volume doesn't merely double; it increases by a factor of eight. Pictographs using areas or volumes can therefore be very misleading.)

EXAMPLE 2   **Pictograph of Incomes and Degrees** *USA Today* published a graph similar to Figure 2-10(a). Figure 2-10(a) is not misleading because the bars have the same width, but it is somewhat too busy and is somewhat difficult to understand.

Figure 2-10(b) is misleading because it depicts the same one-dimensional data with three-dimensional boxes. See the first and last boxes in Figure 2-10(b). Workers with advanced degrees have annual incomes that are approximately 4 times the incomes of those with no high school diplomas, but Figure 2-10(b) exaggerates

continued

this difference by making it appear that workers with advanced degrees have incomes that are roughly *64* times the amounts for workers with no high school diploma. (By making the box for workers with advanced degrees four times wider, four times taller, and four times deeper than the box for those with no diploma, the volumes differ by a factor of 64 instead of a factor of 4.)

In Figure 2-10(c) we use a simple bar graph to depict the data in a fair and objective way that is unencumbered by distracting features. All three parts of Figure 2-10 depict the same data from the U.S. Census Bureau.

Examples 1 and 2 illustrate two of the most common ways graphs can be misleading. Here are two points to keep in mind when critically analyzing graphs:

- Examine the graph to determine whether it is misleading because an axis does not begin at zero, so that differences are exaggerated.

- Examine the graph to determine whether objects of area or volume are used for data that are actually one-dimensional, so that differences are exaggerated.

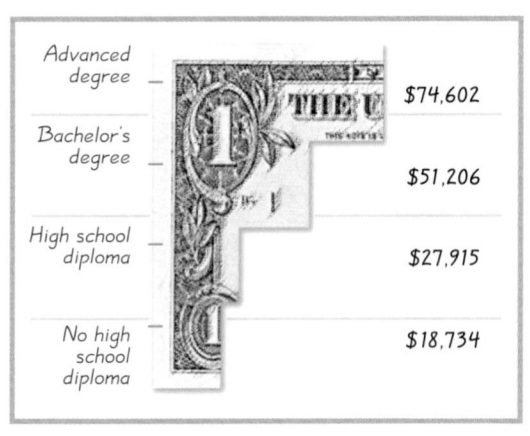

(a)

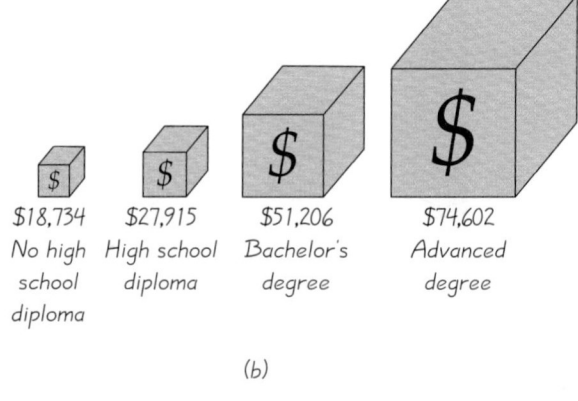

(b)

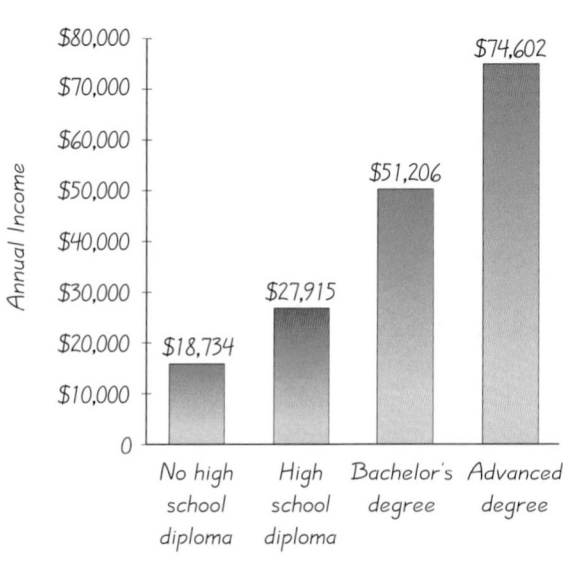

(c)

Figure 2-10   Annual Incomes of Groups with Different Education Levels

2-5   Basic Skills and Concepts

Statistical Literacy and Critical Thinking

1. Dollar Bills The *Washington Post* illustrated diminishing purchasing power of the dollar in five different presidential administrations using five different $1 bills of different sizes. The Eisenhower era was represented by a $1 with purchasing power of $1, and the subsequent administrations were represented with smaller $1 bills corresponding to lower amounts of purchasing power. What is wrong with this illustration?

2. Poll Results America Online (AOL) occasionally conducts online polls in which Internet users can respond to a question. If a graph is constructed to illustrate results from such a poll, and the graph is designed objectively with sound graphing techniques, does the graph provide us with greater understanding of the greater population? Why or why not?

3. Ethics in Statistics Assume that, as a newspaper reporter, you must graph data showing that increased smoking causes an increased risk of lung cancer. Given that people might be helped and lives might be saved by creating a graph that exaggerates the risk of lung cancer, is it ethical to construct such a graph?

4. Areas of Countries In constructing a graph that compares the land areas of the five largest countries, you choose to depict the five areas with squares of different sizes. If the squares are drawn so that the areas are in proportion to the areas of the corresponding countries, is the resulting graph misleading? Why or why not?

In Exercises 5–10, answer the questions about the graphs.

5. Graph of Weights According to data from Gordon, Churchill, Clauser, et al., women have an average (mean) weight of 137 lb or 62 kg, and men have an average (mean) weight of 172 lb or 78 kg. These averages are shown in the accompanying graph. Does the graph depict the data fairly? Why or why not?

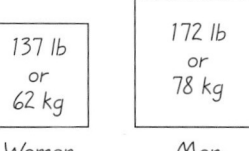

Average Weight

137 lb or 62 kg — Women
172 lb or 78 kg — Men

6. Graph of Teaching Salaries See the accompanying graph that compares teaching salaries of women and men at private colleges and universities (based on data from the U.S. Department of Education). What impression does the graph create? Does the graph depict the data fairly? If not, construct a graph that depicts the data fairly.

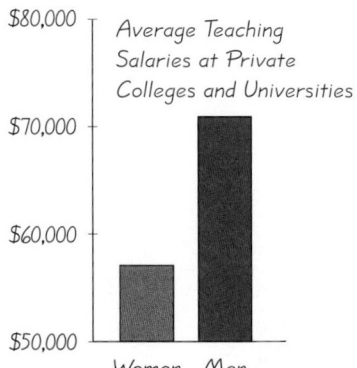

Average Teaching Salaries at Private Colleges and Universities

7. Graph of Incomes The accompanying graph depicts average full-time incomes of women and men aged 18 and over. For a recent year, those incomes were $37,197 for women and $53,059 for men (based on data from the U.S. Census Bureau). Does the graph make a fair comparison of the data? Why or why not? If the graph distorts the data, construct a fair graph.

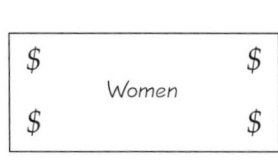

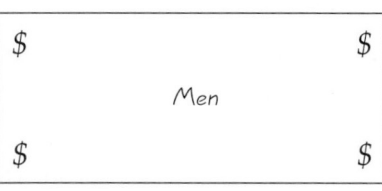

Annual Income

8. Graph of Oil Consumption The accompanying graph uses cylinders to represent barrels of oil consumed by the United States and Japan. Does the graph distort the data or does it depict the data fairly? Why or why not? If the graph distorts the data, construct a graph that depicts the data fairly.

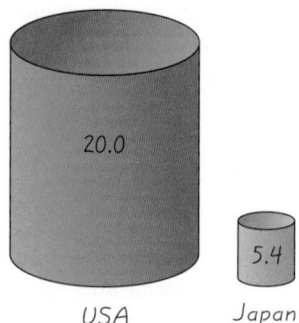

Daily Oil Consumption
(millions of barrels)

9. Braking Distances The accompanying graph shows the braking distances for different cars measured under the same conditions. Describe the ways in which this graph might be deceptive. How much greater is the braking distance of the Acura RL than the braking distance of the Volvo S80? Draw the graph in a way that depicts the data more fairly.

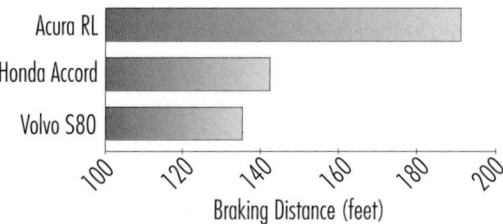

10. Adoptions from China The accompanying bar graph shows the numbers of U.S. adoptions from China in the years 2000 and 2005. What is wrong with this graph? Draw a graph that depicts the data in a fair and objective way.

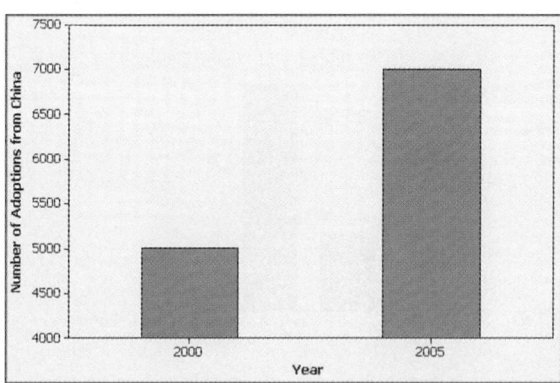

2-5   Beyond the Basics

11. Graphs in the Media A graph similar to the one on the top of the next page appeared in *USA Today*, and it used percentages represented by volumes of portions of someone's head. Are the data presented in a format that makes them easy to understand and compare? Are the data presented in a way that does not mislead? Could the same information be presented in a better way? If so, describe how to construct a graph that better depicts the data.

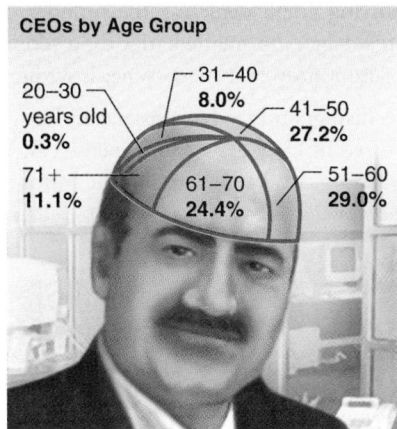

CEOs by Age Group

20–30 years old **0.3%**

31–40 **8.0%**

41–50 **27.2%**

71+ **11.1%**

61–70 **24.4%**

51–60 **29.0%**

SOURCE: Arthur Anderson/Mass Mutual Family Business Survey '97

12. Bar Graph of Undergraduates For a recent year, 38.5% of undergraduates were attending two-year colleges, and the other undergraduates were in four-year colleges (based on data from the U.S. National Center for Education Statistics).

a. Construct a bar graph that is misleading by exaggerating the difference between the two rates.

b. Construct a bar graph that depicts the data objectively.

Review

This chapter focused on methods for organizing, summarizing and graphing data sets. When investigating a data set, the characteristics of center, variation, distribution, outliers, and changing pattern over time are generally very important, and this chapter includes a variety of tools for investigating the distribution of the data. After completing this chapter, you should be able to do the following:

• Construct a frequency distribution or relative frequency distribution to summarize data (Section 2-2).

• Construct a histogram or relative frequency histogram to show the distribution of data (Section 2-3).

• Construct graphs of data using a frequency polygon, dotplot, stemplot, bar graph, multiple bar graph, Pareto chart, pie chart, scatterplot (for paired data), or time-series graph (Section 2-4).

• Critically analyze a graph to determine whether it objectively depicts data or is misleading (Section 2-5).

In addition to constructing frequency distributions and graphs, you should be able to *understand* and *interpret* those results. For example, the Chapter Problem includes Figure 2-1, which summarizes poll results. We should know that the graph is misleading because it uses a vertical scale that does not start at zero, so differences are exaggerated.

Statistical Literacy and Critical Thinking

1. Exploring Data Table 2-2 is a frequency distribution summarizing the pulse rates of females (listed in Table 2-1), and Figure 2-3 is a histogram representing those same pulse rates. When investigating the distribution of that data set, which is more effective: the frequency distribution or the histogram? Why?

2. College Tuition If you want to graph changing tuition costs over the past 20 years, which graph would be better, a histogram or a time-series graph? Why?

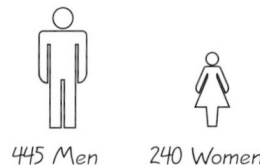

445 Men   240 Women

3. Graph See the accompanying graph depicting the number of men and the number of women who earned associate's degrees in mathematics for a recent year (based on data from the U.S. National Center for Education Statistics). What is wrong with this graph?

4. Normal Distribution A histogram is to be constructed from the durations (in hours) of NASA space shuttle flights listed in Data Set 10 in Appendix B. Without actually constructing that histogram, simply identify two key features of the histogram that would suggest that the data have a *normal distribution.*

Chapter Quick Quiz

1. The first two classes of a frequency distribution are 0–9 and 10–19. What is the class width?

2. The first two classes of a frequency distribution are 0–9 and 10–19. What are the class boundaries of the first class?

3. Can the original 27 values of a data set be identified by knowing that 27 is the frequency for the class of 0–9?

4. True or false: When a die is rolled 600 times, each of the 6 possible outcomes occurs about 100 times as we normally expect, so the frequency distribution summarizing the results is an example of a normal distribution.

5. Fill in the blank: For typical data sets, it is important to investigate center, distribution, outliers, changing patterns of the data over time, and _____.

6. What values are represented by this first row of a stemplot: 5 | 2 2 9?

7. Which graph is best for paired data consisting of the shoe sizes and heights of 30 randomly selected students: histogram, dotplot, scatterplot, Pareto chart, pie chart?

8. True or false: A histogram and a relative frequency histogram constructed from the same data always have the same basic shape, but the vertical scales are different.

9. What characteristic of a data set can be better understood by constructing a histogram?

10. Which graph is best for showing the relative importance of these defect categories for light bulbs: broken glass, broken filament, broken seal, and incorrect wattage label: histogram, dotplot, stemplot, Pareto chart, scatterplot?

Review Exercises

1. Frequency Distribution of Pulse Rates of Males Construct a frequency distribution of the pulse rates of males listed in Table 2-1 on page 53. Use the classes of 50–59, 60–69, and so on. How does the result compare to the frequency distribution for the pulse rates of females as shown in Table 2-2 on page 53?

2. Histogram of Pulse Rates of Males Construct the histogram that corresponds to the frequency distribution from Exercise 1. How does the result compare to the histogram for females (Figure 2-3)?

3. Dotplot of Pulse Rates of Men Construct a dotplot of the pulse rates of males listed in Table 2-1 on page 53 How does the result compare to the dotplot for the pulse rates of females shown in Section 2-4?

4. Stemplot of Pulse Rates of Males Construct a stemplot of the pulse rates of males listed in Table 2-1 on page 53. How does the result compare to the stemplot for the pulse rates of females shown in Section 2-4?

TI 5. Scatterplot of Car Weight and Braking Distance Listed below are the weights (in pounds) and braking distances (in feet) of the first six cars listed in Data Set 16 from Appendix B. Use the weights and braking distances shown below to construct a scatterplot. Based on the result, does there appear to be a relationship between the weight of a car and its braking distance?

Weight (lb)	4035	3315	4115	3650	3565	4030
Braking Distance (ft)	131	136	129	127	146	146

6. Time-Series Graph Listed below are the annual sunspot numbers for a recent sequence of years beginning with 1980. Construct a time-series graph. Is there a trend? If so, what is it?

154.6 140.5 115.9 66.6 45.9 17.9 13.4 29.2 100.2 157.6 142.6 145.7
94.3 54.6 29.9 17.5 8.6 21.5 64.3 93.3 119.6 123.3 123.3 65.9

7. Car Acceleration Times See the accompanying graph illustrating the acceleration times (in seconds) of four different cars. The actual acceleration times are as follows: Volvo XC-90: 7.6 s; Audi Q7: 8.2 s; Volkswagon Passat: 7.0 s; BMW 3 Series: 9.2 s. Does the graph correctly illustrate the acceleration times, or is it somehow misleading? Explain. If the graph is misleading, draw a graph that correctly illustrates the acceleration times.

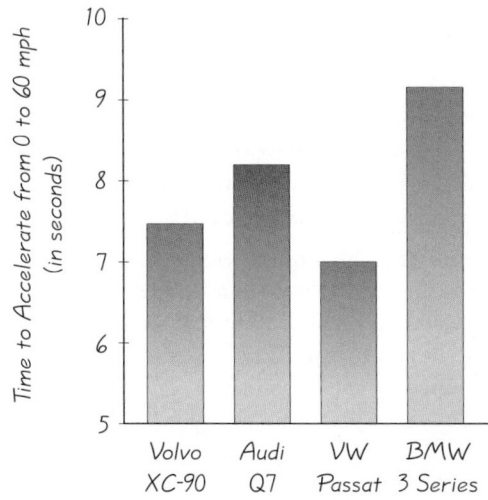

8. Old Faithful Geyser The accompanying table represents a frequency distribution of the duration times (in seconds) of 40 eruptions of the Old Faithful geyser, as listed in Data Set 15 in Appendix B.

a. What is the class width?

b. What are the upper and lower class limits of the first class?

c. What are the upper and lower class boundaries of the first class?

d. Does the distribution of duration times appear to be a normal distribution?

Duration (seconds)	Frequency
100–124	2
125–149	0
150–174	0
175–199	1
200–224	2
225–249	10
250–274	22
275–299	3

Cumulative Review Exercises

In Exercises 1–4, refer to the table in the margin, which summarizes results from a survey of 1733 randomly selected executives (based on data from Korn/Ferry International). Participants responded to this question: "If you could start your career over in a completely different field, would you?"

Response	Relative Frequency
Yes	51%
No	25%
Maybe	24%

1. Frequency Distribution Does the table describe a frequency distribution? Why or why not?

2. Level of Measurement What is the level of measurement of the 1733 individual responses: nominal, ordinal, interval, or ratio? Why?

3. Percentages Given that there are 1733 responses, find the actual number of responses in each category.

4. Sampling Suppose that the results in the table were obtained by mailing a survey to 10,000 executives and recording the 1733 responses that were returned. What is this type of sampling called? Is this type of sample likely to be representative of the population of all executives? Why or why not?

5. Sampling

a. What is a random sample?

b. What is a simple random sample?

c. Assume that the population of the United States is partitioned into 300,000 groups with exactly 1000 subjects in each group. If a computer is used to randomly select one of the groups, is the result a random sample? Simple random sample?

6. Cotinine Levels of Smokers The accompanying frequency distribution summarizes the measured cotinine levels of a simple random sample of 40 smokers (from Data Set 5 in Appendix B).

Cotinine Level	Frequency
0–99	11
100–199	12
200–299	14
300–399	1
400–499	2

a. What is the class width?

b. What are the upper and lower class boundaries of the first class?

c. What is the relative frequency corresponding to the frequency of 11 for the first class?

d. What is the level of measurement of the original cotinine levels: nominal, ordinal, interval, or ratio?

e. Are the measured cotinine levels qualitative data or quantitative data?

TI **7. Histogram** Construct the histogram that represents the data summarized in the table that accompanies Exercise 6. What should be the shape of the histogram in order to conclude that the data have a normal distribution? If using a fairly strict interpretation of a normal distribution, does the histogram suggest that the cotinine levels are normally distributed?

8. Statistics and Parameters The cotinine levels summarized in the table that accompanies Exercise 6 are obtained from a simple random sample of smokers selected from the population of all smokers. If we add the original 40 cotinine levels, then divide the total by 40, we obtain 172.5, which is the average (mean). Is 172.5 a statistic or a parameter? In general, what is the difference between a statistic and a parameter?

TI-83/84 Plus Calculator Project

Manually constructed graphs have a certain primitive charm, but they are generally unsuitable for publications and presentations. Graphs generated by technology are much better for such purposes. Table 2-1 in Section 2-2 lists pulse rates of females and males, but those pulse rates are also listed in Data Set 1 in Appendix B, and they are available as files that can be opened by a TI-83/84 Plus calculator. The female pulse rates are listed in FPULS, which is in the file FHEALTH (female health measurements), and the male pulse rates are listed in MPULS, which is in the file MHEALTH (male health measurements). The data sets FHEALTH and MHEALTH are in the TRIOLA11 APP. Use a TI-83/84 Plus calculator to download the FPULS and MPULS lists, then generate three histograms: (1) a histogram of the pulse rates of females listed in Data Set 1 in Appendix B; (2) a histogram of the pulse rates of males listed in Table 2-1 in Section 2-2; (3) a histogram of the combined list of pulse rates of males and females. If possible, obtain printed copies of the histograms, then compare them. Does it appear that the pulse rates of males and females have similar characteristics? (Later in this book, we will present more formal methods for making such comparisons. See, for example, Section 9-4.)

Hint: You can use the *augment* command to combine the lists of female pulse rates and male pulse rates. Press **2ND** **STAT** to get to the LIST menu, then press ▷ to scroll to **OPS** (operations), then continue to press ▽ select **augment.** You can combine the list FPULS and the list MPULS and store the combined list in L1 with this command: **augment(FPULS, MPULS).** (The entry of FPULS is achieved by pressing **2ND** **STAT** to get the names of the lists; then select FPULS. Do the same for MPULS.) The screen will display the combined list in a format such as {76 72 88 60 72 . . . . You can store the combined list by pressing **STO▸** followed by L1. (To get L1, press **2ND** **1**.)

Data on the Internet

Go to: **http://www.aw.com/triola**

The Internet is host to a wealth of information and much of that information comes from raw data that have been collected or observed. Many Web sites summarize such data using the graphical methods discussed in this chapter. For example, we found the following with just a few clicks:

- Bar graphs at the site of the Centers for Disease Control tell us that the percentage of men and women who report an average of less than 6 hours of sleep per night has increased in each age group over the last two decades.

- A pie chart provided by the National Collegiate Athletic Association (NCAA) shows that an estimated 90.12% of NCAA revenue in 2006–07 came from television and marketing rights fees while only 1.74% came from investments, fees, and services.

The Internet Project for this chapter will further explore graphical representations of data sets found on the Internet. In the process, you will view and collect data sets in the areas of sports, population demographics, and finance, and perform your own graphical analyses.

APPLET PROJECT

The CD included with this book contains applets designed to help visualize various concepts. When conducting polls, it is common to randomly generate the digits of telephone numbers of people to be called. Open the Applets folder on the CD and click on **Start.** Select the menu item of **Random sample.** Enter a minimum value of 0, a maximum value of 9, and 100 for the number of sample values. Construct a frequency distribution of the results. Does the frequency distribution suggest that the digits have been selected as they should?

Cooperative Group Activities

1. In-class activity Table 2-1 in Section 2-2 includes pulse rates of males and females. In class, each student should record his or her pulse rate by counting the number of heartbeats in one minute. Construct a frequency distribution and histogram for the pulse rates of males and construct another frequency distribution and histogram for the pulse rates of females. Compare the results. Is there an obvious difference? Are the results consistent with those found using the data from Table 2-1?

2. Out-of-class activity Search newspapers and magazines to find an example of a graph that is misleading. (See Section 2-5.) Describe how the graph is misleading. Redraw the graph so that it depicts the information correctly.

3. In-class activity Given below are the ages of motorcyclists at the time they were fatally injured in traffic accidents (based on data from the U.S. Department of Transportation). If your objective is to dramatize the dangers of motorcycles for young people, which would be most effective: histogram, Pareto chart, pie chart, dotplot, stemplot, or some other graph? Construct the graph that best meets the objective of dramatizing the dangers of motorcycle driving. Is it okay to deliberately distort data if the objective is one such as saving lives of motorcyclists?

17	38	27	14	18	34	16	42	28
24	40	20	23	31	37	21	30	25
17	28	33	25	23	19	51	18	29

4. Out-of-class activity In each group of three or four students, construct a graph that is effective in addressing this question: Is there a difference between the body mass index (BMI) values for men and for women? (See Data Set 1 in Appendix B.)

5. Out-of-class activity Obtain a copy of *The Visual Display of Quantitative Information,* second edition, by Edward Tufte (Graphics Press, PO Box 430, Cheshire, CT 06410). Find the graph describing Napoleon's march to Moscow and back, and explain why Tufte says that "it may well be the best graphic ever drawn."

6. Out-of-class activity Obtain a copy of *The Visual Display of Quantitative Information,* second edition, by Edward Tufte (Graphics Press, PO Box 430, Cheshire, CT 06410). Find the graph that appeared in *American Education,* and explain why Tufte says that "this may well be the worst graphic ever to find its way into print." Construct a graph that is effective in depicting the same data.

7. Out-of-class activity Find the number of countries that use the metric (SI) system and the number of countries that use the British system (miles, pounds, gallons, etc.). Construct a graph that is effective in depicting the data. What does the graph suggest?

FROM DATA TO DECISION

Do the Academy Awards involve discrimination based on age?

Listed below are the ages of actresses and actors at the times that they won Oscars in the Best Actress and Best Actor categories. The ages are listed in order, beginning with the first Academy Awards ceremony in 1928. (*Notes:* In 1968 there was a tie in the Best Actress category, and the mean of the two ages is used; in 1932 there was a tie in the Best Actor category, and the mean of the two ages is used. These data are suggested by the article "Ages of Oscar-winning Best Actors and Actresses," by Richard Brown and Gretchen Davis, *Mathematics Teacher* magazine. In that article, the year of birth of the award winner was subtracted from the year of the awards ceremony, but the ages listed below are based on the birth date of the winner and the date of the awards ceremony.)

Critical Thinking: Use the methods from this chapter for organizing, summarizing, and graphing data, compare the two data sets. Address these questions: Are there differences between the ages of the Best Actresses and the ages of the Best Actors? Does it appear that actresses and actors are judged strictly on the basis of their artistic abilities? Or does there appear to be discrimination based on age, with the Best Actresses tending to be younger than the Best Actors? Are there any other notable differences?

Best Actresses

22	37	28	63	32	26	31	27	27	28
30	26	29	24	38	25	29	41	30	35
35	33	29	38	54	24	25	46	41	28
40	39	29	27	31	38	29	25	35	60
43	35	34	34	27	37	42	41	36	32
41	33	31	74	33	50	38	61	21	41
26	80	42	29	33	35	45	49	39	34
26	25	33	35	35	28	30	29	61	

Best Actors

44	41	62	52	41	34	34	52	41	37
38	34	32	40	43	56	41	39	49	57
41	38	42	52	51	35	30	39	41	44
49	35	47	31	47	37	57	42	45	42
44	62	43	42	48	49	56	38	60	30
40	42	36	76	39	53	45	36	62	43
51	32	42	54	52	37	38	32	45	60
46	40	36	47	29	43	37	38	45	

NAME:	Bob Sehlinger
JOB:	Publisher
COMPANY:	Menasha Ridge Press

Bob Sehlinger works at Menasha Ridge Press, which publishes, among many other titles, the Unofficial Guide series for John Wiley & Sons (Wiley, Inc.). The Unofficial Guides use statistics extensively to research the experiences that travelers are likely to encounter and to help them make informed decisions that will help them enjoy great vacations.

Q: How do you use statistics in your job and what specific statistical concepts do you use?

A: We use statistics in every facet of the business: expected value analysis for sales forecasting; regression analysis to determine what books to publish in a series, etc., but we're best known for our research in the areas of queuing and evolutionary computations.

The research methodologies used in the *Unofficial Guide* series are ushering in a truly groundbreaking approach to how travel guides are created. Our research designs and the use of technology from the field of operations research have been cited by academe and reviewed in peer journals for quite some time.

We're using a revolutionary team approach and cutting-edge science to provide readers with extremely valuable information not available in other travel series. Our entire organization is guided by individuals with extensive training and experience in research design as well as data collection and analysis.

From the first edition of the *Unofficial Guide to* our research at Walt Disney World, minimizing our readers' wait in lines has been a top priority. We developed and offered our readers field-tested touring plans that allow them to experience as many attractions as possible with the least amount of waiting in line. We field-tested our approach in the park; the group touring without our plans spent an average of $3\frac{1}{2}$ hours more waiting in line and experienced 37% fewer attractions than did those who used our touring plans.

As we add attractions to our list, the number of possible touring plans grows rapidly. The 44 attractions in the Magic Kingdom One-Day Touring Plan for Adults have a staggering 51,090,942,171,709,440,000 possible touring plans. How good are the new touring plans in the *Unofficial Guide?* Our computer program gets typically within about 2% of the optimal touring plan. To put this in perspective, if the hypothetical "perfect" Adult One-Day touring plan took about 10 hours to complete, the *Unofficial* touring plan would take about 10 hours and 12 minutes. Since it would take about 30 years for a really powerful computer to find that "perfect" plan, the extra 12 minutes is a reasonable trade-off.

Q: What background in statistics is required to obtain a job like yours?

A: I work with PhD level statisticians and programmers in developing and executing research designs. I hold an MBA and had a lot of practical experience in operations research before entering publishing, but the main prerequisite in doing the research is knowing enough statistics to see opportunities to use statistics for developing useful information for our readers.

Q: Do you recommend that today's college students study statistics? Why?

A: Absolutely. In a business context, statistics along with accounting and a good grounding in the mathematics of finance are the quantitative cornerstones. Also, statistics are important in virtually every aspect of life.

Q: Which other skills are important for today's college students?

A: Good oral and written expression.

3-1 Review and Preview

3-2 Measures of Center

3-3 Measures of Variation

3-4 Measures of Relative Standing and Boxplots

3

Statistics for Describing, Exploring, and Comparing Data

Do women really talk more than men?

A common belief is that women talk more than men. Is that belief founded in fact, or is it a myth? Do men actually talk more than women? Or do men and women talk about the same amount?

In the book *The Female Brain,* neuro-psychiatrist Louann Brizendine stated that women speak 20,000 words per day, compared to only 7,000 for men. She deleted that statement after complaints from linguistics experts who said that those word counts were not substantiated.

Researchers conducted a study in an attempt to address the issue of words spoken by men and women. Their findings were published in the article "Are Women Really More Talkative Than Men?" (by Mehl, Vazire, Ramirez-Esparza, Slatcher, and Pennebaker, *Science*, Vol. 317, No. 5834). The study involved 396 subjects who each wore a voice recorder that collected samples of conversations over several days. Researchers then analyzed those conversations and counted the numbers of spoken words for each of the subjects. Data Set 8 in Appendix B includes male/female word counts from each of six different sample groups (from results provided by the researchers), but if we combine all of the male word counts and all of the female word counts in Data Set 8, we get two sets of sample data that can be compared. A good way to begin to explore the data is to construct a graph that allows us to visualize the samples. See the relative frequency polygons shown in Figure 3-1. Based on that figure, the samples of word counts from men and women appear to be very close, with no substantial differences.

When comparing the word counts of the sample of men to the word counts of the sample of women, one step is to compare the *means* from the two samples. Shown below are the values of the means and the sample sizes. (Many people are more familiar with the term "average," but that term is not used in statistics; the term "mean" is used instead, and that term is formally defined and discussed in Section 3-2, where we see that a mean is found by adding all of the values and dividing the total by the number of values.)

Figure 3-1 and the sample means give us considerable insight into a comparison of the numbers of words spoken by men and women. In this section we introduce other common statistical methods that are helpful in making comparisons. Using the methods of this chapter and of other chapters, we will determine whether women actually do talk more than men, or whether that is just a myth.

	Males	Females
Sample mean	15,668.5	16,215.0
Sample size	186	210

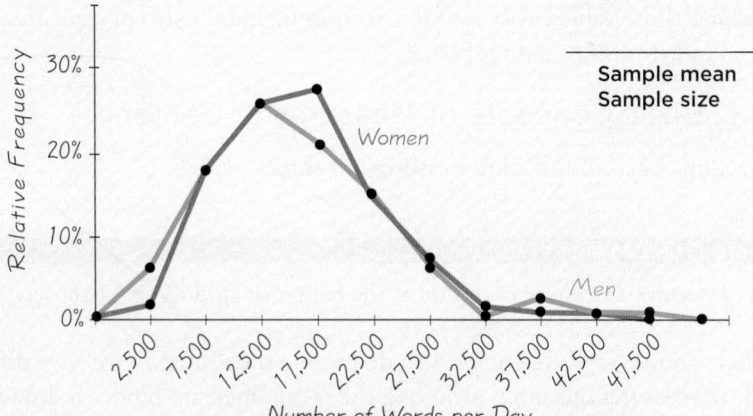

Figure 3-1 Frequency Polygons of Numbers of Words Spoken by Men and Women

Review and Preview

Chapter 1 discussed methods of collecting sample data, and Chapter 2 presented the frequency distribution as a tool for summarizing data. Chapter 2 also presented graphs designed to help us understand some characteristics of the data, including the distribution. We noted in Chapter 2 that when describing, exploring, and comparing data sets, these characteristics are usually extremely important: (1) center; (2) variation; (3) distribution; (4) outliers; and (5) changing characteristics of data over time. In this chapter we introduce important statistics, including the mean, median, and standard deviation. Upon completing this chapter, you should be able to find the mean, median, standard deviation, and variance from a data set, and you should be able to clearly understand and interpret such values. It is especially important to understand values of standard deviation by using tools such as the range rule of thumb.

Critical Thinking and Interpretation: Going Beyond Formulas

In this chapter we present several formulas used to compute basic statistics. Because technology enables us to compute many of these statistics automatically, it is not as important for us to memorize formulas and manually perform complex calculations. Instead, we should focus on understanding and interpreting the values we obtain from them.

The methods and tools presented in Chapter 2 and in this chapter are often called **descriptive statistics,** because they summarize or describe relevant characteristics of data. Later in this book, we will use **inferential statistics** to make inferences, or generalizations, about a population.

 ## Measures of Center

Key Concept In this section we discuss the characteristic of center. In particular, we present measures of center, including *mean* and *median,* as tools for analyzing data. Our focus here is not only to determine the value of each measure of center, but also to interpret those values. Part 1 of this section includes core concepts that should be understood before considering Part 2.

Part 1: Basic Concepts of Measures of Center

This section discusses different measures of center.

 DEFINITION

A **measure of center** is a value at the center or middle of a data set.

There are several different ways to determine the center, so we have different definitions of measures of center, including the mean, median, mode, and midrange. We begin with the mean.

Mean

The (arithmetic) mean is generally the most important of all numerical measurements used to describe data, and it is what most people call an *average.*

⊙ DEFINITION

The **arithmetic mean,** or the **mean,** of a set of data is the measure of center found by adding the data values and dividing the total by the number of data values.

This definition can be expressed as Formula 3-1, in which the Greek letter Σ (uppercase sigma) indicates that the data values should be added. That is, Σx represents the sum of all data values. The symbol n denotes the **sample size,** which is the number of data values.

Formula 3-1

$$\text{mean} = \frac{\Sigma x}{n} \quad \begin{array}{l} \leftarrow \text{ sum of all data values} \\ \leftarrow \text{ number of data values} \end{array}$$

If the data are a *sample* from a population, the mean is denoted by $\bar{x}$ (pronounced "*x*-bar"); if the data are the entire population, the mean is denoted by μ (lowercase Greek mu). (Sample statistics are usually represented by English letters, such as $\bar{x}$, and population parameters are usually represented by Greek letters, such as μ.)

Notation

Σ denotes the *sum* of a set of data values.

x is the *variable* usually used to represent the individual data values.

n represents the *number of data values* in a *sample*.

N represents the *number of data values* in a *population*.

$\bar{x} = \dfrac{\Sigma x}{n}$ is the mean of a set of *sample* values.

$\mu = \dfrac{\Sigma x}{N}$ is the mean of all values in a *population*.

EXAMPLE 1   **Mean** The Chapter Problem refers to word counts from 186 men and 210 women. Find the mean of these first five word counts from men: 27,531; 15,684; 5,638; 27,997; and 25,433.

SOLUTION   The mean is computed by using Formula 3-1. First add the data values, then divide by the number of data values:

$$\bar{x} = \frac{\Sigma x}{n} = \frac{27,531 + 15,684 + 5,638 + 27,997 + 25,433}{5} = \frac{102,283}{5}$$

$$= 20,456.6$$

Since $\bar{x} = 20,456.6$ words, the mean of the first five word counts is 20,456.6 words.

A mean can be found by using a TI-83/84 Plus calculator. See the instructions near the end of this section.

One advantage of the mean is that it is relatively *reliable,* so that when samples are selected from the same population, sample means tend to be more consistent than other measures of center. That is, the means of samples drawn from the same population

---

Average Bob

According to Kevin O'Keefe, author of *The Average American: The Extraordinary Search for the Nation's Most Ordinary Citizen,* Bob Burns is the most average person in the United States. O'Keefe spent 2 years using 140 criteria to identify the single American who is most average. He identified statistics revealing preferences of the majority, and applied them to the many people he encountered. Bob Burns is the only person who satisfied all of the 140 criteria. Bob Burns is 5 ft 8 in. tall, weighs 190 pounds, is 54 years of age, married, has three children, wears glasses, works 40 hours per week, drives an eight-year-old car, has an outdoor grill, mows his own lawn, drinks coffee each day, and walks his dog each evening.

Changing Populations

Included among the five important data set characteristics listed in Chapter 2 is the changing pattern of data over time. Some populations change, and their important statistics change as well. Car seat belt standards haven't changed in 40 years, even though the weights of Americans have increased considerably since then. In 1960, 12.8% of adult Americans were considered obese, compared to 22.6% in 1994.

According to the National Highway Traffic Safety Administration, seat belts must fit a standard crash dummy (designed according to 1960 data) placed in the most forward position, with 4 in. to spare. In theory, 95% of men and 99% of women should fit into seat belts, but those percentages are now lower because of the increases in weight over the last half-century. Some car companies provide seat belt extenders, but some do not.

don't vary as much as the other measures of center. Another advantage of the mean is that it takes every data value into account. However, because the mean is sensitive to every value, just one extreme value can affect it dramatically. Since the mean cannot resist substantial changes caused by extreme values, we say that the mean is not a *resistant* measure of center.

Median

Unlike the mean, the median is a resistant measure of center, because it does not change by large amounts due to the presence of just a few extreme values.

The median can be thought of loosely as a "middle value" in the sense that about half of the values in a data set are below the median and half are above it. The following definition is more precise.

 **DEFINITION**

The **median** of a data set is the measure of center that is the *middle value* when the original data values are arranged in order of increasing (or decreasing) magnitude. The median is often denoted by $\tilde{x}$ (pronounced "*x*-tilde").

To find the median, first *sort* the values (arrange them in order), then follow one of these two procedures:

1. If the number of data values is odd, the median is the number located in the exact middle of the list.

2. If the number of data values is even, the median is found by computing the mean of the two middle numbers.

A TI-83/84 Plus calculator can be used to find a median. See the instructions near the end of this section.

EXAMPLE 2   **Median** Find the median for this sample of data values used in Example 1: 27,531, 15,684, 5,638, 27,997, and 25,433.

SOLUTION   First sort the data values, as shown below:

5,638   15,684   25,433   27,531   27,997

Because the number of data values is an odd number (5), the median is the number located in the exact middle of the sorted list, which is 25,433. The median is therefore 25,433 words. Note that the median of 25,433 is different from the mean of 20,456.6 words found in Example 1.

EXAMPLE 3   **Median** Repeat Example 2 after including the additional data value of 8,077 words. That is, find the median of these word counts: 27,531, 15,684, 5,638, 27,997, 25,433, and 8,077.

SOLUTION   First arrange the values in order:

5,638   8,077   15,684   25,433   27,531   27,997

Because the number of data values is an even number (6), the median is found by computing the mean of the two middle numbers, which are 15,684 and 25,433.

$$\text{Median} = \frac{15{,}684 + 25{,}433}{2} = \frac{41{,}117}{2} = 20{,}558.5$$

The median is 20,558 words.

CAUTION

Never use the term *average* when referring to a measure of center. Use the correct term, such as *mean* or *median*.

Mode

The mode is another measure of center.

DEFINITION

The **mode** of a data set is the value that occurs with the greatest frequency.

A data set can have one mode, more than one mode, or no mode.

- When two data values occur with the same greatest frequency, each one is a mode and the data set is **bimodal.**

- When more than two data values occur with the same greatest frequency, each is a mode and the data set is said to be **multimodal.**

- When no data value is repeated, we say that there is **no mode.**

EXAMPLE 4 **Mode** Find the mode of these word counts:

18,360　18,360　27,531　15,684　5,638　27,997　25,433.

SOLUTION The mode is 18,360 words, because it is the data value with the greatest frequency.

In Example 4 the mode is a single value. Here are two other possible circumstances:

Two modes:　The values of 0, 0, 0, 1, 1, 2, 3, 5, 5, 5 have two modes: 0 and 5.

No mode:　The values of 0, 1, 2, 3, 5 have no mode because no value occurs more than once.

In reality, the mode isn't used much with numerical data. However, the mode is the only measure of center that can be used with data at the nominal level of measurement. (Remember, the nominal level of measurement applies to data that consist of names, labels, or categories only.)

Midrange

Another measure of center is the midrange. Because the midrange uses only the maximum and minimum values, it is too sensitive to those extremes, so the midrange is rarely used. However, the midrange does have three redeeming features: (1) it is very easy to compute; (2) it helps to reinforce the important point that there are several

---

Class Size Paradox

There are at least two ways to obtain the mean class size, and they can have very different results. At one college, if we take the numbers of students in 737 classes, we get a mean of 40 students. But if we were to compile a list of the class sizes for each student and use this list, we would get a mean class size of 147. This large discrepancy is due to the fact that there are many students in large classes, while there are few students in small classes. Without changing the number of classes or faculty, we could reduce the mean class size experienced by students by making all classes about the same size. This would also improve attendance, which is better in smaller classes.

different ways to define the center of a data set; (3) it is sometimes incorrectly used for the median, so confusion can be reduced by clearly defining the midrange along with the median. (See Exercise 3.)

 **DEFINITION**

> The **midrange** of a data set is the measure of center that is the value midway between the maximum and minimum values in the original data set. It is found by adding the maximum data value to the minimum data value and then dividing the sum by 2, as in the following formula:
>
> $$\text{midrange} = \frac{\text{maximum data value} + \text{minimum data value}}{2}$$

EXAMPLE 5   **Midrange** Find the midrange of these values from Example 1: 27,531, 15,684, 5,638, 27,997, and 25,433.

SOLUTION   The midrange is found as follows:

$$\text{midrange} = \frac{\text{maximum data value} + \text{minimum data value}}{2}$$

$$= \frac{27,997 + 5,638}{2} = 16,817.5$$

The midrange is 16,817.5 words.

The term *average* is often used for the mean, but it is sometimes used for other measures of center. To avoid any confusion or ambiguity we use the correct and specific term, such as *mean* or *median*. The term *average* is not used by statisticians and it will not be used throughout the remainder of this book when referring to a specific measure of center.

When calculating measures of center, we often need to round the result. We use the following rule.

Round-Off Rule for the Mean, Median, and Midrange

Carry one more decimal place than is present in the original set of values.

(Because values of the mode are the same as some of the original data values, they can be left as is without any rounding.)

When applying this rule, round only the final answer, *not intermediate values that occur during calculations.* For example, the mean of 2, 3, 5, is 3.333333..., which is rounded to 3.3, which has one more decimal place than the original values of 2, 3, 5. As another example, the mean of 80.4 and 80.6 is 80.50 (one more decimal place than was used for the original values). Because the mode is one or more of the original data values, we do not round values of the mode; we simply use the same original values.

Critical Thinking

Although we can calculate measures of center for a set of sample data, we should always think about whether the results are reasonable. In Section 1-2 we noted that it does not make sense to do numerical calculations with data at the nominal level of measurement, because those data consist of names, labels, or categories only, so statistics such as the mean and median are meaningless. We should also think about the method used to collect the sample data. If the method is not sound, the statistics we obtain may be misleading.

EXAMPLE 6 **Critical Thinking and Measures of Center** For each of the following, identify a major reason why the mean and median are *not* meaningful statistics.

a. Zip codes: 12601, 90210, 02116, 76177, 19102

b. Ranks of stress levels from different jobs: 2, 3, 1, 7, 9

c. Survey respondents are coded as 1 (for Democrat), 2 (for Republican), 3 (for Liberal), 4 (for Conservative), or 5 (for any other political party).

SOLUTION

a. The zip codes don't measure or count anything. The numbers are actually labels for geographic locations.

b. The ranks reflect an ordering, but they don't measure or count anything. The rank of 1 might come from a job that has a stress level substantially greater than the stress level from the job with a rank of 2, so the different numbers don't correspond to the magnitudes of the stress levels.

c. The coded results are numbers that don't measure or count anything. These numbers are simply different ways of expressing names.

Example 6 involved data at the nominal level of measurement that do not justify the use of statistics such as the mean or median. Example 7 involves a more subtle issue.

EXAMPLE 7 **Mean per Capita Personal Income** Per capita personal income is the income that each person would receive if the total national income were divided equally among everyone in the population. Using data from the U.S. Department of Commerce, the mean per capita personal income can be found for each of the 50 states. Some of the values for the latest data available at the time of this writing are:

$$\$29,136 \quad \$35,612 \quad \$30,267 \quad \dots \quad \$36,778$$

The mean of the 50 state means is $33,442. Does it follow that $33,442 is the mean per capita personal income for the entire United States? Why or why not?

continued

SOLUTION   No, $33,442 is not necessarily the mean per capita personal income in the United States. The issue here is that some states have many more people than others. The calculation of the mean for the United States should take into account the number of people in each state. The mean per capita personal income in the United States is actually $34,586, not $33,442. We can't find the mean for the United States population by finding the mean of the 50 state means.

Part 2: Beyond the Basics of Measures of Center

Mean from a Frequency Distribution

When working with data summarized in a frequency distribution, we don't know the exact values falling in a particular class. To make calculations possible, we assume that all sample values in each class are equal to the class midpoint. For example, consider a class interval of 0–9,999 with a frequency of 46 (as in Table 3-1). We assume that all 46 values are equal to 4999.5 (the class midpoint). With the value of 4999.5 repeated 46 times, we have a total of $4999.5 \cdot 46 = 229,977$. We can then add the products from each class to find the total of all sample values, which we then divide by the sum of the frequencies, Σf. Formula 3-2 is used to compute the mean when the sample data are summarized in a frequency distribution. Formula 3-2 is not really a new concept; it is simply a variation of Formula 3-1.

Formula 3-2

First multiply each frequency and class midpoint, then add the products.
$$\downarrow$$

mean from frequency distribution: $$\bar{x} = \frac{\Sigma(f \cdot x)}{\Sigma f}$$
$$\uparrow$$
sum of frequencies

A TI-83/84 Plus calculator can be used to find the mean from a frequency distribution. See the instructions near the end of this section.

The following example illustrates the procedure for finding the mean from a frequency distribution.

Table 3-1   Finding the Mean from a Frequency Distribution

Word Counts from Men	Frequency f	Class Midpoint x	$f \cdot x$
0–9,999	46	4,999.5	229,977.0
10,000–19,999	90	14,999.5	1,349,955.0
20,000–29,999	40	24,999.5	999,980.0
30,000–39,999	7	34,999.5	244,996.5
40,000–49,999	3	44,999.5	134.998.5
Totals:	$\Sigma f = 186$		$\Sigma(f \cdot x) = 2,959,907$

$$\bar{x} = \frac{\Sigma(f \cdot x)}{\Sigma f} = \frac{2,959,907}{186} = 15,913.5$$

(handwritten margin notes): mean? must find class midpoint for each class entry in L1 Freq L2

EXAMPLE 8 **Computing Mean from a Frequency Distribution** The first two columns of Table 3-1 constitute a frequency distribution summarizing the word counts of the 186 men in Data Set 8 from Appendix B. Use the frequency distribution to find the mean.

SOLUTION Table 3-1 illustrates the procedure for using Formula 3-2 when calculating a mean from data summarized in a frequency distribution. The class midpoint values are shown in the third column, and the products $f \cdot x$ are shown in the last column. The calculation using Formula 3-2 is shown at the bottom of Table 3-1. The result is $\bar{x} = 15{,}913.5$ words. If we use the original list of word counts for the 186 men, we get $\bar{x} = 15{,}668.5$ words. The frequency distribution yields an approximation of $\bar{x}$, because it is not based on the exact original list of sample values.

Weighted Mean

When data values are assigned different weights, we can compute a **weighted mean.** Formula 3-3 can be used to compute the weighted mean, w.

Formula 3-3

$$\text{weighted mean: } \bar{x} = \frac{\Sigma(w \cdot x)}{\Sigma w}$$

Formula 3-3 tells us to first multiply each weight w by the corresponding value x, then to add the products, and then finally to divide that total by the sum of the weights Σw.

EXAMPLE 9 **Computing Grade Point Average** In her first semester of college, a student of the author took five courses. Her final grades along with the number of credits for each course were: A (3 credits); A (4 credits); B (3 credits), C (3 credits), and F (1 credit). The grading system assigns quality points to letter grades as follows: A = 4; B = 3; C = 2; D = 1; F = 0. Compute her grade point average.

SOLUTION Use the numbers of credits as weights: $w = 3, 4, 3, 3, 1$. Replace the letter grades of A, A, B, C, and F with the corresponding quality points: $x = 4, 4, 3, 2, 0$. We now use Formula 3-3 as shown below. The result is a first-semester grade point average of 3.07. (Using the preceding round-off rule, the result should be rounded to 3.1, but it is common to round grade point averages with two decimal places.)

$$\bar{x} = \frac{\Sigma(w \cdot x)}{\Sigma w}$$

$$= \frac{(3 \times 4) + (4 \times 4) + (3 \times 3) + (3 \times 2) + (1 \times 0)}{3 + 4 + 3 + 3 + 1}$$

$$= \frac{43}{14} = 3.07$$

Skewness

A comparison of the mean, median, and mode can reveal information about the characteristic of skewness, defined below and illustrated in Figure 3-2.

 DEFINITION

A distribution of data is **skewed** if it is not symmetric and extends more to one side than to the other. (A distribution of data is **symmetric** if the left half of its histogram is roughly a mirror image of its right half.)

Data **skewed to the left** (also called *negatively skewed*) have a longer left tail, and the mean and median are to the left of the mode. Data **skewed to the right** (also called *positively skewed*) have a longer right tail, and the mean and median are to the right of the mode.

Skewed data usually (but not always!) have the mean located farther out in the longer tail than the median. Figure 3-2(a) shows the mean to the left of the median for data skewed to the left, and Figure 3-2(c) shows the mean to the right of the median for data skewed to the right, but those relative positions of the mean and median are not always as shown in the figures. For example, it is possible to have data skewed to the left with a median less than the mean, contrary to the order shown in Figure 3-2(a). For the values of −100, 1.0, 1.5, 1.7, 1.8, 2.0, 3.0, 4.0, 5.0, 50.0, 50.0, 60.0, a histogram shows that the data are skewed to the left, but the mean of 6.7 is *greater than* the median of 2.5, contradicting the order of the mean and median shown in Figure 3-2(a).

The mean and median cannot always be used to identify the shape of the distribution.

In practice, many distributions of data are approximately symmetric and without skewness. Distributions skewed to the right are more common than those skewed to the left because it's often easier to get exceptionally large values than values that are exceptionally small. With annual incomes, for example, it's impossible to get values below zero, but there are a few people who earn millions or billions of dollars in a year. Annual incomes therefore tend to be skewed to the right, as in Figure 3-2(c).

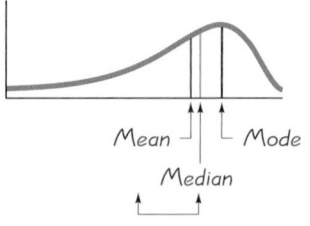

The order of the mean and median may be reversed.

(a) Skewed to the Left (Negatively Skewed): The mean and median are to the *left* of the mode (but their order is not always predictable).

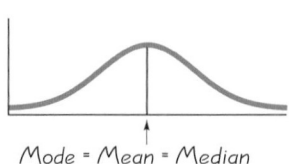

Mode = Mean = Median

(b) Symmetric (Zero Skewness): The mean, median, and mode are the same.

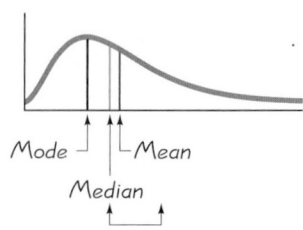

The order of the median and mean may be reversed.

(c) Skewed to the Right (Positively Skewed): The mean and median are to the *right* of the mode (but their order is not always predictable).

Figure 3-2   Skewness

Using the TI-83/84 Plus Calculator to Find Descriptive Statistics from a List of Data

You can use a TI-83/84 Plus calculator to find the mean and median of a set of data. The basic approach is to press **STAT**, select **CALC**, select **1-Var Stats**, and enter the name or label of the list of data values. Although the TI-83/84 Plus calculator does not find the midrange or mode, it has features that are helpful in finding those values. Use the following procedure.

TI-83/84 Plus

```
1-Var Stats
 x̄=15668.52688
 Σx=2914346
 Σx²=5.94498ᴇ10
 Sx=8632.534672
 σx=8609.297659
↓n=186
```

TI-83/84 Plus

```
1-Var Stats
↑n=186
 minX=695
 Q₁=10009
 Med=14290
 Q₃=20565
 maxX=47016
```

1. Enter or open a list of data. You could open any of the lists of data in Appendix B by using the APP named TRIOLA11, you could use any of the lists L1, L2, . . ., L6, or you could use a list of values assigned to a name. We will proceed by assuming that the 186 word counts of males from Data Set 8 in Appendix B have been entered in list L1. (*Hint:* The App named Triola11 has Data Set 8 included in the file WORDS, and the 186 word counts of males are in six individual lists named M1, M2, M3, M4, M5, and M6. Those lists can be combined two at a time into one list which can be stored in L1. To combine lists, use the **augment** command described in the hint for the TI-83/84 Plus Project at the end of Chapter 2.)

2. Press **STAT**, then highlight the top menu item of **CALC** by pressing ▷.

3. We want the first menu item of **1-Var Stats,** and it is the default, so press **ENTER**.

4. The screen should display 1-Var Stats, and we must now identify the list to be used for the calculations. Either enter the name of a list (by pressing **2ND** **STAT** and using the arrow keys to select the desired name) or enter L1 or L2 or L3 or L4 or L5 or L6. If we want to find measures of center for the data listed in L1, the screen should display 1-Var Stats L1.

5. Press **ENTER** to get results such as those shown in the margin. Note that there are more results than can be shown in one screen, so the arrow on the bottom of the first screen indicates that there are more results available, and you can access them by pressing ▽.

6. *Interpret the results:* The displays show that the mean is given by $\bar{x} = 15668.52688$ words, and we should round this result to 15,668.5 words by using one more decimal place than the original data. The median is 14,290.0 words. Because the minimum value (695 words) and the maximum value (47,016 words) are displayed, it is easy to find the midrange, which is $(47{,}016 + 695)/2 = 23{,}855.5$ words. [The midrange is (maximum value + minimum value)/2.]

Using *Sort* to find Mode Although the TI-83/84 Plus calculator does not automatically find the mode from a data set, finding the mode is made easier by using the ability of the calculator to *sort* data (arrange them in order). To sort a list of data, press **STAT**, then select the second item of **SortA** to sort the data in *ascending* order, or select the third item of **SortD** to sort the data in *descending* order. After selecting SortA or SortD, press **ENTER** and enter the name of the list to be sorted, or enter L1, L2, L3 L4, L5, or L6. For example, the command **SortA(L1)** causes the values in the list L1 to be arranged in increasing order. After arranging the data in order, view the data set by scrolling through the list looking for repeated values. [*Helpful hint:* If the data are in a named list, such as AGES, it is probably easier to store the sorted list in L1 for better viewing in a vertical format. To store the values in the list AGES in the list L1, first select the list name of AGES by pressing **2ND** **STAT**, then scroll to select the name of AGES and press **ENTER**. You should now see LAGES. Press **STO▸** **2ND** ① so that the screen will display LAGES ➔ L1. The values in the list AGES will now be in list L1. Press **STAT** **ENTER** to view the values in list L1.]

Table 2-2   Pulse Rates of Females

Pulse Rate	Frequency
60-69	12
70-79	14
80-89	11
90-99	1
100-109	1
110-119	0
120-129	1

TI-83/84 Plus

L1	L2	L3 3
64.5	12	
74.5	14	
84.5	11	
94.5	1	
104.5	1	
114.5	0	
124.5	1	

L3(1)=

Using the TI-83/84 Plus Calculator to Find Descriptive Statistics from a Frequency Distribution

You can use a TI-83/84 Plus calculator to find the mean and median of data represented in the form of a frequency distribution. The basic approach is to enter the *class midpoints* in list L1, then enter the corresponding *class frequencies* in list L2, then use **STAT** and **CALC** and **1-Var Stats** to enter the command **1-Var Stats L1, L2.** Use the following procedure.

1. Given a frequency distribution, enter the class midpoint values in list L1 and enter the corresponding class frequencies in list L2. For example, the frequency distribution of Table 2-2 is reproduced in the margin along with the corresponding TI-83/84 Plus screen display. (Remember, you can enter values in L1 and L2 by pressing **STAT** and **ENTER**.)

2. Now press **STAT**, use the arrow keys to select **CALC,** then select the first menu item of **1-Var Stats**, and press **ENTER**.

3. With 1-Var Stats displayed on the screen, enter **L1, L2** (including the comma) so that the complete command of 1-Var Stats L1, L2 appears on the screen. (To enter L1, L2, use this sequence of keystrokes: **2ND** **1** **,** **2ND** **2**). Press **ENTER** to get the results, which include the mean and median. For the sample data in Table 2-2, the mean is displayed as $\bar{x} = 76.75$ and the median is shown as 74.5.

3-2   Basic Skills and Concepts

Statistical Literacy and Critical Thinking

1. Measures of Center In what sense are the mean, median, mode, and midrange measures of "center"?

2. Average A headline in *USA Today* stated that "Average family income drops 2.3%." What is the role of the term *average* in statistics? Should another term be used in place of *average*?

3. Median In an editorial, the *Poughkeepsie Journal* printed this statement: "The median price—the price exactly in between the highest and lowest—…" Does that statement correctly describe the median? Why or why not?

4. Nominal Data When the Indianapolis Colts recently won the Super Bowl, the numbers on the jerseys of the active players were 29, 41, 50, 58, 79, …, 10 (listed in the alphabetical order of the player's names). Does it make sense to calculate the mean of those numbers? Why or why not?

In Exercises 5–20, find the (a) mean, (b) median, (c) mode, and (d) midrange for the given sample data. Then answer the given questions.

5. Number of English Words A simple random sample of pages from *Merriam-Webster's Collegiate Dictionary, 11th edition,* was obtained. Listed below are the numbers of words defined on those pages. Given that this dictionary has 1459 pages with defined words, estimate the total number of defined words in the dictionary. Is that estimate likely to be an accurate estimate of the number of words in the English language?

51   63   36   43   34   62   73   39   53   79

6. Tests of Child Booster Seats The National Highway Traffic Safety Administration conducted crash tests of child booster seats for cars. Listed below are results from those tests, with the measurements given in hic (standard *head injury condition* units). According to the safety requirement, the hic measurement should be less than 1000 hic. Do the results suggest that all of the child booster seats meet the specified requirement?

774   649   1210   546   431   612

TI **7. Car Crash Costs** The Insurance Institute for Highway Safety conducted tests with crashes of new cars traveling at 6 mi/h. The total cost of the damages was found for a simple random sample of the tested cars and listed below. Do the different measures of center differ very much?

$7448 $4911 $9051 $6374 $4277

TI **8. FICO Scores** The FICO credit rating scores obtained in a simple random sample are listed below. As of this writing, the reported mean FICO score was 678. Do these sample FICO scores appear to be consistent with the reported mean?

714 751 664 789 818 779 698 836 753 834 693 802

TI **9. TV Salaries** Listed below are the top 10 annual salaries (in millions of dollars) of TV personalities (based on data from *OK!* magazine). These salaries correspond to Letterman, Cowell, Sheindlin, Leno, Couric, Lauer, Sawyer, Viera, Sutherland, and Sheen. Given that these are the *top 10* salaries, do we know anything about the salaries of TV personalities in general? Are such *top 10* lists valuable for gaining insight into the larger population?

38 36 35 27 15 13 12 10 9.6 8.4

TI **10. Phenotypes of Peas** Biologists conducted experiments to determine whether a deficiency of carbon dioxide in the soil affects the phenotypes of peas. Listed below are the phenotype codes, where 1 = smooth-yellow, 2 = smooth-green, 3 = wrinkled-yellow, and 4 = wrinkled-green. Can the measures of center be obtained for these values? Do the results make sense?

2 1 1 1 1 1 1 4 1 2 2 1 2 3 3 2 3 1 3 1 3 1 3 2 2

TI **11. Space Shuttle Flights** Listed below are the durations (in hours) of a simple random sample of all flights (as of this writing) of NASA's Space Transport System (space shuttle). The data are from Data Set 10 in Appendix B. Is there a duration time that is very unusual? How might that duration time be explained?

73 95 235 192 165 262 191 376 259 235 381 331 221 244 0

TI **12. Freshman 15** According to the "freshman 15" legend, college freshmen gain 15 pounds (or 6.8 kilograms) during their freshman year. Listed below are the amounts of weight change (in kilograms) for a simple random sample of freshmen included in a study ("Changes in Body Weight and Fat Mass of Men and Women in the First Year of College: A Study of the 'Freshman 15,'" by Hoffman, Policastro, Quick, and Lee, *Journal of American College Health,* Vol. 55, No. 1). Positive values correspond to students who gained weight and negative values correspond to students who lost weight. Do these values appear to support the legend that college students gain 15 pounds (or 6.8 kilograms) during their freshman year? Why or why not?

11 3 0 −2 3 −2 −2 5 −2 7 2 4 1 8 1 0 −5 2

TI **13. Change in MPG Measure** Fuel consumption is commonly measured in miles per gallon. The Environmental Protection Agency designed new fuel consumption tests to be used starting with 2008 car models. Listed below are randomly selected amounts by which the measured MPG ratings *decreased* because of the new 2008 standards. For example, the first car was measured at 16 mi/gal under the old standards and 15 mi/gal under the new 2008 standards, so the amount of the decrease is 1 mi/gal. Would there be much of an error if, instead of retesting all older cars using the new 2008 standards, the mean amount of decrease is subtracted from the measurement obtained with the old standard?

1 2 3 2 4 3 4 2 2 2 2 3 2 2 2 3 2 2 2 2

TI **14. NCAA Football Coach Salaries** Listed below are the annual salaries for a simple random sample of NCAA football coaches (based on data from *USA Today*). How do the mean and median change if the highest salary is omitted?

$150,000 $300,000 $350,147 $232,425 $360,000 $1,231,421 $810,000 $229,000

TI **15. Playing Times of Popular Songs** Listed below are the playing times (in seconds) of songs that were popular at the time of this writing. (The songs are by Timberlake, Furtado, Daughtry, Stefani, Fergie, Akon, Ludacris, Beyonce, Nickelback, Rihanna, Fray, Lavigne, Pink, Mims, Mumidee, and Omarion.) Is there one time that is very different from the others?

448 242 231 246 246 293 280 227 244 213 262 239 213 258 255 257

TI **16. Satellites** Listed below are the numbers of satellites in orbit from different countries. Does one country have an exceptional number of satellites? Can you guess which country has the most satellites?

158 17 15 18 7 3 5 1 8 3 4 2 4 1 2 3 1 1 1 1 1 1 1 1

TI **17. Years to Earn Bachelor's Degree** Listed below are the lengths of time (in years) it took for a random sample of college students to earn bachelor's degrees (based on data from the U.S. National Center for Education Statistics). Based on these results, does it appear that it is common to earn a bachelor's degree in four years?

4 4 4 4 4 4 4.5 4.5 4.5 4.5 4.5 4.5 6 6 8 9 9 13 13 15

TI **18. Car Emissions** Environmental scientists measured the greenhouse gas emissions of a sample of cars. The amounts listed below are in tons (per year), expressed as CO_2 equivalents. Given that the values are a simple random sample selected from Data Set 16 in Appendix B, are these values a simple random sample of cars in use? Why or why not?

7.2 7.1 7.4 7.9 6.5 7.2 8.2 9.3

TI **19. Bankruptcies** Listed below are the numbers of bankruptcy filings in Dutchess County, New York State. The numbers are listed in order for each month of a recent year (based on data from the *Poughkeepsie Journal*). Is there a trend in the data? If so, how might it be explained?

59 85 98 106 120 117 97 95 143 371 14 15

TI **20. Radiation in Baby Teeth** Listed below are amounts of strontium-90 (in millibecquerels or mBq per gram of calcium) in a simple random sample of baby teeth obtained from Pennsylvania residents born after 1979 (based on data from "An Unexpected Rise in Strontium-90 in U.S. Deciduous Teeth in the 1990s," by Mangano, et al., *Science of the Total Environment*). How do the different measures of center compare? What, if anything, does this suggest about the distribution of the data?

155 142 149 130 151 163 151 142 156 133 138 161 128 144 172 137 151 166 147 163

145 116 136 158 114 165 169 145 150 150 150 158 151 145 152 140 170 129 188 156

In Exercises 21–24, find the **mean** *and* **median** *for each of the two samples, then compare the two sets of results.*

TI **21. Cost of Flying** Listed below are costs (in dollars) of roundtrip flights from JFK airport in New York City to San Francisco. (All flights involve one stop and a two-week stay.) The airlines are US Air, Continental, Delta, United, American, Alaska, and Northwest. Does it make much of a difference if the tickets are purchased 30 days in advance or 1 day in advance?

30 Days in Advance: 244   260   264   264   278   318   280
1 Day in Advance: 456   614   567   943   628   1088   536

TI **22. BMI for Miss America** The trend of thinner Miss America winners has generated charges that the contest encourages unhealthy diet habits among young women. Listed below are body mass indexes (BMI) for Miss America winners from two different time periods.

BMI (from the 1920s and 1930s):   20.4   21.9   22.1   22.3   20.3   18.8   18.9   19.4   18.4   19.1
BMI (from recent winners):   19.5   20.3   19.6   20.2   17.8   17.9   19.1   18.8   17.6   16.8

TI **23. Nicotine in Cigarettes** Listed below are the nicotine amounts (in mg per cigarette) for samples of filtered and nonfiltered cigarettes (from Data Set 4 in Appendix B). Do filters appear to be effective in reducing the amount of nicotine?

Nonfiltered: 1.1   1.7   1.7   1.1   1.1   1.4   1.1   1.4   1.0   1.2   1.1   1.1   1.1
1.1   1.1   1.8   1.6   1.1   1.2   1.5   1.3   1.1   1.3   1.1   1.1

Filtered: 0.4   1.0   1.2   0.8   0.8   1.0   1.1   1.1   1.1   0.8   0.8   0.8   0.8
1.0   0.2   1.1   1.0   0.8   1.0   0.9   1.1   1.1   0.6   1.3   1.1

TI **24. Customer Waiting Times** Waiting times (in minutes) of customers at the Jefferson Valley Bank (where all customers enter a single waiting line) and the Bank of Providence (where customers wait in individual lines at three different teller windows) are listed below. Determine whether there is a difference between the two data sets that is not apparent from a comparison of the measures of center. If so, what is it?

Jefferson Valley (single line):   6.5   6.6   6.7   6.8   7.1   7.3   7.4   7.7   7.7   7.7

Providence (individual lines):   4.2   5.4   5.8   6.2   6.7   7.7   7.7   8.5   9.3   10.0

Large Data Sets from Appendix B. *In Exercises 25–28, refer to the indicated data set in Appendix B. Use computer software or a calculator to find the* means *and* medians.

TI **25. Body Temperatures** Use the body temperatures for 12:00 AM on day 2 from Data Set 2 in Appendix B. Do the results support or contradict the common belief that the mean body temperature is 98.6°F?

TI **26. How Long Is a 3/4 in. Screw?** Use the listed lengths of the machine screws from Data Set 19 in Appendix B. The screws are supposed to have a length of 3/4 in. Do the results indicate that the specified length is correct?

TI **27. Home Voltage** Refer to Data Set 13 in Appendix B. Compare the means and medians from the three different sets of measured voltage levels.

TI **28. Movies** Refer to Data Set 9 in Appendix B and consider the gross amounts from two different categories of movies: Movies with R ratings and movies with ratings of PG or PG-13. Do the results appear to support a claim that R-rated movies have greater gross amounts because they appeal to larger audiences than movies rated PG or PG-13?

In Exercises 29–32, find the mean *of the data summarized in the given frequency distribution. Also, compare the* computed means *to the* actual means *obtained by using the original list of data values, which are as follows: (Exercise 29) 21.1 mg; (Exercise 30) 76.3 beats per minute; (Exercise 31) 46.7 mi/h; (Exercise 32) 1.911 lb.*

TI **29.**

Tar (mg) in Nonfiltered Cigarettes	Frequency
10–13	1
14–17	0
18–21	15
22–25	7
26–29	2

TI **30.**

Pulse Rates of Females	Frequency
60–69	12
70–79	14
80–89	11
90–99	1
100–109	1
110–119	0
120–129	1

TI **31. Speeding Tickets** The given frequency distribution describes the speeds of drivers ticketed by the Town of Poughkeepsie police. These drivers were traveling through a 30mi/h speed zone on Creek Road, which passes the author's college. How does the mean speed compare to the posted speed limit of 30mi/h?

Table for Exercise 31

Speed	Frequency
42–45	25
46–49	14
50–53	7
54–57	3
58–61	1

TI **32.**

Weights (lb) of Discarded Plastic	Frequency
0.00–0.99	14
1.00–1.99	20
2.00–2.99	21
3.00–3.99	4
4.00–4.99	2
5.00–5.99	1

33. Weighted Mean A student of the author earned grades of B, C, B, A, and D. Those courses had these corresponding numbers of credit hours: 3, 3, 4, 4, and 1. The grading system assigns quality points to letter grades as follows: A = 4; B = 3; C = 2; D = 1; F = 0. Compute the grade point average (GPA) and round the result with two decimal places. If the Dean's list requires a GPA of 3.00 or greater, did this student make the Dean's list?

34. Weighted Mean A student of the author earned grades of 92, 83, 77, 84, and 82 on her five regular tests. She earned grades of 88 on the final exam and 95 on her class projects. Her combined homework grade was 77. The five regular tests count for 60% of the final grade, the final exam counts for 10%, the project counts for 15%, and homework counts for 15%. What is her weighted mean grade? What letter grade did she earn? (A, B, C, D, or F)

3-2   Beyond the Basics

35. Degrees of Freedom A secondary standard mass is periodically measured and compared to the standard for one kilogram (or 1000 grams). Listed below is a sample of measured masses (in micrograms) that the secondary standard is *below* the true mass of 1000 grams. One of the sample values is missing and is not shown below. The data are from the National Institutes of Standards and Technology, and the mean of the sample is 657.054 micrograms.

a. Find the missing value.

b. We need to create a list of n values that have a specific known mean. We are free to select any values we desire for some of the n values. How many of the n values can be freely assigned before the remaining values are determined? (The result is referred to as the *number of degrees of freedom.*)

$$675.04 \quad 665.10 \quad 631.27 \quad 671.35$$

36. Censored Data As of this writing, there have been 42 different presidents of the United States, and four of them are alive. Listed below are the numbers of years that they lived after their first inauguration, and the four values with the plus signs represent the four presidents who are still alive. (These values are said to be *censored* at the current time that this list was compiled.) What can you conclude about the mean time that a president lives after inauguration?

10 29 26 28 15 23 17 25 0 20 4 1 24 16 12 4 10 17 16 0 7

24 12 4 18 21 11 2 9 36 12 28 3 16 9 25 23 32 30+ 18+ 14+ 6+

37. Trimmed Mean Because the mean is very sensitive to extreme values, we stated that it is not a *resistant* measure of center. The **trimmed mean** is more resistant. To find the 10% trimmed mean for a data set, first arrange the data in order, then delete the bottom 10% of the values and the top 10% of the values, then calculate the mean of the remaining values. For the FICO credit-rating scores in Data Set 24 from Appendix B, find the following. How do the results compare?

a. the mean   **b.** the 10% trimmed mean   **c.** the 20% trimmed mean

38. Harmonic Mean The **harmonic mean** is often used as a measure of center for data sets consisting of rates of change, such as speeds. It is found by dividing the number of values n by the sum of the *reciprocals* of all values, expressed as

$$\frac{n}{\sum \dfrac{1}{x}}$$

(No value can be zero.) The author drove 1163 miles to a conference in Orlando, Florida. For the trip to the conference, the author stopped overnight, and the mean speed from start to finish was 38 mi/h. For the return trip, the author stopped only for food and fuel, and the mean speed from start to finish was 56 mi/h. Can the "average" speed for the combined round trip be found by adding 38 mi/h and 56 mi/h, then dividing that sum by 2? Why or why not? What is the "average" speed for the round trip?

39. Geometric Mean The **geometric mean** is often used in business and economics for finding average rates of change, average rates of growth, or average ratios. Given *n* values (all of which are positive), the geometric mean is the *n*th root of their product. The *average growth factor* for money compounded at annual interest rates of 10%, 5%, and 2% can be found by computing the geometric mean of 1.10, 1.05, and 1.02. Find that average growth factor. What single percentage growth rate would be the same as having three successive growth rates of 10%, 5%, and 2%? Is that result the same as the mean of 10%, 5%, and 2%?

40. Quadratic Mean The **quadratic mean** (or **root mean square**, or **R.M.S.**) is usually used in physical applications. In power distribution systems, for example, voltages and currents are usually referred to in terms of their R.M.S. values. The quadratic mean of a set of values is obtained by squaring each value, adding those squares, dividing the sum by the number of values *n*, and then taking the square root of that result, as indicated below:

$$\text{quadratic mean} = \sqrt{\frac{\Sigma x^2}{n}}$$

Find the R.M.S. of the voltages listed for the generator from Data Set 13 in Appendix B. How does the result compare to the mean? Will the same comparison apply to all other data sets?

41. Median When data are summarized in a frequency distribution, the median can be found by first identifying the *median class* (the class that contains the median). We then assume that the values in that class are evenly distributed and we can interpolate. Letting *n* denote the sum of all class frequencies, and letting *m* denote the sum of the class frequencies that *precede* the median class, the median can be estimated as shown below.

$$(\text{lower limit of median class}) + (\text{class width})\left(\frac{\left(\dfrac{n+1}{2}\right) - (m+1)}{\text{frequency of median class}}\right)$$

Use this procedure to find the median of the frequency distribution given in Exercise 29. How does the result compare to the median of the original list of data, which is 20.0 mg? Which value of the median is better: the value computed for the frequency table or the value of 20.0 mg?

3-3 Measures of Variation

Key Concept In this section we discuss the characteristic of variation. In particular, we present measures of variation, such as the *standard deviation,* as tools for analyzing data. Our focus here is not only to find values of the measures of variation, but also to interpret those values. In addition, we discuss concepts that help us to better understand the standard deviation.

Study Hint: Part 1 of this section presents basic concepts of variation and Part 2 presents additional concepts related to the standard deviation. Although both parts contain several formulas for computation, do not spend too much time memorizing those formulas and doing arithmetic calculations. Instead, make *understanding* and *interpreting* the standard deviation a priority.

Part 1: Basic Concepts of Variation

For a visual illustration of variation, see the accompanying dotplots representing two different samples of IQ scores. Both samples have the same mean of 100, but notice how the top dotplot (based on randomly selected high school students) shows IQ scores that are spread apart much farther than in the bottom dotplot (representing high school students grouped according to grades). This characteristic of spread, or variation, or dispersion, is so important that we develop methods for measuring it with numbers. We begin with the *range*.

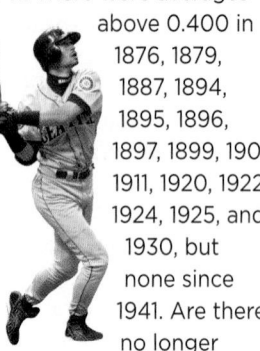

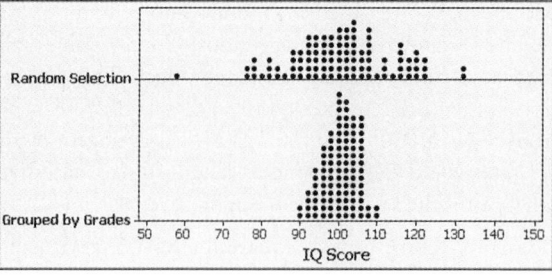

Both samples have the same mean of 100.0.

Range

The first measure of variation we consider is the range.

> **DEFINITION**
>
> The **range** of a set of data values is the difference between the maximum data value and the minimum data value.
>
> $$\text{range} = (\text{maximum data value}) - (\text{minimum data value})$$

Because the range uses only the maximum and the minimum data values, it is very sensitive to extreme values and isn't as useful as other measures of variation that use every data value, such as the standard deviation. However, because the range is so easy to compute and understand, it is used often in statistical process control. (See Section 14-2 for control charts based on the range.)

In general, the range should not be rounded. However, to keep procedures consistent, we round the range using the same round-off rule for all measures of variation discussed in this section.

Round-Off Rule for Measures of Variation

When rounding the value of a measure of variation, carry one more decimal place than is present in the original set of data.

EXAMPLE 1   **Range** As of this writing, India has 1 satellite used for military and intelligence purposes, Japan has 3, and Russia has 14. Find the range of the sample values of 1, 3, and 14.

SOLUTION   The range is found by subtracting the lowest value from the largest value, so we get

$$\text{range} = (\text{maximum value}) - (\text{minimum value}) = 14 - 1 = 13.0$$

The result is shown with one more decimal place than is present in the original data values.

Standard Deviation of a Sample

The *standard deviation* is the measure of variation most commonly used in statistics.

 **DEFINITION**

The **standard deviation** of a set of sample values, denoted by s, is a measure of variation of values about the mean. It is a type of *average deviation* of values from the mean that is calculated by using Formula 3-4 or 3-5. Formula 3-5 is just a different version of Formula 3-4; it is algebraically the same.

Formula 3-4

$$s = \sqrt{\frac{\Sigma(x - \bar{x})^2}{n - 1}}$$ sample standard deviation

Formula 3-5

$$s = \sqrt{\frac{n\Sigma(x^2) - (\Sigma x)^2}{n(n - 1)}}$$ shortcut formula for sample standard deviation (formula used by calculators and computer programs)

Later in this section we describe the reasoning behind these formulas, but for now we recommend that you use Formula 3-4 for a few examples, then learn how to find standard deviation values using your TI-83/84 Plus calculator. The following properties are consequences of the way in which the standard deviation is defined:

- The standard deviation is a measure of variation of all values from the *mean.*

- The value of the standard deviation s is usually positive. It is zero only when all of the data values are the same number. (It is never negative.) Also, larger values of s indicate greater amounts of variation.

- The value of the standard deviation s can increase dramatically with the inclusion of one or more outliers (data values that are very far away from all of the others).

- The units of the standard deviation s (such as minutes, feet, pounds, and so on) are the same as the units of the original data values.

If our goal was to develop skills for manually calculating values of standard deviations, we would focus on Formula 3-5, which simplifies the calculations. However, we prefer to show a calculation using Formula 3-4, because that formula better illustrates that the standard deviation is based on deviations of sample values away from the mean.

EXAMPLE 2 **Using Formula 3-4** Use Formula 3-4 to find the standard deviation of the sample values of 1, 3, and 14 from Example 1.

SOLUTION The left column of Table 3-2 summarizes the general procedure for finding the standard deviation using Formula 3-4, and the right column illustrates that procedure for the sample values 1, 3, and 14. The result shown in Table 3-2 is 7.0, which is rounded to one more decimal place than is present in the original list of sample values (1, 3, 14). Also, the units for the standard deviation are the same as the units of the original data. Because the original data are 1 satellite, 3 satellites, and 14 satellites, the standard deviation is 7.0 satellites.

continued

More Stocks, Less Risk

In their book *Investments*, authors Zvi Bodie, Alex Kane, and Alan Marcus state that "the average standard deviation for returns of portfolios composed of only one stock was 0.554. The average portfolio risk fell rapidly as the number of stocks included in the portfolio increased." They note that with 32 stocks, the standard deviation is 0.325, indicating much less variation and risk. They make the point that with only a few stocks, a portfolio has a high degree of "firm-specific" risk, meaning that the risk is attributable to the few stocks involved. With more than 30 stocks, there is very little firm-specific risk; instead, almost all of the risk is "market risk," attributable to the stock market as a whole. They note that these principles are "just an application of the well-known law of averages."

Table 3-2

General Procedure for Finding Standard Deviation with Formula 3-4	Specific Example using these sample values: 1, 3, 14.
Step 1: Compute the mean $\bar{x}$.	The sum of 1, 3, and 14 is 18, so $$\bar{x} = \frac{\Sigma x}{n} = \frac{1 + 3 + 14}{3} = \frac{18}{3} = 6.0$$
Step 2: Subtract the mean from each individual sample value. (The result is a list of deviations of the form $(x - \bar{x})$.)	Subtract the mean of 6.0 from each sample value to get these deviations away from the mean: $-5, -3, 8$.
Step 3: Square each of the deviations obtained from Step 2. (This produces numbers of the form $(x - \bar{x})^2$.)	The squares of the deviations from Step 2 are: 25, 9, 64.
Step 4: Add all of the squares obtained from Step 3. The result is $\Sigma(x - \bar{x})^2$.	The sum of the squares from Step 3 is $25 + 9 + 64 = 98$.
Step 5: Divide the total from Step 4 by the number $n - 1$, which is 1 less than the total number of sample values present.	With $n = 3$ data values, $n - 1 = 2$, so we divide 98 by 2 to get this result: $\frac{98}{2} = 49$.
Step 6: Find the square root of the result of Step 5. The result is the standard deviation.	The standard deviation is $\sqrt{49} = 7.0$.

EXAMPLE 3 **Using Formula 3-5** Use Formula 3-5 to find the standard deviation of the sample values 1, 3, and 14 from Example 1.

SOLUTION Shown below is the computation of the standard deviation of 1 satellite, 3 satellites, and 14 satellites using Formula 3-5.

$n = 3$ (because there are 3 values in the sample)

$\Sigma x = 18$ (found by adding the sample values: $1 + 3 + 14 = 18$)

$\Sigma x^2 = 206$ (found by adding the squares of the sample values, as in $1^2 + 3^2 + 14^2 = 206$)

Using Formula 3-5, we get

$$s = \sqrt{\frac{n(\Sigma x^2) - (\Sigma x)^2}{n(n-1)}} = \sqrt{\frac{3(206) - (18)^2}{3(3-1)}} = \sqrt{\frac{294}{6}} = 7.0 \text{ satellites}$$

Note that the result is the same as the result in Example 2.

Comparing Variation in Different Samples Table 3-3 shows measures of center and measures of variation for the word counts of the 186 men and 210 women listed in Data Set 8 in Appendix B. From the table we see that the range for men is somewhat larger than the range for women. Table 3-3 also shows that the standard deviation for men is somewhat larger than the standard deviation for women, but *it's a good practice to compare two sample standard deviations only when the sample means are approximately the same.* When comparing variation in samples with very different means, it is better to use the coefficient of variation, which is defined later in this section. We also use the coefficient of variation when we want to compare variation from two samples with different scales or units of values, such as the comparison of variation of heights of men and weights of men (see Example 8, at the end of this section).

Table 3-3   Comparison of Word Counts of
Men and Women

	Men	Women
Mean	15,668.5	16,215.0
Median	14,290.0	15,917.0
Midrange	23,855.5	20,864.5
Range	46,321.0	38,381.0
Standard Deviation	8,632.5	7,301.2

Standard Deviation of a Population

The definition of standard deviation and Formulas 3-4 and 3-5 apply to the standard deviation of *sample* data. A slightly different formula is used to calculate the standard deviation σ (lowercase sigma) of a *population:* Instead of dividing by $n - 1$, we divide by the population size N, as shown here:

$$\text{population standard deviation} \qquad \sigma = \sqrt{\frac{\Sigma(x - \mu)^2}{N}}$$

Because we generally deal with sample data, we will usually use Formula 3-4, in which we divide by $n - 1$. The TI-83/84 Plus calculator gives both the sample standard deviation and the population standard deviation. See the display in the margin. The value identified as Sx is the *sample* standard deviation, and the value identified as σx is the *population* standard deviation.

TI-83/84 Plus

```
1-Var Stats
  x̄=15668.52688
  Σx=2914346
  Σx²=5.94498E10
  Sx=8632.534672
  σx=8609.297659
↓n=186
```

> **CAUTION**
> ....
> When using technology to find the standard deviation of sample data, be sure that you obtain the *sample* standard deviation, not the population standard deviation.

Variance of a Sample and a Population

So far, we have used the term *variation* as a general description of the amount that values vary among themselves. (The terms *dispersion* and *spread* are sometimes used instead of *variation.*) The term *variance* has a specific meaning.

> **DEFINITION**
>
> The **variance** of a set of values is a measure of variation equal to the square of the standard deviation.
>
> Sample variance: s^2 square of the standard deviation s.
>
> Population variance: σ^2 square of the population standard deviation σ.

The sample variance s^2 is an **unbiased estimator** of the population variance σ^2, which means that values of s^2 tend to target the value of σ^2 instead of systematically tending to overestimate or underestimate σ^2. For example, consider an IQ test designed so that the population variance is 225. If you repeat the process of randomly selecting 100 subjects, giving them IQ tests, and calculating the sample variance s^2 in each case, the sample variances that you obtain will tend to center around 225, which is the population variance.

The variance is a statistic used in some statistical methods, such as analysis of variance discussed in Chapter 12. For our present purposes, the variance has this serious disadvantage: *The units of variance are different than the units of the original data set.* For example, if we have data consisting of waiting times in minutes, the units of the variance are min², but what is a square minute? Because the variance uses different units, it is difficult to understand variance as it relates to the original data set. Because of this property, it is better to focus on the standard deviation when trying to develop an understanding of variation, as we do later in this section.

Part 1 of this section introduced basic concepts of variation. The notation we have used is summarized below.

Notation

s = *sample* standard deviation

s^2 = *sample* variance

σ = *population* standard deviation

σ^2 = *population* variance

Note: Articles in professional journals and reports often use SD for standard deviation and VAR for variance.

Part 2: Beyond the Basics of Variation

Using and Understanding Standard Deviation

In this subsection we focus on making sense of the standard deviation, so that it is not some mysterious number devoid of any practical significance.

One crude but simple tool for understanding standard deviation is the **range rule of thumb,** which is based on the principle that for many data sets, the vast majority (such as 95%) of sample values lie within two standard deviations of the mean. We could improve the accuracy of this rule by taking into account such factors as the size of the sample and the distribution, but here we prefer to sacrifice accuracy for the sake of simplicity. Also, we could use three or even four standard deviations instead of two standard deviations, but we want a simple rule that will help us interpret values of standard deviations. Later we study methods that will produce more accurate results.

Range Rule of Thumb

Interpreting a Known Value of the Standard Deviation: We informally defined *usual* values in a data set to be those that are typical and not too extreme. If the standard deviation of a collection of data is known, use it to find rough estimates of the minimum and maximum *usual* sample values as follows:

$$\text{minimum “usual” value} = (\text{mean}) - 2 \times (\text{standard deviation})$$
$$\text{maximum “usual” value} = (\text{mean}) + 2 \times (\text{standard deviation})$$

Estimating a Value of the Standard Deviation s: To roughly estimate the standard deviation from a collection of known sample data, use

$$s \approx \frac{\text{range}}{4}$$

where range = (maximum data value) − (minimum data value).

EXAMPLE 4 **Range Rule of Thumb for Interpreting s** The Wechsler Adult Intelligence Scale involves an IQ test designed so that the mean score is 100 and the standard deviation is 15. Use the range rule of thumb to find the minimum and maximum "usual" IQ scores. Then determine whether an IQ score of 135 would be considered "unusual."

SOLUTION With a mean of 100 and a standard deviation of 15, we use the range rule of thumb to find the minimum and maximum usual IQ scores as follows:

$$\text{minimum "usual" value} = (\text{mean}) - 2 \times (\text{standard deviation})$$
$$= 100 - 2(15) = 70$$
$$\text{maximum "usual" value} = (\text{mean}) + 2 \times (\text{standard deviation})$$
$$= 100 + 2(15) = 130$$

INTERPRETATION Based on these results, we expect that typical IQ scores fall between 70 and 130. Because 135 does not fall within those limits, it would be considered an unusual IQ score.

EXAMPLE 5 **Range Rule of Thumb for Estimating s** Use the range rule of thumb to estimate the standard deviation of the sample of 100 FICO credit rating scores listed in Data Set 24 in Appendix B. Those scores have a minimum of 444 and a maximum of 850.

SOLUTION The range rule of thumb indicates that we can estimate the standard deviation by finding the range and dividing it by 4. With a minimum of 444 and a maximum of 850, the range rule of thumb can be used to estimate the standard deviation s as follows:

$$s \approx \frac{\text{range}}{4} = \frac{850 - 444}{4} = 101.5$$

INTERPRETATION The actual value of the standard deviation is $s = 92.2$. The estimate of 101.5 is off by a fair amount. This illustrates that the range rule of thumb yields a rough estimate that might be off by a considerable amount.

Listed below are properties of the standard deviation.

Properties of the Standard Deviation

• The standard deviation measures the *variation* among data values.

• Values close together have a small standard deviation, but values with much more variation have a larger standard deviation.

• The standard deviation has the same units of measurement (such as minutes or grams or dollars) as the original data values.

• For many data sets, a value is *unusual* if it differs from the mean by more than two standard deviations.

• When comparing variation in two different data sets, compare the standard deviations only if the data sets use the same scale and units and they have means that are approximately the same.

Empirical (or 68–95–99.7) Rule for Data with a Bell-Shaped Distribution

Another concept that is helpful in interpreting the value of a standard deviation is the **empirical rule.** This rule states that *for data sets having a distribution that is approximately bell-shaped,* the following properties apply. (See Figure 3-3.)

• About 68% of all values fall within 1 standard deviation of the mean.

• About 95% of all values fall within 2 standard deviations of the mean.

• About 99.7% of all values fall within 3 standard deviations of the mean.

EXAMPLE 6   **Empirical Rule** IQ scores have a bell-shaped distribution with a mean of 100 and a standard deviation of 15. What percentage of IQ scores are between 70 and 130?

SOLUTION   The key to solving this problem is to recognize that 70 and 130 are each exactly 2 standard deviations away from the mean of 100, as shown below.

$$2 \text{ standard deviations} = 2s = 2(15) = 30$$

Therefore, 2 standard deviations from the mean is

$$100 - 30 = 70$$
$$\text{or} \quad 100 + 30 = 130$$

The empirical rule tells us that about 95% of all values are within 2 standard deviations of the mean, so about 95% of all IQ scores are between 70 and 130.

Figure 3-3
The Empirical Rule

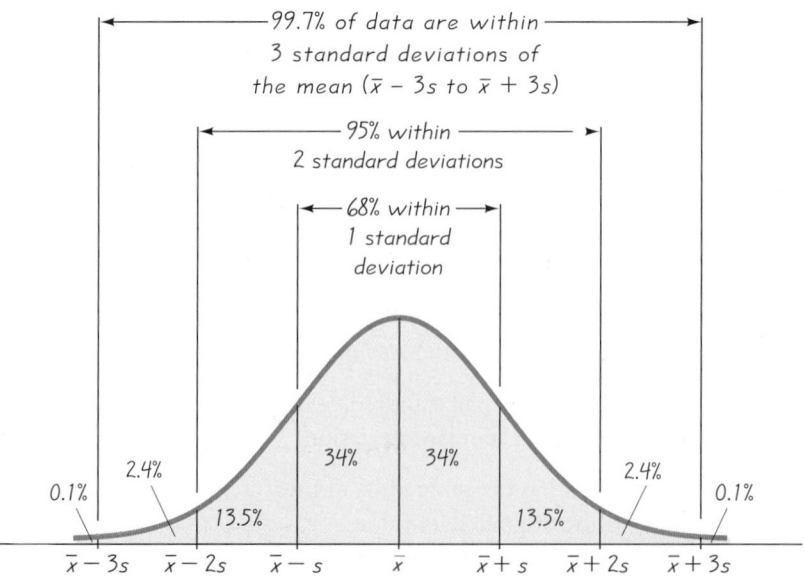

A third concept that is helpful in understanding or interpreting a value of a standard deviation is **Chebyshev's theorem.** The empirical rule applies only to data sets with bell-shaped distributions, but Chebyshev's theorem applies to *any* data set. Unfortunately, results from Chebyshev's theorem are only approximate. Because the results are lower limits ("at least"), Chebyshev's theorem has limited usefulness.

Chebyshev's Theorem

The proportion (or fraction) of any set of data lying within K standard deviations of the mean is always *at least* $1 - 1/K^2$, where K is any positive number greater than 1. For $K = 2$ and $K = 3$, we get the following statements:

- At least 3/4 (or 75%) of all values lie within 2 standard deviations of the mean.
- At least 8/9 (or 89%) of all values lie within 3 standard deviations of the mean.

EXAMPLE 7 **Chebyshev's Theorem** IQ scores have a mean of 100 and a standard deviation of 15. What can we conclude from Chebyshev's theorem?

SOLUTION Applying Chebyshev's theorem with a mean of 100 and a standard deviation of 15, we can reach the following conclusions.

- At least 3/4 (or 75%) of IQ scores are within 2 standard deviations of the mean (between 70 and 130).

- At least 8/9 (or 89%) of all IQ scores are within 3 standard deviations of the mean (between 55 and 145).

When trying to make sense of the standard deviation, we should use one or more of the preceding three concepts. To gain additional insight into the nature of the standard deviation, we now consider the underlying rationale leading to Formula 3-4, which is the basis for its definition. (Recall that Formula 3-5 is simply another version of Formula 3-4.)

Why Is Standard Deviation Defined as in Formula 3-4?

Why do we measure variation using Formula 3-4? In measuring variation in a set of sample data, it makes sense to begin with the individual amounts by which values deviate from the mean. For a particular data value x, the amount of **deviation** is $x - \bar{x}$, which is the difference between the individual x value and the mean. For the values of 1, 3, 14, the mean is 6.0 so the deviations away from the mean are -5, -3, and 8. It would be good to somehow combine those deviations into one number that can serve as a measure of the variation. Simply adding the deviations doesn't work, because the sum will always be zero. To get a statistic that measures variation (instead of always being zero), we need to avoid the canceling out of negative and positive numbers. One approach is to add absolute values, as in $\Sigma|x - \bar{x}|$. If we find the mean of that sum, we get the **mean absolute deviation** (or **MAD**), which is the mean distance of the data from the mean:

$$\text{mean absolute deviation} = \frac{\Sigma|x - \bar{x}|}{n}$$

Because the values of 1, 3, 14 have deviations of -5, -3, and 8, the mean absolute deviation is $(5 + 3 + 8)/3 = 16/3 = 5.3$.

Why Not Use the Mean Absolute Deviation Instead of the Standard Deviation? Computation of the mean absolute deviation uses absolute values, so it uses an operation that is not "algebraic." (The algebraic operations include addition, multiplication, extracting roots, and raising to powers that are integers or fractions, but absolute value is not included among the algebraic operations.) The use of absolute values would create algebraic difficulties in inferential methods of statistics discussed in later chapters. For example, Section 9-3 presents a method for making inferences about the means of two populations, and that method is built around an additive property of variances, but the mean absolute deviation has no such additive property. (Here is a simplified version of the additive property of variances: If you have two independent populations and you randomly select one value from each population and add them, such sums will have a variance equal to the sum of the variances of the two populations.) Also, the mean absolute deviation is *biased*, meaning that when you find mean absolute deviations of samples, you do not tend to target the mean absolute deviation of the population. In contrast, the standard deviation uses only algebraic operations. Because it is based on the square root of a sum of squares, the standard deviation closely parallels distance formulas found in algebra. There are many instances where a statistical procedure is based on a similar sum of squares. Therefore, instead of using absolute values, we square all deviations $(x - \bar{x})$ so that they are nonnegative. This approach leads to the standard deviation. For these reasons, scientific calculators typically include a standard deviation function, but they almost never include the mean absolute deviation.

Why Divide by $n - 1$? After finding all of the individual values of $(x - \bar{x})^2$, we combine them by finding their sum. We then divide by $n - 1$ because there are only $n - 1$ independent values. With a given mean, only $n - 1$ values can be freely assigned any number before the last value is determined. Exercise 37 illustrates that division by $n - 1$ yields a better result than division by n. That exercise shows how division by $n - 1$ causes the sample variance s^2 to target the value of the population variance σ^2, whereas division by n causes the sample variance s^2 to underestimate the value of the population variance σ^2.

Comparing Variation in Different Populations

When comparing variation in two different sets of data, the standard deviations should be compared only if the two sets of data use the same scale and units and they have approximately the same mean. If the means are substantially different, or if the samples use different scales or measurement units, we can use the *coefficient of variation,* defined as follows.

DEFINITION

The **coefficient of variation** (or **CV**) for a set of nonnegative sample or population data, expressed as a percent, describes the standard deviation relative to the mean, and is given by the following:

Sample	Population
$CV = \dfrac{s}{\bar{x}} \cdot 100\%$	$CV = \dfrac{\sigma}{\mu} \cdot 100\%$

EXAMPLE 8   **Heights and Weights of Men** Compare the variation in heights of men to the variation in weights of men, using these sample results obtained from Data Set 1 in Appendix B: for men, the heights yield $\bar{x} = 68.34$ in. and $s = 3.02$ in; the weights yield $\bar{x} = 172.55$ lb and $s = 26.33$ lb. Note that we want to compare variation among *heights* to variation among *weights*.

SOLUTION   We can compare the standard deviations if the same scales and units are used and the two means are approximately equal, but here we have different scales (heights and weights) and different units of measurement (inches and pounds), so we use the coefficients of variation:

$$\text{heights:} \qquad CV = \frac{s}{\bar{x}} \cdot 100\% = \frac{3.02 \text{ in.}}{68.34 \text{ in.}} \cdot 100\% = 4.42\%$$

$$\text{weights:} \qquad CV = \frac{s}{\bar{x}} \cdot 100\% = \frac{26.33 \text{ lb}}{172.55 \text{ lb}} \cdot 100\% = 15.26\%$$

Although the standard deviation of 3.02 in. cannot be compared to the standard deviation of 26.33 lb, we can compare the coefficients of variation, which have no units. We can see that heights (with $CV = 4.42\%$) have considerably less variation than weights (with $CV = 15.26\%$). This makes intuitive sense, because we routinely see that weights among men vary much more than heights. It is very rare to see two adult men with one of them being twice as tall as the other, but it is much more common to see two men with one of them weighing twice as much as the other.

Using the TI-83/84 Plus Calculator to Find Measures of Variation

List of single values   If using the TI-83/84 Plus calculator to find measures of variation from a list of data values, use the same procedures given at the end of Section 3-2. For a list of single values, enter or open the list, then press **STAT**, select **CALC**, press **ENTER**, then enter the label or name, as in 1-Var Stats L1 or 1-Var Stats LAGES. (If you want to enter a list name such as AGES, access the list of variable names by pressing **2ND** **STAT**, then scroll to the name AGES and Press **ENTER**.) Press **ENTER** to obtain the results.

Frequency distribution   If using the TI-83/84 Plus calculator to find measures of variation from a frequency distribution, use the same procedure given at the end of Section 3-2 under the heading of "Using the TI-83/84 Plus Calculator to Find Descriptive Statistics from a Frequency Distribution." That is, enter the class midpoints in list L1, enter the frequencies in list L2, press **STAT**, select **CALC**, press **ENTER**, then enter **1-Var Stats L1, L2** (including the comma).

Interpretation and other measures   The results will include a value indicated by Sx and another value indicated by σx. The Sx value is the *sample* standard deviation, which is the value we usually want. The σx value is the *population* standard deviation, which we rarely want. Also, the range can be easily computed using the minimum and maximum values identified among the results as minX and maxX. To obtain a value of *variance*, simply square the value of the standard deviation. For example, if the calculator displays Sx = 9.0, enter 9.0 and press **x²** **ENTER** to find that the variance is 81.0.

TI-83/84 Plus

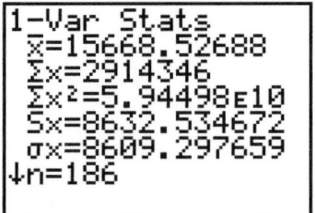

┌── **Sx** denotes the *sample* standard deviation,
│   so $s = 8632.534672$. (This is the value we usually want.)

├── **σx** denotes the *population* standard deviation,
│   so $\sigma = 8609.297659$.

├── From this part of the display, we find the *range* by using the minimum and maxi-
│   mum values: range = maximum − minimum = 47,016 − 695 = 46,321.0.

TI-83/84 Plus

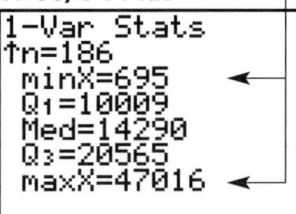

3-3   Basic Skills and Concepts

Statistical Literacy and Critical Thinking

1. Variation and Variance In statistics, how do *variation* and *variance* differ?

2. Correct Statement? In the book *How to Lie with Charts,* it is stated that "the standard deviation is usually shown as plus or minus the difference between the high and the mean, and the low and the mean. For example, if the mean is 1, the high 3, and the low −1, the standard deviation is ±2." Is that statement correct? Why or why not?

3. Comparing Variation Which do you think has more variation: the incomes of a simple random sample of 1000 adults selected from the general population, or the incomes of a simple random sample of 1000 statistics teachers? Why?

4. Unusual Value? The systolic blood pressures of 40 women are given in Data Set 1 in Appendix B. They have a mean of 110.8 mm Hg and a standard deviation of 17.1 mm Hg. The highest systolic blood pressure measurement in this sample is 181 mm Hg. In this context, is a systolic blood pressure of 181 mm Hg "unusual"? Why or why not?

In Exercises 5–20, find the range, variance, *and* standard deviation *for the given sample data. Include appropriate units (such as "minutes") in your results. (The same data were used in Section 3-2 where we found measures of center. Here we find measures of variation.) Then answer the given questions.*

5. Number of English Words *Merriam-Webster's Collegiate Dictionary, 11th edition,* has 1459 pages of defined words. Listed below are the numbers of defined words per page for a simple random sample of those pages. If we use this sample as a basis for estimating the total number of defined words in the dictionary, how does the variation of these numbers affect our confidence in the accuracy of the estimate?

51   63   36   43   34   62   73   39   53   79

6. Tests of Child Booster Seats The National Highway Traffic Safety Administration conducted crash tests of child booster seats for cars. Listed below are results from those tests, with the measurements given in hic (standard *head injury condition* units). According to the safety requirement, the hic measurement should be less than 1000. Do the different child booster seats have much variation among their crash test measurements?

774   649   1210   546   431   612

7. Car Crash Costs The Insurance Institute for Highway Safety conducted tests with crashes of new cars traveling at 6 mi/h. The total cost of the damages for a simple random sample of the tested cars are listed below. Based on these results, is damage of $10,000 *unusual?* Why or why not?

$7448 $4911 $9051 $6374 $4277

8. FICO Scores A simple random sample of FICO credit rating scores is listed below. As of this writing, the mean FICO score was reported to be 678. Based on these results, is a FICO score of 500 *unusual?* Why or why not?

714   751   664   789   818   779   698   836   753   834   693   802

9. TV Salaries Listed below are the top 10 annual salaries (in millions of dollars) of TV personalities (based on data from *OK!* magazine). These salaries correspond to Letterman, Cowell, Sheindlin, Leno, Couric, Lauer, Sawyer, Viera, Sutherland, and Sheen. Given that these are the *top 10* salaries, do we know anything about the variation of salaries of TV personalities in general?

38   36   35   27   15   13   12   10   9.6   8.4

TI **10. Phenotypes of Peas** Biologists conducted an experiment to determine whether a deficiency of carbon dioxide in the soil affects the phenotypes of peas. Listed below are the phenotype codes, where 1 = smooth-yellow, 2 = smooth-green, 3 = wrinkled-yellow, and 4 = wrinkled-green. Can the measures of variation be obtained for these values? Do the results make sense?

2 1 1 1 1 1 1 4 1 2 2 1 2 3 3 3 2 3 1 3 1 3 1 3 2 2

TI **11. Space Shuttle Flights** Listed below are the durations (in hours) of a simple random sample of all flights (as of this writing) of NASA's Space Transport System (space shuttle). The data are from Data Set 10 in Appendix B. Is the lowest duration time *unusual?* Why or why not?

73 95 235 192 165 262 191 376 259 235 381 331 221 244 0

TI **12. Freshman 15** According to the "freshman 15" legend, college freshmen gain 15 pounds (or 6.8 kilograms) during their freshman year. Listed below are the amounts of weight change (in kilograms) for a simple random sample of freshmen included in a study ("Changes in Body Weight and Fat Mass of Men and Women in the First Year of College: A Study of the 'Freshman 15,'" by Hoffman, Policastro, Quick, and Lee, *Journal of American College Health*, Vol. 55, No. 1). Positive values correspond to students who gained weight and negative values correspond to students who lost weight. Is a weight gain of 15 pounds (or 6.8 kg) *unusual?* Why or why not? If 15 pounds (or 6.8 kg) is not unusual, does that support the legend of the "freshman 15"?

11 3 0 −2 3 −2 −2 5 −2 7 2 4 1 8 1 0 −5 2

TI **13. Change in MPG Measure** Fuel consumption is commonly measured in miles per gallon. The Environmental Protection Agency designed new fuel consumption tests to be used starting with 2008 car models. Listed below are randomly selected amounts by which the measured MPG ratings *decreased* because of the new 2008 standards. For example, the first car was measured at 16 mi/gal under the old standards and 15 mi/gal under the new 2008 standards, so the amount of the decrease is 1 mi/gal. Is the decrease of 4 mi/gal *unusual?* Why or why not?

1 2 3 2 4 3 4 2 2 2 2 3 2 2 2 3 2 2 2 2

TI **14. NCAA Football Coach Salaries** Listed below are the annual salaries for a simple random sample of NCAA football coaches (based on data from *USA Today*). How does the standard deviation change if the highest salary is omitted?

$150,000 $300,000 $350,147 $232,425 $360,000 $1,231,421 $810,000 $229,000

TI **15. Playing Times of Popular Songs** Listed below are the playing times (in seconds) of songs that were popular at the time of this writing. (The songs are by Timberlake, Furtado, Daughtry, Stefani, Fergie, Akon, Ludacris, Beyonce, Nickelback, Rihanna, Fray, Lavigne, Pink, Mims, Mumidee, and Omarion.) Does the standard deviation change much if the longest playing time is deleted?

448 242 231 246 246 293 280 227 244 213 262 239 213 258 255 257

TI **16. Satellites** Listed below are the numbers of satellites in orbit from different countries. Based on these results, is it *unusual* for a country to not have any satellites? Why or why not?

158 17 15 18 7 3 5 1 8 3 4 2 4 1 2 3 1 1 1 1 1 1 1 1

TI **17. Years to Earn Bachelor's Degree** Listed below are the lengths of time (in years) it took for a random sample of college students to earn bachelor's degrees (based on data from the U.S. National Center for Education Statistics). Based on these results, is it *unusual* for someone to earn a bachelor's degree in 12 years?

4 4 4 4 4 4 4.5 4.5 4.5 4.5 4.5 4.5 6 6 8 9 9 13 13 15

TI **18. Car Emissions** Environmental scientists measured the greenhouse gas emissions of a sample of cars. The amounts listed below are in tons (per year), expressed as CO_2 equivalents. Is the value of 9.3 tons *unusual?*

7.2 7.1 7.4 7.9 6.5 7.2 8.2 9.3

TI **19. Bankruptcies** Listed below are the numbers of bankruptcy filings in Dutchess County, New York State. The numbers are listed in order for each month of a recent year (based on data from the *Poughkeepsie Journal*). Identify any of the values that are *unusual*.

59 85 98 106 120 117 97 95 143 371 14 15

20. Radiation in Baby Teeth Listed below are amounts of strontium-90 (in millibecquerels or mBq) in a simple random sample of baby teeth obtained from Pennsylvania residents born after 1979 (based on data from "An Unexpected Rise in Strontium-90 in U.S. Deciduous Teeth in the 1990s," by Mangano, et al., *Science of the Total Environment*). Identify any of the values that are *unusual.*

155 142 149 130 151 163 151 142 156 133 138 161 128 144 172 137 151 166 147 163

145 116 136 158 114 165 169 145 150 150 150 158 151 145 152 140 170 129 188 156

Coefficient of Variation. *In Exercises 21–24, find the* **coefficient of variation** *for each of the two sets of data, then compare the variation. (The same data were used in Section 3-2.)*

21. Cost of Flying Listed below are costs (in dollars) of roundtrip flights from JFK airport in New York City to San Francisco. All flights involve one stop and a two-week stay. The airlines are US Air, Continental, Delta, United, American, Alaska, and Northwest.

30 Days in Advance: 244   260   264   264   278   318   280
1 Day in Advance: 456   614   567   943   628   1088   536

22. BMI for Miss America The trend of thinner Miss America winners has generated charges that the contest encourages unhealthy diet habits among young women. Listed below are body mass indexes (BMI) for Miss America winners from two different time periods.

BMI (from the 1920s and 1930s):   20.4 21.9 22.1 22.3 20.3 18.8 18.9 19.4 18.4 19.1

BMI (from recent winners):   19.5 20.3 19.6 20.2 17.8 17.9 19.1 18.8 17.6 16.8

23. Nicotine in Cigarettes Listed below are the nicotine amounts (in mg per cigarette) for samples of filtered and nonfiltered cigarettes (from Data Set 4 in Appendix B).

Nonfiltered: 1.1   1.7   1.7   1.1   1.1   1.4   1.1   1.4   1.0   1.2   1.1   1.1   1.1
                1.1   1.1   1.8   1.6   1.1   1.2   1.5   1.3   1.1   1.3   1.1   1.1

Filtered: 0.4   1.0   1.2   0.8   0.8   1.0   1.1   1.1   1.1   0.8   0.8   0.8   0.8
             1.0   0.2   1.1   1.0   0.8   1.0   0.9   1.1   1.1   0.6   1.3   1.1

24. Customer Waiting Times Waiting times (in minutes) of customers at the Jefferson Valley Bank (where all customers enter a single waiting line) and the Bank of Providence (where customers wait in individual lines at three different teller windows) are listed below.

Jefferson Valley (single line):   6.5   6.6   6.7   6.8   7.1   7.3   7.4   7.7   7.7   7.7
Providence (individual lines):   4.2   5.4   5.8   6.2   6.7   7.7   7.7   8.5   9.3   10.0

Large Data Sets from Appendix B. *In Exercises 25–28, refer to the indicated data set in Appendix B. Use computer software or a calculator to find the range, variance, and standard deviation.*

25. Body Temperatures Use the body temperatures for 12:00 AM on day 2 from Data Set 2 in Appendix B.

26. Machine Screws Use the listed lengths of the machine screws from Data Set 19 in Appendix B.

27. Home Voltage Refer to Data Set 13 in Appendix B. Compare the variation from the three different sets of measured voltage levels.

28. Movies Refer to Data Set 9 in Appendix B and consider the gross amounts from two different categories of movies: those with R ratings, and those with ratings of PG or PG-13. Use the coefficients of variation to determine whether the two categories appear to have the same amount of variation.

Finding Standard Deviation from a Frequency Distribution. *In Exercises 29 and 30, find the standard deviation of sample data summarized in a frequency distribution table by using the formula below, where x represents the class midpoint,* *f represents the class frequency, and n represents the total number of sample values. Also, compare the computed standard deviations to these standard deviations obtained by using Formula 3-4 with the original list of data values: (Exercise 29) 3.2 mg; (Exercise 30) 12.5 beats per minute.*

$$s = \sqrt{\frac{n[\Sigma(f \cdot x^2)] - [\Sigma(f \cdot x)]^2}{n(n-1)}} \quad \text{standard deviation for frequency distribution}$$

29.

Tar (mg) in Nonfiltered Cigarettes	Frequency
10–13	1
14–17	0
18–21	15
22–25	7
26–29	2

30.

Pulse Rates of Females	Frequency
60–69	12
70–79	14
80–89	11
90–99	1
100–109	1
110–119	0
120–129	1

31. Range Rule of Thumb As of this writing, all of the ages of winners of the Miss America Pageant are between 18 years and 24 years. Estimate the standard deviation of those ages.

32. Range Rule of Thumb Use the range rule of thumb to estimate the standard deviation of ages of all instructors at your college.

33. Empirical Rule Heights of women have a bell-shaped distribution with a mean of 161 cm and a standard deviation of 7 cm. Using the empirical rule, what is the approximate percentage of women between

a. 154 cm and 168 cm?   **b.** 147 cm and 175 cm?

34. Empirical Rule The author's Generac generator produces voltage amounts with a mean of 125.0 volts and a standard deviation of 0.3 volt, and the voltages have a bell-shaped distribution. Using the empirical rule, what is the approximate percentage of voltage amounts between

a. 124.4 volts and 125.6 volts?   **b.** 124.1 volts and 125.9 volts?

35. Chebyshev's Theorem Heights of women have a bell-shaped distribution with a mean of 161 cm and a standard deviation of 7 cm. Using Chebyshev's theorem, what do we know about the percentage of women with heights that are within 2 standard deviations of the mean? What are the minimum and maximum heights that are within 2 standard deviations of the mean?

36. Chebyshev's Theorem The author's Generac generator produces voltage amounts with a mean of 125.0 volts and a standard deviation of 0.3 volt. Using Chebyshev's theorem, what do we know about the percentage of voltage amounts that are within 3 standard deviations of the mean? What are the minimum and maximum voltage amounts that are within 3 standard deviations of the mean?

3-3   Beyond the Basics

37. Why Divide by $n - 1$? Let a *population* consist of the values 1, 3, 14. (These are the same values used in Example 1, and they are the numbers of military/intelligence satellites owned by India, Japan, and Russia.) Assume that samples of 2 values are randomly selected *with replacement* from this population. (That is, a selected value is replaced before the second selection is made.)

a. Find the variance σ^2 of the population {1, 3, 14}.

b. After listing the 9 different possible samples of 2 values selected with replacement, find the sample variance s^2 (which includes division by $n - 1$) for each of them, then find the mean of the sample variances s^2.

c. For each of the 9 different possible samples of 2 values selected with replacement, find the variance by treating each sample as if it is a population (using the formula for population variance, which includes division by n), then find the mean of those population variances.

continued

d. Which approach results in values that are better estimates of σ^2: part (b) or part (c)? Why? When computing variances of samples, should you use division by n or $n - 1$?

e. The preceding parts show that s^2 is an unbiased estimator of σ^2. Is s an unbiased estimator of σ?

38. Mean Absolute Deviation Let a population consist of the values of 1, 3, and 14. (These are the same values used in Example 1, and they are the numbers of military/intelligence satellites owned by India, Japan, and Russia.) Show that when samples of size 2 are randomly selected with replacement, the samples have mean absolute deviations that do not center about the value of the mean absolute deviation of the population.

3-4   Measures of Relative Standing and Boxplots

Key Concept In this section we introduce measures of relative standing, which are numbers showing the location of data values relative to the other values within a data set. The most important concept in this section is the z score, which will be used often in following chapters. We also discuss percentiles and quartiles, which are common statistics, as well as a new statistical graph called a boxplot.

Part 1: Basics of z Scores, Percentiles, Quartiles, and Boxplots

z Scores

A z score (or standardized value) is found by converting a value to a standardized scale, as given in the following definition. This definition shows that a z score is the number of standard deviations that a data value is from the mean. We will use z scores extensively in Chapter 6 and later chapters.

> **DEFINITION**
>
> A **z score** (or **standardized value**) is the number of standard deviations that a given value x is above or below the mean. The z score is calculated by using one of the following:
>
> | Sample | | Population |
> |---|---|---|
> | $z = \dfrac{x - \bar{x}}{s}$ | or | $z = \dfrac{x - \mu}{\sigma}$ |

Round-Off Rule for z Scores

Round z scores to two decimal places (such as 2.46).

The round-off rule for z scores is due to the fact that the standard table of z scores (Table A-2 in Appendix A) has z scores with two decimal places. Example 1 illustrates how z scores can be used to compare values, even if they come from different populations.

> **EXAMPLE 1**   **Comparing a Height and a Weight** Example 8 in Section 3-3 used the coefficient of variation to compare the variation among heights of men to the variation among weights of men. We now consider a comparison of two *individual* data values as we try to determine which is more extreme: the 76.2 in. height of a man or the 237.1 lb weight of a man. We obviously cannot compare those two values directly (apples and oranges). Compare those two data values by finding their corresponding z scores. Use these sample results obtained from Data

Set 1 in Appendix B: for men, the heights have mean $\bar{x} = 68.34$ in. and standard deviation $s = 3.02$ in.; the weights have $\bar{x} = 172.55$ lb and $s = 26.33$ lb.

SOLUTION Heights and weights are measured on different scales with different units of measurement, but we can standardize the data values by converting them to z scores:

height of 76.2 in.: $\quad z = \dfrac{x - \bar{x}}{s} = \dfrac{76.2 \text{ in.} - 68.34 \text{ in.}}{3.02 \text{ in.}} = 2.60$

weight of 237.1 lb: $\quad z = \dfrac{x - \bar{x}}{s} = \dfrac{237.1 \text{ lb} - 172.55 \text{ lb}}{26.33 \text{ lb}} = 2.45$

INTERPRETATION The results show that the height of 76.2 in. is 2.60 standard deviations above the mean height, and the weight of 237.1 lb is 2.45 standard deviations above the mean weight. Because the height is more standard deviations above the mean, it is the more extreme value. The height of 76.2 in. is more extreme than the weight of 237.1 lb.

Calculator Hint: When evaluating expressions like those given in Example 1, use *parentheses* to group the numerator. For the height of 76.2 in. in Example 1, an entry of 76.2 ⊖ 68.34 ⊕ 3.02 would yield the *wrong* result of 53.57086093, because that entry would be equivalent to $76.2 - (68.34/3.02)$. (Remember, without parentheses, multiplications and divisions take precedence over additions and subtractions). Instead, use parentheses to group the terms of the numerator as shown below.

Expression	TI-83/84 Plus key sequence	Result
$\dfrac{76.2 - 68.34}{3.02}$	(76.2 ⊖ 68.34) ⊕ 3.02	2.602649007

z Scores, Unusual Values, and Outliers

In Section 3-3 we used the range rule of thumb to conclude that a value is "unusual" if it is more than 2 standard deviations away from the mean. It follows that unusual values have z scores less than -2 or greater than $+2$. (See Figure 3-4.) Using this criterion, we see that the height of 76.2 in. and the weight of 237.1 lb given in Example 1 are unusual because they have z scores greater than 2.

> **Ordinary values:** $-2 \leq z \text{ score} \leq 2$
> **Unusual values:** $z \text{ score} < -2 \quad or \quad z \text{ score} > 2$

The preceding objective criteria can be used to identify unusual values. In Section 2-1 we described outliers as values that are very far away from the vast majority of the other data values, but that description does not provide specific objective criteria for identifying outliers. In this section we provide objective criteria for identifying outliers in the context of boxplots; however, we will continue to consider outliers to be values far away from the vast majority of the other data values. It is important to look for and identify outliers because they can have a substantial effect on statistics (such as the mean and standard deviation), as well as on some of the methods we will consider later.

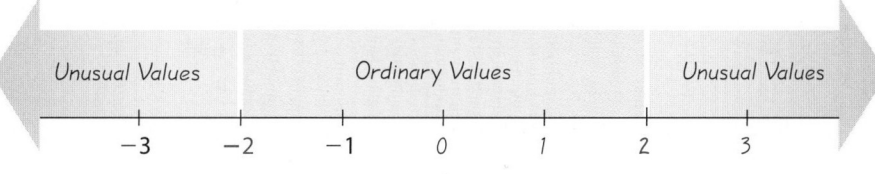

Cost of Laughing Index

There really is a Cost of Laughing Index (CLI), which tracks costs of such items as rubber chickens, Groucho Marx glasses, admissions to comedy clubs, and 13 other leading humor indicators. This is the same basic approach used in developing the Consumer Price Index (CPI), which is based on a weighted average of goods and services purchased by typical consumers. While standard scores and percentiles allow us to compare different values, they ignore any element of time. Index numbers, such as the CLI and CPI, allow us to compare the value of some variable to its value at some base time period. The value of an index number is the current value, divided by the base value, multiplied by 100.

Figure 3-4
Interpreting z Scores
Unusual values are those with z scores less than -2.00 or greater than 2.00.

While considering heights (as in Example 1), note that the height of 61.3 in. converts to $z = -0.68$, as shown below. (We again use $\bar{x} = 68.34$ in. and $s = 3.02$ in.)

$$z = \frac{x - \bar{x}}{s} = \frac{61.3 \text{ in.} - 68.34 \text{ in.}}{3.02 \text{ in.}} = -2.33$$

This height of 61.3 in. illustrates the following principle:

Whenever a data value is less than the mean, its corresponding z score is negative.

z scores are measures of position, in that they describe the location of a value (in terms of standard deviations) relative to the mean. A z score of 2 indicates that a data value is two standard deviations *above* the mean, and a z score of -3 indicates that a value is three standard deviations *below* the mean. Quartiles and percentiles are also measures of position; defined differently than z scores, they are useful for comparing values within the same data set or between different sets of data.

Percentiles

Percentiles are one type of *quantiles*—or *fractiles*—which partition data into groups with roughly the same number of values in each group.

> **DEFINITION**
>
> **Percentiles** are measures of location, denoted $P_1, P_2, \cdots, P_{99}$, which divide a set of data into 100 groups with about 1% of the values in each group

For example, the 50th percentile, denoted P_{50}, has about 50% of the data values below it and about 50% of the data values above it. So the 50th percentile is the same as the median. There is not universal agreement on a single procedure for calculating percentiles, but we will describe two relatively simple procedures for (1) finding the percentile of a data value, and (2) converting a percentile to its corresponding data value. We begin with the first procedure.

Finding the Percentile of a Data Value The process of finding the percentile that corresponds to a particular data value x is given by the following:

$$\text{percentile of value } x = \frac{\text{number of values less than } x}{\text{total number of values}} \cdot 100$$
$$\text{(round the result to the nearest whole number)}$$

> **EXAMPLE 2** **Finding a Percentile: Movie Budgets** Table 3-4 lists the 35 sorted budget amounts (in millions of dollars) from the simple random sample of movies listed in Data Set 9 in Appendix B. Find the percentile for the value of $29 million.

Table 3-4   *Sorted* Movie Budget Amounts (in millions of dollars)

4.5	5	6.5	7	20	20	29	30	35	40
40	41	50	52	60	65	68	68	70	70
70	72	74	75	80	100	113	116	120	125
132	150	160	200	225					

SOLUTION   From the sorted list of budget amounts in Table 3-4, we see that there are 6 budget amounts less than 29, so

$$\text{percentile of } 29 = \frac{6}{35} \cdot 100 = 17 \text{ (rounded to the nearest whole number)}$$

INTERPRETATION   The budget amount of $29 million is the 17th percentile. This can be interpreted loosely as: The budget amount of $29 million separates the lowest 17% of the budget amounts from the highest 83%.

Example 2 shows how to convert from a given sample value to the corresponding percentile. There are several different methods for the reverse procedure of converting a given percentile to the corresponding value in the data set. The procedure we will use is summarized in Figure 3-5 on the next page, which uses the following notation.

Notation

n total number of values in the data set

k percentile being used (Example: For the 25th percentile, $k = 25$.)

L locator that gives the *position* of a value (Example: For the 12th value in the sorted list, $L = 12$.)

P_k kth percentile (Example: P_{25} is the 25th percentile.)

EXAMPLE 3   **Converting a Percentile to a Data Value** Refer to the sorted movie budget amounts in Table 3-4 and use the procedure in Figure 3-5 to find the value of the 90th percentile, P_{90}.

SOLUTION   From Figure 3-5, we see that the sample data are already sorted, so we can proceed to find the value of the locator L. In this computation we use $k = 90$ because we are trying to find the value of the 90th percentile. We use $n = 35$ because there are 35 data values.

$$L = \frac{k}{100} \cdot n = \frac{90}{100} \cdot 35 = 31.5$$

Since $L = 31.5$ is not a whole number, we proceed to the next lower box where we change L by rounding it up from 31.5 to 32. (In this book we typically round off the usual way, but this is one of two cases where we round *up* instead of rounding *off*.) From the last box we see that the value of P_{90} is the 32nd value, counting from the lowest. In Table 3-4, the 32nd value is 150. That is, $P_{90} = \$150$ million. So, about 90% of the movies have budgets below $150 million and about 10% of the movies have budgets above $150 million.

EXAMPLE 4   **Converting a Percentile to a Data Value** Refer to the sorted movie budget amounts listed in Table 3-4. Use Figure 3-5 to find the 60th percentile, denoted by P_{60}.

continued

Figure 3-5

Converting from the _k_th Percentile to the Corresponding Data Value

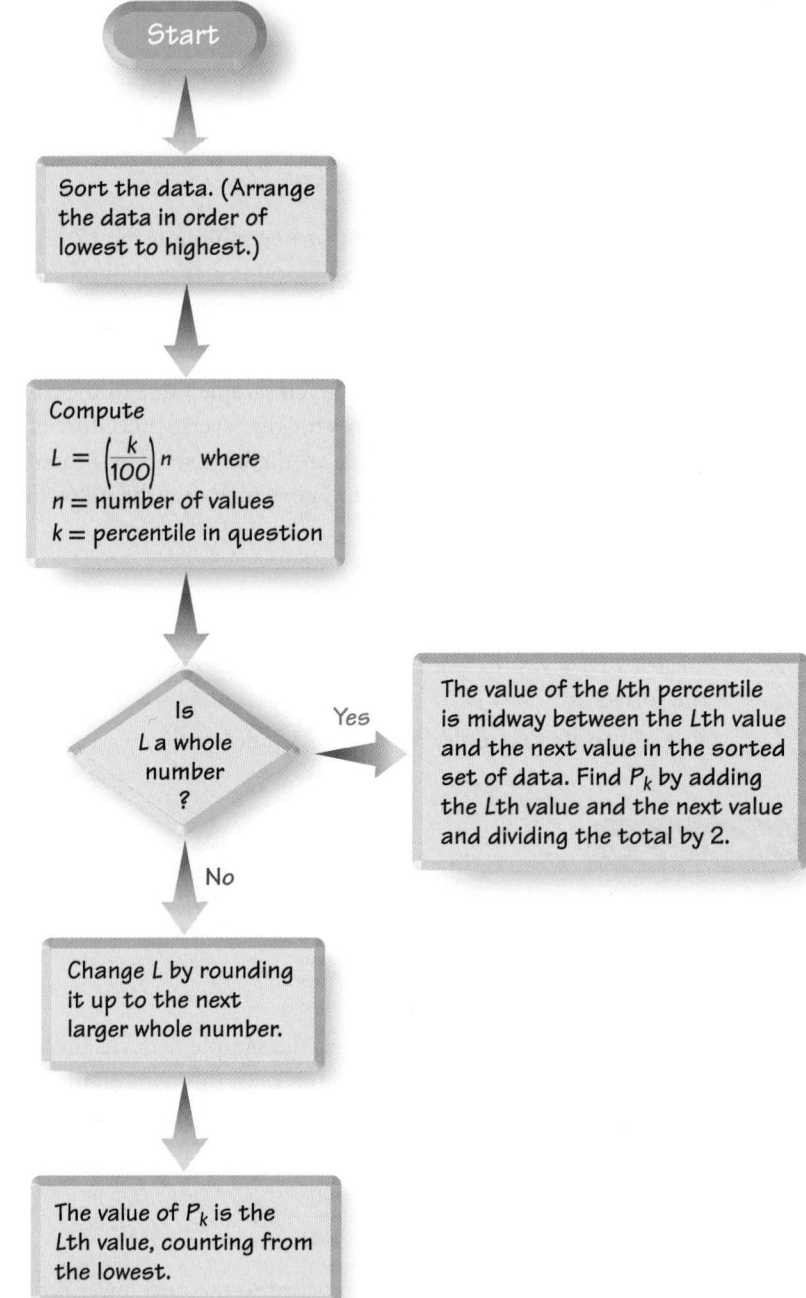

Start

Sort the data. (Arrange the data in order of lowest to highest.)

Compute
$$L = \left(\frac{k}{100}\right)n \quad \text{where}$$
n = number of values
k = percentile in question

Is _L_ a whole number ?

Yes → The value of the _k_th percentile is midway between the _L_th value and the next value in the sorted set of data. Find P_k by adding the _L_th value and the next value and dividing the total by 2.

No ↓

Change _L_ by rounding it up to the next larger whole number.

The value of P_k is the _L_th value, counting from the lowest.

SOLUTION Referring to Figure 3-5, we see that the sample data are already sorted, so we can proceed to compute the value of the locator _L_. In this computation, we use $k = 60$ because we are attempting to find the value of the 60th percentile, and we use $n = 35$ because there are 35 data values.

$$L = \frac{k}{100} \cdot n = \frac{60}{100} \cdot 35 = 21$$

Since $L = 21$ is a whole number, we proceed to the box located at the right. We now see that the value of the 60th percentile is midway between the _L_th (21st) value and the next value in the original set of data. That is, the value of the 60th percentile is midway between the 21st value and the 22nd value. The 21st value is $70 million and the 22nd value is $72 million, so the value midway between them is $71 million. We conclude that the 60th percentile is $P_{60} = \$71$ million.

EXAMPLE 5   **Setting Speed Limits** Listed below are recorded speeds (in mi/h) of randomly selected cars traveling on a section of Highway 405 in Los Angeles (based on data from Sigalert). That section has a posted speed limit of 65 mi/h. Traffic engineers often establish speed limits by using the "85th percentile rule," whereby the speed limit is set so that 85% of drivers are at or below the speed limit.

a. Find the 85th percentile of the listed speeds.

b. Given that speed limits are usually rounded to a multiple of 5, what speed limit is suggested by these data? Explain your choice.

c. Does the existing speed limit on Highway 405 conform to the 85th percentile rule?

68 68 72 73 65 74 73 72 68 65 65 73 66 71 68 74 66 71 65 73
59 75 70 56 66 75 68 75 62 72 60 73 61 75 58 74 60 73 58 75

SOLUTION

a. First we sort the data. Because there are 40 sample values and we want to find the 85th percentile, we use $n = 40$ and $k = 60$. We can now find the location L of the 85th percentile in the *sorted* list:

$$L = \frac{k}{100} \cdot n = \frac{85}{100} \cdot 40 = 34$$

Because $L = 34$ is a whole number, Figure 3-5 indicates that the 85th percentile is located between the 34th and 35th speed in the sorted list. After sorting the listed speeds, the 34th and 35th speeds are both found to be 74 mi/h, so the 85th percentile is 74 mi/h.

b. A speed of 75 mi/h is the multiple of 5 closest to the 85th percentile, but it is probably safer to round down, so that a speed of 70 mi/h is the closest multiple of 5 below the 85th percentile.

c. The existing speed limit of 65 mi/h is below the speed limit determined by the 85th percentile rule, so the existing speed limit does not conform to the 85th percentile rule. (Most California highways have a maximum speed limit of 65 mi/h.)

Quartiles

Just as there are 99 percentiles that divide the data into 100 groups, there are three quartiles that divide the data into four groups.

DEFINITION

Quartiles are measures of location, denoted Q_1, Q_2, and Q_3, which divide a set of data into four groups with about 25% of the values in each group.

Here are descriptions of quartiles that are more accurate than those given in the preceding definition:

Q_1 **(First quartile):**   Separates the bottom 25% of the sorted values from the top 75%. (To be more precise, at least 25% of the sorted values are less than or equal to Q_1, and at least 75% of the values are greater than or equal to Q_1.)

Q_2 **(Second quartile):**   Same as the median; separates the bottom 50% of the sorted values from the top 50%.

Q_3 **(Third quartile):**   Separates the bottom 75% of the sorted values from the top 25%. (To be more precise, at least 75% of the sorted values are less than or equal to Q_3, and at least 25% of the values are greater than or equal to Q_3.)

$$Q_1 = P_{25}$$
$$Q_2 = P_{50}$$
$$Q_3 = P_{75}$$

Finding values of quartiles can be accomplished with the same procedure used for finding percentiles. Simply use the relationships shown in the margin.

EXAMPLE 6   **Finding a Quartile** Refer to the sorted movie budget amounts listed in Table 3-4. Find the value of the first quartile Q_1.

SOLUTION   Finding Q_1 is really the same as finding P_{25}. We proceed to find P_{25} by using the procedure summarized in Figure 3-5. The data are already sorted, and we find the locator L as follows:

$$L = \frac{k}{100} \cdot n = \frac{25}{100} \cdot 35 = 8.75$$

Next, we note that $L = 8.75$ is not a whole number, so we change it by rounding it up to the next larger whole number, getting $L = 9$. The value of P_{25} is the 9th value in the sorted list, so $P_{25} = \$35$ million. The first quartile is given by $Q_1 = \$35$ million.

Just as there is not universal agreement on a procedure for finding percentiles, there is not universal agreement on a single procedure for calculating quartiles, and different computer programs often yield different results. If you use a calculator or computer software for exercises involving quartiles, you may get results that differ slightly from the answers obtained by using the procedures described here.

In earlier sections of this chapter we described several statistics, including the mean, median, mode, range, and standard deviation. Some other statistics are defined using quartiles and percentiles, as in the following:

$$\text{interquartile range (or IQR)} = Q_3 - Q_1$$

$$\text{semi-interquartile range} = \frac{Q_3 - Q_1}{2}$$

$$\text{midquartile} = \frac{Q_3 + Q_1}{2}$$

$$\text{10–90 percentile range} = P_{90} - P_{10}$$

Using the TI-83/84 Plus Calculator to Find Percentiles and Quartiles

Percentiles The TI-83/84 Plus calculator does not have a function for finding percentiles, but determination of percentiles is made easier using the **Sort** function, which arranges the values in order. To sort a list of values, press **STAT**, then select **SortA** (for ascending order) or **SortD** (for descending order). The command SortA(L1) rearranges the entries in list L1 to be in ascending (increasing) order. The command Sort(LAGES) rearranges the entries in the list named AGES to be in ascending (increasing) order. This section's procedure for finding is much easier with a sorted list.

Quartiles Quartiles are included among the descriptive statistics obtained by using the same procedure given at the end of Section 3-2. That is, quartiles can be found by pressing **STAT**, selecting **CALC,** then selecting **1-Var Stats** and entering the label or name of the list as described at the end of Section 3-2. *Important:* The procedure for finding percentiles and quartiles given in this section is simplified, so the results will not always agree with those obtained from a TI-83/84 Plus calculator. Also, the TI-83/84 Plus calculator will identify the second quartile Q_2 as "Med" (median). The TI-83/84 Plus screen in the margin shows that for 186 word counts of men, the quartiles are $Q_1 = 10,009$, $Q_2 = 14,290$, and $Q_3 = 20,565$.

5-Number Summary and Boxplot

The values of the three quartiles are used for the 5-number summary and the construction of boxplot graphs.

> **DEFINITION**
>
> For a set of data, the **5-number summary** consists of the minimum value, the first quartile Q_1, the median (or second quartile Q_2), the third quartile Q_3, and the maximum value.
>
> A **boxplot** (or **box-and-whisker diagram**) is a graph of a data set that consists of a line extending from the minimum value to the maximum value, and a box with lines drawn at the first quartile Q_1, the median, and the third quartile Q_3. (See Figure 3-6 on page 128.)

> **EXAMPLE 7** **Finding a 5-Number Summary** Use the movie budget amounts listed in Table 3-4 to find the 5-number summary.

> **SOLUTION** Because the budget amounts in Table 3-4 are sorted, it is easy to see that the minimum is $4.5 million and the maximum is $225 million. The value of the first quartile is $Q_1 = \$35$ million, as was found in Example 6. Using the procedure from Example 6, we can find that $Q_2 = \$68$ million and $Q_3 = \$113$ million. The 5-number summary is 4.5, 35, 68, 113, 225, all in millions of dollars.
>
> A TI-83/84 Plus calculator can be used to find a 5-number summary. See the display in the margin. Instructions are given near the end of this section.

The 5-number summary is used to construct a boxplot, as in the following procedure.

Procedure for Constructing a Boxplot

1. Find the 5-number summary consisting of the minimum value, Q_1, the median, Q_3, and the maximum value.

2. Construct a scale with values that include the minimum and maximum data values.

3. Construct a box (rectangle) extending from Q_1 to Q_3, and draw a line in the box at the median value.

4. Draw lines extending outward from the box to the minimum and maximum data values.

> **EXAMPLE 8**   **Constructing a Boxplot** Use the movie budget amounts listed in Table 3-4 to construct a boxplot.

> **SOLUTION**   The boxplot uses the 5-number summary found in Example 7: 4.5, 35, 68, 113, 225, all in millions of dollars. Figure 3-6 is the boxplot representing the movie budget amounts listed in Table 3-4.

Figure 3-6

Boxplot of Movie Budget Amounts

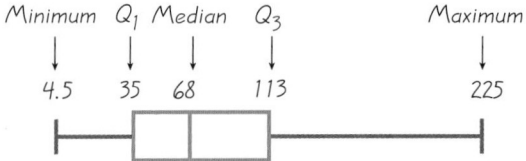

Boxplots give us information about the distribution and spread of the data. Shown below is a boxplot from a data set with a normal (bell-shaped) distribution and a boxplot from a data set with a distribution that is skewed to the right (based on data from *USA Today*).

Normal Distribution: Heights from a Simple Random Sample of Women

Skewed Distribution: Salaries (in thousands of dollars) of NCAA Football Coaches

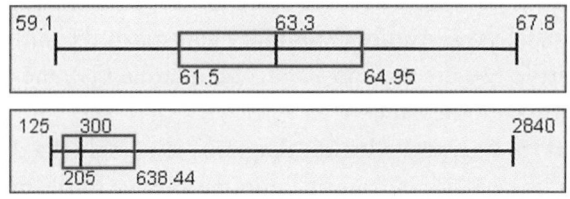

Boxplots don't show as much detailed information as histograms or stemplots, so they might not be the best choice when dealing with a single data set. However, boxplots are often great for comparing two or more data sets. When using two or more boxplots for comparing different data sets, graph the boxplots on the same scale so that comparisons can be easily made.

> **EXAMPLE 9**   **Do Women Really Talk More Than Men?** The Chapter Problem refers to a study in which daily word counts were obtained for a sample of men and a sample of women. The frequency polygons in Figure 3-1 show that the word counts of men and women are not very different. Use Figure 3-1 along with boxplots and sample statistics to address the issue of whether women really do talk more than men.

> **SOLUTION**   The STATDISK-generated boxplots shown below suggest that the numbers of words spoken by men and women are not very different. (Figure 3-1 also suggested that they are not very different.) The summary statistics in Table 3-3 (reproduced here) also suggest that the numbers of words spoken by men and women are not very different. Based on Figure 3-1, the boxplots shown here, and Table 3-3, it appears that women do *not* talk more than men. The common belief that women talk more appears to be an unsubstantiated myth.

Methods discussed later in this book allow us to analyze this issue more formally. We can conduct a *hypothesis test,* which is a formal procedure for addressing claims, such as the claim that women talk more than men. (See Example 4 in Section 9-3, in which a hypothesis test is used to establish that there is not sufficient evidence to justify a statement that men and women have different mean numbers of words spoken in a day.)

STATDISK

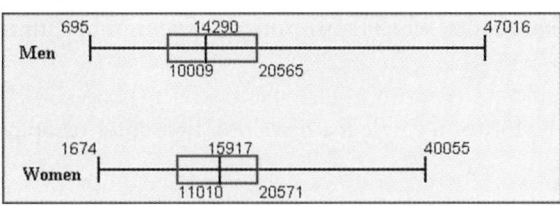

Table 3-3 Comparison of Word Counts of Men and Women

	Men	Women
Mean	15,668.5	16,215.0
Median	14,290.0	15,917.0
Midrange	23,855.5	20,864.5
Range	46,321.0	38,381.0
Standard Deviation	8,632.5	7,301.2

EXAMPLE 10 **Comparing Pulse Rates of Men and Women** Using the pulse rates of the 40 females and the 40 males listed in Data Set 1 in Appendix B, use the same scale to construct boxplots for each of the two data sets. What do the boxplots reveal about the data?

SOLUTION Shown below are STATDISK and TI-83/84 Plus boxplots displayed on the same scale. The top boxplot represents the pulse rates of the females, and the bottom boxplot represents the pulse rates of the males. We can see that the pulse rates of females are generally somewhat greater than those of males. When comparing such data sets, we can now include boxplots among the different tools that allow us to make those comparisons.

STATDISK

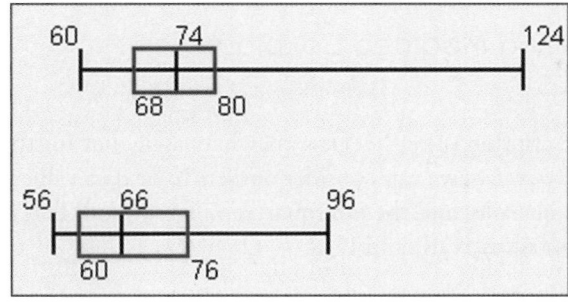

TI-83/84 Plus

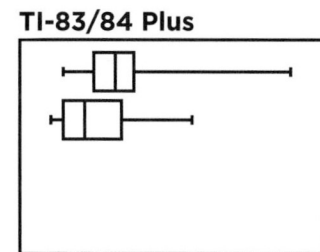

Outliers

When analyzing data, it is important to identify and consider outliers because they can strongly affect values of some important statistics (such as the mean and standard deviation), and they can also strongly affect important methods discussed later in this book. In Section 2-1 we described outliers as sample values that lie very far away from the vast majority of the other values in a set of data, but that description is vague and it does not provide specific objective criteria.

CAUTION

When analyzing data, always identify outliers and consider their effects, which can be substantial.

Using the TI-83/84 Plus Calculator for Five-Number summaries and Boxplots

Five-Number Summaries To obtain a five-number summary from a list of data, use the same procedure for finding descriptive statistics given at the end of Section 3-2. That is, press **STAT**, select **CALC,** then select **1-Var Stats** and enter the label or name of the list of data values. The results do not all fit on one screen, so scroll down by pressing ⊙ to obtain the last five results, which constitute the five-number summary.

One Boxplot To obtain one boxplot for a single list of values, first enter or open the list. Then follow this procedure:

1. Press **2ND** ⌜ Y= ⌝ to select the STAT PLOT feature.
2. Press **ENTER** and turn on Plot1. (Plot2 should be off and Plot3 should be off.)
3. For the basic boxplot described in Part 1 of this section, select the graph type that is in the middle of the second row, as shown in the accompanying screen display. (Selection of the boxplot icon in the beginning of the second row would result in a *modified boxplot,* as described in Part 2 of this section.)
4. For Xlist, enter the label or name of the list of data values, such as L1 or AGES.
5. Press ⌜ ZOOM ⌝ ⌜ 9 ⌝ to select the ZoomStat graphing option, which automatically configures the window so that all of the data are included.

TI-83/84 Plus

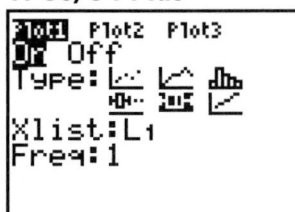

Multiple Boxplots To obtain boxplots for two or three lists of data, follow the above steps, but also turn on Plot2 and/or Plot3 and use them for the additional lists.

Trace Whether displaying one, two, or three boxplots in the window, press the ⌜ TRACE ⌝ key and use the arrow keys to see the actual values of the minimum, maximum, and quartiles.

Part 2: Outliers and Modified Boxplots

Outliers

We noted that the description of outliers is somewhat vague, but for the purposes of constructing *modified boxplots,* we can consider outliers to be data values meeting specific criteria based on quartiles and the interquartile range. (Recall that the interquartile range is often denoted by IQR, and IQR $= Q_3 - Q_1$.)

> **In modified boxplots, a data value is an outlier if it is ...**
> **above Q_3 by an amount greater than 1.5 × IQR**
> or   **below Q_1 by an amount greater than 1.5 × IQR**

Modified Boxplots

The boxplots described earlier are called **skeletal** (or **regular**) **boxplots,** but some statistical software packages provide modified boxplots, which represent outliers as special points. A **modified boxplot** is a boxplot constructed with these modifications: (1) A special symbol (such as an asterisk or point) is used to identify outliers as defined above, and (2) the solid horizontal line extends only as far as the minimum data value that is not an outlier and the maximum data value that is not an outlier. (*Note: Exercises involving modified boxplots are found in the "Beyond the Basics" exercises only.*)

EXAMPLE 11 **Modified Boxplot** Use the pulse rates of females listed in Data Set 1 in Appendix B to construct a modified boxplot.

SOLUTION From the boxplot in Example 10 we see that $Q_1 = 68$ and $Q_3 = 80$. The interquartile range is found as follows: IQR $= Q_3 - Q_1 = 80 - 68 = 12$. Using the criteria for identifying outliers, we look for pulse rates above the third quartile of 80 by an amount that is greater than $1.5 \times$ IQR $= 1.5 \times 12 = 18$, so high outliers are greater than 98. The pulse rates of 104 and 124 satisfy this condition, so those two values are outliers.

Using the criteria for identifying outliers, we also look for pulse rates below the first quartile of 68 by an amount greater than 18 (the value of $1.5 \times$ IQR). Low outliers are below 68 by more than 18, so they are less than 50. From the data set we see that there are no pulse rates of females below 50.

The only outliers of 104 and 124 are clearly identified as the two special points in the TI-83/84 Plus and Minitab display.

MINITAB

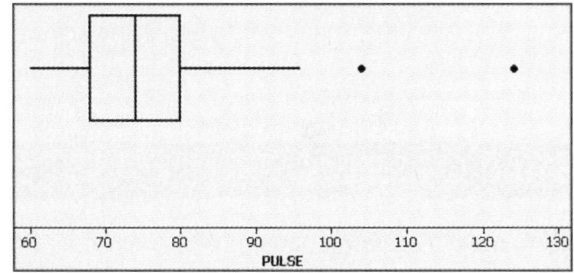

TI-83/84 Plus

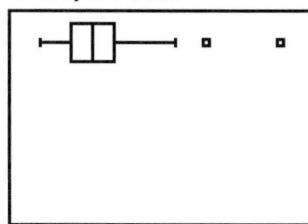

Putting It All Together

We have discussed several basic tools commonly used in statistics. When designing an experiment, analyzing data, reading an article in a professional journal, or doing anything else with data, it is important to consider certain key factors, such as:

- Context of the data
- Source of the data
- Sampling method
- Measures of center
- Measures of variation
- Distribution
- Outliers
- Changing patterns over time
- Conclusions
- Practical implications

This is an excellent checklist, but it should not replace *thinking* about any other relevant factors. It is very possible that some application of statistics requires factors not included in the above list, and it is also possible that some of the factors in the list are not relevant for certain applications.

When comparing the pulse rates of females and males from Data Set 1 in Appendix B, for example, we should understand what the pulse rates represent (pulse rates in beats per minute), the source (the National Center for Health Statistics), the sampling method (simple random sample of health exam subjects), the measures of center (such as $\bar{x} = 69.4$ for males and $\bar{x} = 76.3$ for females), the measures of variation (such as $s = 11.3$ for males and $s = 12.5$ for females), the distribution (histograms that are not substantially different from being bell-shaped), outliers (such as pulse rates of 104 and 124 for females), changing patterns over time (not an issue with the data being considered), conclusions (male pulse rates appear to be lower than female pulse rates), and practical implications (determination of an unusual pulse rate should take the sex of the subject into account).

TI-83/84 Plus

Using the TI-83/84 Plus Calculator for Modified Boxplots

To obtain *modified* boxplots, use the same procedure given at the end of Part 1 of this section, but in Step 3 be sure to select the boxplot icon in the *beginning* of the second row, as shown by the display in the margin. Also, you can select the symbol used for outliers by selecting one of the three available options shown in the line labeled "Mark."

3-4   Basic Skills and Concepts

Statistical Literacy and Critical Thinking

1. z Scores When Reese Witherspoon won an Oscar as Best Actress for the movie *Walk the Line,* her age was converted to a z score of -0.61 when included among the ages of all other Oscar-winning Best Actresses at the time of this writing. Was her age above the mean or below the mean? How many standard deviations away from the mean is her age?

2. z Scores A set of data consists of the heights of presidents of the United States, measured in centimeters. If the height of President Kennedy is converted to a z score, what unit is used for the z score? Centimeters?

3. Boxplots Shown below is a STATDISK-generated boxplot of the durations (in hours) of flights of NASA's Space Shuttle. What do the values of 0, 166, 215, 269, and 423 tell us?

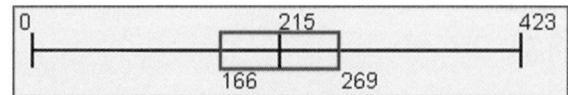

4. Boxplot Comparisons Refer to the two STATDISK-generated boxplots shown below that are drawn on the same scale. One boxplot represents weights of randomly selected men and the other represents weights of randomly selected women. Which boxplot represents women? How do you know? Which boxplot depicts weights with more variation?

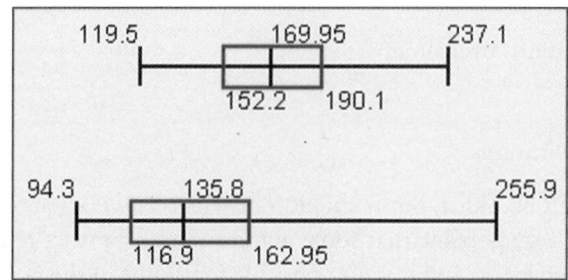

z Scores *In Exercises 5–14, express all z scores with two decimal places.*

5. *z* Score for Helen Mirren's Age As of this writing, the most recent Oscar-winning Best Actress was Helen Mirren, who was 61 at the time of the award. The Oscar-winning Best Actresses have a mean age of 35.8 years and a standard deviation of 11.3 years.

a. What is the difference between Helen Mirren's age and the mean age?

b. How many standard deviations is that (the difference found in part (a))?

c. Convert Helen Mirren's age to a *z* score.

d. If we consider "usual" ages to be those that convert to *z* scores between −2 and 2, is Helen Mirren's age usual or unusual?

6. *z* Score for Philip Seymour Hoffman's Age Philip Seymour Hoffman was 38 years of age when he won a Best Actor Oscar for his role in *Capote*. The Oscar-winning Best Actors have a mean age of 43.8 years and a standard deviation of 8.9 years.

a. What is the difference between Hoffman's age and the mean age?

b. How many standard deviations is that (the difference found in part (a))?

c. Convert Hoffman's age to a *z* score.

d. If we consider "usual" ages to be those that convert to *z* scores between −2 and 2, is Hoffman's age usual or unusual?

7. *z* Score for Old Faithful Eruptions of the Old Faithful geyser have duration times with a mean of 245.0 sec and a standard deviation of 36.4 sec (based on Data Set 15 in Appendix B). One eruption had a duration time of 110 sec.

a. What is the difference between a duration time of 110 sec and the mean?

b. How many standard deviations is that (the difference found in part (a))?

c. Convert the duration time of 110 sec to a *z* score.

d. If we consider "usual" duration times to be those that convert to *z* scores between −2 and 2, is a duration time of 110 sec usual or unusual?

8. *z* Score for World's Tallest Man Bao Xishun is the world's tallest man with a height of 92.95 in. (or 7 ft, 8.95 in.). Men have heights with a mean of 69.6 in. and a standard deviation of 2.8 in.

a. What is the difference between Bao's height and the mean height of men?

b. How many standard deviations is that (the difference found in part (a))?

c. Convert Bao's height to a *z* score.

d. Does Bao's height meet the criterion of being unusual by corresponding to a *z* score that does not fall between −2 and 2?

9. *z* Scores for Body Temperatures Human body temperatures have a mean of 98.20°F and a standard deviation of 0.62°F (based on Data Set 2 in Appendix B). Convert each given temperature to a *z* score and determine whether it is usual or unusual.

a. 101.00°F   **b.** 96.90°F   **c.** 96.98°F

10. *z* Scores for Heights of Women Soldiers The U.S. Army requires women's heights to be between 58 in. and 80 in. Women have heights with a mean of 63.6 in. and a standard deviation of 2.5 in. Find the *z* score corresponding to the minimum height requirement and find the *z* score corresponding to the maximum height requirement. Determine whether the minimum and maximum heights are unusual.

11. *z* Score for Length of Pregnancy A woman wrote to *Dear Abby* and claimed that she gave birth 308 days after a visit from her husband, who was in the Navy. Lengths of pregnancies have a mean of 268 days and a standard deviation of 15 days. Find the *z* score for 308 days. Is such a length unusual? What do you conclude?

12. *z* Score for Blood Count White blood cell counts (in cells per microliter) have a mean of 7.14 and a standard deviation of 2.51 (based on data from the National Center for

Health Statistics). Find the z score corresponding to a person who had a measured white blood cell count of 16.60. Is this level unusually high?

13. Comparing Test Scores Scores on the SAT test have a mean of 1518 and a standard deviation of 325. Scores on the ACT test have a mean of 21.1 and a standard deviation of 4.8. Which is relatively better: a score of 1840 on the SAT test or a score of 26.0 on the ACT test? Why?

14. Comparing Test Scores Scores on the SAT test have a mean of 1518 and a standard deviation of 325. Scores on the ACT test have a mean of 21.1 and a standard deviation of 4.8. Which is relatively better: a score of 1190 on the SAT test or a score of 16.0 on the ACT test? Why?

Percentiles. *In Exercises 15–18, use the given sorted values, which are the numbers of points scored in the Super Bowl for a recent period of 24 years. Find the percentile corresponding to the given number of points.*

36 37 37 39 39 41 43 44 44 47 50 53 54 55 56 56 57 59 61 61 65 69 69 75

15. 47   **16.** 65   **17.** 54   **18.** 41

In Exercises 19–26, use the same list of 24 sorted values given for Exercises 15-18. Find the indicated percentile or quartile.

19. P_{20} **20.** Q_1 **21.** Q_3 **22.** P_{80}

23. P_{50} **24.** P_{75} **25.** P_{25} **26.** P_{95}

27. Boxplot for Super Bowl Points Using the same 24 sorted values given for Exercises 15-18, construct a boxplot and include the values of the 5-number summary.

28. Boxplot for Number of English Words A simple random sample of pages from *Merriam-Webster's Collegiate Dictionary, 11th edition,* was obtained. Listed below are the numbers of defined words on those pages, and they are arranged in order. Construct a boxplot and include the values of the 5-number summary.

34   36   39   43   51   53   62   63   73   79

29. Boxplot for FICO Scores A simple random sample of FICO credit rating scores was obtained, and the sorted scores are listed below. Construct a boxplot and include the values of the 5-number summary.

664   693   698   714   751   753   779   789   802   818   834   836

30. Boxplot for Radiation in Baby Teeth Listed below are sorted amounts of strontium-90 (in millibecquerels or mBq) in a simple random sample of baby teeth obtained from Pennsylvania residents born after 1979 (based on data from "An Unexpected Rise in Strontium-90 in U.S. Deciduous Teeth in the 1990s," by Mangano, et al., *Science of the Total Environment*). Construct a boxplot and include the values of the 5-number summary.

128 130 133 137 138 142 142 144 147 149 151 151 151 155 156 161 163 163 166 172

Boxplots from Larger Data Sets in Appendix B. *In Exercises 31–34, use the given data sets from Appendix B.*

31. Weights of Regular Coke and Diet Coke Use the same scale to construct boxplots for the weights of regular Coke and diet Coke from Data Set 17 in Appendix B. Use the boxplots to compare the two data sets.

32. Boxplots for Weights of Regular Coke and Regular Pepsi Use the same scale to construct boxplots for the weights of regular Coke and regular Pepsi from Data Set 17 in Appendix B. Use the boxplots to compare the two data sets.

33. Boxplots for Weights of Quarters Use the same scale to construct boxplots for the weights of the pre-1964 silver quarters and the post-1964 quarters from Data Set 20 in Appendix B. Use the boxplots to compare the two data sets.

TI 34. Boxplots for Voltage Amounts Use the same scale to construct boxplots for the home voltage amounts and the generator voltage amounts from Data Set 13 in Appendix B. Use the boxplots to compare the two data sets.

3-4   Beyond the Basics

TI 35. Outliers and Modified Boxplot Use the 40 upper leg lengths (cm) listed for females from Data Set 1 in Appendix B. Construct a modified boxplot. Identify any outliers as defined in Part 2 of this section.

TI 36. Outliers and Modified Boxplot Use the gross amounts from movies from Data Set 9 in Appendix B. Construct a modified boxplot. Identify any outliers as defined in Part 2 of this section.

37. Interpolation When finding percentiles using Figure 3-5, if the locator L is not a whole number, we round it up to the next larger whole number. An alternative to this procedure is to *interpolate.* For example, using interpolation with a locator of $L = 23.75$ leads to a value that is 0.75 (or 3/4) of the way between the 23rd and 24th values. Use this method of interpolation to find P_{25} (or Q_1) for the movie budget amounts in Table 3-4 on page 122. How does the result compare to the value that would be found by using Figure 3-5 without interpolation?

38. Deciles and Quintiles For a given data set, there are nine deciles, denoted by $D_1, D_2, \cdots, D_9$, which separate the sorted data into 10 groups, with about 10% of the values in each group. There are also four *quintiles,* which divide the sorted data into 5 groups, with about 20% of the values in each group. (Note the difference between quintiles and quantiles, which were described earlier in this section.)

a. Using the movie budget amounts in Table 3-4 on page 122, find the deciles D_1, D_7, and D_8.

b. Using the movie budget amounts in Table 3-4, find the four quintiles.

Review

In this chapter we discussed various characteristics of data that are generally very important. After completing this chapter, we should be able to do the following:

• Calculate measures of center by finding the mean and median (Section 3-2).

• Calculate measures of variation by finding the standard deviation, variance, and range (Section 3-3).

• *Understand* and *interpret* the standard deviation by using tools such as the range rule of thumb (Section 3-3).

• Compare data values by using z scores, quartiles, or percentiles (Section 3-4).

• Investigate the spread of data by constructing a boxplot (Section 3-4).

Statistical Literacy and Critical Thinking

1. Quality Control Cans of regular Coke are supposed to contain 12 oz of cola. If a quality control engineer finds that the production process results in cans of Coke having a mean of 12 oz, can she conclude that the production process is proceeding as it should? Why or why not?

2. ZIP Codes An article in the *New York Times* noted that the ZIP code of 10021 on the Upper East Side of Manhattan is being split into the three ZIP codes of 10065, 10021, and 10075 (in geographic order from south to north). The ZIP codes of 11 famous residents (including Bill Cosby, Spike Lee, and Tom Wolfe) in the 10021 ZIP code will have these ZIP codes after the change: 10065, 10065, 10065, 10065, 10065, 10021, 10021, 10075, 10075, 10075, 10075. What is wrong with finding the mean and standard deviation of these 11 new ZIP codes?

3. Outlier Nola Ochs recently became the oldest college graduate when she graduated at the age of 95. If her age is included with the ages of 25 typical college students at the times of their graduations, how much of an effect will her age have on the mean, median, standard deviation, and range?

4. Sunspot Numbers The annual sunspot numbers are found for a recent sequence of 24 years. The data are sorted, then it is found that the mean is 81.09, the standard deviation is 50.69, the minimum is 8.6, the first quartile is 29.55, the median is 79.95, the third quartile is 123.3, and the maximum is 157.6. What potentially important characteristic of these annual sunspot numbers is lost when the data are replaced by the sorted values?

Chapter Quick Quiz

1. What is the mean of the sample values 2 cm, 2 cm, 3 cm, 5 cm, and 8 cm?

2. What is the median of the sample values listed in Exercise 1?

3. What is the mode of the sample values listed in Exercise 1?

4. If the standard deviation of a data set is 5.0 ft, what is the variance?

5. If a data set has a mean of 10.0 seconds and a standard deviation of 2.0 seconds, what is the z score corresponding to the time of 4.0 seconds?

6. Fill in the blank: The range, standard deviation, and variance are all measures of _____.

7. What is the symbol used to denote the standard deviation of a sample, and what is the symbol used to denote the standard deviation of a population?

8. What is the symbol used to denote the mean of a sample, and what is the symbol used to denote the mean of a population?

9. Fill in the blank: Approximately _____ percent of the values in a sample are greater than or equal to the 25th percentile.

10. True or false: For any data set, the median is always equal to the 50th percentile.

Review Exercises

(TI) 1. Weights of Steaks A student of the author weighed a simple random sample of Porterhouse steaks, and the results (in ounces) are listed below. The steaks are supposed to be 21 oz because they are listed on the menu as weighing 20 ounces, and they lose an ounce when cooked. Use the listed weights to find the (a) mean; (b) median; (c) mode; (d) midrange; (e) range; (f) standard deviation; (g) variance; (h) Q_1; (i) Q_3.

$$17 \quad 19 \quad 21 \quad 18 \quad 20 \quad 18 \quad 19 \quad 20 \quad 20 \quad 21$$

(TI) 2. Boxplot Using the same weights listed in Exercise 1, construct a boxplot and include the values of the 5-number summary.

(TI) 3. Ergonomics When designing a new thrill ride for an amusement park, the designer must consider the sitting heights of males. Listed below are the sitting heights (in millimeters) obtained from a simple random sample of adult males (based on anthropometric survey data from Gordon, Churchill, et al.). Use the given sitting heights to find the (a) mean; (b) median; (c) mode; (d) midrange; (e) range; (f) standard deviation; (g) variance; (h) Q_1; (i) Q_3.

$$936 \quad 928 \quad 924 \quad 880 \quad 934 \quad 923 \quad 878 \quad 930 \quad 936$$

TI 4. z Score Using the sample data from Exercise 3, find the z score corresponding to the sitting height of 878 mm. Based on the result, is the sitting height of 878 mm unusual? Why or why not?

TI 5. Boxplot Using the same sitting heights listed in Exercise 3, construct a boxplot and include the values of the 5-number summary. Does the boxplot suggest that the data are from a population with a normal (bell-shaped) distribution? Why or why not?

TI 6. Comparing Test Scores SAT scores have a mean of 1518 and a standard deviation of 325. Scores on the ACT test have a mean of 21.1 and a standard deviation of 4.8. Which is relatively better: a score of 1030 on the SAT test or a score of 14.0 on the ACT test? Why?

7. Estimating Mean and Standard Deviation

a. Estimate the mean age of cars driven by students at your college.

b. Use the range rule of thumb to make a rough estimate of the standard deviation of the ages of cars driven by students at your college.

8. Estimating Mean and Standard Deviation

a. Estimate the mean length of time that traffic lights are red.

b. Use the range rule of thumb to make a rough estimate of the standard deviation of the lengths of times that traffic lights are red.

9. Interpreting Standard Deviation Engineers consider the overhead grip reach (in millimeters) of sitting adult women when designing a cockpit for an airliner. Those grip reaches have a mean of 1212 mm and a standard deviation of 51 mm (based on anthropometric survey data from Gordon, Churchill, et al.). Use the range rule of thumb to identify the minimum "usual" grip reach and the maximum "usual" grip reach. Which of those two values is more relevant in this situation? Why?

10. Interpreting Standard Deviation A physician routinely makes physical examinations of children. She is concerned that a three-year-old girl has a height of only 87.8 cm. Heights of three-year-old girls have a mean of 97.5 cm and a standard deviation of 6.9 cm (based on data from the National Health and Nutrition Examination Survey). Use the range rule of thumb to find the maximum and minimum usual heights of three-year-old girls. Based on the result, is the height of 87.8 cm unusual? Should the physician be concerned?

Cumulative Review Exercises

1. Types of Data Refer to the sitting heights listed in Review Exercise 3.

a. Are the sitting heights from a population that is discrete or continuous?

b. What is the level of measurement of the sitting heights? (nominal, ordinal, interval, ratio)

TI 2. Frequency Distribution Use the sitting heights listed in Review Exercise 3 to construct a frequency distribution. Use a class width of 10 mm, and use 870 mm as the lower class limit of the first class.

TI 3. Histogram Use the frequency distribution from Exercise 2 to construct a histogram. Based on the result, does the distribution appear to be uniform, normal (bell-shaped), or skewed?

4. Dotplot Use the sitting heights listed in Review Exercise 3 to construct a dotplot.

5. Stemplot Use the sitting heights listed in Review Exercise 3 to construct a stemplot.

6. a. A set of data is at the nominal level of measurement and you want to obtain a representative data value. Which of the following is most appropriate: mean, median, mode, or midrange? Why?

b. A botanist wants to obtain data about the plants being grown in homes. A sample is obtained by telephoning the first 250 people listed in the local telephone directory. What type of sampling is being used? (random, stratified, systematic, cluster, convenience)

c. An exit poll is conducted by surveying everyone who leaves the polling booth at 50 randomly selected election precincts. What type of sampling is being used? (random, stratified, systematic, cluster, convenience)

d. A manufacturer makes fertilizer sticks to be used for growing plants. A manager finds that the amounts of fertilizer placed in the sticks are not very consistent, so that for some fertilization lasts longer than claimed, while others don't last long enough. She wants to improve quality by making the amounts of fertilizer more consistent. When analyzing the amounts of fertilizer, which of the following statistics is most relevant: mean, median, mode, midrange, standard deviation, first quartile, third quartile? Should the value of that statistic be raised, lowered, or left unchanged?

7. Sampling Shortly after the World Trade Center towers were destroyed, America Online ran a poll of its Internet subscribers and asked this question: "Should the World Trade Center towers be rebuilt?" Among the 1,304,240 responses, 768,731 answered "yes," 286,756 answered "no," and 248,753 said that it was "too soon to decide." Given that this sample is extremely large, can the responses be considered to be representative of the population of the United States? Explain.

8. Sampling What is a simple random sample? What is a voluntary response sample? Which of those two samples is generally better?

9. Observational Study and Experiment What is the difference between an observational study and an experiment?

10. Histogram What is the major flaw in the histogram (below) of the outcomes of 100 rolls of a fair die?

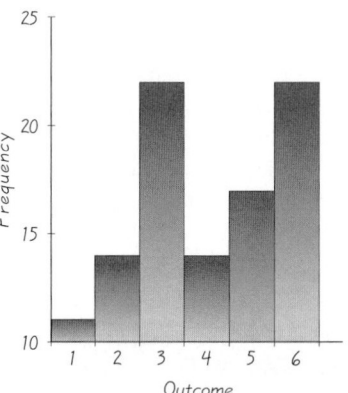

TI-83/84 Plus Calculator Project

When dealing with large data sets, manual entry of data can become quite tedious and time consuming. There are better things to do with your time, such as rotating the tires on your car. Refer to Data Set 13 in Appendix B, which includes measured voltage levels from the author's home, a generator, and an uninterruptible power supply. Instead of manually entering the data, use a TI-83/84 Plus calculator and open the file VOLTAGE in the TRIOLA11 APP. Proceed to generate histograms and find appropriate statistics that allow you to compare the three sets of data. Are there any outliers? Do all three power sources appear to provide electricity with properties that are basically the same? Are there any significant differences? What is a consequence of having voltage that varies too much? Write a brief report including your conclusions and supporting graphs.

Using Statistics to Summarize Data

Go to **http://www.aw.com/triola**

The importance of statistics as a tool to summarize data cannot be underestimated. For example, consider data sets such as the ages of all the students at your school or the annual incomes of every person in the United States. On paper, these data sets would be lengthy lists of numbers, too lengthy to be absorbed and interpreted on their own. In the previous chapter, you learned a variety of graphical tools used to represent such data sets. This chapter focused on the use of numbers or statistics to summarize various aspects of data.

Just as important as being able to summarize data with statistics is the ability to *interpret* such statistics when presented. Given a number such as the arithmetic mean, you need not only to understand what it is telling you about the underlying data, but also what additional statistics you need to put the value of the mean in context.

The Internet Project for this chapter will help you develop these skills using data from such diverse fields as meteorology, entertainment, and health. You will also discover uses for such statistics as the geometric mean that you might not have expected.

APPLET PROJECT

The CD included with this book contains applets designed to help visualize various concepts. Open the Applets folder on the CD and click on **Start.** Select the menu item of **Mean versus median.** Create a set of points that are very close together and then add a point that is far away from the others. What is the effect of the new point on the mean? What is the effect of the new point on the median? Also, create a data set with a median below 2 and a mean between 2 and 4.

Do the Academy Awards involve discrimination based on age?

The *From Data to Decision* project at the end of Chapter 2 listed the ages of actresses and actors at the times that they won Oscars in the Best Actress and Best Actor categories. Refer to those same ages.

Critical Thinking

Use methods from this chapter to compare the two data sets. Are there differences between the ages of the Best Actresses and the ages of the Best Actors? Identify any other notable differences.

Cooperative Group Activities

1. Out-of-class activity Are estimates influenced by anchoring numbers? In the article "Weighing Anchors" in *Omni* magazine, author John Rubin observed that when people estimate a value, their estimate is often "anchored" to (or influenced by) a preceding number, even if that preceding number is totally unrelated to the quantity being estimated. To demonstrate this, he asked people to give a quick estimate of the value of $8 \times 7 \times 6 \times 5 \times 4 \times 3 \times 2 \times 1$. The average answer given was 2250, but when the order of the numbers was reversed, the average became 512. Rubin explained that when we begin calculations with larger numbers (as in $8 \times 7 \times 6$), our estimates tend to be larger. He noted that both 2250 and 512 are far below the correct product, 40,320. The article suggests that irrelevant numbers can play a role in influencing real estate appraisals, estimates of car values, and estimates of the likelihood of nuclear war.

Conduct an experiment to test this theory. Select some subjects and ask them to quickly estimate the value of

$$8 \times 7 \times 6 \times 5 \times 4 \times 3 \times 2 \times 1$$

Then select other subjects and ask them to quickly estimate the value of

$$1 \times 2 \times 3 \times 4 \times 5 \times 6 \times 7 \times 8$$

Record the estimates along with the particular order used. Carefully design the experiment so that conditions are uniform and the two sample groups are selected in a way that minimizes any bias. Don't describe the theory to subjects until after they have provided their estimates. Compare the two sets of sample results by using the methods of this chapter. Provide a printed report that includes the data collected, the detailed methods used, the method of analysis, any relevant graphs and/or statistics, and a statement of conclusions. Include a critique of the experiment, with reasons why the results might not be correct, and describe ways in which the experiment could be improved.

2. Out-of-class activity In each group of three or four students, collect an original data set of values at the interval or ratio level of measurement. Provide the following: (1) a list of sample values, (2) printed computer results of descriptive statistics and graphs, and (3) a written description of the nature of the data, the method of collection, and important characteristics.

3. Out-of-class activity Appendix B includes many real and interesting data sets. In each group of three or four students, select a data set from Appendix B and analyze it using the methods discussed so far in this book. Write a brief report summarizing key conclusions.

4. Out-of-class activity Record the service times of randomly selected customers at a drive-up window of a bank or fast-food restaurant, and describe important characteristics of those times.

5. Out-of-class activity Record the times that cars are parked at a gas pump, and describe important characteristics of those times.

NAME:	Robert S. Holzman, MD
JOB:	Professor of Medicine and Environmental Medicine
COMPANY:	NYU School of Medicine; Hospital Epidemiologist, Bellevue Hospital Center, New York City

D r. Holzman is Professor of Medicine and Environmental Medicine at NYU School of Medicine. He is also a hospital epidemiologist at Bellevue Hospital Center in New York City. As an internist specializing in infectious diseases, he is responsible for the Infection Control Program at Bellevue Hospital. He also teaches medical students and postdoctoral trainees in Clinical Infectious Diseases and Epidemiology.

Q: How do you use statistics in your job and what specific statistical concepts do you use?

A: Much of what I do on a day-to-day basis is applied statistical analysis, including determination of sample size and analysis of clinical trials and laboratory experiments, and the development of regression models for retrospective studies, primarily using logistic regression. I also track hospital infection rates using control charts.

Q: Please describe a specific example of how the use of statistics was helpful in improving a practice or service.

A: A surveillance nurse detected an increase in the isolation of a certain type of bacteria among patients in an intensive care unit 2 months ago. We took action to remedy that increase, and control charts were used to show us that the bacteria levels were returning to their baseline levels.

Q: What background in statistics is required to obtain a job like yours? What other educational requirements are there?

A: To be an academic physician requires college and medical school, followed by at least 5 years of postgraduate training, often more. In many cases students today are combining their medical education with a research-oriented PhD training program. I acquired my own statistical and epidemiologic knowledge through a combination of on-the-job association with statistical professionals, reading statistical texts, and some graduate course work. Today such knowledge is introduced as part of standard training programs, but additional work is still important for mastery.

Q: At your place of work, do you feel job applicants are viewed more favorably if they have studied some statistics?

A: Ability to apply statistical and epidemiologic knowledge to evaluate the medical literature is definitely considered a plus for physicians, even those who work as clinical caregivers. To work as an epidemiologist requires additional statistical study.

Q: Do you recommend that today's college students study statistics? Why?

A: A knowledge of basic probability, data summarization, and the principles of inferential statistics are essential to our understanding of the scientific method and our evaluation of reports of scientific studies. Such reports are found daily in newspapers, and knowledge of what is left out of the report helps temper uncritical acceptance of new "facts."

4-1 Review and Preview

4-2 Basic Concepts of
Probability

4-3 Addition Rule

4-4 Multiplication Rule:
Basics

4-5 Multiplication Rule:
Complements and
Conditional Probability

4-6 Probabilities Through
Simulations

4-7 Counting

4-8 Bayes' Theorem
(on CD-ROM)

4 Probability

Are polygraph instruments effective as "lie detectors"?

A polygraph instrument measures several physical reactions, such as blood pressure, pulse rate, and skin conductivity. Subjects are usually given several questions that must be answered and, based on physical measurements, the polygraph examiner determines whether or not the subject is lying. Errors in test results could lead to an individual being falsely accused of committing a crime or to a candidate being denied a job.

Based on research, the success rates from polygraph tests depend on several factors, including the questions asked, the test subject, the competence of the polygraph examiner, and the polygraph instrument used for the test.

Many experiments have been conducted to evaluate the effectiveness of polygraph devices, but we will consider the data in Table 4-1, which includes results from experiments conducted by researchers Charles R. Honts (Boise State University) and Gordon H. Barland (Department of Defense Polygraph Institute). Table 4-1 summarizes polygraph test results for 98 different subjects. In each case, it was known whether or not the subject lied. So, the table indicates when the polygraph test was correct.

Analyzing the Results

When testing for a condition, such as lying, pregnancy, or disease, the result of the test is either positive or negative. However, sometimes errors occur during the testing process which can yield a *false positive* result or a *false negative* result. For example, a false positive result in a polygraph test would indicate that a subject lied when in fact he or she did not lie. A false negative would indicate that a subject did not lie when in fact he or she lied.

Incorrect Results

- **False positive:** Test *incorrectly* indicates the presence of a condition (such as lying, being pregnant, or having some disease) when the subject does not actually have that condition.
- **False negative:** Test *incorrectly* indicates that subject does not have the condition when the subject actually does have that condition.

Correct Results

- **True positive:** Test *correctly* indicates that the condition is present when it really is present.
- **True negative:** Test *correctly* indicates that the condition is not present when it really is not present.

Measures of Test Reliability

- **Test sensitivity:** The probability of a true positive.
- **Test specificity:** The probability of a true negative.

In this chapter we study the basic principles of *probability* theory. These principles will allow us to address questions related to the reliability (or unreliability) of polygraph tests, such as these: Given the sample results in Table 4-1, what is the probability of a false positive or a false negative? Are those probabilities low enough to support the use of polygraph tests in making judgments about the test subject?

Table 4-1 Results from Experiments with Polygraph Instruments

	Did the Subject Actually Lie?	
	No (Did Not Lie)	Yes (Lied)
Positive test result (Polygraph test indicated that the subject *lied.*)	15 (**false positive**)	42 (**true positive**)
Negative test result (Polygraph test indicated that the subject did *not lie.*)	32 (**true negative**)	9 (**false negative**)

Review and Preview

The previous chapters have been developing some fundamental tools used in the statistical methods to be introduced in later chapters. We have discussed the necessity of sound sampling methods and common measures of characteristics of data, including the mean and standard deviation. The main objective of this chapter is to develop a sound understanding of probability values, because those values constitute the underlying foundation on which the methods of inferential statistics are built. As a simple example, suppose that you have developed a gender-selection procedure and you claim that it greatly increases the likelihood of a baby being a girl. Suppose that independent test results from 100 couples show that your procedure results in 98 girls and only 2 boys. Even though there is a chance of getting 98 girls in 100 births with no special treatment, that chance is so incredibly low that it would be rejected as a reasonable explanation. Instead, it would be generally recognized that the results provide strong support for the claim that the gender-selection technique is effective. This is exactly how statisticians think: They reject explanations based on very low probabilities. Statisticians use the *rare event rule for inferential statistics*.

> **Rare Event Rule for Inferential Statistics**
>
> If, under a given assumption, the probability of a particular observed event is extremely small, we conclude that the assumption is probably not correct.

Although the main objective in this chapter is to develop a sound understanding of probability values that will be used in later chapters of this book, a secondary objective is to develop the basic skills necessary to determine probability values in a variety of other important circumstances.

Basic Concepts of Probability

Key Concept In this section we present three different approaches to finding the *probability* of an event. The most important objective of this section is to learn how to *interpret* probability values, which are expressed as values between 0 and 1. We should know that a small probability, such as 0.001, corresponds to an event that is *unusual*, in the sense that it rarely occurs. We also discuss expressions of *odds* and how probability is used to determine the odds of an event occurring. Although the concepts related to odds are not needed for topics that follow, odds are considered in some everyday situations. For instance, odds are used to determine the likelihood of winning the lottery.

Part 1: Basics of Probability

In considering probability, we deal with procedures (such as taking a polygraph test, rolling a die, answering a multiple-choice test question, or undergoing a test for drug use) that produce outcomes.

⊜ DEFINITION

An **event** is any collection of results or outcomes of a procedure.

A **simple event** is an outcome or an event that cannot be further broken down into simpler components.

The **sample space** for a procedure consists of all possible *simple* events. That is, the sample space consists of all outcomes that cannot be broken down any further.

Example 1 illustrates the concepts defined above.

EXAMPLE 1 In the following display, we use "f" to denote a female baby and "m" to denote a male baby.

Procedure	Example of Event	Complete Sample Space
Single birth	1 female (simple event)	{f, m}
3 births	2 females and 1 male (ffm, fmf, mff are all simple events resulting in 2 females and a male)	{fff, ffm, fmf, fmm, mff, mfm, mmf, mmm}

With one birth, the result of 1 female is a *simple event* because it cannot be broken down any further. With three births, the event of "2 females and 1 male" is *not a simple event* because it can be broken down into simpler events, such as ffm, fmf, or mff. With three births, the *sample space* consists of the 8 simple events listed above. With three births, the outcome of ffm is considered a simple event, because it is an outcome that cannot be broken down any further. We might incorrectly think that ffm can be further broken down into the individual results of f, f, and m, but f, f, and m are not individual outcomes from three births. With three births, there are exactly 8 outcomes that are simple events: fff, ffm, fmf, fmm, mff, mfm, mmf, and mmm.

We first list some basic notation, then we present three different approaches to finding the probability of an event.

Notation for Probabilities

P denotes a probability.

A, B, and *C* denote specific events.

$P(A)$ denotes the probability of event *A* occurring.

1. **Relative Frequency Approximation of Probability** Conduct (or observe) a procedure, and count the number of times that event *A* actually occurs. Based on these actual results, $P(A)$ is *approximated* as follows:

$$P(A) = \frac{\text{number of times } A \text{ occurred}}{\text{number of times the procedure was repeated}}$$

2. **Classical Approach to Probability (Requires Equally Likely Outcomes)**
Assume that a given procedure has *n* different simple events and that *each of*

Probabilities That Challenge Intuition

In certain cases, our subjective estimates of probability values are dramatically different from the actual probabilities. Here is a classical example: If you take a deep breath, there is better than a 99% chance that you will inhale a molecule that was exhaled in dying Caesar's last breath. In that same morbid and unintuitive spirit, if Socrates' fatal cup of hemlock was mostly water, then the next glass of water you drink will likely contain one of those same molecules. Here's another less morbid example that can be verified: In classes of 25 students, there is better than a 50% chance that at least 2 students will share the same birthday (day and month).

Gambling to Win

In the typical state lottery, the "house" has a 65% to 70% advantage, since only 30% to 35% of the money bet is returned as prizes. The house advantage at racetracks is usually around 15%. In casinos, the house advantage is 5.26% for roulette, 1.4% for craps, and 3% to 22% for slot machines.

The house advantage is 5.9% for blackjack, but some professional gamblers can systematically win with a 1% player advantage by using complicated card-counting techniques that require many hours of practice. With one system, the player scans the cards that are shown and subtracts one point for a picture card or a 10 or an ace, and adds one point for the cards 2, 3, 4, 5, 6. Cards of 7 or 8 are ignored. When the count is high and the dealer is deep into the card stack, the deck has disproportionately more high cards, and the odds favor the player more. If a card-counting player were to suddenly change from small bets to large bets, the dealer would recognize the card counting and the player would be ejected. Card counters try to beat this policy by working with a team. When the count is high enough, the player signals an accomplice who enters the game with large bets. A group of MIT students supposedly won millions of dollars by counting cards in blackjack.

those simple events has an equal chance of occurring. If event A can occur in s of these n ways, then

$$P(A) = \frac{\text{number of ways } A \text{ can occur}}{\text{number of different simple events}} = \frac{s}{n}$$

CAUTION

When using the classical approach, always verify that the outcomes are equally likely.

3. Subjective Probabilities $P(A)$, the probability of event A, is *estimated* by using knowledge of the relevant circumstances.

Note that the classical approach requires *equally likely outcomes*. If the outcomes are not equally likely, we must use the relative frequency approximation or we must rely on our knowledge of the circumstances to make an *educated guess*. Figure 4-1 illustrates the three approaches.

When finding probabilities with the relative frequency approach, we obtain an *approximation* instead of an exact value. As the total number of observations increases, the corresponding approximations tend to get closer to the actual probability. This property is stated as a theorem commonly referred to as the *law of large numbers.*

(a)

(b)

(c)

Figure 4-1

Three Approaches to Finding a Probability

(a) **Relative Frequency Approach:** When trying to determine the probability that an individual car crashes in a year, we must examine past results to determine the number of cars in use in a year and the number of them that crashed, then we find the ratio of the number of cars that crashed to the total number of cars. For a recent year, the result is a probability of 0.0480. (See Example 2.)

(b) **Classical Approach:** When trying to determine the probability of winning the grand prize in a lottery by selecting 6 numbers between 1 and 60, each combination has an equal chance of occurring. The probability of winning is 0.0000000200, which can be found by using methods presented later in this chapter.

(c) **Subjective Probability:** When trying to estimate the probability of an astronaut surviving a mission in a space shuttle, experts consider past events along with changes in technologies and conditions to develop an estimate of the probability. As of this writing, that probability has been estimated by NASA scientists as 0.99.

Law of Large Numbers As a procedure is repeated again and again, the relative frequency probability of an event tends to approach the actual probability.

The law of large numbers tells us that relative frequency approximations tend to get better with more observations. This law reflects a simple notion supported by common sense: A probability estimate based on only a few trials can be off by a substantial amount, but with a very large number of trials, the estimate tends to be much more accurate.

Probability and Outcomes That Are Not Equally Likely One common mistake is to incorrectly assume that outcomes are equally likely just because we know nothing about the likelihood of each outcome. When we know nothing about the likelihood of different possible outcomes, we cannot necessarily assume that they are equally likely. For example, we should not conclude that the probability of passing a test is 1/2 or 0.5 (because we either pass the test or do not). The actual probability depends on factors such as the amount of preparation and the difficulty of the test.

EXAMPLE 2 **Probability of a Car Crash** Find the probability that a randomly selected car in the United States will be in a crash this year.

SOLUTION For a recent year, there were 6,511,100 cars that crashed among the 135,670,000 cars registered in the United States (based on data from *Statistical Abstract of the United States*). We can now use the relative frequency approach as follows:

$$P(\text{crash}) = \frac{\text{number of cars that crashed}}{\text{total number of cars}} = \frac{6,511,100}{135,670,000} = 0.0480$$

Note that the classical approach cannot be used since the two outcomes (crash, no crash) are not equally likely.

EXAMPLE 3 **Probability of a Positive Test Result** Refer to Table 4-1 included with the Chapter Problem. Assuming that one of the 98 test results summarized in Table 4-1 is randomly selected, find the probability that it is a positive test result.

SOLUTION The sample space consists of the 98 test results listed in Table 4-1. Among the 98 results, 57 of them are positive test results (found from 42 + 15). Since each test result is equally likely to be selected, we can apply the classical approach as follows:

$$P(\text{positive test result from Table 4-1}) = \frac{\text{number of positive test results}}{\text{total number of results}}$$

$$= \frac{57}{98} = 0.582$$

How Probable?

How do we interpret such terms as *probable, improbable,* or *extremely improbable*? The FAA interprets these terms as follows.

- *Probable:* A probability on the order of 0.00001 or greater for each hour of flight. Such events are expected to occur several times during the operational life of each airplane.

- *Improbable:* A probability on the order of 0.00001 or less. Such events are not expected to occur during the total operational life of a single airplane of a particular type, but may occur during the total operational life of all airplanes of a particular type.

- *Extremely improbable:* A probability on the order of 0.000000001 or less. Such events are so unlikely that they need not be considered to ever occur.

Making Cents of the Lottery

Many people spend large sums of money buying lottery tickets, even though they don't have a realistic sense for their chances of winning. Brother Donald Kelly of Marist College suggests this analogy: Winning the lottery is equivalent to correctly picking the "winning" dime from a stack of dimes that is 21 miles tall! Commercial aircraft typically fly at altitudes of 6 miles, so try to image a stack of dimes more than three times higher than those high-flying jets, then try to imagine selecting the one dime in that stack that represents a winning lottery ticket. Using the methods of this section, find the probability of winning your state's lottery, then determine the height of the corresponding stack of dimes.

EXAMPLE 4 **Genotypes** When studying the affect of heredity on height, we can express each individual genotype, AA, Aa, aA, and aa, on an index card and shuffle the four cards and randomly select one of them. What is the probability that we select a genotype in which the two components are different?

SOLUTION The sample space (AA, Aa, aA, aa) in this case includes equally likely outcomes. Among the 4 outcomes, there are exactly 2 in which the two components are different: Aa and aA. We can use the classical approach to get

$$P(\text{outcome with different components}) = \frac{2}{4} = 0.5$$

EXAMPLE 5 **Probability of a President from Alaska** Find the probability that the next President of the United States is from Alaska.

SOLUTION The sample space consists of two simple events: The next President is from Alaska or is not. If we were to use the relative frequency approach, we would incorrectly conclude that it is impossible for anyone from Alaska to be President, because it has never happened in the past. We cannot use the classical approach because the two possible outcomes are events that are not equally likely. We are left with making a subjective estimate. The population of Alaska is 0.2% of the total United States population, but the remoteness of Alaska presents special challenges to politicians from that state, so an estimated probability of 0.001 is reasonable.

EXAMPLE 6 **Stuck in an Elevator** What is the probability that you will get stuck in the next elevator that you ride?

SOLUTION In the absence of historical data on elevator failures, we cannot use the relative frequency approach. There are two possible outcomes (becoming stuck or not becoming stuck), but they are not equally likely, so we cannot use the classical approach. That leaves us with a subjective estimate. In this case, experience suggests that the probability is quite small. Let's estimate it to be, say, 0.0001 (equivalent to 1 chance in ten thousand). That subjective estimate, based on our general knowledge, is likely to be in the general ballpark of the true probability.

Finding the Total Number of Outcomes In basic probability problems we must be careful to examine the available information and to correctly identify the total number of possible outcomes. In some cases, the total number of possible outcomes is given, but in other cases it must be calculated, as in the next two examples.

EXAMPLE 7 **Gender of Children** Find the probability that when a couple has 3 children, they will have exactly 2 boys. Assume that boys and girls are equally likely and that the gender of any child is not influenced by the gender of any other child.

1st 2nd 3rd

boy-boy-boy

boy-boy-girl

boy-girl-boy

exactly

2 boys

boy-girl-girl

girl-boy-boy

girl-boy-girl

girl-girl-boy

girl-girl-girl

SOLUTION The biggest challenge here is to correctly identify the sample space. It involves more than working only with the numbers 2 and 3 given in the statement of the problem. The sample space consists of 8 different ways that 3 children can occur (see the margin). Those 8 outcomes are equally likely, so we use the classical approach. Of those 8 different possible outcomes, 3 correspond to exactly 2 boys, so

$$P(\text{2 boys in 3 births}) = \frac{3}{8} = 0.375$$

INTERPRETATION There is a 0.375 probability that if a couple has 3 children, exactly 2 will be boys.

EXAMPLE 8 **America Online Survey** The Internet service provider America Online (AOL) asked users this question about Kentucky Fried Chicken (KFC): "Will KFC gain or lose business after eliminating trans fats?" Among the responses received, 1941 said that KFC would gain business, 1260 said that KFC business would remain the same, and 204 said that KFC would lose business. Find the probability that a randomly selected response states that KFC would gain business.

SOLUTION *Hint:* Instead of trying to determine an answer directly from the printed statement, begin by first summarizing the given information in a format that allows you to clearly understand the information. For example, use this format:

1941	gain in business
1260	business remains the same
204	loss in business
3405	total responses

We can now use the relative frequency approach as follows:

$P(\text{response of a gain in business})$

$$= \frac{\text{number who said that KFC would gain business}}{\text{total number of responses}} = \frac{1941}{3405}$$

$$= 0.570$$

INTERPRETATION There is a 0.570 probability that if a response is randomly selected, it was a response of a gain in business. *Important:* Note that the survey involves a voluntary response sample because the AOL users themselves decided whether to respond. Consequently, when interpreting the results of this survey, keep in mind that they do not necessarily reflect the opinions of the general population. The responses reflect only the opinions of those who chose to respond.

Simulations   The statements of the three approaches for finding probabilities and the preceding examples might seem to suggest that we should always use the classical approach when a procedure has equally likely outcomes, but many procedures are so complicated that the classical approach is impractical. In the game of solitaire, for example, the outcomes (hands dealt) are all equally likely, but it is extremely frustrating to try to use the classical approach to find the probability of winning. In such cases we can more easily get good estimates by using the relative frequency approach. Simulations are often helpful when using this approach. A *simulation* of a procedure is a process that behaves in the same ways as the procedure itself, so that similar results are produced. (See Section 4-6 and the Technology Project near the end of this chapter.) For example, it's much easier to use the relative frequency approach for approximating the probability of winning at solitaire—that is, to play the game many times (or to run a computer simulation)—than to perform the complex calculations required with the classical approach.

EXAMPLE 9   **Thanksgiving Day** If a year is selected at random, find the probability that Thanksgiving Day will be (a) on a Wednesday or (b) on a Thursday.

SOLUTION

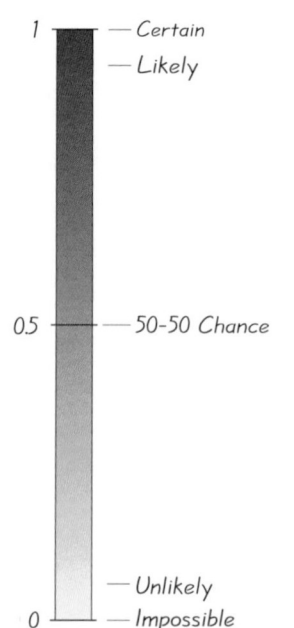

a. Thanksgiving Day always falls on the fourth Thursday in November. It is therefore impossible for Thanksgiving to be on a Wednesday. When an event is impossible, we say that its probability is 0.

b. It is certain that Thanksgiving will be on a Thursday. When an event is certain to occur, we say that its probability is 1.

Because any event imaginable is impossible, certain, or somewhere in between, it follows that the mathematical probability of any event is 0, 1, or a number between 0 and 1 (see Figure 4-2).

1 — Certain
  — Likely

0.5 — 50-50 Chance

  — Unlikely
0 — Impossible

Figure 4-2   Possible Values for Probabilities

CAUTION

Always express a probability as a fraction or decimal number between 0 and 1.

- **The probability of an impossible event is 0.**

- **The probability of an event that is certain to occur is 1.**

- **For any event A, the probability of A is between 0 and 1 inclusive. That is, $0 \leq P(A) \leq 1$.**

In Figure 4-2, the scale of 0 through 1 is shown, and the more familiar and common expressions of likelihood are included.

Complementary Events

Sometimes we need to find the probability that an event A does *not* occur.

DEFINITION

The **complement** of event A, denoted by $\overline{A}$, consists of all outcomes in which event A does *not* occur.

EXAMPLE 10 **Guessing on an SAT Test** A typical question on an SAT test requires the test taker to select one of five possible choices: A, B, C, D, or E. Because only one answer is correct, if you make a random guess, your probability of being correct is 1/5 or 0.2. Find the probability of making a random guess and *not* being correct (or being incorrect).

SOLUTION Because exactly 1 of the 5 responses is correct, it follows that 4 of them are *not* correct, so

$$P(\text{not guessing the correct answer}) = P(\overline{\text{correct}}) = P(\text{incorrect}) = \frac{4}{5} = 0.8$$

INTERPRETATION When guessing for such a multiple-choice question, there is a 0.8 probability of being incorrect. Although test takers are not penalized for wrong guesses, guessing is OK for some questions, especially if you can eliminate any of the choices. In the long run, scores are not affected, but many guesses will tend to result in a low score.

Although it is difficult to develop a universal rule for rounding off probabilities, the following guide will apply to most problems in this text.

Rounding Off Probabilities

When expressing the value of a probability, either give the *exact* fraction or decimal or round off final decimal results to three significant digits. (*Suggestion:* When a probability is not a simple fraction such as 2/3 or 5/9, express it as a decimal so that the number can be better understood.) All digits in a number are significant except for the zeros that are included for proper placement of the decimal point.

EXAMPLE 11 **Rounding Probabilities**

• The probability of 0.04799219 (from Example 2) has seven significant digits (4799219), and it can be rounded to three significant digits as 0.0480. (The zero to the immediate right of the decimal point is *not* significant because it is necessary for correct placement of the decimal point, but the zero at the extreme right is significant because it is not necessary for correct placement of the decimal point.)

• The probability of 1/3 can be left as a fraction, or rounded to 0.333. (Do *not* round to 0.3.)

• The probability of 2/4 (from Example 4) can be expressed as 1/2 or 0.5; because 0.5 is exact, there's no need to express it with three significant digits as 0.500.

• The fraction 1941/3405 (from Example 8) is exact, but its value isn't obvious, so express it as the decimal 0.570.

The mathematical expression of probability as a number between 0 and 1 is fundamental and common in statistical procedures, and we will use it throughout the remainder of this text. A typical computer output, for example, may include a "*P*-value" expression such as "significance less than 0.001." We will discuss the meaning of *P*-values later, but they are essentially probabilities of the type discussed in this section. For now, you should recognize that a probability of 0.001 (equivalent to 1/1000) corresponds to an event so rare that it occurs an average of only once in a thousand trials. Example 12 involves the interpretation of such a small probability value.

> **EXAMPLE 12**   **Unusual Event?** In a clinical experiment of the Salk vaccine for polio, 200,745 children were given a placebo and 201,229 other children were treated with the Salk vaccine. There were 115 cases of polio among those in the placebo group and 33 cases of polio in the treatment group. If we assume that the vaccine has no effect, the probability of getting such test results is found to be "less than 0.001." Is an event with a probability less than 0.001 an *unusual* event? What does that probability imply about the effectiveness of the vaccine?
>
> **SOLUTION**   A probability value less than 0.001 is very small. It indicates that the event will occur fewer than once in a thousand times, so the event is "unusual." The small probability suggests that the test results are not likely to occur if the vaccine has no effect. Consequently, there are two possible explanations for the results of this clinical experiment: (1) The vaccine has no effect and the results occurred by chance; (2) the vaccine has an effect, which explains why the treatment group had a much lower incidence of polio. Because the probability is so small (less than 0.001), the second explanation is more reasonable. We conclude that the vaccine appears to be effective.

The preceding example illustrates the "rare event rule for inferential statistics" given in Section 4-1. Under the assumption of a vaccine with no effect, we find that the probability of the results is extremely small (less than 0.001), so we conclude that the assumption is probably not correct. The preceding example also illustrates the role of probability in making important conclusions about clinical experiments. For now, we should understand that when a probability is small, such as less than 0.001, it indicates that the event is very unlikely to occur.

Part 2: Beyond the Basics of Probability: Odds

Expressions of likelihood are often given as *odds,* such as 50:1 (or "50 to 1"). Because the use of odds makes many calculations difficult, statisticians, mathematicians, and scientists prefer to use probabilities. The advantage of odds is that they make it easier to deal with money transfers associated with gambling, so they tend to be used in casinos, lotteries, and racetracks. Note that in the three definitions that follow, the *actual odds against* and the *actual odds in favor* are calculated with the actual likelihood of some event, but the *payoff odds* describe the relationship between the bet and the amount of the payoff. The actual odds correspond to actual probabilities of outcomes, but the payoff odds are set by racetrack and casino operators. Racetracks and casinos are in business to make a profit, so the payoff odds will not be the same as the actual odds.

DEFINITION

The **actual odds against** event A occurring are the ratio $P(\overline{A})/P(A)$, usually expressed in the form of $a{:}b$ (or "a to b"), where a and b are integers having no common factors.

The **actual odds in favor** of event A occurring are the ratio $P(A)/P(\overline{A})$, which is the reciprocal of the actual odds against that event. If the odds against A are $a{:}b$, then the odds in favor of A are $b{:}a$.

The **payoff odds** against event A occurring are the ratio of net profit (if you win) to the amount bet.

payoff odds against event $A = $ (net profit):(amount bet)

EXAMPLE 13 If you bet \$5 on the number 13 in roulette, your probability of winning is 1/38 and the payoff odds are given by the casino as 35:1.

a. Find the actual odds against the outcome of 13.

b. How much net profit would you make if you win by betting on 13?

c. If the casino was not operating for profit, and the payoff odds were changed to match the actual odds against 13, how much would you win if the outcome were 13?

SOLUTION

a. With $P(13) = 1/38$ and $P(\text{not } 13) = 37/38$, we get

$$\text{actual odds against } 13 = \frac{P(\text{not } 13)}{P(13)} = \frac{37/38}{1/38} = \frac{37}{1} \text{ or } 37{:}1$$

b. Because the payoff odds against 13 are 35:1, we have

$$35{:}1 = (\text{net profit}){:}(\text{amount bet})$$

So there is a \$35 profit for each \$1 bet. For a \$5 bet, the net profit is \$175. The winning bettor would collect \$175 plus the original \$5 bet. That is, the total amount collected would be \$180, for a net profit of \$175.

c. If the casino were not operating for profit, the payoff odds would be equal to the actual odds against the outcome of 13, or 37:1. So there is a net profit of \$37 for each \$1 bet. For a \$5 bet the net profit would be \$185. (The casino makes its profit by paying only \$175 instead of the \$185 that would be paid with a roulette game that is fair instead of favoring the casino.)

4-2   Basic Skills and Concepts

Statistical Literacy and Critical Thinking

1. Interpreting Probability Based on recent results, the probability of someone in the United States being injured while using sports or recreation equipment is 1/500 (based on data from *Statistical Abstract of the United States*). What does it mean when we say that the probability is 1/500? Is such an injury *unusual?*

2. Probability of a Republican President When predicting the chance that we will elect a Republican President in the year 2012, we could reason that there are two possible outcomes (Republican, not Republican), so the probability of a Republican President is 1/2 or 0.5. Is this reasoning correct? Why or why not?

3. Probability and Unusual Events If A denotes some event, what does $\overline{A}$ denote? If $P(A) = 0.995$, what is the value of $P(\overline{A})$? If $P(A) = 0.995$, is $\overline{A}$ *unusual*?

4. Subjective Probability Estimate the probability that the next time you ride in a car, you will *not* be delayed because of some car crash blocking the road.

In Exercises 5–12, express the indicated degree of likelihood as a probability value between 0 and 1.

5. Lottery In one of New York State's instant lottery games, the chances of a win are stated as "4 in 21."

6. Weather A WeatherBug forecast for the author's home was stated as: "Chance of rain: 80%."

7. Testing If you make a random guess for the answer to a true/false test question, there is a 50-50 chance of being correct.

8. Births When a baby is born, there is approximately a 50-50 chance that the baby is a girl.

9. Dice When rolling two dice at the Venetian Casino in Las Vegas, there are 6 chances in 36 that the outcome is a 7.

10. Roulette When playing roulette in the Mirage Casino, you have 18 chances out of 38 of winning if you bet that the outcome is an odd number.

11. Cards It is impossible to get five aces when selecting cards from a shuffled deck.

12. Days When randomly selecting a day of the week, you are certain to select a day containing the letter y.

13. Identifying Probability Values Which of the following values *cannot* be probabilities?

3:1   2/5   5/2   −0.5   0.5   123/321   321/123   0   1

14. Identifying Probability Values

a. What is the probability of an event that is certain to occur?

b. What is the probability of an impossible event?

c. A sample space consists of 10 separate events that are equally likely. What is the probability of each?

d. On a true/false test, what is the probability of answering a question correctly if you make a random guess?

e. On a multiple-choice test with five possible answers for each question, what is the probability of answering a question correctly if you make a random guess?

15. Gender of Children Refer to the list of the eight outcomes that are possible when a couple has three children. (See Example 7.) Find the probability of each event.

a. There is exactly one girl.   **b.** There are exactly two girls.

c. All are girls.

16. Genotypes In Example 4 we noted that a study involved equally likely genotypes represented as AA, Aa, aA, and aa. If one of these genotypes is randomly selected as in Example 4, what is the probability that the outcome is AA? Is obtaining AA unusual?

17. Polygraph Test Refer to the sample data in Table 4-1, which is included with the Chapter Problem.

a. How many responses are summarized in the table?

b. How many times did the polygraph provide a negative test result?

c. If one of the responses is randomly selected, find the probability that it is a negative test result. (Express the answer as a fraction.)

d. Use the rounding method described in this section to express the answer from part (c) as a decimal.

18. Polygraph Test Refer to the sample data in Table 4-1.

a. How many responses were actually lies?

b. If one of the responses is randomly selected, what is the probability that it is a lie? (Express the answer as a fraction.)

c. Use the rounding method described in this section to express the answer from part (b) as a decimal.

19. Polygraph Test Refer to the sample data in Table 4-1. If one of the responses is randomly selected, what is the probability that it is a false positive? (Express the answer as a decimal.) What does this probability suggest about the accuracy of the polygraph test?

20. Polygraph Test Refer to the sample data in Table 4-1. If one of the responses is randomly selected, what is the probability that it is a false negative? (Express the answer as a decimal.) What does this probability suggest about the accuracy of the polygraph test?

21. U.S. Senate The 110th Congress of the United States included 84 male Senators and 16 female Senators. If one of these Senators is randomly selected, what is the probability that a woman is selected? Does this probability agree with a claim that men and women have the same chance of being elected as Senators?

22. Mendelian Genetics When Mendel conducted his famous genetics experiments with peas, one sample of offspring consisted of 428 green peas and 152 yellow peas. Based on those results, estimate the probability of getting an offspring pea that is green. Is the result reasonably close to the expected value of 3/4, as claimed by Mendel?

23. Struck by Lightning In a recent year, 281 of the 290,789,000 people in the United States were struck by lightning. Estimate the probability that a randomly selected person in the United States will be struck by lightning this year. Is a golfer reasoning correctly if he or she is caught out in a thunderstorm and does not seek shelter from lightning during a storm because the probability of being struck is so small?

24. Gender Selection In updated results from a test of MicroSort's XSORT gender-selection technique, 726 births consisted of 668 baby girls and 58 baby boys (based on data from the Genetics & IVF Institute). Based on these results, what is the probability of a girl born to a couple using MicroSort's XSORT method? Does it appear that the technique is effective in increasing the likelihood that a baby will be a girl?

Using Probability to Identify Unusual Events. *In Exercises 25–32, consider an event to be "unusual" if its probability is less than or equal to 0.05. (This is equivalent to the same criterion commonly used in inferential statistics, but the value of 0.05 is not absolutely rigid, and other values such as 0.01 are sometimes used instead.)*

25. Guessing Birthdays On their first date, Kelly asks Mike to guess the date of her birth, not including the year.

a. What is the probability that Mike will guess correctly? (Ignore leap years.)

b. Would it be unusual for him to guess correctly on his first try?

c. If you were Kelly, and Mike did guess correctly on his first try, would you believe his claim that he made a lucky guess, or would you be convinced that he already knew when you were born?

d. If Kelly asks Mike to guess her age, and Mike's guess is too high by 15 years, what is the probability that Mike and Kelly will have a second date?

26. Adverse Effect of Viagra When the drug Viagra was clinically tested, 117 patients reported headaches and 617 did not (based on data from Pfizer, Inc.). Use this sample to estimate the probability that a Viagra user will experience a headache. Is it unusual for a Viagra user to experience headaches? Is the probability high enough to be of concern to Viagra users?

27. Heart Pacemaker Failures Among 8834 cases of heart pacemaker malfunctions, 504 were found to be caused by firmware, which is software programmed into the device (based on data from "Pacemaker and ICD Generator Malfunctions," by Maisel, et al., *Journal of the American Medical Association,* Vol. 295, No. 16). Based on these results, what is the probability that a pacemaker malfunction is caused by firmware? Is a firmware malfunction unusual among pacemaker malfunctions?

28. Bumped from a Flight Among 15,378 Delta airline passengers randomly selected, 3 were bumped from a flight against their wishes (based on data from the U.S. Department of Transportation). Find the probability that a randomly selected passenger is involuntarily bumped. Is such bumping unusual? Does such bumping pose a serious problem for Delta passengers in general? Why or why not?

29. Death Penalty In the last 30 years, death sentence executions in the United States included 795 men and 10 women (based on data from the Associated Press). If an execution is randomly selected, find the probability that the person executed is a woman. Is it unusual for a woman to be executed? How might the discrepancy be explained?

30. Stem Cell Survey Adults were randomly selected for a *Newsweek* poll, and they were asked if they "favor or oppose using federal tax dollars to fund medical research using stem cells obtained from human embryos." Of the adults selected, 481 were in favor, 401 were opposed, and 120 were unsure. Based on these results, find the probability that a randomly selected adult would respond in favor. Is it unusual for an adult to be in favor?

31. Cell Phones in Households In a survey of consumers aged 12 and older conducted by Frank N. Magid Associates, respondents were asked how many cell phones were in use by the household. Among the respondents, 211 answered "none," 288 said "one," 366 said "two," 144 said "three," and 89 responded with four or more. Find the probability that a randomly selected household has four or more cellphones in use. Is it unusual for a household to have four or more cell phones in use?

32. Personal Calls at Work *USA Today* reported on a survey of office workers who were asked how much time they spend on personal phone calls per day. Among the responses, 1065 reported times between 1 and 10 minutes, 240 reported times between 11 and 30 minutes, 14 reported times between 31 and 60 minutes, and 66 said that they do not make personal calls. If a worker is randomly selected, what is the probability the worker does not make personal calls. Is it unusual for a worker to make no personal calls?

Constructing Sample Space. *In Exercises 33–36, construct the indicated sample space and answer the given questions.*

33. Gender of Children: Constructing Sample Space This section included a table summarizing the gender outcomes for a couple planning to have three children.

a. Construct a similar table for a couple planning to have *two* children.

b. Assuming that the outcomes listed in part (a) are equally likely, find the probability of getting two girls.

c. Find the probability of getting exactly one child of each gender.

34. Gender of Children: Constructing Sample Space This section included a table summarizing the gender outcomes for a couple planning to have three children.

a. Construct a similar table for a couple planning to have *four* children.

b. Assuming that the outcomes listed in part (a) are equally likely, find the probability of getting exactly two girls and two boys.

c. Find the probability that the four children are all boys.

35. Genetics: Eye Color Each of two parents has the genotype brown/blue, which consists of the pair of alleles that determine eye color, and each parent contributes one of those alleles to a child. Assume that if the child has at least one brown allele, that color will dominate and the eyes will be brown. (The actual determination of eye color is somewhat more complicated.)

a. List the different possible outcomes. Assume that these outcomes are equally likely.

b. What is the probability that a child of these parents will have the blue/blue genotype?

c. What is the probability that the child will have brown eyes?

36. X-Linked Genetic Disease Men have XY (or YX) chromosomes and women have XX chromosomes. X-linked recessive genetic diseases (such as juvenile retinoschisis) occur when there is a defective X chromosome that occurs *without* a paired X chromosome that is good. In the following, represent a defective X chromosome with lower case x, so a child with the xY or Yx pair of chromosomes will have the disease, while a child with XX or XY or YX or xX or Xx will not have the disease. Each parent contributes one of the chromosomes to the child.

a. If a father has the defective x chromosome and the mother has good XX chromosomes, what is the probability that a son will inherit the disease?

b. If a father has the defective x chromosome and the mother has good XX chromosomes, what is the probability that a daughter will inherit the disease?

c. If a mother has one defective x chromosome and one good X chromosome, and the father has good XY chromosomes, what is the probability that a son will inherit the disease?

d. If a mother has one defective x chromosome and one good X chromosome, and the father has good XY chromosomes, what is the probability that a daughter will inherit the disease?

4-2 Beyond the Basics

Odds. *In Exercises 37–40, answer the given questions that involve odds.*

37. Solitaire Odds A solitaire game was played 500 times. Among the 500 trials, the game was won 77 times. (The results are from the Microsoft solitaire game, and the Vegas rules of "draw 3" with $52 bet and a return of $5 per card are used.) Based on these results, find the odds against winning.

38. Finding Odds in Roulette A roulette wheel has 38 slots. One slot is 0, another is 00, and the others are numbered 1 through 36, respectively. You place a bet that the outcome is an odd number.

a. What is your probability of winning?   **b.** What are the actual odds against winning?

c. When you bet that the outcome is an odd number, the payoff odds are 1:1. How much profit do you make if you bet $18 and win?

d. How much profit would you make on the $18 bet if you could somehow convince the casino to change its payoff odds so that they are the same as the actual odds against winning? (*Recommendation:* Don't actually try to convince any casino of this; their sense of humor is remarkably absent when it comes to things of this sort.)

39. Kentucky Derby Odds When the horse Barbaro won the 132nd Kentucky Derby, a $2 bet that Barbaro would win resulted in a return of $14.20.

a. How much net profit was made from a $2 win bet on Barbaro?

b. What were the payoff odds against a Barbaro win?

c. Based on preliminary wagering before the race, bettors collectively believed that Barbaro had a 57/500 probability of winning. Assuming that 57/500 was the true probability of a Barbaro victory, what were the actual odds against his winning?

d. If the payoff odds were the actual odds found in part (c), how much would a $2 win ticket be worth after the Barbaro win?

40. Finding Probability from Odds If the actual odds against event A are $a:b$, then $P(A) = b/(a + b)$. Find the probability of the horse Cause to Believe winning the 132nd Kentucky Derby, given that the actual odds against his winning that race were 97:3.

41. Relative Risk and Odds Ratio In a clinical trial of 2103 subjects treated with Nasonex, 26 reported headaches. In a control group of 1671 subjects given a placebo, 22

Boys and Girls Are Not Equally Likely

In many probability calculations, good results are obtained by assuming that

boys and girls are equally likely to be born. In reality, a boy is more likely to be born (with probability 0.512) than a girl (with probability 0.488). These results are based on recent data from the National Center for Health Statistics, which showed that the 4,112,856 births in one year included 2,105,458 boys and 2,007,398 girls. Researchers monitor these probabilities for changes that might suggest such factors as changes in the environment and exposure to chemicals.

reported headaches. Denoting the proportion of headaches in the treatment group by p_t and denoting the proportion of headaches in the control (placebo) group by p_c, the *relative risk* is p_t/p_c. The relative risk is a measure of the strength of the effect of the Nasonex treatment. Another such measure is the *odds ratio,* which is the ratio of the odds in favor of a headache for the treatment group to the odds in favor of a headache for the control (placebo) group, found by evaluating the following:

$$\frac{p_t/(1 - p_t)}{p_c/(1 - p_c)}$$

The relative risk and odds ratios are commonly used in medicine and epidemiological studies. Find the relative risk and odds ratio for the headache data. What do the results suggest about the risk of a headache from the Nasonex treatment?

42. Flies on an Orange If two flies land on an orange, find the probability that they are on points that are within the same hemisphere.

43. Points on a Stick Two points along a straight stick are randomly selected. The stick is then broken at those two points. Find the probability that the three resulting pieces can be arranged to form a triangle. (This is possibly the most difficult exercise in this book.)

 ## 4-3 Addition Rule

Key Concept In this section we present the *addition rule* as a device for finding probabilities that can be expressed as $P(A \text{ or } B)$, which denotes the probability that either event A occurs or event B occurs (or they both occur) as the single outcome of a procedure. To find the probability of event A occurring or event B occurring, we begin by finding the total number of ways that A can occur and the number of ways that B can occur, without counting any outcomes more than once.

The key word in this section is "or." Throughout this text we use the *inclusive or,* which means either one or the other or both. (Except for Exercise 41, we will not consider the *exclusive or,* which means either one or the other but not both.)

In the previous section we presented the basics of probability and considered events categorized as *simple* events. In this and the following section we consider *compound events.*

 DEFINITION

A **compound event** is any event combining two or more simple events.

Notation for Addition Rule

$P(A \text{ or } B) = P(\text{in a single trial, event } A \text{ occurs or event } B \text{ occurs or they both occur})$

Understanding the Notation   In this section, $P(A \text{ and } B)$ denotes the probability that A and B both occur in the same trial, but in Section 4-4 we use $P(A \text{ and } B)$ to denote the probability that event A occurs on one trial followed by event B on another trial. The true meaning of $P(A \text{ and } B)$ can therefore be determined only by knowing whether we are referring to one trial that can have outcomes of A and B, or two trials with event A occurring on the first trial and event B occurring on the second trial. The meaning denoted by $P(A \text{ and } B)$ therefore depends upon the context.

In Section 4-2 we considered simple events, such as the probability of getting a false positive when one test result is randomly selected from the 98 test results listed in Table 4-1, reproduced on the next page for convenience. If we randomly select one test result, the probability of a false positive is given by $P(\text{false positive}) = 15/98 = 0.153$. (See Exercise 19 in Section 4-2.) Now let's consider $P(\text{getting a positive test result}$

Table 4-1   Results from Experiments with Polygraph Instruments

	Did the Subject Actually Lie?	
	No (Did Not Lie)	Yes (Lied)
Positive test result	15	42
(Polygraph test indicated that the subject *lied*.)	(false positive)	(true positive)
Negative test result	32	9
(Polygraph test indicated that the subject did *not lie*.)	(true negative)	(false negative)

or a subject who lied) when one of the 98 test results is randomly selected. Refer to Table 4-1 and carefully count the number of subjects who tested positive or lied, but be careful to count subjects once, not twice. Examination of Table 4-1 shows that 66 subjects had positive test results or lied. (*Important:* It is *wrong* to add the 57 subjects with positive test results to the 51 subjects who lied, because this total of 108 counts 42 of the subjects twice.) See the role that the correct total of 66 plays in the following example.

EXAMPLE 1   **Polygraph Test** Refer to Table 4-1. If 1 subject is randomly selected from the 98 subjects given a polygraph test, find the probability of selecting a subject who had a positive test result or lied.

SOLUTION   From Table 4-1 we see that there are 66 subjects who had a positive test result or lied. We obtain that total of 66 by adding the subjects who tested positive to the subjects who lied, being careful to count everyone only once. Dividing the total of 66 by the overall total of 98, we get: P(positive test result or lied) = 66/98 or 0.673.

In Example 1, there are several ways to count the subjects who tested positive or lied. Any of the following would work:

- Color the cells representing subjects who tested positive or lied, then add the numbers in those colored cells, being careful to add each number only once. This approach yields

$$15 + 42 + 9 = 66$$

- Add the 57 subjects who tested positive to the 51 subjects who lied, but the total of 108 involves double-counting of 42 subjects, so compensate for the double-counting by subtracting the overlap consisting of the 42 subjects who were counted twice. This approach yields a result of

$$57 + 51 - 42 = 66$$

- Start with the total of 57 subjects who tested positive, then add those subjects who lied and were not yet included in that total, to get a result of

$$57 + 9 = 66$$

Example 1 illustrates that when finding the probability of an event A or event B, use of the word "or" suggests addition, and the addition must be done without double-counting.

The preceding example suggests a general rule whereby we add the number of outcomes corresponding to each of the events in question:

When finding the probability that event A occurs or event B occurs, find the total of the number of ways A can occur and the number of ways B

Shakespeare's Vocabulary

According to Bradley Efron and Ronald Thisted, Shakespeare's writings included 31,534 different words. They used probability theory to conclude that

Shakespeare probably knew at least another 35,000 words that he didn't use in his writings. The problem of estimating the size of a population is an important problem often encountered in ecology studies, but the result given here is another interesting application. (See "Estimating the Number of Unseen Species: How Many Words Did Shakespeare Know?", in *Biometrika*, Vol. 63, No. 3.)

can occur, but *find that total in such a way that no outcome is counted more than once.*

CAUTION

When using the addition rule, always be careful to avoid counting outcomes more than once.

One way to formalize the rule is to combine the number of ways event *A* can occur with the number of ways event *B* can occur and, if there is any overlap, compensate by subtracting the number of outcomes that are counted twice, as in the following rule.

Formal Addition Rule

$$P(A \text{ or } B) = P(A) + P(B) - P(A \text{ and } B)$$

where $P(A \text{ and } B)$ denotes the probability that *A* and *B* both occur at the same time as an outcome in a trial of a procedure.

Although the formal addition rule is presented as a formula, blind use of formulas is not recommended. It is generally better to *understand* the spirit of the rule and use that understanding, as follows.

Intuitive Addition Rule

To find $P(A \text{ or } B)$, find the sum of the number of ways event *A* can occur and the number of ways event *B* can occur, *adding in such a way that every outcome is counted only once.* $P(A \text{ or } B)$ is equal to that sum, divided by the total number of outcomes in the sample space.

The addition rule is simplified when the events are *disjoint*.

DEFINITION

Events *A* and *B* are **disjoint** (or **mutually exclusive**) if they cannot occur at the same time. (That is, disjoint events do not overlap.)

EXAMPLE 2   **Polygraph Test** Refer to Table 4-1.

a. Consider the procedure of randomly selecting 1 of the 98 subjects included in Table 4-1. Determine whether the following events are disjoint:

A: Getting a subject with a negative test result.

B: Getting a subject who did not lie.

b. Assuming that 1 subject is randomly selected from the 98 that were tested, find the probability of selecting a subject who had a negative test result or did not lie.

SOLUTION

a. In Table 4-1 we see that there are 41 subjects with negative test results and there are 47 subjects who did not lie. The event of getting a subject with a negative test result and getting a subject who did not lie can occur at the same time (because there are 32 subjects who had negative test results and did not lie). Because those events overlap, they can occur at the same time and we say that the events are *not* disjoint.

b. In Table 4-1 we must find the total number of subjects who had negative test results or did not lie, but we must find that total without double-counting. We get a total of 56 (from 32 + 9 + 15). Because 56 subjects had negative test results or did not lie, and because there are 98 total subjects included, we see that

$$P(\text{negative test result or did not lie}) = \frac{56}{98} = 0.571$$

Figure 4-3 shows a Venn diagram that provides a visual illustration of the formal addition rule. In this figure we can see that the probability of *A* or *B* equals the probability of *A* (left circle) plus the probability of *B* (right circle) minus the probability of *A* and *B* (football-shaped middle region). This figure shows that the addition of the areas of the two circles will cause double-counting of the football-shaped middle region. This is the basic concept that underlies the addition rule. Because of the relationship between the addition rule and the Venn diagram shown in Figure 4-3, the notation $P(A \cup B)$ is sometimes used in place of $P(A \text{ or } B)$. Similarly, the notation $P(A \cap B)$ is sometimes used in place of $P(A \text{ and } B)$ so the formal addition rule can be expressed as

$$P(A \cup B) = P(A) + P(B) - P(A \cap B)$$

Whenever *A* and *B* are disjoint, $P(A \text{ and } B)$ becomes zero in the addition rule. Figure 4-4 illustrates that when *A* and *B* are disjoint, we have $P(A \text{ or } B) = P(A) + P(B)$.

We can summarize the key points of this section as follows:

1. To find $P(A \text{ or } B)$, begin by associating use of the word "or" with addition.

2. Consider whether events *A* and *B* are disjoint; that is, can they happen at the same time? If they are not disjoint (that is, they can happen at the same time), be sure to avoid (or at least compensate for) double-counting when adding the relevant probabilities. If you understand the importance of not double-counting when you find $P(A \text{ or } B)$, you don't necessarily have to calculate the value of $P(A) + P(B) - P(A \text{ and } B)$.

Errors made when applying the addition rule often involve double-counting; that is, events that are not disjoint are treated as if they were. One indication of such an error is a total probability that exceeds 1; however, errors involving the addition rule do not always cause the total probability to exceed 1.

Complementary Events

In Section 4-2 we defined the complement of event *A* and denoted it by $\overline{A}$. We said that $\overline{A}$ consists of all the outcomes in which event *A* does *not* occur. Events *A* and $\overline{A}$ must be disjoint, because it is impossible for an event and its complement to occur at the same time. Also, we can be absolutely certain that *A* either does or does not occur, which implies that either *A* or $\overline{A}$ must occur. These observations let us apply the addition rule for disjoint events as follows:

$$P(A \text{ or } \overline{A}) = P(A) + P(\overline{A}) = 1$$

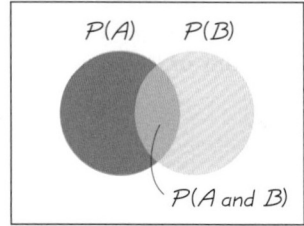

Total Area = 1

P(A)   P(B)

P(A and B)

Figure 4-3 Venn Diagram for Events That Are Not Disjoint

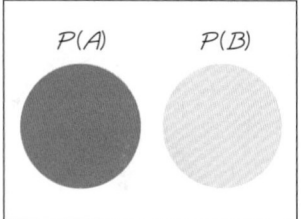

Total Area = 1

P(A)   P(B)

Figure 4-4 Venn Diagram for Disjoint Events

Total Area = 1

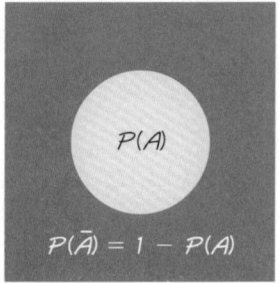

$P(A)$

$P(\overline{A}) = 1 - P(A)$

Figure 4-5   Venn Diagram for the Complement of Event *A*

We justify $P(A \text{ or } \overline{A}) = P(A) + P(\overline{A})$ by noting that A and $\overline{A}$ are disjoint; we justify the total of 1 by our certainty that A either does or does not occur. This result of the addition rule leads to the following three equivalent expressions.

Rule of Complementary Events

$$P(A) + P(\overline{A}) = 1$$
$$P(\overline{A}) = 1 - P(A)$$
$$P(A) = 1 - P(\overline{A})$$

Figure 4-5 visually displays the relationship between $P(A)$ and $P(\overline{A})$.

EXAMPLE 3   FBI data show that 62.4% of murders are cleared by arrests. We can express the probability of a murder being cleared by an arrest as $P(\text{cleared}) = 0.624$. For a randomly selected murder, find $P(\overline{\text{cleared}})$.

SOLUTION   Using the rule of complementary events, we get

$$P(\overline{\text{cleared}}) = 1 - P(\text{cleared}) = 1 - 0.624 = 0.376$$

That is, the probability of a randomly selected murder case *not* being cleared by an arrest is 0.376.

A major advantage of the *rule of complementary events* is that it simplifies certain problems, as we illustrate in Section 4-5.

4-3   Basic Skills and Concepts

Statistical Literacy and Critical Thinking

1. Disjoint Events A single trial of some procedure is conducted and the resulting events are analyzed. In your own words, describe what it means for two events in a single trial to be *disjoint*.

2. Disjoint Events and Complements When considering events resulting from a single trial, if one event is the complement of another event, must those two events be disjoint? Why or why not?

3. Notation Using the context of the addition rule presented in this section and using your own words, describe what $P(A \text{ and } B)$ denotes.

4. Addition Rule When analyzing results from a test of the Microsort gender selection technique developed by the Genetics IVF Institute, a researcher wants to compare the results to those obtained from a coin toss. Consider $P(G \text{ or } H)$, which is the probability of getting a baby girl *or* getting heads from a coin toss. Explain why the addition rule does *not* apply to $P(G \text{ or } H)$.

Determining Whether Events Are Disjoint. *For Exercises 5–12, determine whether the two events are disjoint for a single trial.* **Hint: (***Consider "disjoint" to be equivalent to "separate" or "not overlapping."***)**

5. Randomly selecting a physician at Bellevue Hospital in New York City and getting a surgeon

Randomly selecting a physician at Bellevue Hospital in New York City and getting a female

6. Conducting a Pew Research Center poll and randomly selecting a subject who is a Republican

Conducting a Pew Research Center poll and randomly selecting a subject who is a Democrat

7. Randomly selecting a Corvette from the Chevrolet assembly line and getting one that is free of defects

Randomly selecting a Corvette from the Chevrolet assembly line and getting one with a dead battery

8. Randomly selecting a fruit fly with red eyes

Randomly selecting a fruit fly with sepian (dark brown) eyes

9. Receiving a phone call from a volunteer survey subject who believes that there is solid evidence of global warming

Receiving a phone call from a volunteer survey subject who is opposed to stem cell research

10. Randomly selecting someone treated with the cholesterol-reducing drug Lipitor

Randomly selecting someone in a control group given no medication

11. Randomly selecting a movie with a rating of R

Randomly selecting a movie with a rating of four stars

12. Randomly selecting a college graduate

Randomly selecting someone who is homeless

Finding Complements. *In Exercises 13–16, find the indicated complements.*

13. STATDISK Survey Based on a recent survey of STATDISK users, it is found that $P(M) = 0.05$, where M is the event of getting a Macintosh user when a STATDISK user is randomly selected. If a STATDISK user is randomly selected, what does $P(\overline{M})$ signify? What is its value?

14. Colorblindness Women have a 0.25% rate of red/green color blindness. If a woman is randomly selected, what is the *probability* that she does *not* have red/green color blindness? (*Hint:* The decimal equivalent of 0.25% is 0.0025, not 0.25.)

15. Pew Poll A Pew Research Center poll showed that 79% of Americans believe that it is morally wrong to not report all income on tax returns. What is the probability that an American does not have that belief?

16. Sobriety Checkpoint When the author observed a sobriety checkpoint conducted by the Dutchess County Sheriff Department, he saw that 676 drivers were screened and 6 were arrested for driving while intoxicated. Based on those results, we can estimate that $P(I) = 0.00888$, where I denotes the event of screening a driver and getting someone who is intoxicated. What does $P(\overline{I})$ denote and what is its value?

In Exercises 17–20, use the polygraph test data given in Table 4-1, which is included with the Chapter Problem.

17. Polygraph Test If one of the test subjects is randomly selected, find the probability that the subject had a positive test result or did not lie.

18. Polygraph Test If one of the test subjects is randomly selected, find the probability that the subject did not lie.

19. Polygraph Test If one of the subjects is randomly selected, find the probability that the subject had a true negative test result.

20. Polygraph Test If one of the subjects is randomly selected, find the probability that the subject had a negative test result or lied.

In Exercises 21–26, use the data in the accompanying table, which summarizes challenges by tennis players (based on data reported in USA Today*). The results are from the first U.S. Open that used the Hawk-Eye electronic system for displaying an instant replay used to determine whether the ball is in bounds or out of bounds. In each case, assume that one of the challenges is randomly selected.*

	Was the challenge to the call successful?	
	Yes	No
Men	201	288
Women	126	224

For Exercises 21–26, see the instructions and table on the preceding page.

21. Tennis Instant Replay If S denotes the event of selecting a successful challenge, find $P(\overline{S})$.

22. Tennis Instant Replay If M denotes the event of selecting a challenge made by a man, find $P(\overline{M})$.

23. Tennis Instant Replay Find the probability that the selected challenge was made by a man or was successful.

24. Tennis Instant Replay Find the probability that the selected challenge was made by a woman or was successful.

25. Tennis Instant Replay Find P(challenge was made by a man or was not successful).

26. Tennis Instant Replay Find P(challenge was made by a woman or was not successful).

In Exercises 27–32, refer to the following table summarizing results from a study of people who refused to answer survey questions (based on data from "I Hear You Knocking but You Can't Come In," by Fitzgerald and Fuller, Sociological Methods and Research, Vol. 11, No. 1). In each case, assume that one of the subjects is randomly selected.

	Age					
	18–21	22–29	30–39	40–49	50–59	60 and over
Responded	73	255	245	136	138	202
Refused	11	20	33	16	27	49

27. Survey Refusals What is the probability that the selected person refused to answer? Does that probability value suggest that refusals are a problem for pollsters? Why or why not?

28. Survey Refusals A pharmaceutical company is interested in opinions of the elderly, because they are either receiving Medicare or will receive it soon. What is the probability that the selected subject is someone 60 and over who responded?

29. Survey Refusals What is the probability that the selected person responded or is in the 18–21 age bracket?

30. Survey Refusals What is the probability that the selected person refused to respond or is over 59 years of age?

31. Survey Refusals A market researcher is interested in responses, especially from those between the ages of 22 and 39, because they are the people more likely to make purchases. Find the probability that a selected subject responds or is aged between the ages of 22 and 39.

32. Survey Refusals A market researcher is not interested in refusals or subjects below 22 years of age or over 59. Find the probability that the selected person refused to answer or is below 22 or is older than 59.

In Exercises 33–38, use these results from the "1-Panel-THC" test for marijuana use, which is provided by the company Drug Test Success: Among 143 subjects with positive test results, there are 24 false positive results; among 157 negative results, there are 3 false negative results. (Hint: Construct a table similar to Table 4-1, which is included with the Chapter Problem.)

33. Screening for Marijuana Use

a. How many subjects are included in the study?

b. How many subjects did not use marijuana?

c. What is the probability that a randomly selected subject did not use marijuana?

34. Screening for Marijuana Use If one of the test subjects is randomly selected, find the probability that the subject tested positive or used marijuana.

35. Screening for Marijuana Use If one of the test subjects is randomly selected, find the probability that the subject tested negative or did not use marijuana.

36. Screening for Marijuana Use If one of the test subjects is randomly selected, find the probability that the subject actually used marijuana. Do you think that the result reflects the marijuana use rate in the general population?

37. Screening for Marijuana Use Find the probability of a false positive or false negative. What does the result suggest about the test's accuracy?

38. Screening for Marijuana Use Find the probability of a correct result by finding the probability of a true positive or a true negative. How does this result relate to the result from Exercise 37?

4-3 Beyond the Basics

39. Gender Selection Find $P(G \text{ or } H)$ in Exercise 4, assuming that boys and girls are equally likely.

40. Disjoint Events If events A and B are disjoint and events B and C are disjoint, must events A and C be disjoint? Give an example supporting your answer.

41. Exclusive Or The formal addition rule expressed the probability of A or B as follows: $P(A \text{ or } B) = P(A) + P(B) - P(A \text{ and } B)$. Rewrite the expression for $P(A \text{ or } B)$ assuming that the addition rule uses the *exclusive or* instead of the *inclusive or*. (Recall that the *exclusive or* means either one or the other but not both.)

42. Extending the Addition Rule Extend the formal addition rule to develop an expression for $P(A \text{ or } B \text{ or } C)$. (*Hint:* Draw a Venn diagram.)

43. Complements and the Addition Rule

a. Develop a formula for the probability of not getting either A or B on a single trial. That is, find an expression for $P(\overline{A \text{ or } B})$.

b. Develop a formula for the probability of not getting A or not getting B on a single trial. That is, find an expression for $P(\overline{A} \text{ or } \overline{B})$.

c. Compare the results from parts (a) and (b). Does $P(\overline{A \text{ or } B}) = P(\overline{A} \text{ or } \overline{B})$?

4-4 Multiplication Rule: Basics

Key Concept In Section 4-3 we presented the addition rule for finding $P(A \text{ or } B)$, the probability that a single trial has an outcome of A or B or both. In this section we present the basic multiplication rule, which is used for finding $P(A \text{ and } B)$, the probability that event A occurs in a first trial and event B occurs in a second trial. If the outcome of the first event A somehow affects the probability of the second event B, it is important to adjust the probability of B to reflect the occurrence of event A. The rule for finding $P(A \text{ and } B)$ is called the multiplication rule because it involves the multiplication of the probability of event A and the probability of event B (where, if necessary, the probability of event B is adjusted because of the outcome of event A). In Section 4-3 we associated use of the word "or" with addition. In this section we associate use of the word "and" with multiplication.

Notation

$P(A \text{ and } B) = P(\text{event } A \text{ occurs in a first trial and event } B \text{ occurs in a second trial})$

To illustrate the multiplication rule, let's consider the following example involving test questions used extensively in the analysis and design of standardized tests, such as the SAT, ACT, MCAT (for medicine), and LSAT (for law). For ease of grading, standard tests typically use true/false or multiple-choice questions. Consider a quick quiz in which the first question is a true/false type, while the second question is

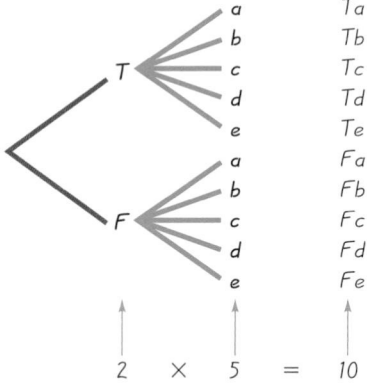

Figure 4-6   Tree Diagram of Test Answers

a multiple-choice type with five possible answers (a, b, c, d, e). We will use the following two questions. Try them!

1. True or false: A pound of feathers is heavier than a pound of gold.

2. Who said that "smoking is one of the leading causes of statistics"?

   a. Philip Morris

   b. Smokey Robinson

   c. Fletcher Knebel

   d. R. J. Reynolds

   e. Virginia Slims

The answers to the two questions are T (for "true") and c. (The first answer is true. Weights of feathers are given in Avoirdupois units, but weights of gold and other precious metals are given in Troy units. An Avoirdupois pound is 453.59 g, which is greater than the 373.24 g in a Troy pound. The second answer is Fletcher Knebel, who was a political columnist and author of books, including *Seven Days in May.*)

One way to find the probability that if someone makes random guesses for both answers, the first answer will be correct *and* the second answer will be correct, is to list the sample space as follows:

$$T,a \quad T,b \quad T,c \quad T,d \quad T,e$$

$$F,a \quad F,b \quad F,c \quad F,d \quad F,e$$

If the answers are random guesses, then the above 10 possible outcomes are equally likely, so

$$P(\text{both correct}) = P(T \text{ and } c) = \frac{1}{10} = 0.1$$

Now note that $P(T \text{ and } c) = 1/10$, $P(T) = 1/2$, and $P(c) = 1/5$, from which we see that

$$\frac{1}{10} = \frac{1}{2} \cdot \frac{1}{5}$$

so that

$$P(T \text{ and } c) = P(T) \times P(c)$$

This suggests that, in general, $P(A \text{ and } B) = P(A) \cdot P(B)$, but let's consider another example before accepting that generalization.

A tree diagram is a picture of the possible outcomes of a procedure, shown as line segments emanating from one starting point. These diagrams are sometimes helpful in determining the number of possible outcomes in a sample space, if the number of possibilities is not too large. The tree diagram shown in Figure 4-6 summarizes the outcomes of the true/false and multiple-choice questions. From Figure 4-6 we see that if both answers are random guesses, all 10 branches are equally likely and the probability of getting the correct pair (T,c) is 1/10. For each response to the first question, there are 5 responses to the second. *The total number of outcomes is 5 taken 2 times, or 10.* The tree diagram in Figure 4-6 therefore provides a visual illustration for using multiplication.

The preceding discussion of the true/false and multiple-choice questions suggests that $P(A \text{ and } B) = P(A) \cdot P(B)$, but Example 1 shows another critical element that should be considered.

EXAMPLE 1 **Polygraph Test** If two of the subjects included in Table 4-1 are randomly selected *without replacement*, find the probability that the first selected person had a positive test result and the second selected person had a negative test result.

Table 4-1   Results from Experiments with Polygraph Instruments

	Did the Subject Actually Lie?	
	No (Did Not Lie)	Yes (Lied)
Positive test result	15	42
(Polygraph test indicated that the subject *lied*.)	(false positive)	(true positive)
Negative test result	32	9
(Polygraph test indicated that the subject did *not lie*.)	(true negative)	(false negative)

SOLUTION First selection:

$$P(\text{positive test result}) = \frac{57}{98}$$

(because there are 57 subjects who tested positive, and the total number of subjects is 98).
Second selection:

$$P(\text{negative test result}) = \frac{41}{97}$$

(after the first selection of a subject with a positive test result, there are 97 subjects remaining, 41 of whom had negative test results).

With $P(\text{first subject has positive test result}) = 57/98$ and $P(\text{second subject has negative test result}) = 41/97$ we have

$$P\left(\begin{array}{l}\text{1st subject has positive test result}\\ \text{and 2nd subject has negative result}\end{array}\right) = \frac{57}{98} \cdot \frac{41}{97} = 0.246$$

The key point is this: *We must adjust the probability of the second event to reflect the outcome of the first event.* Because selection of the second subject is made *without* replacement of the first subject, the second probability must take into account the fact that the first selection removed a subject who tested positive, so only 97 subjects are available for the second selection, and 41 of them had a negative test result.

Example 1 illustrates the important principle that *the probability for the second event B should take into account the fact that the first event A has already occurred.* This principle is often expressed using the following notation.

Notation for Conditional Probability

$P(B|A)$ represents the probability of event B occurring after it is assumed that event A has already occurred. (We can read $B|A$ as "B given A" or as "event B occurring after event A has already occurred.")

For example, playing the California lottery and then playing the New York lottery are *independent* events because the result of the California lottery has absolutely no effect on the probabilities of the outcomes of the New York lottery. In contrast, the event of having your car start and the event of getting to your statistics class on time are *dependent* events, because the outcome of trying to start your car does affect the probability of getting to the statistics class on time.

> ### DEFINITION
>
> Two events A and B are **independent** if the occurrence of one does not affect the *probability* of the occurrence of the other. (Several events are similarly independent if the occurrence of any does not affect the probabilities of the occurrence of the others.) If A and B are not independent, they are said to be **dependent.**

Two events are dependent if the occurrence of one of them affects the *probability* of the occurrence of the other, but this does not necessarily mean that one of the events is a *cause* of the other. See Exercise 9.

Using the preceding notation and definitions, along with the principles illustrated in the preceding examples, we can summarize the key concept of this section as the following *formal multiplication rule*, but it is recommended that you work with the *intuitive multiplication rule*, which is more likely to reflect *understanding* instead of blind use of a formula.

Formal Multiplication Rule

$$P(A \text{ and } B) = P(A) \cdot P(B|A)$$

If A and B are independent events, $P(B|A)$ is the same as $P(B)$. See the following *intuitive multiplication rule*. (Also see Figure 4-7.)

Intuitive Multiplication Rule

When finding the probability that event A occurs in one trial and event B occurs in the next trial, multiply the probability of event A by the probability of event B, but be sure that the probability of event B takes into account the previous occurrence of event A.

Figure 4-7

Applying the Multiplication Rule

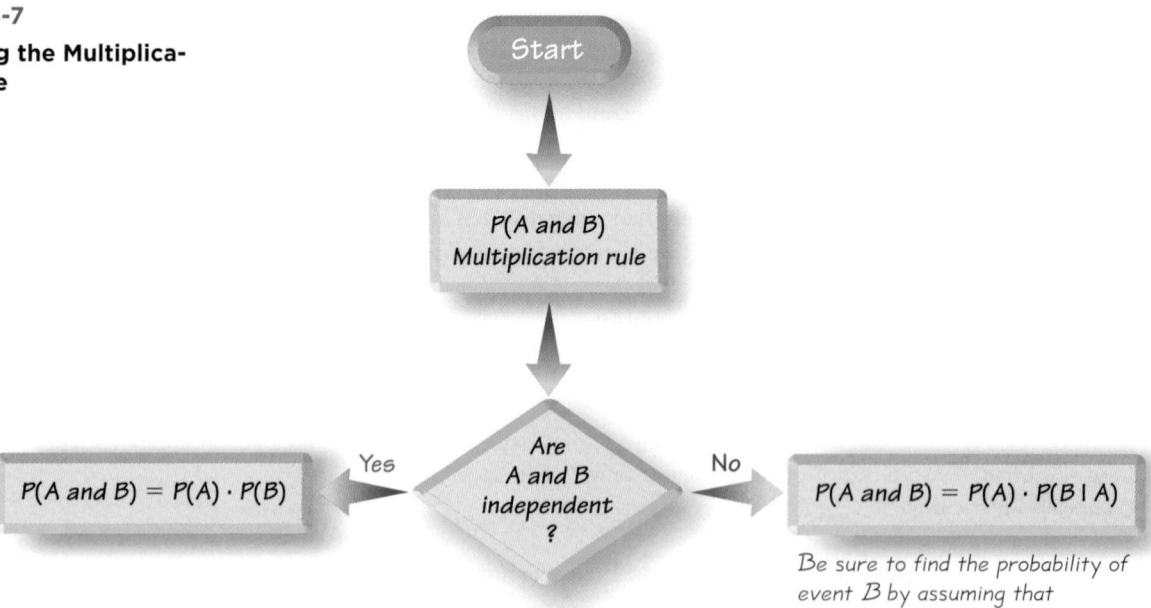

Start

$P(A \text{ and } B)$
Multiplication rule

Are A and B independent ?

Yes → $P(A \text{ and } B) = P(A) \cdot P(B)$

No → $P(A \text{ and } B) = P(A) \cdot P(B \mid A)$

Be sure to find the probability of event B by assuming that event A has already occurred.

CAUTION

··

When applying the multiplication rule, always consider whether the events are independent or dependent, and adjust the calculations accordingly.

In Example 2, we consider two situations: (1) The items are selected *with* replacement; (2) the items are selected *without* replacement. If items are selected with replacement, each selection begins with exactly the same collection of items, but if items are selected without replacement, the collection of items changes after each selection, and we must take those changes into account.

EXAMPLE 2 **Quality Control in Manufacturing** Pacemakers are implanted in patients for the purpose of stimulating pulse rate when the heart cannot do it alone. Each year, there are more than 250,000 pacemakers implanted in the United States. Unfortunately, pacemakers sometimes fail, but the failure rate is low, such as 0.0014 per year (based on data from "Pacemaker and ICD Generator Malfunctions," by Maisel, et al., *Journal of the American Medical Association*, Vol. 295, No. 16). We will consider a small sample of five pacemakers, including three that are good (denoted here by G) and two that are defective (denoted here by D). A medical researcher wants to randomly select two of the pacemakers for further experimentation. Find the probability that the first selected pacemaker is good (G) and the second pacemaker is also good (G). Use each of the following assumptions.

a. Assume that the two random selections are made *with replacement,* so that the first selected pacemaker is replaced before the second selection is made.

b. Assume that the two random selections are made *without replacement,* so that the first selected pacemaker is *not* replaced before the second selection is made.

SOLUTION Before proceeding, it would be helpful to visualize the three good pacemakers and the two defective pacemakers in a way that provides us with greater clarity, as shown below.

$$G \quad G \quad G \quad D \quad D$$

a. If the two pacemakers are randomly selected *with replacement*, the two selections are independent because the second event is not affected by the first outcome. In each of the two selections there are three good (G) pacemakers and two that are defective (D), so we get

$$P(\text{first pacemaker is G and second pacemaker is G}) = \frac{3}{5} \cdot \frac{3}{5} = \frac{9}{25} \text{ or } 0.36$$

b. If the two pacemakers are randomly selected *without replacement*, the two selections are dependent because the probability of the second event is affected by the first outcome. In the first selection, three of the five pacemakers are good (G). After selecting a good pacemaker on the first selection, we are left with four pacemakers including two that are good. We therefore get

$$P(\text{first pacemaker is G and second pacemaker is G}) = \frac{3}{5} \cdot \frac{2}{4} = \frac{6}{20} \text{ or } 0.3$$

continued

Convicted by Probability

A witness described a Los Angeles robber as a Caucasian woman with blond hair in a ponytail who escaped in a yellow car driven by an African-American male with a mustache and beard. Janet and Malcolm Collins fit this description, and they were convicted based on testimony that there is only about 1 chance in 12 million that any couple would have these characteristics. It was estimated that the probability of a yellow car is 1/10, and the other probabilities were estimated to be 1/10, 1/3, 1/10, and 1/1000. The convictions were later overturned when it was noted that no evidence was presented to support the estimated probabilities or the independence of the events. However, because the couple was not randomly selected, a serious error was made in not considering the probability of *other* couples being in the same region with the same characteristics.

INTERPRETATION   Note that in part (b) we adjust the second probability to take into account the selection of a good pacemaker (G) in the first outcome. After selecting G the first time, there would be two Gs among the four pacemakers that remain.

When considering whether to sample with replacement or without replacement, it might seem obvious that a medical researcher would not sample with replacement, as in part (a). However, in statistics we have a special interest in sampling with replacement. (See Section 6-4.)

So far we have discussed two events, but the multiplication rule can be easily extended to several events. In general, the probability of any sequence of independent events is simply the product of their corresponding probabilities. For example, the probability of tossing a coin three times and getting all heads is $0.5 \cdot 0.5 \cdot 0.5 = 0.125$. We can also extend the multiplication rule so that it applies to several dependent events; simply adjust the probabilities as you go along.

Treating Dependent Events as Independent   Part (b) of Example 2 involved selecting items without replacement, and we therefore treated the events as being dependent. However, some calculations are cumbersome, but they can be made manageable by using the common practice of treating events as independent when *small samples* are drawn from *large populations*. In such cases, it is rare to select the same item twice. Here is a common guideline routinely used with applications such as analyses of poll results.

> **Treating Dependent Events as Independent: The 5% Guideline for Cumbersome Calculations**
>
> If calculations are very cumbersome and if a sample size is no more than 5% of the size of the population, treat the selections as being *independent* (even if the selections are made without replacement, so they are technically dependent).

EXAMPLE 3   **Quality Control in Manufacturing** Assume that we have a batch of 100,000 heart pacemakers, including 99,950 that are good (G) and 50 that are defective (D).

a. If two of those 100,000 pacemakers are randomly selected without replacement, find the probability that they are both good.

b. If 20 of those 100,000 pacemakers are randomly selected without replacement, find the probability that they are all good.

SOLUTION   First, note that 5% of 100,000 is $(0.05)(100,000) = 5000$.

a. Even though the sample size of two is no more than 5% of the size of the population of 100,000, we will not use the 5% guideline because the exact calculation is quite easy, as shown on the following page.

P(first pacemaker is good and the second pacemaker is good)

$$= \frac{99,950}{100,000} \cdot \frac{99,949}{99,999} = 0.999$$

b. With 20 pacemakers randomly selected without replacement, the exact calculation becomes quite cumbersome:

$$P(\text{all 20 pacemakers are good}) = \frac{99,950}{100,000} \cdot \frac{99,949}{99,999} \cdot \frac{99,948}{99,998} \cdots \frac{99,931}{99,981}$$

(20 *different* factors)

Because this calculation is extremely cumbersome, we use the 5% guideline by treating the events as independent, even though they are actually dependent. Note that the sample size of 20 is no more than 5% of the population of 100,000, as required. Treating the events as independent, we get the following result, which is easy to calculate.

$$P(\text{all 20 pacemakers are good}) = \frac{99,950}{100,000} \cdot \frac{99,950}{100,000} \cdot \frac{99,950}{100,000} \cdots \frac{99,950}{100,000}$$

$$= \left(\frac{99,950}{100,000}\right)^{20} = 0.990 \quad (20 \; \textit{identical} \; \text{factors})$$

Because the result is rounded to three decimal places, in this case we get the same result that would be obtained by performing the more cumbersome exact calculation with dependent events.

The following example is designed to illustrate the importance of carefully identifying the event being considered. Note that parts (a) and (b) appear to be quite similar, but their solutions are very different.

> **EXAMPLE 4** **Birthdays** Assume that two people are randomly selected and also assume that birthdays occur on the days of the week with equal frequencies.
> **a.** Find the probability that the two people are born on the same day of the week.
> **b.** Find the probability that the two people are both born on Monday.

SOLUTION

a. Because no particular day of the week is specified, the first person can be born on any one of the seven week days. The probability that the second person is born on the same day as the first person is 1/7. The probability that two people are born on the same day of the week is therefore 1/7.

b. The probability that the first person is born on Monday is 1/7 and the probability that the second person is also born on Monday is 1/7. Because the two events are independent, the probability that both people are born on Monday is

$$\frac{1}{7} \cdot \frac{1}{7} = \frac{1}{49}$$

Independent Jet Engines

Soon after departing from Miami, Eastern Airlines Flight 855 had one engine shut down because of a low oil pressure warning light.

As the L-1011 jet turned to Miami for landing, the low pressure warning lights for the other two engines also flashed. Then an engine failed, followed by the failure of the last working engine. The jet descended without power from 13,000 ft to 4000 ft when the crew was able to restart one engine, and the 172 people on board landed safely. With independent jet engines, the probability of all three failing is only 0.0001^3, or about one chance in a trillion. The FAA found that the same mechanic who replaced the oil in all three engines failed to replace the oil plug sealing rings. The use of a single mechanic caused the operation of the engines to become dependent, a situation corrected by requiring that the engines be serviced by different mechanics.

Important Applications of the Multiplication Rule

The following two examples illustrate practical applications of the multiplication rule. Example 5 gives us some insight into *hypothesis testing* (which is introduced in Chapter 8), and Example 6 illustrates the principle of *redundancy*, which is used to increase the reliability of many mechanical and electrical systems.

EXAMPLE 5 **Effectiveness of Gender Selection** A geneticist developed a procedure for increasing the likelihood of female babies. In an initial test, 20 couples use the method and the results consist of 20 females among 20 babies. Assuming that the gender-selection procedure has no effect, find the probability of getting 20 females among 20 babies by chance. Does the resulting probability provide strong evidence to support the geneticist's claim that the procedure is effective in increasing the likelihood that babies will be females?

SOLUTION We want to find P(all 20 babies are female) with the assumption that the procedure has no effect, so that the probability of any individual offspring being a female is 0.5. Because separate pairs of parents were used, we will treat the events as if they are independent. We get this result:

P (all 20 offspring are female)

$= P$(1st is female and 2nd is female and 3rd is female $\cdots$ and 20th is female)

$= P(\text{female}) \cdot P(\text{female}) \cdot \cdots \cdot P(\text{female})$

$= 0.5 \cdot 0.5 \cdot \cdots \cdot 0.5$

$= 0.5^{20} = 0.000000954$

The low probability of 0.000000954 indicates that instead of getting 20 females by chance, a more reasonable explanation is that females appear to be more likely with the gender-selection procedure. Because there is such a small probability (0.000000954) of getting 20 females in 20 births, we do have strong evidence to support the geneticist's claim that the gender-selection procedure is effective in increasing the likelihood that babies will be female.

Calculator Hint: The preceding example required evaluation of 0.5 multiplied by itself 20 times, which is 0.5^{20}. Many calculators have a "power" key labeled x^y, but the TI-83/84 Plus calculator uses ⌃ for the power key. To evaluate 0.5^{20}, enter 0.5 ⌃ 20 **ENTER** to get a result of 0.9536743164**E** 7, which is 0.0000009536743164 when expressed in ordinary notation. In the preceding example, this result was rounded to three significant digits as 0.000000954.

EXAMPLE 6 **Redundancy for Increased Reliability** Modern aircraft engines are now highly reliable. One design feature contributing to that reliability is the use of *redundancy*, whereby critical components are duplicated so that if one fails, the other will work. For example, single-engine aircraft now have two independent electrical systems so that if one electrical system fails, the other can continue to work so that the engine does not fail. For the purposes of this example, we will assume that the probability of an electrical system failure is 0.001.

  a. If the engine in an aircraft has one electrical system, what is the probability that
  it will work?

  b. If the engine in an aircraft has two independent electrical systems, what is
  the probability that the engine can function with a working electrical system?

SOLUTION

a. If the probability of an electrical system failure is 0.001, the probability that it
does *not* fail is 0.999. That is, the probability that the engine can function with a
working electrical system is as follows:

$$P(\text{working electrical system}) = P(\text{electrical system does not fail})$$
$$= 1 - P(\text{electrical system failure}) = 1 - 0.001 = 0.999$$

b. With two independent electrical systems, the engine will function unless *both*
electrical systems fail. The probability that the two independent electrical systems both fail is found by applying the multiplication rule for independent
events as follows.

$$P(\text{both electrical systems fail})$$
$$= P(\text{first electrical system fails } and \text{ the second electrical system fails})$$
$$= 0.001 \times 0.001 = 0.000001$$

There is a 0.000001 probability of both electrical systems failing, so the probability
that the engine can function with a working electrical system is $1 - 0.000001 = 0.999999$.

INTERPRETATION With only one electrical system we can see that there is a
0.001 probability of failure, but with two independent electrical systems, there is
only a 0.000001 probability that the engine will not be able to function with a
working electrical system. With two electrical systems, the chance of a catastrophic
failure drops from 1 in 1000 to 1 in 1,000,000, resulting in a dramatic increase in
safety and reliability. (*Note:* For the purposes of this exercise, we assumed that the
probability of failure of an electrical system is 0.001, but it is actually much lower.
Arjen Romeyn, a transportation safety expert, estimates that the probability of a single engine failure is around 0.0000001 or 0.000000001.)

We can summarize the addition and multiplication rules as follows:

• $P(A \text{ or } B)$: The word "or" suggests addition, and when adding $P(A)$ and $P(B)$,
we must be careful to add in such a way that every outcome is counted only once.

• $P(A \text{ and } B)$: The word "and" suggests multiplication, and when multiplying
$P(A)$ and $P(B)$, we must be careful to be sure that the probability of event B
takes into account the previous occurrence of event A.

4-4 Basic Skills and Concepts

Statistical Literacy and Critical Thinking

1. Independent Events Create your own example of two events that are independent, and
create another example of two other events that are dependent. Do not use examples given in
this section.

2. Notation In your own words, describe what the notation $P(B|A)$ represents.

3. Sample for a Poll There are currently 477,938 adults in Alaska, and they are all included in one big numbered list. The Gallup Organization uses a computer to randomly select 1068 different numbers between 1 and 477,938, and then contacts the corresponding adults for a poll. Are the events of selecting the adults actually independent or dependent? Explain.

4. 5% Guideline Can the events described in Exercise 3 be treated as independent? Explain.

Identifying Events as Independent or Dependent. *In Exercises 5–12, for each given pair of events, classify the two events as* independent *or* dependent. *(If two events are technically dependent but can be treated as if they are independent according to the 5% guideline, consider them to be independent.)*

5. Randomly selecting a TV viewer who is watching *Saturday Night Live*

Randomly selecting a second TV viewer who is watching *Saturday Night Live*

6. Finding that your car radio works

Finding that your car headlights work

7. Wearing plaid shorts with black socks and sandals

Asking someone on a date and getting a positive response

8. Finding that your cell phone works

Finding that your car starts

9. Finding that your television works

Finding that your refrigerator works

10. Finding that your calculator works

Finding that your computer works

11. Randomly selecting a consumer from California

Randomly selecting a consumer who owns a television

12. Randomly selecting a consumer who owns a computer

Randomly selecting a consumer who uses the Internet

Polygraph Test. *In Exercises 13–16, use the sample data in Table 4-1. (See Example 1.)*

13. Polygraph Test If 2 of the 98 test subjects are randomly selected without replacement, find the probability that they both had false positive results. Is it unusual to randomly select 2 subjects without replacement and get 2 results that are both false positive results? Explain.

14. Polygraph Test If 3 of the 98 test subjects are randomly selected without replacement, find the probability that they all had false positive results. Is it unusual to randomly select 3 subjects without replacement and get 3 results that are all false positive results? Explain.

15. Polygraph Test If four of the test subjects are randomly selected without replacement, find the probability that, in each case, the polygraph indicated that the subject lied. Is such an event unusual?

16. Polygraph Test If four of the test subjects are randomly selected without replacement, find the probability that they all had incorrect test results (either false positive or false negative). Is such an event likely?

In Exercises 17–20, use the data in the following table, which summarizes blood groups and Rh types for 100 subjects. These values may vary in different regions according to the ethnicity of the population.

		O	A	B	AB
Type	**Rh⁺**	39	35	8	4
	Rh⁻	6	5	2	1

17. Blood Groups and Types If 2 of the 100 subjects are randomly selected, find the probability that they are both group O and type Rh^+.

a. Assume that the selections are made with replacement.

b. Assume that the selections are made without replacement.

18. Blood Groups and Types If 3 of the 100 subjects are randomly selected, find the probability that they are all group B and type Rh^-.

a. Assume that the selections are made with replacement.

b. Assume that the selections are made without replacement.

19. Universal Blood Donors People with blood that is group O and type Rh^- are considered to be universal donors, because they can give blood to anyone. If 4 of the 100 subjects are randomly selected, find the probability that they are all universal donors.

a. Assume that the selections are made with replacement.

b. Assume that the selections are made without replacement.

20. Universal Recipients People with blood that is group AB and type Rh^+ are considered to be universal recipients, because they can receive blood from anyone. If three of the 100 subjects are randomly selected, find the probability that they are all universal recipients.

a. Assume that the selections are made with replacement.

b. Assume that the selections are made without replacement.

21. Guessing A quick quiz consists of a true/false question followed by a multiple-choice question with four possible answers (a, b, c, d). An unprepared student makes random guesses for both answers.

a. Consider the event of being correct with the first guess and the event of being correct with the second guess. Are those two events independent?

b. What is the probability that both answers are correct?

c. Based on the results, does guessing appear to be a good strategy?

22. Acceptance Sampling With one method of a procedure called *acceptance sampling,* a sample of items is randomly selected without replacement and the entire batch is accepted if every item in the sample is okay. The Telektronics Company manufactured a batch of 400 backup power supply units for computers, and 8 of them are defective. If 3 of the units are randomly selected for testing, what is the probability that the entire batch will be accepted?

23. Poll Confidence Level It is common for public opinion polls to have a "confidence level" of 95%, meaning that there is a 0.95 probability that the poll results are accurate within the claimed margins of error. If each of the following organizations conducts an independent poll, find the probability that all of them are accurate within the claimed margins of error: Gallup, Roper, Yankelovich, Harris, CNN, ABC, CBS, NBC, *New York Times.* Does the result suggest that with a confidence level of 95%, we can expect that almost all polls will be within the claimed margin of error?

24. Voice Identification of Criminal In a case in Riverhead, New York, nine different crime victims listened to voice recordings of five different men. All nine victims identified the same voice as that of the criminal. If the voice identifications were made by random guesses, find the probability that all nine victims would select the same person. Does this constitute reasonable doubt?

25. Testing Effectiveness of Gender-Selection Method Recent developments appear to make it possible for couples to dramatically increase the likelihood that they will conceive a child with the gender of their choice. In a test of a gender-selection method, 3 couples try to have baby girls. If this gender-selection method has no effect, what is the probability that the 3 babies will be all girls? If there are actually 3 girls among 3 children, does this gender-selection method appear to be effective? Why or why not?

26. Testing Effectiveness of Gender Selection Repeat Exercise 25 for these results: Among 10 couples trying to have baby girls, there are 10 girls among the 10 children. If this gender-selection method has no effect, what is the probability that the 10 babies will be all girls? If there are actually 10 girls among 10 children, does this gender-selection method appear to be effective? Why or why not?

27. Redundancy The principle of redundancy is used when system reliability is improved through redundant or backup components. Assume that your alarm clock has a 0.9 probability of working on any given morning.

a. What is the probability that your alarm clock will *not* work on the morning of an important final exam?

b. If you have two such alarm clocks, what is the probability that they both fail on the morning of an important final exam?

c. With one alarm clock, you have a 0.9 probability of being awakened. What is the probability of being awakened if you use two alarm clocks?

d. Does a second alarm clock result in greatly improved reliability?

28. Redundancy The FAA requires that commercial aircraft used for flying in instrument conditions must have two independent radios instead of one. Assume that for a typical flight, the probability of a radio failure is 0.002. What is the probability that a particular flight will be threatened with the failure of both radios? Describe how the second independent radio increases safety in this case.

29. Defective Tires The Wheeling Tire Company produced a batch of 5000 tires that includes exactly 200 that are defective.

a. If 4 tires are randomly selected for installation on a car, what is the probability that they are all good?

b. If 100 tires are randomly selected for shipment to an outlet, what is the probability that they are all good? Should this outlet plan to deal with defective tires returned by consumers?

30. Car Ignition Systems A quality control analyst randomly selects 3 different car ignition systems from a manufacturing process that has just produced 200 systems, including 5 that are defective.

a. Does this selection process involve independent events?

b. What is the probability that all 3 ignition systems are good? (Do not treat the events as independent.)

c. Use the 5% guideline for treating the events as independent, and find the probability that all 3 ignition systems are good.

d. Which answer is better: The answer from part (b) or the answer from part (c)? Why?

4-4   Beyond the Basics

31. System Reliability Refer to the accompanying figure in which surge protectors p and q are used to protect an expensive high-definition television. If there is a surge in the voltage, the surge protector reduces it to a safe level. Assume that each surge protector has a 0.99 probability of working correctly when a voltage surge occurs.

a. If the two surge protectors are arranged in series, what is the probability that a voltage surge will not damage the television? (Do not round the answer.)

b. If the two surge protectors are arranged in parallel, what is the probability that a voltage surge will not damage the television? (Do not round the answer.)

c. Which arrangement should be used for the better protection?

Series Configuration   Parallel Configuration

32. Same Birthdays If 25 people are randomly selected, find the probability that no two of them have the same birthday. Ignore leap years.

33. Drawing Cards Two cards are to be randomly selected without replacement from a shuffled deck. Find the probability of getting an ace on the first card and a spade on the second card.

4-5   Multiplication Rule: Complements and Conditional Probability

Key Concept In Section 4-4 we introduced the basic multiplication rule. In this section we extend our use of the multiplication rule to the following two special applications:

1. **Probability of "at least one":** Find the probability that among several trials, we get *at least one* of some specified event.

2. **Conditional probability:** Find the probability of an event when we have additional information that some other event has already occurred.

We begin with situations in which we want to find the probability that among several trials, *at least one* will result in some specified outcome.

Complements: The Probability of "At Least One"

Let's suppose that we want to find the probability that among 3 children, there is "at least one" girl. In such cases, the meaning of the language must be clearly understood:

- "At least one" is equivalent to "one or more."

- The complement of getting at least one item of a particular type is that you get *no* items of that type. For example, not getting at least 1 girl among 3 children is equivalent to getting no girls (or 3 boys).

We can use the following procedure to find the probability of at least one of some event.

Find the probability of *at least one* of some event by using these steps:

1. **Use the symbol *A* to denote the event of getting *at least one*.**

2. **Let $\overline{A}$ represent the event of getting *none* of the items being considered.**

3. **Calculate the probability that *none* of the outcomes results in the event being considered.**

4. **Subtract the result from 1. That is, evaluate this expression:**

$$P(\text{at least one}) = 1 - P(\text{none}).$$

> **EXAMPLE 1** **Gender of Children** Find the probability of a couple having at least 1 girl among 3 children. Assume that boys and girls are equally likely and that the gender of a child is independent of any other child.

SOLUTION

Step 1: Use a symbol to represent the event desired. In this case, let $A = $ at least 1 of the 3 children is a girl.

continued

Princeton Closes Its ESP Lab

The Princeton Engineering Anomalies Research (PEAR) laboratory recently closed, after it had been in operation since 1975. The purpose of the lab was to conduct studies on extrasensory perception and telekinesis. In one of the lab's experiments, test subjects were asked to think high or think low, then a device would display a random number either above 100 or below 100. The researchers then used statistical methods to determine whether the results differed significantly from what would be expected by chance. The objective was to determine whether the test subjects could use their minds to somehow influence the behavior of the random-generating device.

Because the PEAR lab had been an embarrassment to many members of Princeton's community, they welcomed its closing. After being used for research for 28 years, and after using more than $10 million in funding, the PEAR lab failed to provide results compelling enough to convince anyone that ESP or telekinesis are real phenomena.

Coincidences?

John Adams and Thomas Jefferson (the second and third presidents) both died on July 4, 1826. President Lincoln was assassinated in

Ford's Theater; President Kennedy was assassinated in a Lincoln car made by the Ford Motor Company. Lincoln and Kennedy were both succeeded by vice presidents named Johnson. Fourteen years *before* the sinking of the *Titanic,* a novel described the sinking of the *Titan,* a ship that hit an iceberg; see Martin Gardner's *The Wreck of the Titanic Foretold?* Gardner states, "In most cases of startling coincidences, it is impossible to make even a rough estimate of their probability."

Step 2: Identify the event that is the complement of A.

$$\overline{A} = \textit{not} \text{ getting at least 1 girl among 3 children}$$
$$= \text{all 3 children are boys}$$
$$= \text{boy and boy and boy}$$

Step 3: Find the probability of the complement.

$$P(\overline{A}) = P(\text{boy and boy and boy})$$
$$= \frac{1}{2} \cdot \frac{1}{2} \cdot \frac{1}{2} = \frac{1}{8}$$

Step 4: Find $P(A)$ by evaluating $1 - P(\overline{A})$.

$$P(A) = 1 - P(\overline{A}) = 1 - \frac{1}{8} = \frac{7}{8}$$

INTERPRETATION   There is a 7/8 probability that if a couple has 3 children, at least 1 of them is a girl.

EXAMPLE 2   **Defective Firestone Tires** Assume that the probability of a defective Firestone tire is 0.0003 (based on data from Westgard QC). If the retail outlet CarStuff buys 100 Firestone tires, find the probability that they get at least 1 that is defective. If that probability is high enough, plans must be made to handle defective tires returned by consumers. Should they make those plans?

SOLUTION

Step 1: Use a symbol to represent the event desired. In this case, let $A =$ at least 1 of the 100 tires is defective.

Step 2: Identify the event that is the complement of A.

$$\overline{A} = \textit{not} \text{ getting at least 1 defective tire among 100 tires}$$
$$= \text{all 100 tires are good}$$
$$= \text{good and good and . . . and good (100 times)}$$

Step 3: Find the probability of the complement.

$$P(\overline{A}) = 0.9997 \cdot 0.9997 \cdot 0.9997 \cdots \cdot 0.9997 \text{ (100 factors)}$$
$$= 0.9997^{100} = 0.9704$$

Step 4: Find $P(A)$ by evaluating $1 - P(\overline{A})$.

$$P(A) = 1 - P(\overline{A}) = 1 - 0.9704 = 0.0296$$

INTERPRETATION   There is a 0.0296 probability of at least 1 defective tire among the 100 tires. Because this probability is so low, it is not necessary to make plans for dealing with defective tires returned by consumers.

Conditional Probability

We now consider the second application, which is based on the principle that the probability of an event is often affected by knowledge of circumstances. For example, the probability of a golfer making a hole in one is 1/12,000 (based on past results), but if you learn that the golfer is a professional on tour, the probability is 1/2375 (based on data from *USA Today*). A *conditional probability* of an event is used when the probability is affected by the knowledge of other circumstances, such as the knowledge that a golfer is also a professional on tour.

> **DEFINITION**
>
> A **conditional probability** of an event is a probability obtained with the additional information that some other event has already occurred. $P(B|A)$ denotes the conditional probability of event B occurring, given that event A has already occurred. $P(B|A)$ can be found by dividing the probability of events A and B both occurring by the probability of event A:
>
> $$P(B|A) = \frac{P(A \text{ and } B)}{P(A)}$$

The preceding formula is a formal expression of conditional probability, but blind use of formulas is not recommended. Instead, we recommend the following intuitive approach:

Intuitive Approach to Conditional Probability

The conditional probability of B given A can be found by assuming that event A has occurred, and then calculating the probability that event B will occur.

EXAMPLE 3   **Polygraph Test** Refer to Table 4-1 to find the following:

a. If 1 of the 98 test subjects is randomly selected, find the probability that the subject had a positive test result, given that the subject actually lied. That is, find P(positive test result | subject lied).

b. If 1 of the 98 test subjects is randomly selected, find the probability that the subject actually lied, given that he or she had a positive test result. That is, find P(subject lied | positive test result).

Table 4-1   Results from Experiments with Polygraph Instruments

	Did the Subject Actually Lie?	
	No (Did Not Lie)	Yes (Lied)
Positive test result	15	42
(Polygraph test indicated that the subject *lied*.)	**(false positive)**	**(true positive)**
Negative test result	32	9
(Polygraph test indicated that the subject did *not* lie.)	**(true negative)**	**(false negative)**

SOLUTION

a. *Intuitive Approach to Conditional Probability:* We want P(positive test result | subject lied), the probability of getting someone with a positive test result, *given that the selected*   *continued*

Prosecutor's Fallacy

The *prosecutor's fallacy* is misunderstanding or confusion of two different conditional probabilities: (1) the probability that a defendant is innocent, given that forensic evidence shows a match; (2) the probability that forensics shows a match, given that a person is innocent. The prosecutor's fallacy has led to wrong convictions and imprisonment of some innocent people.

Lucia de Berk is a nurse who was convicted of murder and sentenced to prison in the Netherlands. Hospital administrators observed suspicious deaths that occurred in hospital wards where de Berk had been present. An expert testified that there was only 1 chance in 342 million that her presence was a coincidence. However, mathematician Richard Gill calculated the probability to be closer to 1/50, or possibly as low as 1/5. The court used the probability that the suspicious deaths could have occurred with de Berk present, given that she was innocent. The court should have considered the probability that de Berk is innocent, given that the suspicious deaths occurred when she was present. This error of the prosecutor's fallacy is subtle and can be very difficult to understand and recognize, yet it can lead to the imprisonment of innocent people. (See also the Chapter Problem for Chapter 11.)

subject lied. Here is the key point: If we assume that the selected subject actually lied, we are dealing only with the 51 subjects in the second column of Table 4-1. Among those 51 subjects, 42 had positive test results, so we get this result:

$$P(\text{positive test result} \mid \text{subject lied}) = \frac{42}{51} = 0.824$$

Using the Formula for Conditional Probability: The same result can be found by using the formula for $P(B \mid A)$ given the definition of conditional probability. We use the following notation.

$$P(B \mid A) = P(\text{positive test result} \mid \text{subject lied}) \quad \rightarrow \quad \begin{array}{l} B = \text{positive test result} \\ A = \text{subject lied} \end{array}$$

In the following calculation, we use $P(\text{subject lied and had a positive test result}) = 42/98$ and $P(\text{subject lied}) = 51/98$ to get the following results.

$$P(B \mid A) = \frac{P(A \text{ and } B)}{P(A)}$$

becomes

$$P(\text{positive test result} \mid \text{subject lied}) = \frac{P(\text{subject lied and had a positive test result})}{P(\text{subject lied})}$$

$$= \frac{42/98}{51/98} = 0.824$$

By comparing the intuitive approach to the use of the formula, it should be clear that the intuitive approach is much easier to use, and that it is also less likely to result in errors. The intuitive approach is based on an *understanding* of conditional probability, instead of manipulation of a formula, and understanding is so much better.

b. Here we want $P(\text{subject lied} \mid \text{positive test result})$. This is the probability that the selected subject lied, *given that the subject had a positive test result*. If we assume that the subject had a positive test result, we are dealing with the 57 subjects in the first row of Table 4-1. Among those 57 subjects, 42 lied, so

$$P(\text{subject lied} \mid \text{positive test result}) = \frac{42}{57} = 0.737$$

Again, the same result can be found by applying the formula for conditional probability, but we will leave that for those with a special fondness for manipulations with formulas.

INTERPRETATION   The first result of $P(\text{positive test result} \mid \text{subject lied}) = 0.824$ indicates that a subject who lies has a 0.824 probability of getting a positive test result. The second result of $P(\text{subject lied} \mid \text{positive test result}) = 0.737$ indicates that for a subject who gets a positive test result, there is a 0.737 probability that this subject actually lied.

Confusion of the Inverse

Note that in Example 3, $P(\text{positive test result} \mid \text{subject lied}) \neq P(\text{subject lied} \mid \text{positive test result})$. To incorrectly believe that $P(B \mid A)$ and $P(A \mid B)$ are the same, or to incorrectly use one value for the other, is often called *confusion of the inverse.*

> **EXAMPLE 4**   **Confusion of the Inverse** Consider the probability that it is dark outdoors, given that it is midnight: $P(\text{dark} \mid \text{midnight}) = 1$. (We conveniently ignore the Alaskan winter and other such anomalies.) But the probability that it is midnight, given that it is dark outdoors is almost zero. Because $P(\text{dark} \mid \text{midnight}) = 1$ but $P(\text{midnight} \mid \text{dark})$ is almost zero, we can clearly see that in this case, $P(B \mid A) \neq P(A \mid B)$. Confusion of the inverse occurs when we incorrectly switch those probability values.

Studies have shown that physicians often give very misleading information when they confuse the inverse. Based on real studies, they tended to confuse $P(\text{cancer} \mid \text{positive test result for cancer})$ with $P(\text{positive test result for cancer} \mid \text{cancer})$. About 95% of physicians estimated $P(\text{cancer} \mid \text{positive test result for cancer})$ to be about 10 times too high, with the result that patients were given diagnoses that were very misleading, and patients were unnecessarily distressed by the incorrect information.

4-5   Basic Skills and Concepts

Statistical Literacy and Critical Thinking

1. Interpreting "At Least One" You want to find the probability of getting at least 1 defect when 10 heart pacemakers are randomly selected and tested. What do you know about the exact number of defects if "at least one" of the 10 pacemakers is defective?

2. Notation Use your own words to describe the notation $P(B \mid A)$.

3. Finding Probability A medical researcher wants to find the probability that a heart patient will survive for one year. He reasons that there are two outcomes (survives, does not survive), so the probability is $1/2$. Is he correct? What important information is not included in his reasoning process?

4. Confusion of the Inverse What is confusion of the inverse?

Describing Complements. *In Exercises 5–8, provide a written description of the complement of the given event.*

5. Steroid Testing When the 15 players on the LA Lakers basketball team are tested for steroids, at least one of them tests positive.

6. Quality Control When six defibrillators are purchased by the New York University School of Medicine, all of them are free of defects.

7. X-Linked Disorder When four males are tested for a particular X-linked recessive gene, none of them are found to have the gene.

8. A Hit with the Misses When Brutus asks five different women for a date, at least one of them accepts.

9. Probability of At Least One Girl If a couple plans to have six children, what is the probability that they will have at least one girl? Is that probability high enough for the couple to be very confident that they will get at least one girl in six children?

10. Probability of At Least One Girl If a couple plans to have 8 children (it could happen), what is the probability that there will be at least one girl? If the couple eventually has 8 children and they are all boys, what can the couple conclude?

11. At Least One Correct Answer If you make guesses for four multiple-choice test questions (each with five possible answers), what is the probability of getting at least one correct? If a very lenient instructor says that passing the test occurs if there is at least one correct answer, can you reasonably expect to pass by guessing?

12. At Least One Working Calculator A statistics student plans to use a TI-84 Plus calculator on her final exam. From past experience, she estimates that there is a 0.96 probability that the calculator will work on any given day. Because the final exam is so important, she plans to use redundancy by bringing in two TI-84 Plus calculators. What is the probability that she will be able to complete her exam with a working calculator? Does she really gain much by bringing in the backup calculator? Explain.

13. Probability of a Girl Find the probability of a couple having a baby girl when their fourth child is born, given that the first three children were all girls. Is the result the same as the probability of getting four girls among four children?

14. Credit Risks The FICO (Fair Isaac & Company) score is commonly used as a credit rating. There is a 1% delinquency rate among consumers who have a FICO score above 800. If four consumers with FICO scores above 800 are randomly selected, find the probability that at least one of them becomes delinquent.

15. Car Crashes The probability of a randomly selected car crashing during a year is 0.0480 (based on data from the *Statistical Abstract of the United States*). If a family has four cars, find the probability that at least one of them has a car crash during the year. Is there any reason why the probability might be wrong?

16. Births in China In China, the probability of a baby being a boy is 0.5845. Couples are allowed to have only one child. If relatives give birth to five babies, what is the probability that there is at least one girl? Can that system continue to work indefinitely?

17. Fruit Flies An experiment with fruit flies involves one parent with normal wings and one parent with vestigial wings. When these parents have an offspring, there is a 3/4 probability that the offspring has normal wings and a 1/4 probability of vestigial wings. If the parents give birth to 10 offspring, what is the probability that at least 1 of the offspring has vestigial wings? If researchers need at least one offspring with vestigial wings, can they be reasonably confident of getting one?

18. Solved Robberies According to FBI data, 24.9% of robberies are cleared with arrests. A new detective is assigned to 10 different robberies.

a. What is the probability that at least one of them is cleared with an arrest?

b. What is the probability that the detective clears all 10 robberies with arrests?

c. What should we conclude if the detective clears all 10 robberies with arrests?

19. Polygraph Test Refer to Table 4-1 (included with the Chapter Problem) and assume that 1 of the 98 test subjects is randomly selected. Find the probability of selecting a subject with a positive test result, given that the subject did not lie. Why is this particular case problematic for test subjects?

20. Polygraph Test Refer to Table 4-1 and assume that 1 of the 98 test subjects is randomly selected. Find the probability of selecting a subject with a negative test result, given that the subject lied. What does this result suggest about the polygraph test?

21. Polygraph Test Refer to Table 4-1. Find $P(\text{subject lied} \mid \text{negative test result})$. Compare this result to the result found in Exercise 20. Are $P(\text{subject lied} \mid \text{negative test result})$ and $P(\text{negative test result} \mid \text{subject lied})$ equal?

22. Polygraph Test Refer to Table 4-1.

a. Find $P(\text{negative test result} \mid \text{subject did not lie})$.

b. Find $P(\text{subject did not lie} \mid \text{negative test result})$.

c. Compare the results from parts (a) and (b). Are they equal?

Identical and Fraternal Twins. *In Exercises 23–26, use the data in the following table. Instead of summarizing observed results, the entries reflect the actual probabilities based on births of twins (based on data from the Northern California Twin Registry and the article "Bayesians, Frequentists, and Scientists" by Bradley Efron,* Journal of the American Statistical Association, *Vol. 100, No. 469). Identical twins come from a single egg that splits into two embryos, and fraternal twins* *are from separate fertilized eggs. The table entries reflect the principle that among sets of twins, 1/3 are identical and 2/3 are fraternal. Also, identical twins must be of the same sex and the sexes are equally likely (approximately), and sexes of fraternal twins are equally likely.*

Sexes of Twins

	boy/boy	boy/girl	girl/boy	girl/girl
Identical Twins	5	0	0	5
Fraternal Twins	5	5	5	5

23. Identical Twins

a. After having a sonogram, a pregnant woman learns that she will have twins. What is the probability that she will have identical twins?

b. After studying the sonogram more closely, the physician tells the pregnant woman that she will give birth to twin boys. What is the probability that she will have identical twins? That is, find the probability of identical twins given that the twins consist of two boys.

24. Fraternal Twins

a. After having a sonogram, a pregnant woman learns that she will have twins. What is the probability that she will have fraternal twins?

b. After studying the sonogram more closely, the physician tells the pregnant woman that she will give birth to twins consisting of one boy and one girl. What is the probability that she will have fraternal twins?

25. Fraternal Twins If a pregnant woman is told that she will give birth to fraternal twins, what is the probability that she will have one child of each sex?

26. Fraternal Twins If a pregnant woman is told that she will give birth to fraternal twins, what is the probability that she will give birth to two girls?

27. Redundancy in Alarm Clocks A statistics student wants to ensure that she is not late for an early statistics class because of a malfunctioning alarm clock. Instead of using one alarm clock, she decides to use three. What is the probability that at least one of her alarm clocks works correctly if each individual alarm clock has a 90% chance of working correctly? Does the student really gain much by using three alarm clocks instead of only one? How are the results affected if all of the alarm clocks run on electricity instead of batteries?

28. Acceptance Sampling With one method of the procedure called *acceptance sampling,* a sample of items is randomly selected without replacement, and the entire batch is rejected if there is at least one defect. The Newport Gauge Company has just manufactured a batch of aircraft altimeters, and 3% are defective.

a. If the batch contains 400 altimeters and 2 of them are selected without replacement and tested, what is the probability that the entire batch will be rejected?

b. If the batch contains 4000 altimeters and 100 of them are selected without replacement and tested, what is the probability that the entire batch will be rejected?

29. Using Composite Blood Samples When testing blood samples for HIV infections, the procedure can be made more efficient and less expensive by combining samples of blood specimens. If samples from three people are combined and the mixture tests negative, we know that all three individual samples are negative. Find the probability of a positive result for three samples combined into one mixture, assuming the probability of an individual blood sample testing positive is 0.1 (the probability for the "at-risk" population, based on data from the New York State Health Department).

30. Using Composite Water Samples The Orange County Department of Public Health tests water for contamination due to the presence of *E. coli* (*Escherichia coli*) bacteria. To reduce laboratory costs, water samples from six public swimming areas are combined for one test, and further testing is done only if the combined sample fails. Based on past results, there is a 2% chance of finding *E. coli* bacteria in a public swimming area. Find the probability that a combined sample from six public swimming areas will reveal the presence of *E. coli* bacteria.

4-5   Beyond the Basics

31. Shared Birthdays Find the probability that of 25 randomly selected people,

a. no 2 share the same birthday.

b. at least 2 share the same birthday.

32. Whodunnit? The Atlanta plant of the Medassist Pharmaceutical Company manufactures 400 heart pacemakers, of which 3 are defective. The Baltimore plant of the same company manufactures 800 pacemakers, of which 2 are defective. If 1 of the 1200 pacemakers is randomly selected and is found to be defective, what is the probability that it was manufactured in Atlanta?

33. Roller Coaster The Rock 'n' Roller Coaster at Disney–MGM Studios in Orlando has 2 seats in each of 12 rows. Riders are assigned to seats in the order that they arrive. If you ride this roller coaster once, what is the probability of getting the coveted first row? How many times must you ride in order to have at least a 95% chance of getting a first-row seat at least once?

34. Unseen Coins A statistics professor tosses two coins that cannot be seen by any students. One student asks this question: "Did one of the coins turn up heads?" Given that the professor's response is "yes," find the probability that both coins turned up heads.

4-6   Probabilities Through Simulations

Key Concept In Section 4-2, we briefly discussed simulations. In this section we use simulations as an alternative approach to finding probabilities. The advantage to using simulations is that we can overcome much of the difficulty encountered when using the formal rules discussed in the preceding sections.

We begin by defining a simulation.

DEFINITION

> A **simulation** of a procedure is a process that behaves the same way as the procedure, so that similar results are produced.

Example 1 illustrates the use of a simulation involving births.

EXAMPLE 1   **Gender Selection** In a test of the MicroSort method of gender selection developed by the Genetics & IVF Institute, 127 boys were born among 152 babies born to parents who used the YSORT method for trying to have a baby boy. In order to properly evaluate these results, we need to know the probability of getting at least 127 boys among 152 births, assuming that boys and girls are equally likely. Assuming that male and female births are equally likely, describe a simulation that results in the genders of 152 newborn babies.

SOLUTION   One approach is simply to flip a fair coin 152 times, with heads representing females and tails representing males. Another approach is to use a calculator or computer to randomly generate 152 numbers that are 0s and 1s, with 0 representing a male and 1 representing a female. The numbers must be generated in such a way that they are equally likely. Here are typical results:

0	0	1	0	1	1	1	⋯
↓	↓	↓	↓	↓	↓	↓	⋯
Male	Male	Female	Male	Female	Female	Female	

EXAMPLE 2   **Same Birthdays** Exercise 31 in Section 4-5 refers to the classical birthday problem, in which we find the probability that in a randomly selected group of 25 people, at least 2 share the same birthday. The theoretical solution is somewhat difficult. It isn't practical to survey many different groups of 25 people, so a simulation is a helpful alternative. Describe a simulation that could be used to find the probability that among 25 randomly selected people, at least 2 share the same birthday.

SOLUTION   Begin by representing birthdays by integers from 1 through 365, where 1 = January 1, 2 = January 2, ⋯, 365 = December 31. Then use a calculator or computer program to generate 25 random numbers, each between 1 and 365. Those numbers can then be sorted, so it becomes easy to examine the list to determine whether any 2 of the simulated birth dates are the same. (After sorting, same numbers are adjacent.) We can repeat the process as many times as we like, until we are satisfied that we have a good estimate of the probability. Our estimate of the probability is the number of times we got at least 2 birth dates that are the same, divided by the total number of groups of 25 that were generated. Here are typical results:

20	274	42	93	⋯
↓	↓	↓	↓	⋯
Jan. 20	Oct. 1	Feb. 11	Apr. 3	⋯

There are several ways of obtaining randomly generated numbers from 1 through 365, including the following:

- **A Table of Random Digits:** Refer, for example, to the *CRC Standard Probability and Statistics Tables and Formulae,* which contains a table of 14,000 digits. (In such a table there are many ways to extract numbers from 1 through 365. One way is by referring to the digits in the first three columns and ignoring 000 as well as anything above 365.)

- **TI-83/84 Plus Calculator:** Use the following procedure to generate 25 simulated birthdays, where each birthday is represented by a whole number between 1 and 365.

    1. Press **MATH**, then use the ▷ key to select **PRB,** then use the ▽ key to select the menu item of **randInt** (for random integers).

    2. The format of the **randInt** command requires a minimum number, a maximum number, and the number of values to be generated. To simulate

Probability of an Event That Has Never Occurred

Some events are possible, but are so unlikely that they have never occurred. Here is one such problem of great interest to political scientists: Estimate the probability that your single vote will determine the winner in a U.S. Presidential election. Andrew Gelman, Gary King, and John Boscardin write in the *Journal of the American Statistical Association* (Vol. 93, No. 441) that "the exact value of this probability is of only minor interest, but the number has important implications for understanding the optimal allocation of campaign resources, whether states and voter groups receive their fair share of attention from prospective presidents, and how formal 'rational choice' models of voter behavior might be able to explain why people vote at all." The authors show how the probability value of 1 in 10 million is obtained for close elections.

Monkey Typists

A classical claim is that a monkey randomly hitting a keyboard would eventually produce the complete works of Shakespeare, assuming that it continues to type century after century. The multiplication rule for probability has been used to find such estimates. One result of 1,000,000,000,000, 000,000,000,000,000, 000,000,000 years is considered by some to be too short. In the same spirit, Sir Arthur Eddington wrote this poem: "There once was a brainy baboon, who always breathed down a bassoon. For he said, 'It appears that in billions of years, I shall certainly hit on a tune.'"

25 birthdays, enter 1 for the minimum, 365 for the maximum, and 25 for the number of generated values, so that the command is entered as **randInt(1, 365, 25).** Then press **ENTER**.

3. For ease of viewing, store the list of generated values in L1 by entering this sequence of key strokes: **STO› 2ND 1 ENTER**.

4. It is much easier to inspect the list of values for matches if the list has been arranged in order, so sort the values in list L1 by entering **STAT**, then selecting the menu item of **SortA** (to arrange the values in ascending order). Press **ENTER**. Now identify L1 as the list to be sorted by pressing **2ND 1 ENTER**, so that the screen displays the command of **SortA(L1).** Press **ENTER**, and the screen should provide the message of Done.

5. Inspect list L1 for matches by pressing **STAT ENTER**, so that the entries in L1 are displayed in a single column of values arranged in order. Use the ⌄ key to scroll through the list of values searching for matches. Matches are easy to identify because they will be adjacent equal numbers.

The above steps will result in displays similar to those shown here, although the generated numbers will be different, because they are randomly generated.

TI-83/84 Plus

```
randInt(1,365,25
{79 206 340 133…
Ans→L1
{79 206 340 133…
SortA(L1)
            Done
■
```

TI-83/84 Plus

L1	L2	L3	1
17	------	------	
34			
46			
70			
79			
132			
129			

L1(7)=129

EXAMPLE 3   **Run of Six Heads or Tails** One of the author's favorite class activities is to give this assignment: All students take out a coin and flip it. Students getting heads go home and actually flip a coin 200 times and record the results. Students getting tails make up their own results for 200 coin flips. In the next class, the author could select any student's results and quickly determine whether the results are real or fabricated by using this criterion: If there is a run of six heads or six tails, the results are real, but if there is no such run, the results are fabricated. This is based on the principle that when fabricating results, students almost never include a run of six or more heads or tails, but with 200 actual coin flips, there is a very high probability of getting such a run of at least six heads or tails. This activity is more fun than human beings should be allowed to have. Unfortunately, the calculation for the probability of getting a run of at least six heads or six tails is *extremely* difficult. Fortunately, simulations can let us know whether such runs are likely in 200 coin flips. Without necessarily finding a probability value, simulate 200 actual coin flips, repeat the simulation a few times, then determine whether a run of six heads or tails is likely.

SOLUTION   Let 0 = heads and 1 = tails, then use some simulation technique to generate 200 digits that are all 0s and 1s. Now examine the list. It is easy to quickly determine whether there is a sequence of at least six 0s or six 1s. After repeating the simulation a few times, it will be obvious that a string of six 0s or six 1s will almost always occur, so the probability of getting such a string is very high.

4-6 Basic Skills and Concepts

Statistical Literacy and Critical Thinking

1. Simulating Dice When two dice are rolled, the total is between 2 and 12 inclusive. A student simulates the rolling of two dice by randomly generating numbers between 2 and 12. Does this simulation behave in a way that is similar to actual dice? Why or why not?

2. Simulating Dice Assume that you have access to a computer that can randomly generate whole numbers between any two values. Describe how this computer can be used to simulate the rolling of a pair of dice.

3. Simulating Birthdays A student wants to conduct the simulation described in Example 2, but no calculator or computer is available, so the student uses 365 individual index cards to write the individual numbers between 1 and 365. The student then shuffles the cards, selects one, and records the result. That card is replaced, the cards are again shuffled and a second number is drawn. This process is repeated until 25 birthdays are generated. Does this simulation behave the same way as the process of selecting 25 people and recording their birthdays? Why or why not?

4. Simulating Coin Flips One student conducted the simulation described in Example 3 and stated that the probability of getting a sequence of six 0s or six 1s is 0.977. What is wrong with that statement?

In Exercises 5–8, describe the simulation procedure. (For example, to simulate 10 births, use a random number generator to generate 10 integers between 0 and 1 inclusive, and consider 0 to be a male and 1 to be a female.)

5. Brand Recognition The probability of randomly selecting an adult who recognizes the brand name of McDonald's is 0.95 (based on data from Franchise Advantage). Describe a procedure for using a TI-83/84 Plus calculator to simulate the random selection of 50 adult consumers. Each individual outcome should be an indication of one of two results: (1) The consumer recognizes the brand name of McDonald's; (2) the consumer does not recognize the brand name of McDonald's.

6. Lefties Ten percent of people are left-handed. In a study of dexterity, 15 people are randomly selected. Describe a procedure for using a TI-83/84 Plus calculator to simulate the random selection of 15 people. Each of the 15 outcomes should be an indication of one of two results: (1) Subject is left-handed; (2) subject is not left-handed.

7. Shaquille O'Neal Shaquille O'Neal is a professional basketball star who had a reputation for being a poor free throw shooter. As of this writing, he made 5155 of the 9762 free throws that he attempted, for a success ratio of 0.528. Describe a procedure for using a TI-83/84 Plus calculator to simulate his next free throw. The outcome should be an indication of one of two results: (1) The free throw is made; (2) the free throw is missed.

8. Simulating Hybridization When Mendel conducted his famous hybridization experiments, he used peas with green pods and yellow pods. One experiment involved crossing peas in such a way that 75% of the offspring peas were expected to have green pods, and 25% of the offspring peas were expected to have yellow pods. Describe a procedure for using a TI-83/84 Plus calculator to simulate 20 peas in such a hybridization experiment. Each of the 20 individual outcomes should be an indication of one of two results: (1) The pod is green; (2) the pod is yellow.

In Exercises 9–12, develop a simulation using a TI-83/84 Plus calculator.

TI **9. Simulating Brand Recognition Study** Refer to Exercise 5, which required a description of a simulation.

a. Conduct the simulation and record the number of consumers who recognize the brand name of McDonald's. If possible, obtain a printed copy of the results. Is the proportion of those who recognize McDonald's reasonably close to the value of 0.95?

b. Repeat the simulation until it has been conducted a total of 10 times. In each of the 10 trials, record the proportion of those who recognize McDonald's. Based on the results, do the proportions appear to be very consistent or do they vary widely? Based on the results, would it be *unusual* to randomly select 50 consumers and find that about half of them recognize McDonald's?

10. Simulating Left-Handedness Refer to Exercise 6, which required a description of a simulation.

a. Conduct the simulation and record the number of left-handed people. Is the percentage of left-handed people from the simulation reasonably close to the value of 10%?

b. Repeat the simulation until it has been conducted a total of 10 times. Record the numbers of left-handed people in each case. Based on the results, would it be *unusual* to randomly select 15 people and find that none of them are left-handed?

11. Simulating the Shaq Refer to Exercise 7, which required a description of a simulated free throw by basketball player Shaquille O'Neal.

a. Repeat the simulation five times and record the number of times that the free throw was made. Is the percentage of successful free throws from the simulation reasonably close to the value of 0.528?

b. Repeat part (a) until it has been conducted a total of 10 times. Record the proportion of successful free throws in each case. Based on the results, would it be *unusual* for Shaquille O'Neal to make all of five free throws in a game?

12. Simulating Hybridization Refer to Exercise 8, which required a description of a hybridization simulation.

a. Conduct the simulation and record the number of yellow peas. If possible, obtain a printed copy of the results. Is the percentage of yellow peas from the simulation reasonably close to the value of 25%?

b. Repeat the simulation until it has been conducted a total of 10 times. Record the numbers of peas with yellow pods in each case. Based on the results, do the numbers of peas with yellow pods appear to be very consistent? Based on the results, would it be *unusual* to randomly select 20 such offspring peas and find that none of them have yellow pods?

13. Probability of a Run of Three Use a simulation approach to find the probability that when five consecutive babies are born, there is a run of at least three babies of the same sex. Describe the simulation procedure used, and determine whether such runs are unusual.

14. Probability of a Run of Four Use a simulation approach to find the probability that when six consecutive babies are born, there is a run of at least four babies of the same sex. Describe the simulation procedure used, and determine whether such runs are unusual.

15. Gender-Selection Method As of this writing, the latest results available from the Microsort YSORT method of gender selection consist of 127 boys in 152 births. That is, among 152 sets of parents using the YSORT method for increasing the likelihood of a boy, 127 actually had boys and the other 25 had girls. Assuming that the YSORT method has no effect and that boys and girls are equally likely, simulate 152 births. Is it unusual to get 127 boys in 152 births? What does the result suggest about the YSORT method?

16. Nasonex Treatment Analysis Nasonex is a nasal spray used to treat allergies. In clinical trials, 1671 subjects were given a placebo, and 2 of them developed upper respiratory tract infections. Another 2103 patients were treated with Nasonex and 6 of them developed upper respiratory tract infections. Assume that Nasonex has no effect on upper respiratory tract infections, so that the rate of those infections also applies to Nasonex users. Using the placebo rate of 2/1671, simulate groups of 2103 subjects given the Nasonex treatment, and determine whether a result of 6 upper respiratory tract infections could easily occur. What does that suggest about Nasonex as a cause of upper respiratory tract infections?

4-6   Beyond the Basics

17. Simulating the Monty Hall Problem A problem that once attracted much attention is the *Monty Hall problem,* based on the old television game show *Let's Make a Deal,* hosted by Monty Hall. Suppose you are a contestant who has selected one of three doors after being told that two of them conceal nothing, but that a new red Corvette is behind one of the three. Next, the host opens one of the doors you didn't select and shows that there is nothing behind it. He then offers you the choice of sticking with your first selection or switching to the other unopened door. Should you stick with your first choice or should you switch? Develop a simulation of this game and determine whether you should stick or switch. (According to *Chance* magazine, business schools at such institutions as Harvard and Stanford use this problem to help students deal with decision making.)

18. Simulating Birthdays

a. Develop a simulation for finding the probability that when 50 people are randomly selected, at least 2 of them have the same birth date. Describe the simulation and estimate the probability.

b. Develop a simulation for finding the probability that when 50 people are randomly selected, at least 3 of them have the same birth date. Describe the simulation and estimate the probability.

19. Genetics: Simulating Population Control A classical probability problem involves a king who wanted to increase the proportion of women by decreeing that after a mother gives birth to a son, she is prohibited from having any more children. The king reasons that some families will have just one boy, whereas other families will have a few girls and one boy, so the proportion of girls will be increased. Is his reasoning correct? Will the proportion of girls increase?

4-7   Counting

Key Concept In this section we present methods for counting the number of possible outcomes in a variety of different situations. Probability problems typically require that we know the total number of possible outcomes, but finding that total often requires the methods of this section (because it is not practical to construct a list of the outcomes).

Fundamental Counting Rule

For a sequence of two events in which the first event can occur m ways and the second event can occur n ways, the events together can occur a total of $m \cdot n$ ways.

The fundamental counting rule extends to situations involving more than two events, as illustrated in the following examples.

EXAMPLE 1   **Identity Theft** It's wise not to disclose social security numbers, because they are often used by criminals attempting identity theft. Assume that a criminal is found using your social security number and claims that all of the digits were randomly generated. What is the probability of getting your social security number when randomly generating nine digits? Is the criminal's claim that your number was randomly generated likely to be true?

continued

Too Few Bar Codes

In 1974, a pack of gum was the first item to be scanned

in a super-market. That scan-ning re-quired that the gum be identi-fied with a bar code. Bar codes or Universal Product Codes are used to identify individual items to be purchased. Bar codes used 12 digits that allowed scanners to automatically list and record the price of each item purchased. The use of 12 digits became insufficient as the number of different products increased, so the codes were recently modi-fied to include 13 digits.

Similar problems are encountered when tele-phone area codes are split because there are too many different telephones for one area code in a region. Meth-ods of counting are used to design systems to accom-modate future numbers of units that must be processed or served.

SOLUTION   Each of the 9 digits has 10 possible outcomes: 0, 1, 2, . . . , 9. By applying the fundamental counting rule, we get

$$10 \cdot 10 \cdot 10 \cdot 10 \cdot 10 \cdot 10 \cdot 10 \cdot 10 \cdot 10 = 1{,}000{,}000{,}000$$

Only one of those 1,000,000,000 possibilities corresponds to your social security number, so the probability of randomly generating a social security number and getting yours is 1/1,000,000,000. It is extremely unlikely that a criminal would generate your social security by chance, assuming that only one social security number is generated. (Even if the criminal could generate thousands of social security numbers and try to use them, it is highly unlikely that your number would be generated.) If someone is found using your social security number, it was probably by some other method, such as spying on Internet transactions or searching through your mail or garbage.

EXAMPLE 2   **Chronological Order** Consider the following question given on a history test:

Arrange the following events in chronological order.

a. Boston Tea Party

b. Teapot Dome Scandal

c. The Civil War

The correct answer is a, c, b, but let's assume that a student makes random guesses. Find the probability that this student chooses the correct chronological order.

SOLUTION   Although it is easy to list the six possible arrangements, the funda-mental counting rule gives us another way to approach this problem. When making the random selections, there are 3 possible choices for the first event, 2 remaining choices for the second event, and only 1 choice for the third event, so the total num-ber of possible arrangements is

$$3 \cdot 2 \cdot 1 = 6$$

Because only one of the 6 possible arrangements is correct, the probability of getting the correct chronological order with random guessing is 1/6 or 0.167.

In Example 2, we found that 3 items can be arranged $3 \cdot 2 \cdot 1 = 6$ different ways. This particular solution can be generalized by using the following notation and the *factorial rule*.

TI-83/84 Plus

```
MATH NUM CPX PRB
1:rand
2:nPr
3:nCr
4:!
5:randInt(
6:randNorm(
7:randBin(
```

Notation

The **factorial symbol (!)** denotes the product of decreasing positive whole numbers. For example, $4! = 4 \cdot 3 \cdot 2 \cdot 1 = 24$. By special definition, $0! = 1$.

(Many calculators have a factorial key. On the TI-83/84 Plus calculator, first enter the number, press **MATH** and select **PRB,** and then select menu item 4 and press **ENTER**.)

Factorial Rule

A collection of n different items can be arranged in order $n!$ different ways. (This *factorial rule* reflects the fact that the first item may be selected n different ways, the second item may be selected $n - 1$ ways, and so on.)

Routing problems often involve application of the factorial rule. Verizon wants to route telephone calls through the shortest networks. Federal Express wants to find the shortest routes for its deliveries. American Airlines wants to find the shortest route for returning crew members to their homes.

EXAMPLE 3   **Routes to National Parks** During the summer, you are planning to visit these six national parks: Glacier, Yellowstone, Yosemite, Arches, Zion, and Grand Canyon. You would like to plan the most efficient route and you decide to list all of the possible routes. How many different routes are possible?

SOLUTION   By applying the factorial rule, we know that 6 different parks can be arranged in order 6! different ways. The number of different routes is $6! = 6 \cdot 5 \cdot 4 \cdot 3 \cdot 2 \cdot 1 = 720$. There are 720 different possible routes.

Example 3 is a variation of a classical problem called the *traveling salesman problem.* Because routing problems are so important to so many different companies, and because the number of different routes can be very large, there is a continuing effort to simplify the method of finding the most efficient routes.

According to the factorial rule, n different items can be arranged $n!$ different ways. Sometimes we have n different items, but we need to select *some* of them instead of all of them. For example, if we must conduct surveys in state capitals, but we have time to visit only four capitals, the number of different possible routes is $50 \cdot 49 \cdot 48 \cdot 47 = 5,527,200$. Another way to obtain this same result is to evaluate

$$\frac{50!}{46!} = 50 \cdot 49 \cdot 48 \cdot 47 = 5,527,200$$

In this calculation, note that the factors in the numerator divide out with the factors in the denominator, except for the factors of 50, 49, 48, and 47 that remain. We can generalize this result by noting that if we have n different items available and we want to select r of them, the number of different arrangements possible is $n!/(n - r)!$ as in 50!/46!. This generalization is commonly called the *permutations rule*.

Permutations Rule (When Items Are All Different)

Requirements

1. There are n *different* items available.

2. We select r of the n items (without replacement).

3. We consider rearrangements of the same items to be different sequences. (The permutation of *ABC* is different from *CBA* and is counted separately)

If the preceding requirements are satisfied, the number of *permutations* (or sequences) of r items selected from n different available items (without replacement) is

$$_nP_r = \frac{n!}{(n - r)!}$$

How Many Shuffles?

After conducting extensive research, Harvard mathematician Persi Diaconis found that it takes seven shuffles of a deck of cards to get a complete mixture. The mixture is complete in the sense that all possible arrangements are equally likely. More than seven shuffles will not have a significant effect, and fewer than seven are not enough. Casino dealers rarely shuffle as often as seven times, so the decks are not completely mixed. Some expert card players have been able to take advantage of the incomplete mixtures that result from fewer than seven shuffles.

TI-83/84 Plus The end of this section includes instructions for using a TI-83/84 Plus calculator for applying the permutations rule.

When we use the terms *permutations, arrangements,* or *sequences,* we imply that *order is taken into account,* in the sense that different orderings of the same items are counted separately. The letters *ABC* can be arranged six different ways: *ABC, ACB, BAC, BCA, CAB, CBA.* (Later, we will refer to *combinations,* which do not count such arrangements separately.) In the next example, we are asked to find the total number of different sequences that are possible.

> **EXAMPLE 4**   **Exacta Bet** In horse racing, a bet on an exacta in a race is won by correctly selecting the horses that finish first and second, and you must select those two horses in the correct order. The 132nd running of the Kentucky Derby had a field of 20 horses. If a bettor randomly selects two of those horses for an exacta bet, what is the probability of winning?

> **SOLUTION**   We have $n = 20$ horses available, and we must select $r = 2$ of them without replacement. The number of different sequences of arrangements is found as shown:
>
> $$_nP_r = \frac{n!}{(n-r)!} = \frac{20!}{(20-2)!} = 380$$
>
> There are 380 different possible arrangements of 2 horses selected from the 20 that are available. If one of those arrangements is randomly selected, there is a probability of 1/380 that the winning arrangement is selected.

We sometimes need to find the number of permutations when some of the items are identical to others. The following variation of the permutations rule applies to such cases.

Permutations Rule (When Some Items Are Identical to Others)

Requirements

1. There are *n* items available, and some items are identical to others.

2. We select all of the *n* items (without replacement).

3. We consider rearrangements of distinct items to be different sequences.

If the preceding requirements are satisfied, and if there are n_1 alike, n_2 alike, . . . , n_k alike, the number of *permutations* (or sequences) of all items selected without replacement is

$$\frac{n!}{n_1!n_2!\cdots n_k!}$$

> **EXAMPLE 5**   **Gender Selection** In a preliminary test of the MicroSort gender selection method developed by the Genetics and IVF Institute, 14 couples tried to have baby girls. Analysis of the effectiveness of the MicroSort method is based on a probability value, which in turn is based on numbers of permutations. Let's consider this simple problem: How many ways can 11 girls and 3 boys be arranged in sequence? That is, find the number of permutations of 11 girls and 3 boys.

SOLUTION   We have $n = 14$ babies, with $n_1 = 11$ alike (girls) and $n_2 = 3$ others alike (boys). The number of permutations is computed as follows:

$$\frac{n!}{n_1! \, n_2!} = \frac{14!}{11! \, 3!} = \frac{87,178,291,200}{(39,916,800)(6)} = 364$$

There are 364 different ways to arrange 11 girls and 3 boys.

The preceding example involved n items, each belonging to one of two categories. When there are only two categories, we can stipulate that x of the items are alike and the other $n - x$ items are alike, so the permutations formula simplifies to

$$\frac{n!}{(n - x)! \, x!}$$

This particular result will be used for the discussion of binomial probabilities in Section 5-3.

Combinations Rule

Requirements

1. There are n *different* items available.

2. We select r of the n items (without replacement).

3. We consider rearrangements of the same items to be the same. (The combination *ABC* is the same as *CBA*.)

If the preceding requirements are satisfied, the number of *combinations* of r items selected from n different items is

$$_nC_r = \frac{n!}{(n - r)! \, r!}$$

TI-83/84 Plus The end of this section includes instructions for using a TI-83/84 Plus calculator for applying the combinations rule.

When we intend to select r items from n different items but *do not take order into account,* we are really concerned with possible *combinations* rather than permutations. That is, **when different orderings of the same items are counted separately, we have a permutation problem, but when different orderings of the same items are not counted separately, we have a combination problem** and may apply the combinations rule.

Because choosing between the permutations rule and the combinations rule can be confusing, we provide the following example, which is intended to emphasize the difference between them.

EXAMPLE 6   **Phase I of a Clinical Trial** A clinical test on humans of a new drug is normally done in three phases. Phase I is conducted with a relatively small number of healthy volunteers. Let's assume that we want to treat 8 healthy humans with a new drug, and we have 10 suitable volunteers available.

a. If the subjects are selected and treated *in sequence*, so that the trial is discontinued if anyone displays adverse effects, how many different sequential arrangements are possible if 8 people are selected from the 10 that are available?

b. If 8 subjects are selected from the 10 that are available, and the 8 selected subjects are all treated at the same time, how many different treatment groups are possible?

continued

The Random Secretary

One classical problem of probability goes like this: A secretary addresses 50 different letters and envelopes to 50 different people, but the letters are randomly mixed before being put into envelopes. What is the probability that at least one letter gets into the correct envelope? Although the probability might seem like it should be small, it's actually 0.632. Even with a million letters and a million envelopes, the probability is 0.632. The solution is beyond the scope of this text—way beyond.

Composite Sampling

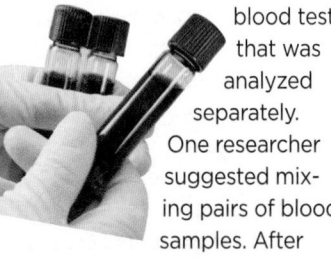

The U.S. Army once tested for syphilis by giving each inductee an individual blood test that was analyzed separately. One researcher suggested mixing pairs of blood samples. After the mixed pairs were tested, syphilitic inductees could be identified by retesting the few blood samples that were in the pairs that tested positive. The total number of analyses was reduced by pairing blood specimens, so why not put them in groups of three or four or more? Probability theory was used to find the most efficient group size, and a general theory was developed for detecting the defects in any population. This technique is known as *composite sampling*.

SOLUTION   Note that in part (a), order is relevant because the subjects are treated sequentially and the trial is discontinued if anyone exhibits a particularly adverse reaction. However, in part (b) the order of selection is irrelevant because all of the subjects are treated at the same time.

a. Because order does count, we want the number of *permutations* of $r = 8$ people selected from the $n = 10$ available people. We get

$$_nP_r = \frac{n!}{(n - r)!} = \frac{10!}{(10 - 8)!} = 1{,}814{,}400$$

b. Because order does *not* count, we want the number of *combinations* of $r = 8$ people selected from the $n = 10$ available people. We get

$$_nC_r = \frac{n!}{(n - r)!\, r!} = \frac{10!}{(10 - 8)!\, 8!} = 45$$

With order taken into account, there are 1,814,400 permutations, but without order taken into account, there are 45 combinations.

EXAMPLE 7   **Florida Lottery** The Florida Lotto game is typical of state lotteries. You must select six different numbers between 1 and 53. You win the jackpot if the same six numbers are drawn in any order. Find the probability of winning the jackpot.

SOLUTION   Because the order of the selected numbers does not matter, you win if you get the correct combination of six numbers. Because there is only one winning combination, the probability of winning the jackpot is 1 divided by the total number of combinations. With $n = 53$ numbers available and with $r = 6$ numbers selected, the number of combinations is

$$_nC_r = \frac{n!}{(n - r)!\, r!} = \frac{53!}{(53 - 6)!\, 6!} = 22{,}957{,}480$$

With 1 winning combination and 22,957,480 different possible combinations, the probability of winning the jackpot is 1/22,957,480.

Five different rules for finding total numbers of outcomes were given in this section. Although not all counting problems can be solved with one of these five rules, they do provide a strong foundation for many real and relevant applications.

Using a TI-83/84 Plus Calculator for Factorials, Permutations, and Combinations

Factorials   First enter the number, then press **MATH**, select **PRB,** then select menu item 4 (which is the ! symbol) and press **ENTER**. For example, 5! is evaluated with **MATH** **PRB** (4) **ENTER**, and the result is 120.

Permutations   When evaluating an expression of the form $_nP_r$, follow these steps:

1. First enter the value of n.
2. Press **MATH**, then select **PRB,** then select $_nP_r$. Press **ENTER**.
3. Now enter the value of r and press **ENTER**.

For example, to evaluate $_9P_3$, first enter 9, then execute Step 2 above, then enter 3 and press **ENTER** to obtain a result of 504. That is, $_9P_3 = 504$.

Combinations When evaluating an expression of the form $_nC_r$, follow these steps:

1. First enter the value of n.
2. Press **MATH**, then select **PRB**, then select $_nC_r$. Press **ENTER**.
3. Now enter the value of r and press **ENTER**.

For example, to evaluate $_9C_3$, first enter 9, then execute Step 2 above, then enter 3 and press **ENTER** to obtain a result of 4. That is, $_9C_3 = 84$.

4-7   Basic Skills and Concepts

Statistical Literacy and Critical Thinking

1. Permutations and Combinations What is the basic difference between a situation requiring application of the permutations rule and one that requires the combinations rule?

2. Combination Lock The typical combination lock uses three numbers between 0 and 49, and they must be selected in the correct sequence. Given the way that these locks work, is the name of "combination" lock correct? Why or why not?

3. Trifecta In horse racing, a trifecta is a bet that the first three finishers in a race are selected, and they are selected in the correct order. Does a trifecta involve combinations or permutations? Explain.

4. Quinela In horse racing, a quinela is a bet that the first two finishers in a race are selected, and they can be selected in any order. Does a quinela involve combinations or permutations? Explain.

Calculating Factorials, Combinations, Permutations. *In Exercises 5–12, evaluate the given expressions and express all results using the usual format for writing numbers (instead of scientific notation).*

5. Factorial Find the number of different ways that five test questions can be arranged in order by evaluating 5!.

6. Factorial Find the number of different ways that the nine players on a baseball team can line up for the National Anthem by evaluating 9!.

7. Blackjack In the game of blackjack played with one deck, a player is initially dealt two cards. Find the number of different two-card initial hands by evaluating $_{52}C_2$.

8. Card Playing Find the number of different possible five-card poker hands by evaluating $_{52}C_5$.

9. Scheduling Routes A manager must select 5 delivery locations from 9 that are available. Find the number of different possible routes by evaluating $_9P_5$.

10. Scheduling Routes A political strategist must visit state capitols, but she has time to visit only 3 of them. Find the number of different possible routes by evaluating $_{50}P_3$.

11. Virginia Lottery The Virginia Win for Life lottery game requires that you select the correct 6 numbers between 1 and 42. Find the number of possible combinations by evaluating $_{42}C_6$.

12. Trifecta Refer to Exercise 3. Find the number of different possible trifecta bets in a race with ten horses by evaluating $_{10}P_3$.

Probability of Winning the Lottery. *Because the California Fantasy 5 lottery is won by selecting the correct five numbers (in any order) between 1 and 39, there are 575,757 different 5-number combinations that could be played, and the probability* *of winning this lottery is 1/575,757. In Exercises 13–16, find the probability of winning the indicated lottery by buying one ticket. In each case, numbers selected are different and order does not matter. Express the result as a fraction.*

13. Lotto Texas Select the six winning numbers from 1, 2, . . . , 54.

14. Florida Lotto Select the six winning numbers from 1, 2, . . . , 53.

15. Florida Fantasy 5 Select the five winning numbers from 1, 2, . . . , 36.

16. Wisconsin Badger Five Answer each of the following.

a. Find the probability of selecting the five winning numbers from 1, 2, . . . , 31.

b. The Wisconsin Badger 5 lottery is won by selecting the correct five numbers from 1, 2, . . . , 31. What is the probability of winning if the rules are changed so that in addition to selecting the correct five numbers, you must now select them in the same order as they are drawn?

17. Identity Theft with Social Security Numbers Identity theft often begins by someone discovering your nine-digit social security number or your credit card number. Answer each of the following. Express probabilities as fractions.

a. What is the probability of randomly generating nine digits and getting *your* social security number.

b. In the past, many teachers posted grades along with the last four digits of the student's social security numbers. If someone already knows the last four digits of your social security number, what is the probability that if they randomly generated the other digits, they would match yours? Is that something to worry about?

18. Identity Theft with Credit Cards Credit card numbers typically have 16 digits, but not all of them are random. Answer the following and express probabilities as fractions.

a. What is the probability of randomly generating 16 digits and getting *your* MasterCard number?

b. Receipts often show the last four digits of a credit card number. If those last four digits are known, what is the probability of randomly generating the other digits of your MasterCard number?

c. Discover cards begin with the digits 6011. If you also know the last four digits of a Discover card, what is the probability of randomly generating the other digits and getting all of them correct? Is this something to worry about?

19. Sampling The Bureau of Fisheries once asked for help in finding the shortest route for getting samples from locations in the Gulf of Mexico. How many routes are possible if samples must be taken at 6 locations from a list of 20 locations?

20. DNA Nucleotides DNA (deoxyribonucleic acid) is made of nucleotides. Each nucleotide can contain any one of these nitrogenous bases: A (adenine), G (guanine), C (cytosine), T (thymine). If one of those four bases (A, G, C, T) must be selected three times to form a linear triplet, how many different triplets are possible? Note that all four bases can be selected for each of the three components of the triplet.

21. Electricity When testing for current in a cable with five color-coded wires, the author used a meter to test two wires at a time. How many different tests are required for every possible pairing of two wires?

22. Scheduling Assignments The starting five players for the Boston Celtics basketball team have agreed to make charity appearances tomorrow night. If you must send three players to a United Way event and the other two to a Heart Fund event, how many different ways can you make the assignments?

23. Computer Design In designing a computer, if a *byte* is defined to be a sequence of 8 bits and each bit must be a 0 or 1, how many different bytes are possible? (A byte is often used to represent an individual character, such as a letter, digit, or punctuation symbol. For example, one coding system represents the letter *A* as 01000001.) Are there enough different bytes for the characters that we typically use, such as lower-case letters, capital letters, digits, punctuation symbols, dollar sign, and so on?

TI 24. Simple Random Sample In Phase I of a clinical trial with gene therapy used for treating HIV, five subjects were treated (based on data from *Medical News Today*). If 20 people were eligible for the Phase I treatment and a simple random sample of five is selected, how many different simple random samples are possible? What is the probability of each simple random sample?

TI 25. Jumble Puzzle Many newspapers carry "Jumble," a puzzle in which the reader must unscramble letters to form words. The letters BUJOM were included in newspapers on the day this exercise was written. How many ways can the letters of BUJOM be arranged? Identify the correct unscrambling, then determine the probability of getting that result by randomly selecting one arrangement of the given letters.

TI 26. Jumble Puzzle Repeat Exercise 25 using these letters: AGGYB.

27. Coca Cola Directors There are 11 members on the board of directors for the Coca Cola Company.

a. If they must elect a chairperson, first vice chairperson, second vice chairperson, and secretary, how many different slates of candidates are possible?

b. If they must form an ethics subcommittee of four members, how many different subcommittees are possible?

TI 28. Safe Combination The author owns a safe in which he stores all of his great ideas for the next edition of this book. The safe combination consists of four numbers between 0 and 99. If another author breaks in and tries to steal these ideas, what is the probability that he or she will get the correct combination on the first attempt? Assume that the numbers are randomly selected. Given the number of possibilities, does it seem feasible to try opening the safe by making random guesses for the combination?

TI 29. MicroSort Gender Selection In a preliminary test of the MicroSort gender-selection method, 14 babies were born and 13 of them were girls.

a. Find the number of different possible sequences of genders that are possible when 14 babies are born.

b. How many ways can 13 girls and 1 boy be arranged in a sequence?

c. If 14 babies are randomly selected, what is the probability that they consist of 13 girls and 1 boy?

d. Does the gender-selection method appear to yield a result that is significantly different from a result that might be expected by random chance?

TI 30. ATM Machine You want to obtain cash by using an ATM machine, but it's dark and you can't see your card when you insert it. The card must be inserted with the front side up and the printing configured so that the beginning of your name enters first.

a. What is the probability of selecting a random position and inserting the card, with the result that the card is inserted correctly?

b. What is the probability of randomly selecting the card's position and finding that it is incorrectly inserted on the first attempt, but it is correctly inserted on the second attempt?

c. How many random selections are required to be absolutely sure that the card works because it is inserted correctly?

TI 31. Designing Experiment Clinical trials of Nasonex involved a group given placebos and another group given treatments of Nasonex. Assume that a preliminary Phase I trial is to be conducted with 10 subjects, including 5 men and 5 women. If 5 of the 10 subjects are randomly selected for the treatment group, find the probability of getting 5 subjects of the same sex. Would there be a problem with having members of the treatment group all of the same sex?

TI 32. Is the Researcher Cheating? You become suspicious when a genetics researcher randomly selects groups of 20 newborn babies and seems to consistently get 10 girls and 10 boys. The researcher claims that it is common to get 10 girls and 10 boys in such cases.

a. If 20 newborn babies are randomly selected, how many different gender sequences are possible?

b. How many different ways can 10 girls and 10 boys be arranged in sequence?

c. What is the probability of getting 10 girls and 10 boys when 20 babies are born?

d. Based on the preceding results, do you agree with the researcher's explanation that it is common to get 10 girls and 10 boys when 20 babies are randomly selected?

33. Powerball As of this writing, the Powerball lottery is run in 29 states. Winning the jackpot requires that you select the correct five numbers between 1 and 55 and, in a separate drawing, you must also select the correct single number between 1 and 42. Find the probability of winning the jackpot.

34. Mega Millions As of this writing, the Mega Millions lottery is run in 12 states. Winning the jackpot requires that you select the correct five numbers between 1 and 56 and, in a separate drawing, you must also select the correct single number between 1 and 46. Find the probability of winning the jackpot.

35. Finding the Number of Area Codes *USA Today* reporter Paul Wiseman described the old rules for the three-digit telephone area codes by writing about "possible area codes with 1 or 0 in the second digit. (Excluded: codes ending in 00 or 11, for toll-free calls, emergency services, and other special uses.)" Codes beginning with 0 or 1 should also be excluded. How many different area codes were possible under these old rules?

36. NCAA Basketball Tournament Each year, 64 college basketball teams compete in the NCAA tournament. Sandbox.com recently offered a prize of $10 million to anyone who could correctly pick the winner in each of the tournament games. (The president of that company also promised that, in addition to the cash prize, he would eat a bucket of worms. Yuck.)

a. How many games are required to get one championship team from the field of 64 teams?

b. If someone makes random guesses for each game of the tournament, find the probability of picking the winner in each game.

c. In an article about the $10 million prize, the *New York Times* wrote that "even a college basketball expert who can pick games at a 70 percent clip has a 1 in _____ chance of getting all the games right." Fill in the blank.

4-7   Beyond the Basics

37. Finding the Number of Computer Variable Names A common computer programming rule is that names of variables must be between 1 and 8 characters long. The first character can be any of the 26 letters, while successive characters can be any of the 26 letters or any of the 10 digits. For example, allowable variable names are A, BBB, and M3477K. How many different variable names are possible?

38. Handshakes and Round Tables

a. Five managers gather for a meeting. If each manager shakes hands with each other manager exactly once, what is the total number of handshakes?

b. If n managers shake hands with each other exactly once, what is the total number of handshakes?

c. How many different ways can five managers be seated at a round table? (Assume that if everyone moves to the right, the seating arrangement is the same.)

d. How many different ways can n managers be seated at a round table?

39. Evaluating Large Factorials Many calculators or computers cannot directly calculate 70! or higher. When n is large, $n!$ can be approximated by $n = 10^K$, where

$$K = (n + 0.5) \log n + 0.39908993 - 0.43429448n.$$

a. You have been hired to visit the capitol of each of the 50 states. How many different routes are possible? Evaluate the answer using the factorial key on a calculator and also by using the approximation given here.

NAME:	Karen De Toro
JOB:	Senior Manager
COMPANY:	Deloitte Consulting LLP

Karen De Toro is Senior Manager at Deloitte Consulting LLP in Chicago, IL. In her role as an actuarial consultant, she uses probability and statistics to determine the cost of life insurance policies for insurance companies.

Q: How do you use statistics in your work?

A: The field of actuarial science is heavily based on statistics. Most actuaries work in the field of insurance. Insurance companies provide a benefit to their customers if certain risky events occur. These events include untimely death, disability, medical problems, loss of property, car accidents, etc. Actuaries are experts in assessing the probabilities of these events occurring and using this information to help insurance companies figure out how much to charge people to provide insurance. We also use this information to determine how much capital an insurance company must have on hand to ensure that they remain solvent, and to help them find ways to use that capital in the most efficient manner.

Q: How critical is statistics in your work?

A: To be an actuary, it is absolutely critical that you have a solid knowledge of statistics. Actuaries are credentialed through a self-study exam process, which requires candidates to exhibit their knowledge of concepts of probability and statistics.

Q: Please cite an example of how your data are used.

A: My area of expertise is life insurance. We use mortality tables that are discrete probability distributions of the risk of death at any particular age. The table essentially gives the probability that an individual of a particular age will die during the year before their next birthday. Our professional organization, the Society of Actuaries, conducts research projects periodically to publish standardized tables using large amounts of data collected on actual deaths. Actuaries in insurance companies do similar studies periodically to see how their own company's experience compares to the published tables. This information is used to determine how much a company should charge for a particular type of life insurance contract and how much money they need to have on hand at any given point in order to pay their customers' claims.

Q: In terms of statistics, what would you recommend for prospective employees?

A: Most entry-level actuaries have degrees in mathematics, actuarial science or statistics. Regardless of the specific degree, entry-level actuaries should generally have two courses in probability and statistics, as well as other advanced math topics.

Q: Do you recommend statistics for today's college students?

A: Definitely. When I got my MBA, it became apparent to me how many fields use statistics. I took some marketing classes in my MBA program, and the marketing methodologies they introduced to us were based in statistical analysis.

Q: Which other skills are important for today's college students?

A: Communications and general business skills are critical, even if you're not planning to enter the business world. These are skills that are applicable in any industry.

Q: Is there anything else that you would like to say that would be helpful to statistics students?

A: There are many fantastic careers and roles you can take on if you have experience with statistics. It is a great skill to have.

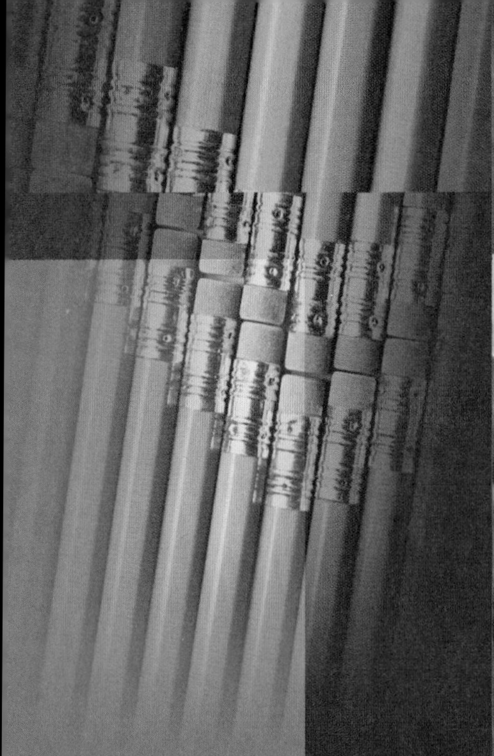

5-1 Review and Preview

5-2 Random Variables

5-3 Binomial Probability Distributions

5-4 Mean, Variance, and Standard Deviation for the Binomial Distribution

5-5 Poisson Probability Distributions

5

Discrete Probability Distributions

Did Mendel's results from plant hybridization experiments contradict his theory?

Gregor Mendel conducted original experiments to study the genetic traits of pea plants. In 1865 he wrote "Experiments in Plant Hybridization," which was published in *Proceedings of the Natural History Society.* Mendel presented a theory that when there are two inheritable traits, one of them will be dominant and the other will be recessive. Each parent contributes one gene to an offspring and, depending on the combination of genes, that offspring could inherit the dominant trait or the recessive trait. Mendel conducted an experiment using pea plants. The pods of pea plants can be green or yellow. When one pea carrying a dominant green gene and a recessive yellow gene is crossed with another pea carrying the same green/yellow genes, the offspring can inherit any one of four combinations of genes, as shown in the table below.

Because green is dominant and yellow is recessive, the offspring pod will be green if either of the two inherited genes is green. The offspring can have a yellow pod only if it inherits the yellow gene from each of the two parents. We can see from the table that when crossing two parents with the green/yellow pair of genes, we expect that 3/4 of the offspring peas should have green pods. That is, $P(\text{green pod}) = 3/4$.

When Mendel conducted his famous hybridization experiments using parent pea plants with the green/yellow combination of genes, he obtained 580 offspring. According to Mendel's theory, 3/4 of the offspring should have green pods, but the actual number of plants with green pods was 428. So the proportion of offspring with green pods to the total number of offspring is $428/580 = 0.738$. Mendel *expected* a proportion of 3/4 or 0.75, but his *actual result* is a proportion of 0.738. In this chapter we will consider the issue of whether the experimental results contradict the theoretical results and, in so doing, we will lay a foundation for *hypothesis testing,* which is introduced in Chapter 8.

Gene from Parent 1		Gene from Parent 2		Offspring Genes		Color of Offspring Pod
green	+	green	→	green/green	→	green
green	+	yellow	→	green/yellow	→	green
yellow	+	green	→	yellow/green	→	green
yellow	+	yellow	→	yellow/yellow	→	yellow

Review and Preview

In this chapter we combine the methods of *descriptive statistics* presented in Chapters 2 and 3 and those of *probability* presented in Chapter 4 to describe and analyze *probability distributions*. Probability distributions describe what will probably happen instead of what actually did happen, and they are often given in the format of a graph, table, or formula. Recall that in Chapter 2 we used observed sample data to construct frequency distributions. In this chapter we use the possible outcomes of a procedure (determined using the methods of Chapter 4) along with the *expected* relative frequencies to construct probability distributions, which serve as models of theoretically perfect frequency distributions. With this knowledge of population outcomes, we are able to find important characteristics, such as the mean and standard deviation, and to compare theoretical probabilities to actual results in order to determine whether outcomes are unusual.

Figure 5-1 provides a visual summary of what we will accomplish in this chapter. Using the methods of Chapters 2 and 3, we would repeatedly roll the die to collect sample data, which could then be described visually (with a histogram or boxplot), or numerically using measures of center (such as the mean) and measures of variation (such as the standard deviation). Using the methods of Chapter 4, we could find the probability of each possible outcome. Then we could construct a probability distribution, which describes the relative frequency table for a die rolled an infinite number of times.

In order to fully understand probability distributions, we must first understand the concept of a random variable, and be able to distinguish between discrete and continuous random variables. In this chapter we focus on *discrete* probability distributions. In particular, we discuss binomial and Poisson probability distributions. We will discuss *continuous* probability distributions in Chapter 6.

The table at the extreme right in Figure 5-1 represents a probability distribution that serves as a model of a theoretically perfect population frequency distribution.

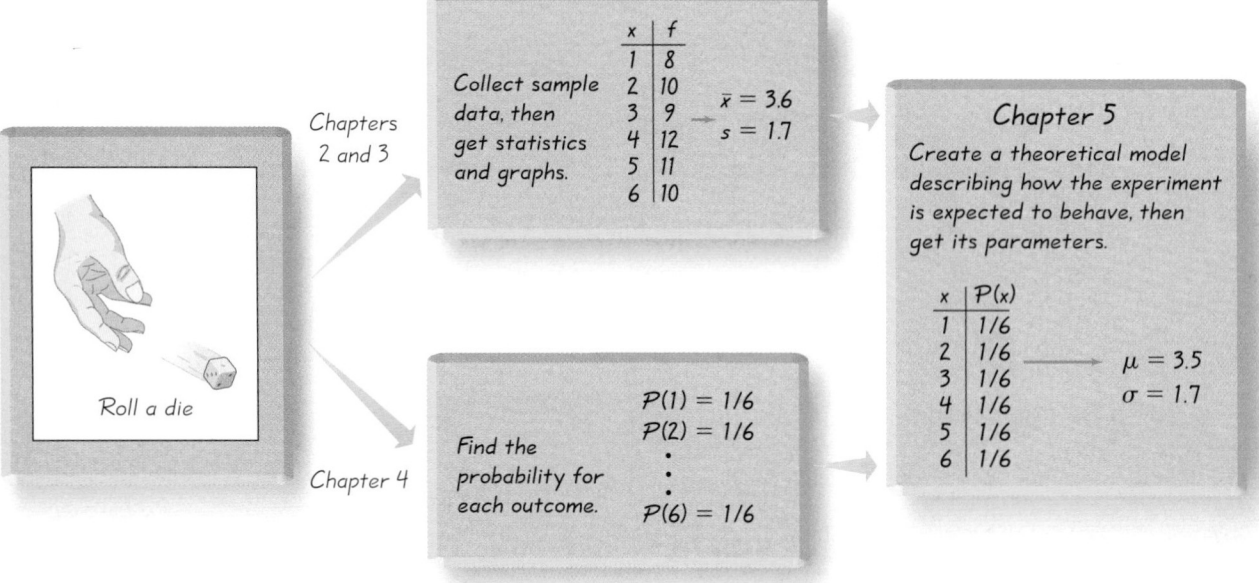

Figure 5-1   Combining Descriptive Methods and Probabilities to Form a Theoretical Model of Behavior

In essence, we can describe the relative frequency table for a die rolled an infinite number of times. With this knowledge of the population of outcomes, we are able to find its important characteristics, such as the mean and standard deviation. The remainder of this book and the very core of inferential statistics are based on some knowledge of probability distributions. We begin by examining the concept of a random variable, and then we consider important distributions that have many real applications.

5-2   Random Variables

Key Concept In this section we consider the concept of random variables and how they relate to probability distributions. We also discuss how to distinguish between discrete random variables and continuous random variables. In addition, we develop formulas for finding the mean, variance, and standard deviation for a probability distribution. Most importantly, we focus on determining whether outcomes are likely to occur by chance or they are unusual (in the sense that they are not likely to occur by chance).

We begin with the related concepts of *random variable* and *probability distribution*.

Table 5-1 Probability Distribution: Probabilities of Numbers of Peas with Green Pods Among 5 Offspring Peas

x (Number of Peas with Green Pods)	P(x)
0	0.001
1	0.015
2	0.088
3	0.264
4	0.396
5	0.237

 **DEFINITION**

A **random variable** is a variable (typically represented by x) that has a single numerical value, determined by chance, for each outcome of a procedure.

A **probability distribution** is a description that gives the probability for each value of the random variable. It is often expressed in the format of a graph, table, or formula.

EXAMPLE 1   **Genetics** Consider the offspring of peas from parents both having the green/yellow combination of pod genes. Under these conditions, the probability that the offspring has a green pod is 3/4 or 0.75. That is, $P(\text{green}) = 0.75$. If five such offspring are obtained, and if we let

$$x = \text{number of peas with green pods among 5 offspring peas}$$

then x is a random variable because its value depends on chance. Table 5-1 is a probability distribution because it gives the probability for each value of the random variable x. (In Section 5-3 we will see how to find the probability values, such as those listed in Table 5-1.)

Note: If a probability value is very small, such as 0.000000123, we can represent it as 0+ in a table, where 0+ indicates that the probability value is a very small positive number. (Representing the small probability as 0 would incorrectly indicate that the event is impossible.)

In Section 1-2 we made a distinction between discrete and continuous data. Random variables may also be discrete or continuous, and the following two definitions are consistent with those given in Section 1-2.

Figure 5-2

Devices Used to Count and Measure Discrete and Continuous Random Variables

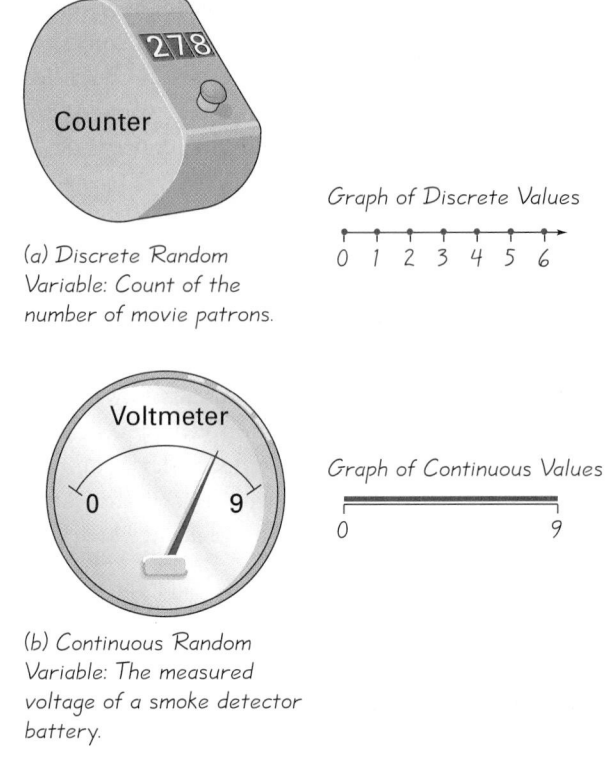

Graph of Discrete Values

0 1 2 3 4 5 6

(a) Discrete Random Variable: Count of the number of movie patrons.

Graph of Continuous Values

0   9

(b) Continuous Random Variable: The measured voltage of a smoke detector battery.

> **DEFINITION**
>
> A **discrete random variable** has either a finite number of values or a countable number of values, where "countable" refers to the fact that there might be infinitely many values, but they can be associated with a counting process, so that the number of values is 0 or 1 or 2 or 3, etc.
>
> A **continuous random variable** has infinitely many values, and those values can be associated with measurements on a continuous scale without gaps or interruptions.

This chapter deals exclusively with discrete random variables, but the following chapters will deal with continuous random variables.

> **EXAMPLE 2**   The following are examples of discrete and continuous random variables.
>
> 1. **Discrete** Let x = the number of eggs that a hen lays in a day. This is a *discrete* random variable because its only possible values are 0, or 1, or 2, and so on. No hen can lay 2.343115 eggs, which would have been possible if the data had come from a continuous scale.
>
> 2. **Discrete** The count of the number of statistics students present in class on a given day is a whole number and is therefore a discrete random variable. The counting device shown in Figure 5-2(a) is capable of indicating only a finite number of values, so it is used to obtain values for a *discrete* random variable.
>
> 3. **Continuous** Let x = the amount of milk a cow produces in one day. This is a *continuous* random variable because it can have any value over a continuous span. During a single day, a cow might yield an amount of milk that can be any value

between 0 gallons and 5 gallons. It would be possible to get 4.123456 gallons, because the cow is not restricted to the discrete amounts of 0, 1, 2, 3, 4, or 5 gallons.

4. Continuous The measure of voltage for a particular smoke detector battery can be any value between 0 volts and 9 volts. It is therefore a continuous random variable. The voltmeter shown in Figure 5-2(b) is capable of indicating values on a continuous scale, so it can be used to obtain values for a *continuous* random variable.

Graphs

There are various ways to graph a probability distribution, but we will consider only the **probability histogram.** Figure 5-3 is a probability histogram. Notice that it is similar to a relative frequency histogram (see Chapter 2), but the vertical scale shows *probabilities* instead of relative frequencies based on actual sample results.

In Figure 5-3, we see that the values of 0, 1, 2, 3, 4, 5 along the horizontal axis are located at the centers of the rectangles. This implies that the rectangles are each 1 unit wide, so the areas of the rectangles are 0.001, 0.015, 0.088, 0.264, 0.396, 0.237. The *areas* of these rectangles are the same as the *probabilities* in Table 5-1. We will see in Chapter 6 and future chapters that such a correspondence between area and probability is very useful in statistics.

Every probability distribution must satisfy each of the following two requirements.

Requirements for a Probability Distribution
1. $\Sigma P(x) = 1$ where *x* assumes all possible values. (The sum of all probabilities must be 1, but values such as 0.999 or 1.001 are acceptable because they result from rounding errors.)
2. $0 \leq P(x) \leq 1$ for every individual value of *x*. (That is, each probability value must be between 0 and 1 inclusive.)

The first requirement comes from the simple fact that the random variable *x* represents all possible events in the entire sample space, so we are certain (with probability 1) that one of the events will occur. In Table 5-1 we see that the sum of the probabilities is 1.001 (due to rounding errors) and that every value $P(x)$ is between 0 and 1. Because Table 5-1 satisfies the above requirements, we confirm that it is a probability distribution. A probability distribution may be described by a table, such as Table 5-1, or a graph, such as Figure 5-3, or a formula.

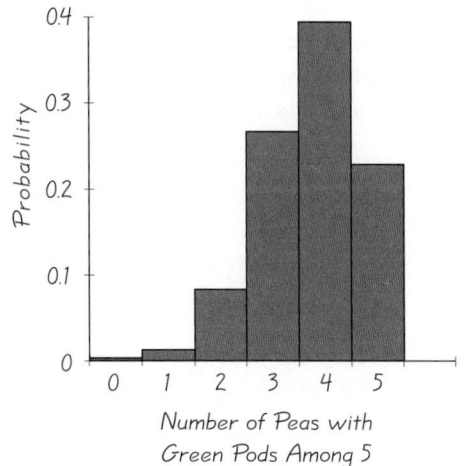

Number of Peas with
Green Pods Among 5

Life Data Analysis

Life data analysis deals with the longevity and failure rates of manufactured products. In one application, it is known that Dell computers have an "infant mortality" rate, whereby the failure rate is highest immediately after the computers are produced. Dell therefore tests or "burns-in" the computers before they are shipped. Dell can optimize profits by using an optimal burn-in time that identifies failures without wasting valuable testing time. Other products, such as cars, have failure rates that increase over time as parts wear out. If General Motors or Dell or any other company were to ignore the use of statistics and life data analysis, it would run the serious risk of going out of business because of factors such as excessive warranty repair costs or the loss of customers who experience unacceptable failure rates.

The *Weibull distribution* is a probability distribution commonly used in life data analysis applications. That distribution is beyond the scope of this book.

Figure 5-3

Probability Histogram

Table 5-2 Cell Phones per Household

x	P(x)
0	0.19
1	0.26
2	0.33
3	0.13

EXAMPLE 3   **Cell Phones** Based on a survey conducted by Frank N. Magid Associates, Table 5-2 lists the probabilities for the number of cell phones in use per household. Does Table 5-2 describe a probability distribution?

SOLUTION   To be a probability distribution, $P(x)$ must satisfy the preceding two requirements. But

$$\Sigma P(x) = P(0) + P(1) + P(2) + P(3)$$
$$= 0.19 + 0.26 + 0.33 + 0.13$$
$$= 0.91 \quad [\text{showing that } \Sigma P(x) \neq 1]$$

Because the first requirement is not satisfied, we conclude that Table 5-2 does *not* describe a probability distribution.

EXAMPLE 4   Does $P(x) = \dfrac{x}{10}$ (where x can be 0, 1, 2, 3, or 4) determine a probability distribution?

SOLUTION   For the given formula we find that $P(0) = 0/10$, $P(1) = 1/10$, $P(2) = 2/10$, $P(3) = 3/10$, and $P(4) = 4/10$, so that

1. $\Sigma P(x) = \dfrac{0}{10} + \dfrac{1}{10} + \dfrac{2}{10} + \dfrac{3}{10} + \dfrac{4}{10} = \dfrac{10}{10} = 1$

2. Each of the $P(x)$ values is between 0 and 1.

Because both requirements are satisfied, the formula given in this example is a probability distribution.

Mean, Variance, and Standard Deviation

In Chapter 2 we described the following characteristics of data (which can be remembered with the mnemonic of CVDOT for "**C**omputer **V**iruses **D**estroy **O**r **T**erminate"): (1) center; (2) variation; (3) distribution; (4) outliers; and (5) time (changing characteristics of data over time). These same characteristics can be used to describe probability distributions. A probability histogram or table can provide insight into the distribution of random variables. The mean is the central or "average" value of the random variable for a procedure repeated an infinite number of times. The variance and standard deviation measure the variation of the random variable. The mean, variance, and standard deviation for a probability distribution can be found by using these formulas:

Formula 5-1 $\mu = \Sigma[x \cdot P(x)]$ Mean for a probability distribution

Formula 5-2 $\sigma^2 = \Sigma[(x - \mu)^2 \cdot P(x)]$ Variance for a probability distribution (easier to understand)

Formula 5-3 $\sigma^2 = \Sigma[x^2 \cdot P(x)] - \mu^2$ Variance for a probability distribution (easier computations)

Formula 5-4 $\sigma = \sqrt{\Sigma[x^2 \cdot P(x)] - \mu^2}$ Standard deviation for a probability distribution

···· EXAMPLE 5

Finding the Mean, Variance, and Standard Deviation
Table 5-1 describes the probability distribution for the number of peas with green pods among 5 offspring peas obtained from parents both having the green/yellow pair of genes. Find the mean, variance, and standard deviation for the probability distribution described in Table 5-1 from Example 1.

SOLUTION

In Table 5-3, the two columns at the left describe the probability distribution given earlier in Table 5-1, and we create the three columns at the right for the purposes of the calculations required.

Using Formulas 5-1 and 5-2 and the table results, we get

Mean: $\quad \mu = \Sigma[x \cdot P(x)] = 3.752 = 3.8 \qquad$ (rounded)

Variance: $\quad \sigma^2 = \Sigma[(x - \mu)^2 \cdot P(x)] = 0.940574 = 0.9$ (rounded)

The standard deviation is the square root of the variance, so

Standard deviation: $\quad \sigma = \sqrt{0.940574} = 0.969832 = 1.0 \qquad$ (rounded)

Table 5-3 Calculating μ, σ, and σ^2 for a Probability Distribution

x	$P(x)$	$x \cdot P(x)$	$(x - \mu)^2 \cdot P(x)$
0	0.001	$0 \cdot 0.001 = 0.000$	$(0 - 3.752)^2 \cdot 0.001 = 0.014078$
1	0.015	$1 \cdot 0.015 = 0.015$	$(1 - 3.752)^2 \cdot 0.015 = 0.113603$
2	0.088	$2 \cdot 0.088 = 0.176$	$(2 - 3.752)^2 \cdot 0.088 = 0.270116$
3	0.264	$3 \cdot 0.264 = 0.792$	$(3 - 3.752)^2 \cdot 0.264 = 0.149293$
4	0.396	$4 \cdot 0.396 = 1.584$	$(4 - 3.752)^2 \cdot 0.396 = 0.024356$
5	0.237	$5 \cdot 0.237 = 1.185$	$(5 - 3.752)^2 \cdot 0.237 = 0.369128$
Total		3.752	0.940574
		↑	↑
		$\mu = \Sigma[x \cdot P(x)]$	$\sigma^2 = \Sigma[(x - \mu)^2 \cdot P(x)]$

INTERPRETATION

The mean number of peas with green pods is 3.8 peas, the variance is 0.9 "peas squared," and the standard deviation is 1.0 pea.

Using the TI-83/84 Plus Calculator

TI-83/84 Plus

```
1-Var Stats
 x̄=3.748251748
 Σx=3.752
 Σx²=15.004
 Sx=
 σx=.9693398892
↓n=1.001
```

The TI-83/84 Plus calculator can be used to find the mean and standard deviation corresponding to Formulas 5-1 and 5-4. Shown here is the screen display that results from the probability distribution summarized in Table 5-1. In this TI-83/84 Plus display, the value shown as $\bar{x}$ is actually the value of the mean μ, and the value shown as σx is the value of the standard deviation σ. Based on the displayed results, we get $\mu = 3.7$ and $\sigma = 1.0$ after rounding. Because of rounding errors, the value of the mean differs slightly from the value found in Example 5. The procedure for obtaining these results is given on page 220.

Rationale for Formulas 5-1 through 5-4

Instead of blindly accepting and using formulas, it is much better to have some understanding of why they work. When computing the mean from a frequency distribution, f represents class frequency and N represents population size. In the expression below, we rewrite the formula for the mean of a frequency table so that it applies to a

How to Choose Lottery Numbers

Many books and suppliers of computer programs claim to be helpful in predicting winning lottery numbers. Some use the theory that particular numbers are "due" (and should be selected) because they haven't been coming up often; others use the theory that some numbers are "cold" (and should be avoided) because they haven't been coming up often; and still others use astrology, numerology, or dreams. Because selections of winning lottery number combinations are independent events, such theories are worthless. A valid approach is to choose numbers that are "rare" in the sense that they are not selected by other people, so that if you win, you will not need to share your jackpot with many others. The combination of 1, 2, 3, 4, 5, 6 is a poor choice because many people tend to select it. In a Florida lottery with a $105 million prize, 52,000 tickets had 1, 2, 3, 4, 5, 6; if that combination had won, the prize would have been only $1000. It's wise to pick combinations not selected by many others. Avoid combinations that form a pattern on the entry card.

population. In the fraction f/N, the value of f is the frequency with which the value x occurs and N is the population size, so f/N is the probability for the value of x. When we replace f/N with $P(x)$, we make the transition from relative frequency based on a limited number of observations to probability based on infinitely many trials.

$$\mu = \frac{\Sigma(f \cdot x)}{N} = \Sigma\left[\frac{f \cdot x}{N}\right] = \Sigma\left[x \cdot \frac{f}{N}\right] = \Sigma[x \cdot P(x)]$$

Similar reasoning enables us to take the variance formula from Chapter 3 and apply it to a random variable for a probability distribution; the result is Formula 5-2. Formula 5-3 is a shortcut version that will always produce the same result as Formula 5-2. Although Formula 5-3 is usually easier to work with, Formula 5-2 is easier to understand directly. Based on Formula 5-2, we can express the standard deviation as

$$\sigma = \sqrt{\Sigma[(x - \mu)^2 \cdot P(x)]}$$

or as the equivalent form given in Formula 5-4.

When applying Formulas 5-1 through 5-4, use this rule for rounding results.

> **Round-off Rule for μ, σ, and σ^2**
>
> **Round results by carrying one more decimal place than the number of decimal places used for the random variable x. If the values of x are integers, round μ, σ, and σ^2 to one decimal place.**

It is sometimes necessary to use a different rounding rule because of special circumstances, such as results that require more decimal places to be meaningful. For example, with four-engine jets the mean number of jet engines working successfully throughout a flight is 3.999714286, which becomes 4.0 when rounded to one more decimal place than the original data. Here, 4.0 would be misleading because it suggests that all jet engines always work successfully. We need more precision to correctly reflect the true mean, such as the precision in the number 3.999714.

Identifying *Unusual* Results with the Range Rule of Thumb

The range rule of thumb (introduced in Section 3-3) may be helpful in interpreting the value of a standard deviation. According to the range rule of thumb, most values should lie within 2 standard deviations of the mean; it is unusual for a value to differ from the mean by more than 2 standard deviations. (The use of 2 standard deviations is not an absolutely rigid value, and other values such as 3 could be used instead.) We can therefore identify "unusual" values by determining that they lie outside of these limits:

Range Rule of Thumb

$$\text{maximum usual value} = \mu + 2\sigma$$
$$\text{minimum usual value} = \mu - 2\sigma$$

CAUTION

Know that the use of the number 2 in the range rule of thumb is somewhat arbitrary, and this rule is a guideline, not an absolutely rigid rule.

Identifying *Unusual* Results with Probabilities

> **Rare Event Rule for Inferential Statistics**
>
> If, under a given assumption (such as the assumption that a coin is fair), the probability of a particular observed event (such as 992 heads in 1000 tosses of a coin) is extremely small, we conclude that the assumption is probably not correct.

Probabilities can be used to apply the rare event rule as follows:

Using Probabilities to Determine When Results Are Unusual

- **Unusually *high* number of successes:** x successes among n trials is an *unusually high* number of successes if the probability of x or more successes is unlikely with a probability of 0.05 or less. This criterion can be expressed as follows: $P(x \text{ or more}) \leq 0.05$.*

- **Unusually *low* number of successes:** x successes among n trials is an *unusually low* number of successes if the probability of x or fewer successes is unlikely with a probability of 0.05 or less. This criterion can be expressed as follows: $P(x \text{ or fewer}) \leq 0.05$.*

*The value 0.05 is not absolutely rigid. Other values, such as 0.01, could be used to distinguish between results that can easily occur by chance and events that are very unlikely to occur by chance.

Study hint: Take time to carefully read and understand the above rare event rule and the following paragraph. The next paragraph illustrates an extremely important approach used often in statistics.

Suppose you were tossing a coin to determine whether it favors heads, and suppose 1000 tosses resulted in 501 heads. This is not evidence that the coin favors heads, because it is very easy to get a result like 501 heads in 1000 tosses just by chance. Yet, the probability of getting *exactly* 501 heads in 1000 tosses is actually quite small: 0.0252. This low probability reflects the fact that with 1000 tosses, *any specific* number of heads will have a very low probability. However, we do not consider 501 heads among 1000 tosses to be *unusual,* because the probability of *501 or more* heads is high: 0.487.

EXAMPLE 6   **Identifying Unusual Results with the Range Rule of Thumb**
In Example 5 we found that for groups of 5 offspring (generated from parents both having the green/yellow pair of genes), the mean number of peas with green pods is 3.8, and the standard deviation is 1.0. Use those results and the range rule of thumb to find the maximum and minimum usual values. Based on the results, determine whether it is unusual to generate 5 offspring peas and find that only 1 of them has a green pod.

SOLUTION   Using the range rule of thumb, we can find the maximum and minimum usual values as follows:

$$\text{maximum usual value:} \quad \mu + 2\sigma = 3.8 + 2(1.0) = 5.8$$
$$\text{minimum usual value:} \quad \mu - 2\sigma = 3.8 - 2(1.0) = 1.8$$

INTERPRETATION   Based on these results, we conclude that for groups of 5 offspring peas, the number of offspring peas with green pods should usually fall between 1.8 and 5.8. If 5 offspring peas are generated as described, it would be unusual to get only 1 with a green pod (because the value of 1 is outside of this range of usual values: 1.8 to 5.8). (In this case, the maximum usual value is actually 5, because that is the largest possible number of peas with green pods.)

EXAMPLE 7   **Identifying Unusual Results with Probabilities** Use *probabilities* to determine whether 1 is an unusually low number of peas with green pods when 5 offspring are generated from parents both having the green/yellow pair of genes.

SOLUTION   To determine whether 1 is an unusually low number of peas with green pods (among 5 offspring), we need to find the probability of getting 1 or fewer peas with green pods. By referring to Table 5-1 on page 211 we can easily get the following results:

$$P(1 \text{ or fewer}) = P(1 \text{ or } 0) = 0.015 + 0.001 = 0.016.$$

INTERPRETATION   Because the probability 0.016 is less than 0.05, we conclude that the result of 1 pea with a green pod is *unusually low*. There is a very small likelihood (0.016) of getting 1 or fewer peas with green pods.

Expected Value

The mean of a discrete random variable is the theoretical mean outcome for infinitely many trials. We can think of that mean as the *expected value* in the sense that it is the average value that we would expect to get if the trials could continue indefinitely. The uses of expected value (also called *expectation,* or *mathematical expectation*) are extensive and varied, and they play an important role in *decision theory.*

DEFINITION

The **expected value** of a discrete random variable is denoted by E, and it represents the mean value of the outcomes. It is obtained by finding the value of $\Sigma[x \cdot P(x)]$

$$E = \Sigma[x \cdot P(x)]$$

CAUTION

An expected value need not be a whole number, even if the different possible values of x might all be whole numbers.

From Formula 5-1 we see that $E = \mu$. That is, the mean of a discrete random variable is the same as its expected value. For example, when generating groups of five offspring peas, the mean number of peas with green pods is 3.8 (see Table 5-3). So, it follows that the expected value of the number of peas with green pods is also 3.8.

Because the concept of expected value is used often in decision theory, the following example involves a real decision.

EXAMPLE 8   **How to Be a Better Bettor** You are considering placing a bet either on the number 7 in roulette or on the "pass line" in the dice game of craps at the Venetian casino in Las Vegas.

a. If you bet $5 on the number 7 in roulette, the probability of losing $5 is 37/38 and the probability of making a net gain of $175 is 1/38. (The prize is $180, including your $5 bet, so the net gain is $175.) Find your expected value if you bet $5 on the number 7 in roulette.

b. If you bet $5 on the pass line in the dice game of craps, the probability of losing $5 is 251/495 and the probability of making a net gain of $5 is 244/495. (If you bet $5 on the Pass Line and win, you are given $10 that includes your bet, so the net gain is $5.) Find your expected value if you bet $5 on the Pass Line.

Which bet is better: A $5 bet on the number 7 in roulette or a $5 bet on the pass line in the dice game? Why?

SOLUTION

a. Roulette The probabilities and payoffs for betting $5 on the number 7 in roulette are summarized in Table 5-4. Table 5-4 also shows that the expected value is $\Sigma[x \cdot P(x)] = -26¢$. That is, for every $5 bet on the number 7, you can expect to lose an average of 26¢.

Table 5-4   Roulette

Event	x	P(x)	x · P(x)
Lose	−$5	37/38	−$4.87
Gain (net)	$175	1/38	$4.61
Total			−$0.26
			(or −26¢)

b. Dice The probabilities and payoffs for betting $5 on the pass line in craps are summarized in Table 5-5. Table 5-5 also shows that the expected value is $\Sigma[x \cdot P(x)] = -8¢$. That is, for every $5 bet on the Pass Line, you can expect to lose an average of 8¢.

Table 5-5   Dice

Event	x	P(x)	x · P(x)
Lose	−$5	251/495	−$2.54
Gain (net)	$5	244/495	$2.46
Total			−$0.08
			(or −8¢)

INTERPRETATION The $5 bet in roulette results in an expected value of −26¢ and the $5 bet in craps results in an expected value of −8¢. The bet in the dice game is better because it has the larger expected value. That is, you are better off losing 8¢ instead of losing 26¢. Even though the roulette game provides an opportunity for a larger payoff, the craps game is better in the long run.

In this section we learned that a random variable has a numerical value associated with each outcome of some random procedure, and a probability distribution has a probability associated with each value of a random variable. We examined methods for finding the mean, variance, and standard deviation for a probability distribution. We saw that the expected value of a random variable is really the same as the mean. Finally, the range rule of thumb or probabilities can be used for determining when outcomes are *unusual*.

Entering the Probability Distribution of Table 5-1

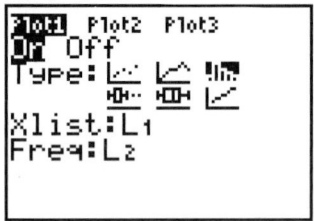

Finding the Mean and Standard Deviation

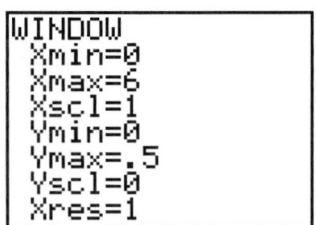

Entries for Selecting Graph

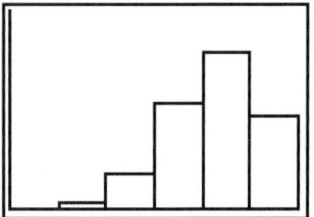

Window Settings for Table 5-1

```
WINDOW
 Xmin=0
 Xmax=6
 Xscl=1
 Ymin=0
 Ymax=.5
 Yscl=0
 Xres=1
```

Probability Histogram for Table 5-1

Using a TI-83 Plus Calculator for Probability Distributions

Entering a Probability Distribution   Let's assume that we have a probability distribution in the format of a table, such as Table 5-1. We can enter that table in a TI-83/84 Plus calculator by using the following procedure.

1. Clear lists L1 and L2 as follows: Press **STAT**, select **4:CrlList**, then enter **L1, L2** (with the comma included), and press **ENTER**.

2. Press **STAT** and **ENTER** to enter the edit mode, and proceed to enter the values of the random variable x in list L1, and enter the corresponding probabilities in list L2. See the first display in the margin for the screen showing the entry of Table 5-1.

Finding the Mean and Standard Deviation   First enter the probability distribution by using the procedure described in Steps 1 and 2 above. Then find the mean and standard deviation by using the following procedure.

1. After completing the entry of the probability distribution as described above, quit the edit mode by pressing **2ND** **MODE** (for "QUIT").

2. Now enter the command of **1-Var Stats L1, L2** by using this sequence of keystrokes: Press **STAT**, then select **CALC** by pressing the ▶ key, and press **ENTER**. The screen should now show 1-Var Stats.

3. Enter **L1, L2** (including the comma), so that the screen shows 1-Var Stats L1, L2. (Enter L1 by pressing **2ND** **1**, press **,**, then enter L2 by pressing **2ND** **2**.)

4. With the command of **1-Var Stats L1, L2** displayed, press **ENTER** to get the results shown in the second screen display in the margin. *Interpreting results:* Note that the value shown as $\bar{x}$ is actually the population mean μ, so $\mu = 3.748251748$, which we round to 3.7. Also, there is no value shown for the sample standard deviation indicated by Sx, but the population standard deviation is indicated by σx, so we know that $\sigma = 0.9693398892$, which we round to 1.0. The value of $n = 1.001$ is the sum of the probabilities in column L2, and the sum is not exactly 1 because of a rounding error in the listed probabilities.

Obtaining a Probability Histogram   Enter the probability distribution as described above, then use the following procedure.

1. After completing the entry of the probability distribution as described above, quit the edit mode by pressing **2ND** **MODE** (for "QUIT").

2. Access the STAT PLOT menu by pressing **2ND** **Y=**.

3. Press **ENTER**, then press **ENTER** again to turn on the first plot (with the other plots turned off). Select the graph type that is the third entry of the first row.

4. For Xlist, enter **L1.**

5. For Freq (frequency), enter **L2.**

6. Press **WINDOW** and adjust the settings. For the probability distribution of Table 5-1, use the settings shown in the screen display, then press **GRAPH** to get the probability histogram shown. You can also press **TRACE** and use the arrow keys to find values corresponding to individual elements of the graph.

5-2   Basic Skills and Concepts

Statistical Literacy and Critical Thinking

1. Random Variable What is a random variable? A friend of the author buys one lottery ticket every week in one year. Over the 52 weeks, she counts the number of times that she won something. In this context, what is the random variable, and what are its possible values?

2. Expected Value A researcher calculates the expected value for the number of girls in three births. He gets a result of 1.5. He then rounds the result to 2, saying that it is not possible to get 1.5 girls when three babies are born. Is this reasoning correct? Explain.

3. Probability Distribution One of the requirements of a probability distribution is that the sum of the probabilities must be 1 (with a small discrepancy allowed for rounding errors). What is the justification for this requirement?

4. Probability Distribution A professional gambler claims that he has loaded a die so that the outcomes of 1, 2, 3, 4, 5, 6 have corresponding probabilities of 0.1, 0.2, 0.3, 0.4, 0.5, and 0.6. Can he actually do what he has claimed? Is a probability distribution described by listing the outcomes along with their corresponding probabilities?

Identifying Discrete and Continuous Random Variables. *In Exercises 5 and 6, identify the given random variable as being* discrete *or* continuous.

5. a. The number of people now driving a car in the United States

b. The weight of the gold stored in Fort Knox

c. The height of the last airplane that departed from JFK Airport in New York City

d. The number of cars in San Francisco that crashed last year

e. The time required to fly from Los Angeles to Shanghai

6. a. The total amount (in ounces) of soft drinks that you consumed in the past year

b. The number of cans of soft drinks that you consumed in the past year

c. The number of movies currently playing in U.S. theaters

d. The running time of a randomly selected movie

e. The cost of making a randomly selected movie

Identifying Probability Distributions. *In Exercises 7–12, determine whether or not a probability distribution is given. If a probability distribution is given, find its mean and standard deviation. If a probability distribution is not given, identify the requirements that are not satisfied.*

TI 7. Genetic Disorder Three males with an X-linked genetic disorder have one child each. The random variable x is the number of children among the three who inherit the X-linked genetic disorder.

x	$P(x)$
0	0.125
1	0.375
2	0.375
3	0.125

TI 8. Caffeine Nation In the accompanying table, the random variable x represents the number of cups or cans of caffeinated beverages consumed by Americans each day (based on data from the National Sleep Foundation).

x	$P(x)$
0	0.22
1	0.16
2	0.21
3	0.16

TI 9. Overbooked Flights Air America has a policy of routinely overbooking flights. The random variable x represents the number of passengers who cannot be boarded because there are more passengers than seats (based on data from an IBM research paper by Lawrence, Hong, and Cherrier).

x	$P(x)$
0	0.051
1	0.141
2	0.274
3	0.331
4	0.187

10. Eye Color Groups of five babies are randomly selected. In each group, the random variable x is the number of babies with green eyes (based on data from a study by Dr. Sorita Soni at Indiana University). (The symbol 0+ denotes a positive probability value that is very small.)

x	$P(x)$
0	0.528
1	0.360
2	0.098
3	0.013
4	0.001
5	0+

11. American Televisions In the accompanying table, the random variable x represents the number of televisions in a household in the United States (based on data from Frank N. Magid Associates).

x	$P(x)$
0	0.02
1	0.15
2	0.29
3	0.26
4	0.16
5	0.12

12. TV Ratings In a study of television ratings, groups of 6 U.S. households are randomly selected. In the accompanying table, the random variable x represents the number of households among 6 that are tuned to *60 Minutes* during the time that the show is broadcast (based on data from Nielsen Media Research).

x	$P(x)$
0	0.539
1	0.351
2	0.095
3	0.014
4	0.001
5	0+
6	0+

Pea Hybridization Experiment. *In Exercises 13–16, refer to the accompanying table, which describes results from eight offspring peas. The random variable x represents the number of offspring peas with green pods.*

13. Mean and Standard Deviation Find the mean and standard deviation for the numbers of peas with green pods.

14. Range Rule of Thumb for Unusual Events Use the range rule of thumb to identify a range of values containing the usual number of peas with green pods. Based on the result, is it unusual to get only one pea with a green pod? Explain.

15. Using Probabilities for Unusual Events

a. Find the probability of getting exactly 7 peas with green pods.

b. Find the probability of getting 7 or more peas with green pods.

c. Which probability is relevant for determining whether 7 is an unusually high number of peas with green pods: the result from part (a) or part (b)?

d. Is 7 an unusually high number of peas with green pods? Why or why not?

16. Using Probabilities for Unusual Events

a. Find the probability of getting exactly 3 peas with green pods.

b. Find the probability of getting 3 or fewer peas with green pods.

c. Which probability is relevant for determining whether 3 is an unusually low number of peas with green pods: the result from part (a) or part (b)?

d. Is 3 an unusually low number of peas with green pods? Why or why not?

Probabilities of Numbers of Peas with Green Pods Among 8 Offspring Peas

x (Number of Peas with Green Pods)	$P(x)$
0	0+
1	0+
2	0.004
3	0.023
4	0.087
5	0.208
6	0.311
7	0.267
8	0.100

17. Baseball World Series Based on past results found in the *Information Please Almanac,* there is a 0.1919 probability that a baseball World Series contest will last four games, a 0.2121 probability that it will last five games, a 0.2222 probability that it will last six games, and a 0.3737 probability that it will last seven games.

a. Does the given information describe a probability distribution?

b. Assuming that the given information describes a probability distribution, find the mean and standard deviation for the numbers of games in World Series contests.

c. Is it unusual for a team to "sweep" by winning in four games? Why or why not?

18. Job Interviews Based on information from MRINetwork, some job applicants are required to have several interviews before a decision is made. The number of required interviews and the corresponding probabilities are: 1 (0.09); 2 (0.31); 3 (0.37); 4 (0.12); 5 (0.05); 6 (0.05).

a. Does the given information describe a probability distribution?

b. Assuming that a probability distribution is described, find its mean and standard deviation.

c. Use the range rule of thumb to identify the range of values for usual numbers of interviews.

d. Is it unusual to have a decision after just one interview? Explain.

19. Bumper Stickers Based on data from CarMax.com, when a car is randomly selected, the number of bumper stickers and the corresponding probabilities are: 0 (0.824); 1 (0.083); 2 (0.039); 3 (0.014); 4 (0.012); 5 (0.008); 6 (0.008); 7 (0.004); 8 (0.004); 9 (0.004).

a. Does the given information describe a probability distribution?

b. Assuming that a probability distribution is described, find its mean and standard deviation.

c. Use the range rule of thumb to identify the range of values for usual numbers of bumper stickers.

d. Is it unusual for a car to have more than one bumper sticker? Explain.

20. Gender Discrimination The Telektronic Company hired 8 employees from a large pool of applicants with an equal number of males and females. If the hiring is done without regard to sex, the numbers of females hired and the corresponding probabilities are as follows: 0 (0.004); 1 (0.031); 2 (0.109); 3 (0.219); 4 (0.273); 5 (0.219); 6 (0.109); 7 (0.031); 8 (0.004).

a. Does the given information describe a probability distribution?

b. Assuming that a probability distribution is described, find its mean and standard deviation.

c. Use the range rule of thumb to identify the range of values for usual numbers of females hired in such groups of eight.

d. If the most recent group of eight newly hired employees does not include any females, does there appear to be discrimination based on sex? Explain.

21. Finding Mean and Standard Deviation Let the random variable x represent the number of girls in a family of three children. Construct a table describing the probability distribution, then find the mean and standard deviation. (*Hint:* List the different possible outcomes.) Is it unusual for a family of three children to consist of three girls?

22. Finding Mean and Standard Deviation Let the random variable x represent the number of girls in a family of four children. Construct a table describing the probability distribution, then find the mean and standard deviation. (*Hint:* List the different possible outcomes.) Is it unusual for a family of four children to consist of four girls?

23. Random Generation of Telephone Numbers A description of a Pew Research Center poll referred to "the random generation of the last two digits of telephone numbers." Each digit has the same chance of being randomly generated. Construct a table representing the probability distribution for digits randomly generated by computer, find its mean and standard deviation, then describe the shape of the probability histogram.

24. Analysis of Leading Digits The analysis of the leading (first) digits of checks led to the conclusion that companies in Brooklyn, New York, were guilty of fraud. For the purposes of this exercise, assume that the leading digits of check amounts are randomly generated by computer.

a. Identify the possible leading digits.

b. Find the mean and standard deviation of such leading digits.

c. Use the range rule of thumb to identify the range of usual values.

d. Can any leading digit be considered unusual? Why or why not?

TI **25. Finding Expected Value for the Illinois Pick 3 Game** In the Illinois Pick 3 lottery game, you pay 50¢ to select a sequence of three digits, such as 233. If you select the same sequence of three digits that are drawn, you win and collect $250.

a. How many different selections are possible?

b. What is the probability of winning?

c. If you win, what is your net profit?

d. Find the expected value.

e. If you bet 50¢ in Illinois' Pick 4 game, the expected value is −25¢. Which bet is better: A 50¢ bet in the Illinois Pick 3 game or a 50¢ bet in the Illinois Pick 4 game? Explain.

TI **26. Expected Value in New Jersey's Pick 4 Game** In New Jersey's Pick 4 lottery game, you pay 50¢ to select a sequence of four digits, such as 1332. If you select the same sequence of four digits that are drawn, you win and collect $2788.

a. How many different selections are possible?

b. What is the probability of winning?

c. If you win, what is your net profit?

d. Find the expected value.

e. If you bet 50¢ in Illinois' Pick 4 game, the expected value is −25¢. Which bet is better: A 50¢ bet in the Illinois Pick 4 game or a 50¢ bet in New Jersey's Pick 4 game? Explain.

TI **27. Expected Value in Roulette** When playing roulette at the Bellagio casino in Las Vegas, a gambler is trying to decide whether to bet $5 on the number 13 or to bet $5 that the outcome is any one of these five possibilities: 0 or 00 or 1 or 2 or 3. From Example 8, we know that the expected value of the $5 bet for a single number is −26¢. For the $5 bet that the outcome is 0 or 00 or 1 or 2 or 3, there is a probability of $5/38$ of making a net profit of $30 and a $33/38$ probability of losing $5.

a. Find the expected value for the $5 bet that the outcome is 0 or 00 or 1 or 2 or 3.

b. Which bet is better: A $5 bet on the number 13 or a $5 bet that the outcome is 0 or 00 or 1 or 2 or 3? Why?

TI **28. Expected Value for *Deal or No Deal*** The television game show *Deal or No Deal* begins with individual suitcases containing the amounts of 1¢, $1, $5, $10, $25, $50, $75, $100, $200, $300, $400, $500, $750, $1000, $5000, $10,000, $25,000, $50,000, $75,000, $100,000, $200,000, $300,000, $400,000, $500,000, $750,000, and $1,000,000. If a player adopts the strategy of choosing the option of "no deal" until one suitcase remains, the payoff is one of the amounts listed, and they are all equally likely.

a. Find the expected value for this strategy.

b. Find the value of the standard deviation.

c. Use the range rule of thumb to identify the range of usual outcomes.

d. Based on the preceding results, is a result of $750,000 or $1,000,000 unusual? Why or why not?

TI **29. Expected Value for Life Insurance** There is a 0.9986 probability that a randomly selected 30-year-old male lives through the year (based on data from the U.S. Department of Health and Human Services). A Fidelity life insurance company charges $161 for insuring that the male will live through the year. If the male does not survive the year, the policy pays out $100,000 as a death benefit.

a. From the perspective of the 30-year-old male, what are the values corresponding to the two events of surviving the year and not surviving?

b. If a 30-year-old male purchases the policy, what is his expected value?

c. Can the insurance company expect to make a profit from many such policies? Why?

TI **30. Expected Value for Life Insurance** There is a 0.9968 probability that a randomly selected 50-year-old female lives through the year (based on data from the U.S. Department of Health and Human Services). A Fidelity life insurance company charges $226 for insuring that the female will live through the year. If she does not survive the year, the policy pays out $50,000 as a death benefit.

a. From the perspective of the 50-year-old female, what are the values corresponding to the two events of surviving the year and not surviving?

b. If a 50-year-old female purchases the policy, what is her expected value?

c. Can the insurance company expect to make a profit from many such policies? Why?

5-2 Beyond the Basics

31. Junk Bonds Kim Hunter has $1000 to invest, and her financial analyst recommends two types of junk bonds. The A bonds have a 6% annual yield with a default rate of 1%. The B bonds have an 8% annual yield with a default rate of 5%. (If the bond defaults, the $1000 is lost.) Which of the two bonds is better? Why? Should she select either bond? Why or why not?

32. Defective Parts: Finding Mean and Standard Deviation The Sky Ranch is a supplier of aircraft parts. Included in stock are eight altimeters that are correctly calibrated and two that are not. Three altimeters are randomly selected without replacement. Let the random variable x represent the number that are not correctly calibrated. Find the mean and standard deviation for the random variable x.

33. Labeling Dice to Get a Uniform Distribution Assume that you have two blank dice, so that you can label the 12 faces with any numbers. Describe how the dice can be labeled so that, when the two dice are rolled, the totals of the two dice are uniformly distributed in such a way that the outcomes of 1, 2, 3, . . . , 12 each have probability 1/12. (See "Can One Load a Set of Dice So That the Sum Is Uniformly Distributed?" by Chen, Rao, and Shreve, *Mathematics Magazine,* Vol. 70, No. 3.)

5-3 Binomial Probability Distributions

Key Concept In this section we focus on one particular category of discrete probability distributions: *binomial* probability distributions. Because binomial probability distributions involve proportions used with methods of inferential statistics discussed later in this book, it is important to understand fundamental properties of this particular class of probability distributions. In this section we present a basic definition of a binomial probability distribution along with notation, and methods for finding probability values. As in other sections, we want to interpret probability values to determine whether events are usual or unusual.

Binomial probability distributions allow us to deal with circumstances in which the outcomes belong to *two* relevant categories, such as acceptable/defective or survived/died. Other requirements are given in the following definition.

> **DEFINITION**
>
> A **binomial probability distribution** results from a procedure that meets all the following requirements:
>
> 1. The procedure has a *fixed number of trials*.
>
> 2. The trials must be *independent*. (The outcome of any individual trial doesn't affect the probabilities in the other trials.)
>
> 3. Each trial must have all outcomes classified into *two categories* (commonly referred to as *success* and *failure*).
>
> 4. The probability of a success remains the same in all trials.

Not At Home

Pollsters cannot simply ignore those who were not at home when they were called the first time. One solution is to make repeated callback attempts until the person can be reached. Alfred Politz and Willard Simmons describe a way to compensate for those missing results without making repeated callbacks. They suggest weighting results based on how often people are not at home. For example, a person at home only two days out of six will have a 2/6 or 1/3 probability of being at home when called the first time. When such a person is reached the first time, his or her results are weighted to count three times as much as someone who is always home. This weighting is a compensation for the other similar people who are home two days out of six and were not at home when called the first time. This clever solution was first presented in 1949.

Independence Requirement   When selecting a sample (such as survey subjects) for some statistical analysis, we usually sample without replacement. Recall that sampling without replacement involves dependent events, which violates the second requirement in the above definition. However, we can often assume independence by applying the following 5% guideline introduced in Section 4-4:

> **Treating Dependent Events as Independent: The 5% Guideline for Cumbersome Calculations**
>
> **If calculations are cumbersome and if a sample size is no more than 5% of the size of the population, treat the selections as being *independent* (even if the selections are made without replacement, so that they are technically dependent).**

If a procedure satisfies the above four requirements, the distribution of the random variable x (number of successes) is called a *binomial probability distribution* (or *binomial distribution*). The following notation is commonly used.

Notation for Binomial Probability Distributions

S and F (success and failure) denote the two possible categories of all outcomes.

$P(S) = p$	(p = probability of a success)
$P(F) = 1 - p = q$	(q = probability of a failure)
n	denotes the fixed number of trials.
x	denotes a specific number of successes in n trials, so x can be any whole number between 0 and n, inclusive.
p	denotes the probability of *success* in *one* of the n trials.
q	denotes the probability of *failure* in *one* of the n trials.
$P(x)$	denotes the probability of getting exactly x successes among the n trials.

The word *success* as used here is arbitrary and does not necessarily represent something good. Either of the two possible categories may be called the success S as long as its probability is identified as p. (The value of q can always be found by subtracting p from 1; if $p = 0.95$, then $q = 1 - 0.95 = 0.05$.)

CAUTION
....

When using a binomial probability distribution, always be sure that x and p both refer to the *same* category being called a success.

> ⋯⋯ **EXAMPLE 1**   **Genetics** Consider an experiment in which 5 offspring peas are generated from 2 parents each having the green/yellow combination of genes for pod color. Recall from the Chapter Problem that the probability an offspring pea will have a green pod is $\frac{3}{4}$ or 0.75. That is, $P(\text{green pod}) = 0.75$. Suppose we want to find the probability that exactly 3 of the 5 offspring peas have a green pod.
>
> **a.** Does this procedure result in a binomial distribution?
>
> **b.** If this procedure does result in a binomial distribution, identify the values of n, x, p, and q.

SOLUTION

a. This procedure does satisfy the requirements for a binomial distribution, as shown below.

 1. The number of trials (5) is fixed.

 2. The 5 trials are independent, because the probability of any offspring pea having a green pod is not affected by the outcome of any other offspring pea.

 3. Each of the 5 trials has two categories of outcomes: The pea has a green pod or it does not.

 4. For each offspring pea, the probability that it has a green pod is 3/4 or 0.75, and that probability remains the same for each of the 5 peas.

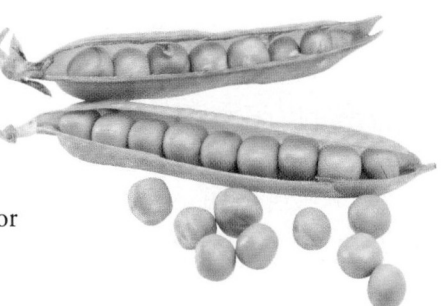

b. Having concluded that the given procedure does result in a binomial distribution, we now proceed to identify the values of n, x, p, and q.

 1. With 5 offspring peas, we have $n = 5$.

 2. We want the probability of exactly 3 peas with green pods, so $x = 3$.

 3. The probability of success (getting a pea with a green pod) for one selection is 0.75, so $p = 0.75$.

 4. The probability of failure (not getting a green pod) is 0.25, so $q = 0.25$.

Again, it is very important to be sure that x and p both refer to the same concept of "success." In this example, we use x to count the number of peas with green pods, so p must be the probability that a pea has a green pod. Therefore, x and p do use the same concept of success (green pod) here.

We now discuss three methods for finding the probabilities corresponding to the random variable x in a binomial distribution. The first method involves calculations using the *binomial probability formula* and is the basis for the other two methods. The second method involves the use of a TI-83/84 Plus calculator, and the third method involves the use of Table A-1. (With technology so widespread, such tables are becoming obsolete.) If you are using computer software or a calculator that automatically produces binomial probabilities, we recommend that you solve one or two exercises using Method 1 to ensure that you understand the basis for the calculations. Understanding is always infinitely better than blind application of formulas.

Method 1: Using the Binomial Probability Formula   In a binomial probability distribution, probabilities can be calculated by using the binomial probability formula.

Formula 5-5 nC_x

$$P(x) = \frac{n!}{(n-x)!\,x!} \cdot p^x \cdot q^{n-x} \quad \text{for } x = 0, 1, 2, \ldots, n$$

where n = number of trials

  x = number of successes among n trials

  p = probability of success in any one trial

  q = probability of failure in any one trial ($q = 1 - p$)

The factorial symbol !, introduced in Section 4-7, denotes the product of decreasing factors. Two examples of factorials are $3! = 3 \cdot 2 \cdot 1 = 6$ and $0! = 1$ (by definition).

EXAMPLE 2   **Genetics** Assuming that the probability of a pea having a green pod is 0.75 (as in the Chapter Problem and Example 1), use the binomial probability formula to find the probability of getting exactly 3 peas with green pods when 5 offspring peas are generated. That is, find $P(3)$ given that $n = 5$, $x = 3$, $p = 0.75$, and $q = 0.25$.

SOLUTION   Using the given values of n, x, p, and q in the binomial probability formula (Formula 5-5), we get

$$P(3) = \frac{5!}{(5-3)!\,3!} \cdot 0.75^3 \cdot 0.25^{5-3}$$

$$= \frac{5!}{2!\,3!} \cdot 0.421875 \cdot 0.0625$$

$$= (10)(0.421875)(0.0625) = 0.263671875$$

The probability of getting exactly 3 peas with green pods among 5 offspring peas is 0.264 (rounded to three significant digits).

Calculation hint: When computing a probability with the binomial probability formula, it's helpful to get a single number for $n!/[(n-x)!\,x!]$, a single number for p^x and a single number for q^{n-x}, then simply multiply the three factors together as shown at the end of the calculation for the preceding example. Don't round too much when you find those three factors; round only at the end.

Method 2: Using Technology The TI-83/84 Plus calculator can be used to find binomial probabilities. The screen display listing binomial probabilities for $n = 5$ and $p = 0.75$, as in Example 2, is given. See that the probability distribution is given as a table.

TI-83/84 PLUS

L1	L2	L3	2
0	9.8E-4	------	
1	.01465		
2	.08789		
3	.26367		
4	.39551		
5	.2373		

L2(7) =

Method 3: Using Table A-1 in Appendix A Table A-1 in Appendix A lists binomial probabilities for select values of n and p. Table A-1 cannot be used for Example 2 because the probability of $p = 0.75$ is not one of the probabilities included. Example 3 illustrates the use of Table A-1.

To use Table A-1, we must first locate n and the desired corresponding value of x. At this stage, one row of numbers should be isolated. Now align that row with the proper probability of p by using the column across the top. The isolated number represents the desired probability. A very small probability, such as 0.000064, is indicated by 0+.

EXAMPLE 3   **McDonald's Brand Recognition** The fast food chain McDonald's has a brand name recognition rate of 95% around the world (based on data from Retail Marketing Group). Assuming that we randomly select 5 people, use Table A-1 to find the following.

a. The probability that exactly 3 of the 5 people recognize McDonald's

b. The probability that the number of people who recognize McDonald's is 3 or fewer

SOLUTION

a. The displayed excerpt from Table A-1 below shows that when $n = 5$ and $p = 0.95$, the probability of $x = 3$ is given by $P(3) = 0.021$.

b. "3 or fewer" successes means that the number of successes is 3 or 2 or 1 or 0.

$$
\begin{aligned}
P(3 \text{ or fewer}) &= P(3 \text{ or } 2 \text{ or } 1 \text{ or } 0) \\
&= P(3) + P(2) + P(1) + P(0) \\
&= 0.021 + 0.001 + 0 + 0 \\
&= 0.022
\end{aligned}
$$

TABLE A-1			Binomial Probabilities						x	P(x)
					p					
n	x	.01		.90	.95	.99	x		x	P(x)
5	0	.951		0+	0+	0+	0		0	0+
	1	.048		0+	0+	0+	1		1	0+
	2	.001		.008	.001	0+	2		2	0.001
	3	0+		.073	.021	.001	3		3	0.021
	4	0+		.328	.204	.048	4		4	0.204
0	5	0+		.590	.774	.951	5		5	0.774

If we wanted to use the binomial probability formula to find $P(3 \text{ or fewer})$, as in part (b) of Example 3, we would need to apply the formula four times to compute four different probabilities, which would then be added. Given this choice between the formula and the table, it makes sense to use the table. Unfortunately, Table A-1 includes only limited values of n as well as limited values of p, so the table doesn't always work.

Given that we now have three different methods for finding binomial probabilities, here is an effective and efficient strategy:

1. Use a TI-83/84 Plus calculator or computer software, if available.

2. If neither the TI-83/84 Plus calculator nor computer software is available, use Table A-1, if possible.

3. If neither the TI-83/84 Plus calculator nor computer software is available and the probabilities can't be found using Table A-1, use the binomial probability formula.

Rationale for the Binomial Probability Formula

The binomial probability formula is the basis for all three methods presented in this section. Instead of accepting and using that formula blindly, let's see why it works.

In Example 2, we used the binomial probability formula to find the probability of getting exactly 3 peas with green pods when 5 offspring peas are generated. With $P(\text{green pod}) = 0.75$, we can use the multiplication rule from Section 4-4 to find the probability that the first 3 peas have green pods while the last 2 peas do not have green pods. We get the following result:

$P(3 \text{ peas with green pods followed by 2 peas with pods that are not green})$

$$
\begin{aligned}
&= 0.75 \cdot 0.75 \cdot 0.75 \cdot 0.25 \cdot 0.25 \\
&= 0.75^3 \cdot 0.25^2 \\
&= 0.0264
\end{aligned}
$$

This result gives a probability of generating 5 offspring in which the first 3 have green pods. However, it does not give the probability of getting exactly 3 peas with green pods because it assumes a particular arrangement for 3 offspring peas with green pods. Other arrangements for generating 3 offspring peas with green pods are possible.

In Section 4-7 we saw that with 3 subjects identical to each other (such as peas with green pods) and 2 other subjects identical to each other (such as peas without green pods), the total number of arrangements, or permutations, is $5!/[(5 - 3)!3!]$ or 10. Each of those 10 different arrangements has a probability of $0.75^3 \cdot 0.25^2$, so the total probability is as follows:

$$P(3 \text{ peas with green pods among } 5) = \frac{5!}{(5 - 3)!3!} \cdot 0.75^3 \cdot 0.25^2$$

This particular result can be generalized as the binomial probability formula (Formula 5-5). That is, the binomial probability formula is a combination of the multiplication rule of probability and the counting rule for the number of arrangements of n items when x of them are identical to each other and the other $n - x$ are identical to each other. (See Exercises 13 and 14.)

The number of outcomes with exactly x successes among n trials

The probability of x successes among n trials for any one particular order

$$P(x) = \frac{n!}{(n - x)!\,x!} \cdot p^x \cdot q^{n-x}$$

Using a TI-83 Plus Calculator for the Binomial Probability Distribution

Finding a Single Probability Value In Example 2, we want to find the probability of exactly 3 green pods when 5 offspring peas are generated. Using the notation introduced in this section, we have $n = 5$, $x = 3$, and $p = 0.75$. That is, given $n = 5$ trials with probability of success $p = 0.75$, we want to find the probability of $x = 3$ successes. This is accomplished with the command **binompdf(5, 0.75, 3),** which is entered with the following procedure.

1. Access the DISTR (distributions) menu by pressing **2ND** **VARS**.
2. The DISTR menu includes the following two items that relate to this section, but we will focus on the first of these two items:
   - **binompdf**   Gives binomial probabilities for *individual* values of x.
   - **binomcdf**   Gives *cumulative* probabilities for values from 0 up to, and including, a desired value of x.

   To find the probability of $x = 3$ successes in $n = 5$ trials with probability $p = 0.75$, use the arrow keys to select **binompdf(,** then enter 5, 0.75, 3 and press **ENTER** to get a result of 0.263671875.

The general format of the binompdf (binomial probability density function) command is **binompdf(n, p, x)** where n, p, and x are as described in this section. (n is the total number of trials, p is the probability of success on a *single* trial, and x is the number of successes among the n trials.)

Binomial Probability Distribution

L1	L2	L3 2
0	9.8E⁻4	------
1	.01465	
2	.08789	
3	.26367	
4	.39551	
5	.2373	
------	------	
L2(7) =		

Finding a Probability Distribution   You can also obtain *all* of the probabilities for all values of x by omitting a specific entry for x, as in **binompdf(n, p).** For example, the command of binompdf(5, 0.75) will yield a list of all of the probability values corresponding to the values of $x = 0$ up to the largest possible value of x (which is the value of n).

Calculator Hint: Store the list of probabilities in list L2 (by pressing **STO** L2), then manually enter the x values of 0, 1, 2, ..., n in L1, so that lists L1 and L2 combine to form a probability distribution, as in the sample screen display shown here.

You could then use the TI-83/84 Plus calculator procedures described in Section 5-2 to find the mean and standard deviation or to generate a probability histogram.

Calculator Hint: If you know that you want to use your TI-83/84 Plus calculator to find a binomial probability, but you cannot remember the correct format of *n, p, x,* use the calculator's HELP feature. It is included with the TI-83/84 Plus Silver Edition, and it is an APP that can be downloaded to a TI-83 Plus or TI-84 Plus calculator. To use HELP, scroll to the binompdf menu item, but press (**+**) *before* pressing **ENTER**. The screen will indicate that the correct format is (**numtrials, p, [,x]**). This is a reminder that the format of the required entry is *n, p,* and *x,* where the entry of *x* is optional.

5-3 Basic Skills and Concepts

Statistical Literacy and Critical Thinking

1. Binomial Probabilities In the United States, 35% of the population has blue eyes (based on data from Dr. P. Sorita Soni at Indiana State University). Suppose you want to find the probability of getting exactly 2 people with blue eyes when 5 people are randomly selected. Why can't the answer be found as follows: Use the multiplication rule to find the probability of getting 2 people with blue eyes followed by 3 people with eyes that are not blue, which is $(0.35)(0.35)(0.65)(0.65)(0.65)$?

2. Notation If we use the binomial probability formula (Formula 5-5) for finding the probability described in Exercise 1, what is wrong with letting *p* denote the probability of getting someone with blue eyes while *x* counts the number of people with eyes that are not blue?

3. Independence A Gallup poll of 1236 adults showed that 12% of the respondents believe that it is bad luck to walk under a ladder. Consider the probability that among 30 randomly selected people from the 1236 who were polled, there are at least 2 who have that belief. Given that the subjects surveyed were selected without replacement, the events are not independent. Can the probability be found by using the binomial probability formula? Why or why not?

4. Notation When using Table A-1 to find the probability of guessing and getting exactly 8 correct answers on a multiple choice test with 10 questions, the result is found to be 0+. What does 0+ indicate? Does 0+ indicate that it is it impossible to get exactly 8 correct answers?

Identifying Binomial Distributions. *In Exercises 5–12, determine whether or not the given procedure results in a binomial distribution. For those that are not binomial, identify at least one requirement that is not satisfied.*

5. Clinical Trial of Lipitor Treating 863 subjects with Lipitor (Atorvastatin) and recording whether there is a "yes" response when they are each asked if they experienced a headache (based on data from Pfizer, Inc.).

6. Clinical Trial of Lipitor Treating 863 subjects with Lipitor (Atorvastatin) and asking each subject "How does your head feel?" (based on data from Pfizer, Inc.).

7. Gender Selection Treating 152 couples with the YSORT gender selection method developed by the Genetics & IVF Institute and recording the ages of the parents.

8. Gender Selection Treating 152 couples with the YSORT gender selection method developed by the Genetics & IVF Institute and recording the gender of each of the 152 babies that are born.

9. Surveying Senators Twenty different Senators are randomly selected from the 100 Senators in the current Congress, and each was asked whether he or she is in favor of abolishing estate taxes.

10. Surveying Governors Fifteen different Governors are randomly selected from the 50 Governors currently in office and the sex of each Governor is recorded.

11. Surveying New Yorkers Five hundred different New York City voters are randomly selected from the population of 2.8 million registered voters, and each is asked if he or she is a Democrat.

12. Surveying Statistics Students Two hundred statistics students are randomly selected and each is asked if he or she owns a TI-84 Plus calculator.

13. Finding Probabilities When Guessing Answers Multiple-choice questions on the SAT test each have 5 possible answers (a, b, c, d, e), one of which is correct. Assume that you guess the answers to 3 such questions.

a. Use the multiplication rule to find the probability that the first 2 guesses are wrong and the third is correct. That is, find $P(WWC)$, where C denotes a correct answer and W denotes a wrong answer.

b. Beginning with WWC, make a complete list of the different possible arrangements of 2 wrong answers and 1 correct answer, then find the probability for each entry in the list.

c. Based on the preceding results, what is the probability of getting exactly 1 correct answer when 3 guesses are made?

14. Finding Probabilities When Guessing Answers A psychology test consists of multiple-choice questions, each having 4 possible answers (a, b, c, d), 1 of which is correct. Assume that you guess the answers to 6 such questions.

a. Use the multiplication rule to find the probability that the first 2 guesses are wrong and the last 4 guesses are correct. That is, find $P(WWCCCC)$, where C denotes a correct answer and W denotes a wrong answer.

b. Beginning with WWCCCC, make a complete list of the different possible arrangements of 2 wrong answers and 4 correct answers, then find the probability for each entry in the list.

c. Based on the preceding results, what is the probability of getting exactly 4 correct answers when 6 guesses are made?

Using Table A-1. *In Exercises 15–20, assume that a procedure yields a binomial distribution with a trial repeated n times. Use Table A-1 to find the probability of x successes given the probability p of success on a given trial.*

15. $n = 2$, $x = 1$, $p = 0.30$ **16.** $n = 5$, $x = 1$, $p = 0.95$

17. $n = 15$, $x = 11$, $p = 0.99$ **18.** $n = 14$, $x = 4$, $p = 0.60$

19. $n = 10$, $x = 2$, $p = 0.05$ **20.** $n = 12$, $x = 12$, $p = 0.70$

Using the Binomial Probability Formula. *In Exercises 21–24, assume that a procedure yields a binomial distribution with a trial repeated n times. Use the binomial probability formula to find the probability of x successes given the probability p of success on a single trial.*

21. $n = 12$, $x = 10$, $p = 3/4$ **22.** $n = 9$, $x = 2$, $p = 0.35$

23. $n = 20$, $x = 4$, $p = 0.15$ **24.** $n = 15$, $x = 13$, $p = 1/3$

Using Calculator Results. *In Exercises 25–28, refer to the accompanying TI-83/84 Plus display. (When blood donors were randomly selected, 45% of them had blood that is Group O (based on data from the Greater New York Blood Program).) The display shows the probabilities obtained by entering the values of $n = 5$ and $p = 0.45$.*

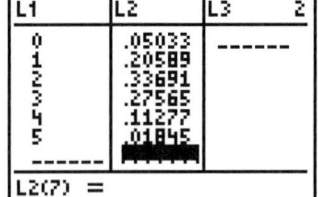

25. Group O Blood Find the probability that at least 1 of the 5 donors has Group O blood. If at least 1 Group O donor is needed, is it reasonable to expect that at least 1 will be obtained?

26. Group O Blood Find the probability that at least 3 of the 5 donors have Group O blood. If at least 3 Group O donors are needed, is it very likely that at least 3 will be obtained?

27. Group O Blood Find the probability that all of the 5 donors have Group O blood. Is it unusual to get 5 Group O donors from five randomly selected donors? Why or why not?

28. Group O Blood Find the probability that at most 2 of the 5 donors have Group O blood.

29. Brand Recognition The brand name of Mrs. Fields (cookies) has a 90% recognition rate (based on data from Franchise Advantage). If Mrs. Fields herself wants to verify that rate by beginning with a small sample of 10 randomly selected consumers, find the probability that exactly 9 of the 10 consumers recognize her brand name. Also find the probability that the number who recognize her brand name is *not* 9.

30. Brand Recognition The brand name of McDonald's has a 95% recognition rate (based on data from Retail Marketing Group). If a McDonald's executive wants to verify that rate by beginning with a small sample of 15 randomly selected consumers, find the probability that exactly 13 of the 15 consumers recognize the McDonald's brand name. Also find the probability that the number who recognize the brand name is *not* 13.

31. Eye Color In the United States, 40% of the population have brown eyes (based on data from Dr. P. Sorita Soni at Indiana University). If 14 people are randomly selected, find the probability that at least 12 of them have brown eyes. Is it unusual to randomly select 14 people and find that at least 12 of them have brown eyes? Why or why not?

32. Credit Rating There is a 1% delinquency rate for consumers with FICO (Fair Isaac & Company) credit rating scores above 800. If the Jefferson Valley Bank provides large loans to 12 people with FICO scores above 800, what is the probability that at least one of them becomes delinquent? Based on that probability, should the bank plan on dealing with a delinquency?

33. Genetics Ten peas are generated from parents having the green/yellow pair of genes, so there is a 0.75 probability that an individual pea will have a green pod. Find the probability that among the 10 offspring peas, at least 9 have green pods. Is it unusual to get at least 9 peas with green pods when 10 offspring peas are generated? Why or why not?

34. Genetics Ten peas are generated from parents having the green/yellow pair of genes, so there is a 0.75 probability that an individual pea will have a green pod. Find the probability that among the 10 offspring peas, at least 1 has a green pod. Why does the usual rule for rounding (with three significant digits) not work in this case?

35. Affirmative Action Programs Researchers conducted a study to determine whether there were significant differences in graduation rates between medical students admitted through special programs (such as affirmative action) and medical students admitted through the regular admissions criteria. It was found that the graduation rate was 94% for the medical students admitted through special programs (based on data from the *Journal of the American Medical Association*).

a. If 10 of the students from the special programs are randomly selected, find the probability that at least 9 of them graduated.

b. Would it be unusual to randomly select 10 students from the special programs and get only 7 that graduate? Why or why not?

36. Slot Machine The author purchased a slot machine configured so that there is a 1/2000 probability of winning the jackpot on any individual trial. Although no one would seriously consider tricking the author, suppose that a guest claims that she played the slot machine 5 times and hit the jackpot twice.

a. Find the probability of exactly 2 jackpots in 5 trials.

b. Find the probability of at least 2 jackpots in 5 trials.

c. Does the guest's claim of hitting 2 jackpots in 5 trials seem valid? Explain.

37. Nielsen Rating The television show *NBC Sunday Night Football* broadcast a game between the Colts and Patriots and received a share of 22, meaning that among the TV sets in use, 22% were tuned to that game (based on data from Nielsen Media Research). An advertiser wants to obtain a second opinion by conducting its own survey, and a pilot survey begins with 20 households having TV sets in use at the time of that same *NBC Sunday Night Football* broadcast.

a. Find the probability that none of the households are tuned to *NBC Sunday Night Football*.

b. Find the probability that at least one household is tuned to *NBC Sunday Night Football*.

c. Find the probability that at most one household is tuned to *NBC Sunday Night Football*.

d. If at most one household is tuned to *NBC Sunday Night Football*, does it appear that the 22% share value is wrong? Why or why not?

38. Composite Sampling A medical testing laboratory saves money by combining blood samples for tests, so that only one test is conducted for several people. The combined sample tests positive if at least one person is infected. If the combined sample tests positive, then individual blood tests are performed. In a test for gonorrhea, blood samples from 30 randomly selected people are combined. Find the probability that the combined sample tests positive with at least one of the 30 people infected. Based on data from the Centers for Disease Control, the probability of a randomly selected person having gonorrhea is 0.00114. Is it likely that such combined samples test positive?

TI 39. Job Survey In a survey of 320 college graduates, 36% reported that they stayed on their first full-time job less than one year (based on data from *USA Today* and Experience.com).

a. If 15 of those survey subjects are randomly selected without replacement for a follow-up survey, find the probability that 5 of them stayed on their first full-time job less than one year.

b. If part (a) is changed so that 20 different survey subjects are selected, explain why the binomial probability formula *cannot* be used.

TI 40. Job Interview Survey In a survey of 150 senior executives, 47% said that the most common job interview mistake is to have little or no knowledge of the company.

a. If 6 of those surveyed executives are randomly selected without replacement for a follow-up survey, find the probability that 3 of them said that the most common job interview mistake is to have little or no knowledge of the company.

b. If part (a) is changed so that 9 of the surveyed executives are to be randomly selected without replacement, explain why the binomial probability formula *cannot* be used.

TI 41. Acceptance Sampling The Medassist Pharmaceutical Company receives large shipments of aspirin tablets and uses this acceptance sampling plan: Randomly select and test 40 tablets, then accept the whole batch if there is only one or none that doesn't meet the required specifications. If one shipment of 5000 aspirin tablets actually has a 3% rate of defects, what is the probability that this whole shipment will be accepted? Will almost all such shipments be accepted, or will many be rejected?

TI 42. Overbooking Flights When someone buys a ticket for an airline flight, there is a 0.0995 probability that the person will not show up for the flight (based on data from an IBM research paper by Lawrence, Hong, and Cherrier). An agent for Air America wants to book 24 persons on an airplane that can seat only 22. If 24 persons are booked, find the probability that not enough seats will be available. Is this probability low enough so that overbooking is not a real concern?

TI 43. Identifying Gender Discrimination After being rejected for employment, Jennifer Summer learns that the Kingston Technology Corporation has hired only 3 women among the last 24 new employees. She also learns that the pool of applicants is very large, with an approximately equal number of qualified men and women. Help her address the charge of gender discrimination by finding the probability of getting 3 or fewer women when 24 people are hired, assuming that there is no discrimination based on gender. Does the resulting probability really support such a charge?

TI 44. Improving Quality The Write Right Company manufactures ballpoint pens and has been experiencing a 6% rate of defective pens. Modifications are made to the manufacturing process in an attempt to improve quality. The manager claims that the modified procedure is better because a test of 60 pens shows that only 1 is defective.

a. Assuming that the 6% rate of defects has not changed, find the probability that among 60 pens, exactly 1 is defective.

b. Assuming that the 6% rate of defects has not changed, find the probability that among 60 pens, none are defective.

c. What probability value should be used for determining whether the modified process results in a defect rate that is less than 6%?

d. What can you conclude about the effectiveness of the modified manufacturing process?

5-3   Beyond the Basics

TI 45. Mendel's Hybridization Experiment The Chapter Problem notes that Mendel obtained 428 peas with green pods when 580 peas were generated. He theorized that the probability of a pea with a green pod is 0.75. If the 0.75 probability value is correct, find the probability of getting 428 peas with green pods among 580 peas. Is that result unusual? Does the result suggest that Mendel's probability value of 0.75 is wrong? Why or why not?

46. Geometric Distribution If a procedure meets all the conditions of a binomial distribution except that the number of trials is not fixed, then the **geometric distribution** can be used. The probability of getting the first success on the xth trial is given by $P(x) = p(1 - p)^{x-1}$ where p is the probability of success on any one trial. Subjects are randomly selected for the National Health and Nutrition Examination Survey conducted by the National Center for Health Statistics, Centers for Disease Control. Find the probability that the first subject to be a universal blood donor (with group O and type Rh$^-$ blood) is the 12th person selected. The probability that someone is a universal donor is 0.06.

47. Hypergeometric Distribution If we sample from a small finite population without replacement, the binomial distribution should not be used because the events are not independent. If sampling is done without replacement and the outcomes belong to one of two types, we can use the **hypergeometric distribution.** If a population has A objects of one type (such as lottery numbers that match the ones you selected), while the remaining B objects are of the other type (such as lottery numbers that you did not select), and if n objects are sampled without replacement (such as 6 lottery numbers), then the probability of getting x objects of type A and $n - x$ objects of type B is

$$P(x) = \frac{A!}{(A - x)!\,x!} \cdot \frac{B!}{(B - n + x)!\,(n - x)!} \div \frac{(A + B)!}{(A + B - n)!\,n!}$$

In the New York State Lotto game, a bettor selects six numbers from 1 to 59 (without repetition), and a winning 6-number combination is later randomly selected. Find the probabilities of the following events and express them in decimal form.

a. You purchase 1 ticket with a 6-number combination and you get all 6 winning numbers.

b. You purchase 1 ticket with a 6-number combination and you get exactly 5 of the winning numbers.

c. You purchase 1 ticket with a 6-number combination and you get exactly 3 of the winning numbers.

d. You purchase 1 ticket with a 6-number combination and you get none of the winning numbers.

48. Multinomial Distribution The binomial distribution applies only to cases involving two types of outcomes, whereas the **multinomial distribution** involves more than two categories. Suppose we have three types of mutually exclusive outcomes denoted by A, B, and C. Let $P(A) = p_1$, $P(B) = p_2$, and $P(C) = p_3$. In n independent trials, the probability of x_1 outcomes of type A, x_2 outcomes of type B, and x_3 outcomes of type C is given by

$$\frac{n!}{(x_1)!(x_2)!(x_3)!} \cdot p_1^{x_1} \cdot p_2^{x_2} \cdot p_3^{x_3}$$

A genetics experiment involves 6 mutually exclusive genotypes identified as A, B, C, D, E, and F, and they are all equally likely. If 20 offspring are tested, find the probability of getting exactly 5 As, 4 Bs, 3 Cs, 2 Ds, 3 Es, and 3 Fs by expanding the above expression so that it applies to 6 types of outcomes instead of only 3.

5-4 Mean, Variance, and Standard Deviation for the Binomial Distribution

Key Concept In this section we consider important characteristics of a binomial distribution, including center, variation, and distribution. That is, given a particular binomial probability distribution, we can find its mean, variance, and standard deviation. In addition to finding these values, a strong emphasis is placed on *interpreting* and *understanding* those values. In particular, we use the range rule of thumb for determining whether events are usual or unusual.

Section 5-2 included Formulas 5-1, 5-3, and 5-4 for finding the mean, variance, and standard deviation from any discrete probability distribution. Because a binomial distribution is a particular type of discrete probability distribution, we could use those same formulas. However, it is much easier to use Formulas 5-6, 5-7, and 5-8 below.

In Formulas 5-6, 5-7, and 5-8, note that $q = 1 - p$. For example, if $p = 0.75$, then $q = 0.25$. (This notation for q was introduced in Section 5-3.)

For Any Discrete Probability Distribution	**For Binomial Distributions**
Formula 5-1 $\mu = \Sigma[x \cdot P(x)]$	Formula 5-6 $\mu = np$
Formula 5-3 $\sigma^2 = \Sigma[x^2 \cdot P(x)] - \mu^2$	Formula 5-7 $\sigma^2 = npq$
Formula 5-4 $\sigma = \sqrt{\Sigma[x^2 \cdot P(x)] - \mu^2}$	Formula 5-8 $\sigma = \sqrt{npq}$

As in earlier sections, finding values for μ and σ is fine, but it is especially important to *interpret* and *understand* those values, so the range rule of thumb can be very helpful. Recall that we can consider values to be unusual if they fall outside of the limits obtained from the following:

Range Rule of Thumb

$$\text{maximum usual value: } \boldsymbol{\mu + 2\sigma}$$

$$\text{minimum usual value: } \boldsymbol{\mu - 2\sigma}$$

EXAMPLE 1   **Genetics** Use Formulas 5-6 and 5-8 to find the mean and standard deviation for the numbers of peas with green pods when groups of 5 offspring peas are generated. Assume that there is a 0.75 probability that an offspring pea has a green pod (as described in the Chapter Problem).

SOLUTION   Using the values $n = 5$, $p = 0.75$, and $q = 0.25$, Formulas 5-6 and 5-8 can be applied as follows:

$$\mu = np = (5)(0.75) = 3.8 \quad \text{(rounded)}$$
$$\sigma = \sqrt{npq} = \sqrt{(5)(0.75)(0.25)} = 1.0 \quad \text{(rounded)}$$

Formula 5-6 for the mean makes sense intuitively. If 75% of peas have green pods and 5 offspring peas are generated, we expect to get around $5 \cdot 0.75 = 3.8$ peas with green pods. This result can be generalized as $\mu = np$. The variance and standard deviation are not so easily justified, and we omit the complicated algebraic manipulations that lead to Formulas 5-7 and 5-8. Instead, refer again to the preceding example and Table 5-3 to verify that for a binomial distribution, Formulas 5-6, 5-7, and 5-8 will produce the same results as Formulas 5-1, 5-3, and 5-4.

EXAMPLE 2   **Genetics** In an actual experiment, Mendel generated 580 offspring peas. He claimed that 75%, or 435, of them would have green pods. The actual experiment resulted in 428 peas with green pods.

a. Assuming that groups of 580 offspring peas are generated, find the mean and standard deviation for the numbers of peas with green pods.

b. Use the range rule of thumb to find the minimum usual number and the maximum usual number of peas with green pods. Based on those numbers, can we conclude that Mendel's actual result of 428 peas with green pods is *unusual*? Does this suggest that Mendel's value of 75% is wrong?

SOLUTION

a. With $n = 580$ offspring peas, with $p = 0.75$, and $q = 0.25$, we can find the mean and standard deviation for the numbers of peas with green pods as follows:

$$\mu = np = (580)(0.75) = 435.0$$

$$\sigma = \sqrt{npq} = \sqrt{(580)(0.75)(0.25)} = 10.4 \quad \text{(rounded)}$$

For groups of 580 offspring peas, the mean number of peas with green pods is 435.0 and the standard deviation is 10.4.

b. We must now interpret the results to determine whether Mendel's actual result of 428 peas is a result that could easily occur by chance, or whether that result is so unlikely that the assumed rate of 75% is wrong. We will use the range rule of thumb as follows:

maximum usual value: $\mu + 2\sigma = 435.0 + 2(10.4) = 455.8$

minimum usual value: $\mu - 2\sigma = 435.0 - 2(10.4) = 414.2$

INTERPRETATION   If Mendel generated many groups of 580 offspring peas and if his 75% rate is correct, the numbers of peas with green pods should usually fall between 414.2 and 455.8. (Calculations with unrounded values yield 414.1 and 455.9.) Mendel actually got 428 peas with green pods, and that value does fall within the range of usual values, so the experimental results are consistent with the 75% rate. The results do not suggest that Mendel's claimed rate of 75% is wrong.

Variation in Statistics Example 2 is a good illustration of the importance of variation in statistics. In a traditional algebra course, we might conclude that 428 is not 75% of 580 simply because 428 does not equal 435 (which is 75% of 580). However, in statistics we recognize that sample results vary. We don't expect to get *exactly* 75% of the peas with green pods. We recognize that as long as the results don't vary too far away from the claimed rate of 75%, they are consistent with that claimed rate of 75%.

In this section we presented easy procedures for finding values of the mean μ and standard deviation σ from a binomial probability distribution. However, it is really important to be able to *interpret* those values by using such devices as the range rule of thumb for identifying a range of usual values.

Reliability and Validity

The reliability of data refers to the consistency with which results occur, whereas the validity of data refers to how well the data measure what they are supposed to measure. The reliability of an IQ test can be judged by comparing scores for the test given on one date to scores for the same test given at another time. To test the validity of an IQ test, we might compare the test scores to another indicator of intelligence, such as academic performance. Many critics charge that IQ tests are reliable, but not valid; they provide consistent results, but don't really measure intelligence.

5-4   Basic Skills and Concepts

Statistical Literacy and Critical Thinking

1. Notation Formula 5-8 shows that the standard deviation σ of values of the random variable x in a binomial probability distribution can be found by evaluating $\sqrt{npq}$. Some books give the expression $\sqrt{np(1 - p)}$. Do these two expressions always give the same result? Explain.

2. Is Anything Wrong? Excel is used to find the mean and standard deviation of a discrete probability distribution and the results are as follows: $\mu = 2.0$ and $\sigma = -3.5$. Can these results be correct? Explain.

3. Variance In a Gallup poll of 1236 adults, it was found that 5% of those polled said that bad luck occurs after breaking a mirror. Based on these results, such randomly selected groups of 1236 adults will have a mean of 61.8 people with that belief, and a standard deviation of 7.7 people. What is the variance? (Express the answer including the appropriate units.)

4. What Is Wrong? A statistics class consists of 10 females and 30 males. Each day, 12 of the students are randomly selected without replacement, and the number of females is counted. Using the methods of this section we get $\mu = 3.0$ females and $\sigma = 1.5$ females, but the value of the standard deviation is wrong. Why don't the methods of this section give the correct results here?

Finding μ, σ, and Unusual Values. *In Exercises 5–8, assume that a procedure yields a binomial distribution with n trials and the probability of success for one trial is p. Use the given values of n and p to find the mean μ and standard deviation σ. Also, use the range rule of thumb to find the minimum usual value $\mu - 2\sigma$ and the maximum usual value $\mu + 2\sigma$.*

5. Guessing on SAT Random guesses are made for 50 SAT multiple choice questions, so $n = 50$ and $p = 0.2$.

6. Gender Selection In an analysis of test results from the YSORT gender selection method, 152 babies are born and it is assumed that boys and girls are equally likely, so $n = 152$ and $p = 0.5$.

7. Drug Test In an analysis of the 1-Panel TCH test for marijuana usage, 300 subjects are tested and the probability of a positive result is 0.48, so $n = 300$ and $p = 0.48$.

8. Gallup Poll A Gallup poll of 1236 adults showed that 14% believe that bad luck follows if your path is crossed by a black cat, so $n = 1236$ and $p = 0.14$.

9. Guessing on an Exam The midterm exam in a nursing course consists of 75 true/false questions. Assume that an unprepared student makes random guesses for each of the answers.

a. Find the mean and standard deviation for the number of correct answers for such students.

b. Would it be unusual for a student to pass this exam by guessing and getting at least 45 correct answers? Why or why not?

10. Guessing Answers The final exam in a sociology course consists of 100 multiple-choice questions. Each question has 5 possible answers, and only 1 of them is correct. An unprepared student makes random guesses for all of the answers.

a. Find the mean and standard deviation for the number of correct answers for such students.

b. Would it be unusual for a student to pass the exam by guessing and getting at least 60 correct answers? Why or why not?

11. Are 16% of M&M's Green? Mars, Inc. claims that 16% of its M&M plain candies are green. A sample of 100 M&Ms is randomly selected.

a. Find the mean and standard deviation for the numbers of green M&Ms in such groups of 100.

b. Data Set 18 in Appendix B consists of a random sample of 100 M&Ms in which 19 are green. Is this result unusual? Does it seem that the claimed rate of 16% is wrong?

12. Are 24% of M&Ms Blue? Mars, Inc., claims that 24% of its M&M plain candies are blue. A sample of 100 M&Ms is randomly selected.

a. Find the mean and standard deviation for the numbers of blue M&Ms in such groups of 100.

b. Data Set 18 in Appendix B consists of a random sample of 100 M&Ms in which 27 are blue. Is this result unusual? Does it seem that the claimed rate of 24% is wrong?

13. Gender Selection In a test of the XSORT method of gender selection, 574 babies are born to couples trying to have baby girls, and 525 of those babies are girls (based on data from the Genetics & IVF Institute).

a. If the gender-selection method has no effect and boys and girls are equally likely, find the mean and standard deviation for the numbers of girls born in groups of 574.

b. Is the result of 525 girls unusual? Does it suggest that the gender-selection method appears to be effective?

14. Gender Selection In a test of the YSORT method of gender selection, 152 babies are born to couples trying to have baby boys, and 127 of those babies are boys (based on data from the Genetics & IVF Institute).

a. If the gender-selection method has no effect and boys and girls are equally likely, find the mean and standard deviation for the numbers of boys born in groups of 152.

b. Is the result of 127 boys unusual? Does it suggest that the gender-selection method appears to be effective?

15. Job Longevity A headline in *USA Today* states that "most stay at first job less than 2 years." That headline is based on an Experience.com poll of 320 college graduates. Among those polled, 78% stayed at their first full-time job less than 2 years.

a. Assuming that 50% is the true percentage of graduates who stay at their first job less than two years, find the mean and standard deviation of the numbers of such graduates in randomly selected groups of 320 graduates.

b. Assuming that the 50% rate in part (a) is correct, find the range of usual values for the numbers of graduates among 320 who stay at their first job less than two years.

c. Find the actual number of surveyed graduates who stayed at their first job less than two years. Use the range of values from part (b) to determine whether that number is unusual. Does the result suggest that the headline is not justified?

d. This statement was given as part of the description of the survey methods used: "Alumni who opted-in to receive communications from Experience were invited to participate in the online poll, and 320 of them completed the survey." What does that statement suggest about the results?

16. Mendelian Genetics When Mendel conducted his famous genetics experiments with plants, one sample of 1064 offspring consisted of 787 plants with long stems and 277 plants with short stems. Mendel theorized that 25% of the offspring plants would have short stems.

a. If Mendel's theory is correct, find the mean and standard deviation for the numbers of plants with short stems in such groups of 1064 offspring plants.

b. Are the actual results unusual? What do the actual results suggest about Mendel's theory?

17. Voting In a past presidential election, the actual voter turnout was 61%. In a survey, 1002 subjects were asked if they voted in the presidential election.

a. Find the mean and standard deviation for the numbers of actual voters in groups of 1002.

b. In the survey of 1002 people, 701 *said* that they voted in the last presidential election (based on data from ICR Research Group). Is this result consistent with the actual voter turnout, or is this result unlikely to occur with an actual voter turnout of 61%? Why or why not?

c. Based on these results, does it appear that accurate voting results can be obtained by asking voters how they acted?

18. Cell Phones and Brain Cancer In a study of 420,095 cell phone users in Denmark, it was found that 135 developed cancer of the brain or nervous system. If we assume that the use of cell phones has no effect on developing such cancer, then the probability of a person having such a cancer is 0.000340.

a. Assuming that cell phones have no effect on developing cancer, find the mean and standard deviation for the numbers of people in groups of 420,095 that can be expected to have cancer of the brain or nervous system.

b. Based on the results from part (a), is it unusual to find that among 420,095 people, there are 135 cases of cancer of the brain or nervous system? Why or why not?

c. What do these results suggest about the publicized concern that cell phones are a health danger because they increase the risk of cancer of the brain or nervous system?

19. Smoking Treatment In a clinical trial of a drug used to help subjects stop smoking, 821 subjects were treated with 1 mg doses of Chantix. That group consisted of 30 subjects who experienced nausea (based on data from Pfizer, Inc.). The probability of nausea for subjects not receiving the treatment was 0.0124.

a. Assuming that Chantix has no effect, so that the probability of nausea was 0.0124, find the mean and standard deviation for the numbers of people in groups of 821 that can be expected to experience nausea.

b. Based on the result from part (a), is it unusual to find that among 821 people, there are 30 who experience nausea? Why or why not?

c. Based on the preceding results, does nausea appear to be an adverse reaction that should be of concern to those who use Chantix?

20. Test of Touch Therapy Nine-year-old Emily Rosa conducted this test: A professional touch therapist put both hands through a cardboard partition and Emily would use a coin flip to randomly select one of the hands. Emily would place her hand just above the hand of the therapist, who was then asked to identify the hand that Emily had selected. The touch therapists believed that they could sense the energy field and identify the hand that Emily had selected. The trial was repeated 280 times. (Based on data from "A Close Look at Therapeutic Touch," by Rosa et al., *Journal of the American Medical Association,* Vol. 279, No. 13.)

a. Assuming that the touch therapists have no special powers and made random guesses, find the mean and standard deviation for the numbers of correct responses in groups of 280 trials.

b. The professional touch therapists identified the correct hand 123 times in the 280 trials. Is that result unusual? What does the result suggest about the ability of touch therapists to select the correct hand by sensing an energy field?

5-4   Beyond the Basics

21. Hypergeometric Distribution As in Exercise 4, assume that a statistics class consists of 10 females and 30 males, and each day, 12 of the students are randomly selected without replacement. Because the sampling is from a small finite population without replacement, the hypergeometric distribution applies. (See Exercise 47 in Section 5-3.) Using the hypergeometric distribution, find the mean and standard deviation for the numbers of girls that are selected on the different days.

22. Acceptable/Defective Products Mario's Pizza Parlor has just opened. Due to a lack of employee training, there is only a 0.8 probability that a pizza will be edible. An order for 5 pizzas has just been placed. What is the minimum number of pizzas that must be made in order to be at least 99% sure that there will be 5 that are edible?

5-5   Poisson Probability Distributions

Key Concept This chapter began by considering discrete probability distributions in general. In Sections 5-3 and 5-4 we discussed the binomial probability distribution, which is one particular type of discrete probability distribution. In this section we introduce the *Poisson distribution,* which is another particular discrete probability distribution. The Poisson distribution is often used for describing the behavior of rare events (with small probabilities).

　　The Poisson distribution is used for describing behavior such as radioactive decay, arrivals of people in a line, eagles nesting in a region, patients arriving at an emergency room, crashes on the Massachusetts Turnpike, and Internet users logging onto a Web site. For example, suppose your local hospital experiences a mean of 2.3 patients arriving at the emergency room on Fridays between 10:00 P.M. and 11:00 P.M. We can use the Poisson distribution to find the probability that for a randomly selected Friday, exactly four patients arrive at the ER between 10:00 P.M. and 11:00 P.M. We use the Poisson distribution, defined as follows.

 **DEFINITION**

The **Poisson distribution** is a discrete probability distribution that applies to occurrences of some event *over a specified interval.* The random variable x is the number of occurrences of the event in an interval. The interval can be time, distance, area, volume, or some similar unit. The probability of the event occurring x times over an interval is given by Formula 5-9.

Formula 5-9

$$P(x) = \frac{\mu^x \cdot e^{-\mu}}{x!} \qquad \text{where } e \approx 2.71828$$

TI-83/84 Plus The end of this section includes instructions for using a TI-83/84 Plus calculator for finding probabilities with a Poisson distribution.

Requirements for the Poisson Distribution

1. The random variable x is the number of occurrences of an event *over some interval.*

2. The occurrences must be *random.*

3. The occurrences must be *independent* of each other.

4. The occurrences must be *uniformly distributed* over the interval being used.

Parameters of the Poisson Distribution

• The mean is μ.

• The standard deviation is $\sigma = \sqrt{\mu}$.

A Poisson distribution differs from a binomial distribution in these fundamental ways:

1. The binomial distribution is affected by the sample size n and the probability p, whereas the Poisson distribution is affected only by the mean μ.

2. In a binomial distribution, the possible values of the random variable x are $0, 1, \ldots, n$, but a Poisson distribution has possible x values of $0, 1, 2, \ldots,$ with no upper limit.

EXAMPLE 1 **Earthquakes** For a recent period of 100 years, there were 93 major earthquakes (measuring at least 6.0 on the Richter scale) in the world (based on data from the *World Almanac and Book of Facts*). Assume that the Poisson distribution is a suitable model.

a. Find the mean number of major earthquakes per year.

b. If $P(x)$ is the probability of x earthquakes in a randomly selected year, find $P(0)$, $P(1)$, $P(2)$, $P(3)$, $P(4)$, $P(5)$, $P(6)$, and $P(7)$.

c. The actual results are as follows: 47 years (0 major earthquakes); 31 years (1 major earthquake); 13 years (2 major earthquakes); 5 years (3 major earthquakes);

Queues

Queuing theory is a branch of mathematics that uses probability and statistics. The study of queues, or waiting lines, is important to businesses such as super-markets, banks, fast-food restaurants, air-lines, and amusement parks. Grand Union supermarkets try to keep checkout lines no longer than three shoppers. Wendy's introduced the "Express Pak" to expedite servicing its numerous drive-through customers. Disney conducts extensive studies of lines at its amusement parks so that it can keep patrons happy and plan for expansion. Bell Laboratories uses queuing theory to optimize telephone network usage, and factories use it to design efficient production lines.

continued

2 years (4 major earthquakes); 0 years (5 major earthquakes); 1 year (6 major earthquakes); 1 year (7 major earthquakes). How do these actual results compare to the probabilities found in part (b)? Does the Poisson distribution appear to be a good model in this case?

SOLUTION

a. The Poisson distribution applies because we are dealing with the occurrences of an event (earthquakes) over some interval (a year). The mean number of earthquakes per year is

$$\mu = \frac{\text{number of earthquakes}}{\text{number of years}} = \frac{93}{100} = 0.93$$

b. Using Formula 5-9, the calculation for $x = 2$ earthquakes in a year is as follows (with μ replaced by 0.93 and e replaced by 2.71828):

$$P(2) = \frac{\mu^x \cdot e^{-\mu}}{x!} = \frac{0.93^2 \cdot 2.71828^{-0.93}}{2!} = \frac{0.8649 \cdot 0.394554}{2} = 0.171$$

The probability of exactly 2 earthquakes in a year is $P(2) = 0.171$. Using the same procedure to find the other probabilities, we get these results: $P(0) = 0.395$, $P(1) = 0.367$, $P(2) = 0.171$, $P(3) = 0.0529$, $P(4) = 0.0123$, $P(5) = 0.00229$, $P(6) = 0.000355$, and $P(7) = 0.0000471$.

c. The probability of $P(0) = 0.395$ from part (b) is the likelihood of getting 0 earthquakes in one year. So in 100 years, the expected number of years with 0 earthquakes is $100 \times 0.395 = 39.5$ years. Using the probabilities from part (b), here are all of the expected frequencies: 39.5, 36.7, 17.1, 5.29, 1.23, 0.229, 0.0355, and 0.00471. These expected frequencies compare reasonably well with the actual frequencies of 47, 31, 13, 5, 2, 0, 1, and 1. Because the expected frequencies agree reasonably well with the actual frequencies, the Poisson distribution is a good model in this case.

Poisson Distribution as an Approximation to the Binomial Distribution

The Poisson distribution is sometimes used to approximate the binomial distribution when n is large and p is small. One rule of thumb is to use such an approximation when the following two requirements are both satisfied.

Requirements for Using the Poisson Distribution as an Approximation to the Binomial

1. $n \geq 100$

2. $np \leq 10$

If both requirements are satisfied and we want to use the Poisson distribution as an approximation to the binomial distribution, we need a value for μ. That value can be calculated by using Formula 5-6 (first presented in Section 5-4):

Formula 5-6

$$\mu = np$$

EXAMPLE 2   **Illinois Pick 3** In the Illinois Pick 3 game, you pay 50¢ to select a sequence of three digits, such as 729. If you play this game once every day, find the probability of winning exactly once in 365 days.

SOLUTION   Because the time interval is 365 days, $n = 365$. Because there is one winning set of numbers among the 1000 that are possible (from 000 to 999), $p = 1/1000$. With $n = 365$ and $p = 1/1000$, the conditions $n \geq 100$ and $np \leq 10$ are both satisfied, so we can use the Poisson distribution as an approximation to the binomial distribution. We first need the value of μ, which is found as follows:

$$\mu = np = 365 \cdot \frac{1}{1000} = 0.365$$

Having found the value of μ, we can now find $P(1)$ by using $x = 1$, $\mu = 0.365$, and $e = 2.71828$, as shown here:

$$P(1) = \frac{\mu^x \cdot e^{-\mu}}{x!} = \frac{0.365^1 \cdot 2.71828^{-0.365}}{1!} = \frac{0.253}{1} = 0.253$$

Using the Poisson distribution as an approximation to the binomial distribution, we find that there is a 0.253 probability of winning exactly once in 365 days. If we use the binomial distribution, we get 0.254, so we can see that the Poisson approximation is quite good here.

Using the TI-83/84 Plus Calculator for the Poisson Distribution

To find an individual probability value from the Poisson distribution, use the following procedure.

1. Access the DISTR (distributions) menu by pressing **2ND** **VARS**.
2. Use the arrow keys to select the menu item of **poissonpdf.**
3. The general format of the **poissonpdf** (Poisson probability density function) command is **poissonpdf(μ x),** where μ is the mean and x is the desired number of successes. For Example 2 in this section, $\mu = 0.365$ and $x = 1$, so enter poissonpdf(0.365, 1) to get a result of 0.2533817776, which can be rounded to 0.253.

Calculator Hint: If you know that you want to use your TI-83/84 Plus calculator to find a probability using the Poisson distribution, but you cannot remember the correct format of μ, x, use the calculator's HELP feature. It is included with the TI-83/84 Plus Silver Edition, and it is an APP that can be downloaded to a TI-83 Plus or TI-84 Plus calculator. To use HELP, scroll to the Poissonpdf menu item, but press **+** *before* pressing **ENTER**. The screen will indicate that the correct format is (μ, x).

5-5   Basic Skills and Concepts

Statistical Literacy and Critical Thinking

1. Poisson Distribution What are the conditions for using the Poisson distribution?

2. Checking Account Last year, the author wrote 126 checks. Let the random variable x represent the number of checks he wrote in one day, and assume that it has a Poisson distribution. What is the mean number of checks written per day? What is the standard deviation? What is the variance?

3. Approximating a Binomial Distribution Assume that we have a binomial distribution with $n = 100$ and $p = 0.1$. It is impossible to get 101 successes in such a binomial distribution, but we *can* compute the probability that $x = 101$ if we use the Poisson distribution to approximate the binomial distribution, and the result is 4.82×10^{-64}. How does that result agree with the impossibility of having $x = 101$ with a binomial distribution?

4. Poisson/Binomial An experiment involves rolling a die 6 times and counting the number of 2s that occur. If we calculate the probability of $x = 0$ occurrences of 2 using the Poisson distribution, we get 0.368, but we get 0.335 if we use the binomial distribution. Which is the correct probability of getting no 2s when a die is rolled 6 times? Why is the other probability wrong?

Using a Poisson Distribution to Find Probability. *In Exercises 5–8, assume that the Poisson distribution applies, and proceed to use the given mean to find the indicated probability.*

TI **5.** If $\mu = 2$, find $P(3)$.   **TI** **6.** If $\mu = 0.3$, find $P(1)$.

TI **7.** If $\mu = 3/4$, find $P(3)$.   **TI** **8.** If $\mu = 1/6$, find $P(0)$.

In Exercises 9–16, use the Poisson distribution to find the indicated probabilities.

TI **9. Motor Vehicle Deaths** Dutchess County, New York, has been experiencing a mean of 35.4 motor vehicle deaths each year.

a. Find the mean number of deaths per day.

b. Find the probability that on a given day, there are more than 2 motor vehicle deaths.

c. Is it unusual to have more than 2 motor vehicle deaths on the same day? Why or why not?

TI **10. Low Birth Weight** A newborn baby is considered to have a low birth weight if it weighs less than 2500 grams. Such babies often require extra care. Dutchess County, New York, has been experiencing a mean of 210.0 cases of low birth weight each year.

a. Find the mean number of low birth weight babies born each day.

b. Find the probability that on a given day, there is more than 1 baby born with a low birth weight.

c. Is it unusual to have more than 1 low birth weight baby born in a day? Why or why not?

TI **11. Radioactive Decay** Radioactive atoms are unstable because they have too much energy. When they release their extra energy, they are said to decay. When studying cesium-137, a nuclear engineer found that over 365 days, 1,000,000 radioactive atoms decayed to 977,287 radioactive atoms.

a. Find the mean number of radioactive atoms that decayed in a day.

b. Find the probability that on a given day, 50 radioactive atoms decayed.

TI **12. Deaths from Horse Kicks** A classical example of the Poisson distribution involves the number of deaths caused by horse kicks to men in the Prussian Army between 1875 and 1894. Data for 14 corps were combined for the 20-year period, and the 280 corps-years included a total of 196 deaths. After finding the mean number of deaths per corps-year, find the probability that a randomly selected corps-year has the following numbers of deaths.

a. 0   **b.** 1   **c.** 2   **d.** 3   **e.** 4

continued

The actual results consisted of these frequencies: 0 deaths (in 144 corps-years);1 death (in 91 corps-years); 2 deaths (in 32 corps-years); 3 deaths (in 11 corps-years); 4 deaths (in 2 corps-years). Compare the actual results to those expected by using the Poisson probabilities. Does the Poisson distribution serve as a good device for predicting the actual results?

TI **13. Homicide Deaths** In one year, there were 116 homicide deaths in Richmond, Virginia (based on "A Classroom Note on the Poisson Distribution: A Model for Homicidal Deaths in Richmond, VA for 1991," by Winston A. Richards in *Mathematics and Computer Education*). For a randomly selected day, find the probability that the number of homicide deaths is

a. 0   **b.** 1   **c.** 2   **d.** 3   **e.** 4

Compare the calculated probabilities to these actual results: 268 days (no homicides); 79 days (1 homicide); 17 days (2 homicides); 1 day (3 homicides); no days with more than 3 homicides.

TI **14. Disease Cluster** Neuroblastoma, a rare form of malignant tumor, occurs in 11 children in a million, so its probability is 0.000011. Four cases of neuroblastoma occurred in Oak Park, Illinois, which had 12,429 children.

a. Assuming that neuroblastoma occurs as usual, find the mean number of cases in groups of 12,429 children.

b. Find the probability that the number of neuroblastoma cases in a group of 12,429 children is 0 or 1.

c. What is the probability of more than one case of neuroblastoma?

d. Does the cluster of four cases appear to be attributable to random chance? Why or why not?

TI **15. Life Insurance** A Fidelity life insurance company charges $226 for a $50,000 life insurance policy for a 50-year-old female. The probability that such a female survives the year is 0.9968 (based on data from the U.S. Department of Health and Human Services). Assume that the company sells 700 such policies to 50-year-old females, so it collects $158,200 in policy payments. The company will make a profit if fewer than four of the 700 women die during the year.

a. What is the mean number of deaths in such groups of 700 females?

b. Find the probability that the company makes a profit from the 700 policies. Is that probability high enough so that the company is almost sure to make a profit?

TI **16. Life Insurance** There is a 0.9986 probability that a randomly selected 30-year-old male lives through the year (based on data from the U.S. Department of Health and Human Services). A Fidelity life insurance company charges $161 for insuring that the male will live through the year. If the male does not survive the year, the policy pays out $100,000 as a death benefit. Assume that the company sells 1300 such policies to 30-year-old males, so it collects $209,300 in policy payments. The company will make a profit if the number of deaths in this group is two or fewer.

a. What is the mean number of deaths in such groups of 1300 males?

b. Use the Poisson distribution to find the probability that the company makes a profit from the 1300 policies.

c. Use the binomial distribution to find the probability that the company makes a profit from the 1300 policies, then compare the result to the result found in part (b).

5-5   Beyond the Basics

TI **17. Poisson Approximation to Binomial Distribution** For a binomial distribution with $n = 10$ and $p = 0.5$, we should not use the Poisson approximation because the conditions $n \geq 100$ and $np \leq 10$ are not both satisfied. Suppose we go way out on a limb and use the Poisson approximation anyway. Are the resulting probabilities unacceptable approximations? Why or why not?

Review

This chapter introduced the concept of a probability distribution, which describes the probability for each value of a random variable. This chapter includes only discrete probability distributions, but the following chapters will include continuous probability distributions. The following key points were discussed:

• A *random variable* has values that are determined by chance.

• A *probability distribution* consists of all values of a random variable, along with their corresponding probabilities. A probability distribution must satisfy two requirements: the sum of all of the probabilities for values of the random variable must be 1, and each probability value must be between 0 and 1 inclusive. This is expressed as $\Sigma P(x) = 1$ and, for each value of x, $0 \le P(x) \le 1$.

• Important characteristics of a *probability distribution* can be explored by constructing a probability histogram and by computing its mean and standard deviation using these formulas:

$$\mu = \Sigma[x \cdot P(x)]$$
$$\sigma = \sqrt{\Sigma[x^2 \cdot P(x)] - \mu^2}$$

• In a *binomial distribution,* there are two categories of outcomes and a fixed number of independent trials with a constant probability. The probability of x successes among n trials can be found by using the binomial probability formula, or Table A-1, or computer software (such as STATDISK, Minitab, or Excel), or a T1-83/84 Plus calculator.

• In a binomial distribution, the mean and standard deviation can be found by calculating the values of $\mu = np$ and $\sigma = \sqrt{npq}$.

• A *Poisson probability distribution* applies to occurrences of some event over a specific interval, and its probabilities can be computed with Formula 5-9.

• *Unusual outcomes:* To distinguish between outcomes that are usual and those that are unusual, we used two different criteria: the range rule of thumb and the use of probabilities.

Using the range rule of thumb to identify unusual values:

maximum usual value = $\mu + 2\sigma$

minimum usual value = $\mu - 2\sigma$

Using probabilities to identify unusual values:

Unusually high number of successes: x successes among n trials is an unusually high number of successes if $P(x$ or more$) \le 0.05$.*

Unusually low number of successes: x successes among n trials is an unusually low number of successes if $P(x$ or fewer$) \le 0.05$.*

*The value of 0.05 is commonly used, but is not absolutely rigid. Other values, such as 0.01, could be used to distinguish between events that can easily occur by chance and events that are very unlikely to occur by chance.

Statistical Literacy and Critical Thinking

1. Random Variable What is a random variable? Is it possible for a discrete random variable to have an infinite number of possible values?

2. Discrete versus Continuous What is the difference between a discrete random variable and a continuous random variable?

3. Binomial Probability Distribution In a binomial probability distribution, the symbols p and q are used to represent probabilities. What is the numerical relationship between p and q?

4. Probability Distributions This chapter described the concept of a discrete probability distribution, and then described the binomial and Poisson probability distributions. Are all discrete probability distributions either binomial or Poisson? Why or why not?

Chapter Quick Quiz

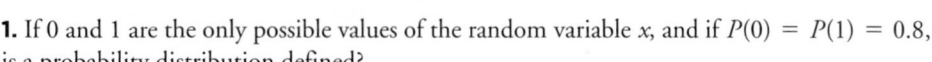

1. If 0 and 1 are the only possible values of the random variable x, and if $P(0) = P(1) = 0.8$, is a probability distribution defined?

2. If 0 and 1 are the only possible values of the random variable x, and if $P(0) = 0.3$ and $P(1) = 0.7$, find the mean of the probability distribution.

3. If boys and girls are equally likely and groups of 400 births are randomly selected, find the mean number of girls in such groups of 400.

4. If boys and girls are equally likely and groups of 400 births are randomly selected, find the standard deviation of the numbers of girls in such groups of 400.

5. A multiple-choice test has 100 questions. For subjects making random guesses for each answer, the mean number of correct answers is 20.0 and the standard deviation of the numbers of correct answers is 4.0. For someone making random guesses for all answers, is it unusual to get 35 correct answers?

In Exercises 6–10, use the following:
A multiple-choice test has 4 questions. For a subject making random guesses for each answer, the probabilities for the number of correct responses are given in the table in the margin. Assume that a subject makes random guesses for each question.

x	$P(x)$
0	0.4096
1	0.4096
2	0.1536
3	0.0256
4	0.0016

6. Is the given probability distribution a binomial probability distribution?

7. Find the probability of getting at least one correct answer.

8. Find the probability of getting all correct answers.

9. Find the probability that the number of correct answers is 2 or 3.

10. Is it unusual to answer all of the questions correctly?

Review Exercises

1. Postponing Death An interesting theory is that dying people have some ability to postpone their death to survive a major holiday. One study involved the analysis of deaths during the time period spanning from the week before Thanksgiving to the week after Thanksgiving. (See "Holidays, Birthdays, and the Postponement of Cancer Death," by Young and Hade, *Journal of the American Medical Association,* Vol. 292, No. 24.) Assume that $n = 8$ such deaths are randomly selected from those that occurred during the time period spanning from the week before Thanksgiving to the week after, and also assume that dying people have no ability to postpone death, so the probability that a death occurred the week *before* Thanksgiving is $p = 0.5$. Construct a table describing the probability distribution, where the random variable x is the number of the deaths (among 8) that occurred the week before Thanksgiving. Express all of the probabilities with three decimal places.

2. Postponing Death Find the mean and standard deviation for the probability distribution described in Exercise 1, then use those values and the range rule of thumb to identify the range of usual values of the random variable. Is it unusual to find that all 8 deaths occurred the week before Thanksgiving? Why or why not?

3. Postponing Death Exercise 1 involves 8 randomly selected deaths during the time period spanning from the week before Thanksgiving to the week after Thanksgiving. Assume now that 20 deaths are randomly selected from that time period.

continued

a. Find the probability that exactly 14 of the deaths occur during the week before Thanksgiving.

b. Is it unusual to have exactly 14 deaths occurring the week before Thanksgiving?

c. If the probability of exactly 14 deaths is very small, does that imply that 14 is an unusually high number of deaths occurring the week before Thanksgiving? Why or why not?

4. Expected Value for *Deal or No Deal* In the television game show *Deal or No Deal*, contestant Elna Hindler had to choose between acceptance of an offer of $193,000 or continuing the game. If she continued to refuse all further offers, she would have won one of these five equally-likely prizes: $75, $300, $75,000, $500,000, and $1,000,000. Find her expected value if she continued the game and refused all further offers. Based on the result, should she accept the offer of $193,000, or should she continue?

5. Expected Value for a *Magazine Sweepstakes* *Reader's Digest* ran a sweepstakes in which prizes were listed along with the chances of winning: $1,000,000 (1 chance in 90,000,000), $100,000 (1 chance in 110,000,000), $25,000 (1 chance in 110,000,000), $5,000 (1 chance in 36,667,000), and $2,500 (1 chance in 27,500,000).

a. Assuming that there is no cost of entering the sweepstakes, find the expected value of the amount won for one entry.

b. Find the expected value if the cost of entering this sweepstakes is the cost of a postage stamp. Is it worth entering this contest?

6. Brand Recognition In a study of brand recognition of Sony, groups of four consumers are interviewed. If x is the number of people in the group who recognize the Sony brand name, then x can be 0, 1, 2, 3, or 4, and the corresponding probabilities are 0.0016, 0.0250, 0.1432, 0.3892, and 0.4096. Does the given information describe a probability distribution? Why or why not?

7. Kentucky Pick 4 In Kentucky's Pick 4 game, you pay $1 to select a sequence of four digits, such as 2283. If you buy only one ticket and win, your prize is $5000 and your net gain is $4999.

a. If you buy one ticket, what is the probability of winning?

b. Construct a table describing the probability distribution corresponding to the purchase of one Pick 4 ticket.

c. If you play this game once every day, find the mean number of wins in years with exactly 365 days.

d. If you play this game once every day, find the probability of winning exactly once in 365 days.

e. Find the expected value for the purchase of one ticket.

8. Reasons for Being Fired "Inability to get along with others" is the reason cited in 17% of worker firings (based on data from Robert Half International, Inc.). Concerned about her company's working conditions, the personnel manager at the Boston Finance Company plans to investigate the five employee firings that occurred over the past year.

a. Assuming that the 17% rate applies, find the probability that at least four of those five employees were fired because of an inability to get along with others.

b. If the personnel manager actually does find that at least four of the firings were due to an inability to get along with others, does this company appear to be very different from other typical companies? Why or why not?

9. Detecting Fraud The Brooklyn District Attorney's office analyzed the leading digits of check amounts in order to identify fraud. The leading digit of 1 is expected to occur 30.1% of the time, according to Benford's law that applies in this case. Among 784 checks issued by a suspect company, there were none with amounts that had a leading digit of 1.

a. For randomly selected checks, there is a 30.1% chance that the leading digit of the check amount is 1. What is the expected number of checks that should have a leading digit of 1?

b. Assume that groups of 784 checks are randomly selected. Find the mean and standard deviation for the numbers of checks with amounts having a leading digit of 1.

c. Use the results from part (b) and the range rule of thumb to find the range of usual values.

d. Given that the 784 actual check amounts had no leading digits of 1, is there very strong evidence that the suspect checks are very different from the expected results? Why or why not?

 **10. World War II Bombs** In analyzing hits by V-1 buzz bombs in World War II, South London was subdivided into 576 regions, each with an area of 0.25 km^2. A total of 535 bombs hit the combined area of 576 regions.

a. What is the mean number of hits per region?

b. If a region is randomly selected, find the probability that it was not hit.

c. Based on the probability in part (b), how many of the 576 regions are expected to have no hits?

d. There were actually 229 regions that were not hit. How does this actual result compare to the result from part (c)?

Cumulative Review Exercises

1. Auditing Checks It is common for professional auditors to analyze checking accounts with a randomly selected sample of checks. Given below are check amounts (in dollars) from a random sample of checks issued by the author.

$$115.00 \quad 188.00 \quad 134.83 \quad 217.60 \quad 142.94$$

a. Find the mean.

b. Find the median.

c. Find the range.

d. Find the standard deviation.

e. Find the variance.

f. Use the range rule of thumb to identify the range of usual values.

g. Based on the result from part (f), are any of the sample values unusual? Why or why not?

h. What is the level of measurement of the data: nominal, ordinal, interval, or ratio?

i. Are the data discrete or continuous?

j. If the sample had consisted of the last five checks that the author issued, what type of sampling would have been used: random, systematic, stratified, cluster, convenience?

k. The checks included in this sample are five of the 134 checks written in the year. Estimate the total value of all checks written in the year.

2. Employee Drug Testing Among companies doing highway or bridge construction, 80% test employees for substance abuse (based on data from the Construction Financial Management Association). A study involves the random selection of 10 such companies.

a. Find the probability that exactly 5 of the 10 companies test for substance abuse.

b. Find the probability that at least half of the companies test for substance abuse.

c. For such groups of 10 companies, find the mean and standard deviation for the number (among 10) that test for substance abuse.

d. Using the results from part (c) and the range rule of thumb, identify the range of usual values.

3. Determining the Effectiveness of an HIV Training Program The New York State Health Department reports a 10% rate of the HIV virus for the "at-risk" population. In one region, an intensive education program is used in an attempt to lower that 10% rate. After running the program, a follow-up study of 150 at-risk individuals is conducted.

a. Assuming that the program has no effect, find the mean and standard deviation for the number of HIV cases in groups of 150 at-risk people.

b. Among the 150 people in the follow-up study, 8% (or 12 people) tested positive for the HIV virus. If the program has no effect, is that rate unusually low? Does this result suggest that the program is effective?

4. Titanic Of the 2223 passengers on board the *Titanic,* 706 survived.

a. If one of the passengers is randomly selected, find the probability that the passenger survived.

b. If two different passengers are randomly selected, find the probability that they both survived.

c. If two different passengers are randomly selected, find the probability that neither of them survived.

5. Energy Consumption Each year, the U.S. Department of Energy publishes an *Annual Energy Review* that includes per capita energy consumption (in millions of Btu) for each of the 50 states. If you calculate the mean of these 50 values, is the result the mean per capita energy consumption for the total population from all 50 states combined? If it is not, explain how you would use those 50 values to calculate the mean per capita energy consumption for the total population from all 50 states combined.

TI-83/84 Plus Calculator Project

United Flight 15 from New York's JFK airport to San Francisco uses a Boeing 757-200 with 182 seats. Because some people with reservations don't show up, United can overbook by accepting more than 182 reservations. If the flight is not overbooked, the airline will lose revenue due to empty seats, but if too many seats are sold and some passengers are denied seats, the airline loses money from the compensation that must be given to the bumped passengers. Assume that there is a 0.0995 probability that a passenger with a reservation will not show up for the flight (based on data from the IBM research paper "Passenger-Based Predictive Modeling of Airline No-Show Rates," by Lawrence, Hong, and Cherrier). Also assume that the airline accepts 200 reservations for the 182 seats that are available.

Find the probability that when 200 reservations are accepted for United Flight 15, there are more passengers showing up than there are seats available. Table A-1 cannot be used and calculations with the binomial probability formula would be extremely time-consuming and tedious. The best approach is to use a T1-83/84 Plus calculator or statistics software. (See Section 5-3 for instructions describing the use of a T1-83/84 Plus calculator.) Is the probability of overbooking small enough so that it does not happen very often, or does it seem too high so that changes must be made to make it lower? Now use trial and error to find the maximum number of reservations that could be accepted so that the probability of having more passengers than seats is 0.05 or less.

INTERNET PROJECT

Probability Distributions and Simulation

Go to: **http://www.aw.com/triola**

Probability distributions are used to predict the outcome of the events they model. For example, if we toss a fair coin, the distribution for the outcome is a probability of 0.5 for heads and 0.5 for tails. If we toss the coin ten consecutive times, we expect five heads and five tails. We might not get this exact result, but in the long run, over hundreds or thousands of tosses, we expect the split between heads and tails to be very close to "50-50."

Proceed to the Internet Project for Chapter 5 where you will find two explorations. In the first exploration you are asked to develop a probability distribution for a simple experiment, and use that distribution to predict the outcome of repeated trial runs of the experiment. In the second exploration, we will analyze a more complicated situation: the paths of rolling marbles as they move in pinball-like fashion through a set of obstacles. In each case, a dynamic visual simulation will allow you to compare the predicted results with a set of experimental outcomes.

The CD included with this book contains applets designed to help visualize various concepts. Open the Applets folder on the CD and double-click on **Start.** Select the menu item of **Binomial Distribution.** Select $n = 10$, $p = 0.4$, and $N = 1000$ for the number of trials. Based on the simulated results, find $P(3)$. Compare that probability to $P(3)$ for a binomial experiment with $n = 10$ and $p = 0.4$, found by using an exact method instead of a simulation. After repeating the simulation several times, comment on how much the estimated value of $P(3)$ varies from simulation to simulation.

FROM DATA TO DECISION

Critical Thinking: Did the jury selection process discriminate?

Rodrigo Partida is an American who is of Mexican ancestry. He was convicted of burglary with intent to commit rape. His conviction took place in Hidalgo County, which is in Texas on the border with Mexico. Hidalgo County had 181,535 people eligible for jury duty, and 79.1% of them were Americans of Mexican ancestry. Among 870 people selected for grand jury duty, 39% (339) were Americans of Mexican ancestry. Partida's conviction was later appealed (*Castaneda v. Partida*) on the basis of the large discrepancy between the 79.1% of the Americans of Mexican ancestry eligible for grand jury duty and the fact that only 39% of such Americans were actually selected.

1. Given that Americans of Mexican ancestry constitute 79.1% of the population of those eligible for jury duty, and given that Partida was convicted by a jury of 12 people with only 58% of them (7 jurors) that were Americans of Mexican ancestry, can we conclude that his jury was selected in a process that discriminates against Americans of Mexican ancestry?

2. Given that Americans of Mexican ancestry constitute 79.1% of the population of 181,535 and, over a period of 11 years, only 339 of the 870 people selected for grand jury duty were Americans of Mexican ancestry, can we conclude that the process of selecting grand jurors discriminated against Americans of Mexican ancestry?

Cooperative Group Activities

1. In-class activity Win $1,000,000! The James Randi Educational Foundation offers a $1,000,000 prize to anyone who can show, "under proper observing conditions, evidence of any paranormal, supernatural, or occult power or event." Divide into groups of three. Select one person who will be tested for extrasensory perception (ESP) by trying to correctly identify a digit randomly selected by another member of the group. Another group member should record the randomly selected digit, the digit guessed by the subject, and whether the guess was correct or wrong. Construct the table for the probability distribution of randomly generated digits, construct the relative frequency table for the random digits that were actually obtained, and construct a relative frequency table for the guesses that were made. After comparing the three tables, what do you conclude? What proportion of guesses are correct? Does it seem that the subject has the ability to select the correct digit significantly more often than would be expected by chance?

2. In-class activity See the preceding activity and *design an experiment* that would be effective in testing someone's claim that they have the ability to identify the color of a card selected from a standard deck of playing cards. Describe the experiment with great detail. Because the prize of $1,000,000 is at stake, we want to be careful to avoid the serious mistake of concluding that the person has the paranormal power when that power is not actually present. There will likely be some chance that the subject could make random guesses and be correct every time, so identify a probability that is reasonable for the event of the subject passing the test with random guesses. Be sure that the test is designed so that this probability is equal to or less than the probability value selected.

3. In-class activity Suppose we want to identify the probability distribution for the number of children born to randomly selected couples. For each student in the class, find the number of brothers and sisters and record the total number of children (including the student) in each family. Construct the relative frequency table for the result obtained. (The values of the random variable x will be 1, 2, 3, . . .) What is wrong with using this relative frequency table as an estimate of the probability distribution for the number of children born to randomly selected couples?

4. Out-of-class activity The analysis of the last digits of data can sometimes reveal whether the data have been collected through actual measurements or reported by the subjects. Refer to an almanac or the Internet and find a collection of data (such as lengths of rivers in the world), then analyze the distribution of last digits to determine whether the values were obtained through actual measurements.

5. Out-of-class activity In Review Exercise 9 it was noted that leading digits of the amounts on checks can be analyzed for fraud. It was also noted that the leading digit of 1 is expected about 30.1% of the time. Obtain a random sample of actual check amounts and record the leading digits. Compare the actual number of checks with amounts that have a leading digit of 1 to the 30.1% rate expected. Do the actual checks conform to the expected rate, or is there a substantial discrepancy? Explain.

NAME:	Sarah Mesnick
JOB:	Behavioral and Molecular Biologist
COMPANY:	Laboratory of Molecular Ecology

*S*arah Mesnick is a behavioral and molecular ecologist, and is a National Research Council postdoctoral fellow. In her work as a marine mammal biologist, she conducts research at sea as well as in the Laboratory of Molecular Ecology. Her research focuses on the social organization and population structure of sperm whales. She received her doctorate in evolutionary biology at the University of Arizona.

Q: What do you do?

A: My research focuses on the relationship between sociality and population structure in sperm whales. We use this information to build better management models for the conservation of this and other endangered marine mammal species.

Q: What concepts of statistics do you use?

A: Currently, I use chi-square and *F*-statistics to examine population structure and regression measures to estimate the degree of relatedness among individuals within whale pods. We use the chi-square and *F*-statistics to determine how many discrete populations of whales are in the Pacific. Discrete populations are managed as independent stocks. The regression analysis of relatedness is used to determine kinship within groups.

Q: Could you cite a specific example illustrating the use of statistics?

A: I'm currently working with tissue samples obtained from three mass strandings of sperm whales. We use genetic markers to determine the degree of relatedness among individuals within the strandings. This is a striking behavior—entire pods swam up onto the beach following a young female calf, stranded, and subsequently all died. We thought that to do something as dramatic as this, the individuals involved must be very closely related. We're finding, however, that they are not. The statistics enable us to determine the probability that two individuals are related given the number of alleles that they share. Also, sperm whales—and many other marine mammal, bird, and turtle species—are injured or killed incidentally in fishing operations. We need to know the size of the population from which these animals are taken. If the population is small, and the incidental kill large, the marine mammal population may be threatened. We use statistics to determine the degree of isolation between putative stocks. If stocks are found to be isolated, we would use this information to prepare management plans specifically designed to conserve the marine mammals of the region. Human activities may need to protect the health of the marine environment and its inhabitants.

Q: How do you approach your research?

A: We try not to have preconceived notions about how the animals are dispersed in their environment. In marine mammals in particular, because they are so difficult to study, there are generally accepted notions about what the animals are doing, yet these have not been critically investigated. In the case of relatedness among individuals within sperm whale groups, they were once thought to be matrilineal and accompanied by a "harem master." With the advent of genetic techniques, and dedicated field work, more open minds and more critical analyses—the statistics come in here—we're able to reassess these notions.

6-1 Review and Preview

6-2 The Standard Normal Distribution

6-3 Applications of Normal Distributions

6-4 Sampling Distributions and Estimators

6-5 The Central Limit Theorem

6-6 Normal as Approximation to Binomial

6-7 Assessing Normality

6 Normal Probability Distributions

254

How do we design airplanes, boats, cars, and homes for safety and comfort?

Ergonomics involves the study of people fitting into their environments. Ergonomics is used in a wide variety of applications such as these: Design a doorway so that most people can walk through it without bending or hitting their head; design a car so that the dashboard is within easy reach of most drivers; design a screw bottle top so that most people have sufficient grip strength to open it; design a manhole cover so that most workers can fit through it. Good ergonomic design results in an environment that is safe, functional, efficient, and comfortable. Bad ergonomic design can result in uncomfortable, unsafe, or possibly fatal conditions. For example, the following real situations illustrate the difficulty in determining safe loads in aircraft and boats.

- "We have an emergency for Air Midwest fifty-four eighty," said pilot Katie Leslie, just before her plane crashed in Charlotte, North Carolina. The crash of the Beech plane killed all of the 21 people on board. In the subsequent investigation, the weight of the passengers was suspected as a factor that contributed to the crash. This prompted the Federal Aviation Administration to order airlines to collect weight information from randomly selected flights, so that the old assumptions about passenger weights could be updated.

- Twenty passengers were killed when the *Ethan Allen* tour boat capsized on New York's Lake George. Based on an assumed mean weight of 140 lb, the boat was certified to carry 50 people. A subsequent investigation showed that most of the passengers weighed more than 200 lb, and the boat should have been certified for a much smaller number of passengers.

- A water taxi sank in Baltimore's Inner Harbor. Among the 25 people on board, 5 died and 16 were injured. An investigation revealed that the safe passenger load for the water taxi was 3500 lb. Assuming a mean passenger weight of 140 lb, the boat was allowed to carry 25 passengers, but the mean of 140 lb was determined 44 years ago when people were not as heavy as they are today. (The mean weight of the 25 passengers aboard the boat that sank was found to be 168 lb.) The National Transportation and Safety Board suggested that the old estimated mean of 140 lb be updated to 174 lb, so the safe load of 3500 lb would now allow only 20 passengers instead of 25.

This chapter introduces the statistical tools that are basic to good ergonomic design. After completing this chapter, we will be able to solve problems in a wide variety of different disciplines, including ergonomics.

6-1 ## Review and Preview

In Chapter 2 we considered the distribution of data, and in Chapter 3 we considered some important measures of data sets, including measures of center and variation. In Chapter 4 we discussed basic principles of probability, and in Chapter 5 we presented the concept of a probability distribution. In Chapter 5 we considered only *discrete* probability distributions, but in this chapter we present *continuous* probability distributions. To illustrate the correspondence between area and probability, we begin with a uniform distribution, but most of this chapter focuses on *normal distributions*. Normal distributions occur often in real applications, and they play an important role in methods of inferential statistics. In this chapter we present concepts of normal distributions that will be used often in the remaining chapters of this text. Several of the statistical methods discussed in later chapters are based on concepts related to the central limit theorem, discussed in Section 6-5. Many other sections require normally distributed populations, and Section 6-7 presents methods for analyzing sample data to determine whether or not the sample appears to be from such a normally distributed population.

DEFINITION

If a continuous random variable has a distribution with a graph that is symmetric and bell-shaped, as in Figure 6-1, and it can be described by the equation given as Formula 6-1, we say that it has a **normal distribution.**

Figure 6-1

The Normal Distribution

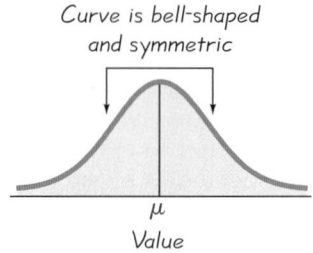

Curve is bell-shaped and symmetric

μ

Value

Formula 6-1

$$y = \frac{e^{-\frac{1}{2}\left(\frac{x-\mu}{\sigma}\right)^2}}{\sigma\sqrt{2\pi}}$$

Formula 6-1 is mathematically challenging, and we include it only to illustrate that any particular normal distribution is determined by two parameters: the mean, μ, and standard deviation, σ. Formula 6-1 is like many an equation with one variable y on the left side and one variable x on the right side. The letters π and e represent the constant values of 3.14159... and 2.71828..., respectively. The symbols μ and σ represent fixed values for the mean and standard deviation, respectively. Once specific values are selected for μ and σ, we can graph Formula 6-1 as we would graph any equation relating x and y; the result is a continuous probability distribution with the same bell shape shown in Figure 6-1. From Formula 6-1 we see that a normal distribution is determined by the fixed values of the mean μ and standard deviation σ. And that's all we need to know about Formula 6-1!

6-2 The Standard Normal Distribution

Key Concept In this section we present the *standard normal distribution*, which has these three properties:

1. Its graph is bell-shaped (as in Figure 6-1).

2. Its mean is equal to 0 (that is, $\mu = 0$).

3. Its standard deviation is equal to 1 (that is, $\sigma = 1$).

In this section we develop the skill to find areas (or probabilities or relative frequencies) corresponding to various regions under the graph of the standard normal distribution. In addition, we find *z*-scores that correspond to areas under the graph.

Uniform Distributions

The focus of this chapter is the concept of a normal probability distribution, but we begin with a *uniform distribution*. The uniform distribution allows us to see two very important properties:

1. The area under the graph of a probability distribution is equal to 1.

2. There is a correspondence between area and probability (or relative frequency), so some probabilities can be found by identifying the corresponding areas.

Chapter 5 considered only discrete probability distributions, but we now consider continuous probability distributions, beginning with the *uniform distribution*.

 DEFINITION

A continuous random variable has a **uniform distribution** if its values are spread *evenly* over the range of possibilities. The graph of a uniform distribution results in a rectangular shape.

EXAMPLE 1 **Home Power Supply** The Newport Power and Light Company provides electricity with voltage levels that are uniformly distributed between 123.0 volts and 125.0 volts. That is, any voltage amount between 123.0 volts and 125.0 volts is possible, and all of the possible values are equally likely. If we randomly select one of the voltage levels and represent its value by the random variable *x*, then *x* has a distribution that can be graphed as in Figure 6-2.

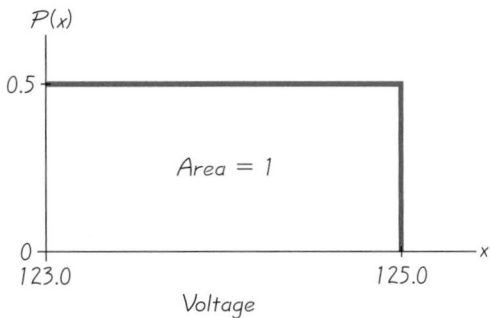

Figure 6-2 Uniform Distribution of Voltage Levels

The Placebo Effect

It has long been believed that placebos actually help some patients. In fact, some formal studies have shown that when given a placebo (a treatment with no medicinal value), many test subjects show some improvement. Estimates of improvement rates have typically ranged between one-third and two-thirds of the patients. However, a more recent study suggests that placebos have no real effect. An article in the *New England Journal of Medicine* (Vol. 334, No. 21) was based on research of 114 medical studies over 50 years. The authors of the article concluded that placebos appear to have some effect only for relieving pain, but not for other physical conditions. They concluded that apart from clinical trials, the use of placebos "cannot be recommended."

The graph of a continuous probability distribution, such as in Figure 6-2, is called a **density curve.** A density curve must satisfy the following two requirements.

Requirements for a Density Curve

1. The total area under the curve must equal 1.

2. Every point on the curve must have a vertical height that is 0 or greater. (That is, the curve cannot fall below the *x*-axis.)

By setting the height of the rectangle in Figure 6-2 to be 0.5, we force the enclosed area to be $2 \times 0.5 = 1$, as required. (In general, the area of the rectangle becomes 1 when we make its height equal to the value of 1/range.) The requirement that the area must equal 1 makes solving probability problems simple, so the following statement is important:

Because the total area under the density curve is equal to 1, there is a correspondence between *area* and *probability*.

EXAMPLE 2   **Voltage Level** Given the uniform distribution illustrated in Figure 6-2, find the probability that a randomly selected voltage level is greater than 124.5 volts.

SOLUTION   The shaded area in Figure 6-3 represents voltage levels that are greater than 124.5 volts. Because the total area under the density curve is equal to 1, there is a correspondence between area and probability. We can find the desired probability by using areas as follows:

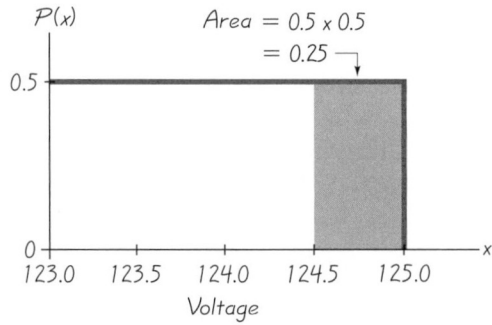

Figure 6-3   Using Area to Find Probability

P(voltage greater than 124.5 volts) = area of shaded region in Figure 6-3

$$= 0.5 \times 0.5$$

$$= 0.25$$

INTERPRETATION   The probability of randomly selecting a voltage level greater than 124.5 volts is 0.25.

Standard Normal Distribution

The density curve of a uniform distribution is a horizontal line, so we can find the area of any rectangular region by applying this formula: Area = width × height. Because the density curve of a normal distribution has a complicated bell shape as shown in Figure 6-1, it is more difficult to find areas. However, the basic principle is the same: *There is a correspondence between area and probability.* In Figure 6-4 we show that for a standard normal distribution, the area under the density curve is equal to 1.

> **DEFINITION**
>
> The **standard normal distribution** is a normal probability distribution with $\mu = 0$ and $\sigma = 1$. The total area under its density curve is equal to 1. (See Figure 6-4.)

It is not easy to find areas in Figure 6-4, so mathematicians have calculated many different areas under the curve, and those areas are included in Table A-2 in Appendix A.

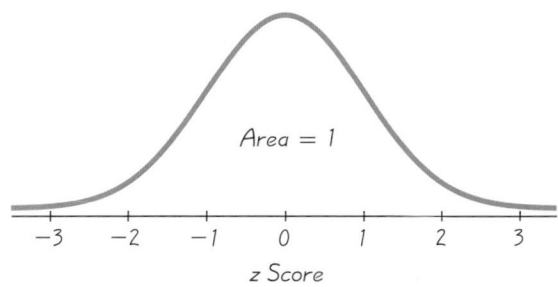

Area = 1

Figure 6-4   Standard Normal Distribution: Bell-Shaped Curve with $\mu = 0$ and $\sigma = 1$

Finding Probabilities When Given *z* Scores

We can find areas (or probabilities) corresponding to different *z* scores by using technology (such as a TI-83/84 Plus calculator) or Table A-2 (in Appendix A and the *Formulas and Tables* insert card). The key features of the different methods are summarized in Table 6-1 on the next page. Because the TI-83/84 Plus calculator gives more accurate results and is not limited to specific values of *z*, **we strongly recommend the use of the TI-83/84 Plus calculator instead of Table A-2.** (When there are discrepancies, answers in Appendix D generally include results based on Table A-2 as well as answers based on the TI-83/84 Plus calculator.)

If using Table A-2, it is essential to understand these points:

1. Table A-2 is designed only for the *standard* normal distribution, which has a mean of 0 and a standard deviation of 1.

2. Table A-2 is on two pages, with one page for *negative z* scores and the other page for *positive z* scores.

3. Each value in the body of the table is a *cumulative area from the left* up to a vertical boundary above a specific *z* score.

4. When working with a graph, avoid confusion between z scores and areas.

> **z score:** **_Distance_ along the horizontal scale of the standard normal distribution; refer to the leftmost column and top row of Table A-2.**
>
> **Area:** **_Region_ under the curve; refer to the values in the body of Table A-2.**

5. The part of the z score denoting hundredths is found across the top row of Table A-2.

CAUTION

When working with a normal distribution, avoid confusion between z scores and areas.

Table 6-1   Methods for Finding Normal Distribution Areas

TI-83/84 Plus Calculator

Gives area bounded on the left and bounded on the right by vertical lines above any specific values.

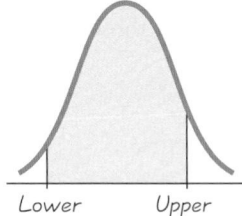

Lower   Upper

TI-83/84   Press **2ND** **VARS** **[2: normal cdf ( ]**, then enter the two z scores separated by a comma, as in (left z score, right z score).

Table A-2

Gives the cumulative area from the left up to a vertical line above a specific value of z.

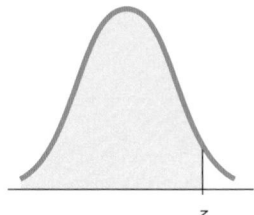

z

Table A-2   The procedure for using Table A-2 is described in the text.

The following example requires that we find the probability associated with a z score less than 1.27. Begin with the z score of 1.27 by locating 1.2 in the left column; next find the value in the adjoining row of probabilities that is directly below 0.07, as shown in the following excerpt from Table A-2.

TABLE A-2	(continued) Cumulative Area from the LEFT									
z	.00	.01	.02	.03	.04	.05	.06	.07	.08	.09
0.0	.5000	.5040	.5080	.5120	.5160	.5199	.5239	.5279	.5319	.5359
0.1	.5398	.5438	.5478	.5517	.5557	.5596	.5636	.5675	.5714	.5753
0.2	.5793	.5832	.5871	.5910	.5948	.5987	.6026	.6064	.6103	.6141
1.0	.8413	.8438	.8461	.8485	.8508	.8531	.8554	.8577	.8599	.8621
1.1	.8643	.8665	.8686	.8708	.8729	.8749	.8770	.8790	.8810	.8830
1.2	.8849	.8869	.8888	.8907	.8925	.8944	.8962	.8980	.8997	.9015
1.3	.9032	.9049	.9066	.9082	.9099	.9115	.9131	.9147	.9162	.9177
1.4	.9192	.9207	.9222	.9236	.9251	.9265	.9279	.9292	.9306	.9319

The area (or probability) value of 0.8980 indicates that there is a probability of 0.8980 of randomly selecting a *z* score less than 1.27. (The following sections will consider cases in which the mean is not 0 or the standard deviation is not 1.)

> **EXAMPLE 3** **Scientific Thermometers** The Precision Scientific Instrument Company manufactures thermometers that are supposed to give readings of 0°C at the freezing point of water. Tests on a large sample of these instruments reveal that at the freezing point of water, some thermometers give readings below 0° (denoted by negative numbers) and some give readings above 0° (denoted by positive numbers). Assume that the mean reading is 0°C and the standard deviation of the readings is 1.00°C. Also assume that the readings are normally distributed. If one thermometer is randomly selected, find the probability that, at the freezing point of water, the reading is less than 1.27°.

SOLUTION The probability distribution of readings is a standard normal distribution, because the readings are normally distributed with $\mu = 0$ and $\sigma = 1$. We need to find the area in Figure 6-5 below $z = 1.27$. The *area* below $z = 1.27$ is equal to the *probability* of randomly selecting a thermometer with a reading less than 1.27°. From Table A-2 we find that this area is 0.8980.

With a TI-83/84 Plus Calculator, select **normalcdf** and enter (−999999, 1.27) as described in the TI-83/84 Plus instructions given near the end of this section. (Because there is no specific lower limit, we use the large negative number of −999999 for the lower limit.) The result is 0.8979576193 as shown in the display in the margin.

TI-83/84 Plus

```
normalcdf(-99999
,1.27)
      .8979576193
```

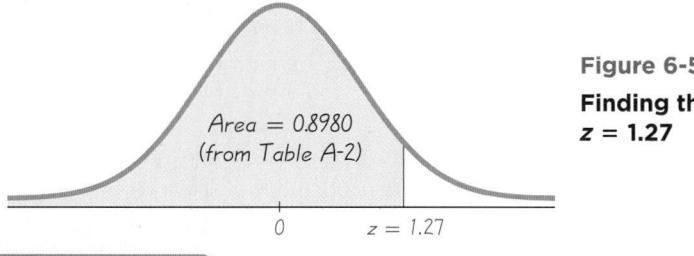

Area = 0.8980
(from Table A-2)

0   z = 1.27

Figure 6-5

Finding the Area Below
z = 1.27

INTERPRETATION The *probability* of randomly selecting a thermometer with a reading less than 1.27° (at the freezing point of water) is equal to the area of 0.8980 shown as the shaded region in Figure 6-5. Another way to interpret this result is to conclude that 89.80% of the thermometers will have readings below 1.27°.

> **EXAMPLE 4** **Scientific Thermometers** Using the thermometers from Example 3, find the probability of randomly selecting one thermometer that reads (at the freezing point of water) above −1.23°.

SOLUTION We again find the desired *probability* by finding a corresponding *area*. We are looking for the area of the region that is shaded in Figure 6-6, but Table A-2 is designed to apply only to cumulative areas from the *left*. Referring to Table A-2 for the page with *negative z* scores, we find that the cumulative area from the left up to $z = -1.23$ is 0.1093 as shown. Because the total area under the curve is 1, we can find the shaded area by subtracting 0.1093 from 1. The result is 0.8907. Even though Table A-2 is designed only for cumulative areas from the left, we can use it to find cumulative areas from the right, as shown in Figure 6-6.

With a TI-83/84 Plus calculator, select **normalcdf** and enter $(-1.23, 999999)$ as described in the TI-83/84 Plus instructions given near the end of this section. (Because there is no specific upper limit, we use the large positive number of 999999 for the upper limit.) The result is 0.8906513833.

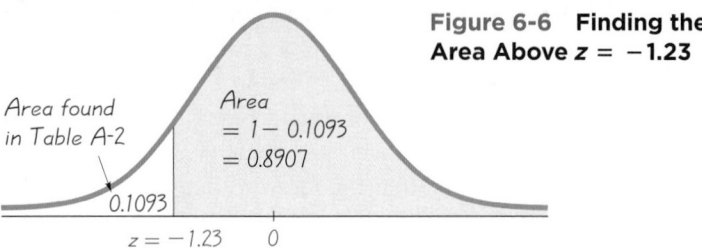

Figure 6-6 Finding the Area Above $z = -1.23$

Area found in Table A-2

Area $= 1 - 0.1093$ $= 0.8907$

0.1093

$z = -1.23$ 0

INTERPRETATION   Because of the correspondence between probability and area, we conclude that the *probability* of randomly selecting a thermometer with a reading above $-1.23°$ at the freezing point of water is 0.8907 (which is the *area* to the right of $z = -1.23$). In other words, 89.07% of the thermometers have readings above $-1.23°$.

Example 4 illustrates a way that Table A-2 can be used indirectly to find a cumulative area from the right. The following example illustrates another way that we can find an area indirectly by using Table A-2.

EXAMPLE 5   **Scientific Thermometers** Make a random selection from the same sample of thermometers from Example 3. Find the probability that the chosen thermometer reads (at the freezing point of water) between $-2.00°$ and $1.50°$.

SOLUTION   We are again dealing with normally distributed values having a mean of $0°$ and a standard deviation of $1°$. The probability of selecting a thermometer that reads between $-2.00°$ and $1.50°$ corresponds to the shaded area in Figure 6-7. Table A-2 cannot be used to find that area directly, but we can use the table to find that $z = -2.00$ corresponds to the area of 0.0228, and $z = 1.50$ corresponds to the area of 0.9332, as shown in the figure. From Figure 6-7 we see that the shaded area is the difference between 0.9332 and 0.0228. The shaded area is therefore $0.9332 - 0.0228 = 0.9104$.

With a TI-83/84 Plus calculator, select **normalcdf** and enter $(-2.00, 1.50)$ as described in the TI-83/84 Plus instructions given near the end of this section. The result is 0.9104427093.

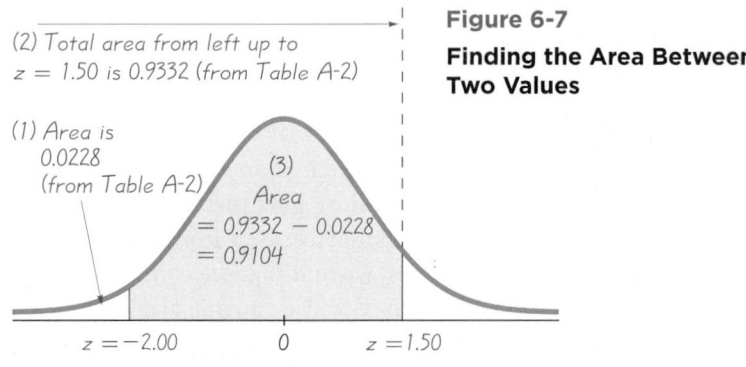

Figure 6-7

Finding the Area Between Two Values

(2) Total area from left up to $z = 1.50$ is 0.9332 (from Table A-2)

(1) Area is 0.0228 (from Table A-2)

(3) Area $= 0.9332 - 0.0228$ $= 0.9104$

$z = -2.00$ 0 $z = 1.50$

INTERPRETATION Using the correspondence between probability and area, we conclude that there is a probability of 0.9104 of randomly selecting one of the thermometers with a reading between $-2.00°$ and $1.50°$ at the freezing point of water. Another way to interpret this result is to state that if many thermometers are selected and tested at the freezing point of water, then 0.9104 (or 91.04%) of them will read between $-2.00°$ and $1.50°$.

Example 5 can be generalized as the following rule: **The area corresponding to the region between two specific z scores can be found by finding the difference between the two areas found in Table A-2.** Figure 6-8 illustrates this general rule. Note that the shaded region B can be found by calculating the *difference* between two areas found from Table A-2: area A and B combined (found in Table A-2 as the area corresponding to z_{Right}) and area A (found in Table A-2 as the area corresponding to z_{Left}). *Study hint:* Don't try to memorize a rule or formula for this case. Focus on understanding how Table A-2 works. If necessary, first draw a graph, shade the desired area, then think of a way to find that area given the condition that Table A-2 provides only cumulative areas from the left.

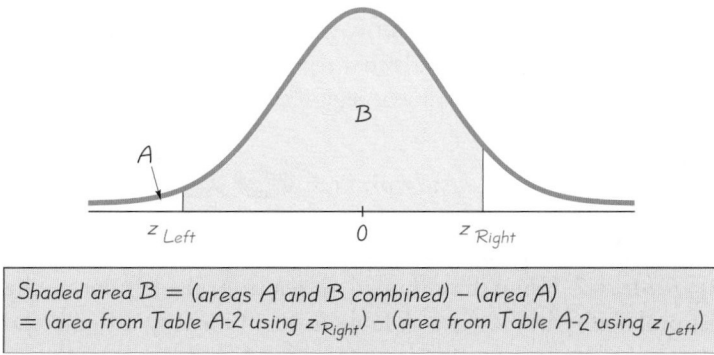

Shaded area B = (areas A and B combined) − (area A)
= (area from Table A-2 using z_{Right}) − (area from Table A-2 using z_{Left})

Figure 6-8   Finding the Area Between Two z Scores

Probabilities such as those in the preceding examples can also be expressed with the following notation.

Notation

$P(a < z < b)$	denotes the probability that the z score is between a and b.
$P(z > a)$	denotes the probability that the z score is greater than a.
$P(z < a)$	denotes the probability that the z score is less than a.

Using this notation, we can express the result of Example 5 as: $P(-2.00 < z < 1.50) = 0.9104$, which states in symbols that the probability of a z score falling between -2.00 and 1.50 is 0.9104. With a continuous probability distribution such as the normal distribution, the probability of getting any single *exact* value is 0. That is, $P(z = a) = 0$. For example, there is a 0 probability of randomly selecting someone and getting a person whose height is exactly 68.12345678 in. In the normal distribution, any single point on the horizontal scale is represented not by a region under the curve, but by a vertical line above the point. For $P(z = 1.50)$ we have a vertical line above $z = 1.50$, but that vertical line by itself contains no area, so $P(z = 1.50) = 0$. With any continuous random variable, the probability of any one exact value is 0, and it follows that $P(a \leq z \leq b) = P(a < z < b)$. It also follows that the probability of getting a z score of *at most b* is equal to the probability of getting a z score *less than b*. It is important to correctly interpret key phrases such as *at most, at least, more than, no more than,* and so on.

Finding *z* Scores from Known Areas

So far in this section, all of the examples involving the standard normal distribution have followed the same format: Given z scores, find areas under the curve. These areas correspond to probabilities. In many cases, we have the reverse: Given the area (or probability), find the corresponding z score. In such cases, we must avoid confusion between z scores and areas. Remember, z scores are *distances* along the horizontal scale, whereas areas (or probabilities) are regions under the curve. (Table A-2 lists z-scores in the left column and across the top row, but areas are found in the *body* of the table.) Also, z scores positioned in the left half of the curve are always negative. If we already know a probability and want to determine the corresponding z score, we find it as follows.

Procedure for Finding a *z* Score from a Known Area

1. Draw a bell-shaped curve and identify the region under the curve that corresponds to the given probability. If that region is not a cumulative region from the left, work instead with a known region that is a cumulative region from the left.

2. *Using a TI-83/84 Plus Calculator:* Press **2ND** **VARS** to get the DISTR (distribution) menu, then select **invNorm**, and then proceed to enter the total area to the *left* of the desired z score.

   Using Table A-2: Using the cumulative area from the left, locate the closest probability in the *body* of Table A-2 and identify the corresponding z score.

> **EXAMPLE 6**   **Scientific Thermometers** Use the same thermometers from Example 3, with temperature readings at the freezing point of water that are normally distributed with a mean of 0°C and a standard deviation of 1.00°C. Find the temperature corresponding to P_{95}, the 95th percentile. That is, find the temperature separating the bottom 95% from the top 5%. See Figure 6-9.

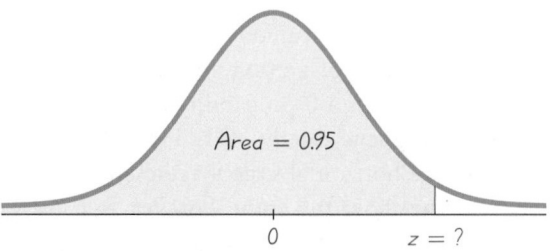

Figure 6-9   Finding the 95th Percentile

Figure 6-9 shows the *z* score that is the 95th percentile, with 95% of the area (or 0.95) below it. Referring to Table A-2, we search for the area of 0.95 *in the body* of the table and then find the corresponding *z* score. In Table A-2 we find the areas of 0.9495 and 0.9505, but there's an asterisk with a special note indicating that 0.9500 corresponds to a *z* score of 1.645. We can now conclude that the *z* score in Figure 6-9 is 1.645, so the 95th percentile is the temperature reading of 1.645°C.

With a TI-83/84 Plus calculator, select **invNorm** and enter (0.95) as described in the TI-83/84 Plus instructions given near the end of this section. The *z* score in Figure 6-9 is found to be 1.644853626.

INTERPRETATION When tested at freezing, 95% of the readings will be less than or equal to 1.645°C, and 5% of them will be greater than or equal to 1.645°C.

Note that in the preceding solution, Table A-2 led to a *z* score of 1.645, which is midway between 1.64 and 1.65. When using Table A-2, we can usually avoid interpolation by simply selecting the closest value. Special cases are listed in the accompanying table because they are often used in a wide variety of applications. (For one of those special cases, the value of $z = 2.576$ gives an area slightly closer to the area of 0.9950, but $z = 2.575$ has the advantage of being the value midway between $z = 2.57$ and $z = 2.58$.) Except in these special cases, we can select the closest value in the table. (If a desired value is midway between two table values, select the larger value.) For *z* scores above 3.49, we can use 0.9999 as an approximation of the cumulative area from the left; for *z* scores below −3.49, we can use 0.0001 as an approximation of the cumulative area from the left.

Table A-2 Special Cases

z Score	Cumulative Area from the Left
1.645	0.9500
−1.645	0.0500
2.575	0.9950
−2.575	0.0050
Above 3.49	0.9999
Below −3.49	0.0001

EXAMPLE 7 **Scientific Thermometers** Using the same thermometers from Example 3, find the temperatures separating the bottom 2.5% and the top 2.5%.

SOLUTION The required *z* scores are shown in Figure 6-10. To find the *z* score located to the left, we search the *body of Table A-2* for an area of 0.025. The result is $z = -1.96$. To find the *z* score located to the right, we search *the body of Table A-2* for an area of 0.975. (Remember that Table A-2 always gives cumulative areas from the *left*.) The result is $z = 1.96$. The values of $z = -1.96$ and $z = 1.96$ separate the bottom 2.5% and the top 2.5%, as shown in Figure 6-10.

If using a TI-83/84 Plus calculator as described near the end of this section, the *z* scores in Figure 6-10 are found to be −1.959963986 and 1.959963986.

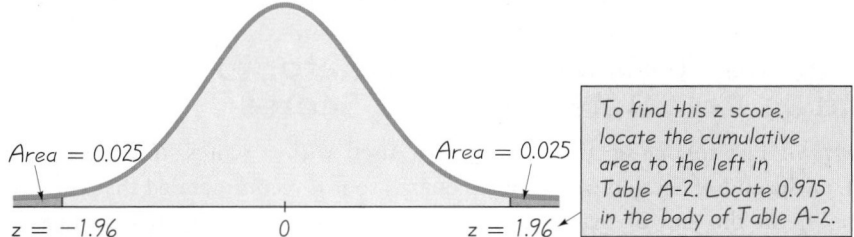

Figure 6-10
Finding z Scores

INTERPRETATION When tested at freezing, 2.5% of the thermometer readings will be equal to or less than −1.96°, and 2.5% of the readings will be equal to or greater than 1.96°. Another interpretation is that at the freezing point of water, 95% of all thermometer readings will fall between −1.96° and 1.96°.

Critical Values For a normal distribution, a **critical value** is a z score on the borderline separating the z scores that are *likely* to occur from those that are *unlikely*. Common critical values are $z = -1.96$ and $z = 1.96$, and they are obtained as shown in Example 7. In Example 7, the values below $z = -1.96$ are not likely to occur, because they occur in only 2.5% of the readings, and the values above $z = 1.96$ are not likely to occur because they also occur in only 2.5% of the readings. The reference to *critical values* is not so important in this chapter, but will become extremely important in the following chapters. The following notation is used for critical z values found by using the standard normal distribution.

Notation

The expression z_α denotes the z score with an area of α to its right. (α is the Greek letter alpha.)

EXAMPLE 8 **Finding z_α** In the expression z_α, let $\alpha = 0.025$ and find the value of $z_{0.025}$.

SOLUTION The notation of $z_{0.025}$ is used to represent the z score with an area of 0.025 to its right. Refer to Figure 6-10 and note that the value of $z = 1.96$ has an area of 0.025 to its right, so $z_{0.025} = 1.96$.

Caution: When using Table A-2 for finding a value of z_α for a particular value of α, note that α is the area to the *right* of z_α, but Table A-2 lists cumulative areas to the *left* of a given z score. To find the value of z_α by using Table A-2, resolve that conflict by using the value of $1 - \alpha$. In Example 8, the value of $z_{0.025}$ can be found by locating the area of 0.9750 in the body of the table.

The examples in this section were created so that the mean of 0 and the standard deviation of 1 coincided exactly with the properties of the standard normal distribution. In reality, it is unusual to find such convenient parameters, because typical normal distributions involve means different from 0 and standard deviations different from 1. In the next section we introduce methods for working with such normal distributions, which are much more realistic and practical.

Using the TI-83/84 Plus Calculator to Find an Area Between Two z Scores

Important: The following procedure can be used with z scores, but the procedure given at the end of Section 6-3 is more general, so it is recommended that you use the procedure at the end of that section. To find the area between two z scores, press **2ND** **VARS** to get to the DISTR (distribution) menu, then select **normalcdf.** You must now enter the lower boundary and the upper, separated by a comma. For the shaded area in Figure 6-7 from Example 5, enter the command **normalcdf(−2.00, 1.50)** to get 0.9104427093.

Hint: If there is no actual lower boundary or no upper boundary, enter a very large value, as shown below.

- Cumulative area to the left (with no actual lower boundary): Enter −999999 for the lower boundary. For the shaded area in Figure 6-5 from Example 3, enter the command **normalcdf(−999999, 1.27)** to get 0.8979576193.

- Cumulative area to the right (with no actual upper boundary): Enter 999999 for the upper boundary. For the shaded area in Figure 6-6 from Example 4, enter the command **normalcdf(−1.23, 999999)** to get 0.8906513833.

Using the TI-83/84 Plus Calculator to Find a z Score Corresponding to a Known Probability

Important: The following procedure can be used with z scores, but the procedure given at the end of Section 5-3 is more general, so it is recommended that you use the procedure at the end of that section. To find a z score corresponding to a known probability, press **2ND** **VARS** to get to the DISTR (distribution) menu, then select **invNorm.** Proceed to enter the *cumulative area to the left* of the desired z score. For the z score in Figure 6-9 from Example 6, enter the command **invNorm(0.95)** to get 1.644853626.

Using the TI-83/84 Plus Calculator to Find a Critical Value z_α

The value z_α is the z score with an area of α to its *right*. Press **2ND** **VARS** to get to the DISTR (distribution) menu, then select **invNorm.** Proceed to enter the *cumulative area to the left* of the desired critical z score, so enter the value of $1 - \alpha$. Example 8 involves finding the critical value $z_{0.025}$, which is the z score with an area of 0.025 to its right. With $\alpha = 0.025$, we get $1 - \alpha = 0.975$, so enter the command **invNorm(0.975)** to get 1.959963986, which can be rounded to 1.96, as in Example 8.

6-2 Basic Skills and Concepts

Statistical Literacy and Critical Thinking

1. Normal Distribution When we refer to a "normal" distribution, does the word "normal" have the same meaning as in ordinary language, or does it have a special meaning in statistics? What exactly is a normal distribution?

2. Normal Distribution A normal distribution is informally described as a probability distribution that is "bell-shaped" when graphed. Describe the "bell shape."

3. Standard Normal Distribution What requirements are necessary for a normal probability distribution to be a *standard* normal probability distribution?

4. Notation What does the notation z_α indicate?

Continuous Uniform Distribution. *In Exercises 5–8, refer to the continuous uniform distribution depicted in Figure 6-2. Assume that a voltage level between 123.0 volts and 125.0 volts is randomly selected, and find the probability that the given voltage level is selected.*

5. Greater than 124.0 volts

6. Less than 123.5 volts

7. Between 123.2 volts and 124.7 volts

8. Between 124.1 volts and 124.5 volts

Standard Normal Distribution. *In Exercises 9–12, find the area of the shaded region. The graph depicts the standard normal distribution with mean 0 and standard deviation 1.*

TI **9.**

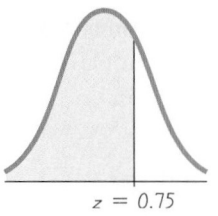

$z = 0.75$

TI **10.**

$z = -0.75$

TI **11.**

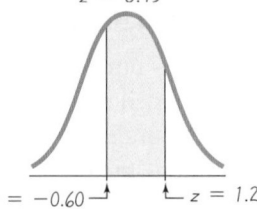

$z = -0.60$ $z = 1.20$

TI **12.**

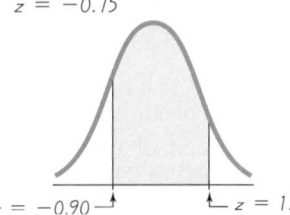

$z = -0.90$ $z = 1.60$

Standard Normal Distribution. *In Exercises 13–16, find the indicated z score. The graph depicts the standard normal distribution with mean 0 and standard deviation 1.*

TI **13.**

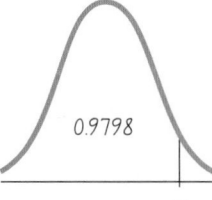

0.9798

z

TI **14.**

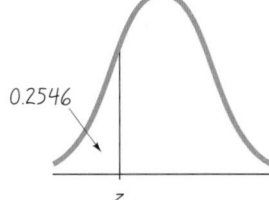

0.2546

z

TI **15.**

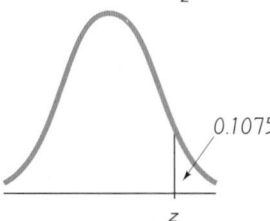

0.1075

z

TI **16.**

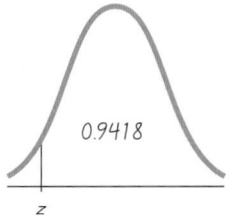

0.9418

z

Standard Normal Distribution. *In Exercises 17–36, assume that thermometer readings are normally distributed with a mean of 0°C and a standard deviation of 1.00°C. A thermometer is randomly selected and tested. In each case, draw a sketch, and find the probability of each reading. (The given values are in Celsius degrees.) If using technology instead of Table A-2, round answers to four decimal places.*

TI **17.** Less than -1.50

TI **18.** Less than -2.75

TI **19.** Less than 1.23

TI **20.** Less than 2.34

TI **21.** Greater than 2.22

TI **22.** Greater than 2.33

TI **23.** Greater than -1.75

TI **24.** Greater than -1.96

TI **25.** Between 0.50 and 1.00

TI **26.** Between 1.00 and 3.00

TI **27.** Between -3.00 and -1.00

TI **28.** Between -1.00 and -0.50

TI **29.** Between -1.20 and 1.95

TI **30.** Between -2.87 and 1.34

TI **31.** Between -2.50 and 5.00

TI **32.** Between -4.50 and 1.00

TI **33.** Less than 3.55

TI **34.** Greater than 3.68

TI **35.** Greater than 0

TI **36.** Less than 0

Basis for the Range Rule of Thumb and the Empirical Rule. *In Exercises 37–40, find the indicated area under the curve of the standard normal distribution, then convert it to a percentage and fill in the blank. The results form the basis for the range rule of thumb and the empirical rule introduced in Section 3-3.*

TI 37. About _____% of the area is between $z = -1$ and $z = 1$ (or within 1 standard deviation of the mean).

TI 38. About _____% of the area is between $z = -2$ and $z = 2$ (or within 2 standard deviations of the mean).

TI 39. About _____% of the area is between $z = -3$ and $z = 3$ (or within 3 standard deviations of the mean).

TI 40. About _____% of the area is between $z = -3.5$ and $z = 3.5$ (or within 3.5 standard deviations of the mean).

Finding Critical Values. *In Exercises 41–44, find the indicated value.*

TI 41. $z_{0.05}$ **TI 42.** $z_{0.01}$

TI 43. $z_{0.10}$ **TI 44.** $z_{0.02}$

Finding Probability. *In Exercises 45–48, assume that the readings on the thermometers are normally distributed with a mean of 0°C and a standard deviation of 1.00°. Find the indicated probability, where z is the reading in degrees.*

TI 45. $P(-1.96 < z < 1.96)$ **TI 46.** $P(z < 1.645)$

TI 47. $P(z < -2.575$ or $z > 2.575)$ **TI 48.** $P(z < -1.96$ or $z > 1.96)$

Finding Temperature Values. *In Exercises 49–52, assume that thermometer readings are normally distributed with a mean of 0°C and a standard deviation of 1.00°C. A thermometer is randomly selected and tested. In each case, draw a sketch, and find the temperature reading corresponding to the given information.*

TI 49. Find P_{95}, the 95th percentile. This is the temperature reading separating the bottom 95% from the top 5%.

TI 50. Find P_1, the 1st percentile. This is the temperature reading separating the bottom 1% from the top 99%.

TI 51. If 2.5% of the thermometers are rejected because they have readings that are too high and another 2.5% are rejected because they have readings that are too low, find the two readings that are cutoff values separating the rejected thermometers from the others.

TI 52. If 0.5% of the thermometers are rejected because they have readings that are too low and another 0.5% are rejected because they have readings that are too high, find the two readings that are cutoff values separating the rejected thermometers from the others.

6-2   Beyond the Basics

TI 53. For a standard normal distribution, find the percentage of data that are

a. within 2 standard deviations of the mean.

b. more than 1 standard deviation away from the mean.

c. more than 1.96 standard deviations away from the mean.

d. between $\mu - 3\sigma$ and $\mu + 3\sigma$.

e. more than 3 standard deviations away from the mean.

TI 54. If a continuous uniform distribution has parameters of $\mu = 0$ and $\sigma = 1$, then the minimum is $-\sqrt{3}$ and the maximum is $\sqrt{3}$.

a. For this distribution, find $P(-1 < x < 1)$.

b. Find $P(-1 < x < 1)$ if you incorrectly assume that the distribution is normal instead of uniform.

c. Compare the results from parts (a) and (b). Does the distribution affect the results very much?

TI 55. Assume that z scores are normally distributed with a mean of 0 and a standard deviation of 1.

a. If $P(z < a) = 0.9599$, find a.

b. If $P(z > b) = 0.9772$, find b.

c. If $P(z > c) = 0.0668$, find c.

d. If $P(-d < z < d) = 0.5878$, find d.

e. If $P(-e < z < e) = 0.0956$, find e.

56. In a continuous uniform distribution,

$$\mu = \frac{\text{minimum} + \text{maximum}}{2} \quad \text{and} \quad \sigma = \frac{\text{range}}{\sqrt{12}}$$

Find the mean and standard deviation for the uniform distribution represented in Figure 6-2.

6-3   Applications of Normal Distributions

Key Concept In this section we introduce real and important applications involving nonstandard normal distributions by extending the procedures presented in Section 6-2. We use a simple conversion (Formula 6-2) that allows us to standardize any normal distribution so that the methods of the preceding section can be used with normal distributions having a mean that is not 0 or a standard deviation that is not 1. Specifically, given some nonstandard normal distribution, we should be able to find probabilities corresponding to values of the variable x, and given some probability value, we should be able to find the corresponding value of the variable x.

To work with a nonstandard normal distribution, we simply standardize values to use the procedures from Section 6-2.

> **If we convert values to standard z-scores using Formula 6-2, then procedures for working with all normal distributions are the same as those for the standard normal distribution.**

Formula 6-2

$$z = \frac{x - \mu}{\sigma} \quad \text{(round } z \text{ scores to 2 decimal places)}$$

The TI-83/84 Plus calculator does not require the above conversion to z scores because probabilities can be found directly. However, if you use Table A-2 to find probabilities, you must first convert values to standard z scores. Regardless of the method you use, you need to clearly understand the above principle, because it is an important foundation for concepts introduced in the following chapters.

Figure 6-11 illustrates the conversion from a nonstandard to a standard normal distribution. The area in any normal distribution bounded by some score x (as in Figure 6-11(a)) is the same as the area bounded by the equivalent z score in the standard normal distribution (as in Figure 6-11(b)). This means that when working with a nonstandard normal distribution, you can use Table A-2 the same way it was used in Section 6-2, as long as you first convert the values to z scores.

Figure 6-11   Converting from a Nonstandard to a Standard Normal Distribution

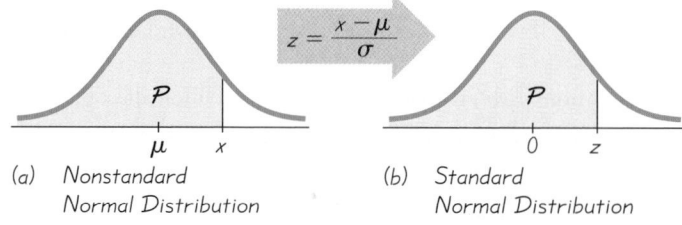

(a)   Nonstandard Normal Distribution   *(b)   Standard Normal Distribution*

Using the TI-83/84 Plus Calculator to Find Areas

See the procedure given near the end of this section.

Using Table A-2 to Find Areas

1. Sketch a normal curve, label the mean and the specific *x* values, then *shade* the region representing the desired probability.

2. For each relevant value *x* that is a boundary for the shaded region, use Formula 6-2 to convert that value to the equivalent *z* score.

3. Refer to Table A-2 to find the area of the shaded region. This area is the desired probability.

The following example applies these three steps to illustrate the relationship between a typical nonstandard normal distribution and the standard normal distribution.

EXAMPLE 1   **Why Do Doorways Have a Height of 6 ft 8 in.?** The typical home doorway has a height of 6 ft 8 in., or 80 in. Because men tend to be taller than women, we will consider only men as we investigate the limitations of that standard doorway height. Given that heights of men are normally distributed with a mean of 69.0 in. and a standard deviation of 2.8 in., find the percentage of men who can fit through the standard doorway without bending or bumping their head. Is that percentage high enough to continue using 80 in. as the standard height? Will a doorway height of 80 in. be sufficient in future years?

SOLUTION

Step 1: See Figure 6-12, which incorporates this information: Men have heights that are normally distributed with a mean of 69.0 in. and a standard deviation of 2.8 in. The shaded region represents the men who can fit through a doorway that has a height of 80 in.

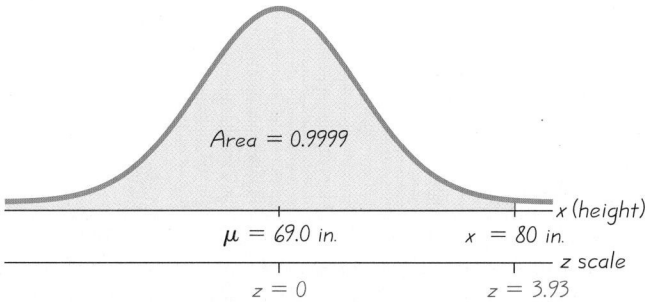

Figure 6-12

Heights (in inches) of Men

Step 2: If using a TI-83/84 Plus calculator with the procedure given near the end of this section, press **2ND** **VARS** and proceed to enter **normalcdf(−999999, 80, 69.0, 2.8)** to get 0.9999572562.

To use Table A-2, we first must use Formula 6-2 to convert from the nonstandard normal distribution to the standard normal distribution. The height of 80 in. is converted to a *z* score as follows:

$$z = \frac{x - \mu}{\sigma} = \frac{80 - 69.0}{2.8} = 3.93$$

Step 3: Referring to Table A-2 and using $z = 3.93$, we find that this *z* score is in the category of "3.50 and up," so the cumulative area to the left of 80 in. is 0.9999 as shown in Figure 6-12.

continued

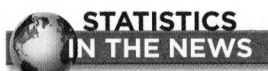
INTERPRETATION   The proportion of men who can fit through the standard doorway height of 80 in. is 0.9999, or 99.99%. Very few men will not be able to fit through the doorway without bending or bumping their head. This percentage is high enough to justify the use of 80 in. as the standard doorway height. However, heights of men and women have been increasing gradually but steadily over the past decades, so the time may come when the standard doorway height of 80 in. may no longer be adequate.

EXAMPLE 2   **Birth Weights** Birth weights in the United States are normally distributed with a mean of 3420 g and a standard deviation of 495 g. The Newport General Hospital requires special treatment for babies that are less than 2450 g (unusually light) or more than 4390 g (unusually heavy). What is the percentage of babies who do not require special treatment because they have birth weights between 2450 g and 4390 g? Under these conditions, do many babies require special treatment?

SOLUTION   Figure 6-13 shows the shaded region representing birth weights between 2450 g and 4390 g. If using a TI-83/84 Plus calculator with the procedure given near the end of this section, press **2ND** **VARS** and proceed to enter **normalcdf(2450, 4390, 3420, 495)** to get 0.949957106. If using Table A-2, we can't find the shaded area directly, but we can find it indirectly by using the same basic procedures presented in Section 6-2, as follows: (1) Find the cumulative area from the left up to 2450; (2) find the cumulative area from the left up to 4390; (3) find the difference between those two areas.

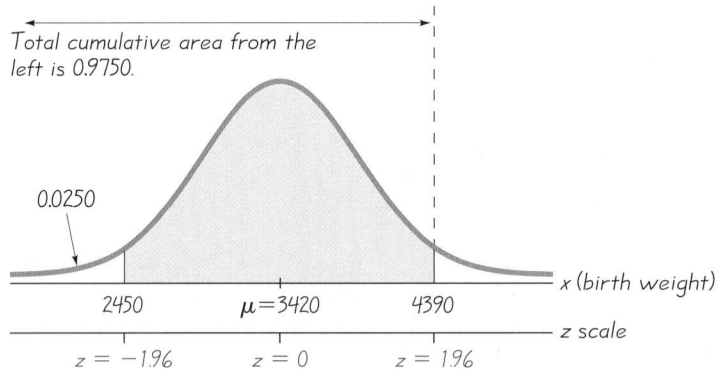

Figure 6-13   Birth Weights

Find the cumulative area up to 2450:

$$z = \frac{x - \mu}{\sigma} = \frac{2450 - 3420}{495} = -1.96$$

Using Table A-2, we find that $z = -1.96$ corresponds to an area of 0.0250, as shown in Figure 6-13.

Find the cumulative area up to 4390:

$$z = \frac{x - \mu}{\sigma} = \frac{4390 - 3420}{495} = 1.96$$

Using Table A-2, we find that $z = 1.96$ corresponds to an area of 0.9750, as shown in Figure 6-13.

Find the shaded area between 2450 and 4390:

$$\text{Shaded area} = 0.9750 - 0.0250 = 0.9500$$

INTERPRETATION Expressing the result as a percentage, we conclude that 95.00% of the babies do not require special treatment because they have birth weights between 2450 g and 4390 g. It follows that 5.00% of the babies do require special treatment because they are unusually light or heavy. The 5.00% rate is probably not too high for typical hospitals.

Finding Values from Known Areas

Here are helpful hints for those cases in which the area (or probability or percentage) is known and we must find the relevant value(s):

1. *Don't confuse z scores and areas.* Remember, z scores are *distances* along the horizontal scale, but areas are *regions* under the normal curve. Table A-2 lists z scores in the left columns and across the top row, but areas are found in the body of the table.

2. *Choose the correct (right/left) side of the graph.* A value separating the top 10% from the others will be located on the right side of the graph, but a value separating the bottom 10% will be located on the left side of the graph.

3. A z score must be *negative* whenever it is located in the *left* half of the normal distribution.

4. Areas (or probabilities) are positive or zero values, but they are never negative.

Graphs are extremely helpful in visualizing, understanding, and successfully working with normal probability distributions, so they should be used whenever possible.

Using the TI-83/84 Plus Calculator for Finding Values
See the procedure near the end of this section.

Using Table A-2 and Formula 6-2 for Finding Values

1. Sketch a normal distribution curve, enter the given probability or percentage in the appropriate region of the graph, and identify the x value(s) being sought.

2. Use Table A-2 to find the z score corresponding to the cumulative left area bounded by x. Refer to the *body* of Table A-2 to find the closest area, then identify the corresponding z score.

3. Using Formula 6-2, enter the values for μ, σ, and the z score found in Step 2, then solve for x. Based on Formula 6-2, we can solve for x as follows:

$$x = \mu + (z \cdot \sigma) \qquad \text{(another form of Formula 6-2)}$$

(If z is located to the left of the mean, be sure that it is a negative number.)

4. Refer to the sketch of the curve to verify that the solution makes sense in the context of the graph and in the context of the problem.

The following example uses the procedure just outlined.

EXAMPLE 3 **Designing Doorway Heights** When designing an environment, one common criterion is to use a design that accommodates 95% of the population. How high should doorways be if 95% of men will fit through without bending or bumping their head? That is, find the 95th percentile of heights of men. Heights of men are normally distributed with a mean of 69.0 in. and a standard deviation of 2.8 in.

SOLUTION

Step 1: Figure 6-14 shows the normal distribution with the height x that we want to identify. The shaded area represents the 95% of men who can fit through the doorway that we are designing.

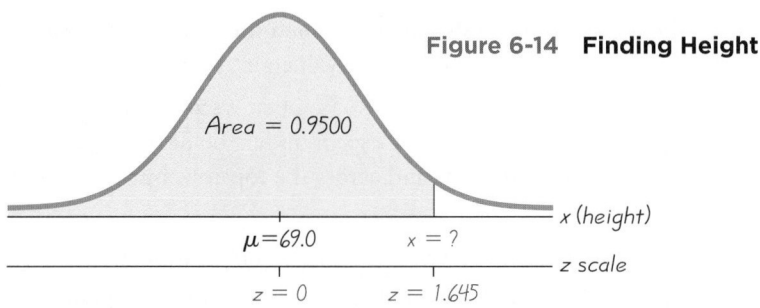

Figure 6-14   **Finding Height**

Step 2: If using a TI-83/84 Plus calculator with the procedure given near the end of this section, press **2ND** **VARS** and proceed to enter **invNorm(0.9500, 69.0, 2.8)** to get 73.60559015.

If using Table A-2, we search for an area of 0.9500 *in the body* of the table. (The area of 0.9500 shown in Figure 6-14 is a cumulative area from the left, and that is exactly the type of area listed in Table A-2.) The area of 0.9500 is between the Table A-2 areas of 0.9495 and 0.9505, but there is an asterisk and footnote indicating that an area of 0.9500 corresponds to $z = 1.645$.

Step 3: With $z = 1.645$, $\mu = 69.0$, and $\sigma = 2.8$, we can solve for x by using Formula 6-2:

$$z = \frac{x - \mu}{\sigma} \quad \text{becomes} \quad 1.645 = \frac{x - 69.0}{2.8}$$

The result of $x = 73.606$ in. can be found directly or by using the following version of Formula 6-2:

$$x = \mu + (z \cdot \sigma) = 69.0 + (1.645 \cdot 2.8) = 73.606$$

Step 4: The solution of $x = 73.6$ in. (rounded) in Figure 6-14 is reasonable because it is greater than the mean of 69.0 in.

INTERPRETATION A doorway height of 73.6 in. (or 6 ft 1.6 in.) would allow 95% of men to fit without bending or bumping their head. It follows that 5% of men would *not* fit through a doorway with a height of 73.6 in. Because so many men walk through doorways so often, this 5% rate is probably not practical.

EXAMPLE 4   **Birth Weights** The Newport General Hospital wants to re-define the minimum and maximum birth weights that require special treatment because they are unusually low or unusually high. After considering relevant factors, a committee recommends special treatment for birth weights in the lowest 3% and the highest 1%. The committee members soon realize that specific birth weights need to be identified. Help this committee by finding the birth weights that separate the lowest 3% and the highest 1%. Birth weights in the United States are normally distributed with a mean of 3420 g and a standard deviation of 495 g.

SOLUTION

Step 1: We begin with the graph shown in Figure 6-15. We have entered the mean of 3420 g, and we have identified the x values separating the lowest 3% and the highest 1%.

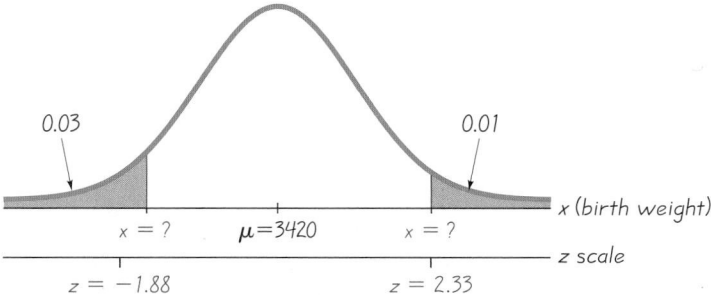

Figure 6-15

Finding the Values Separating the Lowest 3% and the Highest 1%

Step 2: If using a TI-83/84 Plus calculator with the procedure given near the end of this section, press **2ND** **VARS** and proceed to enter **invNorm(0.03, 3420, 495)** to get 2489.007163, which is the leftmost x value. Then enter **invNorm(0.99, 3420, 495)** to get 4571.542199, which is the rightmost x value.

   If using Table A-2, we must use cumulative areas from the left. For the leftmost value of x, the cumulative area from the left is 0.03, so search for an area of 0.03 *in the body* of the table to get $z = -1.88$ (which corresponds to the closest area of 0.0301). For the rightmost value of x, the cumulative area from the left is 0.99, so search for an area of 0.99 *in the body* of the table to get $z = 2.33$ (which corresponds to the closest area of 0.9901).

Step 3: We now solve for the two values of x by using Formula 6-2 directly or by using the following version of Formula 6-2:

Leftmost value of x: $x = \mu + (z \cdot \sigma) = 3420 + (-1.88 \cdot 495) = 2489.4$
Rightmost value of x: $x = \mu + (z \cdot \sigma) = 3420 + (2.33 \cdot 495) = 4573.35$

Step 4: Referring to Figure 6-15, we see that the leftmost value of $x = 2489.4$ g is reasonable because it is less than the mean of 3420 g. Also, the rightmost value of 4573.35 is reasonable because it is above the mean of 3420 g.

INTERPRETATION   The birth weight of 2489 g (rounded) separates the lowest 3% of birth weights, and 4573 g (rounded) separates the highest 1% of birth weights. The hospital now has well-defined criteria for determining whether a newborn baby should be given special treatment for a birth weight that is unusually low or high.

When using the methods of this section with applications involving a normal distribution, it is important to first determine whether you are finding a probability (or area) from a known value of x or finding a value of x from a known probability (or area). Figure 6-16 is a flowchart summarizing the main procedures of this section.

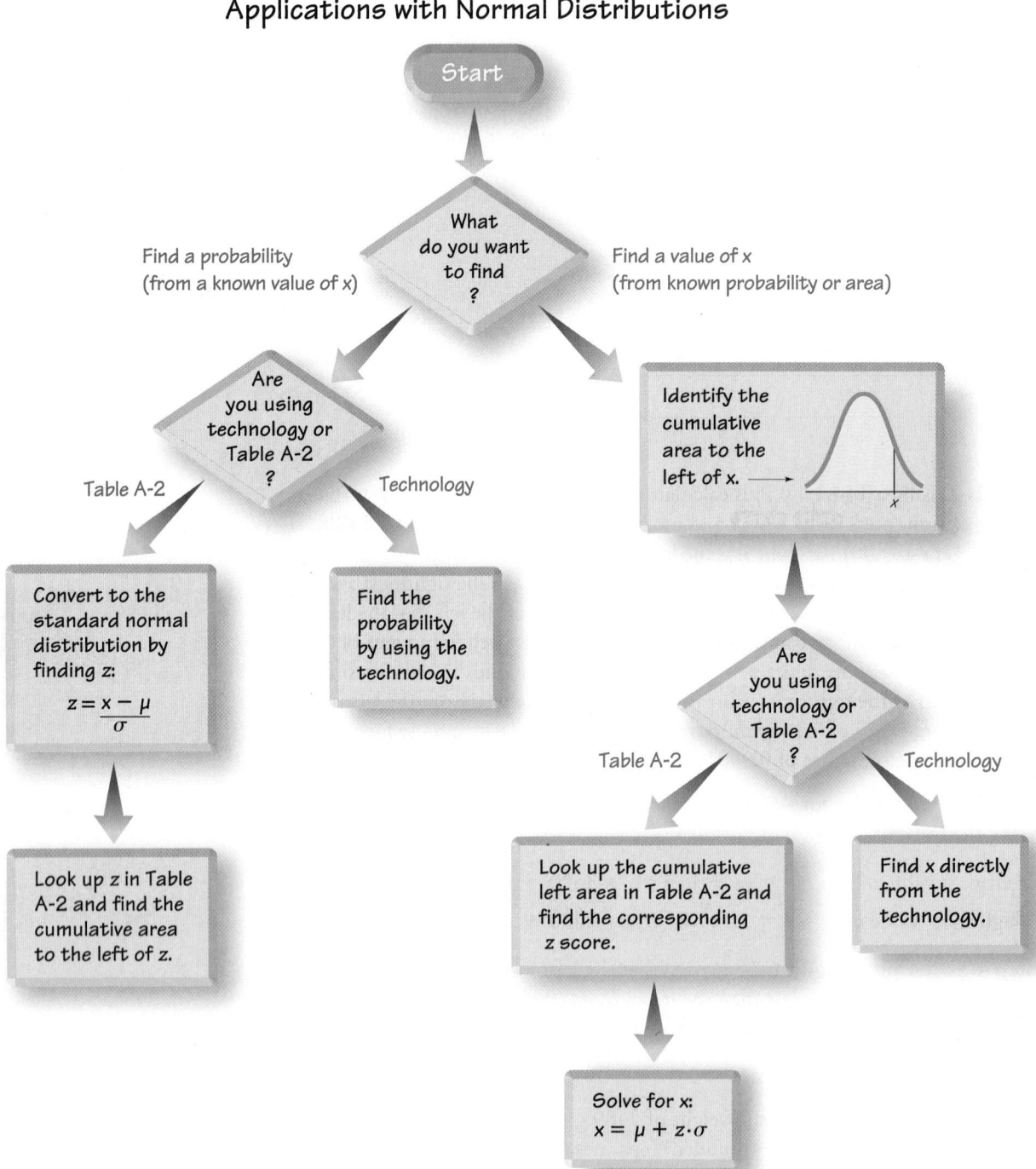

Applications with Normal Distributions

Start

What do you want to find ?

Find a probability (from a known value of x)

Find a value of x (from known probability or area)

Are you using technology or Table A-2 ?

Table A-2

Technology

Convert to the standard normal distribution by finding z:

$$z = \frac{x - \mu}{\sigma}$$

Find the probability by using the technology.

Identify the cumulative area to the left of x.

Look up z in Table A-2 and find the cumulative area to the left of z.

Are you using technology or Table A-2 ?

Table A-2

Technology

Look up the cumulative left area in Table A-2 and find the corresponding z score.

Find x directly from the technology.

Solve for x:

$$x = \mu + z \cdot \sigma$$

Figure 6-16   Procedures for Applications with Normal Distributions

Using a TI-83/84 Plus Calculator to Find Areas

1. Sketch a normal curve, label the mean, label the *x* value at the lower boundary of the desired area, and label the *x* value at the upper boundary of the desired area. Also, *shade* the region representing the desired probability.

2. Press **2ND** **VARS** to get the DISTR (distribution) menu, then select **normalcdf** and proceed to enter the two *x* values, the mean, and the standard deviation, all separated by commas. That is, use this format: **normalcdf(lower bound, upper bound, μ, σ).**

3. Press **ENTER** and the area will be displayed.

Hint: If there is no lower bound or there is no upper bound, enter a very large value, as indicated below.

• Cumulative area from the left (with no actual lower bound): Enter −999999 for the lower bound.

• Cumulative area from the right (with no actual upper bound): Enter 999999 for the upper bound.

Example: The first screen in the margin shows the DISTR (distribution) menu with normalcdf selected. The second screen shows the entries to be used for Example 1. Pressing **ENTER** yields 0.9999572562, which is the area under the normal distribution from the left up to 80, assuming that the mean is 69.0 and the standard deviation is 2.8.

Using a TI-83/84 Plus Calculator to Find Values

1. Sketch a normal distribution curve, enter the given probability or percentage in the appropriate region of the graph, and identify the *x* value being sought.

2. Press **2ND** **VARS** to get the DISTR (distribution) menu, then select **invNorm** and proceed to enter the *cumulative area to the left* of the desired *x* value, the mean, and the standard deviation, all separated by commas. That is, use this format: **invNorm(area to the left, μ, σ).**

3. Press **ENTER** and the value of *x* will be displayed.

Example: The third screen in the margin shows the DISTR (distribution) menu with invNorm selected, and the fourth screen shows the entries for finding the rightmost *x* value in Figure 6-15 from Example 4.

normalcdf on the DISTR Menu

```
DISTR DRAW
1:normalpdf(
2:normalcdf(
3:invNorm(
4:tpdf(
5:tcdf(
6:X²pdf(
7↓X²cdf(
```

Entries for Example 1

```
normalcdf(-99999
9,80,69.0,2.8)
        .9999572562
■
```

Selecting invNorm in the DISTR Menu

```
DISTR DRAW
1:normalpdf(
2:normalcdf(
3:invNorm(
4:tpdf(
5:tcdf(
6:X²pdf(
7↓X²cdf(
```

Finding the Rightmost *x* Value in Example 4

```
invNorm(.99,3420
,495)
        4571.542199
```

6-3 Basic Skills and Concepts

Statistical Literacy and Critical Thinking

1. Normal Distributions What is the difference between a standard normal distribution and a nonstandard normal distribution?

2. IQ Scores The distribution of IQ scores is a nonstandard normal distribution with a mean of 100 and a standard deviation of 15, and a bell-shaped graph is drawn to represent this distribution.

a. What is the area under the curve?

b. What is the value of the median?

c. What is the value of the mode?

3. Normal Distributions The distribution of IQ scores is a nonstandard normal distribution with a mean of 100 and a standard deviation of 15. What are the values of the mean and standard deviation after all IQ scores have been standardized by converting them to z scores using $z = (x - \mu)/\sigma$?

4. Random Digits Computers are often used to randomly generate digits of telephone numbers to be called when conducting a survey. Can the methods of this section be used to find the probability that when one digit is randomly generated, it is less than 5? Why or why not? What is the probability of getting a digit less than 5?

IQ Scores. In Exercises 5–8, find the area of the shaded region. The graphs depict IQ scores of adults, and those scores are normally distributed with a mean of 100 and a standard deviation of 15 (as on the Wechsler test).

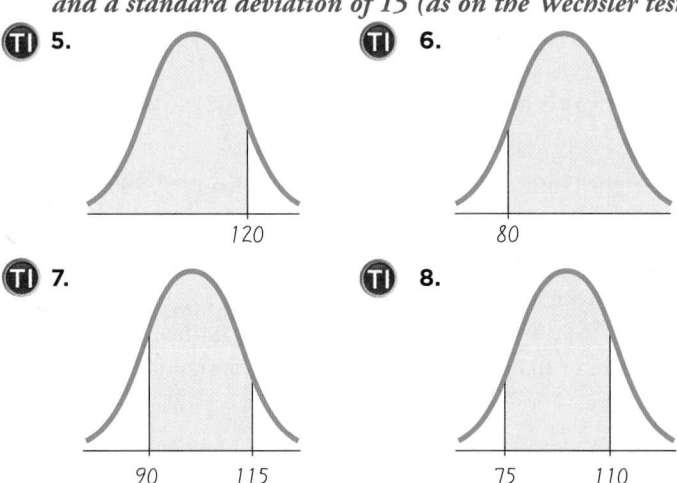

TI 5.   120

TI 6.   80

TI 7.   90   115

TI 8.   75   110

IQ Scores. In Exercises 9–12, find the indicated IQ score. The graphs depict IQ scores of adults, and those scores are normally distributed with a mean of 100 and a standard deviation of 15 (as on the Wechsler test).

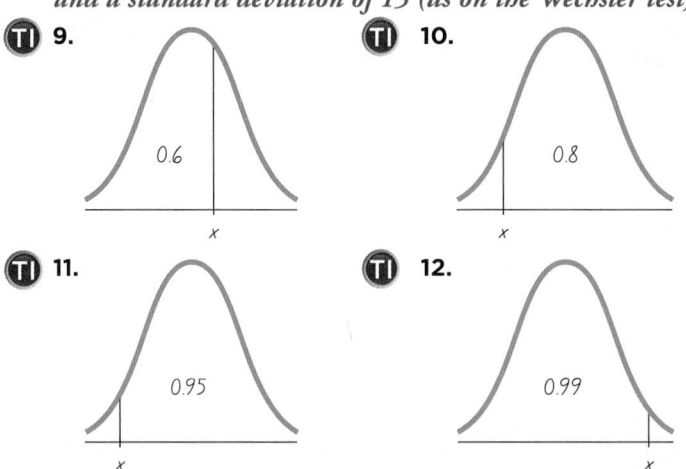

TI 9.   0.6   *x*

TI 10.   0.8   *x*

TI 11.   0.95   *x*

TI 12.   0.99   *x*

IQ Scores. In Exercises 13–20, assume that adults have IQ scores that are normally distributed with a mean of 100 and a standard deviation of 15 (as on the Wechsler test). (Hint: Draw a graph in each case.)

TI 13. Find the probability that a randomly selected adult has an IQ that is less than 115.

TI 14. Find the probability that a randomly selected adult has an IQ greater than 131.5 (the requirement for membership in the Mensa organization).

TI 15. Find the probability that a randomly selected adult has an IQ between 90 and 110 (referred to as the *normal* range).

TI 16. Find the probability that a randomly selected adult has an IQ between 110 and 120 (referred to as *bright normal*).

TI 17. Find P_{30}, which is the IQ score separating the bottom 30% from the top 70%.

TI 18. Find the first quartile Q_1, which is the IQ score separating the bottom 25% from the top 75%.

TI 19. Find the third quartile Q_3, which is the IQ score separating the top 25% from the others.

TI 20. Find the IQ score separating the top 37% from the others.

In Exercises 21–26, use this information (based on data from the National Health Survey):

- *Men's heights are normally distributed with mean 69.0 in. and standard deviation 2.8 in.*
- *Women's heights are normally distributed with mean 63.6 in. and standard deviation 2.5 in.*

TI 21. Doorway Height The Mark VI monorail used at Disney World and the Boeing 757-200 ER airliner have doors with a height of 72 in.

a. What percentage of adult men can fit through the doors without bending?

b. What percentage of adult women can fit through the doors without bending?

c. Does the door design with a height of 72 in. appear to be adequate? Explain.

d. What doorway height would allow 98% of adult men to fit without bending?

TI 22. Doorway Height The Gulfstream 100 is an executive jet that seats six, and it has a doorway height of 51.6 in.

a. What percentage of adult men can fit through the door without bending?

b. What percentage of adult women can fit through the door without bending?

c. Does the door design with a height of 51.6 in. appear to be adequate? Why didn't the engineers design a larger door?

d. What doorway height would allow 60% of men to fit without bending?

TI 23. Tall Clubs International Tall Clubs International is a social organization for tall people. It has a requirement that men must be at least 74 in. tall, and women must be at least 70 in. tall.

a. What percentage of men meet that requirement?

b. What percentage of women meet that requirement?

c. Are the height requirements for men and women fair? Why or why not?

TI 24. Tall Clubs International Tall Clubs International has minimum height requirements for men and women.

a. If the requirements are changed so that the tallest 4% of men are eligible, what is the new minimum height for men?

b. If the requirements are changed so that the tallest 4% of women are eligible, what is the new minimum height for women?

TI 25. U.S. Army Height Requirements for Women The U.S. Army requires women's heights to be between 58 in. and 80 in.

a. Find the percentage of women meeting the height requirement. Are many women being denied the opportunity to join the Army because they are too short or too tall?

b. If the U.S. Army changes the height requirements so that all women are eligible except the shortest 1% and the tallest 2%, what are the new height requirements?

TI 26. Marine Corps Height Requirement for Men The U.S. Marine Corps requires that men have heights between 64 in. and 80 in.

a. Find the percentage of men who meet the height requirements. Are many men denied the opportunity to become a Marine because they do not satisfy the height requirements?

b. If the height requirements are changed so that all men are eligible except the shortest 3% and the tallest 4%, what are the new height requirements?

TI 27. Birth Weights Birth weights in Norway are normally distributed with a mean of 3570 g and a standard deviation of 500 g.

a. If the Ulleval University Hospital in Oslo requires special treatment for newborn babies weighing less than 2700 g, what is the percentage of newborn babies requiring special treatment?

b. If the Ulleval University Hospital officials plan to require special treatment for the lightest 3% of newborn babies, what birth weight separates those requiring special treatment from those who do not?

c. Why is it not practical for the hospital to simply state that babies require special treatment if they are in the bottom 3% of birth weights?

28. Weights of Water Taxi Passengers It was noted in the Chapter Problem that when a water taxi sank in Baltimore's Inner Harbor, an investigation revealed that the safe passenger load for the water taxi was 3500 lb. It was also noted that the mean weight of a passenger was assumed to be 140 lb. Assume a "worst case" scenario in which all of the passengers are adult men. (This could easily occur in a city that hosts conventions in which people of the same gender often travel in groups.) Based on data from the National Health and Nutrition Examination Survey, assume that weights of men are normally distributed with a mean of 172 lb and a standard deviation of 29 lb.

a. If one man is randomly selected, find the probability that he weighs less than 174 lb (the new value suggested by the National Transportation and Safety Board).

b. With a load limit of 3500 lb, how many men passengers are allowed if we assume a mean weight of 140 lb?

c. With a load limit of 3500 lb, how many men passengers are allowed if we use the new mean weight of 174 lb?

d. Why is it necessary to periodically review and revise the number of passengers that are allowed to board?

29. Body Temperatures Based on the sample results in Data Set 2 of Appendix B, assume that human body temperatures are normally distributed with a mean of 98.20°F and a standard deviation of 0.62°F.

a. Bellevue Hospital in New York City uses 100.6°F as the lowest temperature considered to be a fever. What percentage of normal and healthy persons would be considered to have a fever? Does this percentage suggest that a cutoff of 100.6°F is appropriate?

b. Physicians want to select a minimum temperature for requiring further medical tests. What should that temperature be, if we want only 5.0% of healthy people to exceed it? (Such a result is a *false positive*, meaning that the test result is positive, but the subject is not really sick.)

30. Aircraft Seat Width Engineers want to design seats in commercial aircraft so that they are wide enough to fit 99% of all males. (Accommodating 100% of males would require very wide seats that would be much too expensive.) Men have hip breadths that are normally distributed with a mean of 14.4 in. and a standard deviation of 1.0 in. (based on anthropometric survey data from Gordon, Clauser, et al.). Find P_{99}. That is, find the hip breadth for men that separates the smallest 99% from the largest 1%.

31. Lengths of Pregnancies The lengths of pregnancies are normally distributed with a mean of 268 days and a standard deviation of 15 days.

a. One classical use of the normal distribution is inspired by a letter to "Dear Abby" in which a wife claimed to have given birth 308 days after a brief visit from her husband, who was serving in the Navy. Given this information, find the probability of a pregnancy lasting 308 days or longer. What does the result suggest?

b. If we stipulate that a baby is *premature* if the length of pregnancy is in the lowest 4%, find the length that separates premature babies from those who are not premature. Premature babies often require special care, and this result could be helpful to hospital administrators in planning for that care.

32. Sitting Distance A common design requirement is that an item (such as an aircraft or theater seat) must fit the range of people who fall between the 5th percentile for women and the 95th percentile for men. If this requirement is adopted, what is the minimum sitting distance and what is the maximum sitting distance? For the sitting distance, use the buttock-to-knee length. Men have buttock-to-knee lengths that are normally distributed with a mean of 23.5 in. and a standard deviation of 1.1 in. Women have buttock-to-knee lengths that are normally distributed with a mean of 22.7 in. and a standard deviation of 1.0 in.

Large Data Sets. *In Exercises 33 and 34, refer to the data sets in Appendix B and use computer software or a calculator.*

33. Appendix B Data Set: Systolic Blood Pressure Refer to Data Set 1 in Appendix B and use the systolic blood pressure levels for males.

a. Using the systolic blood pressure levels for males, find the mean and standard deviation, and verify that the data have a distribution that is roughly normal.

b. Assuming that systolic blood pressure levels of males are normally distributed, find the 5th percentile and the 95th percentile. (Treat the statistics from part (a) as if they were population parameters.) Such percentiles could be helpful when physicians try to determine whether blood pressure levels are too low or too high.

34. Appendix B Data Set: Duration of Shuttle Flights Refer to Data Set 10 in Appendix B and use the durations (hours) of the NASA shuttle flights.

a. Find the mean and standard deviation, and verify that the data have a distribution that is roughly normal.

b. Treat the statistics from part (a) as if they are population parameters and assume a normal distribution to find the values of the quartiles Q_1, Q_2, and Q_3.

6-3 Beyond the Basics

35. Units of Measurement Heights of women are normally distributed.

a. If heights of individual women are expressed in units of centimeters, what are the units used for the z scores that correspond to individual heights?

b. If heights of all women are converted to z scores, what are the mean, standard deviation, and distribution of these z scores?

36. Using Continuity Correction There are many situations in which a normal distribution can be used as a good approximation to a random variable that has only *discrete* values. In such cases, we can use this *continuity correction:* Represent each whole number by the interval extending from 0.5 below the number to 0.5 above it. Assume that IQ scores are all whole numbers having a distribution that is approximately normal with a mean of 100 and a standard deviation of 15.

a. Without using any correction for continuity, find the probability of randomly selecting someone with an IQ score greater than 103.

b. Using the correction for continuity, find the probability of randomly selecting someone with an IQ score greater than 103.

c. Compare the results from parts (a) and (b).

37. Curving Test Scores A statistics professor gives a test and finds that the scores are normally distributed with a mean of 25 and a standard deviation of 5. She plans to curve the scores.

a. If she curves by adding 50 to each grade, what is the new mean? What is the new standard deviation?

b. Is it fair to curve by adding 50 to each grade? Why or why not?

c. If the grades are curved according to the following scheme (instead of adding 50), find the numerical limits for each letter grade.

A: Top 10%

B: Scores above the bottom 70% and below the top 10%

C: Scores above the bottom 30% and below the top 30%

D: Scores above the bottom 10% and below the top 70%

F: Bottom 10%

d. Which method of curving the grades is fairer: Adding 50 to each grade or using the scheme given in part (c)? Explain.

38. SAT and ACT Tests Scores on the SAT test are normally distributed with a mean of 1518 and a standard deviation of 325. Scores on the ACT test are normally distributed with a mean of 21.1 and a standard deviation of 4.8. Assume that the two tests use different scales to measure the same aptitude.

a. If someone gets a SAT score that is the 67th percentile, find the actual SAT score and the equivalent ACT score.

b. If someone gets a SAT score of 1900, find the equivalent ACT score.

 39. Outliers For the purposes of constructing modified boxplots as described in Section 3-4, outliers were defined as data values that are above Q_3 by an amount greater than $1.5 \times IQR$ or below Q_1 by an amount greater than $1.5 \times IQR$, where IQR is the interquartile range. Using this definition of outliers, find the probability that when a value is randomly selected from a normal distribution, it is an outlier.

6-4   Sampling Distributions and Estimators

Key Concept In this section we consider the concept of a *sampling distribution of a statistic*. Also, we learn some important properties of sampling distributions of the mean, median, variance, standard deviation, range, and proportion. We see that some statistics (such as the mean, variance, and proportion) are unbiased estimators of population parameters, whereas other statistics (such as the median and range) are not.

The following chapters of this book introduce methods for using sample statistics to estimate values of population parameters. Those procedures are based on an understanding of how sample statistics behave, and that behavior is the focus of this section. We begin with the definition of a sampling distribution of a statistic.

 DEFINITION

The **sampling distribution of a statistic** (such as a sample mean or sample proportion) is the distribution of all values of the statistic when all possible samples of the same size n are taken from the same population. (The sampling distribution of a statistic is typically represented as a probability distribution in the format of a table, probability histogram, or formula.)

Sampling Distribution of the Mean

The preceding definition is general, so let's consider the specific sampling distribution of the mean.

 DEFINITION

The **sampling distribution of the mean** is the distribution of sample means, with all samples having the same sample size n taken from the same population. (The sampling distribution of the mean is typically represented as a probability distribution in the format of a table, probability histogram, or formula.)

> ┌─────────────┐
> │ **EXAMPLE 1** │ **Sampling Distribution of the Mean** Consider repeating
> └─────────────┘
> this process: Roll a die 5 times and find the mean $\bar{x}$ of the results. (See Table 6-2
> on the next page.) What do we know about the behavior of all sample means that
> are generated as this process continues indefinitely?

> ┌─────────────┐
> │ **SOLUTION** │   The top portion of Table 6-2 illustrates a process of rolling a die
> └─────────────┘
> 5 times and finding the mean of the results. Table 6-2 shows results from repeating
> this process 10,000 times, but the true sampling distribution of the mean involves
> repeating the process indefinitely. Because the values of 1, 2, 3, 4, 5, 6 are all equally
> likely, the population has a mean of $\mu = 3.5$, and Table 6-2 shows that the 10,000
> sample means have a mean of 3.49. If the process is continued indefinitely, the mean
> of the sample means will be 3.5. Also, Table 6-2 shows that the distribution of the
> sample means is approximately a normal distribution.

> ┌──────────────────┐
> │ **INTERPRETATION** │   Based on the actual sample results shown in the top portion
> └──────────────────┘
> of Table 6-2, we can describe the sampling distribution of the mean by the histogram
> at the top of Table 6-2. The actual sampling distribution would be described by a
> histogram based on all possible samples, not only the 10,000 samples included in
> the histogram, but the number of trials is large enough to suggest that the true sam-
> pling distribution of means is a normal distribution.

The results of Example 1 allow us to observe these two important properties of the
sampling distribution of the mean:

1. The sample means *target* the value of the population mean. (That is, the mean
   of the sample means is the population mean. The expected value of the sample
   mean is equal to the population mean.)

2. The distribution of sample means tends to be a normal distribution. (This will
   be discussed further in the following section, but the distribution tends to become closer to a normal distribution as the sample size increases.)

Sampling Distribution of the Variance

Having discussed the sampling distribution of the mean, we now consider the sampling distribution of the variance.

> ┌─────────────┐
> │ **DEFINITION** │
> └─────────────┘
> The **sampling distribution of the variance** is the distribution of sample
> variances, with all samples having the same sample size n taken from the
> same population. (The sampling distribution of the variance is typically rep-
> resented as a probability distribution in the format of a table, probability
> histogram, or formula.)

Caution: When working with population standard deviations or variances, be sure to
evaluate them correctly. Recall from Section 3-3 that the computations for *population*

Do Boys or Girls Run in the Family?

The author of this book, his siblings, and his siblings' children consist of 11 males and only one female. Is this an example of a phe- nomenon whereby one particular gender runs in a family? This issue was studied by examining a random sample of 8770 households in the United States. The results were reported in the *Chance* magazine article "Does Having Boys or Girls Run in the Family?" by Joseph Rodgers and Debby Doughty. Part of their analysis involves use of the binomial probability distribution. Their conclusion is that "We found no compelling evidence that sex bias runs in the family."

Table 6-2   Specific Results from 10,000 Trials

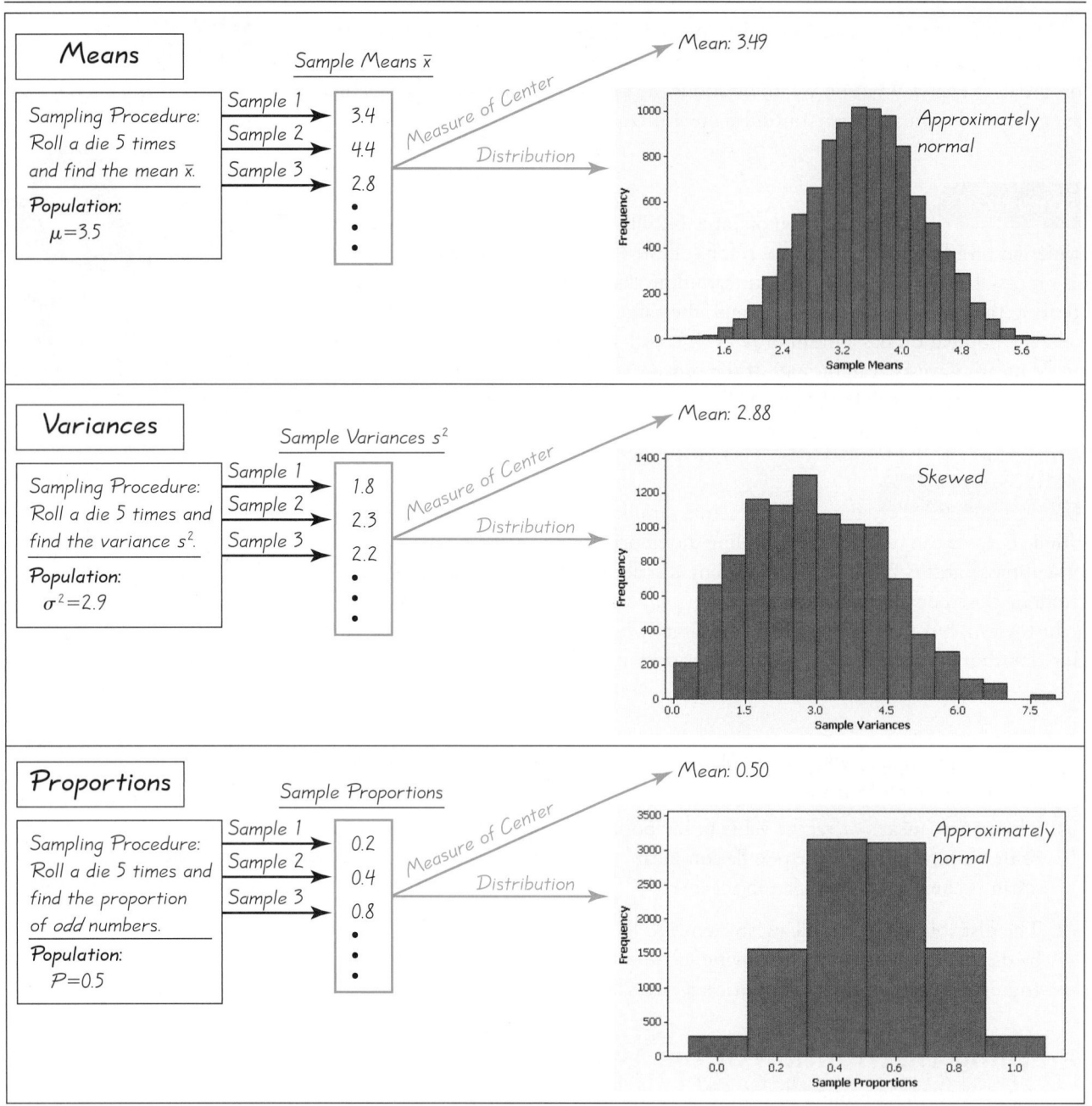

standard deviations or variances involve division by the population size N (not the value of $n - 1$), as shown below.

$$\text{Population standard deviation: } \sigma = \sqrt{\frac{\Sigma(x - \mu)^2}{N}}$$

$$\text{Population variance: } \sigma^2 = \frac{\Sigma(x - \mu)^2}{N}$$

Because the calculations are typically performed with computer software or calculators, be careful to correctly distinguish between the standard deviation of a sample and the standard deviation of a population. Also be careful to distinguish between the variance of a sample and the variance of a population.

EXAMPLE 2 **Sampling Distribution of the Variance** Consider repeating this process: Roll a die 5 times and find the variance s^2 of the results. What do we know about the behavior of all sample variances that are generated as this process continues indefinitely?

SOLUTION The middle portion of Table 6-2 illustrates a process of rolling a die 5 times and finding the variance of the results. Table 6-2 shows results from repeating this process 10,000 times, but the true sampling distribution of the variance involves repeating the process indefinitely. Because the values of 1, 2, 3, 4, 5, 6 are all equally likely, the population has a variance of $\sigma^2 = 2.9$, and Table 6-2 shows that the 10,000 sample variances have a mean of 2.88. If the process is continued indefinitely, the mean of the sample variances will be 2.9. Also, the middle portion of Table 6-2 shows that the distribution of the sample variances is a skewed distribution.

INTERPRETATION Based on the actual sample results shown in the middle portion of Table 6-2, we can describe the sampling distribution of the variance by the histogram in the middle of Table 6-2. The actual sampling distribution would be described by a histogram based on all possible samples, not the 10,000 samples included in the histogram, but the number of trials is large enough to suggest that the true sampling distribution of variances is a distribution skewed to the right.

The results of Example 2 allow us to observe these two important properties of the sampling distribution of the variance:

1. The sample variances *target* the value of the population variance. (That is, the mean of the sample variances is the population variance. The expected value of the sample variance is equal to the population variance.)

2. The distribution of sample variances tends to be a distribution skewed to the right.

Sampling Distribution of Proportion

We now consider the sampling distribution of a proportion.

DEFINITION

The **sampling distribution of the proportion** is the distribution of sample proportions, with all samples having the same sample size n taken from the same population.

We need to distinguish between a population proportion p and some sample proportion, so the following notation is commonly used.

Notation for Proportions

$p = population$ proportion

$\hat{p} = sample$ proportion

EXAMPLE 3   **Sampling Distribution of the Proportion** Consider repeating this process: Roll a die 5 times and find the proportion of *odd* numbers. What do we know about the behavior of all sample proportions that are generated as this process continues indefinitely?

SOLUTION   The bottom portion of Table 6-2 illustrates a process of rolling a die 5 times and finding the proportion of odd numbers. Table 6-2 shows results from repeating this process 10,000 times, but the true sampling distribution of the proportion involves repeating the process indefinitely. Because the values of 1, 2, 3, 4, 5, 6 are all equally likely, the proportion of odd numbers in the population is 0.5, and Table 6-2 shows that the 10,000 sample proportions have a mean of 0.50. If the process is continued indefinitely, the mean of the sample proportions will be 0.5. Also, the bottom portion of Table 6-2 shows that the distribution of the sample proportions is approximately a normal distribution.

INTERPRETATION   Based on the actual sample results shown in the bottom portion of Table 6-2, we can describe the sampling distribution of the proportion by the histogram at the bottom of Table 6-2. The actual sampling distribution would be described by a histogram based on all possible samples, not the 10,000 samples included in the histogram, but the number of trials is large enough to suggest that the true sampling distribution of proportions is a normal distribution.

The results of Example 3 allow us to observe these two important properties of the sampling distribution of the proportion:

1. The sample proportions *target* the value of the population proportion. (That is, the mean of the sample proportions is the population proportion. The expected value of the sample proportion is equal to the population proportion.)

2. The distribution of sample proportions tends to be a normal distribution.

The preceding three examples are based on 10,000 trials and the results are summarized in Table 6-2. Table 6-3 describes the *general* behavior of the sampling distribution of the mean, variance, and proportion, assuming that certain conditions are satisfied. For example, Table 6-3 shows that the sampling distribution of the mean tends to be a normal distribution, but the following section describes conditions that must be satisfied before we can assume that the distribution is normal.

Unbiased Estimators   The preceding three examples show that sample means, variances, and proportions tend to *target* the corresponding population parameters. More formally, we say that sample means, variances, and proportions are *unbiased estimators*. That is, their sampling distributions have a mean that is equal to the mean of the corresponding population parameter. If we want to use a sample statistic (such as a sample proportion from a survey) to estimate a population parameter (such as the population proportion), it is important that the sample statistic used as the estimator *targets* the population parameter instead of being a biased estimator in the sense that it systematically underestimates or overestimates the parameter. The preceding three examples and Table 6-2 involve the mean, variance, and proportion, but here is a summary that includes other statistics.

Table 6-3   General Behavior of Sampling Distributions

Estimators: Unbiased and Biased

Unbiased Estimators

These statistics are unbiased estimators. That is, they target the value of the population parameter:

- Mean $\bar{x}$
- Variance s^2
- Proportion $\hat{p}$

Biased Estimators

These statistics are biased estimators, That is, they do *not* target the population parameter:

- Median
- Range
- Standard deviation s. (*Important Note:* The sample standard deviations do not target the population standard deviation σ, but the bias is relatively small in

large samples, so *s* **is often used to estimate** even though *s* is a biased estimator of σ.)

The preceding three examples all involved rolling a die 5 times, so the number of different possible samples is $6 \times 6 \times 6 \times 6 \times 6 = 7776$. Because there are 7776 different possible samples, it is not practical to manually list all of them. The next example involves a smaller number of different possible samples, so we can list them and we can then describe the sampling distribution of the range in the format of a table for the probability distribution.

EXAMPLE 4   **Sampling Distribution of the Range** Three randomly selected households are surveyed as a pilot project for a larger survey to be conducted later. The numbers of people in the households are 2, 3, and 10 (based on Data Set 22 in Appendix B). Consider the values of 2, 3, and 10 to be a population. Assume that samples of size $n = 2$ are randomly selected with replacement from the population of 2, 3, and 10.

a. List all of the different possible samples, then find the range in each sample.

b. Describe the sampling distribution of the ranges in the format of a table summarizing the probability distribution.

c. Describe the sampling distribution of the ranges in the format of a probability histogram.

d. Based on the results, do the sample ranges target the population range, which is $10 - 2 = 8$?

e. What do these results indicate about the sample range as an estimator of the population range?

SOLUTION

a. In Table 6-4 we list the nine different possible samples of size $n = 2$ selected with replacement from the population of 2, 3, and 10. Table 6-4 also shows the range for each of the nine samples.

b. The nine samples in Table 6-4 are all equally likely, so each sample has a probability of 1/9. The last two columns of Table 6-4 list the values of the range along with the corresponding probabilities, so the last two columns constitute a table summarizing the probability distribution, which can be condensed as shown in Table 6-5. Table 6-5 therefore describes the *sampling distribution* of the sample ranges.

c. Figure 6-17 is the probability histogram based on Table 6-5.

d. The mean of the nine sample ranges is 3.6, but the range of the population is 8. Consequently, the sample ranges do not target the population range.

e. Because the mean of the sample ranges (3.6) does not equal the population range (8), the sample range is a biased estimator of the population range. We can also see that the range is a biased estimator by simply examining Table 6-5 and noting that most of the time, the sample range is well below the population range of 8.

Table 6-4   **Sampling Distribution of the Range**

Sample	Sample Range	Probability
2, 2	0	1/9
2, 3	1	1/9
2, 10	8	1/9
3, 2	1	1/9
3, 3	0	1/9
3, 10	7	1/9
10, 2	8	1/9
10, 3	7	1/9
10, 10	0	1/9

Mean of the sample ranges = 3.6 (rounded)

Table 6-5   **Probability Distribution for the Range**

Sample Range	Probability
0	3/9
1	2/9
7	2/9
8	2/9

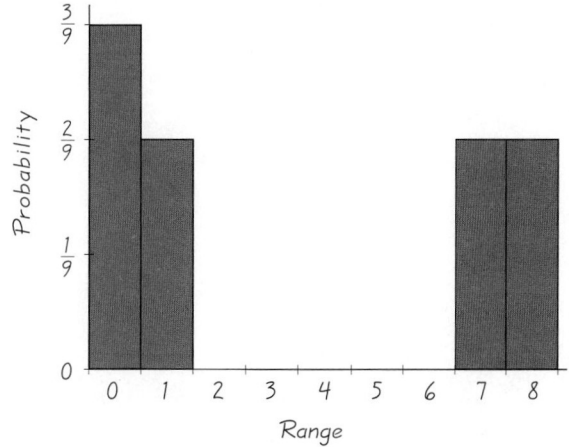

Figure 6-17   Probability Histogram: Sampling Distribution of the Sample Ranges

INTERPRETATION   In this example, we conclude that the sample range is a biased estimator of the population range. This implies that, in general, the sample range should not be used to estimate the value of the population range.

EXAMPLE 5   **Sampling Distribution of the Proportion** In a study of gender selection methods, an analyst considers the process of generating 2 births. When 2 births are randomly selected, the sample space is bb, bg, gb, gg. Those 4 outcomes are equally likely, so the probability of 0 girls is 0.25, the probability of 1 girl is 0.5, and the probability of 2 girls is 0.25. Describe the sampling distribution of the proportion of girls from 2 births as a probability distribution table and also describe it as a probability histogram.

continued

SOLUTION   See the accompanying display. The top table summarizes the probability distribution for the number of girls in 2 births. That top table can be used to construct the probability distribution for the *proportion* of girls in 2 births as shown. The top table can also be used to construct the probability histogram as shown.

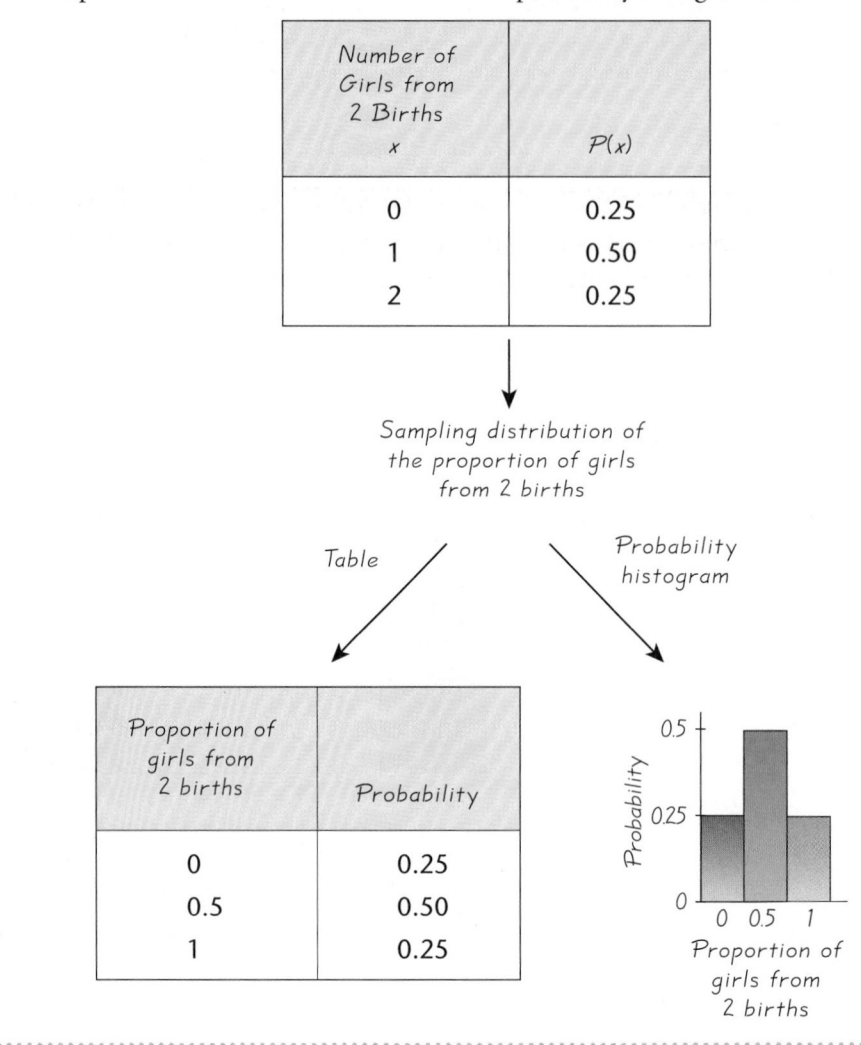

Example 5 shows that a sampling distribution can be described with a table or a graph. Sampling distributions can also be described with a formula (as in Exercise 21), or may be described in some other way, such as this: "The sampling distribution of the sample mean is a normal distribution with $\mu = 100$ and $\sigma = 15$."

Why sample *with* replacement?   All of the examples in this section involved sampling *with replacement*. Sampling *without replacement* would have the very practical advantage of avoiding wasteful duplication whenever the same item is selected more than once. However, we are particularly interested in sampling *with replacement* for these two reasons:

1. When selecting a relatively small sample from a large population, it makes no significant difference whether we sample with replacement or without replacement.

2. Sampling with replacement results in independent events that are unaffected by previous outcomes, and independent events are easier to analyze and result in simpler calculations and formulas.

For the above reasons, we focus on the behavior of samples that are randomly selected *with replacement*. Many of the statistical procedures discussed in the following chapters are based on the assumption that sampling is conducted with replacement.

The key point of this section is to introduce the concept of a sampling distribution of a statistic. Consider the goal of trying to find the mean body temperature of all adults. Because that population is so large, it is not practical to measure the temperature of every adult. Instead, we obtain a sample of body temperatures and use it to estimate the population mean. Data Set 2 in Appendix B includes a sample of 106 such body temperatures. The mean for that sample is $\bar{x} = 98.20°F$. Conclusions that we make about the population mean temperature of all adults require that we understand the behavior of the sampling distribution of all such sample means. Even though it is not practical to obtain every possible sample and we are stuck with just one sample, we can form some very meaningful conclusions about the population of all body temperatures. A major goal of the following sections and chapters is to learn how we can effectively use a sample to form conclusions about a population. In Section 6-5 we consider more details about the sampling distribution of sample means, and in Section 6-6 we consider more details about the sampling distribution of sample proportions.

CAUTION
....

Many methods of statistics require a *simple random sample*. Some samples, such as voluntary response samples or convenience samples, could easily result in very wrong results.

6-4 Basic Skills and Concepts

Statistical Literacy and Critical Thinking

1. Sampling Distribution In your own words describe a sampling distribution.

2. Sampling Distribution Data Set 24 in Appendix B includes a sample of FICO credit rating scores from randomly selected consumers. If we investigate this sample by constructing a histogram and finding the sample mean and standard deviation, are we investigating the sampling distribution of the mean? Why or why not?

3. Unbiased Estimator What does it mean when we say that the sample mean is an unbiased estimator, or that the sample mean "targets" the population mean?

4. Sampling with Replacement Give two reasons why statistical methods tend to be based on the assumption that sampling is conducted *with* replacement, instead of without replacement.

5. Good Sample? You want to estimate the proportion of all U.S. college students who have the profound wisdom to take a statistics course. You obtain a simple random sample of students at New York University. Is the resulting sample proportion a good estimator of the population proportion? Why or why not?

6. Unbiased Estimators Which of the following statistics are unbiased estimators of population parameters?

a. Sample mean used to estimate a population mean

b. Sample median used to estimate a population median

c. Sample proportion used to estimate a population proportion

d. Sample variance used to estimate a population variance

e. Sample standard deviation used to estimate a population standard deviation

f. Sample range used to estimate a population range

7. Sampling Distribution of the Mean Samples of size $n = 1000$ are randomly selected from the population of the last digits of telephone numbers. If the sample mean is found for each sample, what is the distribution of the sample means?

8. Sampling Distribution of the Proportion Samples of size $n = 1000$ are randomly selected from the population of the last digits of telephone numbers, and the proportion of even numbers is found for each sample. What is the distribution of the sample proportions?

In Exercises 9–12, refer to the population and list of samples in Example 4.

9. Sampling Distribution of the Median In Example 4, we assumed that samples of size $n = 2$ are randomly selected without replacement from the population consisting of 2, 3, and 10, where the values are the numbers of people in households. Table 6-4 lists the nine different possible samples.

a. Find the median of each of the nine samples, then summarize the sampling distribution of the medians in the format of a table representing the probability distribution. (*Hint:* Use a format similar to Table 6-5).

b. Compare the population median to the mean of the sample medians.

c. Do the sample medians target the value of the population median? In general, do sample medians make good estimators of population medians? Why or why not?

10. Sampling Distribution of the Standard Deviation Repeat Exercise 9 using standard deviations instead of medians.

11. Sampling Distribution of the Variance Repeat Exercise 9 using variances instead of medians.

12. Sampling Distribution of the Mean Repeat Exercise 9 using means instead of medians.

13. Assassinated Presidents: Sampling Distribution of the Mean The ages (years) of the four U.S. presidents when they were assassinated in office are 56 (Lincoln), 49 (Garfield), 58 (McKinley), and 46 (Kennedy).

a. Assuming that 2 of the ages are randomly selected with replacement, list the 16 different possible samples.

b. Find the mean of each of the 16 samples, then summarize the sampling distribution of the means in the format of a table representing the probability distribution. (Use a format similar to Table 6-5 on page 289).

c. Compare the population mean to the mean of the sample means.

d. Do the sample means target the value of the population mean? In general, do sample means make good estimators of population means? Why or why not?

14. Sampling Distribution of the Median Repeat Exercise 13 using medians instead of means.

15. Sampling Distribution of the Range Repeat Exercise 13 using ranges instead of means.

16. Sampling Distribution of the Variance Repeat Exercise 13 using variances instead of means.

17. Sampling Distribution of Proportion Example 4 referred to three randomly selected households in which the numbers of people are 2, 3, and 10. As in Example 4, consider the values of 2, 3, and 10 to be a population and assume that samples of size $n = 2$ are randomly selected with replacement. Construct a probability distribution table that describes the sampling distribution of the proportion of odd numbers when samples of size $n = 2$ are randomly selected. Does the mean of the sample proportions equal the proportion of odd numbers in the population? Do the sample proportions target the value of the population proportion? Does the sample proportion make a good estimator of the population proportion?

18. Births: Sampling Distribution of Proportion When 3 births are randomly selected, the sample space is bbb, bbg, bgb, bgg, gbb, gbg, ggb, and ggg. Assume that those 8 outcomes are equally likely. Describe the sampling distribution of the *proportion* of girls from 3 births as a probability distribution table. Does the mean of the sample proportions equal the proportion of girls in 3 births? (*Hint:* See Example 5.)

19. Genetics: Sampling Distribution of Proportion A genetics experiment involves a population of fruit flies consisting of 1 male named Mike and 3 females named Anna, Barbara, and Chris. Assume that two fruit flies are randomly selected *with replacement.*

a. After listing the 16 different possible samples, find the proportion of females in each sample, then use a table to describe the sampling distribution of the proportions of females.

b. Find the mean of the sampling distribution.

c. Is the mean of the sampling distribution (from part (b)) equal to the population proportion of females? Does the mean of the sampling distribution of proportions *always* equal the population proportion?

20. Quality Control: Sampling Distribution of Proportion After constructing a new manufacturing machine, 5 prototype integrated circuit chips are produced and it is found that 2 are defective (D) and 3 are acceptable (A). Assume that two of the chips are randomly selected *with replacement* from this population.

a. After identifying the 25 different possible samples, find the proportion of defects in each of them, then use a table to describe the sampling distribution of the proportions of defects.

b. Find the mean of the sampling distribution.

c. Is the mean of the sampling distribution (from part (b)) equal to the population proportion of defects? Does the mean of the sampling distribution of proportions *always* equal the population proportion?

6-4 Beyond the Basics

21. Using a Formula to Describe a Sampling Distribution Example 5 includes a table and graph to describe the sampling distribution of the proportions of girls from 2 births. Consider the formula shown below, and evaluate that formula using sample proportions *x* of 0, 0.5, and 1. Based on the results, does the formula describe the sampling distribution? Why or why not?

$$P(x) = \frac{1}{2(2 - 2x)!(2x)!} \quad \text{where } x = 0, 0.5, 1$$

22. Mean Absolute Deviation Is the mean absolute deviation of a sample a good statistic for estimating the mean absolute deviation of the population? Why or why not? (*Hint:* See Example 4.)

6-5 The Central Limit Theorem

Key Concept In this section we introduce and apply the *central limit theorem.* The central limit theorem tells us that for a population with *any* distribution, the distribution of the sample means approaches a normal distribution as the sample size increases. In other words, if the sample size is large enough, the distribution of sample means can be approximated by a *normal distribution,* even if the original population is not normally distributed. In addition, if the original population has mean μ and standard deviation σ, the mean of the sample means will also be μ, but the standard deviation of the sample means will be $\sigma/\sqrt{n}$, where n is the sample size.

In Section 6-4 we discussed the sampling distribution of $\bar{x}$, and in this section we describe procedures for using that sampling distribution in practical applications. The procedures of this section form the foundation for estimating population parameters and hypothesis testing—topics discussed at length in the following chapters. When selecting a simple random sample of n subjects from a population with mean μ and standard deviation σ, it is essential to know these principles:

1. For a population with any distribution, if $n > 30$, then the sample means have a distribution that can be approximated by a normal distribution with mean μ and standard deviation $\sigma/\sqrt{n}$.

2. If $n \leq 30$ and the original population has a normal distribution, then the sample means have a normal distribution with mean μ and standard deviation $\sigma/\sqrt{n}$.

3. If $n \leq 30$ and the original population does not have a normal distribution, then the methods of this section do not apply.

Here are the key points that form a foundation for the following chapters.

The Central Limit Theorem and the Sampling Distribution of $\bar{x}$

Given

1. The random variable x has a distribution (which may or may not be normal) with mean μ and standard deviation σ.

2. Simple random samples all of the same size n are selected from the population. (The samples are selected so that all possible samples of size n have the same chance of being selected.)

Conclusions

1. The distribution of sample means $\bar{x}$ will, as the sample size increases, approach a *normal* distribution.

2. The mean of all sample means is the population mean μ.

3. The standard deviation of all sample means is $\sigma/\sqrt{n}$.

Practical Rules Commonly Used

1. If the original population is *not normally distributed,* here is a common guideline: For $n > 30$, the distribution of the sample means can be approximated reasonably well by a normal distribution. (There are exceptions, such as populations with very nonnormal distributions requiring sample sizes larger than 30, but such exceptions are relatively rare.) The distribution of sample means gets closer to a normal distribution as the sample size n becomes larger.

2. If the original population is *normally distributed,* then for *any* sample size n, the sample means will be normally distributed.

The central limit theorem involves two different distributions: the distribution of the original population and the distribution of the sample means. As in previous chapters, we use the symbols μ and σ to denote the mean and standard deviation of the original population, but we use the following new notation for the mean and standard deviation of the distribution of sample means.

Notation for the Sampling Distribution of $\bar{x}$

If all possible random samples of size n are selected from a population with mean μ and standard deviation σ, the mean of the sample means is denoted by $\mu_{\bar{x}}$, so

$$\mu_{\bar{x}} = \mu$$

Also, the standard deviation of the sample means is denoted by $\sigma_{\bar{x}}$, so

$$\sigma_{\bar{x}} = \frac{\sigma}{\sqrt{n}}$$

$\sigma_{\bar{x}}$ is called the **standard error of the mean.**

EXAMPLE 1   **Normal, Uniform, and U-Shaped Distributions** Table 6-6 illustrates the central limit theorem. The top dotplots in Table 6-6 show an approximately normal distribution, a uniform distribution, and a distribution with a shape resembling the letter U. In each column, the second dotplot shows the distribution of sample means where $n = 10$, and the bottom dotplot shows the distribution of sample means where $n = 50$. As we proceed down each column of Table 6-6, we can see that the distribution of sample means is approaching the shape of a normal distribution. That characteristic is included among the following observations that we can make from Table 6-6.

- As the sample size increases, the distribution of sample means tends to approach a normal distribution.

- The mean of the sample means is the same as the mean of the original population.

- As the sample size increases, the dotplots become narrower, showing that the standard deviation of the sample means becomes smaller.

The Fuzzy Central Limit Theorem

In *The Cartoon Guide to Statistics,* by Gonick and Smith, the authors describe the Fuzzy Central Limit Theorem as follows: "Data that are influenced by many small and unrelated random effects are approximately normally distributed. This explains why the normal is everywhere: stock market fluctuations, student weights, yearly temperature averages, SAT scores: All are the result of many different effects." People's heights, for example, are the results of hereditary factors, environmental factors, nutrition, health care, geographic region, and other influences which, when combined, produce normally distributed values.

Table 6-6   Sampling Distributions

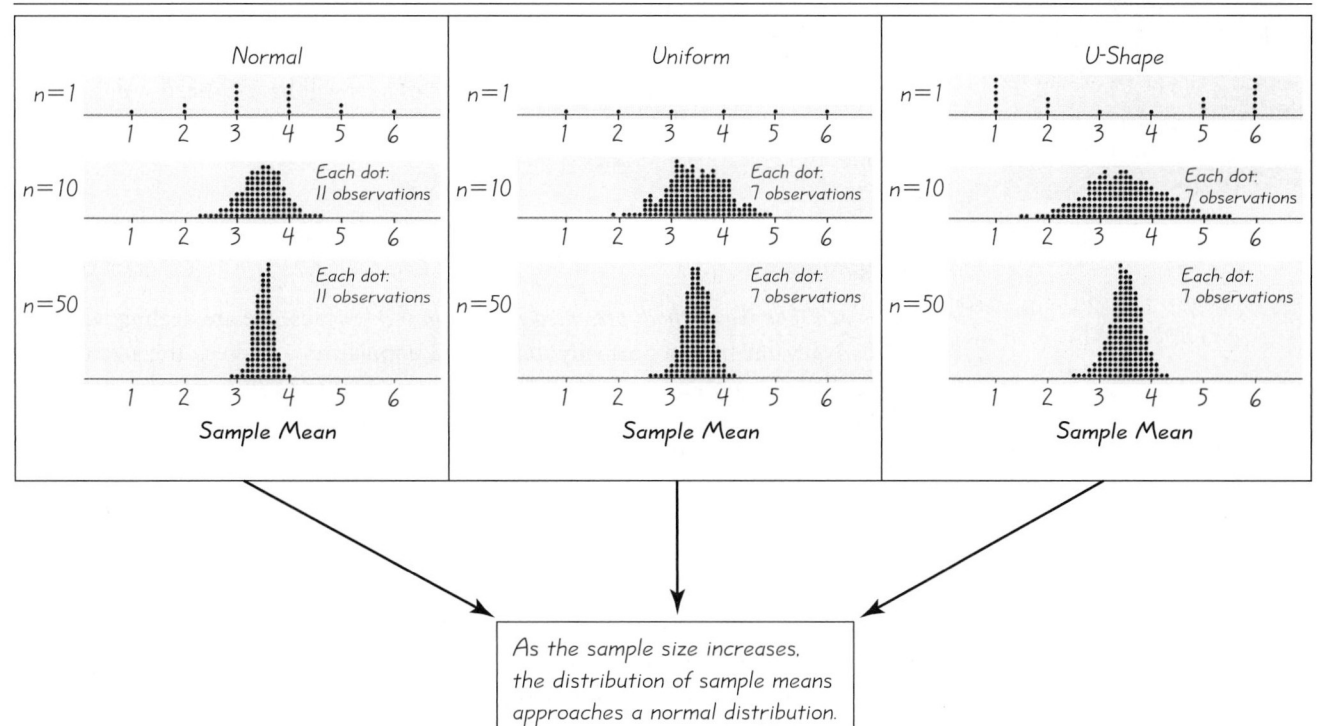

As the sample size increases, the distribution of sample means approaches a normal distribution.

Applying the Central Limit Theorem

Many practical problems can be solved with the central limit theorem. When working with such problems, remember that if the sample size is greater than 30, or if the original population is normally distributed, treat the distribution of sample means as if it were a normal distribution with mean μ and standard deviation $\sigma/\sqrt{n}$.

In Example 2, part (a) involves an *individual* value, but part (b) involves the mean for a *sample* of 20 men, so we must use the central limit theorem in working with the random variable $\bar{x}$. Study this example carefully to understand the fundamental difference between the procedures used in parts (a) and (b).

- *Individual value:* When working with an *individual* value from a normally distributed population, use the methods of Section 6-3. Use $z = \dfrac{x - \mu}{\sigma}$.

- *Sample of values:* When working with a mean for some *sample* (or group), be sure to use the value of $\sigma/\sqrt{n}$ for the standard deviation of the sample means. Use $z = \dfrac{\bar{x} - \mu}{\dfrac{\sigma}{\sqrt{n}}}$.

> **EXAMPLE 2**   **Water Taxi Safety** In the Chapter Problem we noted that some passengers died when a water taxi sank in Baltimore's Inner Harbor. Men are typically heavier than women and children, so when loading a water taxi, let's assume a worst-case scenario in which all passengers are men. Based on data from the National Health and Nutrition Examination Survey, assume that weights of men are normally distributed with a mean of 172 lb and a standard deviation of 29 lb. That is, assume that the population of weights of men is normally distributed with $\mu = 172$ lb and $\sigma = 29$ lb.
>
> **a.** Find the probability that if an *individual* man is randomly selected, his weight will be greater than 175 lb.
>
> **b.** Find the probability that *20 randomly selected men* will have a mean weight that is greater than 175 lb (so that their total weight exceeds the safe capacity of 3500 lb).

> **SOLUTION**

a. *Approach: Use the methods presented in Section 6-3* (because we are dealing with an *individual* value from a normally distributed population). We seek the area of the green-shaded region in Figure 6-18(a). If using a TI-83/84 Plus calculator, press **2ND** **VARS** to get the DISTR menu, then select **normalcdf.** Proceed to enter **normalcdf(175, 999999, 172, 29)** to get the probability of 0.4588035427.

If using Table A-2, we convert the weight of 175 to the corresponding z score:

$$z = \frac{x - \mu}{\sigma} = \frac{175 - 172}{29} = 0.10$$

Use Table A-2 and use $z = 0.10$ to find that the cumulative area to the left of 175 lb is 0.5398. The green-shaded region is therefore $1 - 0.5398 = 0.4602$. The probability of a randomly selected man weighing more than 175 lb is 0.4602.

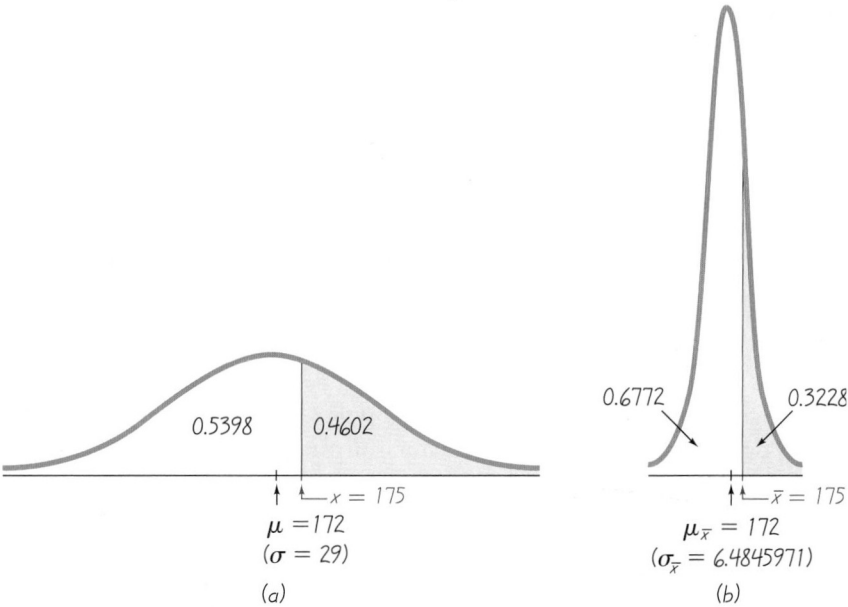

Figure 6-18   Men's Weights
(a) Distribution of Individual Men's Weights; (b) Distribution of Sample Means

b. *Approach: Use the central limit theorem* (because we are dealing with the *mean for a sample* of 20 men, not an individual man). Although the sample size is not greater than 30, we use a normal distribution because the original population of men has a normal distribution, so samples of *any* size will yield means that are normally distributed. Because we are now dealing with a distribution of sample means, we must use the parameters $\mu_{\bar{x}}$ and $\sigma_{\bar{x}}$, which are evaluated as follows:

$$\mu_{\bar{x}} = \mu = 172$$
$$\sigma_{\bar{x}} = \frac{\sigma}{\sqrt{n}} = \frac{29}{\sqrt{20}} = 6.4845971$$

We want to find the green-shaded area shown in Figure 6-18(b). (See how the distribution in Figure 6-18(b) is narrower because the standard deviation is smaller.)

If using a TI-83/84 Plus calculator, be sure to use $29/\sqrt{20} = 6.4845971$ (not 29) for the standard deviation. Press **2ND** **VARS** to get the DISTR menu, then select **normalcdf** and proceed to enter **normalcdf(175, 999999, 172, 6.4845971)** to get the probability of 0.3218131094.

If using Table A-2, we find the relevant z score, which is calculated as follows:

$$z = \frac{\bar{x} - \mu_{\bar{x}}}{\sigma_{\bar{x}}} = \frac{175 - 172}{\dfrac{29}{\sqrt{20}}} = \frac{3}{6.4845971} = 0.46$$

From Table A-2 we find that $z = 0.46$ corresponds to a cumulative left area of 0.6772, so the green-shaded region is $1 - 0.6772 = 0.3228$. The probability that the 20 men have a mean weight greater than 175 lb is 0.3228.   *continued*

INTERPRETATION   There is a 0.4602 probability that an individual man will weigh more than 175 lb, and there is a 0.3228 probability that 20 men will have a mean weight of more than 175 lb. Given that the safe capacity of the water taxi is 3500 lb, there is a fairly good chance (with probability 0.3228) that it will be overweight if is filled with 20 randomly selected men. Given that 21 people have already died, and given the high chance of overloading, it would be wise to limit the number of passengers to some level below 20. The capacity of 20 passengers is just not safe enough.

The calculations used here are exactly the type of calculations used by engineers when they design ski lifts, elevators, escalators, airplanes, and other devices that carry people.

Introduction to Hypothesis Testing

The next two examples present applications of the central limit theorem, but carefully examine the conclusions that are reached. These examples illustrate the type of thinking that is the basis for the important procedure of hypothesis testing (discussed in Chapter 8). These examples use the rare event rule for inferential statistics, first presented in Section 4-1.

Rare Event Rule for Inferential Statistics

If, under a given assumption, the probability of a particular observed event is exceptionally small (such as less than 0.05), we conclude that the assumption is probably not correct.

EXAMPLE 3   **Filling Coke Cans** Cans of regular Coke are labeled to indicate that they contain 12 oz. Data Set 17 in Appendix B lists measured amounts for a sample of Coke cans. The corresponding sample statistics are $n = 36$ and $\bar{x} = 12.19$ oz. If the Coke cans are filled so that $\mu = 12.00$ oz (as labeled) and the population standard deviation is $\sigma = 0.11$ oz (based on the sample results), find the probability that a sample of 36 cans will have a mean of 12.19 oz or greater. Do these results suggest that the Coke cans are filled with an amount greater than 12.00 oz?

SOLUTION   We weren't given the distribution of the population, but because the sample size $n = 36$ exceeds 30, we apply the central limit theorem and conclude that the distribution of sample means is approximately a normal distribution with these parameters:

$$\mu_{\bar{x}} = \mu = 12.00 \qquad\qquad \text{(by assumption)}$$

$$\sigma_{\bar{x}} = \frac{\sigma}{\sqrt{n}} = \frac{0.11}{\sqrt{36}} = 0.018333$$

Figure 6-19 shows the shaded area (see the small region in the right tail of the graph) corresponding to the probability we seek. Having already found the parameters that apply to the distribution shown in Figure 6-19, we can now find the shaded area by using the same procedures developed in Section 6-3.

TI-83/84 Plus

```
normalcdf(12.19,
999999,12.00,0.0
18333)
     1.85020454E-25
```

If using a TI-83/84 Plus calculator, press **2ND** **VARS** and proceed to enter **normalcdf(12.19, 999999, 12.00, 0.018333).** Because there is no actual upper bound, we use the large positive number of 999999 for the upper bound. The result is shown in the margin display. The probability value is expressed in scientific notation, and it is a very small value with 24 zeros following the decimal point. In a case like this, it is common to say that the probability is "less than 0.001."

To use Table A-2, we first find the z score:

$$z = \frac{\bar{x} - \mu_{\bar{x}}}{\sigma_{\bar{x}}} = \frac{12.19 - 12.00}{0.018333} = 10.36$$

Referring to Table A-2, we find that $z = 10.36$ is off the chart. However, for values of z above 3.49, we use 0.9999 for the cumulative left area. We therefore conclude that the shaded region in Figure 6-19 is 0.0001.

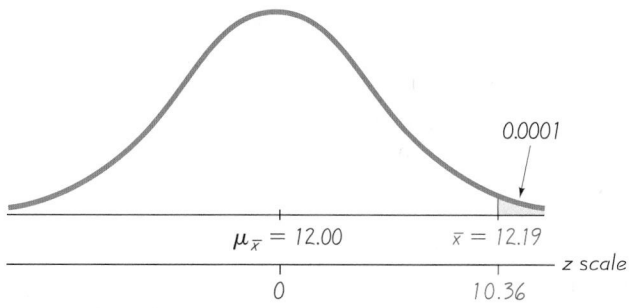

Figure 6-19

Distribution of Amounts of Coke (in ounces)

INTERPRETATION The result shows that if the mean amount in Coke cans is really 12.00 oz, then there is an extremely small probability of getting a sample mean of 12.19 oz or greater when 36 cans are randomly selected. Because we did obtain such a sample mean, there are two possible explanations: Either the population mean really is 12.00 oz and the sample represents a chance event that is extremely rare, or the population mean is actually greater than 12.00 oz and the sample is typical. Since the probability is so low, it seems more reasonable to conclude that the population mean is greater than 12.00 oz. It appears that Coke cans are being filled with more than 12.00 oz. However, the sample mean of 12.19 oz suggests that the mean amount of overfill is very small. It appears that the Coca Cola company has found a way to ensure that very few cans have less than 12 oz while not wasting very much of their product.

EXAMPLE 4 **How Long Is a 3/4 Inch Screw?** It is not totally unreasonable to think that screws labeled as being 3/4 inch in length would have a mean length that is somewhat close to 3/4 in. Data Set 19 in Appendix B includes the lengths of a sample of 50 such screws, with a mean length of 0.7468 in. Assume that the population of all such screws has a standard deviation described by $\sigma = 0.0123$ in. (based on Data Set 19).

a. Assuming that the screws have a mean length of 0.75 in. (or 3/4 inch) as labeled, find the probability that a sample of 50 screws has a mean length of 0.7468 in. or less. (See Figure 6-20.)

continued

b. The probability of getting a sample mean that is "at least as extreme as the given sample mean" is twice the probability found in part (a). Find this probability. (Note that the sample mean of 0.7468 in. misses the labeled mean of 0.75 in. by 0.0032 in., so any other mean is at least as extreme as the sample mean if it is below 0.75 in. by 0.0032 inch or more, or if it is above 0.75 in. by 0.0032 in. or more.)

c. Based on the result in part (b), does it appear that the sample mean misses the labeled mean of 0.75 in. by a significant amount? Explain.

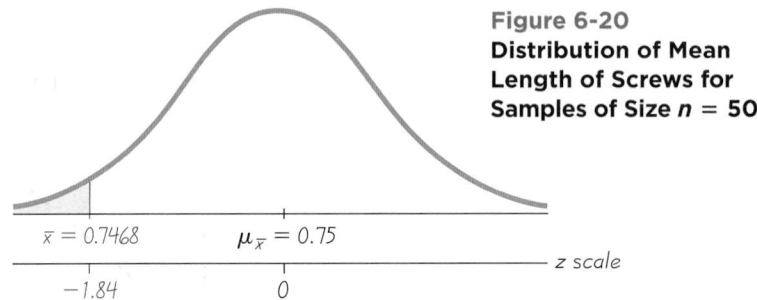

Figure 6-20
Distribution of Mean Length of Screws for Samples of Size *n* = 50

SOLUTION

a. We weren't given the distribution of the population, but because the sample size $n = 50$ exceeds 30, we use the central limit theorem and conclude that the distribution of sample means is a normal distribution with these parameters:

$$\mu_{\bar{x}} = \mu = 0.75 \qquad \text{(by assumption)}$$

$$\sigma_{\bar{x}} = \frac{\sigma}{\sqrt{n}} = \frac{0.0123}{\sqrt{50}} = 0.001739$$

Figure 6-20 shows the shaded area corresponding to the probability that 50 screws have a mean of 0.7468 in. or less. We can find the shaded area by using the same procedures developed in Section 6-3.

If using a TI-83/84 Plus calculator, press **2ND** **VARS** and proceed to enter **normalcdf(− 999999, 0.7468, 0.75, 0.001739).** Because there is no actual lower bound, we use the large negative number of −999999 for the lower bound. The result is 0.0328739288.

To use Table A-2, we first find the *z* score:

$$z = \frac{\bar{x} - \mu_{\bar{x}}}{\sigma_{\bar{x}}} = \frac{0.7468 - 0.75}{0.001739} = -1.84$$

Referring to Table A-2, we find that $z = -1.84$ corresponds to a cumulative left area of 0.0329. The probability of getting a sample mean of 0.7468 in. or less is 0.0329.

b. The probability of getting a sample mean that is "at least as extreme as the given sample mean" is twice the probability found in part (a), so that probability is $2 \times 0.0329 = 0.0658$.

c. The result from part (b) shows that there is a 0.0658 probability of getting a sample mean that is at least as extreme as the given sample mean. Using a 0.05 cutoff probability for distinguishing between usual events and unusual events, we see that the probability of 0.0658 exceeds 0.05, so the sample mean is not unusual. Consequently, we conclude that the given sample mean does not miss the labeled mean of 0.75 in. by a substantial amount. The labeling of 3/4 in. or 0.75 in. appears to be justified.

The reasoning in Examples 3 and 4 is the type of reasoning used in *hypothesis testing*, to be introduced in Chapter 8. For now, we focus on the use of the central limit theorem for finding the indicated probabilities, but we should recognize that this theorem will be used later in developing some very important concepts in statistics.

Correction for a Finite Population

In applying the central limit theorem, our use of $\sigma_{\bar{x}} = \sigma/\sqrt{n}$ assumes that the population has infinitely many members. When we sample with replacement (that is, put back each selected item before making the next selection), the population is effectively infinite. Yet many realistic applications involve sampling without replacement, so successive samples depend on previous outcomes. In manufacturing, quality-control inspectors typically sample items from a finite production run without replacing them. For such a finite population, we may need to adjust $\sigma_{\bar{x}}$. Here is a common rule of thumb:

> **When sampling without replacement and the sample size n is greater than 5% of the finite population size N (that is, $n > 0.05N$), adjust the standard deviation of sample means $\sigma_{\bar{x}}$ by multiplying it by the *finite population correction factor*:**
>
> $$\sqrt{\frac{N-n}{N-1}}$$

Except for Exercises 22 and 23, the examples and exercises in this section assume that the finite population correction factor does *not* apply, because we are sampling with replacement, or the population is infinite, or the sample size doesn't exceed 5% of the population size.

The central limit theorem allows us to use the basic normal distribution methods in a wide variety of different circumstances. In Chapter 7 we will apply the theorem when we use sample data to estimate means of populations. In Chapter 8 we will apply it when we use sample data to test claims made about population means. Table 6-7 summarizes the conditions in which we can and cannot use the normal distribution.

Table 6-7 Distributions of Sample Means

Population (with mean μ and standard deviation σ)	Distribution of Sample Means	Mean of the Sample Means	Standard Deviation of the Sample Means
Normal	Normal (for *any* sample size n)	$\mu_{\bar{x}} = \mu$	$\sigma_{\bar{x}} = \dfrac{\sigma}{\sqrt{n}}$
Not normal with $n > 30$	Normal (approximately)	$\mu_{\bar{x}} = \mu$	$\sigma_{\bar{x}} = \dfrac{\sigma}{\sqrt{n}}$
Not normal with $n \leq 30$	*Not* normal	$\mu_{\bar{x}} = \mu$	$\sigma_{\bar{x}} = \dfrac{\sigma}{\sqrt{n}}$

6-5 Basic Skills and Concepts

Statistical Literacy and Critical Thinking

1. Standard Error of the Mean What is the standard error of the mean?

2. Small Sample If selecting samples of size $n = 2$ from a population with a known mean and standard deviation, what requirement must be satisfied in order to assume that the distribution of the sample means is a normal distribution?

3. Notation What does the notation $\mu_{\bar{x}}$ represent? What does the notation $\sigma_{\bar{x}}$ represent?

4. Distribution of Incomes Assume that we collect a large ($n > 30$) simple random sample of annual incomes of adults in the United States. Because the sample is large, can we approximate the distribution of those incomes with a normal distribution? Why or why not?

Using the Central Limit Theorem. *In Exercises 5–8, assume that SAT scores are normally distributed with mean $\mu = 1518$ and standard deviation $\sigma = 325$ (based on data from the College Board).*

5. a. If 1 SAT score is randomly selected, find the probability that it is less than 1500.

b. If 100 SAT scores are randomly selected, find the probability that they have a mean less than 1500.

6. a. If 1 SAT score is randomly selected, find the probability that it is greater than 1600.

b. If 64 SAT scores are randomly selected, find the probability that they have a mean greater than 1600.

7. a. If 1 SAT score is randomly selected, find the probability that it is between 1550 and 1575.

b. If 25 SAT scores are randomly selected, find the probability that they have a mean between 1550 and 1575.

c. Why can the central limit theorem be used in part (b), even though the sample size does not exceed 30?

8. a. If 1 SAT score is randomly selected, find the probability that it is between 1440 and 1480.

b. If 16 SAT scores are randomly selected, find the probability that they have a mean between 1440 and 1480.

c. Why can the central limit theorem be used in part (b), even though the sample size does not exceed 30?

9. Water Taxi Safety Based on data from the National Health and Nutrition Examination Survey, assume that weights of men are normally distributed with a mean of 172 lb and a standard deviation of 29 lb.

a. Find the probability that if an *individual* man is randomly selected, his weight will be greater than 180 lb.

b. Find the probability that *20 randomly selected men* will have a mean weight that is greater than 180 lb.

c. If 20 men have a mean weight greater than 180 lb, the total weight exceeds the 3500 lb safe capacity of a particular water taxi. Based on the preceding results, is this a safety concern? Why or why not?

10. Mensa Membership in Mensa requires an IQ score above 131.5. Nine candidates take IQ tests, and their summary results indicated that their mean IQ score is 133. (IQ scores are normally distributed with a mean of 100 and a standard deviation of 15.)

a. If 1 person is randomly selected from the general population, find the probability of getting someone with an IQ score of at least 133.

b. If 9 people are randomly selected, find the probability that their mean IQ score is at least 133.

c. Although the summary results are available, the individual IQ test scores have been lost. Can it be concluded that all 9 candidates have IQ scores above 131.5 so that they are all eligible for Mensa membership?

11. Gondola Safety A ski gondola in Vail, Colorado, carries skiers to the top of a mountain. It bears a plaque stating that the maximum capacity is 12 people or 2004 lb. That capacity will be exceeded if 12 people have weights with a mean greater than $2004/12 = 167$ lb. Because men tend to weigh more than women, a "worst case" scenario involves 12 passengers who are all men. Men have weights that are normally distributed with a mean of 172 lb and a standard deviation of 29 lb (based on data from the National Health Survey).

a. Find the probability that if an individual man is randomly selected, his weight will be greater than 167 lb.

b. Find the probability that 12 randomly selected men will have a mean that is greater than 167 lb (so that their total weight is greater than the gondola maximum capacity of 2004 lb).

c. Does the gondola appear to have the correct weight limit? Why or why not?

TI **12. Effect of Diet on Length of Pregnancy** The lengths of pregnancies are normally distributed with a mean of 268 days and a standard deviation of 15 days.

a. If 1 pregnant woman is randomly selected, find the probability that her length of pregnancy is less than 260 days.

b. If 25 randomly selected women are put on a special diet just before they become pregnant, find the probability that their lengths of pregnancy have a mean that is less than 260 days (assuming that the diet has no effect).

c. If the 25 women do have a mean of less than 260 days, does it appear that the diet has an effect on the length of pregnancy, and should the medical supervisors be concerned?

TI **13. Blood Pressure** For women aged 18–24, systolic blood pressures (in mm Hg) are normally distributed with a mean of 114.8 and a standard deviation of 13.1 (based on data from the National Health Survey). Hypertension is commonly defined as a systolic blood pressure above 140.

a. If a woman between the ages of 18 and 24 is randomly selected, find the probability that her systolic blood pressure is greater than 140.

b. If 4 women in that age bracket are randomly selected, find the probability that their mean systolic blood pressure is greater than 140.

c. Given that part (b) involves a sample size that is not larger than 30, why can the central limit theorem be used?

d. If a physician is given a report stating that 4 women have a mean systolic blood pressure below 140, can she conclude that none of the women have hypertension (with a blood pressure greater than 140)?

TI **14. Designing Motorcycle Helmets** Engineers must consider the breadths of male heads when designing motorcycle helmets. Men have head breadths that are normally distributed with a mean of 6.0 in. and a standard deviation of 1.0 in. (based on anthropometric survey data from Gordon, Churchill, et al.).

a. If one male is randomly selected, find the probability that his head breadth is less than 6.2 in.

b. The Safeguard Helmet company plans an initial production run of 100 helmets. Find the probability that 100 randomly selected men have a mean head breadth less than 6.2 in.

c. The production manager sees the result from part (b) and reasons that all helmets should be made for men with head breadths less than 6.2 in., because they would fit all but a few men. What is wrong with that reasoning?

TI **15. Doorway Height** The Boeing 757-200 ER airliner carries 200 passengers and has doors with a height of 72 in. Heights of men are normally distributed with a mean of 69.0 in. and a standard deviation of 2.8 in.

a. If a male passenger is randomly selected, find the probability that he can fit through the doorway without bending.

b. If half of the 200 passengers are men, find the probability that the mean height of the 100 men is less than 72 in.

c. When considering the comfort and safety of passengers, which result is more relevant: The probability from part (a) or the probability from part (b)? Why?

d. When considering the comfort and safety of passengers, why are women ignored in this case?

TI **16. Labeling of M&M Packages** M&M plain candies have a mean weight of 0.8565 g and a standard deviation of 0.0518 g (based on Data Set 18 in Appendix B). The M&M candies used in Data Set 18 came from a package containing 465 candies, and the package label stated that the net weight is 396.9 g. (If every package has 465 candies, the mean weight of the candies must exceed 396.9/465 = 0.8535 g for the net contents to weigh at least 396.9 g.)

continued

a. If 1 M&M plain candy is randomly selected, find the probability that it weighs more than 0.8535 g.

b. If 465 M&M plain candies are randomly selected, find the probability that their mean weight is at least 0.8535 g.

c. Given these results, does it seem that the Mars Company is providing M&M consumers with the amount claimed on the label?

TI 17. Redesign of Ejection Seats When women were allowed to become pilots of fighter jets, engineers needed to redesign the ejection seats because they had been originally designed for men only. The ACES-II ejection seats were designed for men weighing between 140 lb and 211 lb. The weights of women are normally distributed with a mean of 143 lb and a standard deviation of 29 lb (based on data from the National Health Survey).

a. If 1 woman is randomly selected, find the probability that her weight is between 140 lb and 211 lb.

b. If 36 different women are randomly selected, find the probability that their mean weight is between 140 lb and 211 lb.

c. When redesigning the fighter jet ejection seats to better accommodate women, which probability is more relevant: The result from part (a) or the result from part (b)? Why?

TI 18. Vending Machines Currently, quarters have weights that are normally distributed with a mean of 5.670 g and a standard deviation of 0.062 g. A vending machine is configured to accept only those quarters with weights between 5.550 g and 5.790 g.

a. If 280 different quarters are inserted into the vending machine, what is the expected number of rejected quarters?

b. If 280 different quarters are inserted into the vending machine, what is the probability that the mean falls between the limits of 5.550 g and 5.790 g?

c. If you own the vending machine, which result would concern you more? The result from part (a) or the result from part (b)? Why?

TI 19. Filling Pepsi Cans Cans of regular Pepsi are labeled to indicate that they contain 12 oz. Data Set 17 in Appendix B lists measured amounts for a sample of Pepsi cans. The sample statistics are $n = 36$ and $\bar{x} = 12.29$ oz. If the Pepsi cans are filled so that $\mu = 12.00$ oz (as labeled) and the population standard deviation is $\sigma = 0.09$ oz (based on the sample results), find the probability that a sample of 36 cans will have a mean of 12.29 oz or greater. Do these results suggest that the Pepsi cans are filled with an amount greater than 12.00 oz?

TI 20. Body Temperatures Assume that the population of human body temperatures has a mean of 98.6°F, as is commonly believed. Also assume that the population standard deviation is 0.62°F (based on data from University of Maryland researchers). If a sample of size $n = 106$ is randomly selected, find the probability of getting a mean temperature of 98.2°F or lower. (The value of 98.2°F was actually obtained; see the midnight temperatures for Day 2 in Data Set 2 of Appendix B.) Does that probability suggest that the mean body temperature is not 98.6°F?

6-5   Beyond the Basics

TI 21. Doorway Height The Boeing 757-200 ER airliner carries 200 passengers and has doors with a height of 72 in. Heights of men are normally distributed with a mean of 69.0 in. and a standard deviation of 2.8 in.

a. What doorway height would allow 95% of men to enter the aircraft without bending?

b. Assume that half of the 200 passengers are men. What doorway height satisfies the condition that there is a 0.95 probability that this height is greater than the mean height of 100 men?

c. When designing the Boeing 757-200 ER airliner, which result is more relevant: The height from part (a) or the height from part (b)? Why?

TI **22. Correcting for a Finite Population** In a study of Reye's syndrome, 160 children had a mean age of 8.5 years, a standard deviation of 3.96 years, and ages that approximated a normal distribution (based on data from Holtzhauer and others, *American Journal of Diseases of Children*, Vol. 140). Assume that 36 of those children are to be randomly selected for a follow-up study.

a. When considering the distribution of the mean ages for groups of 36 children, should $\sigma_{\bar{x}}$ be corrected by using the finite population correction factor? Explain.

b. Find the probability that the mean age of the follow-up sample group is greater than 10.0 years.

TI **23. Correcting for a Finite Population** The Newport Varsity Club has 210 members. The weights of members have a distribution that is approximately normal with a mean of 163 lb and a standard deviation of 32 lb. The design for a new club building includes an elevator with a capacity limited to 12 passengers.

a. When considering the distribution of the mean weight of 12 passengers, should $\sigma_{\bar{x}}$ be corrected by using the finite population correction factor? Explain.

b. If the elevator is designed to safely carry a load up to 2100 lb, what is the maximum safe mean weight when the elevator has 12 passengers?

c. If the elevator is filled with 12 randomly selected club members, what is the probability that the total load exceeds the safe limit of 2100 lb? Is this probability low enough?

d. What is the maximum number of passengers that should be allowed if we want a 0.999 probability that the elevator will not be overloaded when it is filled with randomly selected club members?

TI **24. Population Parameters** Three randomly selected households are surveyed as a pilot project for a larger survey to be conducted later. The numbers of people in the households are 2, 3, and 10 (based on Data Set 22 in Appendix B). Consider the values of 2, 3, and 10 to be a population. Assume that samples of size $n = 2$ are randomly selected *without* replacement.

a. Find μ and σ.

b. After finding all samples of size $n = 2$ that can be obtained *without* replacement, find the population of all values of $\bar{x}$ by finding the mean of each sample of size $n = 2$.

c. Find the mean $\mu_{\bar{x}}$ and standard deviation $\sigma_{\bar{x}}$ for the population of sample means found in part (b).

d. Verify that

$$\mu_{\bar{x}} = \mu \quad \text{and} \quad \sigma_{\bar{x}} = \frac{\sigma}{\sqrt{n}}\sqrt{\frac{N-n}{N-1}}$$

6-6 Normal as Approximation to Binomial

Key Concept In this section we present a method for using a normal distribution as an approximation to a binomial probability distribution. If the conditions $np \geq 5$ and $nq \geq 5$ are both satisfied, then probabilities from a binomial probability distribution can be approximated reasonably well by using a normal distribution with mean $\mu = np$ and standard deviation $\sigma = \sqrt{npq}$. Because a binomial probability distribution typically uses only whole numbers for the random variable x, while the normal approximation is continuous, we must use a "continuity correction" with a whole number x represented by the interval from $x - 0.5$ to $x + 0.5$. *Note:* Instead of using a normal distribution as an approximation to a binomial probability distribution, most practical applications of the binomial distribution can be handled with a TI-83/84 Plus calculator, but this section introduces the principle that a binomial distribution can be approximated by a normal distribution, and that principle will be used in later chapters.

Section 5-3 stated that a *binomial probability distribution* has (1) a fixed number of trials; (2) trials that are independent; (3) trials that are each classified into two categories commonly referred to as *success* and *failure*; (4) trials with the property that the probability of success remains constant. Also recall this notation:

n = the fixed number of trials.

x = the specific number of successes in n trials

p = probability of *success* in *one* of the n trials.

q = probability of *failure* in *one* of the n trials.

Consider this situation: The author was mailed a survey from Viking River Cruises supposedly sent to "a handful of people." Assume that the survey requested an e-mail address, was sent to 40,000 people, and the percentage of surveys returned with an e-mail address is 3%. Suppose that the true goal of the survey was to acquire a pool of at least 1150 e-mail addresses to be used for aggressive marketing. To find the probability of getting at least 1150 responses with e-mail addresses, we can use the binomial probability distribution with n = 40,000, p = 0.03, and q = 0.97. See the accompanying Minitab display showing a graph of the probability for each number of successes from 1100 to 1300, and notice how the graph appears to be a normal distribution, even though the plotted points are from a binomial distribution. (The other values of x all have probabilities that are very close to zero.) This graph suggests that we can use a normal distribution to approximate the binomial distribution.

MINITAB

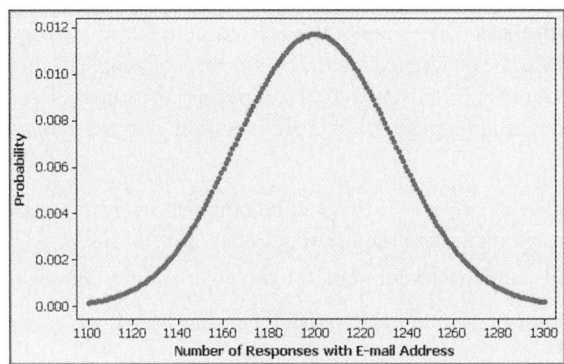

Normal Distribution as an Approximation to the Binomial Distribution

Requirements

1. The sample is a simple random sample of size n from a population in which the proportion of successes is p, or the sample is the result of conducting n independent trials of a binomial experiment in which the probability of success is p.

2. $np \geq 5$ and $nq \geq 5$.

Normal Approximation

If the above requirements are satisfied, then the probability distribution of the random variable x can be approximated by a normal distribution with these parameters:

- $\mu = np$
- $\sigma = \sqrt{npq}$

Continuity Correction

When using the normal approximation, adjust the discrete whole number x by using a *continuity correction*, so that x is represented by the interval from $x - 0.5$ to $x + 0.5$.

Note that the requirements include verification of $np \geq 5$ and $nq \geq 5$. The minimum value of 5 is common, but it is not an absolutely rigid value, and a few textbooks use 10 instead. This requirement is included in the following procedure for using a normal approximation to a binomial distribution:

Procedure for Using a Normal Distribution to Approximate a Binomial Distribution

1. Verify that both of the preceding requirements are satisfied. (If these requirements are not both satisfied, then you must use computer software, or a calculator, or Table A-1, or calculations using the binomial probability formula.)

2. Find the values of the parameters μ and σ by calculating $\mu = np$ and $\sigma = \sqrt{npq}$.

3. Identify the discrete whole number x that is relevant to the binomial probability problem. (For example, if you're trying to find the probability of getting at least 1150 successes among 40,000 trials (as in Example 1), the discrete whole number of concern is $x = 1150$. First focus on the value of 1150 itself, and temporarily ignore whether you want at least 1150, more than 1150, fewer than 1150, at most 1150, or exactly 1150.)

4. Draw a normal distribution centered about μ, then draw a *vertical strip area* centered over x. Mark the left side of the strip with the number equal to $x - 0.5$, and mark the right side with the number equal to $x + 0.5$. (With $x = 1150$, for example, draw a strip from 1149.5 to 1150.5.) *Consider the entire area of the entire strip to represent the probability of the discrete whole number x itself.*

5. Now determine whether the value of x itself should be included in the probability you want. (For example, "at least x" *does* include x itself, but "more than x" *does not* include x itself.) Next, determine whether you want the probability of at least x, at most x, more than x, fewer than x, or exactly x. Shade the area to the right or left of the strip, as appropriate; also shade the interior of the strip *if and only if x itself* is to be included. This total shaded region corresponds to the probability being sought.

6. Using either $x - 0.5$ or $x + 0.5$ in place of x, find the area of the shaded region from Step 5 as follows: First, find the z score: $z = (x - \mu)/\sigma$ (with either $x - 0.5$ or $x + 0.5$ used in place of x). Second, use that z score to find the area to the left of the adjusted value of x. Third, that cumulative left area can now be used to identify the shaded area corresponding to the desired probability.

Voltaire Beats Lottery

In 1729, the philosopher Voltaire became rich by devising a scheme to beat the Paris lottery. The government ran a lottery to repay municipal bonds that had lost some value. The city added large amounts of money with the net effect that the prize values totaled more than the cost of all tickets. Voltaire formed a group that bought all the tickets in the monthly lottery and won for more than a year. A bettor in the New York State Lottery tried to win a share of an exceptionally large prize that grew from a lack of previous winners. He wanted to write a $6,135,756 check that would cover all combinations, but the state declined and said that the nature of the lottery would have been changed.

> **EXAMPLE 1** **Mail Survey** The author was mailed a survey from Viking River Cruises, and the survey included a request for an e-mail address. Assume that the survey was sent to 40,000 people and that for such surveys, the percentage of responses with an e-mail address is 3%. If the true goal of the survey was to acquire a bank of at least 1150 e-mail addresses, find the probability of getting at least 1150 responses with e-mail addresses.

> **SOLUTION** The given problem involves a binomial distribution with a fixed number of trials ($n = 40,000$), which are independent. There are two categories for each survey: a response is obtained with an e-mail address or it is not. The probability of success ($p = 0.03$) presumably remains constant from trial to trial. Calculations with the binomial probability formula are not practical, because we would have to apply it

continued

38,851 times, once for each value of x from 1150 to 40,000 inclusive. We will proceed with the six-step approach of using a normal distribution to approximate the binomial distribution.

Step 1: Requirement check: Although it is unknown how the survey subjects were selected, we will proceed under the assumption that we have a simple random sample.

We must verify that it is reasonable to approximate the binomial distribution by the normal distribution because $np \geq 5$ and $nq \geq 5$. With $n = 40{,}000$, $p = 0.03$, and $q = 1 - p = 0.97$, we verify the required conditions as follows:

$$np = 40{,}000 \cdot 0.03 = 1200 \qquad \text{(Therefore } np \geq 5.)$$

$$nq = 40{,}000 \cdot 0.97 = 38{,}800 \qquad \text{(Therefore } nq \geq 5.)$$

Step 2: We now proceed to find the values for the parameters μ and σ that are needed for the normal distribution. We get the following:

$$\mu = np = 40{,}000 \cdot 0.03 = 1200$$

$$\sigma = \sqrt{npq} = \sqrt{40{,}000 \cdot 0.03 \cdot 0.97} = 34.117444$$

Step 3: We want the probability of at least 1150 responses with e-mail addresses, so $x = 1150$ is the discrete whole number relevant to this example.

Step 4: See Figure 6-21, which shows a normal distribution with mean $\mu = 1200$ and standard deviation $\sigma = 34.117444$. Figure 6-21 also shows the vertical strip from 1149.5 to 1150.5.

Step 5: We want to find the probability of getting *at least 1150* responses with e-mail addresses, so we want to shade the vertical strip representing 1150 as well as the area to its right. The desired area is shaded in green in Figure 6-21.

Step 6: We want the area to the right of 1149.5 in Figure 6-21, so the z score is found by using the values of μ and σ from Step 2 and the boundary value of 1149.5 as follows:

$$z = \frac{x - \mu}{\sigma} = \frac{1149.5 - 1200}{34.117444} = -1.48$$

Using Table A-2, we find that $z = -1.48$ corresponds to an area of 0.0694, so the shaded region in Figure 6-21 is $1 - 0.0694 = 0.9306$.

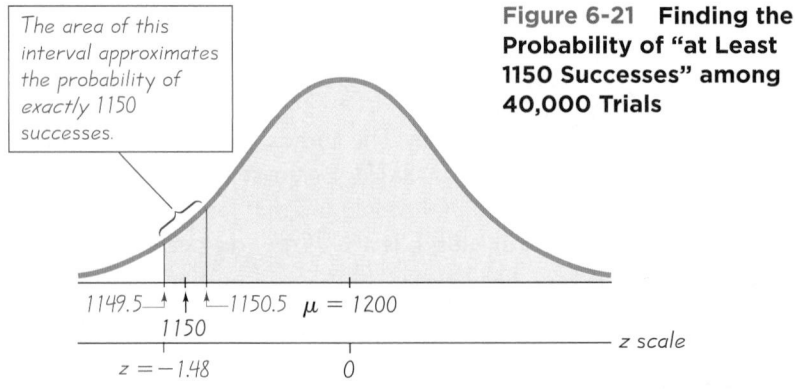

The area of this interval approximates the probability of exactly 1150 successes.

Figure 6-21 Finding the Probability of "at Least 1150 Successes" among 40,000 Trials

INTERPRETATION   There is a 0.9306 probability of getting at least 1150 responses with e-mail addresses among the 40,000 surveys that were mailed. (We get 0.9313 if we use a TI-83/84 Plus calculator for an exact result.) This probability is high enough to conclude that it is very likely that Viking Cruises will attain their goal of at least 1150 responses with e-mail addresses.

If the Viking River Cruises uses a sampling method that does not provide a simple random sample, then the resulting probability of 0.9306 might be very wrong. For example, if they surveyed only past customers, they might be more likely to get a higher response rate, so the preceding calculations might be incorrect. We should never forget the importance of a suitable sampling method.

Continuity Correction

The procedure for using a normal distribution to approximate a binomial distribution includes a *continuity correction,* defined as follows.

> **DEFINITION**
>
> When we use the normal distribution (which is a *continuous* probability distribution) as an approximation to the binomial distribution (which is *discrete*), a **continuity correction** is made to a discrete whole number x in the binomial distribution by representing the discrete whole number x by the *interval* from $x - 0.5$ to $x + 0.5$ (that is, adding and subtracting 0.5).

In the above six-step procedure for using a normal distribution to approximate a binomial distribution, Steps 3 and 4 incorporate the continuity correction. (See Steps 3 and 4 in the solutions to Examples 1 and 2.)

To see examples of continuity corrections, see the common cases illustrated in Figure 6-22. Those cases correspond to the statements in the following list.

Statement	Area
At least 8 (includes 8 and above)	To the *right* of 7.5
More than 8 (doesn't include 8)	To the *right* of 8.5
At most 8 (includes 8 and below)	To the *left* of 8.5
Fewer than 8 (doesn't include 8)	To the *left* of 7.5
Exactly 8	Between 7.5 and 8.5

> **EXAMPLE 2** **Internet Penetration Survey** A recent Pew Research Center survey showed that among 2822 randomly selected adults, 2060 (or 73%) stated that they are Internet users. If the proportion of all adults using the Internet is actually 0.75, find the probability that a random sample of 2822 adults will result in *exactly* 2060 Internet users.

> **SOLUTION** We have $n = 2822$ independent survey subjects, and $x = 2060$ of them are Internet users. We assume that the population proportion is $p = 0.75$, so it follows that $q = 0.25$. We will use a normal distribution to approximate the binomial distribution.
>
> **Step 1:** We begin by checking the requirements. The Pew Research Center has a reputation for sound survey techniques, so it is reasonable to treat the sample as a simple random sample. We now check the requirements that $np \geq 5$ and $nq \geq 5$:
>
> $$np = 2822 \cdot 0.75 = 2116.5 \quad \text{(Therefore } np \geq 5.)$$
>
> $$nq = 2822 \cdot 0.25 = 705.5 \quad \text{(Therefore } nq \geq 5.)$$
> *continued*

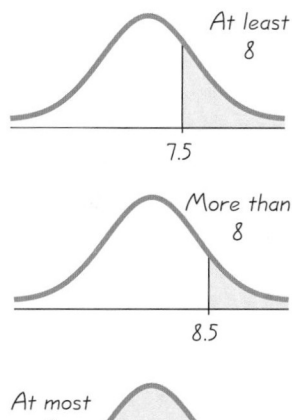

At least 8

7.5

More than 8

8.5

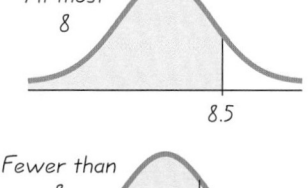

At most 8

8.5

Fewer than 8

7.5

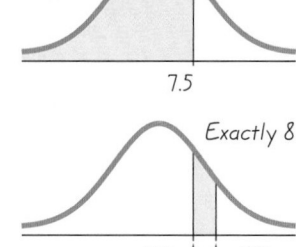

Exactly 8

7.5 8.5

Figure 6-22
Using Continuity Corrections

Step 2: We now find the values for μ and σ. We get the following:

$$\mu = np = 2822 \cdot 0.75 = 2116.5$$
$$\sigma = \sqrt{npq} = \sqrt{2822 \cdot 0.75 \cdot 0.25} = 23.002717$$

Step 3: We want the probability of *exactly* 2060 Internet users, so the discrete whole number relevant to this example is 2060.

Step 4: See Figure 6-23, which is a normal distribution with mean $\mu = 2116.5$ and standard deviation $\sigma = 23.002717$. Also, Figure 6-23 includes a vertical strip from 2059.5 to 2060.5, which represents the probability of exactly 2060 Internet users.

Step 5: Because we want the probability of *exactly* 2060 Internet users, we want the shaded area shown in Figure 6-23.

Step 6: To find the shaded region in Figure 6-23, first find the total area to the left of 2060.5, and then find the total area to the left of 2059.5. Then find the *difference* between those two areas. Let's begin with the total area to the left of 2060.5. If using Table A-2, we must first find the z score corresponding to 2060.5. We get

$$z = \frac{2060.5 - 2116.5}{23.002717} = -2.43$$

We use Table A-2 to find that $z = -2.43$ corresponds to a probability of 0.0075, which is the total area to the left of 2060.5. Now we find the area to the left of 2059.5 by first finding the z score corresponding to 2059.5:

$$z = \frac{2059.5 - 2116.5}{23.002717} = -2.48$$

We use Table A-2 to find that $z = -2.48$ corresponds to a probability of 0.0066, which is the total area to the left of 2059.5. The shaded area is $0.0075 - 0.0066 = 0.0009$. (We also get 0.0009 if we use a TI-83/84 Plus calculator for an exact result.)

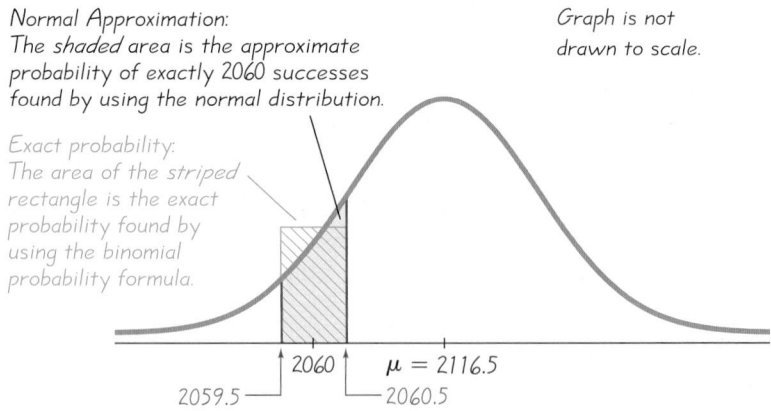

Figure 6-23   Using the Continuity Correction

INTERPRETATION   If we assume that 75% of all adults use the Internet, the probability of getting exactly 2060 Internet users among 2822 randomly selected adults is 0.0009. (Using technology, the probability is 0.000872.) This probability tells us that if the percentage of Internet users in the adult population is 75%, then it is highly unlikely that we will get *exactly* 2060 Internet users when we survey 2822 adults. Actually, when surveying 2822 adults, the probability of *any single number* of Internet users will be very small.

Interpreting Results

When we use a normal distribution as an approximation to a binomial distribution, our ultimate goal is not simply to find a probability number. We often need to make some *judgment* based on the probability value. The following criterion (from Section 5-2) describes the use of probabilities for distinguishing between results that could easily occur by chance and those results that are highly unusual.

Using Probabilities to Determine When Results Are Unusual

- **Unusually high:** *x* successes among *n* trials is an *unusually high* number of successes if *P*(*x* or more) is very small (such as 0.05 or less).

- **Unusually low:** *x* successes among *n* trials is an *unusually low* number of successes if *P*(*x* or fewer) is very small (such as 0.05 or less).

The Role of the Normal Approximation

Almost all practical applications of the binomial probability distribution can now be handled well with computer software or a TI-83/84 Plus calculator. In this section we presented methods for using a normal approximation method instead of software, but, more importantly, we illustrated the principle that under appropriate circumstances, the binomial probability distribution can be approximated by a normal distribution. Later chapters will include procedures based on the use of a normal distribution as an approximation to a binomial distribution, so this section forms a foundation for those important procedures.

Using the TI-83/84 Plus Calculator for Finding Exact Results With the Binomial Distribution

The main focus of this section is the use of the normal distribution as an *approximation* to the binomial distribution, and the TI-83/84 Plus calculator can be used for such approximations. However, the TI-83/84 Plus calculator can often be used to find *exact* values, so that approximations are not necessary. The following procedure can be used to find such exact values.

1. Gain access to the "DISTR" (distributions) menu by pressing **2ND** **VARS**.

2. The DISTR menu includes the options of 0:binompdf and A:binomcdf for a single probability and a cumulative probability.

   - **Single Value** To find the binomial probability for a *single* value of *x*: **binompdf(*n, p, x*)** gives the binomial probability of exactly *x* successes among *n* trials, where each trial has a probability of success *p*.

   - **Cumulative Values** To find a *cumulative* probability for values from 0 up to and including the desired value of *x*: **binomcdf(*n, p, x*)** gives a *cumulative* probability from 0 up to (and including) the desired value of *x*, where *n* is the number of trials and *p* is the probability of success for each trial. (*Hint:* If you want a cumulative probability that is different from 0 up to *x*, you can find it by using binomcdf and adjusting the result. For example, the probability of a value *greater than x* can be found by subtracting the result from 1.)

6-6   Basic Skills and Concepts

Statistical Literacy and Critical Thinking

1. Proportions in Television Nielsen Media Research conducts surveys to determine the proportion of households tuned to television shows. Assume that a different random sample of 5000 households is obtained each week. If the proportion of households tuned to *60 Minutes* is recorded for each Sunday during the course of two years, and the proportions are depicted in a histogram, what is the approximate shape of the histogram? Why?

2. Continuity Correction The Wechsler test is used to measure IQ scores. It is designed so that the mean IQ score is 100 and the standard deviation is 15. It is known that IQ scores have a normal distribution. Assume that we want to find the probability that a randomly selected person has an IQ equal to 107. What is the continuity correction, and how would it be applied in finding that probability?

3. Gender Selection The Genetics & IVF Institute has developed methods for helping couples determine the gender of their children. For comparison, a large sample of randomly selected families with four children is obtained, and the proportion of girls in each family is recorded. Is the normal distribution a good approximation of the distribution of those proportions? Why or why not?

4. μ and σ Multiple choice test questions are commonly used for standardized tests, including the SAT, ACT, and LSAT. When scoring such questions, it is common to compensate for guessing. If a test consists of 100 multiple choice questions, each with possible answers of a, b, c, d, e, and each question has only one correct answer, find μ and σ for the number of correct answers provided by someone who makes random guesses. What do μ and σ measure?

Applying Continuity Correction. *In Exercises 5–12, the given values are discrete. Use the continuity correction and describe the region of the normal distribution that corresponds to the indicated probability. For example, the probability of "more than 20 defective items" corresponds to the area of the normal curve described with this answer: "the area to the right of 20.5."*

5. Probability of more than 8 Senators who are women

6. Probability of at least 2 traffic tickets this year

7. Probability of fewer than 5 passengers who do not show up for a flight

8. Probability that the number of students who are absent is exactly 4

9. Probability of no more than 15 peas with green pods

10. Probability that the number of defective computer power supplies is between 12 and 16 inclusive

11. Probability that the number of job applicants late for interviews is between 5 and 9 inclusive

12. Probability that exactly 24 felony indictments result in convictions

Using Normal Approximation. *In Exercises 13–16, do the following: (a) Find the indicated binomial probability by using Table A-1 in Appendix A. (b) If $np \geq 5$ and $nq \geq 5$, also estimate the indicated probability by using the normal distribution as an approximation to the binomial distribution; if $np < 5$ or $nq < 5$, then state that the normal approximation is not suitable.*

13. With $n = 10$ and $p = 0.5$, find $P(3)$.

14. With $n = 12$ and $p = 0.8$, find $P(9)$.

15. With $n = 8$ and $p = 0.9$, find P(at least 6).

16. With $n = 15$ and $p = 0.4$, find P(fewer than 3).

17. Mail Survey In Example 1 it was noted that the author was mailed a survey from Viking River Cruises, and it included a request for an e-mail address. As in Example 1, assume that the survey was sent to 40,000 people and that for such surveys, the percentage of responses with an e-mail address is 3%. If the goal of the survey was to acquire a bank of at least 1300 e-mail addresses, find the probability of getting at least 1300 responses with e-mail addresses. Is it likely that the goal will be reached?

18. Internet Penetration Survey In Example 2, it was noted that a recent Pew Research Center survey showed that among 2822 randomly selected adults, 2060 (or 73%) stated that they are Internet users. A technology specialist claims that 75% of adults use the Internet, and the results from the survey show a lower percentage because of the random chance variation in surveys. Assuming that the 75% rate is correct, is a result of 2060 Internet users an unusually low number when 2822 adults are randomly selected? Explain.

19. Gender Selection The Genetics & IVF Institute developed its XSORT method to increase the probability of conceiving a girl. Among 574 women using that method, 525 had baby girls. Assuming that the method has no effect so that boys and girls are equally likely, find the probability of getting at least 525 girls among 574 babies. Does the result suggest that the XSORT method is effective? Why or why not?

20. Gender Selection The Genetics & IVF Institute developed its YSORT method to increase the probability of conceiving a boy. Among 152 women using that method, 127 had baby boys. Assuming that the method has no effect so that boys and girls are equally likely, find the probability of getting at least 127 boys among 152 babies. Does the result suggest that the YSORT method is effective? Why or why not?

21. Mendel's Hybridization Experiment When Mendel conducted his famous hybridization experiments, he used peas with green pods and yellow pods. One experiment involved crossing peas in such a way that 25% (or 145) of the 580 offspring peas were expected to have yellow pods. Instead of getting 145 peas with yellow pods, he obtained 152. Assume that Mendel's 25% rate is correct.

a. Find the probability that among the 580 offspring peas, exactly 152 have yellow pods.

b. Find the probability that among the 580 offspring peas, at least 152 have yellow pods.

c. Which result is useful for determining whether Mendel's claimed rate of 25% is incorrect? (Part (a) or part (b)?)

d. Is there strong evidence to suggest that Mendel's rate of 25% is incorrect?

22. Voters Lying? In a survey of 1002 people, 701 said that they voted in a recent presidential election (based on data from ICR Research Group). Voting records show that 61% of eligible voters actually did vote. Given that 61% of eligible voters actually did vote, find the probability that among 1002 randomly selected eligible voters, at least 701 actually did vote. What does the result suggest?

23. Cell Phones and Brain Cancer In a study of 420,095 cell phone users in Denmark, it was found that 135 developed cancer of the brain or nervous system. Assuming that the use of cell phones has no effect on developing such cancers, there is a 0.000340 probability of a person developing cancer of the brain or nervous system. We therefore expect about 143 cases of such cancer in a group of 420,095 randomly selected people. Estimate the probability of 135 or fewer cases of such cancer in a group of 420,095 people. What do these results suggest about media reports that cell phones cause cancer of the brain or nervous system?

24. Employee Hiring There is an 80% chance that a prospective employer will check the educational background of a job applicant (based on data from the Bureau of National Affairs, Inc.). For 100 randomly selected job applicants, find the probability that exactly 85 have their educational backgrounds checked.

25. Universal Donors Six percent of typical people have blood that is group O and type Rh⁻. These people are considered to be universal donors, because they can give blood to anyone. Providence Memorial Hospital is conducting a blood drive because it needs blood from at least 10 universal donors. If 200 volunteers donate blood, what is the probability that the number of universal donors is at least 10? Is the pool of 200 volunteers likely to be sufficient?

26. Acceptance Sampling With the procedure called *acceptance sampling*, a sample of items is randomly selected and the entire batch is either rejected or accepted, depending on the results. The Telektronics Company has just manufactured a large batch of backup power supply units for computers, and 7.5% of them are defective. If the acceptance sampling plan is to randomly select 80 units and accept the whole batch if at most 4 units are defective, what is the probability that the entire batch will be accepted? Based on the result, does the Telektronics Company have quality control problems?

27. M&M Candies: Are 24% Blue? According to Mars (the candy company, not the planet), 24% of M&M plain candies are blue. Data Set 18 in Appendix B shows that among 100 M&Ms chosen, 27 are blue. Assuming that the claimed blue M&Ms rate of 24% is correct, find the probability of randomly selecting 100 M&Ms and getting 27 or more that are blue. Based on the result, is 27 an unusually high number of blue M&Ms when 100 are randomly selected?

28. Detecting Fraud When working for the Brooklyn District Attorney, investigator Robert Burton analyzed the leading digits of amounts on checks from companies that were suspected of fraud. Among 784 checks, 479 had amounts with leading digits of 5, but checks issued in the normal course of honest transactions were expected to have 7.9% of the checks with amounts having leading digits of 5. Is there strong evidence to indicate that the check amounts are significantly different from amounts that are normally expected? Explain?

29. Cholesterol Reducing Drug The probability of flu symptoms for a person not receiving any treatment is 0.019. In a clinical trial of Lipitor (atorvastatin), a drug commonly used to lower cholesterol, 863 patients were given a treatment of 10-mg atorvastatin tablets, and 19 of those patients experienced flu symptoms (based on data from Pfizer, Inc.). Assuming that these tablets have no effect on flu symptoms, estimate the probability that at least 19 of 863 people experience flu symptoms. What do these results suggest about flu symptoms as an adverse reaction to the drug?

30. Polygraph Accuracy Polygraph experiments conducted by researchers Charles R. Honts (Boise State University) and Gordon H. Barland (Department of Defense Polygraph Institute) showed that among 57 polygraph indications of a lie, the truth was told 15 times, so the proportion of *false positive* results among the 57 positive results is 15/57. Assuming that the polygraph makes random guesses, determine whether 15 is an unusually low number of false positive results among the 57 positive results. Does the polygraph appear to be making random guesses? Explain.

31. Overbooking a Boeing 767-300 A Boeing 767-300 aircraft has 213 seats. When someone buys a ticket for a flight, there is a 0.0995 probability that the person will not show up for the flight (based on data from an IBM research paper by Lawrence, Hong, and Cherrier). A ticket agent accepts 236 reservations for a flight that uses a Boeing 767-300. Find the probability that not enough seats will be available. Is this probability low enough so that overbooking is not a real concern?

32. Passenger Load on a Boeing 767-300 An American Airlines Boeing 767-300 aircraft has 213 seats. When fully loaded with passengers, baggage, cargo, and fuel, the pilot must verify that the gross weight is below the maximum allowable limit, and the weight must be properly distributed so that the balance of the aircraft is within safe acceptable limits. When considering the weights of passengers, their weights are estimated according to Federal Aviation Administration rules. Men have a mean weight of 172 lb, whereas women have a mean weight of 143 lb, so disproportionately more male passengers might result in an unsafe overweight situation. Assume that if there are at least 122 men in a roster of 213 passengers, the load must be somehow adjusted. Assume that passengers are booked randomly, and that male passengers and female passengers are equally likely. If the aircraft is full of adults, find the probability that a Boeing 767-300 with 213 passengers has at least 122 men. Based on the result, does it appear that the load must be adjusted often?

6-6   Beyond the Basics

33. Gambling Strategy Marc Taylor plans to place 200 bets of $5 each on a game at the Mirage casino in Las Vegas.

a. One strategy is to bet on the number 7 at roulette. A win pays off with odds of 35:1 and, on any one spin, there is a probability of 1/38 that 7 will be the winning number. Among the 200 bets, what is the minimum number of wins needed for Marc to make a profit? Find the probability that Marc will make a profit.

b. Another strategy is to bet on the pass line in the dice game of craps. A win pays off with odds of 1:1 and, on any one game, there is a probability of 244/495 that he will win. Among the 200 bets, what is the minimum number of wins needed for Marc to make a profit? Find the probability that Marc will make a profit.

c. Based on the preceding results, which game is the better "investment": The roulette game from part (a) or the craps game from part (b)? Why?

 34. Overbooking a Boeing 767-300 A Boeing 767-300 aircraft has 213 seats. When someone buys a ticket for an airline flight, there is a 0.0995 probability that the person will not show up for the flight (based on data from an IBM research paper by Lawrence, Hong, and Cherrier). How many reservations could be accepted for a Boeing 767-300 for there to be at least a 0.95 probability that all reservation holders who show will be accommodated?

 35. Joltin' Joe Assume that a baseball player hits .350, so his probability of a hit is 0.350. (Ignore the complications caused by walks.) Also assume that his hitting attempts are independent of each other.

a. Find the probability of at least 1 hit in 4 tries in a single game.

b. Assuming that this batter gets up to bat 4 times each game, find the probability of getting a total of at least 56 hits in 56 games.

c. Assuming that this batter gets up to bat 4 times each game, find the probability of at least 1 hit in each of 56 consecutive games (Joe DiMaggio's 1941 record).

d. What minimum batting average would be required for the probability in part (c) to be greater than 0.1?

36. Normal Approximation Required This section included the statement that almost all practical applications of the binomial probability distribution can now be handled well with computer software or a TI-83/84 Plus calculator. Using specific computer software or a TI-83/84 Plus calculator, identify a case in which the technology fails so that a normal approximation to a binomial distribution is required.

6-7 Assessing Normality

Key Concept The following chapters describe statistical methods requiring that the data are a simple random sample from a population having a *normal* distribution. In this section we present criteria for determining whether the requirement of a normal distribution is satisfied. The criteria involve (1) visual inspection of a histogram to see if it is roughly bell-shaped; (2) identifying any outliers; (3) constructing a graph called a *normal quantile plot*.

Part 1: Basic Concepts of Assessing Normality

We begin with the definition of a normal quantile plot.

 **DEFINITION**

A **normal quantile plot** (or **normal probability plot**) is a graph of points (x, y) where each x value is from the original set of sample data, and each y value is the corresponding z score that is a quantile value expected from the standard normal distribution.

Procedure for Determining Whether It Is Reasonable to Assume that Sample Data Are From a Normally Distributed Population

1. *Histogram:* Construct a histogram. Reject normality if the histogram departs dramatically from a bell shape.

2. *Outliers:* Identify outliers. Reject normality if there is more than one outlier present. (Just one outlier could be an error or the result of chance variation, but be careful, because even a single outlier can have a dramatic effect on results.)

3. *Normal quantile plot:* If the histogram is basically symmetric and there is at most one outlier, use technology to generate a *normal quantile plot.* Use the following criteria to determine whether or not the distribution is normal. (These criteria can be used loosely for small samples, but they should be used more strictly for large samples.)

Normal Distribution: The population distribution is normal if the pattern of the points is reasonably close to a straight line and the points do not show some systematic pattern that is not a straight-line pattern.

Not a Normal Distribution: The population distribution is *not* normal if either or both of these two conditions applies:

- The points do not lie reasonably close to a straight line.

- The points show some *systematic pattern* that is not a straight-line pattern.

Later in this section we will describe the actual process of constructing a normal quantile plot, but for now we focus on interpreting such a plot.

EXAMPLE 1   **Determining Normality** The accompanying displays show histograms of data along with the corresponding normal quantile plots.

Normal: The first case shows a histogram of IQ scores that is close to being bell-shaped, so the histogram suggests that the IQ scores are from a normal distribution. The corresponding normal quantile plot shows points that are reasonably close to a straight-line pattern, and the points do not show any other systematic pattern that is not a straight line. It is safe to assume that these IQ scores are from a normally distributed population.

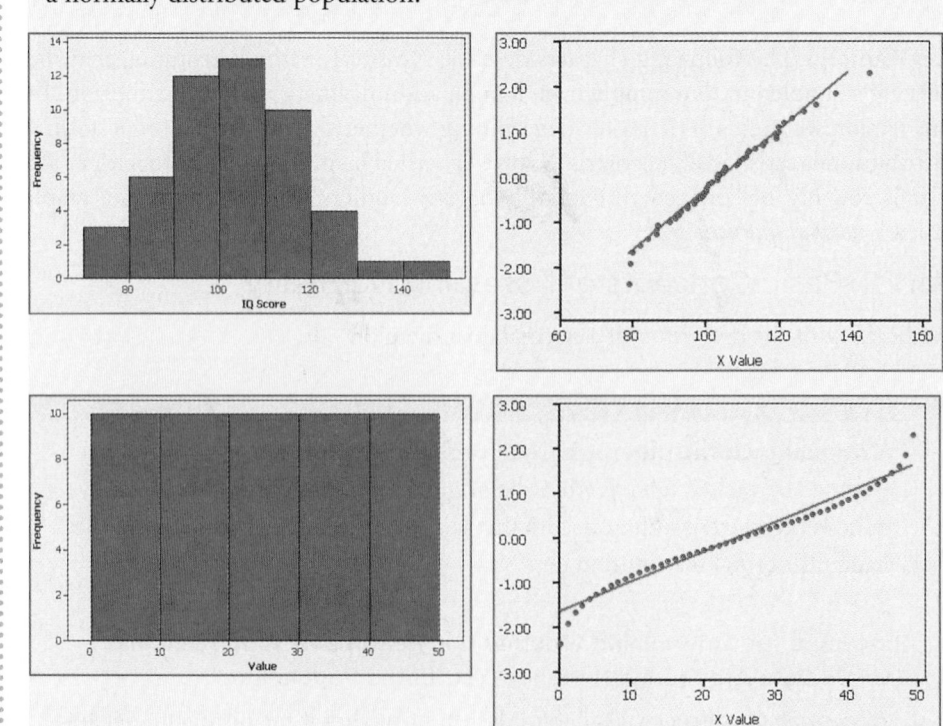

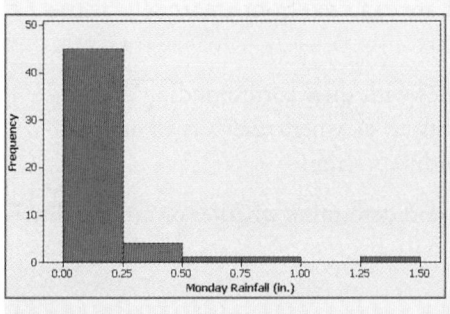
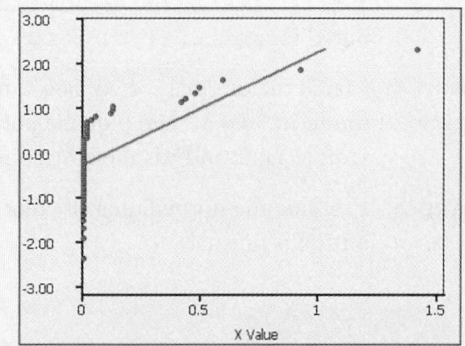

Uniform: The second case shows a histogram of data having a uniform distribution. The corresponding normal quantile plot suggests that the points are not normally distributed because the points show a *systematic pattern that is not a straight-line pattern*. These sample values are not from a population having a normal distribution.

Skewed: The third case shows a histogram of the amounts of rainfall (in inches) in Boston for every Monday during one year. (See Data Set 14 in Appendix B.) The shape of the histogram is skewed, not bell-shaped. The corresponding normal quantile plot shows points that are not at all close to a straight-line pattern. These rainfall amounts are not from a population having a normal distribution.

Here are some important comments about procedures for determining whether data are from a normally distributed population:

- If the requirement of a normal distribution is not too strict, examination of a histogram and consideration of outliers may be all that you need to assess normality.

- Normal quantile plots can be difficult to construct on your own, but they can be generated with a TI-83/84 Plus calculator or suitable computer software, such as STATDISK, SPSS, SAS, Minitab, and Excel.

- In addition to the procedures discussed in this section, there are other more advanced procedures for assessing normality, such as the chi-square goodness-of-fit test, the Kolmogorov-Smirnov test, the Lilliefors test, the Anderson-Darling test, and the Ryan-Joiner test (discussed briefly in Part 2).

Part 2: Beyond the Basics of Assessing Normality

The following is a relatively simple procedure for manually constructing a normal quantile plot, and it is the same procedure used by the TI-83/84 Plus calculator and STATDISK. Some statistical packages use various other approaches, but the interpretation of the graph is basically the same.

Manual Construction of a Normal Quantile Plot

Step 1. First sort the data by arranging the values in order from lowest to highest.

Step 2. With a sample of size n, each value represents a proportion of $1/n$ of the sample. Using the known sample size n, identify the areas of $1/2n$, $3/2n$, and so on. These are the cumulative areas to the left of the corresponding sample values.

Step 3. Use the standard normal distribution (Table A-2 or software or a calculator) to find the z scores corresponding to the cumulative left areas found in

Small Sample

The Children's Defense Fund was organized to promote the welfare of children. The group published *Children Out of School in America,* which reported that in one area, 37.5% of the 16- and 17-year-old children were out of school. This statistic received much press coverage, but it was based on a sample of only 16 children. Another statistic was based on a sample size of only 3 students. (See "Firsthand Report: How Flawed Statistics Can Make an Ugly Picture Look Even Worse," *American School Board Journal*, Vol. 162.)

Step 2. (These are the z scores that are expected from a normally distributed sample.)

Step 4. Match the original sorted data values with their corresponding z scores found in Step 3, then plot the points (x, y), where each x is an original sample value and y is the corresponding z score.

Step 5. Examine the normal quantile plot and determine whether or not the distribution is normal.

EXAMPLE 2 **Movie Lengths** Data Set 9 in Appendix B includes lengths (in minutes) of randomly selected movies. Let's consider only the first 5 movie lengths: 110, 96, 170, 125, 119. With only 5 values, a histogram will not be very helpful in revealing the distribution of the data. Instead, construct a normal quantile plot for these 5 values and determine whether they appear to come from a population that is normally distributed.

SOLUTION   The following steps correspond to those listed in the above procedure for constructing a normal quantile plot.

Step 1. First, sort the data by arranging them in order. We get 96, 110, 119, 125, 170.

Step 2. With a sample of size $n = 5$, each value represents a proportion of $1/5$ of the sample, so we proceed to identify the cumulative areas to the left of the corresponding sample values. The cumulative left areas, which are expressed in general as $1/2n$, $3/2n$, $5/2n$, $7/2n$, and so on, become these specific areas for this example with $n = 5$: $1/10$, $3/10$, $5/10$, $7/10$, and $9/10$. The cumulative left areas expressed in decimal form are 0.1, 0.3, 0.5, 0.7, and 0.9.

Step 3. We now search in the body of Table A-2 for the cumulative left areas of 0.1000, 0.3000, 0.5000, 0.7000, and 0.9000 to find these corresponding z scores: -1.28, -0.52, 0, 0.52, and 1.28.

Step 4. We now pair the original sorted movie lengths with their corresponding z scores. We get these (x, y) coordinates which are plotted in the accompanying TI-83/84 Plus calculator and STATDISK displays: $(96, -1.28)$, $(110, -0.52)$, $(119, 0)$, $(125, 0.52)$, and $(170, 1.28)$.

TI-83/84 Plus

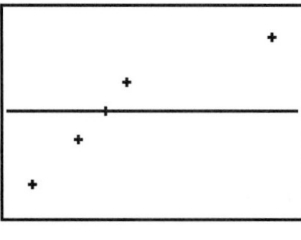

STATDISK

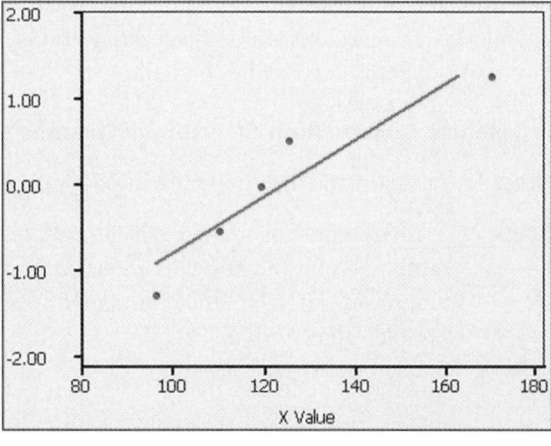

INTERPRETATION We examine the normal quantile plot in the STATDISK display. Because the points appear to lie reasonably close to a straight line and there does not appear to be a systematic pattern that is not a straight-line pattern, we conclude that the sample of five movie lengths appears to come from a normally distributed population.

In the next example, we address the issue of an outlier in a data set.

EXAMPLE 3 **Movie Lengths** Let's repeat Example 2 after changing one of the values so that it becomes an outlier. Change the highest value of 170 min in Example 2 to a length of 1700 min. (The actual longest movie is *Cure for Insomnia* with a length of 5220 min, or 87 hr.) The accompanying TI-83/84 Plus calculator and STATDISK displays show the normal quantile plot of these movie lengths: 110, 96, **1700,** 125, 119. Note how that one outlier affects the graph. This normal quantile plot does *not* result in points with an approximately straight-line pattern. The normal quantile plot suggests that the values of 110, 96, 1700, 125, 119 are from a population with a distribution that is *not* a normal distribution.

TI-83/84 Plus   **STATDISK**

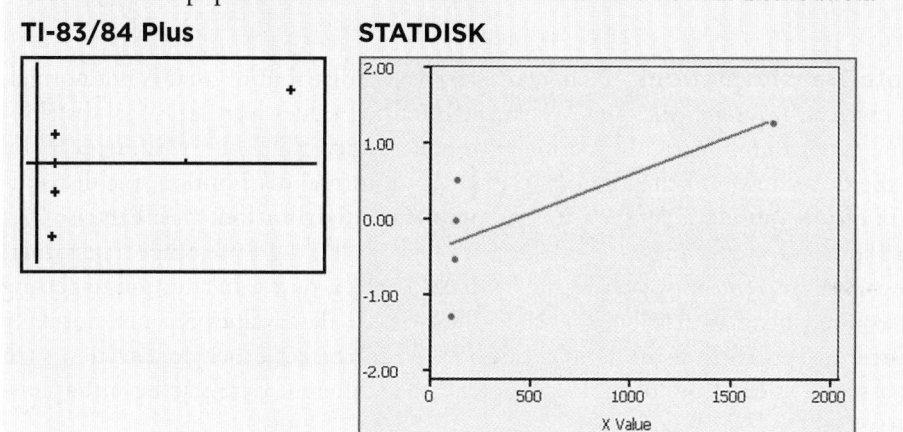

Ryan-Joiner Test   The Ryan-Joiner test is one of several formal tests of normality, each having their own advantages and disadvantages. STATDISK has a feature of **Normality Assessment** that displays a histogram, normal quantile plot, the number of potential outliers, and results from the Ryan-Joiner test. Information about the Ryan-Joiner test is readily available on the Internet.

EXAMPLE 4 **Tobacco in Children's Movies** Data Set 7 in Appendix B includes the times (in seconds) that the use of tobacco was shown in 50 different animated children's movies. Shown below is the STATDISK display summarizing results from the feature of Normality Assessment. All of these results suggest that the sample is *not* from a normally distributed population: (1) The histogram is far from being bell-shaped; (2) the points in the normal quantile plot are far from a straight-line pattern; (3) there appears to be one or more outliers; (4) results from the Ryan-Joiner test indicate that normality should be rejected. The evidence against a normal distribution is strong and consistent.

continued

Is Parachuting Safe?

About 30 people die each year as more than 100,000 people make about 2.25 million parachute jumps. In comparison, a typical year includes about 200 scuba diving fatalities, 7000 drownings, 900 bicycle deaths, 800 lightning deaths, and 1150 deaths from bee stings. Of course, these figures don't necessarily mean that parachuting is safer than bike riding or swimming. A fair comparison should involve fatality rates, not just the total number of deaths.

The author, with much trepidation, made two parachute jumps but quit after missing the spacious drop zone both times. He has also flown in a hang glider, hot air balloon, ultralight, sailplane, and Goodyear blimp.

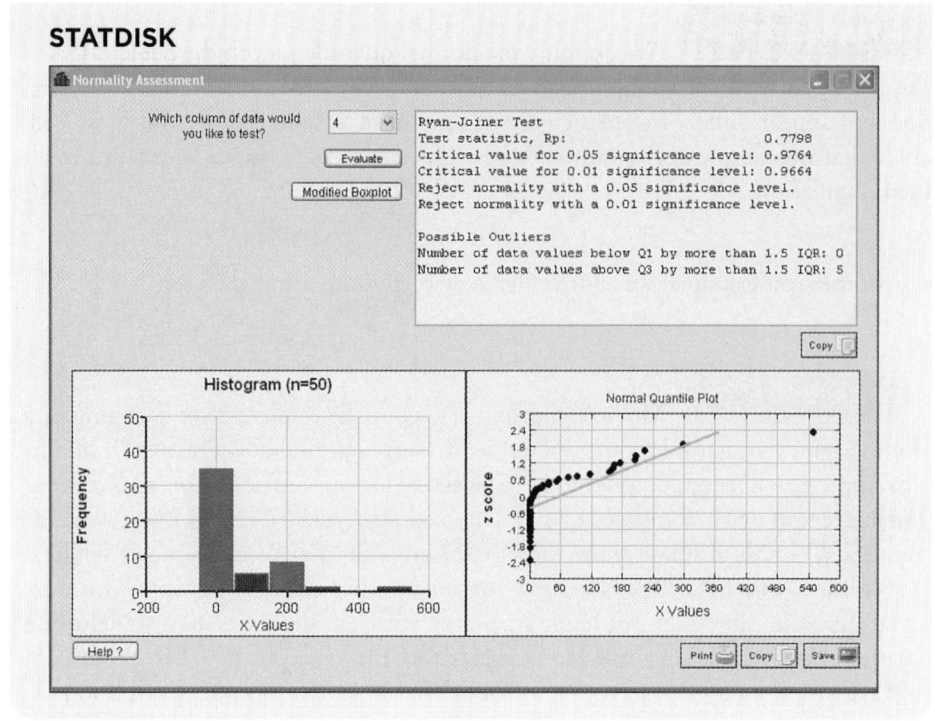

STATDISK

Data Transformations   Many data sets have a distribution that is not normal, but we can *transform* the data so that the modified values have a normal distribution. One common transformation is to replace each value of x with $\log(x + 1)$. If the distribution of the $\log(x + 1)$ values is a normal distribution, the distribution of the x values is referred to as a **lognormal distribution.** (See Exercise 22.) In addition to replacing each x value with $\log(x + 1)$, there are other transformations, such as replacing each x value with $\sqrt{x}$, or $1/x$, or x^2. In addition to getting a required normal distribution when the original data values are not normally distributed, such transformations can be used to correct other deficiencies, such as a requirement (found in later chapters) that different data sets have the same variance.

Using a TI-83/84 Plus Calculator to Generate a Normal Quantile Plot

To generate a normal quantile plot using a TI-83/84 Plus calculator, use the following procedure.

1. First enter the data in list L1, or assign the data to a list name. (Remember that the data sets in Appendix B are available in the APP that is on the CD-ROM included with this book.)

2. Get to the **STAT PLOT** menu by pressing **2ND** **Y=**.

3. Press **ENTER** **ENTER** to turn on Plot 1, and be sure that Plot2 and Plot3 are in the off state.

4. For the "type" item, select the last item in the second row of displayed options, then enter L1 or a list name for the "Data List." Select X for the Data Axis, and select the icon to be used to depict the plotted points. See the accompanying screen display showing the screen for generating a normal quantile plot using the sample values in list L1.

5. Press **ZOOM** **9** to generate the normal quantile plot.

Statistical Literacy and Critical Thinking

1. Normal Quantile Plot What is the purpose of constructing a normal quantile plot?

2. Rejecting Normality Identify two different characteristics of a normal quantile plot, where each characteristic would lead to the conclusion that the data are not from a normally distributed population.

3. Normal Quantile Plot If you select a simple random sample of M&M plain candies and construct a normal quantile plot of their weights, what pattern would you expect in the graph?

4. Criteria for Normality Assume that you have a data set consisting of the ages of all New York City police officers. Examination of a histogram and normal quantile plot are two different ways to assess the normality of that data set. Identify a third way.

Interpreting Normal Quantile Plots. *In Exercises 5–8, examine the normal quantile plot and determine whether it depicts sample data from a population with a normal distribution.*

5. Old Faithful The normal quantile plot represents duration times (in seconds) of Old Faithful eruptions from Data Set 15 in Appendix B.

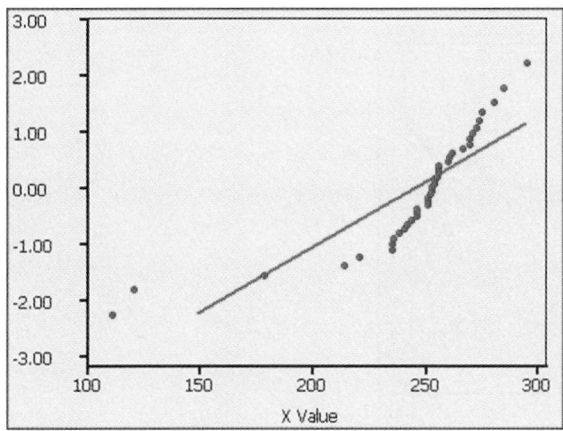

6. Heights of Women The normal quantile plot represents heights of women from Data Set 1 in Appendix B.

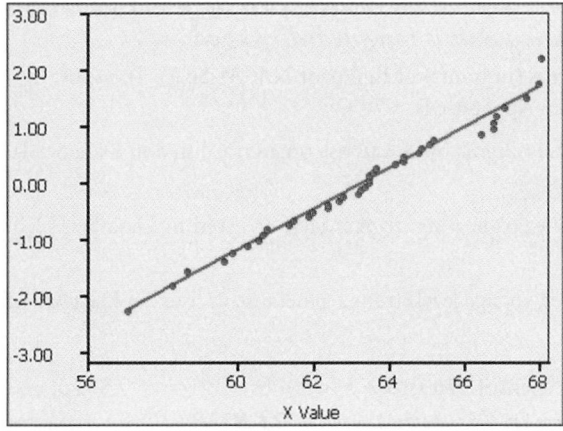

7. Weights of Diet Coke The normal quantile plot represents weights (in pounds) of diet Coke from Data Set 17 in Appendix B.

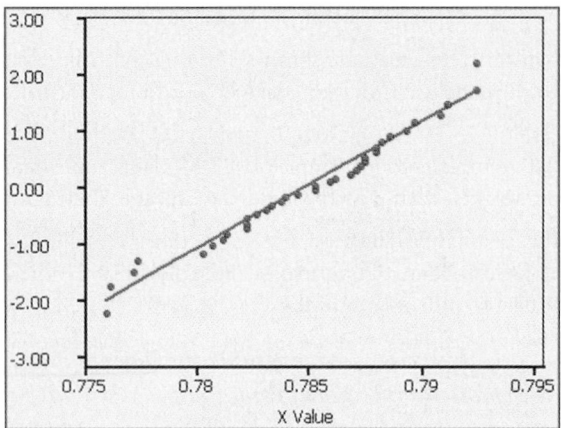

8. Telephone Digits The normal quantile plot represents the last two digits of telephone numbers of survey subjects.

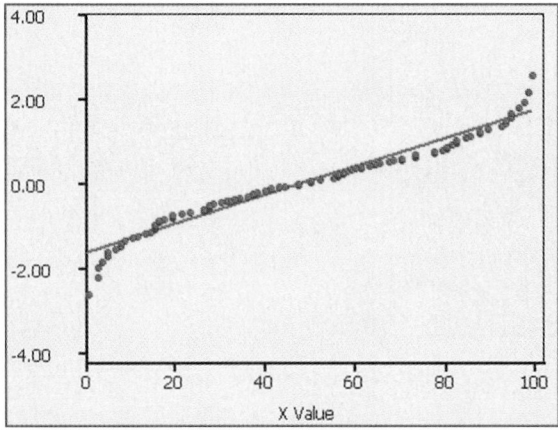

Determining Normality. *In Exercises 9–12, refer to the indicated data set and determine whether the data have a normal distribution. Assume that this requirement is loose in the sense that the population distribution need not be exactly normal, but it must be a distribution that is roughly bell-shaped.*

9. Space Shuttle Flights The lengths (in hours) of flights of NASA's Space Transport System (Shuttle) as listed in Data Set 10 in Appendix B.

10. Astronaut Flights The numbers of flights by NASA astronauts, as listed in Data Set 10 in Appendix B.

11. Heating Degree Days The values of heating degree days, as listed in Data Set 12 in Appendix B.

12. Generator Voltage The measured voltage levels from a generator, as listed in Data Set 13 in Appendix B.

Using Technology to Generate Normal Quantile Plots. *In Exercises 13–16, use the data from the indicated exercise in this section. Use a TI-83/ 84 Plus calculator* *to generate a normal quantile plot. Then determine whether the data come from a normally distributed population.*

TI **13.** Exercise 9   **TI** **14.** Exercise 10

TI **15.** Exercise 11   **TI** **16.** Exercise 12

TI **17. Comparing Data Sets** Using the heights of women and the cholesterol levels of women, as listed in Data Set 1 in Appendix B, analyze each of the two data sets and determine whether each appears to come from a normally distributed population. Compare the results and give a possible explanation for any notable differences between the two distributions.

TI **18. Comparing Data Sets** Using the systolic blood pressure levels and the elbow breadths of women, as listed in Data Set 1 in Appendix B, analyze each of the two data sets and determine whether each appears to come from a normally distributed population. Compare the results and give a possible explanation for any notable differences between the two distributions.

Constructing Normal Quantile Plots. *In Exercises 19 and 20, use the given data values to identify the corresponding z scores that are used for a normal quantile plot. Then construct the normal quantile plot and determine whether the data appear to be from a population with a normal distribution.*

TI **19. Braking Distances** A sample of braking distances (in feet) measured under standard conditions for an Acura RL, Acura TSX, Audi A6, BMW 525i, and Buick LaCrosse: 131, 136, 129, 127, 146.

TI **20. Satellites** A sample of the numbers of satellites in orbit: 158 (United States); 17 (China); 18 (Russia); 15 (Japan); 3 (France); 5 (Germany).

6-7   Beyond the Basics

TI **21. Transformations** The heights (in inches) of men listed in Data Set 1 in Appendix B have a distribution that is approximately normal, so it appears that those heights are from a normally distributed population.

a. If 2 inches is added to each height, are the new heights also normally distributed?

b. If each height is converted from inches to centimeters, are the heights in centimeters also normally distributed?

c. Are the logarithms of normally distributed heights also normally distributed?

TI **22. Lognormal Distribution** The following values are the times (in days) it took for prototype integrated circuits to fail. Test these values for normality, then replace each x value with $\log (x + 1)$ and test the transformed values for normality. What can you conclude?

103	547	106	662	329	510	1169	267	1894	1065
1396	307	362	1091	102	3822	547	725	4337	339

Review

In this chapter we introduced the normal probability distribution—the most important distribution in the study of statistics.

Section 6-2   In Section 6-2 we worked with the standard normal distribution, which is a normal distribution having a mean of 0 and a standard deviation of 1. The total area under the density curve of a normal distribution is 1, so there is a convenient correspondence between areas and probabilities. We presented methods for finding areas (or probabilities) that correspond to standard z scores, and we presented important methods for finding standard z scores that correspond to known areas (or probabilities). Values of areas and z scores can be found using Table A-2 or a TI-83/84 Plus calculator or computer software.

Section 6-3   In Section 6-3 we extended the methods from Section 6-2 so that we could work with any normal distribution, not just the standard normal distribution. We presented the standard score $z = (x - \mu)/\sigma$ for solving problems such as these:

• Given that IQ scores are normally distributed with $\mu = 100$ and $\sigma = 15$, find the probability of randomly selecting someone with an IQ above 90.

• Given that IQ scores are normally distributed with $\mu = 100$ and $\sigma = 15$, find the IQ score separating the bottom 85% from the top 15%.

Section 6-4   In Section 6-4 we introduced the concept of a sampling distribution of a statistic. The sampling distribution of the mean is the probability distribution of sample means, with all samples having the same sample size n. The sampling distribution of the proportion is the probability distribution of sample proportions, with all samples having the same sample size n. In general, the sampling distribution of any statistic is the probability distribution of that statistic.

Section 6-5   In Section 6-5 we presented the following conclusions associated with the central limit theorem:

1. The distribution of sample means $\bar{x}$ will, as the sample size n increases, approach a normal distribution.

2. The mean of the sample means is the population mean μ.

3. The standard deviation of the sample means is $\sigma/\sqrt{n}$.

Section 6-6   In Section 6-6 we noted that a normal distribution can sometimes approximate a binomial probability distribution. If both $np \geq 5$ and $nq \geq 5$, the binomial random variable x is approximately normally distributed with the mean and standard deviation given as $\mu = np$ and $\sigma = \sqrt{npq}$. Because the binomial probability distribution deals with discrete data and the normal distribution deals with continuous data, we apply the continuity correction, which should be used in normal approximations to binomial distributions.

Section 6-7   In Section 6-7 we presented procedures for determining whether sample data appear to come from a population that has a normal distribution. Some of the statistical methods covered later in this book have a loose requirement of a normally distributed population. In such cases, examination of a histogram and outliers might be all that is needed. In other cases, normal quantile plots might be necessary because of factors such as a small sample or a very strict requirement that the population must have a normal distribution.

Statistical Literacy and Critical Thinking

1. Normal Distribution What is a normal distribution? What is a standard normal distribution?

2. Normal Distribution In a study of incomes of individual adults in the United States, it is observed that many people have no income or very small incomes, while there are very few people with extremely large incomes, so a graph of the incomes is skewed instead of being symmetric. A researcher states that because incomes are a normal occurrence, the distribution of incomes is a normal distribution. Is that statement correct? Why or why not?

3. Distribution of Sample Means In each of the past 50 years, a simple random sample of 36 new movies is selected, and the mean of the 36 movie lengths (in minutes) is calculated. What is the approximate distribution of those sample means?

4. Large Sample On one cruise of the ship Queen Elizabeth II, 17% of the passengers became ill from Norovirus. America Online conducted a survey about that incident and received 34,358 responses. Given that the sample is so large, can we conclude that this sample is representative of the population?

Chapter Quiz

1. Find the value of $z_{0.03}$.

2. A process consists of rolling a single die 100 times and finding the mean of the 100 outcomes. If that process is repeated many times, what is the approximate distribution of the resulting means? (uniform, normal, Poisson, binomial)

3. What are the values of μ and σ in the standard normal distribution?

4. For the standard normal distribution, find the area to the right of $z = 1.00$.

5. For the standard normal distribution, find the area between the z scores of -1.50 and 2.50.

In Exercises 6–10, assume that IQ scores are normally distributed with a mean of 100 and a standard deviation of 15.

6. Find the probability that a randomly selected person has an IQ score less than 115.

7. Find the probability that a randomly selected person has an IQ score greater than 118.

8. Find the probability that a randomly selected person has an IQ score between 88 and 112.

9. If 25 people are randomly selected, find the probability that their mean IQ score is less than 103.

10. If 100 people are randomly selected, find the probability that their mean IQ score is greater than 103.

Review Exercises

Heights. *In Exercises 1–4, assume that heights of men are normally distributed with a mean of 69.0 in. and a standard deviation of 2.8 in. Also assume that heights of women are normally distributed with a mean of 63.6 in. and a standard deviation of 2.5 in. (based on data from the National Health Survey).*

1. Bed Length A day bed is 75 in. long.

a. Find the percentage of men with heights that exceed the length of a day bed.

b. Find the percentage of women with heights that exceed the length of a day bed.

c. Based on the preceding results, comment on the length of a day bed.

2. Bed Length In designing a new bed, you want the length of the bed to equal or exceed the height of at least 95% of all men. What is the minimum length of this bed?

3. Designing Caskets The standard casket has an inside length of 78 in.

a. What percentage of men are too tall to fit in a standard casket, and what percentage of women are too tall to fit in a standard casket? Based on those results, does it appear that the standard casket size is adequate?

b. A manufacturer of caskets wants to reduce production costs by making smaller caskets. What inside length would fit all men except the tallest 1%?

4. Heights of Rockettes In order to have a precision dance team with a uniform appearance, height restrictions are placed on the famous Rockette dancers at New York's Radio City Music Hall. Because women have grown taller over the years, a more recent change now requires that a Rockette dancer must have a height between 66.5 in. and 71.5 in. What percentage of women meet this height requirement? Does it appear that Rockettes are taller than typical women?

5. Genetics Experiment In one of Mendel's experiments with plants, 1064 offspring consisted of 787 plants with long stems. According to Mendel's theory, 3/4 of the offspring plants should have long stems. Assuming that Mendel's proportion of 3/4 is correct, find the probability of getting 787 or fewer plants with long stems among 1064 offspring plants.

continued

Based on the result, is 787 offspring plants with long stems unusually low? What does the result imply about Mendel's claimed proportion of 3/4?

6. Sampling Distributions Assume that the following sample statistics were obtained from a simple random sample. Which of the following statements are true?

a. The sample mean $\bar{x}$ targets the population mean μ in the sense that the mean of all sample means is μ.

b. The sample proportion $\hat{p}$ targets the population proportion p in the sense that the mean of all sample proportions is p.

c. The sample variance s^2 targets the population variance σ^2 in the sense that the mean of all sample variances is σ^2.

d. The sample median targets the population median in the sense that the mean of all sample medians is equal to the population median.

e. The sample range targets the population range in the sense that the mean of all sample ranges is equal to the range of the population.

7. High Cholesterol Levels The serum cholesterol levels in men aged 18–24 are normally distributed with a mean of 178.1 and a standard deviation of 40.7. Units are in mg/100 mL, and the data are based on the National Health Survey.

a. If 1 man aged 18–24 is randomly selected, find the probability that his serum cholesterol level is greater than 260, a value considered to be "moderately high."

b. If 1 man aged 18–24 is randomly selected, find the probability that his serum cholesterol level is between 170 and 200.

c. If 9 men aged 18–24 are randomly selected, find the probability that their mean serum cholesterol level is between 170 and 200.

d. The Providence Health Maintenance Organization wants to establish a criterion for recommending dietary changes if cholesterol levels are in the top 3%. What is the cutoff for men aged 18–24?

8. Identifying Gender Discrimination Jennifer Jenson learns that the Newport Temp Agency has hired only 15 women among its last 40 new employees. She also learns that the pool of applicants is very large, with an equal number of qualified men and women. Find the probability that among 40 such applicants, the number of women is 15 or fewer. Based on the result, is there strong evidence to charge that the Newport Temp Agency is discriminating against women?

9. Critical Values

a. Find the standard z score with a cumulative area to its left of 0.6700.

b. Find the standard z score with a cumulative area to its right of 0.9960.

c. Find the value of $z_{0.025}$.

10. Sampling Distributions A large number of simple random samples of size $n = 85$ are obtained from a large population of birth weights having a mean of 3420 g and a standard deviation of 495 g. The sample mean $\bar{x}$ is calculated for each sample.

a. What is the approximate shape of the distribution of the sample means?

b. What is the expected mean of the sample means?

c. What is the expected standard deviation of the sample means?

11. Aircraft Safety Standards Under older Federal Aviation Administration rules, airlines had to estimate the weight of a passenger as 185 lb. (That amount is for an adult traveling in winter, and it includes 20 lb of carry-on baggage.) Current rules require an estimate of 195 lb. Men have weights that are normally distributed with a mean of 172 lb and a standard deviation of 29 lb.

a. If 1 adult male is randomly selected and is assumed to have 20 lb of carry-on baggage, find the probability that his total is greater than 195 lb.

b. If a Boeing 767-300 aircraft is full of 213 adult male passengers and each is assumed to have 20 lb of carry-on baggage, find the probability that the mean passenger weight (including carry-on baggage) is greater than 195 lb. Based on that probability, does a pilot have to be concerned about exceeding this weight limit?

 **12. Assessing Normality** Listed below are the weights (in grams) of a simple random sample of United States one-dollar coins (from Data Set 20 in Appendix B). Do those weights appear to come from a population that has a normal distribution? Why or why not?

8.1008	8.1072	8.0271	8.0813	8.0241	8.0510	7.9817	8.0954	8.0658	8.1238
8.1281	8.0307	8.0719	8.0345	8.0775	8.1384	8.1041	8.0894	8.0538	8.0342

Cumulative Review Exercises

 **1. Salaries of Coaches** Listed below are annual salaries (in thousands of dollars) for a simple random sample of NCAA Division 1-A head football coaches (based on data from the *New York Times*).

235	159	492	530	138	125	128	900	360	212

a. Find the mean $\bar{x}$ and express the result in dollars instead of thousands of dollars.

b. Find the median and express the result in dollars instead of thousands of dollars.

c. Find the standard deviation s and express the result in dollars instead of thousands of dollars.

d. Find the variance s^2 and express the result in appropriate units.

e. Convert the first salary of $235,000 to a z score.

f. What level of measurement (nominal, ordinal, interval, ratio) describes this data set?

g. Are the salaries discrete data or continuous data?

2. Sampling

a. What is a simple random sample?

b. What is a voluntary response sample, and why is it generally unsuitable for statistical purposes?

 **3. Clinical Trial of Nasonex** In a clinical trial of the allergy drug Nasonex, 2103 adult patients were treated with Nasonex and 14 of them developed viral infections.

a. If two different adults are randomly selected from the treatment group, what is the probability that they both developed viral infections?

b. Assuming that the same proportion of viral infections applies to all adults who use Nasonex, find the probability that among 5000 randomly selected adults who use Nasonex, at least 40 develop viral infections.

c. Based on the result from part (b), is 40 an unusually high number of viral infections? Why or why not?

d. Do the given results (14 viral infections among 2103 adult Nasonex users) suggest that viral infections are an adverse reaction to the Nasonex drug? Why or why not?

4. Graph of Car Mileage The accompanying graph depicts the fuel consumption (in miles per gallon) for highway conditions of three cars. Does the graph depict the data fairly, or does it somehow distort the data? Explain.

5. Left-Handedness According to data from the American Medical Association, 10% of us are left-handed.

a. If three people are randomly selected, find the probability that they are all left-handed.

b. If three people are randomly selected, find the probability that at least one of them is left-handed.

c. Why can't we solve the problem in part (b) by using the normal approximation to the binomial distribution?

d. If groups of 50 people are randomly selected, what is the mean number of left-handed people in such groups?

e. If groups of 50 people are randomly selected, what is the standard deviation for the numbers of left-handed people in such groups?

f. Would it be unusual to get 8 left-handed people in a randomly selected group of 50 people? Why or why not?

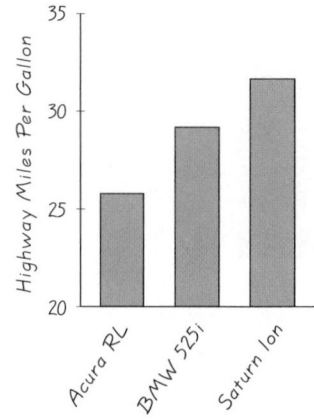

TI-83/84 Plus Calculator Project

Example 2 in Section 6-3 involved birth weights. The solution showed that 95.00% of babies in the United States are born with weights between 2450 g and 4390 g. In that example it was assumed that birth weights are normally distributed with $\mu = 3420$ g and $\sigma = 495$ g. This project describes a different method of solution that is based on a *simulation* technique. We will use the TI-83/84 Plus calculator to randomly generate 500 birth weights from a normally distributed population with $\mu = 3420$ g and $\sigma = 495$ g. Then we will find the percentage of those simulated weights that are between 2450 g and 4390 g. Here is the TI-83/84 Plus procedure:

1. Generate 500 values from a normally distributed population with $\mu = 3420$ g and $\sigma = 495$ g. This can be accomplished by pressing **MATH**, selecting **PRB,** then selecting the menu item of **randNorm** (random values from a normal distribution).

2. The format of the randNorm command requires the mean, followed by the standard deviation, followed by the number of desired values, all separated by commas. To generate 500 values from a normally distributed population with $\mu = 3420$ g and $\sigma = 495$ g, enter the command as **randNorm(3420, 495, 500).**

3. Enter **STO →**L1 to store the data in list L1. (**STO →**L1 can be entered by pressing **STO⟩** **2ND** **1**.)

4. Arrange the data in order by using this procedure: Press **STAT**, select **Sort(A** and enter L1 to obtain the command of **Sort(L1).** (Remember that L1 is entered by pressing **2ND** **1**.)

5. The sorted entries in list L1 can now be examined by pressing **STAT** **ENTER**. Visually examine the entries in list L1 and find the number of simulated birth weights that are between 2450 g and 4390 g, and divide that number by 500 to find the percentage of simulated values that are between 2450 g and 4390 g. Compare the result to the theoretical value of 95.00% that was found in Example 2 in Section 6-3.

6. Write a brief statement describing how the simulation provides an alternative solution to the method used in Example 2 in Section 6-3.

INTERNET PROJECT

Exploring the Central Limit Theorem

Go to: **http://www.aw.com/triola**

The central limit theorem is one of the most important results in statistics. It also may be one of the most surprising. Informally, the central limit theorem says that the normal distribution is everywhere. No matter what probability distribution underlies an experiment, there is a corresponding distribution of means that will be approximately normal in shape.

The best way to both understand and appreciate the central limit theorem is to see it in action. The Internet Project for this chapter will allow you to do just that. You will examine the central limit theorem from both a theoretical and a practical point of view. First, simulations found on the Internet will help you understand the theorem itself. Second, you will see how the theorem is key to such common activities as conducting polls and predicting election outcomes.

APPLET PROJECT

The CD included with this book contains applets designed to help visualize various concepts. Open the Applets folder on the CD and double-click on **Start.** Select the menu item of **Sampling Distributions.** Use the applet to compare the sampling distributions of the mean and the median in terms of center and spread for bell-shaped and skewed distributions. Write a brief report describing your results.

FROM DATA TO DECISION

Critical Thinking: Designing aircraft seating

In this project we consider the issue of determining the "sitting distance" shown in Figure 6-24(a). We define the sitting distance to be the length between the back of the seat cushion and the seat in front. To determine the sitting distance, we must take into account human body measurements. Specifically, we must consider the "buttock-to-knee length," as shown in Figure 6-24(b). Determining the sitting distance for an aircraft is extremely important. If the sitting distance is unnecessarily large, rows of seats might need to be eliminated. It has been estimated that removing a single row of six seats can cost around $8 million over the life of the aircraft. If the sitting distance is too small, passengers might be uncomfortable and might prefer to fly other aircraft, or their safety might be compromised because of their limited mobility.

In determining the sitting distance in our aircraft, we will use previously collected data from measurements of large numbers of people. Results from those measurements are summarized in the given table. We can use the data in the table to determine the required sitting distance, but we must make some hard choices. If we are to accommodate *everyone* in the population, we will have a sitting distance that is so costly in terms of reduced seating that it might not be economically feasible. Some questions we must ask ourselves are: (1) What percentage of the population are we willing to exclude? (2) How much extra room do we want to provide for passenger comfort and safety? Use the available information to determine the sitting distance. Identify the choices and decisions that were made in that determination.

Buttock-to-Knee Length (inches)

	Mean	Standard Deviation	Distribution
Males	23.5 in.	1.1 in.	Normal
Females	22.7 in.	1.0 in.	Normal

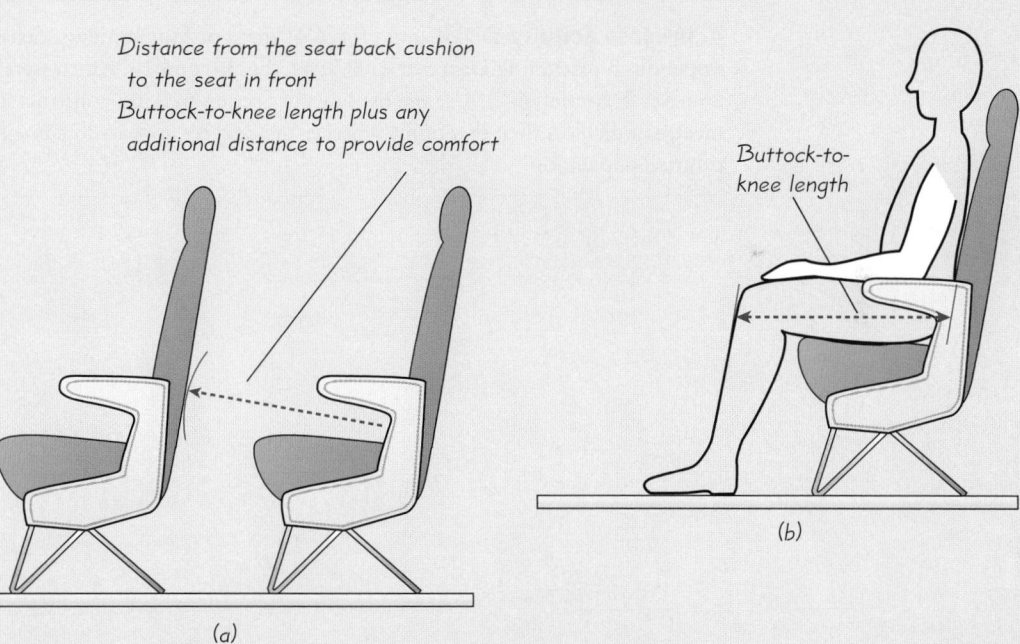

Distance from the seat back cushion to the seat in front
Buttock-to-knee length plus any additional distance to provide comfort

Buttock-to-knee length

(a)

(b)

Figure 6-24 Sitting Distance and Buttock-to-Knee Length

Cooperative Group Activities

1. In-class activity Divide into groups of three or four students and address these issues affecting the design of manhole covers.

• Which of the following is most relevant for determining whether a manhole cover diameter of 24 in. is large enough: weights of men, weights of women, heights of men, heights of women, hip breadths of men, hip breadths of women, shoulder breadths of men, shoulder breadths of women?

• Why are manhole covers usually round? (This was once a popular interview question asked of applicants at IBM, and there are at least three good answers. One good answer is sufficient here.)

2. Out-of-class activity Divide into groups of three or four students. In each group, develop an original procedure to illustrate the central limit theorem. The main objective is to show that when you randomly select samples from a population, the means of those samples tend to be *normally* distributed, regardless of the nature of the population distribution. For this illustration, begin with some population of values that does not have a normal distribution.

3. In-class activity Divide into groups of three or four students. Using a coin to simulate births, each individual group member should simulate 25 births and record the number of simulated girls. Combine all results from the group and record n = total number of births and x = number of girls. Given batches of n births, compute the mean and standard deviation for the number of girls. Is the simulated result usual or unusual? Why?

4. In-class activity Divide into groups of three or four students. Select a set of data from Appendix B (excluding Data Sets 1, 8, 9, 11, 12, 14, and 16, which were used in examples or exercises in Section 6-7). Use the methods of Section 6-7 and construct a histogram and normal quantile plot, then determine whether the data set appears to come from a normally distributed population.

NAME:	Barbara Carvalho
JOB:	Director of the Marist College Poll
NAME:	Lee Miringoff
JOB:	Director of the Marist College Institute for Public Opinion

*B*arbara Carvalho is Director of the Marist College Poll, and Lee Miringoff is Director of the Marist College Institute for Public Opinion. They report on their poll results in many interviews for print and electronic media, including news programs for NBC, CBS, ABC, Fox, and public television. Lee Miringoff appears regularly on NBC's Today show.

Q: What do you do?

A: We do public polling. We survey public issues, approval ratings of public officials in New York City, New York State, and nationwide. We don't do partisan polling for political parties, political candidates, or lobby groups. We are independently funded by Marist College and we have no outside funding that in any way might suggest that we are doing research for any particular group on any one issue.

Q: How do you select survey respondents?

A: For a statewide survey we select respondents in proportion to county voter registrations. Different counties have different refusal rates and if we were to select people at random throughout

the state, we would get an uneven model of what the state looks like. We stratify by county and use random digit dialing so that we get listed and unlisted numbers.

Q: You mentioned refusal rates. Are they a real problem?

A: One of the issues that we deal with extensively is the issue of people who don't respond to surveys. That has been increasing over time and there has been much attention from the survey research community. As a research center we do quite well when compared to others. But when you do face-to-face interviews and have refusal rates of 25% to 50%, there's a real concern to find out who is refusing and why they are not responding, and the impact that has on the representativeness of the studies that we're doing.

Q: Would you recommend a statistics course for students?

A: Absolutely. All numbers are not created equally. Regardless of your field of study or career interests, an ability to critically evaluate research information that is presented to you, to use data to improve services, or to interpret results to develop strategies is a very valuable asset. Surveys, in particular, are everywhere. It is vital that as workers, managers, and citizens we are able to evaluate their accuracy and worth. Statistics cut across disciplines. Students will inevitably find it in their careers at some point.

Q: Do you have any other recommendations for students?

A: It is important for students to take every opportunity to develop their communication and presentation skills. Sharpen not only your ability to speak and write, but also raise your comfort level with new technologies.

7-1 Review and Preview

7-2 Estimating a Population Proportion

7-3 Estimating a Population Mean: σ Known

7-4 Estimating a Population Mean: σ Not Known

7-5 Estimating a Population Variance

7

Estimates and Sample Sizes

How do we interpret a poll about global warming?

Global warming is the increase in the mean temperature of air near the surface of the earth and the increase in the mean temperature of the oceans. Scientists generally agree that global warming is caused by increased amounts of carbon dioxide, methane, ozone, and other gases that result from human activity.

Global warming is believed to be responsible for the retreat of glaciers, the reduction in the Arctic region, and a rise in sea levels. It is feared that continued global warming will result in even higher sea levels, flooding, drought, and more severe weather.

Because global warming appears to have the potential for causing dramatic changes in our environment, it is critical that we recognize that potential. Just how much do we all recognize global warming? In a Pew Research Center poll, respondents were asked "From what you've read and heard, is there solid evidence that the average temperature on earth has been increasing over the past few decades, or not?" In response to that question, 70% of 1501 randomly selected adults in the United States answered "yes." Therefore, among those polled, 70% believe in global warming. Although the subject matter of this poll has great significance, we will focus on the interpretation and analysis of the results. Some important issues that relate to this poll are as follows:

- How can the poll results be used to estimate the percentage of all adults in the United States who believe that the earth is getting warmer?
- How accurate is the result of 70% likely to be?
- Given that only 1501/225,139,000, or 0.0007% of the adult population in the United States were polled, is the sample size too small to be meaningful?
- Does the method of selecting the people to be polled have much of an effect on the results?

We can answer the last question based on the sound sampling methods discussed in Chapter 1. The method of selecting the people to be polled most definitely has an effect on the results. The results are likely to be poor if a convenience sample or some other nonrandom sampling method is used. If the sample is a simple random sample, the results are likely to be good.

Our ability to understand polls and to interpret the results is crucial for our role as citizens. As we consider the topics of this chapter, we will learn more about polls and surveys and how to correctly interpret and present results.

Review and Preview

In Chapters 2 and 3 we used "descriptive statistics" when we summarized data using tools such as graphs, and statistics such as the mean and standard deviation. We use "inferential statistics" when we use sample data to make inferences about population parameters. Two major activities of inferential statistics are (1) to use sample data to estimate values of population parameters (such as a population proportion or population mean), and (2) to test hypotheses or claims made about population parameters. In this chapter we begin working with the true core of inferential statistics as we use sample data to estimate values of population parameters. For example, the Chapter Problem refers to a poll of 1501 adults in the United States, and we see that 70% of them believe that the earth is getting warmer. Based on the sample statistic of 70%, we will estimate the percentage of *all* adults in the United States who believe that the earth is getting warmer. In so doing, we are using the sample results to make an inference about the population.

This chapter focuses on the use of sample data to estimate a population parameter, and Chapter 8 will introduce the basic methods for testing claims (or hypotheses) that have been made about a population parameter.

Because Sections 7-2 and 7-3 use *critical values,* it is helpful to review this notation introduced in Section 6-2: z_α denotes the z score with an area of α to its right. (α is the Greek letter alpha.) See Example 8 in Section 6-2, where it is shown that if $\alpha = 0.025$, the critical value is $z_{0.025} = 1.96$. That is, the critical value of $z_{0.025} = 1.96$ has an area of 0.025 to its right.

Estimating a Population Proportion

Key Concept In this section we present methods for using a *sample proportion* to estimate a *population proportion.* There are three main ideas that we should know and understand in this section.

- The sample proportion is the best *point estimate* of the population proportion.

- We can use a sample proportion to construct a *confidence interval* to estimate the true value of a population proportion, and we should know how to interpret such confidence intervals.

- We should know how to find the sample size necessary to estimate a population proportion.

The concepts presented in this section are used in the following sections and chapters, so it is important to understand this section quite well.

Proportion, Probability, and Percent Although this section focuses on the population proportion p, we can also work with probabilities or percentages. In the Chapter Problem, for example, it was noted that 70% of those polled believe in global warming. The sample statistic of 70% can be expressed in decimal form as 0.70, so the sample proportion is $\hat{p} = 0.70$. (Recall from Section 6-4 that p represents the *population proportion,* and $\hat{p}$ is used to denote the *sample proportion.*)

Point Estimate If we want to estimate a population proportion with a single value, the best estimate is the sample proportion $\hat{p}$. Because $\hat{p}$ consists of a single value, it is called a *point estimate.*

DEFINITION

A **point estimate** is a single value (or point) used to approximate a population parameter.

The sample proportion $\hat{p}$ is the best point estimate of the population proportion p.

We use $\hat{p}$ as the point estimate of p because it is unbiased and it is the most consistent of the estimators that could be used. It is unbiased in the sense that the distribution of sample proportions tends to center about the value of p; that is, sample proportions $\hat{p}$ do not systematically tend to underestimate or overestimate p. (See Section 6-4.) The sample proportion $\hat{p}$ is the most consistent estimator in the sense that the standard deviation of sample proportions tends to be smaller than the standard deviation of any other unbiased estimators.

EXAMPLE 1 **Proportion of Adults Believing in Global Warming** In the Chapter Problem we noted that in a Pew Research Center poll, 70% of 1501 randomly selected adults in the United States believe in global warming, so the sample proportion is $\hat{p} = 0.70$. Find the best point estimate of the proportion of *all* adults in the United States who believe in global warming.

SOLUTION Because the sample proportion is the best point estimate of the population proportion, we conclude that the best point estimate of p is 0.70. When using the sample results to estimate the percentage of all adults in the United States who believe in global warming, the best estimate is 70%.

Why Do We Need Confidence Intervals?

In Example 1 we saw that 0.70 was our *best* point estimate of the population proportion p, but we have no indication of just how *good* our best estimate is. Because a point estimate has the serious flaw of not revealing anything about how good it is, statisticians have cleverly developed another type of estimate. This estimate, called a *confidence interval* or *interval estimate,* consists of a range (or an interval) of values instead of just a single value.

DEFINITION

A **confidence interval** (or **interval estimate**) is a range (or an interval) of values used to estimate the true value of a population parameter. A confidence interval is sometimes abbreviated as CI.

A confidence interval is associated with a confidence level, such as 0.95 (or 95%). The confidence level gives us the success rate of the procedure used to construct the confidence interval. The confidence level is often expressed as the probability or area $1 - \alpha$ (lowercase Greek alpha), where α is the complement of the *confidence level.* For a 0.95 (or 95%) confidence level, $\alpha = 0.05$. For a 0.99 (or 99%) confidence level, $\alpha = 0.01$.

Curbstoning

The glossary for the Census defines *curbstoning as* "the practice by which a census enumerator fabricates a question- naire for a residence without actually visiting it." Curbstoning occurs when a census enu- merator sits on a curbstone (or anywhere else) and fills out survey forms by making up responses. Because data from curbstoning are not real, they can affect the validity of the Census. The extent of curbstoning has been investigated in several studies, and one study showed that about 4% of Census enumerators prac- ticed curbstoning at least some of the time.

The methods of Section 7-2 assume that the sample data have been collected in an appropriate way, so if much of the sample data have been obtained through curbstoning, then the re- sulting confidence interval estimates might be very flawed.

> **DEFINITION**
>
> The **confidence level** is the probability $1 - \alpha$ (often expressed as the equiv- alent percentage value) that the confidence interval actually does contain the population parameter, assuming that the estimation process is repeated a large number of times. (The confidence level is also called the **degree of confidence,** or the **confidence coefficient.**)

The most common choices for the confidence level are 90% (with $\alpha = 0.10$), 95% (with $\alpha = 0.05$), and 99% (with $\alpha = 0.01$). The choice of 95% is most com- mon because it provides a good balance between precision (as reflected in the width of the confidence interval) and reliability (as expressed by the confidence level).

Here's an example of a confidence interval found later (in Example 3), which is based on the sample data of 1501 adults polled, with 70% of them saying that they believe in global warming:

The 0.95 (or 95%) confidence interval estimate of the population propor- tion p is $0.677 < p < 0.723$.

It's common for a media report to include a statement such as this: "Based on a Pew Research Center poll, the proportion of adults believing in global warming is esti- mated to be 70%, with a margin of error of 2 percentage points." (We will discuss the margin of error later in this section.) Note that the confidence level is not mentioned. Although the confidence level should be given when reporting information about a poll, the media usually fail to include it.

Interpreting a Confidence Interval

We must be careful to interpret confidence intervals correctly. There is a correct inter- pretation and many different and creative incorrect interpretations of the confidence interval $0.677 < p < 0.723$.

Correct: "We are 95% confident that the interval from 0.677 to 0.723 actu- ally does contain the true value of the population proportion p." This means that if we were to select many different samples of size 1501 and construct the corresponding confidence intervals, 95% of them would actually contain the value of the population proportion p. (Note that in this correct interpretation, the level of 95% refers to the success rate of the *process* being used to estimate the proportion.)

Incorrect: "There is a 95% chance that the true value of p will fall between 0.677 and 0.723." It would also be incorrect to say that "95% of sample proportions fall between 0.677 and 0.723."

CAUTION

Know the correct interpretation of a confidence interval, as given above.

At any specific point in time, a population has a fixed and constant value p, and a confidence interval constructed from a sample either includes p or does not. Simi- larly, if a baby has just been born and the doctor is about to announce its gender, it's incorrect to say that there is a probability of 0.5 that the baby is a girl; the baby is a girl or is not, and there's no probability involved. A population proportion p is like the baby that has been born—the value of p is fixed, so the confidence interval limits ei- ther contain p or do not, and that is why it's incorrect to say that there is a 95% chance that p will fall between values such as 0.677 and 0.723.

A confidence level of 95% tells us that the *process* we are using will, in the long run, result in confidence interval limits that contain the true population proportion 95% of the time. Suppose that the true proportion of all adults who believe in global warming is $p = 0.75$. Then the confidence interval obtained from the Pew Research Center poll does not contain the population proportion, because the true population proportion 0.75 is not between 0.677 and 0.723. This is illustrated in Figure 7-1. Figure 7-1 shows typical confidence intervals resulting from 20 different samples. With 95% confidence, we expect that 19 out of 20 samples should result in confidence intervals that contain the true value of p, and Figure 7-1 illustrates this with 19 of the confidence intervals containing p, while one confidence interval does not contain p.

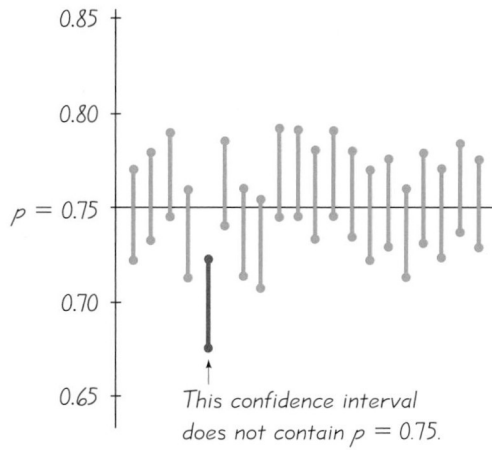

Figure 7-1

Confidence Intervals from 20 Different Samples

CAUTION

Confidence intervals can be used informally to compare different data sets, but *the overlapping of confidence intervals should not be used for making formal and final conclusions about equality of proportions.* (See "On Judging the Significance of Differences by Examining the Overlap Between Confidence Intervals," by Schenker and Gentleman, *American Statistician,* Vol. 55, No. 3.)

Critical Values

The methods of this section (and many of the other statistical methods found in the following chapters) include reference to a standard z score that can be used to distinguish between sample statistics that are likely to occur and those that are unlikely to occur. Such a z score is called a *critical value.* (Critical values were first presented in Section 6-2, and they are formally defined below.) Critical values are based on the following observations:

1. Under certain conditions, the sampling distribution of sample proportions can be approximated by a normal distribution, as shown in Figure 7-2.

2. A z score associated with a sample proportion has a probability of $\alpha/2$ of falling in the right tail of Figure 7-2.

3. The z score separating the right-tail region is commonly denoted by $z_{\alpha/2}$, and is referred to as a *critical value* because it is on the borderline separating z scores from sample proportions that are likely to occur from those that are unlikely to occur.

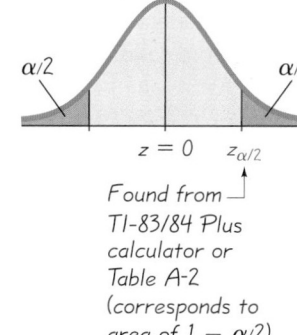

Figure 7-2 Critical Value $z_{\alpha/2}$ in the Standard Normal Distribution

DEFINITION

A **critical value** is the number on the borderline separating sample statistics that are likely to occur from those that are unlikely to occur. The number $z_{\alpha/2}$ is a critical value that is a z score with the property that it separates an area of $\alpha/2$ in the right tail of the standard normal distribution (as in Figure 7-2).

EXAMPLE 2 **Finding a Critical Value** Find the critical value $z_{\alpha/2}$ corresponding to a 95% confidence level.

SOLUTION A 95% confidence level corresponds to $\alpha = 0.05$. Figure 7-3 shows that the area in each of the red-shaded tails is $\alpha/2 = 0.025$. We find $z_{\alpha/2} = 1.96$ by noting that the cumulative area to its left must be $1 - 0.025$, or 0.975.

Using a TI-83/84 Plus Calculator: We want to find the z score with an area of 0.975 to its left, so press **2ND** **VARS** to get to the DISTR (distribution) menu, then select **invNorm.** Proceed to enter 0.975, which is the *cumulative area to the left* of z. Enter the command **invNorm(0.975)** to get 1.959963986, which can be rounded to 1.96.

Using Table A-2: If using Table A-2, we refer to the body of the table (because 0.9750 is an area) and we find that the area of 0.9750 corresponds to $z = 1.96$. For a 95% confidence level, the critical value is therefore $z_{\alpha/2} = 1.96$.

Example 2 showed that a 95% confidence level results in a critical value of $z_{\alpha/2} = 1.96$. This is the most common critical value, and it is listed with two other common values in the table that follows.

Confidence Level	α	Critical Value, $z_{\alpha/2}$
90%	0.10	1.645
95%	0.05	1.96
99%	0.01	2.575

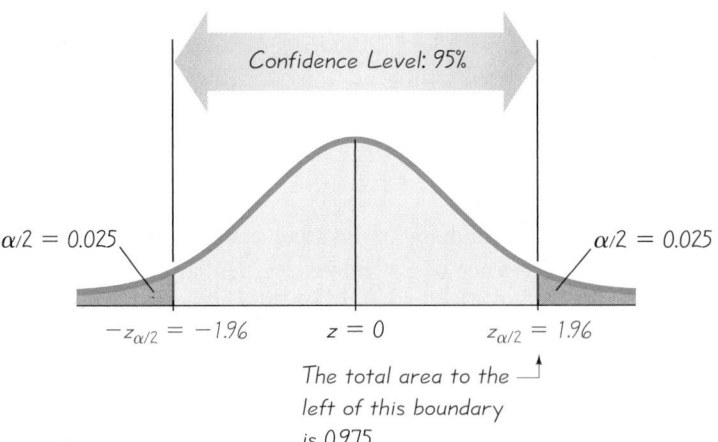

Figure 7-3 **Finding $z_{\alpha/2}$ for a 95% Confidence Level**

Margin of Error

When we collect sample data that result in a sample proportion, such as the Pew Research Center poll given in the Chapter Problem (with 70% of 1501 respondents believing in global warming), we can calculate the sample proportion $\hat{p}$. Because of random variation in samples, the sample proportion is typically different from the population proportion. The difference between the sample proportion and the population proportion can be thought of as an error. We now define the *margin of error E* as follows.

DEFINITION

When data from a simple random sample are used to estimate a population proportion p, the **margin of error,** denoted by E, is the maximum likely difference (with probability $1 - \alpha$, such as 0.95) between the observed sample proportion $\hat{p}$ and the true value of the population proportion p. The margin of error E is also called the *maximum error of the estimate* and can be found by multiplying the critical value and the standard deviation of sample proportions, as shown in Formula 7-1.

Formula 7-1 $E = z_{\alpha/2}\sqrt{\dfrac{\hat{p}\hat{q}}{n}}$ margin of error for proportions

For a 95% confidence level, $\alpha = 0.05$, so there is a probability of 0.05 that the sample proportion will be in error by more than E. This property is generalized in the following box.

Confidence Interval for Estimating a Population Proportion p

Objective

Construct a confidence interval used to estimate a population proportion.

Notation

p = population proportion

$\hat{p}$ = sample proportion

n = number of sample values

E = margin of error

$z_{\alpha/2}$ = z score separating an area of $\alpha/2$ in the right tail of the standard normal distribution

$\hat{q} = 1 - \hat{p}$

Requirements

1. The sample is a simple random sample. (*Caution:* If the sample data have been obtained in a way that is not appropriate, the estimates of the population proportion may be very wrong.)

2. The conditions for the binomial distribution are satisfied. That is, there is a fixed number of trials, the trials are independent, there are two categories of outcomes, and the probabilities remain constant for each trial. (See Section 5-3.)

3. There are at least 5 successes and at least 5 failures. (With the population proportions p and q unknown, we estimate their values using the sample proportion, so this requirement is a way of verifying that $np \geq 5$ and $nq \geq 5$ are both satisfied, so the normal distribution is a suitable approximation to the binomial distribution. There are procedures for dealing with situations in which the normal distribution is not a suitable approximation, as in Exercise 51.)

Confidence Interval

$$\hat{p} - E < p < \hat{p} + E \quad \text{where} \quad E = z_{\alpha/2}\sqrt{\dfrac{\hat{p}\hat{q}}{n}}$$

The confidence interval is often expressed in the following equivalent formats:

$$\hat{p} \pm E$$

or

$$(\hat{p} - E, \hat{p} + E)$$

In Chapter 4, when probabilities were given in decimal form, we rounded to three significant digits. We use that same rounding rule here.

Round-Off Rule for Confidence Interval Estimates of *p*

Round the confidence interval limits for *p* to three significant digits.

We now summarize the procedure for constructing a confidence interval estimate of a population proportion *p*:

Procedure for Constructing a Confidence Interval for *p*

1. Verify that the requirements are satisfied.

2. Refer to Table A-2 or use technology to find the critical value $z_{\alpha/2}$ that corresponds to the desired confidence level.

3. Evaluate the margin of error $E = z_{\alpha/2}\sqrt{\hat{p}\hat{q}/n}$.

4. Using the value of the calculated margin of error E and the value of the sample proportion $\hat{p}$, find the values of the *confidence interval limits* $\hat{p} - E$ and $\hat{p} + E$. Substitute those values in the general format for the confidence interval:

$$\hat{p} - E < p < \hat{p} + E$$

or

$$\hat{p} \pm E$$

or

$$(\hat{p} - E, \hat{p} + E)$$

5. Round the resulting confidence interval limits to three significant digits.

> **EXAMPLE 3**   **Constructing a Confidence Interval: Poll Results** In the Chapter Problem we noted that a Pew Research Center poll of 1501 randomly selected U.S. adults showed that 70% of the respondents believe in global warming. The sample results are $n = 1501$, and $\hat{p} = 0.70$.
>
> **a.** Find the margin of error E that corresponds to a 95% confidence level.
>
> **b.** Find the 95% confidence interval estimate of the population proportion p.
>
> **c.** Based on the results, can we safely conclude that the majority of adults believe in global warming?
>
> **d.** Assuming that you are a newspaper reporter, write a brief statement that accurately describes the results and includes all of the relevant information.

SOLUTION   **REQUIREMENT CHECK** We first verify that the necessary requirements are satisfied. (1) The polling methods used by the Pew Research Center result in samples that can be considered to be simple random samples. (2) The conditions for a binomial experiment are satisfied, because there is a fixed number of trials (1501), the trials are independent (because the response from one person doesn't affect the probability of the response from another person), there are two categories of outcome (subject believes in global warming or does not), and the probability remains constant. Also, with 70% of the respondents believing in global warming, the number who believe is 1051 (or 70% of 1501) and the number who do not believe is 450, so the number of successes (1051) and the number of failures (450) are both at least 5. The check of requirements has been successfully completed. ✓

a. The margin of error is found by using Formula 7-1 with $z_{\alpha/2} = 1.96$ (as found in Example 2), $\hat{p} = 0.70$, $\hat{q} = 0.30$, and $n = 1501$.

$$E = z_{\alpha/2}\sqrt{\frac{\hat{p}\hat{q}}{n}} = 1.96\sqrt{\frac{(0.70)(0.30)}{1501}} = 0.023183$$

b. Constructing the confidence interval is quite easy now that we know the values of $\hat{p}$ and E. We simply substitute those values to obtain this result:

$$\hat{p} - E < p < \hat{p} + E$$
$$0.70 - 0.023183 < p < 0.70 + 0.023183$$
$$0.677 < p < 0.723 \quad \text{(rounded to three significant digits)}$$

TI-83/84 Plus

```
1-PropZInt
(.67702,.72338)
p̂=.7001998668
n=1501
```

This same result could be expressed in the format of 0.70 ± 0.023 or $(0.677, 0.723)$. If we use a TI-83/84 Plus calculator, the confidence interval is given in the format of $(0.677, 0.723)$, as shown in the margin. (See the procedure given at the end of this section.)

   If we want the 95% confidence interval for the true population *percentage,* we could express the result as $67.7\% < p < 72.3\%$.

c. Based on the confidence interval obtained in part (b), it does appear that the proportion of adults who believe in global warming is greater than 0.5 (or 50%), so we can safely conclude that the majority of adults believe in global warming. Because the limits of 0.677 and 0.723 are likely to contain the true population proportion, it appears that the population proportion is a value greater than 0.5.

d. Here is one statement that summarizes the results: 70% of United States adults believe that the earth is getting warmer. That percentage is based on a Pew Research Center poll of 1501 randomly selected adults in the United States. In theory, in 95% of such polls, the percentage should differ by no more than 2.3 percentage points in either direction from the percentage that would be found by interviewing all adults in the United States.

Analyzing Polls Example 3 addresses the poll described in the Chapter Problem. When analyzing results from polls, we should consider the following.

1. The sample should be a simple random sample, not an inappropriate sample (such as a voluntary response sample).

2. The confidence level should be provided. (It is often 95%, but media reports often neglect to identify it.)

3. The sample size should be provided. (It is usually provided by the media, but not always.)

4. Except for relatively rare cases, the quality of the poll results depends on the sampling method and the size of the sample, but the size of the population is usually not a factor.

CAUTION

Never follow the common misconception that poll results are unreliable if the sample size is a small percentage of the population size. The population size is usually not a factor in determining the reliability of a poll.

Determining Sample Size

Suppose we want to collect sample data in order to estimate some population proportion. How do we know *how many* sample items must be obtained? If we solve the formula for the margin of error E (Formula 7-1) for n, we get Formula 7-2. Formula 7-2 requires $\hat{p}$ as an estimate of the population proportion p, but if no such estimate is known (as is often the case), we replace $\hat{p}$ by 0.5 and replace $\hat{q}$ by 0.5, with the result given in Formula 7-3.

Finding the Sample Size Required to Estimate a Population Proportion

Objective

Determine how large the sample should be in order to estimate the population proportion p.

Notation

p = population proportion

$\hat{p}$ = sample proportion

n = number of sample values

E = desired margin of error

$z_{\alpha/2}$ = z score separating an area of $\alpha/2$ in the right tail of the standard normal distribution

$\hat{q} = 1 - \hat{p}$

Requirements

The sample must be a simple random sample of independent subjects.

When an estimate $\hat{p}$ is known: **Formula 7-2** $n = \dfrac{[z_{\alpha/2}]^2 \hat{p}\hat{q}}{E^2}$

When no estimate $\hat{p}$ is known: **Formula 7-3** $n = \dfrac{[z_{\alpha/2}]^2 0.25}{E^2}$

If reasonable estimates of $\hat{p}$ can be made by using previous samples, a pilot study, or someone's expert knowledge, use Formula 7-2. If nothing is known about the value of $\hat{p}$, use Formula 7-3.

Formulas 7-2 and 7-3 are remarkable because they show that the sample size does not depend on the size (N) of the population; the sample size depends on the desired confidence level, the desired margin of error, and sometimes the known estimate of $\hat{p}$. (See Exercise 49 for dealing with cases in which a relatively large sample is selected without replacement from a finite population.)

Round-Off Rule for Determining Sample Size

If the computed sample size n is not a whole number, round the value of n up to the next *larger* whole number.

EXAMPLE 4   **How Many Adults Use the Internet?** The Internet is affecting us all in many different ways, so there are many reasons for estimating the proportion of adults who use it. Assume that a manager for E-Bay wants to determine the current percentage of U.S. adults who now use the Internet. How many adults must be surveyed in order to be 95% confident that the sample percentage is in error by no more than three percentage points?

a. Use this result from a Pew Research Center poll: In 2006, 73% of U.S. adults used the Internet.

b. Assume that we have no prior information suggesting a possible value of the proportion.

SOLUTION

a. The prior study suggests that $\hat{p} = 0.73$, so $\hat{q} = 0.27$ (found from $\hat{q} = 1 - 0.73$). With a 95% confidence level, we have $\alpha = 0.05$, so $z_{\alpha/2} = 1.96$. Also, the margin of error is $E = 0.03$ (the decimal equivalent of "three percentage points"). Because we have an estimated value of $\hat{p}$ we use Formula 7-2 as follows:

$$n = \frac{[z_{\alpha/2}]^2 \hat{p}\hat{q}}{E^2} = \frac{[1.96]^2(0.73)(0.27)}{0.03^2}$$

$$= 841.3104 = 842 \qquad \text{(rounded up)}$$

We must obtain a simple random sample that includes at least 842 adults.

b. As in part (a), we again use $z_{\alpha/2} = 1.96$ and $E = 0.03$, but with no prior knowledge of $\hat{p}$ (or $\hat{q}$), we use Formula 7-3 as follows:

$$n = \frac{[z_{\alpha/2}]^2 \cdot 0.25}{E^2} = \frac{[1.96]^2 \cdot 0.25}{0.03^2}$$

$$= 1067.1111 = 1068 \qquad \text{(rounded up)}$$

INTERPRETATION   To be 95% confident that our sample percentage is within three percentage points of the true percentage for all adults, we should obtain a simple random sample of 1068 adults. By comparing this result to the sample size of 842 found in part (a), we can see that if we have no knowledge of a prior study, a larger sample is required to achieve the same results as when the value of $\hat{p}$ can be estimated.

CAUTION

Try to avoid these two common errors when calculating sample size:

1. Don't make the mistake of using $E = 3$ as the margin of error corresponding to "three percentage points."

2. Be sure to substitute the critical z score for $z_{\alpha/2}$. For example, if you are working with 95% confidence, be sure to replace $z_{\alpha/2}$ with 1.96. Don't make the mistake of replacing $z_{\alpha/2}$ with 0.95 or 0.05.

Finding the Point Estimate and *E* from a Confidence Interval   Sometimes we want to better understand a confidence interval that might have been obtained from a journal article, or generated using computer software or a calculator. If we already know the confidence interval limits, the sample proportion (or the best point estimate) $\hat{p}$ and the margin of error E can be found as follows:

Point estimate of p:

$$\hat{p} = \frac{\text{(upper confidence interval limit)} + \text{(lower confidence interval limit)}}{2}$$

Margin of error:

$$E = \frac{(\text{upper confidence interval limit}) \ - \ (\text{lower confidence interval limit})}{2}$$

EXAMPLE 5   The article "High-Dose Nicotine Patch Therapy," by Dale, Hurt, et al. (*Journal of the American Medical Association,* Vol. 274, No. 17) includes this statement: "Of the 71 subjects, 70% were abstinent from smoking at 8 weeks (95% confidence interval [CI], 58% to 81%)." Use that statement to find the point estimate $\hat{p}$ and the margin of error E.

SOLUTION   From the given statement, we see that the 95% confidence interval is $0.58 < p < 0.81$. The point estimate $\hat{p}$ is the value midway between the upper and lower confidence interval limits, so we get

$$\hat{p} = \frac{(\text{upper confidence limit}) \ + \ (\text{lower confidence limit})}{2}$$

$$= \frac{0.81 + 0.58}{2} = 0.695$$

The margin of error can be found as follows:

$$E = \frac{(\text{upper confidence limit}) \ - \ (\text{lower confidence limit})}{2}$$

$$= \frac{0.81 - 0.58}{2} = 0.115$$

Better-Performing Confidence Intervals

Important note: The exercises for this section are based on the method for constructing a confidence interval as described above, not the confidence intervals described in the following discussion.

The confidence interval described in this section has the format typically presented in introductory statistics courses, but it does not perform as well as some other confidence intervals. The *adjusted Wald confidence interval* performs better in the sense that its probability of containing the true population proportion p is closer to the confidence level that is used. The adjusted Wald confidence interval uses this simple procedure: Add 2 to the number of successes x, add 2 to the number of failures (so that the number of trials n is increased by 4), then find the confidence interval as described in this section. For example, if we use the methods of this section with $x = 10$ and $n = 20$, we get this 95% confidence interval: $0.281 < p < 0.719$. With $x = 10$ and $n = 20$ we use the adjusted Wald confidence interval by letting $x = 12$ and $n = 24$ to get this confidence interval: $0.300 < p < 0.700$. The chance that the confidence interval $0.300 < p < 0.700$ contains p is closer to 95% than the chance that $0.281 < p < 0.719$ contains p.

Another confidence interval that performs better than the one described in this section and the adjusted Wald confidence interval is the *Wilson score confidence interval:*

$$\frac{\hat{p} + \dfrac{z_{\alpha/2}^2}{2n} \pm z_{\alpha/2}\sqrt{\dfrac{\hat{p}\hat{q} + \dfrac{z_{\alpha/2}^2}{4n}}{n}}}{1 + \dfrac{z_{\alpha/2}^2}{n}}$$

(It is easy to see why this approach is not used much in introductory courses.) Using $x = 10$ and $n = 20$, the 95% Wilson score confidence interval is $0.299 < p < 0.701$.

For a discussion of these and other confidence intervals for p, see "Approximation Is Better than 'Exact' for Interval Estimation of Binomial Proportions," by Agresti and Coull, *American Statistician,* Vol. 52, No. 2.

Using the TI-83/84 Plus Calculator for Confidence Interval Estimates of p

Here is the procedure for using the TI-83/84 Plus calculator for the construction of a confidence interval estimate of a population proportion p:

1. You must first identify the number of successes x and the number of trials n. (If you know only the sample size n and the sample proportion $\hat{p}$, the number of successes x can be calculated by using this expression: $x = n\hat{p}$. In Example 3, $n = 1501$ and $\hat{p} = 0.70$, so $x = (1501)(0.70) = 1050.7$. If the resulting value of x is not a whole number, round it to the nearest whole number. The result of 1050.7 is rounded to 1051.)

2. Press **STAT**, then use the ▷ key to select the option of **TESTS.**

3. The menu option of **1-PropZInt** identifies the confidence interval for the case involving one proportion (as described in this section), which is constructed using a critical z score. Select this option by pressing ▽ key to scroll down to **1-PropZInt.**

4. The screen requires an entry for x, an entry for n, and an entry for the C-Level (confidence level, such as 0.95). Enter those values, then scroll down to **Calculate** and press **ENTER**. The first screen in the margin shows the entries corresponding to Example 3. After pressing **ENTER**, the confidence interval will be expressed in a format with the confidence interval limits enclosed within parentheses, such as (.67702, .72338). See the second screen in the margin that corresponds to the results from Example 3. The TI-83/84 Plus confidence interval of (.67702, .72338) can also be expressed as $0.677 < p < 0.723$, as in Example 3.

TI-83/84 Plus

```
1-PropZInt
 x:1051
 n:1501
 C-Level:.95
 Calculate
```

TI-83/84 Plus

```
1-PropZInt
 (.67702,.72338)
 p̂=.7001998668
 n=1501
```

Using Technology for Sample Size

Although the TI-83/84 Plus calculator can be used to calculate the sample size required to estimate a population proportion p, it does not yet include a special function designed specifically for that purpose. STATDISK, a computer software program designed to accompany this textbook, does have a module for determining sample size. STATDISK is on the CD-ROM included with this book or it can be downloaded from www.aw.com/triola. After starting STATDISK, click on **Analysis** on the main menu at the top, then click on the menu item of **Sample Size Determination.** That item has three options, so select **Estimate Proportion,** then proceed to enter the items in the dialog box. Click on **Evaluate** to obtain the required sample size.

7-2   Basic Skills and Concepts

Statistical Literacy and Critical Thinking

1. Poll Results in the Media *USA Today* provided a "snapshot" illustrating poll results from 21,944 subjects. The illustration showed that 43% answered "yes" to this question: "Would you rather have a boring job than no job?" The margin of error was given as ± 1 percentage point. What important feature of the poll was omitted?

2. Margin of Error For the poll described in Exercise 1, describe what is meant by the statement that "the margin of error is ± 1 percentage point."

3. Confidence Interval For the poll described in Exercise 1, we see that 43% of 21,944 people polled answered "yes" to the given question. Given that 43% is the best estimate of the population percentage, why would we need a confidence interval? That is, what additional information does the confidence interval provide?

4. Sampling Suppose the poll results from Exercise 1 were obtained by mailing 100,000 questionnaires and receiving 21,944 responses. Is the result of 43% a good estimate of the population percentage of "yes" responses? Why or why not?

Finding Critical Values. *In Exercises 5–8, find the indicated critical z value.*

5. Find the critical value $z_{\alpha/2}$ that corresponds to a 99% confidence level.

6. Find the critical value $z_{\alpha/2}$ that corresponds to a 99.5% confidence level.

7. Find $z_{\alpha/2}$ for $\alpha = 0.10$.   **8.** Find $z_{\alpha/2}$ for $\alpha = 0.02$.

Expressing Confidence Intervals. *In Exercises 9–12, express the confidence interval using the indicated format.*

9. Express the confidence interval $0.200 < p < 0.500$ in the form of $\hat{p} \pm E$.

10. Express the confidence interval $0.720 < p < 0.780$ in the form of $\hat{p} \pm E$.

11. Express the confidence interval $(0.437, 0.529)$ in the form of $\hat{p} \pm E$.

12. Express the confidence interval 0.222 ± 0.044 in the form of $\hat{p} - E < p < \hat{p} + E$.

Interpreting Confidence Interval Limits. *In Exercises 13–16, use the given confidence interval limits to find the point estimate $\hat{p}$ and the margin of error E.*

13. $(0.320, 0.420)$ **14.** $0.772 < p < 0.776$

15. $0.433 < p < 0.527$ **16.** $0.102 < p < 0.236$

Finding Margin of Error. *In Exercises 17–20, assume that a sample is used to estimate a population proportion p. Find the margin of error E that corresponds to the given statistics and confidence level.*

17. $n = 1000$, $x = 400$, 95% confidence

18. $n = 500$, $x = 220$, 99% confidence

19. 98% confidence; the sample size is 1230, of which 40% are successes.

20. 90% confidence; the sample size is 1780, of which 35% are successes.

Constructing Confidence Intervals. *In Exercises 21–24, use the sample data and confidence level to construct the confidence interval estimate of the population proportion p.*

21. $n = 200$, $x = 40$, 95% confidence   **22.** $n = 2000$, $x = 400$, 95% confidence

23. $n = 1236$, $x = 109$, 99% confidence   **24.** $n = 5200$, $x = 4821$, 99% confidence

Determining Sample Size. *In Exercises 25–28, use the given data to find the minimum sample size required to estimate a population proportion or percentage.*

25. Margin of error: 0.045; confidence level: 95%; $\hat{p}$ and $\hat{q}$ unknown

26. Margin of error: 0.005; confidence level: 99%; $\hat{p}$ and $\hat{q}$ unknown

27. Margin of error: two percentage points; confidence level: 99%; from a prior study, $\hat{p}$ is estimated by the decimal equivalent of 14%.

28. Margin of error: three percentage points; confidence level: 95%; from a prior study, $\hat{p}$ is estimated by the decimal equivalent of 87%.

TI 29. Gender Selection The Genetics and IVF Institute conducted a clinical trial of the XSORT method designed to increase the probability of conceiving a girl. As of this writing, 574 babies were born to parents using the XSORT method, and 525 of them were girls.

a. What is the best point estimate of the population proportion of girls born to parents using the XSORT method?

b. Use the sample data to construct a 95% confidence interval estimate of the percentage of girls born to parents using the XSORT method.

c. Based on the results, does the XSORT method appear to be effective? Why or why not?

TI 30. Gender Selection The Genetics and IVF Institute conducted a clinical trial of the YSORT method designed to increase the probability of conceiving a boy. As of this writing, 152 babies were born to parents using the YSORT method, and 127 of them were boys.

a. What is the best point estimate of the population proportion of boys born to parents using the YSORT method?

b. Use the sample data to construct a 99% confidence interval estimate of the percentage of boys born to parents using the YSORT method.

c. Based on the results, does the YSORT method appear to be effective? Why or why not?

TI 31. Postponing Death An interesting and popular hypothesis is that individuals can temporarily postpone their death to survive a major holiday or important event such as a birthday. In a study of this phenomenon, it was found that in the week before and the week after Thanksgiving, there were 12,000 total deaths, and 6062 of them occurred in the week before Thanksgiving (based on data from "Holidays, Birthdays, and Postponement of Cancer Death," by Young and Hade, *Journal of the American Medical Association,* Vol. 292, No. 24.)

a. What is the best point estimate of the proportion of deaths in the week before Thanksgiving to the total deaths in the week before and the week after Thanksgiving?

b. Construct a 95% confidence interval estimate of the proportion of deaths in the week before Thanksgiving to the total deaths in the week before and the week after Thanksgiving.

c. Based on the result, does there appear to be any indication that people can temporarily postpone their death to survive the Thanksgiving holiday? Why or why not?

TI 32. Medical Malpractice An important issue facing Americans is the large number of medical malpractice lawsuits and the expenses that they generate. In a study of 1228 randomly selected medical malpractice lawsuits, it is found that 856 of them were later dropped or dismissed (based on data from the Physician Insurers Association of America).

a. What is the best point estimate of the proportion of medical malpractice lawsuits that are dropped or dismissed?

b. Construct a 99% confidence interval estimate of the proportion of medical malpractice lawsuits that are dropped or dismissed.

c. Does it appear that the majority of such suits are dropped or dismissed?

TI 33. Mendelian Genetics When Mendel conducted his famous genetics experiments with peas, one sample of offspring consisted of 428 green peas and 152 yellow peas.

a. Find a 95% confidence interval estimate of the percentage of yellow peas.

b. Based on his theory of genetics, Mendel expected that 25% of the offspring peas would be yellow. Given that the percentage of offspring yellow peas is not 25%, do the results contradict Mendel's theory? Why or why not?

TI 34. Misleading Survey Responses In a survey of 1002 people, 701 said that they voted in a recent presidential election (based on data from ICR Research Group). Voting records show that 61% of eligible voters actually did vote.

a. Find a 99% confidence interval estimate of the proportion of people who say that they voted.

b. Are the survey results consistent with the actual voter turnout of 61%? Why or why not?

TI 35. Cell Phones and Cancer A study of 420,095 Danish cell phone users found that 135 of them developed cancer of the brain or nervous system. Prior to this study of cell phone use, the rate of such cancer was found to be 0.0340% for those not using cell phones. The data are from the *Journal of the National Cancer Institute.*

a. Use the sample data to construct a 95% confidence interval estimate of the percentage of cell phone users who develop cancer of the brain or nervous system.

b. Do cell phone users appear to have a rate of cancer of the brain or nervous system that is different from the rate of such cancer among those not using cell phones? Why or why not?

TI 36. Global Warming Poll A Pew Research Center poll included 1708 randomly selected adults who were asked whether "global warming is a problem that requires immediate government action." Results showed that 939 of those surveyed indicated that immediate government action is required. A news reporter wants to determine whether these survey results constitute strong evidence that the majority (more than 50%) of people believe that immediate government action is required.

a. What is the best estimate of the percentage of adults who believe that immediate government action is required?

b. Construct a 99% confidence interval estimate of the proportion of adults believing that immediate government action is required.

c. Is there strong evidence supporting the claim that the majority is in favor of immediate government action? Why or why not?

TI 37. Internet Use In a Pew Research Center poll, 73% of 3011 adults surveyed said that they use the Internet. Construct a 95% confidence interval estimate of the proportion of all adults who use the Internet. Is it correct for a newspaper reporter to write that "3/4 of all adults use the Internet"? Why or why not?

TI 38. Job Interview Mistakes In an Accountemps survey of 150 senior executives, 47% said that the most common job interview mistake is to have little or no knowledge of the company. Construct a 99% confidence interval estimate of the proportion of all senior executives who have that same opinion. Is it possible that exactly half of all senior executives believe that the most common job interview mistake is to have little or no knowledge of the company? Why or why not?

TI 39. AOL Poll After 276 passengers on the *Queen Elizabeth II* cruise ship contracted a norovirus, America Online presented this question on its Internet site: "Would the recent outbreak deter you from taking a cruise?" Among the 34,358 people who responded, 62% answered "yes." Use the sample data to construct a 95% confidence interval estimate of the population of all people who would respond "yes" to that question. Does the confidence interval provide a good estimate of the population proportion? Why or why not?

TI 40. Touch Therapy When she was nine years of age, Emily Rosa did a science fair experiment in which she tested professional touch therapists to see if they could sense her energy field. She flipped a coin to select either her right hand or her left hand, then she placed her selected hand under the corresponding therapist's hand, and she asked the therapists to identify the selected hand without seeing it and without touching it. Among 280 trials, the touch therapists were correct 123 times (based on data in "A Close Look at Therapeutic Touch," *Journal of the American Medical Association,* Vol. 279, No. 13).

a. Given that Emily used a coin toss to select either her right hand or her left hand, what proportion of correct responses would be expected if the touch therapists made random guesses?

b. Using Emily's sample results, what is the best point estimate of the therapist's success rate?

c. Using Emily's sample results, construct a 99% confidence interval estimate of the proportion of correct responses made by touch therapists.

d. What do the results suggest about the ability of touch therapists to select the correct hand by sensing an energy field?

Determining Sample Size. *In Exercises 41–44, find the minimum sample size required to estimate a population proportion or percentage.*

41. Internet Use The use of the Internet is constantly growing. How many randomly selected adults must be surveyed to estimate the percentage of adults in the United States who now use the Internet? Assume that we want to be 99% confident that the sample percentage is within two percentage points of the true population percentage.

a. Assume that nothing is known about the percentage of adults using the Internet.

b. As of this writing, it was estimated that 73% of adults in the United States use the Internet (based on a Pew Research Center poll).

42. Cell Phones As the newly hired manager of a company that provides cell phone service, you want to determine the percentage of adults in your state who live in a household with cell phones and no land-line phones. How many adults must you survey? Assume that you want to be 90% confident that the sample percentage is within four percentage points of the true population percentage.

a. Assume that nothing is known about the percentage of adults who live in a household with cell phones and no land-line phones.

b. Assume that a recent survey suggests that about 8% of adults live in a household with cell phones and no land-line phones (based on data from the National Health Interview Survey).

43. Nitrogen in Tires A campaign was designed to convince car owners that they should fill their tires with nitrogen instead of air. At a cost of about $5 per tire, nitrogen supposedly has the advantage of leaking at a much slower rate than air, so that the ideal tire pressure can be maintained more consistently. Before spending huge sums to advertise the nitrogen, it would be wise to conduct a survey to determine the percentage of car owners who would pay for the nitrogen. How many randomly selected car owners should be surveyed? Assume that we want to be 95% confident that the sample percentage is within three percentage points of the true percentage of all car owners who would be willing to pay for the nitrogen.

44. Name Recognition As this book was being written, former New York City mayor Rudolph Giuliani announced that he was a candidate for the presidency of the United States. If you were a campaign worker and needed to determine the percentage of people that recognized his name, how many people should you have surveyed to estimate that percentage? Assume that you wanted to be 95% confident that the sample percentage was in error by no more than two percentage points, and also assume that a recent survey indicated that Giuliani's name is recognized by 10% of all adults (based on data from a Gallup poll).

Using Appendix B Data Sets. *In Exercises 45–48, use the indicated data set from Appendix B.*

🔵 **45. Green M&M Candies** Refer to Data Set 18 in Appendix B and find the sample proportion of M&Ms that are green. Use that result to construct a 95% confidence interval estimate of the population percentage of M&Ms that are green. Is the result consistent with the 16% rate that is reported by the candy maker Mars? Why or why not?

🔵 **46. Freshman 15 Weight Gain** Refer to Data Set 3 in Appendix B.

a. Based on the sample results, find the best point estimate of the percentage of college students who gain weight in their freshman year.

b. Construct a 95% confidence interval estimate of the percentage of college students who gain weight in their freshman year.

c. Assuming that you are a newspaper reporter, write a statement that describes the results. Include all of the relevant information. (*Hint:* See Example 3 part (d).)

🔵 **47. Precipitation in Boston** Refer to Data Set 14 in Appendix B, and consider days with precipitation values different from 0 to be days with precipitation. Construct a 95% confidence interval estimate of the proportion of Wednesdays with precipitation, and also construct a 95% confidence interval estimate of the proportion of Sundays with precipitation. Compare the results. Does precipitation appear to occur more on either day?

TI **48. Movie Ratings** Refer to Data Set 9 in Appendix B and find the proportion of movies with R ratings. Use that proportion to construct a 95% confidence interval estimate of the proportion of all movies with R ratings. Assuming that the listed movies constitute a simple random sample of all movies, can we conclude that most movies have ratings different from R? Why or why not?

7-2   Beyond the Basics

49. Using Finite Population Correction Factor In this section we presented Formulas 7-2 and 7-3, which are used for determining sample size. In both cases we assumed that the population is infinite or very large and that we are sampling with replacement. When we have a relatively small population with size N and sample without replacement, we modify E to include the *finite population correction factor* shown here, and we can solve for n to obtain the result given here. Use this result to repeat Exercise 43, assuming that we limit our population to the 12,784 car owners living in LaGrange, New York, home of the author. Is the sample size much lower than the sample size required for a population of millions of people?

$$E = z_{\alpha/2}\sqrt{\frac{\hat{p}\hat{q}}{n}}\sqrt{\frac{N-n}{N-1}} \qquad n = \frac{N\hat{p}\hat{q}\,[z_{\alpha/2}]^2}{\hat{p}\hat{q}\,[z_{\alpha/2}]^2 + (N-1)E^2}$$

TI **50. One-Sided Confidence Interval** A *one-sided confidence interval* for p can be expressed as $p < \hat{p} + E$ or $p > \hat{p} - E$, where the margin of error E is modified by replacing $z_{\alpha/2}$ with z_α. If Air America wants to report an on-time performance of at least x percent with 95% confidence, construct the appropriate one-sided confidence interval and then find the percent in question. Assume that a simple random sample of 750 flights results in 630 that are on time.

TI **51. Confidence Interval from Small Sample** Special tables are available for finding confidence intervals for proportions involving small numbers of cases, where the normal distribution approximation cannot be used. For example, given $x = 3$ successes among $n = 8$ trials, the 95% confidence interval found in *Standard Probability and Statistics Tables and Formulae* (CRC Press) is $0.085 < p < 0.755$. Find the confidence interval that would result if you were to incorrectly use the normal distribution as an approximation to the binomial distribution. Are the results reasonably close?

TI **52. Interpreting Confidence Interval Limits** Assume that a coin is modified so that it favors heads, and 100 tosses result in 95 heads. Find the 99% confidence interval estimate of the proportion of heads that will occur with this coin. What is unusual about the results obtained by the methods of this section? Does common sense suggest a modification of the resulting confidence interval?

53. Rule of Three Suppose n trials of a binomial experiment result in no successes. According to the *Rule of Three,* we have 95% confidence that the true population proportion has an upper bound of $3/n$. (See "A Look at the Rule of Three," by Jovanovic and Levy, *American Statistician,* Vol. 51, No. 2.)

a. If n independent trials result in no successes, why can't we find confidence interval limits by using the methods described in this section?

b. If 20 patients are treated with a drug and there are no adverse reactions, what is the 95% upper bound for p, the proportion of all patients who experience adverse reactions to this drug?

54. Poll Accuracy A *New York Times* article about poll results states, "In theory, in 19 cases out of 20, the results from such a poll should differ by no more than one percentage point in either direction from what would have been obtained by interviewing all voters in the United States." Find the sample size suggested by this statement.

7-3 Estimating a Population Mean: σ Known

Key Concept In this section we present methods for estimating a population mean. In addition to knowing the values of the sample data or statistics, we must also know the value of the population standard deviation, σ. Here are three key concepts that should be learned in this section.

1. We should know that the sample mean $\bar{x}$ is the best *point estimate* of the population mean μ.

2. We should learn how to use sample data to construct a *confidence interval* for estimating the value of a population mean, and we should know how to interpret such confidence intervals.

3. We should develop the ability to determine the sample size necessary to estimate a population mean.

Important: The confidence interval described in this section has the requirement that we know the value of the population standard deviation σ, but that value is rarely known in real circumstances. Section 7-4 describes methods for dealing with realistic cases in which σ is not known.

Point Estimate In Section 7-2 we saw that the sample proportion $\hat{p}$ is the best point estimate of the population proportion p. The sample mean $\bar{x}$ is an *unbiased estimator* of the population mean μ, and for many populations, sample means tend to vary less than other measures of center, so the sample mean $\bar{x}$ is usually the best point estimate of the population mean μ.

> **The sample mean $\bar{x}$ is the best point estimate of the population mean.**

Although the sample mean $\bar{x}$ is usually the *best* point estimate of the population mean μ, it does not give us any indication of just how *good* our best estimate is. We get more information from a *confidence interval* (or *interval estimate*), which consists of a range (or an interval) of values instead of just a single value.

Knowledge of σ The listed requirements on the next page include knowledge of the population standard deviation σ, but Section 7-4 presents methods for estimating a population mean without knowledge of the value of σ.

Normality Requirement The requirements on the next page include the property that either the population is normally distributed or $n > 30$. If $n \le 30$, the population need not have a distribution that is exactly normal. The methods of this section are *robust* against departures from normality, which means that these methods are not strongly affected by departures from normality, provided that those departures are not too extreme. We therefore have a loose normality requirement that can be satisfied if there are no outliers and if a histogram of the sample data is not dramatically different from being bell-shaped. (See Section 6-7.)

Sample Size Requirement The normal distribution is used as the distribution of sample means. If the original population is not itself normally distributed, then we say that means of samples with size $n > 30$ have a distribution that can be approximated by a normal distribution. The condition $n > 30$ is a common guideline, but there is no specific minimum sample size that works for all cases.

The minimum sample size actually depends on how much the population distribution departs from a normal distribution. Sample sizes of 15 to 30 are sufficient if the population has a distribution that is not far from normal, but some other populations have distributions that are extremely far from normal and sample sizes greater than 30 might be necessary. In this book we use the simplified criterion of $n > 30$ as justification for treating the distribution of sample means as a normal distribution.

Confidence Level   The confidence interval is associated with a confidence level, such as 0.95 (or 95%). The confidence level gives us the success rate of the procedure used to construct the confidence interval. As in Section 7-2, α is the complement of the confidence level. For a 0.95 (or 95%) confidence level, $\alpha = 0.05$ and $z_{\alpha/2} = 1.96$.

Confidence Interval for Estimating a Population Mean (with σ Known)

Objective

Construct a confidence interval used to estimate a population mean.

Notation

μ = population mean

σ = population standard deviation

$\bar{x}$ = sample mean

n = number of sample values

E = margin of error

$z_{\alpha/2}$ = z score separating an area of $\alpha/2$ in the right tail of the standard normal distribution

Requirements

1. The sample is a simple random sample.

2. The value of the population standard deviation σ is known.

3. Either or both of these conditions is satisfied: The population is normally distributed or $n > 30$.

Confidence Interval

$$\bar{x} - E < \mu < \bar{x} + E \quad \text{where} \quad E = z_{\alpha/2} \cdot \frac{\sigma}{\sqrt{n}}$$

or

$$\bar{x} \pm E$$

or

$$(\bar{x} - E, \bar{x} + E)$$

Procedure for Constructing a Confidence Interval for μ (with Known σ)

1. Verify that the requirements are satisfied.

2. Refer to a TI-83/84 Plus calculator or Table A-2 to find the critical value $z_{\alpha/2}$ that corresponds to the desired confidence level. (For example, if the confidence level is 95%, the critical value is $z_{\alpha/2} = 1.96$.)

3. Evaluate the margin of error $E = z_{\alpha/2} \cdot \sigma/\sqrt{n}$

4. Using the value of the calculated margin of error E and the value of the sample mean $\bar{x}$, find the values of the confidence interval limits: $\bar{x} - E$ and $\bar{x} + E$. Substitute those values in the general format for the confidence interval:

$$\bar{x} - E < \mu < \bar{x} + E$$

or

$$\bar{x} \pm E$$

or

$$(\bar{x} - E, \bar{x} + E)$$

5. Round the resulting values by using the following round-off rule.

Using the TI-83/84 Plus Calculator The TI-83/84 Plus calculator can automatically generate confidence interval estimates of the population mean μ. See the procedure at the end of this section.

> **Round-Off Rule for Confidence Intervals Used to Estimate μ**
>
> 1. When using the *original set of data* to construct a confidence interval, round the confidence interval limits to one more decimal place than is used for the original set of data.
>
> 2. When the original set of data is unknown and only the *summary statistics* $(n, \bar{x}, s)$ are used, round the confidence interval limits to the same number of decimal places used for the sample mean.

Interpreting a Confidence Interval   As in Section 7-2, be careful to interpret confidence intervals correctly. After obtaining a confidence interval estimate of the population mean μ, such as a 95% confidence interval of $164.49 < \mu < 180.61$, there is a correct interpretation and many incorrect interpretations.

Correct: "We are 95% confident that the interval from 164.49 to 180.61 actually does contain the true value of μ." This means that if we were to select many different samples of the same size and construct the corresponding confidence intervals, in the long run 95% of them would actually contain the value of μ. (As in Section 7-2, this correct interpretation refers to the success rate of the *process* being used to estimate the population mean.)

Incorrect: Because μ is a fixed constant, it would be incorrect to say "there is a 95% chance that μ will fall between 164.49 and 180.61." It would also be incorrect to say that "95% of all data values are between 164.49 and 180.61," or that "95% of sample means fall between 164.49 and 180.61." Creative readers can formulate other possible incorrect interpretations.

> **EXAMPLE 1**   **Weights of Men** People have died in boat and aircraft accidents because an obsolete estimate of the mean weight of men was used. In recent decades, the mean weight of men has increased considerably, so we need to update our estimate of that mean so that boats, aircraft, elevators, and other such devices do not become dangerously overloaded. Using the weights of men from Data Set 1 in Appendix B, we obtain these sample statistics for the simple random sample: $n = 40$ and $\bar{x} = 172.55$ lb. Research from several other sources suggests that the population of weights of men has a standard deviation given by $\sigma = 26$ lb.

continued

Estimating Wildlife Population Sizes

The National Forest Management Act protects endangered species, including the northern spotted owl, with the result that the forestry industry was not allowed to cut vast regions of trees in the Pacific Northwest. Biologists and statisticians were asked to analyze the problem, and they concluded that survival rates and population sizes were decreasing for the female owls, known to play an important role in species survival. Biologists and statisticians also studied salmon in the Snake and Columbia Rivers in Washington State, and penguins in New Zealand. In the article "Sampling Wildlife Populations" (*Chance,* Vol. 9, No. 2), authors Bryan Manly and Lyman McDonald comment that in such studies, "biologists gain through the use of modeling skills that are the hallmark of good statistics. Statisticians gain by being introduced to the reality of problems by biologists who know what the crucial issues are."

Captured Tank Serial Numbers Reveal Population Size

During World War II, Allied intelligence specialists wanted to determine the number of tanks Germany was producing. Traditional spy techniques provided unreliable results, but statisticians obtained accurate estimates by analyzing serial numbers on captured tanks. As one example, records show that Germany actually produced 271 tanks in June 1941. The estimate based on serial numbers was 244, but traditional intelligence methods resulted in the extreme estimate of 1550. (See "An Empirical Approach to Economic Intelligence in World War II," by Ruggles and Brodie, *Journal of the American Statistical Association*, Vol. 42.)

TI-83/84 Plus

```
ZInterval
 (164.49,180.61)
 x̄=172.55
 n=40
```

a. Find the best point estimate of the mean weight of the population of all men.

b. Construct a 95% confidence interval estimate of the mean weight of all men.

c. What do the results suggest about the mean weight of 166.3 lb that was used to determine the safe passenger capacity of water vessels in 1960 (as given in the National Transportation and Safety Board safety recommendation M-04-04)?

SOLUTION   **REQUIREMENT CHECK** We must first verify that the requirements are satisfied. (1) The sample is a simple random sample. (2) The value of σ is assumed to be known with $\sigma = 26$ lb. (3) With $n > 30$, we satisfy the requirement that "the population is normally distributed or $n > 30$." The requirements are therefore satisfied. ✓

a. The sample mean of 172.55 lb is the best point estimate of the mean weight for the population of all men.

b. The 0.95 confidence level implies that $\alpha = 0.05$, so $z_{\alpha/2} = 1.96$ (as was shown in Example 2 in Section 7-2). The margin of error E is first calculated as follows. (Extra decimal places are used to minimize rounding errors in the confidence interval.)

$$E = z_{\alpha/2} \cdot \frac{\sigma}{\sqrt{n}} = 1.96 \cdot \frac{26}{\sqrt{40}} = 8.0574835$$

With $\bar{x} = 172.55$ and $E = 8.0574835$, we now construct the confidence interval as follows:

$$\bar{x} - E < \mu < \bar{x} + E$$
$$172.55 - 8.0574835 < \mu < 172.55 + 8.0574835$$
$$164.49 < \mu < 180.61 \quad \text{(rounded to two decimal places as in } \bar{x}\text{)}$$

If we use a TI-83/84 Plus calculator with the procedure given at the end of this section, we get the confidence interval in the format of (164.49, 180.61), as shown in the margin.

c. Based on the confidence interval, it is possible that the mean weight of 166.3 lb used in 1960 could be the mean weight of men today. However, the best point estimate of 172.55 lb suggests that the mean weight of men is now considerably greater than 166.3 lb. Considering that an underestimate of the mean weight of men could result in lives lost through overloaded boats and aircraft, these results strongly suggest that additional data should be collected. (Additional data have been collected, and the assumed mean weight of men has been increased.)

INTERPRETATION   The confidence interval from part (b) could also be expressed as 172.55 ± 8.06 or as (164.49, 180.61). Based on the sample with $n = 40$, $\bar{x} = 172.55$ and σ assumed to be 26, the confidence interval for the population mean μ is 164.49 lb $< \mu <$ 180.61 lb and this interval has a 0.95 confidence level. This means that if we were to select many different simple random samples of 40 men and construct the confidence intervals as we did here, 95% of them would actually contain the value of the population mean μ.

Rationale for the Confidence Interval   The basic idea underlying the construction of confidence intervals relates to this property of the sampling distribution of sample means: If we collect simple random samples of the same size n, the sample means are (at least approximately) normally distributed with mean μ and standard deviation $\sigma/\sqrt{n}$. In the standard score $z = (\bar{x} - \mu_{\bar{x}})/\sigma_{\bar{x}}$, replace $\sigma_{\bar{x}}$ with $\sigma/\sqrt{n}$, replace $\mu_{\bar{x}}$ with μ, then solve for μ to get

$$\mu = \bar{x} - z\frac{\sigma}{\sqrt{n}}$$

In the above equation, use the positive and negative values of z and replace the rightmost term by E. The right-hand side of the equation then yields the confidence interval limits of $\bar{x} - E$ and $\bar{x} + E$ that we are given earlier in this section. For a 95% confidence interval, we let $\alpha = 0.05$, so $z_{\alpha/2} = 1.96$, so there is a 0.95 probability that a sample mean will be within 1.96 standard deviations (or $z_{\alpha/2} \cdot \sigma/\sqrt{n}$ or E) of μ. If the sample mean $\bar{x}$ is within E of the population mean, then μ is between $\bar{x} - E$ and $\bar{x} + E$. That is, $\bar{x} - E < \mu < \bar{x} + E$.

Determining Sample Size Required to Estimate μ

When collecting a simple random sample that will be used to estimate a population mean μ, *how many* sample values must be obtained? For example, suppose we want to estimate the mean weight of airline passengers (an important value for reasons of safety). How many passengers must be randomly selected and weighed? Determining the size of a simple random sample is a very important issue, because samples that are needlessly large waste time and money, and samples that are too small may lead to poor results.

If we use the expression for the margin of error ($E = z_{\alpha/2}\sigma/\sqrt{n}$) and solve for the sample size n, we get Formula 7-4 shown below.

Finding the Sample Size Required to Estimate a Population Mean

Objective

Determine how large a sample should be in order to estimate the population mean μ.

Notation

μ = population mean

σ = population standard deviation

$\bar{x}$ = sample mean

E = desired margin of error

$z_{\alpha/2}$ = z score separating an area of $\alpha/2$ in the right tail of the standard normal distribution

Requirements

The sample must be a simple random sample.

Formula 7-4
$$n = \left[\frac{z_{\alpha/2}\sigma}{E}\right]^2$$

Formula 7-4 is remarkable because it shows that the sample size does not depend on the size (N) of the population; the sample size depends on the desired confidence level, the desired margin of error, and the value of the standard deviation σ. (See Exercise 38 for dealing with cases in which a relatively large sample is selected without replacement from a finite population.)

The sample size must be a whole number, because it represents the number of sample values that must be found. However, Formula 7-4 usually gives a result that is not a whole number, so we use the following round-off rule. (It is based on the principle that when rounding is necessary, the required sample size should be rounded *upward* so that it is at least adequately large as opposed to slightly too small.)

Round-Off Rule for Sample Size *n*

If the computed sample size *n* is not a whole number, round the value of *n* up to the next *larger* whole number.

Dealing with Unknown σ When Finding Sample Size   Formula 7-4 requires that we substitute a known value for the population standard deviation σ, but in reality, it is usually unknown. When determining a required sample size (not constructing a confidence interval), here are some ways that we can work around the problem of not knowing the value of σ:

1. Use the range rule of thumb (see Section 3-3) to estimate the standard deviation as follows: $\sigma \approx$ range/4. (With a sample of 87 or more values randomly selected from a normally distributed population, range/4 will yield a value that is greater than or equal to σ at least 95% of the time. (See "Using the Sample Range as a Basis for Calculating Sample Size in Power Calculations," by Richard Browne, *American Statistician*, Vol. 55, No. 4.)

2. Start the sample collection process without knowing σ and, using the first several values, calculate the sample standard deviation s and use it in place of σ. The estimated value of σ can then be improved as more sample data are obtained, and the sample size can be refined accordingly.

3. Estimate the value of σ by using the results of some other study that was done earlier.

In addition, we can sometimes be creative in our use of other known results. For example, IQ tests are typically designed so that the mean is 100 and the standard deviation is 15. Statistics students have IQ scores with a mean greater than 100 and a standard deviation less than 15 (because they are a more homogeneous group than people randomly selected from the general population). We do not know the specific value of σ for statistics students, but we can play it safe by using $\sigma = 15$. Using a value for σ that is larger than the true value will make the sample size larger than necessary, but using a value for σ that is too small would result in a sample size that is inadequate. *When calculating the sample size n, any errors should always be conservative in the sense that they make n too large instead of too small.*

EXAMPLE 2   **IQ Scores of Statistics Students**   Assume that we want to estimate the mean IQ score for the population of statistics students. How many statistics students must be randomly selected for IQ tests if we want 95% confidence that the sample mean is within 3 IQ points of the population mean?

SOLUTION   For a 95% confidence interval, we have $\alpha = 0.05$, so $z_{\alpha/2} = 1.96$. Because we want the sample mean to be within 3 IQ points of μ, the margin of error is $E = 3$. Also, $\sigma = 15$ (see the discussion that immediately precedes this example). Using Formula 7-4, we get

$$n = \left[\frac{z_{\alpha/2}\sigma}{E} \right]^2 = \left[\frac{1.96 \cdot 15}{3} \right]^2 = 96.04 = 97 \quad \text{(rounded } up\text{)}$$

INTERPRETATION Among the thousands of statistics students, we need to obtain a simple random sample of at least 97 students. Then we need to get their IQ scores. With a simple random sample of only 97 statistics students, we will be 95% confident that the sample mean $\bar{x}$ is within 3 IQ points of the true population mean μ.

If we want a more accurate estimate, we can decrease the margin of error. Halving the margin of error quadruples the sample size, so if you want more accurate results, the sample size must be substantially increased. Because large samples generally require more time and money, there is often a need for a tradeoff between the sample size and the margin of error E.

Using the TI-83/84 Plus Calculator for Confidence Interval Estimates of μ (With σ Known)

Here is the procedure for using a TI-83/84 Plus calculator for the construction of a confidence interval estimate of a population mean μ, *assuming that the population standard deviation σ is known*:

1. Press **STAT**, then use the ▷ key to select the option of **TESTS.**

2. If the population standard deviation σ is known, select the menu option of **ZInterval.** Select this option by pressing ▽ to scroll down. (If σ is not known, use the procedure described in Section 7-4.)

3. The first line, identified as Inpt (for input), gives you the option of using Data or Stats. Use the arrow keys ◁ and ▷ to select either Data or Stats.

   - **Data:** Use this option if you have the sample data values stored as a list.

   - **Stats:** Use this option if you know the values of σ, $\bar{x}$, and n.

4. What happens next depends on the choice made in step 3. If you choose to use Stats, you must enter the values of σ, $\bar{x}$, and n.

   If you choose to use Data, you must enter the known value of σ, the list where the data are stored, and a value for Freq (frequency). Enter 1 for the frequency. (However, if the sample data are stored in two columns as a frequency distribution, the Freq entry should be the list containing frequencies.)

5. Enter a confidence level, such as 0.95, then scroll down to **Calculate** and press **ENTER**. The margin shows the entry screen and results screen corresponding to Example 1. (The List entry of MWT is the list of male weights from the APP named MHEALTH.) The confidence interval will be expressed in a format with the confidence interval limits enclosed within parentheses, as in the second screen display, which corresponds to Example 1. The TI-83/84 Plus confidence interval of (164.49, 180.61) can also be expressed as $164.49 < \mu < 180.61$, as in Example 1.

TI-83/84 Plus

```
ZInterval
 Inpt:DATA Stats
 σ:26
 List:MWT
 Freq:1
 C-Level:.95
 Calculate
```

TI-83/84 Plus

```
ZInterval
 (164.49,180.61)
 x̄=172.55
 n=40
```

Using Technology for Sample Size

Although the TI-83/84 Plus calculator can be used to calculate the sample size required to estimate a population mean μ, it does not yet include a special function designed specifically for that purpose. STATDISK, a computer software program designed to accompany this textbook, does have a module for determining sample size. STATDISK is on the CD-ROM included with this book or it can be downloaded from www.aw.com/triola. After starting STATDISK, click on **Analysis** on the main menu at the top, then click on the menu item of **Sample Size Determination.** That item has three options, so select **Estimate Mean,** then proceed to enter the items in the dialog box. Click on **Evaluate** to obtain the required sample size.

7-3   Basic Skills and Concepts

Statistical Literacy and Critical Thinking

1. Point Estimate In general, what is a *point estimate* of a population parameter? Given a simple random sample of heights from some population, such as the population of all basketball players in the NBA, how would you find the best point estimate of the population mean?

2. Simple Random Sample A design engineer for the Ford Motor Company must estimate the mean leg length of all adults. She obtains a list of the 1275 employees at her facility, then obtains a simple random sample of 50 employees. If she uses this sample to construct a 95% confidence interval to estimate the mean leg length for the population of all adults, will her estimate be good? Why or why not?

3. Confidence Interval Based on the heights of women listed in Data Set 1 in Appendix B, and assuming that heights of women have a standard deviation of $\sigma = 2.5$ in., this 95% confidence interval is obtained: 62.42 in. $< \mu < 63.97$ in. Assuming that you are a newspaper reporter, write a statement that correctly interprets that confidence interval and includes all of the relevant information.

4. Unbiased Estimator One of the features of the sample mean that makes it a good estimator of a population mean μ is that the sample mean is an unbiased estimator. What does it mean for a statistic to be an unbiased estimator of a population parameter?

Finding Critical Values. *In Exercises 5–8, find the indicated critical value $z_{\alpha/2}$.*

5. Find the critical value $z_{\alpha/2}$ that corresponds to a 90% confidence level.

6. Find the critical value $z_{\alpha/2}$ that corresponds to a 98% confidence level.

7. Find $z_{\alpha/2}$ for $\alpha = 0.20$.   **8.** Find $z_{\alpha/2}$ for $\alpha = 0.04$.

Verifying Requirements and Finding the Margin of Error. *In Exercises 9–12, find the margin of error and confidence interval if the necessary requirements are satisfied. If the requirements are not all satisfied, state that the margin of error and confidence interval cannot be calculated using the methods of this section.*

9. Credit Rating FICO (Fair, Isaac, and Company) credit rating scores of a simple random sample of applicants for credit cards: 95% confidence; $n = 50$, $\bar{x} = 677$, and σ is known to be 68.

10. Braking Distances The braking distances of a simple random sample of cars: 95% confidence; $n = 32$, $\bar{x} = 137$ ft, and σ is known to be 7 ft.

11. Rainfall Amounts The amounts of rainfall for a simple random sample of Saturdays in Boston: 99% confidence; $n = 12$, $\bar{x} = 0.133$ in., σ is known to be 0.212 in., and the population is known to have daily rainfall amounts with a distribution that is far from normal.

12. Failure Times The times before failure of integrated circuits used in calculators: 99% confidence; $n = 25$, $\bar{x} = 112$ hours, σ is known to be 18.6 hours, and the distribution of all times before failure is far from normal.

Finding Sample Size. *In Exercises 13–16, use the given information to find the minimum sample size required to estimate an unknown population mean μ.*

13. Credit Rating How many adults must be randomly selected to estimate the mean FICO (credit rating) score of working adults in the United States. We want 95% confidence that the sample mean is within 3 points of the population mean, and the population standard deviation is 68.

14. Braking Distances How many cars must be randomly selected and tested in order to estimate the mean braking distance of registered cars in the United States. We want 99% confidence that the sample mean is within 2 ft of the population mean, and the population standard deviation is known to be 7 ft.

15. Rainfall Amounts How many daily rainfall amounts in Boston must be randomly selected to estimate the mean daily rainfall amount? We want 99% confidence that the sample mean is within 0.010 in. of the population mean, and the population standard deviation is known to be 0.212 in.

16. Failure Times How many integrated circuits must be randomly selected and tested for time to failure in order to estimate the mean time to failure? We want 95% confidence that the sample mean is within 2 hr of the population mean, and the population standard deviation is known to be 18.6 hours.

Interpreting Results. *In Exercises 17–20, refer to the accompanying TI-83/84 Plus calculator display of a 95% confidence interval. The sample display results from using a simple random sample of the amounts of tar (in milligrams) in cigarettes that are all king size, nonfiltered, nonmenthol, and non-light.*

TI-83/84 Plus

```
ZInterval
 (19.853,22.387)
 x̄=21.12
 n=25
```

TI **17.** Identify the value of the point estimate of the population mean μ.

TI **18.** Express the confidence interval in the format of $\bar{x} - E < \mu < \bar{x} + E$.

TI **19.** Express the confidence interval in the format of $\bar{x} \pm E$.

TI **20.** Write a statement that interprets the 95% confidence interval.

TI **21. Weights of Women** Using the simple random sample of weights of women from Data Set 1 in Appendix B, we obtain these sample statistics: $n = 40$ and $\bar{x} = 146.22$ lb. Research from other sources suggests that the population of weights of women has a standard deviation given by $\sigma = 30.86$ lb.

a. Find the best point estimate of the mean weight of all women.

b. Find a 95% confidence interval estimate of the mean weight of all women.

TI **22. NCAA Football Coach Salaries** A simple random sample of 40 salaries of NCAA football coaches has a mean of $415,953. Assume that $\sigma = \$463,364$.

a. Find the best point estimate of the mean salary of all NCAA football coaches.

b. Construct a 95% confidence interval estimate of the mean salary of an NCAA football coach.

c. Does the confidence interval contain the actual population mean of $474,477?

TI **23. Perception of Time** Randomly selected statistics students of the author participated in an experiment to test their ability to determine when 1 min (or 60 seconds) has passed. Forty students yielded a sample mean of 58.3 sec. Assume that $\sigma = 9.5$ sec.

a. Find the best point estimate of the mean time for all statistics students.

b. Construct a 95% confidence interval estimate of the population mean of all statistics students.

c. Based on the results, is it likely that their estimates have a mean that is reasonably close to 60 sec?

TI **24. Red Blood Cell Count** A simple random sample of 50 adults (including males and females) is obtained, and each person's red blood cell count (in cells per microliter) is measured. The sample mean is 4.63. The population standard deviation for red blood cell counts is 0.54.

a. Find the best point estimate of the mean red blood cell count of adults.

b. Construct a 99% confidence interval estimate of the mean red blood cell count of adults.

c. The normal range of red blood cell counts for adults is given by the National Institutes of Health as 4.7 to 6.1 for males and 4.3 to 5.4 for females. What does the confidence interval suggest about these normal ranges?

TI **25. SAT Scores** A simple random sample of 125 SAT scores has a mean of 1522. Assume that SAT scores have a standard deviation of 333.

a. Construct a 95% confidence interval estimate of the mean SAT score.

b. Construct a 99% confidence interval estimate of the mean SAT score.

c. Which of the preceding confidence intervals is wider? Why?

TI **26. Birth Weights** A simple random sample of birth weights in the United States has a mean of 3433 g. The standard deviation of all birth weights is 495 g.

a. Using a sample size of 75, construct a 95% confidence interval estimate of the mean birth weight in the United States.

b. Using a sample size of 75,000, construct a 95% confidence interval estimate of the mean birth weight in the United States.

c. Which of the preceding confidence intervals is wider? Why?

TI 27. Blood Pressure Levels When 14 different second-year medical students at Bellevue Hospital measured the blood pressure of the same person, they obtained the results listed below. Assuming that the population standard deviation is known to be 10 mmHg, construct a 95% confidence interval estimate of the population mean. Ideally, what should the confidence interval be in this situation?

    138   130   135   140   120   125   120   130   130   144   143   140   130   150

TI 28. Telephone Digits Polling organizations typically generate the last digits of telephone numbers so that people with unlisted numbers are included. Listed below are digits randomly generated by STATDISK. Such generated digits are from a population with a standard deviation of 2.87.

a. Use the methods of this section to construct a 95% confidence interval estimate of the mean of all such generated digits.

b. Are the requirements for the methods of this section all satisfied? Does the confidence interval from part (a) serve as a good estimate of the population mean? Explain.

                     1   1   7   0   7   4   5   1   7   6

Large Data Sets from Appendix B. *In Exercises 29 and 30, refer to the data set from Appendix B.*

TI 29. Movie Gross Amounts Refer to Data Set 9 from Appendix B and construct a 95% confidence interval estimate of the mean gross amount for the population of all movies. Assume that the population standard deviation is known to be 100 million dollars.

TI 30. FICO Credit Rating Scores Refer to Data Set 24 in Appendix B and construct the 99% confidence interval estimate of the mean FICO score for the population. Assume that the population standard deviation is 92.2.

Finding Sample Size. *In Exercises 31–36, find the indicated sample size.*

31. Sample Size for Mean IQ of NASA Scientists The Wechsler IQ test is designed so that the mean is 100 and the standard deviation is 15 for the population of normal adults. Find the sample size necessary to estimate the mean IQ score of scientists currently employed by NASA. We want to be 95% confident that our sample mean is within five IQ points of the true mean. The mean for this population is clearly greater than 100. The standard deviation for this population is probably less than 15 because it is a group with less variation than a group randomly selected from the general population; therefore, if we use $\sigma = 15$, we are being conservative by using a value that will make the sample size at least as large as necessary. Assume then that $\sigma = 15$ and determine the required sample size.

32. Sample Size for White Blood Cell Count What sample size is needed to estimate the mean white blood cell count (in cells per microliter) for the population of adults in the United States? Assume that you want 99% confidence that the sample mean is within 0.2 of the population mean. The population standard deviation is 2.5.

33. Sample Size for Atkins Weight Loss Program You want to estimate the mean weight loss of people one year after using the Atkins weight loss program. How many people on that program must be surveyed if we want to be 95% confident that the sample mean weight loss is within 0.25 lb of the true population mean? Assume that the population standard deviation is known to be 10.6 lb (based on data from "Comparison of the Atkins, Ornish, Weight Watchers, and Zone Diets for Weight Loss and Heart Disease Risk Reduction," by Dansinger, et al., *Journal of the American Medical Association,* Vol. 293, No. 1). Is the resulting sample size practical?

34. Grade Point Average A researcher wants to estimate the mean grade point average of all current college students in the United States. She has developed a procedure to standardize scores from colleges using something other than a scale between 0 and 4. How many grade point averages must be obtained so that the sample mean is within 0.1 of the population mean? Assume that a 90% confidence level is desired. Also assume that a pilot study showed that the population standard deviation is estimated to be 0.88.

35. Sample Size Using Range Rule of Thumb You want to estimate the mean amount of annual tuition being paid by current full-time college students in the United States. First use the range rule of thumb to make a rough estimate of the standard deviation of the amounts spent. It is reasonable to assume that tuition amounts range from $0 to about

$40,000. Then use that estimated standard deviation to determine the sample size corresponding to 95% confidence and a $100 margin of error.

36. Sample Size Using Sample Data Refer to Data Set 1 in Appendix B and find the maximum and minimum pulse rates for males, then use those values with the range rule of thumb to estimate σ. How many adult males must you randomly select and test if you want to be 95% confident that the sample mean pulse rate is within 2 beats (per minute) of the true population mean μ? If, instead of using the range rule of thumb, the standard deviation of the male pulse rates in Data Set 1 is used as an estimate of σ, is the required sample size very different? Which sample size is likely to be closer to the correct sample size?

7-3 Beyond the Basics

37. Confidence Interval with Finite Population Correction Factor The standard error of the mean is $\sigma/\sqrt{n}$, provided that the population size is infinite or very large or sampling is with replacement. If the population size N is finite, then the correction factor $\sqrt{(N-n)/(N-1)}$ should be used whenever $n > 0.05N$. The margin of error E is multiplied by this correction factor as shown below. Repeat part (a) of Exercise 25 assuming that the sample is selected without replacement from a population of size 200. How is the confidence interval affected by the additional information about the population size?

$$E = z_{\alpha/2}\frac{\sigma}{\sqrt{n}}\sqrt{\frac{N-n}{N-1}}$$

38. Sample Size with Finite Population Correction Factor The methods of this section assume that sampling is from a population that is very large or infinite, and that we are sampling with replacement. If we have a relatively small population and sample without replacement, we should modify E to include a *finite population correction factor*, so that the margin of error is as shown in Exercise 37, where N is the population size. That expression for the margin of error can be solved for n to yield

$$n = \frac{N\sigma^2(z_{\alpha/2})^2}{(N-1)E^2 + \sigma^2(z_{\alpha/2})^2}$$

Repeat Exercise 32, assuming that a simple random sample is selected without replacement from a population of 500 people. Does the additional information about the population size have much of an effect on the sample size?

7-4 Estimating a Population Mean: σ Not Known

Key Concept In this section we present methods for estimating a population mean when the population standard deviation σ is unknown. With σ unknown, we use the *Student t distribution* (instead of the normal distribution), assuming that the relevant requirements are satisfied. Because σ is typically unknown in real circumstances, the methods of this section are realistic and practical, and they are often used.

As in Section 7-3, the sample mean $\bar{x}$ is the best point estimate (or single-valued estimate) of the population mean μ.

The sample mean $\bar{x}$ is the best point estimate of the population mean μ.

Here is a major point of this section: If σ is not known, but the relevant requirements are satisfied, we use a *Student t distribution* (instead of a normal distribution), as developed by William Gosset (1876–1937). Gosset was a Guinness Brewery employee who needed a distribution that could be used with small samples. The Irish brewery where he worked did not allow the publication of research results, so Gosset published under the pseudonym "Student." (In the interest of research and better serving his readers, the author visited the Guinness Brewery and sampled some of the product. Such commitment!)

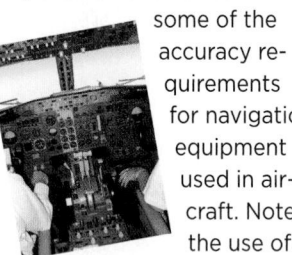
Student *t* Distribution

If a population has a normal distribution, then the distribution of

$$t = \frac{\bar{x} - \mu}{\frac{s}{\sqrt{n}}}$$

is a **Student *t* distribution** for all samples of size *n*. A Student *t* distribution is often referred to simply as a ***t* distribution.**

Because we do not know the value of the population standard deviation σ, we estimate it with the value of the sample standard deviation *s*, but this introduces another source of unreliability, especially with small samples. In order to maintain a desired confidence level, such as 95%, we compensate for this additional unreliability by making the confidence interval wider: We use critical values $t_{\alpha/2}$ (from a Student *t* distribution) that are larger than the critical values of $z_{\alpha/2}$ from the normal distribution. A critical value of $t_{\alpha/2}$ can be found by using technology or Table A-3, but we must first identify the number of *degrees of freedom.*

> ### DEFINITION
>
> The number of **degrees of freedom** for a collection of sample data is the number of sample values that can vary after certain restrictions have been imposed on all data values. The number of degrees of freedom is often abbreviated as **df.**

For example, if 10 students have quiz scores with a mean of 80, we can freely assign values to the first 9 scores, but the 10th score is then determined. The sum of the 10 scores must be 800, so the 10th score must equal 800 minus the sum of the first 9 scores. Because those first 9 scores can be *freely* selected to be any values, we say that there are 9 degrees of freedom available. For the applications of this section, the number of degrees of freedom is simply the sample size minus 1.

$$\textbf{degrees of freedom} = n - 1$$

> **EXAMPLE 1** **Finding a Critical *t* Value** A sample of size $n = 7$ is a simple random sample selected from a normally distributed population. Find the critical value $t_{\alpha/2}$ corresponding to a 95% confidence level.

> **SOLUTION** First, we identify the number of degrees of freedom. Because $n = 7$, the number of degrees of freedom is $n - 1 = 6$. We can now proceed to find the critical *t* value.
>
> **Using the TI-84 Plus Calculator** A critical value $t_{\alpha/2}$ can be found with a TI-84 Plus calculator by using the command **invT.** Press press **2ND** **VARS** to get to the DISTR (distribution) menu, then select **invT.** Proceed to enter the *cumulative area to the left* of *t* followed by a comma followed by the number of degrees of freedom. Enter the command **invT(0.975, 6)** to get 2.446911839, which can be rounded to 2.447.
>
> **Using the TI-83 Plus Calculator** The TI-83 Plus calculator does not have the invT command, but the CD included with this book contains the program TINVRS83, which can be used to find critical values of *t*. If that program is installed, press **PRGM** and select **TINVRS83** and press **ENTER**. You will then be prompted for the required entries.

Using Table A-3 Using Table A-3, we locate the 6th row by referring to the column at the extreme left. A 95% confidence level corresponds to $\alpha = 0.05$, and confidence intervals require that the area α be divided equally between the left and right tails of the distribution (as in Figure 7-4), so we find the column listing values for an *area of 0.05 in two tails*. The value corresponding to the row for 6 degrees of freedom and the column for an area of 0.05 in two tails is 2.447, so $t_{\alpha/2} = 2.447$. (See Figure 7-4.) We could also express this as $t_{0.025} = 2.447$. Such critical values $t_{\alpha/2}$ are used for the margin of error E and confidence interval as shown in the box that follows.

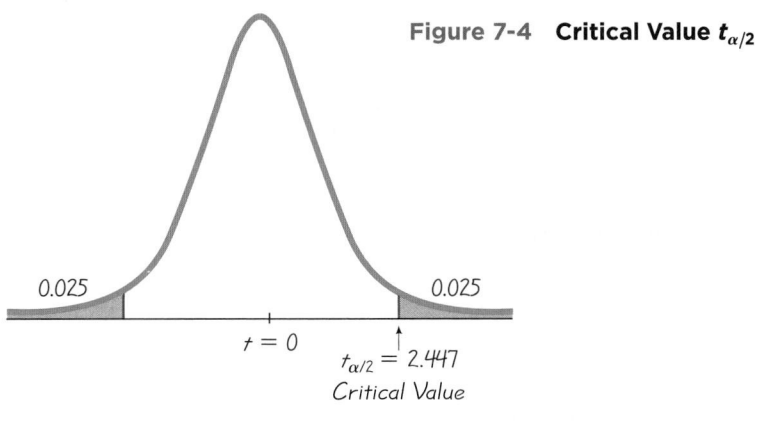

Figure 7-4   Critical Value $t_{\alpha/2}$

0.025   0.025

$t = 0$

$t_{\alpha/2} = 2.447$
Critical Value

Confidence Interval for Estimating a Population Mean (with σ Not Known)

Objective

Construct a confidence interval used to estimate a population mean.

Notation

μ = population mean

$\bar{x}$ = sample mean

s = sample standard deviation

n = number of sample values

E = margin of error

$t_{\alpha/2}$ = critical t value separating an area of $\alpha/2$ in the right tail of the t distribution

Requirements

1. The sample is a simple random sample.

2. Either the sample is from a normally distributed population or $n > 30$.

Confidence Interval

$$\bar{x} - E < \mu < \bar{x} + E \qquad \text{where} \qquad E = t_{\alpha/2}\frac{s}{\sqrt{n}} \qquad (df = n - 1)$$

or

$$\bar{x} \pm E$$

or

$$(\bar{x} - E, \bar{x} + E)$$

Requirements As in Section 7-3, the requirement of a normally distributed population is not a strict requirement, so we can usually consider the population to be normally distributed after using the sample data to confirm that there are no outliers and the histogram has a shape that is not substantially far from a normal distribution.

Also, as in Section 7-3, the requirement that the sample size is $n > 30$ is commonly used as a guideline, but the minimum sample size actually depends on how much the population distribution departs from a normal distribution. (If a population is known to be normally distributed, the distribution of sample means $\bar{x}$ is *exactly* a normal distribution with mean μ and standard deviation $\sigma/\sqrt{n}$; if the population is not normally distributed, large ($n > 30$) samples yield sample means with a distribution that is *approximately* normal with mean μ and standard deviation $\sigma/\sqrt{n}$.)

Procedure for Constructing a Confidence Interval for μ (with σ unknown)

1. Verify that the requirements are satisfied.

2. Using $n - 1$ degrees of freedom, refer to Table A-3 or use technology to find the critical value $t_{\alpha/2}$ that corresponds to the desired confidence level. (For the confidence level, refer to the "Area in Two Tails.")

3. Evaluate the margin of error $E = t_{\alpha/2} \cdot s/\sqrt{n}$.

4. Using the value of the calculated margin of error E and the value of the sample mean $\bar{x}$, find the values of the confidence interval limits: $\bar{x} - E$ and $\bar{x} + E$. Substitute those values in the general format for the confidence interval.

5. Round the resulting confidence interval limits. If using the original set of data, round to one more decimal place than is used for the original set of data. If using summary statistics ($n, \bar{x}, s$), round the confidence interval limits to the same number of decimal places used for the sample mean.

Using the TI-83/84 Plus Calculator The TI-83/84 Plus calculator can automatically generate confidence interval estimates of the population mean μ. See the procedure at the end of this section.

> **EXAMPLE 2**   **Constructing a Confidence Interval: Garlic for Reducing Cholesterol** A common claim is that garlic lowers cholesterol levels. In a test of the effectiveness of garlic, 49 subjects were treated with doses of raw garlic, and their cholesterol levels were measured before and after the treatment. The changes in their levels of LDL cholesterol (in mg/dL) have a mean of 0.4 and a standard deviation of 21.0 (based on data from "Effect of Raw Garlic vs Commercial Garlic Supplements on Plasma Lipid Concentrations in Adults With Moderate Hypercholesterolemia," by Gardner et al., *Archives of Internal Medicine*, Vol. 167). Use the sample statistics of $n = 49$, $\bar{x} = 0.4$, and $s = 21.0$ to construct a 95% confidence interval estimate of the mean net change in LDL cholesterol after the garlic treatment. What does the confidence interval suggest about the effectiveness of garlic in reducing LDL cholesterol?

SOLUTION   **REQUIREMENT CHECK** We must first verify that the requirements are satisfied. (1) The detailed design of the garlic trials justify the assumption that the sample is a simple random sample. (2) The requirement that "the population is normally distributed or $n > 30$" is satisfied because $n = 49$. The requirements are therefore satisfied.

The confidence level of 95% implies that $\alpha = 0.05$. With $n = 49$, the number of degrees of freedom is $n - 1 = 48$. If using Table A-3, we look for the row with 48 degrees of freedom and the column corresponding to $\alpha = 0.05$ in two tails.

Table A-3 does not include 48 degrees of freedom, and the closest number of degrees of freedom is 50, so we can use $t_{\alpha/2} = 2.009$. (If we use a TI-84 Plus calculator, we get the more accurate result of $t_{\alpha/2} = 2.010634722$.)

Using $t_{\alpha/2} = 2.009$, $s = 21.0$, and $n = 49$, we find the margin of error E as follows:

$$E = t_{\alpha/2}\frac{s}{\sqrt{n}} = 2.009 \cdot \frac{21.0}{\sqrt{49}} = 6.027$$

With $\bar{x} = 0.4$ and $E = 6.027$, we construct the confidence interval as follows:

$$\bar{x} - E < \mu < \bar{x} + E$$
$$0.4 - 6.027 < \mu < 0.4 + 6.027$$
$$-5.6 < \mu < 6.4 \quad \text{(rounded to one decimal place,}$$
$$\text{as in the given sample mean)}$$

If we use a TI-83/84 Plus calculator with the procedure given near the end of this section, we get the display shown in the margin. The confidence interval is $(-5.6, 6.4)$ after rounding.

TI-83/84 Plus

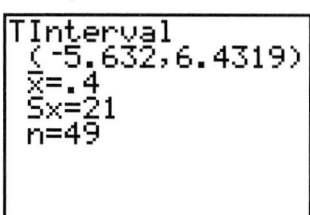

INTERPRETATION   This result could also be expressed in the format of 0.4 ± 6.0 or $(-5.6, 6.4)$. On the basis of the given sample results, we are 95% confident that the limits of -5.6 and 6.4 actually do contain the value of μ, the mean of the changes in LDL cholesterol for the population.

Because the confidence interval limits contain the value of 0, it is very possible that the mean of the changes in LDL cholesterol is equal to 0, suggesting that the garlic treatment did not affect the LDL cholesterol levels. It does not appear that the garlic treatment is effective in lowering LDL cholesterol.

We now list the important properties of the Student t distribution that has been introduced in this section.

Important Properties of the Student t Distribution

1. The Student t distribution is different for different sample sizes. (See Figure 7-5 for the cases $n = 3$ and $n = 12$.)

2. The Student t distribution has the same general symmetric bell shape as the standard normal distribution, but it reflects the greater variability (with wider distributions) that is expected with small samples.

3. The Student t distribution has a mean of $t = 0$ (just as the standard normal distribution has a mean of $z = 0$).

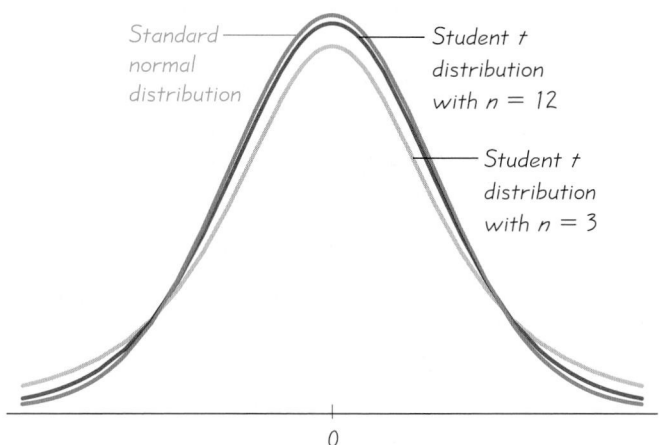

Figure 7-5

Student t Distributions for $n = 3$ and $n = 12$

The Student t distribution has the same general shape and symmetry as the standard normal distribution, but it reflects the greater variability that is expected with small samples.

4. The standard deviation of the Student t distribution varies with the sample size, but it is greater than 1 (unlike the standard normal distribution, which has $\sigma = 1$).

5. As the sample size n gets larger, the Student t distribution gets closer to the standard normal distribution.

Choosing the Appropriate Distribution

It is sometimes difficult to decide whether to use the standard normal z distribution or the Student t distribution. The flowchart in Figure 7-6 and the accompanying Table 7-1 both summarize the key points to consider when constructing confidence intervals for estimating μ, the population mean. In Figure 7-6 or Table 7-1, note that if we have a small sample ($n \leq 30$) drawn from a distribution that differs dramatically from a normal distribution, we can't use the methods described in this chapter. One alternative is to use nonparametric methods (see Chapter 13), and another alternative is to use the computer bootstrap method. In both of those approaches, no assumptions are made about the original population. The bootstrap method is described in the Technology Project at the end of this chapter.

Important: Figure 7-6 and Table 7-1 assume that the sample is a simple random sample. If the sample data have been collected using some inappropriate method, such as a convenience sample or a voluntary response sample, it is very possible that no methods of statistics can be used to find a useful estimate of a population mean.

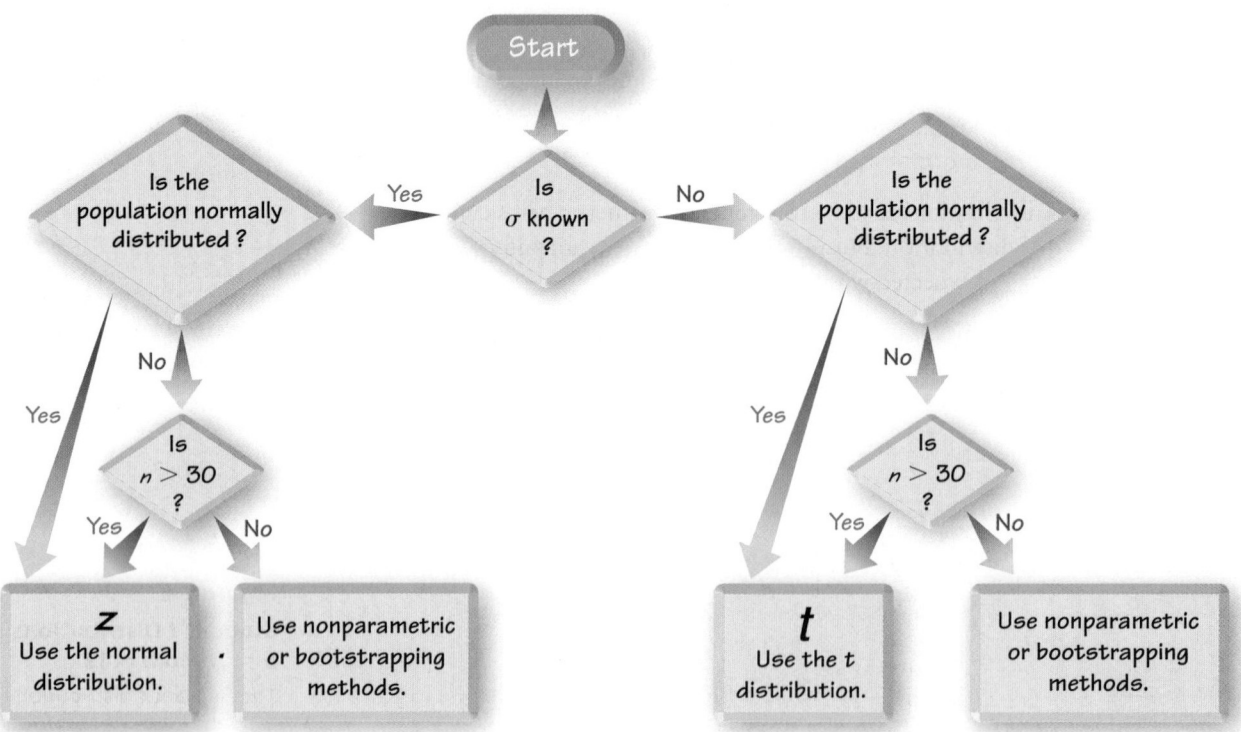

Figure 7-6   Choosing Between *z* and *t*

Table 7-1 Choosing Between *z* and *t*

Method	Conditions
Use normal (*z*) distribution.	σ known and normally distributed population *or* σ known and $n > 30$
Use *t* distribution.	σ not known and normally distributed population *or* σ not known and $n > 30$
Use a nonparametric method or bootstrapping.	Population is not normally distributed and $n \leq 30$.

Notes: 1. **Criteria for deciding whether the population is normally distributed:** The population need not be exactly normal, but it should appear to be somewhat symmetric with one mode and no outliers.
   2. **Sample size $n > 30$:** This is a common guideline, but sample sizes of 15 to 30 are adequate if the population appears to have a distribution that is not far from being normal and there are no outliers. For some population distributions that are extremely far from normal, the sample size might need to be much larger than 30.

The following example focuses on choosing the correct approach.

EXAMPLE 3 **Choosing Distributions** You plan to construct a confidence interval for the population mean μ. Use the given data to determine whether the margin of error E should be calculated using a critical value of $z_{\alpha/2}$ (from the normal distribution), a critical value of $t_{\alpha/2}$ (from a *t* distribution), or neither (so that the methods of Section 7-3 and this section cannot be used).

a. $n = 9, \bar{x} = 75, s = 15$, and the population has a normal distribution.

b. $n = 5, \bar{x} = 20, s = 2$, and the population has a very skewed distribution.

c. $n = 12, \bar{x} = 98.6, \sigma = 0.6$, and the population has a normal distribution. (In reality, σ is rarely known.)

d. $n = 75, \bar{x} = 98.6, \sigma = 0.6$, and the population has a skewed distribution. (In reality, σ is rarely known.)

e. $n = 75, \bar{x} = 98.6, s = 0.6$, and the population has a skewed distribution.

SOLUTION Refer to Figure 7-6 or Table 7-1.

a. Because the population standard deviation σ is not known and the population is normally distributed, the margin of error is calculated using $t_{\alpha/2}$.

b. Because the sample is small ($n \leq 30$) and the population does not have a normal distribution, the margin of error E should not be calculated using a critical value of $z_{\alpha/2}$ or $t_{\alpha/2}$. The methods of Section 7-3 and this section do not apply.

continued

Estimating Crowd Size

There are sophisticated methods of analyzing the size of a crowd. Aerial photographs and measures of people density can be used with reasonably

good accuracy. However, reported crowd size estimates are often simple guesses. After the Boston Red Sox won the World Series for the first time in 86 years, Boston city officials estimated that the celebration parade was attended by 3.2 million fans. Boston police provided an estimate of around 1 million, but it was admittedly based on guesses by police commanders. A photo analysis led to an estimate of around 150,000. Boston University Professor Farouk El-Baz used images from the U.S. Geological Survey to develop an estimate of at most 400,000. MIT physicist Bill Donnelly said that "it's a serious thing if people are just putting out any number. It means other things aren't being vetted that carefully."

Estimating Sugar in Oranges

In Florida, members of the citrus industry make extensive use of statistical methods. One particular application involves the way in which growers are paid for oranges used to make orange juice. An arriving truckload of oranges is first weighed at the receiving plant, then a sample of about a dozen oranges is randomly selected. The sample is weighed and then squeezed, and the amount of sugar in the juice is measured. Based on the sample results, an estimate is made of the total amount of sugar in the entire truckload. Payment for the load of oranges is based on the estimate of the amount of sugar because sweeter oranges are more valuable than those less sweet, even though the amounts of juice may be the same.

c. Because σ is known and the population has a normal distribution, the margin of error is calculated using $z_{\alpha/2}$.

d. Because the sample is large ($n > 30$) and σ is known, the margin of error is calculated using $z_{\alpha/2}$.

e. Because the sample is large ($n > 30$) and σ is not known, the margin of error is calculated using $t_{\alpha/2}$.

EXAMPLE 4   **Confidence Interval for Alcohol in Video Games** Twelve different video games showing substance use were observed. The duration times (in seconds) of alcohol use were recorded, with the times listed below (based on data from "Content and Ratings of Teen-Rated Video Games," by Haninger and Thompson, *Journal of the American Medical Association*, Vol. 291, No. 7). The design of the study justifies the assumption that the sample can be treated as a simple random sample. Use the sample data to construct a 95% confidence interval estimate of μ, the mean duration time that the video showed the use of alcohol.

| 84 | 14 | 583 | 50 | 0 | 57 | 207 | 43 | 178 | 0 | 2 | 57 |

SOLUTION   **REQUIREMENT CHECK** We must first verify that the requirements are satisfied. (1) We can consider the sample to be a simple random sample. (2) When checking the requirement that "the population is normally distributed or $n > 30$," we see that the sample size is $n = 12$, so we must determine whether the data appear to be from a population with a normal distribution. Shown below are a Minitab-generated histogram and a STATDISK-generated normal quantile plot. The histogram does not appear to be bell-shaped, and the points in the normal quantile plot are not reasonably close to a straight-line pattern, so it appears that the times are not from a population having a normal distribution. The requirements are not satisfied. If we were to proceed with the construction of the confidence interval, we would get 1.8 sec $< \mu <$ 210.7 sec, but this result is questionable because it assumes incorrectly that the requirements are satisfied.

INTERPRETATION   Because the requirement that "the population is normally distributed or $n > 30$" is not satisfied, we do not have 95% confidence that the limits of 1.8 sec and 210.7 sec actually do contain the value of the population mean. We should use some other approach for finding the confidence interval limits. For example, the author used bootstrap resampling as described in the Technology Project at the end of this section. The confidence interval of 35.3 sec $< \mu <$ 205.6 sec was obtained.

MINITAB

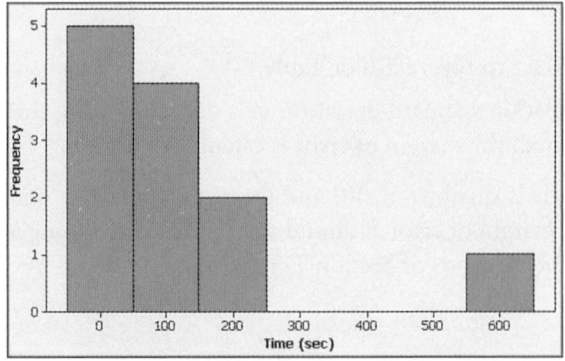

STATDISK

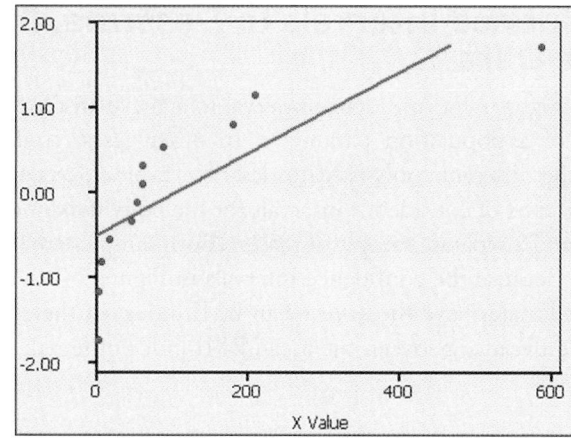

Finding Point Estimate and *E* from a Confidence Interval

Later in this section we will describe how computer software and calculators can be used to find a confidence interval. A typical use requires that you enter a confidence level and sample statistics, and the display shows the confidence interval limits. The sample mean $\bar{x}$ is the value midway between those limits, and the margin of error *E* is one-half the difference between those limits (because the upper limit is $\bar{x} + E$ and the lower limit is $\bar{x} - E$, the distance separating them is $2E$).

Point estimate of μ: $\quad \bar{x} = \dfrac{\text{(upper confidence limit)} + \text{(lower confidence limit)}}{2}$

Margin of error: $\quad\quad E = \dfrac{\text{(upper confidence limit)} - \text{(lower confidence limit)}}{2}$

EXAMPLE 5 **Weights of Garbage** Data Set 22 in Appendix B lists the weights of discarded garbage from a sample of 62 households. The accompanying TI-83/84 Plus calculator screen displays results from using the 62 amounts of total weights (in pounds) to construct a 95% confidence interval estimate of the mean weight of garbage discarded by the population of all households. Use the displayed confidence interval to find the values of the best point estimate $\bar{x}$ and the margin of error *E*.

SOLUTION In the following calculations, results are rounded to three decimal places, which is one additional decimal place beyond the two decimal places used for the original list of weights.

$$\bar{x} = \frac{\text{(upper confidence limit)} + \text{(lower confidence limit)}}{2}$$

$$= \frac{30.607 + 24.28}{2} = 27.444 \text{ lb}$$

$$E = \frac{\text{(upper confidence limit)} - \text{(lower confidence limit)}}{2}$$

$$= \frac{30.607 - 24.28}{2} = 3.164 \text{ lb}$$

TI-83/84 PLUS

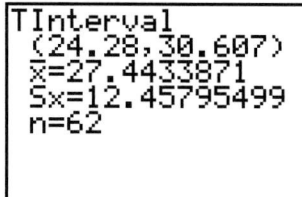

Using Confidence Intervals to Describe, Explore, or Compare Data

In some cases, we might use a confidence interval to achieve an ultimate goal of estimating the value of a population parameter. In other cases, confidence intervals might be among the different tools used to describe, explore, or compare data sets. Figure 7-7 shows graphs of confidence intervals for the body mass indexes (BMI) of a sample of females and a separate sample of males. (Both samples are listed in Data Set 1 in Appendix B.) Because the confidence intervals in Figure 7-7 overlap, it is possible that females and males have the *same* mean BMI index, so there does not appear to be a significant difference between the mean BMI index of females and males.

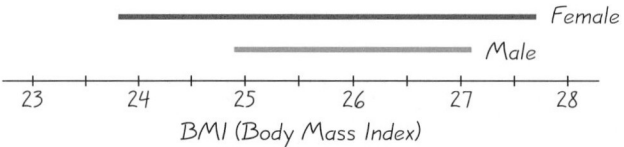

Figure 7-7   BMI Indexes of Females and Males

CAUTION

As in Sections 7-2 and 7-3, confidence intervals can be used *informally* to compare different data sets, but *the overlapping of confidence intervals should not be used for making formal and final conclusions about equality of means.*

Determining Sample Size   Section 7-2 included a subsection describing methods for determining the size of a sample needed to estimate a population proportion, and Section 7-3 included a subsection with methods for determining the size of a sample needed to estimate a population mean. This section does not include a similar subsection. When determining the sample size needed to estimate a population mean, use the procedure described in Section 7-3, which requires an estimated or known value of the population standard deviation.

Using a TI-83 Plus Calculator for Confidence Interval Estimates of μ (With σ Not Known)

Here is the procedure for using a TI-83/84 Plus calculator for the construction of a confidence interval estimate of a population mean μ, assuming that the population standard deviation σ is not known:

1. Press **STAT**, then use the ▷ key to select the option of **TESTS.**
2. If the population standard deviation σ is not known, select the menu option of **TInterval.** Select this option by pressing ▽ to scroll down to **TInterval.** (If σ is known, use the procedure described in Section 7-3.)
3. The first line, identified as Inpt (for input), gives you the option of using Data or Stats. Use the arrow keys ◁ and ▷ to select either Data or Stats.
   - **Data:** Use this option if you have the sample data values stored as a list.
   - **Stats:** Use this option if you know the values of $\bar{x}$, s, and n.

4. What happens next depends on the choice made in step 3. If you choose to use Stats, you must enter the values of $\bar{x}$, s, and n.

If you choose to use Data, you must enter the list where the data are stored, and a value for Freq (frequency). Enter 1 for the frequency. (However, if the sample data are stored in two columns as a frequency distribution, the Freq entry should be the list containing frequencies.)

5. Enter a confidence level, such as 0.95, then scroll down to **Calculate** and press **ENTER**. The first screen in the margin shows the entries corresponding to Example 2. The second screen shows the confidence interval limits enclosed within parentheses. The TI-83/84 Plus confidence interval of $(-5.632, 6.4319)$ can be expressed as $-5.6 < \mu < 6.4$ after rounding, as in Example 2.

TI-83/84 Plus

```
TInterval
 Inpt:Data Stats
 x̄:.4
 Sx:21
 n:49
 C-Level:.95
 Calculate
```

TI-83/84 Plus

```
TInterval
 (-5.632,6.4319)
 x̄=.4
 Sx=21
 n=49
```

7-4 Basic Skills and Concepts

Statistical Literacy and Critical Thinking

1. What's Wrong? A "snapshot" in *USA Today* noted that "Consumers will spend an estimated average of \$483 on merchandise" for back-to-school spending. It was reported that the value is based on a survey of 8453 consumers, and the margin of error is "±1 percentage point." What's wrong with this information?

2. Robust What does it mean when we say that the methods for constructing confidence intervals in this section are *robust* against departures from normality? Are the methods for constructing confidence intervals in this section robust against poor sampling methods?

3. Sampling A national polling organization has been hired to estimate the mean amount of cash carried by adults in the United States. The original sampling plan involved telephone calls placed to 2500 different telephone numbers throughout the United States, but a manager decides to save long-distance telephone expenses by using a simple random sample of 2500 telephone numbers that are all within the state of California. If this sample is used to construct a 95% confidence interval to estimate the population mean, will the estimate be good? Why or why not?

4. Degrees of Freedom A simple random sample of size $n = 5$ is obtained from the population of drivers living in New York City, and the braking reaction time of each driver is measured. The results are to be used for constructing a 95% confidence interval. What is the number of degrees of freedom that should be used for finding the critical value $t_{\alpha/2}$? Give a brief explanation of the number of degrees of freedom.

Using Correct Distribution. *In Exercises 5–12, assume that we want to construct a confidence interval using the given confidence level. Do one of the following, as appropriate: (a) Find the critical value $z_{\alpha/2}$, (b) find the critical value $t_{\alpha/2}$, (c) state that neither the normal nor the t distribution applies.*

TI 5. 95%; $n = 23$; σ is unknown; population appears to be normally distributed.

TI 6. 99%; $n = 25$; σ is known; population appears to be normally distributed.

TI 7. 99%; $n = 6$; σ is unknown; population appears to be very skewed.

TI 8. 95%; $n = 40$; σ is unknown; population appears to be skewed.

TI 9. 90%; $n = 200$; $\sigma = 15.0$; population appears to be skewed.

TI 10. 95%; $n = 9$; σ is unknown; population appears to be very skewed.

TI 11. 99%; $n = 12$; σ is unknown; population appears to be normally distributed.

TI 12. 95%; $n = 38$; σ is unknown; population appears to be skewed.

Finding Confidence Intervals. *In Exercises 13 and 14, use the given confidence level and sample data to find (a) the margin of error and (b) the confidence interval for the population mean μ. Assume that the sample is a simple random sample and the population has a normal distribution.*

TI **13. Hospital Costs** 95% confidence; $n = 20$, $\bar{x} = \$9004$, $s = \$569$ (based on data from hospital costs for car crash victims who wore seat belts, from the U.S. Department of Transportation)

TI **14. Car Pollution** 99% confidence; $n = 7$, $\bar{x} = 0.12$, $s = 0.04$ (original values are nitrogen-oxide emissions in grams/mile, from the Environmental Protection Agency)

Interpreting Display. *In Exercises 15 and 16, use the given data and the corresponding display to express the confidence interval in the format of $\bar{x} - E < \mu < \bar{x} + E$. Also write a statement that interprets the confidence interval.*

15. Weights of Dollar Coins 95% confidence; $n = 20$, $\bar{x} = 8.0710$ g, $s = 0.0411$ g (based on measurements made by the author). See the following SPSS display.

SPSS

			Statistic	Std. Error
Coins	Mean		8.0710	.00919
	95% Confidence Interval for Mean	Lower Bound	8.0518	
		Upper Bound	8.0903	

TI-83/84 PLUS

```
TInterval
 (1.5514,2.2706)
 x̄=1.911
 Sx=1.065
 n=62
```

TI **16. Weights of Plastic Discarded by Households** 99% confidence; $n = 62$, $\bar{x} = 1.911$ lb, $s = 1.065$ lb (based on data from the Garbage Project, University of Arizona). See the TI-83/84 Plus calculator display in the margin.

Constructing Confidence Intervals. *In Exercises 17–30, construct the confidence interval.*

TI **17. Garlic for Reducing Cholesterol** In a test of the effectiveness of garlic for lowering cholesterol, 47 subjects were treated with Garlicin, which is garlic in a processed tablet form. Cholesterol levels were measured before and after the treatment. The changes in their levels of LDL cholesterol (in mg/dL) have a mean of 3.2 and a standard deviation of 18.6 (based on data from "Effect of Raw Garlic vs Commercial Garlic Supplements on Plasma Lipid Concentrations in Adults With Moderate Hypercholesterolemia," by Gardner et al., *Archives of Internal Medicine*, Vol. 167).

a. What is the best point estimate of the population mean net change in LDL cholesterol after the Garlicin treatment?

b. Construct a 95% confidence interval estimate of the mean net change in LDL cholesterol after the Garlicin treatment. What does the confidence interval suggest about the effectiveness of Garlicin in reducing LDL cholesterol?

TI **18. Birth Weights** A random sample of the birth weights of 186 babies has a mean of 3103 g and a standard deviation of 696 g (based on data from "Cognitive Outcomes of Preschool Children with Prenatal Cocaine Exposure," by Singer et al., *Journal of the American Medical Association*, Vol. 291, No. 20). These babies were born to mothers who did not use cocaine during their pregnancies.

a. What is the best point estimate of the mean weight of babies born to mothers who did not use cocaine during their pregnancies?

b. Construct a 95% confidence interval estimate of the mean birth weight for all such babies.

c. Compare the confidence interval from part (b) to this confidence interval obtained from birth weights of babies born to mothers who used cocaine during pregnancy: 2608 g $< \mu <$ 2792 g. Does cocaine use appear to affect the birth weight of a baby?

TI **19. Mean Body Temperature** Data Set 2 in Appendix B includes 106 body temperatures for which $\bar{x} = 98.20°F$ and $s = 0.62°F$.

a. What is the best point estimate of the mean body temperature of all healthy humans?

b. Construct a 99% confidence interval estimate of the mean body temperature of all healthy humans. Do the confidence interval limits contain 98.6°F? What does the sample suggest about the use of 98.6°F as the mean body temperature?

TI **20. Atkins Weight Loss Program** In a test of the Atkins weight loss program, 40 individuals participated in a randomized trial with overweight adults. After 12 months, the mean weight *loss* was found to be 2.1 lb, with a standard deviation of 4.8 lb.

a. What is the best point estimate of the mean weight loss of all overweight adults who follow the Atkins program?

b. Construct a 99% confidence interval estimate of the mean weight loss for all such subjects.

c. Does the Atkins program appear to be effective? Is it practical?

TI **21. Echinacea Treatment** In a study designed to test the effectiveness of echinacea for treating upper respiratory tract infections in children, 337 children were treated with echinacea and 370 other children were given a placebo. The numbers of days of peak severity of symptoms for the echinacea treatment group had a mean of 6.0 days and a standard deviation of 2.3 days. The numbers of days of peak severity of symptoms for the placebo group had a mean of 6.1 days and a standard deviation of 2.4 days (based on data from "Efficacy and Safety of Echinacea in Treating Upper Respiratory Tract Infections in Children," by Taylor et al., *Journal of the American Medical Association,* Vol. 290, No. 21).

a. Construct the 95% confidence interval for the mean number of days of peak severity of symptoms for those who receive echinacea treatment.

b. Construct the 95% confidence interval for the mean number of days of peak severity of symptoms for those who are given a placebo.

c. Compare the two confidence intervals. What do the results suggest about the effectiveness of echinacea?

TI **22. Acupuncture for Migraines** In a study designed to test the effectiveness of acupuncture for treating migraine, 142 subjects were treated with acupuncture and 80 subjects were given a sham treatment. The numbers of migraine attacks for the acupuncture treatment group had a mean of 1.8 and a standard deviation of 1.4. The numbers of migraine attacks for the sham treatment group had a mean of 1.6 and a standard deviation of 1.2.

a. Construct the 95% confidence interval estimate of the mean number of migraine attacks for those treated with acupuncture.

b. Construct the 95% confidence interval estimate of the mean number of migraine attacks for those given a sham treatment.

c. Compare the two confidence intervals. What do the results suggest about the effectiveness of acupuncture?

TI **23. Magnets for Treating Back Pain** In a study designed to test the effectiveness of magnets for treating back pain, 20 patients were given a treatment with magnets and also a sham treatment without magnets. Pain was measured using a standard Visual Analog Scale (VAS). After given the magnet treatments, the 20 patients had VAS scores with a mean of 5.0 and a standard deviation of 2.4. After being given the sham treatments, the 20 patients had VAS scores with a mean of 4.7 and a standard deviation of 2.9.

a. Construct the 95% confidence interval estimate of the mean VAS score for patients given the magnet treatment.

b. Construct the 95% confidence interval estimate of the mean VAS score for patients given a sham treatment.

c. Compare the results. Does the treatment with magnets appear to be effective?

Ⓣ 24. Ages of Oscar Winning Actresses and Actors The ages of the 79 actresses at the time that they won Oscars for the Best Actress category have a mean of 35.8 years and a standard deviation of 11.3 years. The ages of the 79 actors at the time that they won Oscars for the category of Best Actor have a mean of 43.8 years and a standard deviation of 8.9 years. Assume that the samples are simple random samples.

a. Construct the 99% confidence interval estimate of the mean age of actresses at the time that they win Oscars for the Best Actress category.

b. Construct the 99% confidence interval estimate of the mean age of actors at the time that they win Oscars for the Best Actor category.

c. Compare the results.

Ⓣ 25. Monitoring Lead in Air Listed below are measured amounts of lead (in micrograms per cubic meter, or $\mu g/m^3$) in the air. The Environmental Protection Agency (EPA) has established an air quality standard for lead of $1.5 \ \mu g/m^3$. The measurements shown below were recorded at Building 5 of the World Trade Center site on different days immediately following the destruction caused by the terrorist attacks of September 11, 2001. After the collapse of the two World Trade Center buildings, there was considerable concern about the quality of the air. Use the given values to construct a 95% confidence interval estimate of the mean amount of lead in the air. Is there anything about this data set suggesting that the confidence interval might not be very good? Explain.

5.40   1.10   0.42   0.73   0.48   1.10

Ⓣ 26. Estimating Car Pollution In a sample of seven cars, each car was tested for nitrogen-oxide emissions (in grams per mile) and the following results were obtained: 0.06, 0.11, 0.16, 0.15, 0.14, 0.08, 0.15 (based on data from the EPA). Assuming that this sample is representative of the cars in use, construct a 98% confidence interval estimate of the mean amount of nitrogen-oxide emissions for all cars. If the EPA requires that nitrogen-oxide emissions be less than 0.165 g/mi, can we safely conclude that this requirement is being met?

Ⓣ 27. TV Salaries Listed below are the top 10 salaries (in millions of dollars) of television personalities in a recent year (listed in order for Letterman, Cowell, Sheindlin, Leno, Couric, Lauer, Sawyer, Viera, Sutherland, and Sheen, based on data from *OK!* magazine).

a. Use the sample data to construct the 95% confidence interval for the population mean.

b. Do the sample data represent a simple random sample of TV salaries?

c. What is the assumed population? Is the sample representative of the population?

d. Does the confidence interval make sense?

38   36   35   27   15   13   12   10   9.6   8.4

Ⓣ 28. Movie Lengths Listed below are 12 lengths (in minutes) of randomly selected movies from Data Set 9 in Appendix B.

a. Construct a 99% confidence interval estimate of the mean length of all movies.

b. Assuming that it takes 30 min to empty a theater after a movie, clean it, allow time for the next audience to enter, and show previews, what is the minimum time that a theater manager should plan between start times of movies, assuming that this time will be sufficient for typical movies?

110   96   125   94   132   120   136   154   149   94   119   132

Ⓣ 29. Video Games Twelve different video games showing substance use were observed and the duration times of game play (in seconds) are listed below (based on data from "Content and Ratings of Teen-Rated Video Games," by Haninger and Thompson, *Journal of the American Medical Association,* Vol. 291, No. 7). The design of the study justifies the assumption that the sample can be treated as a simple random sample. Use the sample data to construct a 95% confidence interval estimate of μ, the mean duration of game play.

4049   3884   3859   4027   4318   4813   4657   4033   5004   4823   4334   4317

30. Ages of Presidents Listed below are the ages of the Presidents of the United States at the times of their inaugurations. Construct a 99% confidence interval estimate of the mean age of presidents at the times of their inaugurations. What is the population? Does the confidence interval provide a good estimate of the population mean? Why or why not?

42 43 46 46 47 48 49 49 50 51 51 51 51 51 52 52 54 54 54 54 54 55

55 55 55 56 56 56 57 57 57 57 58 60 61 61 61 62 64 64 65 68 69

Appendix B Data Sets. *In Exercises 31 and 32, use the data sets from Appendix B.*

31. Nicotine in Cigarettes Refer to Data Set 4 in Appendix B and assume that the samples are simple random samples obtained from normally distributed populations.

a. Construct a 95% confidence interval estimate of the mean amount of nicotine in cigarettes that are king size, nonfiltered, nonmenthol, and non-light.

b. Construct a 95% confidence interval estimate of the mean amount of nicotine in cigarettes that are 100 mm, filtered, nonmenthol, and non-light.

c. Compare the results. Do filters on cigarettes appear to be effective?

32. Pulse Rates A physician wants to develop criteria for determining whether a patient's pulse rate is atypical, and she wants to determine whether there are significant differences between males and females. Use the sample pulse rates in Data Set 1 from Appendix B.

a. Construct a 95% confidence interval estimate of the mean pulse rate for males.

b. Construct a 95% confidence interval estimate of the mean pulse rate for females.

c. Compare the preceding results. Can we conclude that the population means for males and females are different? Why or why not?

7-4 Beyond the Basics

33. Effect of an Outlier Use the sample data from Exercise 30 to find a 99% confidence interval estimate of the population mean, after changing the first age from 42 years to 422 years. This value is not realistic, but such an error can easily occur during a data entry process. Does the confidence interval change much when 42 years is changed to 422 years? Are confidence interval limits sensitive to outliers? How should you handle outliers when they are found in sample data sets that will be used for the construction of confidence intervals?

34. Alternative Method Figure 7-6 and Table 7-1 summarize the decisions made when choosing between the normal and t distributions. An alternative method included in some textbooks (but almost never used by professional statisticians and almost never included in professional journals) is based on this criterion: Substitute the sample standard deviation s for σ whenever $n > 30$, then proceed as if σ is known. Using this alternative method, repeat Exercise 30. Compare the results to those found in Exercise 30, and comment on the implications of the change in the width of the confidence interval.

35. Finite Population Correction Factor If a simple random sample of size n is selected without replacement from a finite population of size N, and the sample size is more than 5% of the population size ($n > 0.05N$), better results can be obtained by using the finite population correction factor, which involves multiplying the margin of error E by $\sqrt{(N - n)/(N - 1)}$. For the sample of 100 weights of M&M candies in Data Set 18 from Appendix B, we get $\bar{x} = 0.8565$ g and $s = 0.0518$ g. First construct a 95% confidence interval estimate of μ assuming that the population is large, then construct a 95% confidence interval estimate of the mean weight of M&Ms in the full bag from which the sample was taken. The full bag has 465 M&Ms. Compare the results.

36. Confidence Interval for Sample of Size _n_ = 1 When a manned NASA spacecraft lands on Mars, the astronauts encounter a single adult Martian, who is found to be 12.0 ft tall. It is reasonable to assume that the heights of all Martians are normally distributed.

a. The methods of this chapter require information about the variation of a variable. If only one sample value is available, can it give us any information about the variation of the variable?

b. Based on the article "An Effective Confidence Interval for the Mean with Samples of Size One and Two," by Wall, Boen, and Tweedie (*American Statistician,* Vol. 55, No. 2), a 95% confidence interval for μ can be found (using methods not discussed in this book) for a sample of size $n = 1$ randomly selected from a normally distributed population, and it can be expressed as $x \pm 9.68|x|$. Use this result to construct a 95% confidence interval using the single sample value of 12.0 ft, and express it in the format of $\bar{x} - E < \mu < \bar{x} + E$. Based on the result, is it likely that some other randomly selected Martian might be 50 ft tall?

 7-5 Estimating a Population Variance

Key Concept In this section we introduce the chi-square probability distribution so that we can construct confidence interval estimates of a population standard deviation or variance. We also present a method for determining the sample size required to estimate a population standard deviation or variance.

When we considered estimates of proportions and means, we used the normal and Student t distributions. When developing estimates of variances or standard deviations, we use another distribution, referred to as the chi-square distribution. We will examine important features of that distribution before proceeding with the development of confidence intervals.

Chi-Square Distribution

In a normally distributed population with variance σ^2, assume that we randomly select independent samples of size n and, for each sample, compute the sample variance s^2 (which is the square of the sample standard deviation s). The sample statistic $\chi^2 = (n - 1)s^2/\sigma^2$ has a sampling distribution called the **chi-square distribution.**

Chi-Square Distribution	
Formula 7-5	$\chi^2 = \dfrac{(n - 1)s^2}{\sigma^2}$
Where	n = number of sample values
	s^2 = sample variance
	σ^2 = population variance

We denote chi-square by χ^2, pronounced "kigh square." To find critical values of the chi-square distribution, refer to Table A-4. The chi-square distribution is determined by the number of degrees of freedom, and in this chapter we use $n - 1$ degrees of freedom.

degrees of freedom $= n - 1$

In later chapters we will encounter situations in which the degrees of freedom are not $n - 1$, so we should not make the incorrect generalization that the number of degrees of freedom is always $n - 1$.

Properties of the Chi-Square Distribution

1. The chi-square distribution is not symmetric, unlike the normal and Student t distributions (see Figure 7-8). (As the number of degrees of freedom increases, the distribution becomes more symmetric, as Figure 7-9 illustrates.)

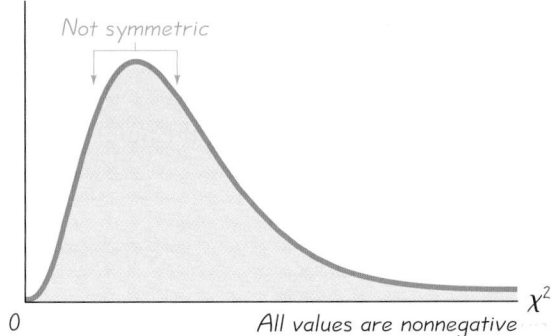

Not symmetric

0   *All values are nonnegative* χ^2

Figure 7-8   Chi-Square Distribution

2. The values of chi-square can be zero or positive, but they cannot be negative (see Figure 7-8).

3. The chi-square distribution is different for each number of degrees of freedom (see Figure 7-9), and the number of degrees of freedom is given by df $= n - 1$. As the number of degrees of freedom increases, the chi-square distribution approaches a normal distribution.

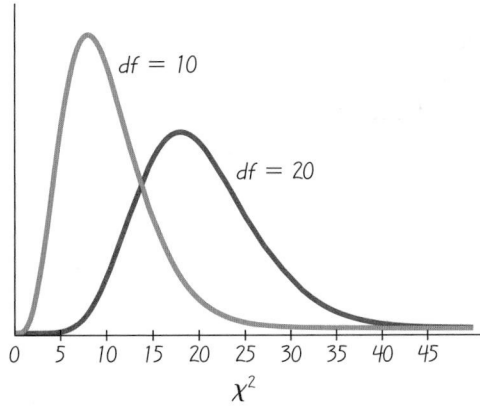

df = 10

df = 20

0 5 10 15 20 25 30 35 40 45
χ^2

**Figure 7-9   Chi-Square Distribution
for df = 10 and df = 20**

Because the chi-square distribution is not symmetric, the confidence interval for σ^2 does not fit a format of $s^2 \pm E$, and we must do separate calculations for the upper and lower confidence interval limits. If using Table A-4 for finding critical values, note the following design feature of that table:

In Table A-4, each critical value of χ^2 corresponds to an area given in the top row of the table, and that area represents the *cumulative area located to the right* of the critical value.

Table A-2 for the standard normal distribution provides cumulative areas from the *left*, but Table A-4 for the chi-square distribution provides cumulative areas from the *right*.

> **EXAMPLE 1**   **Finding Critical Values of χ^2** A simple random sample of ten voltage levels is obtained. Construction of a confidence interval for the population standard deviation σ requires the left and right critical values of χ^2 corresponding to a confidence level of 95% and a sample size of $n = 10$. Find the critical value of χ^2 separating an area of 0.025 in the left tail, and find the critical value of χ^2 separating an area of 0.025 in the right tail.

> **SOLUTION**   With a sample size of $n = 10$, the number of degrees of freedom is df $= n - 1 = 9$. See Figure 7-10.

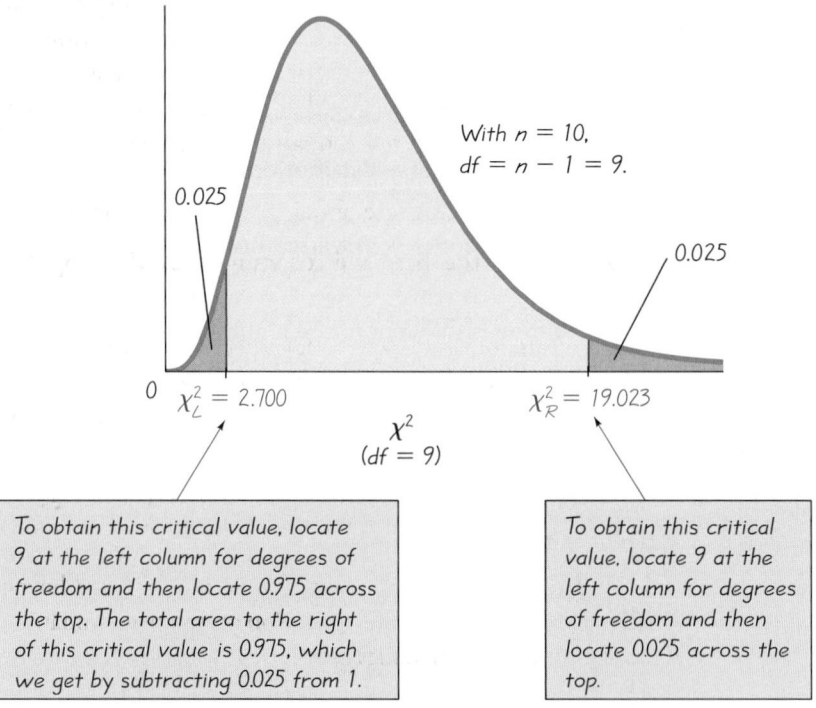

Figure 7-10   Critical Values of the Chi-Square Distribution

If using Table A-4, the critical value to the right ($\chi_R^2 = 19.023$) is obtained in a straightforward manner by locating 9 in the degrees-of-freedom column at the left and 0.025 across the top row. The critical value of $\chi_L^2 = 2.700$ to the left once again corresponds to 9 in the degrees-of-freedom column, but we must locate 0.975 (found by subtracting 0.025 from 1) across the top row because the values in the top row are always *areas to the right* of the critical value. Refer to Figure 7-10 and see that the total area to the right of $\chi_L^2 = 2.700$ is 0.975. Figure 7-10 shows that, for a sample of 10 values taken from a normally distributed population, the chi-square statistic $(n - 1)s^2/\sigma^2$ has a 0.95 probability of falling between the chi-square critical values of 2.700 and 19.023.

Instead of using Table A-4, technology can be used to find critical values of χ^2. A major advantage of technology is that it can be used for any number of degrees of freedom and any confidence level, not just the limited choices included in Table A-4. If using a TI-83/84 Plus calculator, use the program **CHINVR83.**

When obtaining critical values of χ^2 from Table A-4, note that the numbers of degrees of freedom are consecutive integers from 1 to 30, followed by 40, 50, 60, 70, 80, 90, and 100. When a number of degrees of freedom (such as 52) is not found in the table, you can usually use the closest critical value. For example, if the number of degrees of freedom is 52, refer to Table A-4 and use 50 degrees of freedom. (If the number of degrees of freedom is exactly midway between table values, such as 55, simply find the mean of the two χ^2 values.) For numbers of degrees of freedom greater than 100, use the equation given in Exercise 27, or a more extensive table, or use technology.

Estimators of σ^2

In Section 6-4 we showed that sample variances s^2 tend to target (or center on) the value of the population variance σ^2, so we say that s^2 is an *unbiased estimator* of σ^2. That is, sample variances s^2 do not systematically tend to overestimate the value of σ^2, nor do they systematically tend to underestimate σ^2. Instead, they tend to target the value of σ^2 itself. Also, the values of s^2 tend to produce smaller errors by being closer to σ^2 than do other unbiased measures of variation. For these reasons, s^2 is generally used to estimate σ^2. (However, there are other estimators of σ^2 that could be considered better than s^2. For example, even though $(n - 1)s^2/(n + 1)$ is a biased estimator of σ^2, it has the desirable property of minimizing the mean of the squares of the errors and therefore has a better chance of being closer to σ^2. See Exercise 28.)

> **The sample variance s^2 is the best point estimate of the population variance σ^2.**

Because s^2 is an unbiased estimator of σ^2, we might expect that s would be an unbiased estimator of σ, but this is not the case. (See Section 6-4.) If the sample size is large, however, the bias is small, so that we can use s as a reasonably good estimate of σ. Even though it is a biased estimate, s is often used as a point estimate of σ.

> **The sample standard deviation s is commonly used as a point estimate of σ (even though it is a biased estimate).**

Although s^2 is the best point estimate of σ^2, there is no indication of how good it actually is. To compensate for that deficiency, we develop an interval estimate (or confidence interval) that gives us a range of values associated with a confidence level.

Confidence Interval for Estimating a Population Standard Deviation or Variance

Objective

Construct a confidence interval used to estimate a population standard deviation or variance.

Notation

σ = population standard deviation	σ^2 = population variance
s = sample standard deviation	s^2 = sample variance
n = number of sample values	E = margin of error
χ_L^2 = left-tailed critical value of χ^2	χ_R^2 = right-tailed critical value of χ^2

continued

Requirements

1. The sample is a simple random sample.

2. The population must have normally distributed values (even if the sample is large).

Confidence Interval for the Population Variance σ^2

$$\frac{(n-1)s^2}{\chi_R^2} < \sigma^2 < \frac{(n-1)s^2}{\chi_L^2}$$

Confidence Interval for the Population Standard Deviation σ

$$\sqrt{\frac{(n-1)s^2}{\chi_R^2}} < \sigma < \sqrt{\frac{(n-1)s^2}{\chi_L^2}}$$

Requirements   For the methods of this section, departures from normal distributions can lead to gross errors. Consequently, the requirement of a normal distribution is much stricter here than in earlier sections, and we should check the distribution of data by constructing histograms and normal quantile plots, as described in Section 6-7.

Procedure for Constructing a Confidence Interval for σ or σ^2

1. Verify that the requirements are satisfied.

2. Using $n - 1$ degrees of freedom, refer to Table A-4 or use technology to find the critical values χ_R^2 and χ_L^2 that correspond to the desired confidence level.

3. Evaluate the upper and lower confidence interval limits using this format of the confidence interval:

$$\frac{(n-1)s^2}{\chi_R^2} < \sigma^2 < \frac{(n-1)s^2}{\chi_L^2}$$

4. If a confidence interval estimate of σ is desired, take the square root of the upper and lower confidence interval limits and change σ^2 to σ.

5. Round the resulting confidence interval limits. If using the original set of data, round to one more decimal place than is used for the original set of data. If using the sample standard deviation or variance, round the confidence interval limits to the same number of decimal places.

CAUTION

Confidence intervals can be used *informally* to compare the variation in different data sets, but *the overlapping of confidence intervals should not be used for making formal and final conclusions about equality of variances or standard deviations.*

> **EXAMPLE 2**   **Confidence Interval for Home Voltage** The proper operation of typical home appliances requires voltage levels that do not vary much. Listed below are ten voltage levels (in volts) recorded in the author's home on ten different days. (The voltages are from Data Set 13 in Appendix B.) These ten values have a standard deviation of $s = 0.15$ volt. Use the sample data to construct a 95% confidence interval estimate of the standard deviation of all voltage levels.
>
> 123.3   123.5   123.7   123.4   123.6   123.5   123.5   123.4   123.6   123.8

SOLUTION   **REQUIREMENT CHECK** We first verify that the requirements are satisfied. (1) The sample can be treated as a simple random sample. (2) The following displays show a Minitab and a TI-83/84 Plus histogram, and the shape of the histogram is very close to the bell shape of a normal distribution, so the requirement of normality is satisfied. (This check of requirements is Step 1 in the process of finding a confidence interval of σ, so we proceed next with Step 2.) ✓

MINITAB

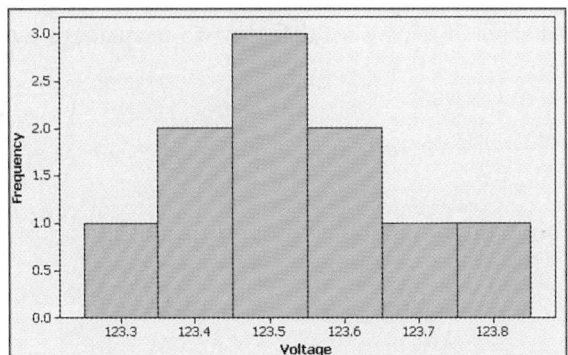

TI-83/84 Plus

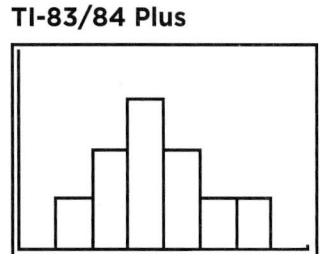

Step 2: The sample size is $n = 10$, so the number of degrees of freedom is given by df $= 10 - 1 = 9$. If we use Table A-4, we refer to the row corresponding to 9 degrees of freedom, and we refer to the columns with areas of 0.975 and 0.025. (For a 95% confidence level, we divide $\alpha = 0.05$ equally between the two tails of the chi-square distribution, and we refer to the values of 0.975 and 0.025 across the top row of Table A-4.) The critical values are $\chi_L^2 = 2.700$ and $\chi_R^2 = 19.023$. (See Example 1.)

Step 3: Using the critical values of 2.700 and 19.023, the sample standard deviation of $s = 0.15$, and the sample size of $n = 10$, we construct the 95% confidence interval by evaluating the following:

$$\frac{(n-1)s^2}{\chi_R^2} < \sigma^2 < \frac{(n-1)s^2}{\chi_L^2}$$

$$\frac{(10-1)(0.15)^2}{19.023} < \sigma^2 < \frac{(10-1)(0.15)^2}{2.700}$$

Step 4: Evaluating the above expression results in $0.010645 < \sigma^2 < 0.075000$. Finding the square root of each part (before rounding), then rounding to two decimal places, yields this 95% confidence interval estimate of the population standard deviation: 0.10 volt $< \sigma <$ 0.27 volt.

INTERPRETATION   Based on this result, we have 95% confidence that the limits of 0.10 volt and 0.27 volt contain the true value of σ. The confidence interval can also be expressed as (0.10, 0.27), but the format of $s \pm E$ *cannot* be used because the confidence interval does not have s at its center.

Rationale for the Confidence Interval We now explain why the confidence intervals for σ and σ^2 have the forms just given. If we obtain simple random samples of size n from a population with variance σ^2, there is a probability of $1 - \alpha$ that the statistic $(n - 1)s^2/\sigma^2$ will fall between the critical values of χ_L^2 and χ_R^2. In other words (and symbols), there is a $1 - \alpha$ probability that both of the following are true:

$$\frac{(n - 1)s^2}{\sigma^2} < \chi_R^2 \quad \text{and} \quad \frac{(n - 1)s^2}{\sigma^2} > \chi_L^2$$

If we multiply both of the preceding inequalities by σ^2 and divide each inequality by the appropriate critical value of χ^2, we see that the two inequalities can be expressed in the equivalent forms:

$$\frac{(n - 1)s^2}{\chi_R^2} < \sigma^2 \quad \text{and} \quad \frac{(n - 1)s^2}{\chi_L^2} > \sigma^2$$

These last two inequalities can be combined into one inequality:

$$\frac{(n - 1)s^2}{\chi_R^2} < \sigma^2 < \frac{(n - 1)s^2}{\chi_L^2}$$

Determining Sample Size   The procedures for finding the sample size necessary to estimate σ^2 are much more complex than the procedures given earlier for means and proportions. Instead of using very complicated procedures, we will use Table 7-2. STATDISK also provides sample sizes. With STATDISK, select **Analysis, Sample Size Determination,** and then **Estimate St Dev.** The TI-83/84 Plus calculator does not provide such sample sizes.

Table 7-2

Sample Size for σ^2		Sample Size for σ	
To be 95% confident that s^2 is within	of the value of σ^2, the sample size n should be at least	To be 95% confident that s is within	of the value of σ, the sample size n should be at least
1%	77,208	1%	19,205
5%	3,149	5%	768
10%	806	10%	192
20%	211	20%	48
30%	98	30%	21
40%	57	40%	12
50%	38	50%	8
To be 99% confident that s^2 is within	of the value of σ^2, the sample size n should be at least	To be 99% confident that s is within	of the value of σ, the sample size n should be at least
1%	133,449	1%	33,218
5%	5,458	5%	1,336
10%	1,402	10%	336
20%	369	20%	85
30%	172	30%	38
40%	101	40%	22
50%	68	50%	14

EXAMPLE 3 **Finding Sample Size for Estimating σ** We want to estimate the standard deviation σ of all voltage levels in a home. We want to be 95% confident that our estimate is within 20% of the true value of σ. How large should the sample be? Assume that the population is normally distributed.

SOLUTION From Table 7-2, we can see that 95% confidence and an error of 20% for σ correspond to a sample of size 48. We should obtain a simple random sample of 48 voltage levels from the population of voltage levels.

Using the TI-83/84 Plus Calculator for Confidence Interval Estimates of σ or σ^2

The TI-83/84 Plus calculator is not yet designed to automatically generate confidence intervals for the population standard deviation σ or the population variance σ^2, but the program **S2INT** can be used. That program was written by Michael Lloyd of Henderson State University, and it is on the CD-ROM included with this book or it can be downloaded from www.aw.com/triola. The program S2INT uses the program ZZINEWT, so that program must also be installed. Use the following procedure.

Entering Data

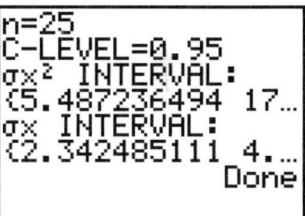

1. Use the enclosed CD-ROM or the Web site www.aw.com/triola to download the programs S2INT and ZZINEWT. (Download instructions can be found at the site www.ti.com, and they typically involve the use of software such as TI-GRAPHLINK or TI CONNECT.)

2. Press **PRGM**, select the program **S2INT**, and proceed to enter the sample variance s^2, the sample size n, and the confidence level (such as 0.95). Press **ENTER**, and wait a while (such as 60 seconds) for the display of the confidence interval limits for σ^2 and σ.

Confidence Interval Results

The accompanying screens result from a sample with $n = 25$ and $s = 3.0$. Note that the program requires entry of the sample variance s^2. The first screen shows the entry of the sample statistics ($s^2 = 3.0^2$ and $n = 25$) and the confidence level of 0.95. The second screen shows the resulting confidence intervals. *Note:* You may have to press **ENTER** and/or press $\triangleright$ to see all of the results. In the second screen, for example, you must press $\triangleright$ to scroll to the right in order to see the value of the upper confidence interval limits. Also, the program initially displayed the confidence interval for σ^2, but the confidence interval for σ can be seen by pressing **ENTER**. Based on the calculator results, the following 95% confidence intervals are obtained:

$$5.5 < \sigma^2 < 17.4$$
$$2.3 < \sigma < 4.2$$

Using Technology for Sample Sizes for Estimating σ or σ^2

The TI-83/84 Plus calculator is not yet designed to automatically generate sample sizes required to estimate the population standard deviation σ or the population variance σ^2, but STATDISK can be used instead. With STATDISK, select **Analysis** from the main menu at the top, then select **Sample Size Determination** and select the option of **Estimate St Dev.** Proceed to enter the confidence level and the percent margin of error (as in Table 7-2), then click **Evaluate.**

7-5   Basic Skills and Concepts

Statistical Literacy and Critical Thinking

1. Interpreting a Confidence Interval Using the weights of the M&M candies listed in Data Set 18 from Appendix B, we use the standard deviation of the sample ($s = 0.05179$ g) to obtain the following 95% confidence interval estimate of the standard deviation of the weights of all M&Ms: 0.0455 g $< \sigma <$ 0.0602 g. Write a statement that correctly interprets that confidence interval.

2. Expressing Confidence Intervals Is the confidence interval given in Exercise 1 equivalent to the expression (0.0455 g, 0.0602 g)? Is the confidence interval given in Exercise 1 equivalent to the expression 0.05285 g $\pm$ 0.00735 g? Why or why not?

3. Valid Confidence Interval? A pollster for the Gallup Organization randomly generates the last two digits of telephone numbers to be called, so the numbers from 00 to 99 are all equally likely. Can the methods of this section be used to construct a confidence interval estimate of the standard deviation of the population of all outcomes? Why or why not?

4. Unbiased Estimators What is an unbiased estimator? Is the sample variance an unbiased estimator of the population variance? Is the sample standard deviation an unbiased estimator of the population standard deviation?

Finding Critical Values. *In Exercises 5–8, find the critical values* χ^2_L *and* χ^2_R *that correspond to the given confidence level and sample size.*

5. 95%; $n = 9$ **6.** 95%; $n = 20$

7. 99%; $n = 81$ **8.** 90%; $n = 51$

Finding Confidence Intervals. *In Exercises 9–12, use the given confidence level and sample data to find a confidence interval for the population standard deviation* σ. *In each case, assume that a simple random sample has been selected from a population that has a normal distribution.*

9. SAT Scores of College Students 95% confidence; $n = 30$, $\bar{x} = 1533$, $s = 333$

10. Speeds of Drivers Ticketed in a 65 mi/h Zone on the Massachusetts Turnpike 95% confidence; $n = 25$, $\bar{x} = 81.0$ mi/h, $s = 2.3$ mi/h.

11. White Blood Cell Count (in Cells per Microliter) 99% confidence; $n = 7$, $\bar{x} = 7.106$, $s = 2.019$

12. Reaction Times of NASCAR Drivers 99% confidence; $n = 8$, $\bar{x} = 1.24$ sec, $s = 0.12$ sec

Determining Sample Size. *In Exercises 13–16, assume that each sample is a simple random sample obtained from a normally distributed population. Use Table 7-2 on page 382 to find the indicated sample size.*

13. Find the minimum sample size needed to be 95% confident that the sample standard deviation s is within 1% of σ. Is this sample size practical in most applications?

14. Find the minimum sample size needed to be 95% confident that the sample standard deviation s is within 30% of σ. Is this sample size practical in most applications?

15. Find the minimum sample size needed to be 99% confident that the sample variance is within 40% of the population variance. Is such a sample size practical in most cases?

16. Find the minimum sample size needed to be 95% confident that the sample variance is within 20% of the population variance.

Finding Confidence Intervals. *In Exercises 17–24, assume that each sample is a simple random sample obtained from a population with a normal distribution.*

17. Birth Weights In a study of the effects of prenatal cocaine use on infants, the following sample data were obtained for weights at birth: $n = 190$, $\bar{x} = 2700$ g, $s = 645$ g (based on data from "Cognitive Outcomes of Preschool Children with Prenatal Cocaine Exposure," by Singer et al., *Journal of the American Medical Association*, Vol. 291, No. 20). Use the sample data to construct a 95% confidence interval estimate of the standard deviation of all birth weights of infants born to mothers who used cocaine during pregnancy. (Because Table A-4 has a maximum of 100 degrees of freedom while we require 189 degrees of freedom, use these critical values obtained from STATDISK: $\chi_L^2 = 152.8222$ and $\chi_R^2 = 228.9638$.) Based on the result, does the standard deviation appear to be different from the standard deviation of 696 g for birth weights of babies born to mothers who did not use cocaine during pregnancy?

TI **18. Weights of M&Ms** Data Set 18 in Appendix B lists 100 weights (in grams) of M&M candies. The minimum weight is 0.696 g and the maximum weight is 1.015 g.

a. Use the range rule of thumb to estimate σ, the standard deviation of weights of all such M&Ms.

b. The 100 weights have a standard deviation of 0.0518 g. Construct a 95% confidence interval estimate of the standard deviation of weights of all M&Ms.

c. Does the confidence interval from part (b) contain the estimated value of σ from part (a)? What do the results suggest about the estimate from part (a)?

TI **19. Movie Lengths** Data Set 9 in Appendix B includes 23 movies with ratings of PG or PG-13, and those movies have lengths (in minutes) with a mean of 120.8 min and a standard deviation of 22.9 min. That same data set also includes 12 movies with R ratings, and those movies have lengths with a mean of 118.1 min and a standard deviation of 20.8 min.

a. Construct a 95% confidence interval estimate of the standard deviation of the lengths of all movies with ratings of PG or PG-13.

b. Construct a 95% confidence interval estimate of the standard deviation of the lengths of all movies with ratings of R.

c. Compare the variation of the lengths of movies with ratings of PG or PG-13 to the variation of the lengths of movies with ratings of R. Does there appear to be a difference?

TI **20. Pulse Rates of Men and Women** Data Set 1 in Appendix B includes 40 pulse rates of men, and those pulse rates have a mean of 69.4 beats per minute and a standard deviation of 11.3 beats per minute. That data set also includes 40 pulse rates of women, and those pulse rates have a mean of 76.3 beats per minute and a standard deviation of 12.5 beats per minute.

a. Construct a 99% confidence interval estimate of the standard deviation of the pulse rates of men.

b. Construct a 99% confidence interval estimate of the standard deviation of the pulse rates of women.

c. Compare the variation of the pulse rates of men and women. Does there appear to be a difference?

TI **21. Video Games** Twelve different video games showing substance use were observed and the duration times of game play (in seconds) are listed below (based on data from "Content and Ratings of Teen-Rated Video Games," by Haninger and Thompson, *Journal of the American Medical Association,* Vol. 291, No. 7). The design of the study justifies the assumption that the sample can be treated as a simple random sample. Use the sample data to construct a 99% confidence interval estimate of σ, the standard deviation of the duration times of game play.

    4049   3884   3859   4027   4318   4813   4657   4033   5004   4823   4334   4317

TI **22. Designing Theater Seats** In the course of designing theater seats, the sitting heights (in mm) of a simple random sample of adult women is obtained, and the results are listed below (based on anthropometric survey data from Gordon, Churchill, et al.). Use the sample data to construct a 95% confidence interval estimate of σ, the standard deviation of sitting heights of all women. Does the confidence interval contain the value of 35 mm, which is believed to be the standard deviation of sitting heights of women?

    849   807   821   859   864   877   772   848   802   807   887   815

TI **23. Monitoring Lead in Air** Listed below are measured amounts of lead (in micrograms per cubic meter, or $\mu g/m^3$) in the air. The EPA has established an air quality standard for lead of 1.5 $\mu g/m^3$. The measurements shown below were recorded at Building 5 of the World Trade Center site on different days immediately following the destruction caused by the terrorist attacks of September 11, 2001. Use the given values to construct a 95% confidence interval estimate of the standard deviation of the amounts of lead in the air. Is there anything about this data set suggesting that the confidence interval might not be very good? Explain.

| 5.40 | 1.10 | 0.42 | 0.73 | 0.48 | 1.10 |

TI **24. a. Comparing Waiting Lines** The listed values are waiting times (in minutes) of customers at the Jefferson Valley Bank, where customers enter a single waiting line that feeds three teller windows. Construct a 95% confidence interval for the population standard deviation σ.

| 6.5 | 6.6 | 6.7 | 6.8 | 7.1 | 7.3 | 7.4 | 7.7 | 7.7 | 7.7 |

b. The listed values are waiting times (in minutes) of customers at the Bank of Providence, where customers may enter any one of three different lines that have formed at three teller windows. Construct a 95% confidence interval for the population standard deviation σ.

| 4.2 | 5.4 | 5.8 | 6.2 | 6.7 | 7.7 | 7.7 | 8.5 | 9.3 | 10.0 |

c. Interpret the results found in parts (a) and (b). Do the confidence intervals suggest a difference in the variation among waiting times? Which arrangement seems better: the single-line system or the multiple-line system?

Using Large Data Sets from Appendix B. *In Exercises 25 and 26, use the data set from Appendix B. Assume that each sample is a simple random sample obtained from a population with a normal distribution.*

TI **25. FICO Credit Rating Scores** Refer to Data Set 24 in Appendix B and use the credit rating scores to construct a 95% confidence interval estimate of the standard deviation of all credit rating scores.

TI **26. Home Energy Consumption** Refer to Data Set 12 in Appendix B and use the sample amounts of home energy consumption (in kWh) to construct a 99% confidence interval estimate of the standard deviation of all energy consumption amounts.

7-5   Beyond the Basics

27. Finding Critical Values In constructing confidence intervals for σ or σ^2, we use Table A-4 to find the critical values χ_L^2 and χ_R^2, but that table applies only to cases in which $n \leq 101$, so the number of degrees of freedom is 100 or smaller. For larger numbers of degrees of freedom, we can approximate χ_L^2 and χ_R^2 by using

$$\chi^2 = \frac{1}{2}\left[\pm z_{\alpha/2} + \sqrt{2k - 1}\right]^2$$

where k is the number of degrees of freedom and $z_{\alpha/2}$ is the critical z score described in Section 7-2. STATDISK was used to find critical values for 189 degrees of freedom with a confidence level of 95%, and those critical values are given in Exercise 17. Use the approximation shown here to find the critical values and compare the results to those found from STATDISK.

28. Finding the Best Estimator Values of s^2 tend to produce smaller errors by being closer to σ^2 than do other unbiased measures of variation. Let's now consider the biased estimator of $(n - 1)s^2/(n + 1)$. Given the population of values {2, 3, 7}, use the value of σ^2, and use the nine different possible samples of size $n = 2$ (for sampling with replacement) for the following.

a. Find s^2 for each of the nine samples, then find the error $s^2 - \sigma^2$ for each sample. Then square those errors. Then find the mean of those squares. The result is the value of the mean square error.

b. Find the value of $(n - 1)s^2/(n + 1)$ for each of the nine samples. Then find the error $(n - 1)s^2/(n + 1) - \sigma^2$ for each sample. Square those errors, then find the mean of those squares. The result is the mean square error.

c. The mean square error can be used to measure how close an estimator comes to the population parameter. Which estimator does a better job by producing the smaller mean square error? Is that estimator biased or unbiased?

Review

In this chapter we introduced basic methods for finding *estimates* of population proportions, means, and variances. This chapter included procedures for finding each of the following:

- point estimate
- confidence interval
- required sample size

We discussed the point estimate (or single-valued estimate) and formed these conclusions:

- Proportion: The best point estimate of p is $\hat{p}$
- Mean: The best point estimate of μ is $\bar{x}$.
- Variation: The value of s is commonly used as a point estimate of σ, even though it is a biased estimate. Also, s^2 is the best point estimate of σ^2.

Because the above point estimates consist of single values, they have the serious shortcoming of not revealing how close to the population parameter that they are likely to be, so confidence intervals (or interval estimates) are commonly used as more informative and useful estimates. We also considered ways of determining the sample sizes necessary to estimate parameters to within given margins of error. This chapter also introduced the Student t and chi-square distributions. We must be careful to use the correct probability distribution for each set of circumstances. This chapter used the following criteria for selecting the appropriate distribution:

Confidence interval for proportion p: Use the *normal* distribution (assuming that the required conditions are satisfied and there are at least 5 successes and at least 5 failures so that the normal distribution can be used to approximate the binomial distribution).

Confidence interval for μ: See Figure 7-6 (page 366) or Table 7-1 (page 367) to choose between the *normal* or t distributions (or conclude that neither applies).

Confidence interval for σ or σ^2:   Use the *chi-square* distribution (assuming that the required conditions are satisfied).

For the confidence interval and sample size procedures in this chapter, it is very important to verify that the requirements are satisfied. If they are not, then we cannot use the methods of this chapter and we may need to use other methods, such as the bootstrap method described in the Technology Project at the end of this chapter, or nonparametric methods, such as those discussed in Chapter 13.

Statistical Literacy and Critical Thinking

1. Estimating Population Parameters Quest Diagnostics is a provider of drug testing for job applicants, and its managers want to estimate the proportion of job applicants who test positive for drugs. In this context, what is a point estimate of that proportion? What is a confidence interval? What is a major advantage of the confidence interval estimate over the point estimate?

2. Interpreting a Confidence Interval Here is a 95% confidence interval estimate of the proportion of all job applicants who test positive when they are tested for drug use: $0.0262 < p < 0.0499$ (based on data from Quest Diagnostics). Write a statement that correctly interprets this confidence interval.

3. Confidence Level What is the confidence level of the confidence interval given in Exercise 2? What is a confidence level in general?

4. Online Poll The Internet service provider AOL periodically conducts polls by posting a survey question on its Web site, and Internet users can respond if they choose to do so. Assume that a survey question asks whether the respondent has a high-definition television in the household and the results are used to construct this 95% confidence interval: $0.232 < p < 0.248$. Can this confidence interval be used to form valid conclusions about the general population? Why or why not?

Chapter Quick Quiz

1. The following 95% confidence interval estimate is obtained for a population mean: $10.0 < \mu < 20.0$. Interpret that confidence interval.

2. With a Democrat and a Republican candidate running for office, a newspaper conducts a poll to determine the proportion of voters who favor the Republican candidate. Based on the poll results, this 95% confidence interval estimate of that proportion is obtained: $0.492 < p < 0.588$. Which of the following statements better describes the results: (1) The Republican is favored by a majority of the voters. (2) The election is too close to call.

3. Find the critical value of $t_{\alpha/2}$ for $n = 20$ and $\alpha = 0.05$.

4. Find the critical value of $z_{\alpha/2}$ for $n = 20$ and $\alpha = 0.10$.

5. Find the sample size required to estimate the percentage of college students who use loans to help fund their tuition. Assume that we want 95% confidence that the proportion from the sample is within two percentage points of the true population percentage.

6. In a poll of 600 randomly selected subjects, 240 answered "yes" when asked if they planned to vote in a state election. What is the best point estimate of the population proportion of all who plan to vote in that election.

7. In a poll of 600 randomly selected subjects, 240 answered "yes" when asked if they planned to vote in a state election. Construct a 95% confidence interval estimate of the proportion of all who plan to vote in that election.

8. In a survey of randomly selected subjects, the mean age of the 36 respondents is 40.0 years and the standard deviation of the ages is 10.0 years. Use these sample results to construct a 95% confidence interval estimate of the mean age of the population from which the sample was selected.

9. Repeat Exercise 8 assuming that the population standard deviation is known to be 10.0 years.

10. Find the sample size required to estimate the mean age of registered drivers in the United States. Assume that we want 95% confidence that the sample mean is within 1/2 year of the true mean age of the population. Also assume that the standard deviation of the population is known to be 12 years.

Review Exercises

TI 1. Reporting Income In a Pew Research Center poll of 745 randomly selected adults, 589 said that it is morally wrong to not report all income on tax returns. Construct a 95% confidence interval estimate of the percentage of all adults who have that belief, and then write a statement interpreting the confidence interval.

2. Determining Sample Size See the survey described in Exercise 1. Assume that you must conduct a new poll to determine the percentage of adults who believe that it is morally wrong to not report all income on tax returns. How many randomly selected adults must you survey if you want 99% confidence that the margin of error is two percentage points? Assume that nothing is known about the percentage that you are trying to estimate.

3. Determining Sample Size See the survey described in Exercise 1. Assume that you must conduct a survey to determine the mean income reported on tax returns, and you have access to actual tax returns. How many randomly selected tax returns must you survey if you want to be 99% confident that the mean of the sample is within $500 of the true population mean? Assume that reported incomes have a standard deviation of $28,785 (based on data from the U.S. Census Bureau). Is the sample size practical?

TI 4. Penny Weights A simple random sample of 37 weights of pennies made after 1983 has a mean of 2.4991 g and a standard deviation of 0.0165 g (based on Data Set 20 in Appendix B). Construct a 99% confidence interval estimate of the mean weight of all such pennies. Design specifications require a population mean of 2.5 g. What does the confidence interval suggest about the manufacturing process?

TI 5. Crash Test Results The National Transportation Safety Administration conducted crash test experiments on five subcompact cars. The head injury data (in hic) recorded from crash test dummies in the driver's seat are as follows: 681, 428, 917, 898, 420. Use these sample results to construct a 95% confidence interval for the mean of head injury measurements from all subcompact cars.

TI 6. Confidence Interval for σ New car design specifications are being considered to control the variation of the head injury measurements. Use the same sample data from Exercise 5 to construct a 95% confidence interval estimate of σ.

TI 7. Cloning Survey A Gallup poll consisted of 1012 randomly selected adults who were asked whether "cloning of humans should or should not be allowed." Results showed that 901 adults surveyed indicated that cloning should not be allowed.

a. Find the best point estimate of the proportion of adults believing that cloning of humans should not be allowed.

b. Construct a 95% confidence interval estimate of the proportion of adults believing that cloning of humans should not be allowed.

c. A news reporter wants to determine whether these survey results constitute strong evidence that the majority (more than 50%) of people are opposed to such cloning. Based on the results, is there strong evidence supporting the claim that the majority is opposed to such cloning? Why or why not?

8. Sample Size You have been hired by a consortium of local car dealers to conduct a survey about the purchases of new and used cars.

a. If you want to estimate the percentage of car owners in your state who purchased new cars (not used), how many adults must you survey if you want 95% confidence that your sample percentage is in error by no more than four percentage points?

b. If you want to estimate the mean amount of money spent by car owners on their last car purchase, how many car owners must you survey if you want 95% confidence that your sample mean is in error by no more than $750? (Based on results from a pilot study, assume that the standard deviation of amounts spent on car purchases is $14,227.)

c. If you plan to obtain the estimates described in parts (a) and (b) with a single survey having several questions, how many people must be surveyed?

TI 9. Discarded Glass Listed below are weights (in pounds) of glass discarded in one week by randomly selected households (based on data from the Garbage Project at the University of Arizona).

a. What is the best point estimate of the mean weight of glass discarded by all households in one week?

b. Construct a 95% confidence interval estimate of the mean weight of glass discarded by all households.

c. Repeat part (b) assuming that the population is normally distributed with a standard deviation known to be 3.108 lb.

| 3.52 | 8.87 | 3.99 | 3.61 | 2.33 | 3.21 | 0.25 | 4.94 |

TI 10. Confidence Intervals for σ and σ^2

a. Use the sample data from Exercise 9 to construct a 95% confidence interval estimate of the population standard deviation.

b. Use the sample data from Exercise 9 to construct a 95% confidence interval estimate of the population variance.

Cumulative Review Exercises

Weights of Supermodels. *Supermodels are sometimes criticized on the grounds that their low weights encourage unhealthy eating habits among young women. In Exercises 1–4, use the following weights (in pounds) of randomly selected supermodels.*

| 125 (Taylor) | 119 (Auermann) | 128 (Schiffer) | 125 (Bundchen) | 119 (Turlington) |
| 127 (Hall) | 105 (Moss) | 123 (Mazza) | 110 (Reilly) | 103 (Barton) |

TI 1. Find the mean, median, and standard deviation.

2. What is the level of measurement of these data (nominal, ordinal, interval, ratio)?

TI 3. Construct a 95% confidence interval for the population mean.

4. Find the sample size necessary to estimate the mean weight of all supermodels so that there is 95% confidence that the sample mean is in error by no more than 2 lb. Assume that a pilot study suggests that the weights of all supermodels have a standard deviation of 7.5 lb.

TI 5. Employment Drug Test If a randomly selected job applicant is given a drug test, there is a 0.038 probability that the applicant will test positive for drug use (based on data from Quest Diagnostics).

a. If a job applicant is randomly selected and given a drug test, what is the probability that the applicant does not test positive for drug use?

b. Find the probability that when two different job applicants are randomly selected and given drug tests, they both test positive for drugs.

c. If 500 job applicants are randomly selected and they are all given drug tests, find the probability that at least 20 of them test positive for drugs.

TI 6. ACT Scores Scores on the ACT test are normally distributed with a mean of 21.1 and a standard deviation of 4.8.

a. If one ACT score is randomly selected, find the probability that it is greater than 20.0.

b. If 25 ACT scores are randomly selected, find the probability that they have a mean greater than 20.

c. Find the ACT score that is the 90th percentile.

7. Sampling What is a simple random sample? What is a voluntary response sample?

8. Range Rule of Thumb Use the range rule of thumb to estimate the standard deviation of grade point averages at a college with a grading system designed so that the lowest and highest possible grade point averages are 0 and 4.

9. Rare Event Rule Find the probability of making random guesses to 12 true/false questions and getting 12 correct answers. If someone did get 12 correct answers, is it possible that they made random guesses? Is it likely that they made random guesses?

10. Sampling Method If you conduct a poll by surveying all of your friends that you see during the next week, which of the following terms best describes the type of sampling used: random, systematic, cluster, convenience, voluntary response? Is the sample likely to be representative of the population?

TI-83/84 Plus Calculator Project

Bootstrap Resampling   The *bootstrap resampling method* can be used to construct confidence intervals for situations in which traditional methods cannot (or should not) be used. Example 4 in Section 7-4 included the following sample of times that different video games showed the use of alcohol (based on data from "Content and Ratings of Teen-Rated

Video Games," by Haninger and Thompson, *Journal of the American Medical Association,* Vol. 291, No. 7).

84	14	583	50	0	57	207	43	178	0	2	57

Example 4 in Section 7-4 showed the histogram and normal quantile plot, and they both suggest that the times are not from a normally distributed population, so methods requiring a normal distribution should not be used.

If we want to use the above sample data for the construction of a confidence interval estimate of the population mean μ, one approach is to use the bootstrap resampling method, which has no requirements about the distribution of the population. This method typically requires a computer to build a bootstrap population by replicating (duplicating) a sample many times. We draw from the sample with replacement, thereby creating an approximation of the original population. In this way, we pull the sample up "by its own bootstraps" to simulate the original population.

Using the sample data given above, proceed to construct a 95% confidence interval estimate of the population mean μ by using the bootstrap method as described in the following steps.

a. Create a new sample of size 12 by selecting 12 values with replacement from the 12 sample values given above. Accomplish this using a TI-83/84 Plus calculator as follows:

i. Press **MATH**, then select **PRB,** then select the menu item of **randInt.**

ii. Press **ENTER**, then proceed to enter (1, 12, 12) so that the command of **randInt(1, 12, 12)** is displayed. This command tells the calculator to randomly select 12 integers between 1 and 12 inclusive, with repetition allowed. A typical result might be this: {3 7 3 2 2 8 9 4 2 5 1 1}.

iii. Consider the 12 randomly generated integers to be *positions* of data values from the original list given above. For example, the randomly generated list of {3 7 3 2 2 8 9 4 2 5 1 1} indicates the 3rd value in the original list, the 7th value, 3rd value, and so on. The list of {3 7 3 2 2 8 9 4 2 5 1 1} therefore corresponds to this list of values from the original data set: {583 207 583 14 14 43 178 50 14 0 84 84}. This constitutes a *resampling* of the original data set.

iv. Now find the mean of the resampled data set. For example, if the resampled data set is {583 207 583 14 14 43 178 50 14 0 84 84}, its mean is 154.5.

b. Repeat the step (a) procedure a total of five times to get a list of 5 sample means. Ideally, we would repeat the process thousands of times. However, in the interest of preserving the sanity of readers, the author has compiled 95 such means, with the *sorted* list of 95 means given below.

c. Combine your five sample means with the 95 sample means in the list below. Be sure to enter them where they belong in the sorted list.

d. Now find the percentiles $P_{2.5}$ and $P_{97.5}$ for the sorted list of 100 means. ($P_{2.5}$ is the 3rd value in the sorted list; $P_{97.5}$ is the 98th value in the sorted list.) Identify the resulting confidence interval by substituting the values for $P_{2.5}$ and $P_{97.5}$ in $P_{2.5} < \mu < P_{97.5}$.

25.8	40.0	40.1	47.9	48.6	48.8	49.3	49.8	52.5	53.5
54.6	55.2	57.6	58.3	58.8	59.8	60.6	61.7	66.5	67.9
69.3	69.6	72.5	72.8	73.8	77.3	77.4	78.7	80.5	80.8
81.4	83.6	84.3	85.8	88.4	89.3	89.6	90.3	93.6	95.1
96.8	98.4	98.8	100.5	102.8	103.4	108.7	109.5	111.1	111.4
112.1	115.5	115.9	116.8	119.0	120.8	121.1	122.3	124.7	126.4
128.1	128.8	130.5	132.3	134.2	134.7	134.8	137.8	139.3	140.9
141.3	142.8	143.8	148.3	151.1	160.7	161.3	161.9	162.4	171.5
178.5	183.8	185.2	187.5	193.0	194.3	194.3	195.5	195.7	199.1
207.6	211.2	219.3	221.6	271.0					

Confidence Intervals

Go to: **http://www.aw.com/triola**

The confidence intervals in this chapter illustrate an important point in the science of statistical estimation. Namely, estimations based on sample data are made with certain degrees of confidence. In the Internet Project for this chapter, you will use confidence intervals to make a statement about the temperature where you live.

After going to this book's Web site, locate the project for this chapter. There you will find instructions on how to use the Internet to find temperature data collected by the weather station nearest your home. With this data in hand, you will construct confidence intervals for temperatures during different time periods and attempt to draw conclusions about temperature change in your area. In addition, you will learn more about the relationship between confidence and probability.

APPLET PROJECT

The CD included with this book contains applets designed to help visualize various concepts. Open the Applets folder on the CD and double-click on **Start.** Select the menu item of **Confidence Intervals for a Proportion.** Using $n = 20$ and $p = 0.7$, click on **Simulate** and find the proportion of the 95% confidence intervals among 100 that contain the population proportion of 0.7. Click on **Simulate** nine more times so that the total number of confidence intervals is 1000. What proportion of the 1000 95% confidence intervals contain $p = 0.7$? Write a brief explanation of the principle illustrated by these results.

Critical Thinking: What do the "Do Not Call" registry survey results tell us?

Surveys have become an important component of American life. They directly affect us in so many ways, including public policy, the television shows we watch, the products we buy, and the political leaders we elect. Because surveys are now such an integral part of our lives, it is important that every citizen has the ability to interpret survey results. Surveys are the focus of this project.

A recent Harris survey of 1961 adults showed that 76% have registered for the "Do Not Call" registry, so that telemarketers do not phone them.

Analyzing the Data

1. Use the survey results to construct a 95% confidence interval estimate of the percentage of all adults on the "Do Not Call" registry.

2. Identify the margin of error for this survey.

3. Explain why it would or would not be okay for a newspaper to make this statement: "Based on results from a recent survey, the majority of adults are not on the 'Do Not Call' registry."

4. Assume that you are a newspaper reporter. Write a description of the survey results for your newspaper.

5. A common criticism of surveys is that they poll only a very small percentage of the population and therefore cannot be accurate. Is a sample of only 1961 adults taken from a population of 225,139,000 adults a sample size that is too small? Write an explanation of why the sample size of 1961 is or is not too small.

6. In reference to another survey, the president of a company wrote to the Associated Press about a nationwide survey of 1223 subjects. Here is what he wrote:

   When you or anyone else attempts to tell me and my associates that 1223 persons account for our opinions and tastes here in America, I get mad as hell! How dare you! When you or anyone else tells me that 1223 people represent America, it is astounding and unfair and should be outlawed.

The writer of that letter then proceeds to claim that because the sample size of 1223 people represents 120 million people, his single letter represents 98,000 (120 million divided by 1223) who share the same views. Do you agree or disagree with this claim? Write a response that either supports or refutes this claim.

Cooperative Group Activities

1. Out-of-class activity Collect sample data, and use the methods of this chapter to construct confidence interval estimates of population parameters. Here are some suggestions for parameters:

• Proportion of students at your college who can raise one eyebrow without raising the other eyebrow.

• Mean age of cars driven by statistics students and/or the mean age of cars driven by faculty.

• Mean length of words in *New York Times* editorials and mean length of words in editorials found in your local newspaper.

• Mean lengths of words in *Time* magazine, *Newsweek* magazine, and *People* magazine.

• Proportion of students at your college who can correctly identify the president, vice president, and secretary of state.

• Proportion of students at your college who are over the age of 18 and are registered to vote.

• Mean age of full-time students at your college.

• Proportion of motor vehicles in your region that are cars.

2. In-class activity Without using any measuring device, each student should draw a line believed to be 3 in. long and another line believed to be 3 cm long. Then use rulers to measure and record the lengths of the lines drawn. Find the means and standard deviations of the two sets of lengths. Use the sample data to construct a confidence interval for the length of the line estimated to be 3 in., then do the same for the length of the line estimated to be 3 cm. Do the confidence interval limits actually contain the correct length? Compare the results. Do the estimates of the 3-in. line appear to be more accurate than those for the 3-cm line?

3. In-class activity Assume that a method of gender selection can affect the probability of a baby being a girl, so that the probability becomes 1/4. Each student should simulate 20 births by drawing 20 cards from a shuffled deck. Replace each card after it has been drawn, then reshuffle. Consider the hearts to be girls and consider all other cards to be boys. After making 20 selections and recording the "genders" of the babies, construct a confidence interval estimate of the proportion of girls. Does the result appear to be effective in identifying the true value of the population proportion? (If decks of cards are not available, use some other way to simulate the births, such as using the random number generator on a calculator or using digits from phone numbers or social security numbers.)

4. Out-of-class activity Groups of three or four students should go to the library and collect a sample consisting of the ages of books (based on copyright dates). Plan and describe the sampling procedure, execute the sampling procedure, then use the results to construct a confidence interval estimate of the mean age of all books in the library.

5. In-class activity Each student should write an estimate of the age of the current President of the United States. All estimates should be collected and the sample mean and standard deviation should be calculated. Then use the sample results to construct a confidence interval. Do the confidence interval limits contain the correct age of the President?

6. In-class activity A class project should be designed to conduct a test in which each student is given a taste of Coke and a taste of Pepsi. The student is then asked to identify which sample is Coke. After all of the results are collected, analyze the claim that the success rate is better than the rate that would be expected with random guesses.

7. In-class activity Each student should estimate the length of the classroom. The values should be based on visual estimates, with no actual measurements being taken. After the estimates have been collected, construct a confidence interval, then measure the length of the room. Does the confidence interval contain the actual length of the classroom? Is there a "collective wisdom," whereby the class mean is approximately equal to the actual room length?

8. In-class activity Divide into groups of three or four. Examine a current magazine such as *Time* or *Newsweek,* and find the proportion of pages that include advertising. Based on the results, construct a 95% confidence interval estimate of the percentage of all such pages that have advertising. Compare results with other groups.

9. In-class activity Divide into groups of two. First find the sample size required to estimate the proportion of times that a coin turns up heads when tossed, assuming that you want 80% confidence that the sample proportion is within 0.08 of the true population proportion. Then toss a coin the required number of times and record your results. What percentage of such confidence intervals should actually contain the true value of the population proportion, which we know is $p = 0.5$? Verify this last result by comparing your confidence interval with the confidence intervals found in other groups.

10. Out-of-class activity Identify a topic of general interest and coordinate with all members of the class to conduct a survey. Instead of conducting a "scientific" survey using sound principles of random selection, use a convenience sample consisting of respondents that are readily available, such as friends, relatives, and other students. Analyze and interpret the results. Identify the population. Identify the shortcomings of using a convenience sample, and try to identify how a sample of subjects randomly selected from the population might be different.

11. Out-of-class activity Each student should find an article in a professional journal that includes a confidence interval of the type discussed in this chapter. Write a brief report describing the confidence interval and its role in the context of the article.

12. Out-of-class activity Obtain a sample and use it to estimate the mean number of hours per week that students at your college devote to studying.

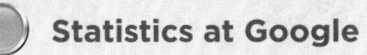

Statistics at Google

NAME: Artem Boytsov

JOB: Senior Software Engineer

COMPANY: Google Inc.

*A*rtem Boytsov is a Senior Software Engineer at Google, Inc. Founded in 1998 by Stanford University students Larry Page and Sergey Brin, Google has become the most widely used Internet search engine. It has become so popular that the word google *is now used as a verb in everyday language. To "google" is to use the Google search engine for finding information on the Internet.*

Q: What concepts of statistics do you use in your work for Google?

A: The statistics tools I use include sampling, normal distribution, Zipf/Pareto distributions, standard deviations, standard errors, correlations, confidence intervals, and conditional probabilities.

Q: How do you use statistics at Google?

A: Statistics and probability theory lie in the core of Google Trends. We use statistics to analyze our users' search behavior. We measure the popularity of terms on the Web to identify factors such as seasonal patterns and correlations between queries.

Q: Please describe at least one specific example illustrating how the use of statistics was successful in improving a product or service.

A: Knowledge of normal distributions, confidence intervals, and error concepts are crucial in analyzing the data. Performance and data quality trade-offs are calculated and made in our product. Further performance improvements were implemented by dismissing insignificant data and exploiting its properties (for example, that query frequencies follow the Zipf distribution).

Q: Is your use of probability and statistics increasing, decreasing, or remaining stable?

A: It is increasing with every feature or internal application of our product.

Q: Please describe how you try to ensure objectivity.

A: We use uniformly selected samples of Google's query log to calculate the probability that a given keyword is present in a random user query. We use conditional probabilities to factor out such factors as Google's overall traffic growth, differences in traffic volume between countries, etc.

Q: How critical do you find your knowledge of statistics for performing your responsibilities?

A: Knowledge of probability theory and statistics is crucial to my job as a Google Trends engineer. Many different projects in Google use statistics, probability, and information theories.

Q: Please cite an example of how your data are used.

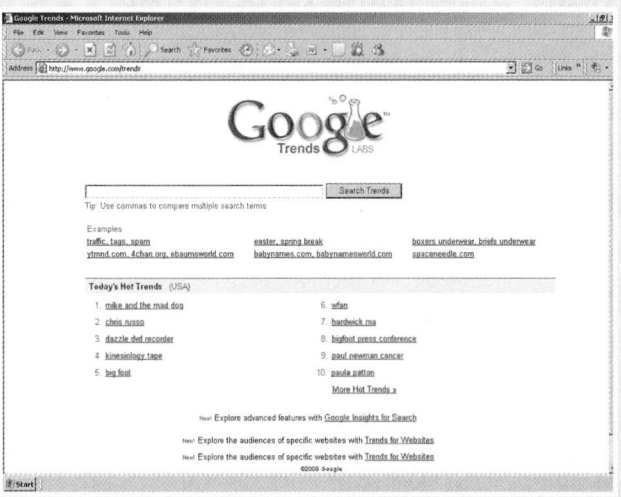

A: A merchant may use Google Trends to plan for fluctuations in seasonal demand. General trend (upward/downward) in popularity helps predict and prepare for such demand changes.

Q: In terms of statistics, what would you recommend for prospective employees?

A: I recommend an introductory course in statistics to anyone. To those pursuing degrees in engineering or finance, I would recommend courses in statistics, probability theory, and information theory. There's a world of information out there, and statistics is your first step to understanding and utilizing this information.

Q: Do you recommend statistics for today's college students?

A: Yes. The ability to analyze data is crucial in the era of information. I face misinterpretation and misuse of data by "statistically illiterate" people quite often, and it's very sad.

8-1 Review and Preview

8-2 Basics of Hypothesis Testing

8-3 Testing a Claim About a Proportion

8-4 Testing a Claim About a Mean: σ Known

8-5 Testing a Claim About a Mean: σ Not Known

8-6 Testing a Claim About a Standard Deviation or Variance

Hypothesis Testing

Does the MicroSort method of gender selection increase the likelihood that a baby will be a girl?

Gender-selection methods are somewhat controversial. Some people believe that use of such methods should be prohibited, regardless of the reason. Others believe that limited use should be allowed for medical reasons, such as to prevent gender-specific hereditary disorders. For example, some couples carry X-linked recessive genes, so that a male child has a 50% chance of inheriting a serious disorder and a female child has no chance of inheriting the disorder. These couples may want to use a gender-selection method to increase the likelihood of having a baby girl so that none of their children inherit the disorder.

Methods of gender selection have been around for many years. In the 1980s, ProCare Industries sold a product called Gender Choice. The product cost only $49.95, but the Food and Drug Administration told the company to stop distributing Gender Choice because there was no evidence to support the claim that it was 80% reliable.

The Genetics & IVF Institute developed a newer gender-selection method called MicroSort. The Microsort XSORT method is designed to increase the likelihood of a baby girl, and the YSORT method is designed to increase the likelihood of a boy. Here is a statement from the MicroSort Web site: "The Genetics & IVF Institute is offering couples the ability to increase the chance of having a child of the desired gender to reduce the probability of X-linked diseases or for family balancing." Stated simply, for a cost exceeding $3000, the Genetics & IVF Institute claims that it can increase the probability of having a baby of the gender that a couple prefers. As of this writing, the MicroSort method is undergoing clinical trials, but these results are available: Among 726 couples who used the XSORT method in trying to have a baby girl, 668 couples did have baby girls, for a success rate of 92.0%. Under normal circumstances with no special treatment, girls occur in 50% of births. (Actually, the current birth rate of girls is 48.79%, but we will use 50% to keep things simple.) These results provide us with an interesting question: Given that 668 out of 726 couples had girls, can we actually support the claim that the XSORT technique is effective in increasing the probability of a girl? Do we now have an effective method of gender selection?

8-1   Review and Preview

In Chapters 2 and 3 we used "descriptive statistics" when we summarized data using tools such as graphs, and statistics such as the mean and standard deviation. Methods of inferential statistics use sample data to make an inference or conclusion about a population. The two main activities of inferential statistics are using sample data to (1) *estimate* a population parameter (such as estimating a population parameter with a confidence interval), and (2) test a hypothesis or claim about a population parameter. In Chapter 7 we presented methods for estimating a population parameter with a confidence interval, and in this chapter we present the method of hypothesis testing.

> **DEFINITION**
>
> In statistics, a **hypothesis** is a claim or statement about a property of a population.
>
> A **hypothesis test** (or **test of significance**) is a procedure for testing a claim about a property of a population.

The main objective of this chapter is to develop the ability to conduct hypothesis tests for claims made about a population proportion p, a population mean μ, or a population standard deviation σ.

Here are examples of hypotheses that can be tested by the procedures we develop in this chapter:

- **Genetics** The Genetics & IVF Institute claims that its XSORT method allows couples to increase the probability of having a baby girl.

- **Business** A newspaper headline makes the claim that most workers get their jobs through networking.

- **Medicine** Medical researchers claim that when people with colds are treated with echinacea, the treatment has no effect.

- **Aircraft Safety** The Federal Aviation Administration claims that the mean weight of an airline passenger (including carry-on baggage) is greater than 185 lb, which it was 20 years ago.

- **Quality Control** When new equipment is used to manufacture aircraft altimeters, the new altimeters are better because the variation in the errors is reduced so that the readings are more consistent. (In many industries, the quality of goods and services can often be improved by reducing variation.)

The formal method of hypothesis testing uses several standard terms and conditions in a systematic procedure.

Study Hint: Start by clearly understanding Example 1 in Section 8-2, then read Sections 8-2 and 8-3 casually to obtain a general idea of their concepts, then study Section 8-2 more carefully to become familiar with the terminology.

> **CAUTION**
>
> When conducting hypothesis tests as described in this chapter and the following chapters, instead of jumping directly to procedures and calculations, be sure to consider the *context* of the data, the *source* of the data, and the *sampling method* used to obtain the sample data. (See Section 1-2.)

8-2 Basics of Hypothesis Testing

Key Concept In this section we present individual components of a hypothesis test. In Part 1 we discuss the basic concepts of hypothesis testing. Because these concepts are used in the following sections and chapters, we should know and understand the following:

- How to identify the null hypothesis and alternative hypothesis from a given claim, and how to express both in symbolic form

- How to calculate the value of the test statistic, given a claim and sample data

- How to identify the critical value(s), given a significance level

- How to identify the *P*-value, given a value of the test statistic

- How to state the conclusion about a claim in simple and nontechnical terms

In Part 2 we discuss the *power* of a hypothesis test.

Part 1: Basic Concepts of Hypothesis Testing

The methods presented in this chapter are based on the rare event rule (Section 4-1) for inferential statistics, so let's review that rule before proceeding.

Rare Event Rule for Inferential Statistics
If, under a given assumption, the probability of a particular observed event is extremely small, we conclude that the assumption is probably not correct.

Following this rule, we test a claim by analyzing sample data in an attempt to distinguish between results that can *easily occur by chance* and results that are *highly unlikely to occur by chance*. We can explain the occurrence of highly unlikely results by saying that either a rare event has indeed occurred or that the underlying assumption is not correct. Let's apply this reasoning in the following example.

> **EXAMPLE 1** **Gender Selection** ProCare Industries, Ltd. provided a product called "Gender Choice," which, according to advertising claims, allowed couples to "increase your chances of having a girl up to 80%." Suppose we conduct an experiment with 100 couples who want to have baby girls, and they all follow the Gender Choice "easy-to-use in-home system" described in the pink package designed for girls. Assuming that Gender Choice has no effect and using only common sense and no formal statistical methods, what should we conclude about the assumption of "no effect" from Gender Choice if 100 couples using Gender Choice have 100 babies consisting of the following?
>
> **a.** 52 girls   **b.** 97 girls

> **SOLUTION**

a. We normally expect around 50 girls in 100 births. The result of 52 girls is close to 50, so we should not conclude that the Gender Choice product is effective. The result of 52 girls could easily occur by chance, so there isn't sufficient evidence to say that Gender Choice is effective, even though the sample proportion of girls is greater than 50%.

continued

Aspirin Not Helpful for Geminis and Libras

Physician Richard Peto submitted an article to *Lancet*, a British medical journal. The article showed that patients had a better chance of surviving a heart attack if they were treated with aspirin within a few hours of their heart attacks. *Lancet* editors asked Peto to break down his results into subgroups to see if recovery worked better or worse for different groups, such as males or females. Peto believed that he was being asked to use too many subgroups, but the editors insisted. Peto then agreed, but he supported his objections by showing that when his patients were categorized by signs of the zodiac, aspirin was useless for Gemini and Libra heart-attack patients, but aspirin is a lifesaver for those born under any other sign. This shows that when conducting multiple hypothesis tests with many different subgroups, there is a very large chance of getting some wrong results.

b. The result of 97 girls in 100 births is extremely unlikely to occur by chance. We could explain the occurrence of 97 girls in one of two ways: Either an *extremely* rare event has occurred by chance, or Gender Choice is effective. The extremely low probability of getting 97 girls suggests that Gender Choice is effective.

In Example 1 we should conclude that the treatment is effective only if we get *significantly* more girls than we would normally expect. Although the outcomes of 52 girls and 97 girls are both greater than 50%, the result of 52 girls is not significant, whereas the result of 97 girls is significant.

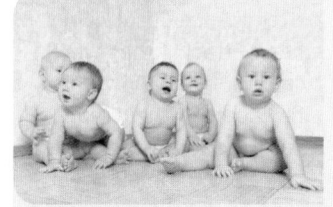

> **EXAMPLE 2**   **Gender Selection** The Chapter Problem includes the latest results from clinical trials of the XSORT method of gender selection. Instead of using the latest available results, we will use these results from preliminary trials of the XSORT method: Among 14 couples using the XSORT method, 13 couples had girls and one couple had a boy. We will proceed to formalize some of the analysis in testing the claim that the XSORT method increases the likelihood of having a girl, but there are two points that can be confusing:

1. Assume $p = 0.5$: Under normal circumstances, with no treatment, girls occur in 50% of births. So $p = 0.5$ and a claim that the XSORT method is effective can be expressed as $p > 0.5$.

2. Instead of P(exactly 13 girls), use P(13 or more girls): When determining whether 13 girls in 14 births is likely to occur by chance, use P(13 *or more* girls). (Stop for a minute and review the subsection of "Using Probabilities to Determine When Results Are Unusual" in Section 5-2.)

Under normal circumstances the proportion of girls is $p = 0.5$, so a claim that the XSORT method is effective can be expressed as $p > 0.5$. We support the claim of $p > 0.5$ only if a result such as 13 girls is unlikely (with a small probability, such as less than or equal to 0.05). Using a normal distribution as an approximation to the binomial distribution (see Section 6-6), we find P(13 or more girls in 14 births) $= 0.0016$. Figure 8-1 shows that with a probability of 0.5, the outcome of 13 girls in 14 births is unusual, so we *reject* random chance as a reasonable explanation. We conclude that the proportion of girls born to couples using the XSORT method is *significantly* greater than the proportion that we expect with random chance. Here are the key components of this example:

• Claim: The XSORT method increases the likelihood of a girl. That is, $p > 0.5$.

• Working assumption: The proportion of girls is $p = 0.5$ (with no effect from the XSORT method).

• The preliminary sample resulted in 13 girls among 14 births, so the sample proportion is $\hat{p} = 13/14 = 0.929$.

• Assuming that $p = 0.5$, we use a normal distribution as an approximation to the binomial distribution to find that P(at least 13 girls in 14 births) $= 0.0016$. (Using Table A-1 or calculations with the binomial probability distribution results in a probability of 0.001.)

• There are two possible explanations for the result of 13 girls in 14 births: Either a random chance event (with the very low probability of 0.0016) has occurred, or the proportion of girls born to couples using the XSORT method is greater than 0.5. Because the probability of getting at least 13 girls by chance is so small (0.0016), we reject random chance as a reasonable explanation. The more reasonable explanation for 13 girls is that the XSORT method is effective in increasing the likelihood of girls. There is sufficient evidence to support a claim that the XSORT method is effective in producing more girls than expected by chance.

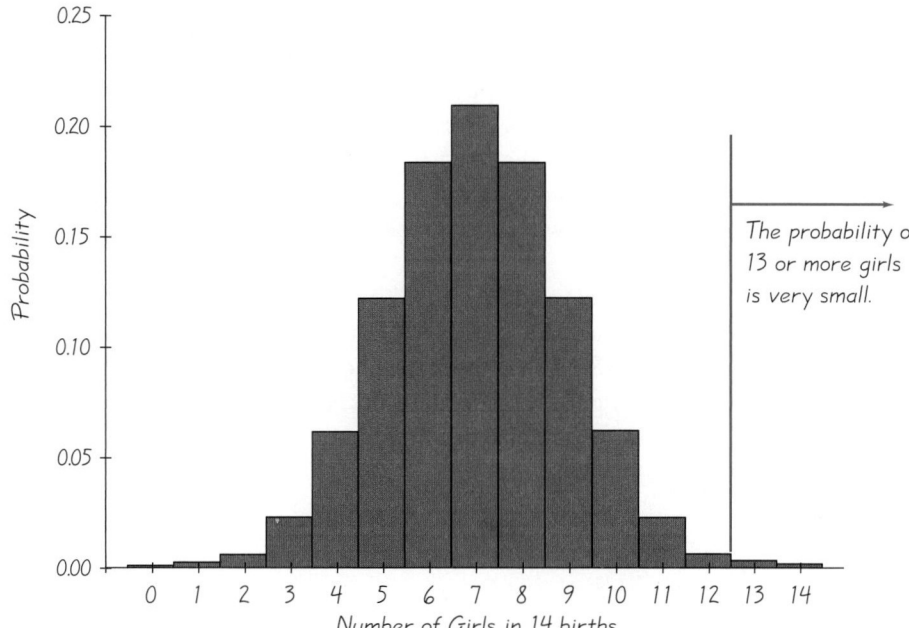

Figure 8-1

Probability Histogram for Numbers of Girls in 14 Births

We now proceed to describe the components of a formal **hypothesis test,** or **test of significance.** Many professional journals will include results from hypothesis tests, and they will use the same components described here.

Working with the Stated Claim: Null and Alternative Hypotheses

- The **null hypothesis** (denoted by H_0) is a statement that the value of a population parameter (such as proportion, mean, or standard deviation) is *equal to* some claimed value. (The term *null* is used to indicate no change or no effect or no difference.) Here is a typical null hypothesis included in this chapter: H_0: $p = 0.5$. We test the null hypothesis directly in the sense that we assume (or pretend) it is true and reach a conclusion to either reject it or fail to reject it.

- The **alternative hypothesis** (denoted by H_1 or H_a or H_A) is the statement that the parameter has a value that somehow differs from the null hypothesis. For the methods of this chapter, the symbolic form of the alternative hypothesis must use one of these symbols: $<, >, \neq$. Here are different examples of alternative hypotheses involving proportions:

$$H_1: p > 0.5 \qquad H_1: p < 0.5 \qquad H_1: p \neq 0.5$$

Note About Always Using the Equal Symbol in H_0: It is now rare, but the symbols $\leq$ and $\geq$ are occasionally used in the null hypothesis H_0. Professional statisticians and professional journals use only the equal symbol for equality. We conduct the hypothesis test by assuming that the proportion, mean, or standard deviation is *equal to* some specified value so that we can work with a single distribution having a specific value.

Note About Forming Your Own Claims (Hypotheses): If you are conducting a study and want to use a hypothesis test to *support* your claim, the claim must be worded so that it becomes the alternative hypothesis (and can be expressed using only the symbols $<$, $>$, or $\neq$). You can never support a claim that some parameter is *equal to* some specified value.

For example, after completing the clinical trials of the XSORT method of gender selection, the Genetics & IVF Institute will want to demonstrate that the method is effective in increasing the likelihood of a girl, so the claim will be stated as $p > 0.5$. In this context of trying to support the goal of the research, the alternative hypothesis is sometimes referred to as the *research hypothesis*. It will be assumed for the purpose of the hypothesis test that $p = 0.5$, but the Genetics & IVF Institute will hope that $p = 0.5$ gets rejected so that $p > 0.5$ is supported. Supporting the alternative hypothesis of $p > 0.5$ will support the claim that the XSORT method is effective.

Note About Identifying H_0 and H_1: Figure 8-2 summarizes the procedures for identifying the null and alternative hypotheses. Next to Figure 8-2 is an example using the claim that "with the XSORT method, the likelihood of having a girl is greater than 0.5." Note that the original statement could become the null hypothesis, it could become the alternative hypothesis, or it might not be either the null hypothesis or the alternative hypothesis.

Figure 8-2

Identifying H_0 and H_1

Example: The claim is that with the XSORT method, the likelihood of having a girl is greater than 0.5. This claim in symbolic form is $p > 0.5$.

If $p > 0.5$ is false, the symbolic form that must be true is $p \leq 0.5$.

H_1: $p > 0.5$
H_0: $p = 0.5$

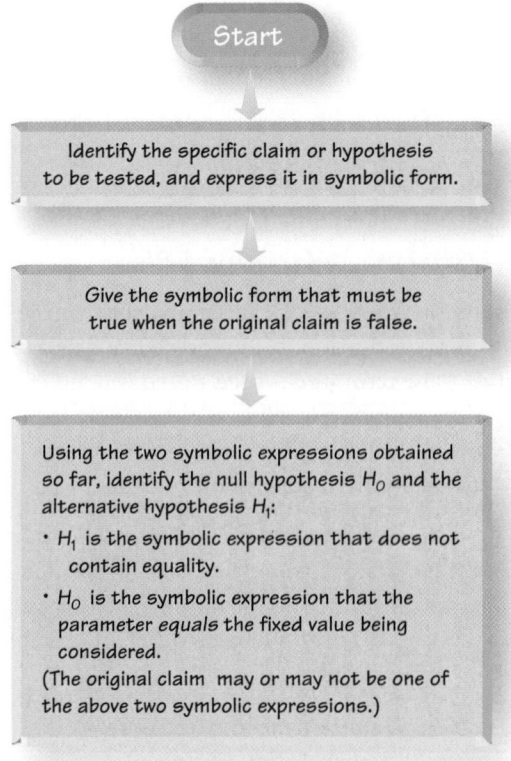

Start

Identify the specific claim or hypothesis to be tested, and express it in symbolic form.

Give the symbolic form that must be true when the original claim is false.

Using the two symbolic expressions obtained so far, identify the null hypothesis H_0 and the alternative hypothesis H_1:

• H_1 is the symbolic expression that does not contain equality.

• H_0 is the symbolic expression that the parameter equals the fixed value being considered.

(The original claim may or may not be one of the above two symbolic expressions.)

EXAMPLE 3 **Identifying the Null and Alternative Hypotheses**
Consider the claim that the mean weight of airline passengers (including carry-on baggage) is at most 195 lb (the current value used by the Federal Aviation Administration). Follow the three-step procedure outlined in Figure 8-2 to identify the null hypothesis and the alternative hypothesis.

SOLUTION Refer to Figure 8-2, which shows the three-step procedure.

Step 1: Express the given claim in symbolic form. The claim that the mean is at most 195 lb is expressed in symbolic form as $\mu \leq 195$ lb.

Step 2: If $\mu \leq 195$ lb is false, then $\mu > 195$ lb must be true.

Step 3: Of the two symbolic expressions $\mu \leq 195$ lb and $\mu > 195$ lb, we see that $\mu > 195$ lb does not contain equality, so we let the alternative hypothesis H_1 be $\mu > 195$ lb. Also, the null hypothesis must be a statement that the mean *equals* 195 lb, so we let H_0 be $\mu = 195$ lb.

Note that in this example, the original claim that the mean is at most 195 lb is neither the alternative hypothesis nor the null hypothesis. (However, we would be able to address the original claim upon completion of a hypothesis test.)

Converting Sample Data to a Test Statistic

The calculations required for a hypothesis test typically involve converting a sample statistic to a *test statistic*.

The **test statistic** is a value used in making a decision about the null hypothesis. It is found by converting the sample statistic (such as the sample proportion $\hat{p}$, the sample mean $\bar{x}$, or the sample standard deviation s) to a score (such as z, t, or χ^2) with the assumption that the null hypothesis is true. In this chapter we use the following test statistics:

Test statistic for proportion
$$z = \frac{\hat{p} - p}{\sqrt{\frac{pq}{n}}}$$

Test statistic for mean
$$z = \frac{\bar{x} - \mu}{\frac{\sigma}{\sqrt{n}}} \quad \text{or} \quad t = \frac{\bar{x} - \mu}{\frac{s}{\sqrt{n}}}$$

Test statistic for standard deviation
$$\chi^2 = \frac{(n-1)s^2}{\sigma^2}$$

The test statistic for a mean uses the normal or Student t distribution, depending on the conditions that are satisfied. For hypothesis tests of a claim about a population mean, this chapter will use the same criteria for using the normal or Student t distributions as described in Section 7-4. (See Figure 7-6 and Table 7-1.)

EXAMPLE 4 **Finding the Value of the Test Statistic** Let's again consider the claim that the XSORT method of gender selection increases the likelihood of having a baby girl. Preliminary results from a test of the XSORT method of gender selection involved 14 couples who gave birth to 13 girls and 1 boy. Use the

continued

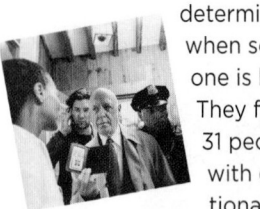

given claim and the preliminary results to calculate the value of the test statistic. Use the format of the test statistic given above, so that a normal distribution is used to approximate a binomial distribution. (There are other exact methods that do not use the normal approximation.)

SOLUTION   From Figure 8-2 and the example displayed next to it, the claim that the XSORT method of gender selection increases the likelihood of having a baby girl results in the following null and alternative hypotheses: H_0: $p = 0.5$ and H_1: $p > 0.5$. We work under the assumption that the null hypothesis is true with $p = 0.5$. The sample proportion of 13 girls in 14 births results in $\hat{p} = 13/14 = 0.929$. Using $p = 0.5$, $\hat{p} = 0.929$, and $n = 14$, we find the value of the test statistic as follows:

$$z = \frac{\hat{p} - p}{\sqrt{\dfrac{pq}{n}}} = \frac{0.929 - 0.5}{\sqrt{\dfrac{(0.5)(0.5)}{14}}} = 3.21$$

INTERPRETATION   We know from previous chapters that a z score of 3.21 is "unusual" (because it is greater than 2). It appears that in addition to being greater than 0.5, the sample proportion of $13/14$ or 0.929 is *significantly* greater than 0.5. Figure 8-3 shows that the sample proportion of 0.929 does fall within the range of values considered to be significant because they are so far above 0.5 that they are not likely to occur by chance (assuming that the population proportion is $p = 0.5$).

Figure 8-3 shows the test statistic of $z = 3.21$, and other components in Figure 8-3 are described as follows.

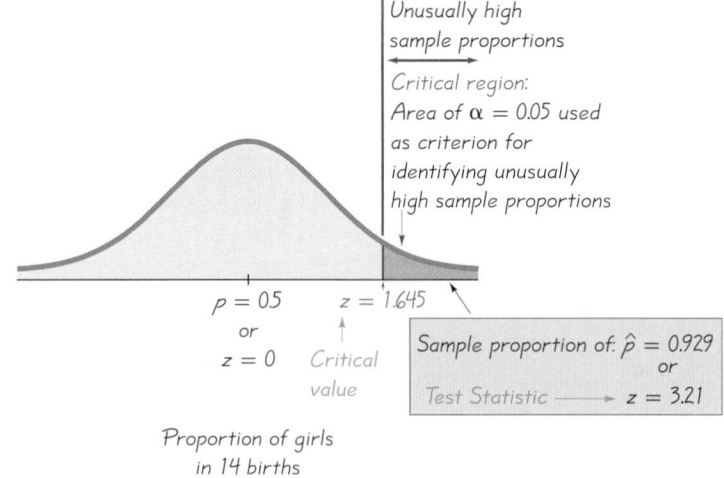

Figure 8-3   Critical Region, Critical Value, Test Statistic

Tools for Assessing the Test Statistic: Critical Region, Significance Level, Critical Value, and *P*-Value

The test statistic alone usually does not give us enough information to make a decision about the claim being tested. The following tools can be used to understand and interpret the test statistic.

- The **critical region** (or **rejection region**) is the set of all values of the test statistic that cause us to reject the null hypothesis. For example, see the red-shaded critical region shown in Figure 8-3.

- The **significance level** (denoted by α) is the probability that the test statistic will fall in the critical region when the null hypothesis is actually true. If the test statistic falls in the critical region, we reject the null hypothesis, so α is the probability of making the mistake of rejecting the null hypothesis when it is true. This is the same α introduced in Section 7-2, where we defined the confidence level for a confidence interval to be the probability $1 - \alpha$. Common choices for α are 0.05, 0.01, and 0.10, with 0.05 being most common.

- A **critical value** is any value that separates the critical region (where we reject the null hypothesis) from the values of the test statistic that do not lead to rejection of the null hypothesis. The critical values depend on the nature of the null hypothesis, the sampling distribution that applies, and the significance level of α. See Figure 8-3 where the critical value of $z = 1.645$ corresponds to a significance level of $\alpha = 0.05$. (See the end of Section 6-2 for a procedure for finding a critical value of z by using a TI-83/84 Plus calculator.)

EXAMPLE 5 **Finding a Critical Value for Critical Region in the Right Tail** Using a significance level of $\alpha = 0.05$, find the critical z value for the alternative hypothesis H_1: $p > 0.5$ (assuming that the normal distribution can be used to approximate the binomial distribution). This alternative hypothesis is used to test the claim that the XSORT method of gender selection is effective, so that baby girls are more likely, with a proportion greater than 0.5.

SOLUTION Refer to Figure 8-3. With H_1: $p > 0.5$, the critical region is in the right tail as shown. With a right-tailed area of 0.05, the critical value is found to be $z = 1.645$ (by using the methods of Section 6-2). If the right-tailed critical region is 0.05, the cumulative area to the left of the critical value is 0.95, and a TI-83/84 Plus calculator or Table A-2 show that the z score corresponding to a cumulative left area of 0.95 is $z = 1.645$. The critical value is $z = 1.645$ as shown in Figure 8-3.

EXAMPLE 6 **Finding Critical Values for a Critical Region in Two Tails** Using a significance level of $\alpha = 0.05$, find the two critical z values for the alternative hypothesis H_1: $p \neq 0.5$ (assuming that the normal distribution can be used to approximate the binomial distribution).

SOLUTION Refer to Figure 8-4(a). With H_1: $p \neq 0.5$, the critical region is in the two tails as shown. If the significance level is 0.05, each of the two tails has an area of 0.025 as shown in Figure 8-4(a). The left critical value of $z = -1.96$ corresponds to a cumulative left area of 0.025. (A TI-83/84 Plus calculator or Table A-2 result in $z = -1.96$ by using the methods of Section 6-2). The rightmost critical value of $z = 1.96$ is found from the cumulative left area of 0.975. (The rightmost critical value is $z_{0.975} = 1.96$.) The two critical values are $z = -1.96$ and $z = 1.96$ as shown in Figure 8-4(a).

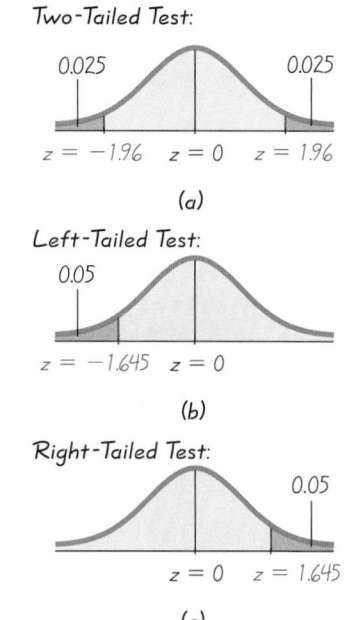

Two-Tailed Test:

0.025   0.025

$z = -1.96$ $z = 0$ $z = 1.96$

(a)

Left-Tailed Test:

0.05

$z = -1.645$ $z = 0$

(b)

Right-Tailed Test:

0.05

$z = 0$ $z = 1.645$

(c)

Figure 8-4 Finding Critical Values

• The **P-value** (or **p-value** or **probability value**) is the probability of getting a value of the test statistic that is *at least as extreme* as the one representing the sample data, assuming that the null hypothesis is true. *P*-values can be found after finding the area beyond the test statistic. The procedure for finding *P*-values is given in Figure 8-5. (When using a TI-83/84 Plus calculator for hypothesis tests, results include the *P*-value.) The procedure can be summarized as follows:

Critical region in the left tail: *P*-value = area to the *left* of the test statistic

Critical region in the right tail: *P*-value = area to the *right* of the test statistic

Critical region in two tails: *P*-value = *twice* the area in the tail beyond the test statistic

The null hypothesis is rejected if the *P*-value is very small, such as 0.05 or less. Here is a memory tool useful for interpreting the *P*-value:

If the *P* is low, the null must go.

If the *P* is high, the null will fly.

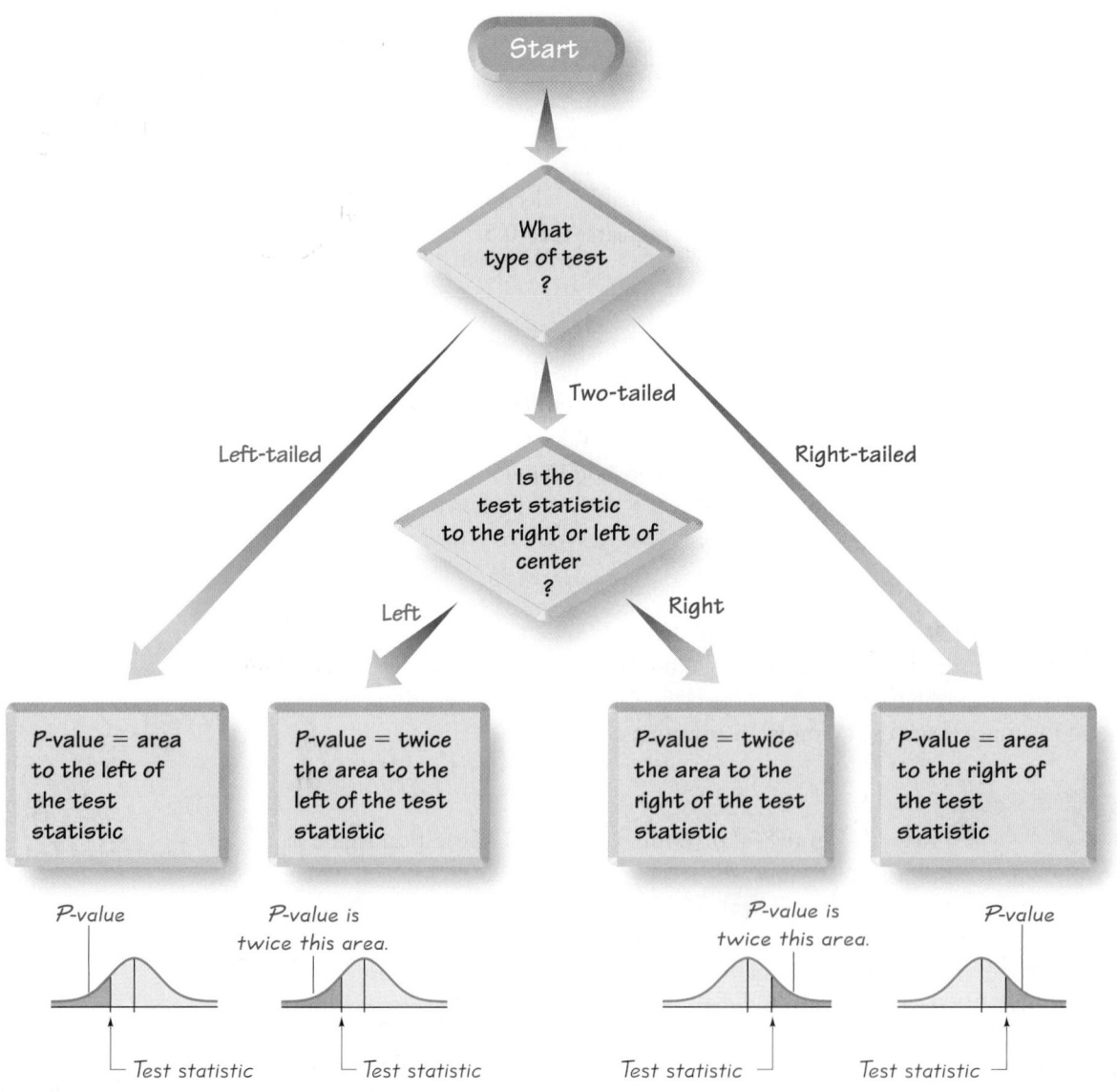

Figure 8-5   Procedure for Finding *P*-Values

CAUTION

Don't confuse a *P*-value with a proportion *p*. Know this distinction:

P-value = probability of getting a test statistic at least as extreme as the one representing sample data

p = population proportion

EXAMPLE 7 Finding a *P*-Value for a Critical Region in the Right Tail

Consider the claim that the XSORT method of gender selection increases the likelihood of having a baby girl, so that $p > 0.5$. Use the test statistic $z = 3.21$ (found from 13 girls in 14 births, as in Example 4). First determine whether the given conditions result in a critical region in the right tail, left tail, or two tails, then use Figure 8-5 to find the *P*-value. Interpret the *P*-value.

SOLUTION

With a claim of $p > 0.5$, the critical region is in the right tail, as shown in Figure 8-3. Using Figure 8-5 to find the *P*-value for a right-tailed test, we see that the *P*-value is the area to the right of the test statistic $z = 3.21$. A TI-83/84 Plus calculator or Table A-2 show that the area to the *right* of $z = 3.21$ is 0.0007, so the *P*-value is 0.0007.

INTERPRETATION

The *P*-value of 0.0007 is very small, and it shows that there is a very small chance of getting the sample results that led to a test statistic of $z = 3.21$. It is very unlikely that we would get 13 (or more) girls in 14 births by chance. This suggests that the XSORT method of gender selection increases the likelihood that a baby will be a girl.

EXAMPLE 8 Finding a *P*-Value for a Critical Region in Two Tails

Consider the claim that with the XSORT method of gender selection, the likelihood of having a baby girl is different from $p = 0.5$, and use the test statistic $z = 3.21$ found from 13 girls in 14 births. First determine whether the given conditions result in a critical region in the right tail, left tail, or two tails, then use Figure 8-5 to find the *P*-value. Interpret the *P*-value.

SOLUTION

The claim that the likelihood of having a baby girl is different from $p = 0.5$ can be expressed as $p \neq 0.5$, so the critical region is in two tails (as in Figure 8-4(a)). Using Figure 8-5 to find the *P*-value for a two-tailed test, we see that the *P*-value is *twice* the area to the right of the test statistic $z = 3.21$. We can use a TI-83/84 Plus calculator or Table A-2 to find that the area to the *right* of $z = 3.21$ is 0.0007. In this case, the *P*-value is twice the area to the right of the test statistic, so we have:

$$P\text{-value} = 2 \times 0.0007 = 0.0014$$

INTERPRETATION

The *P*-value is 0.0014 (or 0.0013 if greater precision is used for the calculations). The small *P*-value of 0.0014 shows that there is a very small chance of getting the sample results that led to a test statistic of $z = 3.21$. This suggests that with the XSORT method of gender selection, the likelihood of having a baby girl is different from 0.5.

Lie Detectors and the Law

Why not require all criminal suspects to take lie detector tests and dispense with trials by jury? The Council of Scientific Affairs of the American Medical Association states, "It is established that classification of guilty can be made with 75% to 97% accuracy, but the rate of false positives is often sufficiently high to preclude use of this (polygraph) test as the sole arbiter of guilt or innocence." A "false positive" is an indication of guilt when the subject is actually innocent. Even with accuracy as high as 97%, the percentage of false positive results can be 50%, so half of the innocent subjects incorrectly appear to be guilty.

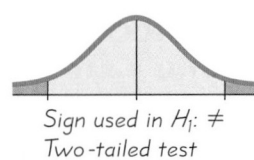

Sign used in H_1: ≠
Two-tailed test

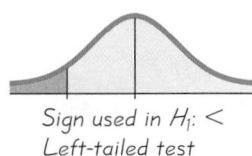

Sign used in H_1: <
Left-tailed test

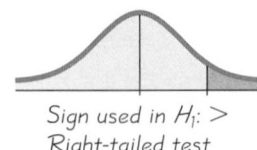

Sign used in H_1: >
Right-tailed test

Figure 8-6 Two-Tailed, Left-Tailed, Right-Tailed Tests

Types of Hypothesis Tests: Two-Tailed, Left-Tailed, Right-Tailed

The *tails* in a distribution are the extreme critical regions bounded by critical values. Determinations of *P*-values and critical values are affected by whether a critical region is in two tails, the left tail, or the right tail. It therefore becomes important to correctly characterize a hypothesis test as two-tailed, left-tailed, or right-tailed.

- **Two-tailed test:** The critical region is in the two extreme regions (tails) under the curve (as in Figure 8-4(a)).
- **Left-tailed test:** The critical region is in the extreme left region (tail) under the curve (as in Figure 8-4(b)).
- **Right-tailed test:** The critical region is in the extreme right region (tail) under the curve (as in Figure 8-4(c)).

Hint: By examining the alternative hypothesis, we can determine whether a test is two-tailed, left-tailed, or right-tailed. The tail will correspond to the critical region containing the values that would conflict significantly with the null hypothesis. A useful check is summarized in Figure 8-6. *Note that the inequality sign in H_1 points in the direction of the critical region.* The symbol ≠ is often expressed in programming languages as < >, and this reminds us that an alternative hypothesis such as $p \neq 0.5$ corresponds to a two-tailed test.

Decisions and Conclusions

The standard procedure of hypothesis testing requires that we directly test the null hypothesis, so our initial conclusion will always be one of the following:

1. Reject the null hypothesis.

2. Fail to reject the null hypothesis.

Decision Criterion The decision to reject or fail to reject the null hypothesis is usually made using either the *P*-value method of testing hypotheses or the traditional method (or classical method). Sometimes, however, the decision is based on confidence intervals. In recent years, use of the *P*-value method has been increasing along with the inclusion of *P*-values in results from software packages. The TI-83/84 Plus calculator provides *P*-values, so the *P*-value method is used for hypothesis tests.

P-value method:	Using the significance level α:
	If *P*-value $\leq \alpha$, *reject H_0*.
	If *P*-value $> \alpha$, *fail to reject H_0*.
Traditional method:	If the test statistic falls within the critical region, *reject H_0*.
	If the test statistic does not fall within the critical region, *fail to reject H_0*.
Another option:	Instead of using a significance level such as $\alpha = 0.05$, simply identify the *P*-value and leave the decision to the reader.
Confidence intervals:	A confidence interval estimate of a population parameter contains the likely values of that parameter.
	If a confidence interval does not include a claimed value of a population parameter, reject that claim.

Wording the Final Conclusion   Figure 8-7 summarizes a procedure for wording the final conclusion in simple, nontechnical terms. Note that only one case leads to wording indicating that the sample data actually *support* the conclusion. If you want to support some claim, state it in such a way that it becomes the alternative hypothesis, and then hope that the null hypothesis gets rejected.

CAUTION

Never conclude a hypothesis test with a statement of "reject the null hypothesis" or "fail to reject the null hypothesis." Always make sense of the conclusion with a statement that uses simple nontechnical wording that addresses the original claim.

Accept/Fail to Reject   A few textbooks continue to say "accept the null hypothesis" instead of "fail to reject the null hypothesis." The term *accept* is somewhat misleading, because it seems to imply incorrectly that the null hypothesis has been proved, but we can never prove a null hypothesis. The phrase *fail to reject* says more correctly that the available evidence isn't strong enough to warrant rejection of the null hypothesis. In this text we use the terminology *fail to reject the null hypothesis,* instead of *accept the null hypothesis.*

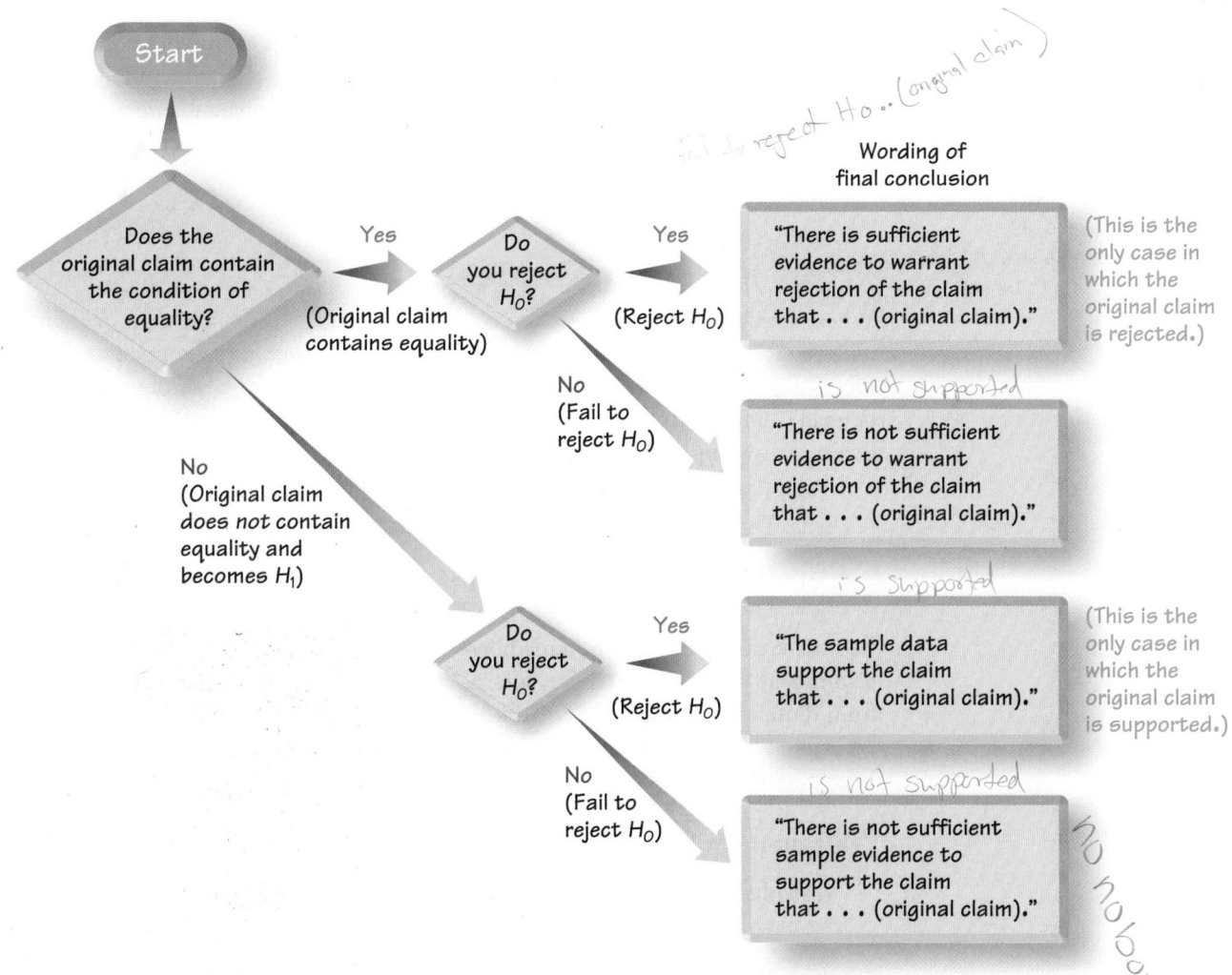

Figure 8-7   Wording of Final Conclusion

Large Sample Size Isn't Good Enough

Biased sample data should not be used for inferences, no matter how large the sample is. For example, in *Women and Love: A Cultural Revolution in Progress,* Shere Hite bases her conclusions on 4500 replies that she received after mailing 100,000 questionnaires to various women's groups. A *random* sample of 4500 subjects would usually provide good results, but Hite's sample is biased. It is criticized for over-representing women who join groups and women who feel strongly about the issues addressed. Because Hite's sample is biased, her inferences are not valid, even though the sample size of 4500 might seem to be sufficiently large.

Multiple Negatives   When stating the final conclusion in nontechnical terms, it is possible to get correct statements with up to three negative terms. (Example: "There is *not* sufficient evidence to warrant *rejection* of the claim of *no* difference between 0.5 and the population proportion.") Such conclusions are confusing, so it is good to restate them in a way that makes them understandable, but care must be taken to not change the meaning. For example, instead of saying that "there is not sufficient evidence to warrant rejection of the claim of no difference between 0.5 and the population proportion," better statements would be these:

- Fail to reject the claim that the population proportion is equal to 0.5.
- Unless stronger evidence is obtained, continue to assume that the population proportion is equal to 0.5.

EXAMPLE 9   **Stating the Final Conclusion**   Suppose a geneticist claims that the XSORT method of gender selection increases the likelihood of a baby girl. This claim of $p > 0.5$ becomes the alternative hypothesis, while the null hypothesis becomes $p = 0.5$. Further suppose that the sample evidence causes us to reject the null hypothesis of $p = 0.5$. State the conclusion in simple, nontechnical terms.

SOLUTION   Refer to Figure 8-7. Because the original claim does not contain equality, it becomes the alternative hypothesis. Because we reject the null hypothesis, the wording of the final conclusion should be as follows: "There is sufficient evidence to support the claim that the XSORT method of gender selection increases the likelihood of a baby girl."

Errors in Hypothesis Tests

When testing a null hypothesis, we arrive at a conclusion of rejecting it or failing to reject it. Such conclusions are sometimes correct and sometimes wrong (even if we do everything correctly). Table 8-1 summarizes the two different types of errors that can be made, along with the two different types of correct decisions. We distinguish between the two types of errors by calling them type I and type II errors.

- **Type I error:** The mistake of rejecting the null hypothesis when it is actually true. The symbol α (alpha) is used to represent the probability of a type I error.

- **Type II error:** The mistake of failing to reject the null hypothesis when it is actually false. The symbol β (beta) is used to represent the probability of a type II error.

Because it can be difficult to remember which error is type I and which is type II, we recommend a mnemonic device, such as "routine for fun." Using only the consonants from those words (**R**ou**T**i**N**e **F**o**R** **F**u**N**), we can easily remember that a type I error is RTN: Reject True Null (hypothesis), whereas a type II error is FRFN: Fail to Reject a False Null (hypothesis).

Notation

α (alpha) = probability of a type I error (the probability of rejecting the null hypothesis when it is true)

β (beta) = probability of a type II error (the probability of failing to reject a null hypothesis when it is false)

Table 8-1 Type I and Type II Errors

		True State of Nature	
		The null hypothesis is true	The null hypothesis is false
Decision	We decide to reject the null hypothesis	**Type I error** (rejecting a true null hypothesis) $P(\text{type I error}) = \alpha$	Correct decision
	We fail to reject the null hypothesis	Correct decision	**Type II error** (failing to reject a false null hypothesis) $P(\text{type II error}) = \beta$

EXAMPLE 10 **Identifying Type I and Type II Errors** Assume that we are conducting a hypothesis test of the claim that a method of gender selection increases the likelihood of a baby girl, so that the probability of a baby girl is $p > 0.5$. Here are the null and alternative hypotheses:

$$H_0: p = 0.5$$
$$H_1: p > 0.5$$

Give statements identifying the following.

a. Type I error   **b.** Type II error

SOLUTION

a. A type I error is the mistake of rejecting a true null hypothesis, so this is a type I error: Conclude that there is sufficient evidence to support $p > 0.5$, when in reality $p = 0.5$. That is, a type I error is made when we conclude that the gender selection method is effective when in reality it has no effect.

b. A type II error is the mistake of failing to reject the null hypothesis when it is false, so this is a type II error: Fail to reject $p = 0.5$ (and therefore fail to support $p > 0.5$) when in reality $p > 0.5$. That is, a type II error is made if we conclude that the gender selection method has no effect, when it really is effective in increasing the likelihood of a baby girl.

Controlling Type I and Type II Errors: One step in our standard procedure for testing hypotheses involves the selection of the significance level α (such as 0.05), which is the probability of a type I error. The values of α, β, and the sample size n are all related, so when you choose or determine any two of them, the third is automatically determined. One common practice is to select the significance level α, then select a sample size that is practical, so the value of β is determined. Generally try to use the largest α that you can tolerate, but for type I errors with more serious consequences, select smaller values of α. Then choose a sample size n as large as is reasonable, based on considerations of time, cost, and other relevant factors. Another common practice is to select α and β, so the required sample size n is automatically determined. (See Example 12 in Part 2 of this section.)

Comprehensive Hypothesis Test In this section we describe the individual components used in a hypothesis test, but the following sections will combine those components in comprehensive procedures. We can test claims about population parameters by using the *P*-value method summarized in Figure 8-8, the traditional method summarized in Figure 8-9, or we can use a confidence interval, as described on page 413.

Figure 8-8   *P*-Value Method

P-Value Method

Start

1. Identify the specific claim or hypothesis to be tested, and put it in symbolic form.

2. Give the symbolic form that must be true when the original claim is false.

3. Of the two symbolic expressions obtained so far, let the alternative hypothesis H_1 be the one not containing equality, so that H_1 uses the symbol $>$ or $<$ or $\neq$. Let the null hypothesis H_0 be the symbolic expression that the parameter *equals* the fixed value being considered.

4. Select the significance level α based on the seriousness of a type 1 error. Make α small if the consequences of rejecting a true H_0 are severe. The values of 0.05 and 0.01 are very common.

5. Identify the statistic that is relevant to this test and determine its sampling distribution (such as normal, *t*, chi-square).

6. Find the test statistic and find the *P*-value (see Figure 8-5). Draw a graph and show the test statistic and *P*-value.

7. Reject H_0 if the *P*-value is less than or equal to the significance level α. Fail to reject H_0 if the *P*-value is greater than α.

8. Restate this previous decision in simple, nontechnical terms, and address the original claim.

Stop

Figure 8-9   Traditional Method

Traditional Method

Start

1. Identify the specific claim or hypothesis to be tested, and put it in symbolic form.

2. Give the symbolic form that must be true when the original claim is false.

3. Of the two symbolic expressions obtained so far, let the alternative hypothesis H_1 be the one not containing equality, so that H_1 uses the symbol $>$ or $<$ or $\neq$. Let the null hypothesis H_0 be the symbolic expression that the parameter *equals* the fixed value being considered.

4. Select the significance level α based on the seriousness of a type 1 error. Make α small if the consequences of rejecting a true H_0 are severe. The values of 0.05 and 0.01 are very common.

5. Identify the statistic that is relevant to this test and determine its sampling distribution (such as normal, *t*, chi-square).

6. Find the test statistic, the critical values, and the critical region. Draw a graph and include the test statistic, critical value(s), and critical region.

7. Reject H_0 if the test statistic is in the critical region. Fail to reject H_0 if the test statistic is not in the critical region.

8. Restate this previous decision in simple, nontechnical terms, and address the original claim.

Stop

Confidence Interval Method

Construct a confidence interval with a confidence level selected as in Table 8-2.
Because a confidence interval estimate of a population parameter contains the likely values of that parameter, reject a claim that the population parameter has a value that is not included in the confidence interval.

Table 8-2	Confidence Level for Confidence Interval		
		Two-Tailed Test	One-Tailed Test
Significance	0.01	99%	98%
Level for	0.05	95%	90%
Hypothesis	0.10	90%	80%
Test			

Confidence Interval Method   For two-tailed hypothesis tests construct a confidence interval with a confidence level of $1 - \alpha$; but for a one-tailed hypothesis test with significance level α, construct a confidence interval with a confidence level of $1 - 2\alpha$. (See Table 8-2 for common cases.) After constructing the confidence interval, use this criterion:

> **A confidence interval estimate of a population parameter contains the likely values of that parameter. We should therefore reject a claim that the population parameter has a value that is not included in the confidence interval.**

CAUTION

In some cases, a conclusion based on a confidence interval may be different from a conclusion based on a hypothesis test. See the comments in the individual sections that follow.

The exercises for this section involve isolated components of hypothesis tests, but the following sections will involve complete and comprehensive hypothesis tests.

Part 2: Beyond the Basics of Hypothesis Testing: The *Power* of a Test

We use β to denote the probability of failing to reject a false null hypothesis, so $P(\text{type II error}) = \beta$. It follows that $1 - \beta$ is the probability of rejecting a false null hypothesis, and statisticians refer to this probability as the *power* of a test, and they often use it to gauge the effectiveness of a hypothesis test in allowing us to recognize that a null hypothesis is false.

> **DEFINITION**
>
> The **power** of a hypothesis test is the probability $(1 - \beta)$ of rejecting a false null hypothesis. The value of the power is computed by using a particular significance level α and a *particular* value of the population parameter that is an alternative to the value assumed true in the null hypothesis.

Note that in the above definition, determination of power requires a particular value that is an alternative to the value assumed in the null hypothesis. Consequently, a hypothesis test can have many different values of power, depending on the particular values of the population parameter chosen as alternatives to the null hypothesis.

EXAMPLE 11   **Power of a Hypothesis Test** Let's again consider these preliminary results from the XSORT method of gender selection: There were 13 girls among the 14 babies born to couples using the XSORT method. If we want to test the claim that girls are more likely ($p > 0.5$) with the XSORT method, we have the following null and alternative hypotheses:

$$H_0\text{: } p = 0.5 \qquad H_1\text{: } p > 0.5$$

Let's use $\alpha = 0.05$. In addition to all of the given test components, we need a particular value of p that is an alternative to the value assumed in the null hypothesis H_0: $p = 0.5$. Using the given test components along with different alternative values of p, we get the following examples of power values. These values of power were found by using Minitab, and exact calculations are used instead of a normal approximation to the binomial distribution.

continued

Specific Alternative Value of p	β	Power of Test (1 − β)
0.6	0.820	0.180
0.7	0.564	0.436
0.8	0.227	0.773
0.9	0.012	0.988

INTERPRETATION   Based on the above list of power values, we see that this hypothesis test has power of 0.180 (or 18.0%) of rejecting H_0: $p = 0.5$ when the population proportion p is actually 0.6. That is, if the true population proportion is actually equal to 0.6, there is an 18.0% chance of making the correct conclusion of rejecting the false null hypothesis that $p = 0.5$. That low power of 18.0% is not good. There is a 0.564 probability of rejecting $p = 0.5$ when the true value of p is actually 0.7. It makes sense that this test is more effective in rejecting the claim of $p = 0.5$ when the population proportion is actually 0.7 than when the population proportion is actually 0.6. (When identifying animals assumed to be horses, there's a better chance of rejecting an elephant as a horse (because of the greater difference) than rejecting a mule as a horse.) In general, increasing the difference between the assumed parameter value and the actual parameter value results in an increase in power, as shown in the above table.

Because the calculations of power are quite complicated, the use of technology is strongly recommended. (In this section, only Exercises 46–48 involve power.)

Power and the Design of Experiments   Just as 0.05 is a common choice for a significance level, a power of at least 0.80 is a common requirement for determining that a hypothesis test is effective. (Some statisticians argue that the power should be higher, such as 0.85 or 0.90.) When designing an experiment, we might consider how much of a difference between the claimed value of a parameter and its true value is an important amount of difference. If testing the effectiveness of the XSORT gender-selection method, a change in the proportion of girls from 0.5 to 0.501 is not very important. A change in the proportion of girls from 0.5 to 0.6 might be important. Such magnitudes of differences affect power. When designing an experiment, a goal of having a power value of at least 0.80 can often be used to determine the minimum required sample size, as in the following example.

EXAMPLE 12   **Finding Sample Size Required to Achieve 80% Power**
Here is a statement similar to one in an article from the *Journal of the American Medical Association:* "The trial design assumed that with a 0.05 significance level, 153 randomly selected subjects would be needed to achieve 80% power to detect a reduction in the coronary heart disease rate from 0.5 to 0.4." Before conducting the experiment, the researchers selected a significance level of 0.05 and a power of at least 0.80. They also decided that a reduction in the proportion of coronary heart disease from 0.5 to 0.4 is an important difference that they wanted to detect (by correctly rejecting the false null hypothesis). Using a significance level of 0.05, power of 0.80, and the alternative proportion of 0.4, technology such as Minitab is used to find that the required minimum sample size is 153. The researchers can then proceed by obtaining a sample of at least 153 randomly selected subjects. Due to factors such as dropout rates, the researchers are likely to need somewhat more than 153 subjects. (See Exercise 48.)

8-2   Basic Skills and Concepts

Statistical Literacy and Critical Thinking

1. Hypothesis Test In reporting on an *Elle*/MSNBC.COM survey of 61,647 people, *Elle* magazine stated that "just 20% of bosses are good communicators." Without performing formal calculations, do the sample results appear to support the claim that less than 50% of people believe that bosses are good communicators? What can you conclude after learning that the survey results were obtained over the Internet from people who chose to respond?

2. Interpreting *P*-Value When the clinical trial of the XSORT method of gender selection is completed, a formal hypothesis test will be conducted with the alternative hypothesis of $p > 0.5$, which corresponds to the claim that the XSORT method increases the likelihood of having a girl, so that the proportion of girls is greater than 0.5. If you are responsible for developing the XSORT method and you want to show its effectiveness, which of the following *P*-values would you prefer: 0.999, 0.5, 0.95, 0.05, 0.01, 0.001? Why?

3. Proving that the Mean Equals 325 mg Bottles of Bayer aspirin are labeled with a statement that the tablets each contain 325 mg of aspirin. A quality control manager claims that a large sample of data can be used to support the claim that the mean amount of aspirin in the tablets is equal to 325 mg, as the label indicates. Can a hypothesis test be used to support that claim? Why or why not?

4. Supporting a Claim In preliminary results from couples using the Gender Choice method of gender selection to increase the likelihood of having a baby girl, 20 couples used the Gender Choice method with the result that 8 of them had baby girls and 12 had baby boys. Given that the sample proportion of girls is 8/20 or 0.4, can the sample data support the claim that the proportion of girls is greater than 0.5? Can any sample proportion less than 0.5 be used to support a claim that the population proportion is greater than 0.5?

Stating Conclusions About Claims. *In Exercises 5–8, make a decision about the given claim. Use only the rare event rule stated in Section 8-2, and make subjective estimates to determine whether events are likely. For example, if the claim is that a coin favors heads and sample results consist of 11 heads in 20 flips, conclude that there is not sufficient evidence to support the claim that the coin favors heads (because it is easy to get 11 heads in 20 flips by chance with a fair coin).*

5. Claim: A coin favors heads when tossed, and there are 90 heads in 100 tosses.

6. Claim: The proportion of households with telephones is greater than the proportion of 0.35 found in the year 1920. A recent simple random sample of 2480 households results in a proportion of 0.955 households with telephones (based on data from the U.S. Census Bureau).

7. Claim: The mean pulse rate (in beats per minute) of students of the author is less than 75. A simple random sample of students has a mean pulse rate of 74.4.

8. Claim: Movie patrons have IQ scores with a standard deviation that is less than the standard deviation of 15 for the general population. A simple random sample of 40 movie patrons results in IQ scores with a standard deviation of 14.8.

Identifying H_0 and H_1 *In Exercises 9–16, examine the given statement, then express the null hypothesis H_0 and alternative hypothesis H_1 in symbolic form. Be sure to use the correct symbol (μ, p, σ) for the indicated parameter.*

9. The mean annual income of employees who took a statistics course is greater than $60,000.

10. The proportion of people aged 18 to 25 who currently use illicit drugs is equal to 0.20 (or 20%).

11. The standard deviation of human body temperatures is equal to 0.62°F.

12. The majority of college students have credit cards.

13. The standard deviation of duration times (in seconds) of the Old Faithful geyser is less than 40 sec.

14. The standard deviation of daily rainfall amounts in San Francisco is 0.66 cm.

15. The proportion of homes with fire extinguishers is 0.80.

16. The mean weight of plastic discarded by households in one week is less than 1 kg.

Finding Critical Values. *In Exercises 17–24, assume that the normal distribution applies and find the critical z values.*

17. Two-tailed test; $\alpha = 0.01$.

18. Two-tailed test; $\alpha = 0.10$.

19. Right-tailed test; $\alpha = 0.02$.

20. Left-tailed test; $\alpha = 0.10$.

21. $\alpha = 0.05$; H_1 is $p \neq 98.6°F$

22. $\alpha = 0.01$; H_1 is $p > 0.5$.

23. $\alpha = 0.005$; H_1 is $p < 5280$ ft.

24. $\alpha = 0.005$; H_1 is $p \neq 45$ mm

Finding Test Statistics. *In Exercises 25–28, find the value of the test statistic z using*

$$z = \frac{\hat{p} - p}{\sqrt{\dfrac{pq}{n}}}$$

25. Genetics Experiment The claim is that the proportion of peas with yellow pods is equal to 0.25 (or 25%). The sample statistics from one of Mendel's experiments include 580 peas with 152 of them having yellow pods.

26. Carbon Monoxide Detectors The claim is that less than 1/2 of adults in the United States have carbon monoxide detectors. A KRC Research survey of 1005 adults resulted in 462 who have carbon monoxide detectors.

27. Italian Food The claim is that more than 25% of adults prefer Italian food as their favorite ethnic food. A Harris Interactive survey of 1122 adults resulted in 314 who say that Italian food is their favorite ethnic food.

28. Seat Belts The claim is that more than 75% of adults always wear a seat belt in the front seat. A Harris Poll of 1012 adults resulted in 870 who say that they always wear a seat belt in the front seat.

Finding *P*-values. *In Exercises 29–36, use the given information to find the P-value. (Hint: Follow the procedure summarized in Figure 8-5.) Also, use a 0.05 significance level and state the conclusion about the null hypothesis (reject the null hypothesis or fail to reject the null hypothesis).*

29. The test statistic in a left-tailed test is $z = -1.25$.

30. The test statistic in a right-tailed test is $z = 2.50$.

31. The test statistic in a two-tailed test is $z = 1.75$.

32. The test statistic in a two-tailed test is $z = -0.55$.

33. With H_1: $p \neq 0.707$, the test statistic is $z = -2.75$.

34. With H_1: $p \neq 3/4$, the test statistic is $z = 0.35$.

35. With H_1: $p > 1/4$, the test statistic is $z = 2.30$.

36. With H_1: $p < 0.777$, the test statistic is $z = -2.95$.

Stating Conclusions. *In Exercises 37–40, state the final conclusion in simple non-technical terms. Be sure to address the original claim. (Hint: See Figure 8-7.)*

37. Original claim: The percentage of blue M&Ms is greater than 5%.

Initial conclusion: Fail to reject the null hypothesis.

38. Original claim: The percentage of on-time U.S. airline flights is less than 75%.

Initial conclusion: Reject the null hypothesis.

39. Original claim: The percentage of Americans who know their credit score is equal to 20%.

Initial conclusion: Fail to reject the null hypothesis.

40. Original claim: The percentage of Americans who believe in heaven is equal to 90%.

Initial conclusion: Reject the null hypothesis.

Identifying Type I and Type II Errors. *In Exercises 41–44, identify the type I error and the type II error that correspond to the given hypothesis.*

41. The percentage of nonsmokers exposed to secondhand smoke is equal to 41%.

42. The percentage of Americans who believe that life exists only on earth is equal to 20%.

43. The percentage of college students who consume alcohol is greater than 70%.

44. The percentage of households with at least two cell phones is less than 60%.

8-2 Beyond the Basics

45. Significance Level

a. If a null hypothesis is rejected with a significance level of 0.05, is it also rejected with a significance level of 0.01? Why or why not?

b. If a null hypothesis is rejected with a significance level of 0.01, is it also rejected with a significance level of 0.05? Why or why not?

46. Interpreting Power Chantix tablets are used as an aid to help people stop smoking. In a clinical trial, 129 subjects were treated with Chantix twice a day for 12 weeks, and 16 subjects experienced abdominal pain (based on data from Pfizer, Inc.). If someone claims that more than 8% of Chantix users experience abdominal pain, that claim is supported with a hypothesis test conducted with a 0.05 significance level. Using 0.18 as an alternative value of p, the power of the test is 0.96. Interpret this value of the power of the test.

47. Calculating Power Consider a hypothesis test of the claim that the MicroSort method of gender selection is effective in increasing the likelihood of having a baby girl ($p > 0.5$). Assume that a significance level of $\alpha = 0.05$ is used, and the sample is a simple random sample of size $n = 64$.

a. Assuming that the true population proportion is 0.65, find the power of the test, which is the probability of rejecting the null hypothesis when it is false. (*Hint:* With a 0.05 significance level, the critical value is $z = 1.645$, so any test statistic in the right tail of the accompanying top graph is in the rejection region where the claim is supported. Find the sample proportion $\hat{p}$ in the top graph, and use it to find the power shown in the bottom graph.)

b. Explain why the red shaded region of the bottom graph corresponds to the power of the test.

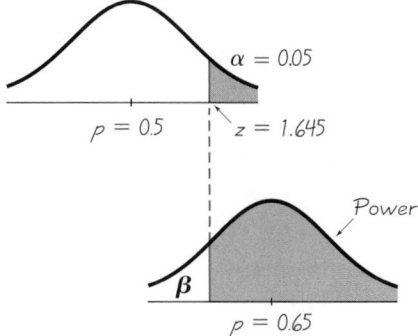

48. Finding Sample Size to Achieve Power Researchers plan to conduct a test of a gender selection method. They plan to use the alternative hypothesis of H_1: $p > 0.5$ and a significance level of $\alpha = 0.05$. Find the sample size required to achieve at least 80% power in detecting an increase in p from 0.5 to 0.55. (*This is a very difficult exercise. Hint: See Exercise 47.*)

8-3 Testing a Claim About a Proportion

Key Concept In Section 8-2 we presented the individual components of a hypothesis test. In this section we present complete procedures for testing a hypothesis (or claim) made about a population proportion. We illustrate hypothesis testing with the *P*-value method, the traditional method, and the use of confidence intervals. In addition to testing claims about population proportions, we can use the same procedures for testing claims about probabilities or the decimal equivalents of percents.

The following are examples of the types of claims we will be able to test:

- **Genetics** The Genetics & IVF Institute claims that its XSORT method allows couples to increase the probability of having a baby girl, so that the proportion of girls with this method is greater than 0.5.

- **Medicine** Pregnant women can correctly guess the sex of their babies so that they are correct more than 50% of the time.

- **Entertainment** Among the television sets in use during a recent Super Bowl game, 64% were tuned to the Super Bowl.

Two common methods for testing a claim about a population proportion are (1) to use a normal distribution as an approximation to the binomial distribution, and (2) to use an exact method based on the binomial probability distribution. Part 1 of this section uses the approximate method with the normal distribution, and Part 2 of this section briefly describes the exact method.

Part 1: Basic Methods of Testing Claims About a Population Proportion p

The following box includes the key elements used for testing a claim about a population proportion.

Requirements

Objective

Test a claim about a population proportion using a formal method of hypothesis testing.

Notation

n = sample size or number of trials

$\hat{p} = \dfrac{x}{n}$ (*sample* proportion)

p = population proportion (based on the claim, p is the value used in the null hypothesis)

$q = 1 - p$

Requirements

1. The sample observations are a simple random sample.

2. The conditions for a *binomial distribution* are satisfied. (There are a fixed number of independent trials having constant probabilities, and each trial has two outcome categories of "success" and "failure.")

3. The conditions $np \geq 5$ and $nq \geq 5$ are both satisfied, so **the binomial distribution of sample proportions can be approximated by a normal distribution with $\mu = np$ and $\sigma = \sqrt{npq}$** (as described in Section 6-6). Note that p is the *assumed* proportion used in the claim, not the sample proportion.

Test Statistic for Testing a Claim About a Proportion

$$z = \frac{\hat{p} - p}{\sqrt{\dfrac{pq}{n}}}$$

P-values: Use the standard normal distribution (Table A-2) and refer to Figure 8-5.

Critical values: Use the standard normal distribution (Table A-2).

TI-83/84 Plus Calculator The TI-83/84 Plus calculator can be used to automatically generate the test statistic and *P*-value. See the procedure at the end of this section.

CAUTION

Reminder: Don't confuse a *P*-value with a proportion *p*. **P-value** = probability of getting a test statistic at least as extreme as the one representing sample data, but p = population proportion.

The above test statistic does not include a correction for continuity (as described in Section 6-6), because its effect tends to be very small with large samples.

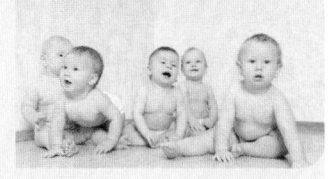

EXAMPLE 1 **Testing the Effectiveness of the MicroSort Method of Gender Selection** The Chapter Problem described these results from trials of the XSORT method of gender selection developed by the Genetics & IVF Institute: Among 726 babies born to couples using the XSORT method in an attempt to have a baby girl, 668 of the babies were girls and the others were boys. Use these results with a 0.05 significance level to test the claim that among babies born to couples using the XSORT method, the proportion of girls is greater than the value of 0.5 that is expected with no treatment. Here is a summary of the claim and the sample data:

Claim: With the XSORT method, the proportion of girls is greater than 0.5. That is, $p > 0.5$.

Sample data: $n = 726$ and $\hat{p} = \dfrac{668}{726} = 0.920$

Before starting the hypothesis test, verify that the necessary requirements are satisfied.

REQUIREMENT CHECK We first check the three requirements.

1. It is not likely that the subjects in the clinical trial are a simple random sample, but a selection bias is not really an issue here, because a couple wishing to have a baby girl can't affect the sex of their baby without an effective treatment. Volunteer couples are self-selected, but that does not affect the results in this situation.

2. There is a fixed number (726) of independent trials with two categories (the baby is either a girl or boy).

continued

3. The requirements $np \geq 5$ and $nq \geq 5$ are both satisfied with $n = 726$, $p = 0.5$, and $q = 0.5$. (We get $np = (726)(0.5) = 363 \geq 5$ and $nq = (726)(0.5) = 363 \geq 5$.)

The three requirements are satisfied.

P-Value Method

Figure 8-8 on page 412 lists the steps for using the *P*-value method. Using those steps from Figure 8-8, we can test the claim in Example 1 as follows.

Step 1. The original claim in symbolic form is $p > 0.5$.

Step 2. The opposite of the original claim is $p \leq 0.5$.

Step 3. Of the preceding two symbolic expressions, the expression $p > 0.5$ does not contain equality, so it becomes the alternative hypothesis. The null hypothesis is the statement that p equals the fixed value of 0.5. We can therefore express H_0 and H_1 as follows:

$$H_0: p = 0.5$$
$$H_1: p > 0.5$$

Step 4. We use the significance level of $\alpha = 0.05$, which is a very common choice.

Step 5. Because we are testing a claim about a population proportion p, the sample statistic $\hat{p}$ is relevant to this test. The sampling distribution of sample proportions $\hat{p}$ can be approximated by a normal distribution in this case.

Step 6. The test statistic $z = 22.63$ is calculated as follows:

$$z = \frac{\hat{p} - p}{\sqrt{\frac{pq}{n}}} = \frac{0.920 - 0.5}{\sqrt{\frac{(0.5)(0.5)}{726}}} = 22.63$$

We now find the *P*-value by using the following procedure, which is shown in Figure 8-5:

Left-tailed test:	*P*-value = area to left of test statistic z
Right-tailed test:	*P*-value = area to right of test statistic z
Two-tailed test:	*P*-value = *twice* the area of the extreme region bounded by the test statistic z

Because the hypothesis test we are considering is right-tailed with a test statistic of $z = 22.63$, the *P*-value is the area to the right of $z = 22.63$. Referring to Table A-2, we see that for values of $z = 3.50$ and higher, we use 0.0001 for the cumulative area to the *right* of the test statistic. The *P*-value is therefore 0.0001. Figure 8-10 shows the test statistic and *P*-value for this example.

TI-83/84 Plus

```
1-PropZTest
prop>.5
z=22.63922338
p=0
p=.9201101928
n=726
```

TI-83/84 Plus If we use a TI-83/84 Plus calculator with the procedure given near the end of this section, we obtain the display shown in the margin. The test statistic is shown as $z = 22.63922338$, or $z = 22.64$ when rounded. (This test statistic uses the unrounded sample proportion of 0.9201101928, and it is more accurate than the one shown above.) The TI-83/84 Plus display also shows that the *P*-value is 0, which indicates that the actual *P*-value is a positive number so small that it can be considered to be 0 for all practical purposes.

Step 7. Because the *P*-value is less than or equal to the significance level of $\alpha = 0.05$, we reject the null hypothesis.

Step 8. We conclude that there is sufficient sample evidence to support the claim that among babies born to couples using the XSORT method, the proportion

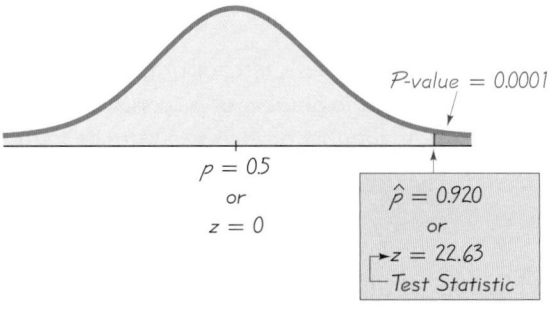

Figure 8-10   *P*-Value Method

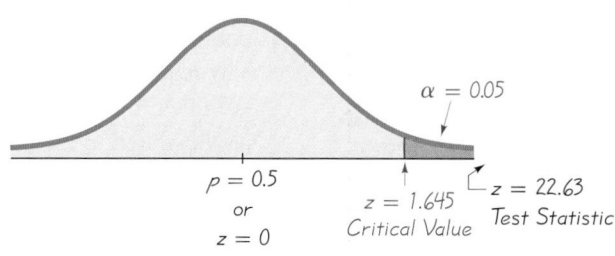

Figure 8-11   Traditional Method

of girls is greater than 0.5. (See Figure 8-7 for help with wording this final conclusion.) It does appear that the XSORT method is effective.

Traditional Method

The traditional method of testing hypotheses is summarized in Figure 8-9. When using the traditional method with the claim given in Example 1, Steps 1 through 5 are the same as in Steps 1 through 5 for the *P*-value method, as shown above. We continue with Step 6 of the traditional method.

Step 6. The test statistic is computed to be $z = 22.63$ as shown for the preceding *P*-value method. With the traditional method, we now find the critical value (instead of the *P*-value). This is a right-tailed test, so the area of the critical region is an area of $\alpha = 0.05$ in the right tail. Referring to Table A-2 and applying the methods of Section 6-2, we find that the critical value of $z = 1.645$ is at the boundary of the critical region. See Figure 8-11.

Step 7. Because the test statistic falls within the critical region, we reject the null hypothesis.

Step 8. We conclude that there is sufficient sample evidence to support the claim that among babies born to couples using the XSORT method, the proportion of girls is greater than 0.5. It does appear that the XSORT method is effective.

Confidence Interval Method

The claim of $p > 0.5$ can be tested with a 0.05 significance level by constructing a 90% confidence interval (as shown in Table 8-2 on page 412). (In general, for *two-tailed* hypothesis tests construct a confidence interval with a confidence level corresponding to the significance level, but for *one-tailed* hypothesis tests use a confidence level corresponding to *twice* the significance level, as in Table 8-2.)

The 90% confidence interval estimate of the population proportion p is found using the sample data consisting of $n = 726$ and $\hat{p} = 668/726 = 0.920$. Using the methods of Section 7-2 we get: $0.904 < p < 0.937$. That entire interval is above 0.5. Because we are 90% confident that the limits of 0.904 and 0.937 contain the true value of p, we have sufficient evidence to support the claim that $p > 0.5$, so the conclusion is the same as with the *P*-value method and the traditional method.

> **CAUTION**
>
> ....
>
> When testing claims about a population proportion, the traditional method and the *P*-value method are equivalent in the sense that they always yield the same results, but the confidence interval method is not equivalent to them and may result in a different conclusion. (Both the traditional method and *P*-value method use the same standard deviation based on the *claimed proportion p*, but the confidence interval uses an estimated standard deviation based on the *sample proportion $\hat{p}$*.) Here is a good strategy: Use a confidence interval to *estimate* a population proportion, but use the *P*-value method or traditional method for *testing a claim* about a proportion.

Finding the Number of Successes *x*

Computer software and calculators designed for hypothesis tests of proportions usually require input consisting of the sample size n and the number of successes x, but the sample proportion is often given instead of x. The number of successes x can be found as illustrated in Example 2. Note that x must be rounded to the nearest whole number.

EXAMPLE 2   **Finding the Number of Successes *x*** A study addressed the issue of whether pregnant women can correctly guess the sex of their baby. Among 104 recruited subjects, 55% correctly guessed the sex of the baby (based on data from "Are Women Carrying 'Basketballs' Really Having Boys? Testing Pregnancy Folklore," by Perry, DiPietro, and Constigan, *Birth,* Vol. 26, No. 3). How many of the 104 women made correct guesses?

SOLUTION   The number of women who made correct guesses is 55% of 104, or $0.55 \times 104 = 57.2$. The product 0.55×104 is 57.2, but the number of women who guessed correctly must be a whole number, so we round the product to the nearest whole number of 57.

Although a media report about this study used "55%," the more precise percentage of 54.8% is obtained by using the actual number of correct guesses (57) and the sample size (104). When conducting the hypothesis test, better results can be obtained by using the sample proportion of 0.548 (instead of 0.55).

EXAMPLE 3   **Can a Pregnant Woman Predict the Sex of Her Baby?** Example 2 referred to a study in which 57 out of 104 pregnant women correctly guessed the sex of their babies. Use these sample data to test the claim that the success rate of such guesses is no different from the 50% success rate expected with random chance guesses. Use a 0.05 significance level.

SOLUTION   **REQUIREMENT CHECK** (1) Given that the subjects were recruited and given the other conditions described in the study, it is reasonable to treat the sample as a simple random sample. (2) There is a fixed number (104) of independent trials with two categories (the mother correctly guessed the sex of her baby or did not). (3) The requirements $np \geq 5$ and $nq \geq 5$ are both satisfied with $n = 104$, $p = 0.5$, and $q = 0.5$. We get $np = (104)(0.5) = 52 \geq 5$ and $nq = (104)(0.5) = 52 \geq 5$. The three requirements are all satisfied.

We proceed to conduct the hypothesis test using the *P*-value method summarized in Figure 8-8.

Step 1: The original claim is that the success rate is no different from 50%. We express this in symbolic form as $p = 0.50$.

Step 2: The opposite of the original claim is $p \neq 0.50$.

Step 3: Because $p \neq 0.50$ does not contain equality, it becomes H_1. We get

$$H_0: p = 0.50 \qquad \text{(null hypothesis and original claim)}$$

$$H_1: p \neq 0.50 \qquad \text{(alternative hypothesis)}$$

Step 4: The significance level is $\alpha = 0.05$.

Step 5: Because the claim involves the proportion p, the statistic relevant to this test is the sample proportion $\hat{p}$, and the sampling distribution of sample proportions can be approximated by the normal distribution.

Step 6: The test statistic $z = 0.98$ is calculated as follows:

$$z = \frac{\hat{p} - p}{\sqrt{\dfrac{pq}{n}}} = \frac{\dfrac{57}{104} - 0.50}{\sqrt{\dfrac{(0.50)(0.50)}{104}}} = 0.98$$

Refer to Figure 8-5 for the procedure for finding the P-value. Figure 8-5 shows that for this two-tailed test with the test statistic located to the right of the center (because $z = 0.98$ is positive), the P-value is *twice* the area to the right of the test statistic. Using Table A-2, we see that $z = 0.98$ has an area of 0.8365 to its left. So the area to the right of $z = 0.98$ is $1 - 0.8365 = 0.1635$, which we double to get 0.3270. (The TI-83/84 Plus calculator provides a more accurate P-value of 0.3268.)

Step 7: Because the P-value of 0.3270 is greater than the significance level of 0.05, we fail to reject the null hypothesis.

INTERPRETATION Methods of hypothesis testing never allow us to support a claim of equality, so we cannot conclude that pregnant women have a success rate equal to 50% when they guess the sex of their babies. Here is the correct conclusion: There is not sufficient evidence to warrant rejection of the claim that women who guess the sex of their babies have a success rate equal to 50%.

Traditional Method: If we were to repeat Example 3 using the traditional method of testing hypotheses, we would see that in Step 6, the critical values are found to be $z = -1.96$ and $z = 1.96$. In Step 7, we would fail to reject the null hypothesis because the test statistic of $z = 0.98$ would not fall within the critical region. We would reach the same conclusion given in Example 3.

Confidence Interval Method: If we were to repeat the preceding example using the confidence interval method, we would obtain this 95% confidence interval: $0.452 < p < 0.644$. Because the confidence interval limits do contain the value of 0.5, the success rate could be 50%, so there is not sufficient evidence to reject the 50% rate. In this case, the P-value method, traditional method, and confidence interval method all lead to the same conclusion.

Part 2: Exact Method for Testing Claims About a Population Proportion p

Instead of using the normal distribution as an *approximation* to the binomial distribution, we can get *exact* results by using the binomial probability distribution itself. Binomial probabilities are a nuisance to calculate manually, but technology makes this approach quite simple. Also, this exact approach does not require that $np \geq 5$ and $nq \geq 5$, so we have a method that applies when that requirement is not satisfied. To test hypotheses using the exact binomial distribution, use the binomial probability distribution with the P-value method, use the value of p assumed in the null hypothesis, and find P-values as follows:

Left-tailed test: The P-value is the probability of getting x or fewer successes among the n trials.

Right-tailed test: The P-value is the probability of getting x or more successes among the n trials.

Process of Drug Approval

Gaining FDA approval for a new drug is expensive and time consuming. Here are the different stages of getting approval for a new drug:

- **Phase I study:** The safety of the drug is tested with a small (20–100) group of volunteers.
- **Phase II:** The drug is tested for effectiveness in randomized trials involving a larger (100–300) group of subjects. This phase often has subjects randomly assigned to either a treatment group or a placebo group.
- **Phase III:** The goal is to better understand the effectiveness of the drug as well as its adverse reactions. This phase typically involves 1,000–3,000 subjects, and it might require several years of testing.

Lisa Gibbs wrote in *Money* magazine that "the (drug) industry points out that for every 5,000 treatments tested, only 5 make it to clinical trials and only 1 ends up in drugstores." Total cost estimates vary from a low of $40 million to as much as $1.5 billion.

Home Field Advantage

In the *Chance* magazine article "Predicting Professional Sports Game Outcomes from Intermediate Game Scores," authors Harris Cooper, Kristina DeNeve, and Frederick Mosteller used statistics to analyze two common beliefs: Teams have an advantage when they play at home, and only the last quarter of professional basketball games really counts. Using a random sample of hundreds of games, they found that for the four top sports, the home team wins about 58.6% of games. Also, basketball teams ahead after 3 quarters go on to win about 4 out of 5 times, but baseball teams ahead after 7 innings go on to win about 19 out of 20 times. The statistical methods of analysis included the chi-square distribution applied to a contingency table.

Two-tailed test: If $\hat{p} > p$, the *P*-value is twice the probability of getting *x* or more successes;

if $\hat{p} < p$, the *P*-value is twice the probability of getting *x* or fewer successes.

EXAMPLE 4   **Using the Exact Method** Repeat Example 3 using exact binomial probabilities instead of the normal distribution. That is, test the claim that when pregnant women guess the sex of their babies, they have a 50% success rate. Use the sample data consisting of 104 guesses, of which 57 are correct. Use a 0.05 significance level.

SOLUTION   **REQUIREMENT CHECK** We need to check only the first two requirements listed near the beginning of this section, but those requirements were checked in Example 3, so we can proceed with the solution. ✔

As in Example 3, the null and alternative hypotheses are as follows:

$$H_0: p = 0.50 \qquad \text{(null hypothesis and original claim)}$$
$$H_1: p \neq 0.50 \qquad \text{(alternative hypothesis)}$$

Instead of calculating the test statistic and *P*-value as in Example 3, we use technology to find probabilities in a binomial distribution with $p = 0.50$. Because this is a two-tailed test with $\hat{p} > p$ (or $57/104 > 0.50$), the *P*-value is *twice* the probability of getting 57 or more successes among 104 trials, assuming that $p = 0.50$. See the accompanying STATDISK display of exact probabilities from the binomial distribution. This STATDISK display shows that the probability of 57 or more successes is 0.1887920, so the *P*-value is $2 \times 0.1887920 = 0.377584$. The *P*-value of 0.377584 is high (greater than 0.05), which shows that the 57 correct guesses in 104 trials can be easily explained by chance. Because the *P*-value is greater than the significance level of 0.05, fail to reject the null hypothesis and reach the same conclusion obtained in Example 3.

STATDISK

Num Trials, n:	104		Evaluate
Success Prob, p:	0.5		

Mean:	52.0000
St Dev:	5.0990
Variance:	26.0000

x	P(x)	P(x or fewer)	P(x or greater)
52	0.0780512	0.5390256	0.5390256
53	0.0765785	0.6156041	0.4609744
54	0.0723241	0.6879282	0.3843959
55	0.0657492	0.7536775	0.3120718
56	0.0575306	0.8112080	0.2463225
57	0.0484468	0.8596548	0.1887920
58	0.0392586	0.8989134	0.1403452
59	0.0306084	0.9295219	0.1010866
60	0.0229563	0.9524782	0.0704781
61	0.0165586	0.9690368	0.0475218
62	0.0114842	0.9805210	0.0309632
63	0.0076561	0.9881772	0.0194790
64	0.0049047	0.9930819	0.0118228

TI-83/84 Plus: In Example 3 we used STATDISK because the required probability of 57 or more successes could be identified directly from the STATDISK display. The TI-83/84 Plus command of **binomcdf(n, p, x)** yields the probability of x or fewer successes among n outcomes in which the probability of success is p. For Example 3, the command of **binomcdf(104, 0.5, 56)** yields the probability of 56 or fewer correct guesses, which is 0.8112080451, as shown in the accompanying screen display. The probability of 57 or more correct guesses is therefore $1 - 0.8112080451 = 0.1887919549$. The P-value is twice that result, so the P-value is 0.3775839098.

In Example 3, we obtained a P-value of 0.3270, but the exact method of Example 4 provides a more accurate P-value of 0.377584. The normal approximation to the binomial distribution is usually taught in introductory statistics courses, but technology is changing the way statistical methods are used. The time may come when the exact method eliminates the need for the normal approximation to the binomial distribution for testing claims about population proportions.

Rationale for the Test Statistic: The test statistic used in Part 1 of this section is justified by noting that when using the normal distribution to approximate a binomial distribution, we use $\mu = np$ and $\sigma = \sqrt{npq}$ to get

$$z = \frac{x - \mu}{\sigma} = \frac{x - np}{\sqrt{npq}}$$

We used the above expression in Section 6-6 along with a correction for continuity, but when testing claims about a population proportion, we make two modifications. First, we don't use the correction for continuity because its effect is usually very small for the large samples we are considering. Second, instead of using the above expression to find the test statistic, we use an equivalent expression obtained by dividing the numerator and denominator by n, and we replace x/n by the symbol $\hat{p}$ to get the test statistic we are using. The end result is that the test statistic is simply the same standard score (from Section 3-4) of $z = (x - \mu)/\sigma$, but modified for the binomial notation.

Using a TI-83/84 Plus Calculator for Testing a Claim About a Proportion

Requirements: When using a TI-83/84 Plus calculator for testing a claim about a population proportion p, you must know the sample size n and the number of successes x. If you do not know the value of x, but you know the value of the sample size n and the sample proportion $\hat{p}$, calculate the value of x as follows: $x = n \cdot \hat{p}$, then round the result to the nearest whole number, as illustrated in Example 2.

Because the TI-83/84 Plus calculator provides the test statistic and P-value (but not the critical value), we use the P-value method of testing hypotheses. Here is the procedure for using a TI-83/84 Plus calculator for testing a claim about a population proportion:

1. Press **STAT**, then press **▶** twice to select **TESTS.**

2. Select the menu item of **1-PropZTest** (for a hypothesis test of a claim involving one proportion).

3. For the first required entry indicated by $p0$, enter the claimed value of the population proportion. It is the same value used in the statement of the null and alternative hypotheses.

4. For second required entry of x, enter the number of successes. (See the above comment about calculating x when only the sample size n and sample proportion $\hat{p}$ are known.)

5. For the third required entry of n, enter the sample size.

TI-83/84 Plus

```
binomcdf(104,0.5
,56)
        .8112080451
```

Entries for Example 3

```
1-PropZTest
 P0:.5
 x:57
 n:104
 prop≠P0 <P0 >P0
 Calculate Draw
```

Results for Example 3

```
1-PropZTest
 prop≠.5
 z=.9805806757
 P=.3267995636
 p̂=.5480769231
 n=104
```

6. For the next line, select the type of test (two-tailed, left-tailed, right-tailed). Select the option that uses the same symbol as the alternative hypothesis. For a two-tailed case select the option including ≠, for a left-tailed test select the option of <p0, and for a right-tailed test select the option of >p0.

7. Use the arrow keys to scroll down to the bottom line, select **Calculate,** then press **ENTER**.

See Example 3, which requires that we test the claim of $p = 0.5$ using the sample data consisting of $n = 104$ and $x = 57$. The first accompanying screen display shows the required entries and the second screen shows the results.

Interpreting the results: See the displayed results and note that the TI-83/84 Plus calculator provides the test statistic of $z = 0.98$ and the P-value of 0.3268 when rounded. Because the P-value is greater than the significance level of 0.05, we fail to reject the null hypothesis. We conclude that there is not sufficient sample evidence to warrant rejection of the claim women who guess the sex of their babies have a success rate equal to 50%.

8-3   Basic Skills and Concepts

Statistical Literacy and Critical Thinking

1. Sample Proportion In a Harris poll, adults were asked if they are in favor of abolishing the penny. Among the responses, 1261 answered "no," 491 answered "yes," and 384 had no opinion. What is the sample proportion of *yes* responses, and what notation is used to represent it?

2. Online Poll America Online conducted a survey in which Internet users were asked to respond to this question: Do you want to live to be 100?" Among 5266 responses, 3042 were responses of "yes." Is it valid to use these sample results for testing the claim that the majority of the general population wants to live to be 100? Why or why not?

3. Interpreting *P*-Value In 280 trials with professional touch therapists, correct responses to a question were obtained 123 times. The P-value of 0.979 is obtained when testing the claim that $p > 0.5$ (the proportion of correct responses is greater than the proportion of 0.5 that would be expected with random chance). What is the value of the sample proportion? Based on the P-value of 0.979, what should we conclude about the claim that $p > 0.5$?

4. Notation and *P*-Value

a. Refer to Exercise 3 and distinguish between the value of p and the P-value.

b. We previously stated that we can easily remember how to interpret P-values with this: "If the P is low, the null must go. If the P is high, the null will fly." What does this mean?

In Exercises 5–8, identify the indicated values or interpret the given display. Use the normal distribution as an approximation to the binomial distribution (as described in Part 1 of this section).

TI **5. College Applications Online** A recent study showed that 53% of college applications were submitted online (based on data from the National Association of College Admissions Counseling). Assume that this result is based on a simple random sample of 1000 college applications, with 530 submitted online. Use a 0.01 significance level to test the claim that among all college applications the percentage submitted online is equal to 50%.

a. What is the test statistic?   **b.** What are the critical values?

c. What is the P-value?   **d.** What is the conclusion?

e. Can a hypothesis test be used to "prove" that the percentage of college applications submitted online is equal to 50%, as claimed?

TI **6. Driving and Texting** In a survey, 1864 out of 2246 randomly selected adults in the United States said that texting while driving should be illegal (based on data from Zogby

International). Consider a hypothesis test that uses a 0.05 significance level to test the claim that more than 80% of adults believe that texting while driving should be illegal.

a. What is the test statistic?   **b.** What is the critical value?

c. What is the *P*-value?   **d.** What is the conclusion?

7. Driving and Cell Phones In a survey, 1640 out of 2246 randomly selected adults in the United States said that they use cell phones while driving (based on data from Zogby International). When testing the claim that the proportion of adults who use cell phones while driving is equal to 75%, the TI-83/84 Plus calculator display in the margin is obtained. Use the results from the display with a 0.05 significance level to test the stated claim.

8. Percentage of Arrests A survey of 750 people aged 14 or older showed that 35 of them were arrested within the last year (based on FBI data). Minitab was used to test the claim that fewer than 5% of people aged 14 or older were arrested within the last year. Use the results from the Minitab display and use a 0.01 significance level to test the stated claim.

TI-83/84 PLUS

```
1-PropZTest
 prop≠.75
 z=-2.168475396
 p=.0301224268
 p̂=.7301869991
 n=2246
```

MINITAB

```
Test of p = 0.05 vs p < 0.05

                          99% Upper
Sample   X     N   Sample p   Bound   Z-Value   P-Value
1        35   750  0.046667  0.064584  -0.42     0.338
```

Testing Claims About Proportions. *In Exercises 9–32, test the given claim. Identify the null hypothesis, alternative hypothesis, test statistic, P-value or critical value(s), conclusion about the null hypothesis, and final conclusion that addresses the original claim. Use the P-value method unless your instructor specifies otherwise. Use the normal distribution as an approximation to the binomial distribution (as described in Part 1 of this section).*

9. Reporting Income In a Pew Research Center poll of 745 randomly selected adults, 589 said that it is morally wrong to not report all income on tax returns. Use a 0.01 significance level to test the claim that 75% of adults say that it is morally wrong to not report all income on tax returns.

10. Voting for the Winner In a presidential election, 308 out of 611 voters surveyed said that they voted for the candidate who won (based on data from ICR Survey Research Group). Use a 0.01 significance level to test the claim that among all voters, the percentage who believe that they voted for the winning candidate is equal to 43%, which is the actual percentage of votes for the winning candidate. What does the result suggest about voter perceptions?

11. Tennis Instant Replay The Hawk-Eye electronic system is used in tennis for displaying an instant replay that shows whether a ball is in bounds or out of bounds. In the first U.S. Open that used the Hawk-Eye system, players could challenge calls made by referees. The Hawk-Eye system was then used to confirm or overturn the referee's call. Players made 839 challenges, and 327 of those challenges were successful with the call overturned (based on data reported in *USA Today*). Use a 0.01 significance level to test the claim that the proportion of challenges that are successful is greater than 1/3. What do the results suggest about the quality of the calls made by the referees?

12. Screening for Marijuana Usage The company Drug Test Success provides a "1-Panel-THC" test for marijuana usage. Among 300 tested subjects, results from 27 subjects were wrong (either a false positive or a false negative). Use a 0.05 significance level to test the claim that less than 10% of the test results are wrong. Does the test appear to be good for most purposes?

13. Clinical Trial of Tamiflu Clinical trials involved treating flu patients with Tamiflu, which is a medicine intended to attack the influenza virus and stop it from causing flu symptoms. Among 724 patients treated with Tamiflu, 72 experienced nausea as an adverse reaction. Use a 0.05 significance level to test the claim that the rate of nausea is greater than the 6% rate experienced by flu patients given a placebo. Does nausea appear to be a concern for those given the Tamiflu treatment?

14. Postponing Death An interesting and popular hypothesis is that individuals can temporarily postpone their death to survive a major holiday or important event such as a birthday. In a study of this phenomenon, it was found that there were 6062 deaths in the week before Thanksgiving, and 5938 deaths the week after Thanksgiving (based on data from "Holidays, Birthdays, and Postponement of Cancer Death," by Young and Hade, *Journal of the American Medical Association,* Vol. 292, No. 24). If people can postpone their deaths until after Thanksgiving, then the proportion of deaths in the week before should be less than 0.5. Use a 0.05 significance level to test the claim that the proportion of deaths in the week before Thanksgiving is less than 0.5. Based on the result, does there appear to be any indication that people can temporarily postpone their death to survive the Thanksgiving holiday?

15. Cell Phones and Cancer In a study of 420,095 Danish cell phone users, 135 subjects developed cancer of the brain or nervous system (based on data from the *Journal of the National Cancer Institute* as reported in *USA Today*). Test the claim of a once popular belief that such cancers are affected by cell phone use. That is, test the claim that cell phone users develop cancer of the brain or nervous system at a rate that is different from the rate of 0.0340% for people who do not use cell phones. Because this issue has such great importance, use a 0.005 significance level. Should cell phone users be concerned about cancer of the brain or nervous system?

16. Predicting Sex of Baby Example 3 in this section included a hypothesis test involving pregnant women and their ability to predict the sex of their babies. In the same study, 45 of the pregnant women had more than 12 years of education, and 32 of them made correct predictions. Use these results to test the claim that women with more than 12 years of education have a proportion of correct predictions that is greater than the 0.5 proportion expected with random guesses. Use a 0.01 significance level. Do these women appear to have an ability to correctly predict the sex of their babies?

17. Cheating Gas Pumps When testing gas pumps in Michigan for accuracy, fuel-quality enforcement specialists tested pumps and found that 1299 of them were not pumping accurately (within 3.3 oz when 5 gal is pumped), and 5686 pumps were accurate. Use a 0.01 significance level to test the claim of an industry representative that less than 20% of Michigan gas pumps are inaccurate. From the perspective of the consumer, does that rate appear to be low enough?

18. Gender Selection for Boys The Genetics and IVF Institute conducted a clinical trial of the YSORT method designed to increase the probability that a baby is a boy. As of this writing, among the babies born to parents using the YSORT method, 172 were boys and 39 were girls. Use the sample data with a 0.01 significance level to test the claim that with this method, the probability of a baby being a boy is greater than 0.5. Does the YSORT method of gender selection appear to work?

19. Lie Detectors Trials in an experiment with a polygraph include 98 results that include 24 cases of wrong results and 74 cases of correct results (based on data from experiments conducted by researchers Charles R. Honts of Boise State University and Gordon H. Barland of the Department of Defense Polygraph Institute). Use a 0.05 significance level to test the claim that such polygraph results are correct less than 80% of the time. Based on the results, should polygraph test results be prohibited as evidence in trials?

20. Stem Cell Survey Adults were randomly selected for a *Newsweek* poll. They were asked if they "favor or oppose using federal tax dollars to fund medical research using stem cells obtained from human embryos." Of those polled, 481 were in favor, 401 were opposed, and 120 were unsure. A politician claims that people don't really understand the stem cell issue and their responses to such questions are random responses equivalent to a coin toss. Exclude the 120 subjects who said that they were unsure, and use a 0.01 significance level to test the claim that the proportion of subjects who respond in favor is equal to 0.5. What does the result suggest about the politician's claim?

21. Nielsen Share A recently televised broadcast of *60 Minutes* had a 15 share, meaning that among 5000 monitored households with TV sets in use, 15% of them were tuned to *60 Minutes.*

Use a 0.01 significance level to test the claim of an advertiser that among the households with TV sets in use, less than 20% were tuned to *60 Minutes*.

TI 22. New Sheriff in Town In recent years, the Town of Newport experienced an arrest rate of 25% for robberies (based on FBI data). The new sheriff compiles records showing that among 30 recent robberies, the arrest rate is 30%, so she claims that her arrest rate is greater than the 25% rate in the past. Is there sufficient evidence to support her claim that the arrest rate is greater than 25%?

TI 23. Job Interview Mistakes In an Accountemps survey of 150 senior executives, 47.3% said that the most common job interview mistake is to have little or no knowledge of the company. Test the claim that in the population of all senior executives, 50% say that the most common job interview mistake is to have little or no knowledge of the company. What important lesson is learned from this survey?

TI 24. Smoking and College Education A survey showed that among 785 randomly selected subjects who completed four years of college, 18.3% smoke and 81.7% do not smoke (based on data from the American Medical Association). Use a 0.01 significance level to test the claim that the rate of smoking among those with four years of college is less than the 27% rate for the general population. Why would college graduates smoke at a lower rate than others?

TI 25. Internet Use When 3011 adults were surveyed in a Pew Research Center poll, 73% said that they use the Internet. Is it okay for a newspaper reporter to write that "3/4 of all adults use the Internet"? Why or why not?

TI 26. Global Warming As part of a Pew Research Center poll, subjects were asked if there is solid evidence that the earth is getting warmer. Among 1501 respondents, 20% said that there is not such evidence. Use a 0.01 significance level to test the claim that less than 25% of the population believes that there is not solid evidence that the earth is getting warmer. What is a possible consequence of a situation in which too many people incorrectly believe that there is not evidence of global warming during a time when global warming is occurring?

TI 27. Predicting Sex of Baby Example 3 in this section included a hypothesis test involving pregnant women and their ability to correctly predict the sex of their baby. In the same study, 59 of the pregnant women had 12 years of education or less, and it was reported that 43% of them correctly predicted the sex of their baby. Use a 0.05 significance level to test the claim that these women have no ability to predict the sex of their baby, and the results are not significantly different from those that would be expected with random guesses. What do you conclude?

TI 28. Bias in Jury Selection In the case of *Casteneda v. Partida*, it was found that during a period of 11 years in Hidalgo County, Texas, 870 people were selected for grand jury duty, and 39% of them were Americans of Mexican ancestry. Among the people eligible for grand jury duty, 79.1% were Americans of Mexican ancestry. Use a 0.01 significance level to test the claim that the selection process is biased against Americans of Mexican ancestry. Does the jury selection system appear to be fair?

TI 29. Scream A survey of 61,647 people included several questions about office relationships. Of the respondents, 26% reported that bosses scream at employees. Use a 0.05 significance level to test the claim that more than 1/4 of people say that bosses scream at employees. How is the conclusion affected after learning that the survey is an *Elle*/MSNBC.COM survey in which Internet users chose whether to respond?

TI 30. Is Nessie Real? This question was posted on the America Online Web site: Do you believe the Loch Ness monster exists? Among 21,346 responses, 64% were "yes." Use a 0.01 significance level to test the claim that most people believe that the Loch Ness monster exists. How is the conclusion affected by the fact that Internet users who saw the question could decide whether to respond?

TI 31. Finding a Job Through Networking In a survey of 703 randomly selected workers, 61% got their jobs through networking (based on data from Taylor Nelson Sofres Research).

Use the sample data with a 0.05 significance level to test the claim that most workers (more than 50%) get their jobs through networking. What does the result suggest about the strategy for finding a job after graduation?

TI **32. Mendel's Genetics Experiments** When Gregor Mendel conducted his famous hybridization experiments with peas, one such experiment resulted in 580 offspring peas, with 26.2% of them having yellow pods. According to Mendel's theory, 1/4 of the offspring peas should have yellow pods. Use a 0.05 significance level to test the claim that the proportion of peas with yellow pods is equal to 1/4.

Large Data Sets. *In Exercises 33–36, use the Data Set from Appendix B to test the given claim.*

TI **33. M&Ms** Refer to Data Set 18 in Appendix B and find the sample proportion of M&Ms that are red. Use that result to test the claim of Mars, Inc., that 20% of its plain M&M candies are red.

TI **34. Freshman 15** Data Set 3 in Appendix B includes results from a study described in "Changes in Body Weight and Fat Mass of Men and Women in the First Year of College: A Study of the 'Freshman 15,'" by Hoffman, Policastro, Quick, and Lee, *Journal of American College Health*, Vol. 55, No. 1. Refer to that data set and find the proportion of men included in the study. Use a 0.05 significance level to test the claim that when subjects were selected for the study, they were selected from a population in which the percentage of males is equal to 50%.

TI **35. Bears** Refer to Data Set 6 in Appendix B and find the proportion of male bears included in the study. Use a 0.05 significance level to test the claim that when the bears were selected, they were selected from a population in which the percentage of males is equal to 50%.

TI **36. Movies** According to the *Information Please* almanac, the percentage of movies with ratings of R has been 55% during a recent period of 33 years. Refer to Data Set 9 in Appendix B and find the proportion of movies with ratings of R. Use a 0.01 significance level to test the claim that the movies in Data Set 9 are from a population in which 55% of the movies have R ratings.

8-3   Beyond the Basics

TI **37. Exact Method** Repeat Exercise 36 using the exact method with the binomial distribution, as described in Part 2 of this section.

TI **38. Using Confidence Intervals to Test Hypotheses** When analyzing the last digits of telephone numbers in Port Jefferson, it is found that among 1000 randomly selected digits, 119 are zeros. If the digits are randomly selected, the proportion of zeros should be 0.1.

a. Use the traditional method with a 0.05 significance level to test the claim that the proportion of zeros equals 0.1.

b. Use the *P*-value method with a 0.05 significance level to test the claim that the proportion of zeros equals 0.1.

c. Use the sample data to construct a 95% confidence interval estimate of the proportion of zeros. What does the confidence interval suggest about the claim that the proportion of zeros equals 0.1?

d. Compare the results from the traditional method, the *P*-value method, and the confidence interval method. Do they all lead to the same conclusion?

39. Coping with No Successes In a simple random sample of 50 plain M&M candies, it is found that none of them are blue. We want to use a 0.01 significance level to test the claim of Mars, Inc., that the proportion of M&M candies that are blue is equal to 0.10. Can the methods of this section be used? If so, test the claim. If not, explain why not.

40. Power For a hypothesis test with a specified significance level α, the probability of a type I error is α, whereas the probability β of a type II error depends on the particular value of p that is used as an alternative to the null hypothesis.

a. Using an alternative hypothesis of $p < 0.4$, a sample size of $n = 50$, and assuming that the true value of p is 0.25, find the power of the test. See Exercise 47 in Section 8-2. (*Hint:* Use the values $p = 0.25$ and $pq/n = (0.25)(0.75)/50$.)

b. Find the value of β, the probability of making a type II error.

c. Given the conditions cited in part (a), what do the results indicate about the effectiveness of the hypothesis test?

 # Testing a Claim About a Mean: σ Known

Key Concept In this section we discuss hypothesis testing methods for claims made about a population mean, assuming that the population standard deviation is a known value. The following section presents methods for testing a claim about a mean when σ is not known. Here we use the normal distribution with the same components of hypothesis tests that were introduced in Section 8-2.

The requirements, test statistic, critical values, and *P*-value are summarized as follows:

Testing Claims About a Population Mean (with σ Known)

Objective

Test a claim about a population mean (with σ known) by using a formal method of hypothesis testing.

Notation

$n =$ sample size

$\bar{x} = sample$ mean

$\mu_{\bar{x}} = population$ mean of all sample means from samples of size n (this value is based on the claim and is used in the null hypothesis)

$\sigma =$ known value of the population standard deviation

Requirements

1. The sample is a simple random sample.

2. The value of the population standard deviation σ is known.

3. Either or both of these conditions is satisfied: The population is normally distributed or $n > 30$.

Test Statistic for Testing a Claim About a Mean (with σ Known)

$$z = \frac{\bar{x} - \mu_{\bar{x}}}{\frac{\sigma}{\sqrt{n}}}$$

P-values: Use the standard normal distribution (Table A-2) and refer to Figure 8-5.

Critical values: Use the standard normal distribution (Table A-2).

TI-83/84 Plus Calculator The TI-83/84 Plus calculator can be used to automatically generate the test statistic and *P*-value. See the procedure at the end of this section.

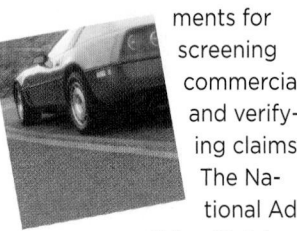
Knowledge of σ   The listed requirements include knowledge of the population standard deviation σ, but Section 8-5 presents methods for testing claims about a mean when σ is not known. In reality, the value of σ is usually unknown, so the methods of Section 8-5 are used much more often than the methods of this section.

Normality Requirement   The requirements include the property that either the population is normally distributed or $n > 30$. If $n \le 30$, we can consider the normality requirement to be satisfied if there are no outliers and if a histogram of the sample data is not dramatically different from being bell-shaped. (The methods of this section are *robust* against departures from normality, which means that these methods are not strongly affected by departures from normality, provided that those departures are not too extreme.) However, the methods of this section often yield very poor results from samples that are not simple random samples.

Sample Size Requirement   The normal distribution is used as the distribution of sample means. If the original population is not itself normally distributed, we use the condition $n > 30$ for justifying use of the normal distribution, but there is no specific minimum sample size that works for all cases. Sample sizes of 15 to 30 are sufficient if the population has a distribution that is not far from normal, but some other populations have distributions that are extremely far from normal and sample sizes greater than 30 might be necessary. In this book we use the simplified criterion of $n > 30$ as justification for treating the distribution of sample means as a normal distribution.

> **EXAMPLE 1**   **Overloading Boats: *P*-Value Method** People have died in boat accidents because an obsolete estimate of the mean weight of men was used. Using the weights of the simple random sample of men from Data Set 1 in Appendix B, we obtain these sample statistics: $n = 40$ and $\bar{x} = 172.55$ lb. Research from several other sources suggests that the population of weights of men has a standard deviation given by $\sigma = 26$ lb. Use these results to test the claim that men have a mean weight greater than 166.3 lb, which was the weight in the National Transportation and Safety Board's recommendation M-04-04. Use a 0.05 significance level, and use the *P*-value method outlined in Figure 8-8.

> **SOLUTION**   **REQUIREMENT CHECK** (1) The sample is a simple random sample. (2) The value of σ is known (26 lb). (3) The sample size is $n = 40$, which is greater than 30. The requirements are satisfied.

We follow the *P*-value procedure summarized in Figure 8-8.

Step 1: The claim that men have a mean weight greater than 166.3 lb is expressed in symbolic form as $\mu > 166.3$ lb.

Step 2: The alternative (in symbolic form) to the original claim is $\mu \le 166.3$ lb.

Step 3: Because the statement $\mu > 166.3$ lb does not contain the condition of equality, it becomes the alternative hypothesis. The null hypothesis is the statement that $\mu = 166.3$ lb. (See Figure 8-2 for the procedure used to identify the null hypothesis H_0 and the alternative hypothesis H_1.)

$$H_0: \mu = 166.3 \text{ lb} \qquad \text{(null hypothesis)}$$

$$H_1: \mu > 166.3 \text{ lb} \qquad \text{(alternative hypothesis and original claim)}$$

Step 4: As specified in the statement of the problem, the significance level is $\alpha = 0.05$.

Step 5: Because the claim is made about the *population mean μ*, the sample statistic most relevant to this test is the *sample mean* $\bar{x} = 172.55$ lb. Because σ is assumed to be known (26 lb) and the sample size is greater than 30, the central limit theorem indicates that the distribution of sample means can be approximated by a *normal* distribution.

Step 6: The test statistic is calculated as follows:

$$z = \frac{\bar{x} - \mu_{\bar{x}}}{\frac{\sigma}{\sqrt{n}}} = \frac{172.55 - 166.3}{\frac{26}{\sqrt{40}}} = 1.52$$

Using this test statistic of $z = 1.52$, we now proceed to find the *P*-value. See Figure 8-5 for the flowchart summarizing the procedure for finding *P*-values. This is a right-tailed test, so the *P*-value is the area to the *right* of $z = 1.52$, which is 0.0643. (Table A-2 shows that the area to the *left* of $z = 1.52$ is 0.9357, so the area to the right of $z = 1.52$ is $1 - 0.9357 = 0.0643$.) The *P*-value is 0.0643, as shown in Figure 8-12.

TI-83/84 Plus The TI-83/84 Plus calculator provides *P*-values, but not a critical value, so the *P*-value method of hypothesis testing applies. In this example, if we use the TI-83/84 Plus calculator with the procedure given near the end of this section, we obtain the display shown in the margin. This display shows the test statistic of $z = 1.520325798$ and the *P*-value of 0.0642145783. This *P*-value is 0.0642 when rounded, and it is more accurate than the *P*-value obtained by using Table A-2.

Step 7: Because the *P*-value of 0.0643 is greater than the significance level of $\alpha = 0.05$, we fail to reject the null hypothesis.

INTERPRETATION   The *P*-value of 0.0643 tells us that if men have a mean weight given by $\mu = 166.3$ lb, there is a good chance (0.0643) of getting a sample mean of 172.55 lb. A sample mean such as 172.55 lb could easily occur by chance. There is not sufficient evidence to support a conclusion that the population mean is greater than 166.3 lb, as in the National Transportation and Safety Board's recommendation.

TI-83/84 Plus

```
Z-Test
 μ>166.3
 z=1.520325798
 p=.0642145783
 x̄=172.55
 n=40
```

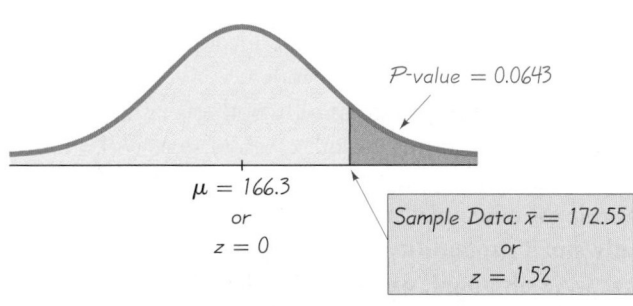

Figure 8-12   *P*-Value Method: Testing the Claim that $\mu > 166.3$ lb

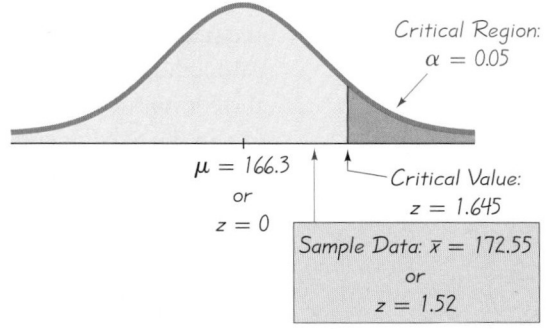

Figure 8-13   Traditional Method: Testing the Claim that $\mu > 166.3$ lb

EXAMPLE 2   **Overloading Boats: Traditional Method** If the traditional method of testing hypotheses is used for Example 1, the first five steps would be the same. In Step 6 we find the critical value of $z = 1.645$ instead of finding the *P*-value. The critical value of $z = 1.645$ is the value separating an area of 0.05 (the significance level) in the right tail of the standard normal distribution (see Table A-2).

continued

We again fail to reject the null hypothesis because the test statistic of $z = 1.52$ does not fall in the critical region, as shown in Figure 8-13. The final conclusion is the same as in Example 1.

EXAMPLE 3   **Overloading Boats: Confidence Interval Method** We can use a confidence interval for testing a claim about μ when σ is known. For a one-tailed hypothesis test with a 0.05 significance level, we construct a 90% confidence interval (as summarized in Table 8-2 on page 412). If we use the sample data in Example 1 with $\sigma = 26$ lb, we can test the claim that $\mu > 166.3$ lb using the methods of Section 7-3 to construct this 90% confidence interval:

$$165.8 \text{ lb} < \mu < 179.3 \text{ lb}$$

Because that confidence interval contains 166.3 lb, we cannot support a claim that μ is greater than 166.3 lb. See Figure 8-14, which illustrates this point: Because the confidence interval from 165.8 lb to 179.3 lb is likely to contain the true value of μ, we cannot support a claim that the value of μ is greater than 166.3 lb. It is very possible that μ has a value that is at or below 166.3 lb.

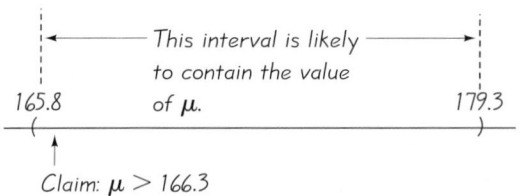

Figure 8-14   Confidence Interval Method: Testing the Claim that $\mu > 166.3$ lb

In Section 8-3 we saw that when testing a claim about a population proportion, the traditional method and P-value method are equivalent, but the confidence interval method is somewhat different. When testing a claim about a population mean, there is no such difference, and all three methods are equivalent.

In the remainder of the text, we will apply methods of hypothesis testing to other circumstances. It is easy to become entangled in a complex web of steps without ever understanding the underlying rationale of hypothesis testing. The key to that understanding lies in the rare event rule for inferential statistics: **If, under a given assumption, there is an exceptionally small probability of getting sample results at least as extreme as the results that were obtained, we conclude that the assumption is probably not correct.** When testing a claim, we make an assumption (null hypothesis) of equality. We then compare the assumption and the sample results to form one of the following conclusions:

- If the sample results (or more extreme results) can easily occur when the assumption (null hypothesis) is true, we attribute the relatively small discrepancy between the assumption and the sample results to chance.
- If the sample results (or more extreme results) cannot easily occur when the assumption (null hypothesis) is true, we explain the relatively large discrepancy between the assumption and the sample results by concluding that the assumption is not true, so we reject the assumption.

Using a TI-83/84 Plus Calculator for Testing a Claim About a Mean (With σ Known)

When using a TI-83/84 Plus calculator for testing a claim about a population mean μ *and the value of the population standard deviation σ is known*, you have the option of using either the summary statistics of n and $\bar{x}$, or you can use a list of original sample data values.

Because the TI-83/84 Plus calculator provides the test statistic and P-value (but not the critical value), we use the P-value method of testing hypotheses. Here is the procedure:

1. Press **STAT**, then press ▷ twice to select **TESTS.**

2. Select the first menu item of **Z-Test.**

3. The first line, identified as Inpt (for input), gives you the option of using Data or Stats. Use the arrow keys ◁ and ▷ to select either Data or Stats.

   • **Data:** Use this option if you have the sample data values stored as a list.

   • **Stats:** Use this option if you know the values of $\bar{x}$ and n.

4. For the value indicated by $\mu0$, enter the claimed value of the population mean. This is the same value used in the null and alternative hypotheses.

5. For the value indicated by σ, enter the known value of the *population* standard deviation. (If only the sample standard deviation s is known, use the procedure given at the end of the following section.)

6. If you selected Stats in Step 3, proceed to enter the values of the sample mean $\bar{x}$ and the sample size n.

   If you selected Data in Step 3, proceed to enter the name of the list where the data are stored (such as L1 or a named list such as AGES), and enter 1 for the frequency. (However, if the sample data are stored in two columns as a frequency distribution, the Freq entry should be the list containing frequencies).

7. For the next line, select the type of test (two-tailed, left-tailed, right-tailed). Select the option that uses the same symbol as the alternative hypothesis. For a two-tailed case select the option including ≠, for a left-tailed test select the option of $<\mu0$, and for a right-tailed test select the option of $>\mu0$.

8. Use the arrow keys to scroll down to the bottom line, select **Calculate,** then press **ENTER**.

Entries for Example 1

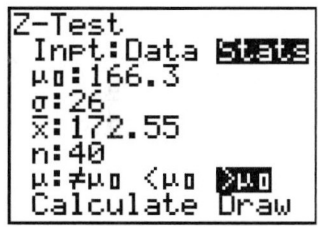

Example 1 involved the claim of $\mu > 166.3$, it is assumed that σ = 26 lb, and sample data are given as $n = 40$ and $\bar{x} = 172.55$ lb. Shown here are the screen displays for the required entries and the generated results.

Interpreting the results: See the displayed results and note that the TI-83/84 Plus calculator provides the test statistic of $z = 1.520325798$ and the P-value of 0.0642145783. Because the P-value is greater than the significance level of 0.05, we fail to reject the null hypothesis and conclude that there is not sufficient sample evidence to support the claim that the population mean is greater than 166.3 lb.

Results for Example 1

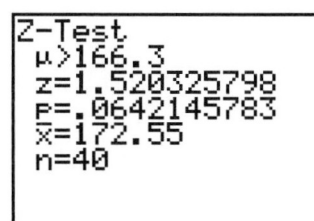

8-4   Basic Skills and Concepts

Statistical Literacy and Critical Thinking

1. Identifying Requirements Data Set 4 in Appendix B lists the amounts of nicotine (in milligrams per cigarette) in 25 different king size cigarettes. If we want to use that sample to test the claim that all king size cigarettes have a mean of 1.5 mg of nicotine, identify the requirements that must be satisfied.

2. Verifying Normality Because the amounts of nicotine in king size cigarettes listed in Data Set 4 in Appendix B constitute a sample of size $n = 25$, we must satisfy the requirement that the population is normally distributed. How do we verify that a population is normally distributed?

3. Confidence Interval If you want to construct a confidence interval to be used for testing the claim that college students have a mean IQ score that is greater than 100, and you want the test conducted with a 0.01 significance level, what confidence level should be used for the confidence interval?

4. Practical Significance A hypothesis test that the Zone diet is effective (when used for one year) results in this conclusion: There is sufficient evidence to support the claim that the mean weight change is less than 0 (so there is a loss of weight). The sample of 40 subjects had a mean weight loss of 2.1 lb (based on data from "Comparison of the Atkins, Ornish, Weight Watchers, and Zone Diets for Weight Loss and Heart Disease Reduction," by Dansinger, et al., *Journal of the American Medical Association,* Vol. 293, No. 1). Does the weight loss of 2.1 pounds have statistical significance? Does the weight loss of 2.1 pounds have practical significance? Explain.

Testing Hypotheses. *In Exercises 5–18, test the given claim. Identify the null hypothesis, alternative hypothesis, test statistic, P-value or critical value(s), conclusion about the null hypothesis, and final conclusion that addresses the original claim. Use the P-value method unless your instructor specifies otherwise.*

TI-83/84 PLUS

```
Z-Test
 μ≠5
 z=1.341572341
 P=.1797348219
 x̄=5.07
 n=40
```

TI 5. Wrist Breadth of Women A jewelry designer claims that women have wrist breadths with a mean equal to 5 cm. A simple random sample of the wrist breadths of 40 women has a mean of 5.07 cm (based on Data Set 1 in Appendix B). Assume that the population standard deviation is 0.33 cm. Use the accompanying TI-83/84 Plus display to test the designer's claim.

6. Weights of Pennies The U.S. Mint has a specification that pennies have a mean weight of 2.5 g. Assume that weights of pennies have a standard deviation of 0.0165 g and use the accompanying Minitab display to test the claim that the sample is from a population with a mean that is less than 2.5 g. These Minitab results were obtained using the 37 weights of post-1983 pennies listed in Data Set 20 in Appendix B.

```
Test of mu = 2.5 vs < 2.5. Assumed s.d. = 0.0165
                  95% Upper
  N     Mean    StDev    Bound      Z       P
  37  2.49910  0.01648  2.50356   -0.33   0.370
```

TI 7. Writing a Hit Song In the manual "How to Have a Number One the Easy Way," by KLF Publications, it is stated that a song "must be no longer than three minutes and thirty seconds" (or 210 seconds). A simple random sample of 40 current hit songs results in a mean length of 252.5 sec. (The songs are by Timberlake, Furtado, Daughtry, Stefani, Fergie, Akon, Ludacris, etc.) Assume that the standard deviation of song lengths is 54.5 sec. Use a 0.05 significance level to test the claim that the sample is from a population of songs with a mean greater than 210 sec. What do these results suggest about the advice given in the manual?

TI 8. Red Blood Cell Count A simple random sample of 50 adults is obtained, and each person's red blood cell count (in cells per microliter) is measured. The sample mean is 5.23. The population standard deviation for red blood cell counts is 0.54. Use a 0.01 significance level to test the claim that the sample is from a population with a mean less than 5.4, which is a value often used for the upper limit of the range of normal values. What do the results suggest about the sample group?

TI 9. M&M Weights A simple random sample of the weights of 19 green M&Ms has a mean of 0.8635 g (as in Data Set 18 in Appendix B). Assume that σ is known to be 0.0565 g. Use a 0.05 significance level to test the claim that the mean weight of all green M&Ms is equal to 0.8535 g, which is the mean weight required so that M&Ms have the weight printed on the package label. Do green M&Ms appear to have weights consistent with the package label?

TI 10. Human Body Temperature Data Set 2 in Appendix B includes a sample of 106 body temperatures with a mean of 98.20°F. Assume that σ is known to be 0.62°F. Use a 0.05 significance level to test the claim that the mean body temperature of the population is equal to 98.6°F, as is commonly believed. Is there sufficient evidence to conclude that the common belief is wrong?

TI 11. Is the Diet Practical? When 40 people used the Weight Watchers diet for one year, their mean weight *loss* was 3.0 lb (based on data from "Comparison of the Atkins, Ornish, Weight Watchers, and Zone Diets for Weight Loss and Heart Disease Reduction," by Dansinger, et al., *Journal of the American Medical Association,* Vol. 293, No. 1). Assume that the standard deviation of all such weight changes is $\sigma = 4.9$ lb and use a 0.01 significance level to test the claim that the mean weight loss is greater than 0. Based on these results, does the diet appear to be effective? Does the diet appear to have practical significance?

TI 12. Loaded Die When a fair die is rolled many times, the outcomes of 1, 2, 3, 4, 5, and 6 are equally likely, so the mean of the outcomes should be 3.5. The author drilled holes into a die and loaded it by inserting lead weights, then rolled it 16 times to obtain a mean of 2.9375. Assume that the standard deviation of the outcomes is 1.7078, which is the standard deviation for a fair die. Use a 0.05 significance level to test the claim that outcomes from the loaded die have a mean different from the value of 3.5 expected with a fair die. Is there anything about the sample data suggesting that the methods of this section should not be used?

TI 13. Sitting Height A student of the author measured the sitting heights of 36 male classmate friends, and she obtained a mean of 92.8 cm. The population of males has sitting heights with a mean of 91.4 cm and a standard deviation of 3.6 cm (based on anthropometric survey data from Gordon, Churchill, et al.). Use a 0.05 significance level to test the claim that males at her college have a mean sitting height different from 91.4 cm. Is there anything about the sample data suggesting that the methods of this section should not be used?

TI 14. Weights of Bears The health of the bear population in Yellowstone National Park is monitored by periodic measurements taken from anesthetized bears. A sample of 54 bears has a mean weight of 182.9 lb. Assuming that σ is known to be 121.8 lb, use a 0.05 significance level to test the claim that the population mean of all such bear weights is greater than 150 lb.

TI 15. NCAA Football Coach Salaries A simple random sample of 40 salaries of NCAA football coaches in the NCAA has a mean of $415,953. The standard deviation of all salaries of NCAA football coaches is $463,364. Use a 0.05 significance level to test the claim that the mean salary of a football coach in the NCAA is less than $500,000.

TI 16. Cans of Coke A simple random sample of 36 cans of regular Coke has a mean volume of 12.19 oz (based on Data Set 17 in Appendix B). Assume that the standard deviation of all cans of regular Coke is 0.11 oz. Use a 0.01 significance level to test the claim that cans of regular Coke have volumes with a mean of 12 oz, as stated on the label. If there is a difference, is it substantial?

TI 17. Juiced Baseballs Tests of older baseballs showed that when dropped 24 ft onto a concrete surface, they bounced an average of 235.8 cm. In a test of 40 new baseballs, the bounce heights had a mean of 235.4 cm. Assume that the standard deviation of bounce heights is 4.5 cm (based on data from Brookhaven National Laboratory and *USA Today*). Use a 0.05 significance level to test the claim that the new baseballs have bounce heights with a mean different from 235.8 cm. Are the new baseballs different?

TI 18. Garbage The totals of the individual weights of garbage discarded by 62 households in one week have a mean of 27.443 lb (based on Data Set 22 in Appendix B). Assume that the standard deviation of the weights is 12.458 lb. Use a 0.05 significance level to test the claim that the population of households has a mean less than 30 lb, which is the maximum amount that can be handled by the current waste removal system. Is there any cause for concern?

Using Raw Data. *In Exercises 19 and 20, test the given claim. Identify the null hypothesis, alternative hypothesis, test statistic, P-value or critical value(s), conclusion about the null hypothesis, and final conclusion that addresses the original claim. Use the P-value method unless your instructor specifies otherwise.*

TI 19. FICO Credit Scores A simple random sample of FICO credit rating scores is obtained, and the scores are listed below. As of this writing, the mean FICO score was reported to be 678. Assuming the the standard deviation of all FICO scores is known to be 58.3, use a 0.05 significance level to test the claim that these sample FICO scores come from a population with a mean equal to 678.

    714 751 664 789 818 779 698 836 753 834 693 802

TI 20. California Speeding Listed below are recorded speeds (in mi/h) of randomly selected cars traveling on a section of Highway 405 in Los Angeles (based on data from Sigalert). That part of the highway has a posted speed limit of 65 mi/h. Assume that the standard deviation of speeds is 5.7 mi/h and use a 0.01 significance level to test the claim that the sample is from a population with a mean that is greater than 65 mi/h.

    68 68 72 73 65 74 73 72 68 65 65 73 66 71 68 74 66 71 65 73

    59 75 70 56 66 75 68 75 62 72 60 73 61 75 58 74 60 73 58 75

Large Data Sets from Appendix B. *In Exercises 21 and 22, use the data set from Appendix B to test the given claim. Identify the null hypothesis, alternative hypothesis, test statistic, P-value or critical value(s), conclusion about the null hypothesis, and final conclusion that addresses the original claim. Use the P-value method unless your instructor specifies otherwise.*

TI **21. Do the Screws Have a Length of 3/4 in.?** A simple random sample of 50 stainless steel sheet metal screws is obtained from those suppled by Crown Bolt, Inc., and the length of each screw is measured using a vernier caliper. The lengths are listed in Data Set 19 of Appendix B. Assume that the standard deviation of all such lengths is 0.012 in., and use a 0.05 significance level to test the claim that the screws have a mean length equal to 3/4 in. (or 0.75 in.), as indicated on the package labels. Do the screw lengths appear to be consistent with the package label?

TI **22. Power Supply** Data Set 13 in Appendix B lists measured voltage amounts supplied directly to the author's home. The Central Hudson power supply company states that it has a target power supply of 120 volts. Using those home voltage amounts and assuming that the standard deviation of all such voltage amounts is 0.24 V, test the claim that the mean is 120 volts. Use a 0.01 significance level.

8-4   Beyond the Basics

23. Interpreting Power For Example 1 in this section, the hypothesis test has power of 0.2296 (or 0.2281 using technology) of supporting the claim that $\mu > 166.3$ lb when the actual population mean is 170 lb.

a. Interpret the given value of the power.

b. Identify the value of β, and interpret that value.

TI **24. Calculating Power of a Test** For Example 1 in this section, find the power of the test in supporting the claim that $\mu > 166.3$ lb when the actual population mean is 180 lb. Also find β, the probability of a type II error. Is the test effective in supporting the claim that $\mu > 166.3$ lb when the true population mean is 180 lb?

8-5   Testing a Claim About a Mean: σ Not Known

Key Concept In Section 8-4 we discussed methods for testing a claim about a population mean, but that section is based on the unrealistic assumption that the value of the population standard deviation σ is somehow known. In this section we present methods for testing a claim about a population mean, but we do not require that σ is known. The methods of this section are referred to as a *t* test because they use the Student *t* distribution that was introduced in Section 7-4. The requirements, test statistic, P-value, and critical values are summarized as follows.

Testing Claims About a Population Mean (with σ Not Known)

Objective

Test a claim about a population mean (with σ not known) by using a formal method of hypothesis testing.

Notation

n = sample size

$\bar{x}$ = *sample* mean

$\mu_{\bar{x}}$ = *population* mean of all sample means from samples of size n (this value is based on the claim and is used in the null hypothesis)

Requirements

1. The sample is a simple random sample.

2. The value of the population standard deviation σ is *not* known.

3. Either or both of these conditions is satisfied: The population is normally distributed or $n > 30$.

Test Statistic for Testing a Claim About a Mean (with σ Not Known)

$$t = \frac{\bar{x} - \mu_{\bar{x}}}{\frac{s}{\sqrt{n}}}$$

(Round t to three decimal places, as in Table A-3.)

***P*-values and Critical values:** Use Table A-3 and use df $= n - 1$ for the number of degrees of freedom. (See Figure 8-5 for *P*-value procedures.)

TI-83/84 Plus Calculator The TI-83/84 Plus calculator can be used to automatically generate the test statistic and *P*-value. See the procedure at the end of this section.

Normality Requirement This *t* test is *robust* against a departure from normality, meaning that the test works reasonably well if the departure from normality is not too extreme. We can usually satisfy this normality requirement by verifying that there are no outliers and that the histogram has a shape that is not very far from a normal distribution.

Sample Size We use the simplified criterion of $n > 30$ as justification for treating the distribution of sample means as a normal distribution, but the minimum sample size actually depends on how much the population distribution departs from a normal distribution.

Here are the important properties of the Student *t* distribution:

Important Properties of the Student *t* Distribution

1. The Student *t* distribution is different for different sample sizes (see Figure 7-5 in Section 7-4).

2. The Student *t* distribution has the same general bell shape as the standard normal distribution; its wider shape reflects the greater variability that is expected when *s* is used to estimate σ.

3. The Student *t* distribution has a mean of $t = 0$ (just as the standard normal distribution has a mean of $z = 0$).

4. The standard deviation of the Student *t* distribution varies with the sample size and is greater than 1 (unlike the standard normal distribution, which has $\sigma = 1$).

5. As the sample size *n* gets larger, the Student *t* distribution gets closer to the standard normal distribution.

Choosing the Correct Method

When testing a claim about a population mean, first be sure that the sample data have been collected with an appropriate sampling method, such as a simple random sample (otherwise, it is very possible that no methods of statistics apply). If we have a simple random sample, a hypothesis test of a claim about μ might use the Student *t* distribution, the normal distribution, or it might require nonparametric methods or bootstrap resampling techniques. (Nonparametric methods, which do not require a particular distribution, are discussed in Chapter 13; the bootstrap resampling technique

Meta-Analysis

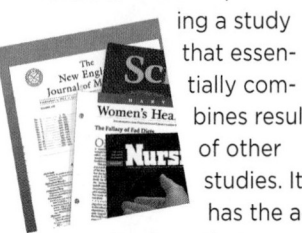

The term *meta-analysis* refers to a technique of doing a study that essentially combines results of other studies. It has the advantage that separate smaller samples can be combined into one big sample, making the collective results more meaningful. It also has the advantage of using work that has already been done. Meta-analysis has the disadvantage of being only as good as the studies that are used. If the previous studies are flawed, the "garbage in, garbage out" phenomenon can occur. The use of meta-analysis is currently popular in medical research and psychological research. As an example, a study of migraine headache treatments was based on data from 46 other studies. (See "Meta-Analysis of Migraine Headache Treatments: Combining Information from Heterogeneous Designs," by Dominici et al., *Journal of the American Statistical Association,* Vol. 94, No. 445.)

TI-83/84 Plus

```
T-Test
 μ>166.3
 t=1.501271202
 p=.0706695554
 x̄=172.55
 Sx=26.33
 n=40
```

is described in the Technology Project at the end of Chapter 7.) See pages 366–367 where Figure 7-6 and Table 7-1 both summarize the decisions to be made in choosing between the normal and Student *t* distributions. The Student *t* distribution is used under these conditions:

> **To test a claim about a population mean, use the Student *t* distribution when the sample is a simple random sample, σ is *not* known, and either or both of these conditions is satisfied:**
>
> **The population is normally distributed or $n > 30$.**

EXAMPLE 1 **Overloading Boats: Traditional Method** In Example 1 of the preceding section, we noted that people have died in boat accidents because an obsolete estimate of the mean weight of men was used. Using the weights of the simple random sample of men from Data Set 1 in Appendix B, we obtain these sample statistics: $n = 40$ and $\bar{x} = 172.55$ lb, $s = 26.33$ lb. Do not assume that the value of σ is known. Use these results to test the claim that men have a mean weight greater than 166.3 lb, which was the weight in the National Transportation and Safety Board's recommendation M-04-04. Use a 0.05 significance level.

SOLUTION **REQUIREMENT CHECK** (1) The sample is a simple random sample. (2) The value of the population standard deviation is not known. (3) The sample size is $n = 40$, which is greater than 30. The requirements are satisfied.

Step 1: The claim that men have a mean weight greater than 166.3 lb is expressed in symbolic form as $\mu > 166.3$ lb.

Step 2: The alternative (in symbolic form) to the original claim is $\mu \leq 166.3$ lb.

Step 3: Because the statement $\mu > 166.3$ lb does not contain the condition of equality, it becomes the alternative hypothesis. The null hypothesis is the statement that $\mu = 166.3$ lb.

$$H_0: \mu = 166.3 \text{ lb} \quad \text{(null hypothesis)}$$
$$H_1: \mu > 166.3 \text{ lb} \quad \text{(alternative hypothesis and original claim)}$$

Step 4: As specified in the statement of the problem, the significance level is $\alpha = 0.05$.

Step 5: Because the claim is made about the *population mean μ*, the sample statistic most relevant to this test is the *sample mean $\bar{x} = 172.55$* lb.

Step 6: The test statistic is calculated as follows:

$$t = \frac{\bar{x} - \mu_{\bar{x}}}{\frac{s}{\sqrt{n}}} = \frac{172.55 - 166.3}{\frac{26.33}{\sqrt{40}}} = 1.501$$

TI-83/84 Plus *P*-value Method The TI-83/84 Plus calculator provides *P*-values, but not critical values, so the *P*-value method of hypothesis testing applies. In this example, if we use the TI-83/84 Plus calculator with the procedure given near the end of this section, we obtain the display shown in the margin. This display shows the test statistic of $t = 1.501271202$ and the *P*-value of 0.0706695554.

Step 7: Because the P-value of 0.0706695554 is greater than the significance level of 0.05, we fail to reject the null hypothesis.

INTERPRETATION Because we fail to reject the null hypothesis, we conclude that there is not sufficient evidence to support a conclusion that the population mean is greater than 166.3 lb, as in the National Transportation and Safety Board's recommendation.

Traditional Method If we were using the traditional method in Example 1, we would need to find the critical value. The critical value of 1.685 can be found from the command of **invT(0.95, 39)** on the TI-84 Plus calculator, or the program **TINVERS83** on the TI-83 Plus calculator, or it can be found from Table A-3. Because the test statistic of $t = 1.501$ does not fall in the critical region bounded by the critical value of $t = 1.685$ as shown in Figure 8-15, we would again fail to reject the null hypothesis and we obtain the same interpretation given in Example 1.

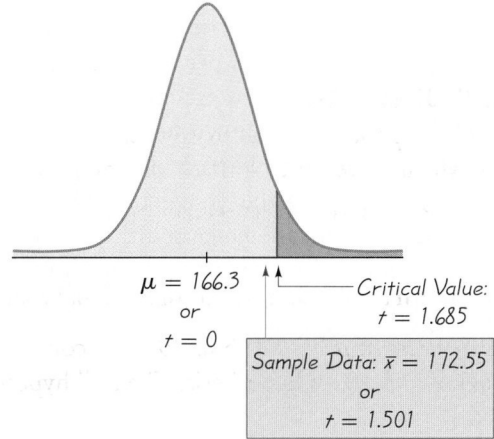

Figure 8-15

Traditional Method: Testing the Claim that $\mu >$ 166.3 lb

Confidence Interval Method We can use a confidence interval for testing a claim about μ when σ is not known. For a two-tailed hypothesis test with a 0.05 significance level, we construct a 95% confidence interval, but for a one-tailed hypothesis test with a 0.05 significance level we construct a 90% confidence interval (as described in Table 8-2 on page 412.)

In Section 8-3 we saw that when testing a claim about a population proportion, the traditional method and P-value method are equivalent, but the confidence interval method is somewhat different. When testing a claim about a population mean, there is no such difference, and all three methods are equivalent.

EXAMPLE 2 **Confidence Interval Method** The sample data from Example 1 result in the statistics $n = 40$, $\bar{x} = 172.55$ lb, $s = 26.33$ lb, and σ is not known. Using a 0.05 significance level, test the claim that $\mu > 166.3$ by applying the confidence interval method.

Lefties Die Sooner?

A study by psychologists Diane Halpern and Stanley Coren received considerable media attention and generated considerable interest when it concluded that left-handed people don't live as long as right-handed people. Based on their study, it appeared that left-handed people live an average of nine years less than righties. The Halpern/Coren study has been criticized for using flawed data. They used secondhand data by surveying relatives about people who had recently died. The myth of lefties dying younger became folklore that has survived many years. However, more recent studies show that left-handed people do *not* have shorter lives than those who are right-handed.

TI-83/84 Plus

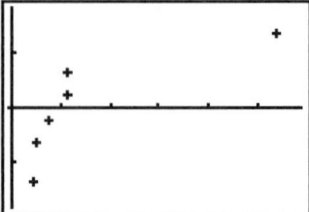

SOLUTION   **REQUIREMENT CHECK** The requirements have already been verified in Example 1.

Using the methods described in Section 7-4, we construct this 90% confidence interval:

$$165.54 \text{ lb} < \mu < 179.56 \text{ lb}$$

Because the assumed value of $\mu = 166.3$ lb is contained within the confidence interval, we cannot reject the null hypothesis that $\mu = 166.3$ lb. Based on the 40 sample values given in the example, we do not have sufficient evidence to support the claim that the mean weight is greater than 166.3 lb. Based on the confidence interval, the true value of μ is likely to be any value between 165.54 lb and 179.56 lb, including 166.3 lb.

The next example involves a small sample with a list of the original sample values. Because the sample is small (30 or fewer), use of the t test requires that we verify that the sample appears to be from a population with a distribution that is not too far from a normal distribution.

EXAMPLE 3   **Small Sample: Monitoring Lead in Air** Listed below are measured amounts of lead (in micrograms per cubic meter, or $\mu g/m^3$) in the air. The Environmental Protection Agency has established an air quality standard for lead of 1.5 $\mu g/m^3$. The measurements shown below constitute a simple random sample of measurements recorded at Building 5 of the World Trade Center site on different days immediately following the destruction caused by the terrorist attacks of September 11, 2001. After the collapse of the two World Trade Center buildings, there was considerable concern about the quality of the air. Use a 0.05 significance level to test the claim that the sample is from a population with a mean greater than the EPA standard of 1.5 $\mu g/m^3$.

5.40   1.10   0.42   0.73   0.48   1.10

SOLUTION   **REQUIREMENT CHECK** (1) The sample is a simple random sample. (2) The value of σ is not known. (3) Because the sample size of $n = 6$ is not greater than 30, we must verify that the sample appears to be from a population having a normal distribution. The accompanying TI-83/84 Plus and STATDISK normal quantile plots show that the points do not lie reasonably close to a straight line, so the normality requirement is very questionable. Also, the value of 5.40 appears to be an outlier. Formal hypothesis tests of normality also suggest that the sample data are not from a population with a normal distribution. The necessary requirements are not satisfied, so use of the methods of this section may yield poor results.

STATDISK

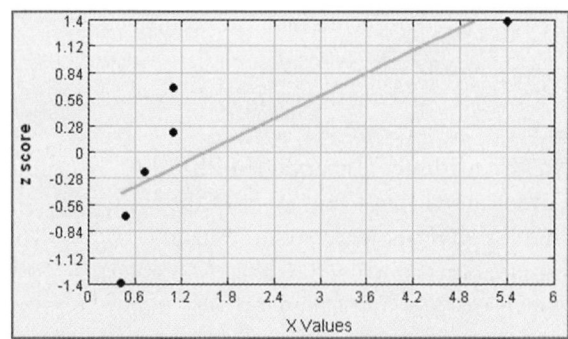

If we were to proceed with the *t* test, we would find that the test statistic is $t = 0.049$, the critical value is $t = 2.015$, and the *P*-value > 0.10. With technology, the *P*-value is 0.4814. We should fail to reject the null hypothesis of $\mu = 1.5$, so there is not sufficient evidence to support the claim that $\mu > 1.5$. But given that the requirements are not satisfied, this conclusion is questionable.

Using a TI-83/84 Plus Calculator for Testing a Claim About a Mean (With σ *Not* Known)

When using a TI-83/84 Plus calculator for testing a claim about a population mean μ and the value of the population standard deviation σ is *not* known, you have the option of using either the summary statistics of $\bar{x}$, s, n, or you can use a list of original sample data values.

Because the TI-83/84 Plus calculator provides the test statistic and P-value (but not the critical value), we use the P-value method of testing hypotheses. Here is the procedure:

1. Press **STAT**, then press ▷ twice to select **TESTS.**

2. Select the menu item of **T-Test.**

3. The first line, identified as Inpt (for input), gives you the option of using Data or Stats. Use the arrow keys ◁ and ▷ to select either Data or Stats.

   • **Data:** Use this option if you have the sample data values stored as a list.

   • **Stats:** Use this option if you know the values of $\bar{x}$, s, n.

4. For the value indicated by $\mu 0$, enter the claimed value of the population mean. This is the same value used in the null and alternative hypotheses.

5. If you selected Stats in Step 3, proceed to enter the values of the sample mean $\bar{x}$, the sample standard deviation s, and the sample size n.

   If you selected Data in Step 3, proceed to enter the list where the data are stored (such as L1 or a named list such as AGES), and enter 1 for the frequency. (However, if the sample data are stored in two columns as a frequency distribution, the Freq entry should be the list containing frequencies).

6. For the next line, select the type of test (two-tailed, left-tailed, right-tailed). Select the option that uses the same symbol as the alternative hypothesis. For a two-tailed case select the option including ≠, for a left-tailed test select the option of $<\mu 0$, and for a right-tailed test select the option of $>\mu 0$.

7. Use the arrow keys to scroll down to the bottom line, select **Calculate,** then press **ENTER**.

Example 1 involved the claim of $\mu > 166.3$ lb, and the sample statistics were given as $n = 40$, $\bar{x} = 172.55$, and $s = 26.33$. The entry screen for Example 1 is shown, followed by the screen showing the results.

Interpreting the results: See the displayed results and note that the TI-83/84 Plus calculator provides the test statistic of $t = 1.501271202$ and the *P*-value of 0.0706695554. Because the *P*-value is greater than the significance level of 0.05, we fail to reject the null hypothesis and conclude that there is not sufficient sample evidence to support the claim that the mean is greater than 166.3 lb.

Entries for Example 1

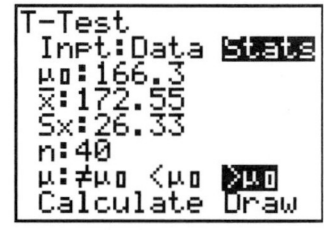

Results for Example 1

```
T-Test
 μ>166.3
 t=1.501271202
 P=.0706695554
 x̄=172.55
 Sx=26.33
 n=40
```

8-5   Basic Skills and Concepts

Statistical Literacy and Critical Thinking

1. Normality Requirement Given a simple random sample of 20 speeds of cars on Highway 405 in California, you want to test the claim that the sample values are from a population with a mean greater than the posted speed limit of 65 mi/h. Is it necessary to determine whether the sample is from a normally distributed population? If so, what methods can be used to make that determination?

2. df In statistics, what does *df* denote? If a simple random sample of 20 speeds of cars on California Highway 405 is to be used to test the claim that the sample values are from a population with a mean greater than the posted speed limit of 65 mi/h, what is the specific value of df?

3. *t* Test What is a *t* test? Why is the letter *t* used?

4. Reality Check Unlike the preceding section, this section does not include a requirement that the value of the population standard deviation must be known. Which section is more likely to apply in realistic situations: this section or the preceding section? Why?

Using Correct Distribution. *In Exercises 5–8, determine whether the hypothesis test involves a sampling distribution of means that is a normal distribution, Student t distribution, or neither. (Hint: See Figure 7-6 and Table 7-1.)*

5. Claim about IQ scores of statistics instructors: $\mu > 100$. Sample data: $n = 15$, $\bar{x} = 118$, $s = 11$. The sample data appear to come from a normally distributed population with unknown μ and σ.

6. Claim about FICO credit scores of adults: $\mu = 678$. Sample data: $n = 12$, $\bar{x} = 719$, $s = 92$. The sample data appear to come from a population with a distribution that is not normal, and σ is unknown.

7. Claim about daily rainfall amounts in Boston: $\mu < 0.20$ in. Sample data: $n = 19$, $\bar{x} = 0.10$ in, $s = 0.26$ in. The sample data appear to come from a population with a distribution that is very far from normal, and σ is unknown.

8. Claim about daily rainfall amounts in Boston: $\mu < 0.20$ in. Sample data: $n = 52$, $\bar{x} = 0.10$ in, $s = 0.26$ in. The sample data appear to come from a population with a distribution that is very far from normal, and σ is known.

Testing Hypotheses. *In Exercises 9–24, assume that a simple random sample has been selected from a normally distributed population and test the given claim. Unless specified by your instructor, use either the traditional method or P-value method for testing hypotheses. Identify the null and alternative hypotheses, test statistic, P-value (or range of P-values), critical value(s), and state the final conclusion that addresses the original claim.*

9. Writing a Hit Song In the KLF Publications manual "How to Have a Number One the Easy Way," it is stated that a song "must be no longer than three minutes and thirty seconds" (or 210 seconds). A simple random sample of 40 current hit songs results in a mean length of 252.5 sec and a standard deviation of 54.5 sec. (The songs are by Timberlake, Furtado, Daughtry, Stefani, Fergie, Akon, Ludacris, etc.) Use a 0.05 significance level and the accompanying Minitab display to test the claim that the sample is from a population of songs with a mean greater than 210 sec. What do these results suggest about the advice given in the manual?

MINITAB

One-Sample T
Test of mu = 210 vs > 210

N	Mean	StDev	SE Mean	95% Lower Bound	T	P
40	252.50	54.50	8.62	237.98	4.93	0.000

TI **10. Red Blood Cell Count** A simple random sample of 50 adults is obtained, and each person's red blood cell count (in cells per microliter) is measured. The sample mean is 5.23 and the sample standard deviation is 0.54. Use a 0.01 significance level and the accompanying TI-83/84 Plus display to test the claim that the sample is from a population with a mean less than 5.4, which is a value often used for the upper limit of the range of normal values. What do the results suggest about the sample group?

TI **11. Cigarette Tar** A simple random sample of 25 filtered 100 mm cigarettes is obtained, and the tar content of each cigarette is measured. The sample has a mean of 13.2 mg and a standard deviation of 3.7 mg (based on Data Set 4 in Appendix B). Use a 0.05 significance level to test the claim that the mean tar content of filtered 100 mm cigarettes is less than 21.1 mg, which is the mean for unfiltered king size cigarettes. What do the results suggest about the effectiveness of the filters?

TI **12. Is the Diet Practical?** When 40 people used the Weight Watchers diet for one year, their mean weight *loss* was 3.0 lb and the standard deviation was 4.9 lb (based on data from "Comparison of the Atkins, Ornish, Weight Watchers, and Zone Diets for Weight Loss and Heart Disease Reduction," by Dansinger, et al., *Journal of the American Medical Association*, Vol. 293, No. 1). Use a 0.01 significance level to test the claim that the mean weight loss is greater than 0 lb. Based on these results, does the diet appear to be effective? Does the diet appear to have practical significance?

TI **13. Weights of Pennies** The U.S. Mint has a specification that pennies have a mean weight of 2.5 g. Data Set 20 in Appendix B lists the weights (in grams) of 37 pennies manufactured after 1983. Those pennies have a mean weight of 2.49910 g and a standard deviation of 0.01648 g. Use a 0.05 significance level to test the claim that this sample is from a population with a mean weight equal to 2.5 g. Do the pennies appear to conform to the specifications of the U.S. Mint?

TI **14. Analysis of Pennies** In an analysis investigating the usefulness of pennies, the cents portions of 100 randomly selected credit card charges are recorded. The sample has a mean of 47.6 cents and a standard deviation of 33.5 cents. If the amounts from 0 cents to 99 cents are all equally likely, the mean is expected to be 49.5 cents. Use a 0.01 significance level to test the claim that the sample is from a population with a mean equal to 49.5 cents. What does the result suggest about the cents portions of credit card charges?

TI **15. Time Required for Bachelor's Degree** Researchers collected a simple random sample of the times that 81 college students required to earn their bachelor's degrees. The sample has a mean of 4.8 years and a standard deviation of 2.2 years (based on data from the National Center for Education Statistics). Use a 0.05 significance level to test the claim that the mean time for all college students is greater than 4.5 years.

TI **16. Uninterruptible Power Supply (UPS)** Data Set 13 in Appendix B lists measured voltage amounts obtained from the author's back-up UPS (APC model CS 350). According to the manufacturer, the normal output voltage is 120 volts. The 40 measured voltage amounts from Data Set 13 have a mean of 123.59 volts and a standard deviation of 0.31 volt. Use a 0.05 significance level to test the claim that the sample is from a population with a mean equal to 120 volts.

TI **17. Analysis of Pennies** In an analysis investigating the usefulness of pennies, the cents portions of 100 randomly selected checks are recorded. The sample has a mean of 23.8 cents and a standard deviation of 32.0 cents. If the amounts from 0 cents to 99 cents are all equally likely, the mean is expected to be 49.5 cents. Use a 0.01 significance level to test the claim that the sample is from a population with a mean less than 49.5 cents. What does the result suggest about the cents portions of the checks?

TI **18. California Speeding** A simple random sample of 40 recorded speeds (in mi/h) is obtained from cars traveling on a section of Highway 405 in Los Angeles. The sample has a mean of 68.4 mi/h and a standard deviation of 5.7 mi/h (based on data from Sigalert). Use a 0.05 significance level to test the claim that the mean speed of all cars is greater than the posted speed limit of 65 mi/h.

TI **19. Car Emissions** Data Set 16 in Appendix B lists the measured greenhouse gas emissions from 32 different cars. The sample has a mean of 7.78 tons and a standard deviation of

1.08 tons. (The amounts are in tons per year, expressed as CO_2 equivalents.) Use a 0.05 significance level to test the claim that all cars have a mean greenhouse gas emission of 8.00 tons.

20. Heights of Supermodels The heights are measured for the simple random sample of supermodels Crawford, Bundchen, Pestova, Christenson, Hume, Moss, Campbell, Schiffer, and Taylor. They have a mean height of 70.0 in. and a standard deviation of 1.5 in. Use a 0.01 significance level to test the claim that supermodels have heights with a mean that is greater than the mean height of 63.6 in. for women in the general population. Given that there are only nine heights represented, can we really conclude that supermodels are taller than the typical woman?

21. Tests of Child Booster Seats The National Highway Traffic Safety Administration conducted crash tests of child booster seats for cars. Listed below are results from those tests, with the measurements given in hic (standard *head injury condition* units). The safety requirement is that the hic measurement should be less than 1000 hic. Use a 0.01 significance level to test the claim that the sample is from a population with a mean less than 1000 hic. Do the results suggest that all of the child booster seats meet the specified requirement?

$$774 \quad 649 \quad 1210 \quad 546 \quad 431 \quad 612$$

22. Number of English Words A simple random sample of pages from *Merriam-Webster's Collegiate Dictionary, 11th edition,* is obtained. Listed below are the numbers of words defined on those pages. Given that this dictionary has 1459 pages with defined words, the claim that there are more than 70,000 defined words is the same as the claim that the mean number of defined words on a page is greater than 48.0. Use a 0.05 significance level to test the claim that the mean number of defined words on a page is greater than 48.0. What does the result suggest about the claim that there are more than 70,000 defined words in the dictionary?

$$51 \quad 63 \quad 36 \quad 43 \quad 34 \quad 62 \quad 73 \quad 39 \quad 53 \quad 79$$

23. Car Crash Costs The Insurance Institute for Highway Safety conducted tests with crashes of new cars traveling at 6 mi/h. The total cost of the damage was found. Results are listed below for a simple random sample of the tested cars. Use a 0.05 significance level to test the claim that when tested under the same standard conditions, the damage costs for the population of cars has a mean of $5000.

$$\$7448 \quad \$4911 \quad \$9051 \quad \$6374 \quad \$4277$$

24. BMI for Miss America The trend of thinner Miss America winners has generated charges that the contest encourages unhealthy diet habits among young women. Listed below are body mass indexes (BMI) for recent Miss America winners. Use a 0.01 significance level to test the claim that recent Miss America winners are from a population with a mean BMI less than 20.16, which was the BMI for winners from the 1920s and 1930s. Do recent winners appear to be significantly different from those in the 1920s and 1930s?

$$19.5 \quad 20.3 \quad 19.6 \quad 20.2 \quad 17.8 \quad 17.9 \quad 19.1 \quad 18.8 \quad 17.6 \quad 16.8$$

Large Data Sets. *In Exercises 25–28, use the Data Set from Appendix B to test the given claim.*

25. Do the Screws Have a Length of 3/4 in.? A simple random sample of 50 stainless steel sheet metal screws is obtained from those supplied by Crown Bolt, Inc., and the length of each screw is measured using a vernier caliper. The lengths are listed in Data Set 19 of Appendix B. Use a 0.05 significance level to test the claim that the screws have a mean length equal to 3/4 in. (or 0.75 in.), as indicated on the package labels. Do the screw lengths appear to be consistent with the package label?

26. Power Supply Data Set 13 in Appendix B lists measured voltage amounts supplied directly to the author's home. The Central Hudson power supply company states that it has a target power supply of 120 volts. Using those home voltage amounts, test the claim that the mean is 120 volts. Use a 0.01 significance level.

27. Is 98.6°F Wrong? Data Set 2 in Appendix B includes measured human body temperatures. Use the temperatures listed for 12 AM on day 2 to test the common belief that the mean body temperature is 98.6°F. Does that common belief appear to be wrong?

TI **28. FICO Credit Scores** Data Set 24 in Appendix B includes a simple random sample of FICO credit rating scores. As of this writing, the mean FICO score was reported to be 678. Use a 0.05 significance level to test the claim that the sample of FICO scores comes from a population with a mean equal to 678.

8-5   Beyond the Basics

TI **29. Alternative Method** When testing a claim about the population mean μ using a simple random sample from a normally distributed population with unknown σ, an alternative method (not used in this book) is to use the methods of this section if the sample is small ($n \leq 30$), but if the sample is large ($n > 30$) substitute s for σ and proceed as if σ is known (as in Section 8-4). A sample of 32 IQ scores has $\bar{x} = 105.3$ and $s = 15.0$. Use a 0.05 significance level to test the claim that the sample is from a population with a mean equal to 100. Use the alternative method and compare the results to those obtained by using the method of this section. Does the alternative method always yield the same conclusion as the t test?

30. Using the Wrong Distribution When testing a claim about a population mean with a simple random sample selected from a normally distributed population with unknown σ, the Student t distribution should be used for finding critical values and/or a P-value. If the standard normal distribution is incorrectly used instead, does that mistake make you more or less likely to reject the null hypothesis, or does it not make a difference? Explain.

31. Finding Critical t Values When finding critical values, we sometimes need significance levels other than those available in Table A-3. Some computer programs approximate critical t values by calculating

$$t = \sqrt{df \cdot (e^{A^2/df} - 1)}$$

where $df = n - 1$, $e = 2.718$, $A = z(8 \cdot df + 3)/(8 \cdot df + 1)$, and z is the critical z score. Use this approximation to find the critical t score corresponding to $n = 75$ and a significance level of 0.05 in a right-tailed case. Compare the results to the critical t value of 1.666 found from a TI-83/84 Plus calculator.

32. Interpreting Power For Example 1 in this section, the hypothesis test has power of 0.2203 of supporting the claim that $\mu > 166.3$ lb when the actual population mean is 170 lb.

a. Interpret the given value of the power.

b. Identify the value of β, and interpret that value.

TI **33. Calculating Power of a Test** For Example 1 in this section, find the power of the test in supporting the claim that $\mu > 166.3$ lb when the actual population mean is 180 lb. Also find β, the probability of a type II error. Is the test effective in supporting the claim that $\mu > 166.3$ lb when the true population mean is 180 lb?

8-6   Testing a Claim About a Standard Deviation or Variance

Key Concept In this section we introduce methods for testing a claim made about a population standard deviation σ or population variance σ^2. The methods of this section use the chi-square distribution that was first introduced in Section 7-5. The assumptions, test statistic, P-value, and critical values are summarized as follows.

Testing Claims About σ or σ^2

Objective

Test a claim about a population standard deviation σ (or population variance σ^2) using a formal method of hypothesis testing.

Notation

n = sample size

s = *sample* standard deviation

s^2 = *sample* variance

σ = claimed value of the *population* standard deviation

σ^2 = claimed value of the *population* variance

Requirements

1. The sample is a simple random sample.

2. The population has a normal distribution. (This is a much stricter requirement than the requirement of a normal distribution when testing claims about means, as in Sections 8-4 and 8-5.)

Test Statistic for Testing a Claim About σ or σ^2

$$\chi^2 = \frac{(n-1)s^2}{\sigma^2}$$ (round to three decimal places, as in Table A-4)

P-values and Critical Values: Use Table A-4 with df = $n - 1$ for the number of degrees of freedom. (Table A-4 is based on *cumulative areas from the right*.)

CAUTION

The χ^2 test of this section is not *robust* against a departure from normality, meaning that the test does not work well if the population has a distribution that is far from normal. The condition of a normally distributed population is therefore a much stricter requirement in this section than it was in Sections 8-4 and 8-5.

The chi-square distribution was introduced in Section 7-5, where we noted the following important properties.

Properties of the Chi-Square Distribution

1. All values of χ^2 are nonnegative, and the distribution is not symmetric (see Figure 8-16).

2. There is a different χ^2 distribution for each number of degrees of freedom (see Figure 8-17).

3. The critical values are found in Table A-4 using

$$\text{degrees of freedom} = n - 1$$

Table A-4 is based on cumulative areas from the *right* (unlike the entries in Table A-2, which are cumulative areas from the left). Critical values are found in Table A-4 by first locating the row corresponding to the appropriate number of degrees of freedom (where df $= n - 1$). Next, the significance level α is used to determine the correct column. The following examples are based on a significance level of $\alpha = 0.05$, but any other significance level can be used in a similar manner.

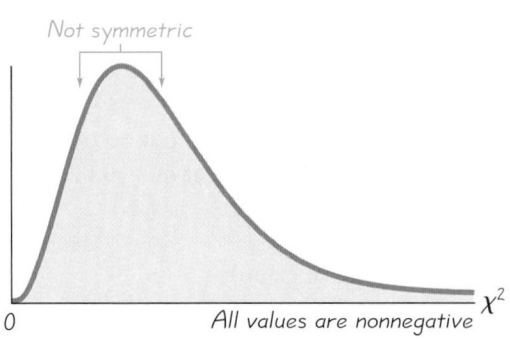

Figure 8-16 **Properties of the Chi-Square Distribution**

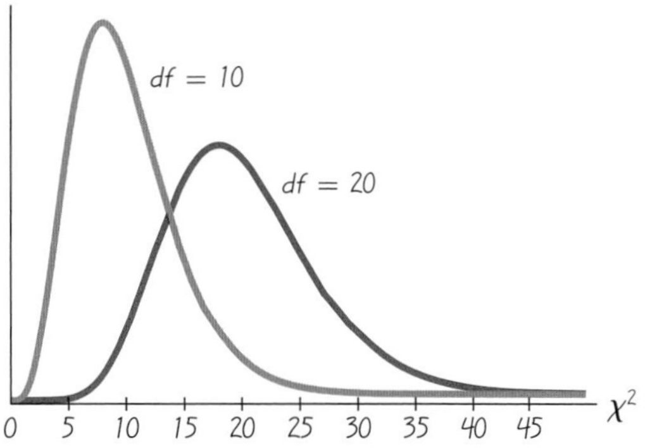

Figure 8-17 **Chi-Square Distribution for df = 10 and df = 20**

Right-tailed test:	Because the area to the *right* of the critical value is 0.05, locate 0.05 at the top of Table A-4.
Left-tailed test:	With a left-tailed area of 0.05, the area to the *right* of the critical value is 0.95, so locate 0.95 at the top of Table A-4.
Two-tailed test:	Unlike the normal and Student t distributions, the critical values in this χ^2 test will be two different positive values (instead of something like ± 1.96). Divide a significance level of 0.05 between the left and right tails, so the areas to the *right* of the two critical values are 0.975 and 0.025, respectively. Locate 0.975 and 0.025 at the top of Table A-4. (See Figure 7-10 and Example 1 on page 378.)

> **EXAMPLE 1** **Quality Control of Coins: Traditional Method** A common goal in business and industry is to improve the quality of goods or services by reducing *variation*. Quality control engineers want to ensure that a product has an acceptable mean, but they also want to produce items of *consistent* quality so that there will be few defects. If weights of coins have a specified mean but too much variation, some will have weights that are too low or too high, so that vending machines will not work correctly (unlike the stellar performance that they now provide). Consider the simple random sample of the 37 weights of post-1983 pennies listed in Data Set 20 in Appendix B. Those 37 weights have a mean of 2.49910 g and a standard deviation of 0.01648 g. U.S. Mint specifications require that pennies be manufactured so that the mean weight is 2.500 g. A hypothesis test will verify that the sample appears to come from a population with a mean of 2.500 g as required, but use a 0.05 significance level to test the claim that the population of weights has a *standard deviation* less than the specification of 0.0230 g.

SOLUTION **REQUIREMENT CHECK** (1) The sample is a simple random sample. (2) Based on the following TI-83/84 Plus calculator histogram and normal quantile plot, the sample does appear to come from a population having a normal distribution. The histogram is close to being bell-shaped. The points on the normal quantile plot are quite close to a straight-line pattern, and there is no other pattern. There are no outliers. The departure from an exact normal distribution is relatively minor. Both requirements are satisfied. ✓

"How Statistics Can Help Save Failing Hearts"

A *New York Times* article by David Leonhardt featured the headline of "How Statistics Can Help Save Failing Hearts." Leonhardt writes that patients have the best chance of recovery if their clogged arteries are opened within two hours of a heart attack. In 2005, the U.S. Department of Health and Human Services began posting hospital data on its Web site www.hospitalcompare.hhs.gov, and it included the percentage of heart attack patients who received treatment for blocked arteries within two hours of arrival at the hospital. Not wanting to be embarrassed by poor data, doctors and hospitals are reducing the time it takes to unblock those clogged arteries. Leonhardt writes about the University of California, San Francisco Medical Center, which cut its time in half from almost three hours to about 90 minutes. Effective use of simple statistics can save lives.

TI-83/84 Plus Histogram

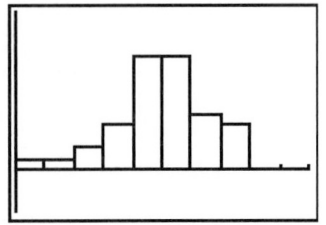

Normal Quantile Plot

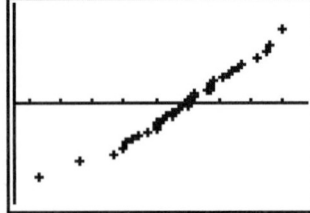

We will use the traditional method of testing hypotheses as outlined in Figure 8-9.

Step 1: The claim is expressed in symbolic form as $\sigma < 0.0230$ g.

Step 2: If the original claim is false, then $\sigma \geq 0.0230$ g.

Step 3: The expression $\sigma < 0.0230$ g does not contain equality, so it becomes the alternative hypothesis. The null hypothesis is the statement that $\sigma = 0.0230$ g.

$$H_0: \sigma = 0.0230 \text{ g}$$

$$H_1: \sigma < 0.0230 \text{ g} \qquad \text{(original claim)}$$

Step 4: The significance level is $\alpha = 0.05$.

Step 5: Because the claim is made about σ, we use the chi-square distribution.

Step 6: The test statistic is

$$\chi^2 = \frac{(n-1)s^2}{\sigma^2} = \frac{(37-1)(0.01648)^2}{0.0230^2} = 18.483$$

The critical value from Table A-4 corresponds to 36 degrees of freedom and an "area to the right" of 0.95 (based on the significance level of 0.05 for a left-tailed test). Table A-4 does not include 36 degrees of freedom, but Table A-4 shows that the critical value is between 18.493 and 26.509. (Using technology, the critical value is 23.269.) See Figure 8-18.

Step 7: Because the test statistic is in the critical region, reject the null hypothesis.

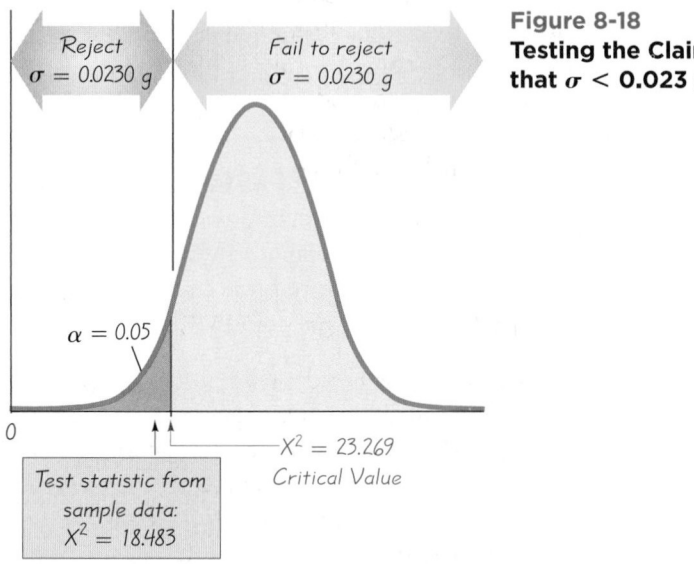

Figure 8-18
Testing the Claim that $\sigma < 0.023$ g

INTERPRETATION   There is sufficient evidence to support the claim that the standard deviation of weights is less than 0.0230 g. It appears that the variation is less than 0.0230 g as specified, so the manufacturing process is acceptable.

P-Value Method

Example 1 can also be solved with the *P*-value approach summarized in Figures 8-5 and 8-8. When using Table A-4, we usually cannot find *exact P*-values because that chi-square distribution table includes only selected values of α and selected numbers of degrees of freedom, but exact *P*-values are easily found using technology.

EXAMPLE 2 **Quality Control of Coins: *P*-Value Method** Repeat Example 1 using the *P*-value method of testing hypotheses.

SOLUTION If technology is used for the preceding example, the *P*-value is 0.0069. Because the *P*-value is less than the significance level of 0.05, we reject the null hypothesis and arrive at the same conclusion given in Example 1.

Confidence Interval Method

Example 1 can also be solved with the confidence interval method of testing hypotheses.

EXAMPLE 3 **Quality Control of Coins: Confidence Interval Method** Repeat the hypothesis test in Example 1 by constructing a suitable confidence interval.

SOLUTION Because the hypothesis test is left-tailed and the significance level is 0.05, we should construct a 90% confidence interval (as indicated in Table 8-2 on page 412). Using the methods described in Section 7-5, we can use the sample data ($n = 37$, $s = 0.01648$ g) to construct this 90% confidence interval: 0.01385 g $< \sigma < 0.02050$ g. Based on this confidence interval, we can support the claim that σ is less than 0.0230 g (because all values of the confidence interval are less than 0.0230 g). We reach the same conclusion found with the traditional and *P*-value methods.

Using a TI-83/84 Plus Calculator for Testing a Claim About a Standard Deviation or Variance
Method 1: Using the Program S2TEST

The TI-83/84 Plus calculator is not designed to automatically generate results for a hypothesis test about a population standard deviation σ or a population variance σ^2, but the program **S2TEST** can be used for this hypothesis test. That program was written by Michael Lloyd of Henderson State University, and it can be downloaded from the CD-ROM included with this book, or it can be downloaded from the Web site www.aw.com/triola. Use the following procedure.

1. Use the enclosed CD-ROM or the Web site www.aw.com/triola to download the program S2INT. (Download instructions can be found at the site www.ti.com, and they typically involve the use of software such as TI-GRAPHLINK or TI CONNECT.)
2. Press **PRGM**, select the program **S2TEST**, then press **ENTER**.
3. Proceed to enter the value of the population variance σ^2. (The value of the population standard deviation σ is used in the null and alternative hypotheses.) Be sure to enter the value of σ^2 when prompted, then press **ENTER**.
4. Next, there will be a prompt for the entry of the sample variance s^2. Enter the square of the sample standard deviation, then press **ENTER**.
5. There will be a prompt for the entry of the sample size n, so enter that value and press **ENTER**.

6. Now select the type of test as follows:
   - For a two-tailed hypothesis test, select the first option that includes the symbol $\neq$.
   - For a left-tailed test, select the option that includes the symbol $<$.
   - For a right-tailed test, select the option that includes the symbol $>$.
7. Press **ENTER** and the test statistic and *P*-value will be provided.

If using this procedure with Example 1, the test statistic of 18.4825225 will be displayed along with a *P*-value of 0.0068732909.

Method 2: Manual Calculations with *P*-value

Manually calculate the value of the test statistic and number of degrees of freedom ($n - 1$). After finding the test statistic and number of degrees of freedom, the *P*-value can be found from the TI-83/84 calculator as follows:

1. Press **2ND** **VARS** to get to the DISTR (distribution) menu, then select χ^2**cdf.**
2. The χ^2cdf command requires three items separated by commas in this format:

   (lower bound, upper bound, degrees of freedom)

   Find the *P*-value by using the number of degrees of freedom and the test statistic as shown below.

   Left-tailed test: The *P*-value is the result of χ^2cdf (-999999, test statistic, degrees of freedom).

   Right-tailed test: The *P*-value is the result of χ^2cdf (test statistic, 999999, degrees of freedom).

   Two-tailed test: The *P*-value is the smaller of the following two results:
   - Twice the value obtained by entering
     χ^2cdf (-999999, test statistic, degrees of freedom)
   - Twice the value obtained by entering
     χ^2cdf (test statistic, 999999, degrees of freedom)

   The example in this section involved a left-tailed test with the test statistic of $\chi^2 = 118.483$ and degrees of freedom of $n - 1 = 36$. Using the above procedure, the *P*-value is the result of χ^2cdf (-999999, 18.483, 36), which is 0.0069 when rounded.

   STATDISK does have the ability to conduct hypothesis tests involving claims about a population standard deviation σ or a population variance σ^2. Select **Analysis** from the main menu at the top, then select **Hypothesis Testing** and proceed to the case identified as **St Dev One Sample.**

8-6   Basic Skills and Concepts

Statistical Literacy and Critical Thinking

1. Normality Requirement Hypothesis tests of claims about the population mean or population standard deviation both require a simple random sample from a normally distributed population. How does the normality requirement for a hypothesis test of a claim about a standard deviation differ from the normality requirement for a hypothesis test of a claim about a mean?

2. Confidence Interval Method of Hypothesis Testing There is a claim that the lengths of men's hands have a standard deviation less than 200 mm. You plan to test that claim with a 0.01 significance level by constructing a confidence interval. What level of confidence should be used for the confidence interval? Will the conclusion based on the confidence interval be the same as the conclusion based on a hypothesis test that uses the traditional method or the *P*-value method?

3. Requirements There is a claim that daily rainfall amounts in Boston have a standard deviation equal to 0.25 in. Sample data show that daily rainfall amounts are from a population with a distribution that is very far from normal. Can the use of a very large sample compensate for the lack of normality, so that the methods of this section can be used for the hypothesis test?

4. Testing a Claim About a Variance There is a claim that men have foot breadths with a variance equal to 36 mm^2. Is a hypothesis test of the claim that the variance is equal to 36 mm^2 equivalent to a test of the claim that the standard deviation is equal to 6 mm?

(TI) **Finding Test Components.** *In Exercises 5–8, find the test statistic and critical value(s). Also, use Table A-4 to find limits containing the P-value, then determine whether there is sufficient evidence to support the given alternative hypothesis.*

5. Birth Weights $H_1: \sigma \neq 696$ g, $\alpha = 0.05$, $n = 25$, $s = 645$ g.

6. Supermodel Weights $H_1: \sigma < 29$ lb, $\alpha = 0.05$, $n = 8$, $s = 7.5$ lb.

7. Customer Waiting Times $H_1: \sigma > 3.5$ minutes, $\alpha = 0.01$, $n = 15$, $s = 4.8$ minutes.

8. Precipitation Amounts $H_1: \sigma \neq 0.25$, $\alpha = 0.01$, $n = 26$, $s = 0.18$.

Testing Claims About Variation. *In Exercises 9–20, test the given claim. Assume that a simple random sample is selected from a normally distributed population. Use either the P-value method or the traditional method of testing hypotheses unless your instructor indicates otherwise.*

(TI) **9. Weights of Pennies** The examples in this section involved the claim that post-1983 pennies have weights with a standard deviation less than 0.0230 g. Data Set 20 in Appendix B includes the weights of a simple random sample of pre-1983 pennies, and that sample has a standard deviation of 0.03910 g. Use a 0.05 significance level to test the claim that pre-1983 pennies have weights with a standard deviation greater than 0.0230 g. Based on these results and those found in Example 1, does it appear that weights of pre-1983 pennies vary more than those of post-1983 pennies?

(TI) **10. Pulse Rates of Men** A simple random sample of 40 men results in a standard deviation of 11.3 beats per minute (based on Data Set 1 in Appendix B). The normal range of pulse rates of adults is typically given as 60 to 100 beats per minute. If the range rule of thumb is applied to that normal range, the result is a standard deviation of 10 beats per minute. Use the sample results with a 0.05 significance level to test the claim that pulse rates of men have a standard deviation greater than 10 beats per minute.

(TI) **11. Cigarette Tar** A simple random sample of 25 filtered 100 mm cigarettes is obtained, and the tar content of each cigarette is measured. The sample has a standard deviation of 3.7 mg (based on Data Set 4 in Appendix B). Use a 0.05 significance level to test the claim that the tar content of filtered 100 mm cigarettes has a standard deviation different from 3.2 mg, which is the standard deviation for unfiltered king size cigarettes.

(TI) **12. Weight Loss from Diet** When 40 people used the Weight Watchers diet for one year, their weight *losses* had a standard deviation of 4.9 lb (based on data from "Comparison of the Atkins, Ornish, Weight Watchers, and Zone Diets for Weight Loss and Heart Disease Reduction," by Dansinger, et al., *Journal of the American Medical Association,* Vol. 293, No. 1). Use a 0.01 significance level to test the claim that the amounts of weight loss have a standard deviation equal to 6.0 lb, which appears to be the standard deviation for the amounts of weight loss with the Zone diet.

(TI) **13. Heights of Supermodels** The heights are measured for the simple random sample of supermodels Crawford, Bundchen, Pestova, Christenson, Hume, Moss, Campbell, Schiffer, and Taylor. Those heights have a mean of 70.0 in. and a standard deviation of 1.5 in. Use a 0.05 significance level to test the claim that supermodels have heights with a standard deviation less than 2.5 in., which is the standard deviation for heights of women from the general population. What does the conclusion reveal about heights of supermodels?

(TI) **14. Statistics Test Scores** Tests in the author's statistics classes have scores with a standard deviation equal to 14.1. One of his last classes had 27 test scores with a standard deviation of 9.3. Use a 0.01 significance level to test the claim that this class has less variation than other past classes. Does a lower standard deviation suggest that this last class is doing better?

TI 15. Pulse Rates of Women A simple random sample of pulse rates of 40 women results in a standard deviation of 12.5 beats per minute (based on Data Set 1 in Appendix B). The normal range of pulse rates of adults is typically given as 60 to 100 beats per minute. If the range rule of thumb is applied to that normal range, the result is a standard deviation of 10 beats per minute. Use the sample results with a 0.05 significance level to test the claim that pulse rates of women have a standard deviation equal to 10 beats per minute.

TI 16. Analysis of Pennies In an analysis investigating the usefulness of pennies, the cents portions of 100 randomly selected credit card charges are recorded, and they have a mean of 47.6 cents and a standard deviation of 33.5 cents. If the amounts from 0 cents to 99 cents are all equally likely, the mean is expected to be 49.5 cents and the population standard deviation is expected to be 28.866 cents. Use a 0.01 significance level to test the claim that the sample is from a population with a standard deviation equal to 28.866 cents. If the amounts from 0 cents to 99 cents are all equally likely, is the requirement of a normal distribution satisfied? If not, how does that affect the conclusion?

TI 17. BMI for Miss America Listed below are body mass indexes (BMI) for recent Miss America winners. Use a 0.01 significance level to test the claim that recent Miss America winners are from a population with a standard deviation of 1.34, which was the standard deviation of BMI for winners from the 1920s and 1930s. Do recent winners appear to have variation that is different from that of the 1920s and 1930s?

| 19.5 | 20.3 | 19.6 | 20.2 | 17.8 | 17.9 | 19.1 | 18.8 | 17.6 | 16.8 |

TI 18. Vitamin Supplement and Birth Weight Listed below are birth weights (in kilograms) of male babies born to mothers on a special vitamin supplement (based on data from the New York State Department of Health). Test the claim that this sample comes from a population with a standard deviation equal to 0.470 kg, which is the standard deviation for male birth weights in general. Use a 0.05 significance level. Does the vitamin supplement appear to affect the variation among birth weights?

| 3.73 | 4.37 | 3.73 | 4.33 | 3.39 | 3.68 | 4.68 | 3.52 |
| 3.02 | 4.09 | 2.47 | 4.13 | 4.47 | 3.22 | 3.43 | 2.54 |

TI 19. Aircraft Altimeters The Skytek Avionics company uses a new production method to manufacture aircraft altimeters. A simple random sample of new altimeters resulted in errors listed below. Use a 0.05 level of significance to test the claim that the new production method has errors with a standard deviation greater than 32.2 ft, which was the standard deviation for the old production method. If it appears that the standard deviation is greater, does the new production method appear to be better or worse than the old method? Should the company take any action?

| −42 | 78 | −22 | −72 | −45 | 15 | 17 | 51 | −5 | −53 | −9 | −109 |

TI 20. Playing Times of Popular Songs Listed below are the playing times (in seconds) of songs that were popular at the time of this writing. (The songs are by Timberlake, Furtado, Daughtry, Stefani, Fergie, Akon, Ludacris, Beyonce, Nickelback, Rihanna, Fray, Lavigne, Pink, Mims, Mumidee, and Omarion.) Use a 0.05 significance level to test the claim that the songs are from a population with a standard deviation less than one minute.

| 448 | 242 | 231 | 246 | 246 | 293 | 280 | 227 | 244 | 213 | 262 | 239 | 213 | 258 | 255 | 257 |

8-6   Beyond the Basics

21. Finding Critical Values of χ^2 For large numbers of degrees of freedom, we can approximate critical values of χ^2 as follows:

$$\chi^2 = \frac{1}{2}\left(z + \sqrt{2k - 1}\right)^2$$

Here k is the number of degrees of freedom and z is the critical value, found in Table A-2. For example, if we want to approximate the two critical values of χ^2 in a two-tailed hypothesis test with $\alpha = 0.01$ and a sample size of 100, we let $k = 99$ with $z = -2.575$ followed by $k = 99$ and $z = 2.575$. Use this approximation to estimate the critical values of χ^2 in a two-tailed hypothesis test with $n = 100$ and $\alpha = 0.01$. Use this approach to find the critical values for Exercise 16.

22. Finding Critical Values of χ^2 Repeat Exercise 21 using this approximation (with k and z as described in Exercise 21):

$$\chi^2 = k\left(1 - \frac{2}{9k} + z\sqrt{\frac{2}{9k}}\right)^3$$

Review

Two major activities of statistics are estimating population parameters (as with confidence intervals) and hypothesis testing. In this chapter we presented basic methods for testing claims about a population proportion, population mean, or population standard deviation (or variance).

In Section 8-2 we presented the fundamental components of a hypothesis test: null hypothesis, alternative hypothesis, test statistic, critical region, significance level, critical value, P-value, type I error, and type II error. We also discussed two-tailed tests, left-tailed tests, right-tailed tests, and the statement of conclusions. We used those components in identifying three different methods for testing hypotheses:

1. The P-value method (summarized in Figure 8-8)

2. The traditional method (summarized in Figure 8-9)

3. Confidence intervals (discussed in Chapter 7)

In Sections 8-3 through 8-6 we discussed specific methods for dealing with different parameters. Because it is so important to be correct in selecting the distribution and test statistic, we provide Table 8-3, which summarizes the hypothesis testing procedures of this chapter.

Table 8-3 Hypothesis Tests

Parameter	Requirements: Simple Random Sample and . . .	Distribution and Test Statistic	Critical and P-Values
Proportion	$np \geq 5$ and $nq \geq 5$	Normal: $z = \dfrac{\hat{p} - p}{\sqrt{\dfrac{pq}{n}}}$	Table A-2
Mean	σ known and normally distributed population or σ known and $n > 30$	Normal: $z = \dfrac{\bar{x} - \mu_{\bar{x}}}{\dfrac{\sigma}{\sqrt{n}}}$	Table A-2
	σ not known and normally distributed population or σ not known and $n > 30$	Student t: $t = \dfrac{\bar{x} - \mu_{\bar{x}}}{\dfrac{s}{\sqrt{n}}}$	Table A-3
	Population not normally distributed and $n \leq 30$	Use a nonparametric method or bootstrapping	
Standard Deviation or Variance	Population normally distributed	Chi-square: $\chi^2 = \dfrac{(n-1)s^2}{\sigma^2}$	Table A-4

Statistical Literacy and Critical Thinking

1. Interpreting *P*-Values Using 52 rainfall amounts for Sundays in Boston, a test of the claim that $\mu > 0$ in. results in a *P*-value of 0.0091. What does the *P*-value suggest about the claim? In general, what does the following memory aid suggest about the interpretation of *P*-values: "If the *P* is low, the null must go. If the *P* is high, the null will fly."

2. Practical Significance A very large simple random sample consists of the differences between the heights of the first-born male child and the second-born male child. With $n = 295{,}362$, $\bar{x} = 0.019$ in., and $s = 3.91$ in., a test of the claim that $\mu > 0$ in. results in a *P*-value of 0.0041. Is there statistical significance? Is there practical significance? Explain.

3. Voluntary Response Sample Some magazines and newspapers conduct polls in which the sample results are a voluntary response sample. What is a voluntary response sample? In general, can such a voluntary response sample be used with a hypothesis test to make a valid conclusion about the larger population?

4. Robust What does it mean when we say that a particular method of hypothesis testing is *robust* against departures from normality? Is the *t* test of a population mean robust against departures from normality? Is the χ^2 test of a population standard deviation robust against departures from normality?

Chapter Quick Quiz

1. Identify the null and alternative hypotheses resulting from the claim that the proportion of males is greater than 0.5. Express those hypotheses in symbolic form.

2. If a population has a normal distribution, which distribution is used to test the claim that $\mu < 98.6$, given a sample of 25 values with a sample mean of 98.2 and a sample standard deviation of 0.62? (normal, *t*, chi-square, binomial, uniform)

3. If a population has a normal distribution, which distribution is used to test the claim that a population has a standard deviation equal to 0.75, given a sample of 25 values with a sample mean of 98.2 and a sample standard deviation of 0.62? (normal, *t*, chi-square, binomial, uniform)

4. True or false: In hypothesis testing, it is never valid to form a conclusion of supporting the null hypothesis.

5. Find the *P*-value in a test of the claim that a population mean is equal to 100, given that the test statistic is $z = 1.50$.

6. Find the test statistic obtained when testing the claim that $p = 0.4$ when given sample data consisting of $x = 30$ successes among $n = 100$ trials.

7. Find the critical value(s) obtained when using a 0.05 significance level for testing the claim that $\mu = 100$ when given sample data consisting of $\bar{x} = 90$, $s = 10$, and $n = 20$.

8. Find the *P*-value obtained when testing the claim that $p = 0.75$ when given sample data resulting in a test statistic of $z = 1.20$.

9. What is the final conclusion obtained when testing the claim that $p > 0.25$, given that the *P*-value is 0.5555?

10. True or false: If correct methods of hypothesis testing are used with a large simple random sample, the conclusion will always be correct.

Review Exercises

TI **1. Rate of Smoking** A simple random sample of 1088 adults between the ages of 18 and 44 is conducted. It is found that 261 of the 1088 adults smoke (based on data from the National Health Interview Survey). Use a 0.05 significance level to test the claim that less than 1/4 of such adults smoke.

TI 2. Graduation Rate A simple random sample is conducted of 1486 college students who are seeking bachelor's degrees, and it includes 802 who earned bachelor's degrees within five years. Use a 0.01 significance level to test the claim that most college students earn bachelor's degrees within five years.

TI 3. Weights of Cars When planning for construction of a parkway, engineers must consider the weights of cars to be sure that the road surface is strong enough. A simple random sample of 32 cars yields a mean of 3605.3 lb and a standard deviation of 501.7 lb (based on Data Set 16 in Appendix B). Use a 0.01 significance level to test the claim that the mean weight of cars is less than 3700 lb. When considering weights of cars for the purpose of constructing a road that is strong enough, is it the mean that is most relevant? If not, what weight is most relevant?

TI 4. Weights of Cars Repeat Exercise 3 by assuming that weights of cars have a standard deviation known to be 520 lb.

TI 5. Herb Consumption Among 30,617 randomly selected adults, 5787 consumed herbs within the past 12 months (based on data from "Use of Herbs Among Adults Based on Evidence-Based Indications: Findings From the National Health Survey," by Bardia, et al., *Mayo Clinic Proceedings,* Vol. 82, No. 5). Use a 0.01 significance level to test the claim that fewer than 20% of adults consumed herbs within the past 12 months.

TI 6. Are Thinner Aluminum Cans Weaker? The axial load of an aluminum can is the maximum weight that the sides can support before collapsing. The axial load is an important measure, because the top lids are pressed onto the sides with pressures that vary between 158 lb and 165 lb. Pepsi experimented with thinner aluminum cans, and a random sample of 175 of the thinner cans has axial loads with a mean of 267.1 lb and a standard deviation of 22.1 lb. Use a 0.01 significance level to test the claim that the thinner cans have a mean axial load that is less than 281.8 lb, which is the mean axial load of the thicker cans that were then in use. Do the thinner cans appear to be strong enough so that they are not crushed when the top lids are pressed onto the sides?

TI 7. Random Generation of Data The TI-83/84 Plus calculator can be used to generate random data from a normally distributed population. The command **randNorm(74, 12.5, 100)** generates 100 values from a normally distributed population with $\mu = 74$ and $\sigma = 12.5$ (for pulse rates of women). One such generated sample of 100 values has a mean of 74.4 and a standard deviation of 11.7. Assume that σ is known to be 12.5 and use a 0.05 significance level to test the claim that the sample actually does come from a population with a mean equal to 74. Based on the results, does it appear that the calculator's random number generator is working correctly?

TI 8. Random Generation of Data Repeat Exercise 7 without making the assumption that the population standard deviation is known.

TI 9. Random Generation of Data Use the sample results from Exercise 7 to test the claim that the generated values are from a population with a standard deviation equal to 12.5. Use a 0.05 significance level.

TI 10. Weights of Cars A simple random sample of 32 cars yields a mean weight of 3605.3 lb, a standard deviation of 501.7 lb, and the sample weights appear to be from a normally distributed population (based on Data Set 16 in Appendix B). Use a 0.01 significance level to test the claim that the standard deviation of the weights of cars is less than 520 lb.

Cumulative Review Exercises

TI 1. Olympic Winners Listed below are the winning times (in seconds) of women in the 100-meter dash for consecutive summer Olympic games, listed in order by year. Assume that the times are sample data from a larger population. Find the values of the indicated statistics.

| 11.07 | 11.08 | 11.06 | 10.97 | 10.54 | 10.82 | 10.94 | 10.75 | 10.93 |

a. Mean.   **b.** Median.   **c.** Standard deviation.   **d.** Variance.   **e.** Range.

2. Olympic Winners Exercise 1 lists the winning times (in seconds) of women in the 100-meter dash for consecutive summer Olympic games.

a. What is the level of measurement of the data? (nominal, ordinal, interval, ratio)

b. Are the values discrete or continuous?

c. Do the values constitute a simple random sample?

d. What important characteristic of the data is not considered when finding the sample statistics indicated in Exercise 1?

e. Which of the following graphs is most helpful in understanding important characteristics of the data: stemplot, boxplot, histogram, pie chart, time series plot, Pareto chart?

3. Confidence Interval for Olympic Winners Use the sample values given in Exercise 1 to construct a 95% confidence interval estimate of the population mean. Assume that the population has a normal distribution. Can the result be used to estimate winning times in the future? Why or why not?

4. Hypothesis Test for Olympic Winners Use the sample values given in Exercise 1 to test the claim that the mean winning time is less than 11 sec. Use a 0.05 significance level. What can we conclude about winning times in the future?

5. Histogram Minitab is used to construct a histogram of the weights of a simple random sample of M&M candies, and the result is shown on the next page.

a. Does the sample appear to be from a population with a normal distribution?

b. How many sample values are represented in the histogram?

c. What is the class width used in the histogram?

d. Use the histogram to estimate the mean weight.

e. Can the histogram be used to identify the exact values in the original list of sample data?

MINITAB

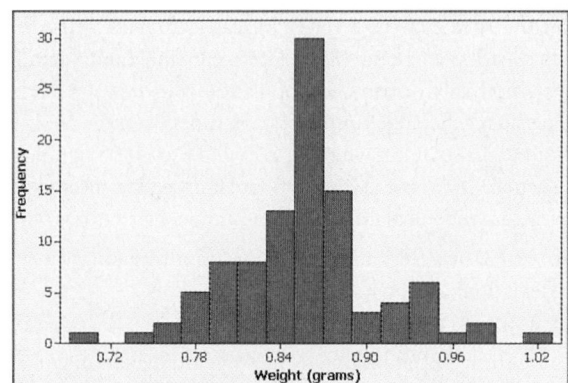

6. Histogram Minitab is used to generate a histogram of the outcomes from 100 rolls of a die. What is wrong with the graph?

MINITAB

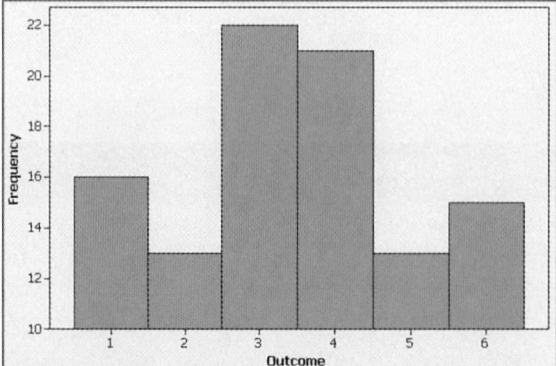

7. Frequency Distribution Refer to the histogram provided for Exercise 6 and construct the frequency distribution that summarizes the outcomes. Then find the mean of the 100 outcomes.

8. Probability in Hypothesis Tests A hypothesis test is conducted with a 0.05 significance level, so that there is a 0.05 probability of making the mistake of rejecting a true null hypothesis. If two different independent hypothesis tests are conducted with a 0.05 significance level, what is the probability that both conclusions are mistakes of rejecting a true null hypothesis?

9. Sitting Eye Heights When designing a movie theater with stadium seating, engineers decide to consider the sitting eye heights of women. Those heights have a mean of 739 mm and a standard deviation of 33 mm and they are normally distributed (based on anthropometric survey data from Gordon, Churchill, Clauser).

a. For a randomly selected woman, what is the *probability* that she has a sitting eye height less than 700 mm?

b. What *percentage* of women have a sitting eye height greater than 750 mm?

c. For 50 randomly selected women, what is the probability that their mean sitting height is less than 730 mm?

d. For sitting heights of women, find the value of P_{90}, which is the 90th percentile.

10. Sitting Eye Heights: Testing for Normality Listed below is a simple random sample of sitting eye heights (in mm) of men (based on anthropometric survey data from Gordon, Churchill, Clauser). Determine whether these sample heights appear to come from a population with a normal distribution. Explain.

773 771 821 815 765 811 764 761 778 838 801 808 778 803 740 761 734 803 844 790

TI-83/84 Plus Calculator Project

Is the mean weight of discarded plastic less than 2 lb?

This project involves a large data set and a simulation method as a different way to test hypotheses.

a. Using the sample weights of discarded plastic from Data Set 22 in Appendix B, test the claim that the mean weight of plastic discarded in a week is less than 2 lb. Use a 0.05 significance level.

b. The basic idea underlying a hypothesis test is the rare event rule for inferential statistics first introduced in Chapter 4. Using that rule, we need to determine whether a sample mean is "unusual" or if it can easily occur by chance. Make that determination using simulations. Repeatedly generate samples of 62 weights from a normally distributed population having the assumed mean of 2 lb (as in the null hypothesis). (For the standard deviation, use $s = 1.06548$ found from Data Set 22.) Then, based on the sample means that are found, determine whether a sample mean such as the one found from Data Set 22 is "unusual" or can easily occur by chance. If a sample mean such as the one found from Data Set 22 (or a lower sample mean) is found in 5% or fewer of the simulated samples, conclude that the sample mean from Data Set 22 cannot easily occur by chance, so there is sufficient evidence to support the claim that the mean is less than 2 lb. Generate enough samples to be reasonably confident in determining that a sample mean such as the one obtained from Data Set 22 (or any lower sample mean) is unusual or can easily occur by chance. Record all of your results.

For part (b), use the TI-83/84 Plus calculator to repeat the process of randomly generating data from a normally distributed population with a mean of 2 and a standard deviation of 1.06548. Here is the procedure:

1. First clear list L1 by pressing **STAT**, then select **ClrList** and enter L1 and press **ENTER**.

2. Now press **MATH**, then select **PRB** and select the menu item of **randNorm(** and press **ENTER**.

3. Proceed to enter the mean, standard deviation, and sample size. To generate 62 weights from a normally distributed population with a mean of 2 and a standard deviation of 1.06548, enter **randNorm(2, 1.06548, 62).** Press **ENTER**.

4. Store the 62 simulated weights in list L1. (This can be accomplished by pressing **STO)**  **2ND** **1** **ENTER**.)

5. Now find the mean of the sample values in L1 as follows: Press **STAT**, select **CALC**, select the first menu item of **1-VAR Stats,** press **ENTER**, enter L1, then press **ENTER**.

Repeat these five steps often enough to determine whether it is "unusual" (with a probability of 0.05 or less) to obtain the sample mean weight of plastic from Data Set 22 (or any lower sample mean). After making that determination, what do you conclude about the claim that the mean weight of discarded plastic is less than 2 lb? Write a summary of your results and conclusions.

INTERNET PROJECT

Hypothesis Testing

Go to: **http://www.aw.com/triola**

This chapter introduced the methodology behind hypothesis testing, an essential technique of inferential statistics. This Internet Project will have you conduct tests using a variety of data sets in different areas of study. For each subject, you will be asked to

- collect data available on the Internet.
- formulate a null and alternative hypothesis based on a given question.

- conduct a hypothesis test at a specified level of significance.
- summarize your conclusion.

Locate the Internet Project for this chapter. There you will find guided investigations in the fields of education, economics, and sports, and a classical example from the physical sciences.

APPLET PROJECT

The CD included with this book contains applets designed to help visualize various concepts. Open the Applets folder on the CD and double-click on **Start**. Select the menu item of **Hypothesis test for a proportion.**

Conduct the simulations based on the hypothesis test in Examples 1 and 2 from Section 8-3. Write a brief explanation of what the applet does, and what the results show. Include a printout of the results.

FROM DATA TO DECISION

Critical Thinking: Analyzing poll results

On the day that this project was written, AOL conducted an Internet poll and asked users to respond to this question: "Where do you live?" If a user chose to respond, he or she could choose either "urban area" or "rural area." Results included 51,318 responses of "urban area" and 37,888 responses of "rural area."

Analyzing the Results

a. Use the given poll results with a 0.01 significance level to test the claim that most people live in urban areas.

b. Which of the following best describes the sample: Convenience, simple random sample, random sample,

voluntary response sample, stratified sample, cluster sample?

c. Considering the sampling method used, does the hypothesis test from Part (a) appear to be valid?

d. Given that the size of the sample is extremely large, can this sample size compensate for a poor method of sampling?

e. What valid conclusion can be made from the sample results? Based on the sample results, can we conclude anything about the whole population?

Cooperative Group Activities

1. In-class activity Without using any measuring device, each student should draw a line believed to be 3 in. long and another line believed to be 3 cm long. Then use rulers to measure and record the lengths of the lines drawn. Find the means and standard deviations of the two sets of lengths. Test the claim that the lines estimated to be 3 in. have a mean length that is equal to 3 in. Test the claim that the lines estimated to be 3 cm have a mean length that is equal to 3 cm. Compare the results. Do the estimates of the 3-in. line appear to be more accurate than those for the 3-cm line?

2. In-class activity Assume that a method of gender selection can affect the probability of a baby being a girl, so that the probability becomes 1/4. Each student should simulate 20 births by drawing 20 cards from a shuffled deck. Replace each card after it has been drawn, then reshuffle. Consider the hearts to be girls and consider all other cards to be boys. After making 20 selections and recording the "genders" of the babies, use a 0.10 significance level to test the claim that the proportion of girls is equal to 1/4. How many students are expected to get results leading to the wrong conclusion that the proportion is not 1/4? How does that relate to the probability of a type I error? Does this procedure appear to be effective in identifying the effectiveness of the gender-selection method? (If decks of cards are not available, use some other way to simulate the births, such as using the random number generator on a calculator or using digits from phone numbers or social security numbers.)

3. Out-of-class activity Groups of three or four students should go to the library and collect a sample consisting of the ages of books (based on copyright dates). Plan and describe the sampling plan, execute the sampling procedure, then use the results to estimate the mean of the ages of all books in the library.

4. In-class activity Each student should write an estimate of the age of the current President of the United States. All estimates should be collected and the sample mean and standard deviation should be calculated. Then test the hypothesis that the mean of all such estimates is equal to the actual current age of the President.

5. In-class activity A class project should be designed to conduct a test in which each student is given a taste of Coke and a taste of Pepsi. The student is then asked to identify which sample is Coke. After all of the results are collected, test the claim that the success rate is better than the rate that would be expected with random guesses.

6. In-class activity Each student should estimate the length of the classroom. The values should be based on visual estimates, with no actual measurements being taken. After the estimates have been collected, measure the length of the room, then test the claim that the sample mean is equal to the actual length of the classroom. Is there a "collective wisdom," whereby the class mean is approximately equal to the actual room length?

7. Out-of-class activity Using a wristwatch that is reasonably accurate, set the time to be exact. Use a radio station or telephone time report which states that "at the tone, the time is . . . " If you cannot set the time to the nearest second, record the error for the watch you are using. Now compare the time on your watch to the time on others. Record the errors with positive signs for watches that are ahead of the actual time and negative signs for those watches that are behind the actual time. Use the data to test the claim that the mean error of all wristwatches is equal to 0. Do we collectively run on time, or are we early or late? Also test the claim that the standard deviation of errors is less than 1 min. What are the practical implications of a standard deviation that is excessively large?

8. In-class activity In a group of three or four people, conduct an ESP experiment by selecting one of the group members as the subject. Draw a circle on one small piece of paper and draw a square on another sheet of the same size. Repeat this experiment 20 times: Randomly select the circle or the square and place it in the subject's hand behind his or her back so that it cannot be seen, then ask the subject to identify the shape (without looking at it); record whether the response is correct. Test the claim that the subject has ESP because the proportion of correct responses is greater than 0.5.

9. In-class activity After dividing into groups with sizes between 10 and 20 people, each group member should record the number of heartbeats in a minute. After calculating $\bar{x}$ and s, each group should proceed to test the claim that the mean is greater than 60, which is the author's result. (When people exercise, they tend to have lower pulse rates, and the author runs five miles a few times each week. What a guy.)

10. Out-of-class activity As part of a Gallup poll, subjects were asked "Are you in favor of the death penalty for persons convicted of murder?" Sixty-five percent of the respondents said that they were in favor, while 27% were against and 8% had no opinion. Use the methods of Section 7-2 to determine the sample size necessary to estimate the proportion of students at your college who are in favor. The class should agree on a confidence level and margin of error. Then divide the sample size by the number of students in the class, and conduct the survey by having each class member ask the appropriate number of students at the college. Analyze the results to determine whether the students differ significantly from the results found in the Gallup poll.

11. Out-of-class activity Each student should find an article in a professional journal that includes a hypothesis test of the type discussed in this chapter. Write a brief report describing the hypothesis test and its role in the context of the article.

12. In-class activity Adult males have sitting heights with a mean of 91.4 cm and a standard deviation of 3.6 cm, and adult females have sitting heights with a mean of 85.2 cm and a standard deviation of 3.5 cm (based on anthropometric survey data from Gordon, Churchill, et al.). Collect sample data in class by measuring sitting heights and conduct appropriate hypothesis tests to determine whether there are significant differences from the population parameters.

STATISTICS FOR CONSUMERS

NAME:	Michael Saccucci
JOB:	Director of Statistics and Quality Management
COMPANY:	Consumers Union

Michael Saccucci is Director of Statistics and Quality Management for Consumers Union, which tests products and services and provides ratings and recommendations to consumers in Consumer Reports *magazine. The author met with Mike Saccucci and toured the product-testing facilities. He was extremely impressed with the involvement of statisticians, the extreme care taken in the design of experiments, and the careful and effective use of statistical analyses of test results.*

Q: What statistical concepts and procedures do you use at Consumers Union?

A: We use any number of statistical procedures, many of which are discussed in this text. For example, in a study to evaluate the quality and safety of chicken, we developed a complex sampling scheme so that the different manufacturers were fairly represented. In a study of sunscreens, we used the normal distribution to help determine the appropriate number of replicates necessary to fairly evaluate the products. Depending on the type of test, we might have to construct a completely randomized design, a randomized block design, or some other type of experimental design to ensure that our results are accurate and unbiased. During the analysis phase, we might use any number of techniques, such as analysis of variance, regression, time series, categorical, or nonparametric analysis.

I think I have one of the most interesting jobs. I never know what to expect on a given day. One day I may be sitting in on the training session for wine tasting to learn about the testing procedures. On another day, I may be discussing various ways to test paints. On most days, though, I spend a significant amount of time using a computer to help design an upcoming study or to sift through large amounts of data that will ultimately be used as the basis for product ratings.

Q: What steps do you take to ensure objectivity in your testing procedures?

A: It is Consumers Union's policy that all tests must be performed in an objective, scientific manner, and with due regard for the safety of test personnel. We go to great lengths to adhere to this policy. For example, we don't accept any type of outside advertising in our publications. We employ anonymous shoppers located throughout the United States to purchase our test samples in ways normally available to consumers. We don't accept free samples from anyone, including vendors. And we don't test unsolicited samples sent from a manufacturer. In addition, technicians use randomized experimental designs to ensure that our testing is done with scientific integrity and objectivity. When practical, tested items are blind coded so that the testers do not know which brands they are evaluating.

Q: Are the ratings and recommendations in *Consumer Reports* magazine based on statistical significance alone?

A: No. The information we provide must be useful to consumers. Our technicians perform a variety of tests to evaluate a product's performance. These tests are designed to simulate conditions of predictable consumer usage. If it turns out that there is a statistically significant, but not meaningful difference in test results, we would not rate one brand over another. In testing water sealants, for example, we might find that there is a statistically significant difference between the amounts of water that seep through two different brands of sealant. However, if the difference amounts to a few drops of water, we would rate the products similarly for that attribute.

Q: Do you feel that job applicants are viewed more favorably if they have studied some statistics?

A: Given the level of innumeracy that exists today, I believe that a basic knowledge of statistics would be viewed favorably for just about any field of study. This would be especially true of quantitative fields, such as the sciences, engineering, and business. It is extremely important for everyone to have an understanding of statistics in order to effectively process the huge amounts of information that we are presented with each day in our professional and personal lives. A focus on statistical thinking would be especially useful.

463

9-1 Review and Preview

9-2 Inferences About Two Proportions

9-3 Inferences About Two Means: Independent Samples

9-4 Inferences from Dependent Samples

9-5 Comparing Variation in Two Samples

9

Inferences from Two Samples

Is the "Freshman 15" real, or is it a myth?

There is a popular belief that college students typically gain 15 lb (or 6.8 kg) during their freshman year. This 15 lb weight gain has been deemed the "Freshman 15." Reasonable explanations for this phenomenon include the new stresses of college life (not including a statistics class, which is just plain fun), new eating habits, increased levels of alcohol consumption, less free time for physical activities, cafeteria food with an abundance of fat and carbohydrates, the new freedom to choose among a variety of foods (including sumptuous pizzas that are just a phone call away), and a lack of sleep that results in lower levels of leptin, which helps regulate appetite and metabolism. But is the Freshman 15 real, or is it a myth that has been perpetuated through anecdotal evidence and/or flawed data?

Several studies have focused on the credibility of the Freshman 15 belief. We will consider results from one reputable study with results published in the article "Changes in Body Weight and Fat Mass of Men and Women in the First Year of College: A Study of the 'Freshman 15'," by Daniel Hoffman, Peggy Policastro, Virginia Quick, and Soo-Kyung Lee, *Journal of American College Health,* Vol. 55, No. 1. The authors of that article have provided the data from their study, and much of it is listed in Data Set 3 in Appendix B. If you examine the weights in Data Set 3, you should note the following:

- The weights in Data Set 3 are in *kilograms,* not pounds, and 15 lb is equivalent to 6.8 kg. The "Freshman 15 (pounds)" is equivalent to the "Freshman 6.8 kilograms."

- Data Set 3 includes two weights for each of the 67 study subjects. Each subject was weighed in September of the freshman year, and again in April of the freshman year. These two measurements were made at the beginning and end of the seven months of campus life that passed between the measurements. It is important to recognize that each individual pair of before and after measurements is from the same student, so the lists of 67 before weights and 67 after weights constitute *paired* data from the 67 subjects in the study.

- Because the "Freshman 15" refers to weight gained, we will use weight changes in this format:

  (April weight) − (September weight)

  If a student does gain 15 lb, the value of (April weight) − (September weight) is 15 lb, or 6.8 kg. (A negative weight "gain" indicates that the student lost weight.)

- The published article about the Freshman 15 study includes some limitations, including these:

  1. All subjects volunteered for the study.
  2. All of the subjects were attending Rutgers, the State University of New Jersey.

The "Freshman 15" constitutes a *claim* made about the population of college students. If we use μ_d to denote the mean of the (April weight) − (September weight) differences for college students during their freshman year, the "Freshman 15" is the claim that $\mu_d = 15$ lb or $\mu_d = 6.8$ kg. Because the sample weights are measured in kilograms, we will consider the claim to be $\mu_d = 6.8$ kg. Later in this chapter, a formal hypothesis test will be used to test this claim. We will then be able to reach one of two possible conclusions: Either there is sufficient evidence to warrant rejection of the claim that $\mu_d = 6.8$ kg (so the "Freshman 15" is rejected), or we will conclude that there is not sufficient evidence to warrant rejection of the claim that $\mu_d = 6.8$ kg (so the "Freshman 15" cannot be rejected). We will then be able to determine whether or not the Freshman 15 is a myth.

9-1   Review and Preview

In Chapters 7 and 8 we introduced methods of *inferential statistics*. In Chapter 7 we presented methods of constructing confidence interval estimates of population parameters. In Chapter 8 we presented methods of testing claims made about population parameters. Chapters 7 and 8 both involved methods for dealing with a sample from a single population. The objective of this chapter is to extend the methods for *estimating* values of population parameters and the methods for *testing hypotheses* to situations involving *two* sets of sample data instead of just one. The following are examples typical of those found in this chapter, which presents methods for using sample data from two populations so that inferences can be made about those populations.

- Test the claim that when college students are weighed at the beginning and end of their freshman year, the differences show a mean weight gain of 15 pounds (as in the "Freshman 15" belief).

- Test the claim that the proportion of children who contract polio is less for children given the Salk vaccine than for children given a placebo.

- Test the claim that subjects treated with Lipitor have a mean cholesterol level that is lower than the mean cholesterol level for subjects given a placebo.

Because there are many studies involving a comparison of *two* samples, the methods of this chapter apply to a wide variety of real situations.

9-2   Inferences About Two Proportions

Key Concept In this section we present methods for (1) testing a claim made about the two population proportions and (2) constructing a confidence interval estimate of the difference between the two population proportions. This section is based on proportions, but we can use the same methods for dealing with probabilities or the decimal equivalents of percentages.

Objectives

Test a claim about two population proportions or construct a confidence interval estimate of the difference between two population proportions.

Notation for Two Proportions

For population 1 we let

$p_1 = $ *population* proportion

$n_1 = $ size of the sample

$x_1 = $ number of successes in the sample

$\hat{p}_1 = \dfrac{x_1}{n_1}$ (*sample* proportion)

$\hat{q}_1 = 1 - \hat{p}_1$ (complement of $\hat{p}_1$)

The corresponding notations p_2, n_2, x_2, $\hat{p}_2$, and $\hat{q}_2$ apply to population 2.

Pooled Sample Proportion

The **pooled sample proportion** is denoted by $\bar{p}$ and is given by:

$$\bar{p} = \frac{x_1 + x_2}{n_1 + n_2}$$

$$\bar{q} = 1 - \bar{p}.$$

Requirements

1. The sample proportions are from two simple random samples that are *independent*. (Samples are *independent* if the sample values selected from one population are not related to or somehow naturally paired or matched with the sample values selected from the other population.)

2. For each of the two samples, the number of successes is at least 5 and the number of failures is at least 5. (That is, $np \geq 5$ and $nq \geq 5$ for each of the two samples).

Test Statistic for Two Proportions (with H_0: $p_1 = p_2$)

$$z = \frac{(\hat{p}_1 - \hat{p}_2) - (p_1 - p_2)}{\sqrt{\dfrac{\bar{p}\,\bar{q}}{n_1} + \dfrac{\bar{p}\,\bar{q}}{n_2}}} \qquad \text{where } p_1 - p_2 = 0 \text{ (assumed in the null hypothesis)}$$

$$\hat{p}_1 = \frac{x_1}{n_1} \qquad \text{and} \qquad \hat{p}_2 = \frac{x_2}{n_2} \qquad \text{(sample proportions)}$$

$$\bar{p} = \frac{x_1 + x_2}{n_1 + n_2} \qquad (\textit{pooled} \text{ sample proportion}) \qquad \text{and} \qquad \bar{q} = 1 - \bar{p}$$

P-value: Use Table A-2. (Use the computed value of the test statistic z and find the P-value by following the procedure summarized in Figure 8-5.)

Critical values: Use Table A-2. (Based on the significance level α, find critical values by using the same procedures introduced in Section 8-2.)

Confidence Interval Estimate of $p_1 - p_2$

The confidence interval estimate of the difference $p_1 - p_2$ is:

$$(\hat{p}_1 - \hat{p}_2) - E < (p_1 - p_2) < (\hat{p}_1 - \hat{p}_2) + E$$

where the margin of error E is given by $E = z_{\alpha/2}\sqrt{\dfrac{\hat{p}_1\hat{q}_1}{n_1} + \dfrac{\hat{p}_2\hat{q}_2}{n_2}}$

Rounding: Round the confidence interval limits to three significant digits.

    TI-83/84 Plus Calculator The TI-83/84 Plus calculator can automatically provide the test statistic and P-value for a hypothesis test, and it can also generate a confidence interval. See the procedures at the end of this section.

Hypothesis Tests

For tests of hypotheses made about two population proportions, we consider only tests having a null hypothesis of $p_1 = p_2$. (For claims that the difference between p_1 and p_2 is equal to a nonzero constant, see Exercise 39.) The following example will help clarify the roles of x_1, n_1, $\hat{p}_1$, $\bar{p}$, and so on. Note that under the assumption of equal proportions, the best estimate of the common proportion is obtained by pooling both samples into one big sample, so that $\bar{p}$ is the estimator of the common population proportion.

EXAMPLE 1 **Do Airbags Save Lives?** The table below lists results from a simple random sample of front-seat occupants involved in car crashes (based on data from "Who Wants Airbags?" by Meyer and Finney, *Chance*, Vol. 18, No. 2). Use a 0.05 significance level to test the claim that the fatality rate of occupants is lower for those in cars equipped with airbags.

	Airbag Available	No Airbag Available
Occupant Fatalities	41	52
Total number of occupants	11,541	9,853

SOLUTION **REQUIREMENTS CHECK** We first verify that the two necessary requirements are satisfied. (1) The data are from two simple random samples, and the two samples are independent of each other. (2) The airbag group includes 41 occupants who were killed and 11,500 occupants who were not killed, so the number of successes is at least 5 and the number of failures is at least 5. The second group includes 52 occupants who were killed and 9801 who were not killed, so the number of successes is at least 5 and the number of failures is at least 5. The requirements are satisfied.

We will use the *P*-value method of hypothesis testing, as summarized in Figure 8-8. In the following steps we stipulate that the group with airbags is Sample 1, and the group without airbags is Sample 2.

Step 1: The claim that the fatality rate is lower for those with airbags can be expressed as $p_1 < p_2$.

Step 2: If $p_1 < p_2$ is false, then $p_1 \geq p_2$.

Step 3: Because the claim of $p_1 < p_2$ does not contain equality, it becomes the alternative hypothesis. The null hypothesis is the statement of equality, so we have

$$H_0: p_1 = p_2 \qquad H_1: p_1 < p_2 \text{ (original claim)}$$

Step 4: The significance level is $\alpha = 0.05$.

Step 5: We will use the normal distribution (with the test statistic given earlier in this section) as an approximation to the binomial distribution. We estimate the common value of p_1 and p_2 with the pooled sample estimate $\bar{p}$ calculated as shown below, with extra decimal places used to minimize rounding errors in later calculations.

$$\bar{p} = \frac{x_1 + x_2}{n_1 + n_2} = \frac{41 + 52}{11,541 + 9,853} = 0.004347$$

With $\bar{p} = 0.004347$, it follows that $\bar{q} = 1 - 0.004347 = 0.995653$.

Step 6: We can now find the value of the test statistic:

$$z = \frac{(\hat{p}_1 - \hat{p}_2) - (p_1 - p_2)}{\sqrt{\dfrac{\bar{p}\,\bar{q}}{n_1} + \dfrac{\bar{p}\,\bar{q}}{n_2}}}$$

$$= \frac{\left(\dfrac{41}{11,541} - \dfrac{52}{9,853}\right) - 0}{\sqrt{\dfrac{(0.004347)(0.995653)}{11,541} + \dfrac{(0.004347)(0.995653)}{9,853}}}$$

$$= -1.91$$

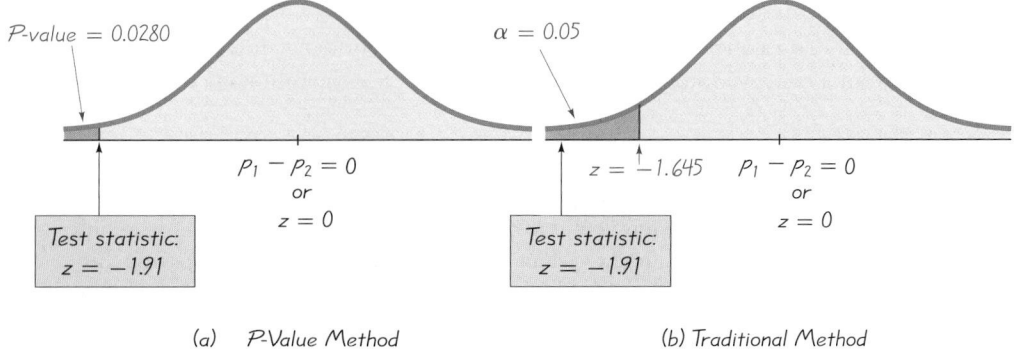

Figure 9-1   **Testing the Claim of a Lower Fatality Rate With Airbags**

TI-83/84 Plus Using a TI-83/84 Plus calculator with the instructions given at the end of this section, we obtain the display shown in the margin. The results include the test statistic $z = -1.91164305$ and the P-value of 0.0279609289.

The P-value in this left-tailed test could also be found by using the TI-83/84 Plus calculator or Table A-2 for identifying the area to the left of the test statistic as shown in Figure 9-1 (a).

Step 7: Because the P-value of 0.0280 is less than the significance level of $\alpha = 0.05$, we reject the null hypothesis of $p_1 = p_2$.

TI-83/84 Plus

```
2-PropZTest
 p1<p2
 z=-1.91164305
 P=.0279609289
 p̂1=.0035525518
 p̂2=.0052775804
↓p̂=.0043470132
```

INTERPRETATION   We must address the original claim that the fatality rate is lower for occupants in cars equipped with airbags. Because we reject the null hypothesis, we conclude that there is sufficient evidence to support the claim that the proportion of accident fatalities for occupants in cars with airbags is less than the proportion of fatalities for occupants in cars without airbags. (See Figure 8-7 for help in wording the final conclusion.) Based on these results, it appears that airbags are effective in saving lives.

The sample data used in this example are only part of the data given in the article cited in the statement of the problem. If all of the available data are used, the test statistic becomes $z = -57.76$, and the P-value is very close to 0, so using all of the data provides even more compelling evidence of the effectiveness of airbags in saving lives.

Traditional Method of Testing Hypotheses

The traditional approach can also be used for Example 1. In Step 6, instead of finding the P-value, find the critical value. With a significance level of $\alpha = 0.05$ in a left-tailed test based on the normal distribution, we use a TI-83/84 Plus calculator or refer to Table A-2 and find that an area of $\alpha = 0.05$ in the left tail corresponds to the critical value of $z = -1.645$. See Figure 9-1(b) where we can see that the test statistic of $z = -1.91$ does fall in the critical region bounded by the critical value of $z = -1.645$. We again reject the null hypothesis. The conclusions are the same as in Example 1.

Confidence Intervals

Using the format given earlier in this section, we can construct a confidence interval estimate of the difference between population proportions ($p_1 - p_2$). If a confidence interval estimate of $p_1 - p_2$ does not include 0, we have evidence suggesting that p_1 and p_2 have different values. The confidence interval uses a standard deviation based on estimated values of the population proportions, whereas a hypothesis test uses a

Polio Experiment

In 1954 an experiment was conducted to test the effectiveness of the Salk vaccine as protection against the devastating effects of polio. Approximately 200,000 children were injected with an ineffective salt solution, and 200,000 other children were injected with the vaccine. The experiment was "double blind" because the children being injected didn't know whether they were given the real vaccine or the placebo, and the doctors giving the injections and evaluating the results didn't know either. Only 33 of the 200,000 vaccinated children later developed paralytic polio, whereas 115 of the 200,000 injected with the salt solution later developed paralytic polio. Statistical analysis of these and other results led to the conclusion that the Salk vaccine was indeed effective against paralytic polio.

TI-83/84 Plus

```
2-PropZInt
 (-.0032,-2E-4)
 p̂1=.0035525518
 p̂2=.0052775804
 n1=11541
 n2=9853
```

standard deviation based on the assumption that the two population proportions are equal. Consequently, a conclusion based on a confidence interval might be different from a conclusion based on a hypothesis test. See the following caution.

CAUTION

When testing a claim about two population proportions, the *P*-value method and the traditional method are equivalent, but they are *not* equivalent to the confidence interval method. If you want to test a claim about two population proportions, use the *P*-value method or traditional method; if you want to estimate the difference between two population proportions, use a confidence interval.

Also, *don't test for equality of two population proportions by determining whether there is an overlap between two individual confidence interval estimates of the two individual population proportions.* When compared to the confidence interval estimate of $p_1 - p_2$, the analysis of overlap between two individual confidence intervals is more conservative (by rejecting equality less often), and it has less power (because it is less likely to reject $p_1 = p_2$ when in reality $p_1 \neq p_2$). (See "On Judging the Significance of Differences by Examining the Overlap Between Confidence Intervals," by Schenker and Gentleman, *American Statistician,* Vol. 55, No. 3.) See Exercise 37.

> **EXAMPLE 2** **Confidence Interval for Airbags** Use the sample data given in Example 1 to construct a 90% confidence interval estimate of the difference between the two population proportions. (As shown in Table 8-2 on page 412, the confidence level of 90% is comparable to the significance level of $\alpha = 0.05$ used in the preceding left-tailed hypothesis test.) What does the result suggest about the effectiveness of airbags in an accident?

> **SOLUTION** **REQUIREMENTS CHECK** We are using the same data from Example 1, and the same requirement check applies here. So, the requirements are satisfied.

With a 90% confidence level, $z_{\alpha/2} = 1.645$ (from Table A-2). We first calculate the value of the margin of error *E* as shown.

$$E = z_{\alpha/2}\sqrt{\frac{\hat{p}_1\hat{q}_1}{n_1} + \frac{\hat{p}_2\hat{q}_2}{n_2}} = 1.645\sqrt{\frac{\left(\frac{41}{11{,}541}\right)\left(\frac{11{,}500}{11{,}541}\right)}{11{,}541} + \frac{\left(\frac{52}{9{,}853}\right)\left(\frac{9801}{9{,}853}\right)}{9{,}853}}$$

$$= 0.001507$$

With $\hat{p}_1 = 41/11{,}541 = 0.003553$, $\hat{p}_2 = 52/9{,}853 = 0.005278$, and $E = 0.001507$, the confidence interval is evaluated as follows, with the confidence interval limits rounded to three significant digits:

$$(\hat{p}_1 - \hat{p}_2) - E < (p_1 - p_2) < (\hat{p}_1 - \hat{p}_2) + E$$
$$(0.003553 - 0.005278) - 0.001507 < (p_1 - p_2) < (0.003553 - 0.005278) + 0.001507$$
$$-0.00323 < (p_1 - p_2) < -0.000218$$

TI-83/84 Plus Using a TI-83/84 Plus calculator with the instructions given at the end of this section, we obtain the display shown in the margin. The results show that the confidence interval is $(-0.0032, -0.0002)$, where the value of -0.0002 is the value of $^-2\text{E}^-4$ in scientific notation. That confidence interval can also be expressed in the format of $-0.0032 < (p_1 - p_2) < -0.0002$. By automatically generating the confidence interval limits, the TI-83/84 Plus calculator saves us much effort and gives us more time to focus on the interpretation of the results.

> INTERPRETATION The confidence interval limits do not contain 0, implying that there is a significant difference between the two proportions. The confidence interval suggests that the fatality rate is lower for occupants in cars with air bags than for occupants in cars without air bags. The confidence interval also provides an estimate of the amount of the difference between the two fatality rates.

Rationale: Why Do the Procedures of This Section Work? The test statistic given for hypothesis tests is justified by the following:

With $n_1 p_1 \geq 5$ and $n_1 q_1 \geq 5$, the distribution of $\hat{p}_1$ can be approximated by a normal distribution with mean p_1, standard deviation $\sqrt{p_1 q_1 / n_1}$, and variance $p_1 q_1 / n_1$ (based on Sections 6-6 and 7-2). They also apply to the second sample. Because $\hat{p}_1$ and $\hat{p}_2$ are each approximated by a normal distribution, the difference $\hat{p}_1 - \hat{p}_2$ will also be approximated by a normal distribution with mean $p_1 - p_2$ and variance

$$\sigma^2_{(\hat{p}_1 - \hat{p}_2)} = \sigma^2_{\hat{p}_1} + \sigma^2_{\hat{p}_2} = \frac{p_1 q_1}{n_1} + \frac{p_2 q_2}{n_2}$$

(The above result is based on this property: The variance of the *differences* between two independent random variables is the *sum* of their individual variances.) The pooled estimate of the common value of p_1 and p_2 is $\bar{p} = (x_1 + x_2)/(n_1 + n_2)$. If we replace p_1 and p_2 by $\bar{p}$ and replace q_1 and q_2 by $\bar{q} = 1 - \bar{p}$, the above variance leads to the following standard deviation:

$$\sigma_{(\hat{p}_1 - \hat{p}_2)} = \sqrt{\frac{\bar{p}\,\bar{q}}{n_1} + \frac{\bar{p}\,\bar{q}}{n_2}}$$

We now know that the distribution of $p_1 - p_2$ is approximately normal, with mean $p_1 - p_2$ and standard deviation as shown above, so the z test statistic has the form given earlier.

The form of the confidence interval requires an expression for the variance different from the one given above. When constructing a confidence interval estimate of the difference between two proportions, we don't assume that the two proportions are equal, and we estimate the standard deviation as

$$\sqrt{\frac{\hat{p}_1 \hat{q}_1}{n_1} + \frac{\hat{p}_2 \hat{q}_2}{n_2}}$$

In the test statistic

$$z = \frac{(p_1 - p_2) - (p_1 - p_2)}{\sqrt{\dfrac{\hat{p}_1 \hat{q}_1}{n_1} + \dfrac{\hat{p}_2 \hat{q}_2}{n_2}}}$$

use the positive and negative values of z (for two tails) and solve for $p_1 - p_2$. The results are the limits of the confidence interval given earlier.

Using a TI-83/84 Plus Calculator for Inferences About Two Proportions

Requirements: When using a TI-83/84 Plus calculator for inferences about two population proportions, you must know the sample size n and the number of successes x for the samples from each of the two populations. That is, you must know the values of x_1, n_1, x_2, and n_2. For either sample, if you do not know the value of x, but you know the value of the sample size n and the sample proportion $\hat{p}$, calculate the value of x as follows: $x = n \cdot \hat{p}$, then round the result to the nearest whole number. For example, if we know that $n = 200$ and the success rate is 45%, then x is 45% of 200, which is $0.45 \times 200 = 90$.

Death Penalty as Deterrent

A common argument supporting the death penalty is that it discourages others from committing murder. Jeffrey Grogger of the University of California analyzed daily homicide data in California for a four-year period during which executions were frequent. Among his conclusions published in the *Journal of the American Statistical Association* (Vol. 85, No. 410): "The analyses conducted consistently indicate that these data provide no support for the hypothesis that executions deter murder in the short term." This is a major social policy issue, and the efforts of people such as Professor Grogger help to dispel misconceptions so that we have accurate information with which to address such issues.

Entries for Example 1

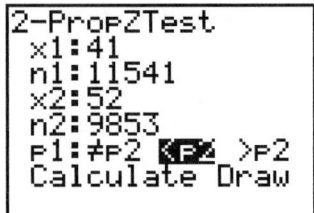

Results for Example 1

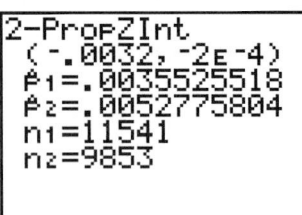

Hypothesis Testing   Given below is the procedure for using a TI-83/84 Plus calculator for testing a claim about two population proportions. *Because the TI-83/84 Plus calculator provides the test statistic and P-value (but not the critical value), we use the P-value method of testing hypotheses.*

1. Press **STAT**, then press ▷ twice to select **TESTS.**
2. Select the menu item of **2-PropZTest** (for a hypothesis test of a claim involving two proportions).
3. Enter the values of x_1, n_1, x_2, and n_2. All of those values must be whole numbers.
4. For the next line, select the type of test (two-tailed, left-tailed, right-tailed) by selecting the expression of the alternative hypothesis:
$$p_1 \neq p_2 \qquad p_1 < p_2 \qquad p_1 > p_2$$
5. Use the arrow keys to scroll down to the bottom line, select **Calculate,** then press **ENTER**.

The two accompanying screen displays correspond to the hypothesis test illustrated in Example 1. The first screen shows the required entries and the second screen shows the results. We can see from the results screen that the *P*-value is 0.0279609289, or 0.0280 when rounded. Because the *P*-value is less than the significance level of 0.05, we reject the null hypothesis of $p_1 = p_2$ and we conclude that there is sufficient evidence to support the claim that the proportion of accident fatalities in cars with airbags is less than the proportion of fatalities in cars without airbags.

Confidence Intervals   Here is the procedure for using a TI-83/84 Plus calculator for finding a confidence interval estimate of the difference ($p_1 - p_2$) between two population proportions:

1. Press **STAT**, then press ▷ twice to select **TESTS.**
2. Select the menu item of **2-PropZInt** (for a confidence interval involving two proportions).
3. Enter the values of x_1, n_1, x_2, and n_2. All of those values must be whole numbers.
4. For the next line labeled C-Level, enter the confidence level. Enter 0.95 for a 95% confidence level.
5. Use the arrow keys to scroll down to the bottom line, select **Calculate,** then press **ENTER**. The result will consist of confidence interval limits enclosed within parentheses.

Example 2 Confidence Interval

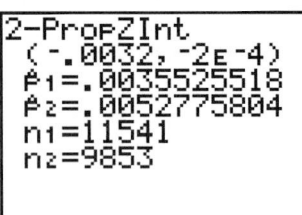

Example 2 illustrates the construction of a 90% confidence interval estimate of the difference $p_1 - p_2$ based on the same sample data given in Example 1. The accompanying display shows the confidence interval for Example 2. Note that the TI-83/84 Plus calculator expresses the confidence interval in the format of $(-0.0032, {}^{-}2E{}^{-}4)$. That confidence interval can also be expressed as $-0.0032 < (p_1 - p_2) < -0.0002$. Remember that the expression ${}^{-}2E{}^{-}4$ is in scientific notation and can be expressed in standard form as -0.0002. (See the discussion of scientific notation in Section 1-6.)

9-2   Basic Skills and Concepts

Statistical Literacy and Critical Thinking

1. Verifying Requirements   A student of the author surveyed her friends and found that among 20 males, 4 smoke and among 30 female friends, 6 smoke. Give *two* reasons why these results should not be used for a hypothesis test of the claim that the proportions of male smokers and female smokers are equal.

2. Interpreting Confidence Interval   In clinical trials of the drug Zocor, some subjects were treated with Zocor and others were given a placebo. The 95% confidence interval estimate of the difference between the proportions of subjects who experienced headaches is $-0.0518 < p_1 - p_2 < 0.0194$ (based on data from Merck & Co., Inc.). Write a statement interpreting that confidence interval.

3. Notation In clinical trials of the drug Zocor, 1583 subjects were treated with Zocor and 15 of them experienced headaches. A placebo is used for 157 other subjects, and 8 of them experienced headaches (based on data from Merck & Co., Inc.). We plan to conduct a hypothesis test involving a claim about the proportions of headaches of subjects treated with Zocor to subjects given a placebo. Identify the values of $\hat{p}_1$, $\hat{p}_2$, and $\bar{p}$. Also, what do the symbols p_1 and p_2 represent?

4. Equivalence of Methods Given a simple random sample of men and a simple random sample of women, we want to use a 0.05 significance level to test the claim that the percentage of men who smoke is equal to the percentage of women who smoke. One approach is to use the P-value method of hypothesis testing, a second approach is to use the traditional method of hypothesis testing, and a third approach is to base the conclusion on the 95% confidence interval estimate of $p_1 - p_2$. Will all three approaches always result in the same conclusion? Explain.

Finding Number of Successes. *In Exercises 5 and 6, find the number of successes x suggested by the given statement.*

5. Heart Pacemakers From an article in *Journal of the American Medical Association:* Among 8834 malfunctioning pacemakers, in 15.8% the malfunctions were due to batteries.

6. Drug Clinical Trial From Pfizer: Among 129 subjects who took Chantix as an aid to stop smoking, 12.4% experienced nausea.

Calculations for Testing Claims. *In Exercises 7 and 8, assume that you plan to use a significance level of $\alpha = 0.05$ to test the claim that $p_1 = p_2$. Use the given sample sizes and numbers of successes to find (a) the pooled estimate $\bar{p}$, (b) the z test statistic, (c) the critical z values, and (d) the P-value.*

TI 7. Online College Applications The numbers of online applications from simple random samples of college applications for 2003 and for the current year are given below (based on data from the National Association of College Admission Counseling).

	2003	Current Year
Number of applications in sample	36	27
Number of online applications in sample	13	14

TI 8. Drug Clinical Trial Chantix is a drug used as an aid to stop smoking. The numbers of subjects experiencing insomnia for each of two treatment groups in a clinical trial of the drug Chantix are given below (based on data from Pfizer):

	Chantix Treatment	Placebo
Number in group	129	805
Number experiencing insomnia	19	13

Calculations for Confidence Intervals. *In Exercises 9 and 10, assume that you plan to construct a 95% confidence interval using the data from the indicated exercise. Find (a) the margin of error E, and (b) the 95% confidence interval.*

TI 9. Exercise 7 **TI 10.** Exercise 8

Interpreting Displays. *In Exercises 11 and 12, conduct the hypothesis test by using the results from the given displays.*

11. Clinical Trials of Lipitor Lipitor is a drug used to control cholesterol. In clinical trials of Lipitor, 94 subjects were treated with Lipitor and 270 subjects were given a placebo. Among those treated with Lipitor, 7 developed infections. Among those given a placebo, 27 developed infections. Use a 0.05 significance level to test the claim that the rate of infections was the same for those treated with Lipitor and those given a placebo.

12. Bednets to Reduce Malaria In a randomized controlled trial in Kenya, insecticide-treated bednets were tested as a way to reduce malaria. Among 343 infants using bednets, 15 developed malaria. Among 294 infants not using bednets, 27 developed malaria (based on data from "Sustainability of Reductions in Malaria Transmission and Infant Mortality in Western Kenya with Use of Insecticide-Treated Bednets," by Lindblade, et al., *Journal of the American*

Author as a Witness

The author was asked to testify in New York State Supreme Court by a former student who was contesting a lost reelection to the office of Dutchess County Clerk. The author testified by using statistics to show that the voting behavior in one contested district was significantly different from the behavior in all other districts. When the opposing attorney asked about results of a confidence interval, he asked if the 5% error (from a 95% confidence level) could be added to the three percentage point margin of error to get a total error of 8%, thereby indicating that he did not understand the basic concept of a confidence interval. The judge cited the author's testimony, upheld the claim of the former student, and ordered a new election in the contested district. That judgment was later overturned by the appellate court on the grounds that the ballot irregularities should have been contested before the election, not after.

TI-83/84 PLUS

```
2-PropZTest
 P1≠P2
 z=-.7326279116
 P=.4637852543
 p̂1=.0744680851
 p̂2=.1
↓p̂=.0934065934
```

Medical Association, Vol. 291, No. 21). Use a 0.01 significance level to test the claim that the incidence of malaria is lower for infants using bednets. Do the bednets appear to be effective?

MINITAB

```
Difference = p (1) - p (2)
Estimate for difference:  -0.0481050
99% upper bound for difference:  -0.00125315
Test for difference = 0 (vs < 0):  Z = -2.44  P-Value = 0.007
```

13. Drug Use in College In a 1993 survey of 560 college students, 171 said that they used illegal drugs during the previous year. In a recent survey of 720 college students, 263 said that they used illegal drugs during the previous year (based on data from the National Center for Addiction and Substance Abuse at Columbia University). Use a 0.05 significance level to test the claim that the proportion of college students using illegal drugs in 1993 was less than it is now.

14. Drug Use in College Using the sample data from Exercise 13, construct the confidence interval corresponding to the hypothesis test conducted with a 0.05 significance level. What conclusion does the confidence interval suggest?

15. Are Seat Belts Effective? A simple random sample of front-seat occupants involved in car crashes is obtained. Among 2823 occupants not wearing seat belts, 31 were killed. Among 7765 occupants wearing seat belts, 16 were killed (based on data from "Who Wants Airbags?" by Meyer and Finney, *Chance,* Vol. 18, No. 2). Construct a 90% confidence interval estimate of the difference between the fatality rates for those not wearing seat belts and those wearing seat belts. What does the result suggest about the effectiveness of seat belts?

16. Are Seat Belts Effective? Use the sample data in Exercise 15 with a 0.05 significance level to test the claim that the fatality rate is higher for those not wearing seat belts.

17. Morality and Marriage A Pew Research Center poll asked randomly selected subjects if they agreed with the statement that "It is morally wrong for married people to have an affair." Among the 386 women surveyed, 347 agreed with the statement. Among the 359 men surveyed, 305 agreed with the statement. Use a 0.05 significance level to test the claim that the percentage of women who agree is different from the percentage of men who agree. Does there appear to be a difference in the way women and men feel about this issue?

18. Morality and Marriage Using the sample data from Exercise 17, construct the confidence interval corresponding to the hypothesis test conducted with a 0.05 significance level. What conclusion does the confidence interval suggest?

19. Raising the Roof in Baseball In a recent baseball World Series, the Houston Astros were ordered to keep the roof of their stadium open. The Houston team claimed that this would make them lose a home-field advantage, because the noise from fans would be less effective. During the regular season, Houston won 36 of 53 games played with the roof closed, and they won 15 of 26 games played with the roof open. Treat these results as simple random samples, and use a 0.05 significance level to test the claim that the proportion of wins at home is higher with a closed roof than with an open roof. Does the closed roof appear to be an advantage?

20. Raising the Roof in Baseball Using the sample data from Exercise 19, construct the confidence interval corresponding to the hypothesis test conducted with a 0.05 significance level. What conclusion does the confidence interval suggest?

21. Is Echinacea Effective for Colds? Rhino viruses typically cause common colds. In a test of the effectiveness of echinacea, 40 of the 45 subjects treated with echinacea developed rhinovirus infections. In a placebo group, 88 of the 103 subjects developed rhinovirus infections (based on data from "An Evaluation of Echinacea Angustifolia in Experimental Rhinovirus Infections," by Turner, et al., *New England Journal of Medicine,* Vol. 353, No. 4). Construct a 95% confidence interval estimate of the difference between the two rates of infection. Does echinacea appear to have any effect on the infection rate?

22. Is Echinacea Effective for Colds? Use the data from Exercise 21 to test the claim that the echinacea treatment has an effect. If you were a physician, would you recommend echinacea?

23. Sick Cruise Ship In one trip of the Royal Caribbean cruise ship *Freedom of the Seas,* 338 of the 3823 passengers became ill with a Norovirus. At about the same time, 276 of the 1652 passengers on the *Queen Elizabeth II* cruise ship became ill with a Norovirus. Treat the sample results as simple random samples from large populations, and use a 0.01 significance level to test the claim that the rate of Norovirus illness on the *Freedom of the Seas* is less than the rate on the *Queen Elizabeth II*. Based on the result, does it appear that when a Norovirus outbreak occurs on a cruise ship, the proportion of infected passengers can vary considerably?

24. Sick Cruise Ship Using the sample data from Exercise 23, construct the confidence interval corresponding to the hypothesis test conducted with a 0.01 significance level. What conclusion does the confidence interval suggest?

25. Tennis Challenges When the Hawk-Eye instant replay system for tennis was introduced at the U.S. Open, men challenged 489 referee calls, and 201 of them were successfully upheld by the Hawk-Eye system. Women challenged 350 referee calls, and 126 of them were successfully upheld by the Hawk-Eye system (based on data from *USA Today*). Construct a 99% confidence interval estimate of the difference between the success rates for challenges made by men and women. What does the confidence interval suggest about the success rates of the men and women tennis players?

26. Tennis Challenges Using the data from Exercise 25, test the claim that men and women tennis players have different success rates when challenging calls. Use a 0.01 significance level.

27. Are the Radiation Effects the Same for Men and Women? Among 2739 female atom bomb survivors, 1397 developed thyroid diseases. Among 1352 male atom bomb survivors, 436 developed thyroid diseases (based on data from "Radiation Dose-Response Relationships for Thyroid Nodules and Autoimmune Thyroid Diseases in Hiroshima and Nagasaki Atomic Bomb Survivors 55–58 Years After Radiation Exposure," by Imaizumi, et al., *Journal of the American Medical Association,* Vol. 295, No. 9). Use a 0.01 significance level to test the claim that the female survivors and male survivors have different rates of thyroid diseases.

28. Are the Radiation Effects the Same for Men and Women? Using the sample data from Exercise 27, construct the confidence interval corresponding to the hypothesis test conducted with a 0.01 significance level. What conclusion does the confidence interval suggest?

29. Global Warming Survey A Pew Research Center Poll asked subjects "Is there solid evidence that the earth is getting warmer?" 69% of 731 male respondents answered "yes," and 70% of 770 female respondents answered "yes." Construct a 90% confidence interval estimate of the difference between the proportions of "yes" responses from males and females. What do you conclude from the result?

30. Global Warming Survey Use the sample data in Exercise 29 with a 0.05 significance level to test the claim that the percentage of males who answer "yes" is less than the percentage of females who answer "yes."

31. Tax Returns and Campaign Funds Tax returns include an option of designating $3 for presidential election campaigns, and it does not cost the taxpayer anything to make that designation. In a simple random sample of 250 tax returns from 1976, 27.6% of the returns designated the $3 for the campaign. In a simple random sample of 300 recent tax returns, 7.3% of the returns designated the $3 for the campaign (based on data from *USA Today*). Use a 0.01 significance level to test the claim that the percentage of returns designating the $3 for the campaign was greater in 1976 than it is now.

32. Tax Returns and Campaign Funds Using the sample data from Exercise 31, construct the confidence interval corresponding to the hypothesis test conducted with a 0.01 significance level. What conclusion does the confidence interval suggest?

33. Adverse Effects of Viagra In an experiment, 16% of 734 subjects treated with Viagra experienced headaches. In the same experiment, 4% of 725 subjects given a placebo experienced headaches (based on data from Pfizer). Use a 0.01 significance level to test the claim that the proportion of headaches is greater for those treated with Viagra. Do headaches appear to be a concern for those who take Viagra?

34. Adverse Effects of Viagra Using the sample data from Exercise 33, construct the confidence interval corresponding to the hypothesis test conducted with a 0.01 significance level. What conclusion does the confidence interval suggest?

35. Employee Perceptions A total of 61,647 people responded to an *Elle*/MSNBC.COM survey. It was reported that 50% of the respondents were women and 50% men. Of the women, 27% said that female bosses are harshly critical; of the men, 25% said that female bosses are harshly critical. Construct a 95% confidence interval estimate of the difference between the proportions of women and men who said that female bosses are harshly critical. How is the result affected by the fact that the respondents chose whether to participate in the survey?

TI 36. Employee Perceptions Use the sample data in Exercise 35 with a 0.05 significance level to test the claim that the percentage of women who say that female bosses are harshly critical is greater than the percentage of men. Does the significance level of 0.05 used in this exercise correspond to the 95% confidence level use for the preceding exercise? Considering the sampling method, is the hypothesis test valid?

9-2   Beyond the Basics

TI 37. Interpreting Overlap of Confidence Intervals In the article "On Judging the Significance of Differences by Examining the Overlap Between Confidence Intervals," by Schenker and Gentleman (*American Statistician,* Vol. 55, No. 3), the authors consider sample data in this statement: "Independent simple random samples, each of size 200, have been drawn, and 112 people in the first sample have the attribute, whereas 88 people in the second sample have the attribute."

a. Use the methods of this section to construct a 95% confidence interval estimate of the difference $p_1 - p_2$. What does the result suggest about the equality of p_1 and p_2?

b. Use the methods of Section 7-2 to construct individual 95% confidence interval estimates for each of the two population proportions. After comparing the overlap between the two confidence intervals, what do you conclude about the equality of p_1 and p_2?

c. Use a 0.05 significance level to test the claim that the two population proportions are equal. What do you conclude?

d. Based on the preceding results, what should you conclude about equality of p_1 and p_2? Which of the three preceding methods is least effective in testing for equality of p_1 and p_2?

TI 38. Equivalence of Hypothesis Test and Confidence Interval Two different simple random samples are drawn from two different populations. The first sample consists of 20 people with 10 having a common attribute. The second sample consists of 2000 people with 1404 of them having the same common attribute. Compare the results from a hypothesis test of $p_1 = p_2$ (with a 0.05 significance level) and a 95% confidence interval estimate of $p_1 - p_2$.

39. Testing for Constant Difference To test the null hypothesis that the difference between two population proportions is equal to a nonzero constant c, use the test statistic

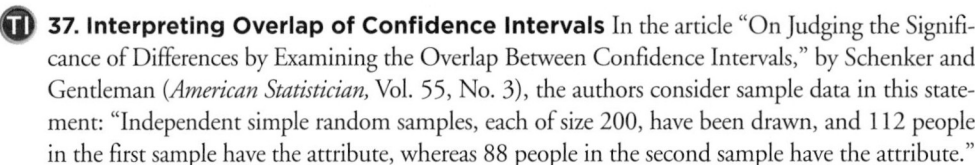

$$z = \frac{(\hat{p}_1 - \hat{p}_2) - c}{\sqrt{\dfrac{\hat{p}_1\hat{q}_1}{n_1} + \dfrac{\hat{p}_2\hat{q}_2}{n_2}}}$$

As long as n_1 and n_2 are both large, the sampling distribution of the test statistic z will be approximately the standard normal distribution. Refer to Exercise 27 and use a 0.01 significance level to test the claim that the rate of thyroid disease among female atom bomb survivors is equal to 15 percentage points more than that for male atom bomb survivors.

40. Determining Sample Size The sample size needed to estimate the difference between two population proportions to within a margin of error E with a confidence level of $1 - \alpha$ can be found by using the following expression.

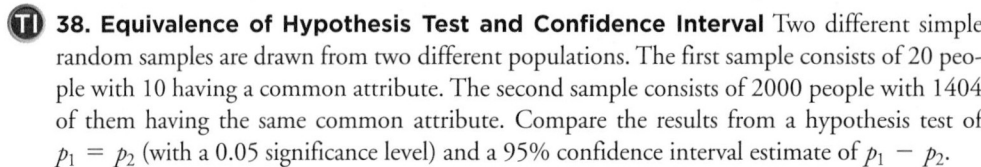

$$E = z_{\alpha/2}\sqrt{\frac{p_1 q_1}{n_1} + \frac{p_2 q_2}{n_2}}$$

In the above formula, replace n_1 and n_2 by n (assuming that both samples have the same size) and replace each of p_1, q_1, p_2, and q_2 by 0.5 (because their values are not known). Then solve for n.

Use this approach to find the size of each sample if you want to estimate the difference between the proportions of men and women who have their own computers. Assume that you want 95% confidence that your error is no more than 0.03.

Inferences About Two Means: Independent Samples

9-3

Key Concept In this section we present methods for using sample data from two independent samples to test hypotheses made about two population means or to construct confidence interval estimates of the difference between two population means. In Part 1 we discuss situations in which the standard deviations of the two populations are unknown and are not assumed to be equal. In Part 2 we discuss two other situations: (1) The two population standard deviations are both known; (2) the two population standard deviations are unknown but are assumed to be equal. Because σ is typically unknown in real situations, most attention should be given to the methods described in Part 1.

Part 1: Independent Samples with σ_1 and σ_2 Unknown and Not Assumed Equal

This section involves two *independent* samples, and the following section deals with samples that are *dependent*, so it is important to know the difference between independent samples and dependent samples.

> **DEFINITION**
>
> Two samples are **independent** if the sample values from one population are not related to or somehow naturally paired or matched with the sample values from the other population.
>
> Two samples are **dependent** if the sample values are *paired*. (That is, each pair of sample values consists of two measurements from the same subject (such as before/after data), or each pair of sample values consists of matched pairs (such as husband/wife data), where the matching is based on some inherent relationship.)

EXAMPLE 1 **Independent Samples** University of Arizona psychologists conducted a study in which 210 women and 186 men wore microphones so that the numbers of words that they spoke could be recorded. The sample word counts for men and the sample word counts for women are two independent samples, because the subjects were not paired or matched in any way.

EXAMPLE 2 **Dependent Samples** Rutgers University researchers conducted a study in which 67 students were weighed in September of their freshman year and again in April of their freshman year. The two samples are dependent, because each September weight is paired with the April weight for the same student.

Using Statistics to Identify Thieves

Methods of statistics can be used to determine that an employee is stealing, and they can also be used to estimate the amount stolen. The following are some of the indicators that have been used. For comparable time periods, samples of sales have means that are significantly different. The mean sale amount decreases significantly. There is a significant increase in the proportion of "no sale" register openings. There is a significant decrease in the ratio of cash receipts to checks. Methods of hypothesis testing can be used to identify such indicators. (See "How To Catch a Thief," by Manly and Thomson, *Chance,* Vol. 11, No. 4.)

> EXAMPLE 3 **Clinical Experiment** In an experiment designed to study the effectiveness of treatments for viral croup, 46 children were treated with low humidity and 46 other children were treated with high humidity. The Westley Croup Score was used to assess the results after one hour. Both samples have the same number of subjects and the sample scores can be listed in adjacent columns of the same length; however, the scores are from two different groups of subjects. So, the samples are independent. (See "Controlled Delivery of High vs Low Humidity vs Mist Therapy for Croup Emergency Departments," by Scolnik, et al., *Journal of the American Medical Association,* Vol. 295, No. 11).

The following box summarizes key elements of a hypothesis test of a claim about two independent population means and a confidence interval estimate of the difference between the means from two independent populations.

Objectives

Test a claim about two independent population means or construct a confidence interval estimate of the difference between two independent population means.

Notation

For population 1 we let

$\mu_1 = $ *population* mean $\qquad\qquad \bar{x}_1 = $ *sample* mean

$\sigma_1 = $ *population* standard deviation $\qquad\qquad s_1 = $ *sample* standard deviation

$n_1 = $ size of the first sample

The corresponding notations μ_2, σ_2, $\bar{x}_2$, s_2, and n_2 apply to population 2.

Requirements

1. σ_1 and σ_2 are unknown and it is not assumed that σ_1 and σ_2 are equal.

2. The two samples are *independent.*

3. Both samples are *simple random samples.*

4. Either or both of these conditions is satisfied: The two sample sizes are both *large* (with $n_1 > 30$ and $n_2 > 30$) or both samples come from populations having normal distributions. (These methods are robust against departures from normality, so for small samples, the normality requirement is loose in the sense that the procedures perform well as long as there are no outliers and departures from normality are not too extreme.)

Hypothesis Test Statistic for Two Means: Independent Samples

$$t = \frac{(\bar{x}_1 - \bar{x}_2) - (\mu_1 - \mu_2)}{\sqrt{\dfrac{s_1^2}{n_1} + \dfrac{s_2^2}{n_2}}} \qquad \text{(where } \mu_1 - \mu_2 \text{ is often assumed to be 0)}$$

continued

Degrees of Freedom: When finding critical values or P-values, use the following for determining the number of degrees of freedom, denoted by df. (Although these two methods typically result in different numbers of degrees of freedom, the conclusion of a hypothesis test is rarely affected by the choice.)

1. In this book we use this simple and conservative estimate: df = smaller of $n_1 - 1$ and $n_2 - 1$.

2. The TI-83/84 Plus calculator uses the more accurate but more difficult estimate given in Formula 9-1. (We will not use Formula 9-1 for the examples and exercises in this book.)

Formula 9-1

$$df = \frac{(A + B)^2}{\dfrac{A^2}{n_1 - 1} + \dfrac{B^2}{n_2 - 1}}$$

where $A = \dfrac{s_1^2}{n_1}$ and $B = \dfrac{s_2^2}{n_2}$

P-values: Refer to the t distribution in Table A-3. Use the procedure summarized in Figure 8-5.

Critical values: Refer to the t distribution in Table A-3.

Confidence Interval Estimate of $\mu_1 - \mu_2$: Independent Samples

The confidence interval estimate of the difference $\mu_1 - \mu_2$ is

$$(\bar{x}_1 - \bar{x}_2) - E < (\mu_1 - \mu_2) < (\bar{x}_1 - \bar{x}_2) + E$$

where $E = t_{\alpha/2}\sqrt{\dfrac{s_1^2}{n_1} + \dfrac{s_2^2}{n_2}}$

and the number of degrees of freedom df is as described above for hypothesis tests. (In this book, we use df = smaller of $n_1 - 1$ and $n_2 - 1$.)

TI-83/84 Plus Calculator The TI-83/84 Plus calculator can automatically provide the test statistic and P-value for a hypothesis test, and it can also generate a confidence interval. See the procedures at the end of this section.

CAUTION
....
Before conducting a hypothesis test, consider the context of the data, the source of the data, the sampling method, and explore the data with graphs and descriptive statistics. Be sure to verify that the requirements are satisfied.

Equivalence of Methods The P-value method of hypothesis testing, the traditional method of hypothesis testing, and confidence intervals all use the same distribution and standard error, so they are equivalent in the sense that they result in the same conclusions. A null hypothesis of $\mu_1 = \mu_2$ (or $\mu_1 - \mu_2 = 0$) can be tested using the P-value method, the traditional method, or by determining whether the confidence interval includes 0.

EXAMPLE 4 **Are Men and Women Equal Talkers?** A headline in *USA Today* proclaimed that "Men, women are equal talkers." That headline referred to a study of the numbers of words that samples of men and women spoke in a day. Given below are the results from the study, which are included in Data Set 8 in Appendix B (based on "Are Women Really More Talkative Than Men?" by Mehl,

continued

Do Real Estate Agents Get You the Best Price?

When a real estate agent sells a home, does he or she get the best price for the seller? This question was addressed by Steven Levitt and Stephen Dubner in *Freak-onomics*. They collected data from thousands of homes near Chicago, including homes owned by the agents themselves. Here is what they write: "There's one way to find out: measure the difference between the sales data for houses that belong to real-estate agents themselves and the houses they sold on behalf of clients. Using the data from the sales of those 100,000 Chicago homes, and controlling for any number of variables—location, age and quality of the house, aesthetics, and so on—it turns out that a real-estate agent keeps her own home on the market an average of ten days longer and sells it for an extra 3-plus percent, or $10,000 on a $300,000 house." A conclusion such as this can be obtained by using the methods of this section.

et al., *Science*, Vol. 317, No. 5834). Use a 0.05 significance level to test the claim that men and women speak the same mean number of words in a day. Does there appear to be a difference?

Number of Words Spoken in a Day

Men	Women
$n_1 = 186$	$n_2 = 210$
$\bar{x}_1 = 15{,}668.5$	$\bar{x}_2 = 16{,}215.0$
$s_1 = 8632.5$	$s_2 = 7301.2$

SOLUTION   **REQUIREMENTS CHECK** (1) The values of the two population standard deviations are not known and we are not making an assumption that they are equal. (2) The two samples are independent because the word counts for the sample of men are in no way matched or paired with the word counts for the sample of women. (3) We assume that the samples are simple random samples. (The article in *Science* magazine describes the sample design.) (4) Both samples are large, so it is not necessary to verify that each sample appears to come from a population with a normal distribution, but the accompanying TI-83/84 Plus calculator display of the histogram for the word counts (in thousands) for men shows that the distribution is not substantially far from being a normal distribution. The histogram for the word counts for the women is very similar. The requirements are satisfied.

TI-83/84 Plus

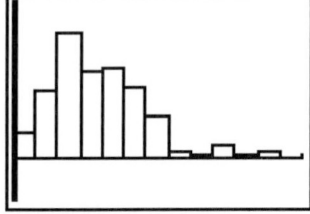

We now proceed with the hypothesis test. We use the *P*-value method summarized in Figure 8-8 on page 412.

Step 1: The claim that men and women have the same mean can be expressed as $\mu_1 = \mu_2$.

Step 2: If the original claim is false, then $\mu_1 \neq \mu_2$.

Step 3: The alternative hypothesis is the expression not containing equality, and the null hypothesis is an expression of equality, so we have

$$H_0: \mu_1 = \mu_2 \text{ (original claim)} \qquad H_1: \mu_1 \neq \mu_2$$

We now proceed with the assumption that $\mu_1 = \mu_2$, or $\mu_1 - \mu_2 = 0$.

Step 4: The significance level is $\alpha = 0.05$.

Step 5: Because we have two independent samples and we are testing a claim about the two population means, we use a *t* distribution with the test statistic given earlier in this section.

Step 6: The test statistic is calculated as follows:

$$t = \frac{(\bar{x}_1 - \bar{x}_2) - (\mu_1 - \mu_2)}{\sqrt{\frac{s_1^2}{n_1} + \frac{s_2^2}{n_2}}} = \frac{(15{,}668.5 - 16{,}215.0) - 0}{\sqrt{\frac{8632.5^2}{186} + \frac{7301.2^2}{210}}} = -0.676$$

TI-83/84 Plus Using a TI-83/84 Plus calculator with the instructions given at the end of this section, we obtain the display shown in the margin. The calculator automatically computes the value of the test statistic $t = -0.676$ (rounded) and the *P*-value 0.4997738726. (Because the TI-83/84 Plus calculator provides *P*-values and not critical values, the *P*-value method of testing hypotheses is used.)

Step 7: Because the *P*-value of 0.4997738726 is greater than the significance level of 0.05, we fail to reject the null hypothesis of $\mu_1 = \mu_2$ (or $\mu_1 - \mu_2 = 0$).

If we had used the traditional method of testing hypotheses, we would find the critical values of $t = \pm 1.972$, and we would fail to reject the null hypothesis because the test statistic does not fall in the critical region, as shown in Figure 9-2.

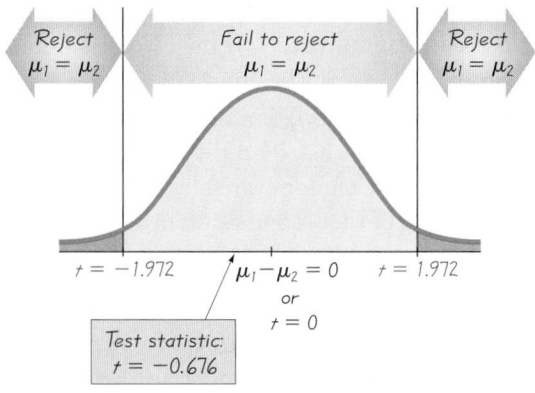

Figure 9-2 **Testing the Claim of Equal Means for Men and Women**

INTERPRETATION There is not sufficient evidence to warrant rejection of the claim that men and women speak the same mean number of words in a day. There does not appear to be a significant difference between the two means.

EXAMPLE 5 **Confidence Interval for Word Counts from Men and Women** Using the sample data given in Example 4, construct a 95% confidence interval estimate of the difference between the mean number of words spoken by men and the mean number of words spoken by women.

SOLUTION **REQUIREMENTS CHECK** Because we are using the same data from Example 4, the same requirement check applies here, so the requirements are satisfied. ✓

We first find the value of the margin of error *E*. We use the same critical value of $t_{\alpha/2} = 1.972$ found in Example 4. (A more accurate critical value is 1.966.)

$$E = t_{\alpha/2}\sqrt{\frac{s_1^2}{n_1} + \frac{s_2^2}{n_2}} = 1.972\sqrt{\frac{8632.5^2}{186} + \frac{7301.2^2}{210}} = 1595.4$$

TI-83/84 Plus

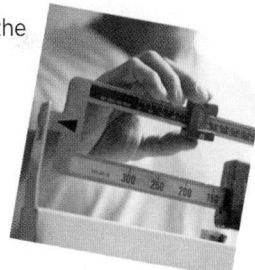

```
2-SampTTest
 μ1≠μ2
 t=-.6755202804
 P=.4997738726
 df=364.2590079
 x̄1=15668.5
↓x̄2=16215
```

Expensive Diet Pill

There are many past examples in which ineffective treatments were marketed for substantial profits. Capsules of "Fat Trapper" and "Exercise in a Bottle," manufactured by the Enforma Natural Products company, were advertised as being effective treatments for weight reduction. Advertisements claimed that after taking the capsules, fat would be blocked and calories would be burned, even without exercise. Because the Federal Trade Commission identified claims that appeared to be unsubstantiated, the company was fined $10 million for deceptive advertising.

The effectiveness of such treatments can be determined with experiments in which one group of randomly selected subjects is given the treatment, while another group of randomly selected subjects is given a placebo. The resulting weight losses can be compared using statistical methods, such as those described in this section.

continued

Super Bowls

Students were invited to a Super Bowl game and half of them were given large 4-liter snack bowls while the other half were given smaller 2-liter bowls. Those using the large bowls consumed 56% more than those using the smaller bowls. (See "Super Bowls: Serving Bowl Size and Food Consumption," by Wansink and Cheney, *Journal of the American Medical Association,* Vol. 293, No. 14.)

A separate study showed that there is "a significant increase in fatal motor vehicle crashes during the hours following the Super Bowl telecast in the United States." Researchers analyzed 20,377 deaths on 27 Super Bowl Sundays and 54 other Sundays used as controls. They found a 41% increase in fatalities after Super Bowl games. (See "Do Fatal Crashes Increase Following a Super Bowl Telecast?" by Redelmeier and Stewart, *Chance,* Vol. 18, No. 1.)

Using $E = 1595.4$ and $\bar{x}_1 = 15{,}668.5$ and $\bar{x}_2 = 16{,}215.0$, we now find the desired confidence interval as follows:

$$(\bar{x}_1 - \bar{x}_2) - E < (\mu_1 - \mu_2) < (\bar{x}_1 - \bar{x}_2) + E$$
$$-2141.9 < (\mu_1 - \mu_2) < 1048.9$$

If we use the TI-83/84 Plus calculator to obtain more accurate results, we get the confidence interval of $-2137 < (\mu_1 - \mu_2) < 1044.4$, so we can see that the above confidence interval is quite good.

INTERPRETATION   We are 95% confident that the limits of -2141.9 words and 1048.9 words actually do contain the difference between the two population means. Because those limits do contain 0, this confidence interval suggests that it is very possible that the two population means are equal, so there is not a significant difference between the two means.

Rationale: Why Do the Test Statistic and Confidence Interval Have the Particular Forms We Have Presented?   If the given assumptions are satisfied, the sampling distribution of $\bar{x}_1 - \bar{x}_2$ can be approximated by a t distribution with mean equal to $\mu_1 - \mu_2$ and standard deviation equal to $\sqrt{s_1^2/n_1 + s_2^2/n_2}$. This last expression for the standard deviation is based on the property that the variance of the *differences* between two independent random variables equals the variance of the first random variable *plus* the variance of the second random variable.

Part 2: Alternative Methods

Part 1 of this section dealt with situations in which the two population standard deviations are unknown and are not assumed to be equal. In Part 2 we address two other situations: (1) The two population standard deviations are both known; (2) the two population standard deviations are unknown but are assumed to be equal. We now describe the procedures for these alternative situations.

Alternative Method When σ_1 and σ_2 Are Known

In reality, the population standard deviations σ_1 and σ_2 are almost never known, but if they are known, the test statistic and confidence interval are based on the normal distribution instead of the t distribution. See the summary box below.

Inferences About Means of Two Independent Populations, With σ_1 and σ_2 Known

Requirements

1. The two population standard deviations σ_1 and σ_2 are both known.

2. The two samples are *independent.*

3. Both samples are *simple random samples.*

4. Either or both of these conditions is satisfied: The two sample sizes are both *large* (with $n_1 > 30$ and $n_2 > 30$) or both samples come from populations having normal distributions. (For small samples, the normality requirement is loose in the sense that the procedures perform well as long as there are no outliers and departures from normality are not too extreme.)

Hypothesis Test

Test statistic:

$$z = \frac{(\bar{x}_1 - \bar{x}_2) - (\mu_1 - \mu_2)}{\sqrt{\dfrac{\sigma_1^2}{n_1} + \dfrac{\sigma_2^2}{n_2}}}$$

P-values and **critical values:** Refer to Table A-2.

Confidence Interval Estimate of $\mu_1 - \mu_2$

Confidence interval:

$$(\bar{x}_1 - \bar{x}_2) - E < (\mu_1 - \mu_2) < (\bar{x}_1 - \bar{x}_2) + E$$

where

$$E = z_{\alpha/2}\sqrt{\frac{\sigma_1^2}{n_1} + \frac{\sigma_2^2}{n_2}}$$

Figure 9-3 summarizes the methods for inferences about two independent population means.

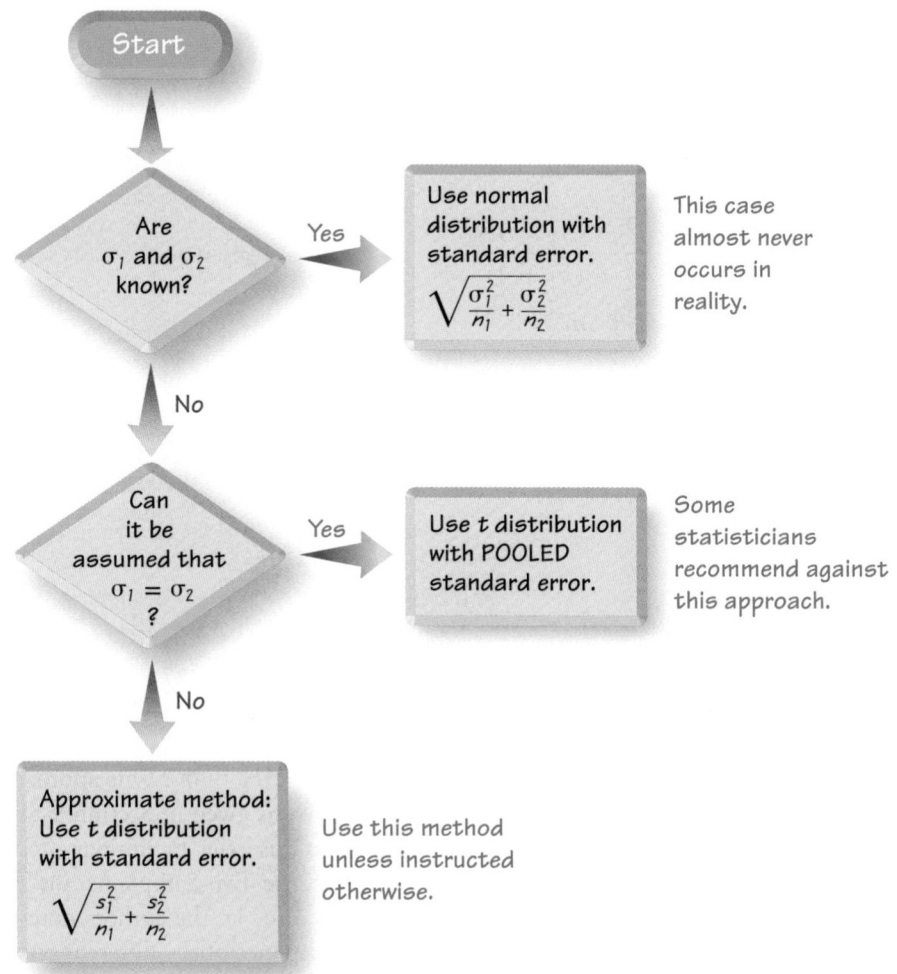

Figure 9-3 · **Methods for Inferences About Two Independent Means**

Better Results with Smaller Class Size

An experiment at the State University of New York at Stony Brook found that students did significantly

better in classes limited to 35 students than in large classes with 150 to 200 students. For a calculus course, failure rates were 19% for the small classes compared to 50% for the large classes. The percentages of A's were 24% for the small classes and 3% for the large classes. These results suggest that students benefit from smaller classes, which allow for more direct interaction between students and teachers.

Alternative Method: Assume That $\sigma_1 = \sigma_2$ and *Pool* the Sample Variances

Even when the specific values of σ_1 and σ_2 are not known, if it can be assumed that they have the *same* value, the sample variances s_1^2 and s_2^2 can be *pooled* to obtain an estimate of the common population variance σ^2. The **pooled estimate of σ^2** is denoted by s_p^2 and is a weighted average of s_1^2 and s_2^2, which is included in the following box.

Inferences About Means of Two Independent Populations, Assuming That $\sigma_1 = \sigma_2$

Requirements

1. The two population standard deviations are not known, but they are assumed to be equal. That is, $\sigma_1 = \sigma_2$.

2. The two samples are *independent*.

3. Both samples are *simple random samples*.

4. Either or both of these conditions is satisfied: The two sample sizes are both *large* (with $n_1 > 30$ and $n_2 > 30$) or both samples come from populations having normal distributions. (For small samples, the normality requirement is loose in the sense that the procedures perform well as long as there are no outliers and departures from normality are not too extreme.)

Hypothesis Test

Test statistic:

$$t = \frac{(\bar{x}_1 - \bar{x}_2) - (\mu_1 - \mu_2)}{\sqrt{\dfrac{s_p^2}{n_1} + \dfrac{s_p^2}{n_2}}}$$

where

$$s_p^2 = \frac{(n_1 - 1)s_1^2 + (n_2 - 1)s_2^2}{(n_1 - 1) + (n_2 - 1)} \qquad (Pooled \text{ variance})$$

and the number of degrees of freedom is given by df $= n_1 + n_2 - 2$.

Confidence Interval Estimate of $\mu_1 - \mu_2$

Confidence interval:

$$(\bar{x}_1 - \bar{x}_2) - E < (\mu_1 - \mu_2) < (\bar{x}_1 - \bar{x}_2) + E$$

where

$$E = t_{\alpha/2}\sqrt{\frac{s_p^2}{n_1} + \frac{s_p^2}{n_2}}$$

and s_p^2 is as given in the above test statistic and the number of degrees of freedom is df $= n_1 + n_2 - 2$.

If we want to use this method, how do we determine that $\sigma_1 = \sigma_2$? One approach is to use a hypothesis test of the null hypothesis $\sigma_1 = \sigma_2$, as given in Section 9-5, but that approach is not recommended and we will not use the preliminary test of $\sigma_1 = \sigma_2$. In the article "Homogeneity of Variance in the Two-Sample Means Test" (by Moser and Stevens, *American Statistician,* Vol. 46, No. 1), the authors note that we rarely know that $\sigma_1 = \sigma_2$. They analyze the performance of the different tests by considering sample sizes and powers of the tests. They conclude that more effort should be spent learning the method given in Part 1, and less emphasis should be placed on the method based on the assumption of $\sigma_1 = \sigma_2$. Unless instructed otherwise, we use the following strategy, which is consistent with the recommendations in the article by Moser and Stevens:

> **Assume that σ_1 and σ_2 are unknown, do *not* assume that $\sigma_1 = \sigma_2$, and use the test statistic and confidence interval given in Part 1 of this section. (See Figure 9-3.)**

Why Don't We Just Eliminate the Method of Pooling Sample Variances?

If we use randomness to assign subjects to treatment and placebo groups, we know that the samples are drawn from the same population. So if we conduct a hypothesis test assuming that two population means are equal, it is not unreasonable to also assume that the samples are from populations with the same standard deviations (but we should still check that assumption). The advantage of this alternative method of pooling sample variances is that the number of degrees of freedom is a little higher, so hypothesis tests have more power, and confidence intervals are a little narrower. Consequently, statisticians sometimes use this method of pooling, and that is why we include it in this subsection.

Using the TI-83/84 Plus Calculator for Inferences About Two Means (With Independent Samples)

1. Press **STAT**, then press **▷** twice to select **TESTS.**

2. Select one of the following as appropriate:

   2-SampTTest Hypothesis test (with σ_1 and σ_2 *not* known)
   2-SampTInt Confidence interval (with σ_1 and σ_2 *not* known)
   2-SampZTest Hypothesis test (with σ_1 and σ_2 known)
   2-SampZInt Confidence interval (with σ_1 and σ_2 known)

3. Select **Data** if the two sets of sample data are already stored in the calculator as lists, or select **Stats** if you know the summary statistics of $\bar{x}_1$, s_1, n_1, $\bar{x}_2$, s_2, and n_2.

4. If you selected **Data** in step 3, proceed to enter the list names or labels, such as L1 and L2. If you selected **Stats** in step 3, proceed to enter the values of the statistics for each sample.

5. The next step depends on whether you are conducting a hypothesis test or constructing a confidence interval.
   - *Hypothesis test*: If conducting a hypothesis test, select the type of test (two-tailed, left-tailed, or right-tailed). Select the format corresponding to the alternative hypothesis ($\mu_1 \neq \mu_2$, $\mu_1 < \mu_2$, or $\mu_1 > \mu_2$).
   - *Confidence interval*: If creating a confidence interval, you must enter a confidence level, such as 0.95.

6. If using the t distribution (by selecting either 2-SampTTest or 2-SampTInt in step 2), you will see a line displaying **Pooled: No Yes.** It is recommended that you select **No** so that the sample variances are *not* pooled. (Select **Yes** only if you are making the assumption that the two population variances are equal.)

7. Use the arrow keys to scroll down to the bottom line, select **Calculate,** then press **ENTER**.

The accompanying screen displays correspond to the hypothesis test illustrated in Example 4 and the construction of a confidence interval illustrated in Example 5. Note the P-value of 0.4998 (rounded) in the hypothesis test. Also note that the confidence interval limits can be rounded and expressed in this format: $-2137 < (\mu_1 - \mu_2) < 1044.4$.

Hypothesis Test from Example 4

```
2-SampTTest
 µ1≠µ2
 t=-.6755202804
 P=.4997738726
 df=364.2590079
 x̄1=15668.5
↓x̄2=16215
```

Confidence Interval from Example 5

```
2-SampTInt
 (-2137,1044.4)
 df=364.2590079
 x̄1=15668.5
 x̄2=16215
 Sx1=8632.5
↓Sx2=7301.2
```

9-3   Basic Skills and Concepts

Statistical Literacy and Critical Thinking

1. Interpreting Confidence Intervals If the pulse rates of men and women from Data Set 1 in Appendix B are used to construct a 95% confidence interval for the difference between the two population means, the result is $-12.2 < \mu_1 - \mu_2 < -1.6$, where pulse rates of men correspond to population 1 and pulse rates of women correspond to population 2. Express the confidence interval with pulse rates of women being population 1 and pulse rates of men being population 2.

2. Interpreting Confidence Intervals What does the confidence interval in Exercise 1 suggest about the pulse rates of men and women?

3. Significance Level and Confidence Level Assume that you want to use a 0.01 significance level to test the claim that the mean pulse rate of men is less than the mean pulse rate of women. What *confidence level* should be used if you want to test that claim using a confidence interval?

4. Degrees of Freedom Assume that you want to use a 0.01 significance level to test the claim that the mean pulse rate of women is greater than the mean pulse rate of men using the sample data from Data Set 1 in Appendix B. Both samples have 40 values. If we use df = smaller of $n_1 - 1$ and $n_2 - 1$, we get df = 39, and the corresponding critical value is $t = 2.426$. If we calculate df using Formula 9-1, we get df = 77.2, and the corresponding critical value is 2.376. How is using a critical value of $t = 2.426$ "more conservative" than using the critical value of 2.376?

Independent and Dependent Samples. *In Exercises 5–8, determine whether the samples are independent or dependent.*

5. Blood Pressure Data Set 1 in Appendix B includes systolic blood pressure measurements from each of 40 randomly selected men and 40 randomly selected women.

6. Home Sales Data Set 23 in Appendix B includes the list price and selling price for each of 40 randomly selected homes.

7. Reducing Cholesterol To test the effectiveness of Lipitor, cholesterol levels are measured in 250 subjects before and after Lipitor treatments.

8. Voltage On each of 40 different days, the author measured the voltage supplied to his home and he also measured the voltage produced by his gasoline-powered generator. (The data are listed in Data Set 13 in Appendix B.) One sample consists of the voltages in his home and the second sample consists of the voltages produced by the generator.

In Exercises 9–32, assume that the two samples are independent simple random samples selected from normally distributed populations. Do not assume that the population standard deviations are equal, unless your instructor stipulates otherwise.

TI 9. Hypothesis Test of Effectiveness of Humidity in Treating Croup In a randomized controlled trial conducted with children suffering from viral croup, 46 children were treated with low humidity while 46 other children were treated with high humidity. Researchers used the Westley Croup Score to assess the results after one hour. The low humidity group had a mean score of 0.98 with a standard deviation of 1.22 while the high humidity group had a mean score of 1.09 with a standard deviation of 1.11 (based on data from "Controlled Delivery of High vs Low Humidity vs Mist Therapy for Croup Emergency Departments," by Scolnik, et al., *Journal of the American Medical Association*, Vol. 295, No. 11). Use a 0.05 significance level to test the claim that the two groups are from populations with the same mean. What does the result suggest about the common treatment of humidity?

TI 10. Confidence Interval for Effectiveness of Humidity in Treating Croup Refer to the sample data given in Exercise 9 and construct a 95% confidence interval estimate of the difference between the mean Westley Croup Score of children treated with low humidity and the mean score of children treated with high humidity. What does the confidence interval suggest about humidity as a treatment for croup?

11. Confidence Interval for Cigarette Tar The mean tar content of a simple random sample of 25 unfiltered king size cigarettes is 21.1 mg, with a standard deviation of 3.2 mg. The mean tar content of a simple random sample of 25 filtered 100 mm cigarettes is 13.2 mg with a standard deviation of 3.7 mg (based on data from Data Set 4 in Appendix B). Construct a 90% confidence interval estimate of the difference between the mean tar content of unfiltered king size cigarettes and the mean tar content of filtered 100 mm cigarettes. Does the result suggest that 100 mm filtered cigarettes have less tar than unfiltered king size cigarettes?

12. Hypothesis Test for Cigarette Tar Refer to the sample data in Exercise 11 and use a 0.05 significance level to test the claim that unfiltered king size cigarettes have a mean tar content greater than that of filtered 100 mm cigarettes. What does the result suggest about the effectiveness of cigarette filters?

13. Hypothesis Test for Checks and Charges The author collected a simple random sample of the cents portions from 100 checks and from 100 credit card charges. The cents portions of the checks have a mean of 23.8 cents and a standard deviation of 32.0 cents. The cents portions of the credit charges have a mean of 47.6 cents and a standard deviation of 33.5 cents. Use a 0.05 significance level to test the claim that the cents portions of the check amounts have a mean that is less than the mean of the cents portions of the credit card charges. Give one reason that might explain a difference.

14. Confidence Interval for Checks and Charges Refer to the sample data given in Exercise 13 and construct a 90% confidence interval for the difference between the mean of the cents portions from checks and the mean of the cents portions from credit card charges. What does the confidence interval suggest about the means of those amounts?

15. Hypothesis Test for Heights of Supermodels The heights are measured for the simple random sample of supermodels Crawford, Bundchen, Pestova, Christenson, Hume, Moss, Campbell, Schiffer, and Taylor. They have a mean of 70.0 in. and a standard deviation of 1.5 in. Data Set 1 in Appendix B lists the heights of 40 women who are not supermodels, and they have heights with a mean of 63.2 in. and a standard deviation of 2.7 in. Use a 0.01 significance level to test the claim that the mean height of supermodels is greater than the mean height of women who are not supermodels.

16. Confidence Interval for Heights of Supermodels Use the sample data from Exercise 15 to construct a 98% confidence interval for the difference between the mean height of supermodels and the mean height of women who are not supermodels. What does the result suggest about those two means?

17. Confidence Interval for Braking Distances of Cars A simple random sample of 13 four-cylinder cars is obtained, and the braking distances are measured. The mean braking distance is 137.5 ft and the standard deviation is 5.8 ft. A simple random sample of 12 six-cylinder cars is obtained and the braking distances have a mean of 136.3 ft with a standard deviation of 9.7 ft (based on Data Set 16 in Appendix B). Construct a 90% confidence interval estimate of the difference between the mean braking distance of four-cylinder cars and the mean braking distance of six-cylinder cars. Does there appear to be a difference between the two means?

18. Hypothesis Test for Braking Distances of Cars Refer to the sample data given in Exercise 17 and use a 0.05 significance level to test the claim that the mean braking distance of four-cylinder cars is greater than the mean braking distance of six-cylinder cars.

19. Hypothesis Test for Cigarette Nicotine Scientists collect a simple random sample of 25 menthol cigarettes and 25 nonmenthol cigarettes. Both samples consist of cigarettes that are filtered, 100 mm long, and non-light. The menthol cigarettes have a mean nicotine amount of 0.87 mg and a standard deviation of 0.24 mg. The nonmenthol cigarettes have a mean nicotine amount of 0.92 mg and a standard deviation of 0.25 mg. Use a 0.05 significance level to test the claim that menthol cigarettes and nonmenthol cigarettes have different amounts of nicotine. Does menthol appear to have an effect on the nicotine content?

TI 20. Confidence Interval for Cigarette Nicotine Refer to the sample data in Exercise 19 and construct a 95% confidence interval estimate of the difference between the mean nicotine amount in menthol cigarettes and the mean nicotine amount in nonmenthol cigarettes. What does the result suggest about the effect of menthol?

TI 21. Hypothesis Test for Mortgage Payments Simple random samples of high-interest (8.9%) mortgages and low-interest (6.3%) mortgages were obtained. For the 40 high-interest mortgages, the borrowers had a mean FICO credit score of 594.8 and a standard deviation of 12.2. For the 40 low-interest mortgages, the borrowers had a mean FICO credit score of 785.2 and a standard deviation of 16.3 (based on data from *USA Today*). Use a 0.01 significance level to test the claim that the mean FICO score of borrowers with high-interest mortgages is lower than the mean FICO score of borrowers with low-interest mortgages. Does the FICO credit rating score appear to affect mortgage payments? If so, how?

TI 22. Confidence Interval for Mortgage Payments Use the sample data from Exercise 21 to construct a 98% confidence interval estimate of the difference between the mean FICO credit score of borrowers with high interest rates and the mean FICO credit score of borrowers with low interest rates. What does the result suggest about the FICO credit rating score of a borrower and the interest rate that is paid?

TI 23. Hypothesis Test for Discrimination The Revenue Commissioners in Ireland conducted a contest for promotion. Statistics from the ages of the unsuccessful and successful applicants are given below (based on data from "Debating the Use of Statistical Evidence in Allegations of Age Discrimination," by Barry and Boland, *American Statistician*, Vol. 58, No. 2). Some of the applicants who were unsuccessful in getting the promotion charged that the competition involved discrimination based on age. Treat the data as samples from larger populations and use a 0.05 significance level to test the claim that the unsuccessful applicants are from a population with a greater mean age than the mean age of successful applicants. Based on the result, does there appear to be discrimination based on age?

> Ages of unsuccessful applicants $n = 23$, $\bar{x} = 47.0$ years, $s = 7.2$ years
>
> Ages of successful applicants $n = 30$, $\bar{x} = 43.9$ years, $s = 5.9$ years

TI 24. Confidence Interval for Discrimination Using the sample data from Exercise 23, construct a 90% confidence interval estimate of the difference between the mean age of unsuccessful applicants and the mean age of successful applicants. What does the result suggest about discrimination based on age?

TI 25. Hypothesis Test for Effect of Marijuana Use on College Students Many studies have been conducted to test the effects of marijuana use on mental abilities. In one such study, groups of light and heavy users of marijuana in college were tested for memory recall, with the results given below (based on data from "The Residual Cognitive Effects of Heavy Marijuana Use in College Students," by Pope and Yurgelun-Todd, *Journal of the American Medical Association*, Vol. 275, No. 7). Use a 0.01 significance level to test the claim that the population of heavy marijuana users has a lower mean than the light users. Should marijuana use be of concern to college students?

> Items sorted correctly by light marijuana users: $n = 64$, $\bar{x} = 53.3$, $s = 3.6$
>
> Items sorted correctly by heavy marijuana users: $n = 65$, $\bar{x} = 51.3$, $s = 4.5$

TI 26. Confidence Interval for Effects of Marijuana Use on College Students Refer to the sample data used in Exercise 25 and construct a 98% confidence interval for the difference between the two population means. Does the confidence interval include zero? What does the confidence interval suggest about the equality of the two population means?

TI 27. Hypothesis Test for Magnet Treatment of Pain People spend huge sums of money (currently around $5 billion annually) for the purchase of magnets used to treat a wide variety of pains. Researchers conducted a study to determine whether magnets are effective in treating back pain. Pain was measured using the visual analog scale, and the results given below are among the results obtained in the study (based on data from "Bipolar Permanent Magnets for the Treatment of Chronic Lower Back Pain: A Pilot Study," by Collacott, Zimmerman,

White, and Rindone, *Journal of the American Medical Association,* Vol. 283, No. 10). Use a 0.05 significance level to test the claim that those treated with magnets have a greater mean reduction in pain than those given a sham treatment (similar to a placebo). Does it appear that magnets are effective in treating back pain? Is it valid to argue that magnets might appear to be effective if the sample sizes are larger?

Reduction in pain level after magnet treatment: $n = 20$, $\bar{x} = 0.49$, $s = 0.96$

Reduction in pain level after sham treatment: $n = 20$, $\bar{x} = 0.44$, $s = 1.4$

TI 28. Confidence Interval for Magnet Treatment of Pain Refer to the sample data from Exercise 27 and construct a 90% confidence interval estimate of the difference between the mean reduction in pain for those treated with magnets and the mean reduction in pain for those given a sham treatment. Based on the result, does it appear that the magnets are effective in reducing pain?

TI 29. BMI for Miss America The trend of thinner Miss America winners has generated charges that the contest encourages unhealthy diet habits among young women. Listed below are body mass indexes (BMI) for Miss America winners from two different time periods. Consider the listed values to be simple random samples selected from larger populations.

a. Use a 0.05 significance level to test the claim that recent winners have a lower mean BMI than winners from the 1920s and 1930s.

b. Construct a 90% confidence interval for the difference between the mean BMI of recent winners and the mean BMI of winners from the 1920s and 1930s.

BMI (from recent winners):   19.5 20.3 19.6 20.2 17.8 17.9 19.1 18.8 17.6 16.8

BMI (from the 1920s and 1930s):   20.4 21.9 22.1 22.3 20.3 18.8 18.9 19.4 18.4 19.1

TI 30. Radiation in Baby Teeth Listed below are amounts of strontium-90 (in millibecquerels or mBq per gram of calcium) in a simple random sample of baby teeth obtained from Pennsylvania residents and New York residents born after 1979 (based on data from "An Unexpected Rise in Strontium-90 in U.S. Deciduous Teeth in the 1990s," by Mangano, et al., *Science of the Total Environment*).

a. Use a 0.05 significance level to test the claim that the mean amount of strontium-90 from Pennsylvania residents is greater than the mean amount from New York residents.

b. Construct a 90% confidence interval of the difference between the mean amount of strontium-90 from Pennsylvania residents and the mean amount from New York residents

Pennsylvania: 155 142 149 130 151 163 151 142 156 133 138 161

New York: 133 140 142 131 134 129 128 140 140 140 137 143

TI 31. Longevity Listed below are the numbers of years that popes and British monarchs (since 1690) lived after their election or coronation (based on data from *Computer-Interactive Data Analysis*, by Lunn and McNeil, John Wiley & Sons). Treat the values as simple random samples from a larger population.

a. Use a 0.01 significance level to test the claim that the mean longevity for popes is less than the mean for British monarchs after coronation.

b. Construct a 98% confidence interval of the difference between the mean longevity for popes and the mean longevity for British monarchs. What does the result suggest about those two means?

Popes: 2   9   21   3   6   10   18   11   6   25   23   6   2   15   32
                25   11   8   17   19   5   15   0   26

Kings and Queens: 17   6   13   12   13   33   59   10   7   63   9   25   36   15

TI 32. Sex and Blood Cell Counts White blood cell counts are helpful for assessing liver disease, radiation, bone marrow failure, and infectious diseases. Listed below are white blood cell counts found in simple random samples of males and females (based on data from the Third National Health and Nutrition Examination Survey).

a. Use a 0.01 significance level to test the claim that females and males have different mean white blood cell counts.

b. Construct a 99% confidence interval of the difference between the mean white blood cell count of females and males. Based on the result, does there appear to be a difference?

Female:	8.90	6.50	9.45	7.65	6.40	5.15	16.60	5.75	11.60
	5.90	9.30	8.55	10.80	4.85	4.90	8.75	6.90	9.75
	4.05	9.05	5.05	6.40	4.05	7.60	4.95	3.00	9.10

Male:	5.25	5.95	10.05	5.45	5.30	5.55	6.85	6.65	6.30
	6.40	7.85	7.70	5.30	6.50	4.55	7.10	8.00	4.70
	4.40	4.90	10.75	11.00	9.60				

Large Data Sets. *In Exercises 33–36, use the indicated Data Sets from Appendix B. Assume that the two samples are independent simple random samples selected from normally distributed populations. Do not assume that the population standard deviations are equal,*

TI **33. Movie Income** Refer to Data Set 9 in Appendix B. Use the amounts of money grossed by movies with ratings of PG or PG-13 as one sample, and use the amounts of money grossed by movies with R ratings.

a. Use a 0.01 significance level to test the claim that movies with ratings of PG or PG-13 have a higher mean gross amount than movies with R ratings.

b. Construct a 98% confidence interval estimate of the difference between the mean amount of money grossed by movies with ratings of PG or PG-13 and the mean amount of money grossed by movies with R ratings. What does the confidence interval suggest about movies as an investment?

TI **34. Word Counts** Refer to Data Set 8 in Appendix B. Use the word counts for male and female psychology students recruited in Mexico (see the columns labeled 3M and 3F).

a. Use a 0.05 significance level to test the claim that male and female psychology students speak the same mean number of words in a day.

b. Construct a 95% confidence interval estimate of the difference between the mean number of words spoken in a day by male and female psychology students in Mexico. Do the confidence interval limits include 0, and what does that suggest about the two means?

TI **35. Voltage** Refer to Data Set 13 in Appendix B. Use a 0.05 significance level to test the claim that the sample of home voltages and the sample of generator voltages are from populations with the same mean. If there is a statistically significant difference, does that difference have practical significance?

TI **36. Weights of Coke** Refer to Data Set 17 in Appendix B and test the claim that because they contain the same amount of cola, the mean weight of cola in cans of regular Coke is the same as the mean weight of cola in cans of Diet Coke. If there is a difference in the mean weights, identify the most likely explanation for that difference.

Pooling. *In Exercises 37–40, assume that the two samples are independent simple random samples selected from normally distributed populations. Also assume that the population standard deviations are equal ($\sigma_1 = \sigma_2$), so that the standard error of the differences between means is obtained by pooling the sample variances as described in Part 2 of this section.*

TI **37. Hypothesis Test with Pooling** Repeat Exercise 9 with the additional assumption that $\sigma_1 = \sigma_2$. How are the results affected by this additional assumption?

TI **38. Confidence Interval with Pooling** Repeat Exercise 10 with the additional assumption that $\sigma_1 = \sigma_2$. How are the results affected by this additional assumption?

TI **39. Confidence Interval with Pooling** Repeat Exercise 11 with the additional assumption that $\sigma_1 = \sigma_2$. How are the results affected by this additional assumption?

TI **40. Hypothesis Test with Pooling** Repeat Exercise 12 with the additional assumption that $\sigma_1 = \sigma_2$. How are the results affected by this additional assumption?

9-3   Beyond the Basics

 41. Effects of an Outlier Refer to Exercise 31 and create an outlier by changing the first value listed for kings and queens from 17 years to 1700 years. After making that change, describe the effects of the outlier on the hypothesis test and confidence interval. Does the outlier have a dramatic effect on the results?

 42. Effects of Units of Measurement How are the results of Exercise 31 affected if all of the longevity times are converted from years to months? In general, does the choice of the scale affect the conclusions about equality of the two population means, and does the choice of scale affect the confidence interval?

43. Effect of No Variation in Sample An experiment was conducted to test the effects of alcohol. Researchers measured the breath alcohol levels for a treatment group of people who drank ethanol and another group given a placebo. The results are given in the accompanying table. Use a 0.05 significance level to test the claim that the two sample groups come from populations with the same mean. The given results are based on data from "Effects of Alcohol Intoxication on Risk Taking, Strategy, and Error Rate in Visuomotor Performance," by Streufert, et al., *Journal of Applied Psychology,* Vol. 77, No. 4.

$$\text{Treatment Group:} \quad n_1 = 22 \quad \bar{x}_1 = 0.049 \quad s_1 = 0.015$$
$$\text{Placebo Group:} \quad n_2 = 22 \quad \bar{x}_2 = 0.000 \quad s_2 = 0.000$$

 44. Calculating Degrees of Freedom How is the number of degrees of freedom for Exercises 9 and 10 affected if Formula 9-1 is used instead of selecting the smaller of $n_1 - 1$ and $n_2 - 1$? If Formula 9-1 is used for the number of degrees of freedom instead of the smaller of $n_1 - 1$ and $n_2 - 1$, how are the hypothesis test and the confidence interval affected? In what sense is "df = smaller of $n_1 - 1$ and $n_2 - 1$" a more conservative estimate of the number of degrees of freedom than the estimate obtained with Formula 9-1?

9-4   Inferences from Dependent Samples

Key Concept In this section we present methods for testing hypotheses and constructing confidence intervals involving the mean of the differences of the values from two dependent populations.

With dependent samples, there is some relationship whereby each value in one sample is paired with a corresponding value in the other sample. Here are two typical examples of dependent samples:

- Each pair of sample values consists of two measurements from the same subject. *Example:* The weight of a freshman student was 64 kg in September and 68 kg in April.

- Each pair of sample values consists of a matched pair. *Example:* The body mass index (BMI) of a husband is 25.1 and the BMI of his wife is 19.7.

Because the hypothesis test and confidence interval use the same distribution and standard error, they are *equivalent* in the sense that they result in the same conclusions. Consequently, the null hypothesis that the mean difference equals 0 can be tested by determining whether the confidence interval includes 0.

There are no exact procedures for dealing with dependent samples, but the *t* distribution serves as a reasonably good approximation, so the following methods are commonly used.

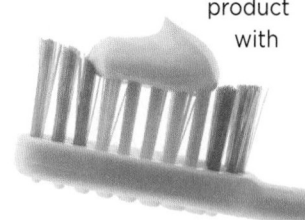

Objectives

Test a claim about the mean of the differences from dependent samples or construct a confidence interval estimate of the mean of the differences from dependent samples.

Notation for Dependent Samples

d = individual difference between the two values in a single matched pair

μ_d = mean value of the differences d for the *population* of all pairs of data

$\overline{d}$ = mean value of the differences d for the paired *sample* data

s_d = standard deviation of the differences d for the paired *sample* data

n = number of *pairs* of data

Requirements

1. The sample data are dependent.

2. The samples are simple random samples.

3. Either or both of these conditions is satisfied: The number of pairs of sample data is large ($n > 30$) or the pairs of values have differences that are from a population having a distribution that is approximately normal. (These methods are robust against departures for normality, so for small samples, the normality requirement is loose in the sense that the procedures perform well as long as there are no outliers and departures from normality are not too extreme.)

Hypothesis Test Statistic for Dependent Samples

$$t = \frac{\overline{d} - \mu_d}{\dfrac{s_d}{\sqrt{n}}}$$

where degrees of freedom $= n - 1$.

P-values and **Critical values:**   Table A-3 (t distribution)

Confidence Intervals for Dependent Samples

$$\overline{d} - E < \mu_d < \overline{d} + E$$

where

$$E = t_{\alpha/2}\frac{s_d}{\sqrt{n}}$$

Critical values of $t_{\alpha/2}$: Use Table A-3 with $n - 1$ degrees of freedom.

TI-83/84 Plus Calculator The TI-83/84 Plus calculator can automatically provide the test statistic and *P*-value for a hypothesis test, and it can also generate a confidence interval. See the procedures at the end of this section.

EXAMPLE 1   **Hypothesis Test of Claimed Freshman Weight Gain** Data Set 3 in Appendix B includes measured weights of college students in September and April of their freshman year. Table 9-1 lists a small portion of those sample values. (Here we use only a small portion of the available data so that we can better illustrate the method of hypothesis testing.) Use the sample data in Table 9-1 with a 0.05 significance level to test the claim that for the population of students, the mean change in weight from September to April is equal to 0 kg.

Table 9-1 Weight (kg) Measurements of Students in Their Freshman Year

April weight	66	52	68	69	71
September weight	67	53	64	71	70
Difference d = (April weight) − (September weight)	−1	−1	4	−2	1

SOLUTION   **REQUIREMENTS CHECK** We address the three requirements listed earlier in this section. (1) The samples are dependent because the values are paired, with each pair measured from the same student. (2) Instead of being a simple random sample of selected students, all subjects volunteered for the study, so the second requirement is not satisfied. This limitation is cited in the journal article describing the results of the study. We will proceed as if the requirement of a simple random sample is satisfied; see the comments in the *interpretation* that follows the solution. (3) The number of pairs is not large, so we should check for normality of the differences and we should check for outliers. Inspection of the differences shows that there are no outliers, and the accompanying TI-83/84 Plus calculator display shows the histogram with a distribution that is not substantially far from being normal. (A normal quantile plot also suggests that the differences are from a population with a distribution that is approximately normal.) The requirements are satisfied.

TI-83/84 Plus

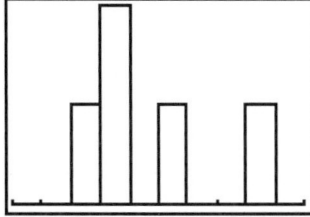

Let's express the amounts of weight gained from September to April by considering differences in this format: (April weight) − (September weight). If we use μ_d (where the subscript d denotes "difference") to denote the mean of the "April − September" differences in weight of college students during their freshman year, the claim is that $\mu_d = 0$ kg.

We will follow the same basic method of hypothesis testing that was introduced in Chapter 8, but we use the test statistic for dependent samples that was given earlier in this section.

Step 1: The claim is that $\mu_d = 0$ kg. (That is, the mean weight gain is equal to 0 kg.)

Step 2: If the original claim is not true, we have $\mu_d \neq 0$ kg.

Step 3: The null hypothesis must express equality and the alternative hypothesis cannot include equality, so we have

$$H_0: \mu_d = 0 \text{ kg (original claim)} \qquad H_1: \mu_d \neq 0 \text{ kg}$$

Step 4: The significance level is $\alpha = 0.05$.

Step 5: We use the Student t distribution.

Step 6: Before finding the value of the test statistic, we must first find the values of $\overline{d}$, and s_d. Refer to Table 9-1 and use the differences of −1, −1, 4, −2, and 1 to find these sample statistics: $\overline{d} = 0.2$ and $s_d = 2.4$. Using these sample statistics
continued

Twins in Twinsburg

During the first weekend in August of each year, Twinsburg, Ohio celebrates its annual "Twins Days in Twinsburg" festival. Thousands of twins from around the world have attended this festival in the past. Scientists saw the festival as an opportunity to study identical twins. Because they have the same basic genetic structure, identical twins are ideal for studying the different effects of heredity and environment on a variety of traits, such as male baldness, heart disease, and deafness—traits that were recently studied at one Twinsburg festival. A study of twins showed that myopia (near-sightedness) is strongly affected by hereditary factors, not by environmental factors such as watching television, surfing the Internet, or playing computer or video games.

TI-83/84 Plus

```
T-Test
 μ≠0
 t=.1873171623
 p=.8605297156
 x̄=.2
 Sx=2.387467277
 n=5
```

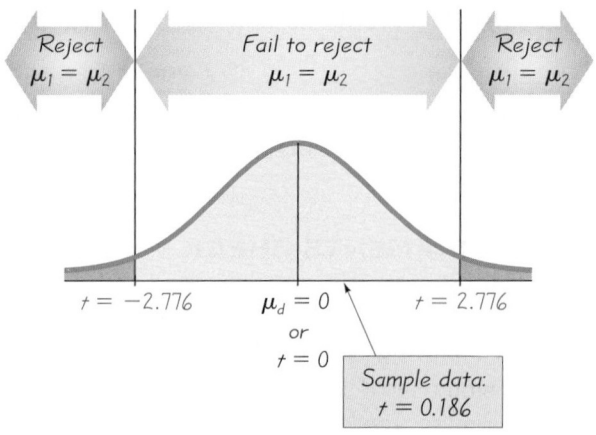

Figure 9-4   Distribution of Differences d Found from Paired Sample Data

and the assumption of the hypothesis test that $\mu_d = 0$ kg, we can now find the value of the test statistic. (The TI-83/84 Plus calculator uses more decimal places and provides the more accurate test statistic of $t = 0.187$.)

$$t = \frac{\overline{d} - \mu_d}{\frac{s_d}{\sqrt{n}}} = \frac{0.2 - 0}{\frac{2.4}{\sqrt{5}}} = 0.186$$

TI-83/84 Plus Using a TI-83/84 Plus calculator with the instructions given at the end of this section, we obtain the display shown in the margin. The calculator automatically computes the value of the test statistic $t = 0.187$ (rounded) and the P-value 0.8605 (rounded). (Because the TI-83/84 Plus calculator provides P-values and not critical values, the P-value method of testing hypotheses is used.)

Step 7: Because the P-value of 0.8605 is greater than the significance level of 0.05, we fail to reject the null hypothesis.

If we had used the traditional method of testing hypotheses, we would find the critical values of $t = \pm 2.776$, and we would fail to reject the null hypothesis because the test statistic does not fall in the critical region, as shown in Figure 9-4.

INTERPRETATION   We conclude that there is not sufficient evidence to warrant rejection of the claim that for the population of students, the mean change in weight from September to April is equal to 0 kg. Based on the sample results listed in Table 9-1, there does not appear to be a significant weight gain from September to April.

The conclusion should be qualified with the limitations noted in the article about the study. The requirement of a simple random sample is not satisfied, because only Rutgers students were used. Also, the study subjects are volunteers, so there is a potential for a self-selection bias. In the article describing the study, the authors cited these limitations and stated that "Researchers should conduct additional studies to better characterize dietary or activity patterns that predict weight gain among young adults who enter college or enter the workforce during this critical period in their lives."

Example 2 uses the P-value method with all 67 pairs of data (from Data Set 3 in Appendix B) instead of the 5 pairs of data shown in Table 9-1.

TI-83/84 Plus

```
T-Test
μ≠0
t=2.482154809
p=.0156113186
x̄=1.179104478
Sx=3.888309491
n=67
```

EXAMPLE 2 **Hypothesis Test of Claimed Freshman Weight Gain**
Example 1 used only the five pairs of sample values listed in Table 9-1, but Data Set 3 in
Appendix B includes results from 67 subjects. If we repeat Example 1 using the
TI-83/84 Plus calculator with all 67 pairs of sample data, we obtain the results in the
margin. This display shows that with the 67 pairs of sample data, the test statistic is
$t = 2.482$ and the P-value is 0.0156. Because the P-value is less than the significance
level of 0.05, we now reject the null hypothesis. We now conclude that there is sufficient
evidence to warrant rejection of the claim that the mean difference is equal to 0 kg.

Examples 1 and 2 illustrate the method of testing hypotheses. Examples 3 and 4
illustrate the construction of confidence intervals.

EXAMPLE 3 **Confidence Interval for Estimating the Mean Weight
Change** Using the same paired sample data in Table 9-1, construct a 95% confidence interval estimate of μ_d, which is the mean of the "April–September" weight
differences of college students in their freshman year.

SOLUTION **REQUIREMENTS CHECK** The solution for Example 1
includes verification that the requirements are satisfied. ✓
We use the values of $\bar{d} = 0.2$, $s_d = 2.4$, $n = 5$, and $t_{\alpha/2} = 2.776$ (found from
Table A-3 with $n - 1 = 4$ degrees of freedom and an area of 0.05 in two tails). We
first find the value of the margin of error E.

$$E = t_{\alpha/2}\frac{s_d}{\sqrt{n}} = 2.776 \cdot \frac{2.4}{\sqrt{5}} = 3.0$$

We now find the confidence interval.

$$\bar{d} - E < \mu_d < \bar{d} + E$$
$$0.2 - 3.0 < \mu_d < 0.2 + 3.0$$
$$-2.8 < \mu_d < 3.2$$

TI-83/84 Plus

```
TInterval
 (-2.764,3.1644)
x̄=.2
Sx=2.387467277
n=5
```

TI-83/84 Plus Using a TI-83/84 Plus calculator with the instructions given at the
end of this section, we obtain the display shown in the margin. The confidence interval format of $(-2.764, 3.1644)$ can be expressed as $-2.8 < \mu_d < 3.2$, where the
confidence interval limits are rounded to include one more decimal place than in the
original list of data values.

INTERPRETATION We have 95% confidence that the limits of -2.8 kg and
3.2 kg contain the true value of the mean weight change from September to April.
In the long run, 95% of such samples will lead to confidence interval limits that actually do contain the true population mean of the differences. Note that the confidence interval includes the value of 0 kg, so it is very possible that the mean of the
weight changes is equal to 0 kg.

EXAMPLE 4 **Confidence Interval for Estimating the Mean Weight
Change** Data Set 3 in Appendix B includes results from 67 subjects. If we repeat
Example 3 using the TI-83/84 Plus calculator with all 67 pairs of sample data, we
obtain the following result.

95% Confidence interval: $0.23067 < \mu_d < 2.1275$

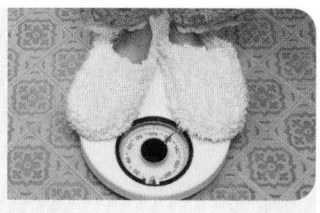

continued

> **INTERPRETATION** This confidence interval suggests that the mean weight gain is likely to be between 0.2 kg and 2.1 kg. This confidence interval does not include the value of 0 kg, so the larger data set suggests that the typical college student does gain some weight during the freshman year, and the mean amount of the weight gains is estimated to be between 0.2 kg and 2.1 kg.

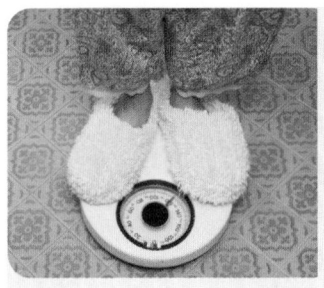

TI-83/84 Plus

```
T-Test
 μ≠6.8
 t=-11.83265191
 p=5.72725ε-18
 x̄=1.179104478
 Sx=3.888309491
 n=67
```

EXAMPLE 5 **Is the *Freshman 15* a Myth?** The Chapter Problem describes the urban legend known as the *Freshman 15,* which is the common belief that students gain an average of 15 lb (or 6.8 kg) during their freshman year. Let's again express the amounts of weight gained from September to April by considering the sample values in this format: (April weight) − (September weight). (In this format, positive differences represent *gains* in weight, and negative differences represent *losses* of weight. Based on this format, the *Freshman 15* claim is that the mean of the differences is 15 lb or 6.8 kg.) If we use μ_d to denote the mean of the "April − September" differences in weight of college students during their freshman year, the "Freshman 15" is the claim that $\mu_d = 15$ lb or $\mu_d = 6.8$ kg. If we test $\mu_d = 6.8$ kg using a 0.05 significance level with all 67 subjects from Data Set 3 in Appendix B, we get the TI-83/84 Plus calculator results displayed in the margin.

The TI-83/84 Plus calculator display shows that the test statistic is $t = -11.833$ and the *P*-value is 0.000 (rounded to three decimal places). Because the *P*-value is less than the significance level of 0.05, we reject the null hypothesis. There is sufficient evidence to warrant rejection of the claim that the mean weight change is equal to 6.8 kg (or 15 pounds).

The confidence interval from Example 4 shows that the mean weight gain is likely to be between 0.2 kg and 2.1 kg (or between 0.4 lb and 4.6 lb), so the claim of a mean weight gain of 15 lb appears to be unfounded. These results suggest that the *Freshman 15* is a myth. This conclusion should again be qualified with the limitations of the study. Only Rutgers students were used, and study subjects volunteered instead of being randomly selected. However, the findings from this study are generally consistent with those from other similar studies, so the *Freshman 15* does appear to be a myth. Based on Data Set 3 in Appendix B, it appears that students do gain some weight during their freshman year, but the mean weight gain is much more modest than the 15 pounds claimed in the *Freshman 15* myth.

Experimental Design Suppose we want to conduct an experiment to compare the effectiveness of two different types of fertilizer (one organic and one chemical). The fertilizers are to be used on 20 plots of land with equal area, but varying soil quality. To make a fair comparison, we should divide each of the 20 plots in half so that one half is treated with organic fertilizer and the other half is treated with chemical fertilizer, creating dependent samples. The yields can then be matched by the plots they share, resulting in paired data. The advantage to using paired data is that we reduce extraneous variation, which could occur if each plot were treated with one type of fertilizer rather than both—that is, if the samples were independent. This strategy for designing an experiment can be generalized by the following design principle:

> **When designing an experiment or planning an observational study, using dependent samples with paired data is generally better than using two independent samples.**

Using a TI-83/84 Plus Calculator for Inferences from Dependent Samples

Caution: For inferences from dependent samples (matched pairs), do *not* use the menu items of **2-SampTTest** or **2-SampTInt,** because they are designed for *independent samples.* Instead, use the following procedure, which begins by creating a list of the differences between the values in each pair of sample data.

1. Clear lists L1, L2, and L3 by pressing **STAT**, selecting the menu item of **ClrList,** and entering L1, L2, L3 (with the commas included).

2. Enter the data for the first variable in list L1 and also enter the data for the second variable in list L2.

3. Create a list of the differences and store the list in L3 by clearing the screen and entering **L1 − L2 → L3.** This expression can be entered with the following sequence of keystrokes: **2ND** **1** **−** **2ND** **2** **STO▸** **2ND** **3** **ENTER**.

4. Press **STAT**, then press **▷** twice to select **TESTS.**

5. Select one of the following:

   T-Test   for a hypothesis test
   TInterval   for a confidence interval

6. For the Inpt option, select **Data.**

7. Proceed to enter the required items.

   • For the List entry, enter **L3** (by pressing **2ND** **3**).

   • If conducting a hypothesis test, enter 0 for the value of μ_0, which corresponds to a null hypothesis of $\mu_d = 0$. That is, the test is conducted by assuming that the mean difference is equal to 0.

   • For the Freq (frequency) value, enter 1.

   • For a confidence interval, enter the confidence level, such as 0.95. For a hypothesis test, select the type of test (two-tailed, left-tailed, right-tailed).

8. Use the arrow keys to scroll down to the bottom line, select **Calculate,** then press **ENTER**.

9-4   Basic Skills and Concepts

Statistical Literacy and Critical Thinking

TI **1. Notation** Listed below are the time intervals (in minutes) before and after eruptions of the Old Faithful geyser. Find the values of $\overline{d}$ and s_d. In general, what does μ_d represent?

| Time interval before eruption | 98 | 92 | 95 | 87 | 96 |
| Time interval after eruption | 92 | 95 | 92 | 100 | 90 |

2. Clinical Test The drug Dozenol is tested on 40 male subjects recruited from New York and 40 female subjects recruited from California. The researcher pairs the 40 male subjects and the 40 female subjects. Can the methods of this section be used to analyze the results? Why or why not?

3. Paired Pulse Rates and Cholesterol Levels Using Data Set 1 in Appendix B, a researcher pairs pulse rates and cholesterol levels for the 40 women. Can the methods of this section be used to construct a confidence interval? Why or why not?

4. Confidence Intervals Example 4 showed that the 67 *dependent* April and September weight measurements from Data Set 3 in Appendix B result in this 95% confidence interval:

0.2 kg $< \mu_d <$ 2.1 kg. If the same data are treated as two *independent* samples, the result is this 95% confidence interval: -2.7 kg $< \mu_1 - \mu_2 <$ 5.0 kg. What is the fundamental difference between interpretations of these two confidence intervals?

Calculations with Paired Sample Data. *In Exercises 5 and 6, assume that you want to use a 0.05 significance level to test the claim that the paired sample data come from a population for which the mean difference is $\mu_d = 0$. Find (a) $\bar{d}$, (b) s_d, (c) the t test statistic, and (d) the critical values.*

5. Car Mileage Listed below are measured fuel consumption amounts (in miles/gal) from a sample of cars (Acura RL, Acura TSX, Audi A6, BMW 525i) taken from Data Set 16 in Appendix B.

City fuel consumption	18	22	21	21
Highway fuel consumption	26	31	29	29

6. Forecast Temperatures Listed below are predicted high temperatures that were forecast before different days (based on Data Set 11 in Appendix B).

Predicted high temperature forecast three days ahead	79	86	79	83	80
Predicted high temperature forecast five days ahead	80	80	79	80	79

7. Confidence Interval Using the sample paired data in Exercise 5, construct a 95% confidence interval for the population mean of all differences, in this format: (city fuel consumption) $-$ (highway fuel consumption).

8. Confidence Interval Using the sample paired data in Exercise 6, construct a 99% confidence interval for the population mean of all differences, in this format: (high temperature predicted three days ahead) $-$ (high temperature predicted five days ahead).

In Exercises 9–20, assume that the paired sample data are simple random samples and that the differences have a distribution that is approximately normal.

9. Does BMI Change During Freshman Year? Listed below are body mass indices (BMI) of the same students included in Table 9-1 on page 493. The BMI of each student was measured in September and April of the freshman year (based on data from "Changes in Body Weight and Fat Mass of Men and Women in the First Year of College: A Study of the 'Freshman 15'," by Hoffman, Policastro, Quick, and Lee, *Journal of American College Health*, Vol. 55, No. 1). Use a 0.05 significance level to test the claim that the mean change in BMI for all students is equal to 0. Does BMI appear to change during freshman year?

April BMI	20.15	19.24	20.77	23.85	21.32
September BMI	20.68	19.48	19.59	24.57	20.96

10. Confidence Interval for BMI Changes Use the same paired data from Exercise 9 to construct a 95% confidence interval estimate of the change in BMI during freshman year. Does the confidence interval include 0, and what does that suggest about BMI during freshman year?

11. Are Best Actresses Younger than Best Actors? Listed below are ages of actresses and actors at the times that they won Oscars. The data are paired according to the years that they won. Use a 0.05 significance level to test the common belief that best actresses are younger than best actors. Does the result suggest a problem in our culture?

Best Actresses	28	32	27	27	26	24	25	29	41	40	27	42	33	21	35
Best Actors	62	41	52	41	34	40	56	41	39	49	48	56	42	62	29

12. Are Flights Cheaper When Scheduled Earlier? Listed below are the costs (in dollars) of flights from New York (JFK) to San Francisco for US Air, Continental, Delta, United, American, Alaska, and Northwest. Use a 0.01 significance level to test the claim that flights scheduled one day in advance cost more than flights scheduled 30 days in advance. What strategy appears to be effective in saving money when flying?

Flight scheduled one day in advance	456	614	628	1088	943	567	536
Flight scheduled 30 days in advance	244	260	264	264	278	318	280

13. Does Your Body Temperature Change During the Day? Listed below are body temperatures (in °F) of subjects measured at 8:00 AM and at 12:00 AM (from University of Maryland physicians listed in Data Set 2 in Appendix B). Construct a 95% confidence interval estimate of the difference between the 8:00 AM temperatures and the 12:00 AM temperatures. Is body temperature basically the same at both times?

8:00 AM	97.0	96.2	97.6	96.4	97.8	99.2
12:00 AM	98.0	98.6	98.8	98.0	98.6	97.6

14. Is Blood Pressure the Same for Both Arms? Listed below are systolic blood pressure measurements (mm Hg) taken from the right and left arms of the same woman (based on data from "Consistency of Blood Pressure Differences Between the Left and Right Arms," by Eguchi, et al., *Archives of Internal Medicine,* Vol. 167). Use a 0.05 significance level to test for a difference between the measurements from the two arms. What do you conclude?

Right arm	102	101	94	79	79
Left arm	175	169	182	146	144

15. Is Friday the 13th Unlucky? Researchers collected data on the numbers of hospital admissions resulting from motor vehicle crashes, and results are given below for Fridays on the 6th of a month and Fridays on the following 13th of the same month (based on data from "Is Friday the 13th Bad for Your Health?" by Scanlon, et al., *British Medical Journal,* Vol. 307, as listed in the *Data and Story Line* online resource of data sets). Use a 0.05 significance level to test the claim that when the 13th day of a month falls on a Friday, the numbers of hospital admissions from motor vehicle crashes are not affected.

Friday the 6th:	9	6	11	11	3	5
Friday the 13th:	13	12	14	10	4	12

16. Tobacco and Alcohol in Children's Movies Listed below are times (seconds) that animated Disney movies showed the use of tobacco and alcohol. (See Data Set 7 in Appendix B.) Use a 0.05 significance level to test the claim that the mean of the differences is greater than 0 sec, so that more time is devoted to showing tobacco than alcohol. For animated children's movies, how much time should be spent showing the use of tobacco and alcohol?

Tobacco use (sec)	176	51	0	299	74	2	23	205	6	155
Alcohol use (sec)	88	33	113	51	0	3	46	73	5	74

17. Car Repair Costs Listed below are the costs (in dollars) of repairing the front ends and rear ends of different cars when they were damaged in controlled low-speed crash tests (based on data from the Insurance Institute for Highway Safety). The cars are Toyota, Mazda, Volvo, Saturn, Subaru, Hyundai, Honda, Volkswagen, and Nissan. Construct a 95% confidence interval of the mean of the differences between front repair costs and rear repair costs. Is there a difference?

Front repair cost	936	978	2252	1032	3911	4312	3469	2598	4535
Rear repair cost	1480	1202	802	3191	1122	739	2769	3375	1787

18. Self-Reported and Measured Male Heights As part of the National Health and Nutrition Examination Survey, the Department of Health and Human Services obtained self-reported heights and measured heights for males aged 12–16. All measurement are in inches. Listed below are sample results.

a. Is there sufficient evidence to support the claim that there is a difference between self-reported heights and measured heights of males aged 12–16? Use a 0.05 significance level.

b. Construct a 95% confidence interval estimate of the mean difference between reported heights and measured heights. Interpret the resulting confidence interval, and comment on the implications of whether the confidence interval limits contain 0.

Reported height	68	71	63	70	71	60	65	64	54	63	66	72
Measured height	67.9	69.9	64.9	68.3	70.3	60.6	64.5	67.0	55.6	74.2	65.0	70.8

19. Car Fuel Consumption Ratings Listed below are combined city–highway fuel consumption ratings (in miles/gal) for different cars measured under both the old rating system and a new rating system introduced in 2008 (based on data from *USA Today*). The new ratings were implemented in response to complaints that the old ratings were too high. Use a 0.01 significance level to test the claim the old ratings are higher than the new ratings.

Old rating	16	18	27	17	33	28	33	18	24	19	18	27	22	18	20	29	19	27	20	21
New rating	15	16	24	15	29	25	29	16	22	17	16	24	20	16	18	26	17	25	18	19

20. Heights of Winners and Runners-Up Listed below are the heights (in inches) of candidates who won presidential elections and the heights of the candidates who were runners up. The data are in chronological order, so the corresponding heights from the two lists are matched. For candidates who won more than once, only the heights from the first election are included, and no elections before 1900 are included.

a. A well-known theory is that winning candidates tend to be taller than the corresponding losing candidates. Use a 0.05 significance level to test that theory. Does height appear to be an important factor in winning the presidency?

b. If you plan to test the claim in part (a) by using a confidence interval, what confidence level should be used? Construct a confidence interval using that confidence level, then interpret the result.

Won Presidency								Runner-Up							
71	74.5	74	73	69.5	71.5	75	72	73	74	68	69.5	72	71	72	71.5
70.5	69	74	70	71	72	70	67	70	68	71	72	70	72	72	72

Large Data Sets. *In Exercises 21–24, use the indicated Data Sets from Appendix B. Assume that the paired sample data are simple random samples and the differences have a distribution that is approximately normal.*

21. Voltage Refer to the voltages listed in Data Set 13 in Appendix B.

a. The list of home voltages were measured from the author's home, and the list of UPS voltages were measured from the author's uninterruptible power supply with voltage supplied by the same power company on the same day. Use a 0.05 significance level to test the claim that these paired sample values have differences that are from a population with a mean of 0 volts. What do you conclude?

b. Why should the methods of this section *not* be used with the home voltages and the generator voltages?

22. Repeat Exercise 9 using the BMI measurements from all 67 subjects listed in Data Set 3 in Appendix B.

23. Paper or Plastic? Refer to Data Set 22 in Appendix B. Construct a 95% confidence interval estimate of the mean of the differences between weights of discarded paper and weights of discarded plastic. Which seems to weigh more: discarded paper or discarded plastic?

24. Glass and Food Refer to Data Set 22 in Appendix B. Construct a 95% confidence interval estimate of the mean of the differences between weights of discarded glass and weights of discarded food. Which seems to weigh more: discarded glass or discarded food? Which creates more of an environmental problem: discarded glass or discarded food? Why?

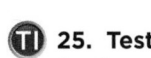

9-4 Beyond the Basics

TI **25. Testing Reaction Times** Students of the author were tested for reaction times (in thousandths of a second) using their right and left hands. (Each value is the elapsed time between the release of a strip of paper and the instant that it is caught by the subject.) Results from five of the students are included in the graph below. Use a 0.05 significance level to test the claim that there is no difference between the reaction times of the right and left hands.

MINITAB

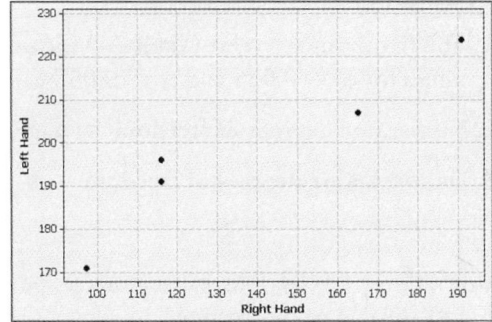

26. Effects of an Outlier and Units of Measurement

a. When using the methods of this section, can an outlier have a dramatic effect on the hypothesis test and confidence interval?

b. The examples in this section used weights measured in kilograms. If we convert all sample weights to pounds, will the change in the units affect the hypothesis tests? Are confidence intervals affected by such a change in units? How?

Comparing Variation in Two Samples

Key Concept In this section we present the F test for comparing two population variances (or standard deviations). The F test (named for statistician Sir Ronald Fisher) uses the F distribution introduced in this section. The F test requires that both populations have normal distributions, and this test is *very* sensitive to departures from normal distributions. Part 1 describes the F test procedure, and Part 2 gives a brief description of two alternative methods for comparing variation in two samples.

Part 1: F Test for Comparing Variances

Recall that a sample variance s^2 is the square of the sample standard deviation s. In this section we designate the *larger* of the two sample variances as s_1^2 (so that computations are easier). The smaller sample variance is denoted as s_2^2.

Objective

Test a claim about two population standard deviations or variances.

Notation for Hypothesis Tests with Two Variances or Standard Deviations

$s_1^2 = $ *larger* of the two sample variances

$n_1 = $ size of the sample with the *larger* variance

$\sigma_1^2 = $ variance of the population from which the sample with the *larger* variance was drawn

The symbols s_2^2, n_2, and σ_2^2 are used for the other sample and population.

continued

Requirements

1. The two populations are *independent*.

2. The two samples are simple random samples.

3. The two populations are each *normally distributed*. (This *F* test is *not robust,* meaning that it performs

poorly if one or both of the populations has a distribution that is not normal. The requirement of normal distributions is therefore quite strict for this *F* test.)

Test Statistic for Hypothesis Tests with Two Variances

$F = \dfrac{s_1^2}{s_2^2}$ (where s_1^2 is the *larger* of the two sample variances)

Critical values: Use Table A-5 to find critical *F* values that are determined by the following:

1. The significance level α (Table A-5 includes critical values for $\alpha = 0.025$ and $\alpha = 0.05$.)

2. **Numerator degrees of freedom $= n_1 - 1$**

3. **Denominator degrees of freedom $= n_2 - 1$**

TI-83/84 Plus Calculator The TI-83/84 Plus calculator can be used for hypothesis tests. See the procedure at the end of this section.

F Distribution For two normally distributed populations with equal variances ($\sigma_1^2 = \sigma_2^2$), the sampling distribution of the test statistic $F = s_1^2/s_2^2$ is the **F distribution** shown in Figure 9-5 (provided that we have not yet imposed the stipulation that the larger sample variance is s_1^2). If you repeat the process of selecting samples from two normally distributed populations with equal variances, the distribution of the ratio s_1^2/s_2^2 is the *F* distribution.

In Figure 9-5, note these properties of the *F* distribution:

• The *F* distribution is not symmetric.

• Values of the *F* distribution cannot be negative.

• The exact shape of the *F* distribution depends on the two different degrees of freedom.

Finding Critical F Values Because the TI-83/84 Plus calculator provides *P*-values, we don't need to find critical values. To find a critical *F* value corresponding to a 0.05 significance level, refer to Table A-5 and use the right-tail area of 0.025 or 0.05, depending on the type of test:

• *Two-tailed* test: Use Table A-5 with 0.025 in the right tail. (We have $\alpha = 0.05$ divided between the two tails, so the area in the right tail is 0.025.)

• *One-tailed* test: Use Table A-5 with $\alpha = 0.05$ in the right tail.

Figure 9-5

F Distribution

There is a different *F* distribution for each different pair of degrees of freedom for the numerator and denominator.

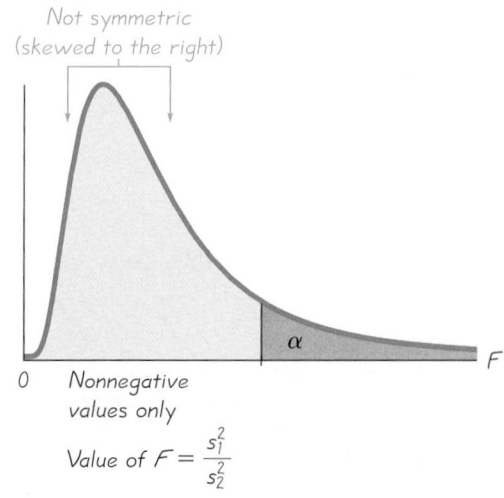

Not symmetric (skewed to the right)

α

F

0   Nonnegative values only

Value of $F = \dfrac{s_1^2}{s_2^2}$

Find the critical value of F in the column with the number $n_1 - 1$ and the row with the number $n_2 - 1$. Because we are stipulating that the larger sample variance is s_1^2, all one-tailed tests will be right-tailed and all two-tailed tests will require that we find only the critical value located to the right. (We have no need to find a critical value at the left tail, which can be somewhat tricky. See Exercise 23.)

Table A-5 provides critical values for select sample sizes and significance levels, but the TI-83/84 Plus calculator provides P-values for any sample sizes and significance levels.

Interpreting the F Test Statistic   If the two populations really do have equal variances, then the ratio s_1^2/s_2^2 tends to be close to 1 because s_1^2 and s_2^2 tend to be close in value. But if the two populations have radically different variances, s_1^2 and s_2^2 tend to be very different numbers. If we let s_1^2 be the larger sample variance, then the ratio s_1^2/s_2^2 will be a large number whenever s_1^2 and s_2^2 are far apart in value. Consequently, a value of F near 1 will be evidence in favor of $\sigma_1^2 = \sigma_2^2$, but a large value of F will be evidence against the conclusion of equality of the population variances.

> *Large* values of F are evidence against $\sigma_1^2 = \sigma_2^2$.

Claims About Standard Deviations   The F test statistic applies to a claim made about two variances, but we can also use it for claims about two population standard deviations. Any claim about two population standard deviations can be restated in terms of the corresponding variances.

Explore the Data!   Because the F test requirement of normal distributions is so important and so strict, we should begin by examining the distributions of the samples with histograms, boxplots, and normal quantile plots, and we should search for outliers. See the requirement check in the following example.

> **EXAMPLE 1**   **Comparing Variation in Weights of Quarters**   Data Set 20 in Appendix B includes weights (in grams) of quarters made before 1964 and weights of quarters made after 1964. Sample statistics are listed below. When designing coin vending machines, we must consider the standard deviations of pre-1964 quarters and post-1964 quarters. Use a 0.05 significance level to test the claim that the weights of pre-1964 quarters and the weights of post-1964 quarters are from populations with the same standard deviation.
>
> | Pre-1964 Quarters | Post-1964 Quarters |
> |---|---|
> | $n = 40$ | $n = 40$ |
> | $s = 0.08700$ g | $s = 0.06194$ g |

> **SOLUTION**   **REQUIREMENTS CHECK** (1) The two populations are clearly independent of each other. Quarters made before 1964 are not at all related to those made after 1964. The quarters are not matched or paired in any way. (2) The two samples are simple random samples selected from coins in circulation. (3) The two samples appear to be from populations having normal distributions, based on the STATDISK histograms and normal quantile plots shown below. Also, there are no outliers. The requirements are satisfied. ✅

continued

PRE-1964 QUARTERS

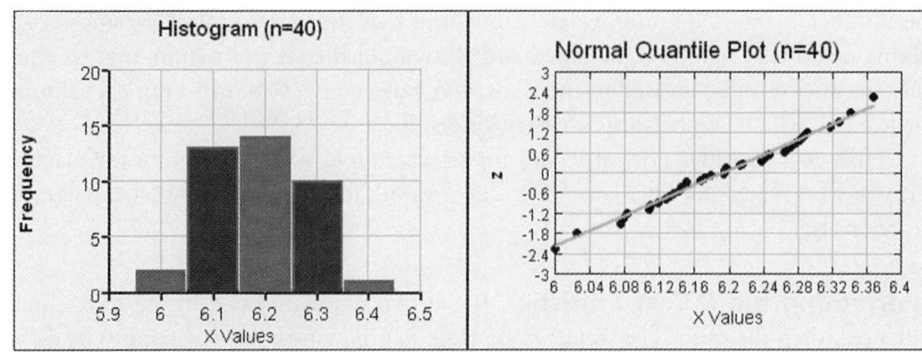

POST-1964 QUARTERS

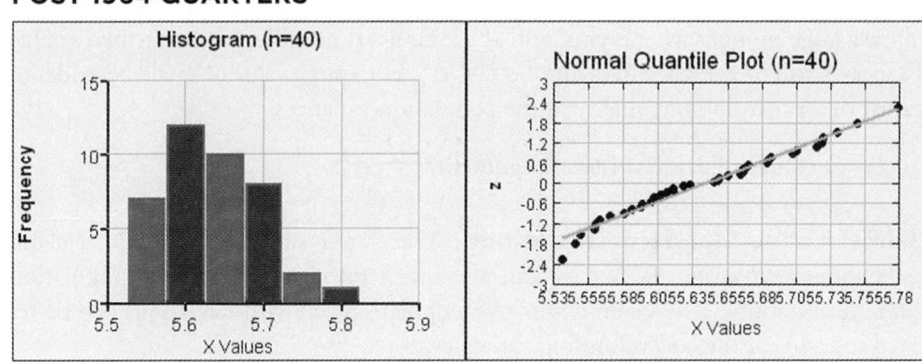

Instead of using the sample standard deviations to test the claim of equal population standard deviations, we use the sample variances to test the claim of equal population variances, but we can state conclusions in terms of standard deviations. Because we stipulate in this section that the larger variance is denoted by s_1^2, we let $s_1^2 = 0.08700^2$ and $s_2^2 = 0.06194^2$. We now proceed to use the traditional method of testing hypotheses as outlined in Figure 8-9.

Step 1: The claim of equal standard deviations is equivalent to a claim of equal variances, which we express symbolically as $\sigma_1^2 = \sigma_2^2$.

Step 2: If the original claim is false, then $\sigma_1^2 \neq \sigma_2^2$.

Step 3: Because the null hypothesis is the statement of equality and because the alternative hypothesis cannot contain equality, we have

$$H_0: \sigma_1^2 = \sigma_2^2 \text{ (original claim)} \qquad H_1: \sigma_1^2 \neq \sigma_2^2$$

Step 4: The significance level is $\alpha = 0.05$.

Step 5: Because this test involves two population variances, we use the F distribution.

Step 6: The test statistic is

$$F = \frac{s_1^2}{s_2^2} = \frac{0.08700^2}{0.06194^2} = 1.9729$$

TI-83/84 Plus

```
2-SampFTest
 σ1≠σ2
 F=1.972859254
 p=.0367533905
 Sx1=.087
 Sx2=.06194
↓n1=40
```

TI-83/84 Plus Using a TI-83/84 Plus calculator with the instructions given at the end of this section, we obtain the display shown in the margin. The calculator automatically computes the value of the test statistic $F = 1.9729$ (rounded) and the P-value 0.0368 (rounded). (Because the TI-83/84 Plus calculator provides P-values and not critical values, we use the P-value method of testing hypotheses.)

Step 7: Because the *P*-value of 0.0368 is less than the significance level of 0.05, we reject the null hypothesis of equal variances. There is sufficient evidence to warrant rejection of the claim of equal standard deviations.

   If we had used the traditional method of testing hypotheses, we would need to find only the rightmost critical value of $F = 1.8907$, as shown in Figure 9-6. Because the test statistic falls in the critical region, we reject the null hypothesis of equal variances, and we conclude that there is sufficient evidence to reject the claim of equal standard deviations.

INTERPRETATION   There is sufficient evidence to warrant rejection of the claim that the two standard deviations are equal. The variation among weights of quarters made after 1964 is significantly different from the variation among weights of quarters made before 1964.

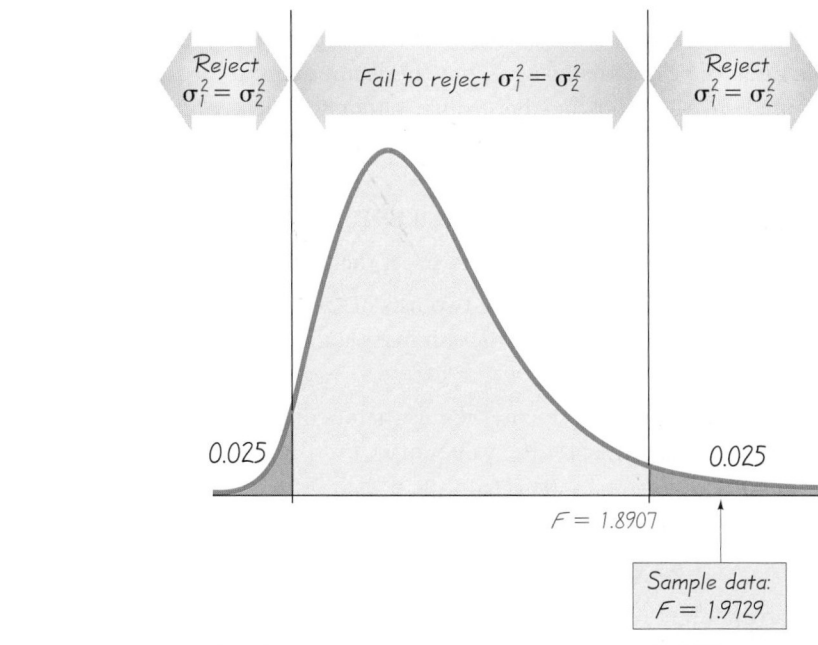

Figure 9-6
Distribution of s_1^2/s_2^2 for Weights of Pre-1964 Quarters and Post-1964 Quarters

Confidence Intervals   Exercise 24 deals with the construction of confidence intervals.

Part 2: Alternative Methods

Part 1 of this section presents the *F* test for comparing variances. Because that test is so sensitive to departures from normality, we now briefly describe two alternative methods that are not so sensitive to departures from normality.

Count Five   The *count five* method is a relatively simple alternative to the *F* test, and it does not require normally distributed populations. (See "A Quick, Compact, Two-Sample Dispersion Test: Count Five," by McGrath and Yeh, *American Statistician*, Vol. 59, No. 1.) If the two sample sizes are equal, and if one sample has at least five of the largest mean absolute deviations (MAD), then we conclude that its population has a larger variance. See Exercise 21 for the specific procedure.

Levene-Brown-Forsythe Test   The *Levene-Brown-Forsythe test* (or modified Levene's test) is another alternative to the *F* test, and it is much more robust. This test begins with a transformation of each set of sample values. Within the first sample, replace each *x* value with |*x* − median|, and do the same for the second sample. Using the transformed values, conduct a *t* test of equality of means for independent samples, as described in Part 1 of Section 9-3. Because the transformed values are now deviations, the *t* test for equality of means is actually a test comparing variation in the two samples. See Exercise 22.

In addition to the count five test and the Levene-Brown-Forsythe test, there are other alternatives to the *F* test, as well as adjustments that improve the performance of the *F* test. See "Fixing the *F* Test for Equal Variances," by Shoemaker, *American Statistician,* Vol. 57, No. 2.

Using a TI-83/84 Plus Calculator for Hypothesis Tests With Two Variances or Two Standard Deviations

Entries for Example 1

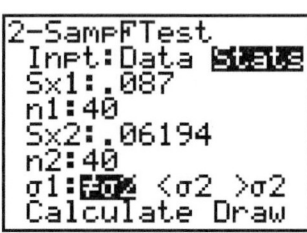

The TI-83/84 Plus calculator is designed for hypothesis tests (but not confidence intervals) involving two variances or two standard deviations. In order to simplify the hypothesis-testing procedure, the method described in this section requires that the larger sample variance be designated as s_1^2, but this is not necessary if a TI-83/84 Plus calculator is used. In the following procedure, either sample can be designated as sample 1. (However, the answers in Appendix D are based on the assumption that $s_1^2 > s_2^2$.)

1. Press **STAT**, then press ▷ twice to select **TESTS**.

2. Using the ▽ arrow key, scroll down to select the menu item of **2-SampFTest.**

3. For the Inpt entry, select **Data** if the two lists of sample data are already stored in the calculator, or select **Stats** if the summary statistics of s_1, n_1, s_2, n_2 are all known.

4. If **Data** was selected in Step 3, enter the list names or labels, such as L1 and L2. If **Stats** was selected, enter the values of s_1, n_1, s_2, and n_2.

5. Select the type of test (two-tailed, left-tailed, right-tailed) by selecting the format of the alternative hypothesis:

$$\sigma_1 \neq \sigma_2 \qquad \sigma_1 < \sigma_2 \qquad \sigma_1 > \sigma_2$$

Results for Example 1

```
2-SampFTest
 σ1≠σ2
 F=1.972859254
 p=.0367533905
 Sx1=.087
 Sx2=.06194
↓n1=40
```

6. Use the arrow keys to scroll down to the bottom line, select **Calculate,** then press **ENTER**.

If this procedure is used with the sample statistics given in the example from this section, the entry screen and the screen of results will be as shown in the accompanying displays. Note that the *P*-value of 0.0368 (rounded) is provided, so the *P*-value approach to hypothesis testing is being used. Because this *P*-value is less than the significance level of 0.05, we reject the null hypothesis and get the same conclusion given in Example 1.

9-5   Basic Skills and Concepts

Statistical Literacy and Critical Thinking

1. Interpreting *F* When testing the claim that two different simple random samples of heights of men are from populations having the same standard deviation, the author obtained the *F* test statistic of 1.010 (based on data from the National Health and Nutrition Examination Survey). What does the value of the *F* test statistic reveal about the sample data?

2. *F* Distribution The author repeated the process of selecting two different random samples of heights of men (from data obtained through the National Health and Nutrition Examination

Survey). In each case, the ratio s_1^2/s_2^2 was recorded without the stipulation that s_1 is the larger of the two standard deviations. Identify two different properties of the distribution of values of that ratio.

3. Robust What does it mean when we say that the F test described in this section is not *robust* against departures from normality? Name two alternatives that are more robust against departures from normality.

4. Testing Normality Given that the F test is not robust against departures from normality, it becomes necessary to verify that the two samples are from populations having distributions that are quite close to normal distributions. Assume that you want to test the claim of equal standard deviations using the samples of cholesterol levels of men and women listed in Data Set 1 in Appendix B. What are some methods that can be used to test for normality?

Hypothesis Test of Equal Variances. *In Exercises 5 and 6, test the given claim. Use a significance level of $\alpha = 0.05$ and assume that all populations are normally distributed.*

5. Zinc Treatment Claim: Weights of babies born to mothers given placebos vary more than weights of babies born to mothers given zinc supplements (based on data from "The Effect of Zinc Supplementation on Pregnancy Outcome," by Goldenberg, et al., *Journal of the American Medical Association,* Vol. 274, No. 6). Sample results are summarized below.

Placebo group: $n = 16, \bar{x} = 3088$ g, $s = 728$ g

Treatment group: $n = 16, \bar{x} = 3214$ g, $s = 669$ g

6. Weights of Pennies Claim: Weights of pre-1983 pennies and weights of post-1983 pennies have the same amount of variation. (The results are based on Data Set 20 in Appendix B.)

Weights of pre-1983 pennies: $n = 35, \bar{x} = 3.07478$ g, $s = 0.03910$ g

Weights of post-1983 pennies: $n = 37, \bar{x} = 2.49910$ g, $s = 0.01648$ g

7. Interpreting Display from Loads on Cans The axial load (in pounds) of a cola can is the maximum load that can be applied to the top before the can is crushed. When testing the claim that axial loads of cola cans with wall thickness of 0.0111 in. have the same standard deviation as the axial loads of cola cans with wall thickness of 0.0109 in., we obtain the accompanying TI-83/84 Plus calculator display. (The original data are listed in Data Set 21 in Appendix B.) Using the display and a 0.01 significance level, test the claim that the two samples are from populations with the same standard deviation.

TI-83/84 Plus

```
2-SampFTest
  σ1≠σ2
  F=1.577726851
  P=.0027873066
  Sx1=27.7746881
  Sx2=22.1122677
↓x̄1=281.8057143
```

8. Interpreting Display for Student and Faculty Car Ages Students at the author's college randomly selected samples of student cars and faculty cars and recorded their ages based on the registration stickers. See the following Excel display of the results. What is the P-value for a hypothesis test of equal standard deviations? Is there sufficient evidence to support the claim that the ages of faculty cars and the ages of student cars have different amounts of variation?

EXCEL

F-Test Two-Sample for Variances		
	Variable 1	Variable 2
Mean	7.890022573	5.990152447
Variance	13.46556454	13.31957893
Observations	217	152
df	216	151
F	1.010960227	
P(F<=f) one-tail	0.474614068	
F Critical one-tail	1.284757846	

Hypothesis Tests of Claims About Variation. *In Exercises 9–18, test the given claim. Assume that both samples are independent simple random samples from populations having normal distributions.*

9. Baseline Characteristics In journal articles about clinical experiments, it is common to include *baseline characteristics* of the different treatment groups so that they can be compared. In an article about the effects of different diets, a table of baseline characteristics showed that 40 subjects treated with the Atkins diet had a mean age of 47 years with a standard deviation of 12 years. Also, 40 subjects treated with the Zone diet had a mean age of 51 years with a standard deviation of 9 years. Use a 0.05 significance level to test the claim that subjects from both treatment groups have ages with the same amount of variation. How are comparisons of treatments affected if the treatment groups have different characteristics?

10. Braking Distances of Cars A random sample of 13 four-cylinder cars is obtained, and the braking distances are measured and found to have a mean of 137.5 ft and a standard deviation of 5.8 ft. A random sample of 12 six-cylinder cars is obtained and the braking distances have a mean of 136.3 ft and a standard deviation of 9.7 ft (based on Data Set 16 in Appendix B). Use a 0.05 significance level to test the claim that braking distances of four-cylinder cars and braking distances of six-cylinder cars have the same standard deviation.

11. Testing Effects of Alcohol Researchers conducted an experiment to test the effects of alcohol. The errors were recorded in a test of visual and motor skills for a treatment group of 22 people who drank ethanol and another group of 22 people given a placebo. The errors for the treatment group have a standard deviation of 2.20, and the errors for the placebo group have a standard deviation of 0.72 (based on data from "Effects of Alcohol Intoxication on Risk Taking, Strategy, and Error Rate in Visuomotor Performance," by Streufert, et al., *Journal of Applied Psychology,* Vol. 77, No. 4). Use a 0.05 significance level to test the claim that the treatment group has errors that vary more than the errors of the placebo group.

12. Home Size and Selling Price Using the sample data from Data Set 23 in Appendix B, 21 homes with living areas under 2000 ft^2 have selling prices with a standard deviation of \$32,159.73. There are 19 homes with living areas greater than 2000 ft^2 and they have selling prices with a standard deviation of \$66,628.50. Use a 0.05 significance level to test the claim of a real estate agent that homes larger than 2000 ft^2 have selling prices that vary more than the smaller homes.

13. Magnet Treatment of Pain Researchers conducted a study to determine whether magnets are effective in treating back pain, with results given below (based on data from "Bipolar Permanent Magnets for the Treatment of Chronic Lower Back Pain: A Pilot Study," by Collacott, Zimmerman, White, and Rindone, *Journal of the American Medical Association,* Vol. 283, No. 10). The values represent measurements of pain using the visual analog scale. Use a 0.05 significance level to test the claim that those given a sham treatment (similar to a placebo) have pain reductions that vary more than the pain reductions for those treated with magnets.

Reduction in pain level after sham treatment: $n = 20, \bar{x} = 0.44, s = 1.4$

Reduction in pain level after magnet treatment: $n = 20, \bar{x} = 0.49, s = 0.96$

14. Effects of Marijuana Use on College Students In a study of the effects of marijuana use, light and heavy users of marijuana in college were tested for memory recall, with the results given below (based on data from "The Residual Cognitive Effects of Heavy Marijuana Use in College Students," by Pope and Yurgelun-Todd, *Journal of the American Medical Association,* Vol. 275, No. 7). Use a 0.05 significance level to test the claim that the population of heavy marijuana users has a standard deviation different from that of light users.

Items sorted correctly by light marijuana users: $n = 64, \bar{x} = 53.3, s = 3.6$

Items sorted correctly by heavy marijuana users: $n = 65, \bar{x} = 51.3, s = 4.5$

15. Radiation in Baby Teeth Listed below are amounts of strontium-90 (in millibecquerels or mBq per gram of calcium) in a simple random sample of baby teeth obtained from

Pennsylvania residents and New York residents born after 1979 (based on data from "An Unexpected Rise in Strontium-90 in U.S. Deciduous Teeth in the 1990s," by Mangano, et al., *Science of the Total Environment*). Use a 0.05 significance level to test the claim that amounts of Strontium-90 from Pennsylvania residents vary more than amounts from New York residents.

Pennsylvania: 155 142 149 130 151 163 151 142 156 133 138 161

New York: 133 140 142 131 134 129 128 140 140 140 137 143

16. BMI for Miss America Listed below are body mass indexes (BMI) for Miss America winners from two different time periods. Use a 0.05 significance level to test the claim that winners from both time periods have BMI values with the same amount of variation.

BMI (from recent winners): 19.5 20.3 19.6 20.2 17.8 17.9 19.1 18.8 17.6 16.8

BMI (from the 1920s and 1930s): 20.4 21.9 22.1 22.3 20.3 18.8 18.9 19.4 18.4 19.1

17. Discrimination The Revenue Commissioners in Ireland conducted a contest for promotion. Ages of the unsuccessful and successful applicants are given below (based on data from "Debating the Use of Statistical Evidence in Allegations of Age Discrimination," by Barry and Boland, *American Statistician*, Vol. 58, No. 2). Use a 0.05 significance level to test the claim that both samples are from populations having the same standard deviation.

Unsuccessful applicants: 34 37 37 38 41 42 43 44 44 45 45 45
46 48 49 53 53 54 54 55 56 57 60

Successful applicants: 27 33 36 37 38 38 39 42 42 43 43 44
44 44 45 45 45 45 46 46 47 47 48 48
49 49 51 51 52 54

18. Platelet Counts Listed below are samples of platelet counts (number per mm^3) from randomly selected men and women (based on data from the National Health and Nutrition Examination Survey). Low platelet counts may result in excessive bleeding, while high platelet counts increase the risk of thrombosis. Use a 0.05 significance level to test the claim that men and women have platelet counts with the same standard deviation.

Female: 224.0 364.5 468.0 323.5 306.5 264.5 233.0 254.5 463.0
282.5 307.5 360.5 315.0 284.0 259.5 259.5 369.0 471.0
198.0 390.0 269.5 344.5 386.5 256.0 226.0 259.0 271.5

Male: 264.5 360.0 384.5 171.0 328.5 267.0 238.0 251.0 321.5
282.5 291.5 164.0 199.5 220.0 245.0 266.0 369.0 210.5
234.0 244.5 365.5 265.0 225.0

Large Data Sets. *In Exercises 19 and 20, use the indicated Data Sets from Appendix B. Assume that both samples are independent simple random samples from populations having normal distributions.*

19. Freshman 15 Study Use the sample weights (in kg) of male and female college students measured in April of their freshman year, as listed in Data Set 3 in Appendix B. Use a 0.05 significance level to test the claim that near the end of the freshman year, weights of male college students vary more than weights of female college students.

20. Heights Use the samples of heights of men and women listed in Data Set 1 in Appendix B and use a 0.05 significance level to test the claim that heights of men vary more than heights of women.

9-5   Beyond the Basics

21. Count Five Test for Comparing Variation in Two Populations Use the original weights of pre-1964 quarters and post-1964 quarters listed in Data Set 20 in Appendix B. Instead of using the F test as in Example 1 in this section, use the following procedure for a "count five" test of equal variation. What do you conclude?

a. For the first sample, find the absolute deviation of each value. The absolute deviation of a sample value x is $|x - \bar{x}|$. Sort the absolute deviation values. Do the same for the second sample.

b. Let c_1 be the count of the number of absolute deviation values in the first sample that are greater than the largest absolute deviation value in the other sample. Also, let c_2 be the count of the number of absolute deviation values in the second sample that are greater than the largest absolute deviation value in the other sample. (One of these counts will always be zero.)

c. If the sample sizes are equal ($n_1 = n_2$), use a critical value of 5. If $n_1 \neq n_2$, calculate the critical value shown below.

$$\frac{\log(\alpha/2)}{\log\left(\dfrac{n_1}{n_1 + n_2}\right)}$$

d. If $c_1 \geq$ critical value, then conclude that $\sigma_1^2 > \sigma_2^2$. If $c_2 \geq$ critical value, then conclude that $\sigma_2^2 > \sigma_1^2$. Otherwise, fail to reject the null hypothesis of $\sigma_1^2 = \sigma_2^2$.

22. Levene-Brown-Forsythe Test for Comparing Variation in Two Populations Repeat Example 1 in this section using the Levene-Brown-Forsythe test. What do you conclude?

23. Finding Lower Critical F Values For hypothesis tests that were two-tailed, the methods of Part 1 require that we need to find only the upper critical value. Let's denote that value by F_R, where the subscript indicates the critical value for the right tail. The lower critical value F_L (for the left tail) can be found as follows: First interchange the degrees of freedom, then take the reciprocal of the resulting F value found in Table A-5. Assuming a significance level of 0.05, find the critical values F_L and F_R for a two-tailed hypothesis test with a sample of size 10 and another sample of size 21.

24. Constructing Confidence Intervals In addition to testing claims involving σ_1^2 and σ_2^2, we can also construct confidence interval estimates of the ratio σ_1^2/σ_2^2 using the following:

$$\left(\frac{s_1^2}{s_2^2} \cdot \frac{1}{F_R}\right) < \frac{\sigma_1^2}{\sigma_2^2} < \left(\frac{s_1^2}{s_2^2} \cdot \frac{1}{F_L}\right)$$

Here F_L and F_R are as described in Exercise 23. Refer to Data Set 18 in Appendix B, and construct a 95% confidence interval estimate for the ratio of the standard deviation of the weights of red M&Ms to the standard deviation of the weights of yellow M&Ms. Do the confidence interval limits include 1, and what can you conclude from whether confidence interval limits include 1?

Review

Two main activities of inferential statistics are (1) constructing confidence interval estimates of population parameters, and (2) using methods of hypothesis testing to test claims about population parameters. In Chapters 7 and 8 we discussed the estimation of population parameters and methods of testing hypotheses made about population parameters, but Chapters 7 and 8 considered only cases involving a single population. In this chapter we considered two samples drawn from two populations. This chapter presented methods for constructing confidence interval estimates and testing hypotheses for two population proportions (Section 9-2), for the means of two independent populations (Section 9-3), for the mean difference from two dependent populations (Section 9-4), and for two population standard deviations or variances (Section 9-5).

Statistical Literacy and Critical Thinking

1. Robust What does it mean when we say that some methods in this chapter are *robust* against departures from normality? Which method of this chapter is not robust against departures from normality?

2. Ginormous The word *ginormous* was added to the Merriam-Webster Dictionary at the time this exercise was written. AOL conducted an online poll in which Internet users were asked "What do you think of the word 'ginormous'?" Among the Internet users who chose to respond, 12,908 gave the word a thumbs up, while 12,224 other Internet users gave it a thumbs down. What do these results tell us about how the general population feels about the word *ginormous?* What methods of statistics can be used with the sample data for inferences about the general population? Explain.

3. Independent or Dependent Samples? A nutritionist selects a simple random sample of 50 cans of Coke and another simple random sample of 50 cans of Pepsi. The cans are arranged as 50 pairs, then the sugar content of each can is measured. Are the two samples (Coke and Pepsi) independent or dependent? Explain.

4. Comparing Ages An employee of the U.S. Department of Labor obtains the mean age of men and the mean age of women for each of the 50 states. She then uses those means to construct a confidence interval estimate of the difference between the mean age of men in the United States and the mean age of women in the United States. Why is that procedure *not* valid?

Chapter Quick Quiz

1. Identify the null and alternative hypotheses resulting from the claim that the proportion of male teachers in California is greater than the proportion of male teachers in Texas.

2. Find the value of the pooled proportion $\bar{p}$ obtained when testing the claim that $p_1 = p_2$ with the sample data $x_1 = 20$, $n_1 = 50$ and $x_2 = 55$, $n_2 = 100$.

3. Find the value of the test statistic resulting from the hypothesis test described in Exercise 2.

4. When testing the claim that $p_1 = p_2$, a test statistic of $z = -2.05$ is obtained. Find the P-value.

5. When testing the claim that $\mu_1 > \mu_2$, a P-value of 0.0001 is obtained. What is the final conclusion?

6. Identify the null and alternative hypotheses resulting from the claim that when comparing heights of husbands to the heights of their wives, the mean of the differences is equal to zero. Express those hypotheses in symbolic form.

7. Identify the null and alternative hypotheses resulting from the claim that the mean age of voters in California is less than the mean age of voters in Iowa.

8. Which distribution is used to test the claim that the standard deviation of the ages of Florida voters is equal to the standard deviation of New York voters? (normal, t, chi-square, F, binomial)

9. When testing the claim that two populations have different means, the P-value of 0.0009 is obtained. What should you conclude?

10. True or false: When testing a claim about the means of two independent populations, the alternative hypothesis can never contain the condition of equality.

Review Exercises

TI **1. Carpal Tunnel Syndrome Treatments** Carpal tunnel syndrome is a common wrist complaint resulting from a compressed nerve, and it is often caused by repetitive wrist movements. In a randomized controlled trial, among 73 patients treated with surgery and evaluated one year later, 67 were found to have successful treatments. Among 83 patients treated with splints and evaluated one year later, 60 were found to have successful treatments (based on data from "Splinting vs Surgery in the Treatment of Carpal Tunnel Syndrome," by Gerritsen, et al., *Journal of the American Medical Association*, Vol. 288, No. 10). In a journal article about the trial, authors claimed that "treatment with open carpal tunnel release surgery resulted in better outcomes than treatment with wrist splinting for patients with CTS (carpal tunnel syndrome)." Use a 0.01 significance level to test that claim. What treatment strategy is suggested by the results?

2. Effects of Cocaine on Children Researchers conducted a study to assess the effects that occur when children are exposed to cocaine before birth. Children were tested at age 4 for object assembly skill, which was described as "a task requiring visual-spatial skills related to mathematical competence." The 190 children born to cocaine users had a mean of 7.3 and a standard deviation of 3.0. The 186 children not exposed to cocaine had a mean score of 8.2 with a standard deviation of 3.0. (The data are based on "Cognitive Outcomes of Preschool Children with Prenatal Cocaine Exposure," by Singer, et al., *Journal of the American Medical Association*, Vol. 291, No. 20.) Use a 0.05 significance level to test the claim that prenatal cocaine exposure is associated with lower scores of four-year-old children on the test of object assembly.

3. Historical Data Set In 1908, "Student" (William Gosset) published the article "The Probable Error of a Mean" (*Biometrika*, Vol. 6, No. 1). He included the data listed below for two different types of straw seed (regular and kiln dried) that were used on adjacent plots of land. The listed values are the yields of straw in cwt per acre, and the yields are paired by the plot of land that they share.

a. Using a 0.05 significance level, test the claim that there is no difference between the yields from the two types of seed.

b. Construct a 95% confidence interval estimate of the mean difference between the yields from the two types of seed.

c. Does it appear that either type of seed is better?

Regular	19.25	22.75	23	23	22.5	19.75	24.5	15.5	18	14.25	17
Kiln dried	25	24	24	28	22.5	19.5	22.25	16	17.25	15.75	17.25

4. Effect of Blinding Among 13,200 submitted abstracts that were blindly evaluated (with authors and institutions not identified), 26.7% were accepted for publication. Among 13,433 abstracts that were not blindly evaluated, 29.0% were accepted (based on data from "Effect of Blinded Peer Review on Abstract Acceptance," by Ross, et al., *Journal of the American Medical Association*, Vol. 295, No. 14). Use a 0.01 significance level to test the claim that the acceptance rate is the same with or without blinding. How might the results be explained?

5. Comparing Readability of J. K. Rowling and Leo Tolstoy Listed below are Flesch Reading Ease scores taken from randomly selected pages in J. K. Rowling's *Harry Potter and the Sorcerer's Stone* and Leo Tolstoy's *War and Peace*. (Higher scores indicate writing that is easier to read.) Use a 0.05 significance level to test the claim that *Harry Potter and the Sorcerer's Stone* is easier to read than *War and Peace*. Is the result as expected?

Rowling: 85.3 84.3 79.5 82.5 80.2 84.6 79.2 70.9 78.6 86.2 74.0 83.7

Tolstoy: 69.4 64.2 71.4 71.6 68.5 51.9 72.2 74.4 52.8 58.4 65.4 73.6

6. Before/After Drug Effects Captopril is a drug designed to lower systolic blood pressure. When subjects were tested with this drug, their systolic blood pressure readings (in mm Hg) were measured before and after drug treatment, with the results given in the following table (based on data from "Essential Hypertension: Effect of an Oral Inhibitor of Angiotensin-Converting Enzyme," by MacGregor, et al., *British Medical Journal*, Vol. 2).

a. Use the sample data to construct a 99% confidence interval for the mean difference between the before and after readings.

b. Is there sufficient evidence to support the claim that captopril is effective in lowering systolic blood pressure?

Subject	A	B	C	D	E	F	G	H	I	J	K	L
Before	200	174	198	170	179	182	193	209	185	155	169	210
After	191	170	177	167	159	151	176	183	159	145	146	177

7. Smoking and Gender A simple random sample of 280 men included 71 who smoke, and a simple random sample of 340 women included 68 who smoke (based on data from the National Health and Nutrition Examination Survey). Use a 0.05 significance level to test the claim that the proportion of men who smoke is greater than the proportion of women who smoke.

8. Income and Education A simple random sample of 80 workers with high school diplomas is obtained, and the annual incomes have a mean of $37,622 and a standard deviation of $14,115. Another simple random sample of 39 workers with bachelor's degrees is obtained, and the annual incomes have a mean of $77,689, with a standard deviation of $24,227. Use a 0.01 significance level to test the claim that workers with a high school diploma have a lower mean annual income than workers with a bachelor's degree. Does solving this exercise contribute to a higher income?

9. Comparing Variation Using the sample data from Exercise 8 and a 0.05 significance level, test the claim that the two samples are from populations with the same standard deviation.

10. Comparing Variation The *baseline characteristics* of different treatment groups are often included in journal articles. In a study, 84 subjects in the treatment group had Mini-Mental State Examination scores with a mean of 18.6 and a standard deviation of 5.9. On the same exam, 69 subjects in the control group had a mean score of 17.5 with a standard deviation of 5.2 (based on data from "Effectiveness of Collaborative Care for Older Adults With Alzheimer Disease in Primary Care," by Callahan, et al., *Journal of the American Medical Association,* Vol. 295, No. 18). Use a 0.05 significance level to test the claim that the two samples are from populations with the same amount of variation.

Cumulative Review Exercises

1. Word Counts Listed below are the numbers of words (in thousands) males and females in randomly selected *couples* spoke in a day (based on data from "Are Women Really More Talkative Than Men?" by Mehl, Vazire, Ramirez-Esparza, Slatcher, and Pennebaker, *Science,* Vol. 317, No. 5834).

Male	9	25	16	21	15	8	14	19	8	14
Female	9	12	38	28	21	16	34	20	18	21

a. Are the two samples independent or dependent? Why?

b. Find the mean, median, mode, range, and standard deviation of the word counts for males. Express results with the appropriate units.

c. What is the level of measurement of the sample data? (nominal, ordinal, interval, ratio)

2. Word Counts Use the sample data from couples listed in Exercise 1, and use a 0.05 significance level to test the claim that among couples, females are more talkative than males.

3. Word Counts Refer to the sample data listed in Exercise 1. Assume that instead of being couples, the males and females have no relationships with each other, so the values are not paired. Use a 0.05 significance level to test the claim that the two samples are from populations with the same mean.

4. Confidence Interval for Word Counts Use the word counts for males from Exercise 1 and construct a 95% confidence interval estimate of the number of words males in couple relationships speak in a day.

5. Constructing a Frequency Distribution Frequency distributions are generally used for data sets larger than the samples in Exercise 1, but construct a frequency distribution summarizing the word counts for males. Use a class width of 4 and use 6 for the lower limit of the first class.

6. Normal Distribution Assume that the numbers of words males speak in a day are normally distributed with a mean of 15,000 words and a standard deviation of 6000 words.

a. If a male is randomly selected, find the probability that he speaks more than 17,000 words in a day.

b. If 9 males are randomly selected, find the probability that the mean number of words they speak in a day is greater than 17,000 words.

c. Find P_{90}.

7. Sample Size for Survey The Ford Motor Company is considering the name *Chameleon* for a new model of hybrid car. The marketing division wants to conduct a survey to estimate the percentage of car owners who answer "yes" when asked if the name *Chameleon* creates a positive image. How many car owners must be surveyed in order to be 90% confident that the sample percentage is in error by no more than 2.5 percentage points?

8. Discrimination Survey In a survey of executives, respondents were asked if they have witnessed gender discrimination within their company. Among the respondents, 126 said that they have witnessed such discrimination, and 205 said that they have not (based on data from Ladders.com). Use the sample results to construct a 95% confidence interval estimate of the percentage of executives who have witnessed gender discrimination within their company.

9. Working Students Assume that 50% of full-time college students have jobs (based on data from the Department of Education and *USA Today*). Also assume that a simple random sample of 50 full-time college students is obtained.

a. For simple random samples of groups of 50 full-time college students, what is the mean of the numbers who have jobs?

b. For simple random samples of groups of 50 full-time college students, what is the standard deviation of the numbers who have jobs?

c. Find the probability that among 50 randomly selected full-time college students, at least 20 have jobs.

10. Firearm Rejections For a recent year, 1.6% of the applications for transfer of firearms were rejected (based on data from the U.S. Bureau of Justice Statistics). If 20 such applications are randomly selected, find the probability that none of them are rejected. Is such an event unusual? Why or why not?

TI-83/84 Plus Calculator Project

The TI-83/84 Plus calculator is capable of generating normally distributed data drawn from a population with a specified mean and standard deviation. Use this procedure:

1. Clear lists L1 and L2. (Press **STAT**, select **ClrList,** and enter L1, L2 with the commas included.)

2. Press **MATH** and use ▷ to select the top menu item of **PRB.**

3. Select the menu item of **randNorm.**

4. The format for this command is **randNorm(μ, σ, n)**, where μ is the population mean, σ is the population standard deviation, and n is the sample size.

5. Press **ENTER** and the random data will be generated. **STO▶** can be used to store the data in a list, such as L1.

Use the above procedure to generate two sets of sample data that represent simulated IQ scores, as described below:

IQ scores of treatment group: Generate 10 sample values from a normally distributed population with mean 100 and standard deviation 15.

IQ scores of placebo group: Generate 12 sample values from a normally distributed population with mean 100 and standard deviation 15.

We can see from the way the data are generated that both data sets actually come from the same population, so there should be no difference between the two sample means.

a. After generating the two data sets, use a 0.10 significance level to test the claim that the two samples come from populations with the same mean.

b. If this experiment is repeated many times, what is the expected percentage of trials leading to the conclusion that the two population means are different? How does this relate to a type I error?

c. If your generated data should lead to the conclusion that the two population means are different, would this conclusion be correct or incorrect in reality? How do you know?

d. If part (a) is repeated 20 times, what is the probability that none of the hypothesis tests leads to rejection of the null hypothesis?

e. Repeat part (a) 20 times. How often was the null hypothesis of equal means rejected? Is this the result you expected?

INTERNET PROJECT

Comparing Populations

Go to: **http://www.aw.com/triola**

The previous chapter showed you methods for testing hypotheses about a single population. This chapter expanded on those ideas, allowing you to test hypotheses about the relationships between two populations. In a similar fashion, the Internet Project for this chapter differs from that of the previous chapter in that you will need data for two populations or groups to conduct investigations.

In this Internet Project you will find several hypothesis-testing problems involving multiple populations. In these problems, you will analyze salary fairness, population demographics, and a traditional superstition. In each case you will formulate the problem as a hypothesis test, collect relevant data, then conduct and summarize the appropriate test.

APPLET PROJECT

Open the Applets folder on the CD and double-click on **Start.** Select the menu item of **Simulate the probability of a head with an unfair coin [$P(H) = 0.2$].** Obtain simulated results from 100 flips. Then select the menu item of **Simulate the probability of a head with a fair coin.** Obtain simulated results from 100 flips.

Use the methods of this section to test for equality of the proportion of heads with the unfair coin and the proportion of heads with the fair coin. Repeat both simulations using 1000 flips, then repeat the hypothesis test. What can you conclude?

FROM DATA TO DECISION

Critical Thinking: Do Academy Awards involve age discrimination?

Listed below are the ages of actresses and actors at the times that they won Oscars for the categories of Best Actress and Best Actor. The ages are listed in chronological order by row, so that corresponding locations in the two tables are from the same year. (*Notes:* In 1968 there was a tie in the Best Actress category, and the mean of the two ages is used; in 1932 there was a tie in the Best Actor category, and the mean of the two ages is used. These data are suggested by the article "Ages of Oscar-winning Best Actors and Actresses," by Richard Brown and Gretchen Davis, *Mathematics Teacher* magazine. In that article, the year of birth of the award winner was subtracted from the year of the awards ceremony, but the ages in the tables below are based on the birth date of the winner and the date of the awards ceremony.)

Analyzing the Results

1. First *explore* the data using suitable statistics and graphs. Use the results to make informal comparisons.

2. Determine whether there are significant differences between the ages of the Best Actresses and the ages of the Best Actors. Use appropriate hypothesis tests. Describe the methods used and the conclusions reached.

3. Discuss cultural implications of the results. Does it appear that actresses and actors are judged strictly on the basis of their artistic abilities? Or does there appear to be discrimination based on age, with the Best Actresses tending to be younger than the Best Actors? Are there any other notable differences?

Best Actresses

22	37	28	63	32	26	31	27	27	28
30	26	29	24	38	25	29	41	30	35
35	33	29	38	54	24	25	46	41	28
40	39	29	27	31	38	29	25	35	60
43	35	34	34	27	37	42	41	36	32
41	33	31	74	33	50	38	61	21	41
26	80	42	29	33	35	45	49	39	34
26	25	33	35	35	28	30	29	61	

Best Actors

44	41	62	52	41	34	34	52	41	37
38	34	32	40	43	56	41	39	49	57
41	38	42	52	51	35	30	39	41	44
49	35	47	31	47	37	57	42	45	42
44	62	43	42	48	49	56	38	60	30
40	42	36	76	39	53	45	36	62	43
51	32	42	54	52	37	38	32	45	60
46	40	36	47	29	43	37	38	45	

Cooperative Group Activities

1. Out-of-class activity Survey married couples and record the number of credit cards each person has. Analyze the paired data to determine whether husbands have more credit cards, wives have more credit cards, or they both have about the same number of credit cards. Try to identify reasons for any discrepancy.

2. Out-of-class activity Measure and record the height of the husband and the height of the wife from each of several different married couples. Estimate the mean of the differences between heights of husbands and the heights of their wives. Compare the result to the difference between the mean height of men and the mean height of women included in Data Set 1 in Appendix B. Do the results suggest that height is a factor when people select marriage partners?

3. Out-of-class activity Are estimates influenced by anchoring numbers? Refer to the related Chapter 3 Cooperative Group Activity. In Chapter 3 we noted that, according to author John Rubin, when people must estimate a value, their estimate is often "anchored" to (or influenced by) a preceding number. In that Chapter 3 activity, some subjects were asked to quickly estimate the value of $8 \times 7 \times 6 \times 5 \times 4 \times 3 \times 2 \times 1$, and others were asked to quickly estimate the value of $1 \times 2 \times 3 \times 4 \times 5 \times 6 \times 7 \times 8$. In Chapter 3, we could compare the two sets of results by using statistics (such as the mean) and graphs (such as boxplots). The methods of Chapter 9 now allow us to compare the results with a formal hypothesis test. Specifically, collect your own sample data and test the claim that when we begin with larger numbers (as in $8 \times 7 \times 6$), our estimates tend to be larger.

4. In-class activity Divide into groups according to gender, with about 10 or 12 students in each group. Each group member should record his or her pulse rate by counting the number of heartbeats in 1 minute, and the group statistics ($n, \bar{x}, s$) should be calculated. The groups should test the null hypothesis of no difference between their mean pulse rate and the mean of the pulse rates for the population from which subjects of the same gender were selected for Data Set 1 in Appendix B.

5. Out-of-class activity Randomly select a sample of male students and a sample of female students and ask each selected person a yes/no question, such as whether they support a death penalty for people convicted of murder, or whether they believe that the federal government should fund stem cell research. Record the response, the gender of the respondent, and the gender of the person asking the question. Use a formal hypothesis test to determine whether there is a difference between the proportions of *yes* responses from males and females. Also, determine whether the responses appear to be influenced by the gender of the interviewer.

6. Out-of-class activity Use a watch to record the waiting times of a sample of McDonald's customers and the waiting times of a sample of Burger King customers. Use a hypothesis test to determine whether there is a significant difference.

7. Out-of-class activity Construct a short survey of just a few questions, including a question asking the subject to report his or her height. After the subject has completed the survey, measure the subject's height (without shoes) using an accurate measuring system. Record the gender, reported height, and measured height of each subject. Do male subjects appear to exaggerate their heights? Do female subjects appear to exaggerate their heights? Do the errors for males appear to have the same mean as the errors for females?

8. In-class activity Without using any measuring device, ask each student to draw a line believed to be 3 in. long and another line believed to be 3 cm long. Then use rulers to measure and record the lengths of the lines drawn. Record the errors along with the genders of the students making the estimates. Test the claim that when estimating the length of a 3 in. line, the mean error from males is equal to the mean error from females. Also, do the results show that we have a better understanding of the British system of measurement (inches) than the SI system (centimeters)?

9. In-class activity Use a ruler as a device for measuring reaction time. One person should suspend the ruler by holding it at the top while the subject holds his or her thumb and forefinger at the bottom edge, ready to catch the ruler when it is released. Record the distance that the ruler falls before it is caught. Convert that distance to the time (in seconds) that it took the subject to react and catch the ruler. (If the distance is measured in inches, use $t = \sqrt{d/192}$. If the distance is measured in centimeters, use $t = \sqrt{d/487.68}$.) Test each subject once with the dominant hand and once with the other hand, and record the paired data. Does there appear to be a difference between the mean of the reaction times using the dominant hand and the mean from the other hand? Do males and females appear to have different mean reaction times?

10. Out-of-class activity Obtain simple random samples of cars in the student and faculty parking lots, and test the claim that students and faculty have the same proportions of foreign cars.

11. Out-of-class activity Obtain simple random samples of cars in parking lots of a discount store and an upscale department store, and test the claim that cars are newer in the parking lot of the upscale department store.

12. Out-of-class activity Obtain sample data and test the claim that husbands are older than their wives.

13. Out-of-class activity Obtain sample to test the claim that in the college library, science books have a mean age that is less than the mean age of English books.

14. Out-of-class activity Obtain sample data and test the claim that when people report their heights, they tend to provide values that are greater than their actual heights.

15. Out-of-class activity Conduct experiments and collect data to test the claim that there are no differences in taste between ordinary tap water and different brands of bottled water.

16. Out-of-class activity Collect sample data and test the claim that people who exercise tend to have pulse rates that are lower than those who do not exercise.

17. Out-of-class activity Collect sample data and test the claim that the proportion of female students who smoke is equal to the proportion of male students who smoke.

| NAME: | Mark T. Lycett |
| JOB: | Faculty and Research, Dept. of Anthropology, University of Chicago |

| NAME: | Kathleen Morrison |
| JOB: | Faculty and Research, Dept. of Anthropology, University of Chicago |

Mark T. Lycett and Kathleen Morrison are both on the faculty of the Department of Anthropology at the University of Chicago. Dr. Lycett's research deals with issues of economic, social, and political transformation associated with Spanish colonialism in the southwestern United States, and Dr. Morrison's research in southern India deals with problems of agricultural change, imperialism, and regional economic organization.

Q: How important is the use of statistics in archaeology?

A: It would be impossible to conduct archaeological research without at least a working knowledge of basic statistics.

Q: What concepts of statistics do you use?

A: Archaeologists make extensive use of both descriptive and inferential statistics on a daily basis. Exploratory data analysis using a variety of graphical and numerical summaries is increasingly common in modern archaeology. Archaeological problems routinely include studies of association for categorical variables, hypothesis testing for both 2-sample and *k*-sample data, correlation and regression problems, and a suite of nonparametric approaches.

Q: Please give a specific example illustrating the use of statistics in your work.

A: We have explored the size distribution of ancient grass pollen grains to investigate changes in agriculture in both the New and Old Worlds during the first centuries of European colonial expansion. Although almost all important crops are grasses with morphologically similar pollen, New World staple crops (corn) have much larger pollen grains than wild grasses, and Old World crops (principally wheat, barley, and rice) are intermediate in size. By studying the size distribution of reference samples of these staple crops as well as fossil grass pollen from archaeological contexts, we have been able to specify the range of crops introduced and grown at colonial period sites in New Mexico and India.

Our data have been used to make inferences about the number and kind of archaeological sites that existed in our study areas; to reconstruct ancient patterns of vegetation, agriculture, and economy; and to study the effects of colonialism and imperialism on local social, economic, and religious practices.

Q: Is your use of probability and statistics increasing, decreasing, or remaining stable?

A: Both the number and variety of statistical applications in archaeology is increasing, particularly as more sophisticated spatial databases become available through the widespread use of Geographic Information Systems technology.

Q: In terms of statistics, what would you recommend for prospective employees?

A: When we were college students, we understood that statistics would be a part of our professional lives, but we never imagined the degree to which we would use it on a daily basis. Undergraduates interested in archaeology should begin with an introductory course in probability and statistics. Those with professional or academic goals should consider more advanced undergraduate or graduate level course work in quantitative data analysis.

519

10-1 Review and Preview

10-2 Correlation

10-3 Regression

10-4 Variation and Prediction Intervals

10-5 Multiple Regression

10-6 Modeling

10

Correlation and Regression

Can we predict the cost of subway fare from the price of a slice of pizza?

In 1964, Eric Bram, a typical New York City teenager, noticed that the cost of a slice of cheese pizza was the same as the cost of a subway ride. Over the years, he noticed that those two costs seemed to increase by about the same amounts. In 1980, when the cost of a slice of pizza increased, he told the *New York Times* that the cost of subway fare would increase. His prediction proved to be correct.

In the recent *New York Times* article "Will Subway Fares Rise? Check at Your Pizza Place," reporter Clyde Haberman wrote that in New York City, the subway fare and the cost of a slice of pizza "have run remarkably parallel for decades." A random sample of costs (in dollars) of pizza and subway fares are listed in Table 10-1. Table 10-1 also includes values of the Consumer Price Index (CPI) for the New York metropolitan region, with the index of 100 assigned to the base period

there is a correlation between the cost of a slice of pizza and the cost of a subway fare. Because an informal conclusion based on an inspection of the scatterplot is largely subjective, we must use other tools for addressing questions such as:

- If there is a correlation between two variables, how can it be described? Is there an *equation* that can be used to predict the cost of a subway fare given the cost of a slice of pizza?
- If we can predict the cost of a subway fare, how accurate is that prediction likely to be?
- Is there also a correlation between the CPI and the cost of a subway fare, and if so, is the CPI better for predicting the cost of a subway fare?

These questions will be addressed in this chapter.

Table 10-1 Cost of a Slice of Pizza, Subway Fare, and the CPI

Year	1960	1973	1986	1995	2002	2003
Cost of Pizza	0.15	0.35	1.00	1.25	1.75	2.00
Subway Fare	0.15	0.35	1.00	1.35	1.50	2.00
CPI	30.2	48.3	112.3	162.2	191.9	197.8

from 1982 to 1984. The Consumer Price Index reflects the costs of a standard collection of goods and services, including such items as a gallon of milk and a loaf of bread.

From Table 10-1, we see that the paired pizza/subway fare costs are approximately the same for the given years. As a first step, we should examine the data visually. Recall from Section 2-4 that a scatterplot is a plot of (x, y) paired data. The pattern of the plotted data points is often helpful in determining whether there is a *correlation*, or association, between the two variables. The scatterplot shown in Figure 10-1 suggests that

TI-83/84 Plus

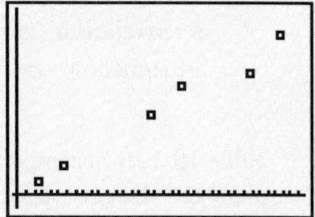

Figure 10-1 Scatterplot of Pizza Costs and Subway Costs

10-1   Review and Preview

In Chapter 9 we presented methods for making inferences from two samples. In Section 9-4 we considered two dependent samples, with each value of one sample somehow paired with a value from the other sample. In Section 9-4 we considered the differences between the paired values, and we illustrated the use of hypothesis tests for claims about the population of differences. We also illustrated the construction of confidence interval estimates of the mean of all such differences. In this chapter we again consider paired sample data, but the objective is fundamentally different from that of Section 9-4. In this chapter we introduce methods for determining whether a *correlation,* or association, between two variables exists and whether the correlation is linear. For linear correlations, we can identify an equation that best fits the data and we can use that equation to predict the value of one variable given the value of the other variable. In this chapter, we also present methods for analyzing differences between predicted values and actual values. In addition, we consider methods for identifying linear equations for correlations among three or more variables. We conclude the chapter with some basic methods for developing a mathematical model that can be used to describe nonlinear correlations between two variables.

10-2   Correlation

Key Concept In Part 1 of this section we introduce the *linear correlation coefficient r,* which is a numerical measure of the strength of the association between two variables representing quantitative data. Using paired sample data (sometimes called **bivariate data**), we find the value of r (usually using technology), then we use that value to conclude that there is (or is not) a linear correlation between the two variables. In this section we consider only *linear* relationships, which means that when graphed, the points approximate a straight-line pattern. In Part 2, we discuss methods of hypothesis testing for correlation.

Part 1: Basic Concepts of Correlation

We begin with the basic definition of *correlation,* a term commonly used in the context of an association between two variables.

 DEFINITION

A **correlation** exists between two variables when the values of one variable are somehow associated with the values of the other variable.

Table 10-1, for example, includes paired sample data consisting of costs of a slice of pizza and the corresponding costs of a subway fare in New York City. We will determine whether there is a correlation between the variable x (cost of a slice of pizza) and the variable y (cost of a subway fare).

Exploring the Data

Before doing any formal statistical analyses, we should use a scatterplot to explore the data visually. We can examine the scatterplot for any distinct patterns and for any outliers, which are points far away from all the other points. If the plotted points show a distinct pattern, we can conclude that there is a correlation between the two variables in a sample of paired data.

Figure 10-2 shows four scatterplots with different characteristics. The scatterplot in Figure 10-2(a) shows a distinct straight-line, or linear, pattern. We say that there is a *positive* correlation between x and y, since as the x-values increase, the corresponding y-values increase. The scatterplot in Figure 10-2(b) shows a distinct linear pattern. We say that there is a *negative* correlation between x and y, since as the x-values increase, the corresponding y-values decrease. The scatterplot in Figure 10-2(c) shows no distinct pattern and suggests that there is no correlation between x and y. The scatterplot in Figure 10-2(d) shows a distinct pattern suggesting a correlation between x and y, but the pattern is not that of a straight line.

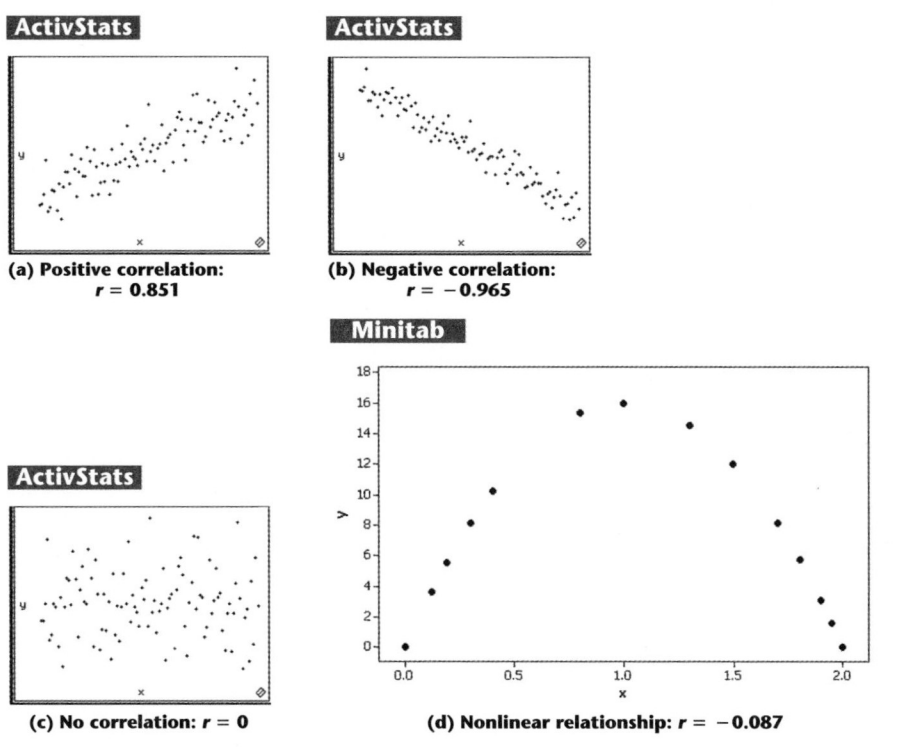

(a) Positive correlation:
$r = 0.851$

(b) Negative correlation:
$r = -0.965$

(c) No correlation: $r = 0$

(d) Nonlinear relationship: $r = -0.087$

Figure 10-2   Scatterplots

Linear Correlation Coefficient

Because conclusions based on visual examinations of scatterplots are largely subjective, we need more objective measures. We use the linear correlation coefficient r, which is useful for detecting straight-line patterns.

> **DEFINITION**
>
> The **linear correlation coefficient** r measures the strength of the linear correlation between the paired quantitative x- and y-values in a *sample*. (Its value is computed by using Formula 10-1 or Formula 10-2, included in the box on page 524. The linear correlation coefficient is sometimes referred to as the **Pearson product moment correlation coefficient** in honor of Karl Pearson (1857–1936), who originally developed it.)

Because the linear correlation coefficient r is calculated using sample data, it is a sample statistic used to measure the strength of the linear correlation between x and y. If we had every pair of population values for x and y, the result of Formula 10-1 or Formula 10-2 would be a population parameter, represented by ρ (Greek letter rho).

Speeding Out-of-Towners Ticketed More?

Are police more likely to issue a ticket to a speeding driver who is out-of-town than to a local driver? George Mason University researchers Michael Makowsky and Thomas Stratmann addressed this question by examining more than 60,000 warnings and tickets issued by Massachusetts police in one year. They found that out-of-town drivers from Massachusetts were 10% more likely to be ticketed than local drivers, and the 10% figure rose to 20% for out-of-state drivers. They also found a statistical association between a town's finances and speeding tickets. When compared to local drivers, out-of-town drivers had a 37% greater chance of being ticketed when speeding in a town in which voters had rejected a proposition to raise taxes more than the 2.5% amount allowed by state law. Such analyses can be conducted using methods of correlation and regression.

Objective

Determine whether there is a linear correlation between two variables.

Notation for the Linear Correlation Coefficient

n = number of pairs of sample data

Σ denotes addition of the items indicated.

Σx = sum of all x-values.

Σx^2 indicates that each x-value should be squared and then those squares added.

$(\Sigma x)^2$ indicates that the x-values should be added and the total then squared. It is extremely important to avoid confusing Σx^2 and $(\Sigma x)^2$.

Σxy indicates that each x-value should first be multiplied by its corresponding y-value. After obtaining all such products, find their sum.

r = linear correlation coefficient for *sample* data.

ρ = linear correlation coefficient for a *population* of paired data.

Requirements

Given any collection of sample paired quantitative data, the linear correlation coefficient r can always be computed, but the following requirements should be satisfied when using the sample data to make a conclusion about correlation in the population.

1. The sample of paired (x, y) data is a simple random sample of quantitative data. (It is important that the sample data have not been collected using some inappropriate method, such as using a voluntary response sample.)

2. Visual examination of the scatterplot must confirm that the points approximate a straight-line pattern.

3. Because results can be strongly affected by the presence of outliers, any outliers must be removed if they

are known to be errors. The effects of any other outliers should be considered by calculating r with and without the outliers included.

Note: Requirements 2 and 3 above are simplified attempts at checking this formal requirement:

The pairs of (x, y) data must have a **bivariate normal distribution.** Normal distributions are discussed in Chapter 6, but this assumption basically requires that for any fixed value of x, the corresponding values of y have a distribution that is approximately normal, and for any fixed value of y, the values of x have a distribution that is approximately normal. This requirement is usually difficult to check, so for now, we will use Requirements 2 and 3 as listed above.

Formulas for Calculating *r*

Formula 10-1

$$r = \frac{n(\Sigma xy) - (\Sigma x)(\Sigma y)}{\sqrt{n(\Sigma x^2) - (\Sigma x)^2}\sqrt{n(\Sigma y^2) - (\Sigma y)^2}}$$

Formula 10-1 is a shortcut formula that simplifies manual calculations, but r is usually calculated with computer software or a calculator.

Formula 10-2

$$r = \frac{\sum(z_x z_y)}{n - 1}$$

where z_x is the z score for the sample value x and z_y is the z score for the sample value y.

Interpreting the Linear Correlation Coefficient *r*

- *Using Computer Software to Interpret r:* If the P-value computed from r is less than or equal to the significance level, conclude that there is a linear correlation. Otherwise, there is not sufficient evidence to support the conclusion of a linear correlation.

- *Using Table A-6 to Interpret r:* If the absolute value of r, denoted $|r|$, exceeds the value in Table A-6, conclude that there is a linear correlation. Otherwise, there is not sufficient evidence to support the conclusion of a linear correlation.

TI-83/84 Plus Calculator: The TI-83/84 Plus calculator is designed to generate the value of the linear correlation coefficient r. See the instructions at the end of this section.

CAUTION

Know that the methods of this section apply to a *linear* correlation. If you conclude that there does not appear to be linear correlation, know that it is possible that there might be some other association that is not linear.

Rounding the Linear Correlation Coefficient r

Round the linear correlation coefficient r to three decimal places (so that its value can be directly compared to critical values in Table A-6). If manually calculating r and other statistics in this chapter, rounding in the middle of a calculation often creates substantial errors, so try using your calculator's memory to store intermediate results and round only the final result.

Properties of the Linear Correlation Coefficient r

1. The value of r is always between -1 and 1 inclusive. That is,
$$-1 \leq r \leq 1$$

2. *If all values of either variable are converted to a different scale, the value of r does not change.*

3. *The value of r is not affected by the choice of x or y.* Interchange all x- and y-values and the value of r will not change.

4. *r measures the strength of a linear relationship.* It is not designed to measure the strength of a relationship that is not linear (as in Figure 10-2(d)).

5. *r is very sensitive to outliers* in the sense that a single outlier can dramatically affect its value.

Calculating the Linear Correlation Coefficient r

The following three examples illustrate three different methods for finding the value of the linear correlation coefficient r, but you need to use only one method. *The use of computer software (as in Example 1) is strongly recommended.* If manual calculations are absolutely necessary, Formula 10-1 is recommended (as in Example 2). If a better understanding of r is desired, Formula 10-2 is recommended (as in Example 3).

Table 10-2   Costs of a Slice of Pizza and Subway Fare (in dollars)

Cost of Pizza	0.15	0.35	1.00	1.25	1.75	2.00
Subway Fare	0.15	0.35	1.00	1.35	1.50	2.00

EXAMPLE 1   **Finding r Using Computer Software** The paired pizza/subway fare costs from Table 10-1 are shown here in Table 10-2. Use computer software with these paired sample values to find the value of the linear correlation coefficient r for the paired sample data.

SOLUTION   **REQUIREMENT CHECK** We can always calculate the linear correlation coefficient r from paired quantitative data, but we should check the requirements if we want to use that value for making a conclusion about correlation. (1) The data are a simple random sample of quantitative data. (2) The plotted points in the scatterplot in Figure 10-1 do approximate a straight-line pattern. (3) The

continued

TI-83/84 Plus

```
LinRegTTest
 y=a+bx
 β≠0 and ρ≠0
↑b=.9450213806
 s=.1229869984
 r²=.9757704494
 r=.9878109381
```

scatterplot in Figure 10-1 also shows that there are no outliers. The requirements are satisfied.

If using computer software or a calculator, the value of r will be automatically calculated. For example, see the bottom line of the TI-83/84 Plus calculator display in the margin, which shows that $r = 0.988$ (rounded).

EXAMPLE 2   **Finding r Using Formula 10-1** Use Formula 10-1 to find the value of the linear correlation coefficient r for the paired pizza/subway fare costs given in Table 10-2.

SOLUTION   **REQUIREMENT CHECK** See the discussion of the requirement check in Example 1. The same comments apply here.

Using Formula 10-1, the value of r is calculated as shown below. Note that the variable x is used for the pizza costs, and the variable y is used for subway fare costs. Because there are six pairs of data, $n = 6$. Other required values are computed in Table 10-3.

Table 10-3   Calculating r with Formula 10-1

x (Pizza)	y (Subway)	x^2	y^2	xy
0.15	0.15	0.0225	0.0225	0.0225
0.35	0.35	0.1225	0.1225	0.1225
1.00	1.00	1.0000	1.0000	1.0000
1.25	1.35	1.5625	1.8225	1.6875
1.75	1.50	3.0625	2.2500	2.6250
2.00	2.00	4.0000	4.0000	4.0000
$\Sigma x = 6.50$	$\Sigma y = 6.35$	$\Sigma x^2 = 9.77$	$\Sigma y^2 = 9.2175$	$\Sigma xy = 9.4575$

Using the values in Table 10-3 and Formula 10-1, r is calculated as follows:

$$r = \frac{n\Sigma xy - (\Sigma x)(\Sigma y)}{\sqrt{n(\Sigma x^2) - (\Sigma x)^2}\sqrt{n(\Sigma y^2) - (\Sigma y)^2}}$$

$$= \frac{6(9.4575) - (6.50)(6.35)}{\sqrt{6(9.77) - (6.50)^2}\sqrt{6(9.2175) - (6.35)^2}}$$

$$= \frac{15.47}{\sqrt{16.37}\sqrt{14.9825}} = 0.988$$

EXAMPLE 3   **Finding r Using Formula 10-2** Use Formula 10-2 to find the value of the linear correlation coefficient r for the paired pizza/subway fare costs given in Table 10-2.

SOLUTION   **REQUIREMENT CHECK** See the discussion of the requirement check in Example 1. The same comments apply here.

If manual calculations are absolutely necessary, Formula 10-1 is much easier than Formula 10-2, but Formula 10-2 has the advantage of making it easier to *understand* how r works. (See the *rationale* for r discussed later in this section.) As in Example 2, the variable x is used for the pizza costs, and the variable y is used for subway fare costs. In Formula 10-2, each sample value is replaced by its corresponding z score. For example, the pizza costs have a mean of $\bar{x} = 1.083333$ and a standard deviation of $s_x = 0.738693$, so the first pizza cost of 0.15 results in this z score:

$$z_x = \frac{x - \bar{x}}{s_x} = \frac{0.15 - 1.083333}{0.738693} = -1.26349$$

The above calculation shows that the first pizza cost of $x = 0.15$ is converted to the z score of -1.26349. Table 10-4 lists the z scores for all of the pizza costs (see the third column) and the z scores for all of the subway fare costs (see the fourth column). (The subway fare costs have a mean of $\bar{y} = 1.058333$ and a standard deviation of $s_y = 0.706694$.) The last column of Table 10-4 lists the products $z_x \cdot z_y$.

Table 10-4   Calculating r with Formula 10-2

x (Pizza)	y (Subway)	z_x	z_y	$z_x \cdot z_y$
0.15	0.15	−1.26349	−1.28533	1.62400
0.35	0.35	−0.99274	−1.00232	0.99504
1.00	1.00	−0.11281	−0.08254	0.00931
1.25	1.35	0.22562	0.41272	0.09312
1.75	1.50	0.90250	0.62498	0.56404
2.00	2.00	1.24093	1.33250	1.65354
				$\Sigma(z_x z_y) = 4.93905$

Using $\Sigma(z_x \cdot z_y) = 4.93905$ from Table 10-4, the value of r is calculated by using Formula 10-2 as shown below.

$$r = \frac{\Sigma(z_x z_y)}{n - 1} = \frac{4.93905}{5} = 0.988$$

Interpreting the Linear Correlation Coefficient r

After calculating the linear correlation coefficient r, we must interpret its meaning. Using the criteria given in the preceding box, we can base our interpretation on a P-value or a critical value from Table A-6.

> **Using the TI-83/84 Plus Calculator: If the computed P-value is less than or equal to the significance level, conclude that there is a linear correlation. Otherwise, there is not sufficient evidence to support the conclusion of a linear correlation.**

> **Using Table A-6: If the absolute value of the computed value of r exceeds the value in Table A-6, conclude that there is a linear correlation. Otherwise, there is not sufficient evidence to support the conclusion of a linear correlation.**

If using Table A-6, we conclude that there is a linear correlation if $|r|$ exceeds the value found in Table A-6. This is equivalent to the condition that r is either greater than the value from Table A-6 or less than the negative of the value from Table A-6. It might be helpful to think of critical values from Table A-6 as being both positive and negative. For the pizza/subway fare data, Table A-6 yields $r = 0.811$ (for six pairs of data and a 0.05 significance level). So we can compare the computed value of $r = 0.988$ to the values of ± 0.811 as shown in Figure 10-3 on the next page. (Exercise

answers in Appendix C include critical values in a format such as ± 0.811.) Figures such as Figure 10-3 are helpful in visualizing and *understanding* the relationship between the computed r and the critical values from Table A-6, and they follow the same general pattern of graphs included in Chapters 8 and 9.

Figure 10-3
Critical Values from Table A-6 and the Computed Value of r

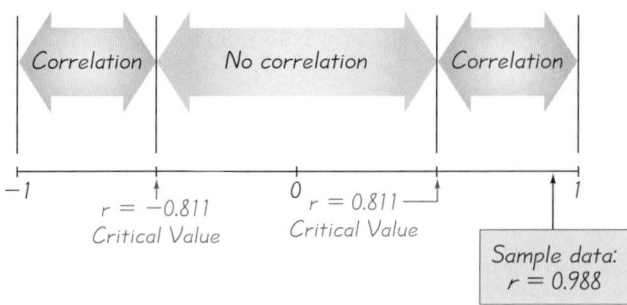

EXAMPLE 4   **Interpreting r** Interpret the value of $r = 0.988$ found in Examples 1, 2, and 3 (based on the pizza/subway fare costs listed in Table 10-2). Use the significance level of 0.05. Is there sufficient evidence to support a claim that there is a linear correlation between the costs of a slice of pizza and subway fares?

SOLUTION   **REQUIREMENT CHECK** The requirement check in Example 1 also applies here.

We can base our conclusion about correlation on either the *P*-value obtained from computer software or the critical value found in Table A-6.

TI-83/84 Plus

Using Computer Software to Interpret r: If the computed *P*-value is less than or equal to the significance level, conclude that there is a linear correlation. Otherwise, there is not sufficient evidence to support the conclusion of a linear correlation.

The fifth line of the display in the margin shows a *P*-value of 2.2195436E⁻4, which is 0.000222 when expressed in standard form and rounded. This *P*-value is less than the significance level of 0.05, so we conclude that *there is sufficient evidence to support the conclusion that there is a linear correlation between the costs of pizza and subway fares.*

Using Table A-6 to Interpret r: If $|r|$ exceeds the value in Table A-6, conclude that there is a linear correlation. Otherwise, there is not sufficient evidence to support the conclusion of a linear correlation.

If we refer to Table A-6 with $n = 6$ pairs of sample pizza/subway fare data, we obtain the critical value of 0.811 for $\alpha = 0.05$. (Critical values and the role of α are described in Chapters 7 and 8. With 6 pairs of data and no linear correlation between costs of pizza and subway fares, there is a 5% chance that $|r|$ will exceed 0.811.) Because $|0.988|$ exceeds the value of 0.811 from Table A-6, we conclude that there is a linear correlation. (Instead of using the condition with absolute value, we could construct a graph such as Figure 10-3, which shows the relationship between the computed r and the critical values from Table A-6.) *There is sufficient evidence to support the conclusion that there is a linear correlation between the costs of pizza and subway fares.*

EXAMPLE 5 **Interpreting *r*** Using a 0.05 significance level, interpret the value of *r* = 0.117 found using the 62 pairs of weights of discarded paper and glass listed in Data Set 22 in Appendix B. When the paired data are used with computer software, the *P*-value is found to be 0.364. Is there sufficient evidence to support a claim of a linear correlation between the weights of discarded paper and glass?

SOLUTION **REQUIREMENT CHECK** (1) The sample is a simple random sample of quantitative data. (2) A scatterplot shows that the points approximate a straight-line pattern (even though the points are not very close to the straight line that they approximate). (3) There are no outliers that are far away from almost all of the other pairs of data. ✓

*Using the TI-83/84 Plus Calculator to Interpret *r*:* The *P*-value obtained from the TI-83/84 Plus calculator is 0.364. Because the *P*-value is *not* less than or equal to 0.05, we conclude that there is not sufficient evidence to support a claim of a linear correlation between weights of discarded paper and glass.

*Using Table A-6 to Interpret *r*:* If we refer to Table A-6 with *n* = 62 pairs of sample data, we obtain the critical value of 0.254 (approximately) for $\alpha = 0.05$. Because $|0.117|$ does *not* exceed the value of 0.254 from Table A-6, we conclude that there is not sufficient evidence to support a claim of a linear correlation between weights of discarded paper and glass.

Interpreting *r*: Explained Variation

If we conclude that there is a linear correlation between *x* and *y*, we can find a linear equation that expresses *y* in terms of *x*, and that equation can be used to predict values of *y* for given values of *x*. In Section 10-3 we will describe a procedure for finding such equations and show how to predict values of *y* when given values of *x*. But a predicted value of *y* will not necessarily be the exact result, because in addition to *x*, there are other factors affecting *y*, such as random variation and other characteristics not included in the study. In Section 10-4 we will present a rationale and more details about this important principle:

The value of r^2 is the proportion of the variation in *y* that is explained by the linear relationship between *x* and *y*.

EXAMPLE 6 **Explained Variation** Using the pizza/subway fare costs in Table 10-2, we have found that the linear correlation coefficient is *r* = 0.988. What proportion of the variation in the subway fare can be explained by the variation in the costs of a slice of pizza?

SOLUTION With *r* = 0.988, we get $r^2 = 0.976$.

INTERPRETATION We conclude that 0.976 (or about 98%) of the variation in the cost of a subway fares can be explained by the linear relationship between the costs of pizza and subway fares. This implies that about 2% of the variation in costs of subway fares cannot be explained by the costs of pizza.

Palm Reading

Some people believe that the length of their palm's lifeline can be used to predict longevity. In a letter published in the *Journal of the American Medical Association,* authors M. E. Wilson and L. E. Mather refuted that belief with a study of cadavers. Ages at death were recorded, along with the lengths of palm lifelines. The authors concluded that there is no correlation between age at death and length of lifeline. Palmistry lost, hands down.

Common Errors Involving Correlation

We now identify three of the most common sources of errors made in interpreting results involving correlation:

1. *A common error is to conclude that correlation implies causality.* Using the sample data in Table 10-2, we can conclude that there is a correlation between the costs of pizza and subway fares, but we cannot conclude that increases in pizza costs *cause* increases in subway fares. Both costs might be affected by some other variable lurking in the background. (A **lurking variable** is one that affects the variables being studied, but is not included in the study.)

2. *Another error arises with data based on averages.* Averages suppress individual variation and may inflate the correlation coefficient. One study produced a 0.4 linear correlation coefficient for paired data relating income and education among individuals, but the linear correlation coefficient became 0.7 when regional averages were used.

3. *A third error involves the property of linearity.* If there is no linear correlation, there might be some other correlation that is not linear, as in Figure 10-2(d). (Figure 10-2(d) is a scatterplot that depicts the relationship between distance above ground and time elapsed for an object thrown upward.)

CAUTION

Know that *correlation does not imply causality.*

Part 2: Formal Hypothesis Test (Requires Coverage of Chapter 8)

A formal hypothesis test is commonly used to determine whether there is a significant linear correlation between two variables. The following box contains key elements of the hypothesis test.

Hypothesis Test for Correlation (Using Test Statistic *r*)

Notation

n = number of pairs of sample data
r = linear correlation coefficient for a *sample* of paired data.

ρ = linear correlation coefficient for a *population* of paired data.

Requirements

The requirements are the same as those given in the preceding box.

Hypotheses

$$H_0: \rho = 0 \quad \text{(There is no linear correlation.)}$$

$$H_1: \rho \neq 0 \quad \text{(There is a linear correlation.)}$$

Test Statistic: *r*

Critical values: Refer to Table A-6.

Conclusion

- If $|r|$ > critical value from Table A-6, reject H_0 and conclude that there is sufficient evidence to support the claim of a linear correlation.

- If $|r|$ ≤ critical value, fail to reject H_0 and conclude that there is not sufficient evidence to support the claim of a linear correlation.

EXAMPLE 7 **Hypothesis Test with Pizza/Subway Fare Costs** Use the paired pizza/subway fare data in Table 10-2 to test the claim that there is a linear correlation between the costs of a slice of pizza and the subway fares. Use a 0.05 significance level.

SOLUTION **REQUIREMENT CHECK** The solution in Example 1 already includes verification that the requirements are satisfied.

To claim that there is a linear correlation is to claim that the population linear correlation coefficient ρ is different from 0. We therefore have the following hypotheses:

$$H_0: \rho = 0 \quad \text{(There is no linear correlation.)}$$

$$H_1: \rho \neq 0 \quad \text{(There is a linear correlation.)}$$

The test statistic is $r = 0.988$ (from Examples 1, 2, and 3). The critical value of $r = 0.811$ is found in Table A-6 with $n = 6$ and $\alpha = 0.05$. Because $|0.988| > 0.811$, we reject $H_0: \rho = 0$. (Rejecting "no linear correlation" indicates that there is a linear correlation.)

INTERPRETATION We conclude that there is sufficient evidence to support the claim of a linear correlation between costs of a slice of pizza and subway fares.

P-Value Method for a Hypothesis Test for Correlation

The preceding method of hypothesis testing involves relatively simple calculations. Computer software packages typically use a *P*-value method based on a *t* test. The key components of the *t* test are as follows.

Hypothesis Test for Correlation (Using *P*-Value from a *t* Test)

Hypotheses

$$H_0: \rho = 0 \quad \text{(There is no linear correlation.)}$$

$$H_1: \rho \neq 0 \quad \text{(There is a linear correlation.)}$$

Test Statistic

$$t = \frac{r}{\sqrt{\dfrac{1 - r^2}{n - 2}}}$$

P-value: Use the TI-83/84 Plus calculator or use Table A-3 with $n - 2$ degrees of freedom to find the *P*-value corresponding to the test statistic *t*.

Conclusion

- If the *P*-value is less than or equal to the significance level, reject H_0 and conclude that there is sufficient evidence to support the claim of a linear correlation.

- If the *P*-value is greater than the significance level, fail to reject H_0 and conclude that there is not sufficient evidence to support the claim of a linear correlation.

EXAMPLE 8 **Hypothesis Test with Pizza/Subway Fare Costs** Use the paired pizza/subway fare data in Table 10-2 and use the *P*-value method to test the claim that there is a linear correlation between the costs of a slice of pizza and subway fares. Use a 0.05 significance level.

SOLUTION   **REQUIREMENT CHECK** The solution in Example 1 already includes verification that the requirements are satisfied. ✅

To claim that there is a linear correlation is to claim that the population linear correlation coefficient ρ is different from 0. We therefore have the following hypotheses:

$$H_0: \rho = 0 \quad \text{(There is no linear correlation.)}$$

$$H_1: \rho \neq 0 \quad \text{(There is a linear correlation.)}$$

The linear correlation coefficient is $r = 0.988$ (from Examples 1, 2, and 3) and $n = 6$ (because there are six pairs of sample data), so the test statistic is

$$t = \frac{r}{\sqrt{\dfrac{1 - r^2}{n - 2}}} = \frac{0.988}{\sqrt{\dfrac{1 - 0.988^2}{6 - 2}}} = 12.793$$

The TI-83/84 Plus calculator uses more precision to obtain the more accurate test statistic of $t = 12.692$. The TI-83/84 Plus calculator shows that the *P*-value is 0.00022. Because the *P*-value is less than the significance level of 0.05, we reject H_0.

INTERPRETATION   We conclude that there is sufficient evidence to support the claim of a linear correlation between costs of a slice of pizza and subway fares.

One-Tailed Tests: Examples 7 and 8 illustrate a two-tailed hypothesis test. The examples and exercises in this section will generally involve only two-tailed tests, but one-tailed tests can occur with a claim of a positive linear correlation or a claim of a negative linear correlation. In such cases, the hypotheses will be as shown here.

Claim of *Negative* Correlation	Claim of *Positive* Correlation
(Left-tailed test)	**(Right-tailed test)**
$H_0: \rho = 0$	$H_0: \rho = 0$
$H_1: \rho < 0$	$H_1: \rho > 0$

For these one-tailed tests, the *P*-value method can be used as in earlier chapters.

Rationale: We have presented Formulas 10-1 and 10-2 for calculating *r* and have illustrated their use. Those formulas are given again below, along with some other formulas that are "equivalent" in the sense that they all produce the same values.

Formula 10-1

$$r = \frac{n\Sigma xy - (\Sigma x)(\Sigma y)}{\sqrt{n(\Sigma x^2) - (\Sigma x)^2} \sqrt{n(\Sigma y^2) - (\Sigma y)^2}}$$

Formula 10-2

$$r = \frac{\sum(z_x z_y)}{n-1}$$

$$r = \frac{\sum(x-\bar{x})(y-\bar{y})}{(n-1)s_x s_y} \qquad r = \frac{\sum\left[\dfrac{(x-\bar{x})}{s_x}\dfrac{(y-\bar{y})}{s_y}\right]}{n-1} \qquad r = \frac{s_{xy}}{\sqrt{s_{xx}}\sqrt{s_{yy}}}$$

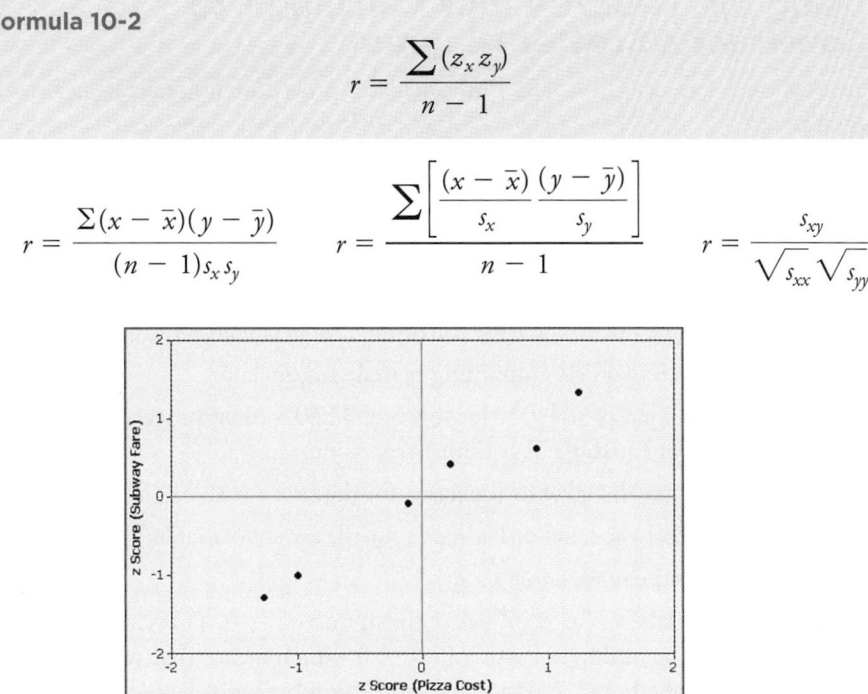

Figure 10-4   Scatterplot of z Scores from Pizza/Subway Data

We will use Formula 10-2 to help us understand the reasoning that underlies the development of the linear correlation coefficient. Because Formula 10-2 uses z scores, the value of $\sum(z_x z_y)$ does not depend on the scale that is used. Figure 10-1 shows the scatterplot of the original pizza/subway fare data, and Figure 10-4 shows the scatterplot of the z scores from the paired pizza/subway data. Compare Figure 10-1 to Figure 10-4 and see that they are essentially the same scatterplots with different scales. The red lines in Figure 10-4 form the same coordinate axes that we have all come to know and love from earlier mathematics courses. The red lines partition Figure 10-4 into four quadrants. If the points of the scatterplot approximate an up-hill line (as in the figure), individual values of the product $z_x \cdot z_y$ tend to be positive (because most of the points are found in the first and third quadrants, where the values of z_x and z_y are either both positive or both negative), so $\sum(z_x z_y)$ tends to be positive. If the points of the scatterplot approximate a downhill line, most of the points are in the second and fourth quadrants, where $z_x \cdot z_y$ are opposite in sign, so $\sum(z_x z_y)$ tends to be negative. Points that follow no linear pattern tend to be scattered among the four quadrants, so the value of $\sum(z_x z_y)$ tends to be close to 0. We can therefore use $\sum(z_x z_y)$ as a measure of how the points are configured among the four quadrants. A large positive sum suggests that the points are predominantly in the first and third quadrants (corresponding to a positive linear correlation), a large negative sum suggests that the points are predominantly in the second and fourth quadrants (corresponding to a negative linear correlation), and a sum near 0 suggests that the points are scattered among the four quadrants (with no linear correlation). We divide $\sum(z_x z_y)$ by $n-1$ to get a type of average instead of a statistic that becomes larger simply because there are more data values. (The reasons for dividing by $n-1$ instead of n are essentially the same reasons that relate to the standard deviation.) The end result is Formula 10-2, which can be algebraically manipulated into any of the other expressions for r.

Entering Paired Data

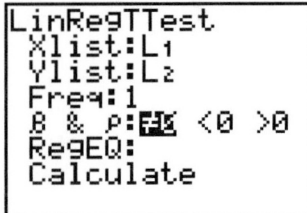

```
LinRegTTest
 Xlist:L1
 Ylist:L2
 Freq:1
 B & ρ:≠0 <0 >0
 RegEQ:
 Calculate
```

Initial Results

```
LinRegTTest
 y=a+bx
 B≠0 and ρ≠0
 t=12.69203165
 p=2.2195436ε-4
 df=4
↓a=.034560171
```

Results After Scrolling Down

```
LinRegTTest
 y=a+bx
 B≠0 and ρ≠0
↑b=.9450213806
 s=.1229869984
 r²=.9757704494
 r=.9878109381
```

Using the TI-83/84 Plus Calculator to Construct a Scatter Diagram

See Section 2-4 for the TI-83/84 Plus calculator procedure for generating a scatter diagram.

Using a TI-83/84 Plus Calculator for Correlation

Important: The following procedure applies to this section and Sections 10-3 and 10-4. This procedure requires that you have paired sample data in two different lists. The paired data can be stored in lists such as L1 and L2, or they can be stored in named lists, such as those obtained from the APP containing the data sets from Appendix B.

1. Enter or open two lists containing paired sample data.
2. Press **STAT**, then press ▷ twice to select **TESTS,** then proceed to select the menu item of **LinRegTTest** (linear regression *t* test).
3. For Xlist, enter the label or list name for the first variable, such as L1.
4. For Ylist, enter the label or list name for the second variable, such as L2.
5. For Freq (frequency), enter 1.
6. For the row labeled "β & ρ" select the option of $\neq 0$. This choice corresponds to a test of the null hypothesis H_0: $\rho = 0$, which means that we are testing to determine whether there is a linear correlation between the two variables.
7. Scroll down to **Calculate** and press the **ENTER** key.
8. There are too many results to fit on one TI-83/84 Plus calculator screen, so press ▽ to scroll down for more results.

For example, if you enter the paired sample data in Table 10-1 in lists L1 and L2, the first screen in the margin shows the required entries. The second screen shows the initial results, and the third screen shows the results obtained by scrolling down.

Interpretation of results: The accompanying screen displays show $r = 0.988$ (rounded), $t = 12.692$ (rounded), and *P*-value = 0.000222 (rounded and expressed in standard form). Because the *P*-value is less than 0.05, we reject the null hypothesis of $\rho = 0$ and conclude that there is a linear correlation between the two variables. The values of *a, b,* and *s* will be discussed in the following two sections.

10-2   Basic Skills and Concepts

Statistical Literacy and Critical Thinking

1. Notation For each of several randomly selected years, the total number of points scored in the Super Bowl football game and the total number of new cars sold in the United States are recorded. For this sample of paired data, what does *r* represent? What does ρ represent? Without doing any research or calculations, estimate the value of *r*.

2. Correlation and Causality What is meant by the statement that correlation does not imply causality?

3. Cause of Global Warming If we find that there is a linear correlation between the concentration of carbon dioxide (CO_2) in our atmosphere and the global temperature, does that indicate that changes in the concentration of carbon dioxide cause changes in the global temperature? Why or why not?

4. Weight Loss and Correlation In a test of the Weight Watchers weight loss program, weights of 40 subjects are recorded before and after the program. Assume that the before/after weights result in $r = 0.876$. Is there sufficient evidence to support a claim of a linear correlation between the before/after weights? Does the value of *r* indicate that the program is effective in reducing weight? Why or why not?

Interpreting *r*. In Exercises 5–8, use a significance level of α = 0.05.

5. Discarded Garbage and Household Size In a study conducted by University of Arizona researchers, the total weight (in lb) of garbage discarded in one week and the household size were found for 62 households. A TI-83/84 Plus calculator was used to find that the value of the linear correlation coefficient is 0.758. Is there sufficient evidence to support the claim that there is a linear correlation between the weight of discarded garbage and the household size? Explain.

6. Heights of Mothers and Daughters The heights (in inches) of a sample of eight mother/daughter pairs of subjects were measured. Using a TI-83/84 Plus calculator with the paired mother/daughter heights, the linear correlation coefficient is found to be 0.693 (based on data from the National Health Examination Survey). Is there sufficient evidence to support the claim that there is a linear correlation between the heights of mothers and the heights of their daughters? Explain.

7. Height and Pulse Rate The heights (in inches) and pulse rates (in beats per minute) for a sample of 40 women were measured. Using a TI-83/84 Plus calculator with the paired height/pulse data, the linear correlation coefficient is found to be 0.202 (based on data from the National Health Examination Survey). Is there sufficient evidence to support the claim that there is a linear correlation between the heights and pulse rates of women? Explain.

8. Supermodel Height and Weight The heights and weights of a sample of 9 supermodels were measured. Using a TI-83/84 Plus calculator, the linear correlation coefficient of the 9 pairs of measurements is found to be 0.360. (The supermodels are Alves, Avermann, Hilton, Dyer, Turlington, Hall, Campbell, Mazza, and Hume.) Is there sufficient evidence to support the claim that there is a linear correlation between the heights and weights of supermodels? Explain.

Importance of Graphing. *Exercises 9 and 10 provide two data sets from "Graphs in Statistical Analysis," by F. J. Anscombe,* **The American Statistician,** *Vol. 27. For each exercise,*

a. Construct a scatterplot.

b. Find the value of the linear correlation coefficient *r*, then determine whether there is sufficient evidence to support the claim of a linear correlation between the two variables.

c. Identify the feature of the data that would be missed if part (b) was completed without constructing the scatterplot.

9.

x	10	8	13	9	11	14	6	4	12	7	5
y	9.14	8.14	8.74	8.77	9.26	8.10	6.13	3.10	9.13	7.26	4.74

10.

x	10	8	13	9	11	14	6	4	12	7	5
y	7.46	6.77	12.74	7.11	7.81	8.84	6.08	5.39	8.15	6.42	5.73

11. Effects of an Outlier Refer to the accompanying Minitab-generated scatterplot.

a. Examine the pattern of all 10 points and subjectively determine whether there appears to be a correlation between *x* and *y*.

b. After identifying the 10 pairs of coordinates corresponding to the 10 points, find the value of the correlation coefficient *r* and determine whether there is a linear correlation.

c. Now remove the point with coordinates (10, 10) and repeat parts (a) and (b).

d. What do you conclude about the possible effect from a single pair of values?

12. Effects of Clusters Refer to the following Minitab-generated scatterplot on the top of the next page. The four points in the lower left corner are measurements from women, and the four points in the upper right corner are from men.

a. Examine the pattern of the four points in the lower left corner (from women) only, and subjectively determine whether there appears to be a correlation between *x* and *y* for women.

b. Examine the pattern of the four points in the upper right corner (from men) only, and subjectively determine whether there appears to be a correlation between *x* and *y* for men.

c. Find the linear correlation coefficient using only the four points in the lower left corner (for women). Will the four points in the upper left corner (for men) have the same linear correlation coefficient?

MINITAB

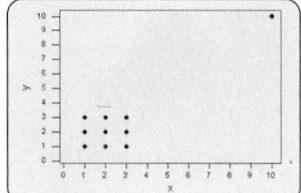

MINITAB

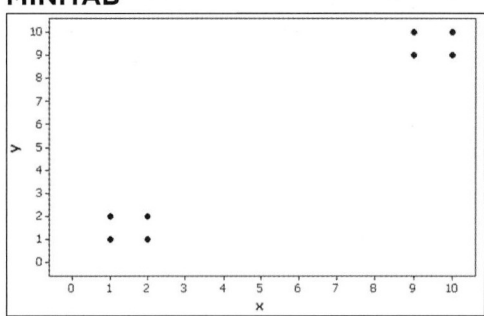

d. Find the value of the linear correlation coefficient using all eight points. What does that value suggest about the relationship between x and y?

e. Based on the preceding results, what do you conclude? Should the data from women and the data from men be considered together, or do they appear to represent two different and distinct populations that should be analyzed separately?

Testing for a Linear Correlation. *In Exercises 13–28, construct a scatterplot, find the value of the linear correlation coefficient r, and find the critical values of r from Table A-6 using* $\alpha = 0.05$. *Determine whether there is sufficient evidence to support a claim of a linear correlation between the two variables. (Save your work because the same data sets will be used in Section 10-3 exercises.)*

T1 **13. CPI and Pizza** The paired values of the Consumer Price Index (CPI) and the cost of a slice of pizza from Table 10-1 in the Chapter Problem are listed below. Is there a linear correlation between the CPI and the cost of a slice of pizza?

CPI	30.2	48.3	112.3	162.2	191.9	197.8
Cost of Pizza	0.15	0.35	1.00	1.25	1.75	2.00

T1 **14. CPI and Subway Fare** The paired values of the Consumer Price Index (CPI) and the cost of subway fare from Table 10-1 in the Chapter Problem are listed below. Is there a linear correlation between the CPI and subway fare?

CPI	30.2	48.3	112.3	162.2	191.9	197.8
Subway Fare	0.15	0.35	1.00	1.35	1.50	2.00

T1 **15. Blood Pressure Measurements** Listed below are systolic blood pressure measurements (in mm Hg) obtained from the same woman (based on data from "Consistency of Blood Pressure Differences Between the Left and Right Arms," by Eguchi, et al., *Archives of Internal Medicine,* Vol. 167). Is there sufficient evidence to conclude that there is a linear correlation between right and left arm systolic blood pressure measurements?

Right Arm	102	101	94	79	79
Left Arm	175	169	182	146	144

T1 **16. Heights of Presidents and Runners-Up** Theories have been developed about the heights of winning candidates for the U.S. presidency and the heights of candidates who were runners-up. Listed below are heights (in inches) from recent presidential elections. Is there a linear correlation between the heights of candidates who won and the heights of the candidates who were runners-up?

Winner	69.5	73	73	74	74.5	74.5	71	71
Runner-Up	72	69.5	70	68	74	74	73	76

T1 **17. Measuring Seals from Photos** Listed below are the overhead widths (in cm) of seals measured from photographs and the weights (in kg) of the seals (based on "Mass Estimation of Weddell Seals Using Techniques of Photogrammetry," by R. Garrott of Montana State

University.) The purpose of the study was to determine if weights of seals could be determined from overhead photographs. Is there sufficient evidence to conclude that there is a linear correlation between overhead widths of seals from photographs and the weights of the seals?

Overhead Width	7.2	7.4	9.8	9.4	8.8	8.4
Weight	116	154	245	202	200	191

18. Casino Size and Revenue Listed below are sizes (in thousands of square feet) and revenue (in millions of dollars) from casinos in Atlantic City (based on data from the *New York Times*). Is there sufficient evidence to conclude that there is a linear correlation between size and revenue of casinos?

Size	160	227	140	144	161	147	141
Revenue	189	157	140	127	123	106	101

19. Air Fares Listed below are costs (in dollars) of air fares for different airlines from New York City (JFK) to San Francisco. The costs are based on tickets purchased 30 days in advance and one day in advance, and the airlines are US Air, Continental, Delta, United, American, Alaska, and Northwest. Is there sufficient evidence to conclude that there is a linear correlation between costs of tickets purchased 30 days in advance and those purchased one day in advance?

30 Days	244	260	264	264	278	318	280
One Day	456	614	567	943	628	1088	536

20. Commuters and Parking Spaces Listed below are the numbers of commuters and the numbers of parking spaces at different Metro-North railroad stations (based on data from Metro-North). Is there a linear correlation between the numbers of commuters and the numbers of parking spaces?

Commuters	3453	1350	1126	3120	2641	277	579	2532
Parking Spaces	1653	676	294	950	1216	179	466	1454

21. Car Repair Costs Listed below are repair costs (in dollars) for cars crashed at 6 mi/h in full-front crash tests and the same cars crashed at 6 mi/h in full-rear crash tests (based on data from the Insurance Institute for Highway Safety). The cars are the Toyota Camry, Mazda 6, Volvo S40, Saturn Aura, Subaru Legacy, Hyundai Sonata, and Honda Accord. Is there sufficient evidence to conclude that there is a linear correlation between the repair costs from full-front crashes and full-rear crashes?

Front	936	978	2252	1032	3911	4312	3469
Rear	1480	1202	802	3191	1122	739	2767

22. New Car Mileage Ratings Listed below are combined city–highway fuel economy ratings (in mi/gal) for different cars. The old ratings are based on tests used before 2008 and the new ratings are based on tests that went into effect in 2008. Is there sufficient evidence to conclude that there is a linear correlation between the old ratings and the new ratings?

Old	16	27	17	33	28	24	18	22	20	29	21
New	15	24	15	29	25	22	16	20	18	26	19

23. Global Warming Concerns about global warming have led to studies of the relationship between global temperature and the concentration of carbon dioxide (CO_2). Listed below are concentrations (in parts per million) of CO_2 and temperatures (in °C) for different years (based on data from the Earth Policy Institute). Is there a linear correlation between temperature and concentration of CO_2?

CO_2	314	317	320	326	331	339	346	354	361	369
Temperature	13.9	14.0	13.9	14.1	14.0	14.3	14.1	14.5	14.5	14.4

24. Costs of Televisions Listed below are prices (in dollars) and quality rating scores of rear-projection televisions (based on data from *Consumer Reports*). All of the televisions have screen sizes of 55 in. or 56 in. Is there sufficient evidence to conclude that there is a linear correlation between the price and the quality rating score of rear-projection televisions? Does it appear that as the price increases, the quality score also increases? Do the results suggest that as you pay more, you get better quality?

Price	2300	1800	2500	2700	2000	1700	1500	2700
Quality Score	74	73	70	66	63	62	52	68

25. Baseball Listed below are baseball team statistics consisting of the proportions of wins and the result of this difference: Difference = (number of runs scored) − (number of runs allowed). The statistics are from a recent year, and the teams are NY (Yankees), Toronto, Boston, Cleveland, Texas, Houston, San Francisco, and Kansas City. Is there sufficient evidence to conclude that there is a linear correlation between the proportion of wins and the above difference?

Difference	163	55	−5	88	51	16	−214
Wins	0.599	0.537	0.531	0.481	0.494	0.506	0.383

26. Crickets and Temperature One classic application of correlation involves the association between the temperature and the number of times a cricket chirps in a minute. Listed below are the numbers of chirps in 1 min and the corresponding temperatures in °F (based on data from *The Song of Insects* by George W. Pierce, Harvard University Press). Is there a linear correlation between the number of chirps in 1 min and the temperature?

Chirps in 1 min	882	1188	1104	864	1200	1032	960	900
Temperature (°F)	69.7	93.3	84.3	76.3	88.6	82.6	71.6	79.6

27. Brain Size and Intelligence Listed below are brain sizes (in cm³) and Wechsler IQ scores of subjects (based on data from StatLib and "Brain Size, Head Size, and Intelligence Quotient in Monozygotic Twins," by Tramo, et al., *Neurology*, Vol. 50, No. 5). Is there sufficient evidence to conclude that there is a linear correlation between brain size and IQ score? Does it appear that people with larger brains are more intelligent?

Brain Size	965	1029	1030	1285	1049	1077	1037	1068	1176	1105
IQ	90	85	86	102	103	97	124	125	102	114

28. Ages of Best Actresses and Actors Listed below are ages of actresses and actors at the times that they won Oscars. Corresponding ages are matched so that they are from the same year. Is there sufficient evidence to conclude that there is a linear correlation between ages of best actresses and best actors?

Best Actresses

26	80	42	29	33	35	45	49	39	34
26	25	33	35	35	28	30	29	61	

Best Actors

51	32	42	54	52	37	38	32	45	60
46	40	36	47	29	43	37	38	45	

Large Data Sets. *In Exercises 29–32, use the data from Appendix B to construct a scatterplot, find the value of the linear correlation coefficient r, and find the critical values of r from Table A-6 using α = 0.05. Determine whether there is sufficient evidence to support the claim of a linear correlation between the two variables. (Save your work because the same data sets will be used in Section 10-3 exercises.)*

29. Movie Budgets and Gross Refer to Data Set 9 in Appendix B and use the paired data consisting of movie budget amounts and the amounts that the movies grossed.

TI **30. Car Weight and Braking Distance** Refer to Data Set 16 in Appendix B and use the weights of cars and the corresponding braking distances.

TI **31. Word Counts of Men and Women** Refer to Data Set 8 in Appendix B and use the word counts measured from men and women in couple relationships listed in the first two columns of Data Set 8.

TI **32. Cigarette Tar and Nicotine** Refer to Data Set 4 in Appendix B and use the tar and nicotine data from king size cigarettes.

Identifying Correlation Errors. *In Exercises 33–36, describe the error in the stated conclusion. (See the list of common errors included in this section.)*

33. *Given:* There is a linear correlation between the number of cigarettes smoked each day and the pulse rate, so that more smoking is associated with a higher pulse rate.

Conclusion: Smoking causes an increase in the pulse rate.

34. *Given:* There is a linear correlation between annual personal income and years of education.

Conclusion: More education causes a person's income to rise.

35. *Given:* There is a linear correlation between state average commuting times and state average commuting costs.

Conclusion: There is a linear correlation between individual commuting times and individual commuting costs.

36. *Given:* The linear correlation coefficient for the IQ test scores and head circumferences of test subjects is very close to 0.

Conclusion: IQ scores and head circumferences are not related in any way.

10-2 Beyond the Basics

TI **37. Transformed Data** In addition to testing for a linear correlation between x and y, we can often use *transformations* of data to explore other relationships. For example, we might replace each x value by x^2 and use the methods of this section to determine whether there is a linear correlation between y and x^2. Given the paired data in the accompanying table, construct the scatterplot and then test for a linear correlation between y and each of the following. Which case results in the largest value of r?

a. x **b.** x^2 **c.** $\log x$ **d.** $\sqrt{x}$ **e.** $1/x$

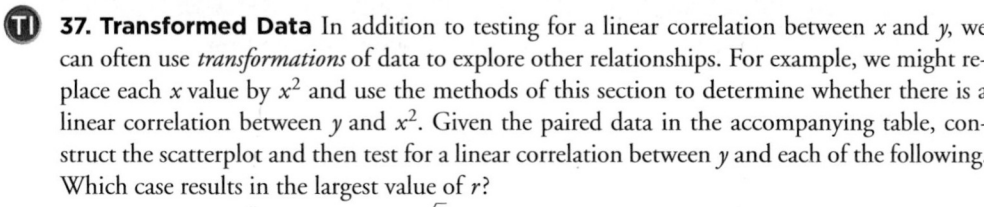

x	1	2	3	4	5	8
y	0	0.3	0.5	0.6	0.7	0.9

TI **38. Finding Critical *r*-Values** The critical r values of Table A-6 are found by using the formula

$$r = \frac{t}{\sqrt{t^2 + n - 2}}$$

where the t value is found from Table A-3 assuming a two-tailed case with $n - 2$ degrees of freedom. Table A-6 lists the results for selected values of n and α. Use the formula for r given here and Table A-3 (with $n - 2$ degrees of freedom) to find the critical r values for the given cases.

a. $H_1: \rho \neq 0$, $n = 47$, $\alpha = 0.05$

b. $H_1: \rho \neq 0$, $n = 102$, $\alpha = 0.01$

c. $H_1: \rho < 0$, $n = 40$, $\alpha = 0.05$

d. $H_1: \rho > 0$, $n = 72$, $\alpha = 0.01$

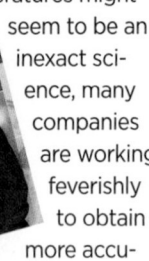

10-3 Regression

Key Concept In Section 10-2, we presented methods for finding the value of the linear correlation coefficient r and for determining whether there is a linear correlation between two variables. In Part 1 of this section, we find the equation of the straight line that best fits the paired sample data. That equation algebraically describes the relationship between the two variables. The best-fitting straight line is called the *regression line*, and its equation is called the *regression equation*. We can graph the regression equation on a scatterplot to visually determine how well it fits the data. We also present methods for using the regression equation to make predictions. In Part 2 we discuss marginal change, influential points, and residual plots as a tool for analyzing correlation and regression results.

Part 1: Basic Concepts of Regression

Two variables are sometimes related in a *deterministic* way, meaning that given a value for one variable, the value of the other variable is exactly determined without any error, as in the equation $y = 12x$ for converting a distance x from feet to inches. Such equations are considered in algebra courses, but statistics courses focus on *probabilistic* models, which are equations with a variable that is not determined completely by the other variable. For example, the height of a child cannot be determined completely by the height of the father and/or mother. Sir Francis Galton (1822–1911) studied the phenomenon of heredity and showed that when tall or short couples have children, the heights of those children tend to *regress*, or revert to the more typical mean height for people of the same gender. We continue to use Galton's "regression" terminology, even though our data do not involve the same height phenomena studied by Galton.

 DEFINITION

Given a collection of paired sample data, the **regression equation**

$$\hat{y} = b_0 + b_1 x$$

algebraically describes the relationship between the two variables x and y.
The graph of the regression equation is called the **regression line** (or *line of best fit*, or *least-squares line*).

The regression equation expresses a relationship between x (called the **explanatory variable,** or **predictor variable,** or **independent variable**) and $\hat{y}$ (called the **response variable,** or **dependent variable**). The preceding definition shows that in statistics, the typical equation of a straight line $y = mx + b$ is expressed in the form $\hat{y} = b_0 + b_1 x$, where b_0 is the y-intercept and b_1 is the slope.

The slope b_1 and y-intercept b_0 can also be found using the following formulas.

$$b_1 = \frac{n(\Sigma xy) - (\Sigma x)(\Sigma y)}{n(\Sigma x^2) - (\Sigma x)^2} \qquad b_0 = \frac{(\Sigma y)(\Sigma x^2) - (\Sigma x)(\Sigma xy)}{n(\Sigma x^2) - (\Sigma x)^2}$$

The values of b_1 and b_0 can be easily found by using any one of the many computer programs and calculators designed to provide those values. (See the TI-83/84 Plus calculator instructions at the end of this section.) Once we have evaluated b_1 and b_0, we can identify the equation of the estimated regression line, which has the following special property: *The regression line fits the sample points best.* (The specific criterion used to determine which line fits "best" is the least-squares property, which will be described later.)

Objective

Find the equation of a regression line.

Notation for the Equation of a Regression Line

	Population Parameter	Sample Statistic
y-intercept of regression equation	β_0	b_0
Slope of regression equation	β_1	b_1
Equation of the regression line	$y = \beta_0 + \beta_1 x$	$\hat{y} = b_0 + b_1 x$

Requirements

1. The sample of paired (x, y) data is a *random* sample of quantitative data.

2. Visual examination of the scatterplot shows that the points approximate a straight-line pattern.

3. Outliers can have a strong effect on the regression equation, so remove any outliers if they are known to be errors. Consider the effects of any outliers that are not known errors.

Note: Requirements 2 and 3 above are simplified attempts at checking these formal requirements for regression analysis:

- For each fixed value of x, the corresponding values of y have a normal distribution.

- For the different fixed values of x, the distributions of the corresponding y-values all have the same standard deviation. (This is violated if part of the scatterplot shows points very close to the regression line while another portion of the scatterplot shows points that are much farther away from the regression line. See the discussion of residual plots in Part 2 of this section.)

- For the different fixed values of x, the distributions of the corresponding y-values have means that lie along the same straight line.

The methods of this section are not seriously affected if departures from normal distributions and equal standard deviations are not too extreme.

Formulas for finding the slope b_1 and y-intercept b_0 in the regression equation $\hat{y} = b_0 + b_1 x$

Formula 10-3

Slope: $\quad b_1 = r \dfrac{s_y}{s_x}$

where r is the linear correlation coefficient, s_y is the standard deviation of the y values, and s_x is the standard deviation of the x values

Formula 10-4

y-intercept: $\quad b_0 = \bar{y} - b_1 \bar{x}$

TI-83/84 Plus Calculator: The TI-83/84 Plus calculator is designed to generate the value of the slope and y-intercept. See the end of this section.

Rounding the Slope b_1 and the y-Intercept b_0

Round b_1 and b_0 to three significant digits. It's difficult to provide a simple universal rule for rounding values of b_1 and b_0, but this rule will work for most situations in this book. (Depending on how you round, this book's answers to examples and exercises may be slightly different from your answers.)

EXAMPLE 1 **Using Technology to Find the Regression Equation** Refer to the sample data given in Table 10-1 in the Chapter Problem. Use technology to find the equation of the regression line in which the explanatory variable (or x variable) is the cost of a slice of pizza and the response variable (or y variable) is the corresponding cost of a subway fare.

SOLUTION **REQUIREMENT CHECK** (1) The data are assumed to be a simple random sample. (2) Figure 10-1 is a scatterplot showing a pattern of points that does appear to be a straight-line pattern. (3) There are no outliers. The requirements are satisfied. ✓

Using software: The use of computer software or a calculator is recommended for finding the equation of a regression line. Shown below are the results from STATDISK, Minitab, Excel, and the TI-83/84 Plus calculator. Note that Minitab actually provides the equation, while STATDISK, Excel, and the TI-83/84 Plus calculator list the values of the y-intercept and the slope. All of these technologies show that the regression equation can be expressed as $\hat{y} = 0.0346 + 0.945x$, where $\hat{y}$ is the predicted cost of a subway fare and x is the cost of a slice of pizza.

STATDISK

```
Regression Results:
Y= b0 + b1x:
Y Intercept, b0:        0.0345602
Slope, b1:              0.9450214
```

EXCEL

	Coefficients	Standard Error
Intercept	0.034560171	0.095012806
X Variable 1	0.945021381	0.074457849

MINITAB

```
Regression Analysis: Subway versus Pizza

The regression equation is
Subway = 0.0346 + 0.945 Pizza
```

TI-83/84 PLUS

```
LinRegTTest
 y=a+bx
 ß≠0 and ρ≠0
↑a=.034560171
 b=.9450213806
 s=.1229869984
↓r²=.9757704494
```

We should know that the regression equation is an *estimate* of the true regression equation. This estimate is based on one particular set of sample data, but another sample drawn from the same population would probably lead to a slightly different equation.

EXAMPLE 2 **Using Manual Calculations to Find the Regression Equation** Refer to the sample data given in Table 10-1 in the Chapter Problem. Use Formulas 10-3 and 10-4 to find the equation of the regression line in which the explanatory variable (or x variable) is the cost of a slice of pizza and the response variable (or y variable) is the corresponding cost of a subway fare.

SOLUTION **REQUIREMENT CHECK** The requirements are verified in Example 1.

We begin by finding the slope b_1 with Formula 10-3 as follows (with extra digits included for greater accuracy).

$$b_1 = r\frac{s_y}{s_x} = 0.987811 \cdot \frac{0.706694}{0.738693} = 0.945 \quad \text{(rounded to three significant digits)}$$

After finding the slope b_1, we can now use Formula 10-4 to find the y-intercept as follows.

$$b_0 = \bar{y} - b_1\bar{x} = 1.058333 - (0.945)(1.083333) = 0.0346 \quad \text{(rounded to three significant digits)}$$

Using these results for b_1 and b_0, we can now express the regression equation as $\hat{y} = 0.0346 + 0.945x$, where $\hat{y}$ is the predicted cost of a subway fare and x is the cost of a slice of pizza.

INTERPRETATION   As in Example 1, the regression equation is an *estimate* of the true regression equation $y = \beta_0 + \beta_1 x$, and other sample data would probably result in a different equation.

EXAMPLE 3   **Graphing the Regression Line** Graph the regression equation $\hat{y} = 0.0346 + 0.945x$ (found in Examples 1 and 2) on the scatterplot of the pizza/subway fare data and examine the graph to subjectively determine how well the regression line fits the data.

SOLUTION   Shown here is the TI-83/84 Plus calculator display of the scatterplot with the graph of the regression line included. We can see that the regression line fits the data quite well.

TI-83/84 Plus

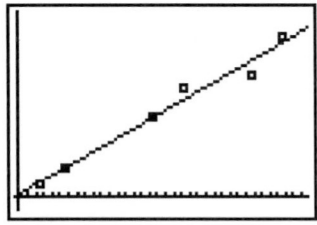

Using the Regression Equation for Predictions

Regression equations are often useful for *predicting* the value of one variable, given some specific value of the other variable. When making predictions, we should consider the following:

1. Use the regression equation for predictions only if the graph of the regression line on the scatterplot confirms that the regression line fits the points reasonably well.

2. Use the regression equation for predictions only if the linear correlation coefficient r indicates that there is a linear correlation between the two variables (as described in Section 10-2).

3. Use the regression line for predictions only if the data do not go much beyond the scope of the available sample data. (Predicting too far beyond the scope of the available sample data is called *extrapolation,* and it could result in bad predictions.)

4. If the regression equation does not appear to be useful for making predictions, the best predicted value of a variable is its point estimate, which is its sample mean.

Strategy for Predicting Values of Y

Is the regression equation a good model?
- The regression line graphed in the scatterplot shows that the line fits the points well.
- r indicates that there is a linear correlation.
- The prediction is not much beyond the scope of the available sample data.

Yes.
The regression equation is a good model.

No.
The regression equation is *not* a good model.

Substitute the given value of x into the regression equation $\hat{y} = b_0 + b_1x$.

Regardless of the value of x, the best predicted value of y is the value of $\bar{y}$ (the mean of the y values).

Figure 10-5   Recommended Strategy for Predicting Values of *y*

Figure 10-5 summarizes a strategy for predicting values of a variable y when given some value of x. Figure 10-5 shows that if the regression equation is a good model, then we substitute the value of x into the regression equation to find the predicted value of y. However, if the regression equation is not a good model, the best predicted value of y is simply $\bar{y}$, the mean of the y values. Remember, this strategy applies to *linear* patterns of points in a scatterplot. If the scatterplot shows a pattern that is not a straight-line pattern, other methods apply, as described in Section 10-6.

EXAMPLE 4   Predicting Subway Fare Table 10-5 includes the pizza/subway fare costs from the Chapter Problem, as well as the total number of runs scored in the baseball World Series for six different years. As of this writing, the cost of a slice of pizza in New York City was $2.25, and 33 runs were scored in the last World Series.

a. Use the pizza/subway fare data from Table 10-5 to predict the cost of a subway fare given that a slice of pizza costs $2.25.

b. Use the runs/subway fare data from Table 10-5 to predict the subway fare in a year in which 33 runs were scored in the World Series.

Table 10-5 Costs of a Slice of Pizza, Total Number of Runs Scored in World Series, and Subway Fare

Year	1960	1973	1986	1995	2002	2003
Cost of Pizza	0.15	0.35	1.00	1.25	1.75	2.00
Runs Scored in World Series	82	45	59	42	85	38
Subway Fare	0.15	0.35	1.00	1.35	1.50	2.00

SOLUTION Shown below are key points in the solutions for parts (a) and (b). Note that in part (a), the pizza/subway fare data result in a good regression model, so the predicted subway fare is found through substitution of $x = \$2.25$ into the regression equation. However, part (b) shows that the baseball/subway data do not result in a good regression model, so the predicted subway fare is $\bar{y}$, the mean of the subway fares.

(a) Using pizza/subway fare data to predict subway fare when pizza costs $2.25:

The regression line fits the points well, as shown here.

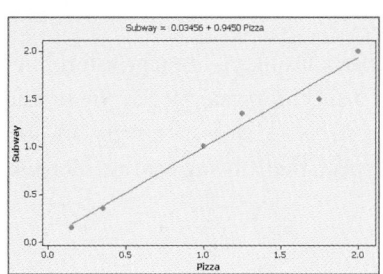

$r = 0.988$, which suggests that there is a linear correlation between pizza costs and subway fares. (The P-value is 0.00022.)

The pizza cost of $2.25 is not too far beyond the scope of the available data.

↓

Because the regression equation
$\hat{y} = 0.0346 + 0.945x$
is a good model, substitute
$x = \$2.25$ to get a predicted subway
fare of $\hat{y} = \$2.16$.

(b) Using runs/subway fare data to predict subway fare when 33 runs are scored in the World Series:

The regression line does *not* fit the points well, as shown here.

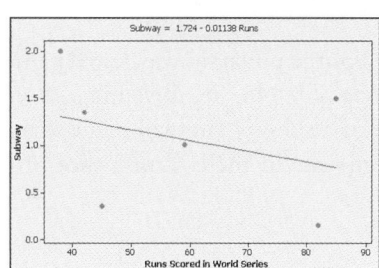

$r = -0.332$, which suggests that there is *not* a linear correlation between Word Series runs and subway fares. (The P-value is 0.520.)

33 runs is not too far beyond the scope of the available data.

↓

Because the regression equation is *not* a good model, the best predicted subway fare is $\bar{y} = \$1.06$.

INTERPRETATION Note the key point of this example: Use the regression equation for predictions only if it is a good model. If the regression equation is not a good model, use the predicted value of $\bar{y}$.

continued

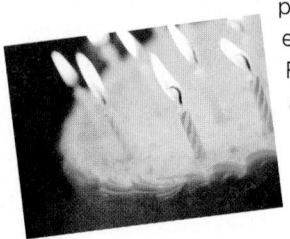
In part (a), the predicted subway fare is not likely to be the inconvenient amount of $2.16. A more likely fare would be $2.25 (which is $2.16 rounded up to the nearest multiple of 25 cents). In part (b), the predicted value of $1.06 ignores the pattern of rising subway fares over time. Given the past pattern of subway fares, a better predicted value for part (b) is the latest fare of $2.00.

Part 2: Beyond the Basics of Regression

In Part 2 we consider the concept of marginal change, which is helpful in interpreting a regression equation; then we consider the effects of outliers and special points called *influential points*. We also discuss *residuals*.

Interpreting the Regression Equation: Marginal Change

We can use the regression equation to see the effect on one variable when the other variable changes by some specific amount.

 DEFINITION

In working with two variables related by a regression equation, the **marginal change** in a variable is the amount that it changes when the other variable changes by exactly one unit. The slope b_1 in the regression equation represents the marginal change in y that occurs when x changes by one unit.

For the pizza/subway fare data from the Chapter Problem, the regression line has a slope of 0.945. So, if we increase x (the cost of a slice of pizza) by $1, the predicted cost of a subway fare will increase by $0.945, or 94.5¢. That is, for every additional $1 increase in the cost of a slice of pizza, we expect the subway fare to increase by 94.5¢.

Outliers and Influential Points

A correlation/regression analysis of bivariate (paired) data should include an investigation of *outliers* and *influential points,* defined as follows.

 DEFINITION

In a scatterplot, an **outlier** is a point lying far away from the other data points.

Paired sample data may include one or more **influential points,** which are points that strongly affect the graph of the regression line.

To determine whether a point is an outlier, examine the scatterplot to see if the point is far away from the others. Here's how to determine whether a point is an influential point: First graph the regression line resulting from the data with the point included, then graph the regression line resulting from the data with the point excluded. If the graph changes by a considerable amount, the point is influential. Influential points are often found by identifying those outliers that are *horizontally* far away from the other points.

EXAMPLE 5 **Influential Point** Consider the pizza/subway fare data from the Chapter Problem. The scatterplot located to the left below shows the regression line. If we include this additional pair of data: $x = 2.00$, $y = -20.00$ (pizza is still $2.00 per slice, but the subway fare is $-$20.00, which means that people are paid $20 to ride the subway), this additional point would be an influential point because the graph of the regression line would change considerably, as shown by the regression line located to the right below. Compare the two graphs and you will see clearly that the addition of that one pair of values has a very dramatic effect on the regression line, so that additional point is an influential point. The additional point is also an outlier because it is far from the other points.

PIZZA/SUBWAY DATA FROM THE CHAPTER PROBLEM

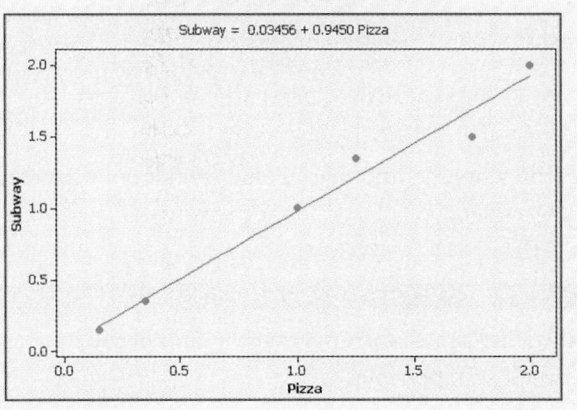

PIZZA/SUBWAY DATA WITH AN INFLUENTIAL POINT

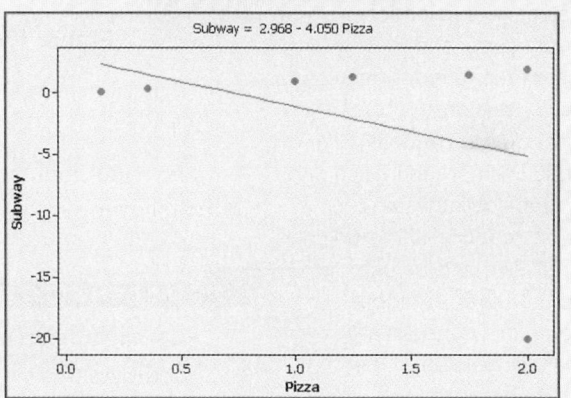

Residuals and the Least-Squares Property

We have stated that the regression equation represents the straight line that "best" fits the data. The criterion to determine the line that is better than all others is based on the vertical distances between the original data points and the regression line. Such distances are called *residuals*.

>  **DEFINITION**
>
> For a pair of sample x and y values, the **residual** is the difference between the *observed* sample value of y and the y-value that is *predicted* by using the regression equation. That is,
>
> $$\text{residual} = \text{observed } y - \text{predicted } y = y - \hat{y}$$

This definition has not yet won any prizes for simplicity, but you can easily understand residuals by referring to Figure 10-6 on the next page, which corresponds to the paired sample data shown in the margin. In Figure 10-6, the residuals are represented by the dashed lines.

x	1	2	4	5
y	4	24	8	32

Consider the sample point with coordinates of (5, 32). If we substitute $x = 5$ into the regression equation $\hat{y} = 5 + 4x$, we get a *predicted* value of $\hat{y} = 25$. But the actual *observed* sample value is $y = 32$. The difference $y - \hat{y} = 32 - 25 = 7$ is a residual.

Predicting Condo Prices

A massive study involved 99,491 sales of condominiums and cooperatives in Manhattan. The study used 41 different variables used to predict the value  of the condo or co-op. The variables include condition of the unit, the neighborhood, age, size, and whether there are doormen. Some conclusions: With all factors equal, a condo is worth 15.5% more than a co-op; a fireplace increases the value of a condo 9.69% and it increases the value of a co-op 11.36%; an additional bedroom in a condo increases the value by 7.11% and it increases the value in a co-op by 18.53%. This use of statistical methods allows buyers and sellers to estimate value with much greater accuracy. Methods of multiple regression (Section 10-5) are used when there is more than one predictor variable, as in this study. (Based on data from "So How Much Is That . . . Worth," by Dennis Hevesi, *New York Times*.)

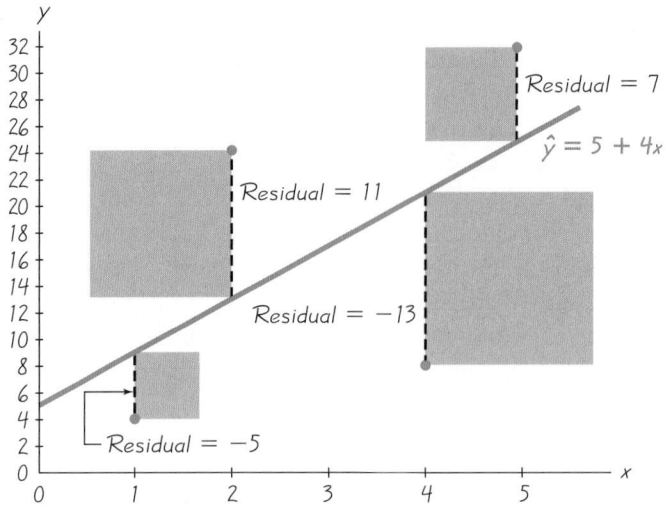

Figure 10-6   Residuals and Squares of Residuals

The regression equation represents the line that "best" fits the points according to the following *least-squares property.*

DEFINITION

A straight line satisfies the **least-squares property** if the sum of the squares of the residuals is the smallest sum possible.

From Figure 10-6, we see that the residuals are -5, 11, -13, and 7, so the sum of their squares is

$$(-5)^2 + 11^2 + (-13)^2 + 7^2 = 364$$

We can visualize the least-squares property by referring to Figure 10-6, where the squares of the residuals are represented by the red-square areas. The sum of the red-square areas is 364, which is the smallest sum possible. Use any other straight line, and the red squares will combine to produce an area larger than the combined red area of 364.

Fortunately, we need not deal directly with the least-squares property when we want to find the equation of the regression line. Calculus has been used to build the least-squares property into Formulas 10-3 and 10-4. Because the derivations of these formulas require calculus, we don't include the derivations in this text.

Residual Plots

In this section and the preceding section we listed simplified requirements for the effective analyses of correlation and regression results. We noted that we should always begin with a scatterplot, and we should verify that the pattern of points is approximately a straight-line pattern. We should also consider outliers. A *residual plot* can be another helpful tool for analyzing correlation and regression results and for checking the requirements necessary for making inferences about correlation and regression.

> ### ⊜ DEFINITION
>
> A **residual plot** is a scatterplot of the (x, y) values after each of the y-coordinate values has been replaced by the residual value $y - \hat{y}$ (where $\hat{y}$ denotes the predicted value of y). That is, a residual plot is a graph of the points $(x, y - \hat{y})$.

To construct a residual plot, use the same x-axis as the scatterplot, but use a vertical axis of residual values. Draw a horizontal reference line through the residual value of 0, then plot the paired values of $(x, y - \hat{y})$. Because the manual construction of residual plots can be tedious, the use of computer software is strongly recommended. When analyzing a residual plot, look for a pattern in the way the points are configured, and use these criteria:

- The residual plot should not have an obvious pattern that is not a straight-line pattern. (This confirms that a scatterplot of the sample data is a straight-line pattern and not some other pattern that is not a straight line.)

- The residual plot should not become thicker (or thinner) when viewed from left to right. (This confirms the requirement that for the different fixed values of x, the distributions of the corresponding y-values all have the same standard deviation.)

EXAMPLE 6 **Residual Plot** The pizza/subway fare data from the Chapter Problem are used to obtain the accompanying residual plots. The first sample x value of 0.15 for the cost of a slice of pizza is substituted into the regression equation of $\hat{y} = 0.0346 + 0.945x$ (found in Examples 1 and 2). The result is the predicted value of $\hat{y} = 0.17635$. For the first value of $x = 0.15$, the corresponding y-value is 0.15, so the value of the residual is $y - \hat{y} = 0.15 - 0.17635 = -0.02635$. Using the x value of 0.15 and the residual of -0.02635, we get the coordinates of the point $(0.15, -0.02635)$, which is the leftmost point in the residual plot shown here. This residual plot becomes thicker, suggesting that the regression equation might not be a good model.

MINITAB

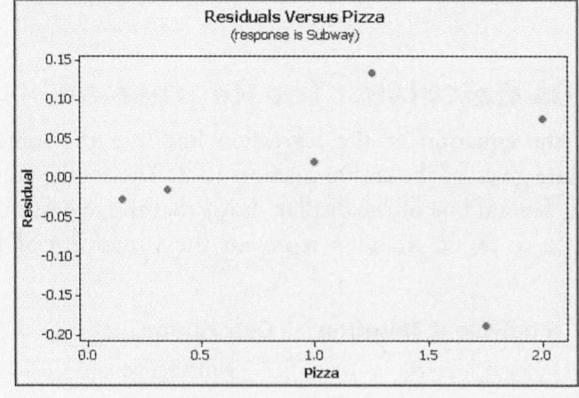

TI-83/84 Plus

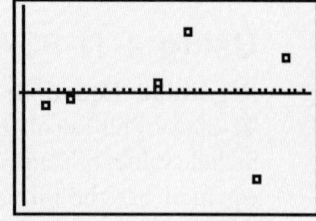

See the three residual plots on the next page. The leftmost residual plot suggests that the regression equation is a good model. The middle residual plot shows a distinct pattern, suggesting that the sample data do not follow a straight-line pattern as required. The rightmost residual plot becomes thicker, which suggests that the requirement of equal standard deviations is violated.

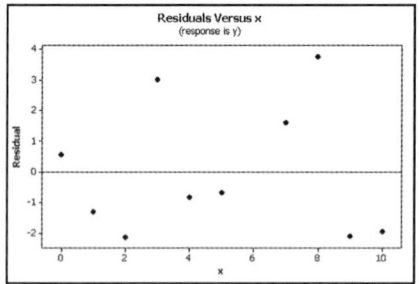

Residual Plot Suggesting that the Regression Equation is a Good Model

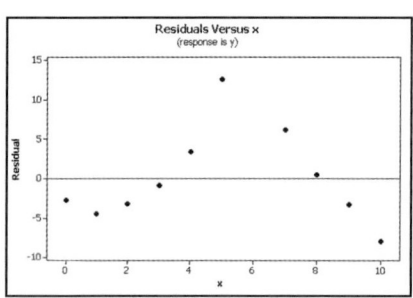

Residual Plot with an Obvious Pattern, Suggesting that the Regression Equation Is Not a Good Model

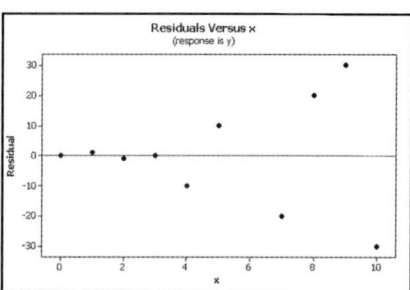

Regression Plot that Becomes Thicker, Suggesting that the Regression Equation Is Not a Good Model

Complete Regression Analysis

In Part 1 of this section, we identified simplified criteria for determining whether a regression equation is a good model. A more complete and thorough analysis can be implemented with the following steps.

1. Construct a scatterplot and verify that the pattern of the points is approximately a straight-line pattern without outliers. (If there are outliers, consider their effects by comparing results that include the outliers to results that exclude the outliers.)

2. Construct a residual plot and verify that there is no pattern (other than a straight-line pattern) and also verify that the residual plot does not become thicker (or thinner).

3. Use a histogram and/or normal quantile plot to confirm that the values of the residuals have a distribution that is approximately normal.

4. Consider any effects of a pattern over time.

Using a TI-83/84 Plus Calculator for Regression

Regression Equation: To find the equation of the regression line, use the same TI-83/84 Plus calculator procedure given at the end of Section 10-2. The results will include values of a and b, and the second line of the display shows that the regression equation has the format of $y = a + bx$, so a and b represent the y-intercept and slope, respectively.

TI-83/84 Plus Calculator	Equivalent Notation	Description
a	b_0	y-intercept
b	b_1	slope

Graph of Scatterplot and Regression Line: To obtain a TI-83/84 Plus graph of a scatterplot with the regression line included, use this procedure:

1. Generate the scatterplot using the procedure described in Section 2-4.

2. Press $\boxed{\text{Y=}}$ and complete the equation for Y1. For example, see the accompanying screen display for the entry of the regression equation corresponding to the sample data for pizza costs and subway fares from Table 10-1 ($\hat{y} = 0.0346 + 0.945x$). The variable x can be entered by pressing $\boxed{\text{ALPHA}}$ $\boxed{\text{STO}}$ (because the alphabet character x is located above the $\boxed{\text{STO}}$ key).

3. Press the $\boxed{\text{GRAPH}}$ key and the same screen will show the scatterplot with the regression line included.

Graphing Regression Line

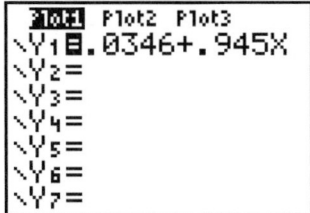

10-3   Basic Skills and Concepts

Statistical Literacy and Critical Thinking

1. Notation and Terminology A physician measured the weights and cholesterol levels of a random sample of men. The regression equation is $\hat{y} = -116 + 2.44x$, where x represents weight (in pounds). What does the symbol $\hat{y}$ represent? What does the predictor variable represent? What does the response variable represent?

2. Best-Fitting Line In what sense is the regression line the straight line that "best" fits the points in a scatterplot?

3. Correlation and Slope Formula 10-3 shows that the slope of a regression line can be found by evaluating $r \cdot s_y/s_x$. What do we know about the graph of the regression line if r is a positive value? What do we know about the graph of the regression line if r is a negative value?

4. Notation What is the difference between the regression equation $\hat{y} = b_0 + b_1x$ and the regression equation $y = \beta_0 + \beta_1x$?

Making Predictions. *In Exercises 5–8, use the given data to find the best predicted value of the response variable. Be sure to follow the prediction procedure summarized in Figure 10-5.*

5. Discarded Garbage and Household Size In a study conducted by University of Arizona researchers, the total weight (in pounds) of garbage discarded in one week and the household size were recorded for 62 households. The linear correlation coefficient is $r = 0.759$ and the regression equation is $\hat{y} = 0.445 + 0.119x$, where x represents the total weight of discarded garbage. The mean of the 62 garbage weights is 27.4 lb and the 62 households have a mean size of 3.71 people. What is the best predicted number of people in a household that discards 50 lb of garbage?

6. Heights of Mothers and Daughters A sample of eight mother/daughter pairs of subjects was obtained, and their heights (in inches) were measured. The linear correlation coefficient is 0.693 and the regression equation is $\hat{y} = 69.0 - 0.0849x$, where x represents the height of the mother (based on data from the National Health Examination Survey). The mean height of the mothers is 63.1 in. and the mean height of the daughters is 63.3 in. Find the best predicted height of a daughter given that the mother has a height of 60 in.

7. Height and Pulse Rate A sample of 40 women is obtained, and their heights (in inches) and pulse rates (in beats per minute) are measured. The linear correlation coefficient is 0.202 and the equation of the regression line is $\hat{y} = 18.2 + 0.920x$, where x represents height (based on data from the National Health Examination Survey). The mean of the 40 heights is 63.2 in. and the mean of the 40 pulse rates is 76.3 beats per minute. Find the best predicted pulse rate of a woman who is 70 in. tall.

8. Supermodel Heights and Weights Heights (in inches) and weights (in pounds) are obtained from a random sample of nine supermodels (Alves, Avermann, Hilton, Dyer, Turlington, Hall, Campbell, Mazza, and Hume). The linear correlation coefficient is 0.360 and the equation of the regression line is $\hat{y} = 31.8 + 1.23x$, where x represents height. The mean of the nine heights is 69.3 in. and the mean of the nine weights is 117 lb. What is the best predicted weight of a supermodel with a height of 72 in.?

Finding the Equation of the Regression Line. *In Exercises 9 and 10, use the given data to find the equation of the regression line. Examine the scatterplot and identify a characteristic of the data that is ignored by the regression line.*

TI **9.**

x	10	8	13	9	11	14	6	4	12	7	5
y	9.14	8.14	8.74	8.77	9.26	8.10	6.13	3.10	9.13	7.26	4.74

TI **10.**

x	10	8	13	9	11	14	6	4	12	7	5
y	7.46	6.77	12.74	7.11	7.81	8.84	6.08	5.39	8.15	6.42	5.73

TI **11. Effects of an Outlier** Refer to the Minitab-generated scatterplot given in Exercise 11 of Section 10-2.

a. Using the pairs of values for all 10 points, find the equation of the regression line.

b. After removing the point with coordinates (10, 10), use the pairs of values for the remaining nine points and find the equation of the regression line.

c. Compare the results from parts (a) and (b).

TI **12. Effects of Clusters** Refer to the Minitab-generated scatterplot given in Exercise 12 of Section 10-2.

a. Using the pairs of values for all 8 points, find the equation of the regression line.

b. Using only the pairs of values for the four points in the lower left corner, find the equation of the regression line.

c. Using only the pairs of values for the four points in the upper right corner, find the equation of the regression line.

d. Compare the results from parts (a), (b), and (c).

Finding the Equation of the Regression Line and Making Predictions. *Exercises 13–28 use the same data sets as Exercises 13–28 in Section 10-2. In each case, find the regression equation, letting the first variable be the predictor (x) variable. Find the indicated predicted value by following the prediction procedure summarized in Figure 10-5.*

TI **13. CPI and Pizza** Find the best predicted cost of a slice of pizza when the Consumer Price Index (CPI) is 182.5 (in the year 2000).

CPI	30.2	48.3	112.3	162.2	191.9	197.8
Cost of Pizza	0.15	0.35	1.00	1.25	1.75	2.00

TI **14. CPI and Subway Fare** Find the best predicted cost of subway fare when the Consumer Price Index (CPI) is 182.5 (in the year 2000).

CPI	30.2	48.3	112.3	162.2	191.9	197.8
Subway Fare	0.15	0.35	1.00	1.35	1.50	2.00

TI **15. Blood Pressure Measurements** Find the best predicted systolic blood pressure in the left arm given that the systolic blood pressure in the right arm is 100 mm Hg.

Right Arm	102	101	94	79	79
Left Arm	175	169	182	146	144

TI **16. Heights of Presidents and Runners-Up** Find the best predicted height of runner-up Goldwater, given that the height of the winning presidential candidate Johnson is 75 in. Is the predicted height of Goldwater close to his actual height of 72 in.?

Winner	69.5	73	73	74	74.5	74.5	71	71
Runner-Up	72	69.5	70	68	74	74	73	76

TI **17. Measuring Seals from Photos** Find the best predicted weight (in kg) of a seal if the overhead width measured from the photograph is 9.0 cm.

Overhead Width	7.2	7.4	9.8	9.4	8.8	8.4
Weight	116	154	245	202	200	191

TI **18. Casino Size and Revenue** Find the best predicted amount of revenue (in millions of dollars) given that the Trump Plaza casino has a size of 87 thousand ft². How does the result compare to the actual revenue of $65.1 million?

Size	160	227	140	144	161	147	141
Revenue	189	157	140	127	123	106	101

TI **19. Air Fares** Find the best predicted cost of a ticket purchased one day in advance, given that the cost of the ticket is $300 if purchased 30 days in advance of the flight.

30 Days	244	260	264	264	278	318	280
One Day	456	614	567	943	628	1088	536

TI **20. Commuters and Parking Spaces** The Metro-North Station of Greenwich, CT has 2804 commuters. Find the best predicted number of parking spots at that station. Is the predicted value close to the actual value of 1274?

Commuters	3453	1350	1126	3120	2641	277	579	2532
Parking Spots	1653	676	294	950	1216	179	466	1454

TI **21. Car Repair Costs** Find the best predicted repair cost from a full-rear crash for a Volkswagon Passat, given that its repair cost from a full-front crash is $4594. How does the result compare to the $982 actual repair cost from a full-rear crash?

Front	936	978	2252	1032	3911	4312	3469
Rear	1480	1202	802	3191	1122	739	2767

TI **22. New Car Mileage Ratings** Find the best predicted new mileage rating of a Jeep Grand Cherokee given that the old rating is 19 mi/gal. Is the predicted value close to the actual value of 17 mi/gal?

Old	16	27	17	33	28	24	18	22	20	29	21
New	15	24	15	29	25	22	16	20	18	26	19

TI **23. Global Warming** Find the best predicted temperature for a recent year in which the concentration (in parts per million) of CO_2 is 370.9. Is the predicted temperature close to the actual temperature of 14.5° (Celsius)?

CO_2	314	317	320	326	331	339	346	354	361	369
Temperature	13.9	14.0	13.9	14.1	14.0	14.3	14.1	14.5	14.5	14.4

TI **24. Costs of Televisions** Find the best predicted quality score of a Hitachi television with a price of $1900. Is the predicted quality score close to the actual quality score of 56?

Price	2300	1800	2500	2700	2000	1700	1500	2700
Quality Score	74	73	70	66	63	62	52	68

TI **25. Baseball** Listed below are statistics from seven baseball teams. The statistics consist of the proportions of wins and the result of this difference: Difference = (number of runs scored) − (number of runs allowed) for a recent year. Find the best predicted winning proportion for San Diego, which has a difference of 52 runs. Is the predicted proportion close to the actual proportion of 0.543?

Difference	163	55	−5	88	51	16	−214
Wins	0.599	0.537	0.531	0.481	0.494	0.506	0.383

TI **26. Crickets and Temperature** Find the best predicted temperature (in °F) at a time when a cricket chirps 3000 times in one minute. What is wrong with this predicted value?

Chirps in 1 min	882	1188	1104	864	1200	1032	960	900
Temperature (°F)	69.7	93.3	84.3	76.3	88.6	82.6	71.6	79.6

TI **27. Brain Size and Intelligence** Find the best predicted IQ score of someone with a brain size of 1275 cm^3.

Brain Size	965	1029	1030	1285	1049	1077	1037	1068	1176	1105
IQ	90	85	86	102	103	97	124	125	102	114

TI **28. Ages of Best Actresses and Actors** Find the best predicted age of the Best Actor at the time that the age of the Best Actress is 75 years.

Best Actresses

26	80	42	29	33	35	45	49	39	34
26	25	33	35	35	28	30	29	61	

Best Actors

51	32	42	54	52	37	38	32	45	60
46	40	36	47	29	43	37	38	45	

Large Data Sets. *Exercises 29–32 use the same Appendix B data sets as Exercises 29–32 in Section 10-2. In each case, find the regression equation, letting the first variable be the predictor (x) variable. Find the indicated predicted values following the prediction procedure summarized in Figure 10-5.*

TI **29. Movie Budgets and Gross** Refer to Data Set 9 in Appendix B and use the paired data consisting of movie budget amounts and the amounts that the movies grossed. Find the best predicted amount that a movie will gross if its budget is $120 million.

TI **30. Car Weight and Braking Distance** Refer to Data Set 16 in Appendix B and use the weights of cars and the corresponding braking distances. Find the best predicted braking distance for a car that weighs 4000 lb.

TI **31. Word Counts of Men and Women** Refer to Data Set 8 in Appendix B and use the word counts measured for men and women from the couples listed in the first two columns of

Data Set 8. Find the best predicted word count of a woman given that her male partner speaks 6000 words in a day.

TI **32. Cigarette Tar and Nicotine** Refer to Data Set 4 in Appendix B and use the tar and nicotine data from king size cigarettes. Find the best predicted amount of nicotine in a king size cigarette with 10 mg of tar.

10-3 Beyond the Basics

33. Equivalent Hypothesis Tests Explain why a test of the null hypothesis $H_0: \rho = 0$ is equivalent to a test of the null hypothesis $H_0: \beta_1 = 0$ where ρ is the linear correlation coefficient for a population of paired data, and β_1 is the slope of the regression line for that same population.

TI **34. Testing Least-Squares Property** According to the least-squares property, the regression line minimizes the sum of the squares of the residuals. Refer to Data Set 1 in Appendix B and use the paired data consisting of the first six pulse rates and the first six systolic blood pressures of males.

a. Find the equation of the regression line.

b. Identify the residuals, and find the sum of squares of the residuals.

c. Show that the equation $\hat{y} = 70 + 0.5x$ results in a larger sum of squares of residuals.

TI **35. Residual Plot** Refer to Data Set 1 in Appendix B and use the paired data consisting of the first six pulse rates and the first six systolic blood pressures of males. Construct the residual plot. Does the residual plot suggest that the regression equation is a bad model? Why or why not? Does the scatter diagram suggest that the regression equation is a bad model? Why or why not?

TI **36. Using Logarithms to Transform Data** If a scatterplot reveals a nonlinear (not a straight line) pattern that you recognize as another type of curve, you may be able to apply the methods of this section. For the data given in the margin, find the linear equation ($\hat{y} = b_0 + b_1 x$) that best fits the sample data, and find the logarithmic equation ($\hat{y} = a + b \ln x$) that best fits the sample data. (*Hint:* Begin by replacing each x-value with $\ln x$.) Which of these two equations fits the data better? Why?

x	2	48	377	4215
y	1	4	6	10

10-4 Variation and Prediction Intervals

Key Concept In Section 10-3 we presented a method for using a regression equation to find a predicted value of y. In this section we present a method for constructing a *prediction interval*, which is an interval estimate of a predicted value of y. (Interval estimates of parameters are *confidence intervals*, but interval estimates of variables are called *prediction intervals*.)

Explained and Unexplained Variation

We first examine measures of *deviation* and *variation* for a pair of (x, y) values. Let's consider the specific case shown in Figure 10-7. Imagine a sample of paired (x, y) data that includes the specific values of $(5, 19)$. Assume that we use this sample of paired data to find the following results:

 • There is sufficient evidence to support the claim of a linear correlation between x and y.

 • The equation of the regression line is $\hat{y} = 3 + 2x$.

Super Bowl as Stock Market Predictor

The "Super Bowl omen" states that a Super Bowl victory by a team with NFL origins is followed by a year in which the New York Stock Exchange index rises; otherwise, it falls. (In 1970, the NFL and AFL merged into the current NFL.) After the first 29 Super Bowl games, the prediction was correct 90% of the time, but it has been much less successful in recent years. As of this writing, it has been correct in 31 of 40 Super Bowl games, for a 78% success rate. Forecasting and predicting are important goals of statistics and investment advisors, but common sense suggests that no one should base investments on the outcome of one football game. Other indicators used to forecast stock market performance include rising skirt hemlines, aspirin sales, limousines on Wall Street, orders for cardboard boxes, sales of beer versus wine, and elevator traffic at the New York Stock Exchange.

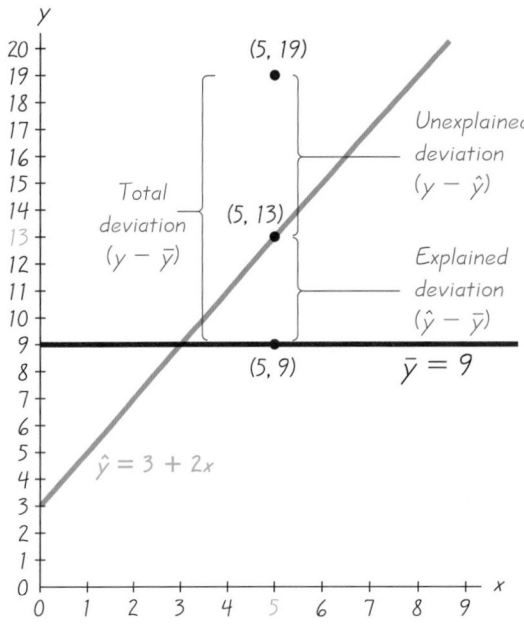

Figure 10-7

Unexplained, Explained, and Total Deviation

- The mean of the y-values is given by $\bar{y} = 9$.
- One of the pairs of sample data is $x = 5$ and $y = 19$.
- The point $(5, 13)$ is one of the points on the regression line, because substituting $x = 5$ into the regression equation of $\hat{y} = 3 + 2x$ yields $\hat{y} = 13$.

Figure 10-7 shows that the point $(5, 13)$ lies on the regression line, but the point $(5, 19)$ from the original data set does not lie on the regression line. If we completely ignore correlation and regression concepts and want to predict a value of y given a value of x and a collection of paired (x, y) data, our best guess would be the mean $\bar{y}$. But in this case there is a linear correlation between x and y, so a better way to predict the value of y when $x = 5$ is to substitute $x = 5$ into the regression equation to get $\hat{y} = 13$. We can explain the discrepancy between $\bar{y} = 9$ and $\hat{y} = 13$ by noting that there is a linear relationship best described by the regression line. Consequently, when $x = 5$, the predicted value of y is 13, not the mean value of 9. For $x = 5$, the predicted value of y is 13, but the observed sample value of y is actually 19. The discrepancy between $\hat{y} = 13$ and $y = 19$ cannot be explained by the regression line, and it is called an *unexplained deviation,* or a *residual.* This unexplained deviation can be expressed in symbols as $y - \hat{y}$.

As in Section 3-3 where we defined the standard deviation, we again consider a *deviation* to be a difference between a value and the mean. (In this case, the mean is $\bar{y} = 9$.) Examine Figure 10-7 carefully and note these specific deviations from $\bar{y} = 9$:

Total deviation (from $\bar{y} = 9$) of the point $(5, 19) = y - \bar{y} = 19 - 9 = 10$

Explained deviation (from $\bar{y} = 9$) of the point $(5, 19) = \hat{y} - \bar{y} = 13 - 9 = 4$

Unexplained deviation (from $\bar{y} = 9$) of the point $(5, 19) = y - \hat{y} = 19 - 13 = 6$

These deviations from the mean are generalized and formally defined as follows.

DEFINITION

Assume that we have a collection of paired data containing the sample point (x, y), that $\hat{y}$ is the predicted value of y (obtained by using the regression equation), and that the mean of the sample y-values is $\bar{y}$.

The **total deviation** of (x, y) is the vertical distance $y - \bar{y}$, which is the distance between the point (x, y) and the horizontal line passing through the sample mean $\bar{y}$.

The **explained deviation** is the vertical distance $\hat{y} - \bar{y}$, which is the distance between the predicted y-value and the horizontal line passing through the sample mean $\bar{y}$.

The **unexplained deviation** is the vertical distance $y - \hat{y}$, which is the vertical distance between the point (x, y) and the regression line. (The distance $y - \hat{y}$ is also called a *residual,* as defined in Section 10-3.)

We can see the following relationship in Figure 10-7:

$$(\text{total deviation}) = (\text{explained deviation}) + (\text{unexplained deviation})$$
$$(y - \bar{y}) \quad = \quad (\hat{y} - \bar{y}) \quad + \quad (y - \hat{y})$$

The above expression involves deviations away from the mean, and it applies to any one particular point (x, y). If we sum the squares of deviations using all points (x, y), we get amounts of *variation.* The same relationship applies to the sums of squares shown in Formula 10-5, even though the above expression is not algebraically equivalent to Formula 10-5. In Formula 10-5, the **total variation** is the sum of the squares of the total deviation values, the **explained variation** is the sum of the squares of the explained deviation values, and the **unexplained variation** is the sum of the squares of the unexplained deviation values.

> **Formula 10-5**
>
> $$(\text{total variation}) = (\text{explained variation}) + (\text{unexplained variation})$$
> or $\quad \Sigma(y - \bar{y})^2 \quad = \quad \Sigma(\hat{y} - \bar{y})^2 \quad + \quad \Sigma(y - \hat{y})^2$

In Section 10-2 we saw that the linear correlation coefficient r can be used to find the proportion of the total variation in y that can be explained by the linear correlation. This statement was made in Section 10-2:

> **The value of r^2 is the proportion of the variation in y that is explained by the linear relationship between x and y.**

This statement about the explained variation is formalized with the following definition.

DEFINITION

The **coefficient of determination** is the amount of the variation in y that is explained by the regression line. It is computed as

$$r^2 = \frac{\text{explained variation}}{\text{total variation}}$$

TI-83/84 Plus

```
LinRegTTest
 y=a+bx
 β≠0 and ρ≠0
↑b=.9450213806
 s=.1229869984
 r²=.9757704494
 r=.9878109381
```

Using the TI-83/84 Plus The TI-83/84 Plus calculator automatically provides the value of r^2 along with the value of r. See the display in the margin showing the value of r on the bottom line and the value of r^2 on the line above the bottom line.

We can find r^2 by using technology, or we can simply square the linear correlation coefficient r.

EXAMPLE 1 **Pizza/Subway Fare Costs: Finding the Coefficient of Determination** In Section 10-2 we used the paired pizza/subway fare costs from the Chapter Problem to find that $r = 0.988$. Find the coefficient of determination. Also, find the percentage of the total variation in y (subway fare) that can be explained by the linear relationship between the cost of a slice of pizza and the cost of a subway fare.

continued

SOLUTION The coefficient of determination is $r^2 = 0.988^2 = 0.976$. Because r^2 is the proportion of total variation that is explained, we conclude that 97.6% of the total variation in subway fares can be explained by the cost of a slice of pizza. This means that 2.4% of the total variation in cost of subway fares can be explained by factors other than the cost of a slice of pizza. But remember that these results are estimates based on the given sample data. Other sample data will likely result in different estimates.

Prediction Intervals

In Section 10-3 we used the sample data from Table 10-1 to find the regression equation $\hat{y} = 0.0346 + 0.945x$, where $\hat{y}$ represents the predicted cost of a subway fare and x represents the cost of a slice of pizza. We then used that equation to predict the cost of a subway fare, given that the cost of a slice of pizza is $x = \$2.25$, and we found that the best predicted cost of a subway fare is $2.16. (See Example 4 in Section 10-3.) Because the predicted cost of a subway fare of $2.16 is a single value, it is referred to as a *point estimate.* In Chapter 7 we saw that point estimates have the serious disadvantage of not giving us any information about how accurate they might be. Here, we know that $2.16 is the best predicted value, but we don't know anything about the accuracy of that value. In Chapter 7 we developed confidence interval estimates to overcome that disadvantage, and in this section we follow the same approach. We will use a *prediction interval.*

DEFINITION

A **prediction interval** is an interval estimate of a predicted value of y.

An interval estimate of a *parameter* (such as the mean of all subway fares) is referred to as a *confidence interval,* but an interval estimate of a *variable* (such as the predicted subway fare) is called a *prediction interval.*

The development of a prediction interval requires a measure of the spread of sample points about the regression line. Recall that the unexplained deviation (or residual) is the vertical distance between a sample point and the regression line, as illustrated in Figure 10-7. The *standard error of estimate* is a collective measure of the spread of the sample points about the regression line, and it is formally defined as follows.

DEFINITION

The **standard error of estimate,** denoted by s_e, is a measure of the differences (or distances) between the observed sample y-values and the predicted values $\hat{y}$ that are obtained using the regression equation. It is given as

$$s_e = \sqrt{\frac{\Sigma(y - \hat{y})^2}{n - 2}} \qquad \text{(where } \hat{y} \text{ is the predicted } y\text{-value)}$$

or as the following equivalent formula:

Formula 10-6 $\qquad s_e = \sqrt{\dfrac{\Sigma y^2 - b_0 \Sigma y - b_1 \Sigma xy}{n - 2}}$

STATDISK, Minitab, Excel, and the TI-83/84 Plus calculator are all designed to automatically compute the value of s_e. (See the TI-83/84 Plus calculator instructions at the end of this section.)

The development of the standard error of estimate s_e closely parallels that of the ordinary standard deviation introduced in Section 3-3. Just as the standard deviation is a measure of how values deviate from their mean, the standard error of estimate s_e is a measure of how sample data points deviate from their regression line. The reasoning behind dividing by $n - 2$ is similar to the reasoning that led to division by $n - 1$ for the ordinary standard deviation. It is important to note that relatively smaller values of s_e reflect points that stay close to the regression line, and relatively larger values occur with points farther away from the regression line.

Formula 10-6 is algebraically equivalent to the other equation in the definition, but Formula 10-6 is generally easier to work with because it doesn't require that we compute each of the predicted values $\hat{y}$ by substitution in the regression equation. However, Formula 10-6 does require that we find the y-intercept b_0 and the slope b_1 of the estimated regression line.

EXAMPLE 2   **Pizza/Subway Fare Costs: Finding s_e** Use Formula 10-6 to find the standard error of estimate s_e for the paired pizza/subway fare data listed in Table 10-1 in the Chapter Problem.

SOLUTION   Using the sample data in Table 10-1, we find these values:

$$n = 6 \qquad \Sigma y^2 = 9.2175 \qquad \Sigma y = 6.35 \qquad \Sigma xy = 9.4575$$

In Section 10-3 we used the pizza/subway fare sample data to find the y-intercept and the slope of the regression line. Those values are given here with extra decimal places for greater precision.

$$b_0 = 0.034560171 \qquad b_1 = 0.94502138$$

We can now use these values in Formula 10-6 to find the standard error of estimate s_e.

$$s_e = \sqrt{\frac{\Sigma y^2 - b_0 \Sigma y - b_1 \Sigma xy}{n - 2}}$$

$$= \sqrt{\frac{9.2175 - (0.034560171)(6.35) - (0.94502138)(9.4575)}{6 - 2}}$$

$$= 0.12298700 = 0.123 \qquad \text{(rounded)}$$

We can measure the spread of the sample points about the regression line with the standard error of estimate $s_e = 0.123$. The standard error of estimate s_e can be used to construct interval estimates that will help us see if our point estimates of y are dependable. Assume that for each fixed value of x, the corresponding sample values of y are normally distributed about the regression line, and those normal distributions have the same variance. The following interval estimate applies to an *individual* y-value. (For a confidence interval used to predict the *mean* of all y-values for some given x-value, see Exercise 26.)

Prediction Interval for an Individual *y*

Given the fixed value x_0, the prediction interval for an individual y is

$$\hat{y} - E < y < \hat{y} + E$$

continued

where the margin of error E is

$$E = t_{\alpha/2}s_e\sqrt{1 + \frac{1}{n} + \frac{n(x_0 - \bar{x})^2}{n(\Sigma x^2) - (\Sigma x)^2}}$$

and x_0 represents the given value of x, $t_{\alpha/2}$ has $n - 2$ degrees of freedom, and s_e is found from Formula 10-6.

EXAMPLE 3   **Pizza/Subway Fare Costs: Finding a Prediction Interval**
For the paired pizza/subway fare costs from the Chapter Problem, we have found that for a pizza cost of $2.25, the best predicted cost of a subway fare is $2.16. Construct a 95% prediction interval for the cost of a subway fare, given that a slice of pizza costs $2.25 (so that $x = 2.25$).

SOLUTION   From Section 10-2 we know that $r = 0.988$, so that there is sufficient evidence to support a claim of a linear correlation (at the 0.05 significance level), and the regression equation is $\hat{y} = 0.0346 + 0.945x$. In Example 2 we found that $s_e = 0.12298700$. We obtain the following statistics from the pizza/subway fare data in Table 10-1:

$$n = 6 \qquad \bar{x} = 1.0833333 \qquad \Sigma x = 6.50 \qquad \Sigma x^2 = 9.77$$

From Table A-3 we find $t_{\alpha/2} = 2.776$ (based on $6 - 2 = 4$ degrees of freedom with $\alpha = 0.05$ in two tails). We first calculate the margin of error E by letting $x_0 = 2.25$ (because we want the prediction interval of the subway fare given that the pizza cost is $x = 2.25$).

$$E = t_{\alpha/2}s_e\sqrt{1 + \frac{1}{n} + \frac{n(x_0 - \bar{x})^2}{n(\Sigma x^2) - (\Sigma x)^2}}$$

$$= (2.776)(0.12298700)\sqrt{1 + \frac{1}{6} + \frac{6(2.25 - 1.0833333)^2}{6(9.77) - (6.50)^2}}$$

$$= (2.776)(0.12298700)(1.2905606) = 0.441 \qquad \text{(rounded)}$$

With $\hat{y} = 2.16$ and $E = 0.441$, we get the prediction interval as follows:

$$\hat{y} - E < y < \hat{y} + E$$
$$2.16 - 0.441 < y < 2.16 + 0.441$$
$$1.72 < y < 2.60$$

INTERPRETATION   If the cost of a slice of pizza is $2.25, we have 95% confidence that the cost of a subway fare is between $1.72 and $2.60. That's a fairly large range of possible values, and one major factor contributing to the large range is that the sample size is very small with $n = 6$.

In addition to knowing that if a slice of pizza costs $2.25, the predicted cost of a subway fare is $2.16, we now have a sense of the reliability of that estimate. The 95% prediction interval found in this example shows that the actual cost of a subway fare can vary substantially from the predicted value of $2.16.

Using the TI-83/84 Plus Calculator to Construct a Prediction Interval

When using a given value of x to construct a prediction interval $\hat{y} - E < y < \hat{y} + E$, you must calculate the limits of $\hat{y} - E$ and $\hat{y} + E$. Use this procedure:

1. Using the given value of x, find the corresponding value of $\hat{y}$ as follows: Identify the regression equation (by using the procedure at the end of Section 10-3), then substitute the given value of x into that equation. The result is the value $\hat{y}$, which is the value of y that is predicted from the given value of x.

2. Find the value of the margin of error E by using the components listed below (as illustrated in Example 3).

   $t_{\alpha/2}$ found from Table A-3 by using $n - 2$ degrees of freedom

   s_e provided by the TI-83/84 Plus calculator by using the procedure given at the end of Section 10-2

   x_0 given value of x used for the prediction

   n number of pairs of data

   Σx sum of the x values

   Σx^2 sum of the squares of the x values

   Press **STAT**, select **CALC**, **1-VarStats,** enter L1, press **ENTER**.

3. Using the value of $\hat{y}$ from Step 1 and the value of E from Step 2, proceed to find the values of the limits $\hat{y} - E$ and $\hat{y} + E$. Substitute those values to obtain the prediction interval $\hat{y} - E < y < \hat{y} + E$.

Although the prediction interval is not completely generated, the TI-83/84 Plus calculator is quite helpful in finding it.

10-4 Basic Skills and Concepts

Statistical Literacy and Critical Thinking

1. s_e Notation Assume that you have paired values consisting of heights (in inches) and weights (in lb) from 40 randomly selected men (as in Data Set 1 in Appendix B), and that you plan to use a height of 70 in. to predict weight. In your own words, describe what s_e represents.

2. Prediction Interval Using the heights and weights described in Exercise 1, a height of 70 in. is used to find that the predicted weight is 180 lb. In your own words, describe a prediction interval in this situation.

3. Prediction Interval Using the heights and weights described in Exercise 1, a height of 70 in. is used to find that the predicted weight is 180 lb. What is the major advantage of using a prediction interval instead of the predicted weight of 180 lb? Why is the terminology of *prediction interval* used instead of *confidence interval*?

4. Coefficient of Determination Using the heights and weights described in Exercise 1, the linear correlation coefficient r is 0.522. Find the value of the coefficient of determination. What practical information does the coefficient of determination provide?

Interpreting the Coefficient of Determination. In Exercises 5–8, use the value of the linear correlation coefficient r to find the coefficient of determination and the percentage of the total variation that can be explained by the linear relationship between the two variables from the Appendix B data sets.

5. $r = 0.873$ ($x =$ tar in menthol cigarettes, $y =$ nicotine in menthol cigarettes)

6. $r = 0.744$ ($x =$ movie budget, $y =$ movie gross)

7. $r = -0.865$ ($x =$ car weight, $y =$ city fuel consumption in mi/gal)

8. $r = -0.488$ ($x =$ age of home, $y =$ home selling price)

Interpreting a Computer Display. *In Exercises 9–12, refer to the TI-83/84 Plus calculator displays obtained by using the paired data consisting of weights (in lb) of 32 cars and their highway fuel consumption amounts (in mi/gal), as listed in Data Set 16 in Appendix B. The weights were used for the x variable.*

TI-83/84 Plus

```
LinRegTTest
 y=a+bx
 β≠0 and ρ≠0
 t=-7.467705745
 p=2.5307076ε-8
 df=30
↓a=50.50155398
```

TI-83/84 Plus

```
LinRegTTest
 y=a+bx
 β≠0 and ρ≠0
↑b=-.0058685076
 s=2.194981373
 r²=.650213605
 r=-.8063582362
```

(TI) 9. Testing for Correlation Use the information provided in the displays to determine the value of the linear correlation coefficient. Given that there are 32 pairs of data, is there sufficient evidence to support a claim of a linear correlation between the weights of cars and their highway fuel consumption amounts?

(TI) 10. Identifying Total Variation What percentage of the total variation in highway fuel consumption can be explained by the linear correlation between weight and highway fuel consumption?

(TI) 11. Predicting Highway Fuel Consumption If a car weighs 4000 lb, what is the single value that is the best predicted amount of highway fuel consumption? (Assume that there is a linear correlation between weight and highway fuel consumption.)

(TI) 12. Finding Prediction Interval For a car weighing 4000 lb, identify the 95% prediction interval estimate of the amount of highway fuel consumption, and write a statement interpreting that interval.

Finding Measures of Variation. *In Exercises 13–16, find the (a) explained variation, (b) unexplained variation, (c) total variation, (d) coefficient of determination, and (e) standard error of estimate s_e. In each case, there is sufficient evidence to support a claim of a linear correlation so that it is reasonable to use the regression equation when making predictions. (Results from these exercises are used in Exercises 17–20.)*

(TI) 13. CPI and Pizza The Consumer Price Index (CPI) and the cost of a slice of pizza from Table 10-1 in the Chapter Problem are listed below.

CPI	30.2	48.3	112.3	162.2	191.9	197.8
Cost of Pizza	0.15	0.35	1.00	1.25	1.75	2.00

(TI) 14. CPI and Subway Fare The Consumer Price Index (CPI) and the cost of subway fare from Table 10-1 in the Chapter Problem are listed below.

CPI	30.2	48.3	112.3	162.2	191.9	197.8
Subway Fare	0.15	0.35	1.00	1.35	1.50	2.00

(TI) 15. Measuring Seals from Photos Listed below are the overhead widths (in cm) of seals measured from photographs and the weights of the seals (in kg). (The data are based on "Mass Estimation of Weddell Seals Using Techniques of Photogrammetry," by R. Garrott of Montana State University.)

Overhead Width	7.2	7.4	9.8	9.4	8.8	8.4
Weight	116	154	245	202	200	191

TI **16. Global Warming** Listed below are concentrations (in parts per million) of CO_2 and temperatures (in °C) for different years (based on data from the Earth Policy Institute).

CO_2	314	317	320	326	331	339	346	354	361	369
Temperature	13.9	14.0	13.9	14.1	14.0	14.3	14.1	14.5	14.5	14.4

TI **17. Effect of Variation on Prediction Interval** Refer to the data given in Exercise 13.

a. Find the predicted cost of a slice of pizza for the year 2001, when the CPI was 187.1.

b. Find a 95% prediction interval estimate of the cost of a slice of pizza when the CPI was 187.1.

TI **18. Finding Predicted Value and Prediction Interval** Refer to Exercise 14.

a. Find the predicted cost of subway fare for the year 2001, when the CPI was 187.1.

b. Find a 95% prediction interval estimate of the cost of subway fare when the CPI was 187.1.

TI **19. Finding Predicted Value and Prediction Interval** Refer to the data given in Exercise 15.

a. Find the predicted weight (in kg) of a seal given that the width from an overhead photograph is 9.0 cm.

b. Find a 95% prediction interval estimate of the weight (in kilograms) of a seal given that the width from an overhead photograph is 9.0 cm.

TI **20. Finding Predicted Value and Prediction Interval** Refer to the data described in Exercise 16.

a. Find the predicted temperature (in °C) when the CO_2 concentration is 370.9 parts per million.

b. Find a 99% prediction interval estimate of the temperature (in °C) when the CO_2 concentration is 370.9 parts per million.

Finding a Prediction Interval. *In Exercises 21–24, refer to the pizza/subway sample data from the Chapter Problem. Let x represent the cost of a slice of pizza and let y represent the corresponding subway fare. Use the given pizza cost and the given confidence level to construct a prediction interval estimate of the subway fare. (See Example 3 in this section.)*

TI **21.** Cost of a slice of pizza: $2.10; 99% confidence

TI **22.** Cost of a slice of pizza: $2.10; 90% confidence

TI **23.** Cost of a slice of pizza: $0.50; 95% confidence

TI **24.** Cost of a slice of pizza: $0.75; 99% confidence

10-4 Beyond the Basics

25. Confidence Intervals for β_0 and β_1 Confidence intervals for the y-intercept β_0 and slope β_1 for a regression line ($y = \beta_0 + \beta_1 x$) can be found by evaluating the limits in the intervals below.

$$b_0 - E < \beta_0 < b_0 + E$$

where

$$E = t_{\alpha/2} s_e \sqrt{\frac{1}{n} + \frac{\bar{x}^2}{\sum x^2 - \frac{(\sum x)^2}{n}}}$$

continued

Predictors for Success in College

When a college accepts a new student, it would like to have some positive indication that the

student will be successful in his or her studies. College admissions deans consider SAT scores, standard achievement tests, rank in class, difficulty of high school courses, high school grades, and extracurricular activities. In a study of characteristics that make good predictors of success in college, it was found that class rank and scores on standard achievement tests are better predictors than SAT scores. A multiple regression equation with college grade-point average predicted by class rank and achievement test score was not improved by including another variable for SAT score. This particular study suggests that SAT scores should not be included among the admissions criteria, but supporters argue that SAT scores are useful for comparing students from different geographic locations and high school backgrounds.

$$b_1 - E < \beta_1 < b_1 + E$$

where

$$E = t_{\alpha/2} \cdot \frac{s_e}{\sqrt{\Sigma x^2 - \dfrac{(\Sigma x)^2}{n}}}$$

The y-intercept b_0 and the slope b_1 are found from the sample data and $t_{\alpha/2}$ is found from Table A-3 by using $n - 2$ degrees of freedom. Using the pizza/subway data from the Chapter Problem, find the 95% confidence interval estimates of β_0 and β_1.

26. Confidence Interval for Mean Predicted Value Example 3 in this section illustrated the procedure for finding a prediction interval for an *individual* value of y. When using a specific value x_0 for predicting the *mean* of all values of y, the confidence interval is as follows:

$$\hat{y} - E < \bar{y} < \hat{y} + E$$

where

$$E = t_{\alpha/2} \cdot s_e \sqrt{\frac{1}{n} + \frac{n(x_0 - \bar{x})^2}{n(\Sigma x^2) - (\Sigma x)^2}}$$

The critical value $t_{\alpha/2}$ is found with $n - 2$ degrees of freedom. Use the pizza/subway data from the Chapter Problem to find a 95% confidence interval estimate of the mean subway fare given that the price of a slice of pizza is $2.25.

 10-5 Multiple Regression

Key Concept The preceding sections of this chapter apply to a linear correlation between *two* variables. In this section we present a method for analyzing a linear relationship with *more than two* variables. We focus on three key elements: (1) the multiple regression equation, (2) the value of adjusted R^2, and (3) the P-value. Because of the complex nature of the necessary calculations, manual calculations are impractical and a threat to mental health, so in this section we emphasize the use and interpretation of results from statistical software or a TI-83/84 Plus calculator.

Part 1: Basic Concepts of a Multiple Regression Equation

As in the preceding sections of this chapter, we will consider *linear* relationships only. The following *multiple regression equation* describes linear relationships involving more than two variables.

> **DEFINITION**
>
> A **multiple regression equation** expresses a linear relationship between a response variable y and two or more predictor variables $(x_1, x_2, \ldots, x_k)$. The general form of a multiple regression equation obtained from sample data is
>
> $$\hat{y} = b_0 + b_1 x_1 + b_2 x_2 + \cdots + b_k x_k.$$

We will use the following notation, which follows naturally from the notation used in Section 10-3.

Finding a Multiple Regression Equation

Objective

Use sample data to find a multiple regression equation that is useful for predicting values of the response variable y.

Notation

$\hat{y} = b_0 + b_1 x_1 + b_2 x_2 + \cdots + b_k x_k$ (General form of the multiple regression equation)

$n = $ sample size

$k = $ number of *predictor* variables. (The predictor variables are also called *independent variables* or x variables.)

$\hat{y} = $ predicted value of y (computed using the multiple regression equation)

$x_1, x_2, \ldots, x_k$ are the predictor variables

$\beta_0, \beta_1, \beta_2, \ldots, \beta_k$ are the parameters for the multiple regression equation

$y = \beta_0 + \beta_1 x_1 + \beta_2 x_2 + \cdots + \beta_k x_k$ found from the *population* of data.

$b_0, b_1, b_2, \ldots, b_k$ are the *sample estimates* of the parameters $\beta_0, \beta_1, \beta_2, \ldots, \beta_k$.

Requirements

For any specific set of x values, the regression equation is associated with a random error often denoted by ε. We assume that such errors are normally distributed with a mean of 0 and a standard deviation of σ, and that the random errors are independent. These assumptions are difficult to check. We assume throughout this section that the necessary requirements are satisfied.

Procedure for Finding a Multiple Regression Equation

Manual calculations are not practical, so computer software such as STATDISK, Minitab, Excel, or a TI-83/84 Plus calculator must be used. (See the TI-83/84 Plus calculator instructions at the end of this section.)

Because we use computer software or a calculator to find multiple regression equations, we ignore the actual calculations and focus instead on *interpreting* computer displays.

> **EXAMPLE 1**   **Heights of Mothers, Fathers, and Daughters** Table 10-6 on the next page includes a random sample of heights of mothers, fathers, and their daughters (based on data from the National Health and Nutrition Examination). Find the multiple regression equation in which the response (y) variable is the height of a daughter and the predictor (x) variables are the height of the mother and height of the father.

SOLUTION   The TI-83/84 Plus calculator results are shown below.

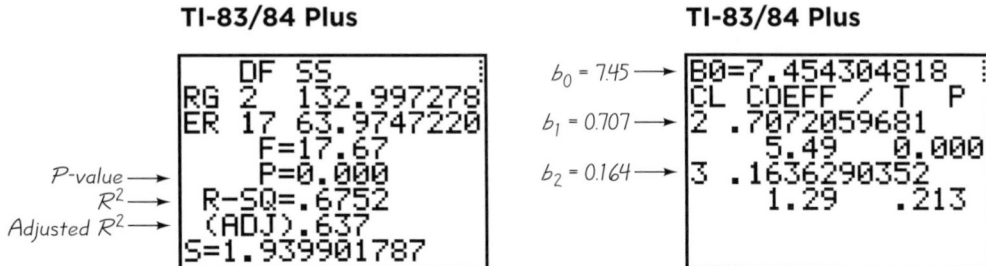

continued

Table 10-6 Heights (in inches) of Mothers, Fathers, and their Daughters

Height of Mother	Height of Father	Height of Daughter
63	64	58.6
67	65	64.7
64	67	65.3
60	72	61.0
65	72	65.4
67	72	67.4
59	67	60.9
60	71	63.1
58	66	60.0
72	75	71.1
63	69	62.2
67	70	67.2
62	69	63.4
69	62	68.4
63	66	62.2
64	76	64.7
63	69	59.6
64	68	61.0
60	66	64.0
65	68	65.4

Interpreting TI-83/84 Plus calculator results: The two screens on page 565 might not win many awards for clarity of presentation, but they do contain the key elements that we need in this section. The important results from the first screen are listed below. In the second screen, see that CL (column) is the top heading for the *column number,* and each column number is followed by the coefficient used in the multiple regression equation, a *t* test statistic value, and a *P*-value for the variable corresponding to the column. Listed below are the key elements that should be extracted from the second screen.

First screen: *P*-value $= 0.000$

$R^2 = 0.6752$

Adjusted $R^2 = 0.637$

Second screen: $b_0 = 7.45$ (rounded)

$b_1 = 0.707$ [Coefficient for column 2 (rounded)]

$b_2 = 0.164$ [Coefficient for column 3 (rounded)]

From the display, we see that the multiple regression equation is

Height $= 7.45 + 0.707$ Mother $+ 0.164$ Father

Using our notation presented earlier in this section, we could write this equation as

$$\hat{y} = 7.45 + 0.707x_1 + 0.164x_2$$

where $\hat{y}$ is the predicted height of a daughter, x_1 is the height of the mother, and x_2 is the height of the father.

If a multiple regression equation fits the sample data well, it can be used for predictions. For example, if we determine that the multiple regression equation in Example 1 is suitable for predictions, and if we have a mother 63 in. tall and a father 69 in. tall, we can predict the height of the daughter by substituting those values into the regression equation to get a predicted height of 63.4 in. for their daughter.

R^2 and Adjusted R^2

R^2 denotes the **multiple coefficient of determination,** which is a measure of how well the multiple regression equation fits the sample data. A perfect fit would result in $R^2 = 1$, and a very good fit results in a value near 1. A very poor fit results in a value of R^2 close to 0. The value of $R^2 = 0.6752$ in the TI-83/84 Plus display for Example 1 indicates that 67.52% of the variation in heights of daughters can be explained by the heights of the mothers and fathers. However, the multiple coefficient of determination R^2 has a serious flaw: As more variables are included, R^2 increases. (R^2 could remain the same, but it usually increases.) The largest R^2 is obtained by simply including *all* of the available variables, but the best multiple regression equation does not necessarily use all of the available variables. Because of that flaw, comparison of different multiple regression equations is better accomplished with the *adjusted coefficient of determination,* which is R^2 adjusted for the number of variables and the sample size.

DEFINITION

The **adjusted coefficient of determination** is the multiple coefficient of determination R^2 modified to account for the number of variables and the sample size. (It is calculated by using Formula 10-7.)

Formula 10-7

$$\text{adjusted } R^2 = 1 - \frac{(n-1)}{[n - (k+1)]}(1 - R^2)$$

where

$n =$ sample size

$k =$ number of predictor (x) variables

The preceding display for Example 1 shows the adjusted coefficient of determination as (ADJ).637. If we use Formula 10-7 with the R^2 value of 0.6752, $n = 20$ and $k = 2$, we find that the adjusted R^2 value is 0.637, confirming the displayed value of (ADJ).637. When we compare this multiple regression equation to others, it is better to use the adjusted R^2 of 0.637.

P-Value

The P-value is a measure of the overall significance of the multiple regression equation. The displayed P-value in Example 1 of 0.000 (rounded to three decimal places) is small, indicating that the multiple regression equation has good overall significance and is usable for predictions. That is, it makes sense to predict heights of daughters based on heights of mothers and fathers. Like the adjusted R^2, this P-value is a good measure of how well the equation fits the sample data. The value of 0.000 results from a test of the null hypothesis that $\beta_1 = \beta_2 = 0$. Rejection of $\beta_1 = \beta_2 = 0$ implies that at least one of β_1 and β_2 is not 0, indicating that this regression equation is effective in predicting heights of daughters. A complete analysis of the Minitab results might include other important elements, such as the significance of the individual coefficients, but we will limit our discussion to the three key components—multiple regression equation, adjusted R^2, and P-value.

Finding the Best Multiple Regression Equation

When trying to find the best multiple regression equation, we should not necessarily include all of the available predictor variables. Finding the best multiple regression equation requires a good dose of judgment, and there is no exact and automatic procedure that can be used to find the best multiple regression equation. *Determination of the best multiple regression equation is often quite difficult, and is beyond the scope of this book,* but the following guidelines should provide some help.

Guidelines for Finding the Best Multiple Regression Equation

1. *Use common sense and practical considerations to include or exclude variables.* For example, when trying to find a good multiple regression equation for predicting the height of a daughter, we should exclude the height of the physician who delivered the daughter, because that height is obviously irrelevant.

2. *Consider the P-value.* Select an equation having overall significance, as determined by the P-value found in the computer display.

3. *Consider equations with high values of adjusted R^2, and try to include only a few variables.* Instead of including almost every available variable, try to include relatively few predictor (x) variables. Use these guidelines:

   • Select an equation having a value of adjusted R^2 with this property: If an additional predictor variable is included, the value of adjusted R^2 does not increase by a substantial amount.

   • For a particular number of predictor (x) variables, select the equation with the largest value of adjusted R^2.

   • In excluding predictor (x) variables that don't have much of an effect on the response (y) variable, it might be helpful to find the linear correlation coefficient r for each pair of variables being considered. If two predictor values have a very high linear correlation coefficient, there is no need to include them both, and we should exclude the variable with the lower value of r.

NBA Salaries and Performance

Researcher Matthew Weeks investigated the correlation between NBA salaries and basketball game statistics. In addition to salary (S), he considered minutes played (M), assists (A), rebounds (R), and points scored (P), and he used data from 30 players. The multiple regression equation is $S = -0.716 - 0.0756M - 0.425A + 0.0536R + 0.742P$ with $R^2 = 0.458$. Because of a high correlation between minutes played (M) and points scored (P), and because points scored had a higher correlation with salary, the variable of minutes played was removed from the multiple regression equation. Also, the variables of assists (A) and rebounds (R) were not found to be significant, so they were removed as well. The single variable of points scored appeared to be the best choice for predicting NBA salaries, but the predictions were found to be not very accurate because of other variables not considered, such as popularity of the player.

EXAMPLE 2 **Predicting Household Size from Discarded Garbage** Data Set 22 in Appendix B includes the household size (number of people) and the individual weights of discarded metal, paper, plastic, glass, food, yard waste, textile materials, weights of all other items, and the total discarded weight obtained from 62 households. An objective of the study was to determine whether population counts could be made by measuring the discarded garbage. Considering some or all of the predictor variables, find the multiple regression equation that is *best* for predicting household size. Is the best multiple regression equation a *good* equation for predicting household size?

SOLUTION We first exclude the variable of total weight, because it is simply the sum of the weights of the other categories. Table 10-7 includes a few of the many other possibilities for choices of predictor variables. From Table 10-7 we see that the combination of predictor variables of plastic and glass and metal and other appears to be best because that combination has the highest adjusted R^2 of 0.672 and the lowest P-value of 0.000. If we were to consider every other possible combination of predictor variables, we would find that plastic/glass/metal/other is the best combination. But we might question whether the adjusted R^2 of 0.672 is sufficiently greater than the adjusted R^2 of 0.556 found using plastic as the only predictor variable. Given the difficulties involved in separating and weighing the different categories of garbage, it seems that we would be better off using plastic as the only predictor variable. The best regression equation is therefore $\hat{y} = 1.08 + 1.38x$, where $\hat{y}$ is the predicted household size and x is the weight of discarded plastic. Because this regression equation has an adjusted R^2 of 0.556 and a low P-value of 0.000, it is good for predicting household size.

Table 10-7 Searching for the Best Multiple Regression Equation

Predictor Variables	Adjusted R^2	P-Value	
Yard	0.005	0.256	← **Not best:** P-value is too high.
Plastic and Glass and Metal and Other	**0.672**	0.000	← **Best:** Highest adjusted R^2 and lowest P-value.
Plastic	0.556	0.000	← **Best if using only one predictor variable.**
Plastic and Food	0.551	0.000	← **Not best:** The adjusted R^2 value is not highest.

The preceding guidelines for finding the best multiple regression equation are based on the adjusted R^2 and the P-value, but we could also conduct individual hypothesis tests based on values of the regression coefficients. Consider the regression coefficient of β_1. A test of the null hypothesis $\beta_1 = 0$ can tell us whether the corresponding predictor variable should be included in the regression equation. Rejection of $\beta_1 = 0$ suggests that β_1 has a nonzero value and is therefore helpful for predicting the value of the response variable. Procedures for such tests are described in Exercise 17.

Part 2: Dummy Variables and Logistic Regression

In this section, all variables have been continuous in nature. The height of a daughter can be any value over a continuous range of heights, so it is a good example of a continuous variable. However, many applications involve a **dichotomous variable,** which is a variable with only *two* possible discrete values (such as male/female or dead/alive or cured/not cured). A common procedure is to represent the two possible discrete values by 0 and 1, where 0 represents a "failure" (such as death) and 1 represents a success. A dichotomous variable with the two possible values of 0 and 1 is called a *dummy variable.*

Procedures of analysis differ dramatically, depending on whether the dummy variable is a predictor (x) variable or the response (y) variable. If we include a dummy variable as another *predictor* (x) variable, we can use the methods of this section, as illustrated in Example 3.

Table 10-8 Heights (in inches) of Mothers, Fathers, and their Children

Height of Mother	Height of Father	Sex of Child	Height of Child
66	70	M	62.5
66	64	M	69.1
64	68	M	67.1
66	74	M	71.1
64	62	M	67.4
64	67	M	64.9
62	72	M	66.5
62	72	M	66.5
63	71	M	67.5
65	71	M	71.9
63	64	F	58.6
64	67	F	65.3
65	72	F	65.4
59	67	F	60.9
58	66	F	60.0
63	69	F	62.2
62	69	F	63.4
63	66	F	62.2
63	69	F	59.6
60	66	F	64.0

EXAMPLE 3 **Using a Dummy Variable** Use the data in Table 10-8 and, for a predictor variable, use the dummy variable of sex (coded as 0 = female, 1 = male). (The data in Table 10-8 are based on data from the National Health and Nutrition Examination.) Given that a mother is 63 in. tall and a father is 69 in. tall, find the multiple regression equation and use it to predict the height of (a) a daughter and (b) a son.

SOLUTION Using the methods of this section and a TI-83/84 Plus calculator, we get this regression equation:

$$\text{Height of Child} = 25.6 + 0.377(\text{Height of Mother}) \\ + 0.195(\text{Height of Father}) + 4.15(\text{Sex})$$

where the value of the dummy variable is 0 for a daughter and 1 for a son.

Icing the Kicker

Just as a kicker in football is about to attempt a field goal, it is a common strategy for the opposing coach to call a time-out to "ice" the kicker. The theory is that the kicker has time to think and become nervous and less confident, but does the practice actually work? In "The Cold-Foot Effect" by Scott M. Berry in *Chance* magazine, the author wrote about his statistical analysis of results from two NFL seasons. He uses a logistic regression model with variables such as wind, clouds, precipitation, temperature, the pressure of making the kick, and whether a time-out was called prior to the kick. He writes that "the conclusion from the model is that icing the kicker works—it is likely icing the kicker reduces the probability of a successful kick."

a. To find the predicted height of a *daughter,* we substitute 0 for the sex variable, and we also substitute 63 in. for the mother's height and 69 in. for the father's height. The result is a predicted height of 62.8 in. for a daughter.

b. To find the predicted height of a *son,* we substitute 1 for the sex variable, and we also substitute 63 in. for the mother's height and 69 in. for the father's height. The result is a predicted height of 67.0 in. for a son.

   The coefficient of 4.15 in the regression equation shows that when given the height of a mother and the height of a father, a son will have a predicted height that is 4.15 in. more than the height of a daughter.

Logistic Regression   In Example 3, we could use the methods of this section because the dummy variable of sex is a *predictor* variable. If the dummy variable is the response (*y*) variable, we cannot use the methods of this section, and we should use a different method known as **logistic regression.** Example 4 illustrates the method of logistic regression.

EXAMPLE 4   **Logistic Regression**   Let a sample data set consist of the heights, weights, waist sizes, and pulse rates of women and men as listed in Data Set 1 in Appendix B. Let the *response y* variable represent gender (0 = female, 1 = male). Using the 80 values of *y* and the combined list of corresponding heights, weights, waist sizes, and pulse rates, we can use logistic regression to obtain this model:

$$\ln\left(\frac{p}{1-p}\right) = -41.8193 + 0.679195(\text{HT}) - 0.0106791(\text{WT})$$

$$+ 0.0375373(\text{WAIST}) - 0.0606805(\text{PULSE})$$

In the above expression, *p* represents a probability. A value of $p = 0$ indicates that the person is a female and $p = 1$ indicates a male. A value of $p = 0.2$ shows that there is a 0.2 probability of the person being a male, so it follows that there is a 0.8 chance that the person is a female. If we use the above model and substitute a height of 72 in., a weight of 200 lb, a waist circumference of 90 cm, and a pulse rate of 85 beats per minute, we can solve for *p* to get $p = 0.960$, indicating that such a large person is very likely to be a male. In contrast, a small person with a height of 60 in., a weight of 90 lb, a waist size of 68 cm, and a pulse rate of 85 beats per minute results in a value of $p = 0.00962$, indicating that such a small person is very likely to be a female.

This book does not include detailed procedures for using logistic regression, but several books are devoted to this topic, and several other textbooks include detailed information about this method.

   When we discussed regression in Section 10-3, we listed four common errors that should be avoided when using regression equations to make predictions. These same errors should be avoided when using multiple regression equations. Be especially careful about concluding that a cause–effect association exists.

Using the TI-83/84 Plus Calculator for Multiple Regression

The TI-83/84 Plus program A2MULREG must be downloaded to your calculator from the CD-ROM included with this book. On the CD-ROM, select the *software* folder, then select *TI-83 TI-84 Plus Programs.* After the program A2MULREG has been downloaded to your calculator, use the following procedure.

   1. Enter the sample data in matrix D, with the first column containing the values of the dependent (*y*) variable. (A matrix is a rectangular array of data, and it

consists of multiple rows and columns.) Here are two basic ways to enter the values in matrix D:

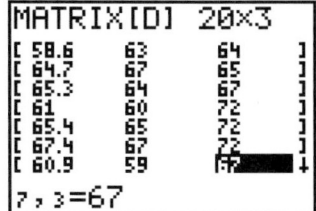

- *Manual entry of sample data:* To manually enter the sample data in matrix D, press **2ND** and the **x⁻¹** key to get the **Matrix** menu. Select **EDIT,** scroll down for **[D],** then press **ENTER**. Begin by entering the dimension of the matrix in the format of rows by columns. For the number of rows, enter the number of sample values listed for each variable. For the number of columns, enter the total number of variables (including the dependent variable). For example, to enter all of the data in Table 10-6 as matrix [D], enter 20 for the number of rows (because there are 20 sample values listed for each variable) and enter 3 for the number of columns (because there are 3 different variables). See the accompanying screen showing the creation of matrix [D] for the data in Table 10-1. After entering the dimension of the matrix, enter the first value, press **ENTER**, then proceed to enter the other values followed by the **ENTER** key after each value. Remember, the first column of matrix [D] must contain the sample values for the dependent (y) variable.

- *Using existing lists:* If the data are already stored as individual lists, those lists can be combined and stored in matrix [D]. To combine lists into a matrix, press **2ND** and the **x⁻¹** key to get the **Matrix** menu. Select the top menu item of **MATH,** then select the item of **List→matr.** Now enter the list names (with the first list corresponding to the dependent y variable), and end with the matrix name [D], all separated by commas. Enter the matrix name of [D] by opening the MATRIX menu, scrolling to [D], and pressing **ENTER**. This is all likely to appear confusing, so consider the objective of doing multiple regression with the sample data in Table 10-6. Assume that the heights of mothers in the first column have been entered as the list named MOM, the heights of fathers in the second column have been entered as the list named DAD, and the heights of the daughters in the third column have been entered as the list named CHILD. Here is the command that puts these lists into matrix [D] with the first column of the matrix representing the dependent (y) variable: **List→matr(CHILD, MOM, DAD, [D]).** It sure beats manual entry of all of the sample data.

2. Press **PRGM**, select **A2MULREG** and press **ENTER** three times, then select **MULT REGRESSION** and press **ENTER**.

3. When prompted, enter the number of *independent* variables, then enter the column numbers of the independent variables that you want to include. (Remember, the y–variable is the dependent variable, and it is not one of the independent variables.)

4. When finished with the entry process, the screen will display some of the results, then press **ENTER** to get the remaining results. See the sample screen displays included in Example 1.

5. When finished, press **ENTER**, then select the **QUIT** option.

10-5   Basic Skills and Concepts

Statistical Literacy and Critical Thinking

1. Notation In multiple regression equations, what do b_1 and β_1 denote? How do they differ?

2. Adjusted Coefficient of Determination

a. When comparing different multiple regression equations for predicting the selling price of a 1960 Corvette, why is the adjusted R^2 a better measure than R^2?

b. When using the sample data in Table 10-6, the single predictor variable of the mother's height results in an adjusted R^2 value of 0.623, and the two predictor variables (mother's height and father's height) result in an adjusted R^2 value of 0.637. Given that the use of the two predictor variables results in the larger value of adjusted R^2, why is the regression equation with the single predictor variable better?

3. Predicting Eye Color A geneticist wants to develop a method for predicting the eye color of a baby, given the eye color of each parent. Can the methods of this section be used? Why or why not?

4. Response and Predictor Variables The regression equation $\hat{y} = -3528 + 1.02x_1 - 1.94x_2$ is obtained using sample data consisting of home selling prices (based on Data Set 23 in Appendix B). In that equation, $\hat{y}$ represents the predicted selling price, x_1 represents the list price, and x_2 represents the annual tax amount. Identify the response variables and predictor variables. In general, how do a response variable and a predictor variable differ?

TI-83/84 Plus

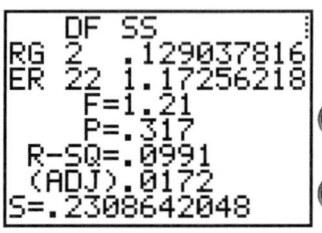

TI-83/84 Plus

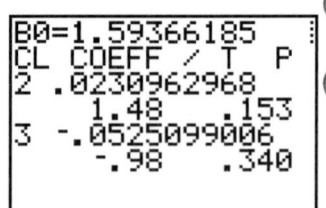

Interpreting a Computer Display. *In Exercises 5–8, refer to the TI-83/84 Plus calculator displays and answer the given questions or identify the indicated items. The displays are based on the measured amounts of tar, carbon monoxide (CO), and nicotine in a sample of 25 king size cigarettes listed in Data Set 4 in Appendix B.*

5. Cigarette Measurements Identify the multiple regression equation that expresses the amount of nicotine in terms of the amount of tar and carbon monoxide (CO).

6. Cigarette Measurements Identify the following:

a. The P-value corresponding to the overall significance of the multiple regression equation

b. The value of the multiple coefficient of determination R^2

c. The adjusted value of R^2

7. Cigarette Measurements Should the multiple regression equation be used for predicting the amount of nicotine based on the amounts of tar and CO? Why or why not?

8. Cigarette Measurements A cigarette has 26 mg of tar and 15 mg of CO. Use the multiple regression equation to determine the predicted amount of nicotine. Is the result likely to be a good predicted value? Why or why not?

Home Selling Prices: Finding the Best Multiple Regression Equation. *In Exercises 9–12, refer to the accompanying table, which was obtained using data from homes sold (from Data Set 23 in Appendix B). The response (y) variable is selling price (in dollars). The predictor (x) variables are LP (list price in dollars), LA (living area of the home in square feet), and LOT (lot size in acres).*

Predictor (x) Variables	P-value	R^2	Adjusted R^2	Regression Equation
LP, LA, Lot	0.000	0.990	0.989	$\hat{y} = 1120 + 0.972\ LP + 0.281\ LA + 465\ LOT$
LP, LA	0.000	0.990	0.989	$\hat{y} = -40.5 + 0.985\ LP - 0.985\ LA$
LP, LOT	0.000	0.990	0.989	$\hat{y} = 1004 + 0.974\ LP + 429\ LOT$
LA, LOT	0.000	0.815	0.805	$\hat{y} = 111,309 + 98.2\ LA + 17,269\ LOT$
LP	0.000	0.990	0.990	$\hat{y} = 99.2 + 0.979\ LP$
LA	0.000	0.643	0.633	$\hat{y} = 133,936 + 101\ LA$
LOT	0.003	0.215	0.194	$\hat{y} = 310,191 + 19,217\ LOT$

9. If only one predictor (x) variable is used to predict the selling price of a home, which single variable is best? Why?

10. If exactly two predictor (x) variables are to be used to predict the selling price of a home, which two variables should be chosen? Why?

11. Which regression equation is best for predicting the selling price? Why?

12. A home is for sale with a list price of $400,000, it has a living area of 3000 square feet, and it is on a 2 acre lot. What is the best predicted value of the selling price? Is that predicted selling price likely to be a good estimate? Is that predicted value likely to be very accurate?

Appendix B Data Sets. *In Exercises 13–16, refer to the indicated data set in Appendix B.*

TI **13. Predicting Nicotine in Cigarettes** Refer to Data Set 4 in Appendix B and use the tar, nicotine, and CO amounts for the cigarettes that are 100 mm long, filtered, nonmenthol, and non-light (the last set of measurements). Find the best regression equation for predicting the amount of nicotine in a cigarette. Why is it best? Is the best regression equation a good regression equation for predicting the nicotine content? Why or why not?

TI **14. Predicting Movie Gross Amount** Refer to Data Set 9 in Appendix B and find the best regression equation with movie gross amount (in millions of dollars) as the response (y) variable. Ignore the MPAA ratings. Why is this equation best? Is this "best" equation good for predicting the amount of money that a movie will gross? Does the combination of predictor variables make sense?

TI **15. Car Mileage** Refer to Data Set 16 in Appendix B and find the best regression equation with highway fuel consumption (in mi/gal) as the response (y) variable. Because the car's weight, length, and engine displacement are all easy to measure, use only those variables as the possible predictor variables. Is the "best" equation good for predicting the highway fuel consumption?

TI **16. Old Faithful** Refer to Data Set 15 in Appendix B and determine the best regression equation that expresses the response variable (y) of time interval after an eruption in terms of one or more of the variables of duration, time interval before the eruption, and height of the eruption. Explain your choice.

10-5 Beyond the Basics

TI **17. Testing Hypotheses About Regression Coefficients** If the coefficient β_1 has a nonzero value, then it is helpful in predicting the value of the response variable. If $\beta_1 = 0$, it is not helpful in predicting the value of the response variable and can be eliminated from the regression equation. To test the claim that $\beta_1 = 0$, use the test statistic $t = (b_1 - 0)/s_{b_1}$. Critical values or P-values can be found using the t distribution with $n - (k + 1)$ degrees of freedom, where k is the number of predictor (x) variables and n is the number of observations in the sample. The test statistic t and P-value can be found using TI-83/84 Plus results. For example, the displays in Example 1 show that for β_1, $t = 5.49$ and the P-value is 0.000. Use the sample data in Table 10-6 and the Minitab display in Example 1 to test the claim that $\beta_1 = 0$. Also test the claim that $\beta_2 = 0$. What do the results imply about the regression equation?

TI **18. Confidence Interval for a Regression Coefficient** A confidence interval for the regression coefficient β_1 is expressed as

$$b_1 - E < \beta_1 < b_1 + E$$

where

$$E = t_{\alpha/2} s_{b_1}$$

The critical t score is found using $n - (k + 1)$ degrees of freedom, where k, n, and s_{b_1} are as described in Exercise 17. Use the sample data in Table 10-6 and the TI-83/84 Plus calculator display in Example 1 to construct 95% confidence interval estimates of β_1 (the coefficient for the variable representing height of the mother) and β_2 (the coefficient for the variable representing height of the father). Does either confidence interval include 0, suggesting that the variable be eliminated from the regression equation?

TI **19. Dummy Variable** Refer to Data Set 6 in Appendix B and use the sex, age, and weight of the bears. For sex, let 0 represent female and let 1 represent male. (In Data Set 6, males are already represented by 1, but for females change the sex values from 2 to 0.) Letting the response (y) variable represent weight, use the variable of age and the dummy variable of sex to find the multiple regression equation. Use the equation to find the predicted weight of a bear with the characteristics given below. Does sex appear to have much of an effect on the weight of a bear?

a. Female bear that is 20 years of age

b. Male bear that is 20 years of age

 Modeling

Key Concept The previous sections of this chapter deal with *linear* relationships only, but this section introduces some basic concepts of finding a *nonlinear* function that fits sample data. We refer to such a function as a **mathematical model.** A mathematical model is simply a mathematical function that "fits" or describes real-world data. Instead of using randomly selected sample data, we will consider data collected periodically over time or some other basic unit of measurement. There are some powerful statistical methods that we could discuss (such as *time series*), but the main objective of this section is to describe briefly how technology can be used to find a good mathematical model.

The following are some generic models as listed in a menu from the TI-83/84 Plus calculator (press **STAT,** then select **CALC**):

Linear: $y = a + bx$ Quadratic: $y = ax^2 + bx + c$

Logarithmic: $y = a + b \ln x$ Exponential: $y = ab^x$

Power: $y = ax^b$

The particular model you select depends on the nature of the sample data, and a scatterplot can be very helpful in making that determination. The illustrations that follow are graphs of some common models displayed on a TI-83/84 Plus calculator.

TI-83/84 PLUS

Linear: $y = 1 + 2x$ Logarithmic: $y = 1 + 2 \ln x$ Power: $y = 3x^{2.5}$

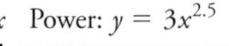

Quadratic: $y = x^2 - 8x + 18$ Exponential: $y = 2^x$

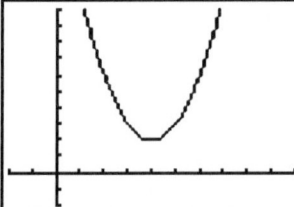 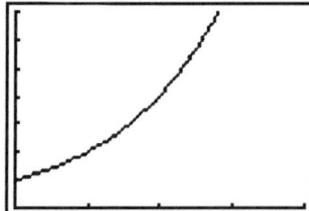

Here are three basic rules for developing a good mathematical model:

1. *Look for a pattern in the graph.* Use the sample data to construct a graph (such as a scatterplot). Then compare the basic pattern to the known generic graphs of linear, quadratic, logarithmic, exponential, and power functions. (See which of the TI-83/84 Plus calculator graphs shown here is closest to the graph of the sample data.)

2. *Find and compare values of R^2.* For each model being considered, use computer software or a TI-83/84 Plus calculator to find the value of the coefficient of determination R^2. Values of R^2 can be interpreted here the same way that they were interpreted in Section 10-5: Select functions that result in larger values of R^2, because such larger values correspond to functions that better fit the observed points. However, don't place much importance on small differences, such as the difference between $R^2 = 0.984$ and $R^2 = 0.989$. (Another measurement used to assess the quality of a model is the sum of squares of the residuals. See Exercise 19.)

3. *Think.* Use common sense. Don't use a model that leads to predicted values that are unrealistic. Use the model to calculate future values, past values, and values for missing data, then determine whether the results are realistic and make sense.

> **EXAMPLE 1** **Finding the Best Population Model** Table 10-9 lists the population of the United States for different years. Find a mathematical model for the population size, then predict the size of the U.S. population in the year 2020.

Table 10-9   Population (in millions) of the United States

Year	1800	1820	1840	1860	1880	1900	1920	1940	1960	1980	2000
Coded year	1	2	3	4	5	6	7	8	9	10	11
Population	5	10	17	31	50	76	106	132	179	227	281

> **SOLUTION** First, we "code" the year values by using 1, 2, 3 . . . , instead of 1800, 1820, 1840. . . The reason for this coding is to use values of x that are much smaller and much less likely to cause computational difficulties.

1. *Look for a pattern in the graph.* Examine the pattern of the data values in the TI-83/84 Plus display (shown in the margin) and compare that pattern to the generic models shown earlier in this section. The pattern of those points is clearly not a straight line, so we rule out a linear model. Good candidates for the model appear to be the quadratic, exponential, and power functions.

2. *Find and compare values of R^2.* The following displays show the TI-83/84 Plus results based on the quadratic, exponential, and power models. Comparing the values of the coefficient R^2, it appears that the quadratic model is best because it has the highest value of 0.9992. However, the other displayed values are also quite high. If we select the quadratic function as the best model, we conclude that the equation $y = 2.77x^2 - 6.00x + 10.01$ best describes the relationship between the year x (coded with $x = 1$ representing 1800, $x = 2$ representing 1820, and so on) and the population y (in millions).

TI-83/84 PLUS

```
QuadReg
 y=ax²+bx+c
 a=2.766899767
 b=-6.002797203
 c=10.01212121
 R²=.9991688446
```

TI-83/84 PLUS

```
ExpReg
 y=a*b^x
 a=5.236195756
 b=1.48297613
 r²=.9631105179
 r=.9813819429
```

TI-83/84 PLUS

```
PwrReg
 y=a*x^b
 a=3.353115397
 b=1.766059823
 r²=.976406226
 r=.9881326966
```

continued

Clinical Trial Cut Short

What do you do when you're testing a new treatment and, before your study ends, you find that it is clearly effective?  You should cut the study short and inform all participants of the treatment's effectiveness. This happened when hydroxyurea was tested as a treatment for sickle cell anemia. The study was scheduled to last about 40 months, but the effectiveness of the treatment became obvious and the study was stopped after 36 months. (See "Trial Halted as Sickle Cell Treatment Proves Itself," by Charles Marwick, *Journal of the American Medical Association,* Vol. 273, No. 8.)

TI-83/84 PLUS

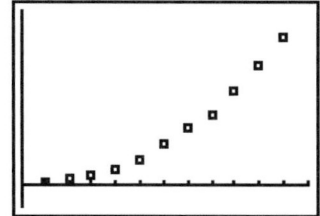

To predict the U.S. population for the year 2020, first note that the year 2020 is coded as $x = 12$ (see Table 10-9). Substituting $x = 12$ into the quadratic model of $y = 2.77x^2 - 6.00x + 10.01$ results in $y = 337$, which indicates that the U.S. population is estimated to be 337 million in the year 2020.

3. *Think.* The forecast result of 337 million in 2020 seems reasonable. (As of this writing, the latest U.S. Bureau of the Census projection is that the population in 2020 will be 336 million.) However, there is considerable danger in making estimates for times that are beyond the scope of the available data. For example, the quadratic model suggests that in 1492, the U.S. population was 671 million—an absurd result. The quadratic model appears to be good for the available data (1800–2000), but other models might be better if it is absolutely necessary to make future population estimates.

EXAMPLE 2   **Interpreting R^2** In Example 1, we obtained the value of $R^2 = 0.9992$ for the quadratic model. Interpret that value as it relates to the predictor variable of year and the response variable of population size.

SOLUTION   In the context of the year/population data from Table 10-9, the value of $R^2 = 0.9992$ can be interpreted as follows: 99.92% of the variation in the population size can be explained by the quadratic regression equation (given in Example 1) that relates year and population size.

In "Modeling the U.S. Population" (*AMATYC Review,* Vol. 20, No. 2), Sheldon Gordon uses more data than Table 10-9, and he uses much more advanced techniques to find better population models. In that article, he makes this important point:

> **"The best choice (of a model) depends on the set of data being analyzed and requires an exercise in judgment, not just computation."**

Using the TI-83/84 Plus Calculator for Modeling

To use TI-83/84 Plus for creating mathematical models such as those discussed in this section, use the following procedure.

1. First enter or open the sample data. The data can be manually entered in lists L1 and L2, or existing named lists can be opened by using their names.
2. Turn on the diagnostics feature by pressing **2ND** **0** to get the **CATALOG** menu. Use the arrow key ⌄ to scroll down to **DiagnosticsOn** and press the **ENTER** key twice.
3. Press **STAT** and select **CALC.**
4. Select the desired model from the list of available options, then press **ENTER**.
5. Enter the two list names or labels, using a comma to separate them. Press **ENTER** when finished.

The display will be similar to those given in this section. The format of the equation will be given along with the coefficients used in the equation. The value of R^2 is also included for many of the models.

10-6 Basic Skills and Concepts

Statistical Literacy and Critical Thinking

1. Claimed Value of R^2 When using data consisting of the number of motor vehicles produced in the United States for each year of the last 30 years, an analyst claims that he obtained a value of $R^2 = 1$. What does that value indicate about the data? Do you believe the analyst's claim? Why or why not?

2. Super Bowl and R^2 When using the numbers of points scored in each Super Bowl from 1980 to the last Super Bowl at the time that this exercise was written, we obtain the following values of R^2 for the different models: Linear: 0.002; quadratic: 0.082; logarithmic: 0.003; exponential: 0.005; power: 0.001. Based on these results, which model is best? Is the best model a good model? What do the results suggest about predicting the number of points scored in a future Super Bowl game?

3. Interpreting R^2 In Exercise 2, the quadratic model results in $R^2 = 0.082$. Identify the percentage of the variation in Super Bowl points that can be explained by the quadratic model relating the variable of year and the variable of points scored. (*Hint:* See Example 2.) What does the result suggest about the usefulness of the quadratic model?

4. Projections In this section we found that for population values from the year 1800 to the year 2000, the best model is described by $y = 2.77x^2 - 6.00x + 10.01$, where the population value of y is in millions. What is wrong with using this model to project the population size for the year 2999?

Finding the Best Model. *In Exercises 5–16, construct a scatterplot and identify the mathematical model that best fits the given data. Assume that the model is to be used only for the scope of the given data, and consider only linear, quadratic, logarithmic, exponential, and power models.*

TI **5.** The table lists the amounts of weekly salary increases y (in dollars) specified in a labor contract negotiated with employees of the Telektronic corporation.

Year	1	2	3	4	5
Increase (y)	10	12	14	16	18

TI **6.** The table lists the value y (in dollars) of $100 deposited in a certificate of deposit at MetLife Bank.

Year	1	2	3	4	5	6	7
Value	100	105	110.25	115.76	121.55	127.63	134.01

TI **7.** The table lists the distance d (in ft) above the ground for an object dropped in a vacuum from a height of 500 ft. The time t (in sec) is the time after the object has been released.

t	1	2	3	4	5
d	484	436	356	244	100

8. The table lists the cost y (in dollars) of purchasing a volume of topsoil, where the volume of topsoil is a cube with each side having a length of x ft.

x	1	2	3	4	5
Cost	1.10	8.80	29.70	70.40	137.50

TI **9. Subway Fare** Use the year/subway fare data in Table 10-1 from the Chapter Problem. Let x represent the year, with 1960 coded as 1, 1973 coded as 14, and so on. Let y represent the subway fare. Does the best model appear to be a good model? Why or why not? Using the best model, find the projected subway fare in the year 2020.

TI **10. Deaths from Motor Vehicle Crashes** Listed below are the numbers of deaths in the United States resulting from motor vehicle crashes. Using the best model and the second-best model, find the projected number of such deaths for the year 2010. Are the two estimates very different?

Year	1975	1980	1985	1990	1995	2000	2005
Deaths	44,525	51,091	43,825	44,599	41,817	41,945	43,443

TI **11. Manatee Deaths from Boats** Listed below are the numbers of Florida manatee deaths resulting from encounters with watercraft for each year beginning with 1980 (based on data from Florida Fish and Wildlife Conservation). Is the best model much better than all of the others? Find the projected number of such deaths for 2006. The actual number of deaths in 2006 was 92. How does the actual number of manatee deaths compare to the projected number of deaths?

16 24 20 15 34 33 33 39 43 50 47 53 38 35 49 42 60 54 67 82 78 81 95 73 69 80

TI **12. Manatee Deaths from Natural Causes** Listed below are the numbers of Florida manatee deaths resulting from natural causes for each year beginning with 1980 (based on data from Florida Fish and Wildlife Conservation). Is the best model a very good model? Why or why not? Find the projected number of such deaths for 2006. The actual number of natural deaths in 2006 was 81. How does the actual number of natural deaths compare to the projected number of natural deaths?

5 9 41 6 24 19 1 10 15 18 21 13 20 22 33 35 101 42 12 37 37 34 59 102 25 88

TI **13. Physics Experiment** An experiment in a physics class involves dropping a golf ball and recording the distance (in m) it falls for different times (in sec) after it was released. The data are given in the table below. Project the distance for a time of 12 sec, given that the golf ball is dropped from a building that is 50 m tall.

Time	0	0.5	1	1.5	2	2.5	3
Distance	0	1.2	4.9	11.0	19.5	30.5	44.0

TI **14. Stock Market** Listed below in order by row are the annual high values of the Dow Jones Industrial Average for each year beginning with 1980. What is the best predicted value for the year 2006? Given that the actual high value in 2006 was 12,464, how good was the predicted value? What does the pattern suggest about the stock market for investment purposes?

1000 1024 1071 1287 1287 1553 1956 2722 2184 2791 3000 3169 3413 3794 3978 5216 6561 8259 9374 11,568 11,401 11,350 10,635 10,454 10,855 10,941

TI **15. Global Warming** Listed below are the global mean temperatures (in degrees °C) of the earth's surface for the years 1950, 1955, 1960, 1965, 1970, 1975, 1980, 1985, 1990, 1995, 2000, and 2005. Find the predicted temperature for the year 2010.

13.8   13.9   14.0   13.9   14.1   14.0   14.3   14.1   14.5   14.5   14.4   14.8

TI **16. Carbon Dioxide** Listed below are concentrations of carbon dioxide (in parts per million) in the earth's atmosphere for the years 1950, 1955, 1960, 1965, 1970, 1975, 1980, 1985, 1990, 1995, 2000, and 2005. Find the predicted concentration of carbon dioxide for the year 2010.

311   314   317   320   326   331   339   346   354   361   369   381

10-6 Beyond the Basics

TI **17. Moore's Law** In 1965, Intel cofounder Gordon Moore initiated what has since become known as *Moore's law:* the number of transistors per square inch on integrated circuits will double approximately every 18 months. The table below lists the number of transistors (in thousands) for different years.

Year	1971	1974	1978	1982	1985	1989	1993	1997	2000	2002	2003	2007
Transistors	2.3	5	29	120	275	1180	3100	7500	42,000	220,000	410,000	789,000

a. Assuming that Moore's law is correct and transistors double every 18 months, which mathematical model best describes this law: linear, quadratic, logarithmic, exponential, power? What specific function describes Moore's law?

b. Which mathematical model best fits the listed sample data?

c. Compare the results from parts (a) and (b). Does Moore's law appear to be working reasonably well?

 **18. Population in 2050** As of this writing, the U.S. Bureau of the Census used its own model to predict a population of 420 million for the United States in 2050. Use the data in Table 10-9 to find the value of R^2 and the 2050 projected population for the linear, quadratic, logarithmic, exponential, and power models. Do any of the models yield a projected population close to 420 million in 2050?

 **19. Sum of Squares Criterion** In addition to the value of R^2, another measurement used to assess the quality of a model is the *sum of squares of the residuals.* Recall from Section 10-3 that a residual is the difference between an observed y value and the value of y predicted from the model, which is denoted as $\hat{y}$. Better models have smaller sums of squares. Refer to the data in Table 10-9.

a. Find $\Sigma(y - \hat{y})^2$, the sum of squares of the residuals resulting from the linear model.

b. Find the sum of squares of residuals resulting from the quadratic model.

c. Verify that according to the sum of squares criterion, the quadratic model is better than the linear model.

Review

This chapter presents basic methods for investigating correlations between variables.

• In Section 10-2 we presented methods for using scatterplots and the linear correlation coefficient r to determine whether there is sufficient evidence to support a claim of a linear correlation between two variables.

• In Section 10-3 we presented methods for finding the equation of the regression line that best fits the paired data. When the regression line fits the data reasonably well, the regression equation can be used to predict the value of a variable, given some value of the other variable.

• In Section 10-4 we introduced the concept of total variation, with components of explained and unexplained variation. The coefficient of determination r^2 gives us the proportion of the variation in the response variable (y) that can be explained by the linear correlation between x and y. We discussed methods for constructing prediction intervals, which are helpful in judging the accuracy of predicted values.

• In Section 10-5 we presented methods for finding a multiple regression equation, which expresses the relationship of a response variable to two or more predictor variables. We also described methods for finding the value of the multiple coefficient of determination R^2, the adjusted R^2, and a P-value for the overall significance of the equation. Those values are helpful for comparing different multiple regression equations as well as finding the best multiple regression equation. Because of the nature of the calculations involved in this section, the methods are based on the interpretation of results from computer software.

• In Section 10-6 we presented basic methods for finding a mathematical model, which is a function that can be used to describe a relationship between two variables. Unlike the preceding sections of this chapter, Section 10-6 included several nonlinear functions.

Statistical Literacy and Critical Thinking

1. Matched Pairs Section 10-2 deals with correlation and Section 9-4 deals with inferences from matched pairs. Given that both sections deal with matched pairs of sample data, what is the basic difference between the goals of those two sections?

2. Correlation Using measurements from 54 bears, it is found that the linear correlation between the chest sizes (distance around the chest) and the weights of the bears is $r = 0.963$ (based on Data Set 6 in Appendix B). Is there sufficient evidence to support the claim of a linear correlation between chest size and weight? If so, does that imply that a larger chest size in a bear is the cause of a larger weight?

3. Interpreting _r_ A jeweler at Tiffany & Company computes the value of the linear correlation coefficient for pairs of sample data consisting of Tiffany prices for gold wedding rings and the corresponding prices at a discount store. She obtains a value of $r = 1$ and concludes that the prices at both companies are the same. Is she correct? Why or why not?

4. Interpreting _r_ A research scientist for the Telektronic company obtains paired data consisting of the cost of manufacturing memory chips of different sizes and the amount of memory that can be stored on those chips. After finding that $r = 0$, she concludes that there is no relationship between those two variables. Is that conclusion correct? Why or why not?

Chapter Quick Quiz

1. Using 10 pairs of sample data, if you compute the value of the linear correlation coefficient r and obtain a result of 2.650, what should you conclude?

2. Using 10 pairs of sample data, if you compute the value of the linear correlation coefficient r and obtain a result of 0.989, what should you conclude?

3. True or false: If sample data result in a linear correlation coefficient of $r = -0.999$, the points are quite close to a straight-line pattern that is downhill (when viewed from left to right).

4. Using 10 pairs of sample data, the value of $r = 0.099$ is found. What should you conclude?

5. True or false: If there is no linear correlation between two variables, then the two variables are not related in any way.

6. Find the critical values of r for a test of the claim that there is a linear correlation between two variables, given that the sample consists of 15 pairs of data and the significance level is 0.05.

7. A scatterplot shows that 20 points fit a perfect straight-line pattern that falls from left to right. What is the value of the linear correlation coefficient?

8. If sample data result in the regression equation of $\hat{y} = -5 + 2x$ and a linear correlation coefficient of $r = 0.999$, find the best predicted value of y for $x = 10$.

9. If sample data result in a linear correlation coefficient of $r = 0.400$, what proportion of the variation in y is explained by the linear relationship between x and y?

10. True or false: If 50 pairs of sample data are used to find $r = 0.999$ where x measures salt consumption and y measures blood pressure, then we can conclude that higher salt consumption causes a rise in blood pressure.

Review Exercises

 1. Body Temperature The table on the top of the next page lists the body temperatures (in °F) of subjects measured at 8:00 AM and later at midnight (based on Data Set 2 in Appendix B).

a. Construct a scatterplot. What does the scatterplot suggest about a linear correlation between 8:00 AM body temperatures and midnight body temperatures?

b. Find the value of the linear correlation coefficient and determine whether there is sufficient evidence to support a claim of a linear correlation between body temperatures measured at 8:00 AM and again at midnight.

c. Letting y represent the midnight temperatures and letting x represent the 8:00 AM temperatures, find the regression equation.

d. Based on the given sample data, what is the best predicted midnight body temperature of someone with a body temperature of 98.3°F measured at 8:00 AM?

8:00 AM	98.2	97.7	97.3	97.5	97.1	98.6
Midnight	97.4	99.4	98.4	98.6	98.4	98.5

2. Height and Weight Shown below are select Minitab results obtained using the heights in inches) and weights (in lb) of 40 randomly selected males (based on Data Set 1 in Appendix B).

a. Determine whether there is sufficient evidence to support a claim of a linear correlation between heights and weights of males.

b. What percentage of the variation in weights of males can be explained by the linear correlation between height and weight?

c. Letting y represent weights of males and letting x represent heights of males, identify the regression equation.

d. Find the best predicted weight of a male who is 72 in. tall.

```
Pearson correlation of HT and WT = 0.522
P-Value = 0.001

The regression equation is
WT = - 139 + 4.55 HT
```

TI **3. Length and Weight** Listed below are the body lengths (in inches) and weights (in lb) of randomly selected bears.

a. Construct a scatterplot. What does the scatterplot suggest about a linear correlation between lengths and weights of bears?

b. Find the value of the linear correlation coefficient and determine whether there is sufficient evidence to support a claim of a linear correlation between lengths of bears and their weights.

c. Letting y represent weights of bears and letting x represent their weights, find the regression equation.

d. Based on the given sample data, what is the best predicted weight of a bear with a length of 72.0 in.?

Length	40	64	65	49	47
Weight	65	356	316	94	86

Predicting Height. *The table below lists upper leg lengths, arm circumferences, and heights of randomly selected males (based Data Set 1 in Appendix B). All measurements are in centimeters. Use these data for Exercises 4 and 5.*

Leg	40.9	43.1	38.0	41.0	46.0
Arm	33.7	30.3	32.8	31.0	36.2
Height	166	178	160	174	173

TI **4. a.** Construct a scatterplot of the leg/height paired data. What does the scatterplot suggest about a linear correlation between upper leg length and height?

b. Find the value of the linear correlation coefficient and determine whether there is sufficient evidence to support a claim of a linear correlation between upper leg length and height of males.

c. Letting y represent the heights of males and letting x represent the upper leg lengths of males, find the regression equation.

d. Based on the given sample data, what is the best predicted height of a male with an upper leg length of 45 cm?

TI **5.** Use a TI-83/84 Plus calculator to find the multiple regression equation of the form $\hat{y} = b_0 + b_1x_1 + b_2x_2$, where the response variable y represents heights, x_1 represents upper leg lengths, and x_2 represents arm circumferences of males. Identify the value of the multiple coefficient of determination R^2, the adjusted R^2, and the P-value representing the overall significance of the multiple regression equation. Use a 0.05 significance level and determine whether the regression equation can be used to predict the height of a male when given his upper leg length and arm circumference.

Cumulative Review Exercises

Heights of Males. *Listed below are randomly selected heights (in inches) of males from 1877 and from a recent National Health and Nutrition Examination Survey. (The 1877 data are from "Peirce and Bowditch: An American Contribution to Correlation and Regression," by Rovine and Anderson,* **The American Statistician,** *Vol. 58, No. 3.) Use the data for Exercises 1–6.*

Heights from 1877	71	62	64	68	68	67	65	65	66	66
Recent Heights	63	66	68	72	73	62	71	69	69	68

TI **1.** Find the mean, median, and standard deviation for each of the two samples.

TI **2.** Use a 0.05 significance level to test the claim that males in 1877 had a mean height that is less than the mean height of males today.

TI **3.** Use a 0.05 significance level to test the claim that heights of men from 1877 have a mean less than 69.1 in., which is the mean height given for men today (based on anthropometric data from Gordon, Churchill, et al.).

TI **4.** Construct a 95% confidence interval estimate of the mean height of males in 1877.

TI **5.** Construct a 95% confidence interval estimate of the difference between the mean height of males now and the mean height of males in 1877. (Use the recent heights as the first sample.) Does the confidence interval include 0? What does that tell us about the two population means?

6. Why would it not make sense to use the data in a test for a linear correlation between heights from 1877 and current heights?

7. a. What is the difference between a *statistic* and a *parameter*?

b. What is a simple random sample?

c. What is a *voluntary response sample,* and why are such samples generally unsuitable for using methods of statistics to make inferences about populations?

8. Body mass index measurements of adults are normally distributed with a mean of 26 and a standard deviation of 5 (based on Data Set 1 in Appendix B). Is a body mass index of 40 an outlier? Why or why not?

TI **9.** Body mass index measurements of adults are normally distributed with a mean of 26 and a standard deviation of 5 (based on Data Set 1 in Appendix B).

a. Find the probability of randomly selecting a person with a body mass index greater than 28.

b. If 16 people are randomly selected, find the probability that their mean body mass index is greater than 28.

10. According to a study conducted by Dr. P. Sorita Soni at Indiana University, 12% of the population have green eyes. If four people are randomly selected for a study of eye pigmentation, find the probability that all of them have green eyes. If a researcher is hired to randomly select the study subjects and she returns with four subjects all having green eyes, what would you conclude?

TI-83/84 Plus Calculator Project

The table below summarizes key statistics for each baseball team for a recent year.

a. Using the paired data consisting of the proportions of wins and the numbers of runs scored, find the linear correlation coefficient r and determine whether there is sufficient evidence to support a claim of a linear correlation between those two variables. Then find the regression equation with the response variable y representing the proportions of wins and the predictor variable x representing the numbers of runs scored.

b. Using the paired data consisting of the proportions of wins and the numbers of runs allowed, find the linear correlation coefficient r and determine whether there is sufficient evidence to support a claim of a linear correlation between those two variables. Then, find the regression equation with the response variable y representing the proportions of wins and the predictor variable x representing the numbers of runs allowed.

c. Use the paired data consisting of the proportions of wins and these differences: (Runs scored) − (runs allowed). Find the linear correlation coefficient r and determine whether there is sufficient evidence to support a claim of a linear correlation between those two variables. Then find the regression equation with the response variable y representing the proportions of wins and the predictor variable x representing the differences of (runs scored) − (runs allowed).

d. Compare the preceding results. Which appears to be more effective for winning baseball games: a strong defense or a strong offense? Explain.

e. Find the regression equation with the response variable y representing the winning percentage and the two predictor variables of runs scored and runs allowed. Does that equation appear to be useful for predicting a team's proportion of wins based on the number of runs scored and the number of runs allowed? Explain.

f. Using the paired data consisting of the numbers of runs scored and the numbers of runs allowed, find the linear correlation coefficient r and determine whether there is sufficient evidence to support a claim of a linear correlation between those two variables. What does the result suggest about the offensive strengths and the defensive strengths of the different teams?

Team	Wins	Losses	Proportion of Wins	Runs Scored	Runs Allowed	(Runs Scored) − (Runs Allowed)
N. Y.(AL)	97	65	0.599	930	767	163
Toronto	87	75	0.537	809	754	55
Boston	86	76	0.531	820	825	−5
Baltimore	70	92	0.432	768	899	−131
Tampa Bay	61	101	0.377	689	856	−167
Minnesota	96	66	0.593	801	683	118
Detroit	95	67	0.586	822	675	147
Chi. (AL)	90	72	0.556	868	794	74
Cleveland	78	84	0.481	870	782	88
Kansas City	62	100	0.383	757	971	−214
Oakland	93	69	0.574	771	727	44
L. A. (AL)	89	73	0.549	766	732	34
Texas	80	82	0.494	835	784	51
Seattle	78	84	0.481	756	792	−36
N. Y. (NL)	97	65	0.599	834	731	103
Philadelphia	85	77	0.525	865	812	53
Atlanta	79	83	0.488	849	805	44
Florida	78	84	0.481	758	772	−14
Washington	71	91	0.438	746	872	−126
St. Louis	83	78	0.516	781	762	19
Houston	82	80	0.506	735	719	16
Cincinnati	80	82	0.494	749	801	−52
Milwaukee	75	87	0.463	730	833	−103
Pittsburgh	67	95	0.414	691	797	−106
Chi. (NL)	66	96	0.407	716	834	−118
San Diego	88	74	0.543	731	679	52
L. A. (NL)	88	74	0.543	820	751	69
S. F.	76	85	0.472	746	790	−44
Arizona	76	86	0.469	773	788	−15
Colorado	76	86	0.469	813	812	1

INTERNET PROJECT

Linear Regression

Go to: **http://www.aw.com/triola**

The linear correlation coefficient is a tool that is used to measure the strength of the linear relationship between two sets of measurements. From a strictly computational point of view, the correlation coefficient may be found for any two data sets of paired values, regardless of what the data values represent. For this reason, certain questions should be asked whenever a correlation is being investigated. Is it reasonable to expect a linear correlation? Could a perceived correlation be caused by a third quantity related to each of the variables being studied?

The Internet Project for this chapter will guide you to several sets of paired data in the fields of sports, medicine, and economics. You will then apply the methods of this chapter, computing correlation coefficients and determining regression lines, while considering the true relationships between the variables involved.

APPLET PROJECT

Open the Applets folder on the CD and double-click on **Start.** Select the menu item of **Correlation by eye,** Use the applet to develop a skill in estimating the value of the linear correlation coefficient r by visually examining a scatterplot. Try to guess the value of r for 10 different data sets. Try to create a data set with a value of r that is approximately 0.9. Try to create a data set with a value of r that is close to 0.

Also use the menu item of **Regression by eye.** Try to move the green line so that it is the regression line. Repeat this until you can identify the regression line reasonably well.

Critical Thinking: Is the pain medicine Duragesic effective in reducing pain?

Listed below are measures of pain intensity before and after using the proprietary drug Duragesic (based on data from Janssen Pharmaceutical Products, L.P.) The data are listed in order by row, and corresponding measures are from the same subject before and after treatment.

For example, the first subject had a measure of 1.2 before treatment and a measure of 0.4 after treatment. Each pair of measurements is from one subject, and the intensity of pain was measured using the standard visual analog score.

Pain Intensity Before Duragesic Treatment

1.2	1.3	1.5	1.6	8.0	3.4	3.5	2.8	2.6	2.2
3.0	7.1	2.3	2.1	3.4	6.4	5.0	4.2	2.8	3.9
5.2	6.9	6.9	5.0	5.5	6.0	5.5	8.6	9.4	10.0
7.6									

Pain Intensity After Duragesic Treatment

0.4	1.4	1.8	2.9	6.0	1.4	0.7	3.9	0.9	1.8
0.9	9.3	8.0	6.8	2.3	0.4	0.7	1.2	4.5	2.0
1.6	2.0	2.0	6.8	6.6	4.1	4.6	2.9	5.4	4.8
4.1									

Analyzing the Results

1. Use the given data to construct a scatterplot, then use the methods of Section 10-2 to test for a linear correlation between the pain intensity before and after treatment. If there does appear to be a linear correlation, does it follow that the drug treatment is effective?

2. Use the given data to find the equation of the regression line. Let the response (y) variable be the pain intensity after treatment. What would be the equation of the regression line for a treatment having absolutely no effect?

3. The methods of Section 9-3 can be used to test the claim that two populations have the same mean. Identify the specific claim that the treatment is effective, then use the methods of Section 9-3 to test that claim. The methods of Section 9-3 are based on the requirement that the samples are independent. Are they independent in this case?

4. The methods of Section 9-4 can be used to test a claim about matched data. Identify the specific claim that the treatment is effective, then use the methods of Section 9-4 to test that claim.

5. Which of the preceding results is best for determining whether the drug treatment is effective in reducing pain? Based on the preceding results, does the drug appear to be effective?

Cooperative Group Activities

1. In-class activity Divide into groups of 8 to 12 people. For each group member, measure the person's height and also measure his or her navel height, which is the height from the floor to the navel. Is there a correlation between height and navel height? If so, find the regression equation with height expressed in terms of navel height. According to an old theory, the average person's ratio of height to navel height is the golden ratio: $(1 + \sqrt{5})/2 \approx 1.6$. Does this theory appear to be reasonably accurate?

2. In-class activity Divide into groups of 8 to 12 people. For each group member, measure height and arm span. For the arm span, the subject should stand with arms extended, like the wings on an airplane. It's easy to mark the height and arm span on a chalkboard, then measure the distances there. Using the paired sample data, is there a correlation between height and arm span? If so, find the regression equation with height expressed in terms of arm span. Can arm span be used as a reasonably good predictor of height?

3. In-class activity Divide into groups of 8 to 12 people. For each group member, use a string and ruler to measure head circumference and forearm length. Is there a relationship between these two variables? If so, what is it?

4. In-class activity Use a ruler as a device for measuring reaction time. One person should suspend the ruler by holding it at the top while the subject holds his or her thumb and forefinger at the bottom edge ready to catch the ruler when it is released. Record the distance that the ruler falls before it is caught. Convert that distance to the time (in seconds) that it took the subject to react and catch the ruler. (If the distance is measured in inches, use $t = \sqrt{d/192}$. If the distance is measured in centimeters, use $t = \sqrt{d/487.68}$.) Test each subject once with the right hand and once with the left hand, and record the paired data. Test for a correlation. Find the equation of the regression line. Does the equation of the regression line suggest that the dominant hand has a faster reaction time?

5. In-class activity Divide into groups of 8 to 12 people. Record the pulse rate of each group member by counting the number of heart beats in 1 min. Then measure and record each person's height. Is there a relationship between pulse rate and height? If so, what is it?

6. In-class activity Collect data from each student consisting of the number of credit cards and the number of keys that the student has in his or her possession. Is there a correlation? If so, what is it? Try to identify at least one reasonable explanation for the presence or absence of a correlation.

7. In-class activity Divide into groups of three or four people. Appendix B includes many data sets not yet included in examples or exercises in this chapter. Search Appendix B for a pair of variables of interest, then investigate correlation and regression. State your conclusions and try to identify practical applications.

8. Out-of-class activity Divide into groups of three or four people. Investigate the relationship between two variables by collecting your own paired sample data and using the methods of this chapter to determine whether there is a significant linear correlation. Also identify the regression equation and describe a procedure for predicting values of one of the variables when given values of the other variable. Suggested topics:

• Is there a relationship between taste and cost of different brands of chocolate chip cookies (or colas)? Taste can be measured on some number scale, such as 1 to 10.

• Is there a relationship between salaries of professional baseball (or basketball, or football) players and their season achievements?

• Is there a relationship between the lengths of men's (or women's) feet and their heights?

• Is there a relationship between student grade-point averages and the amount of television watched? If so, what is it?

• Is there a relationship between hours studied each week and grade point average? If so, what is it?

NAME:	Mark D. Haskell
JOB:	Director, Forecasting and Analysis
COMPANY:	Walt Disney World Resort

As Director of Forecasting and Analysis for Walt Disney World Resort, Mark D. Haskell leads a team of people responsible for planning and forecasting values such as attendance, hotel occupancy, and projected revenue. By analyzing various factors, Mark and his team help Disney continue to work to ensure that each guest has an enjoyable and memorable experience at Walt Disney World Resort.

Q: What do you do in your work?

A: I lead a team of people responsible for planning and forecasting such metrics as theme park attendance, occupancy at each of our resort hotels, and the revenue the Walt Disney World will realize from these key business drives.

Q: How do you use statistics and what specific statistical concepts do you use?

A: Statistics is central to the forecasting process. Many of our forecasting tools are based upon multiple regression techniques, with some of those models more complex than others. We also use very basic statistical concepts on a daily basis, whether reporting the "mean absolute percent error" for our forecasts, understanding measures of central tendency, distributions, and sampling techniques when reviewing marketing research, or running correlations to understand how different variables align with our key business drives. There are many approaches that can be used in creating high quality forecasts, but statistics is a basic building block for almost all of those approaches.

Q: Describe a specific example of how the use of statistics was useful in improving a product or service.

A: My team recently used correlation analysis to help us understand what sources of data would be most helpful in predicting attendance and spending at one of our retail centers. Based upon that work, we developed a regression model that helps company leaders understand revenue potential, determine staffing needs, set operating hours, identify new product opportunities, and identify capital investment needs, to name just a few.

Q: What background in statistics is required to obtain a job like yours?

A: I have a Masters Degree in Economics, specializing in quantitative analysis methods. Generally some form of advanced degree with emphasis in statistical analysis would be required to succeed in my role.

Q: Do you feel job applicants at your company are viewed more favorably if they have studied some statistics?

A: Some level of experience with statistics is required for roles on the Forecasting and Analysis team. There are many other roles at Walt Disney World that would look favorably on those who have studied statistics.

Q: Do you recommend that today's college students study statistics? Why?

A: Absolutely. In a business world that is fascinated with numbers and data, statistics is a key to being able to properly analyze and summarize vast quantities of data. Even if you aren't responsible for conducting the analysis, you need a basic understanding to properly use the information for decision making. You need to learn how to use statistics properly, or you risk having those with a better understanding of statistics use them against you.

Q: What other skills are important for today's college students?

A: Communication skills, both verbal and written. There is tremendous value in having people who can analyze complex information, then simplify it and clearly communicate it for easy consumption.

11-1 Review and Preview

11-2 Goodness-of-Fit

11-3 Contingency Tables

11-4 McNemar's Test for Matched Pairs

11

Goodness-of-Fit and Contingency Tables

Is the nurse a serial killer?

Three alert nurses at the Veteran's Affairs Medical Center in Northampton, Massachusetts noticed an unusually high number of deaths at times when another nurse, Kristen Gilbert, was working. Those same nurses later noticed missing supplies of the drug epinephrine, which is a synthetic adrenaline that stimulates the heart. They reported their growing concerns, and an investigation followed. Kristen Gilbert was arrested and charged with four counts of murder and two counts of attempted murder. When seeking a grand jury indictment, prosecutors provided a key piece of evidence consisting of a two-way table showing the numbers of shifts with deaths when Gilbert was working. See Table 11-1.

Table 11-1   Two-Way Table with Deaths When Gilbert Was Working

	Shifts with a death	Shifts without a death
Gilbert was working	40	217
Gilbert was not working	34	1350

The numbers in Table 11-1 might be better understood with a graph, such as Figure 11-1, which shows the death rates during shifts when Gilbert was working and when she was not working. Figure 11-1 seems to make it clear that shifts when Gilbert was working had a much higher death rate than shifts when she was not working, but we need to determine whether those results are statistically significant.

George Cobb, a leading statistician and statistics educator, became involved in the Gilbert case at the request of an attorney for the defense. Cobb wrote a report stating that the data in Table 11-1 should have been presented to the grand jury (as it was) for purposes of indictment, but that it should not be presented at the actual trial. He noted that the data in Table 11-1 are based on observations and do not show that Gilbert actually *caused* deaths. Also, Table 11-1 includes information about many other deaths that were not relevant to the trial. The judge ruled that the data in Table 11-1 could not be used at the trial. Kristen Gilbert was convicted on other evidence and is now serving a sentence of life in prison, without the possibility of parole.

This chapter will include methods for analyzing data in tables, such as Table 11-1. We will analyze Table 11-1 to see what conclusions could be presented to the grand jury that provided the indictment.

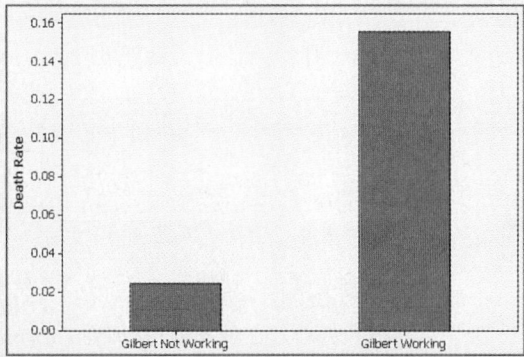

Figure 11-1   Bar Graph of Death Rates with Gilbert Working and Not Working

11-1 Review and Preview

We began a study of inferential statistics in Chapter 7 when we presented methods for estimating a parameter for a single population and in Chapter 8 when we presented methods of testing claims about a single population. In Chapter 9 we extended those methods to situations involving two populations. In Chapter 10 we considered methods of correlation and regression using paired sample data. In this chapter we use statistical methods for analyzing categorical (or qualitative, or attribute) data that can be separated into different cells. We consider hypothesis tests of a claim that the observed frequency counts agree with some claimed distribution. We also consider contingency tables (or two-way frequency tables), which consist of frequency counts arranged in a table with at least two rows and two columns. We conclude this chapter by considering two-way tables involving data consisting of matched pairs.

The methods of this chapter use the same χ^2 (chi-square) distribution that was first introduced in Section 7-5. See Section 7-5 for a quick review of properties of the χ^2 distribution.

11-2 Goodness-of-Fit

Key Concept In this section we consider sample data consisting of observed frequency counts arranged in a single row or column (called a one-way frequency table). We will use a hypothesis test for the claim that the observed frequency counts agree with some claimed distribution, so that there is a *good fit* of the observed data with the claimed distribution.

Because we test for how well an observed frequency distribution fits some specified theoretical distribution, the method of this section is called a *goodness-of-fit test*.

 DEFINITION

A **goodness-of-fit test** is used to test the hypothesis that an observed frequency distribution fits (or conforms to) some claimed distribution.

Objective

Conduct a goodness-of-fit test.

Notation

O represents the *observed frequency* of an outcome, found by tabulating the sample data.

E represents the *expected frequency* of an outcome, found by assuming that the distribution is as claimed.

k represents the *number of different categories* or outcomes.

n represents the total *number of trials* (or observed sample values).

Requirements

1. The data have been randomly selected.

2. The sample data consist of frequency counts for each of the different categories.

3. For each category, the *expected* frequency is at least 5. (The expected frequency for a category is the frequency that would occur if the data actually have the distribution that is being claimed. There is no requirement that the *observed* frequency for each category must be at least 5.)

Test Statistic for Goodness-of-Fit Tests

$$\chi^2 = \sum \frac{(O - E)^2}{E}$$

Critical Values

1. Critical values are found in Table A-4 by using $k - 1$ degrees of freedom, where k is the number of categories.

2. Goodness-of-fit hypothesis tests are always *right-tailed*.

P-Values

P-values are typically provided by computer software, or a range of *P*-values can be found from Table A-4.

TI-83/84 Plus Calculator: The TI-83/84 Plus calculator generates *P*-values, so the *P*-value method of hypothesis testing becomes easy. See the instructions at the end of this section.

Finding Expected Frequencies

Conducting a goodness-of-fit test requires that we identify the observed frequencies, then determine the frequencies expected with the claimed distribution. Table 11-2 on the next page includes observed frequencies with a sum of 80, so $n = 80$. If we assume that the 80 digits were obtained from a population in which all digits are equally likely, then we *expect* that each digit should occur in 1/10 of the 80 trials, so each of the 10 expected frequencies is given by $E = 8$. In general, if we are assuming that all of the expected frequencies are equal, each expected frequency is $E = n/k$, where n is the total number of observations and k is the number of categories. In other cases in which the expected frequencies are not all equal, we can often find the expected frequency for each category by multiplying the sum of all observed frequencies and the probability p for the category, so $E = np$. We summarize these two procedures here.

• **Expected frequencies are equal: $E = n/k$.**

• **Expected frequencies are not all equal: $E = np$ for each individual category.**

As good as these two preceding formulas for E might be, it is better to use an informal approach. Just ask, "How can the observed frequencies be split up among the different categories so that there is perfect agreement with the claimed distribution?" Also, note that the *observed* frequencies must all be whole numbers because they represent actual counts, but the *expected* frequencies need not be whole numbers. For example, when rolling a single die 33 times, the expected frequency for each possible outcome is $33/6 = 5.5$. The expected frequency for rolling a 3 is 5.5, even though it is impossible to have the outcome of 3 occur exactly 5.5 times.

We know that sample frequencies typically deviate somewhat from the values we theoretically expect, so we now present the key question: Are the differences between the actual *observed* values O and the theoretically *expected* values E statistically significant? We need a measure of the discrepancy between the O and E values, so we use the test statistic given with the requirements and critical values. (Later, we will explain how this test statistic was developed, but you can see that it has differences of $O - E$ as a key component.)

The χ^2 test statistic is based on differences between the observed and expected values. If the observed and expected values are *close,* the χ^2 test statistic will be small and the *P*-value will be large. If the observed and expected frequencies are *not close,*

Figure 11-2

Relationships Among the χ^2 Test Statistic, *P*-Value, and Goodness-of-Fit

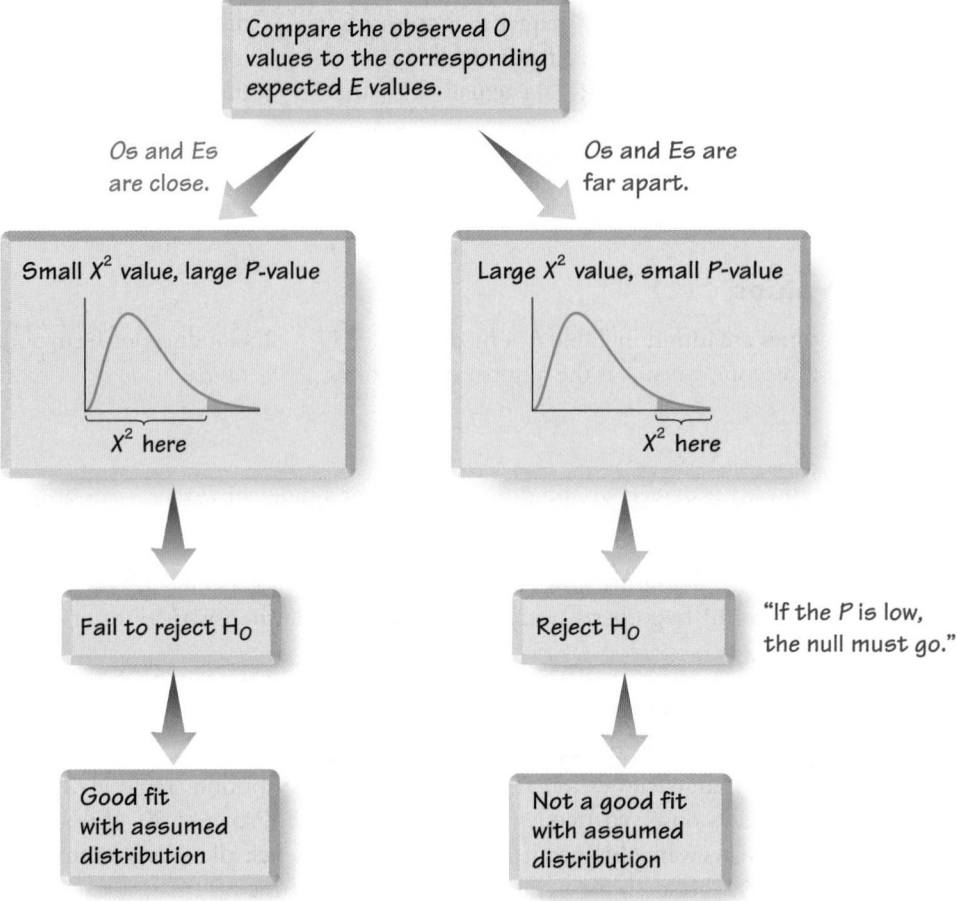

the χ^2 test statistic will be large and the *P*-value will be small. Figure 11-2 summarizes this relationship. The hypothesis tests of this section are always right-tailed, because the critical value and critical region are located at the extreme right of the distribution. If confused, just remember this:

"If the *P* is low, the null must go."

(If the *P*-value is small, reject the null hypothesis that the distribution is as claimed.)

Once we know how to find the value of the test statistic and the critical value, we can test hypotheses by using the same general procedures introduced in Chapter 8.

Table 11-2   Last Digits of Weights

Last Digit	Frequency
0	7
1	14
2	6
3	10
4	8
5	4
6	5
7	6
8	12
9	8

EXAMPLE 1   **Last Digits of Weights** Data Set 1 in Appendix B includes weights from 40 randomly selected adult males and 40 randomly selected adult females. Those weights were obtained as part of the National Health Examination Survey. When obtaining weights of subjects, it is extremely important to actually weigh individuals instead of asking them to report their weights. By analyzing the *last digits* of weights, researchers can verify that weights were obtained through actual measurements instead of being reported. When people report weights, they typically round to a whole number, so reported weights tend to have many last digits consisting of 0. In contrast, if people are actually weighed with a scale having precision to the nearest 0.1 pound, the weights tend to have last digits that are uniformly distributed, with 0, 1, 2, . . . , 9 all occurring with roughly the same frequencies. Table 11-2 shows the frequency distribution of the last digits from the

80 weights listed in Data Set 1 in Appendix B. (For example, the weight of 201.5 lb has a last digit of 5, and this is one of the data values included in Table 11-2.)

Test the claim that the sample is from a population of weights in which the last digits do *not* occur with the same frequency. Based on the results, what can we conclude about the procedure used to obtain the weights?

SOLUTION   **REQUIREMENT CHECK** (1) The data come from randomly selected subjects. (2) The data do consist of frequency counts, as shown in Table 11-2. (3) With 80 sample values and 10 categories that are claimed to be equally likely, each expected frequency is 8, so each expected frequency does satisfy the requirement of being a value of at least 5. All of the requirements are satisfied.

The claim that the digits do not occur with the same frequency is equivalent to the claim that the relative frequencies or probabilities of the 10 cells ($p_0, p_1, \ldots, p_9$) are not all equal. We will use the traditional method for testing hypotheses (see Figure 8-9).

Step 1: The original claim is that the digits do not occur with the same frequency. That is, at least one of the probabilities $p_0, p_1, \ldots, p_9$ is different from the others.

Step 2: If the original claim is false, then all of the probabilities are the same. That is, $p_0 = p_1 = p_2 = p_3 = p_4 = p_5 = p_6 = p_7 = p_8 = p_9$.

Step 3: The null hypothesis must contain the condition of equality, so we have

$$H_0: p_0 = p_1 = p_2 = p_3 = p_4 = p_5 = p_6 = p_7 = p_8 = p_9$$
$$H_1: \text{At least one of the probabilities is different from the others.}$$

Step 4: No significance level was specified, so we select $\alpha = 0.05$.

Step 5: Because we are testing a claim about the distribution of the last digits being a uniform distribution, we use the goodness-of-fit test described in this section. The χ^2 distribution is used with the test statistic given earlier.

Step 6: The observed frequencies O are listed in Table 11-2. Each corresponding expected frequency E is equal to 8 (because the 80 digits would be uniformly distributed among the 10 categories).

***P*-Value Method Using the TI-83/84 Plus** The TI-84 Plus calculator includes the goodness-of-fit test, and a goodness-of-fit program can be downloaded to the TI-83 Plus calculator. (See the instructions at the end of this section.) The display from the TI-84 Plus calculator is shown in the margin, and we can see that the test statistic is $\chi^2 = 11.250$ and the *P*-value is 0.259 (rounded). The test statistic can also be calculated manually as shown in Table 11-3.

Traditional Method The traditional method of testing hypotheses can be used by finding the critical value of $\chi^2 = 16.919$ (found in Table A-4 with $\alpha = 0.05$ in the right tail and degrees of freedom $k - 1 = 9$). The test statistic and critical value are shown in Figure 11-3, where we can see that the test statistic does not fall in the critical region.

Step 7: Because the *P*-value of 0.259 is greater than the significance level of 0.05, we fail to reject the null hypothesis. (If we use the traditional method of testing hypothesis, we fail to reject the null hypothesis because the test statistic does not fall in the critical region.)

Step 8: There is not sufficient evidence to support the claim that the last digits do not occur with the same relative frequency.

INTERPRETATION   This goodness-of-fit test suggests that the last digits provide a reasonably good fit with the claimed distribution of equally likely frequencies. Instead of asking the subjects how much they weigh, it appears that their weights were actually measured as they should have been.

Which Car Seats Are Safest?

Many people believe that the back seat of a car is the safest place to sit, but is it? University of Buffalo researchers analyzed more than 60,000 fatal car crashes and found that the middle back seat is the safest place to sit in a car. They found that sitting in that seat makes a passenger 86% more likely to survive than those who sit in the front seats, and they are 25% more likely to survive than those sitting in either of the back seats nearest the windows. An analysis of seat belt use showed that when not wearing a seat belt in the back seat, passengers are three times more likely to die in a crash than those wearing seat belts in that same seat. Passengers concerned with safety should sit in the middle back seat wearing a seat belt.

TI-83/84 Plus

```
X²GOF-Test
 X²=11.25
 P=.2589612558
 df=9
 CNTRB={.125 4.…
```

Example 1 involves a situation in which the claimed frequencies for the different categories are all equal. The methods of this section can also be used when the hypothesized probabilities (or frequencies) are different, as shown in Example 2.

Table 11-3   Calculating the χ^2 Test Statistic for the Last Digits of Weights

Last Digit	Observed Frequency O	Expected Frequency E	$O - E$	$(O - E)^2$	$\dfrac{(O - E)^2}{E}$
0	7	8	−1	1	0.125
1	14	8	6	36	4.500
2	6	8	−2	4	0.500
3	10	8	2	4	0.500
4	8	8	0	0	0.000
5	4	8	−4	16	2.000
6	5	8	−3	9	1.125
7	6	8	−2	4	0.500
8	12	8	4	16	2.000
9	8	8	0	0	0.000

$$\chi^2 = \sum \frac{(O - E)^2}{E} = 11.250$$

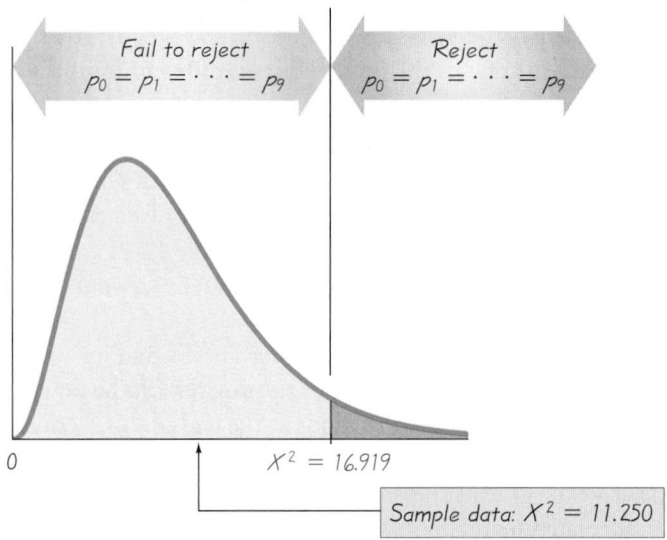

Figure 11-3   Test of $p_0 = p_1 = p_2 = p_3 = p_4 = p_5 = p_6 = p_7 = p_8 = p_9$

> **EXAMPLE 2**   **World Series Games** Table 11-4 lists the numbers of games played in the baseball World Series, as of this writing. That table also includes the expected proportions for the numbers of games in a World Series, assuming that in each series, both teams have about the same chance of winning. Use a 0.05 significance level to test the claim that the actual numbers of games fit the distribution indicated by the probabilities.

Table 11-4   Numbers of Games in World Series Contests

Games played	4	5	6	7
Actual World Series contests	19	21	22	37
Expected proportion	2/16	4/16	5/16	5/16

SOLUTION   **REQUIREMENT CHECK** (1) We begin by noting that the observed numbers of games are not randomly selected from a larger population. However, we treat them as a random sample for the purpose of determining whether they are typical results that might be obtained from such a random sample. (2) The data do consist of frequency counts. (3) Each expected frequency is at least 5, as will be shown later in this solution. All of the requirements are satisfied.

Step 1: The original claim is that the actual numbers of games fit the distribution indicated by the expected proportions. Using subscripts corresponding to the number of games, we can express this claim as $p_4 = 2/16$ and $p_5 = 4/16$ and $p_6 = 5/16$ and $p_7 = 5/16$.

Step 2: If the original claim is false, then at least one of the proportions does not have the value as claimed.

Step 3: The null hypothesis must contain the condition of equality, so we have

H_0: $p_4 = 2/16$ and $p_5 = 4/16$ and $p_6 = 5/16$ and $p_7 = 5/16$.

H_1: At least one of the proportions is not equal to the given claimed value.

Step 4: The significance level is $\alpha = 0.05$.

Step 5: Because we are testing a claim that the distribution of numbers of games in World Series contests is as claimed, we use the goodness-of-fit test described in this section. The χ^2 distribution is used with the test statistic given earlier.

P-Value Method Using the TI-83/84 Plus The display from the TI-84 Plus calculator is shown in the margin and we can see that the test statistic is $\chi^2 = 7.885$ (rounded) and the P-value is 0.0485 (rounded). The test statistic can also be calculated manually as shown in Table 11-5.

Traditional Method The traditional method of testing hypotheses can be used by finding the critical value of $\chi^2 = 7.815$ (found in Table A-4 with $\alpha = 0.05$ in the right tail and degrees of freedom $k - 1 = 3$).

Step 7: Because the P-value of 0.0485 is less than the significance level of 0.05, we reject the null hypothesis. (If we use the traditional method of testing hypothesis, we reject the null hypothesis because the test statistic does not fall in the critical region.)

TI-83/84 Plus

```
X²GOF-Test
 X²=7.884848485
 P=.0484524636
 df=3
 CNTRB={3.54671…
```

Table 11-5   **Calculating the χ^2 Test Statistic for the Numbers of World Series Games**

Number of Games	Observed Frequency O	Expected Frequency E = np	O − E	(O − E)²	(O − E)² / E
4	19	$99 \cdot \dfrac{2}{16} = 12.3750$	6.6250	43.8906	3.5467
5	21	$99 \cdot \dfrac{4}{16} = 24.7500$	−3.7500	14.0625	0.5682
6	22	$99 \cdot \dfrac{5}{16} = 30.9375$	−8.9375	79.8789	2.5819
7	37	$99 \cdot \dfrac{5}{16} = 30.9375$	6.0625	36.7539	1.1880

$$\chi^2 = \sum \frac{(O - E)^2}{E} = 7.885$$

Step 8: There is sufficient evidence to warrant rejection of the claim that actual numbers of games in World Series contests fit the distribution indicated by the expected proportions given in Table 11-4.

INTERPRETATION   This goodness-of-fit test suggests that the numbers of games in World Series contests do not fit the distribution expected from probability calculations. Different media reports have noted that seven-game series occur much more than expected. The results in Table 11-4 show that seven-game series occurred 37% of the time, but they were expected to occur only 31% of the time. (A *USA Today* headline stated that "Seven-game series defy odds.") So far, no reasonable explanations have been provided for the discrepancy.

In Figure 11-4 we graph the expected proportions of 2/16, 4/16, 5/16, and 5/16 along with the observed proportions of 19/99, 21/99, 22/99, and 37/99, so that we can visualize the discrepancy between the distribution that was claimed and the frequencies that were observed. The points along the red line represent the expected proportions, and the points along the green line represent the observed proportions. Figure 11-4 shows disagreement between the expected proportions (red line) and the observed proportions (green line), and the hypothesis test in Example 2 shows that the discrepancy is statistically significant.

Figure 11-4

Observed and Expected Proportions in the Numbers of World Series Games

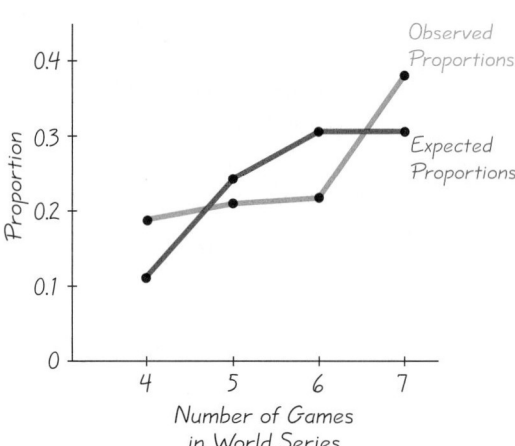

Rationale for the Test Statistic: Examples 1 and 2 show that the χ^2 test statistic is a measure of the discrepancy between observed and expected frequencies. Simply summing the differences between observed and expected values does not result in an effective measure because that sum is always 0. Squaring the $O - E$ values provides a better statistic. (The reasons for squaring the $O - E$ values are essentially the same as the reasons for squaring the $x - \bar{x}$ values in the formula for standard deviation.) The value of $\Sigma(O - E)^2$ measures only the magnitude of the differences, but we need to find the magnitude of the differences relative to what was expected. This relative magnitude is found through division by the expected frequencies, as in the test statistic.

The theoretical distribution of $\Sigma(O - E)^2/E$ is a discrete distribution because the number of possible values is finite. The distribution can be approximated by a chi-square distribution, which is continuous. This approximation is generally considered acceptable, provided that all expected values E are at least 5. (There are ways of circumventing the problem of an expected frequency that is less than 5, such as combining categories so that all expected frequencies are at least 5. Also, there are other methods that can be used when not all expected frequencies are at least 5.)

The number of degrees of freedom reflects the fact that we can freely assign frequencies to $k - 1$ categories before the frequency for every category is determined. (Although we say that we can "freely" assign frequencies to $k - 1$ categories, we cannot have negative frequencies nor can we have frequencies so large that their sum exceeds the total of the observed frequencies for all categories combined.)

Using the TI-83/84 Plus Calculator for Testing Goodness-of-Fit

The TI-84 Plus calculator can execute the methods of this section, but the TI-83 Plus calculator requires Michael Lloyd's program X2GOF that must be downloaded from the CD-ROM included with this book or the Web site www.aw.com/triola. Here are the procedures.

TI-84 Plus Calculator

1. Enter the observed frequencies in list L1 and enter the expected frequencies in list L2.
2. Press **STAT**, press **▷** twice to select **TESTS,** then select χ^2**GOF-Test.**
3. For the Observed entry, enter L1, and for the Expected entry, enter L2.
4. Scroll to the line labeled df and enter the number of degrees of freedom, which is 1 less than the number of categories.
5. Scroll down to **Calculate** and press **ENTER**.

Results will include the χ^2 test statistic and the P-value.

TI-83 Plus Calculator

1. Download the program X2GOF from the CD-ROM or the Web site www.aw.com/triola.
2. Enter the observed frequencies in list L1 and enter the expected frequencies in list L2.
3. Press **PRGM** and select **X2GOF** and press **ENTER**.
4. At the prompt of OBS LIST, enter L1, which contains the list of observed frequencies. Press **ENTER**.
5. At the prompt of EXP LIST, enter L2, which contains the list of expected frequencies. Press **ENTER**.

Results will be displayed. The value of X2 is the χ^2 test statistic. The value of P is the P-value. The accompanying screen display results from using the program X2GOF on the TI-83 Plus calculator with the observed and expected frequencies from Example 2. If you compare the TI-83 Plus results to the TI-84 Plus results shown in Example 2, you can see that the results are similar.

TI-83/84 Plus

```
OBS LIST=L₁
EXP LIST=L₂
X2=7.884848485
P=.0484524636
DF=3
COMP LIST:
{3.546717172 .5…
```

11-2   Basic Skills and Concepts

Statistical Literacy and Critical Thinking

1. Goodness-of-Fit A *New York Times*/CBS News Poll typically involves the selection of random digits to be used for telephone numbers. The *New York Times* states that "within each (telephone) exchange, random digits were added to form a complete telephone number, thus permitting access to listed and unlisted numbers." When such digits are randomly generated, what is the distribution of those digits? Given such randomly generated digits, what is a test for "goodness-of-fit"?

2. Interpreting Values of χ^2 When generating random digits as in Exercise 1, we can test the generated digits for goodness-of-fit with the distribution in which all of the digits are equally likely. What does an exceptionally large value of the χ^2 test statistic suggest about the goodness-of-fit? What does an exceptionally small value of the χ^2 test statistic (such as 0.002) suggest about the goodness-of-fit?

3. Observed/Expected Frequencies A wedding caterer randomly selects clients from the past few years and records the months in which the wedding receptions were held. The results are listed below (based on data from *The Amazing Almanac*). Assume that you want to test the claim that weddings occur in different months with the same frequency. Briefly describe what O and E represent, then find the values of O and E.

Month	Jan.	Feb.	March	April	May	June	July	Aug.	Sept.	Oct.	Nov.	Dec.
Number	5	8	7	9	13	17	11	10	10	12	8	10

4. *P*-Value When using the data from Exercise 3 to conduct a hypothesis test of the claim that weddings occur in the 12 months with equal frequency, we obtain the *P*-value of 0.477. What does that *P*-value tell us about the sample data? What conclusion should be made?

In Exercises 5–20, conduct the hypothesis test and provide the test statistic, critical value and/or P-value, and state the conclusion.

TI **5. Testing a Slot Machine** The author purchased a slot machine (Bally Model 809), and tested it by playing it 1197 times. There are 10 different categories of outcome, including no win, win jackpot, win with three bells, and so on. When testing the claim that the observed outcomes agree with the expected frequencies, the author obtained a test statistic of $\chi^2 = 8.185$. Use a 0.05 significance level to test the claim that the actual outcomes agree with the expected frequencies. Does the slot machine appear to be functioning as expected?

TI **6. Grade and Seating Location** Do "A" students tend to sit in a particular part of the classroom? The author recorded the locations of the students who received grades of A, with these results: 17 sat in the front, 9 sat in the middle, and 5 sat in the back of the classroom. When testing the assumption that the "A" students are distributed evenly throughout the room, the author obtained the test statistic of $\chi^2 = 7.226$. If using a 0.05 significance level, is there sufficient evidence to support the claim that the "A" students are not evenly distributed throughout the classroom? If so, does that mean you can increase your likelihood of getting an A by sitting in the front of the room?

TI **7. Pennies from Checks** When considering effects from eliminating the penny as a unit of currency in the United States, the author randomly selected 100 checks and recorded the cents portions of those checks. The table below lists those cents portions categorized according to the indicated values. Use a 0.05 significance level to test the claim that the four categories are equally likely. The author expected that many checks for whole dollar amounts would result in a disproportionately high frequency for the first category, but do the results support that expectation?

Cents portion of check	0–24	25–49	50–74	75–99
Number	61	17	10	12

TI **8. Flat Tire and Missed Class** A classic tale involves four carpooling students who missed a test and gave as an excuse a flat tire. On the makeup test, the instructor asked the students to identify the particular tire that went flat. If they really didn't have a flat tire, would they be able to identify the same tire? The author asked 41 other students to identify the tire they would select. The results are listed in the following table (except for one student who selected the spare). Use a 0.05 significance level to test the author's claim that the results fit a uniform distribution. What does the result suggest about the ability of the four students to select the same tire when they really didn't have a flat?

Tire	Left front	Right front	Left rear	Right rear
Number selected	11	15	8	6

9. Pennies from Credit Card Purchases When considering effects from eliminating the penny as a unit of currency in the United States, the author randomly selected the amounts from 100 credit card purchases and recorded the cents portions of those amounts. The table below lists those cents portions categorized according to the indicated values. Use a 0.05 significance level to test the claim that the four categories are equally likely. The author expected that many credit card purchases for whole dollar amounts would result in a disproportionately high frequency for the first category, but do the results support that expectation?

Cents portion	0–24	25–49	50–74	75–99
Number	33	16	23	28

10. Occupational Injuries Randomly selected nonfatal occupational injuries and illnesses are categorized according to the day of the week that they first occurred, and the results are listed below (based on data from the Bureau of Labor Statistics). Use a 0.05 significance level to test the claim that such injuries and illnesses occur with equal frequency on the different days of the week.

Day	Mon	Tues	Wed	Thurs	Fri
Number	23	23	21	21	19

11. Loaded Die The author drilled a hole in a die and filled it with a lead weight, then proceeded to roll it 200 times. Here are the observed frequencies for the outcomes of 1, 2, 3, 4, 5, and 6, respectively: 27, 31, 42, 40, 28, 32. Use a 0.05 significance level to test the claim that the outcomes are not equally likely. Does it appear that the loaded die behaves differently than a fair die?

12. Births Records of randomly selected births were obtained and categorized according to the day of the week that they occurred (based on data from the National Center for Health Statistics). Because babies are unfamiliar with our schedule of weekdays, a reasonable claim is that births occur on the different days with equal frequency. Use a 0.01 significance level to test that claim. Can you provide an explanation for the result?

Day	Sun	Mon	Tues	Wed	Thurs	Fri	Sat
Number of births	77	110	124	122	120	123	97

13. Kentucky Derby The table below lists the frequency of wins for different post positions in the Kentucky Derby horse race. A post position of 1 is closest to the inside rail, so that horse has the shortest distance to run. (Because the number of horses varies from year to year, only the first ten post positions are included.) Use a 0.05 significance level to test the claim that the likelihood of winning is the same for the different post positions. Based on the result, should bettors consider the post position of a horse racing in the Kentucky Derby?

Post Position	1	2	3	4	5	6	7	8	9	10
Wins	19	14	11	14	14	7	8	11	5	11

14. Measuring Weights Example 1 in this section is based on the principle that when certain quantities are measured, the last digits tend to be uniformly distributed, but if they are estimated or reported, the last digits tend to have disproportionately more 0s or 5s. The last digits of the September weights in Data Set 3 in Appendix B are summarized in the table below. Use a 0.05 significance level to test the claim that the last digits of 0, 1, 2, ..., 9 occur with the same frequency. Based on the observed digits, what can be inferred about the procedure used to obtain the weights?

Last digit	0	1	2	3	4	5	6	7	8	9
Number	7	5	6	7	14	5	5	8	6	4

15. UFO Sightings Cases of UFO sightings are randomly selected and categorized according to month, with the results listed in the table below (based on data from Larry Hatch). Use a 0.05 significance level to test the claim that UFO sightings occur in the different months with equal frequency. Is there any reasonable explanation for the two months that have the highest frequencies?

Month	Jan.	Feb.	March	April	May	June	July	Aug.	Sept.	Oct.	Nov.	Dec.
Number	1239	1111	1428	1276	1102	1225	2233	2012	1680	1994	1648	1125

16. Violent Crimes Cases of violent crimes are randomly selected and categorized by month, with the results shown in the table below (based on data from the FBI). Use a 0.01 significance level to test the claim that the rate of violent crime is the same for each month. Can you explain the result?

Month	Jan.	Feb.	March	April	May	June	July	Aug.	Sept.	Oct.	Nov.	Dec.
Number	786	704	835	826	900	868	920	901	856	862	783	797

17. Genetics The Advanced Placement Biology class at Mount Pearl Senior High School conducted genetics experiments with fruit flies, and the results in the following table are based on the results that they obtained. Use a 0.05 significance level to test the claim that the observed frequencies agree with the proportions that were expected according to principles of genetics.

Characteristic	Red eye/ normal wing	Sepia eye/ normal wing	Red eye/ vestigial wing	Sepia eye/ vestigial wing
Frequency	59	15	2	4
Expected proportion	9/16	3/16	3/16	1/16

18. Do World War II Bomb Hits Fit a Poisson Distribution? In analyzing hits by V-1 buzz bombs in World War II, South London was subdivided into regions, each with an area of 0.25 km^2. Shown below is a table of actual frequencies of hits and the frequencies expected with the Poisson distribution. (The Poisson distribution is described in Section 5-5.) Use the values listed and a 0.05 significance level to test the claim that the actual frequencies fit a Poisson distribution.

Number of bomb hits	0	1	2	3	4 or more
Actual number of regions	229	211	93	35	8
Expected number of regions (from Poisson distribution)	227.5	211.4	97.9	30.5	8.7

19. M&M Candies Mars, Inc. claims that its M&M plain candies are distributed with the following color percentages: 16% green, 20% orange, 14% yellow, 24% blue, 13% red, and 13% brown. Refer to Data Set 18 in Appendix B and use the sample data to test the claim that the color distribution is as claimed by Mars, Inc. Use a 0.05 significance level.

20. Bias in Clinical Trials? Researchers investigated the issue of race and equality of access to clinical trials. The table below shows the population distribution and the numbers of participants in clinical trials involving lung cancer (based on data from "Participation in Cancer Clinical Trials," by Murthy, Krumholz, and Gross, *Journal of the American Medical Association*, Vol. 291, No. 22). Use a 0.01 significance level to test the claim that the distribution of clinical trial participants fits well with the population distribution. Is there a race/ethnic group that appears to be very underrepresented?

Race/ethnicity	White non-Hispanic	Hispanic	Black	Asian/Pacific Islander	American Indian/ Alaskan Native
Distribution of Population	75.6%	9.1%	10.8%	3.8%	0.7%
Number in Lung Cancer Clinical Trials	3855	60	316	54	12

Benford's Law. *According to Benford's law, a variety of different data sets include numbers with leading (first) digits that follow the distribution shown in the table below. In Exercises 21–24, test for goodness-of-fit with Benford's law.*

Leading Digit	1	2	3	4	5	6	7	8	9
Benford's law: distribution of leading digits	30.1%	17.6%	12.5%	9.7%	7.9%	6.7%	5.8%	5.1%	4.6%

TI **21. Detecting Fraud** When working for the Brooklyn District Attorney, investigator Robert Burton analyzed the leading digits of the amounts from 784 checks issued by seven suspect companies. The frequencies were found to be 0, 15, 0, 76, 479, 183, 8, 23, and 0, and those digits correspond to the leading digits of 1, 2, 3, 4, 5, 6, 7, 8, and 9, respectively. If the observed frequencies are substantially different from the frequencies expected with Benford's law, the check amounts appear to result from fraud. Use a 0.01 significance level to test for goodness-of-fit with Benford's law. Does it appear that the checks are the result of fraud?

TI **22. Author's Check Amounts** Exercise 21 lists the observed frequencies of leading digits from amounts on checks from seven suspect companies. Here are the observed frequencies of the leading digits from the amounts on checks written by the author: 68, 40, 18, 19, 8, 20, 6, 9, 12. (Those observed frequencies correspond to the leading digits of 1, 2, 3, 4, 5, 6, 7, 8, and 9, respectively.) Using a 0.05 significance level, test the claim that these leading digits are from a population of leading digits that conform to Benford's law. Do the author's check amounts appear to be legitimate?

TI **23. Political Contributions** Amounts of recent political contributions are randomly selected, and the leading digits are found to have frequencies of 52, 40, 23, 20, 21, 9, 8, 9, and 30. (Those observed frequencies correspond to the leading digits of 1, 2, 3, 4, 5, 6, 7, 8, and 9, respectively, and they are based on data from "Breaking the (Benford) Law: Statistical Fraud Detection in Campaign Finance," by Cho and Gaines, *American Statistician,* Vol. 61, No. 3.) Using a 0.01 significance level, test the observed frequencies for goodness-of-fit with Benford's law. Does it appear that the political campaign contributions are legitimate?

TI **24. Check Amounts** In the trial of *State of Arizona vs. Wayne James Nelson,* the defendant was accused of issuing checks to a vendor that did not really exist. The amounts of the checks are listed below in order by row. When testing for goodness-of-fit with the proportions expected with Benford's law, it is necessary to combine categories because not all expected values are at least 5. Use one category with leading digits of 1, a second category with leading digits of 2, 3, 4, 5, and a third category with leading digits of 6, 7, 8, 9. Using a 0.01 significance level, is there sufficient evidence to conclude that the leading digits on the checks do not conform to Benford's law?

$ 1,927.48	$27,902.31	$86,241.90	$72,117.46	$81,321.75	$97,473.96
$93,249.11	$89,658.16	$87,776.89	$92,105.83	$79,949.16	$87,602.93
$96,879.27	$91,806.47	$84,991.67	$90,831.83	$93,766.67	$88,336.72
$94,639.49	$83,709.26	$96,412.21	$88,432.86	$71,552.16	

11-2   Beyond the Basics

TI **25. Testing Effects of Outliers** In conducting a test for the goodness-of-fit as described in this section, does an outlier have much of an effect on the value of the χ^2 test statistic? Test for the effect of an outlier in Example 1 after changing the first frequency in Table 11-2 from 7 to 70. Describe the general effect of an outlier.

TI **26. Testing Goodness-of-Fit with a Normal Distribution** Refer to Data Set 21 in Appendix B for the axial loads (in pounds) of the aluminum cans that are 0.0109 in. thick.

Axial load	Less than 239.5	239.5–259.5	259.5–279.5	More than 279.5
Frequency				

a. Enter the observed frequencies in the above table.

b. Assuming a normal distribution with mean and standard deviation given by the sample mean and standard deviation, use the methods of Chapter 6 to find the probability of a randomly selected axial load belonging to each class.

c. Using the probabilities found in part (b), find the expected frequency for each category.

d. Use a 0.01 significance level to test the claim that the axial loads were randomly selected from a normally distributed population. Does the goodness-of-fit test suggest that the data are from a normally distributed population?

11-3 Contingency Tables

Key Concept In this section we consider *contingency tables* (or *two-way frequency tables*), which include frequency counts for categorical data arranged in a table with at least two rows and at least two columns. In Part 1 of this section, we present a method for conducting a hypothesis test of the null hypothesis that the row and column variables are independent of each other. This test of independence is used in real applications quite often. In Part 2, we will use the same method for a test of homogeneity, whereby we test the claim that different populations have the same proportion of some characteristics.

Part 1: Basic Concepts of Testing for Independence

In this section we use standard statistical methods to analyze frequency counts in a contingency table (or two-way frequency table). We begin with the definition of a contingency table.

> **DEFINITION**
>
> A **contingency table** (or **two-way frequency table**) is a table in which frequencies correspond to two variables. (One variable is used to categorize rows, and a second variable is used to categorize columns.)

EXAMPLE 1   **Contingency Table from Echinacea Experiment** Table 11-6 is a contingency table with two rows and three columns. The cells of the table contain frequencies. The row variable identifies whether the subjects became infected, and the column variable identifies the treatment group (placebo, 20% extract group, or 60% extract group).

Table 11-6   Results from Experiment with Echinacea

	Treatment Group		
	Placebo	Echinacea: 20% extract	Echinacea: 60% extract
Infected	88	48	42
Not infected	15	4	10

We will now consider a hypothesis test of independence between the row and column variables in a contingency table. We first define a *test of independence*.

 DEFINITION

A **test of independence** tests the null hypothesis that in a contingency table, the row and column variables are independent.

Objective

Conduct a hypothesis test for independence between the row variable and column variable in a contingency table.

Notation

O represents the *observed frequency* in a cell of a contingency table.

E represents the *expected frequency* in a cell, found by assuming that the row and column variables are independent.

r represents the number of rows in a contingency table (not including labels).

c represents the number of columns in a contingency table (not including labels).

Requirements

1. The sample data are randomly selected.

2. The sample data are represented as frequency counts in a two-way table.

3. For every cell in the contingency table, the *expected* frequency E is at least 5. (There is no requirement that every *observed* frequency must be at least 5. Also, there is no requirement that the population must have a normal distribution or any other specific distribution.)

Null and Alternative Hypotheses

The null and alternative hypotheses are as follows:

H_0: The row and column variables are *independent*.

H_1: The row and column variables are dependent.

Test Statistic for a Test of Independence

$$\chi^2 = \sum \frac{(O - E)^2}{E}$$

where O is the observed frequency in a cell and E is the expected frequency found by evaluating

$$E = \frac{(\text{row total})\,(\text{column total})}{(\text{grand total})}$$

Critical Values

1. The critical values are found in Table A-4 using

degrees of freedom $= (r - 1)(c - 1)$

where r is the number of rows and c is the number of columns.

2. Tests of independence with a contingency table are always *right-tailed*.

P-Values

P-values are typically provided by computer software, or a range of *P*-values can be found from Table A-4.

TI-83/84 Plus Calculator: The TI-83/84 Plus calculator generates *P*-values, so the *P*-value method of hypothesis testing becomes easy. See the instructions at the end of this section

The test statistic allows us to measure the amount of disagreement between the frequencies actually observed and those that we would theoretically expect when the two variables are independent. Large values of the χ^2 test statistic are in the rightmost region of the chi-square distribution, and they reflect significant differences between observed and expected frequencies. The distribution of the test statistic χ^2 can be approximated by the chi-square distribution, provided that all expected frequencies are at least 5. The number of degrees of freedom $(r - 1)(c - 1)$ reflects the fact that because we know the total of all frequencies in a contingency table, we can freely assign frequencies to only $r - 1$ rows and $c - 1$ columns before the frequency for every cell is determined. (However, we cannot have negative frequencies or frequencies so large that any row (or column) sum exceeds the total of the observed frequencies for that row (or column).)

Finding Expected Values *E*

The test statistic χ^2 is found by using the values of O (observed frequencies) and the values of E (expected frequencies). The expected frequency E can be found for a cell by simply multiplying the total of the row frequencies by the total of the column frequencies, then dividing by the grand total of all frequencies, as shown in Example 2.

> **EXAMPLE 2**   **Finding Expected Frequency** Refer to Table 11-6 and find the expected frequency for the first cell, where the observed frequency is 88.

SOLUTION   The first cell lies in the first row (with a total frequency of 178) and the first column (with total frequency of 103). The "grand total" is the sum of all frequencies in the table, which is 207. The expected frequency of the first cell is

$$E = \frac{(\text{row total})(\text{column total})}{(\text{grand total})} = \frac{(178)(103)}{207} = 88.570$$

INTERPRETATION   We know that the first cell has an observed frequency of $O = 88$ and an expected frequency of $E = 88.570$. We can interpret the expected value by stating that if we assume that getting an infection is independent of the treatment, then we expect to find that 88.570 of the subjects would be given a placebo and would get an infection. There is a discrepancy between $O = 88$ and $E = 88.570$, and such discrepancies are key components of the test statistic.

To better understand expected frequencies, pretend that we know only the row and column totals, as in Table 11-7, and that we must fill in the cell expected frequencies by assuming independence (or no relationship) between the row and column variables. In the first row, 178 of the 207 subjects got infections, so $P(\text{infection}) = 178/207$. In the first column, 103 of the 207 subjects were given a placebo, so $P(\text{placebo}) = 103/207$. Because we are assuming independence between getting an infection and the treatment group, the multiplication rule for independent events $[P(A \text{ and } B) = P(A) \cdot P(B)]$ is expressed as

$$P(\text{infection and placebo}) = P(\text{infection}) \cdot P(\text{placebo})$$

$$= \frac{178}{207} \cdot \frac{103}{207}$$

Table 11-7 Results from Experiment with Echinacea

| | Treatment Group | | | Row totals: |
	Placebo	Echinacea: 20% extract	Echinacea: 60% extract	
Infected				178
Not infected				29
Column totals:	103	52	52	Grand total: 207

We can now find the *expected value* for the first cell by multiplying the probability for that cell by the total number of subjects, as shown here:

$$E = n \cdot p = 207 \left[\frac{178}{207} \cdot \frac{103}{207} \right] = 88.570$$

The form of this product suggests a general way to obtain the expected frequency of a cell:

$$\text{Expected frequency } E = (\text{grand total}) \cdot \frac{(\text{row total})}{(\text{grand total})} \cdot \frac{(\text{column total})}{(\text{grand total})}$$

This expression can be simplified to

$$E = \frac{(\text{row total}) \cdot (\text{column total})}{(\text{grand total})}$$

We can now proceed to conduct a hypothesis test of independence, as in Example 3.

EXAMPLE 3 **Does Echinacea Have an Effect on Colds?** Common colds are typically caused by a rhinovirus. In a test of the effectiveness of echinacea, some test subjects were treated with echinacea extracted with 20% ethanol, some were treated with echinacea extracted with 60% ethanol, and others were given a placebo. All of the test subjects were then exposed to rhinovirus. Results are summarized in Table 11-6 (based on data from "An Evaluation of *Echinacea angustifolia* in Experimental Rhinovirus Infections," by Turner, et al., *New England Journal of Medicine,* Vol. 353, No. 4). Use a 0.05 significance level to test the claim that getting an infection (cold) is independent of the treatment group. What does the result indicate about the effectiveness of echinacea as a treatment for colds?

SOLUTION **REQUIREMENT CHECK** (1) The subjects were recruited and were randomly assigned to the different treatment groups. (2) The results are expressed as frequency counts in Table 11-6. (3) The expected frequencies are all at least 5. (The expected frequencies are 88.570, 44.715, 44.715, 14.430, 7.285, and 7.285.) The requirements are satisfied.

The null hypothesis and alternative hypothesis are as follows:

H_0: Getting an infection is independent of the treatment.

H_1: Getting an infection and the treatment are dependent.

The significance level is $\alpha = 0.05$.

Because the data are in the form of a contingency table, we use the χ^2 distribution with this test statistic:

$$\chi^2 = \sum \frac{(O - E)^2}{E} = \frac{(88 - 88.570)^2}{88.570} + \cdots + \frac{(10 - 7.285)^2}{7.285}$$

$$= 2.925$$

TI-83/84 Plus

```
X²-Test
 X²=2.925496946
 P=.231598856
 df=2
```

P-**Value Method Using the TI-83/84 Plus** The display from the TI-83/84 Plus calculator is shown here, and we can see that the test statistic is automatically calculated as $\chi^2 = 2.925$ (rounded). The display also shows the *P*-value of 0.232 (rounded).

Traditional Method The traditional method of testing hypotheses can be used by finding the critical value of $\chi^2 = 5.991$. That value is found in Table A-4 with $\alpha = 0.05$ in the right tail and degrees of freedom given by $(r - 1)(c - 1) = (2 - 1)(3 - 1) = 2$.

Because the *P*-value of 0.232 is greater than the significance level of 0.05, we fail to reject the null hypothesis of independence between getting an infection and treatment. (If we use the traditional method of testing hypothesis, we fail to reject the null hypothesis of independence because the test statistic does not fall in the critical region, as shown in Figure 11-5)

INTERPRETATION It appears that getting an infection is independent of the treatment group. This suggests that echinacea is not an effective treatment for colds.

Figure 11-5

Test of Independence for the Echinacea Data

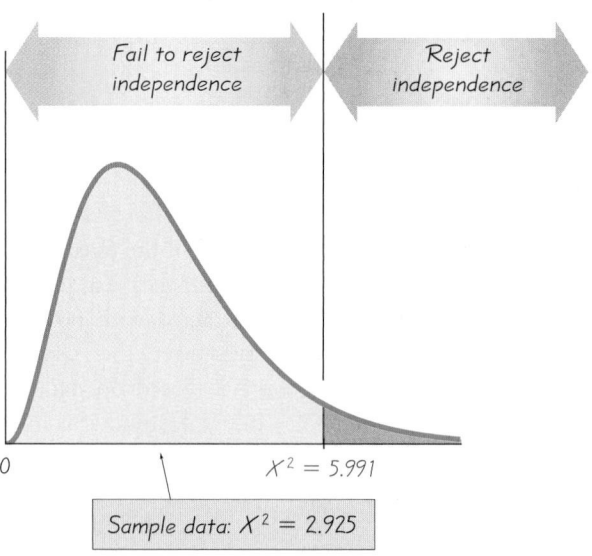

Sample data: $\chi^2 = 2.925$

EXAMPLE 4 **Is the Nurse a Serial Killer?** Table 11-1 provided with the Chapter Problem consists of a contingency table with a row variable (whether Kristen Gilbert was on duty) and a column variable (whether the shift included a death). Test the claim that whether Gilbert was on duty for a shift is independent of whether a patient died during the shift. Because this is such a serious analysis, use a significance level of 0.01. What does the result suggest about the charge that Gilbert killed patients?

SOLUTION **REQUIREMENT CHECK** (1) The data in Table 11-1 can be treated as random data for the purpose of determining whether such random data could easily occur by chance. (2) The sample data are represented as frequency counts in a two-way table. (3) Each expected frequency is at least 5. (The expected frequencies are 11.589, 245.411, 62.411, and 1321.589.) The requirements are satisfied. ✓

The null hypothesis and alternative hypothesis are as follows:

H_0: Whether Gilbert was working is independent of whether there was a death during the shift.

H_1: Whether Gilbert was working and whether there was a death during the shift are dependent.

The TI-83/84 Plus calculator results include the test statistic of $\chi^2 = 86.481$ (rounded). The displayed P-value is $1.410899E{-}20$, but that number is in scientific notation and it becomes 0.000 when written in standard form and rounded. (See Section 1-6 for a discussion of scientific notation.) Because the P-value is less than the significance level of 0.01, we reject the null hypothesis of independence. There is sufficient evidence to warrant rejection of independence between the row and column variables.

TI-83/84 Plus Results

```
χ²-Test
 χ²=86.48094995
 P=1.410899E-20
 df=1
```

INTERPRETATION We reject independence between whether Gilbert was working and whether a patient died during a shift. It appears that there is an association between Gilbert working and patients dying. (Note that this does not show that Gilbert *caused* the deaths, so this is not evidence that could be used at her trial, but it was evidence that led investigators to pursue other evidence that eventually led to her conviction for murder.)

As in Section 11-2, if observed and expected frequencies are close, the χ^2 test statistic will be small and the P-value will be large. If observed and expected frequencies are not close, the χ^2 test statistic will be large and the P-value will be small. These relationships are summarized and illustrated in Figure 11-6 on the next page.

Part 2: Test of Homogeneity and the Fisher Exact Test
Test of Homogeneity

In Part 1 of this section, we focused on the test of independence between the row and column variables in a contingency table. In Part 1, the sample data are from one population, and individual sample results are categorized with the row and column variables. However, we sometimes obtain samples drawn from *different* populations, and we want to determine whether those populations have the same proportions of the characteristics being considered. The *test of homogeneity* can be used in such cases. (The word *homogeneous* means "having the same quality," and in this context, we are testing to determine whether the proportions are the same.)

DEFINITION

In a **test of homogeneity,** we test the claim that *different populations* have the same proportions of some characteristics.

Figure 11-6

Relationships Among Key Components in Test of Independence

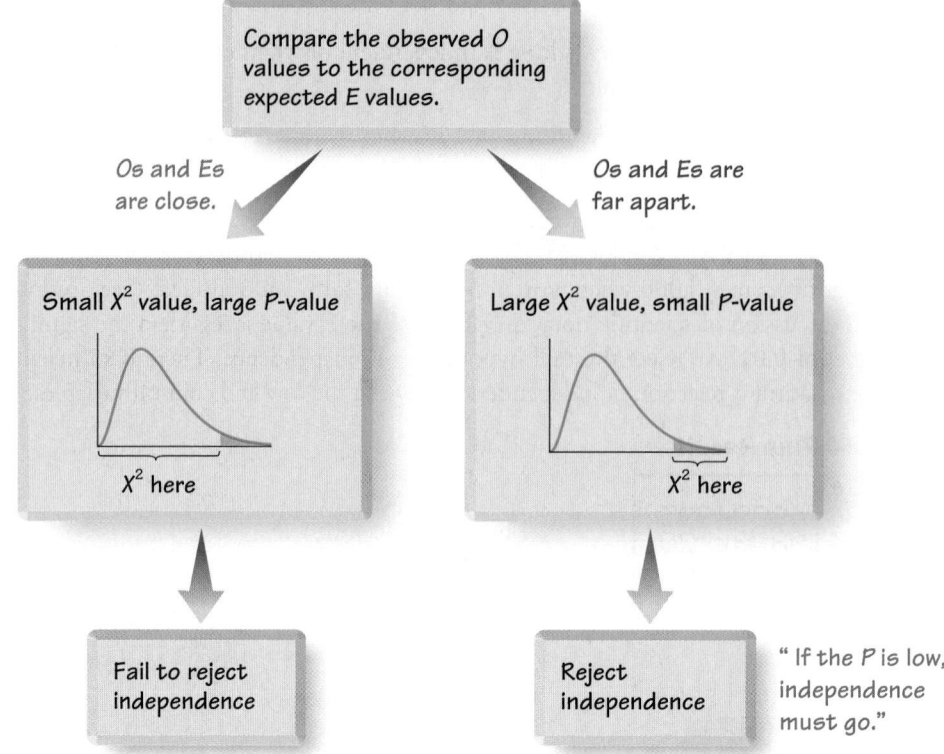

Compare the observed O values to the corresponding expected E values.

Os and Es are close.

Os and Es are far apart.

Small X^2 value, large P-value

X^2 here

Large X^2 value, small P-value

X^2 here

Fail to reject independence

Reject independence

"If the P is low, independence must go."

In conducting a test of homogeneity, we can use the same notation, requirements, test statistic, critical value, and procedures presented in Part 1 of this section, with one exception: Instead of testing the null hypothesis of *independence* between the row and column variables, we test the null hypothesis that *the different populations have the same proportions of some characteristics.*

EXAMPLE 5   **Influence of Gender** Does a pollster's gender have an effect on poll responses by men? A *U.S. News & World Report* article about polls stated: "On sensitive issues, people tend to give 'acceptable' rather than honest responses; their answers may depend on the gender or race of the interviewer." To support that claim, data were provided for an Eagleton Institute poll in which surveyed men were asked if they agreed with this statement: "Abortion is a private matter that should be left to the woman to decide without government intervention." We will analyze the effect of gender on male survey subjects only. Table 11-8 is based on the responses of surveyed men. Assume that the survey was designed so that male interviewers were instructed to obtain 800 responses from male subjects, and female interviewers were instructed to obtain 400 responses from male subjects. Using a 0.05 significance level, test the claim that the proportions of agree/disagree responses are the same for the subjects interviewed by men and the subjects interviewed by women.

Table 11-8   Gender and Survey Responses

	Gender of Interviewer	
	Man	Woman
Men who agree	560	308
Men who disagree	240	92

SOLUTION

REQUIREMENT CHECK (1) The data are random. (2) The sample data are represented as frequency counts in a two-way table. (3) The expected frequencies (shown in the accompanying Minitab display as 578.67, 289.33, 221.33, and 110.67) are all at least 5. All of the requirements are satisfied.

Because this is a test of homogeneity, we test the claim that the proportions of agree/disagree responses are the same for the subjects interviewed by males and the subjects interviewed by females. We have two separate populations (subjects interviewed by men and subjects interviewed by women), and we test for homogeneity with these hypotheses:

H_0: The proportions of agree/disagree responses are the same for the subjects interviewed by men and the subjects interviewed by women.

H_1: The proportions are different.

The significance level is $\alpha = 0.05$. We use the same χ^2 test statistic described earlier, and it is calculated using the same procedure. Instead of listing the details of that calculation, we provide the TI-83/84 Plus calculator display for the data in Table 11-8.

TI-83 Plus

```
X²-Test
 X²=6.529343179
 P=.0106109132
 df=1
```

The display includes the test statistic of $\chi^2 = 6.529$ (rounded) and the P-value of 0.0106. Using the P-value approach to hypothesis testing, we reject the null hypothesis of equal (homogeneous) proportions (because the P-value of 0.0106 is less than 0.05). There is sufficient evidence to warrant rejection of the claim that the proportions are the same.

INTERPRETATION
It appears that response and the gender of the interviewer are dependent. Although this statistical analysis cannot be used to justify any statement about causality, it does appear that men are influenced by the gender of the interviewer.

Fisher Exact Test

The procedures for testing hypotheses with contingency tables with two rows and two columns (2 × 2) have the requirement that every cell must have an expected frequency of at least 5. This requirement is necessary for the χ^2 distribution to be a suitable approximation to the exact distribution of the χ^2 test statistic. The *Fisher exact test* is often used for a 2 × 2 contingency table with one or more expected frequencies that are below 5. The Fisher exact test provides an *exact P*-value and does not require an approximation technique. Because the calculations are quite complex, it's a good idea to use computer software when using the Fisher exact test. STATDISK and Minitab both have the ability to perform the Fisher exact test.

Entering Matrix [A] for Table 11-1

Entries for χ^2 Test

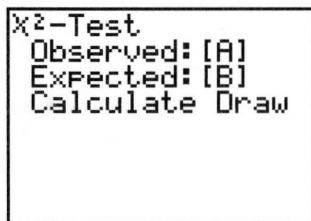

TI-83/84 Plus Results

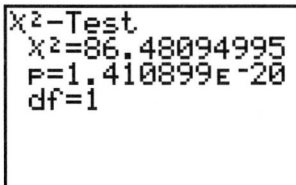

Using the TI-83/84 Plus Calculator for Contingency Tables

The TI-83/84 Plus calculator includes the procedure identified as χ^2-**Test,** and it is designed for the methods of this section. We can easily find the test statistic and the *P*-value. Because the *P*-value is generated, we can use the *P*-value method of hypothesis testing, and there is no need to find critical values from Table A-4. Use the following procedure.

1. Enter the observed frequencies as a *matrix* A, which is a rectangular arrangement of values. The matrix consists of rows and columns, and it should contain the same data in the same configuration as the original contingency table. Press **2ND** **x⁻¹** to get the **MATRIX** menu.

2. Use the arrow key **▷** to select the top menu item of **EDIT,** and press **ENTER** to select the default matrix [A].

3. Enter the dimensions of matrix [A] in the format of $r \times c$, where r is the number of rows and c is the number of columns. Press **ENTER** after entering each dimension.

4. Enter all of the observed frequencies in matrix [A]. See the accompanying TI-83/84 Plus screen that shows the entry of the matrix representing the data in Table 11-1.

5. Press **STAT**, select **TESTS,** and then select the menu item of χ^2-**Test.** Be sure that the observed matrix is the one you entered, such as matrix [A]. (There is no need to enter expected frequencies in matrix [B]; the expected frequencies will be calculated and entered into matrix [B] automatically.) Scroll down to **Calculate** and press **ENTER** to get the test statistic and *P*-value. See the accompanying screen display showing the results in Example 4. The results include the test statistic of $\chi^2 = 86.481$ (rounded) and the *P*-value of 0.000 (expressed in standard form and rounded).

11-3	Basic Skills and Concepts

Statistical Literacy and Critical Thinking

1. Polio Vaccine Results of a test of the Salk vaccine against polio are summarized in the table below. If we test the claim that getting paralytic polio is independent of whether the child was treated with the Salk vaccine or was given a placebo, the TI-83/84 Plus calculator provides a *P*-value of 1.732517E−11, which is in scientific notation. Write the *P*-value in a standard form that is not in scientific notation. Based on the *P*-value, what conclusion should we make? Does the vaccine appear to be effective?

	Paralytic polio	No paralytic polio
Salk vaccine	33	200,712
Placebo	115	201,114

2. Cause and Effect Based on the data in the table provided with Exercise 1, can we conclude that the Salk vaccine causes a decrease in the rate of paralytic polio? Why or why not?

3. Interpreting *P*-Value Refer to the *P*-value given in Exercise 1. Interpret that *P*-value by completing this statement: The *P*-value is the probability of _____.

4. Right-Tailed Test Why are the hypothesis tests described in this section always right-tailed, as in Example 1?

In Exercises 5 and 6, test the given claim using the displayed software results.

(TI) 5. Home Field Advantage Winning team data were collected for teams in different sports, with the results given in the accompanying table (based on data from "Predicting Professional

Sports Game Outcomes from Intermediate Game Scores," by Copper, DeNeve, and Mosteller, *Chance*, Vol. 5, No. 3–4). The TI-83/84 Plus results are also displayed. Use a 0.05 level of significance to test the claim that home/visitor wins are independent of the sport.

	Basketball	Baseball	Hockey	Football
Home team wins	127	53	50	57
Visiting team wins	71	47	43	42

TI-83/84 PLUS

```
X²-Test
 X²=4.737208763
 P=.1920828463
 df=3
```

6. Crime and Strangers The Minitab display shows results from the table below, which lists data obtained from randomly selected crime victims (based on data from the U.S. Department of Justice). What can we conclude?

	Homicide	Robbery	Assault
Criminal was a stranger	12	379	727
Criminal was acquaintance or relative	39	106	642

MINITAB

Chi-Sq = 119.330, DF = 2, P-Value = 0.000

In Exercises 7–22, test the given claim.

7. Instant Replay in Tennis The table below summarizes challenges made by tennis players in the first U.S. Open that used the Hawk-Eye electronic instant replay system. Use a 0.05 significance level to test the claim that success in challenges is independent of the gender of the player. Does either gender appear to be more successful?

	Was the challenge to the call successful?	
	Yes	No
Men	201	288
Women	126	224

8. Open Roof or Closed Roof? In a recent baseball World Series, the Houston Astros wanted to close the roof on their domed stadium so that fans could make noise and give the team a better advantage at home. However, the Astros were ordered to keep the roof open, unless weather conditions justified closing it. But does the closed roof really help the Astros? The table below shows the results from home games during the season leading up to the World Series. Use a 0.05 significance level to test for independence between wins and whether the roof is open or closed. Does it appear that a closed roof really gives the Astros an advantage?

	Win	Loss
Closed roof	36	17
Open roof	15	11

9. Testing a Lie Detector The table below includes results from polygraph (lie detector) experiments conducted by researchers Charles R. Honts (Boise State University) and Gordon H. Barland (Department of Defense Polygraph Institute). In each case, it was known if the subject lied or did not lie, so the table indicates when the polygraph test was correct. Use a 0.05 significance level to test the claim that whether a subject lies is independent of the polygraph test indication. Do the results suggest that polygraphs are effective in distinguishing between truths and lies?

	Did the Subject Actually Lie?	
	No (Did Not Lie)	Yes (Lied)
Polygraph test indicated that the subject *lied*.	15	42
Polygraph test indicated that the subject did *not lie*.	32	9

10. Clinical Trial of Chantix Chantix is a drug used as an aid for those who want to stop smoking. The adverse reaction of nausea has been studied in clinical trials, and the table below summarizes results (based on data from Pfizer). Use a 0.01 significance level to test the claim that nausea is independent of whether the subject took a placebo or Chantix. Does nausea appear to be a concern for those using Chantix?

	Placebo	Chantix
Nausea	10	30
No nausea	795	791

11. Amalgam Tooth Fillings The table below shows results from a study in which some patients were treated with amalgam restorations and others were treated with composite restorations that do not contain mercury (based on data from "Neuropsychological and Renal Effects of Dental Amalgam in Children," by Bellinger, et al., *Journal of the American Medical Association*, Vol. 295, No. 15). Use a 0.05 significance level to test for independence between the type of restoration and the presence of any adverse health conditions. Do amalgam restorations appear to affect health conditions?

	Amalgam	Composite
Adverse health condition reported	135	145
No adverse health condition reported	132	122

12. Amalgam Tooth Fillings In recent years, concerns have been expressed about adverse health effects from amalgam dental restorations, which include mercury. The table below shows results from a study in which some patients were treated with amalgam restorations and others were treated with composite restorations that do not contain mercury (based on data from "Neuropsychological and Renal Effects of Dental Amalgam in Children," by Bellinger, et al., *Journal of the American Medical Association*, Vol. 295, No. 15). Use a 0.05 significance level to test for independence between the type of restoration and sensory disorders. Do amalgam restorations appear to affect sensory disorders?

	Amalgam	Composite
Sensory disorder	36	28
No sensory disorder	231	239

13. Is Sentence Independent of Plea? Many people believe that criminals who plead guilty tend to get lighter sentences than those who are convicted in trials. The accompanying table summarizes randomly selected sample data for San Francisco defendants in burglary cases (based on data from "Does It Pay to Plead Guilty? Differential Sentencing and the Functioning of the Criminal Courts," by Brereton and Casper, *Law and Society Review*, Vol. 16, No. 1). All of the subjects had prior prison sentences. Use a 0.05 significance level to test the claim that the sentence (sent to prison or not sent to prison) is independent of the plea. If you were an attorney defending a guilty defendant, would these results suggest that you should encourage a guilty plea?

	Guilty Plea	Not Guilty Plea
Sent to prison	392	58
Not sent to prison	564	14

14. Is the Vaccine Effective? In a *USA Today* article about an experimental vaccine for children, the following statement was presented: "In a trial involving 1602 children, only 14 (1%) of the 1070 who received the vaccine developed the flu, compared with 95 (18%) of the 532 who got a placebo." The data are shown in the table below. Use a 0.05 significance level to test for independence between the variable of treatment (vaccine or placebo) and the variable representing flu (developed flu, did not develop flu). Does the vaccine appear to be effective?

	Developed Flu?	
	Yes	No
Vaccine treatment	14	1056
Placebo	95	437

15. Which Treatment Is Better? A randomized controlled trial was designed to compare the effectiveness of splinting versus surgery in the treatment of carpal tunnel syndrome. Results are given in the table below (based on data from "Splinting vs. Surgery in the Treatment of Carpal Tunnel Syndrome," by Gerritsen, et al., *Journal of the American Medical Association*, Vol. 288, No. 10). The results are based on evaluations made one year after the treatment. Using a 0.01 significance level, test the claim that success is independent of the type of treatment. What do the results suggest about treating carpal tunnel syndrome?

	Successful Treatment	Unsuccessful Treatment
Splint treatment	60	23
Surgery treatment	67	6

16. Norovirus on Cruise Ships The *Queen Elizabeth II* cruise ship and Royal Caribbean's *Freedom of the Seas* cruise ship both experienced outbreaks of norovirus within two months of each other. Results are shown in the table below. Use a 0.05 significance level to test the claim that getting norovirus is independent of the ship. Based on these results, does it appear that an outbreak of norovirus has the same effect on different ships?

	Norovirus	No norovirus
Queen Elizabeth II	276	1376
Freedom of the Seas	338	3485

17. Global Warming Survey A Pew Research poll was conducted to investigate opinions about global warming. The respondents who answered yes when asked if there is solid evidence that the earth is getting warmer were then asked to select a cause of global warming. The results are given in the table below. Use a 0.05 significance level to test the claim that the sex of the respondent is independent of the choice for the cause of global warming. Do men and women appear to agree, or is there a substantial difference?

	Human activity	Natural patterns	Don't know or refused to answer
Male	314	146	44
Female	308	162	46

18. Global Warming Survey A Pew Research poll was conducted to investigate opinions about global warming. The respondents who answered yes when asked if there is solid evidence that the earth is getting warmer were then asked to select a cause of global warming. The results for two age brackets are given in the table below. Use a 0.01 significance level to test the claim that the age bracket is independent of the choice for the cause of global warming. Do respondents from both age brackets appear to agree, or is there a substantial difference?

	Human activity	Natural patterns	Don't know or refused to answer
Under 30	108	41	7
65 and over	121	71	43

19. Clinical Trial of Campral Campral is a drug used to help patients continue their abstinence from the use of alcohol. Adverse reactions of Campral have been studied in clinical trials, and the table below summarizes results for digestive system effects among patients from different treatment groups (based on data from Forest Pharmaceuticals, Inc.). Use a 0.01 significance level to test the claim that experiencing an adverse reaction in the digestive system is independent of the treatment group. Does Campral treatment appear to have an effect on the digestive system?

	Placebo	Campral 1332 mg	Campral 1998 mg
Adverse effect on digestive system	344	89	8
No effect on digestive system	1362	774	71

20. Is Seat Belt Use Independent of Cigarette Smoking? A study of seat belt users and nonusers yielded the randomly selected sample data summarized in the given table (based on data from "What Kinds of People Do Not Use Seat Belts?" by Helsing and Comstock, *American Journal of Public Health,* Vol. 67, No. 11). Test the claim that the amount of smoking is independent of seat belt use. A plausible theory is that people who smoke more are less concerned about their health and safety and are therefore less inclined to wear seat belts. Is this theory supported by the sample data?

	Number of Cigarettes Smoked per Day			
	0	1–14	15–34	35 and over
Wear seat belts	175	20	42	6
Don't wear seat belts	149	17	41	9

21. Clinical Trial of Lipitor Lipitor is the trade name of the drug atorvastatin, which is used to reduce cholesterol in patients. (This is the largest-selling drug in the world, with $13 billion in sales for a recent year.) Adverse reactions have been studied in clinical trials, and the table below summarizes results for infections in patients from different treatment groups (based on data from Parke-Davis). Use a 0.05 significance level to test the claim that getting an infection is independent of the treatment. Does the atorvastatin treatment appear to have an effect on infections?

	Placebo	Atorvastatin 10 mg	Atorvastatin 40 mg	Atorvastatin 80 mg
Infection	27	89	8	7
No infection	243	774	71	87

22. Injuries and Motorcycle Helmet Color A case-control (or retrospective) study was conducted to investigate a relationship between the colors of helmets worn by motorcycle drivers and whether they are injured or killed in a crash. Results are given in the table below (based on data from "Motorcycle Rider Conspicuity and Crash Related Injury: Case-Control Study," by Wells, et al., *BMJ USA,* Vol. 4). Test the claim that injuries are independent of helmet color. Should motorcycle drivers choose helmets with a particular color? If so, which color appears best?

	Color of Helmet				
	Black	White	Yellow/Orange	Red	Blue
Controls (not injured)	491	377	31	170	55
Cases (injured or killed)	213	112	8	70	26

11-3   Beyond the Basics

23. Test of Homogeneity Table 11-8 summarizes data for male survey subjects, but the table below summarizes data for a sample of women (based on data from an Eagleton Institute poll). Using a 0.01 significance level, and assuming that the sample sizes of 800 men and 400 women are predetermined, test the claim that the proportions of agree/disagree responses are the same for the subjects interviewed by men and the subjects interviewed by women. Does it appear that the gender of the interviewer affected the responses of women?

	Gender of Interviewer	
	Man	Woman
Women who agree	512	336
Women who disagree	288	64

24. Using Yates' Correction for Continuity The chi-square distribution is continuous, whereas the test statistic used in this section is discrete. Some statisticians use *Yates' correction for continuity* in cells with an expected frequency of less than 10 or in all cells of a contingency table with two rows and two columns. With Yates' correction, we replace

$$\sum \frac{(O - E)^2}{E} \quad \text{with} \quad \sum \frac{(|O - E| - 0.5)^2}{E}$$

Given the contingency table in Exercise 7, find the value of the χ^2 test statistic with and without Yates' correction. What effect does Yates' correction have?

25. Equivalent Tests A χ^2 test involving a 2 × 2 table is equivalent to the test for the difference between two proportions, as described in Section 9-2. Using the table in Exercise 7, verify that the χ^2 test statistic and the z test statistic (found from the test of equality of two proportions) are related as follows: $z^2 = \chi^2$. Also show that the critical values have that same relationship.

11-4   McNemar's Test for Matched Pairs

Key Concept The methods in Section 11-3 for analyzing two-way tables are based on *independent* data. For 2 × 2 tables consisting of frequency counts that result from *matched pairs,* the frequency counts within each matched pair are not independent and, for such cases, we can use McNemar's test for matched pairs. In this section we present the method of using McNemar's test for testing the null hypothesis that the frequencies from the discordant (different) categories occur in the same proportion.

Table 11-9 shows a general format for summarizing results from data consisting of frequency counts from matched pairs. Table 11-9 refers to two different treatments (such as two different eye drop solutions) applied to two different parts of each subject (such as left eye and right eye). It's a bit difficult to correctly read a table such as Table 11-9. The total number of subjects is $a + b + c + d$, and each of those subjects yields results from each of two parts of a matched pair. If $a = 100$, then 100 subjects were cured with both treatments. If $b = 50$ in Table 11-9, then each of 50 subjects had no cure with treatment X but they were each cured with treatment Y. Remember, the entries in Table 11-9 are frequency counts of *subjects,* not the total number of individual components in the matched pairs. If 500 people have each eye treated with two different ointments, the value of $a + b + c + d$ is 500 (the number of subjects), not 1000 (the number of treated eyes).

Table 11-9   2 × 2 Table with Frequency Counts from Matched Pairs

		Treatment X	
		Cured	Not Cured
	Cured	a	b
Treatment Y			
	Not cured	c	d

Because the frequency counts in Table 11-9 result from *matched* pairs, the data are not independent and we cannot use the methods from Section 11-3. Instead, we use McNemar's test.

DEFINITION

McNemar's test uses frequency counts from *matched pairs* of nominal data from two categories to test the null hypothesis that for a 2×2 table such as Table 11-9, the frequencies b and c occur in the same proportion.

Objective

Test for a difference in proportions by using McNemar's test for matched pairs.

Notation

a, b, c, and d represent the frequency counts from a 2×2 table consisting of frequency counts from *matched pairs*. (The total number of subjects is $a + b + c + d$.)

Requirements

1. The sample data have been randomly selected.

2. The sample data consist of *matched pairs* of frequency counts.

3. The data are at the nominal level of measurement, and each observation can be classified two ways: (1) According to the category distinguishing values with each matched pair (such as left eye and right eye), and (2) according to another category with two possible values (such as cured/not cured).

4. For tables such as Table 11-9, the frequencies are such that $b + c \geq 10$.

Null and Alternative Hypotheses

H_0: The proportions of the frequencies b and c (as in Table 11-9) are the same.

H_1: The proportions of the frequencies b and c (as in Table 11-9) are different.

Test Statistic (for testing the null hypothesis that for tables such as Table 11-9, the frequencies b and c occur in the same proportion):

$$\chi^2 = \frac{(|b - c| - 1)^2}{b + c}$$

where the frequencies of b and c are obtained from the 2×2 table with a format similar to Table 11-9. (The frequencies b and c must come from "discordant" (or different) pairs, as described later in this section.)

Critical Values

1. The critical region is located in the *right tail only*.

2. The critical values are found in Table A-4 by using **degrees of freedom = 1.**

P-Values

P-values can be found using a TI-83/84 Plus calculator, or a range of *P*-values can be found from Table A-4.

> **EXAMPLE 1**   **Are Hip Protectors Effective?** A randomized controlled trial was designed to test the effectiveness of hip protectors in preventing hip fractures in the elderly. Nursing home residents each wore protection on one hip, but not the other. Results are summarized in Table 11-10 (based on data from "Efficacy of Hip Protector to Prevent Hip Fracture in Nursing Home Residents," by Kiel, et al., *Journal of the American Medical Association,* Vol. 298, No. 4). Using a 0.05 significance level, apply McNemar's test to test the null hypothesis that the following two proportions are the same:
>
> - The proportion of subjects with no hip fracture on the protected hip and a hip fracture on the unprotected hip.
>
> - The proportion of subjects with a hip fracture on the protected hip and no hip fracture on the unprotected hip.
>
> Based on the results, do the hip protectors appear to be effective in preventing hip fractures?

SOLUTION   **REQUIREMENT CHECK** (1) The data are from randomly selected subjects. (2) The data consist of matched pairs of frequency counts. (3) The data are at the nominal level of measurement and each observation can be categorized according to two variables. (One variable has values of "hip protection was worn" and "hip protection was not worn," and the other variable has values of "hip was fractured" and "hip was not fractured.") (4) For Table 11-10, $b = 10$ and $c = 15$, so that $b + c = 25$, which is at least 10. All of the requirements are satisfied. ✓

Although Table 11-10 might appear to be a 2×2 contingency table, we cannot use the procedures of Section 11-3 because the data come from *matched pairs* (instead of being independent). Instead, we use McNemar's test.

After comparing the frequency counts in Table 11-9 to those given in Table 11-10, we see that $b = 10$ and $c = 15$, so the test statistic can be calculated as follows:

$$\chi^2 = \frac{(|b - c| - 1)^2}{b + c} = \frac{(|10 - 15| - 1)^2}{10 + 15} = 0.640$$

P-Value Method Using the TI-83/84 Plus Using a TI-83/84 Plus calculator with the instructions given at the end of this section, we can find the *P*-value of 0.424. Because the *P*-value of 0.424 is greater than the significance level of 0.05, we fail to reject the null hypothesis.

Traditional Method The traditional method of testing hypotheses can be used by finding the critical value of $\chi^2 = 3.841$. That value is found in Table A-4 with $\alpha = 0.05$ in the right tail and 1 degree of freedom. Because the test statistic of $\chi^2 = 0.640$ is not in the right-tailed critical region bounded by the critical value of $\chi^2 = 3.841$, we fail to reject the null hypothesis of independence.

INTERPRETATION   The proportion of hip fractures with the protectors worn is not significantly different from the proportion of hip fractures without the protectors worn. The hip protectors do not appear to be effective in preventing hip fractures.

Table 11-10   Randomized Controlled Trial of Hip Protectors

		No Hip Protector Worn	
		No Hip Fracture	Hip Fracture
Hip Protector Worn	No Hip Fracture	309	10
	Hip Fracture	15	2

Note that in the calculation of the test statistic in Example 1, we did not use the 309 subjects with no fractured hips, nor did we use the frequency of 2 representing subjects with both hips fractured. We used only those subjects with a fracture in one hip but not in the other. That is, we are using only the results from the categories that are *different*. Such pairs of different categories are referred to as *discordant pairs*.

> **DEFINITION**
>
> **Discordant pairs** of results come from matched pairs of results in which the two categories are different (as in the frequencies b and c in Table 11-9).

When trying to determine whether hip protectors are effective, we are not helped by any subjects with no fractures, and we are not helped by any subjects with both hips fractured. The differences are reflected in the discordant results from the subjects with one hip fractured while the other hip is not fractured. Consequently, the test statistic includes only the two frequencies that result from the two discordant (or different) pairs of categories.

CAUTION

When applying McNemar's test, be careful to use only the frequencies from the pairs of categories that are *different*. Do not blindly use the frequencies in the upper right and lower left corners, because they do not necessarily represent the discordant pairs. If Table 11-10 were reconfigured as shown below, it would be inconsistent in its format, but it would be technically correct in summarizing the same results as Table 11-10; however, blind use of the frequencies of 2 and 309 would result in the *wrong* test statistic.

		No Hip Protector Worn	
		No Hip Fracture	Hip Fracture
	Hip Fracture	15	2
Hip Protector Worn			
	No Hip Fracture	309	10

In this reconfigured table, the discordant pairs of frequencies are these:

Hip fracture/No hip fracture: 15
No hip fracture/Hip fracture: 10

With this reconfigured table, we should again use the frequencies of 15 and 10 (as in Example 1), not 2 and 309. In a more perfect world, all such 2×2 tables would be configured with a consistent format, and we would be much less likely to use the wrong frequencies.

In addition to comparing treatments given to matched pairs (as in Example 1), McNemar's test is often used to test a null hypothesis of no change in before/after types of experiments. (See Exercises 5–12.)

Using the TI-83/84 Plus Calculator for McNemar's Test

The TI-83/84 Plus calculator does not include a procedure designed for McNemar's test, but it can be used to find the P-value for the test. Use the following procedure.

1. Use the methods of this section to calculate the χ^2 test statistic.

2. Press **2ND** **VARS** to select the **DISTR** (distribution) menu.

3. Select the menu item of χ^2**cdf(** and proceed to enter the following three values separated by commas:

    i. The χ^2 test statistic

    ii. The number 999999

    iii. The number 1

(*Note:* The format for the χ^2 cdf command is to enter the lower boundary of χ^2, the upper boundary of χ^2, and the number of degrees of freedom. Because we want the area to the right of the χ^2 test statistic, we enter the value of 999999 for the upper boundary, which essentially means that there is no upper boundary.) For Example 1, the entries should appear as follows:

$$\chi^2\text{cdf}(0.640, 999999, 1).$$

4. Press **ENTER** and the *P*-value for the test will be displayed. The entry of χ^2cdf(0.640, 999999, 1) results in a *P*-value of 0.4237107972, or 0.424 when rounded.

11-4   Basic Skills and Concepts

Statistical Literacy and Critical Thinking

1. McNemar's Test The table below summarizes results from a study in which 186 students in an introductory statistics course were each given algebra problems in two different formats: a symbolic format and a verbal format (based on data from "Changing Student's Perspectives of McNemar's Test of Change," by Levin and Serlin, *Journal of Statistics Education,* Vol. 8, No. 2). Assume that the data are randomly selected. Using only an examination of the table entries, does either format appear to be better? If so, which one? Why?

		Verbal Format	
		Mastery	Nonmastery
	Mastery	74	31
Symbolic Format			
	Nonmastery	33	48

2. Discordant Pairs Refer to the table in Exercise 1. Identify the discordant pairs of results.

3. Discordant Pairs Refer to the data in Exercise 1. Explain why McNemar's test ignores the frequencies of 74 and 48.

4. Requirement Check Refer to the data in Exercise 1. Identify which requirements are satisfied for McNemar's test.

In Exercises 5–12, refer to the following table. The table summarizes results from an experiment in which subjects were first classified as smokers or nonsmokers, then they were given a treatment, then later they were again classified as smokers or nonsmokers (based on data from Pfizer Pharmaceuticals in clinical trials of Chantix).

		Before Treatment	
		Smoke	Don't Smoke
	Smoke	460	4
After treatment			
	Don't smoke	361	192

5. Sample Size How many subjects are included in the experiment?

6. Treatment Effectiveness How many subjects changed their smoking status after the treatment?

7. Treatment Ineffectiveness How many subjects appear to be unaffected by the treatment one way or the other?

8. Why Not *t* Test? Section 9-4 presented procedures for data consisting of matched pairs. Why can't we use the procedures of Section 9-4 for the analysis of the results summarized in the table?

9. Discordant Pairs Which of the following pairs of before/after results are *discordant?*

a. smoke/smoke

b. smoke/don't smoke

c. don't smoke/smoke

d. don't smoke/don't smoke

10. Test Statistic Using the appropriate frequencies, find the value of the test statistic.

11. Critical Value Using a 0.01 significance level, find the critical value.

12. Conclusion Based on the preceding results, what do you conclude? How does the conclusion make sense in terms of the original sample results?

TI **13. Testing Hip Protectors** Example 1 in this section used results from subjects who used hip protection at least 80% of the time. Results from a larger data set were obtained from the same study, and the results are shown in the table below (based on data from "Efficacy of Hip Protector to Prevent Hip Fracture in Nursing Home Residents," by Kiel, et al., *Journal of the American Medical Association,* Vol. 298, No. 4). Use a 0.05 significance level to test the effectiveness of the hip protectors.

		No Hip Protector Worn	
		No Hip Fracture	Hip Fracture
Hip Protector Worn	No Hip Fracture	1004	17
	Hip Fracture	21	0

TI **14. Predicting Measles Immunity** Pregnant women were tested for immunity to the rubella virus, and they were also tested for immunity to measles, with results given in the following table (based on data from "Does Rubella Predict Measles Immunity? A Serosurvey of Pregnant Women," by Kennedy, et al., *Infectious Diseases in Obstetrics and Gynecology,* Vol. 2006). Use a 0.05 significance level to apply McNemar's test. What does the result tell us? If a woman is likely to become pregnant and she is found to have rubella immunity, should she also be tested for measles immunity?

		Measles	
		Immune	Not Immune
Rubella	Immune	780	62
	Not Immune	10	7

TI **15. Treating Athlete's Foot** Randomly selected subjects are inflicted with tinea pedis (athlete's foot) on each of their feet. One foot is treated with a fungicide solution while the other foot is given a placebo. The results are given in the accompanying table. Using a 0.05 significance level, test the effectiveness of the treatment.

		Fungicide Treatment	
		Cure	No Cure
Placebo	Cure	5	12
	No cure	22	55

16. Treating Athlete's Foot Repeat Exercise 15 after changing the frequency of 22 to 66.

 17. PET/CT Compared to MRI In the article "Whole-Body Dual-Modality PET/CT and Whole Body MRI for Tumor Staging in Oncology" (Antoch, et al., *Journal of the American Medical Association,* Vol. 290, No. 24), the authors cite the importance of accurately identifying the stage of a tumor. Accurate staging is critical for determining appropriate therapy. The article discusses a study involving the accuracy of positron emission tomography (PET) and computed tomography (CT) compared to magnetic resonance imaging (MRI). Using the data in the given table for 50 tumors analyzed with both technologies, does there appear to be a difference in accuracy? Does either technology appear to be better?

		PET/CT	
		Correct	Incorrect
MRI	Correct	36	1
	Incorrect	11	2

 18. Testing a Treatment In the article "Eradication of Small Intestinal Bacterial Overgrowth Reduces Symptoms of Irritable Bowel Syndrome" (Pimentel, Chow, and Lin, *American Journal of Gastroenterology,* Vol. 95, No. 12), the authors include a discussion of whether antibiotic treatment of bacteria overgrowth reduces intestinal complaints. McNemar's test was used to analyze results for those subjects with eradication of bacterial overgrowth. Using the data in the given table, does the treatment appear to be effective against abdominal pain?

		Abdominal Pain Before Treatment?	
		Yes	No
Abdominal pain after treatment?	Yes	11	1
	No	14	3

11-4 Beyond the Basics

19. Correction for Continuity The test statistic given in this section includes a correction for continuity. The test statistic given below does not include the correction for continuity, and it is sometimes used as the test statistic for McNemar's test. Refer to Exercise 18 and find the value of the test statistic using the expression given below, and compare the result to the one found in the exercise.

$$\chi^2 = \frac{(b - c)^2}{b + c}$$

20. Using Common Sense Consider the table given in Exercise 17. The frequencies of 36 and 2 are not included in the computations, but how are your conclusions modified if those two frequencies are changed to 8000 and 7000 respectively?

21. Small Sample Case The requirements for McNemar's test include the condition that $b + c \geq 10$ so that the distribution of the test statistic can be approximated by the chi-square distribution. Refer to the table on the next page. McNemar's test should not be used because the condition of $b + c \geq 10$ is not satisfied since $b = 2$ and $c = 6$. Instead, use the binomial distribution to find the probability that among 8 equally likely outcomes, the results consist of 6 items in one category and 2 in the other category, or the results are more extreme. That is, use a probability of 0.5 to find the probability that among $n = 8$ trials, the number of successes x is 6 or 7 or 8. Double that probability to find the *P*-value for this test. Compare the result to the *P*-value of 0.289 that results from using the chi-square approximation, even though the condition of $b + c \geq 10$ is violated. What do you conclude about the two treatments?

		Treatment with Pedacream	
		Cured	Not Cured
Treatment with Fungacream	Cured	12	2
	Not cured	6	20

Review

The three sections of this chapter all involve applications of the χ^2 distribution to categorical data consisting of frequency counts. In Section 11-2 we described methods for using frequency counts from different categories for testing goodness-of-fit with some claimed distribution. The test statistic given below is used in a right-tailed test in which the χ^2 distribution has $k - 1$ degrees of freedom, where k is the number of categories. This test requires that each of the expected frequencies must be at least 5.

$$\text{Test statistic is } \chi^2 = \sum \frac{(O - E)^2}{E}$$

In Section 11-3 we described methods for testing claims involving contingency tables (or two-way frequency tables), which have at least two rows and two columns. Contingency tables incorporate two variables: One variable is used for determining the row that describes a sample value, and the second variable is used for determining the column that describes a sample value. We conduct a test of independence between the row and column variables by using the test statistic given below. This test statistic is used in a right-tailed test in which the χ^2 distribution has the number of degrees of freedom given by $(r - 1)(c - 1)$, where r is the number of rows and c is the number of columns. This test requires that each of the expected frequencies must be at least 5.

$$\text{Test statistic is } \chi^2 = \sum \frac{(O - E)^2}{E}$$

In Section 11-4 we introduced McNemar's test for testing the null hypothesis that a sample of matched pairs of data comes from a population in which the discordant (different) pairs occur in the same proportion. The test statistic is given below. The frequencies of b and c must come from "discordant" pairs. This test statistic is used in a right-tailed test in which the χ^2 distribution has 1 degree of freedom.

$$\text{Test statistic is } \chi^2 = \frac{(|b - c| - 1)^2}{b + c}$$

Statistical Literacy and Critical Thinking

1. Categorical Data In what sense are the data in the table below *categorical* data? (The data are from Pfizer, Inc.)

	Celebrex	Ibuprofen	Placebo
Nausea	145	23	78
No Nausea	4001	322	1786

2. Terminology Refer to the table given in Exercise 1. Why is that table referred to as a *two-way* table?

3. Cause/Effect Refer to the table given in Exercise 1. After analysis of the data in such a table, can we ever conclude that a treatment of Celebrex and/or Ibuprofen *causes* nausea? Why or why not?

4. Observed and Expected Frequencies Refer to the table given in Exercise 1. The cell with the observed frequency of 145 has an expected frequency of 160.490. Describe what that expected frequency represents.

Chapter Quick Quiz

Questions 1–4 refer to the sample data in the following table (based on data from the Dutchess County STOP-DWI Program). The table summarizes results from randomly selected fatal car crashes in which the driver had a blood-alcohol level greater than 0.10.

Day	Sun	Mon	Tues	Wed	Thurs	Fri	Sat
Number	40	24	25	28	29	32	38

1. What are the null and alternative hypotheses corresponding to a test of the claim that fatal DWI crashes occur equally on the different days of the week?

2. When testing the claim in Question 1, what are the observed and expected frequencies for Sunday?

3. If using a 0.05 significance level for a test of the claim that the proportions of DWI fatalities are the same for the different days of the week, what is the critical value?

4. Given that the *P*-value for the hypothesis test is 0.2840, what do you conclude?

5. When testing the null hypothesis of independence between the row and column variables in a contingency table, is the test two-tailed, left-tailed, or right-tailed?

6. What distribution is used for testing the null hypothesis that the row and column variables in a contingency table are independent? (normal, *t*, *F*, chi-square, uniform)

Questions 7–10 refer to the sample data in the following table (based on data from a Gallup poll). The table summarizes results from a survey of workers and senior-level bosses who were asked if it was seriously unethical to monitor employee e-mail.

	Yes	No
Workers	192	244
Bosses	40	81

7. If using the given sample data for a hypothesis test, what are the appropriate null and alternative hypotheses?

8. If testing the null hypothesis with a 0.05 significance level, find the critical value.

9. Given that the *P*-value for the hypothesis test is 0.0302, what do you conclude when using a 0.05 significance level?

10. Given that the *P*-value for the hypothesis test is 0.0302, what do you conclude when using a 0.01 significance level?

Review Exercises

 1. Testing for Adverse Reactions The table on the next page summarizes results from a clinical trial (based on data from Pfizer, Inc). Use a 0.05 significance level to test the claim that experiencing nausea is independent of whether a subject is treated with Celebrex, Ibuprofen, or a placebo. Does the adverse reaction of nausea appear to be about the same for the different treatments?

	Celebrex	Ibuprofen	Placebo
Nausea	145	23	78
No Nausea	4001	322	1786

TI **2. Lightning Deaths** Listed below are the numbers of deaths from lightning on the different days of the week. The deaths were recorded for a recent period of 35 years (based on data from the National Oceanic and Atmospheric Administration). Use a 0.01 significance level to test the claim that deaths from lightning occur on the different days with the same frequency. Can you provide an explanation for the result?

Day	Sun	Mon	Tues	Wed	Thurs	Fri	Sat
Number of deaths	574	445	429	473	428	422	467

TI **3. Participation in Clinical Trials by Race** Researchers conducted a study to investigate racial disparity in clinical trials of cancer. Among the randomly selected participants, 644 were white, 23 were Hispanic, 69 were black, 14 were Asian/Pacific Islander, and 2 were American Indian/Alaskan Native. The proportions of the U.S. population of the same groups are 0.757, 0.091, 0.108, 0.038, and 0.007, respectively. (Based on data from "Participation in Clinical Trials," by Murthy, Krumholz, and Gross, *Journal of the American Medical Association,* Vol. 291, No. 22.) Use a 0.05 significance level to test the claim that the participants fit the same distribution as the U.S. population. Why is it important to have proportionate representation in such clinical trials?

TI **4. Effectiveness of Treatment** A clinical trial tested the effectiveness of bupropion hydrochloride in helping people who want to stop smoking. Results of abstinence from smoking 52 weeks after the treatment are summarized in the table below (based on data from "A Double-Blind, Placebo-Controlled, Randomized Trial of Bupropion for Smoking Cessation in Primary Care," by Fossatti, et al., *Archives of Internal Medicine,* Vol. 167, No. 16). Use a 0.05 significance level to test the claim that whether a subject smokes is independent of whether the subject was treated with bupropion hydrochloride or a placebo. Does the bupropion hydrochloride treatment appear to be better than a placebo? Is the bupropion hydrochloride treatment highly effective?

	Bupropion Hydrochloride	Placebo
Smoking	299	167
Not Smoking	101	26

TI **5. McNemar's Test** Parents and their children were surveyed in a study of children's respiratory systems. They were asked if the children coughed early in the morning, and results are shown in the table below (based on data from "Cigarette Smoking and Children's Respiratory Symptoms: Validity of Questionnaire Method," by Bland, et al., *Revue d'Epidemiologie et Sante Publique,* Vol. 27). Use a 0.05 significance level to test the claim that the following proportions are the same: (1) the proportion of cases in which the child indicated no cough while the parent indicated coughing; (2) the proportion of cases in which the child indicated coughing while the parent indicated no coughing. What do the results tell us?

		Child Response	
		Cough	No Cough
Parent Response	Cough	29	104
	No Cough	172	5097

Cumulative Review Exercises

1. Cleanliness The American Society for Microbiology and the Soap and Detergent Association released survey results indicating that among 3065 men observed in public restrooms, 2023 of them washed their hands, and among 3011 women observed, 2650 washed their hands (based on data from *USA Today*).

a. Is the study an experiment or an observational study?

b. Are the given numbers discrete or continuous?

c. Are the given numbers statistics or parameters?

d. Is there anything about the study that might make the results questionable?

2. Cleanliness Refer to the results given in Exercise 1 and use a 0.05 significance level to test the claim that the proportion of men who wash their hands is equal to the proportion of women who wash their hands. Is there a significant difference?

3. Cleanliness Refer to the results given in Exercise 1. Construct a two-way frequency table and use a 0.05 significance level to test the claim that hand washing is independent of gender.

4. Golf Scores Listed below are first round and fourth round golf scores of randomly selected golfers in a Professional Golf Association Championship (based on data from the *New York Times*). Find the mean, median, range, and standard deviation for the first round scores, then find those same statistics for the fourth round scores. Compare the results.

First round	71	68	75	72	74	67
Fourth round	69	69	69	72	70	73

5. Golf Scores Refer to the sample data given in Exercise 4. Use a 0.05 significance level to test for a linear correlation between the first round scores and the fourth round scores.

6. Golf Scores Using only the first round golf scores given in Exercise 4, construct a 95% confidence interval estimate of the mean first round golf score for all golfers. Interpret the result.

7. Wise Action for Job Applicants In an Accountemps survey of 150 randomly selected senior executives, 88% said that sending a thank-you note after a job interview increases the applicant's chances of being hired (based on data from *USA Today*). Construct a 95% confidence interval estimate of the percentage of all senior executives who believe that a thank-you note is helpful. What very practical advice can be gained from these results?

8. Testing a Claim Refer to the sample results given in Exercise 7 and use a 0.01 significance level to test the claim that more than 75% of all senior executives believe that a thank-you note after a job interview increases the applicant's chances of being hired.

9. Ergonomics When designing the cockpit of a single-engine aircraft, engineers must consider the upper leg lengths of men. Those lengths are normally distributed with a mean of 42.6 cm and a standard deviation of 2.9 cm (based on Data Set 1 in Appendix B).

a. If one man is randomly selected, find the probability that his upper leg length is greater than 45 cm.

b. If 16 men are randomly selected, find the probability that their mean upper leg length is greater than 45 cm.

c. When designing the aircraft cockpit, which result is more meaningful: the result from part (a) or the result from part (b)? Why?

10. Tall Women The probability of randomly selecting a woman who is more than 5 feet tall is 0.925 (based on data from the National Health and Nutrition Examination Survey). Find the probability of randomly selecting five women and finding that all of them are more than 5 feet tall. Is it unusual to randomly select five women and find that all of them are more than 5 feet tall? Why or why not?

TI-83/84 Plus Calculator Project

A TI-83/84 Plus calculator can be used to generate equally likely random digits between 0 and 9 inclusive. This project will test the randomness of such generated values to determine whether the calculator appears to be working correctly. We will use the TI-83/84 Plus calculator to randomly generate 500 digits between 0 and 9 inclusive, then we will test those digits to determine whether they appear to be from a population with a uniform distribution. Use the procedure given below.

1. Press **MATH**, then use the arrow key **▷** to select the top menu item of **PRB.**
2. Select the menu item of **randInt** (for random integer).
3. The format for the randInt command is this: (lower limit, upper limit, number of values). To generate 500 random digits between 0 and 9 inclusive, enter the command of **randInt(0, 9, 500).** (If this command creates a memory error message, delete some or all of the lists that have been created. Press **2ND** **+** to get the **MEM** (memory) menu, then select **Mem Mgmt/Del,** then select **List** and proceed to delete unneeded lists.)
4. Store the results in list L1 by pressing **STO▷** **2ND** **1**.
5. Sort the results. (Press **STAT**, select **SortA,** enter L1, then press **ENTER**).
6. Use the sorted values in list L1 to enter the frequencies in the following table.

Digit	0	1	2	3	4	5	6	7	8	9
Frequency										

7. Use a 0.05 significance level with the sample data from Step 6 to test the claim that the sample digits come from a population with a uniform distribution (so that all digits are equally likely). Does the calculator's random number generator appear to be working as it should?

INTERNET PROJECT

Contingency Tables

Go to: **http://www.aw.com/triola**

An important characteristic of tests of independence with contingency tables is that the data collected need not be quantitative in nature. A contingency table summarizes observations by the categories or labels of the rows and columns. As a result, characteristics such as gender, race, and political party all become fair game for formal hypothesis testing procedures. In the Internet Project for this chapter you will find links to a variety of demographic data. With these data sets, you will conduct tests in areas as diverse as academics, politics, and the entertainment industry. In each test, you will draw conclusions related to the independence of interesting characteristics.

APPLET PROJECT

Open the Applets folder on the CD and double-click on **Start.** Select the menu item of **Random numbers.** Randomly generate 100 whole numbers between 0 and 9 inclusive. Construct a frequency distribution of the results, then use the methods of this chapter to test the claim that the whole numbers between 0 and 9 are equally likely.

FROM DATA TO DECISION

Critical Thinking: Was the law of "women and children first" followed in the sinking of the *Titanic?*

One of the most notable sea disasters occurred with the sinking of the *Titanic* on Monday, April 15, 1912. The table below summarizes the fate of the passengers and crew. A common rule of the sea is that when a ship is threatened with sinking, women and children are the first to be saved.

Fate of Passengers and Crew on the *Titanic*

	Men	Women	Boys	Girls
Survived	332	318	29	27
Died	1360	104	35	18

Analyzing the Results

If we examine the data, we see that 19.6% of the men (332 out of 1692) survived, 75.4% of the women (318 out of 422) survived, 45.3% of the boys (29 out of 64) survived, and 60% of the girls (27 out of 45) survived. There do appear to be differences, but are the differences really *significant?*

First construct a bar graph showing the percentage of survivors in each of the four categories (men, women, boys, girls). What does the graph suggest?

Next, treat the 2223 people aboard the *Titanic* as a *sample*. We could take the position that the *Titanic* data in the above table constitute a *population* and therefore should not be treated as a sample, so that methods of inferential statistics do not apply. But let's stipulate that the data in the table are sample data randomly selected from the population of all theoretical people who would find themselves in the same conditions. Realistically, no other people will actually find themselves in the same conditions, but we will make that assumption for the purposes of this discussion and analysis. We can then determine whether the observed differences have statistical significance. Use one or more formal hypothesis tests to investigate the claim that although some men survived while some women and children died, the rule of "women and children first" was essentially followed. Identify the hypothesis test(s) used and interpret the results by addressing the claim that when the *Titanic* sank on its maiden voyage, the rule of "women and children first" was essentially followed.

Cooperative Group Activities

1. Out-of-class activity Divide into groups of four or five students. The instructions for Exercises 21–24 in Section 11-2 noted that according to Benford's law, a variety of different data sets include numbers with leading (first) digits that follow the distribution shown in the table below. Collect original data and use the methods of Section 11-2 to support or refute the claim that the data conform reasonably well to Benford's law. Here are some possibilities that might be considered: (1) amounts on the checks that you wrote; (2) prices of stocks; (3) populations of counties in the United States; (4) numbers on street addresses; (5) lengths of rivers in the world.

Leading Digit	1	2	3	4	5	6	7	8	9
Benford's law:	30.1%	17.6%	12.5%	9.7%	7.9%	6.7%	5.8%	5.1%	4.6%

2. Out-of-class activity Divide into groups of four or five students and collect past results from a state lottery. Such results are often available on Web sites for individual state lotteries. Use the methods of Section 11-2 to test that the numbers are selected in such a way that all possible outcomes are equally likely.

3. Out-of-class activity Divide into groups of four or five students. Each group member should survey at least 15 male students and 15 female students at the same college by asking two questions: (1) Which political party does the subject favor most? (2) If the subject were to make up an absence excuse of a flat tire, which tire would he or she say went flat if the instructor asked? (See Exercise 8 in Section 11-2.) Ask the subject to write the two responses on an index card, and also record the gender of the subject and whether the subject wrote with the right or left hand. Use the methods of this chapter to analyze the data collected. Include these tests:

• The four possible choices for a flat tire are selected with equal frequency.

• The tire identified as being flat is independent of the gender of the subject.

• Political party choice is independent of the gender of the subject.

• Political party choice is independent of whether the subject is right- or left-handed.

• The tire identified as being flat is independent of whether the subject is right- or left-handed.

• Gender is independent of whether the subject is right- or left-handed.

• Political party choice is independent of the tire identified as being flat.

4. Out-of-class activity Divide into groups of four or five students. Each group member should select about 15 other students and first ask them to "randomly" select four digits each. After the four digits have been recorded, ask each subject to write the last four digits of his or her social security number. Take the "random" sample results and mix them into one big sample, then mix the social security digits into a second big sample. Using the "random" sample set, test the claim that students select digits randomly. Then use the social security digits to test the claim that they come from a population of random digits. Compare the results. Does it appear that students can randomly select digits? Are they likely to select any digits more often than others? Are they likely to select any digits less often than others? Do the last digits of social security numbers appear to be randomly selected?

5. In-class activity Divide into groups of three or four students. Each group should be given a die along with the instruction that it should be tested for "fairness." Is the die fair or is it biased? Describe the analysis and results.

6. Out-of-class activity Divide into groups of two or three students. The analysis of last digits of data can sometimes reveal whether values are the results of actual measurements or whether they are reported estimates. Refer to an almanac and find the lengths of rivers in the world, then analyze the last digits to determine whether those lengths appear to be actual measurements or whether they appear to be reported estimates. (Instead of lengths of rivers, you could use heights of mountains, heights of the tallest buildings, lengths of bridges, and so on.)

NAME:	Jackie Macmullan
JOB:	Associate Editor, Sports Columnist
COMPANY:	*Boston Globe*

A previous edition of this book included an interview with Bill James, a recognized baseball expert who specializes in the analysis of baseball statistics and identifying the best game strategies based on past results. The Boston Red Sox hired Bill James as an advisor, and he is credited with some of the changes that led to the first World Series victory by the Red Sox in 86 years. Another Bostonian who makes extensive use of statistics in the world of sports is Jackie Macmullan, who is Associate Editor and sports columnist for the Boston Globe.

Q: How do you use statistics when writing about sports?

A: Sports is all about statistics, really. Especially baseball. Yesterday I wrote about why C.C. Sabathia won the Cy Young award over Josh Beckett. Beckett had more wins but lost to Sabathia because of "deeper numbers," which are the underlying statistics. The big one in this case was quality starts—a pretty good indicator of how proficient you were on the mound (Sabathia's 25 vs. Beckett's 20). In many of my stories on sports, and baseball in particular, the use of statistics gives readers a deeper understanding of the game and its players.

Q: Please describe a specific example illustrating the use of statistics.

A: For many years, as I covered the NBA as a beat writer for the Boston Celtics, I kept a binder full of information on every aspect of the game. After each game, I would record everything—points

per game, turnovers, assists, etc., for each player on the team. I would also keep track of the opposing team and what they did—any player who scored 10 or more points, for example. I would break down all of the action by quarters, noting where the Celtics turned the game around. These meticulous records kept me on top of every player and their tendencies. I relied on that book for a tremendous amount of obscure little things, too, which I think are what makes a strong sportswriter.

Q: Is your use of probability and statistics increasing, decreasing, or remaining stable?

A: Now, everyone is so sophisticated in their use of statistics. Every college and professional team has their own statistical team that records all their data as they play. Every single team in every single sport has its own scouting system that breaks every statistic down. Here's an example: There was a company hired by the Chicago Bulls that would break down everything Michael Jordan did. Things like, "when Jordan gets ball on left block, shoots 36% of time, on right, 31% of time" so you'd know where he'd prefer to shoot from and his rate of success in every location. Data told you players may like to do a certain thing more, but weren't necessarily successful at it. This really influences other teams and strategies, because they know everything these players do and how good they are at doing it.

Q: How critical do you find your knowledge of statistics for performing your responsibilities?

A: Using statistics really enhances my writing — people don't keep their own stats now, but if you do keep them yourself it takes time, effort, and concentration, and the data become more valuable to you. Collecting data like this provides information that you didn't necessarily know you were looking for. For example, from my NBA binder, I would often find that three quarters of the way through the season, I had learned about player tendencies that I didn't see on my own, or even realize I was recording.

Q: In terms of statistics, what would you recommend for prospective employees?

A: An introductory course would be fine. You really have to know how to crunch numbers to do my job effectively. Some of the best articles are written by people who make these unique statistical observations.

12-1 Review and Preview

12-2 One-Way ANOVA

12-3 Two-Way ANOVA

12 Analysis of Variance

Do crash test dummies tell us that larger cars are safer?

The safety of cars is determined from a variety of tests. In one test, cars are crashed into a fixed barrier at 35 mi/h with a crash test dummy in the driver's seat. Unlike the many dummies we have all seen in driver's seats, these dummies are not human, and they are designed to record the damage resulting from the frontal crash. Table 12-1 lists chest deceleration measurements (in g, where g is a force of gravity). Larger values indicate greater amounts of deceleration, which are likely to result in greater injuries to drivers.

The data in Table 12-1 are current as of this writing. The small cars are the Honda Civic, Ford Focus, Chevrolet Aveo, Volkswagon Jetta, Toyota Corolla, Kia Spectra, Saturn Ion, Mazda 3, Subaru Imprezza, and Suzuki Forenza. The medium cars are the Honda Accord, Volkswagon Passat, Pontiac Grand Prix, Toyota Camry, Volvo S40, Nissan Altima, Chevrolet Malibu, Ford Fusion, Saturn Aura, and Chrysler Sebring. The large cars are the Toyota Avalon, Dodge Charger, Ford Five Hundred, Hyundai Azera, Chrysler 300, Buick Lucerne, Mercury Grand Marquis, Cadillac STS, Lincoln MKZ, and Saab 9-5.

The TI-83/84 Plus calculator boxplots are shown for small cars (top boxplot), medium cars (middle boxplot), and large cars (bottom boxplot). Because of their relative positions, the boxplots suggest that larger cars are safer because they have lower chest deceleration measurements. But are those visual differences *significant*?

Table 12-1 shows that the small cars have the largest mean chest deceleration measurement of 44.7 g, and the large cars have the smallest mean chest deceleration measurement of 39.0 g. The values of the sample means suggest that larger cars are safer than smaller cars. But are those numerical differences *significant*?

Here we want to compare the means from three independent samples. In this chapter we present a common method for comparing three or more sample means. We will determine whether the differences among the means from the data in Table 12-1 are significant. We will then determine whether large cars are safer in the sense that they have lower chest deceleration measurements.

Table 12-1 Chest Deceleration Measurements (in g) from Car Crash Tests

Small Cars	44	43	44	54	38	43	42	45	44	50	→ $\bar{x}$ = 44.7 g
Medium Cars	41	49	43	41	47	42	37	43	44	34	→ $\bar{x}$ = 42.1 g
Large Cars	32	37	38	45	37	33	38	45	43	42	→ $\bar{x}$ = 39.0 g

TI-83/84 Plus

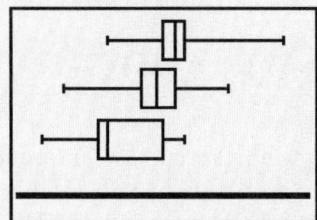

Review and Preview

In Section 9-3 we presented methods for comparing the means from two independent samples. In this chapter we will learn how to test for equality of three or more population means by using the method of one-way analysis of variance. The term *one-way* is used because the sample data are separated into groups according to *one* characteristic. Instead of referring to the main objective of testing for equal means, the term *analysis of variance* refers to the *method* we use, which is based on an analysis of sample variances.

We will also learn how to compare populations separated into categories using *two* characteristics (or factors), such as gender and eye color. Because the sample data are categorized according to two different factors, the method is referred to as *two-way* analysis of variance.

F Distribution

The analysis of variance (ANOVA) methods of this chapter require the *F* distribution, which was first introduced in Section 9-5. In Section 9-5 we noted that the *F* distribution has the following properties (see Figure 12-1):

1. The *F* distribution is not symmetric.

2. Values of the *F* distribution cannot be negative.

3. The exact shape of the *F* distribution depends on the two different degrees of freedom.

Critical *F* values are given in Table A-5.

Figure 12-1

_F_ Distribution

There is a different *F* distribution for each different pair of degrees of freedom for numerator and denominator.

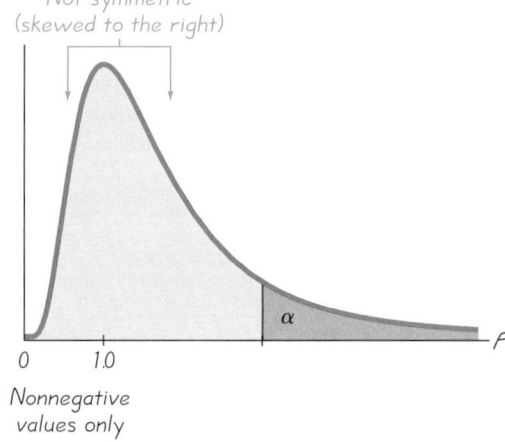

Not symmetric (skewed to the right)

α

F

0   1.0

Nonnegative values only

 One-Way ANOVA

Key Concept In this section we introduce the method of *one-way analysis of variance*, which is used for tests of hypotheses that three or more population means are all equal, as in $H_0: \mu_1 = \mu_2 = \mu_3$. Because the calculations are very complicated, we emphasize the interpretation of results obtained by using software or a TI-83/84 Plus calculator. Here is a recommended study strategy:

1. Understand that a small *P*-value (such as 0.05 or less) leads to rejection of the null hypothesis of equal means. ("If the *P* (value) is low, the null must go.") With a large *P*-value (such as greater than 0.05), fail to reject the null hypothesis of equal means.

2. Develop an understanding of the underlying rationale by studying the examples in this section.

Part 1: Basics of One-Way Analysis of Variance

When testing for equality of three or more population means, use the method of one-way analysis of variance.

DEFINITION

One-way analysis of variance (ANOVA) is a method of testing the equality of three or more population means by analyzing sample variances. One-way analysis of variance is used with data categorized with *one* **treatment** (or **factor**), which is a characteristic that allows us to distinguish the different populations from one another.

The term *treatment* is used because early applications of analysis of variance involved agricultural experiments in which different plots of farmland were treated with different fertilizers, seed types, insecticides, and so on. Table 12-1 uses the one "treatment" (or factor) of size of a car. That factor has three different categories: small, medium, and large.

Using One-Way Analysis of Variance for Testing Equality of Three or More Population Means

Objective

Test a claim that three or more populations have the same mean.

Requirements

1. The populations have distributions that are approximately normal. (This is a loose requirement, because the method works well unless a population has a distribution that is very far from normal. If a population does have a distribution that is far from normal, use the Kruskal-Wallis test described in Section 13-5.)

2. The populations have the same variance σ^2 (or standard deviation σ). (This is a loose requirement, because the method works well unless the population variances differ by large amounts. Statistician George E. P. Box showed that as long as the sample sizes are equal (or nearly equal), the variances can differ by amounts that make the largest up to nine times the smallest and the results of ANOVA will continue to be essentially reliable.)

3. The samples are simple random samples of quantitative data.

4. The samples are independent of each other. (The samples are not matched or paired in any way.)

5. The different samples are from populations that are categorized in only one way.

Procedure for Testing H_0: $\mu_1 = \mu_2 = \mu_3 = \cdots$

1. Use a TI-83/84 Plus calculator, STATDISK, Minitab, or Excel, to obtain results.

2. Identify the *P*-value from the display.

3. Form a conclusion based on these criteria:

   • If the *P*-value $\leq \alpha$, reject the null hypothesis of equal means and conclude that at least one of the population means is different from the others.

   • If the *P*-value $> \alpha$, fail to reject the null hypothesis of equal means.

EXAMPLE 1   **Car Crash Test Measurements** Use the chest deceleration measurements listed in Table 12-1 and a significance level of $\alpha = 0.05$ to test the claim that the three samples come from populations with means that are all equal.

SOLUTION   **REQUIREMENT CHECK** (1) Based on the three samples listed in Table 12-1, the three populations appear to have distributions that are approximately normal, as indicated by normal quantile plots. (2) The three samples in Table 12-1 have standard deviations of 4.4 g, 4.4 g, and 4.6 g, so the three population variances appear to be about the same. (3) The samples are simple random samples of cars selected by the author. (4) The samples are independent of each other; the cars are not matched in any way. (5) The three samples are from populations categorized according to the single factor of size (small, medium, large). The requirements are satisfied. ✓

The null hypothesis and the alternative hypothesis are as follows:

$H_0: \mu_1 = \mu_2 = \mu_3.$

H_1: At least one of the means is different from the others.

The significance level is $\alpha = 0.05$.

Step 1: Use technology to obtain ANOVA results, such as one of those shown.

TI-83/84 PLUS

```
One-way ANOVA
 F=4.094413408
 P=.0279863448
 Factor
  df=2
  SS=162.866667
↓ MS=81.4333333
```

```
One-way ANOVA
↑ MS=81.4333333
 Error
  df=27
  SS=537
  MS=19.8888889
 Sxp=4.45969605
```

STATDISK

Source:	DF:	SS:	MS:	Test Stat, F:	Critical F:	P-Value:
Treatment:	2	162.866667	81.433333	4.094413	3.354128	0.027986
Error:	27	537.00	19.888889			
Total:	29	699.866667				

Reject the Null Hypothesis
Reject equality of means

MINITAB

One-way ANOVA: Small, Medium, Large

Source	DF	SS	MS	F	P
Factor	2	162.9	81.4	4.09	0.028
Error	27	537.0	19.9		
Total	29	699.9			

S = 4.460   R-Sq = 23.27%   R-Sq(adj) = 17.59%

EXCEL

ANOVA						
Source of Variation	SS	df	MS	F	P-value	F crit
Between Groups	162.8667	2	81.43333	4.094413	0.027986	3.354131
Within Groups	537	27	19.88889			
Total	699.8667	29				

Step 2: The displays all show that the P-value is 0.028 when rounded.

Step 3: Because the P-value of 0.028 is less than the significance level of $\alpha = 0.05$, we reject the null hypothesis of equal means.

INTERPRETATION There is sufficient evidence to warrant rejection of the claim that the three samples come from populations with means that are all equal. Based on the samples of measurements listed in Table 12-1, we conclude that those values come from populations having means that are not all the same. On the basis of this ANOVA test, we cannot conclude that any particular mean is different from the others, but we can informally note that the sample mean is smallest for the large cars. Because small measurements correspond to less trauma experienced by the crash test dummies, it appears that the large cars are safest, but this conclusion is not formally justified by this ANOVA test.

CAUTION

When we conclude that there is sufficient evidence to reject the claim of equal population means, we cannot conclude from ANOVA that any particular mean is different from the others. (There are several other methods that can be used to identify the specific means that are different, and some of them are discussed in Part 2 of this section.)

How is the *P*-Value Related to the Test Statistic? *Larger* values of the test statistic result in *smaller P*-values, so the ANOVA test is right-tailed. Figure 12-2 shows the relationship between the *F* test statistic and the *P*-value. Assuming that the populations have the same variance σ^2 (as required for the test), the *F* test statistic is the ratio of these two estimates of σ^2: (1) variation *between* samples (based on variation among sample means); and (2) variation *within* samples (based on the sample variances).

Test Statistic for One-Way ANOVA: $F = \dfrac{\text{variance between samples}}{\text{variance within samples}}$

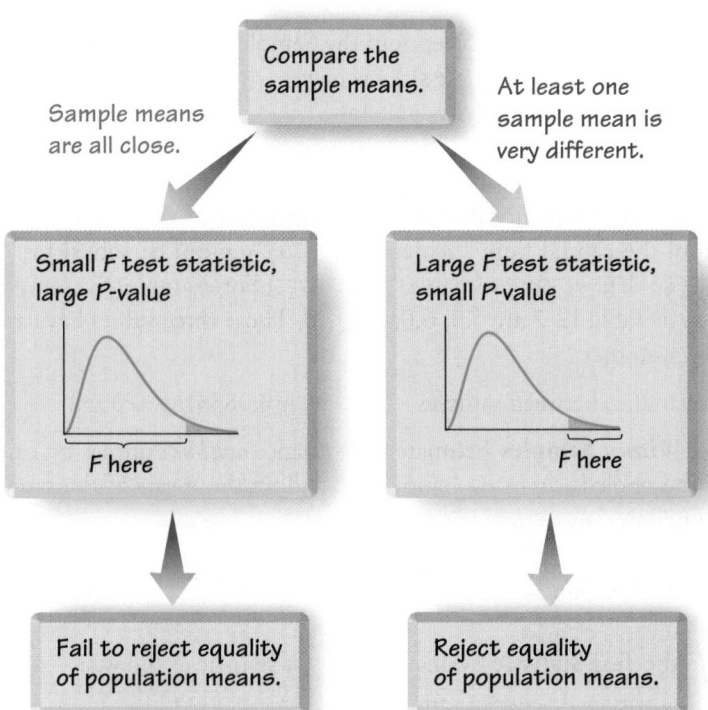

Figure 12-2

Relationship Between the *F* Test Statistic and the *P*-Value

The numerator of the F test statistic measures variation between sample means. The estimate of variance in the denominator depends only on the sample variances and is not affected by differences among the sample means. Consequently, sample means that are close in value result in a small F test statistic and a large P-value, so we conclude that there are no significant differences among the sample means. Sample means that are very far apart in value result in a large F test statistic and a small P-value, so we reject the claim of equal means.

Why Can't We Just Test Two Samples at a Time? If we want to test for equality among three or more population means, why do we need a new procedure when we can test for equality of two means using the methods presented in Section 9-3? For example, if we want to use the sample data from Table 12-1 to test the claim that the three populations have the same mean, why not simply pair them off and test two at a time by testing $H_0: \mu_1 = \mu_2$, $H_0: \mu_2 = \mu_3$, and $H_0: \mu_1 = \mu_3$? For the data in Table 12-1, the approach of testing equality of two means at a time requires three different hypothesis tests. If we use a 0.05 significance level for each of those three hypothesis tests, the actual overall confidence level could be as low as 0.95^3 (or 0.857). In general, as we increase the number of individual tests of significance, we increase the risk of finding a difference by chance alone (instead of a real difference in the means). The risk of a type I error—finding a difference in one of the pairs when no such difference actually exists—is far too high. The method of analysis of variance helps us avoid that particular pitfall (rejecting a true null hypothesis) by using *one test* for equality of several means, instead of several tests that each compare two means at a time.

> **CAUTION**
> ....
> When testing for equality of three or more populations, use analysis of variance. Do *not* use multiple hypothesis tests with two samples at a time.

Part 2: Calculations and Identifying Means That Are Different
Calculations with Equal Sample Sizes *n*

Let's consider Table 12-2. Compare Data Set A to Data Set B. Note that Data Set A is the same as Data Set B with this exception: the Sample 1 values each differ by 10. If the data sets all have the same sample size (as in $n = 4$ for Table 12-2), the following calculations aren't too difficult, as shown below.

Variance *Between* Samples Find the variance *between* samples by evaluating $ns_{\bar{x}}^2$, where $s_{\bar{x}}^2$ is the variance of the sample means and n is the size of each of the samples. That is, consider the sample means to be an ordinary set of values and calculate the variance. (From the central limit theorem, $\sigma_{\bar{x}} = \sigma/\sqrt{n}$ can be solved for σ to get $\sigma = \sqrt{n} \cdot \sigma_{\bar{x}}$, so that we can estimate σ^2 with $ns_{\bar{x}}^2$.) For example, the sample means for Data Set A in Table 12-2 are 5.5, 6.0, and 6.0. Those three values have a variance of $s_{\bar{x}}^2 = 0.0833$, so that

$$\text{variance between samples} = ns_{\bar{x}}^2 = 4(0.0833) = 0.3332$$

Variance *Within* Samples Estimate the variance *within* samples by calculating s_p^2, which is the pooled variance obtained by finding the mean of the sample variances. The sample variances in Table 12-2 are 3.0, 2.0, and 2.0, so that

$$\text{variance within samples} = s_p^2 = \frac{3.0 + 2.0 + 2.0}{3} = 2.3333$$

Calculate the Test Statistic Evaluate the F test statistic as follows:

$$F = \frac{\text{variance between samples}}{\text{variance within samples}} = \frac{ns_{\bar{x}}^2}{s_p^2} = \frac{0.3332}{2.3333} = 0.1428$$

Table 12-2 **Effect of a Mean on the *F* Test Statistic**

	A add 10			**B**		
Sample 1	Sample 2	Sample 3		Sample 1	Sample 2	Sample 3
7	6	4		17	6	4
3	5	7		13	5	7
6	5	6		16	5	6
6	8	7		16	8	7
$n_1 = 4$	$n_2 = 4$	$n_3 = 4$		$n_1 = 4$	$n_2 = 4$	$n_3 = 4$
$\bar{x}_1 = 5.5$	$\bar{x}_2 = 6.0$	$\bar{x}_3 = 6.0$		$\bar{x}_1 = 15.5$	$\bar{x}_2 = 6.0$	$\bar{x}_3 = 6.0$
$s_1^2 = 3.0$	$s_2^2 = 2.0$	$s_3^2 = 2.0$		$s_1^2 = 3.0$	$s_2^2 = 2.0$	$s_3^2 = 2.0$

Variance between samples	$ns_{\bar{x}}^2 = 4(0.0833) = 0.3332$	$ns_{\bar{x}}^2 = 4(30.0833) = 120.3332$
Variance within samples	$s_p^2 = \dfrac{3.0 + 2.0 + 2.0}{3} = 2.3333$	$s_p^2 = \dfrac{3.0 + 2.0 + 2.0}{3} = 2.3333$
F test statistic	$F = \dfrac{ns_{\bar{x}}^2}{s_p^2} = \dfrac{0.3332}{2.3333} = \mathbf{0.1428}$	$F = \dfrac{ns_{\bar{x}}^2}{s_p^2} = \dfrac{120.3332}{2.3333} = \mathbf{51.5721}$
P-value	*P*-value = 0.8688	*P*-value = 0.0000118

The critical value of *F* is found by assuming a right-tailed test, because large values of *F* correspond to significant differences among means. With *k* samples each having *n* values, the numbers of degrees of freedom are as follows.

Degrees of Freedom: (*k* = number of samples and *n* = sample size)

$$\text{numerator degrees of freedom} = k - 1$$
$$\text{denominator degrees of freedom} = k(n - 1)$$

For Data Set A in Table 12-2, $k = 3$ and $n = 4$, so the degrees of freedom are 2 for the numerator and $3(4 - 1) = 9$ for the denominator. With $\alpha = 0.05$, 2 degrees of freedom for the numerator, and 9 degrees of freedom for the denominator, the critical *F* value from Table A-5 is 4.2565. If we were to use the traditional method of hypothesis testing with Data Set A in Table 12-2, we would see that this right-tailed test has a test statistic of $F = 0.1428$ and a critical value of $F = 4.2565$, so the test statistic is not in the critical region. We therefore fail to reject the null hypothesis of equal means.

To really see how the method of analysis of variance works, consider both collections of sample data in Table 12-2. Note that the three samples in Data Set A are identical to the three samples in Data Set B, except that each value in Sample 1 of Data Set B is 10 more than the corresponding value in Data Set A. The three sample means in A are very close, but there are substantial differences in B. However, the three sample variances in A are identical to those in B.

Adding 10 to each data value in the first sample of Table 12-2 has a dramatic effect on the test statistic, with *F* changing from 0.1428 to 51.5721. Adding 10 to each data value in the first sample also has a dramatic effect on the *P*-value, which changes from 0.8688 (not significant) to 0.0000118 (significant). Note that the variance between samples in A is 0.3332, but for B it is 120.3332 (indicating that the sample means in B are farther apart). Note also that the variance within samples is 2.3333 in both parts, because the variance *within* a sample isn't affected when we add a constant to every sample value. *The change in the F test statistic and the P-value is attributable only to the change in $\bar{x}_1$.* This illustrates the key point underlying the method of one-way analysis of variance: **The *F* test statistic is very sensitive to sample *means*, even though it is obtained through two different estimates of the common population *variance.*

Adding 10 to each value of the first sample causes the three sample means to grow farther apart, with the result that the *F* test statistic increases and the *P*-value decreases.

Calculations with Unequal Sample Sizes

While the calculations required for cases with equal sample sizes are reasonable, they become more complicated when the sample sizes are not all the same. The same basic reasoning applies because we calculate an *F* test statistic that is the ratio of two different estimates of the common population variance σ^2, but those estimates involve *weighted* measures that take the sample sizes into account, as shown below.

$$F = \frac{\text{variance between samples}}{\text{variance within samples}} = \frac{\left[\dfrac{\sum n_i (\bar{x}_i - \bar{\bar{x}})^2}{k - 1}\right]}{\left[\dfrac{\sum (n_i - 1)s_i^2}{\sum (n_i - 1)}\right]}$$

where

$\bar{\bar{x}}$ = mean of all sample values combined

k = number of population means being compared

n_i = number of values in the *i*th sample

$\bar{x}_i$ = mean of values in the *i*th sample

s_i^2 = variance of values in the *i*th sample

The factor of n_i is included so that larger samples carry more weight. The denominator of the test statistic is simply the mean of the sample variances, but it is a weighted mean based on the sample sizes.

Because calculating this test statistic can lead to large rounding errors, the various software packages typically use a different (but equivalent) expression that involves SS (for sum of squares) and MS (for mean square) notation. Although the following notation and components are complicated and involved, the basic idea is the same: The test statistic *F* is a ratio with a numerator reflecting variation *between* the means of the samples and a denominator reflecting variation *within* the samples. If the populations have equal means, the *F* ratio tends to be small, but if the population means are not equal, the *F* ratio tends to be significantly large. Key components in our ANOVA method are described as follows.

SS(total), or total sum of squares, is a measure of the total variation (around $\bar{\bar{x}}$) in all of the sample data combined.

Formula 12-1

$$SS(\text{total}) = \sum (x - \bar{\bar{x}})^2$$

SS(total) can be broken down into the components of SS(treatment) and SS(error), described as follows.

SS(treatment), also referred to as SS(factor), SS(between groups), or SS(between samples), is a measure of the variation *between* the sample means.

> **Formula 12-2**
>
> $$\text{SS(treatment)} = n_1(\bar{x}_1 - \bar{\bar{x}})^2 + n_2(\bar{x}_2 - \bar{\bar{x}})^2 + \cdots + n_k(\bar{x}_k - \bar{\bar{x}})^2$$
> $$= \Sigma n_i(\bar{x}_i - \bar{\bar{x}})^2$$

If the population means $(\mu_1, \mu_2, \ldots, \mu_k)$ are equal, then the sample means $\bar{x}_1, \bar{x}_2, \ldots, \bar{x}_k$ will all tend to be close together and also close to $\bar{\bar{x}}$. The result will be a relatively small value of SS(treatment). If the population means are not all equal, however, then at least one of $\bar{x}_1, \bar{x}_2, \ldots, \bar{x}_k$ will tend to be far apart from the others and also far apart from $\bar{\bar{x}}$. The result will be a relatively large value of SS(treatment).

SS(error), also referred to as SS(within groups) or SS(within samples), is a sum of squares representing the variation that is assumed to be common to all the populations being considered.

> **Formula 12-3**
>
> $$\text{SS(error)} = (n_1 - 1)s_1^2 + (n_2 - 1)s_2^2 + \cdots + (n_k - 1)s_k^2$$
> $$= \Sigma(n_i - 1)s_i^2$$

Given the preceding expressions for SS(total), SS(treatment), and SS(error), the following relationship will always hold.

> **Formula 12-4**
>
> $$\text{SS(total)} = \text{SS(treatment)} + \text{SS(error)}$$

SS(treatment) and SS(error) are both sums of squares, and if we divide each by its corresponding number of degrees of freedom, we get *mean* squares. Some of the following expressions for mean squares include the notation N:

$$N = \textbf{total number of values in all samples combined}$$

MS(treatment) is a mean square for treatment, obtained as follows:

> **Formula 12-5**
>
> $$\text{MS(treatment)} = \frac{\text{SS(treatment)}}{k - 1}$$

MS(error) is a mean square for error, obtained as follows:

> **Formula 12-6**
>
> $$\text{MS(error)} = \frac{\text{SS(error)}}{N - k}$$

MS(total) is a mean square for the total variation, obtained as follows:

Formula 12-7

$$MS(total) = \frac{SS(total)}{N - 1}$$

Test Statistic for ANOVA with Unequal Sample Sizes

In testing the null hypothesis H_0: $\mu_1 = \mu_2 = \cdots = \mu_k$ against the alternative hypothesis H_1: The means are not all equal, the test statistic

Formula 12-8

$$F = \frac{MS(treatment)}{MS(error)}$$

has an F distribution (when the null hypothesis H_0 is true) with degrees of freedom given by

$$\text{numerator degrees of freedom} = k - 1$$
$$\text{denominator degrees of freedom} = N - k$$

This test statistic is essentially the same as the one given earlier, and its interpretation is also the same as described earlier. The denominator depends only on the sample variances that measure variation within the treatments and is not affected by the differences among the sample means. In contrast, the numerator is affected by differences among the sample means. If the differences among the sample means are excessively large, they will cause the numerator to be excessively large, so F will also be excessively large. Consequently, very large values of F suggest unequal means, and the ANOVA test is therefore right-tailed.

Designing the Experiment   With one-way (or single-factor) analysis of variance, we use one factor as the basis for partitioning the data into different categories. If we conclude that the differences among the means are significant, we can't be absolutely sure that the differences can be explained by the factor being used. It is possible that the variation of some other unknown factor is responsible. One way to reduce the effect of the extraneous factors is to design the experiment so that it has a **completely randomized design,** in which each element is given the same chance of belonging to the different categories, or treatments. For example, you might assign subjects to two different treatment groups and a placebo group through a process of random selection equivalent to picking slips of paper from a bowl. Another way to reduce the effect of extraneous factors is to use a **rigorously controlled design,** in which elements are carefully chosen so that all other factors have no variability. In general, good results require that the experiment be carefully designed and executed.

Identifying Which Means Are Different

After conducting an analysis of variance test, we might conclude that there is sufficient evidence to reject a claim of equal population means, but we cannot conclude from ANOVA that any *particular* means are different from the others. There are several formal and informal procedures that can be used to identify the specific means that are different. Here are two *informal* methods for comparing means:

- Construct boxplots of the data sets to see if one or more of the data sets is very different from the others.

- Construct confidence interval estimates of the means from the data sets, then compare those confidence intervals to see if one or more of them does not overlap with the others.

There are several formal procedures for identifying which means are different. Some of the tests, called **range tests,** allow us to identify subsets of means that are not significantly different from each other. Other tests, called **multiple comparison tests,** use pairs of means, but they make adjustments to overcome the problem of having a significance level that increases as the number of individual tests increases. There is no consensus on which test is best, but some of the more common tests are the Duncan test, Student-Newman-Keuls test (or SNK test), Tukey test (or Tukey honestly significant difference test), Scheffé test, Dunnett test, least significant difference test, and the Bonferroni test. Let's consider the Bonferroni test to see an example of a multiple comparison test. Here is the procedure:

Bonferroni Multiple Comparison Test

Step 1. Do a separate t test for each pair of samples, but make the adjustments described in the following steps.

Step 2. For an estimate of the variance σ^2 that is common to all of the involved populations, use the value of MS(error), which uses all of the available sample data. The value of MS(error) is typically obtained when conducting the analysis of variance test. Using the value of MS(error), calculate the value of the test statistic t, as shown below. The particular test statistic calculated below is based on the choice of Sample 1 and Sample 2; change the subscripts and use another pair of samples until all of the different possible pairs of samples have been tested.

$$t = \frac{\bar{x}_1 - \bar{x}_2}{\sqrt{MS(error) \cdot \left(\dfrac{1}{n_1} + \dfrac{1}{n_2} \right)}}$$

Step 3. After calculating the value of the test statistic t for a particular pair of samples, find either the critical t value or the P-value, but make the following adjustment so that the overall significance level does not increase.

   P-value: Use the test statistic t with df $= N - k$, where N is the total number of sample values and k is the number of samples, and find the P-value the usual way, but adjust the P-value by multiplying it by the number of different possible pairings of two samples. (For example, with three samples, there are three different possible pairings, so adjust the P-value by multiplying it by 3.)

   Critical value: When finding the critical value, adjust the significance level α by dividing it by the number of different possible pairings of two samples. (For example, with three samples, there are three different possible pairings, so adjust the significance level by dividing it by 3.)

Note that in Step 3 of the preceding Bonferroni procedure, either an individual test is conducted with a much lower significance level, or the P-value is greatly increased. Rejection of equality of means therefore requires differences that are much farther apart. This adjustment in Step 3 compensates for the fact that we are doing several tests instead of only one test.

EXAMPLE 2　**Bonferroni Test** Example 1 in this section used analysis of variance with the sample data in Table 12-1. We concluded that there is sufficient evidence to warrant rejection of the claim of equal means. Use the Bonferroni test with a 0.05 significance level to identify which mean is different from the others.

SOLUTION　The Bonferroni test requires a separate t test for each different possible pair of samples. Here are the null hypotheses to be tested:

$$H_0: \mu_1 = \mu_2 \qquad H_0: \mu_1 = \mu_3 \qquad H_0: \mu_2 = \mu_3$$

We begin with $H_0: \mu_1 = \mu_2$. From Table 12-1 we see that $\bar{x}_1 = 44.7$, $n_1 = 10$, $\bar{x}_2 = 42.1$, and $n_2 = 10$. From the technology results shown in Example 1 we also know that MS(error) = 19.888889. We can now evaluate the test statistic:

$$t = \frac{\bar{x}_1 - \bar{x}_2}{\sqrt{MS(error) \cdot \left(\frac{1}{n_1} + \frac{1}{n_2}\right)}}$$

$$= \frac{44.7 - 42.1}{\sqrt{19.888889 \cdot \left(\frac{1}{10} + \frac{1}{10}\right)}} = 1.303626224$$

The number of degrees of freedom is df = $N - k = 30 - 3 = 27$. With a test statistic of $t = 1.303626244$ and with df = 27, the two-tailed P-value is 0.203368, but we adjust this P-value by multiplying it by 3 (the number of different possible pairs of samples) to get a final P-value of 0.610. Because this P-value is not small (less than 0.05), we fail to reject the null hypothesis. It appears that Samples 1 and 2 do not have significantly different means.

Instead of continuing with separate hypothesis tests for the other two pairings, see the SPSS display showing all of the Bonferroni test results. (The first row of numerical results corresponds to the results found here; see the value of 0.610, which is calculated here.) The display shows that the mean for Sample 1 (small cars) is significantly different from the mean for Sample 3 (large cars). Based on the Bonferroni test, it appears that the measurements from small cars have a mean that is significantly different from the mean for large cars.

SPSS BONFERRONI RESULTS

Dependent Variable: Chest
Tukey HSD

(I) VAR00002	(J) VAR00002	Mean Difference (I-J)	Std. Error	Sig.	95% Confidence Interval	
					Lower Bound	Upper Bound
1	2	2.600	1.994	.405	-2.35	7.55
	3	5.700*	1.994	.021	.75	10.65
2	1	-2.600	1.994	.405	-7.55	2.35
	3	3.100	1.994	.282	-1.85	8.05
3	1	-5.700*	1.994	.021	-10.65	-.75
	2	-3.100	1.994	.282	-8.05	1.85

*. The mean difference is significant at the .05 level.

Using the TI-83/84 Plus Calculator for One-Way Analysis of Variance

To use the TI-83/84 Plus calculator for one-way analysis of variance, use the following procedure.

1. The sample data must be available as lists L1, L2, L3, ... or as named lists that were created or downloaded, such as those available from the APP named TRIOLA11.

2. Press **STAT**, select the top menu item of **TESTS,** and continue to press ⌄ to select the menu item of **ANOVA.**

3. The format for the ANOVA command is to enter the list labels or names, all separated by commas. See the examples below.

TI-83/84 Plus command	Lists of sample data
ANOVA(L1, L2, L3)	L1, L2, L3
ANOVA(SMKR, ETS, NOETS)	SMKR, ETS, NOETS

4. Press **ENTER** and the results will be displayed.

5. The results do not all fit in a single window, so press the arrow key ⌄ to scroll down for more of the results. However, the top of the first window contains the most important results, which are the F test statistic and the P-value. See the sample results included in Example 1.

12-2 Basic Skills and Concepts

Statistical Literacy and Critical Thinking

1. ANOVA Listed below are skull breadths obtained from skulls of Egyptian males from three different epochs (based on data from *Ancient Races of the Thebaid,* by Thomson and Randall-Maciver). Assume that we plan to use an analysis of variance test with a 0.05 significance level to test the claim that the different epochs have the same mean.

a. In this context, what characteristic of the data indicates that we should use *one-way* analysis of variance?

b. If the objective is to test the claim that the three epochs have the same mean, why is the method referred to as analysis of *variance?*

4000 B.C.	131	138	125	129	132	135	132	134	138
1850 B.C.	129	134	136	137	137	129	136	138	134
150 A.D.	128	138	136	139	141	142	137	145	137

2. Why One Test? Refer to the sample data given in Exercise 1. If we want to test for equality of the three means, why don't we use three separate hypothesis tests for $\mu_1 = \mu_2, \mu_2 = \mu_3$, and $\mu_1 = \mu_3$?

3. Interpreting *P*-Value If we use a 0.05 significance level in analysis of variance with the sample data given in Exercise 1, we get a *P*-value of 0.031. What should we conclude?

4. Which Mean Is Different? Refer to the sample data given in Exercise 1. Given that the three sample means are 132.7, 134.4, and 138.1, can we use analysis of variance to conclude that the mean skull breadth from 150 A.D. is different from the means in 400 B.C. and 1850 B.C.? Why or why not?

In Exercises 5–16, use analysis of variance for the indicated test.

5. Readability Measures Samples of pages were randomly selected from *The Bear and the Dragon* by Tom Clancy, *Harry Potter and the Sorcerer's Stone* by J. K. Rowling, and *War and Peace* by Leo Tolstoy. The Flesch Reading Ease scores were obtained from each page, and the TI-83/84 Plus calculator results from analysis of variance are given here. Use a 0.05 significance level to test the claim that the three books have the same mean Flesch Reading Ease score.

TI-83/84 PLUS

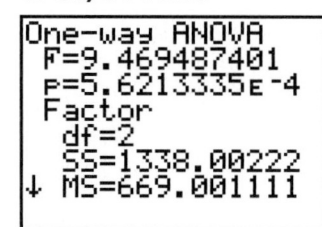

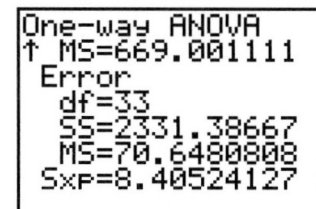

6. Words Per Sentence Samples of pages were randomly selected from the same three books identified in Exercise 5. The mean number of words per sentence was computed for each page, and the analysis of variance results from Minitab are shown below. Using a 0.05 significance level, test the claim that the three books have the same mean number of words per sentence.

MINITAB

Source	DF	SS	MS	F	P
Factor	2	266.5	133.3	3.91	0.030
Error	33	1123.7	34.1		
Total	35	1390.3			

7. Weight Loss from Different Diets A study of the Atkins, Zone, Weight Watchers, and Ornish weight loss programs involved 160 subjects. Each program was followed by 40 subjects. The subjects were weighed before starting the weight loss program and again one year after being on the program. The ANOVA results from Excel are given below (based on data from "Comparison of the Atkins, Ornish, Weight Watchers, and Zone Diets for Weight Loss and Heart Disease Risk Reduction," by Dansinger, et al., *Journal of the American Medical Association,* Vol. 293, No. 1). Use a 0.05 significance level to test the claim that the mean weight loss is the same for the diets. Given that the mean amounts of weight loss after one year are 2.1 lb, 3.2 lb, 3.0 lb, and 3.3 lb for the four diets, do the diets appear to be effective?

EXCEL

ANOVA

Source of Variation	SS	df	MS	F	P-value	F crit
Between Groups	35.99984	3	11.99995	0.35206	0.787709	2.662569
Within Groups	5317.256	156	34.08497			
Total	5353.256	159				

8. Weights of M&Ms Using the weights of M&Ms (in g) from the six different color categories listed in Data Set 18 in Appendix B, the STATDISK results from analysis of variance using a 0.05 significance level are shown below. Identify the test statistic, critical value, and *P*-value. What do you conclude?

STATDISK

Source:	DF:	SS:	MS:	Test Stat, F:	Critical F:	P-Value:
Treatment:	5	0.006115	0.001223	0.443039	2.311272	0.81734
Error:	94	0.259466	0.00276			
Total:	99	0.265581				

9. Movie Gross Amounts If we use the amounts (in millions of dollars) grossed by movies in categories with PG, PG-13, and R ratings, we obtain the SPSS analysis of variance results shown below. The original sample data are listed in Data Set 9 in Appendix B. Use a 0.05 significance level to test the claim that PG movies, PG-13 movies, and R movies have the same mean gross amount.

SPSS

Gross

	Sum of Squares	df	Mean Square	F	Sig.
Between Groups	84801.867	2	42400.933	5.313	.010
Within Groups	255396.1	32	7981.129		
Total	340198.0	34			

10. Voltage Amounts Data Set 13 in Appendix B lists voltage amounts measured from electricity supplied directly to the author's home, an independent Generac generator (model PP 5000), and an uninterruptible power supply (APC model CS 350) connected to the author's home power supply. The results are shown below for analysis of variance obtained using JMP software. Use a 0.05 significance level to test the claim that the three power supplies have the same mean voltage. Can electrical appliances be expected to behave the same way when run from the three different power sources?

JMP

Source	DF	Sum of Squares	Mean Square	F Ratio	Prob > F
Column 2	2	28.816667	14.4083	183.0126	<.0001*
Error	117	9.211250	0.0787		
C. Total	119	38.027917			

11. Head Injury in a Car Crash Listed below are head injury data from crash test dummies used in the same cars from the Chapter Problem. These measurements are in hic, which denotes a standard head injury criterion. Use a 0.05 significance level to test the null hypothesis that the different car categories have the same mean. Do these data suggest that larger cars are safer?

Small Cars	290	406	371	544	374	501	376	499	479	475
Medium Cars	245	502	474	505	393	264	368	510	296	349
Large Cars	342	216	335	698	216	169	608	432	510	332

12. Femur Injury in a Car Crash Listed below are measured loads (in lb) on the left femur of crash test dummies used in the same cars listed in the Chapter Problem. Use a 0.05 significance level to test the null hypothesis that the different car categories have the same mean. Do these data suggest that larger cars are safer?

Small Cars	548	782	1188	707	324	320	634	501	274	437
Medium Cars	194	280	1076	411	617	133	719	656	874	445
Large Cars	215	937	953	1636	937	472	882	562	656	433

13. Triathlon Times Jeff Parent is a statistics instructor who participates in triathlons. Listed below are times (in minutes and seconds) he recorded while riding a bicycle for five laps through each mile of a 3-mile loop. Use a 0.05 significance level to test the claim that it takes the same time to ride each of the miles. Does one of the miles appear to have a hill?

Mile 1	3:15	3:24	3:23	3:22	3:21
Mile 2	3:19	3:22	3:21	3:17	3:19
Mile 3	3:34	3:31	3:29	3:31	3:29

14. Car Emissions Listed below are measured amounts of greenhouse gas emissions from cars in three different categories (from Data Set 16 in Appendix B). The measurements are in tons per year, expressed as CO_2 equivalents. Use a 0.05 significance level to test the claim that the different car categories have the same mean amount of greenhouse gas emissions. Based on the results, does the number of cylinders appear to affect the amount of greenhouse gas emissions?

Four cylinder	7.2	7.9	6.8	7.4	6.5	6.6	6.7	6.5	6.5	7.1	6.7	5.5	7.3
Six cylinder	8.7	7.7	7.7	8.7	8.2	9.0	9.3	7.4	7.0	7.2	7.2	8.2	
Eight cylinder	9.3	9.3	9.3	8.6	8.7	9.3	9.3						

In Exercises 15 and 16, use the data set from Appendix B.

15. Nicotine in Cigarettes Refer to Data Set 4 in Appendix B and use the amounts of nicotine (mg per cigarette) in the king size cigarettes, the 100 mm menthol cigarettes, and the 100 mm nonmenthol cigarettes. The king size cigarettes are nonfiltered, nonmenthol, and non-light. The 100 mm menthol cigarettes are filtered and non-light. The 100 mm nonmenthol cigarettes are filtered and non-light. Use a 0.05 significance level to test the claim that the three categories of cigarettes yield the same mean amount of nicotine. Given that only the king size cigarettes are not filtered, do the filters appear to make a difference?

16. Tar in Cigarettes Refer to Data Set 4 in Appendix B and use the amounts of tar (mg per cigarette) in the three categories of cigarettes described in Exercise 15. Use a 0.05 significance level to test the claim that the three categories of cigarettes yield the same mean amount of tar. Given that only the king size cigarettes are not filtered, do the filters appear to make a difference?

12-2   Beyond the Basics

17. Using the Tukey Test This section included a display of the Bonferroni test results from Table 12-1 included with the Chapter Problem. Shown here is the SPSS-generated display of results from the Tukey test using the same data. Compare the Tukey test results to those from the Bonferroni test.

SPSS

Dependent Variable: Chest
Tukey HSD

(I) VAR00002	(J) VAR00002	Mean Difference (I-J)	Std. Error	Sig.	95% Confidence Interval Lower Bound	95% Confidence Interval Upper Bound
1	2	2.600	1.994	.405	-2.35	7.55
	3	5.700*	1.994	.021	.75	10.65
2	1	-2.600	1.994	.405	-7.55	2.35
	3	3.100	1.994	.282	-1.85	8.05
3	1	-5.700*	1.994	.021	-10.65	-.75
	2	-3.100	1.994	.282	-8.05	1.85

*. The mean difference is significant at the .05 level.

18. Using the Bonferroni Test Shown below are partial results from using the Bonferroni test with the sample data from Exercise 14. Assume that a 0.05 significance level is being used.

a. What do the displayed results tell us?

b. Use the Bonferroni test procedure to test for a significant difference between the mean amount of greenhouse gas emissions from six-cylinder cars and the mean from eight-cylinder cars. Identify the test statistic and either the P-value or critical values. What do the results indicate?

SPSS

(I) Sample	(J) Sample	Mean Difference (I-J)	Std. Error	Sig.	95% Confidence Interval Lower Bound	95% Confidence Interval Upper Bound
1	2	-1.2019*	.2494	.000	-1.836	-.568
	3	-2.2912*	.2920	.000	-3.033	-1.549

12-3   Two-Way ANOVA

Key Concept In this section we introduce the method of *two-way analysis of variance,* which is used with data partitioned into categories according to *two* factors. The method of this section requires that we begin by testing for an *interaction* between the two factors. Then we test to determine whether the row factor has an effect and we also test to determine whether the column factor has an effect.

Table 12-3 is an example of data categorized with *two* factors:

1. Type: One factor is the row variable of type of car (foreign, domestic).

2. Size: The second factor is the column variable of size of the car (small, medium, large).

The subcategories in Table 12-3 are often called *cells,* so Table 12-3 has six cells containing three values each.

In analyzing the sample data in Table 12-3, we have already discussed one-way analysis of variance for a single factor, so it might seem reasonable to simply proceed with one-way ANOVA for the factor of site and another one-way ANOVA for the factor of treatment, but that approach wastes information and totally ignores a very important feature: the possible effect of an interaction between the two factors.

Table 12-3   Chest Deceleration Measurements (in g) from Car Crash Tests

	Size of Car		
	Small	Medium	Large
Foreign	44	41	32
	54	49	45
	43	47	42
Domestic	43	43	37
	44	37	38
	42	34	33

 DEFINITION

> There is an **interaction** between two factors if the effect of one of the fac-
> tors changes for different categories of the other factor.

As an example of an *interaction* between two factors, consider food pairings.
Peanut butter and jelly interact well, but ketchup and ice cream interact in a way that
results in a bad taste, so we rarely see someone eating ice cream topped with ketchup.
Physicians must be careful to avoid prescribing drugs with interactions that produce
adverse effects. It was found that the antifungal drug Nizoral (ketoconazole) inter-
acted with the antihistamine drug Seldane (terfenadine) in such a way that Seldane
was not metabolized properly, causing abnormal heart rhythms in some patients. Seldane was subsequently removed from the market. Just think of an interaction effect as
an effect due to the combination of the two factors.

Exploring Data   Let's explore the data in Table 12-3 by calculating the mean for each
cell and by constructing a graph. The individual cell means are shown in Table 12-4.
Those means vary from a low of 36.0 to a high of 47.0, so they appear to vary considerably. Figure 12-3 is an *interaction graph*, which shows graphs of those means, and
that figure has two very notable features:

Table 12-4   Means (in g) of Cells from Table 12-3

	Size of Car		
	Small	Medium	Large
Foreign	47.0	45.7	39.7
Domestic	43.0	38.0	36.0

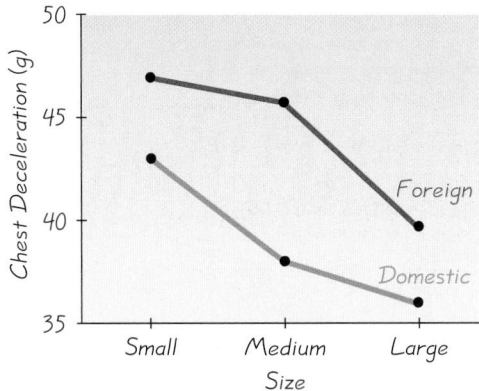

- **Larger means:** Because the line segments representing foreign cars are *higher*
  than the line segments for domestic cars, it appears that foreign cars have consistently *larger* measures of chest deceleration.

- **Interaction:** Because the line segments representing foreign cars appear to be
  approximately *parallel* to the line segments for domestic cars, it appears that

Figure 12-3

**Interaction Graph
of Means of Cells
in Table 12-3**

foreign and domestic cars behave the same for the different car size categories, so there does not appear to be an interaction effect.

In general, if a graph such as Figure 12-3 results in line segments that are approximately *parallel,* we have evidence that there is *not an interaction* between the row and column variables. If the line segments from foreign and domestic cars were far from parallel, we would have evidence of an interaction between site and treatment. These observations based on Table 12-4 and Figure 12-3 are largely subjective, so we will proceed with the more objective method of two-way analysis of variance.

TI-83/84 Plus In using the method of two-way analysis of variance, we consider three possible tests: (1) test for an interaction between the two factors; (2) test for an effect from the row factor; (3) test for an effect from the column factor. The calculations are quite involved, so *we will assume that a TI-83/84 Plus calculator or software is being used.* The end of this section shows the procedure for using a TI-83/84 Plus calculator for two-way analysis of variance.

The next page shows the requirements and basic procedure for two-way analysis of variance (ANOVA). The procedure is also summarized in Figure 12-4.

Figure 12-4

Procedure for Two-Way Analysis of Variance

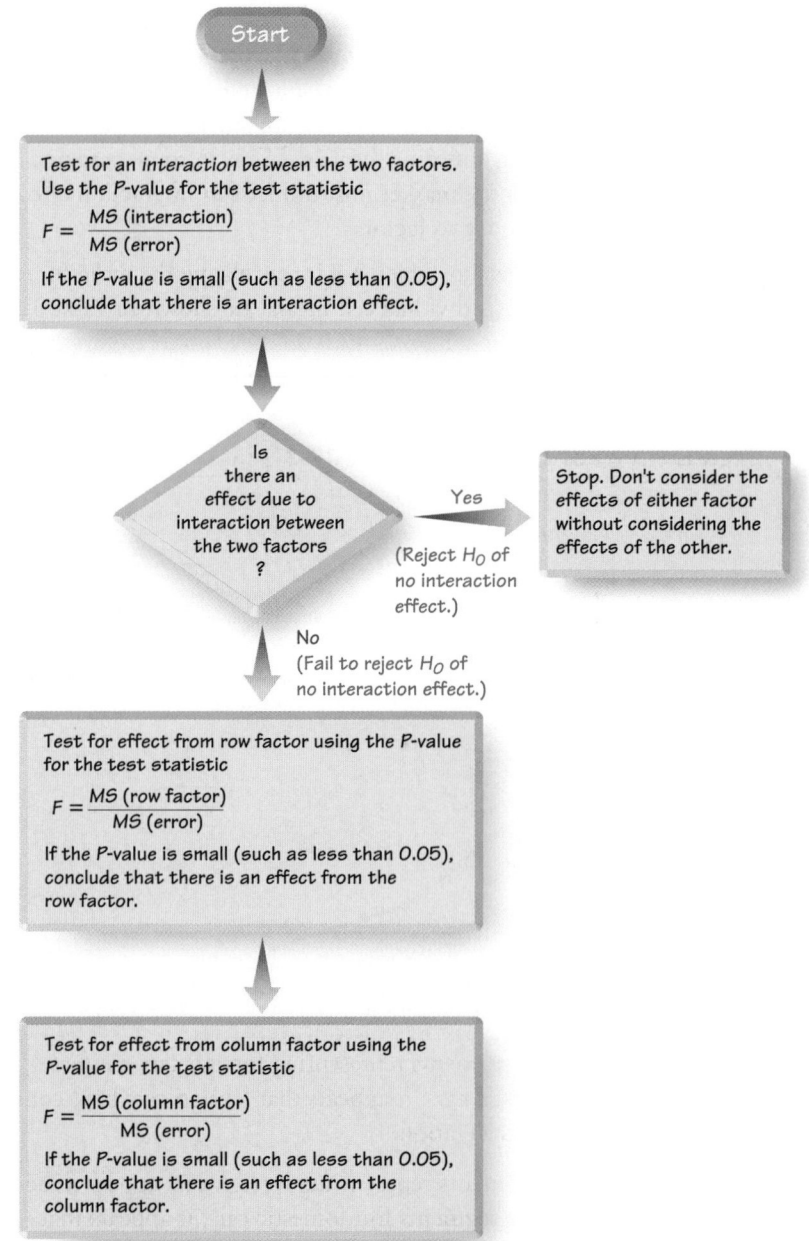

Objective

With sample data categorized with a row variable and a column variable, use two-way analysis of variance to test for an interaction effect, an effect from the row factor, and an effect from the column factor.

Requirements

1. For each cell, the sample values come from a population with a distribution that is approximately normal. (This procedure is robust against reasonable departures from normal distributions.)

2. The populations have the same variance σ^2 (or standard deviation σ). (This procedure is robust against reasonable departures from the requirement of equal variances.)

3. The samples are simple random samples of quantitative data.

4. The samples are independent of each other. (This procedure does not apply to samples that are not independent.)

5. The sample values are categorized two ways. (This is the basis for the name of the method: *two-way* analysis of variance.)

6. All of the cells have the same number of sample values. (This is called a *balanced* design.)

Procedure for Two-Way ANOVA (See Figure 12-4)

Step 1: *Interaction Effect:* In two-way analysis of variance, begin by testing the null hypothesis that there is no interaction between the two factors. Use technology to find the P-value corresponding to the following test statistic:

$$F = \frac{MS(\text{interaction})}{MS(\text{error})}$$

Conclusion:

- If the P-value corresponding to the above test statistic is small (such as less than or equal to 0.05), reject the null hypothesis of no interaction. Conclude that there is an interaction effect.

- If the P-value is large (such as greater than 0.05), fail to reject the null hypothesis of no interaction between the two factors. Conclude that there is no interaction effect.

Step 2: *Row/Column Effects:* If we conclude that there is an interaction effect, then we should stop now; we should not proceed with the two additional tests. (If there is an interaction between factors, we shouldn't consider the effects of either factor without considering those of the other.)

If we conclude that there is no interaction effect, then we should proceed with the following two hypothesis tests.

Row Factor: For the row factor, test the null hypothesis H_0: There are no effects from the row factor (that is, the row means are equal). Find the P-value corresponding to the test statistic $F = MS(\text{row})/MS(\text{error})$.

Conclusion:

- If the P-value corresponding to the test statistic is small (such as less than or equal to 0.05), reject the null hypothesis of no effect from the row factor. Conclude that there is an effect from the row factor.

- If the P-value is large (such as greater than 0.05), fail to reject the null hypothesis of no effect from the row factor. Conclude that there is no effect from the row factor.

Column Factor: For the column factor, test the null hypothesis H_0: There are no effects from the column factor (that is, the column means are equal). Find the P-value corresponding to the test statistic $F = MS(\text{column})/MS(\text{error})$.

Conclusion:

- If the P-value corresponding to the test statistic is small (such as less than or equal to 0.05), reject the null hypothesis of no effect from the column factor. Conclude that there is an effect from the column factor.

- If the P-value is large (such as greater than 0.05), fail to reject the null hypothesis of no effect from the column factor. Conclude that there is no effect from the column factor.

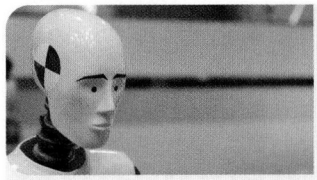

EXAMPLE 1 **Car Crash Test Measurements** Given the chest deceleration measurements in Table 12-3, use two-way analysis of variance to test for an interaction effect, an effect from the row factor of type of car (foreign, domestic), and an effect from the column factor of car size (small, medium, large). Use a 0.05 significance level.

SOLUTION **REQUIREMENT CHECK** (1) For each cell, the sample values appear to be from a normally distributed population, as indicated by normal quantile plots. (2) The variances of the cells are 37.0, 1.0, 17.3, 21.0, 46.3, and 7.0, so they vary considerably, but there are only three sample values in each cell, so we need extreme differences to reject equal variances. The test is robust against departures from equal variances, but we might have some reservations about this requirement. We will assume that this requirement is satisfied. (3) The samples are simple random samples of cars selected by the author. (4) The samples are independent of each other; the cars are not matched in any way. (5) The sample values are categorized in two ways (whether the car is foreign or domestic, and whether the car is small, medium, or large). (6) All of the cells have the same number (three) of sample values. The requirements are satisfied.

The calculations are quite involved, so we use a computer software or a TI-83/84 Plus calculator. The Minitab two-way analysis of variance display for the data in Table 12-3 is shown here. The corresponding TI-83/84 Plus calculator results are shown below the Minitab results.

MINITAB

Source	DF	SS	MS	F	P
Type	1	117.556	117.556	5.44	0.038
Size	2	154.778	77.389	3.58	0.060
Interaction	2	14.778	7.389	0.34	0.717
Error	12	259.333	21.611		
Total	17	546.444			

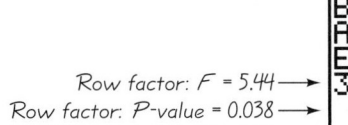

TI-83/84 Plus

```
     DF    SS
A   1   117.555555
B   2   154.777777
AB  2   14.7777777
ER  12  259.333333
3       F(A)=5.44
          P=.038
        F(B)=3.58
```

Row factor: *F* = 5.44 →
Row factor: *P*-value = 0.038 →
Column factor: *F* = 3.58 →

TI-83/84 Plus

```
B       P=.060
    F(AB)=.34
        P=.717

S=4.648775227
```

← *Column factor: P-value = 0.060*
← *Interaction: F = 0.34*
← *Interaction: P-value = 0.717*

Step 1: Interaction Effect: We begin by testing the null hypothesis that there is no interaction between the two factors. Using Minitab for the data in Table 12-3, we get the results shown in the preceding Minitab display, and we find the following test statistic:

$$F = \frac{\text{MS(interaction)}}{\text{MS(error)}} = \frac{7.389}{21.611} = 0.34$$

Interpretation: The corresponding *P*-value is 0.717, so we fail to reject the null hypothesis of no interaction between the two factors. It does not appear that the chest deceleration measurements are affected by an interaction between size of the car (small, medium, large) and the type of car (foreign, domestic). There does not appear to be an interaction effect.

Step 2: Row/Column Effects: Because there does not appear to be an interaction effect, we proceed to test for effects from the row and column factors. The two hypothesis tests use these null hypotheses:

H_0: There are no effects from the row factor (that is, the row means are equal).
H_0: There are no effects from the column factor (that is, the column means are equal).

Row Factor: For the row factor (type), we refer to the preceding Minitab display of results to find the *P*-value corresponding to the following test statistic:

$$F = \frac{\text{MS(type)}}{\text{MS(error)}} = \frac{117.556}{21.611} = 5.44$$

Conclusion: The corresponding *P*-value is 0.038. Because that *P*-value is less than the significance level of 0.05, we reject the null hypothesis of no effects from the type of car. That is, chest deceleration measurements do appear to be affected by whether the car is foreign or domestic.

Column Factor: For the column factor (size), we refer to the preceding Minitab display of results to find the *P*-value corresponding to the following test statistic:

$$F = \frac{\text{MS(size)}}{\text{MS(error)}} = \frac{77.389}{21.611} = 3.58$$

Conclusion: The corresponding *P*-value is 0.060. Because that *P*-value is greater than the significance level of 0.05, we fail to reject the null hypothesis of no effects from size. That is, chest deceleration measurements do not appear to be affected by whether the car is small, medium, or large.

Based on the sample data in Table 12-3, we conclude that chest deceleration measurements appear to be affected by whether the car is foreign or domestic, but those measurements do not appear to be affected by the size of the car.

CAUTION

Two-way analysis of variance is not one-way analysis of variance done twice. Be sure to test for an *interaction* between the two factors.

Special Case: One Observation per Cell and No Interaction Table 12-3 contains 3 observations per cell. If our sample data consist of only one observation per cell, we lose MS(interaction), SS(interaction), and df(interaction) because those values are based on sample variances computed for each individual cell. If there is only one observation per cell, there is no variation within individual cells and those sample variances cannot be calculated, so we use the following procedure.

Procedure for Two-Way Analysis of Variance with One Observation Per Cell

If it seems reasonable to assume (based on knowledge about the circumstances) that there is no interaction between the two factors, make that assumption and then proceed as before to test the following two hypotheses separately:

H_0: There are no effects from the row factor.

H_0: There are no effects from the column factor.

EXAMPLE 2 **One Observation Per Cell: Car Crash Test Measurements** Table 12-5 below has only one observation per cell. (Table 12-5 is obtained from Table 12-3.) The Minitab and TI-83/84 Plus calculator results from Table 12-5 are shown below the table. Use a 0.05 significance level to test for an effect from the row factor of type of car (foreign, domestic) and also test for an effect from the column factor of car size. Assume that there is no effect from an interaction between type of car and car size.

continued

Table 12-5   **One Observation Per Cell: Chest Deceleration Measurements**

	Size of Car		
	Small	Medium	Large
Foreign	44	41	32
Domestic	43	43	37

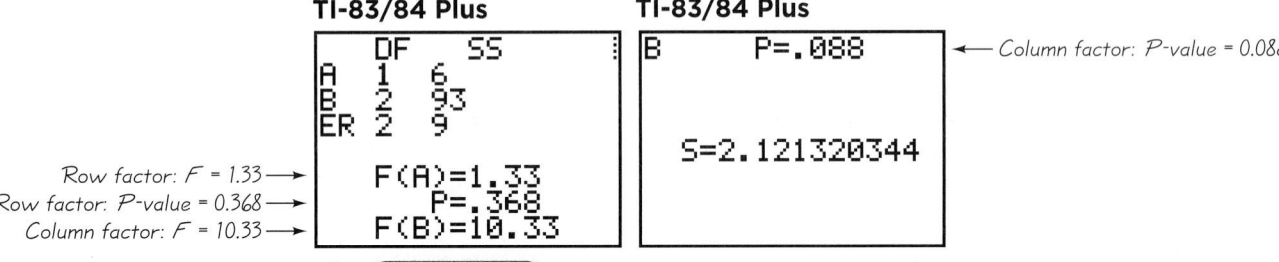

MINITAB

```
Source  DF   SS    MS     F      P
Type    1    6    6.0   1.33   0.368
Size    2   93   46.5  10.33   0.088
Error   2    9    4.5
Total   5  108
```

TI-83/84 Plus

```
      DF    SS
A     1     6
B     2    93
ER    2     9

   F(A)=1.33
      P=.368
   F(B)=10.33
```

Row factor: *F* = 1.33 →
Row factor: *P-value* = 0.368 →
Column factor: *F* = 10.33 →

TI-83/84 Plus

```
B        P=.088


   S=2.121320344
```

← Column factor: *P-value* = 0.088

SOLUTION   **Row factor:** We first use the results from the Minitab display to test the null hypothesis of no effects from the row factor of type of car (foreign or domestic).

$$F = \frac{MS(\text{type})}{MS(\text{error})} = \frac{6}{4.5} = 1.33$$

This test statistic is not significant, because the corresponding *P*-value is 0.368. We fail to reject the null hypothesis; it appears that the chest deceleration measurements are not affected by whether the car is foreign or domestic.

Column factor: We now use the Minitab display to test the null hypothesis of no effect from the column factor of car size. The test statistic is

$$F = \frac{MS(\text{size})}{MS(\text{error})} = \frac{46.5}{4.5} = 10.33$$

This test statistic is not significant because the corresponding *P*-value is 0.088. We fail to reject the null hypothesis, so it appears that the chest deceleration measurements are not affected by the size of the car.

In this section we have briefly discussed an important branch of statistics. We have emphasized the interpretation of computer displays while omitting the manual calculations and formulas, which are quite complex.

Using the TI-83/84 Plus Calculator for Two-Way Analysis of Variance

The two-way analysis of variance procedure requires that we first create a matrix [D] containing the sample data, and we must use the program A1ANOVA that is on the CD-ROM included with this book. On the CD-ROM, select the **Software** folder, then select **TI-83/TI-84 Plus Programs.** After the program A1ANOVA has been downloaded to the calculator, use the following procedure.

1. You can manually enter the data into matrix [D], or you can use lists if they are already available. Here are the two basic procedures for entering the sample data into matrix [D]:

- *Manual entry of sample data:* Enter the sample data into matrix [D], with the first column containing the sample values, the second column containing the corresponding row numbers, and the third column containing the corresponding column numbers. To manually enter the sample data into matrix D, first get to the **Matrix** menu by pressing **2ND** **x⁻¹**. Select **EDIT,** scroll down to select **[D],** then press **ENTER**. Begin by entering the dimension of the matrix in the format of rows by columns. For the number of rows, enter the total number of sample values. For the number of columns, enter 3. For example, to enter the data in Table 12-3 as matrix [D], enter 18 for the number of rows (because there are 18 sample values) and enter 3 for the number of columns. Proceed to enter the sample data and the corresponding row and column numbers. See the accompanying screen showing the creation of matrix [D] for the data in Table 12-3. (Only the top seven rows of the matrix are shown.)

- *Using existing lists:* If the data are already stored as an individual list, it can be entered into matrix [D]. Store the list in L1, then create a list L2 consisting of the corresponding row numbers, and also create a list L3 consisting of the corresponding column numbers. To combine lists L1, L2, L3 into matrix [D], first select the **Matrix** menu by pressing **2ND** **x⁻¹**. Select the top menu item of **MATH,** then select **List→matr.** Now enter L1, L2, L3, and end with the matrix name [D], all separated by commas. Enter the matrix name of [D] by opening the MATRIX menu, scrolling to [D], and pressing enter. Here is the command that puts lists L1, L2, L3 into matrix [D]:

List→matr(L1, L2, L3, [D]).

2. After entering the sample data into matrix [D] as described in Step 1, we can proceed to run the two-way analysis of variance program. Press **PRGM**, select **A1ANOVA** and press **ENTER** **ENTER**, then select **RAN BLOCK DESI** (randomized block design) and press **ENTER**.

3. A message will be displayed beginning with "DATA IN N*3 MAT [D]." Press **ENTER**.

4. Select the option of **CONTINUE** and press **ENTER**.

5. The program will work with the data that were previously stored in matrix [D], and the results will be displayed. The results will not all fit in a single window, so press **ENTER** to see the remaining results, as in the results shown in Example 1 and Example 2.

Interpreting a TI-83/84 Plus Calculator Display The TI-83/84 Plus calculator display is not in a format that is easy to read, so focus on the following key components.

Interaction:	Look for **F(AB)=**. The *F* test statistic is the value shown, and the *P*-value for the interaction is on the line below this test statistic.
Effect from row factor:	Look for **F(A)=**. The value shown is the *F* test statistic for the effect from the row factor, and the *P*-value is on the line below this test statistic.
Effect from column factor:	Look for **F(B)=**. The value shown is the *F* test statistic for the effect from the column factor, and the *P*-value is on the line below this test statistic.

Entering Data into Matrix [D]

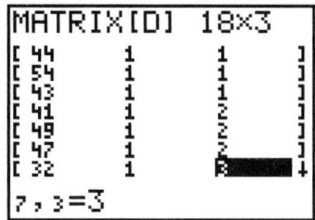

12-3 Basic Skills and Concepts

Statistical Literacy and Critical Thinking

1. Two-Way ANOVA Researchers randomly select and weigh men and women (as in Data Set 1 in Appendix B). Their weights are entered in the table below, so that each cell includes five weights. What characteristic of the data suggests that the appropriate method of analysis is *two-way* analysis of variance? That is, what is "two-way" about the data?

	Age		
	Under 30	30–40	Over 40
Female			
Male			

2. Two-Way ANOVA If the weights are entered in the table described in Exercise 1, what can we determine by using the method of two-way analysis of variance?

3. Balanced Design If the weights are entered in the table described in Exercise 1, is the result a *balanced design?* Why or why not?

4. Interaction Shown below is a Minitab-generated interaction plot representing weights of poplar trees grown at different sites (site 1 and site 2) with different treatments (none, fertilizer, irrigation, fertilizer and irrigation). What does this graph suggest about the interaction between the two factors?

MINITAB

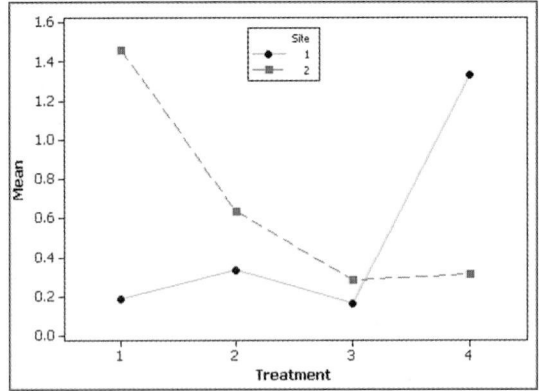

Interpreting a Computer Display. *Exercises 5–7 use the given Minitab display, which results from the heights of 32 randomly selected men and 32 randomly selected women listed in Data Set 1 in Appendix B. The row variable of sex has two values (male, female) and the column variable of age consists of two age brackets (below 30, above 30). Use a 0.05 significance level for the hypothesis test.*

MINITAB

Source	DF	SS	MS	F	P
Sex	1	439.426	439.426	53.11	0.000
Age	1	24.379	24.379	2.95	0.091
Interaction	1	1.995	1.995	0.24	0.625
Error	60	496.479	8.275		
Total	63	962.280			

5. Interaction Effect Test the null hypothesis that heights are not affected by an interaction between sex and age bracket. What do you conclude?

6. Effect of Sex Assume that heights are not affected by an interaction between sex and age bracket. Is there sufficient evidence to support the claim that sex has an effect on height?

7. Effect of Age Bracket Assume that heights are not affected by an interaction between sex and age bracket. Is there sufficient evidence to support the claim that age bracket has an effect on height?

Interpreting a Computer Display. *In Exercises 8–10, use the TI-83/84 Plus calculator display on the next page, which results from the head injury measurements from car crash dummies listed on the next page. The measurements are in hic (head injury criterion) units, and they are from the same cars used for Table 12-3. Use a 0.05 significance level to test the given claim.*

	Size of Car		
	Small	Medium	Large
Foreign	290	245	342
	544	502	698
	501	393	332
Domestic	406	474	216
	371	368	335
	376	349	169

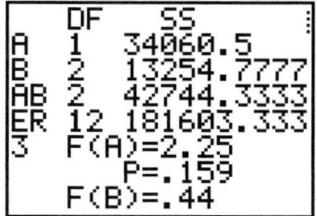

TI-83/84 Plus

8. Interaction Effect Test the null hypothesis that head injury measurements are not affected by an interaction between the type of car (foreign, domestic) and size of the car (small, medium, large). What do you conclude?

9. Effect from Type Assume that head injury measurements are not affected by an interaction between type of car (foreign, domestic) and size of car (small, medium, large). Is there sufficient evidence to support the claim that the type of car has an effect on head injury measurements?

10. Effect of Size Assume that head injury measurements are not affected by an interaction between type of car (foreign, domestic) and size of car (small, medium, large). Is there sufficient evidence to support the claim that size of the car (small, medium, large) has an effect on head injury measurements?

Interpreting a Computer Display. *In Exercises 11–13, use the TI-83/84 Plus calculator display, which results from measures of self-esteem listed in the table below. The data are from Richard Lowry and are based on a student project at Vassar College supervised by Jannay Morrow. The objective of the project was to study how levels of self-esteem in subjects relate to their perceived self-esteem in other target people who were described in writing. Self-esteem levels were measured using the Coopersmith Self-Esteem Inventory, and the test here works well even though the data are at the ordinal level of measurement. Use a 0.05 significance level to test the given claim.*

TI-83/84 Plus

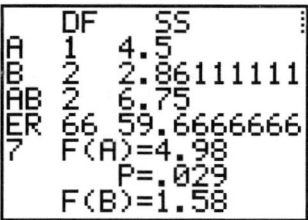

		Subject's Self-Esteem					
		Low		Medium		High	
Target's Self-Esteem	Low	4	4	3	3	3	1
		3	5	3	4	3	3
		4	4	4	2	3	5
		5	4	4	4	3	2
		2	4	1	2	3	3
		4	2	2	3	3	3
	High	2	2	4	3	3	2
		4	2	1	2	3	2
		2	3	1	3	3	4
		2	4	2	4	3	4
		2	2	3	1	4	3
		2	3	1	4	3	4

TI-83/84 Plus

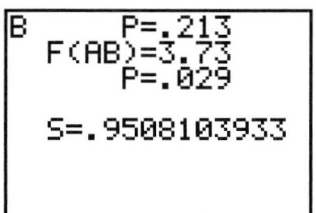

11. Interaction Effect Test the null hypothesis that measurements of self-esteem are not affected by an interaction between the subject's self-esteem and the target's self esteem. What do you conclude?

12. Effect from Target Assume that self-esteem measurements are not affected by an interaction between subject self-esteem and target self-esteem. Is there sufficient evidence to support the claim that the category of the target (low, high) has an effect on measures of self-esteem?

13. Effect of Subject Assume that self-esteem measurements are not affected by an interaction between subject self-esteem and target self-esteem. Is there sufficient evidence to support the claim that the self-esteem of the subject (low, medium, high) has an effect on the measurements of self-esteem?

In Exercises 14 and 15, use a TI-83/84 Plus calculator to obtain the results from two-way analysis of variance.

14. Cholesterol Levels The following table lists measured cholesterol levels from Data Set 1 in Appendix B. Are cholesterol levels affected by an interaction between sex and age? Are cholesterol levels affected by sex? Are cholesterol levels affected by age?

	Age		
	Under 30	30–50	Over 50
Male	265 303 1252 230 957	702 277 176 416 120	75 189 288 578 31
Female	325 112 62 301 223	146 173 149 462 94	254 384 318 600 309

15. Pancake Experiment Listed below are ratings of pancakes made by experts (based on data from Minitab). Different pancakes were made with and without a supplement, and different amounts of whey were used. Are the ratings affected by an interaction between the use of the supplement and the amount of whey? Are ratings affected by use of the supplement? Are ratings affected by the amount of whey?

	Whey			
	0%	10%	20%	30%
No Supplement	4.4 4.5 4.3	4.6 4.5 4.8	4.5 4.8 4.8	4.6 4.7 5.1
Supplement	3.3 3.2 3.1	3.8 3.7 3.6	5.0 5.3 4.8	5.4 5.6 5.3

One Observation Per Cell. *In Exercises 16 and 17, refer to the indicated data and use a 0.05 significance level for the hypothesis test.*

16. Cholesterol Levels Refer to the sample data in Exercise 14 and use only the first entry in each cell. Assume that there is no effect on cholesterol levels from an interaction between age bracket and sex. Is there sufficient evidence to support the claim that cholesterol levels are affected by sex? Is there sufficient evidence to support the claim that cholesterol levels are affected by age bracket?

17. Pancake Experiment List Refer to the sample data in Exercise 15 and use only the first entry in each cell. Assume that there is no effect on rating from an interaction between the use of the supplement and the amount of whey. Is there sufficient evidence to support the claim that ratings are affected by the use of the supplement? Is there sufficient evidence to support the claim that ratings are affected by the amount of whey?

12-3   Beyond the Basics

18. Transformations of Data Example 1 illustrated the use of two-way ANOVA to analyze the sample data in Table 12-3. How are the results affected in each of the following cases?

a. The same constant is added to each sample value.

b. Each sample value is multiplied by the same nonzero constant.

c. The format of the table is transposed, so that the row and column factors are interchanged.

d. The first sample value in the first cell is changed so that it becomes an outlier.

Review

One-Way Analysis of Variance   In Section 12-2 we presented the method of one-way analysis of variance, which is a method used to test for equality of three or more population means. (The requirements and procedure are listed in Section 12-2.) Because of the complex nature of the required calculations, we focused on the interpretation of P-values obtained using technology. When using one-way analysis of variance for testing equality of three or more population means, we use this decision criterion:

• If the P-value is small (such as 0.05 or less), reject the null hypothesis of equal population means and conclude that at least one of the population means is different from the others.

• If the P-value is large (such as greater than 0.05), fail to reject the null hypothesis of equal population means. Conclude that there is not sufficient evidence to warrant rejection of equal population means.

Two-Way Analysis of Variance   In Section 12-3 we presented the method of two-way analysis of variance, which is used with data categorized according to two different factors. One factor is used to arrange the sample data in different rows, while the other factor is used to arrange the sample data in different columns. The procedure for two-way analysis of variance is summarized here:

1. Interaction Test for an interaction between the two factors.

• If the *P*-value for the interaction is small (such as 0.05 or less), there appears to be an interaction effect, and we should stop here and not proceed with the following two tests.

• If the *P*-value is large (such as greater than 0.05), there does not appear to be an interaction effect, and we should proceed with the following two tests.

2. Row Factor Test for an effect from the factor used to arrange the sample data in different rows.

• If the *P*-value for the row factor is small (such as 0.05 or less), there appears to be an effect from the row factor.

• If the *P*-value for the row factor is large (such as greater than 0.05), there does not appear to be an effect from the row factor.

3. Column Factor Test for an effect from the factor used to arrange the sample data in different columns.

• If the *P*-value for the column factor is small (such as 0.05 or less), there appears to be an effect from the column factor.

• If the *P*-value for the column factor is large (such as greater than 0.05), there does not appear to be an effect from the column factor.

In Section 12-3 we also considered the use of two-way analysis of variance for the special case in which there is only one observation per cell.

Statistical Literacy and Critical Thinking

1. Independent Sample Data Data Set 13 in Appendix B lists measured voltage amounts from a gasoline-powered generator, a home's electrical system, and an uninterruptible backup power supply that is powered with the same home's electrical system. If the sample voltage amounts are measured from the three sources at precisely the same times, are the data independent? Should we use one-way analysis of variance? Why or why not?

2. One Way ANOVA and Two-Way ANOVA What is the main difference between one-way analysis of variance and two-way analysis of variance?

3. Movie Data Data Set 9 in Appendix B lists data from 35 movies. If more movies are included and the MPAA ratings (PG, PG-13, R) are categorized according to movie running time (under 2 hours, over 2 hours) and budget (under $75 million, over $75 million), can the methods of this chapter be used to test for effects of movie running time and budget bracket? Why or why not?

4. Car Crash Tests Both sections of this chapter used results from car crash tests. (See the sample data in Table 12-1 and Table 12-3.) If costs of future car crash tests are reduced by using only the least expensive cars in size categories of small, medium, and large, can inferences be made about the safety of all cars? Why or why not?

Chapter Quick Quiz

1. What is one-way analysis of variance used for?

2. Are one-way analysis of variance tests left-tailed, right-tailed, or two-tailed?

3. In one-way analysis of variance tests, do larger test statistics result in larger *P*-values, smaller *P*-values, or *P*-values that are unrelated to the value of the test statistic?

4. The following display results from using Minitab for one-way analysis with sample data consisting of measured grade level reading scores from 12 randomly selected pages in three different books (by Rowling, Clancy, and Tolstoy). What is the value of the test statistic?

MINITAB

Source	DF	SS	MS	F	P
Factor	2	68.19	34.09	8.98	0.001
Error	33	125.31	3.80		
Total	35	193.50			

5. Using the same one-way analysis test and results given in Exercise 4, what is the null hypothesis and what do you conclude about it?

6. Using the same one-way analysis test and results given in Exercise 4, what is the final conclusion?

7. What is the fundamental difference between one-way analysis of variance and two-way analysis of variance?

8. Given below is a Minitab display resulting from two-way analysis of variance with sample data consisting of 18 different student estimates of the length of a classroom. The values are categorized according to sex and major (math, business, liberal arts). What do you conclude about an interaction between sex and major?

MINITAB

Source	DF	SS	MS	F	P
Sex	1	29.389	29.3889	0.78	0.395
Major	2	10.111	5.0556	0.13	0.876
Interaction	2	14.111	7.0556	0.19	0.832
Error	12	453.333	37.7778		
Total	17	506.944			

9. Using the same results given in Exercise 8, does it appear that the length estimates are affected by the sex of the subject?

10. Using the same results given in Exercise 8, does it appear that the length estimates are affected by the subject's major?

Review Exercises

1. Baseline Characteristics Experiments and clinical trials with different treatment groups commonly include information about the characteristics of those groups. In a study of four different weight loss programs, each program had 40 subjects. The means and standard deviations of the ages in each group are as follows: Atkins ($\bar{x} = 47$ years, $s = 12$ years); Zone ($\bar{x} = 51$ years, $s = 9$ years); Weight Watchers ($\bar{x} = 49$ years, $s = 10$ years); Ornish ($\bar{x} = 49$ years, $s = 12$ years). These statistics are listed along with a *P*-value of 0.41. These results are from "Comparison of the Atkins, Ornish, Weight Watchers, and Zone Diets for Weight Loss and Heart Disease Risk Reduction," by Dansinger, et al., *Journal of the American Medical Association,* Vol. 293, No. 1.

a. How many variables are used to categorize the sample data consisting of ages?

b. What specific method is used to find the *P*-value of 0.41?

c. What does the *P*-value of 0.41 indicate about the baseline characteristic of age?

d. What would a small *P*-value (such as 0.001) indicate about the ages, and how would that affect the results of the study?

TI-83/84 Plus

```
One-way ANOVA
 F=54.69558779
 p=1.441117E-10
 Factor
  df=2
  SS=6166725.45
↓ MS=3083362.72
```

2. Car Weight Listed below are the weights (in lb) of cars in three different categories (from Data Set 16 in Appendix B). The TI-83/84 Plus calculator results are also shown. Use a 0.05 significance level to test the claim that the different car categories have the same mean weight. Do cars with more cylinders weigh more?

Four cylinder	3315 3565 3135 3190 2760 3195 2980 2875 3060 3235 2865 2595 3465
Six cylinder	4035 4115 3650 4030 3710 4095 4020 3915 3745 3475 3600 3630
Eight cylinder	4105 4170 4180 3860 4205 4415 4180

Interpreting a Computer Display. *In Exercises 3–5, use the TI-83/84 Plus calculator display on the next page, which results from the values listed on the next page. The car types are foreign and domestic. The values are loads (in lb) on the left femur of crash test dummies in the same cars from Example 1 in Section 12-3.*

	Size of Car		
	Small	Medium	Large
Foreign	548	194	215
	707	280	1636
	320	617	433
Domestic	782	1076	937
	1188	719	953
	634	445	472

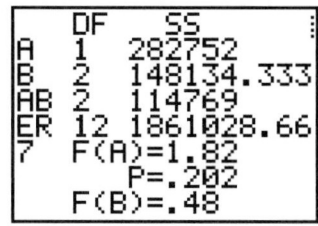

TI-83/84 Plus

3. Interaction Effect Test the null hypothesis that loads on the left femur are not affected by an interaction between the type of car and the size of the car.

4. Effect from Type Assume that left femur loads are not affected by an interaction between type of car (foreign, domestic) and size of car (small, medium, large). Is there sufficient evidence to support the claim that the type of car has an effect on left femur load measurements?

5. Effect of Size Assume that left femur loads are not affected by an interaction between type of car (foreign, domestic) and size of car (small, medium, large). Is there sufficient evidence to support the claim that size of the car (small, medium, large) has an effect on left femur load measurements?

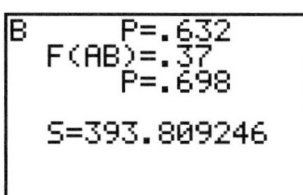

TI-83/84 Plus

6. Carbon Monoxide in Cigarettes Listed below are amounts of carbon monoxide (mg per cigarette) in samples of king size cigarettes, 100 mm menthol cigarettes, and 100 mm nonmenthol cigarettes (from Data Set 4 in Appendix B). The king size cigarettes are nonfiltered, nonmenthol, and non-light. The 100 mm menthol cigarettes are filtered and non-light. The 100 mm non-menthol cigarettes are filtered and non-light. Use a 0.05 significance level to test the claim that the three categories of cigarettes yield the same mean amount of carbon monoxide. Given that only the king size cigarettes are not filtered, do the filters appear to make a difference?

King Size 16 16 16 16 16 17 16 15 16 14 16 16 16 16 16 14 16 16 14
18 15 16 14 16 16

Menthol 15 17 19 9 17 17 15 17 15 17 17 15 17 17 18 11 18 3 17
14 15 22 16 7 9

Nonmenthol 4 19 17 18 18 13 17 15 15 12 18 17 18 16 3 18 15 18 15
17 15 15 7 16 14

7. Smoking, Body Temperature, Gender The table below lists body temperatures obtained from randomly selected subjects (based on Data Set 2 in Appendix B). The temperatures are categorized according to gender and whether the subject smokes. Using a 0.05 significance level, test for an interaction between gender and smoking, test for an effect from gender, and test for an effect from smoking. What do you conclude?

	Smokes	Does not smoke
Male	98.4 98.4 99.4 98.6	98.0 98.0 98.8 97.0
Female	98.8 98.0 98.7 98.4	97.7 98.0 98.2 99.1

8. Longevity The table below lists the numbers of years that U.S. presidents, popes, and British monarchs (since 1690) lived after their inauguration, election, or coronation, respectively. As of this writing, the last president is Gerald Ford, the last pope is John Paul II, and the last British monarch is George VI. Determine whether the survival times for the three groups differ. (Table based on data from *Computer-Interactive Data Analysis,* by Lunn and McNeil, John Wiley & Sons.)

Presidents 10 29 26 28 15 23 17 25 0 20 4 1 24 16 12 4 10 17 16
0 7 24 12 4 18 21 11 2 9 36 12 28 3 16 9 25 23 32

Popes 2 9 21 3 6 10 18 11 6 25 23 6 2 15 32 25 11 8 17
19 5 15 0 26

Monarchs 17 6 13 12 13 33 59 10 7 63 9 25 36 15

Cumulative Review Exercises

In Exercises 1–4, refer to the numbers of years that U.S. presidents, popes, and British monarchs lived after their inauguration, election, or coronation respectively. The data are listed in the table in Review Exercise 8.

TI **1. a.** Find the mean for each of the three groups.

b. Find the standard deviation for each of the three groups.

TI **2.** Test the claim that there is a difference between the mean for presidents and the mean for British monarchs.

TI **3.** Use the longevity times for presidents and determine whether they appear to come from a population having a normal distribution. Explain why the distribution does or does not appear to be normal.

TI **4.** Use the longevity times for presidents and construct a 95% confidence interval estimate of the population mean.

TI **5. Body Temperatures** Listed below are body temperatures (in °F) randomly selected from the larger sample in Data Set 2 of Appendix B.

> 98.3   96.5   97.3   98.9   98.2   97.6   98.4   98.0   97.6

a. What is the level of measurement of these temperatures?

b. Are body temperatures discrete data or continuous data?

c. What is the sample mean?   **d.** What is the sample median?

e. What is the sample range?   **f.** What is the sample standard deviation?

g. What is the sample variance?

TI **6. Movie Ratings: Frequency Distribution** Listed below are viewer ratings of movies (from Data Set 9 in Appendix B). Construct a frequency distribution with a class width of 1.0, and use 1.0 for the lower limit of the first class.

> 6.7 2.2 7.6 8.0 7.4 5.8 4.8 7.7 8.3 5.7 6.5 8.6 1.9 7.8 5.8 2.0 7.8 8.4
> 7.5 2.7 6.0 6.3 7.0 6.6 2.0 7.8 8.0 6.4 8.5 6.6 5.3 6.8 8.1 6.7 7.3

TI **7. Movie Ratings: Histogram** Use the frequency distribution from the preceding exercise to construct the corresponding histogram. Based on the result, does it appear that the viewer ratings are from a population with a normal distribution? Why or why not?

TI **8. Movie Ratings and MPAA Ratings** Listed below are samples of movie viewer ratings categorized according to MPAA ratings of PG, PG-13, and R (based on Data Set 9 in Appendix B). First, informally compare the three sample means, then conduct a formal test of the claim that the three samples are from populations with the same mean. Use a 0.05 significance level.

PG   5.8 6.4 7.7
PG–13 2.0 2.2 2.7 4.8 5.3 6.0 6.5 6.7 6.8 7.0 7.3 7.4 7.5 7.8 7.8 7.8 8.0 8.1 8.4 8.5
R   1.9 2.0 5.7 5.8 6.3 6.6 6.6 6.7 7.6 8.0 8.3 8.6

TI **9. Garbage** Listed below are the amounts (in lb) of paper and plastic discarded by randomly selected households (from Data Set 22 in Appendix B).

a. Test for a linear correlation between the amounts of discarded paper and plastic.

b. Find the equation of the regression line. Let the variable x represent amounts of discarded paper and let y represent amounts of discarded plastic.

c. Is the regression equation found in part (b) a good model for predicting an amount of discarded plastic when given some specific amount of discarded paper? Why or why not?

Paper   9.55   11.42   11.36   10.99   9.09
Plastic   2.19   3.05   2.44   1.44   1.28

TI **10. Firearm Injuries** The table below lists numbers of firearm injuries arranged according to circumstances and whether the firearm was a handgun or a rifle or shotgun (based on data from "Hospitalization Charges, Costs, and Income for Firearm-Related Injuries at a University Trauma Center," by Kizer, et al., *Journal of the American Medical Association,* Vol. 273, No. 22). Use a 0.05 significance level to test the claim that the injury category is independent of the type of weapon.

	Unintentional	Self-Inflicted	Assault
Handgun	31	35	162
Rifle or Shotgun	13	7	67

TI-83/84 Plus Calculator Project

Refer to Data Set 8 in Appendix B. Use a TI-83/84 Plus calculator for the following.

a. Test the null hypothesis that the six samples of word counts for *males* (columns 1, 3, 5, 7, 9, 11) are from populations with the same mean. Print the results and write a brief summary of your conclusions.

b. Test the null hypothesis that the six samples of word counts for *females* (columns 2, 4, 6, 8, 10, 12) are from populations with the same mean. Print the results and write a brief summary of your conclusions.

c. If we want to compare the number of words spoken by men to the number of words spoken by women, does it make sense to combine the six columns of word counts for males and combine the six columns of word counts for females, then compare the two samples? Why or why not?

INTERNET PROJECT

Analysis of Variance

Go to: **http://www.aw.com/triola**

Follow the link to the Internet Project for this chapter. The project provides the background for experiments in areas as varied as athletic performance, consumer product labeling, and biology of the human body. In each case, the associated data will lend itself naturally to groupings ideal for application of this chapter's techniques. You will formulate the appropriate hypotheses, then conduct and summarize ANOVA tests.

APPLET PROJECT

Open the Applets folder on the CD and double-click on **Start.**

a. Select the menu item of **Random numbers** and generate 10 random whole numbers between 0 and 9 inclusive.

b. Repeat part (a).

c. Select the menu item of **Random numbers** and generate 10 random whole numbers between 5 and 14 inclusive.

d. Use one-way analysis of variance to test the claim that the three samples are from populations with the same mean. Be sure to check the requirements for one-way analysis of variance. What can you conclude?

Cooperative Group Activities

1. Out-of-class activity The *World Almanac and Book of Facts* includes a section called "Noted Personalities," with subsections composed of architects, artists, business leaders, military leaders, philosophers, political leaders, scientists, writers, entertainers, and others. Design and conduct an observational study that begins with choosing samples from select groups, followed by a comparison of life spans of people from the different groups. Do any particular groups appear to have life spans that are different from the other groups? Can you explain such differences?

2. In-class activity Ask each student in the class to estimate the length of the classroom. Specify that the length is the distance between the chalkboard and the opposite wall. On the same sheet of paper, each student should also write his or her gender (male/female) and major.

Then divide into groups of three or four, and use the data from the entire class to address these questions:

• Is there a significant difference between the mean estimate for males and the mean estimate for females?

• Is there sufficient evidence to reject equality of the mean estimates for different majors? Describe how the majors were categorized.

• Does an interaction between gender and major have an effect on the estimated length?

• Does gender appear to have an effect on estimated length?

• Does major appear to have an effect on estimated length?

3. Out-of-class activity Divide into groups of three or four students. Each group should survey other students at the same college by asking them to identify their major and gender. You might include other factors, such as employment (none, part-time, full-time) and age (under 21, 21–30, over 30). For each surveyed subject, determine the accuracy of the time on his or her wristwatch. First set your own watch to the correct time using an accurate and reliable source ("At the tone, the time is . . ."). For watches that are ahead of the correct time, record positive times. For watches that are behind the correct time, record negative times. Use the sample data to address questions such as these:

• Does gender appear to have an effect on the accuracy of the wristwatch?

• Does major have an effect on wristwatch accuracy?

• Does an interaction between gender and major have an effect on wristwatch accuracy?

4. Out-of-class activity Divide into groups of three or four students. Each student should go to a different fast-food restaurant (McDonald's, Burger King, Wendy's) and randomly select customers as they enter a line to place an order. Record the time from entering the line to picking up the completed order. Use a sample of at least 10 different customers. Use the same day of the week and the same time of day. Test for equality of the service times.

FROM DATA TO DECISION

Critical Thinking: Is Lipitor effective in lowering LDL cholesterol?

With current sales of Lipitor exceeding $13 billion each year, it has become the best selling drug ever. The author asked Pfizer for original data from clinical drug trials of Lipitor, but Pfizer declined to provide the data. The data shown below are based on results given in a Parke-Davis memo from David G. Orloff, M.D., the medical team leader in the clinical trials. The data below refer to atorvastatin, and Lipitor is the trade name of atorvastatin. LDL cholesterol is considered the bad cholesterol, so a subject's condition is generally improved if the LDL cholesterol is lowered. The changes in LDL cholesterol listed in the table are measured in mg/dL. Note that when compared to baseline values, negative values in the following data indicate that the LDL cholesterol has been lowered.

Changes in LDL Cholesterol from Baseline Values (a negative value represents a decrease)

Placebo Group: −3 5 6 −2
−7 8 5 −6 −1 7 −4 3

Group treated with 10 mg of atorvastatin: −28 −27 −23 −25 −27 −29 −22 −22 −26 −23 −23 −22 −24 −21 −25 −26 −23 −24 −23 −22 −22 −20 −29 −29 −27 −24 −28 −26 −22 −26 −23 −26 −25 −29 −27 −27 −23

Group treated with 20 mg of atorvastatin: −28 −32 −29 −39 −31 −35 −25 −36 −35 −26 −29 −34 −30

Group treated with 80 mg of atorvastatin: −42 −41 −38 −42 −41 −41 −40 −44 −32 −37 −41 −37 −34 −31

Analyzing the Results

Analyze the data. Does it appear that atorvastatin treatment has an effect? If atorvastatin treatment does have an effect, is it the desired effect? Does it appear that larger doses of atorvastatin treatment result in greater beneficial effects? Write a brief report summarizing your findings and include specific statistical tests and results.

NAME:	Jeffrey Foy
JOB:	Toxicologist
COMPANY:	Cabot Corporation

Jeffrey Foy, a toxicologist working for the Cabot Corporation, is responsible for the hazard evaluation of the chemicals Cabot Corporation produces. It is his job to understand how the company's products may affect humans or the environment and help decide on the best ways to protect both.

Q: What do you do in your job?

A: My responsibilities include arranging and evaluating toxicological studies, writing material safety data sheets, and helping our research and development groups produce materials that are safe for both people and the environment or to understand what potential hazards the materials might have.

Q: What concepts of statistics do you use?

A: The primary concept I use is hypothesis testing (probability testing).

Q: How do you use statistics on the job?

A: I use statistics daily. Statistical methods have been and are used in two ways in my work. First, statistics is used to help determine how I design my experiments. Second, statistics is used to determine if the data generated are significant or sometimes if they're even good enough to use.

Studies that I am involved in can cost from as little as $1000 to as much as $500,000 or more, and if you don't properly determine how you are going to evaluate the data, you could cost your company a great deal of time and money. If the experiment is done properly, then we move on to analyze the data. The data from the studies we perform are used to assess any potential health effects our products may have on our workers, customers, or the environment. The results are used to determine how chemicals can be sold or handled. When performing experiments at a testing laboratory or drug company you want to determine if your materials have an effect, whether desired (a drug curing a disease) or undesired (that same drug being toxic). Statistics plays an enormous role in our evaluation of the significance of the effects.

Q: Please describe one specific example illustrating how the use of statistics was successful in improving a product or service.

A: A toxicology study costing about $300,000 was recently conducted. The data from the study were to be used to help determine if a particular chemical caused any effects in the subjects studied. After the study was performed, flaws were found in both the data and statistics used. It took an additional two years to properly review the data and finish the health evaluation. If the proper methods and endpoints had been chosen, then the additional time and money may not have been necessary. It was the understanding of the data and correct statistical evaluation that helped prevent the failure and potential repeat of the study.

13-1 Review and Preview

13-2 Sign Test

13-3 Wilcoxon Signed-Ranks Test for Matched Pairs

13-4 Wilcoxon Rank-Sum Test for Two Independent Samples

13-5 Kruskal-Wallis Test

13-6 Rank Correlation

13-7 Runs Test for Randomness

13

Nonparametric Statistics

Are the "best" universities the most difficult to get into?

Table 13-1 lists scores and ranks from randomly selected national universities. The universities are Stanford, Rochester, Syracuse, Columbia, Georgetown, Tufts, Wake Forest, and Johns Hopkins. The first score is a measure of overall quality (with higher scores indicating higher quality), and these scores are based on factors including assessments by administrators, faculty resources, graduation rates, and student retention. The selectivity ranks are based on factors including standard test scores of applicants, the acceptance rate, and the proportion of students in the top 10% of their graduating high school class. Lower selectivity ranks correspond to schools that are more selective. The ranks are based on the colleges included in the sample, not the population of all colleges. The scores and ranks are based on data provided by *U.S. News and World Report* magazine.

Are the colleges with the highest overall quality scores also the colleges that are most difficult to get into? That is, is there an association between the overall quality score and the selectivity rank? If so, does that association indicate that *higher* overall quality scores are associated with *lower* selectivity ranks?

We should always begin by exploring the data, and the accompanying TI-83/84 Plus scatterplot shows that there does appear to be a relationship between the two variables. Examining the scatterplot from left to right, there appears to be a downward trend, but it is not clear whether the pattern is strong enough to justify the conclusion of a correlation between the two variables. We need a more objective measure of correlation.

Determination of a correlation between two variables was discussed in Chapter 10, but that method is a parametric method in the sense that there is a requirement of a normal distribution. Because the selectivity values are ranks from 1 through 8, they violate the normal distribution requirement, so the methods of Section 10-2 should not be used. Instead, we will use nonparametric methods when a requirement of a normal distribution is not satisfied. In Section 13-6 we use ranks of data to determine whether there is a correlation between two variables. This method can be applied to the sample data in Table 13-1.

TI-83/84 PLUS

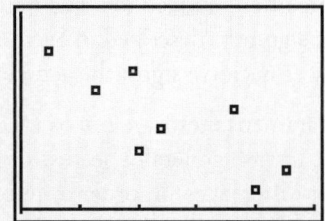

Table 13-1 Overall Quality Scores and Selectivity Ranks of National Universities

Overall quality	95	63	55	90	74	70	69	86
Selectivity rank	2	6	8	1	4	3	7	5

Review and Preview

13-1

In the preceding chapters, we presented a variety of different methods of inferential statistics. Many of those methods require normally distributed populations and are based on sampling from a population with specific parameters, such as the mean μ, standard deviation σ, or population proportion p. The main objective of this chapter is to introduce methods of *nonparametric* statistics, which do not have the stricter requirements of corresponding parametric methods.

> ### DEFINITION
>
> **Parametric tests** have requirements about the nature or shape of the populations involved; **nonparametric tests** do not require that samples come from populations with normal distributions or any other particular distributions. Consequently, nonparametric tests of hypotheses are often called **distribution-free tests.**

The term *nonparametric* suggests that the test is not based on a parameter, but there are some nonparametric tests that do depend on a parameter such as the median. The nonparametric tests do not, however, require a particular distribution, so they are sometimes referred to as *distribution-free* tests. The following are major advantages and disadvantages of nonparametric methods.

Advantages of Nonparametric Methods

1. Nonparametric methods can be applied to a wide variety of situations because they do not have the more rigid requirements of the corresponding parametric methods. In particular, nonparametric methods do not require normally distributed populations.

2. Unlike parametric methods, nonparametric methods can often be applied to categorical data, such as the genders of survey respondents.

Disadvantages of Nonparametric Methods

1. Nonparametric methods tend to waste information because exact numerical data are often reduced to a qualitative form. For example, with the nonparametric sign test (described in Section 13-2), weight losses by dieters are recorded simply as negative signs; the actual magnitudes of the weight losses are ignored.

2. Nonparametric tests are not as efficient as parametric tests, so with a nonparametric test we generally need stronger evidence (such as a larger sample or greater differences) in order to reject a null hypothesis.

When the requirements of population distributions are satisfied, nonparametric tests are generally less efficient than their parametric counterparts, but the reduced efficiency can be compensated for by an increased sample size. For example, in Section 13-6 we present a concept called *rank correlation,* which has an efficiency rating of 0.91 when compared to the linear correlation presented in Chapter 10. This means that with all other things being equal, nonparametric rank correlation requires 100 sample observations to achieve the same results as 91 sample observations analyzed through parametric linear correlation, assuming the stricter requirements for using the parametric method are met. Table 13-2 lists the nonparametric methods covered in this chapter, along with the corresponding parametric method and **efficiency** rating. Table 13-2 shows that several nonparametric tests have efficiency ratings above 0.90, so the lower efficiency might not be an important factor in choosing between

Table 13-2 *Efficiency: Comparison of Parametric and Nonparametric Tests*

Application	Parametric Test	Nonparametric Test	Efficiency Rating of Nonparametric Test with Normal Population
Matched pairs of sample data	*t* test or *z* test	Sign test	0.63
		Wilcoxon signed-ranks test	0.95
Two independent samples	*t* test or *z* test	Wilcoxon rank-sum test	0.95
Several independent samples	Analysis of variance (*F* test)	Kruskal-Wallis test	0.95
Correlation	Linear correlation	Rank correlation test	0.91
Randomness	No parametric test	Runs test	No basis for comparison

parametric and nonparametric methods. However, because parametric tests do have higher efficiency ratings than their nonparametric counterparts, it's generally better to use the parametric tests when their required assumptions are satisfied.

Ranks

In Sections 13-3 through 13-6 we use methods based on ranks, which are defined below.

 **DEFINITION**

Data are *sorted* when they are arranged according to some criterion, such as smallest to largest or best to worst. A **rank** is a number assigned to an individual sample item according to its order in the sorted list. The first item is assigned a rank of 1, the second item is assigned a rank of 2, and so on.

Handling ties in ranks: If a tie in ranks occurs, the usual procedure is to find the mean of the ranks involved and then assign this mean rank to each of the tied items, as in the following example.

EXAMPLE 1 The numbers 4, 5, 5, 5, 10, 11, 12, and 12 are given ranks of 1, 3, 3, 3, 5, 6, 7.5, and 7.5, respectively. The table below illustrates the procedure for handling ties.

Sorted Data	Preliminary Ranking	Rank
4	1	**1**
5	2	**3**
5	3 } Mean is 3.	**3**
5	4	**3**
10	5	**5**
11	6	**6**
12	7 } Mean is 7.5.	**7.5**
12	8	**7.5**

 ## 13-2 Sign Test

Key Concept In this section we discuss how to apply the *sign test* procedure, which can be used with claims about data consisting of matched pairs, claims involving nominal data, or claims about the median of a population. The sign test involves converting data values to plus and minus signs, then testing for disproportionately more of either sign.

DEFINITION

The **sign test** is a nonparametric (distribution-free) test that uses plus and minus signs to test different claims, including:

1. Claims involving matched pairs of sample data
2. Claims involving nominal data
3. Claims about the median of a single population

Basic Concept of the Sign Test   The basic idea underlying the sign test is to analyze the frequencies of the plus and minus signs to determine whether they are significantly different. For example, consider the results of clinical trials of the MicroSort method of gender selection. Among 726 couples who used the XSORT method in trying to have a baby girl, 668 couples did have baby girls. Is 668 girls in 726 births *significant?* Common sense probably suggests that 668 girls in 726 births is significant, but what about 365 girls in 726 births? Or 400 girls in 726 births? The sign test allows us to determine when such results are significant. Figure 13-1 summarizes the sign test procedure.

For consistency and simplicity, we will use a test statistic based on the number of times that the ***less frequent*** sign occurs.

Sign Test

Objective

Test a claim about matched pairs of data, or test a claim about nominal data with two categories, or test a claim about the median of a population using the sign test.

Notation

x = the number of times the *less frequent* sign occurs n = the total number of positive and negative signs combined (after discarding any zeros)

Requirements

The sample data are a simple random sample.
Note: There is *no* requirement that the sample data come from a population with a particular distribution, such as a normal distribution.

Test Statistic

x (the number of times the less frequent sign occurs)

P-Values

P-values can be found using the TI-83/84 Plus calculator.

TI-83/84 Plus   First discard any zeros. Identify the value of x, which is the number of times the less frequent sign occurs, then identify the value of n (the total number of positive and negative signs combined). The TI-83/84 Plus calculator can then be used to find the *P*-value. *Because the TI-83/84 Plus calculator provides an exact P-value in almost all cases, we use the P-value method of testing hypotheses.* Instructions are given at the end of this section.

When using the TI-83/84 Plus calculator, the basic idea is to assume that positive and negative signs are equally likely to occur, so that each has a probability of 0.5. Then use the **binomcdf** function to find the probability of getting x or a smaller frequency among n signs. That probability can then be used to find the *P*-value. We should be careful to avoid the mistake of supporting an alternative hypothesis when the sample data contradict that alternative hypothesis, as in Example 1 which follows.

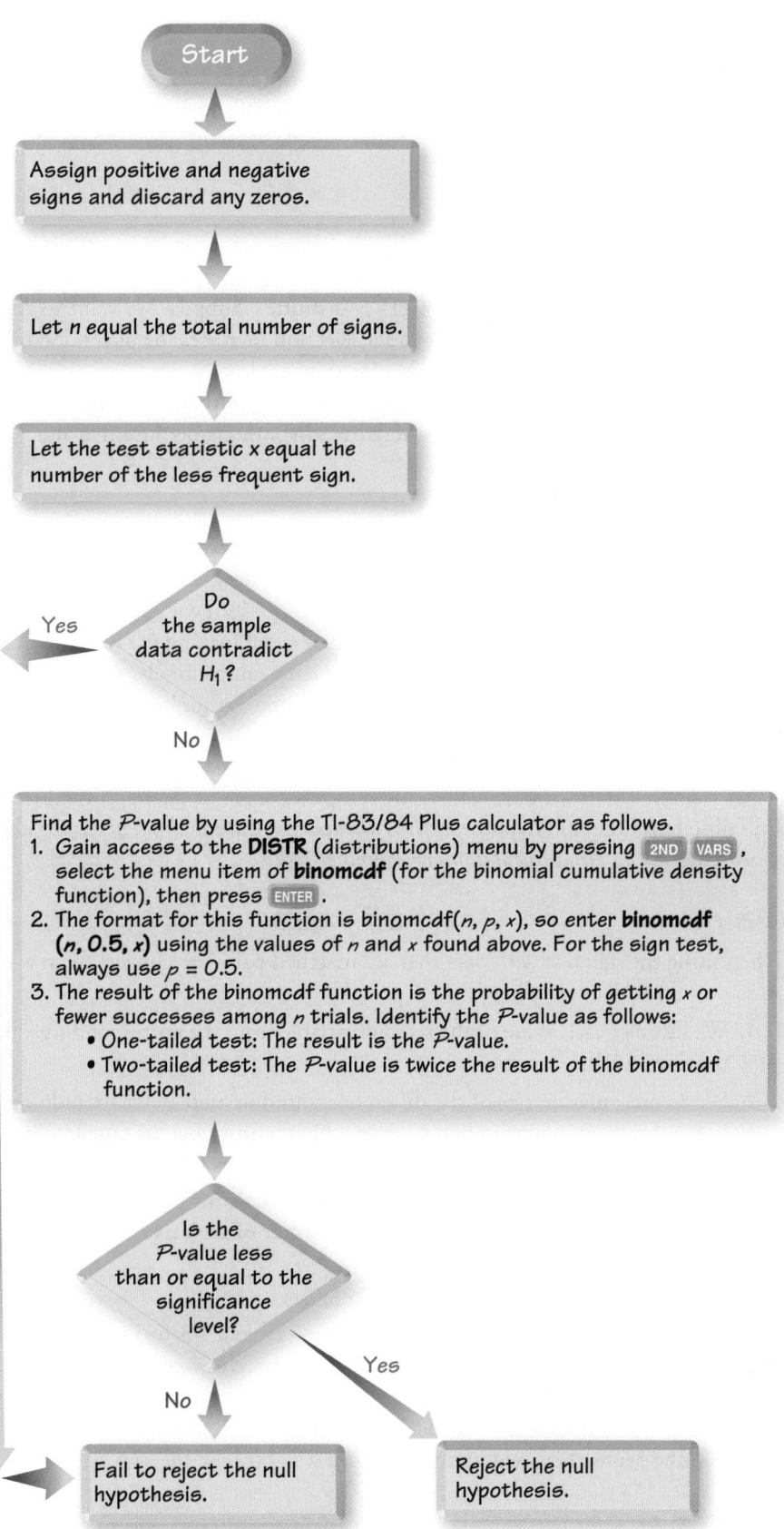

Figure 13-1 Sign Test Procedure

Class Attendance and Grades

In a study of 424 under-graduates at the University of Michigan, it was found that students with the worst attendance records tended to get the lowest grades. (Is anybody surprised?) Those who were absent less than 10% of the time tended to receive grades of B or above. The study also showed that students who sit in the front of the class tend to get significantly better grades.

Gender Gap in Drug Testing

A study of the relationship between heart attacks and doses of aspirin involved 22,000 male physicians. This study, like many others, excluded women. The General Accounting Office recently criticized the National Institutes of Health for not including both sexes in many studies because results of medical tests on males do not necessarily apply to females. For example, women's hearts are different from men's in many important ways. When forming conclusions based on sample results, we should be wary of an inference that extends to a population larger than the one from which the sample was drawn.

CAUTION

When applying the sign test in a one-tailed test, we need to be very careful to avoid making the wrong conclusion when one sign occurs significantly more often than the other, but the sample data contradict the alternative hypothesis. See the following example.

> **EXAMPLE 1**   **Data Contradicting the Alternative Hypothesis** Among 726 couples who used the XSORT method of gender selection, 58 had boys (based on data from the Genetics & IVF Institute). Suppose we want to test the claim that the XSORT method of gender selection increases the likelihood of baby *boys* so that the probability of a boy is $p > 0.5$.
>
> The claim of $p > 0.5$ becomes the alternative hypothesis, but the sample proportion of 58/726 *contradicts* the alternative hypothesis because it is not greater than 0.5.

> **INTERPRETATION**   An alternative hypothesis can never be supported with data that contradict it. It should be obvious that we could never support a claim that $p > 0.5$ with a sample proportion of 58/726 (or 0.0799), which is *less than* 0.5.
>
> When testing a claim, we should be careful to avoid making the fundamental mistake of thinking that a claim is supported because the sample results are significant; the sample results must be significant in the same direction as the alternative hypothesis.

Figure 13-1 summarizes the procedure for the sign test and includes this check: Do the sample data contradict H_1? If the sample data are in the opposite direction of H_1, fail to reject the null hypothesis. *It is always important to think about the data and to avoid relying on blind calculations or computer results.*

Claims Involving Matched Pairs

When using the sign test with data that are matched pairs, we convert the raw data to plus and minus signs as follows:

1. We subtract each value of the second variable from the corresponding value of the first variable.

2. We record only the *sign* of the difference found in Step 1. We exclude *ties:* that is, we throw out any matched pairs in which both values are equal.

The main concept underlying this use of the sign test is as follows:

> **If the two sets of data have equal medians, the number of positive signs should be approximately equal to the number of negative signs.**

> **EXAMPLE 2**   **Freshman Weight Gain** Table 13-3 includes some of the weights listed in Data Set 3 in Appendix B. Those weights were measured from college students in September and April of their freshman year. Use the sample data in Table 13-3 with a 0.05 significance level to test the claim that there is no difference between the September weights and the April weights. Use the sign test.

Table 13-3   **Weight (kg) Measurements of Students in Their Freshman Year**

September weight	67	53	64	74	67	70	55	74	62	57
April weight	66	52	68	77	67	71	60	82	65	58
Sign of difference	+	+	−	−	0	−	−	−	−	−

SOLUTION   **REQUIREMENT CHECK** The only requirement of the sign test is that the sample data are a simple random sample. Instead of being a simple random sample of selected students, all subjects volunteered for the study, so the requirement is not satisfied. This limitation is cited in the journal article describing the results of the study. We will proceed as if the requirement of a simple random sample is satisfied. ✓

If there is no difference between the April weights and the corresponding September weights, the numbers of positive and negative signs should be approximately equal. In Table 13-3 we have 7 negative signs, 2 positive signs, and 1 difference of 0. The sign test tells us whether or not the numbers of positive and negative signs are approximately equal.

The null hypothesis is the claim of no difference between the April weights and the September weights, and the alternative hypothesis is the claim that there is a difference.

H_0: There is no difference. (The median of the differences is equal to 0.)

H_1: There is a difference. (The median of the differences is not equal to 0.)

Following Figure 13-1, we let $n = 9$ (the total number of signs) and we let $x = 2$.

The sample data do not contradict H_1, because there is a difference between the 2 positive signs and the 7 negative signs. The sample data show a difference, and we need to continue with the test to determine whether that difference is significant.

TI-83/84 Plus Figure 13-1 shows that we should find the *P*-value by using the binomcdf function. With $x = 2$ and $n = 9$, we enter **binomcdf(9, 0.5, 2)** to get a probability of 0.08984375, but this value is doubled because we have a two-tailed test. The *P*-value is $2(0.08984375) = 0.1797$ (rounded). That relatively large *P*-value shows that it is easy to get 2 or fewer positive signs among 9. Because the *P*-value of 0.1797 is greater than the significance level of 0.05, we fail to reject the null hypothesis.

There is not sufficient evidence to warrant rejection of the claim that the median of the differences is equal to 0.

INTERPRETATION   We conclude that the September and April weights do not appear to be different. (If we use the parametric *t* test for matched pairs (Section 9-4), we conclude that the mean difference is not zero, so the September weights and April weights appear to be different.)

The conclusion should be qualified with the limitations noted in the article about the study. Only Rutgers students were used, and study subjects were volunteers instead of being a simple random sample.

Claims Involving Nominal Data with Two Categories

In Chapter 1 we defined nominal data to be data that consist of names, labels, or categories only. The nature of nominal data limits the calculations that are possible, but we can identify the *proportion* of the sample data that belong to a particular category, and we can test claims about the corresponding population proportion *p*. The following example uses nominal data consisting of genders (girls/boys). The sign test is used by representing girls with positive (+) signs and boys with negative (−) signs. (Those signs are chosen arbitrarily, honest.)

EXAMPLE 3   **Gender Selection** The Genetics and IVF Institute conducted a clinical trial of its methods for gender selection. As of this writing, 668 of 726 babies born to parents using the XSORT method of gender selection were girls. Use the sign test and a 0.05 significance level to test the claim that this method of gender selection is effective in increasing the likelihood of a baby girl.

SOLUTION   **REQUIREMENT CHECK** The only requirement is that the sample data are a simple random sample. Based on the design of this experiment, we can assume that the sample data are a simple random sample. ✓

Let p denote the population proportion of baby girls. The claim that girls are more likely with the XSORT method can be expressed as $p > 0.5$, so the null and alternative hypotheses are as follows:

$$H_0: \quad p = 0.5 \quad \text{(the proportion of girls is equal to 0.5)}$$
$$H_1: \quad p > 0.5 \quad \text{(girls are more likely)}$$

Denoting girls by the positive sign $(+)$ and boys by the negative sign $(-)$, we have 668 positive signs and 58 negative signs. Using the sign test procedure summarized in Figure 13-1, we let the test statistic x be the smaller of 668 and 58, so $x = 58$ boys. *Instead of trying to determine whether 668 girls is high enough to be significant, we proceed with the equivalent task of trying to determine whether 58 boys is low enough to be significant, so we treat the test as a* left-tailed *test.*

The sample data do not contradict the alternative hypothesis because the sample proportion of girls is 668/726, which is greater than 0.5, as in the above alternative hypothesis.

We let $x = 58$ (the less frequent sign) and $n = 726$ (the total number of signs). If we use the TI-83/84 Plus calculator with the procedure given at the end of this section, we use the command **binomcdf(726, 0.5, 58)** to get a *P*-value of 0. (The actual probability is a positive number so small that it can be considered to be 0 for all practical purposes.) The binomcdf function gives us the probability of *x or fewer* outcomes, so we know that if boys and girls are equally likely, it is extremely unlikely that we would get 58 or fewer boys (or, equivalently, it is extremely unlikely that we would get 668 or more girls). Because the *P*-value is less than the significance level of 0.05, we reject the null hypothesis that the proportion of girls is equal to 0.5. There is sufficient sample evidence to support the claim that girls are more likely with the XSORT method.

INTERPRETATION   The XSORT method of gender selection does appear to be effective in increasing the likelihood of a girl.

Claims About the Median of a Single Population

The next example illustrates the procedure for using the sign test in testing a claim about the median of a single population. See how the negative and positive signs are based on the claimed value of the median.

EXAMPLE 4   **Body Temperatures** Data Set 2 in Appendix B includes measured body temperatures of adults. Use the 106 temperatures listed for 12 A.M. on Day 2 with the sign test to test the claim that the median is less than 98.6°F. Of the 106 subjects, 68 had temperatures below 98.6°F, 23 had temperatures above 98.6°F, and 15 had temperatures equal to 98.6°F.

SOLUTION   **REQUIREMENT CHECK** The only requirement is that the sample data are a simple random sample. Based on the design of this experiment, we assume that the sample data are a simple random sample. ✓

The claim that the median is less than 98.6°F is the alternative hypothesis, while the null hypothesis is the claim that the median is equal to 98.6°F.

H_0: Median is equal to 98.6°F.   (median $=$ 98.6°F)

H_1: Median is less than 98.6°F.   (median $<$ 98.6°F)

Following the procedure outlined in Figure 13-1, we use the negative sign ($-$) to denote each temperature that is below 98.6°F, and we use the positive sign ($+$) to denote each temperature that is above 98.6°F. Note that we discard the 15 data values of 98.6 since they result in differences of zero. We therefore have 68 negative signs and 23 positive signs, so $n = 91$ and $x = 23$ (the number of the less frequent sign). The sample data do not contradict the alternative hypothesis, because most of the 91 temperatures are below 98.6°F.

If we use the TI-83/84 Plus calculator with the procedure given at the end of this section, we use the command **binomcdf(91, 0.5, 23)** to get the *P*-value of 0.00000126. Because the *P*-value is so small, we reject the null hypothesis that the median is equal to 98.6°F.

INTERPRETATION There is sufficient sample evidence to support the claim that the median body temperature of healthy adults is less than 98.6°F. It is not equal to 98.6, as is commonly believed.

In Example 4, the sign test of the claim that the median is below 98.6°F results in a *P*-value of 0.00000126. However, a parametric test of the claim that $\mu < 98.6$°F results in a test statistic of $t = -6.611$ with a *P*-value of 0.000000000813. Because the *P*-value from the sign test is not as low as the *P*-value from the parametric test, we see that the sign test isn't as sensitive as the parametric test. Both tests lead to rejection of the null hypothesis, but the sign test doesn't consider the sample data to be as extreme, partly because the sign test uses only information about the *direction* of the data, ignoring the *magnitudes* of the data values. The next section introduces the Wilcoxon signed-ranks test, which largely overcomes that disadvantage.

Using the TI-83/84 Plus Calculator for the Sign Test

The TI-83/84 Plus calculator does not have a built-in function dedicated to the sign test, but it is easy to use the calculator's **binomcdf** function to find the *P*-value for a sign test. The binomcdf function uses the binomial probability distribution for finding the probability of *x* or fewer successes.

1. Assign positive and negative signs and discard any zeros. Let the test statistic *x* equal the count of the less frequent sign, and let *n* equal the number of positive and negative signs combined.

2. Verify that the sample data do not contradict the alternative hypothesis. (If the sense of the sample data is in conflict with the alternative hypothesis, skip the remaining steps and fail to reject the null hypothesis.)

3. Gain access to the **DISTR** (distributions) menu by pressing **2ND** **VARS**.

4. Scroll down to select the menu item of **binomcdf** (for the binomial cumulative density function), then press **ENTER**.

5. The format for this function is binomcdf(*n, p, x*), so enter **binomcdf(*n*, 0.5, *x*),** using the values of *n* and *x* found in Step 1. For the sign test, always use $p = 0.5$.

6. The result of Step 5 is the probability of getting *x* or fewer successes among *n* trials. *Double this value for two-tailed tests.* The final result is the *P*-value for the sign test.

7. If the *P*-value is less than or equal to the significance level, reject the null hypothesis. If the *P*-value is greater than the significance level, fail to reject the null hypothesis.

13-2   Basic Skills and Concepts

Statistical Literacy and Critical Thinking

1. Nonparametric Test The Genetics and IVF Institute conducted a clinical trial of its methods for gender selection. As of this writing, 172 of 211 babies born to parents using the YSORT method were boys. If the sign test is used, why is it considered to be a "nonparametric" test or a "distribution-free" test?

2. Sign Test For the hypothesis test referred to in Exercise 1, what is it about the procedure that causes us to refer to it as the "sign" test?

3. Sign Test Procedure The Genetics and IVF Institute conducted a clinical trial of its YSORT gender-selection method. The results consisted of 172 boys and 39 girls. Assume that you must conduct a test of the claim that the YSORT method increases the likelihood of *girls*. In what sense is the alternative hypothesis contradicted by the data? Why isn't it necessary to actually conduct the test?

4. Efficiency of the Sign Test Refer to Table 13-2 on page 667 and identify the efficiency of the sign test. What does that value tell us about the sign test?

In Exercises 5–8, assume that matched pairs of data result in the given number of signs when the value of the second variable is subtracted from the corresponding value of the first variable. Use the sign test with a 0.05 significance level to test the null hypothesis of no difference.

5. Positive signs: 13; negative signs: 1; ties: 0 (from a preliminary test of the MicroSort method of gender selection)

6. Positive signs: 5; negative signs: 7; ties: 1 (from a class project testing for the difference between reported and measured heights of males)

7. Positive signs: 360; negative signs: 374; ties: 22 (from a Gallup poll of Internet users who were asked if they make travel plans through the Internet)

8. Positive signs: 512; negative signs: 327; ties: 0 (from challenges to referee calls in the U.S. Open tennis tournament)

In Exercises 9–12, use the sign test for the data consisting of matched pairs.

9. Oscar Winners Listed below are ages of actresses and actors at the times that they won Oscars. The data are paired according to the years that they won. Use a 0.05 significance level to test the claim that there is no difference between the ages of best actresses and the ages of best actors at the time that the awards were presented.

Best Actresses	28	32	27	27	26	24	25	29	41	40	27	42	33	21	35
Best Actors	62	41	52	41	34	40	56	41	39	49	48	56	42	62	29

10. Airline Fares Listed below are the costs (in dollars) of flights from New York (JFK) to San Francisco for US Air, Continental, Delta, United, American, Alaska, and Northwest. Use a 0.05 significance level to test the claim that there is no difference in cost between flights scheduled one day in advance and those scheduled 30 days in advance. What appears to be a wise scheduling strategy?

Flight scheduled one day in advance	456	614	628	1088	943	567	536
Flight scheduled 30 days in advance	244	260	264	264	278	318	280

11. Tobacco and Alcohol in Children's Movies Listed below are times (in sec) that animated Disney movies showed the use of tobacco and alcohol. (See Data Set 7 in Appendix B.) Use a 0.05 significance level to test the claim that for a typical animated movie, the time spent depicting the use of alcohol is less than the time spent depicting the use of tobacco.

Tobacco use (sec)	176	51	0	299	74	2	23	205	6	155
Alcohol use (sec)	88	33	113	51	0	3	46	73	5	74

12. Car Fuel Consumption Ratings Listed below are combined city-highway fuel consumption ratings (in miles/gal) for different cars measured under the old rating system and a new rating system introduced in 2008 (based on data from *USA Today*). The new ratings were implemented in response to complaints that the old ratings were too high. Use a 0.01 significance level to test the claim the old ratings are higher than the new ratings.

| Old rating | 16 | 18 | 27 | 17 | 33 | 28 | 33 | 18 | 24 | 19 | 18 | 27 | 22 | 18 | 20 | 29 | 19 | 27 | 20 | 21 |
| New rating | 15 | 16 | 24 | 15 | 29 | 25 | 29 | 16 | 22 | 17 | 16 | 24 | 20 | 16 | 18 | 26 | 17 | 25 | 18 | 19 |

In Exercises 13–16, use the sign test for the claim involving nominal data.

13. Gender Selection The Genetics and IVF Institute conducted a clinical trial of its methods for gender selection. As of this writing, 172 of 211 babies born to parents using the YSORT method were boys. Use a 0.01 significance level to test the claim that the YSORT method is effective in increasing the likelihood of a boy.

14. Cheating Gas Pumps When testing gas pumps in Michigan for accuracy, fuel-quality enforcement specialists tested pumps and found that 1299 were not accurate (within 3.3 oz when 5 gal is pumped), and 5686 were accurate. Use a 0.01 significance level to test the claim of an industry representative that less than half of Michigan gas pumps are inaccurate.

15. Predicting Sex of Baby A study addressed the issue of whether women have the ability to predict the sex of their babies. Among 104 recruited subjects, 55% correctly guessed the sex of the baby (based on data from "Are Women Carrying 'Basketballs' Really Having Boys? Testing Pregnancy Folklore," by Perry, DiPietro, and Constigan, *Birth,* Vol. 26, No. 3). Use a 0.05 significance level to test the claim that women do not have the ability to predict the sex of their babies.

16. Stem Cell Survey Adults randomly selected for a *Newsweek* poll were asked if they "favor or oppose using federal tax dollars to fund medical research using stem cells obtained from human embryos." Of the subjects surveyed, 481 were in favor, 401 were opposed, and 120 were unsure. A politician claims that people don't really understand the stem cell issue and their responses to such questions are random responses equivalent to a coin flip. Use a 0.01 significance level to test the claim that the proportion of subjects who respond in favor is equal to 0.5. What does the result suggest about the politician's claim?

Appendix B Data Sets. *In Exercises 17–20, refer to the indicated data set in Appendix B and use the sign test for the claim about the median of a population.*

17. Testing for Median Weight of Quarters Refer to Data Set 20 in Appendix B for the weights (in g) of randomly selected quarters that were minted after 1964. The quarters are supposed to have a median weight of 5.670 g. Use a 0.01 significance level to test the claim that the median is equal to 5.670 g. Do quarters appear to be minted according to specifications?

18. Voltage Levels Refer to Data Set 13 in Appendix B for the home voltage levels. The power company (Central Hudson) supplying the power states that the target voltage is 120 V. Use a 0.01 significance level to test the claim that the median voltage is equal to 120 V.

19. Coke Contents Refer to Data Set 17 in Appendix B for the amounts (in oz) in cans of regular Coke. The cans are labeled to indicate that the contents are 12 oz of Coke. Use a 0.05 significance level to test the claim that cans of Coke are filled so that the median amount is 12 oz. If the median is not 12 oz, are consumers being cheated?

20. Lengths of Sheet Metal Screws Refer to Data Set 19 in Appendix B for the lengths (in inches) of a simple random sample of 50 stainless steel sheet metal screws obtained from those supplied by Crown Bolt, Inc. The screws are packaged with a label indicating 3/4 in. length. Use a 0.05 significance level to test the claim that the screws have a median equal to 3/4 in. (or 0.75 in.). Do the screws appear to have lengths consistent with the label?

13-2 Beyond the Basics

21. Procedures for Handling Ties In the sign test procedure described in this section, we exclude ties (represented by 0 instead of a sign of $+$ or $-$). A second approach is to treat half of the 0s as positive signs and half as negative signs. (If the number of 0s is odd, exclude one so that they can be divided equally.) With a third approach, in two-tailed tests make half of the 0s positive and half negative; in one-tailed tests make all 0s either positive or negative, whichever supports the null hypothesis. Repeat Example 4 using the second and third approaches to handling ties. Do the different approaches lead to very different results?

13-3   Wilcoxon Signed-Ranks Test for Matched Pairs

Key Concept In this section we introduce the *Wilcoxon signed-ranks test,* which involves the conversion of the sample data to ranks. This test can be used for the two different applications described in the following definition.

> **DEFINITION**
>
> The **Wilcoxon signed-ranks test** is a nonparametric test that uses ranks for these applications:
> 1. Test a null hypothesis that the population of matched pairs has differences with a median equal to zero.
> 2. Test a null hypothesis that a single population has a claimed value of the median.

Claims Involving Matched Pairs

The sign test (Section 13-2) can be used with matched pairs, but the sign test uses only the signs of the differences. By using ranks, the Wilcoxon signed-ranks test takes the magnitudes of the differences into account. Because the Wilcoxon signed-ranks test incorporates and uses more information than the sign test, it tends to yield conclusions that better reflect the true nature of the data.

Wilcoxon Signed-Ranks Test

Objective

Use the Wilcoxon signed-ranks test with matched pairs for the following null and alternative hypotheses:

H_0: The matched pairs have differences that come from a population with a median equal to zero.

H_1: The matched pairs have differences that come from a population with a nonzero median.

Notation

$T =$ the smaller of the following two sums:

1. The sum of the positive ranks of the nonzero differences d

2. The absolute value of the sum of the negative ranks of the nonzero differences d

Requirements

1. The data consist of matched pairs that are a simple random sample.

2. The population of differences (found from the pairs of data) has a distribution that is approximately *symmetric,* meaning that the left half of its histogram is roughly a mirror image of its right half.

Note: There is *no* requirement that the data have a normal distribution.

Test Statistic

If $n \leq 30$, the test statistic is T.

If $n > 30$, the test statistic is

$$z = \frac{T - \dfrac{n(n+1)}{4}}{\sqrt{\dfrac{n(n+1)(2n+1)}{24}}}$$

Critical Values

1. If $n \leq 30$, the critical T value is found in Table A-7.   2. If $n > 30$, the critical z values are found in Table A-2.

Wilcoxon Signed-Ranks Procedure To see how the following steps are applied, refer to the sample data listed in Table 13-4.

Step 1: For each pair of data, find the difference d by subtracting the second value from the first value. Discard any pairs that have a difference of 0.

EXAMPLE: The third row of Table 13-4 lists the differences found by subtracting the April weights from the corresponding September weights.

Step 2: *Ignore the signs of the differences,* then sort the differences from lowest to highest and replace the differences by the corresponding rank value (as described in Section 13-1). When differences have the same numerical value, assign to them the mean of the ranks involved in the tie.

EXAMPLE: The fourth row of Table 13-4 shows the ranks of the values of $|d|$. Consider the d values of 1, 1, -1, -1. If we ignore their signs, they are tied for the rank values of 1, 2, 3, 4, so they are each assigned a rank of 2.5, which is the mean of the ranks involved in the tie (or the mean of 1, 2, 3, 4).

Step 3: Attach to each rank the sign of the difference from which it came. That is, insert the signs that were ignored in Step 2.

EXAMPLE: The bottom row of Table 13-4 lists the same ranks found in the fourth row, but the signs of the differences shown in the third row are inserted.

Step 4: Find the sum of the ranks that are positive. Also find the absolute value of the sum of the negative ranks.

EXAMPLE: The bottom row of Table 13-4 lists the signed ranks. The sum of the positive ranks is $2.5 + 2.5 = 5$. The sum of the negative ranks is $(-7) + (-5.5) + (-2.5) + (-8) + (-9) + (-5.5) + (-2.5) = -40$, and the absolute value of this sum is 40. The two rank sums are 5 and 40.

Step 5: Let T be the *smaller* of the two sums found in Step 4. Either sum could be used, but for a simplified procedure we arbitrarily select the smaller of the two sums.

EXAMPLE: The data in Table 13-4 result in the rank sums of 5 and 40, so 5 is the smaller of those two sums.

Step 6: Let n be the number of pairs of data for which the difference d is not 0.

EXAMPLE: The data in Table 13-4 have 9 differences that are not 0, so $n = 9$.

Step 7: Determine the test statistic and critical values based on the sample size, as shown in the preceding box.

EXAMPLE: For the data in Table 13-4 the test statistic is $T = 5$. The sample size is $n = 9$, so the critical value is found in Table A-7. Using a 0.05 significance level with a two-tailed test, the critical value from Table A-7 is 6.

Table 13-4   Weights (in kg) of Students in Their Freshman Year

September weight	67	53	64	74	67	70	55	74	62	57
April weight	66	52	68	77	67	71	60	82	65	58
d (difference)	1	1	-4	-3	0	-1	-5	-8	-3	-1
Rank of $|d|$	2.5	2.5	7	5.5		2.5	8	9	5.5	2.5
Signed rank	2.5	2.5	-7	-5.5		-2.5	-8	-9	-5.5	-2.5

Step 8: When forming the conclusion, reject the null hypothesis if the sample data lead to a test statistic that is in the critical region—that is, the test statistic is less than or equal to the critical value(s). Otherwise, fail to reject the null hypothesis.

EXAMPLE: If the test statistic is T (instead of z), reject the null hypothesis if T is less than or equal to the critical value. Fail to reject the null hypothesis if T is greater than the critical value. Since $T = 5$ and the critical value is 6, we reject the null hypothesis.

> **EXAMPLE 1**   **Freshman Weight Gain** The first two rows of Table 13-4 include some of the weights from Data Set 3 in Appendix B. Those weights were measured from college students in September and April of their freshman year. Use the sample data in the first two rows of Table 13-4 to test the claim that there is no difference between the September weights and the April weights. Use the Wilcoxon signed-ranks test with a 0.05 significance level.

TI-83/84 PLUS

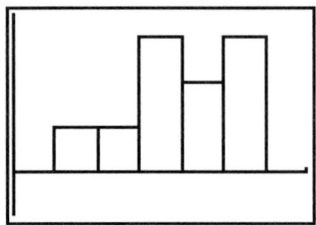

SOLUTION   **REQUIREMENT CHECK** (1) The data should be a simple random sample. Instead of being a simple random sample of selected students, all subjects volunteered, and this is discussed in the journal article describing the results of the study. We will proceed as if the requirement of a simple random sample is satisfied. (2) The TI-83/84 Plus histogram of the differences in the third row of Table 13-4 is shown here. The left side of the graph should be roughly a mirror image of the right side, which does not appear to be the case. But with only 9 differences, the difference between the left and right sides is not too extreme, so we will consider this requirement to be satisfied.

The null hypothesis is the claim of no difference between the April weights and the September weights, and the alternative hypothesis is the claim that there is a difference.

H_0: There is no difference.   (The median of the differences is equal to 0.)

H_1: There is a difference.   (The median of the differences is not equal to 0.)

Test Statistic: Because we are using the Wilcoxon signed-ranks test, the test statistic is calculated by using the eight-step procedure presented earlier in this section. Those steps include examples illustrating the calculation of the test statistic with the sample data in Table 13-4, and the result is the test statistic of $T = 5$.

Critical Value: The sample size is $n = 9$, so the critical value is found in Table A-7. Using a 0.05 significance level with a two-tailed test, the critical value from Table A-7 is found to be 6.

Conclusion: Table A-7 includes a note stating that we should reject the null hypothesis if the test statistic T is less than or equal to the critical value. Because the test statistic of $T = 5$ is less than the critical value of 6, we reject the null hypothesis.

INTERPRETATION   We conclude that the September and April weights do *not* appear to be about the same. The large number of negative differences indicates that most students gained weight during their freshman year. The conclusion should be qualified with the limitations noted in the article about the study. Only Rutgers students were used, and study subjects were volunteers instead of being a simple random sample.

In Example 1, if we use the parametric t test for matched pairs (Section 9-4), we conclude that the mean difference is not zero, so the September weights and April weights appear to be different, as in Example 1. However, the sign test in Section 13-2 led to the conclusion of no difference. By using only positive and negative signs, the sign test did not use the magnitudes of the differences, but the Wilcoxon signed-ranks test was more sensitive to those magnitudes through its use of ranks.

Claims About the Median of a Single Population

The Wilcoxon signed-ranks test can also be used to test a claim that a single population has some claimed value of the median. The preceding procedures can be used with one simple adjustment:

> **When finding the differences d in Step 1, subtract the claimed value of the median from each of the sample values (instead of finding the difference between each matched pair of sample values).**

This adjustment allows us to treat the sample of individual values as a sample consisting of matched pairs, so we can use the same procedure described earlier in this section.

EXAMPLE 2   **Body Temperatures** Data Set 2 in Appendix B includes measured body temperatures of adults. Use the 106 temperatures listed for 12 A.M. on Day 2 with the Wilcoxon signed-ranks test to test the claim that the median is less than 98.6°F.

SOLUTION   **REQUIREMENT CHECK** (1) By pairing each individual sample value with the median of 98.6°F, we satisfy the requirement of having matched pairs. The design of the experiment that led to the data in Data Set 2 justifies treating the sample as a simple random sample. (2) The requirement of an approximately symmetric distribution of differences is satisfied, because a histogram of those differences is approximately symmetric. ✓

If we use the TI-83/84 Plus calculator as described at the end of this section, we find that $n = 91$ and $T = 661$, and we also get the results in the accompanying screen display. From the test statistic $z = 5.66772127$, we get a P-value of 0.000 (rounded), so we reject the null hypothesis that the population of differences between body temperatures and the claimed median of 98.6°F is zero. There is sufficient evidence to support the claim that the median body temperature is less than 98.6°F. This is the same conclusion that results from the sign test, as in Example 4 in Section 13-2.

Results from Program SRTEST

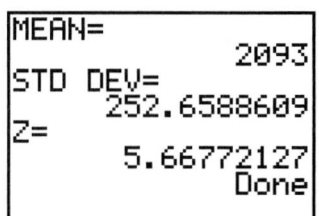

```
MEAN=
                   2093
STD DEV=
           252.6588609
Z=
             5.66772127
                   Done
```

Rationale: In Example 1, the unsigned ranks of 1 through 9 have a total of 45, so if there are no significant differences, each of the two signed-rank totals should be around $45 \div 2$, or 22.5. That is, the negative ranks and positive ranks should split up as 22.5–22.5 or something close, such as 21–24. The table of critical values shows that at the 0.05 significance level with 9 pairs of data, a split of 6–39 represents a significant departure from the null hypothesis, and any split that is farther apart (such as 5–40 or 2–43) will also represent a significant departure from the null hypothesis. Conversely, splits like 7–38 do not represent significant departures from a 22.5–22.5 split, and they would not justify rejecting the null hypothesis. The Wilcoxon signed-ranks test is based on the lower rank total, so instead of analyzing both numbers constituting the split, we consider only the lower number.

The sum of all the ranks $1 + 2 + 3 + \cdots + n$ is equal to $n(n + 1)/2$. If this rank sum is to be divided equally between two categories (positive and negative), each of the two totals should be near $n(n + 1)/4$, which is half of $n(n + 1)/2$. Recognition of this principle helps us understand the test statistic used when $n > 30$.

Using the TI-83/84 Plus Calculator for the Wilcoxon Signed-Ranks Test

The TI-83/84 Plus calculator is not programmed for the Wilcoxon signed-ranks test, but the program SRTEST (by Michael Lloyd) can be used as follows.

1. First download and install the program SRTEST from the CD-ROM or the Web site www.aw.com/triola. (If using the CD-ROM, select the **Software** folder, select **TI-83 TI-84 Plus Programs,** select **Michael Lloyd Programs,** then select **programs.** The program SRTEST should now be available for download.)

2. Also download the program ZZRANK, which is used by the program SRTEST.

3. Create a list of differences between values in the matched pairs. (The first set of values can be entered in list L1, the second set of values can be entered in list L2, then the differences can be stored in list L3 by entering L1 − L2 → L3, where the **STO** key is used for the arrow.)

4. Press the **PRGM** key and select **SRTEST.** Press the **ENTER** key twice.

5. When given the prompt of DATA=, enter the list containing the differences, such as L3, then press **ENTER**.

6. The screen will display the sample size indicated by N, the sum of the positive ranks indicated by $T+$, and the sum of the negative ranks indicated by $T-$.

7. Press **ENTER** again to see the mean and standard deviation.

8. Press **ENTER** once again to see the z score.

9. Using the calculator results:
   - If $n \leq 30$, let the test statistic T equal the smaller of the two rank sums found in Step 6. Find the critical T value from Table A-7.
   - If $n > 30$, let the test statistic z equal the value given in Step 8. Find the critical z values from Table A-2, or use the z test statistic to find the P-value.

See Example 2, where the program SRTEST yields $n = 91$ and $T = 661$. (The other rank sum is given as 3525.) The SRTEST results also include the test statistic $z = 5.66772127$ shown in the screen display included with Example 2. Because $n > 30$, that test statistic can be used to find the P-value of 0.000 (rounded).

13-3 Basic Skills and Concepts

Statistical Literacy and Critical Thinking

1. Wilcoxon Signed-Ranks Test and the Sign Test The same sample data are used for the sign test in Example 2 in Section 13-2 and Example 1 for the Wilcoxon signed-ranks test in this section. Why do the two tests result in different conclusions? Which conclusion is likely to be better? Why?

2. Fraternal Twins In a study of fraternal twins of different genders, researchers measured the heights of each twin. The data show that for each matched pair, the height of the male is greater than the height of his twin sister. What is the value of T?

3. Sample Size and Critical Value In 1908, William Gosset published the article "The Probable Error of a Mean" under the pseudonym of "Student" (*Biometrika,* Vol. 6, No. 1). He included the data listed below for yields from two different types of seed (regular and kiln dried) that were used on adjacent plots of land. The listed values are the yields of straw in cwt per acre, where cwt represents 100 lb. If the Wilcoxon signed-ranks test is used to test the claim that there is no difference between the yields from the two types of seed, what is the sample size *n*? If the significance level is 0.05, what is the critical value?

Regular	19.25	22.75	23	23	22.5	19.75	24.5	15.5	18	14.25	17
Kiln dried	25	24	24	28	22.5	19.5	22.25	16	17.25	15.75	17.25

4. Ranks Using the matched data listed in Exercise 3, the differences are as follows: −5.75, −1.25, −1, −5, 0, 0.25, 2.25, −0.5, 0.75, −1.5, and −0.25. List the corresponding ranks of those differences after discarding the 0 and ignoring their signs.

Using the Wilcoxon Signed-Ranks Test. *In Exercises 5 and 6, refer to the given paired sample data and use the Wilcoxon signed-ranks test to test the claim that the matched pairs have differences that come from a population with a median equal to zero.*

TI **5. Is Friday the 13th Unlucky?** Researchers collected data on the numbers of hospital admissions resulting from motor vehicle crashes. The results given below are for Fridays on the 6th of a month and Fridays on the 13th of the same month (based on data from "Is Friday the 13th Bad for Your Health?" by Scanlon, et al., *BMJ,* Vol. 307, as listed in the *Data and Story Line* online resource of data sets). Use a 0.05 significance level to test the claim that when the 13th day of a month falls on a Friday, the numbers of hospital admissions from motor vehicle crashes are not affected.

Friday the 6th	9	6	11	11	3	5
Friday the 13th	13	12	14	10	4	12

TI **6. Weather Forecasts?** Listed below are actual high temperatures and the high temperatures forecast one day in advance (based on Data Set 11 in Appendix B). Use a 0.05 significance level to test the claim that the population of differences has a median of zero. What do the results suggest about the accuracy of the predictions?

Actual high temperature	80	77	81	85	73	73	80	72
High temperature forecast one day before	78	75	81	85	76	75	79	74

Using the Wilcoxon Signed-Ranks Test. *In Exercises 7–10, refer to the sample data for the given exercises in Section 13-2. Use the Wilcoxon signed-ranks test to test the claim that the matched pairs have differences that come from a population with a median equal to zero. Use a 0.05 significance level.*

TI **7.** Exercise 9   **TI** **8.** Exercise 10

TI **9.** Exercise 11   **TI** **10.** Exercise 12

Appendix B Data Sets. *In Exercises 11–14, refer to the sample data for the given exercises in Section 13-2. Use the Wilcoxon signed-ranks test for the claim about the median of a population.*

TI **11.** Exercise 17   **TI** **12.** Exercise 18

TI **13.** Exercise 19   **TI** **14.** Exercise 20

Ethics in Survey Reports

The American Association for Public Opinion Research developed a voluntary code of ethics to be used in news reports of survey results. This code requires that the following be included: (1) identification of the survey sponsor, (2) date the survey was conducted, (3) size of the sample, (4) nature of the population sampled, (5) type of survey used, (6) exact wording of survey questions. Surveys funded by the U.S. government are subject to a prescreening that assesses the risk to those surveyed, the scientific merit of the survey, and the guarantee of the subject's consent to participate.

13-3   Beyond the Basics

15. Rank Sums

a. If we have sample paired data with 50 nonzero differences and the sum of the positive ranks is 300, find the absolute value of the sum of the negative ranks.

b. If we have sample paired data with n nonzero differences and one of the two rank sums is k, find an expression for the other rank sum.

13-4 Wilcoxon Rank-Sum Test for Two Independent Samples

Key Concept In this section we introduce the *Wilcoxon rank-sum test*, which uses ranks of values from two independent samples to test the null hypothesis that the two populations have equal medians. The Wilcoxon rank-sum test is equivalent to the **Mann-Whitney** U **test** (see Exercise 13), which is included in some other textbooks and software packages (such as Minitab). The basic idea underlying the Wilcoxon rank-sum test is this: If two samples are drawn from identical populations and the individual values are all ranked as one combined collection of values, then the high and low ranks should fall evenly between the two samples. If the low ranks are found predominantly in one sample and the high ranks are found predominantly in the other sample, we suspect that the two populations have different medians.

Unlike the parametric tests in Section 9-3, the Wilcoxon rank-sum test does *not* require normally distributed populations and it can be used with data at the ordinal level of measurement, such as data consisting of ranks. In Table 13-2 we noted that the Wilcoxon rank-sum test has a 0.95 efficiency rating when compared to the parametric test. Because this test has such a high efficiency rating and involves easier calculations, it is often preferred over the parametric test, even when the requirement of normality is satisfied.

CAUTION

Don't confuse the Wilcoxon rank-sum test for two *independent* samples with the Wilcoxon signed-ranks test for matched pairs. Use Internal Revenue Service as the mnemonic for IRS to remind us of "**I**ndependent: **R**ank **S**um."

 DEFINITION

The **Wilcoxon rank-sum test** is a nonparametric test that uses ranks of sample data from two independent populations to test the null hypothesis that the two independent samples come from populations with equal medians. (The alternative hypothesis is the claim that the two populations have different medians, or that the first population has a median greater than the median of the second population, or that the first population has a median less than the median of the second population.)

Wilcoxon Rank-Sum Test

Objective

Use the Wilcoxon rank-sum test with samples from two independent populations for the following null and alternative hypotheses:

H_0: The two samples come from populations with equal medians.

H_1: The median of the first population is different from (or greater than, or less than) the median from the second population.

Notation

n_1 = size of Sample 1

n_2 = size of Sample 2

R_1 = sum of ranks for Sample 1

R_2 = sum of ranks for Sample 2

R = same as R_1 (sum of ranks for Sample 1)

μ_R = mean of the sample R values that is expected when the two populations have equal medians

σ_R = standard deviation of the sample R values that is expected with two populations having equal medians

Requirements

1. There are two independent simple random samples.

2. Each of the two samples has more than 10 values. (For samples with 10 or fewer values, special tables are available in reference books, such as *CRC Standard*

Probability and Statistics Tables and Formulae, published by CRC Press.)

Note: There is *no* requirement that the two populations have a normal distribution or any other particular distribution.

Test Statistic

$$z = \frac{R - \mu_R}{\sigma_R}$$

where

$$\mu_R = \frac{n_1(n_1 + n_2 + 1)}{2}$$

$$\sigma_R = \sqrt{\frac{n_1 n_2(n_1 + n_2 + 1)}{12}}$$

n_1 = size of the sample from which the rank sum R is found

n_2 = size of the other sample

R = sum of ranks of the sample with size n_1

Critical Values

Critical values can be found in Table A-2 (because the test statistic is based on the normal distribution).

P-Values P-values can be found using the z test statistic and Table A-2.

Procedure for Finding the Value of the Test Statistic To see how the following steps are applied, refer to the sample data listed in Table 13-5 below.

1. Temporarily combine the two samples into one big sample, then replace each sample value with its rank. (The lowest value gets a rank of 1, the next lowest value gets a rank of 2, and so on. If values are tied, assign to them the mean of the ranks involved in the tie. See Section 13-1 for a description of ranks and the procedure for handling ties.)

EXAMPLE: In Table 13-5, the ranks of the combined data set are shown in parentheses. The rank of 1 is assigned to the lowest value of 122. The rank of 2 is assigned to the next lowest value of 127. The rank of 3 is assigned to the next lowest value of 128. The rank of 4.5 is assigned to the values of 129 and 129, since they are tied for the ranks of 4 and 5.

Table 13-5

Braking Distance (ft)	
4-Cylinders	6-Cylinders
136 (12.5)	131 (7.5)
146 (23)	129 (4.5)
139 (15.5)	127 (2)
131 (7.5)	146 (23)
137 (14)	155 (25)
144 (20)	122 (1)
133 (9.5)	143 (18)
144 (20)	133 (9.5)
129 (4.5)	128 (3)
144 (20)	146 (23)
130 (6)	139 (15.5)
140 (17)	136 (12.5)
135 (11)	
$n_1 = 13$	$n_2 = 12$
$R_1 = 180.5$	$R_2 = 144.5$

2. Find the sum of the ranks for either one of the two samples.

EXAMPLE: In Table 13-5, the sum of the ranks from the first sample is 180.5. (That is, $12.5 + 23 + \cdots + 11 = 180.5$.)

3. Calculate the value of the z test statistic as shown in the preceding box, where either sample can be used as "Sample 1." (If both sample sizes are greater than 10, then the sampling distribution of R is approximately normal with mean μ_R and standard deviation σ_R, and the test statistic is as shown in the preceding box.)

EXAMPLE: Calculations of μ_R and σ_R and z are shown in Example 1, which follows.

EXAMPLE 1 **Braking Distances of Cars** Table 13-5 lists the braking distances (in ft) of samples of 4-cylinder cars and 6-cylinder cars (from Data Set 16 in Appendix B). Use a 0.05 significance level to test the claim that 4-cylinder cars and 6-cylinder cars have the same median braking distance.

SOLUTION **REQUIREMENT CHECK** (1) The sample data are two independent simple random samples. (2) The sample sizes are 13 and 12, so both sample sizes are greater than 10. The requirements are satisfied. ✓

The null and alternative hypotheses are as follows:

H_0: The braking distances of 4-cylinder cars and 6-cylinder cars have the same median.

H_1: The braking distances of 4-cylinder cars and 6-cylinder cars have different medians.

Rank all 25 braking distances combined, beginning with a rank of 1 (assigned to the lowest value of 122). The ranks corresponding to the individual sample values are shown in parentheses in Table 13-5. R denotes the sum of the ranks for the sample we choose as Sample 1. If we choose the 4-cylinder cars as Sample 1, we get

$$R = 12.5 + 23 + \cdots + 11 = 180.5$$

Because there are 13 cars with 4 cylinders, we have $n_1 = 13$. Also, $n_2 = 12$ because there are 12 cars with 6 cylinders. We can now find the values of μ_R, σ_R, and the test statistic z.

$$\mu_R = \frac{n_1(n_1 + n_2 + 1)}{2} = \frac{13(13 + 12 + 1)}{2} = 169$$

$$\sigma_R = \sqrt{\frac{n_1 n_2(n_1 + n_2 + 1)}{12}} = \sqrt{\frac{(13)(12)(13 + 12 + 1)}{12}} = 18.385$$

$$z = \frac{R - \mu_R}{\sigma_R} = \frac{180.5 - 169}{18.385} = 0.63$$

The test is two-tailed because a large positive value of z would indicate that disproportionately more higher ranks are found in Sample 1, and a large negative value of z would indicate that disproportionately more lower ranks are found in Sample 1. In either case, we would have strong evidence against the claim that the two samples come from populations with equal medians.

The significance of the test statistic z can be treated as in previous chapters. We are testing (with $\alpha = 0.05$) the hypothesis that the two populations have equal medians, so we have a two-tailed test with critical values $z = \pm 1.96$. The test statistic of

$z = 0.63$ does *not* fall within the critical region, so we fail to reject the null hypothesis that braking distances of 4-cylinder cars and 6-cylinder cars have the same median.

INTERPRETATION There is not sufficient evidence to warrant rejection of the claim that 4-cylinder cars and 6-cylinder cars have the same median braking distance. Based on the available sample data, it appears that 4-cylinder cars and 6-cylinder cars have braking distances with about the same median.

In Example 1, if we interchange the two sets of sample values and consider the sample of 6-cylinder cars to be the first sample, then $R = 144.5$, $\mu_R = 156$, $\sigma_R = 18.385$, and $z = -0.63$, so the conclusion is exactly the same.

Using the TI-83/84 Plus Calculator for the Wilcoxon Rank-Sum Test

The TI-83/84 Plus calculator is not programmed for the Wilcoxon rank-sum test, but the program RSTEST can be used. Use the following procedure.

1. First download and install the program RSTEST (rank-sum test) from the CD-ROM or the Web site www.aw.com/triola. (If using the CD-ROM, select the **Software** folder, select **TI-83 TI-84 Plus Programs,** select **Michael Lloyd Programs,** then select **programs.** The program RSTEST should now be available for download.)

2. Also download the program ZZRANK, which is used by the program RSTEST.

3. Enter the two sets of sample data as lists in L1 and L2 (or use named lists).

4. Press **PRGM** and select RSTEST and press **ENTER** **ENTER**.

5. When given the prompt of GROUP A=, enter L1 (or the name of a list) and press **ENTER**.

6. When given the prompt of GROUP B=, enter L2 (or the name of a list) and press **ENTER**.

7. Press **ENTER** once again to get the test statistic z.

8. Find the critical values by referring to Table A-2.

9. Using the calculator results: The program RSTEST will provide a test statistic z, rank sum R, mean, and standard deviation by using calculations based on the sample with the fewer number of values (instead of using the first sample). Consequently, the program RSTEST might provide the same test statistic that would be calculated with the methods of this section, or it might provide a test statistic with an opposite sign. *Use the test statistic provided by RSTEST and the conclusion will be the same as the conclusion reached by using the methods of this section.*

For the sample data in Table 13-5 included with Example 1, enter the first column of sample values in L1 and enter the second column of values in L2. If you run the program RSTEST as described above, you get the two accompanying screen displays. The first screen has a heading of "GROUP B USED," which indicates that the calculations are based on the second column of values. The first screen shows that for the second column of sample values, the rank sum is 144.5 (as shown in Table 13-5). Because the second column of values is used, the mean is given as 156 (instead of 169 found by using the first column), and the standard deviation is 18.385 (rounded). The second screen shows the test statistic of $z = -0.63$ (rounded). That test statistic leads to the same conclusions given in Example 1. (The values of "TEST STAT U" and "MEAN U" given in the second screen can be ignored. They are values used in the Mann-Whitney U test.)

First Screen from RSTEST

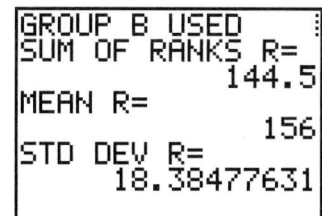

Second Screen from RSTEST

13-4 Basic Skills and Concepts

Statistical Literacy and Critical Thinking

1. Parametric and Nonparametric Tests Listed below are randomly selected values from STATDISK and the TI-83/84 Plus calculator. Both samples are obtained by selecting a uniform distribution of whole numbers between 1 and 1000 inclusive. When trying to test for a difference between the population of such random numbers from each of the two sources, which test should *not* be used: the parametric *t* test or the Wilcoxon rank-sum test? Why?

STATDISK:	606 561 834 421 481 716 272 703 556 94 172 161
TI-83/84 Plus	929 71 678 798 485 768 853 990 297 392 904 644

2. Independent Samples Are the two samples in Exercise 1 independent or dependent? Explain.

3. What Are We Testing? Refer to the sample data in Exercise 1.

a. Given that each sample was generated from a uniform distribution of whole numbers between 1 and 1000, what is the median of each population?

b. Suppose the Wilcoxon rank-sum test is used for the data in Exercise 1. Are we testing that each of the two samples is from a population with the median identified in part (a)?

4. Efficiency Refer to Table 13-2 on page 667 and identify the efficiency of the Wilcoxon rank-sum test. What does that value tell us about the test?

Using the Wilcoxon Rank-Sum Test. *In Exercises 5–8, use the Wilcoxon rank-sum test.*

5. Discrimination Based on Age The Revenue Commissioners in Ireland conducted a contest for promotion. The ages of the unsuccessful and successful applicants are given below (based on data from "Debating the Use of Statistical Evidence in Allegations of Age Discrimination," by Barry and Boland, *The American Statistician,* Vol. 58, No. 2). Some of the applicants who were unsuccessful in getting the promotion charged that the competition involved discrimination based on age. Use a 0.05 significance level to test the claim that the unsuccessful applicants are from a population with the same median age as the successful applicants. Based on the result, does there appear to be discrimination based on age?

Ages of Unsuccessful Applicants	Ages of Successful Applicants
34 37 37 38 41 42 43 44 44 45	27 33 36 37 38 38 39 42 42 43
45 45 46 48 49 53 53 54 54 55	43 44 44 44 45 45 45 45 46 46
56 57 60	47 47 48 48 49 49 51 51 52 54

6. Radiation in Baby Teeth Listed below are amounts of strontium-90 (in millibecquerels or mBq per gram of calcium) in a simple random sample of baby teeth obtained from Pennsylvania residents and New York residents born after 1979 (based on data from "An Unexpected Rise in strontium-90 in U.S. Deciduous Teeth in the 1990s," by Mangano, et al., *Science of the Total Environment*). Use a 0.05 significance level to test the claim that the median amount of strontium-90 from Pennsylvania residents is the same as the median from New York residents.

Pennsylvania	155 142 149 130 151 163 151 142 156 133 138 161
New York	133 140 142 131 134 129 128 140 140 140 137 143

7. Popes and Monarchs Listed below are the numbers of years that U.S. presidents, popes (since 1690), and British monarchs lived after they were inaugurated, elected, or coronated. As of this writing, the last president is Gerald Ford and the last pope is John Paul II. (The times are based on data from *Computer-Interactive Data Analysis,* by Lunn and McNeil, John Wiley & Sons.) Use a 0.05 significance level to test the claim that the two samples of longevity data from popes and monarchs are from populations with the same median.

Presidents	10 29 26 28 15 23 17 25 0 20 4 1 24 16 12 4 10 17 16 0 7 24 12 4
	18 21 11 2 9 36 12 28 3 16 9 25 23 32
Popes	2 9 21 3 6 10 18 11 6 25 23 6 2 15 32 25 11 8 17 19 5 15 0 26
Monarchs	17 6 13 12 13 33 59 10 7 63 9 25 36 15

TI **8. Presidents and Popes** Refer to the longevity data for U.S. presidents and popes in Exercise 7. Use a 0.05 significance level to test the claim that the two samples are from populations with the same median.

Appendix B Data Sets. *In Exercises 9–12, refer to the indicated data set in Appendix B and use the Wilcoxon rank-sum test.*

TI **9. Cigarettes** Refer to Data Set 4 in Appendix B for the amounts of nicotine (in mg per cigarette) in the sample of king size cigarettes, which are nonfiltered, nonmenthol, and non-light, and for the amounts of nicotine in the 100 mm cigarettes, which are filtered, nonmenthol, and non-light. Use a 0.01 significance level to test the claim that the median amount of nicotine in nonfiltered king size cigarettes is greater than the median amount of nicotine in 100 mm filtered cigarettes.

TI **10. Cigarettes** Refer to Data Set 4 in Appendix B for the amounts of tar in the sample of king size cigarettes, which are nonfiltered, nonmenthol, and non-light, and for the amounts of tar in the 100 mm cigarettes, which are filtered, nonmenthol, and non-light. Use a 0.01 significance level to test the claim that the median amount of tar in nonfiltered king size cigarettes is greater than the median amount of tar in 100 mm filtered cigarettes.

TI **11. Movie Gross** Refer to Data Set 9 in Appendix B. Use the amounts of money grossed by movies with ratings of PG or PG-13 as one sample, and use the amounts of money grossed by movies with *R* ratings as a second sample. Use a 0.05 significance level to test the claim that movies with ratings of PG or PG-13 have a higher median gross amount than movies with *R* ratings.

TI **12. Body Mass Index** Refer to Data Set 1 in Appendix B for the body mass index (BMI) measurements of random samples of men and women. Use a 0.05 significance level to test the claim that men and women have different median BMI values.

13-4   Beyond the Basics

13. Using the Mann-Whitney *U* Test The Mann-Whitney *U* test is equivalent to the Wilcoxon rank-sum test for independent samples in the sense that they both apply to the same situations and always lead to the same conclusions. In the Mann-Whitney *U* test we calculate

$$z = \frac{U - \dfrac{n_1 n_2}{2}}{\sqrt{\dfrac{n_1 n_2 (n_1 + n_2 + 1)}{12}}}$$

where

$$U = n_1 n_2 + \frac{n_1 (n_1 + 1)}{2} - R$$

Use the braking distance measurements listed in Table 13-5 on page 684 to find the *z* test statistic for the Mann-Whitney *U* test. Compare this value to the *z* test statistic found using the Wilcoxon rank-sum test.

14. Finding Critical Values Assume that we have two treatments (A and B) that produce quantitative results, and we have only two observations for treatment A and two observations for treatment B. We cannot use the test statistic given in this section because both sample sizes do not exceed 10.

	Rank			Rank sum for treatment A
1	2	3	4	
A	A	B	B	3

a. Complete the accompanying table by listing the five rows corresponding to the other five cases, and enter the corresponding rank sums for treatment A.

b. List the possible values of *R* and their corresponding probabilities. (Assume that the rows of the table from part (a) are equally likely.)

c. Is it possible, at the 0.10 significance level, to reject the null hypothesis that there is no difference between treatments A and B? Explain.

Polls and Psychologists

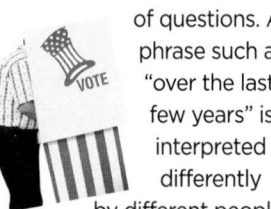

Poll results can be dramatically affected by the wording of questions. A phrase such as "over the last few years" is interpreted differently by different people. Over the last few years (actually, since 1980), survey researchers and psychologists have been working together to improve surveys by decreasing bias and increasing accuracy. In one case, psychologists studied the finding that 10 to 15 percent of those surveyed say they voted in the last election when they did not. They experimented with theories of faulty memory, a desire to be viewed as responsible, and a tendency of those who usually vote to say that they voted in the most recent election, even if they did not. Only the last theory was actually found to be part of the problem.

 Kruskal-Wallis Test

Key Concept In this section we introduce the *Kruskal-Wallis test,* which uses ranks of data from three or more independent simple random samples to test the null hypothesis that the samples come from populations with equal medians.

In Section 12-2 we used one-way analysis of variance (ANOVA) to test the null hypothesis that three or more populations have the same *mean,* but ANOVA requires that all of the involved populations have normal distributions. The Kruskal-Wallis test for equal *medians* does not require normal distributions.

 DEFINITION

The **Kruskal-Wallis test** (also called the *H* **test**) is a nonparametric test that uses ranks of simple random samples from three or more independent populations to test the null hypothesis that the populations have the same median. (The alternative hypothesis is the claim that the populations have medians that are not all equal.)

In applying the Kruskal-Wallis test, we compute the *test statistic H, which has a distribution that can be approximated by the chi-square distribution as long as each sample has at least five observations.* When we use the chi-square distribution in this context, the number of degrees of freedom is $k - 1$, where k is the number of samples. (For a quick review of the key features of the chi-square distribution, see Section 7-5.)

The *H* test statistic is basically a measure of the variance of the rank sums $R_1, R_2, \ldots,$ R_k. If the ranks are distributed evenly among the sample groups, then *H* should be a relatively small number. If the samples are very different, then the ranks will be excessively low in some groups and high in others, with the net effect that *H* will be large. Consequently, only large values of *H* lead to rejection of the null hypothesis that the samples come from identical populations. *The Kruskal-Wallis test is therefore a right-tailed test.*

Kruskal-Wallis Test

Objective

Use the Kruskal-Wallis test with simple random samples from three or more independent populations for the following null and alternative hypotheses:

H_0: The samples come from populations with equal medians.

H_1: The samples come from populations with medians that are not all equal.

Notation

N = total number of observations in all samples combined
k = number of different samples
R_1 = sum of ranks for Sample 1
n_1 = number of observations in Sample 1

For Sample 2, the sum of ranks is R_2 and the number of observations is n_2, and similar notation is used for the other samples.

Requirements

1. We have at least three independent simple random samples.

2. Each sample has at least five observations. (If samples have fewer than five observations, refer to special tables of critical values, such as *CRC Standard Probability and Statistics Tables and Formulae,* published by CRC Press.)

Note: There is *no* requirement that the populations have a normal distribution or any other particular distribution.

Test Statistic

$$H = \frac{12}{N(N+1)}\left(\frac{R_1^2}{n_1} + \frac{R_2^2}{n_2} + \cdots + \frac{R_k^2}{n_k}\right) - 3(N+1)$$

Critical Values

1. The test is *right-tailed.*

2. df $= k - 1$. (Because the test statistic H can be approximated by a chi-square distribution, critical values can be found in Table A-4 with $k - 1$ degrees of freedom, where k is the number of different samples.)

Procedure for Finding the Value of the H Test Statistic To see how the following steps are applied, refer to the sample data in Table 13-6.

1. Temporarily combine all samples into one big sample and assign a rank to each sample value. (Sort the values from lowest to highest, and in cases of ties, assign to each observation the mean of the ranks involved.)

EXAMPLE: In Table 13-6, the numbers in parentheses are the ranks of the combined data set. The rank of 1 is assigned to the lowest value of 32, the rank of 2 is assigned to the next lowest value of 33, and so on. In the case of ties, each of the tied values is assigned the mean of the ranks involved in the tie.

2. For each sample, find the sum of the ranks and find the sample size.

EXAMPLE: In Table 13-6, the sum of the ranks from the first sample is 203.5, it is 152.5 for the second sample, and it is 109 for the third sample.

3. Calculate H using the results of Step 2 and the notation and test statistic given in the preceding box.

EXAMPLE: The test statistic is computed in Example 1.

Table 13-6   Chest Deceleration Measurements (in g) from Car Crash Tests

Small		Medium		Large	
44	(21.5)	41	(10.5)	32	(1)
43	(17)	49	(28)	37	(5)
44	(21.5)	43	(17)	38	(8)
54	(30)	41	(10.5)	45	(25)
38	(8)	47	(27)	37	(5)
43	(17)	42	(13)	33	(2)
42	(13)	37	(5)	38	(8)
45	(25)	43	(17)	45	(25)
44	(21.5)	44	(21.5)	43	(17)
50	(29)	34	(3)	42	(13)
$n_1 = 10$		$n_2 = 10$		$n_3 = 10$	
$R_1 = 203.5$		$R_2 = 152.5$		$R_3 = 109$	

> **EXAMPLE 1**
>
> **Car Crash Test Measurements** The chest deceleration measurements resulting from car crash tests are listed in Table 13-6. The data are from the Chapter Problem in Chapter 12. Test the claim that the three samples come from populations with medians that are all equal. Use a 0.05 significance level.

continued

SOLUTION   **REQUIREMENT CHECK** (1) Each of the three samples is a simple random independent sample. (2) Each sample size is at least 5. The requirements are satisfied. ✓

The null and alternative hypotheses are as follows:

H_0:   The populations of chest deceleration measurements from the three categories have the same median.

H_1:   The populations of chest deceleration measurements from the three categories have medians that are not all the same.

Test Statistic In determining the value of the H test statistic, we first rank all of the data, then we find the sum of the ranks for each category. In Table 13-6, ranks are shown in parentheses next to the original sample values. Next we find the sample size, n, and sum of ranks, R, for each sample. Those values are shown at the bottom of Table 13-6. Because the total number of observations is 30, we have $N = 30$. We can now evaluate the test statistic as follows:

$$H = \frac{12}{N(N + 1)} \left(\frac{R_1^2}{n_1} + \frac{R_2^2}{n_2} + \cdots + \frac{R_k^2}{n_k} \right) - 3(N + 1)$$

$$= \frac{12}{30(30 + 1)} \left(\frac{203.5^2}{10} + \frac{152.5^2}{10} + \frac{109^2}{10} \right) - 3(30 + 1)$$

$$= 5.774$$

Critical Value Because each sample has at least five observations, the distribution of H is approximately a chi-square distribution with $k - 1$ degrees of freedom. The number of samples is $k = 3$, so we have $3 - 1 = 2$ degrees of freedom. Refer to Table A-4 to find the critical value of 5.991, which corresponds to 2 degrees of freedom and a 0.05 significance level (with an area of 0.05 in the right tail). The critical value of 5.991 is shown in Figure 13-2. (The chi-square distribution has the general shape shown in Figure 13-2 whenever the number of degrees of freedom is 1 or 2.)

Figure 13-2 shows that the H test statistic 5.774 is not in the critical region bounded by 5.991, so we fail to reject the null hypothesis of equal medians. (In Section 12-2, we used analysis of variance to reject the null hypothesis of equal *means*.)

INTERPRETATION   There is not sufficient evidence to reject the claim that chest deceleration measurements from small cars, medium cars, and large cars all have equal medians. The medians do not appear to be different.

Figure 13-2

Chi-Square Distribution for Example 1

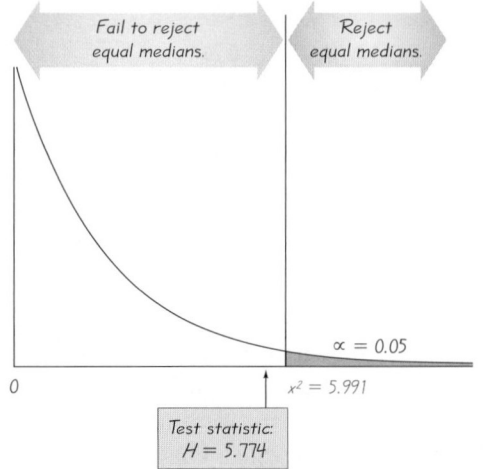

Rationale: The Kruskal-Wallis H test statistic is the rank version of the F test statistic used in the analysis of variance discussed in Chapter 12. When we deal with ranks R instead of original values x, many components are predetermined. For example, the sum of all ranks can be expressed as $N(N + 1)/2$, where N is the total number of values in all samples combined. The expression

$$H = \frac{12}{N(N + 1)} \Sigma n_i (\bar{R}_i - \bar{\bar{R}})^2$$

where

$$\bar{R}_i = \frac{R_i}{n_i} \quad \text{and} \quad \bar{\bar{R}} = \frac{\Sigma R_i}{\Sigma n_i}$$

combines weighted variances of ranks to produce the H test statistic given here. This expression for H is algebraically equivalent to the expression for H given earlier as the test statistic.

Using the TI-83/84 Plus Calculator for the Kruskal-Wallis Test

The TI-83/84 Plus calculator is not programmed for the Kruskal-Wallis test, but the program KWTEST can be used. Use the following procedure.

1. First download and install the program KWTEST (Kruskal-Wallis test) from the CD-ROM or the Web site www.aw.com/triola. (If using the CD-ROM, select the **Software** folder, select **TI-83 TI-84 Plus Programs,** select **Michael Lloyd Programs,** then select **programs.** The program KWTEST should now be available for download.)

2. Also download the program ZZRANK, which is used by the program KWTEST.

3. Enter the lists of sample data in separate columns of matrix [A]. The accompanying screen display shows a portion of the matrix corresponding to the sample data in Table 13-6 used for Example 1. (To create matrix [A]: Press **2ND** **x⁻¹** to get the **MATRIX** menu, then press $\boxed{)}\boxed{)}$ to scroll to **EDIT,** then press **ENTER** to select the first item of [A]. Enter the dimensions of the matrix, where the first dimension is the sample size of the *largest* sample and the second dimension is the number of different sample lists. *Note:* If the samples have different sizes and one of the data values is zero, add some convenient constant to all of the sample values so that no zeros are present.)

4. Press **PRGM**, select **KWTEST,** then press **ENTER**. The value of the H test statistic and the number of degrees of freedom will be provided. Refer to Table A-4 to find the critical value.

If the program KWTEST is used with the sample data in Table 13-6, the display will be as shown in the second screen displayed here. The test statistic of $H = 5.774$ (rounded) and DF = 2 are both provided. These are the same values given in Example 1.

Entering Lists in Matrix [A]

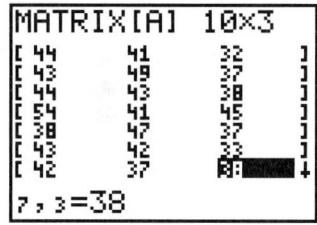

Kruskal-Wallis Test Results for Example 1

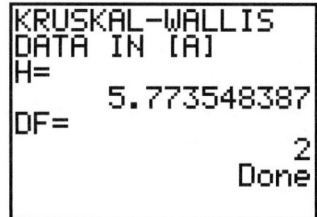

13-5   Basic Skills and Concepts

Statistical Literacy and Critical Thinking

1. Independent Samples Listed on top of the next page are skull breadths obtained from skulls of Egyptian males from three different epochs (based on data from *Ancient Races of the Thebaid,* by Thomson and Randall-Maciver). The Kruskal-Wallis test of equal medians requires independent samples. Are the listed samples independent? Why or why not?

4000 B.C.	125	129	131	132	132	134	135	138	138
1850 B.C.	129	129	134	134	136	137	137	138	136
150 A.D.	128	136	137	137	138	139	141	142	145

2. Simple Random Samples If the Kruskal-Wallis test is to be used with the data in Exercise 1, the samples should be simple random samples. What is a simple random sample?

3. Ranks Refer to the sample data listed in Exercise 1 and assume that the Kruskal-Wallis test is to be used to test the null hypothesis of equal medians. After ranking all of the sample values, find the value of R_1, which is the sum of the ranks for the first sample.

4. Efficiency Refer to Table 13-2 on page 667 and identify the efficiency of the Kruskal-Wallis test. What does that value tell us about the test?

Using the Kruskal-Wallis Test. *In Exercises 5–10, use the Kruskal-Wallis test.*

5. Archaeology Refer to the three samples of skull breadths listed in Exercise 1 and use a 0.05 significance level to test the claim that the samples are from populations with the same median. Changes in head shape over time suggest that interbreeding occurred with immigrant populations. Is interbreeding of cultures suggested by the data?

6. Laboratory Testing of Flammability of Children's Sleepwear Flammability tests were conducted on children's sleepwear. Pieces of fabric were burned under controlled conditions using the Vertical Semirestrained Test. After the burning stopped, the length of the charred portion was measured and recorded. Results are given in the margin for the same fabric tested at different laboratories. Did the different laboratories obtain the same results?

7. Head Injury in a Car Crash Listed below are head injury data from crash test dummies. (The data are from the same cars used in the Chapter Problem for Chapter 12.) These measurements are in hic, which denotes a standard head injury criterion. Use a 0.05 significance level to test the null hypothesis that the different car categories have the same median. Do these data suggest that larger cars are safer?

Small Cars	290	406	371	544	374	501	376	499	479	475
Medium Cars	245	502	474	505	393	264	368	510	296	349
Large Cars	342	216	335	698	216	169	608	432	510	332

8. Femur Injury in a Car Crash Listed below are measured loads (in lb) on the left femur of crash test dummies. (The data are from the same cars used in the Chapter Problem for Chapter 12.) Use a 0.05 significance level to test the null hypothesis that the different car categories have the same median. Do these data suggest that larger cars are safer?

Small Cars	548	782	1188	707	324	320	634	501	274	437
Medium Cars	194	280	1076	411	617	133	719	656	874	445
Large Cars	215	937	953	1636	937	472	882	562	656	433

9. Car Emissions Listed below are measured amounts of greenhouse gas emissions from cars in three different categories (from Data Set 16 in Appendix B). The measurements are in tons per year, expressed as CO_2 equivalents. Use a 0.05 significance level to test the claim that the different car categories have the same median amount of greenhouse gas emissions. Based on the results, does the number of cylinders appear to affect the amount of greenhouse gas emissions?

Four cylinder	7.2	7.9	6.8	7.4	6.5	6.6	6.7	6.5	6.5	7.1	6.7	5.5	7.3
Six cylinder	8.7	7.7	7.7	8.7	8.2	9.0	9.3	7.4	7.0	7.2	7.2	8.2	
Eight cylinder	9.3	9.3	9.3	8.6	8.7	9.3	9.3						

10. Car Fuel Consumption Listed below are the highway fuel consumption amounts (in mi/gal) from cars in three different categories (from Data Set 16 in Appendix B). Use a

Data for Exercise 6

Laboratory

1	2	3	4	5
2.9	2.7	3.3	3.3	4.1
3.1	3.4	3.3	3.2	4.1
3.1	3.6	3.5	3.4	3.7
3.7	3.2	3.5	2.7	4.2
3.1	4.0	2.8	2.7	3.1
4.2	4.1	2.8	3.3	3.5
3.7	3.8	3.2	2.9	2.8
3.9	3.8	2.8	3.2	
3.1	4.3	3.8	2.9	
3.0	3.4	3.5		
2.9	3.3			

0.05 significance level to test the claim that the different car categories have the same median highway fuel consumption. Based on the results, does the number of cylinders appear to affect highway fuel consumption?

Four cylinder	31	30	32	30	32	34	32	34	32	29	32	38	31
Six cylinder	26	29	29	27	27	25	25	30	37	32	31	27	
Eight cylinder	25	25	25	27	25	25	25						

Appendix B Data Sets. *In Exercises 11 and 12, use the Kruskal-Wallis test with the data set from Appendix B.*

 11. Nicotine in Cigarettes Refer to Data Set 4 in Appendix B and use the amounts of nicotine (mg per cigarette) in the king size cigarettes, the 100 mm menthol cigarettes, and the 100 mm nonmenthol cigarettes. The king size cigarettes are nonfiltered, nonmenthol, and non-light. The 100 mm menthol cigarettes are filtered and non-light. The 100 mm nonmenthol cigarettes are filtered and non-light. Use a 0.05 significance level to test the claim that the three categories of cigarettes yield the same median amount of nicotine. Given that only the king size cigarettes are not filtered, do the filters appear to make a difference?

 12. Tar in Cigarettes Refer to Data Set 4 in Appendix B and use the amounts of tar (mg per cigarette) in the three categories of cigarettes described in Exercise 11. Use a 0.05 significance level to test the claim that the three categories of cigarettes yield the same median amount of tar. Given that only the king size cigarettes are not filtered, do the filters appear to make a difference?

13-5 Beyond the Basics

13. Correcting the *H* Test Statistic for Ties In using the Kruskal-Wallis test, there is a correction factor that should be applied whenever there are many ties: Divide *H* by

$$1 - \frac{\sum T}{N^3 - N}$$

For each individual group of tied observations in the combined set of all sample data, calculate $T = t^3 - t$, where *t* is the number of observations that are tied within the individual group. Find *t* for each group of tied values, then compute the value of *T* for each group, then add the *T* values to get $\sum T$. The value of *N* is the total number of observations in all samples combined. Use this procedure to find the corrected value of *H* for Example 1. Does the corrected value of *H* differ substantially from the value found in Example 1?

13-6 Rank Correlation

Key Concept In this section we describe the nonparametric method of the *rank correlation test,* which uses paired data to test for an association between two variables. In Chapter 10 we used paired sample data to compute values for the linear correlation coefficient *r,* but in this section we use *ranks* as the basis for computing the rank correlation coefficient r_s. As in Chapter 10, we begin an analysis of paired data by constructing a scatterplot so that we can identify any patterns in the data.

 **DEFINITION**

> The **rank correlation test** (or **Spearman's rank correlation test**) is a nonparametric test that uses ranks of sample data consisting of matched pairs. It is used to test for an association between two variables.

We use the notation r_s for the rank correlation coefficient so that we don't confuse it with the linear correlation coefficient r. The subscript s does not refer to a standard deviation; it is used in honor of Charles Spearman (1863–1945), who originated the rank correlation approach. In fact, r_s is often called **Spearman's rank correlation coefficient.** Key components of the rank correlation test are given in the following box, and the rank correlation procedure is summarized in Figure 13-3.

Rank Correlation

Objective

Compute the rank correlation coefficient r_s and use it to test for an association between two variables. The null and alternative hypotheses are as follows:

H_0: $\rho_s = 0$ (There is *no* correlation between the two variables.)

H_1: $\rho_s \neq 0$ (There is a correlation between the two variables.)

Notation

r_s = rank correlation coefficient for sample paired data (r_s is a sample statistic)

ρ_s = rank correlation coefficient for all the population data (ρ_s is a population parameter)

n = number of pairs of sample data

d = difference between ranks for the two values within a pair

Requirements

The paired data are a simple random sample. *Note:* Unlike the parametric methods of Section 10-2, there is *no* requirement that the sample pairs of data have a bivariate normal distribution (as described in Section 10-2). There is *no* requirement of a normal distribution for any population.

Test Statistic

No ties: After converting the data in each sample to ranks, if there are no ties among ranks for the first variable and there are no ties among ranks for the second variable, the exact value of the test statistic can be calculated using this formula:

$$r_s = 1 - \frac{6\Sigma d^2}{n(n^2 - 1)}$$

Ties: After converting the data in each sample to ranks, if either variable has ties among its ranks, the exact value of the test statistic r_s can be found by using Formula 10-1 with the ranks:

$$r_s = \frac{n\Sigma xy - (\Sigma x)(\Sigma y)}{\sqrt{n(\Sigma x^2) - (\Sigma x)^2}\sqrt{n(\Sigma y^2) - (\Sigma y)^2}}$$

Critical Values

1. If $n \leq 30$, critical values are found in Table A-8.

2. If $n > 30$, critical values of r_s are found using Formula 13-1.

Formula 13-1 $r_s = \dfrac{\pm z}{\sqrt{n - 1}}$ (critical values when $n > 30$)

where the value of z corresponds to the significance level. (For example, if $\alpha = 0.05$, $z = 1.96$.)

Advantages: Rank correlation has these advantages over the parametric methods discussed in Chapter 10:

1. The nonparametric method of rank correlation can be used in a wider variety of circumstances than the parametric method of linear correlation. With rank correlation, we can analyze paired data that are ranks or can be converted to ranks. For example, if two judges rank 30 different gymnasts, we can use rank correlation, but not linear correlation. Unlike the parametric methods of Chapter 10, the method of rank correlation does *not* require a normal distribution for any population.

2. Rank correlation can be used to detect some (not all) relationships that are not linear.

Disadvantage: A disadvantage of rank correlation is its efficiency rating of 0.91, as described in Section 13-1. This efficiency rating shows that with all other circumstances being equal, the nonparametric approach of rank correlation requires 100 pairs of sample data to achieve the same results as only 91 pairs of sample observations analyzed through the parametric approach, assuming that the stricter requirements of the parametric approach are met.

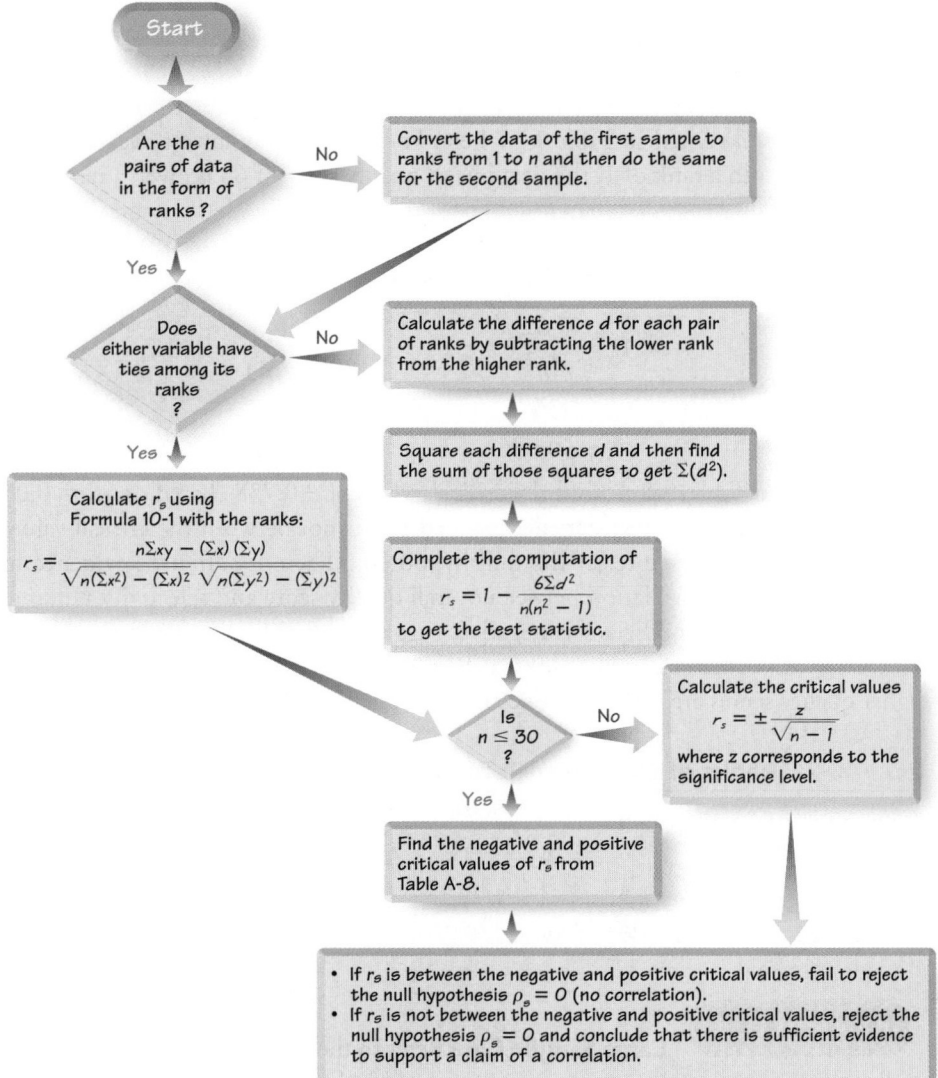

Direct Link Between Smoking and Cancer

When we find a statistical correlation between two variables, we must be extremely careful to avoid the mistake of concluding that there is a cause-effect link. The tobacco industry has consistently emphasized that correlation does not imply causality. However, Dr. David Sidransky of Johns Hopkins University now says that "we have such strong molecular proof that we can take an individual cancer and potentially, based on the patterns of genetic change, determine whether cigarette smoking was the cause of that cancer." Based on his findings, he also said that "the smoker had a much higher incidence of the mutation, but the second thing that nailed it was the very distinct pattern of mutations … so we had the smoking gun." Although statistical methods cannot prove that smoking *causes* cancer, such proof can be established with physical evidence of the type described by Dr. Sidransky.

Figure 13-3   Rank Correlation Procedure for Testing H_0: $\rho_s = 0$

· · · · · · · · **EXAMPLE 1** **Are the Best Universities the Most Difficult to Get Into?**
Table 13-1 lists overall quality scores and selectivity rankings of a sample of national universities (based on data from *U.S. News and World Report*). Find the value of the rank correlation coefficient and use it to determine whether there is a correlation between the overall quality scores and the selectivity rankings. Use a 0.05 significance level. Based on the result, does it appear that national universities with higher overall quality scores are more difficult to get into?

SOLUTION **REQUIREMENT CHECK** The only requirement is that the paired data are a simple random sample. The colleges included are a simple random sample from those available.

The selectivity data consist of ranks that are not normally distributed. So, we use the rank correlation coefficient to test for a relationship between overall quality score and selectivity rank.

The null and alternative hypotheses are as follows:

$$H_0: \rho_s = 0$$
$$H_1: \rho_s \neq 0$$

Following the procedure of Figure 13-3, we begin by converting the data in Table 13-1 into the ranks listed in Table 13-7 here. For example, the lowest overall quality score of 55 is given a rank of 1, the next lowest quality score of 63 is given a rank of 2, and so on. In Table 13-7 we see that neither of the two variables has ties among ranks, so the exact value of the test statistic can be calculated as shown below. We use $n = 8$ (for 8 pairs of data) and $\sum d^2 = 156$ (as shown in Table 13-7) to get

$$r_s = 1 - \frac{6\sum d^2}{n(n^2 - 1)} = 1 - \frac{6(156)}{8(8^2 - 1)}$$

$$= 1 - \frac{936}{504} = -0.857$$

Now we refer to Table A-8 to find the critical values of ± 0.738 (based on $\alpha = 0.05$ and $n = 8$). Because the test statistic $r_s = -0.857$ is not between the critical values of -0.738 and 0.738, we reject the null hypothesis. There is sufficient evidence to support a claim of a correlation between overall quality score and selectivity ranking. The rank correlation coefficient is negative, suggesting that higher quality scores are associated with lower selectivity ranks. It does appear that national universities with higher quality scores are more selective and are more difficult to get into.

Table 13-7   Ranks of Data from Table 13-1

Overall Quality	8	2	1	7	5	4	3	6
Selectivity Rank	2	6	8	1	4	3	7	5
Difference d	6	4	7	6	1	1	4	1
d^2	36	16	49	36	1	1	16	1 $\rightarrow \sum d^2 = 156$

EXAMPLE 2 **Large Sample Case** Refer to the measured systolic and diastolic blood pressure measurements of 40 randomly selected females in Data Set 1 in Appendix B and use a 0.05 significance level to test the claim that among women, there is a correlation between systolic blood pressure and diastolic blood pressure.

SOLUTION   **REQUIREMENT CHECK** The data are a simple random sample. ✓

Test Statistic The value of the rank correlation coefficient is $r_s = 0.780$, which can be found using computer software or a TI-83/84 Plus calculator.

Critical Values Because there are 40 pairs of data, we have $n = 40$. Because n exceeds 30, we find the critical values from Formula 13-1 instead of Table A-8. With $\alpha = 0.05$ in two tails, we let $z = 1.96$ to get the critical values of -0.314 and 0.314, as shown below.

$$r_s = \frac{\pm 1.96}{\sqrt{40 - 1}} = \pm 0.314$$

The test statistic of $r_s = 0.780$ is not between the critical values of -0.314 and 0.314, so we reject the null hypothesis of $\rho_s = 0$. There is sufficient evidence to support the claim that among women, there is a correlation between systolic blood pressure and diastolic blood pressure.

The next example illustrates the principle that rank correlation can sometimes be used to detect relationships that are not linear.

EXAMPLE 3   **Detecting a Nonlinear Pattern** An experiment involves a growing population of bacteria. Table 13-8 lists randomly selected times (in hr) after the experiment is begun, and the number of bacteria present. Use a 0.05 significance level to test the claim that there is a correlation between time and population size.

SOLUTION   **REQUIREMENT CHECK** The data are a simple random sample. ✓

The null and alternative hypotheses are as follows:

$$H_0: \rho_s = 0 \quad \text{(no correlation)}$$
$$H_1: \rho_s \neq 0 \quad \text{(correlation)}$$

We follow the rank correlation procedure summarized in Figure 13-3. The original values are not ranks, so we convert them to ranks and enter the results in Table 13-9. (Section 13-1 describes the procedure for converting scores into ranks.) There are no ties among the ranks for the times, nor are there ties among the ranks for population size, so we proceed by calculating the differences, d, and squaring

Table 13-8   Number of Bacteria in a Growing Population

Time (hrs)	6	107	109	125	126	128	133	143	177	606
Population Size	2	3	4	10	16	29	35	38	41	45

Table 13-9   Ranks from Table 13-8

Ranks of Times	1	2	3	4	5	6	7	8	9	10
Ranks of Populations	1	2	3	4	5	6	7	8	9	10
Difference d	0	0	0	0	0	0	0	0	0	0
d^2	0	0	0	0	0	0	0	0	0	0

them. Next we find the sum of the d^2 values, which is 0. We now calculate the value of the test statistic:

$$r_s = 1 - \frac{6\Sigma d^2}{n(n^2 - 1)} = 1 - \frac{6(0)}{10(10^2 - 1)}$$

$$= 1 - \frac{0}{990} = 1$$

Since $n = 10$, we use Table A-8 to get the critical values of ± 0.648. Finally, the test statistic of $r_s = 1$ is not between -0.648 and 0.648, so we reject the null hypothesis of $\rho_s = 0$. There is sufficient evidence to conclude that there is a correlation between time and population size.

In Example 3, if we test for a linear correlation using the methods of Section 10-2, we get a test statistic of $r = 0.621$ and critical values of -0.632 and 0.632, so we conclude that there is not sufficient evidence to support a claim of a linear correlation between time and population size. If we examine the accompanying TI-83/84 Plus scatterplot, we can see that the pattern of points is not a straight-line pattern. Example 3 illustrates this advantage of the nonparametric approach over the parametric approach: *With rank correlation, we can sometimes detect relationships that are not linear.*

TI-83/84 PLUS

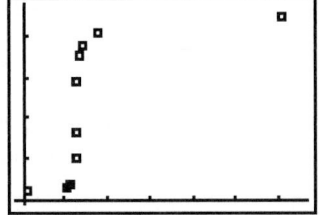

Using the TI-83/84 Plus Calculator for Rank Correlation

The TI-83/84 Plus calculator is not designed to calculate the rank correlation coefficient from original sample values, but we can use it for that purpose. The basic idea is to replace the original sample data with ranks, then proceed as if we were calculating the linear correlation coefficient as in Section 10–2.

1. If the original sample values are not already ranks, replace each of the original sample values by its corresponding rank. [Enter the values for one of the variables in list L1, then use the command **SortA(L1)** to sort those values. The ranks can be easily identified from the sorted list of sample values. Then use the same process to identify the ranks for the second sample.]

2. Press **STAT**, select the top menu item of **TESTS,** then select the menu item of **LinRegTTest.** For the entry labeled Xlist, enter the list containing the ranks for the first variable. For the entry labeled Ylist, enter the list containing the ranks for the second variable.

3. For the entry labeled Freq (frequency), enter 1.

4. For the row labeled β & ρ, choose $\neq 0.$

5. Scroll down to **Calculate** and press **ENTER**.

6. Press ⌄ to scroll down to identify the value of r. Because the calculation of r is done using ranks, the value shown as r is actually the rank correlation coefficient r_s. *Caution: Ignore the displayed P-value* because it is calculated with a t distribution instead of the correct distribution for Spearman's rank r.

Using the data in Table 13-1, we first find the ranks listed in Table 13-7, then we obtain $r_s = -0.857$ (rounded). The conclusions are the same as those given in Example 1.

13-6   Basic Skills and Concepts

Statistical Literacy and Critical Thinking

1. Regression Suppose the methods of this section are used with paired sample data, and the conclusion is that there is sufficient evidence to support the claim of a correlation between the two variables. Can we use the methods of Section 10-3 to find the regression equation that can be used for predictions? Why or why not?

2. Ranks, Differences, and r_s The table below lists the values of new cars sold by dealers and the values of clothes sold by clothing stores in five recent years (based on data from the U.S. Census Bureau). All values are in billions of dollars. Answer the following without using computer software or a calculator.

a. Identify the ranks corresponding to each of the variables.

b. Identify the differences d.

c. What is the value of Σd^2?

d. What is the value of r_s?

New Cars	56.8	58.7	59.4	61.8	63.5	67.5
Clothing	111.8	118.2	119.4	123.0	127.4	136.8

3. Notation Refer to the paired sample data given in Exercise 2. In that context, what is the difference between r_s and ρ_s? Why is the subscript s used? Does the subscript s represent the same standard deviation s introduced in Section 3-3?

4. Efficiency Refer to Table 13-2 on page 667 and identify the efficiency of the rank correlation test. What does that value tell us about the test?

In Exercises 5 and 6, use the scatterplot to find the value of the rank correlation coefficient r_s, and the critical values corresponding to a 0.05 significance level used to test the null hypothesis of $\rho_s = 0$. Determine whether there is a correlation.

5. Distance/Time Data for a Dropped Object

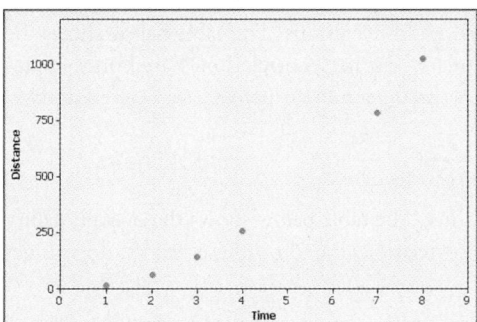

6. Altitude/Time Data for a Descending Aircraft

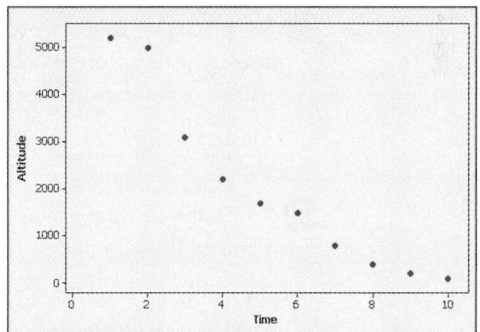

Finding Critical Values. *In Exercises 7 and 8, find the critical values(s) r_s using either Table A-8 or Formula 13-1, as appropriate. Assume that the null hypothesis is $\rho_s = 0$ so the test is two-tailed. Also, n denotes the number of pairs of data.*

7. a. $n = 15$, $\alpha = 0.05$ **8. a.** $n = 9$, $\alpha = 0.01$
b. $n = 24$, $\alpha = 0.01$ **b.** $n = 16$, $\alpha = 0.05$
c. $n = 100$, $\alpha = 0.05$ **c.** $n = 37$, $\alpha = 0.05$
d. $n = 65$, $\alpha = 0.01$ **d.** $n = 82$, $\alpha = 0.01$

Testing for Rank Correlation. *In Exercises 9–16, use the rank correlation coefficient to test for a correlation between the two variables. Use a significance level of $\alpha = 0.05$.*

9. Judges of Marching Bands Two judges ranked seven bands in the Texas state finals competition of marching bands (Coppell, Keller, Grapevine, Dickinson, Poteet, Fossil Ridge,

Heritage), and their rankings are listed below (based on data from the University Interscholastic League). Test for a correlation between the two judges. Do the judges appear to rank about the same or are they very different?

Band	Cpl	Klr	Grp	Dck	Ptt	FR	Her
First Judge	1	3	4	7	5	6	2
Second Judge	6	4	5	1	3	2	7

10. Judges of Marching Bands In the same competition described in Exercise 9, a third judge ranked the bands with the results shown below. Test for a correlation between the first and third judges. Do the judges appear to rank about the same or are they very different?

Band	Cpl	Klr	Grp	Dck	Ptt	FR	Her
First Judge	1	3	4	7	5	6	2
Third Judge	3	4	1	5	7	6	2

11. Ranking DWI Judges Judges in Bernalillo County in New Mexico were ranked for their DWI conviction rates and their recidivism rates, where recidivism refers to a subsequent DWI arrest for a person previously charged with DWI. The results for judges Gentry, Ashanti, Niemczyk, Baca, Clinton, Gomez, Barnhart, Walton, Nakamura, Kavanaugh, Brown, and Barela are shown below (based on data from Steven Flint of the DWI Resource Center). Test for a correlation between conviction rate and recidivism rate. Do conviction rates appear to be related to recidivism rates?

Conviction	1	2	3	4	5	6	7	8	9	10	11	12
Recidivism	6	2	10	4	12	9	8	7	1	5	3	11

12. Plasma TVs *Consumer Reports* magazine tested large plasma TVs. The table below shows the rankings of TVs by overall quality score and cost. High values are given low ranks, so the TV with a quality rank of 1 is the TV with the highest quality score, and a TV given a cost rank of 1 is the most expensive TV. Test for a correlation. Based on these results, can you expect to get higher quality by purchasing a more expensive plasma TV?

Quality	1	2	3	4	5	6	7	8	9	10
Cost	2	3	6	1	10	4	9	5	8	7

13. LCD TVs *Consumer Reports* magazine tested LCD televisions. The table below shows the overall quality score and cost in hundreds of dollars. Test for a correlation. Based on these results, can you expect to get higher quality by purchasing a more expensive LCD television?

Quality	74	71	68	65	63	62	59	57	57	53	51
Cost	27	30	38	23	20	13	27	23	14	13	20

14. Paint *Consumer Reports* magazine tested paints. The table below shows the overall quality score and cost in dollars per gallon. Test for a correlation. Based on these results, do you get better quality paint by paying more?

Quality	90	87	87	86	86	86	82	81	78	62	61	59	23
Cost	27	32	34	30	20	19	19	36	15	39	24	25	15

15. Measuring Seals from Photos Listed below are the overhead widths (in cm) of seals measured from photographs and the weights of the seals (in kg). The data are based on "Mass Estimation of Weddell Seals Using Techniques of Photogrammetry," by R. Garrott of Montana State University. The purpose of the study was to determine if weights of seals could be determined from overhead photographs. Is there sufficient evidence to conclude that there is a correlation between overhead widths of seals from photographs and the weights of the seals?

Overhead Width	7.2	7.4	9.8	9.4	8.8	8.4
Weight	116	154	245	202	200	191

16. Crickets and Temperature The association between the temperature and the number of times a cricket chirps in 1 min was studied. Listed below are the numbers of chirps in 1 min and the corresponding temperatures in degrees Fahrenheit (based on data from *The Song of Insects* by George W. Pierce, Harvard University Press). Is there sufficient evidence to conclude that there is a relationship between the number of chirps in 1 min and the temperature?

Chirps in 1 min	882	1188	1104	864	1200	1032	960	900
Temperature (°F)	69.7	93.3	84.3	76.3	88.6	82.6	71.6	79.6

Appendix B Data Sets. *In Exercises 17 and 18, use the data sets from Appendix B to test for rank correlation with a 0.05 significance level.*

 17. Word Counts of Men and Women Refer to Data Set 8 in Appendix B and use the word counts measured from men and women from the couples included in the data set. Those word counts are listed in the first two columns of Data Set 8.

 18. Cigarette Tar and Nicotine Refer to Data Set 4 in Appendix B and use the tar and nicotine data from king size cigarettes.

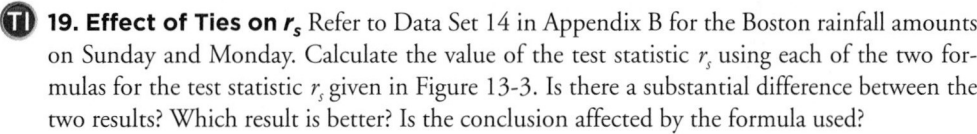

13-6   Beyond the Basics

 19. Effect of Ties on r_s Refer to Data Set 14 in Appendix B for the Boston rainfall amounts on Sunday and Monday. Calculate the value of the test statistic r_s using each of the two formulas for the test statistic r_s given in Figure 13-3. Is there a substantial difference between the two results? Which result is better? Is the conclusion affected by the formula used?

13-7 Runs Test for Randomness

Key Concept In this section we introduce the *runs test for randomness,* which can be used to determine whether the sample data in a sequence are in a random order. This test is based on sample data that have two characteristics, and it analyzes *runs* of those characteristics to determine whether the runs appear to result from some random process, or whether the runs suggest that the order of the data is not random.

 **DEFINITION**

> A **run** is a sequence of data having the same characteristic; the sequence is preceded and followed by data with a different characteristic or by no data at all.
>
> The **runs test** uses the number of runs in a sequence of sample data to test for randomness in the order of the data.

Fundamental Principle of the Runs Test

The fundamental principle of the runs test can be briefly stated as follows:

Reject randomness if the number of runs is very low or very high.

- Example: The sequence of genders FFFFFMMMMM is not random because it has only 2 runs, so the number of runs is very *low.*

- Example: The sequence of genders FMFMFMFMFM is not random because there are 10 runs, which is very *high.*

CAUTION

The runs test for randomness is based on the *order* in which the data occur; it is *not* based on the *frequency* of the data. For example, a sequence of 3 men and 20 women might appear to be random, but the issue of whether 3 men and 20 women constitute a *biased* sample (with disproportionately more women) is *not* addressed by the runs test.

Is the iPod Random Shuffle Really Random?

In *The Guardian,* Steven Levy wrote about an interview with Steve Jobs, CEO of Apple, in which he presented Jobs with this dilemma: "I have a situation with my iPod. The shuffle function just doesn't seem random. Some artists come up way too much and some don't come up at all." According to Jeff Robbin, a head of the iTunes development team, "It is absolutely unequivocally random." Mathematician John Allen Paulos said that "We often interpret and impose patterns on events that are random." Levy goes on to state that when we think that the iPod shuffle is not random, the problem is in our perceptions. Our minds perceive patterns and trends that don't really exist. We often hear runs of consecutive songs by the same artist and think that this is not random, but with true randomness, such consecutive runs are much more likely than we would expect.

The incorrect perception of nonrandomness caused Apple to introduce a "smart shuffle" feature in a new version of iTunes. This feature allows users to control multiple consecutive songs by the same artist. With this feature, consecutive runs by the same artist would be avoided. According to Steve Jobs, "We're making it less random to make it feel more random."

The exact criteria for determining whether a number of runs is very high or low are found in the box on the next page, which summarizes the key elements of the runs test for randomness. The procedure for the runs test for randomness is also summarized in Figure 13-4.

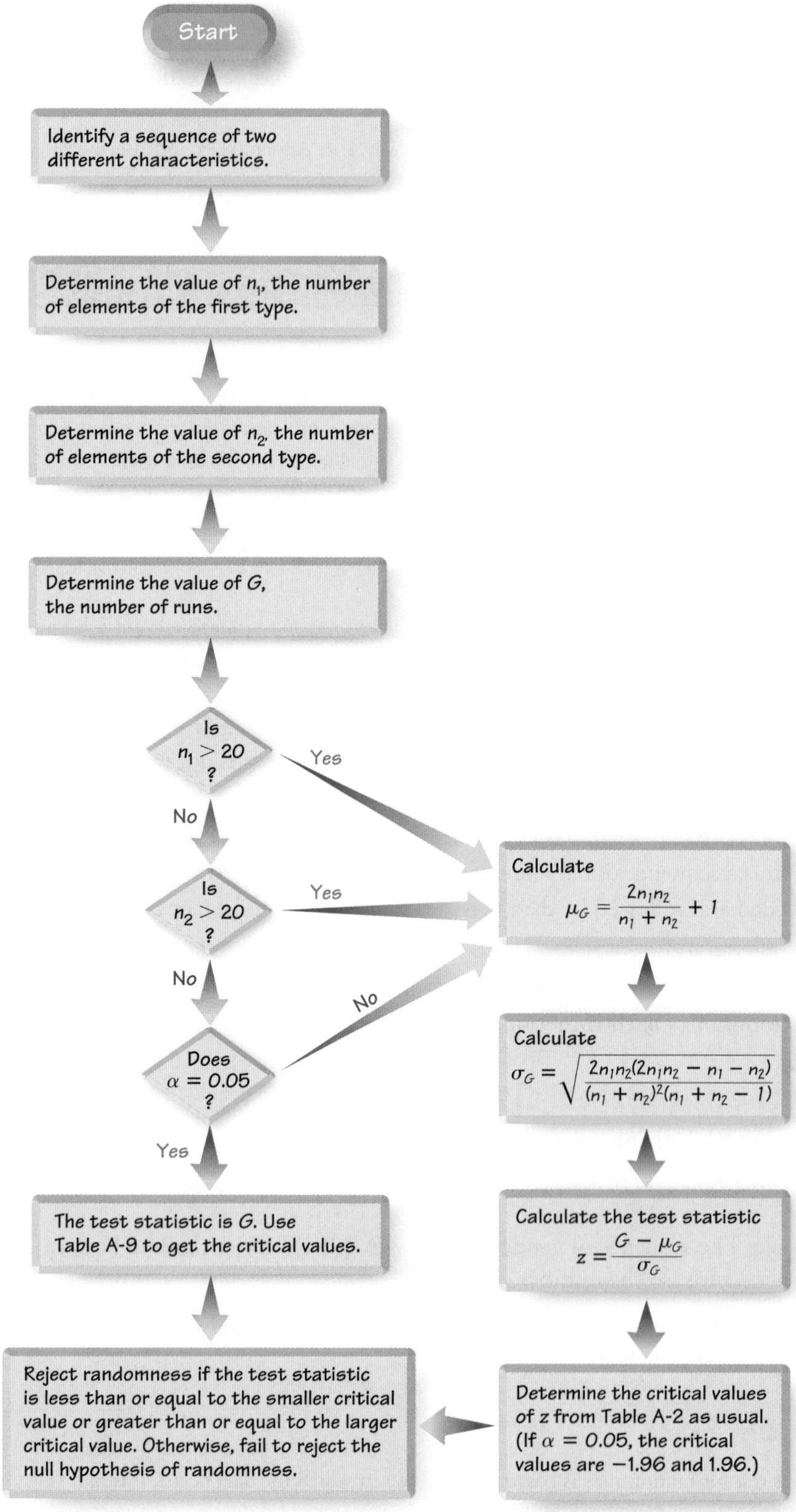

Start

Identify a sequence of two different characteristics.

Determine the value of n_1, the number of elements of the first type.

Determine the value of n_2, the number of elements of the second type.

Determine the value of G, the number of runs.

Is $n_1 > 20$? — Yes →

No ↓

Is $n_2 > 20$? — Yes →

No ↓

Does $\alpha = 0.05$? — No →

Yes ↓

Calculate
$$\mu_G = \frac{2n_1 n_2}{n_1 + n_2} + 1$$

Calculate
$$\sigma_G = \sqrt{\frac{2n_1 n_2 (2n_1 n_2 - n_1 - n_2)}{(n_1 + n_2)^2 (n_1 + n_2 - 1)}}$$

Calculate the test statistic
$$z = \frac{G - \mu_G}{\sigma_G}$$

Determine the critical values of z from Table A-2 as usual. (If $\alpha = 0.05$, the critical values are −1.96 and 1.96.)

The test statistic is G. Use Table A-9 to get the critical values.

Reject randomness if the test statistic is less than or equal to the smaller critical value or greater than or equal to the larger critical value. Otherwise, fail to reject the null hypothesis of randomness.

Figure 13-4   Procedure for Runs Test for Randomness

Runs Test for Randomness

Objective

Apply the runs test for randomness to a *sequence* of sample data to test for randomness in the *order* of the data. Use the following null and alternative hypotheses.

H_0: The data are in a random sequence.

H_1: The data are in a sequence that is not random.

Notation

n_1 = number of elements in the sequence that have one particular characteristic. (The characteristic chosen for n_1 is arbitrary.)

n_2 = number of elements in the sequence that have the other characteristic

G = number of runs

Requirements

1. The sample data are arranged according to some ordering scheme, such as the order in which the sample values were obtained.

2. Each data value can be categorized into one of *two* separate categories (such as male/female).

Test Statistic and Critical Values

For Small Samples and $\alpha = 0.05$: If $n_1 \leq 20$ and $n_2 \leq 20$ and the significance level is $\alpha = 0.05$, the test statistic and critical values are as follows:

Test statistic: G

Critical values: Use Table A-9.

Decision criterion: Reject randomness if the number of runs G is

- less than or equal to the smaller critical value found in Table A-9.

or • greater than or equal to the larger critical value found in Table A-9.

For Large Samples or $\alpha \neq 0.05$: If $n_1 > 20$ or $n_2 > 20$ or $\alpha \neq 0.05$, the test statistic and critical values are as follows:

Test statistic:

$$z = \frac{G - \mu_G}{\sigma_G}$$

where

$$\mu_G = \frac{2n_1 n_2}{n_1 + n_2} + 1$$

and

$$\sigma_G = \sqrt{\frac{(2n_1 n_2)(2n_1 n_2 - n_1 - n_2)}{(n_1 + n_2)^2 (n_1 + n_2 - 1)}}$$

Critical values of z: Use Table A-2.

Decision criterion: Reject randomness if the test statistic z is

- less than or equal to the negative critical z score (such as -1.96).

or • greater than or equal to the positive critical z score (such as 1.96).

Sports Hot Streaks

It is a common belief that athletes often have "hot streaks"—that is, brief periods of extraordinary success. Stanford University psychologist Amos Tversky and other researchers used statistics to analyze the thousands of shots taken by the Philadelphia 76ers for one full season and half of another. They found that the number of "hot streaks" was no different than you would expect from random trials with the outcome of each trial independent of any preceding results. That is, the probability of a hit doesn't depend on the preceding hit or miss.

EXAMPLE 1   **Small Sample: Genders of Study Participants** Listed below are the genders of the first 15 subjects participating in the "Freshman 15" study with results listed in Data Set 3 in Appendix B. Use a 0.05 significance level to test for randomness in the sequence of genders.

$$\text{M M M M F M F F F F F M M F F}$$

SOLUTION   **REQUIREMENT CHECK** (1) The data are arranged in order. (2) Each data value is categorized into one of two separate categories (male/female). The requirements are satisfied.

We will follow the procedure summarized in Figure 13-4. The sequence of two characteristics (male/female) has been identified. We must now find the values of n_1, n_2, and the number of runs G. The sequence is shown below with spacing used to better identify the separate runs.

$$\underbrace{\text{M M M M}}_{\text{1st run}} \quad \underbrace{\text{F}}_{\text{2nd run}} \quad \underbrace{\text{M}}_{\text{3rd run}} \quad \underbrace{\text{F F F F F}}_{\text{4th run}} \quad \underbrace{\text{M M}}_{\text{5th run}} \quad \underbrace{\text{F F}}_{\text{6th run}}$$

We can see that there are 7 males and 8 females, and the number of runs is 6. We therefore have

$$n_1 = \text{total number of males} = 7$$
$$n_2 = \text{total number of females} = 8$$
$$G = \text{number of runs} = 6$$

Because $n_1 \leq 20$ and $n_2 \leq 20$ and $\alpha = 0.05$, the test statistic is $G = 6$ (the number of runs), and we refer to Table A-9 to find the critical values of 4 and 13. Because $G = 6$ is neither less than or equal to the critical value of 4, nor is it greater than or equal to the critical value of 13, *we do not reject randomness*. There is not sufficient evidence to reject randomness in the sequence of genders. It appears that the sequence of genders is random.

Numerical Data: Randomness Above and Below the Mean or Median

In Example 1 we tested for randomness in the sequence of data that clearly fit into two categories. We can also test for randomness in the way numerical data fluctuate above or below a mean or median. To test for randomness above and below the median, for example, use the sample data to find the value of the median, then replace each individual value with the letter A if it is *above* the median, and replace it with B if it is *below* the median. Delete any values that are equal to the median. It is helpful to write the As and Bs directly above or below the numbers they represent because this makes checking easier and also reduces the chance of having the wrong number of letters. After finding the sequence of A and B letters, we can proceed to apply the runs test as described earlier. Economists use the runs test for randomness above and below the median to identify trends or cycles. An upward economic trend would contain a predominance of Bs at the beginning and As at the end, so the number of runs would be small. A downward trend would have As dominating at the beginning and Bs at the end, with a low number of runs. A cyclical pattern would yield a sequence that systematically changes, so the number of runs would tend to be large.

EXAMPLE 2 **Randomness Above and Below the Median** Listed below is the sequence of annual motor vehicle deaths in the U.S. over a consecutive ten-year period (latest data available as of this writing). Use a 0.05 significance level to test for randomness above and below the median. What does the result suggest about motor vehicle deaths?

42,065 42,013 41,501 41,717 41,945 42,196 43,005 42,884 42,836 43,443
  ↓   ↓   ↓   ↓   ↓   ↓   ↓   ↓   ↓   ↓
  B   B   B   B   B   A   A   A   A   A

SOLUTION **REQUIREMENT CHECK** (1) The data are arranged in order. (2) Each data value is categorized into one of two separate categories (below the median or above the median). The requirements are satisfied. ✓

The median of the listed sample values is 42,130.5. We denote a value below the median of 42,130.5 by B (below) and we denote a value above the median by A (above). If there had been any values equal to the median of 42,130.5, they would have been deleted. The sequence of Bs and As is shown below the sample values. That sequence has 5 Bs, so $n_1 = 5$. The sequence has 5 As, so $n_2 = 5$. There are 2 runs, so $G = 2$.

Because $n_1 \leq 20$ and $n_2 \leq 20$ and $\alpha = 0.05$, the test statistic is $G = 2$ (the number of runs), and we refer to Table A-9 to find the critical values of 2 and 10. Because $G = 2$ is less than or equal to the critical value of 2, *we reject the null hypothesis of randomness*. Because all of the values below the median are in the beginning of the sequence and all of the values above the median are at the end of the sequence, it appears that there is an upward trend in the numbers of motor vehicle deaths.

EXAMPLE 3 **Large Sample: Global Warming** Listed below are the global mean temperatures (in °C) of the earth's surface (based on data from the Goddard Institute for Space Studies). The temperatures each represent one year, and they are listed in order by row. Use a 0.05 significance level to test for randomness above and below the mean. What does the result suggest about the earth's temperature?

13.88 13.88 14.00 13.96 13.59 13.77 13.75 13.55 13.77 14.04 13.78 13.44 13.60 13.61 13.68
13.68 13.73 13.85 13.79 13.76 13.95 13.95 13.70 13.64 13.58 13.75 13.85 13.60 13.70 13.69
13.79 13.74 13.67 13.72 13.98 14.06 13.80 13.54 13.67 13.91 13.85 13.95 13.91 13.84 13.89
13.85 14.04 13.95 14.00 13.78 13.97 14.03 14.04 13.89 14.05 13.92 14.01 14.12 14.15 13.98
14.14 14.11 14.10 14.06 14.11 13.99 14.01 14.12 13.97 13.91 13.83 13.98 14.03 14.12 13.91
13.91 13.82 14.08 14.10 14.05 13.98 14.10 14.05 14.03 13.65 13.75 13.93 13.98 13.91 14.00
14.04 13.90 13.95 14.18 13.94 13.98 13.79 14.16 14.07 14.13 14.27 14.40 14.10 14.34 14.16
14.13 14.19 14.35 14.42 14.28 14.49 14.44 14.16 14.18 14.31 14.47 14.36 14.40 14.71 14.44
14.41 14.56 14.70 14.64 14.60 14.77

SOLUTION **REQUIREMENT CHECK** (1) The data are arranged in order. (2) Each data value is categorized into one of two separate categories (below the mean or above the mean). The requirements are satisfied. ✓

The null and alternative hypotheses are as follows:

H_0: The sequence is random.

H_1: The sequence is not random.

continued

The mean of the 126 temperatures is 13.998°C. The test statistic is obtained by first finding the number of temperatures below the mean and the number of temperatures above the mean. Examination of the sequence results in these values:

$$n_1 = \text{number of temperatures below the mean} = 68$$
$$n_2 = \text{number of temperatures above the mean} = 58$$
$$G = \text{number of runs} = 32$$

Since $n_1 > 20$, we need to calculate the test statistic z. We must first evaluate μ_G and σ_G as follows:

$$\mu_G = \frac{2n_1 n_2}{n_1 + n_2} + 1 = \frac{2(68)(58)}{68 + 58} + 1 = 63.6032$$

$$\sigma_G = \sqrt{\frac{(2n_1 n_2)(2n_1 n_2 - n_1 - n_2)}{(n_1 + n_2)^2(n_1 + n_2 - 1)}}$$

$$= \sqrt{\frac{(2)(68)(58)[2(68)(58) - 68 - 58]}{(68 + 58)^2(68 + 58 - 1)}} = 5.55450$$

We now find the test statistic:

$$z = \frac{G - \mu_G}{\sigma_G} = \frac{32 - 63.6032}{5.55450} = -5.69$$

Because the significance level is $\alpha = 0.05$ and we have a two tailed test, the critical values are $z = -1.96$ and $z = 1.96$. The test statistic of $z = -5.69$ falls within the critical region, so we reject the null hypothesis of randomness. The given sequence does not appear to be random.

INTERPRETATION This hypothesis test shows that the sequence of global temperatures over the past 126 years is not random. The accompanying Minitab graph shows that there is an upward trend. Claims of global warming appear to be supported by the data.

MINITAB

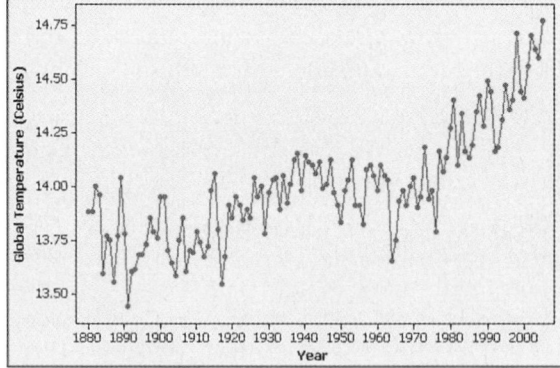

Using Technology for the Runs Test for Randomness

The TI-83/84 Plus calculator is not programmed for the runs test for randomness, but STATDISK can be used instead. STATDISK is a program included with the CD-ROM that is packaged with this book. Load STATDISK, then use the following procedure.

1. Determine the values of n_1, n_2, and the number of runs G as described in this section.

2. Select **Analysis** from the main menu bar, then select **Runs Test**.

3. Proceed to enter the required data in the dialog box.

4. Click **Evaluate** and the results will be displayed. The results will include the test statistic (G or z as appropriate), critical values, and conclusion.

Shown below are the STATDISK results from Example 3, including the test statistic z and the critical values.

STATDISK

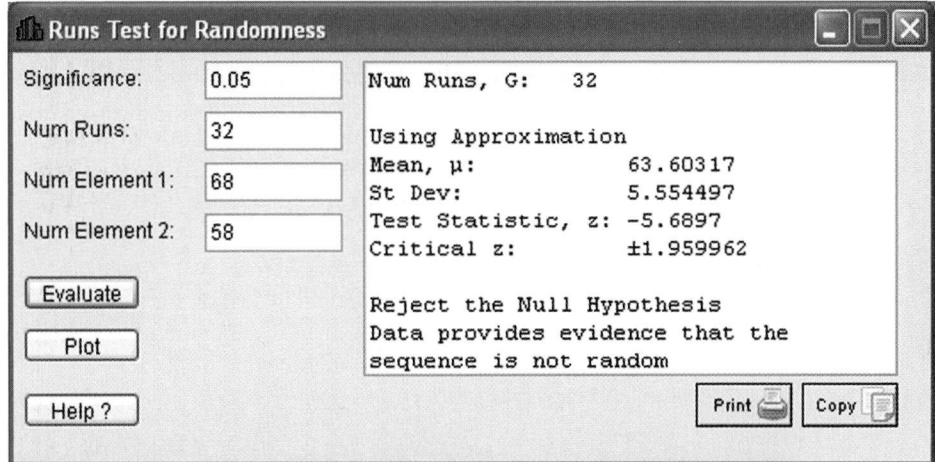

Runs Test for Randomness	
Significance: 0.05	Num Runs, G: 32
Num Runs: 32	Using Approximation
Num Element 1: 68	Mean, μ: 63.60317
Num Element 2: 58	St Dev: 5.554497
	Test Statistic, z: -5.6897
Evaluate	Critical z: ±1.959962
Plot	Reject the Null Hypothesis
	Data provides evidence that the
Help ?	sequence is not random

13-7   Basic Skills and Concepts

Statistical Literacy and Critical Thinking

1. Testing for Bias The last 103 baseball seasons (as of this writing) ended with 61 World Series wins by American League teams, compared to 42 wins by National League teams. Can the runs test be used to show that the American League is better because disproportionately more World Series contests are won by American League teams?

2. Notation Listed below are the genders of the first 25 subjects listed in Data Set 3 in Appendix B. Use that sequence to identify the values of n_1, n_2, and G that would be used in the runs test for randomness.

M M M M M M M M M M M M M M M M M M F F F M F M M M

3. Runs Test If the runs test is used with the sequence of genders of all 107 subjects listed in Data Set 3 in Appendix B, we fail to reject the null hypothesis that the sequence is random. Does it follow that the subjects have been selected in a way that is suitable for statistical purposes?

4. Sequential Data If results from the 107 subjects listed in Data Set 3 in Appendix B are rearranged so that all of the males are placed at the beginning of the list and all of the females are placed at the end of the list, can the runs test be used to determine whether the genders of the subjects are in a random order?

Using the Runs Test for Randomness. *In Exercises 5–10, use the runs test with a significance level of $\alpha = 0.05$. (All data are listed in order by row.)*

5. Oscar Winners Listed below are the genders of the younger winner in the categories of Best Actor and Best Actress for recent and consecutive years. Do the genders of the younger winners appear to occur randomly?

F   F   F   M   M   F   F   F   F   F   F   M   M   M   M   M

6. Cell Phone Subscriptions Listed below are the numbers of cell phone subscriptions (in thousands) in the United States for 11 recent years. Shown below the numbers are letters indicating whether the number is below (B) the mean or above (A) the mean, which is 63,526.2 thousand. Test for randomness of the numbers below and above the mean. Does there appear to be a trend?

Year	1985	1987	1989	1991	1993	1995	1997	1999	2001	2003	2005
Number	340	1231	3509	7557	16,009	33,786	55,312	86,047	128,375	158,722	207,900
	↓	↓	↓	↓	↓	↓	↓	↓	↓	↓	↓
	B	B	B	B	B	B	B	A	A	A	A

7. Testing for Randomness of Presidential Election Winners The political parties of the winning candidates for a recent sequence of presidential elections are listed below. D denotes Democratic party and R denotes Republican party. Does it appear that we elect Democrat and Republican candidates in a random sequence?

R R D R D R R R R D D R R R D D D

D D R R D D R R D R R R D D R R

8. Odd and Even Digits in Pi A *New York Times* article about the calculation of decimal places of π noted that "mathematicians are pretty sure that the digits of π are indistinguishable from any random sequence." Given below are the first 30 decimal places of π. Test for randomness of odd (O) and even (E) digits.

1 4 1 5 9 2 6 5 3 5 8 9 7 9 3 2 3 8 4 6 2 6 4 3 3 8 3 2 7 9

9. Draft Lottery In 1970, a lottery was used to determine who would be drafted into the U.S. Army. The 366 dates in the year were placed in individual capsules, they were mixed, then capsules were selected to identify birth dates of men to be drafted first. The first 30 results are listed below. Test for randomness before and after the middle of the year, which is July 1.

Sept. 14 Apr. 24 Dec. 30 Feb. 14 Oct. 18 Sept. 6 Oct. 26 Sept. 7 Nov. 22 Dec. 6 Aug. 31 Dec. 7 July 8 Apr. 11 July 12 Dec. 29 Jan. 15 Sept. 26 Nov. 1 June 4 Aug. 10 June 26 July 24 Oct. 5 Feb. 19 Dec. 14 July 21 June 5 Mar. 2 Mar. 31

10. Temperatures Listed below are the high temperatures (in °F) near the author's home on consecutive days beginning with September 1 of a recent year (from Data Set 11 in Appendix B). The mean of these high temperatures is 73.8°F. Test for randomness above and below the mean.

80 77 81 85 73 73 80 72 83 81 75 78 80 71 73 78 75 63

63 70 77 82 81 76 77 76 74 66 66 62 71 68 66 71 58

Runs Test with Large Samples. *In Exercises 11–14, use the runs test with a significance level of $\alpha = 0.05$. (All data are listed in order by row.)*

11. Testing for Randomness of Super Bowl Victories Listed below are the conference designations of teams that won the Super Bowl, where N denotes a team from the NFC and A denotes a team from the AFC. Do the results suggest that either conference is superior?

N N A A A N A A A A A N A A A N N A N N N N N N N N N N N N N A N A A N A A N A

12. Baseball World Series Victories Test the claim that the sequence of World Series wins by American League and National League teams is random. Given below are recent results, with American League and National League teams represented by A and N, respectively.

A N A N N N A A A A N A A A A N A N N A A N N A A A A N A N
N A A A A A N A N A N A N A N A A A A A A A N N A N A N N A A N
N N A N A N A N A A A N N A A N N N N A A A A N A N A N A N A A A
N A N A A A N A N A A N A

13. Stock Market: Testing for Randomness Above and Below the Median Listed below are the annual high values of the Dow Jones Industrial Average for a recent sequence of years. Test for randomness below and above the median. What does the result suggest about the stock market as an investment consideration?

969 995 943 985 969 842 951 1036 1052 892 882 1015 1000 908 898
1000 1024 1071 1287 1287 1553 1956 2722 2184 2791 3000 3169 3413 3794 3978
5216 6561 8259 9374 11568 11401 11350 10635 10454 10855 10941 12464 14198

14. Global Warming Listed below are carbon dioxide concentrations (in parts per million) at the earth's surface for 50 recent and consecutive years (based on data from the Goddard Institute for Space Studies). Test for randomness above and below the mean.

314.3 314.8 315.3 316.0 316.9 317.6 318.5 319.0 319.5 320.1 321.3 322.1 323.1 324.6 325.7
326.3 327.5 329.6 330.3 331.2 332.2 333.9 335.5 336.9 338.7 340.0 341.1 342.8 344.4 345.9
347.1 349.0 351.4 352.9 354.2 355.6 356.4 357.1 358.9 360.9 362.6 363.8 366.6 368.3 369.5
371.0 373.1 375.6 377.4 379.6

13-7   Beyond the Basics

15. Finding Critical Values

a. Using all of the elements A, A, A, B, B, B, B, B, B, list the 84 different possible sequences.

b. Find the number of runs for each of the 84 sequences.

c. Use the results from parts (a) and (b) to find your own critical values for *G*.

d. Compare your results to those given in Table A-9.

Review

This chapter introduced six different nonparametric tests, which are also called distribution-free tests because they do not require that the populations have a particular distribution, such as a normal distribution. Nonparametric tests are not as efficient as parametric tests, so we generally need stronger evidence before we reject a null hypothesis.

Table 13-10 lists the nonparametric tests presented in this chapter, along with their functions. The table also lists the corresponding parametric tests.

Table 13-10   Summary of Nonparametric Tests

Nonparametric Test	Function	Parametric Test
Sign test (Section 13-2)	Test for claimed value of median with one sample	*z* test or *t* test (Sections 8-4, 8-5)
	Test for differences between matched pairs	*t* test (Section 9-4)
	Test for claimed value of a proportion	*z* test (Section 8-3)
Wilcoxon signed-ranks test (Section 13-3)	Test for differences between matched pairs	*t* test (Section 9-4)
Wilcoxon rank-sum test (Section 13-4)	Test for difference between two independent samples	*t* test or *z* test (Section 9-3)
Kruskal-Wallis test (Section 13-5)	Test that more than two independent populations have the same median	Analysis of variance (Section 12-2)
Rank correlation (Section 13-6)	Test for relationship between two variables	Linear correlation (Section 10-2)
Runs test (Section 13-7)	Test for randomness of sample data	No parametric test

Statistical Literacy and Critical Thinking

1. Nonparametric Test What is a *nonparametric* test? What is a parametric test?

2. Distribution-Free Test What is the difference between a nonparametric test and a distribution-free test?

3. Rank Many nonparametric tests are based on ranks. What is a rank? What are the ranks of the following lengths (in hr) of NASA Space Shuttle Transport System flights (from Data Set 10 in Appendix B): 54, 54, 192, 169, 122?

4. Efficiency Nonparametric tests are typically not as *efficient* as a corresponding parametric test, provided the necessary requirements are satisfied. What does the efficiency measure? If nonparametric tests are less efficient than parametric tests, why should we use them?

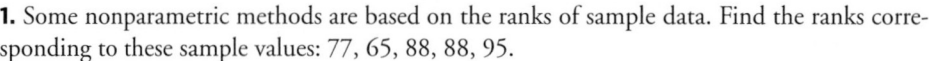

Chapter Quick Quiz

1. Some nonparametric methods are based on the ranks of sample data. Find the ranks corresponding to these sample values: 77, 65, 88, 88, 95.

2. What is the purpose of the runs test for randomness?

3. Identify an advantage of using rank correlation instead of linear correlation.

4. What does it mean when we say that a nonparametric test is less *efficient* when compared to a corresponding parametric method?

5. Which of the following terms is sometimes used instead of "nonparametric" test: normality test; abnormality test; distribution-free test; last testament?

6. True or false: A major advantage of using rank correlation is that it can detect any pattern of the points, even if the pattern is not that of a straight line.

7. Given the following configurations of sample data, identify the one that *cannot* be used with the sign test: matched pairs; four independent samples; data at the nominal level of measurement.

8. Given the following configurations of sample data, identify the one that can be used with the Wilcoxon rank-sum test: one sample of individual values; two independent samples; four independent samples; matched pairs.

9. Given the following configurations of sample data, identify the one that can be used with the Wilcoxon signed-ranks test: one sample of individual values; two independent samples; four independent samples; matched pairs.

10. Given the following configurations of sample data, identify the one that can be used with the Kruskal-Wallis test: one sample of individual values; two independent samples; four independent samples; matched pairs.

Review Exercises

Using Nonparametric Tests. *In Exercises 1–10, use a 0.05 significance level with the indicated test. If no particular test is specified, use the appropriate nonparametric test from this chapter.*

1. World Series The last 103 baseball seasons (as of this writing) ended with 61 World Series wins by American League teams, compared to 42 wins by National League teams. Use the sign test with a 0.05 significance level to test the claim that in each World Series, the American League team has a 0.5 probability of winning.

2. Body Temperatures Listed below are measured body temperatures (in °F) of randomly selected subjects (from Data Set 2 in Appendix B). Use the sign test with a 0.05 significance level to test the claim that body temperatures have a median equal to 98.6°F.

  98.0 98.0 97.0 97.7 98.2 98.4 96.5 98.8 97.4 98.6

3. Body Temperatures Repeat Exercise 2 using the Wilcoxon signed-ranks test.

4. Randomness Refer to the body temperatures listed in Exercise 2. Use only the decimal parts of the temperatures (0, 0, 0, 7, and so on) to test the claim that the sequence of odd and even digits is random.

5. Student and *U.S. News and World Report* Rankings of Colleges Each year, *U.S. News and World Report* publishes rankings of colleges based on statistics such as admission rates, graduation rates, class size, faculty–student ratio, faculty salaries, and peer ratings of administrators. Economists Christopher Avery, Mark Glickman, Caroline Minter Hoxby, and Andrew Metrick took an alternative approach of analyzing the college choices of 3240 high-achieving school seniors. They examined the colleges that offered admission along with the colleges that the students chose to attend. The table below lists rankings for a small sample of colleges. Find the value of the rank correlation coefficient and use it to determine whether there is a correlation between the student rankings and the rankings of the magazine.

Student Ranks	1	2	3	4	5	6	7	8
U.S. News and World Report Ranks	1	2	5	4	7	6	3	8

6. Weather Forecasts Listed below are actual high temperatures and the high temperatures forecast five days in advance (based on Data Set 11 in Appendix B). Use a 0.05 significance level with the sign test to test the claim that the population of differences has a median of zero. What do the results suggest about the accuracy of the forecasts?

Actual high temperature	78 80 71 73 78 75 63 63 70 77
High temperature forecast five days earlier	79 74 75 76 78 76 75 77 71 74

7. Weather Forecasts Repeat the preceding exercise using the Wilcoxon signed-ranks test.

8. Femur Injury in a Car Crash Listed below are measured loads (in lb) on the right femur of crash test dummies. (The data are from the same cars used in the Chapter Problem for Chapter 12.) Test the claim that the different car categories have the same median. Do these data suggest that larger cars are safer?

Small Cars	63 1001 1261 1048 307 925 491 917 750 1008
Medium Cars	257 905 756 547 461 787 677 1023 1444 632
Large Cars	752 669 740 1202 669 1290 554 683 1023 786

9. Falling Temperatures Listed below are low temperatures (in °F) from the first half of September and the second half of September (from Data Set 11 in Appendix B). Use the Wilcoxon rank-sum test to test the claim that temperatures in the first half of September and temperatures in the second half of September have the same median. The region near the author's home where the temperatures were recorded is known to become cooler as the fall season progresses. Do the sample data support that trend?

September 1–15	54 54 55 60 64 51 59 61 68 62 53 52 56 56 54
September 16–30	64 62 55 48 40 47 49 53 51 54 58 48 61 57 53

10. Temperatures and Randomness Refer to the temperatures listed in Exercise 9 and consider them to be one consecutive sequence of 30 temperatures. Test for randomness of even and odd temperatures.

Cumulative Review Exercises

In Exercises 1–5, use the data in the table below, which are cholesterol levels (in mg per dL of blood) and corresponding weights (in lb) for randomly selected adult women from Data Set 1 in Appendix B.

Cholesterol (mg)	264 181 267 384 98 62 126 89 531 130
Weight (lb)	114.8 149.3 107.8 160.1 127.1 123.1 111.7 156.3 218.8 110.2

1. Finding Statistics Find the mean, median, range, standard deviation, and variance of the cholesterol levels. Given that the cholesterol levels are in mg, express the results using the appropriate units.

2. Scatterplot Construct a scatterplot of the paired cholesterol/weight data.

3. Linear Correlation Use a 0.05 significance level to test for a linear correlation between cholesterol level and weight.

4. Regression Find the equation of the regression line for the cholesterol/weight data. What is the best predicted weight of a woman with a cholesterol level of 100 mg?

5. Rank Correlation Use a 0.05 significance level with rank correlation to test for a correlation between cholesterol level and weight.

In Exercises 6–9, use the Flesch-Kincaid Grade Level measurements from randomly selected pages in Tom Clancy's **The Bear and the Dragon,** *J. K. Rowling's* **Harry Potter and the Sorcerer's Stone,** *and Leo Tolstoy's* **War and Peace.** *Those measurements reflect the reading level of the sampled text.*

Clancy	8.8 5.4 5.0 7.6 9.0 3.2 12.0 4.9 4.6 5.0 6.4 6.1
Rowling	5.2 3.7 6.1 4.9 4.4 3.2 5.6 6.9 5.7 4.1 6.7 4.4
Tolstoy	8.6 9.8 6.1 5.9 7.7 10.9 8.2 7.2 11.0 11.5 8.4 5.9

TI **6. ANOVA** Use a 0.05 significance level with analysis of variance to test the claim that the three books have the same mean Flesch-Kincaid Grade Level. Do the three books appear to be written with the same reading level?

TI **7. Kruskal-Wallis Test** Use a 0.05 significance level with the Kruskal-Wallis test to test the claim that the samples of Flesch-Kincaid Grade Level measurements are from books with the same median reading level.

TI **8. *t* Test** Use a 0.05 significance level with a *t* test to test the claim that the samples of Flesch-Kincaid Grade Level scores from Clancy and Rowling have the same mean.

TI **9. Wilcoxon Rank-Sum Test** Use a 0.05 significance level with the Wilcoxon rank-sum test to test the claim that the samples of Flesch-Kincaid Grade Level scores from Clancy and Rowling are from populations having the same median.

10. Cell Phones and Crashes: Analyzing Newspaper Report In an article from the Associated Press, it was reported that researchers "randomly selected 100 New York motorists who had been in an accident and 100 who had not been in an accident. Of those in accidents, 13.7 percent owned a cellular phone, while just 10.6 percent of the accident-free drivers had a phone in the car." What is wrong with these results?

TI-83/84 Plus Calculator Project

Randomly generated digits are commonly used for a variety of different applications. Pollsters typically select telephone numbers by using random digits, so that unlisted numbers can be included. Simulations, such as those discussed in Section 4-6, often require the random generation of digits. This project involves the random generation of digits 0, 1, 2, . . ., 9 in such a way that they are all equally likely.

1. Listed below are data sets that were generated by using the random features of different technologies. (Excel, STATDISK, and Minitab are statistical software packages.) Use the appropriate nonparametric method from this chapter to determine whether the randomly generated digits listed below come from identical populations. Identify the method used, identify your results, and state your conclusions.

TI-83/84 Plus:	5 8 4 9 5 4 0 1 4 2 4 3 1 4 9 7 5 3 7 5
Excel:	3 1 5 9 8 8 4 8 1 2 0 0 1 2 0 2 3 5 3 9
STATDISK:	1 7 8 3 0 0 4 9 7 5 6 5 9 6 1 1 5 5 8 5
Minitab:	1 2 5 1 1 3 1 8 8 5 3 5 5 3 5 3 5 2 3 3

2. Use a TI-83/84 Plus calculator to generate your own list of 100 randomly selected digits. Generate the data by pressing **MATH**, selecting the top menu item of **PROB,** then selecting **randInt.** The format for the randInt command is this:

randInt(min. integer, max. integer, number of generated integers)

Use the command **randInt(0, 9, 100)** to generate 100 randomly selected digits, where each digit is between 0 and 9 inclusive. (*Hint:* Use the **STO▶** key to store the generated digits in list L1.) Test for randomness of the generated digits. Describe the method used and the conclusion reached.

INTERNET PROJECT

Nonparametric Tests

Go to: **http://www.aw.com/triola.**

This chapter introduced hypothesis-testing methods of the nonparametric or distribution-free variety. Nonparametric methods allow you to test hypotheses without making assumptions regarding

the underlying distribution of the population being sampled.

In the Internet Project you will apply nonparametric tests to a new set of data as well as to data analyzed in earlier projects. Further, you will examine random number sequences that arise both theoretically and in the sciences.

APPLET PROJECT

Open the Applets folder on the CD and double-click on **Start**. Select the menu item of **Random numbers**. Generate 50 random numbers between 0 and 1 inclusive. Apply the runs test for randomness to the results. What can you conclude?

FROM DATA TO DECISION

Critical Thinking: Was the draft lottery random?

In 1970, a lottery was used to determine who would be drafted into the U.S. Army. The 366 dates in the year were placed in individual capsules. First, the 31 January capsules were placed in a box; then the 29 February capsules were added and the two months were mixed. Then the 31 March capsules were added and the three months were mixed. This process continued until all months were included. The first capsule selected was September 14, so men born on that date were drafted first. The accompanying list shows the 366 dates in the order of selection.

Analyzing the Results

a. Use the runs test to test the sequence for randomness above and below the median of 183.5.

b. Use the Kruskal-Wallis test to test the claim that the 12 months had priority numbers drawn from the same population.

c. Calculate the 12 monthly means. Then plot those 12 means on a graph. (The horizontal scale lists the 12 months, and the vertical scale ranges from 100 to 260.) Note any pattern suggesting that the original priority numbers were not randomly selected.

d. Based on the results from parts (a), (b), and (c), decide whether this particular draft lottery was fair. Write a statement explaining why you believe that it was or was not fair. If you decided that this lottery was unfair, describe a process for selecting lottery numbers that would have been fair.

Jan	305	159	251	215	101	224	306	199	194	325	329	221	318	238	017	121
	235	140	058	280	186	337	118	059	052	092	355	077	349	164	211	
Feb	086	144	297	210	214	347	091	181	338	216	150	068	152	004	089	212
	189	292	025	302	363	290	057	236	179	365	205	299	285			
Mar	108	029	267	275	293	139	122	213	317	323	136	300	259	354	169	166
	033	332	200	239	334	265	256	258	343	170	268	223	362	217	030	
Apr	032	271	083	081	269	253	147	312	219	218	014	346	124	231	273	148
	260	090	336	345	062	316	252	002	351	340	074	262	191	208		
May	330	298	040	276	364	155	035	321	197	065	037	133	295	178	130	055
	112	278	075	183	250	326	319	031	361	357	296	308	226	103	313	
Jun	249	228	301	020	028	110	085	366	335	206	134	272	069	356	180	274
	073	341	104	360	060	247	109	358	137	022	064	222	353	209		
Jul	093	350	115	279	188	327	050	013	277	284	248	015	042	331	322	120
	098	190	227	187	027	153	172	023	067	303	289	088	270	287	193	
Aug	111	045	261	145	054	114	168	048	106	021	324	142	307	198	102	044
	154	141	311	344	291	339	116	036	286	245	352	167	061	333	011	
Sep	225	161	049	232	082	006	008	184	263	071	158	242	175	001	113	207
	255	246	177	063	204	160	119	195	149	018	233	257	151	315		
Oct	359	125	244	202	024	087	234	283	342	220	237	072	138	294	171	254
	288	005	241	192	243	117	201	196	176	007	264	094	229	038	079	
Nov	019	034	348	266	310	076	051	097	080	282	046	066	126	127	131	107
	143	146	203	185	156	009	182	230	132	309	047	281	099	174		
Dec	129	328	157	165	056	010	012	105	043	041	039	314	163	026	320	096
	304	128	240	135	070	053	162	095	084	173	078	123	016	003	100	

Cooperative Group Activities

1. In-class activity Use the existing seating arrangement in your class and apply the runs test to determine whether the students are arranged randomly according to gender. After recording the seating arrangement, analysis can be done in subgroups of three or four students.

2. In-class activity Divide into groups of 8 to 12 people. For each group member, *measure* his or her height and *measure* his or her arm span. For the arm span, the subject should stand with arms extended, like the wings on an airplane. It's easy to mark the height and arm span on a chalkboard, then measure the distances there. Divide the following tasks among subgroups of three or four people.

a. Use rank correlation with the paired sample data to determine whether there is a correlation between height and arm span.

b. Use the sign test to test for a difference between the two variables.

c. Use the Wilcoxon signed-ranks test to test for a difference between the two variables.

3. In-class activity Do Activity 2 using pulse rate instead of arm span. Measure pulse rates by counting the number of heartbeats in 1 min.

4. Out-of-class activity Divide into groups of three or four students. Investigate the relationship between two variables by collecting your own paired sample data and using the methods of Section 13-6 to determine whether there is a correlation. Suggested topics:

• Is there a relationship between taste and cost of different brands of chocolate chip cookies (or colas)? (Taste can be measured on some number scale, such as 1 to 10.)

• Is there a relationship between salaries of professional baseball (or basketball or football) players and their season achievements?

• Rates versus weights: Is there a relationship between car fuel-consumption rates and car weights?

• Is there a relationship between the lengths of men's (or women's) feet and their heights?

• Is there a relationship between student gradepoint averages and the amount of television watched?

• Is there a relationship between heights of fathers (or mothers) and heights of their first sons (or daughters)?

5. Out-of-class activity See this chapter's "From Data to Decision" project, which involves analysis of the 1970 lottery used for drafting men into the U.S. Army. Because the 1970 results raised concerns about the randomness of selecting draft priority numbers, design a new procedure for generating the 366 priority numbers. Use your procedure to generate the 366 numbers and test your results using the techniques suggested in parts (a), (b), and (c) of the "From Data to Decision" project. How do your results compare to those obtained in 1970? Does your random selection process appear to be better than the one used in 1970? Write a report that clearly describes the process you designed. Also include your analyses and conclusions.

6. Out-of-class activity Divide into groups of three or four. Survey students by asking them to identify their major and gender. For each surveyed subject, determine the accuracy of the time on his or her wristwatch. First set your own watch to the correct time using an accurate and reliable source ("At the tone, the time is . . ."). For watches that are ahead of the correct time, record positive times. For watches that are behind the correct time, record negative times. Use the sample data to address these questions:

• Do the errors appear to be the same for both genders?

• Do the errors appear to be the same for the different majors?

7. In-class activity Divide into groups of 8 to 12 people. For each group member, measure the person's height and also measure his or her navel height, which is the height from the floor to the navel. Use the rank correlation coefficient to determine whether there is a correlation between height and navel height.

8. In-class activity Divide into groups of three or four people. Appendix B includes many data sets not yet addressed by the methods of this chapter. Search Appendix B for variables of interest, then investigate using appropriate methods of nonparametric statistics. State your conclusions and try to identify practical applications.

9. Out-of-class activity Divide into groups of three or four, with at least one member of each group having an iPod. Establish two categories of songs, such as those by males or females, then test the sequence of iPod songs for randomness.

NAME:	Joanna Burger
JOB:	Professor
COMPANY:	Rutgers University

*J*oanna Burger is a Distinguished Professor of Biology at Rutgers University, and she is a member of the Environmental and Occupational Health Sciences Institute. Joanna teaches, does research, and serves on many national and international environmental committees dealing with endangered species, contaminants in wildlife, the effects of chemicals on animal behavior, and the effects of people on ecosystems.

Q: What concepts of statistics do you use in your work?

A: I use a variety of statistical approaches including both parametric and nonparametric methods. Without a firm understanding of statistics I would not be able to test whether environmental factors affect reproductive success. I use statistics to test hypotheses that I generate by observing animals within their natural environments. While observation leads to hypotheses, only with the use of well-designed experiments and statistical tests is it possible to answer the questions. For research and teaching in the field of ecology, animal behavior, and ecotoxicology, knowledge of statistics is essential to obtain a good job, and hold it.

Q: Could you give a specific example of how you have used statistics in the past?

A: Statistics is very helpful in identifying which factors are important in influencing animal behavior. Birds nest in particular

habitats, but we questioned whether they nest randomly or select very specific nest sites. This is important because conservation requires knowing what animals need in order to create, protect, and/or manage that habitat. I tested the hypothesis that common terns were selecting particular salt marsh islands. By comparing statistically a wide range of environmental factors (such as island height, island size, and vegetation type and density) on all islands, with the same set of factors on islands used for nesting by terns, we could show that terns actually selected a very specific set of characteristics. Although there were over 250 islands in the bay where this study was conducted, only 36 met the criteria used by the terns. The characteristics they were selecting actually resulted in their choosing islands that were high enough to avoid summer storm tides, but low enough so that predators could not survive over the winter. Islands that are high enough to avoid winter storm tides can have viable predator populations, such as fox and raccoons, which will eat the eggs and chicks of the terns.

Q: Is a knowledge of statistics critical for your work?

A: A firm understanding of statistics is absolutely essential to conducting research with humans and animals. Using hypothesis testing, and multiple regression analysis, it is possible to begin to identify and evaluate the factors that affect behavior, such as fishing and consumption behavior of people, foraging behavior of shorebirds, and nesting behavior of seabirds.

Q: In terms of statistics, what would you recommend for prospective employees?

A: Anyone who wants to be in the field of conservation biology, ecotoxicology, animal behavior, or ecology needs a wide range of statistical skills. Two or three courses would be best, including general statistics, regression, and nonparametrics. The nature of each problem and the characteristics of the data will determine the statistics that are required, and one should not be limited by a lack of statistical knowledge.

14-1 Review and Preview

14-2 Control Charts for Variation and Mean

14-3 Control Charts for Attributes

14 Statistical Process Control

Global Warming: Is the earth's temperature out of control?

Principles of statistical process control are routinely used by businesses to monitor the quality of the goods they produce and the services they provide. This chapter will include some of the typical business applications, but in this Chapter Problem we consider the process of monitoring temperatures of the earth. Table 14-1 lists the global mean temperature (in °C) of the earth for each year from 1880, with projections used for the last four years. Table 14-1 is arranged so that the temperatures for each decade are listed in order in a separate row. For example, the temperatures for 1880, 1881, and 1882 are 13.88, 13.88, and 14.00, respectively. The last two columns show the mean for each decade and the range for each decade.

We will analyze the data in Table 14-1 as we would analyze any business process, such as the filling of cans of Coke, or the times required to repair defective iPods. By involving the issue of global warming, we emphasize that applications of statistical process control go beyond business statistics.

Many scientists believe that we are experiencing global warming caused by human activity, such as creating carbon dioxide by burning fossil fuels. Many others believe that there really is no global warming. We will see how methods of statistics can be used to monitor the earth's temperature. Specifically, we will determine whether the earth's temperature results from a process that is out of control, or whether the earth's temperature is behaving as it should.

Table 14-1   Annual Temperatures (°C) of the Earth

											$\bar{x}$	Range
1880s	13.88	13.88	14.00	13.96	13.59	13.77	13.75	13.55	13.77	14.04	13.819	0.490
1890s	13.78	13.44	13.60	13.61	13.68	13.68	13.73	13.85	13.79	13.76	13.692	0.410
1900s	13.95	13.95	13.70	13.64	13.58	13.75	13.85	13.60	13.70	13.69	13.741	0.370
1910s	13.79	13.74	13.67	13.72	13.98	14.06	13.80	13.54	13.67	13.91	13.788	0.520
1920s	13.85	13.95	13.91	13.84	13.89	13.85	14.04	13.95	14.00	13.78	13.906	0.260
1930s	13.97	14.03	14.04	13.89	14.05	13.92	14.01	14.12	14.15	13.98	14.016	0.260
1940s	14.14	14.11	14.10	14.06	14.11	13.99	14.01	14.12	13.97	13.91	14.052	0.230
1950s	13.83	13.98	14.03	14.12	13.91	13.91	13.82	14.08	14.10	14.05	13.983	0.300
1960s	13.98	14.10	14.05	14.03	13.65	13.75	13.93	13.98	13.91	14.00	13.938	0.450
1970s	14.04	13.90	13.95	14.18	13.94	13.98	13.79	14.16	14.07	14.13	14.014	0.390
1980s	14.27	14.40	14.10	14.34	14.16	14.13	14.19	14.35	14.42	14.28	14.264	0.320
1990s	14.49	14.44	14.16	14.18	14.31	14.47	14.36	14.40	14.71	14.44	14.396	0.550
2000s	14.41	14.56	14.70	14.64	14.60	14.77	14.64	14.66	14.68	14.70	14.636	0.360

Review and Preview

In Section 2-1 we noted that an important characteristic of data is a changing pattern over time. Some populations change over time so that values of parameters change. The main objective of this chapter is to learn how to construct and interpret *control charts* that can be used to monitor changing characteristics of data over time. That knowledge will better prepare us for work with businesses trying to improve the quality of their goods and services.

Minitab and other software packages include programs for automatically generating charts of the type discussed in this chapter, and we will include examples of such displays. Control charts are good examples of visual tools that allow us to *see* and *understand* some property of data that would be difficult or impossible to understand without graphs. The world needs more people who can construct and interpret graphs, such as the control charts described in this chapter.

Control Charts for Variation and Mean

Key Concept In this section we construct run charts, R charts, and $\bar{x}$ charts so that we can monitor characteristics of data over time. We can use such charts to determine whether a process is statistically stable (or within statistical control).

The following definition formally describes the type of data that will be considered in this chapter.

  DEFINITION

Process data are data arranged according to some time sequence. They are measurements of a characteristic of goods or services that result from some combination of equipment, people, materials, methods, and conditions.

EXAMPLE 1   **Earth's Temperatures as Process Data** Table 14-1 includes process data consisting of the earth's mean temperature for each year of the past 13 decades. Because the data in Table 14-1 are arranged according to the time at which they were selected, they are process data.

It is important to recognize this point:

Characteristics of process data can change over time.

Companies have gone bankrupt because they unknowingly allowed manufacturing processes to deteriorate without constant monitoring. They suffered from a failure to monitor process data.

Run Charts

There are various methods that can be used to monitor a process to ensure that desired characteristics don't change—analysis of a *run chart* is one such method.

DEFINITION

A **run chart** is a sequential plot of *individual* data values over time. One axis (usually the vertical axis) is used for the data values, and the other axis (usually the horizontal axis) is used for the time sequence.

EXAMPLE 2 **Run Chart of Earth's Temperatures** Treating the 130 mean temperatures of the earth in Table 14-1 as a string of consecutive measurements, construct a run chart using a vertical axis for the temperatures and a horizontal axis to identify the chronological order of the sample data, beginning with the first year of 1880.

SOLUTION Figure 14-1 includes Minitab and TI-83/84 Plus run charts for the data in Table 14-1. The vertical scale accommodates the minimum and maximum temperature values of 13.44°C and 14.77°C, respectively. The horizontal scale is designed to include the 130 values arranged in sequence by year. The first point represents the first value of 13.88°C, and so on.

MINITAB

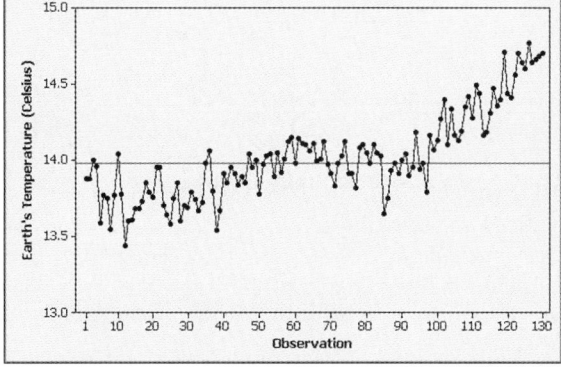

TI-83/84 PLUS

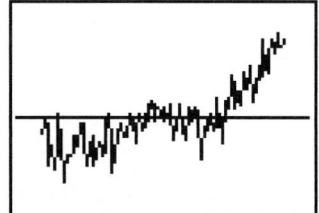

Figure 14-1 Run Charts of Individual Temperatures in Table 14-1

In Figure 14-1, the horizontal scale identifies the sample number, so the number 20 indicates the 20th temperature. The vertical scale represents the temperature of the earth. Now examine Figure 14-1 and try to identify any *patterns*. From Figure 14-1 we see that as time progresses from left to right, the heights of the points appear to increase in value. If this pattern continues, rising temperatures will cause melting of large ice formations and widespread flooding, as well as substantial climate changes. Figure 14-1 is evidence of global warming, which threatens us in many different ways.

DEFINITION

A process is **statistically stable** (or **within statistical control**) if it has only natural variation, with no patterns, cycles, or unusual points.

Interpreting Run Charts Only when a process is *statistically stable* can its data be treated as if they came from a population with a constant mean, standard deviation,

The Flynn Effect: Upward Trend in IQ Scores

A run chart or control chart of IQ scores would reveal that they exhibit an upward trend, because IQ scores have been steadily increasing since they began to be used about 70 years ago.  The trend is worldwide, and it is the same for different types of IQ tests, even those that rely heavily on abstract and nonverbal reasoning with minimal cultural influence. This upward trend has been named the *Flynn effect,* because political scientist James R. Flynn discovered the trend in his studies of U.S. military recruits. The amount of the increase is quite substantial: Based on a current mean IQ score of 100, it is estimated that the mean IQ in 1920 would be about 77. The typical student of today is therefore brilliant when compared to his or her great-grandparents. So far, there is no generally accepted explanation for the Flynn effect.

Costly Assignable Variation

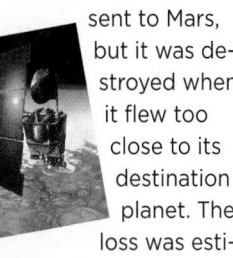

The Mars Climate Orbiter was launched by NASA and sent to Mars, but it was destroyed when it flew too close to its destination planet. The loss was estimated at $125 million. The cause of the crash was found to be confusion between the use of units used for calculations. Acceleration data were provided in the English units of pounds of force, but the Jet Propulsion Laboratory assumed that those units were in metric "newtons" instead of pounds. The thrusters of the spacecraft subsequently provided wrong amounts of force in adjusting the position of the spacecraft. The errors caused by the discrepancy were fairly small at first, but the cumulative error over months of the spacecraft's journey proved to be fatal to its success.

In 1962, the rocket carrying the Mariner 1 satellite was destroyed by ground controllers when it went off course due to a missing minus sign in a computer program.

MINITAB

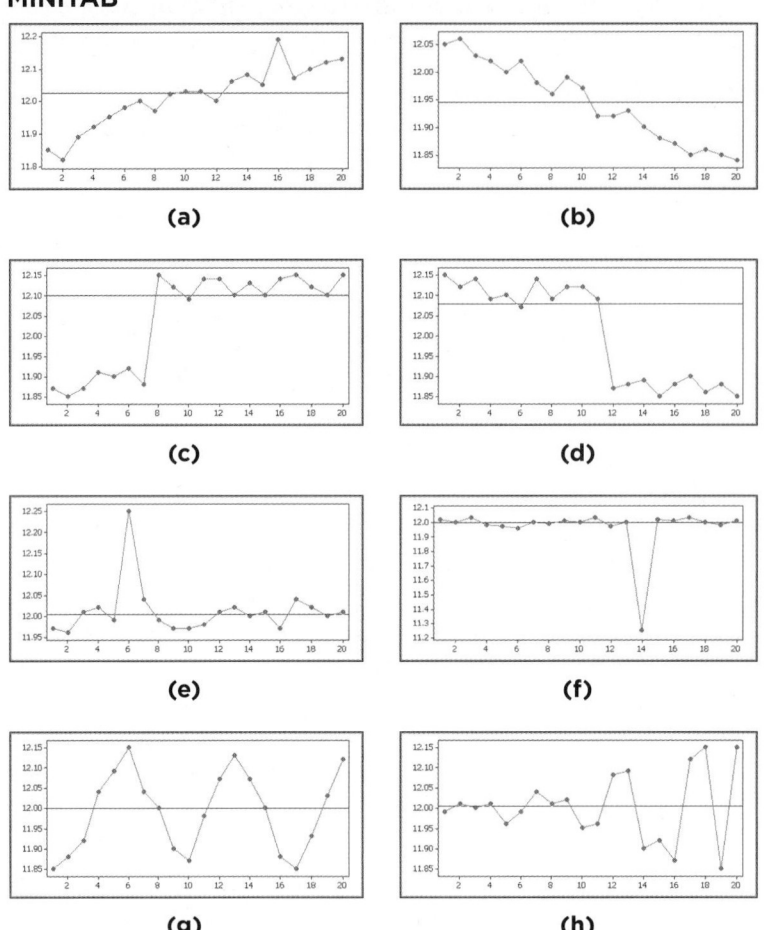

Figure 14-2   Processes That Are Not Statistically Stable

distribution, and other characteristics. Figure 14-2 consists of run charts illustrating typical patterns showing ways in which the process of filling 12-oz cola cans may not be statistically stable.

- **Figure 14-2(a):** There is an obvious *upward trend* that corresponds to values that are increasing over time. If the filling process were to follow this type of pattern, the cans would be filled with more and more cola until they began to overflow, eventually leaving the employees swimming in cola.

- **Figure 14-2(b):** There is an obvious *downward trend* that corresponds to steadily decreasing values. The cans would be filled with less and less cola until they were extremely underfilled. Such a process would require a complete reworking of the cans in order to get them full enough for distribution to consumers.

- **Figure 14-2(c):** There is an *upward shift*. A run chart such as this one might result from an adjustment to the filling process, making all subsequent values higher.

- **Figure 14-2(d):** There is a *downward shift*—the first few values are relatively stable, and then something happened so that the last several values are relatively stable, but at a much lower level.

- **Figure 14-2(e):** The process is stable except for one *exceptionally high value*. The cause of that unusual value should be investigated. Perhaps the cans became temporarily stuck and one particular can was filled twice instead of once.

- **Figure 14-2(f):** There is an *exceptionally low value.* The cause of that unusually low value should be investigated. Perhaps the filling nozzle became temporarily clogged.

- **Figure 14-2(g):** There is a *cyclical pattern* (or repeating cycle). This pattern is clearly nonrandom and therefore reveals a statistically unstable process. Perhaps periodic overadjustments are being made to the machinery, with the effect that some desired value is continually being chased but never quite captured.

- **Figure 14-2(h):** The *variation is increasing over time.* This is a common problem in quality control. The net effect is that products vary more and more until almost all of them are worthless. For example, some cola cans will be overflowing with wasted cola, and some will be underfilled and unsuitable for distribution to consumers.

Many different methods of quality control attempt to *reduce variation* in the product or service. For example, Ford became concerned with variation when it found that its transmissions required significantly more warranty repairs than the same type of transmissions made by Mazda in Japan. A study showed that the Mazda transmissions had substantially less variation in the gearboxes; that is, crucial gearbox measurements varied much less in the Mazda transmissions. Although the Ford transmissions were built within the allowable limits, the Mazda transmissions were more reliable because of their lower variation. Variation in a process can result from two types of causes.

**DEFINITION**

Random variation is due to chance; it is the type of variation inherent in any process that is not capable of producing every good or service exactly the same way every time.

Assignable variation results from causes that can be identified (such factors as defective machinery, untrained employees, and so on).

Later in the chapter we will consider ways to distinguish between assignable variation and random variation.

The run chart is one tool for monitoring the stability of a process. We will now consider *control charts,* which are also useful for monitoring the stability of a process.

Control Chart for Monitoring Variation: The *R* Chart

In the article "The State of Statistical Process Control As We Proceed into the 21st Century" (Stoumbos, Reynolds, Ryan, and Woodall, *Journal of the American Statistical Association,* Vol. 95, No. 451), the authors state that "control charts are among the most important and widely used tools in statistics. Their applications have now moved far beyond manufacturing into engineering, environmental science, biology, genetics, epidemiology, medicine, finance, and even law enforcement and athletics." We begin with the definition of a control chart.

**DEFINITION**

A **control chart** of a process characteristic (such as mean or variation) consists of values plotted sequentially over time, and it includes a **centerline** as well as a **lower control limit** (LCL) and an **upper control limit** (UCL). The centerline represents a central value of the characteristic measurements, whereas the control limits are boundaries used to separate and identify any points considered to be *unusual.*

We will assume that the population standard deviation σ is not known as we now consider two of several different types of *control charts:*

1. *R* charts (or range charts) used to monitor variation

2. $\bar{x}$ charts used to monitor means

When using control charts to monitor a process, it is common to consider *R* charts and $\bar{x}$ charts together, because a statistically unstable process may be the result of increasing *variation,* changing *means,* or both.

An **R chart** (or **range chart**) is a plot of the sample ranges instead of individual sample values, and it is used to monitor the *variation* in a process. (It might make more sense to use standard deviations, but range charts are quite effective for cases in which the size of the samples (or subgroups) is 10 or fewer. If the samples all have a size greater than 10, the use of an *s* chart is recommended instead of an *R* chart. (See Exercise 21.)) In addition to plotting the range values, we include a centerline located at $\bar{R}$, which denotes the mean of all sample ranges, as well as another line for the lower control limit and a third line for the upper control limit. The following is a summary of notation and the components of the *R* chart.

Monitoring Process Variation: Control Chart for *R*

Objective

Construct a control chart for *R* (or an "*R* chart") that can be used to determine whether the *variation* of process data is within statistical control.

Requirements

1. The data are process data consisting of a sequence of samples all of the same size *n*.

2. The distribution of the process data is essentially normal.

3. The individual sample data values are independent.

Notation

n = size of each sample, or *subgroup*

$\bar{R}$ = mean of the sample ranges (that is, the sum of the sample ranges divided by the number of samples)

Graph

Points plotted: Sample ranges (one point for each sample or subgroup)

Centerline: $\bar{R}$ (the mean of the sample ranges)

Upper control limit (UCL): $D_4\bar{R}$ (where D_4 is found in Table 14-2)

Lower control limit (LCL): $D_3\bar{R}$ (where D_3 is found in Table 14-2)

The values of D_4 and D_3 were computed by quality-control experts, and they are intended to simplify calculations. The upper and lower control limits of $D_4\bar{R}$ and $D_3\bar{R}$ are values that are roughly equivalent to 99.7% confidence interval limits. It is therefore highly unlikely that values from a statistically stable process would fall beyond those limits. If a value does fall beyond the control limits, it's very likely that the process is not statistically stable.

Table 14-2   Control Chart Constants

n: Number of Observations in Subgroup	$\bar{x}$		s		R	
	A_2	A_3	B_3	B_4	D_3	D_4
2	1.880	2.659	0.000	3.267	0.000	3.267
3	1.023	1.954	0.000	2.568	0.000	2.574
4	0.729	1.628	0.000	2.266	0.000	2.282
5	0.577	1.427	0.000	2.089	0.000	2.114
6	0.483	1.287	0.030	1.970	0.000	2.004
7	0.419	1.182	0.118	1.882	0.076	1.924
8	0.373	1.099	0.185	1.815	0.136	1.864
9	0.337	1.032	0.239	1.761	0.184	1.816
10	0.308	0.975	0.284	1.716	0.223	1.777
11	0.285	0.927	0.321	1.679	0.256	1.744
12	0.266	0.886	0.354	1.646	0.283	1.717
13	0.249	0.850	0.382	1.618	0.307	1.693
14	0.235	0.817	0.406	1.594	0.328	1.672
15	0.223	0.789	0.428	1.572	0.347	1.653
16	0.212	0.763	0.448	1.552	0.363	1.637
17	0.203	0.739	0.466	1.534	0.378	1.622
18	0.194	0.718	0.482	1.518	0.391	1.608
19	0.187	0.698	0.497	1.503	0.403	1.597
20	0.180	0.680	0.510	1.490	0.415	1.585
21	0.173	0.663	0.523	1.477	0.425	1.575
22	0.167	0.647	0.534	1.466	0.434	1.566
23	0.162	0.633	0.545	1.455	0.443	1.557
24	0.157	0.619	0.555	1.445	0.451	1.548
25	0.153	0.606	0.565	1.435	0.459	1.541

Source: Adapted from *ASTM Manual on the Presentation of Data and Control Chart Analysis,* © 1976 ASTM, pp. 134–136. Reprinted with permission of American Society for Testing and Materials.

Bribery Detected with Control Charts

Control charts were used to help convict a person who bribed Florida jai alai players to lose. (See "Using Control Charts to Corroborate Bribery in Jai Alai," by Charnes and Gitlow, *The American Statistician,* Vol. 49, No. 4.) An auditor for one jai alai facility noticed that abnormally large sums of money were wagered for certain types of bets, and some contestants didn't win as much as expected when those bets were made. *R* charts and $\bar{x}$ charts were used in court as evidence of highly unusual patterns of betting. Examination of the control charts clearly shows points well beyond the upper control limit, indicating that the process of betting was out of statistical control. The statistician was able to identify a date at which assignable variation appeared to stop, and prosecutors knew that it was the date of the suspect's arrest.

EXAMPLE 3   *R* **Chart of Earth's Temperatures** Refer to the temperatures of the earth listed in Table 14-1. Using the samples of size $n = 10$ for each decade, construct a control chart for *R*.

SOLUTION   Note that Table 14-1 shows the sample ranges in the last column. $\bar{R}$ is the mean of those 13 sample ranges, so its value is found as follows:

$$\bar{R} = \frac{0.49 + 0.41 + \cdots + 0.36}{13} = 0.3777$$

continued

The centerline for our R chart is therefore located at $\overline{R} = 0.3777$. To find the upper and lower control limits, we must first find the values of D_3 and D_4. Referring to Table 14-2 for $n = 10$, we get $D_4 = 1.777$ and $D_3 = 0.223$, so the control limits are as follows:

$$\text{Upper control limit: } D_4\overline{R} = (1.777)(0.3777) = 0.6712$$
$$\text{Lower control limit: } D_3\overline{R} = (0.223)(0.3777) = 0.0842$$

Using a centerline value of $\overline{R} = 0.3777$ and control limits of 0.6712 and 0.0842, we now proceed to plot the 13 sample ranges as 13 individual points. The result is shown in the Minitab display and the TI-83/84 Plus display.

MINITAB

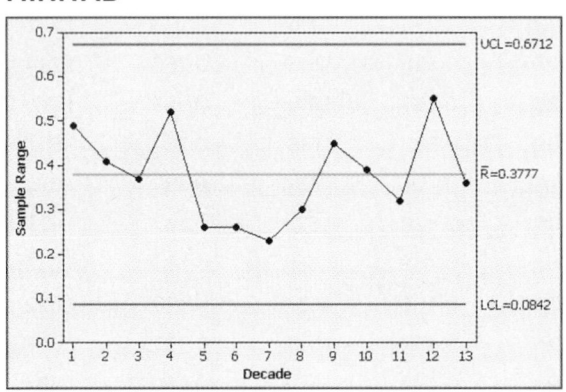

TI-83/84 PLUS

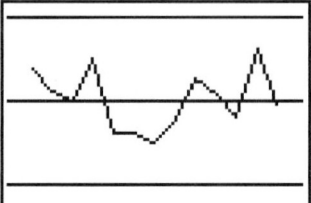

Interpreting Control Charts

When interpreting control charts, the following caution is important:

CAUTION

Upper and lower control limits of a control chart are based on the *actual* behavior of the process, not the *desired* behavior. Upper and lower control limits are totally unrelated to any process *specifications* that may have been decreed by the manufacturer.

When investigating the quality of some process, there are typically two key questions that need to be addressed:

1. Based on the current behavior of the process, can we conclude that the process is within statistical control?

2. Do the process goods or services meet design specifications?

The methods of this chapter are intended to address the first question, but not the second. That is, we are focusing on the behavior of the process with the objective of determining whether the process is within statistical control.

Also, we should clearly understand the specific criteria for determining whether a process is in statistical control (that is, whether it is statistically stable). So far, we have noted that a process is not statistically stable if its pattern resembles any of the patterns shown in Figure 14-2. This criterion is included with some others in the following list.

**Criteria for Determining When a Process Is Not Statistically Stable
(Out of Statistical Control)**

1. There is a pattern, trend, or cycle that is obviously not random (such as those depicted in Figure 14-2).

2. There is a point lying outside of the region between the upper and lower control limits. (That is, there is a point above the upper control limit or below the lower control limit.)

3. *Run of 8 Rule:* There are eight consecutive points all above or all below the centerline. (With a statistically stable process, there is a 0.5 probability that a point will be above or below the centerline, so it is very unlikely that eight consecutive points will all be above the centerline or all below it.)

We will use only the three out-of-control criteria listed above, but some businesses use additional criteria such as these:

• There are six consecutive points all increasing or all decreasing.

• There are 14 consecutive points all alternating between up and down (such as up, down, up, down, and so on).

• Two out of three consecutive points are beyond control limits that are 2 standard deviations away from the centerline.

• Four out of five consecutive points are beyond control limits that are 1 standard deviation away from the centerline.

EXAMPLE 4   **Interpreting *R* Chart of Earth's Temperatures** Examine the *R* chart shown in the Minitab display for Example 3 and determine whether the process variation is within statistical control.

SOLUTION   We can interpret control charts for *R* by applying the three out-of-control criteria just listed. Applying the three criteria to the Minitab display of the *R* chart, we conclude that variation in this process is within statistical control. (1) There is no obvious trend or pattern that is not random. (2) No point lies outside of the region between the upper and lower control limits. (3) There are not eight consecutive points all above or all below the centerline.

INTERPRETATION   We conclude that the *variation* (not necessarily the mean) of the process is within statistical control.

Control Chart for Monitoring Means: The $\bar{x}$ Chart

An $\bar{x}$ **chart** is a plot of the sample means, and it is used to monitor the *center* in a process. In addition to plotting the sample means, we include a centerline located at $\bar{\bar{x}}$, which denotes the mean of all sample means (equal to the mean of all sample values combined), as well as another line for the lower control limit and a third line for the upper control limit. Using the approach common in business and industry, the centerline and control limits are based on ranges instead of standard deviations. (See Exercise 22 for an $\bar{x}$ chart based on standard deviations.)

Monitoring Process Mean: Control Chart for $\bar{x}$

Objective

Construct a control chart for $\bar{x}$ (or an $\bar{x}$ chart) that can be used to determine whether the *center* of process data is within statistical control.

Requirements

1. The data are process data consisting of a sequence of samples all of the same size n.

2. The distribution of the process data is essentially normal.

3. The individual sample data values are independent.

Notation

n = size of each sample, or *subgroup*

$\bar{\bar{x}}$ = mean of all sample means (equal to the mean of all sample values combined)

Graph

Points plotted: Sample means

Centerline: $\bar{\bar{x}}$ = mean of all sample means

Upper control limit (UCL): $\bar{\bar{x}} + A_2\bar{R}$ (where A_2 is found in Table 14-2)

Lower control limit (LCL): $\bar{\bar{x}} - A_2\bar{R}$ (where A_2 is found in Table 14-2)

> **EXAMPLE 5** $\bar{x}$ **Chart of Earth's Temperatures** Refer to the earth's temperatures in Table 14-1. Using samples of size $n = 10$ for each decade, construct a control chart for $\bar{x}$. Based on the control chart for $\bar{x}$, only, determine whether the process mean is within statistical control.

SOLUTION   Before plotting the 13 points corresponding to the 13 values of $\bar{x}$, we must first find the values for the centerline and control limits. We get

$$\bar{\bar{x}} = \frac{13.819 + 13.692 + \cdots + 14.636}{13} = 14.019$$

$$\bar{R} = \frac{0.49 + 0.41 + \cdots + 0.36}{13} = 0.3777$$

Referring to Table 14-2, we find that for $n = 10$, $A_2 = 0.308$. Knowing the values of $\bar{x}$, A_2, and $\bar{R}$, we can now evaluate the control limits.

Upper control limit: $\bar{\bar{x}} + A_2\bar{R} = 14.019 + (0.308)(0.3777) = 14.135$

Lower control limit: $\bar{\bar{x}} - A_2\bar{R} = 14.019 - (0.308)(0.3777) = 13.903$

The resulting control chart for $\bar{x}$ will be as shown in the Minitab and TI-83/84 Plus calculator displays on the next page.

INTERPRETATION   Examination of the $\bar{x}$ chart shows that the process mean is out of statistical control because at least one of the three out-of-control criteria is not satisfied. Specifically, the first criterion is violated because there is a trend of values that are increasing over time, and the second criterion is violated because there are points lying beyond the control limits.

MINITAB

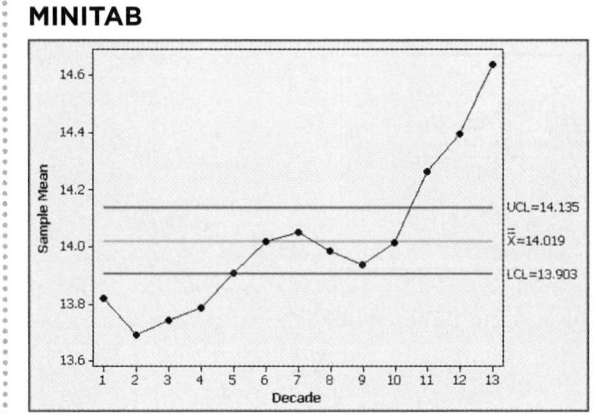

TI-83/84 PLUS

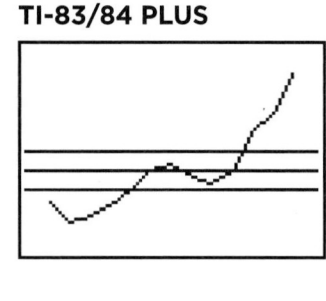

By analyzing the temperature data in Table 14-1 with a run chart, an R chart, and an $\bar{x}$ chart, we can see that the process is out of statistical control because of the pattern of increasing values over time. Based on the data from the past 130 years, it appears that the earth is warming. Many reputable scientists attribute the cause to man-made conditions, including a greenhouse effect caused by greenhouse gases, such as carbon dioxide, methane, and nitrous oxide. If that is the case, it would be wise to take corrective action so that the process is brought within statistical control and the pattern of increasing temperatures is discontinued.

Using a TI-83/84 Plus Calculator for Run Charts, R Charts, and $\bar{x}$ Charts

This section includes examples of a TI-83/84 Plus–generated run chart, R chart, and $\bar{x}$ chart. This section also includes examples of Minitab charts. Comparisons should lead to the obvious conclusion that Minitab is better in the sense that it generates charts that include more information and are easier to read. Consequently, if choosing between a TI-83/84 Plus calculator and Minitab for the purpose of creating a run chart, R chart, or $\bar{x}$ chart, Minitab is the better choice. (Along with some other advantages, Minitab automatically produces the upper and lower control limits as well as the centerline, so there is no need to manually determine them.) The following instructions describe the use of the TI-83/84 Plus calculator.

Run Chart

1. To construct a run chart using a TI-83/84 Plus calculator, first enter all of the sample data in list L1.

2. In list L2 enter the consecutive integers 1, 2, 3, . . . (stopping when L2 has the same length as L1).

3. Find the mean of all of the sample data. (Press **STAT**, select **CALC,** press **ENTER**, enter L1, and press **ENTER** again.)

4. Set up Plot 1 as shown in the accompanying screen display. (Press **2ND** $\boxed{Y=}$ to get to the **STAT PLOTS** menu.) Note that list L2 is used for the Xlist entry, and list L1 is used for the Ylist entry.

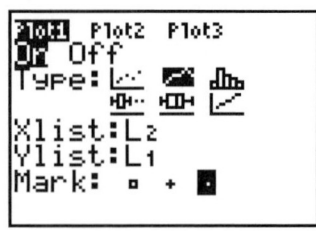

5. Press $\boxed{Y=}$ and enter the mean found in Step 3. For the run chart in Example 2, the mean is 14.019, so enter that value as shown in the second screen.

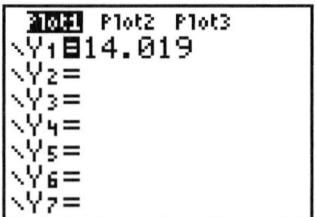

6. Press **2ND** $\boxed{ZOOM}$ and select **AxesOff.**

7. Press $\boxed{ZOOM}$ $\boxed{9}$ to select the **Zoomstat** menu item, which automatically configures the display window so that all of the data points are shown.

8. After the run chart is displayed, change back to AxesOn. Press **2ND** $\boxed{ZOOM}$, use the arrow keys to highlight **AxesON,** then press **ENTER**.

R Chart

1. In list L1, enter the sample *ranges* (instead of the original list of sample data).

2. In list L2 enter the consecutive integers 1, 2, 3, . . . (stopping when L2 has the same length as L1).

3. Find the mean of all of the sample ranges, and find the values used for the upper and lower control limits (as described in this section).

4. Set up **Plot1** as shown in Step 4 of the above procedure for generating a run chart.

5. Press $\boxed{Y=}$ and enter the mean of all of the sample ranges (as shown in Step 5 for the above procedure for generating a run chart.) For **Y2=** enter the value used for the upper control limit, and for **Y3=** enter the value used for the lower control limit.

6. Press **2ND** $\boxed{ZOOM}$ and select **AxesOff.**

7. Press $\boxed{ZOOM}$ $\boxed{9}$ to select the **Zoomstat** menu item, which automatically configures the display window so that all of the data points are shown. If the lines for the upper and lower control limits are not both shown, press $\boxed{WINDOW}$ and change the values used for Ymin and Ymax.

8. After the run chart is displayed, change back to AxesOn. Press **2ND** ⟨ZOOM⟩, use the arrow keys to highlight **AxesON,** then press **ENTER**.

$\bar{x}$ Chart

1. In list L1, enter the sample *means* (instead of the original list of sample data).

2. In list L2 enter the consecutive integers 1, 2, 3, . . . (stopping when L2 has the same length as L1).

3. Find the mean of all of the sample means, and find the values used for the upper and lower control limits (as described in this section).

4. Set up **Plot1** as shown in Step 4 of the above procedure for generating a run chart.

5. Press ⟨Y=⟩ and enter the mean of all of the sample ranges (as shown in Step 5 for the above procedure for generating a run chart.) For **Y2=** enter the value used for the upper control limit, and for **Y3=** enter the value used for the lower control limit.

6. Press **2ND** ⟨ZOOM⟩ and select **AxesOff.**

7. Press ⟨ZOOM⟩ ⟨9⟩ to select the **Zoomstat** menu item, which automatically configures the display window so that all of the data points are shown. If the lines for the upper and lower control limits are not both shown, press ⟨WINDOW⟩ and change the values used for Ymin and Ymax.

8. After the run chart is displayed, change back to **AxesOn.** Press **2ND** ⟨ZOOM⟩, use the arrow keys to highlight **AxesON,** then press **ENTER**.

14-2 Basic Skills and Concepts

Statistical Literacy and Critical Thinking

1. Product Specs Consider process data consisting of the amounts of Coke (in oz) in randomly selected cans of regular Coke. Recent $\bar{x}$ and R control charts show that the process of filling cans of Coke is within statistical control. Does being within statistical control indicate that cans of Coke labeled 12 ounces actually have amounts of Coke that are reasonably close to 12 oz? Why or why not?

2. Notation Consider process data consisting of the amounts of Coke (in oz) in randomly selected cans of regular Coke. The process is to be monitored with $\bar{x}$ and R control charts based on samples of 50 cans randomly selected each day for 20 consecutive days of production. In this context, what do $\bar{\bar{x}}$, $\bar{R}$, UCL, and LCL denote?

3. Process Data Variation Consider process data consisting of the amounts of Coke (in oz) in randomly selected cans of regular Coke. That process is currently within statistical control, yet the amounts of Coke vary. In this context, what is *random variation?* Give an example of assignable variation.

4. Lake Mead Elevations Shown below are an $\bar{x}$ chart (top) and an R chart (bottom) obtained using the monthly elevations of Lake Mead at Hoover Dam (based on data from the U.S. Department of the Interior). The elevations are in feet above sea level. The control charts are based on the 12 monthly elevations for each of the 71 consecutive and recent years available as of this writing. What do the control charts tell us about Lake Mead?

MINITAB

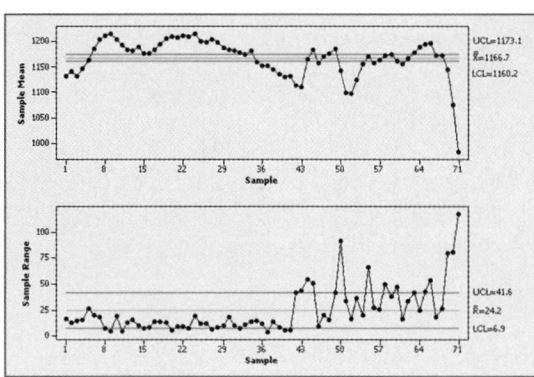

Interpreting Run Charts. *In Exercises 5–8, examine the run chart from a process of filling 12-oz cans of cola and do the following: (a) Determine whether the process is within statistical control; (b) if the process is not within statistical control, identify reasons why it is not; (c) apart from being within statistical control, does the process appear to be behaving as it should?*

5.

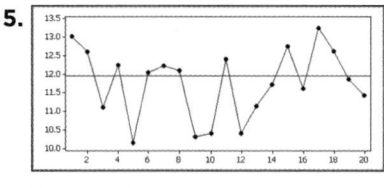

6.

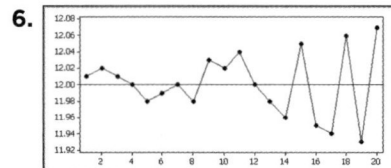

7.

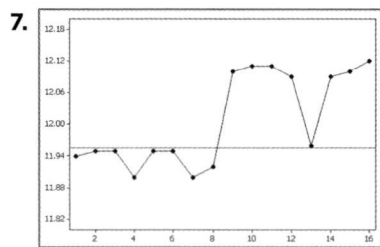

8.

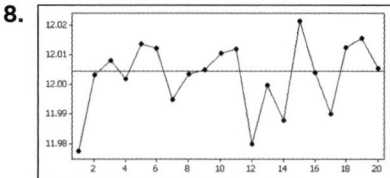

Atmospheric Carbon Dioxide. *In Exercises 9–12, use the data in the following table, which lists carbon dioxide concentrations (in parts per million) for each year from 1880 to 2009, with projected values used for the last four years. Atmospheric carbon dioxide is believed to be the result of human activity and a major contributor to the greenhouse effect that is at least partly responsible for global warming.*

Atmospheric Carbon Dioxide Concentrations (in parts per million)

1880s	290.7	291.0	291.2	291.4	291.6	291.9	292.1	292.3	292.6	292.9
1890s	293.2	293.5	293.8	294.1	294.3	294.6	294.9	295.2	295.5	295.8
1900s	295.6	295.3	295.1	294.8	295.9	296.9	297.5	298.1	298.6	299.2
1910s	299.4	299.6	299.9	300.1	300.3	300.5	300.7	300.9	301.1	301.2
1920s	301.4	301.6	302.3	302.9	303.6	304.2	304.9	305.5	305.6	305.8
1930s	305.9	306.1	306.2	306.3	306.5	306.6	306.8	306.9	307.1	307.3
1940s	307.4	307.6	307.7	307.9	308.4	308.9	309.3	309.8	310.3	310.8
1950s	311.3	311.7	312.2	312.7	313.2	313.7	314.3	314.8	315.3	316.0
1960s	316.9	317.6	318.5	319.0	319.5	320.1	321.3	322.1	323.1	324.6
1970s	325.7	326.3	327.5	329.6	330.3	331.2	332.2	333.9	335.5	336.9
1980s	338.7	340.0	341.1	342.8	344.4	345.9	347.1	349.0	351.4	352.9
1990s	354.2	355.6	356.4	357.1	358.9	360.9	362.6	363.8	366.6	368.3
2000s	369.5	371.0	373.1	375.6	377.4	379.6	379.6	381.2	382.8	384.4

9. Carbon Dioxide: Notation After finding the values of the mean and range for each decade, find the values of $\bar{\bar{x}}$, and $\bar{R}$. Also find the values of LCL and UCL for an R chart, and find the values of LCL and UCL for an $\bar{x}$ chart.

10. Carbon Dioxide: Run Chart Construct a run chart for the 130 values. Does there appear to be a pattern suggesting that the process is not within statistical control? What are the practical implications of the run chart?

11. Carbon Dioxide: R Chart Let each subgroup consist of the 10 values within a decade. Construct an R chart and determine whether the process variation is within statistical control. If it is not, identify which of the three out-of-control criteria lead to rejection of statistically stable variation.

12. Carbon Dioxide: $\bar{x}$ Chart Let each subgroup consist of the 10 values within a decade. Construct an $\bar{x}$ chart and determine whether the process mean is within statistical control. If it is not, identify which of the three out-of-control criteria lead to rejection of a statistically stable mean.

Energy Consumption. *In Exercises 13–16, refer to Data Set 12 in Appendix B and use the amounts of electricity consumed (in kWh) in the author's home. Let each subgroup consist of the six amounts within the same year, so that there are eight subgroups with six amounts in each subgroup.*

13. Energy Consumption: Notation After finding the values of the mean and range for each year, find the values of $\bar{\bar{x}}$, and $\bar{R}$. Then find the values of LCL and UCL for an R chart and for an $\bar{x}$ chart.

14. Energy Consumption: R Chart Let each subgroup consist of the 6 values within a year. Construct an R chart and determine whether the process variation is within statistical control. If it is not, identify which of the three out-of-control criteria lead to rejection of statistically stable variation.

15. Energy Consumption: $\bar{x}$ Chart Let each subgroup consist of the 6 values within a year. Construct an $\bar{x}$ chart and determine whether the process mean is within statistical control. If it is not, identify which of the three out-of-control criteria lead to rejection of a statistically stable mean.

16. Energy Consumption: Run Chart Construct a run chart for the 48 values. Does there appear to be a pattern suggesting that the process is not within statistical control?

Energy Consumption. *In Exercises 17–20, refer to Data Set 13 in Appendix B and use the measured voltage amounts for the power supplied directly to the author's home. Let each subgroup consist of the five amounts within the business days of a week, so the first five voltages constitute the first subgroup, the second five voltages constitute the second subgroup, and so on. The result is eight subgroups with five values each.*

17. Home Voltage: Notation After finding the values of the mean and range for each subgroup, find the values of $\bar{\bar{x}}$, and $\bar{R}$. Then find the values of LCL and UCL for an R chart and for an $\bar{x}$ chart.

18. Home Voltage: $\bar{x}$ Chart Using subgroups of five voltage amounts, construct an $\bar{x}$ chart and determine whether the process mean is within statistical control. If it is not, identify which of the three out-of-control criteria lead to rejection of a statistically stable mean.

19. Home Voltage: Run Chart Construct a run chart for the 40 voltage amounts. Does there appear to be a pattern suggesting that the process is not within statistical control?

20. Home Voltage: R Chart Using subgroups of five voltage amounts, construct an R chart and determine whether the process variation is within statistical control. If it is not, identify which of the three out-of-control criteria lead to rejection of statistically stable variation.

14-2 Beyond the Basics

21. s Chart In this section we described control charts for R and $\bar{x}$ based on ranges. Control charts for monitoring variation and center (mean) can also be based on standard deviations. An *s* chart for monitoring variation is made by plotting sample standard deviations with a centerline at $\bar{s}$ (the mean of the sample standard deviations) and control limits at $B_4\bar{s}$ and $B_3\bar{s}$, where B_4 and B_3 are found in Table 14-2 on page 723. Construct an *s* chart for the data of Table 14-1. Compare the result to the R chart given in Example 3.

22. $\bar{x}$ Chart Based on Standard Deviations An $\bar{x}$ chart based on standard deviations (instead of ranges) is made by plotting sample means with a centerline at $\bar{\bar{x}}$ and control limits at $\bar{\bar{x}} + A_3\bar{s}$ and $\bar{\bar{x}} - A_3\bar{s}$, where A_3 is found in Table 14-2 on page 723 and $\bar{s}$ is the mean of the sample standard deviations. Use the data in Table 14-1 to construct an $\bar{x}$ chart based on standard deviations. Compare the result to the $\bar{x}$ chart based on sample ranges (as in Example 5).

Control Charts for Attributes

Key Concept In this section we present a method for constructing a control chart to monitor the proportion p for some *attribute,* such as whether a service or manufactured item is defective or nonconforming. (A good or a service is nonconforming if it doesn't meet specifications or requirements; nonconforming goods are sometimes discarded, repaired, or called "seconds" and sold at reduced prices.) The control chart is interpreted using the same three criteria from Section 14-2 to determine whether the process is statistically stable. As in Section 14-2, we select samples of size n at regular time intervals and plot points in a sequential graph with a centerline and control limits. (There are ways to deal with samples of different sizes, but we don't consider them here.).

> **DEFINITION**
>
> A **control chart for p** (or **p chart**) is a graph of proportions of some attribute (such as whether products are defective) plotted sequentially over time, and it includes a centerline, a lower control limit (LCL), and an upper control limit (UCL).

The notation and control chart values are as summarized in the following box. In this box, the attribute of "defective" can be replaced by any other relevant attribute (so that each sample item belongs to one of two distinct categories).

Monitoring a Process Attribute: Control Chart for *p*

Objective

Construct a control chart for *p* (or a "*p* chart") that can be used to determine whether the proportion of some attribute (such as whether products are defective) from process data is within statistical control.

Requirements

1. The data are process data consisting of a sequence of samples all of the same size *n*.

2. Each sample item belongs to one of two categories (such as defective or not defective).

3. The individual sample data values are independent.

Notation

$\bar{p}$ = pooled estimate of the proportion of defective items in the process

$$= \frac{\text{total number of defects found among all items sampled}}{\text{total number of items sampled}}$$

$\bar{q}$ = pooled estimate of the proportion of process items that are *not* defective

$$= 1 - \bar{p}$$

n = size of each sample or subgroup

Graph

Points plotted: Proportions from the individual samples of size *n*

Centerline: $\bar{p}$

Upper control limit: $\bar{p} + 3\sqrt{\dfrac{\bar{p}\,\bar{q}}{n}}$ (Use 1 if this result is greater than 1.)

Lower control limit: $\bar{p} - 3\sqrt{\dfrac{\bar{p}\,\bar{q}}{n}}$ (Use 0 if this result is negative.)

CAUTION

Upper and lower control limits of a control chart for a proportion *p* are based on the *actual* behavior of the process, not the *desired* behavior. Upper and lower control limits are totally unrelated to any process *specifications* that may have been decreed by the manufacturer.

We use $\bar{p}$ for the centerline because it is the best estimate of the proportion of defects from the process. The expressions for the control limits correspond to 99.7% confidence interval limits as described in Section 7-2.

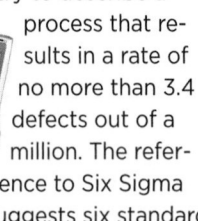

EXAMPLE 1 **Defective Heart Defibrillators** The Guidant Corporation manufactures implantable heart defibrillators. Families of people who have died using these devices are suing the company. According to *USA Today*, "Guidant did not alert doctors when it knew 150 of every 100,000 Prizm 2DR defibrillators might malfunction each year." Because lives could be lost, it is important to monitor the manufacturing process of implantable heart defibrillators.

Consider a manufacturing process that includes careful testing of each defibrillator. Listed below are the numbers of defective defibrillators in successive batches of 10,000. Construct a control chart for the proportion p of defective defibrillators and determine whether the process is within statistical control. If not, identify which of the three out-of-control criteria apply.

Defects 15 12 14 14 11 16 17 11 16 7 8 5 7 6 6 8 5 7 8 7

SOLUTION The centerline for the control chart is located by the value of $\bar{p}$:

$$\bar{p} = \frac{\text{total number of defects from all samples combined}}{\text{total number of defibrillators sampled}}$$

$$= \frac{15 + 12 + 14 + \cdots + 7}{20 \cdot 10{,}000} = \frac{200}{200{,}000} = 0.001$$

Because $\bar{p} = 0.001$, it follows that $\bar{q} = 1 - \bar{p} = 0.999$. Using $\bar{p} = 0.001$, $\bar{q} = 0.999$, and $n = 10{,}000$, we find the positions of the centerline and the control limits as follows:

Centerline: $\bar{p} = 0.001$

Upper control limit:

$$\bar{p} + 3\sqrt{\frac{\bar{p}\,\bar{q}}{n}} = 0.001 + 3\sqrt{\frac{(0.001)(0.999)}{10{,}000}} = 0.001948$$

Lower control limit:

$$\bar{p} - 3\sqrt{\frac{\bar{p}\,\bar{q}}{n}} = 0.001 - 3\sqrt{\frac{(0.001)(0.999)}{10{,}000}} = 0.000052$$

Having found the values for the centerline and the control limits, we can proceed to plot the proportions of defective defibrillators. The Minitab and TI-83/84 Plus control charts for p are shown here.

MINITAB

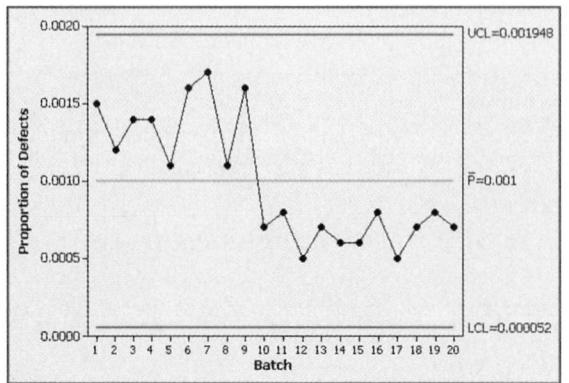

TI-83/84 PLUS

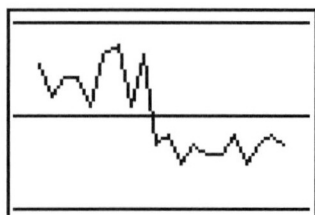

Six Sigma in Industry

Six Sigma is the term used in industry to describe a process that results in a rate of no more than 3.4 defects out of a million. The reference to Six Sigma suggests six standard deviations away from the center of a normal distribution, but the assumption of a perfectly stable process is replaced with the assumption of a process that drifts slightly, so the defect rate is no more than 3 or 4 defects per million.

Started around 1985 at Motorola, Six Sigma programs now attempt to improve quality and increase profits by reducing variation in processes. Motorola saved more than $940 million in three years. Allied Signal reported a savings of $1.5 billion. GE, Polaroid, Ford, Honeywell, Sony, and Texas Instruments are other major companies that have adopted the Six Sigma goal.

INTERPRETATION We can interpret the control chart for p by considering the three out-of-control criteria listed in Section 14-2. Using those criteria, we conclude that this process is out of statistical control for this reason: There appears to be a downward trend. Also, there are 8 consecutive points lying above the centerline, and there are also 8 consecutive points lying below the centerline. Although the process is out of statistical control, it appears to have been somehow *improved,* because the proportion of defects has dropped. The company would be wise to investigate the process so that the cause of the lowered rate of defects can be understood and continued in the future.

Using a TI-83/84 Plus Calculator for Control Charts for p

When constructing a control chart for p, Minitab automatically determines the upper and lower control limits as well as the centerline, and it provides a display that is easy to read. The use of Minitab is therefore recommended. The following instructions describe the use of the TI-83/84 Plus calculator.

1. To construct a control chart for p using a TI-83/84 Plus calculator, first enter all of the sample proportions in list L1.

2. In list L2 enter the consecutive integers 1, 2, 3, . . . (stopping when L2 has the same length as L1).

3. Find the three values to be used for the centerline, upper control limit, and lower control limit (as described in this section).

4. Set up Plot1 as shown in the accompanying screen display. (Press **2ND** ⌐Y=⌐ to get to the **STAT PLOTS** menu.) Note that list L2 is used for the Xlist entry, and list L1 is used for the Ylist entry.

5. Press ⌐Y=⌐ and use three separate lines to enter the three values used for the centerline, the upper control limit, and the lower control limit (from Step 3).

6. Press **2ND** ⌐ZOOM⌐ and select **AxesOff.**

7. Press ⌐ZOOM⌐ ⌐9⌐ to select the **Zoomstat** menu item, which automatically configures the display window so that all of the data points are shown. If the lines for the upper and lower control limits are not both shown, press ⌐WINDOW⌐ and change the values used for Ymin and Ymax.

8. After the p chart is displayed, change back to AxesOn. (Press **2ND** ⌐ZOOM⌐, use the arrow keys to highlight **AxesON,** then press **ENTER**.)

14-3 Basic Skills and Concepts

Statistical Literacy and Critical Thinking

1. Monitoring Aspirin The labels on a bottle of Bayer aspirin indicate that the tablets contain 325 mg of aspirin. Suppose manufacturing specifications require that tablets have between 315 mg and 335 mg of aspirin, so a tablet is considered to be a defect if the amount of aspirin is not within those limits. If the proportion of defects is monitored with a p chart and is found to be within statistical control, can we conclude that almost all of the tablets meet the manufacturing specifications? Why or why not?

2. Notation Assume that Bayer aspirin tablets are monitored to ensure that the proportions of defects are within statistical control. A quality control inspector randomly selects samples with 100 tablets in each sample. If the numbers of defects for the first five samples are 2, 1, 0, 4, and 3, find the value of $\bar{p}$.

3. Control Limits Refer to Exercise 2 and find the values of the upper and lower control limits. Does either of those values need to be adjusted in some way? Explain.

4. Interpreting a Control Chart Assume that the proportion of defective Bayer aspirin tablets is monitored with a control chart and we conclude that the process is not within statistical control because there is a downward pattern that is not random. Should the downward pattern be corrected? What should the company do?

Determining Whether a Process Is in Control. *In Exercises 5–8, examine the given control chart for p and determine whether the process is within statistical control. If it is not, identify which of the three out-of-control criteria apply.*

5.

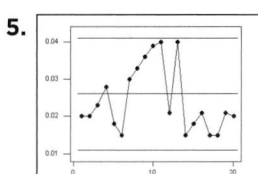

6.

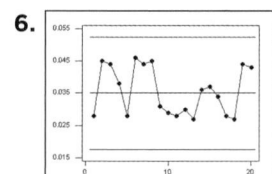

7.

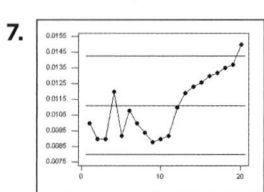

8.

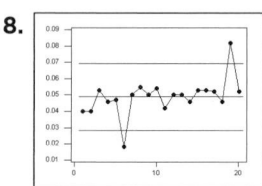

Constructing Control Charts for *p*. *In Exercises 9–14, use the given process data to construct a control chart for p. In each case, use the three out-of-control criteria listed in Section 14-2 and determine whether the process is within statistical control. If it is not, identify which of the three out-of-control criteria apply.*

9. *p* **Chart for Defective Defibrillators** Consider a process that includes careful testing of each manufactured defibrillator (as in Example 1). Listed below are the numbers of defective defibrillators in successive batches of 10,000. Construct a control chart for the proportion *p* of defective defibrillators and determine whether the process is within statistical control. If not, identify which of the three out-of-control criteria apply.

Defects: 20 14 22 27 12 12 18 23 25 19 24 28 21 25 17 19 17 22 15 20

10. *p* **Chart for Defective Defibrillators** Repeat Exercise 9 assuming that the size of each batch is 100 instead of 10,000. Compare the control chart to the one found for Exercise 9. Comment on the general quality of the manufacturing process described in Exercise 9 compared to the manufacturing process described in this exercise.

11. *p* **Chart for College Enrollment** In each of 15 recent and consecutive years, 1000 high school completers were randomly selected and the number who enrolled in college was determined, with the results listed below. Does the *p* chart indicate that such college enrollments are high enough? (The values are based on data from the U.S. National Center for Education Statistics, and they are the most recent values available at the time of this writing.)

601 625 619 626 619 619 650 670 656 629 633 618 652 639 667

12. *p* **Chart for Violent Crimes** In each of 15 recent and consecutive years, 100,000 people in the United States were randomly selected and the number who were victims of violent crime was determined, with the results listed below. Does the rate of violent crime appear to exhibit acceptable behavior? (The values are based on data from the U.S. Department of Justice, and they are the most recent values available at the time of this writing.)

730 758 758 747 714 685 637 611 568 523 507 505 474 476 486

13. p Chart for Voters In each of 23 recent and consecutive years of national elections, 1000 people of voting age in the United States were randomly selected and the number who voted was determined, with the results listed below. Odd-numbered years correspond to years of presidential elections. Comment on the voting behavior of the population. (The values are based on data from the U.S. Census Bureau, and they are the most recent values available at the time of this writing.)

585 454 577 454 550 436 506 357 488 345 475 377
474 336 449 336 513 365 459 331 471 347 514

14. p Chart for Birth Rate In each of 20 recent and consecutive years, 10,000 people were randomly selected and the numbers of births they generated were found, with the results given below. How might the results be explained? (The listed values are based on data from the U.S. Department of Health and Human Services, and they are the most recent values available at the time of this writing.)

157 160 164 167 167 162 158 154 150 146 144 142 143 142 144 141 139 141 140 140

14-3   Beyond the Basics

15. Constructing an np Chart A variation of the control chart for p is the **np chart** in which the *actual numbers* of defects are plotted instead of the *proportions* of defects. The np chart will have a centerline value of $n\bar{p}$, and the control limits will have values of $n\bar{p} + 3\sqrt{n\bar{p}\,\bar{q}}$ and $n\bar{p} - 3\sqrt{n\bar{p}\,\bar{q}}$. The p chart and the np chart differ only in the scale of values used for the vertical axis. Construct the np chart for Example 1 in this section. Compare the np chart to the control chart for p given in this section.

Review

In this chapter we introduced run charts and control charts, which are commonly used to monitor process data with the objective of maintaining or improving the quality of goods or services. Process data were defined to be data arranged according to some time sequence. Control charts have a centerline, an upper control limit, and a lower control limit. A process is statistically stable (or within statistical control) if it has only natural variation, with no patterns, cycles, or unusual points. Decisions about statistical stability are based on how a process is actually behaving, not on how we might like it to behave because of such factors as manufacturer specifications. The following graphs were described:

• *Run chart:* a sequential plot of *individual* data values over time

• *R chart:* a control chart that uses ranges in an attempt to monitor the *variation* in a process

• *x̄ chart:* a control chart used to determine whether the process *mean* is within statistical control

• *p chart:* a control chart used to monitor the proportion of some process *attribute*, such as whether items are defective

Statistical Literacy and Critical Thinking

1. Statistical Process Control The title of this chapter is "Statistical Process Control." What does that mean?

2. Monitoring a Process over Time Lipitor is a drug designed to lower cholesterol levels. With current sales of Lipitor exceeding $13 billion each year, it has become the best selling drug ever. One dosage level of Lipitor is provided with tablets containing 10 mg of the drug atorvastatin. If the manufacturing process is set up to produce tablets containing between 9.5 mg and 10.5 mg of atorvastatin, why is it important to monitor the manufacturing process over time? What would be a possible adverse consequence of a Lipitor manufacturing process that is not monitored?

3. Control Charts Refer to the process described in Exercise 2. When monitoring the amount of atorvastatin in Lipitor tablets, why is it important to use an $\bar{x}$ chart and an R chart together?

4. Interpreting Control Charts Refer to the process described in Exercise 2. If $\bar{x}$ and R charts show that the process is within statistical control, can we conclude that almost all of the tablets contain between 9.5 mg and 10.5 mg of atorvastatin? Why or why not?

Chapter Quick Quiz

1. What are *process* data?

2. What is the difference between *random variation* and *assignable variation?*

3. Identify three specific criteria for determining when a process is out of statistical control.

4. What is the difference between an R chart and an $\bar{x}$ chart?

In Exercises 5–8, use the following two control charts that result from testing newly manufactured aircraft altimeters. The original sample values are errors obtained when the altimeters are tested at an altitude of 1000 ft.

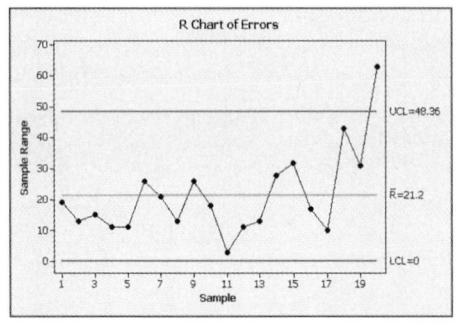 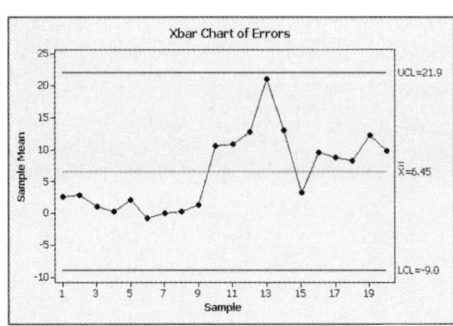

5. Is the process variation within statistical control? Why or why not?

6. What is the value of $\bar{R}$? In general, how is a value of $\bar{R}$ obtained?

7. Is the process mean within statistical control? Why or why not?

8. What is the value of $\bar{\bar{x}}$? In general, how is a value of $\bar{\bar{x}}$ found?

9. What is a p chart?

10. True or false: A manufacturing process results in a greater proportion of defects whenever a p chart of defective items reveals an out-of-control condition.

Review Exercises

Constructing Control Charts for Aluminum Cans. *Exercises 1–5 refer to the axial loads of aluminum cans with a thickness of 0.0109 in. listed in Data Set 21 in Appendix B. Each day, seven cans were randomly selected and tested, and Data Set 21 shows results for 25 consecutive days of manufacturing. The axial load is* *the maximum load (in lb) that a can withstands before collapsing from the pressure applied to the top. (The top lids are pressed into place with pressures that vary between 158 lb and 165 lb.)*

1. Run Chart Construct a run chart for the 21 axial loads from the first three days, as listed below. Based on the result, does there appear to be a pattern suggesting that the process is not within statistical control?

Day 1	270	273	258	204	254	228	282
Day 2	278	201	264	265	223	274	230
Day 3	250	275	281	271	263	277	275

2. R Chart Using subgroups of size $n = 7$ corresponding to the rows of the table for the first three days given in Exercise 1, find the values of $\bar{R}$, the lower control limit, and the upper control limit that would be used to construct an R chart. (Do not construct the R chart.)

3. R Chart If we use subgroups of size $n = 7$ corresponding to the rows of the table, and if we use all of the data from 25 days of production, we get the R chart shown below. Interpret that R chart.

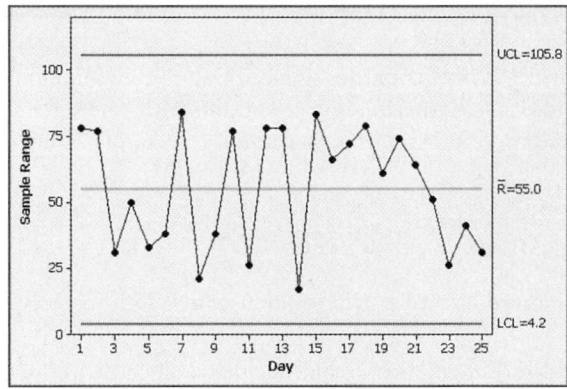

4. $\bar{x}$ Chart Using subgroups of size $n = 7$ corresponding to the rows of the table for the first three days given in Exercise 1, find the values of $\bar{\bar{x}}$, the lower control limit, and the upper control limit that would be used to construct an $\bar{x}$ chart. (Do not construct the chart.)

5. $\bar{x}$ Chart If we use subgroups of size $n = 7$ corresponding to the rows of the table, and if we use all of the data from 25 days of production, we get the $\bar{x}$ chart shown below. Interpret that $\bar{x}$ chart.

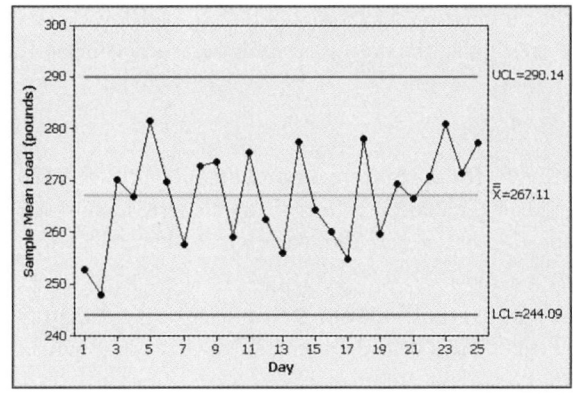

TI **6. Control Chart for Homicides** In each of 15 recent and consecutive years, 1,000,000 people in the United States were randomly selected and the number who were homicide victims was determined, with the results listed below. Are the proportions of homicides within statistical control? What does the control chart suggest about the trend? (The values are based on data from the U.S. Department of Justice, and they are the most recent values available at the time of this writing.)

85   87   94   98   93   95   90   82   74   68   63   57   55   56   56

TI **7. Control Chart for Defects** The Acton Pharmaceutical Company manufactures antacid tablets that are supposed to contain 750 mg of calcium carbonate, similar to the Tums tablets manufactured by GlaxoSmithKline. Each day, 100 tablets are randomly selected and the amount of calcium carbonate is measured. A tablet is considered defective if it has obvious physical deformities or the amount of calcium carbonate is not between 735 mg and 765 mg. The numbers of defects are listed below for consecutive production days. Construct an appropriate control chart and determine whether the process is within statistical control. If not, identify which criteria lead to rejection of statistical stability.

Defects   5   4   3   6   7   5   9   2   10   3   2   5   12   7   1

Cumulative Review Exercises

TI **1. Temperature and Carbon Dioxide** Listed below are concentrations of carbon dioxide (in parts per million) of the earth's atmosphere and the earth's mean temperature (in °C) for each of ten recent and consecutive years. (The last few pairs of values are projections.)

| **Carbon Dioxide** | 369.5 | 371.0 | 373.1 | 375.6 | 377.4 | 379.6 | 379.6 | 381.2 | 382.8 | 384.4 |
| **Temperature** | 14.41 | 14.56 | 14.70 | 14.64 | 14.60 | 14.77 | 14.64 | 14.66 | 14.68 | 14.70 |

a. Use a 0.05 significance level to test for a linear correlation between carbon dioxide and temperature.

b. If there is a linear correlation, can we conclude that an increase in carbon dioxide concentration causes an increase in the earth's temperature? Why or why not?

c. Find the equation of the straight line that best fits the sample data. Let x represent carbon dioxide concentration.

d. What is the best predicted temperature for a year in which the concentration of carbon dioxide was 290.7 parts per million? How close is the predicted temperature to the actual temperature of 13.88°C?

TI **2. Control Chart for Defective Seat Belts** The Flint Accessory Corporation manufactures seat belts for cars. Federal specifications require that the webbing must have a breaking strength of at least 5000 lb. During each week of production, 200 belts are randomly selected and tested for breaking strength. A belt is considered defective if it breaks before reaching the force of 5000 lb. The numbers of defects for a sequence of 10 weeks are listed below. Use a control chart for p to verify that the process is within statistical control. If it is not in control, explain why it is not.

2   3   1   5   9   1   12   15   2   17

TI **3. Confidence Interval for Defective Seat Belts** Refer to the data in Exercise 2 and, using all of the data from the 2000 seat belts that were tested, construct a 95% confidence interval for the proportion of defects. Also, write a statement that interprets the confidence interval.

TI **4. Hypothesis Test for Defective Seat Belts** Refer to the data in Exercise 2 and, using all of the data from the 2000 seat belts that were tested, use a 0.05 significance level to test the claim that the rate of defects is greater than 3%.

5. Using Probability in Control Charts When interpreting control charts, one of the three out-of-control criteria is that there are eight consecutive points all above or all below the centerline. For a statistically stable process, there is a 0.5 probability that a point will be above the centerline and a 0.5 probability that a point will be below the centerline. In each of the following, assume that sample values are independent and the process is statistically stable.

a. Find the probability that when eight consecutive points are randomly selected, they are all above the centerline.

b. Find the probability that when eight consecutive points are randomly selected, they are all below the centerline.

c. Find the probability that when eight consecutive points are randomly selected, they are all above or all below the centerline.

TI 6. Designing Motorcycle Helmets Engineers must consider the breadths of male heads when designing motorcycle helmets. Men have head breadths that are normally distributed with a mean of 6.0 in. and a standard deviation of 1.0 in. (based on anthropometric survey data from Gordon, Churchill, et al.).

a. What percentage of men have head breadths greater than 7.0 in.? If helmets are made to accommodate only males with head breadths less than 7.0 in., would those helmets exclude too many males?

b. If, due to financial constraints, the helmets are designed to fit all men except those with head breadths that are in the smallest 5% or largest 5%, find the minimum and maximum head breadths that the helmets will fit.

7. AOL Poll America Online conducted an online poll in which Internet users were asked, "Have you ever seen a UFO?" There were 37,021 Internet users who responded with "yes," and there were 80,806 Internet users who responded with "no." Based on the results, can we conclude that most people have not seen a UFO? Why or why not?

TI 8. Normality Assessment The author conducted a survey of his students and requested the value of the coins that each student possessed. Some of the results (in cents) are listed below. Do these values appear to be from a population having a normal distribution? Why or why not?

0	100	0	23	185	0	0	43	35	250	178	10	90	0	200
0	40	73	20	500	0	0	35	130	62	5	0	0	0	0

TI 9. Sample Statistics Refer to the sample values listed in Exercise 8 and find the mean, median, and standard deviation.

TI 10. Claimed Proportion Use the sample values listed in Exercise 8 to test the claim that most students have some change in their possession. Use a 0.05 significance level.

TI-83/84 Plus Calculator Project

a. Use the TI-83/84 Plus calculator to simulate the following process for 20 days: Each day, 200 pacemakers are manufactured with a 5% rate of defects. The proportion of defects is recorded for each of the 20 days. The pacemakers for one day are simulated by randomly generating 200 integers, where each is between 1 and 100. Consider an outcome of 1, 2, 3, 4, or 5 to be a defect, with 6 through 100 being acceptable. This corresponds to a 5% rate of defects.

Here are the TI-83/84 Plus instructions: Press **MATH**, select **PRB,** then select the menu item of **randInt.** The format for the randInt command is to enter the minimum integer, the maximum integer, and the number of generated integers, all separated by commas. Use the command

randInt(1, 100, 200)

to obtain 200 randomly generated integers, where each is between 1 and 100 inclusive. Considering outcomes of 1, 2, 3, 4, and 5 to be defects, record the number of defects for one day. Repeat this procedure until results for 20 days have been simulated. *Hint:* This procedure is easier if you store the random integers in a list and then sort them.

b. Using the numbers of defects from 20 different days, construct a p chart for the proportion of defective pacemakers, and determine whether the process is within statistical control. Since we know the process is actually stable with $p = 0.05$, the conclusion that it is not stable would be a type I error; that is, we would have a false positive signal, causing us to believe that the process needed to be adjusted when in fact it should be left alone.

c. The result from part(a) is a simulation of 20 days. Now simulate another 10 days of manufacturing pacemakers, but modify these last 10 days so that the defect rate is 10% instead of 5%. (Consider outcomes of 1, 2, 3, 4, 5, 6, 7, 8, 9, and 10 to be defects.)

d. Combine the data generated from parts(a) and (c) to represent a total of 30 days of sample results. Construct a p chart for this combined data set. Is the process out of control? If we concluded that the process was not out of control, we would be making a type II error; that is, we would believe that the process was okay when in fact it should be repaired or adjusted to correct the shift to the 10% rate of defects.

INTERNET PROJECT

Control Charts

Go to: **http://www.aw.com/triola**

This chapter introduces different charting techniques used to summarize and study data associated with a process along with methods for analyzing the stability of that process. With the exception of the run chart, individual data points are not needed to construct a chart. For example, the R chart is constructed from sample ranges while the p chart is based on sample proportions.

This is an important point, as data collected from third-party sources are often given in terms of summarizing statistics.

Locate the Internet Project dealing with control charts. There you will be directed to data sets and sources of data for use in constructing control charts. From the resulting charts you will be asked to interpret and discuss trends in the underlying processes.

APPLET PROJECT

Open the Applets folder on the CD and double-click on **Start.** Select the menu item of **Simulating the probability of rolling a 6.**

a. Simulate 50 rolls of a die, and record the proportion of 6s.

b. Repeat part (a) 19 more times.

c. Use the methods of this chapter to determine whether or not the proportions of 6s are within statistical control. What can you conclude?

FROM DATA TO DECISION

Critical Thinking: Are the axial loads within statistical control? Is the process of manufacturing cans proceeding as it should?

Review Exercises 1–5 used process data from a New York company that manufactures 0.0109-in.-thick aluminum cans for a major beverage supplier. Refer to Data Set 21 in Appendix B and conduct an analysis of the process data for the cans that are 0.0111 in. thick. The values in the data set are the measured axial loads of cans, and the top lids are pressed into place with pressures that vary between 158 lb and 165 lb.

Analyzing the Results

Based on the given process data, should the company take any corrective action? Write a report summarizing your conclusions. Address not only the issue of statistical stability, but also the ability of the cans to withstand the pressures applied when the top lids are pressed into place. Also compare the behavior of the 0.0111-in. cans to the behavior of the 0.0109-in. cans and recommend which thickness should be used.

Cooperative Group Activities

1. Out-of-class activity Collect your own process data and analyze them using the methods of this chapter. It would be ideal to collect data from a real manufacturing process, but that may be difficult to accomplish. If so, consider using a simulation or referring to published data, such as those found in an almanac. Here are some suggestions:

• Shoot five basketball foul shots (or shoot five crumpled sheets of paper into a wastebasket) and record the number of shots made; then repeat this procedure 20 times, and use a p chart to test for statistical stability in the proportion of shots made.

• Measure your pulse rate by counting the number of times your heart beats in 1 min. Measure your pulse rate four times each hour for several hours, then construct appropriate control charts. What factors contribute to random variation? Assignable variation?

• Go through newspapers for the past 12 weeks and record the closing of the Dow Jones Industrial Average (DJIA) for each business day. Use run and control charts to explore the statistical stability of the DJIA. Identify at least one practical consequence of having this process statistically stable, and identify at least one practical consequence of having this process out of statistical control.

• Find the marriage rate per 10,000 population for several years. (See the *Information Please Almanac* or the *Statistical Abstract of the United States.*) Assume that in each year 10,000 people were randomly selected and surveyed to determine whether they were married. Use a *p* chart to test for statistical stability of the marriage rate. (Other possible rates: death, accident fatality, crime.)

Obtain a printed copy of computer results, and write a report summarizing your conclusions.

2. In-class activity If the instructor can distribute the numbers of absences for each class meeting, groups of three or four students can analyze them for statistical stability and make recommendations based on the conclusions.

3. Out-of-class activity Conduct research to identify *Deming's funnel experiment,* then use a funnel and marbles to collect data for the different rules for adjusting the funnel location. Construct appropriate control charts for the different rules of funnel adjustment. What does the funnel experiment illustrate? What do you conclude?

NAME:	Andrew Chang
JOB:	Marketing Strategy Manager
COMPANY:	AirTran Airways

Andrew Chang is the Manager of Marketing Strategy for AirTran Airways. He oversees the frequent flier program, marketing research, and marketing analysis.

Q: How do you use statistics on the job?

A: We use statistics almost every day. We perform several campaigns where we test different messages and subjects to determine what combinations of variables optimize response rates and therefore revenue. We also use multivariate regressions to create media mix models in order to plan and forecast for the types of media channels to use for a given market or season.

Q: Describe at least one specific example illustrating how the use of statistics was successful in improving a product or service.

A: Every week, we run several tests on our weekly e-mail communications to determine how to maximize open rates, maximize click-through rates, and decrease un-subscriptions. We use simple Chi-square tests to determine which versions performed better in each category and we use the results for the next week's mailing.

Q: Is your use of probability and statistics increasing, decreasing, or remaining stable?

A: Our use of probability and statistics is certainly increasing, and for almost every new person we hire in the group, knowledge and experience of statistics is required.

Q: How critical do you find your knowledge of statistics for performing your responsibilities?

A: It is extremely critical for this job. With fierce and smart competition, we are forced to try and think one-step ahead. Using statistics gives us a framework on how to think, analyze, and solve problems, as opposed to just guessing what customers like or don't like.

Q: In terms of statistics, what would you recommend for prospective employees?

A: I would certainly recommend at least an introductory course of statistics to everyone. Whether you are going to perform your own analyses or not, you have to be able to question, validate, and understand the work of others.

Q: Are job applicants in your field viewed more favorably if they have studied some statistics?

A: Yes. In some positions, knowledge and experience with statistics are required.

Q: While a college student, did you expect to be using statistics on the job?

A: Most people in college certainly do not expect to use statistics on the job, and I was one of them. However, once I learned about its use in marketing, I became addicted to it and could not wait to use it in whatever job I took.

Q: Do you recommend statistics for today's college students?

A: Absolutely. I would require it. In today's competitive environment, it is almost irresponsible to make decisions based on feeling and emotion alone.

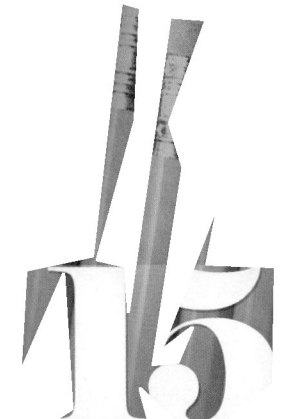

 Projects

Key Concept A final project is an excellent activity that can be a valuable and rewarding experience for students in the introductory statistics course. This final project provides students with the opportunity to use principles of statistics in a real and interesting application. This section provides a suggested format for such a project.

Group Project vs. Individual Project Although different topics could be assigned to individuals, group projects tend to be more effective because they help develop the interpersonal skills that are so necessary in today's working environment. One study showed that the "inability to get along with others" is the main reason for firing employees, so a group project can be very helpful in preparing students for their future work environments. Groups of three, four, or five students work well. The professor should select groups with consideration given to important factors such as past class performance and attendance.

Oral Report A 10- to 15-minute class presentation should involve all group members in a coordinated effort to clearly describe the important components of the study. Students typically have some reluctance to speak in public, so a brief oral report can be very helpful in building confidence. The oral report is an activity that can better prepare students for future professional activities.

Written Report The main objective of the project is not to produce a written document equivalent to a term paper. However, a brief written report should be submitted, and it should include the following components:

1. List of data collected along with a description of how the data were obtained.

2. Description of the method of analysis

3. Relevant graphs and/or statistics, including computer or TI-83/84 Plus displays

4. Statement of conclusions

5. Reasons why the results might not be correct, along with a description of ways in which the study could be improved, given sufficient time and money

Large Classes or Online Classes: Posters or PowerPoint Some classes are too large for individual or group projects. Online classes are not able to meet as a group. For such classes, reports of individual or small group projects can be presented through posters similar to those found at conference poster sessions. Posters or Power-Point presentations summarizing important elements of a project can be submitted to professors for evaluation.

Project Topics   The "Cooperative Group Activities" listed near the end of each chapter include more than 100 suggestions for projects. The following comments about a survey can be another excellent source of project topics.

Survey   A survey can be an excellent source of data that can be used in a statistics project. The sample survey below collects information that can be used to address questions such as these:

1. When people "randomly" select digits (as in Question 2), are the results actually random?

2. Do the last four digits of social security numbers appear to be random?

3. Do males and females carry different amounts of change?

4. Do males and females have different numbers of credit cards?

5. Is there a difference in pulse rates between those who exercise and those who do not?

6. Is there a difference in pulse rates between those who smoke and those who do not?

7. Is there a relationship between exercise and smoking?

8. Is there a relationship between eye color and exercise?

9. Is there a relationship between exercise and the number of hours worked each week?

10. Is there a correlation between height and pulse rate?

Survey

1. _____ Female _____ Male

2. Randomly select four digits and enter them here:
_____ _____ _____ _____

3. Eye color: _____

4. Enter your height in inches: _____

5. What is the total value of all coins now in your possession? _____

6. How many keys are in your possession at this time?
_____

7. How many credit cards are in your possession at this time? _____

8. Enter the last four digits of your social security number. (For reasons of security, you can rearrange the order of these digits.) _____ _____ _____
_____ _____

9. Record your pulse rate by counting the number of heartbeats for 1 minute: _____

10. Do you exercise vigorously (such as running, swimming, cycling, tennis, basketball, etc.) for at least 20 minutes at least twice a week? _____ Yes _____ No

11. How many credit hours of courses are you taking this semester? _____

12. Are you currently employed? _____ Yes _____ No
If yes, how many hours do you work each week?
_____

13. During the past 12 months, have you been the driver of a car that was involved in an accident?
_____ Yes _____ No

14. Do you smoke? _____ Yes _____ No

15. _____ Left-handed _____ Right-handed
_____ Ambidextrous

15-2 Procedures

Key Concept This section describes a general approach for the statistical analysis of data.

Context, Source, Sampling Method   Instead of mindlessly plugging data into some particular statistical procedure, we should begin with some basic considerations, including these:

1. Clearly identify the *context* of the data (as discussed in Section 1-2).

2. Consider the *source* of the data and determine whether that source presents any issues of bias that might affect the validity of the data (as discussed in Section 1-2).

3. Consider the *sampling method* to ensure that it is the type of sampling likely to result in data that are representative of the population. Be especially wary of voluntary response samples.

Exploring, Comparing, Describing   After collecting data, first consider exploring, describing, and comparing data. Start with an appropriate *graph* for the data. (For example, with sample data consisting of single values, construct a histogram, normal quantile plot, and boxplot. With paired data, construct a scatterplot.) Using the basic tools described in Chapters 2 and 3, consider the following:

1. *Center:* Find the mean and median, which are measures of center giving us an indication of where the middle of the data set is located.

2. *Variation:* Find the range and standard deviation, which are measures of the amount that the sample values vary among themselves.

3. *Distribution:* Construct a histogram to see the shape of the distribution of the data. Also construct a normal quantile plot and determine if the data are from a population having a normal distribution.

4. *Outliers:* Identify any sample values that lie very far away from the vast majority of the other sample values. If there are outliers, try to determine whether they are errors that should be corrected. If the outliers are correct values, study their effects by repeating the analysis with the outliers excluded.

5. *Time:* Determine if the population is stable or if its characteristics are changing over time.

Inferences: Estimating Parameters and Hypothesis Testing   When trying to use sample data for making inferences about a population, it is often difficult to choose the best procedure. Figure 15-1 includes the major methods included in this book, along with a scheme for determining which of those methods should be used. Figure 15-1 applies to a fixed population. If the data are from a process that may change over time, construct a control chart (see Chapter 14) to determine whether the process is statistically stable. Figure 15-1 applies to process data only if the process is statistically stable. In addition to the procedures identified in Figure 15-1, there are many other methods that might be more suitable for a particular statistical analysis. Consult your friendly professional statistician for help with other methods.

Conclusions and Practical Implications   After completing the statistical analysis, we should state conclusions in a way that is clear to those unfamiliar with statistics and its terminology, and we should carefully avoid making statements not justified by the statistical analysis (such as using a correlation to conclude that one variable is the *cause* of the other). Also, we should identify practical implications of the results.

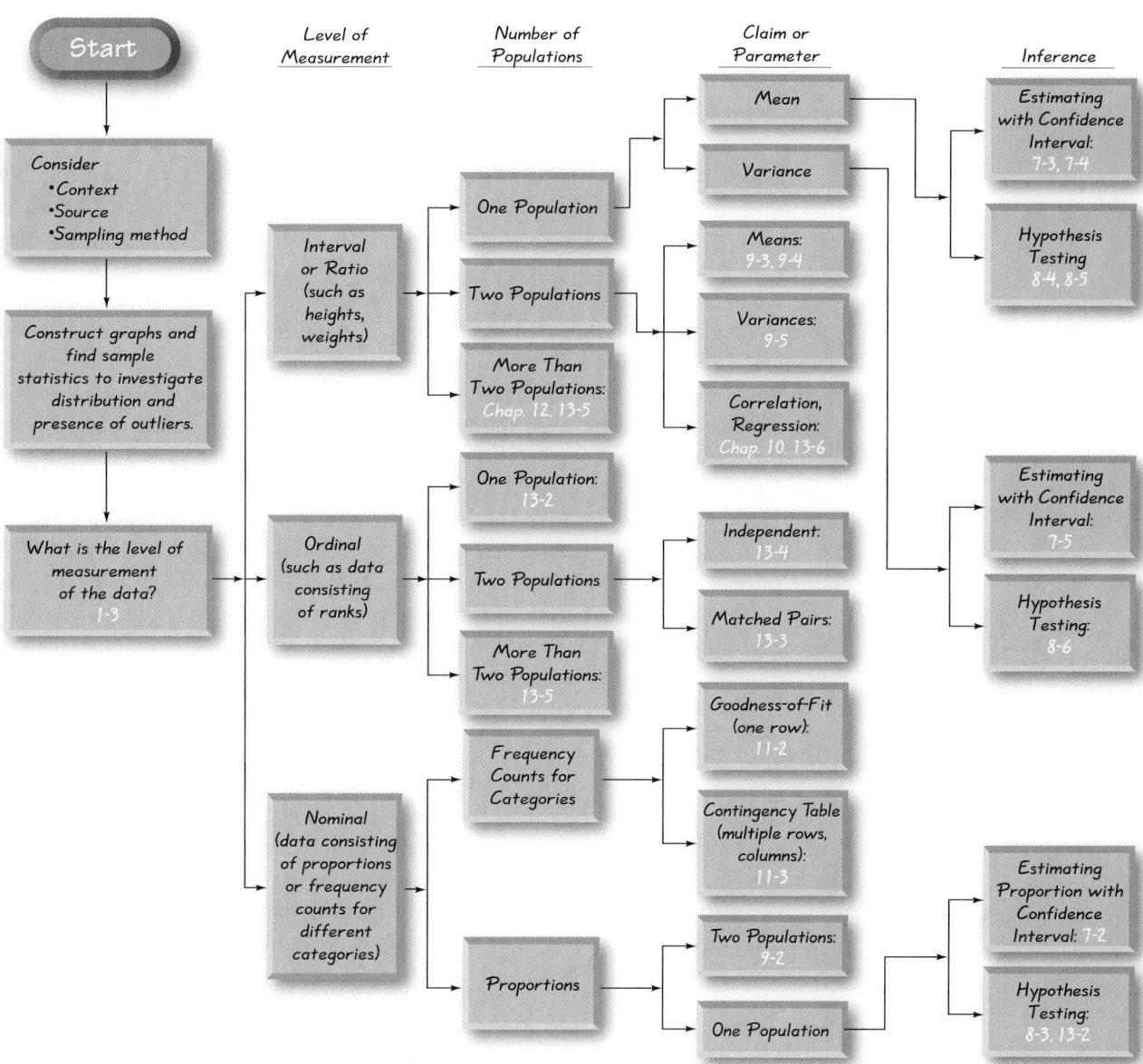

Figure 15-1   Selecting the Appropriate Procedure

Perspectives

Key Concept No single introductory statistics course can make anyone an expert statistician. The introductory course has a limited scope, and many important topics are not included.

Successful completion of an introductory statistics course results in benefits that extend far beyond the attainment of credit toward a college degree. You will have improved job marketability. You will be better prepared to critically analyze reports in the media and professional journals. You will understand the basic concepts of probability and chance. You will know that in attempting to gain insight into a set of data, it is important to consider the context of the data, the source of the data, and the sampling methods used. You will know that, given sample data, you should investigate measures of center (such as mean and median), measures of variation (such as range and standard deviation), the distribution (via a frequency distribution or graph), the presence of outliers, and whether the population is stable or is changing over time. You will know and understand the importance of estimating population parameters (such as mean, standard deviation, and proportion), as well as testing claims made about population parameters.

Throughout this text we have emphasized the importance of good sampling methods. You should recognize that a bad sample may be beyond repair by even the most expert statisticians using the most sophisticated techniques. There are many mail, magazine, and telephone call-in surveys that allow respondents to be "self-selected." The results of such surveys are generally worthless when judged according to the criteria of sound statistical methodology. Keep this in mind when you encounter voluntary response (self-selected) surveys, so that you don't let them affect your beliefs and decisions. You should also recognize, however, that many surveys and polls obtain very good results, even though the sample sizes might seem to be relatively small. Although many people refuse to believe it, a nationwide survey of only 1200 voters can provide good results if the sampling is carefully planned and executed.

Throughout this text we have emphasized the *interpretation* of results. Computers and calculators are quite good at yielding results, but such results generally require careful interpretation. We should recognize that a result is not automatically valid simply because it was computer-generated. Computers don't think, and they are quite capable of providing results that are quite ridiculous when considered in the context of the real world. We should always apply the most important and indispensable tool in all of statistics: *common sense!*

The Educated Person   Once upon a time, a person was considered to be educated if he or she could simply read. That time has long passed. Today, an educated person is capable of critical thinking, possesses an intellectual curiosity, and can communicate effectively both orally and in writing. An educated person can relate to all other people, including those from different cultures, as well as those who might not be so educated. An educated person has statistical literacy and an ability to think statistically. Successful completion of an introductory statistics course is a great achievement of a truly educated person.

Appendices

Appendix A: Tables

Appendix B: Data Sets

Appendix C: Bibliography of Books and Web sites

Appendix D: Answers to Odd-Numbered Exercises, plus Answers to All end-of-chapter Statistical Literacy and Critical Thinking Exercises, Chapter Quick Quizzes, Chapter Review, and Cumulative Review Exercises

Appendix A: Tables

Table A-1   Binomial Probabilities

Table A-2   Standard Normal Distribution

Table A-3 t Distribution

Table A-4   Chi-Square (χ^2) Distribution

Table A-5 F Distribution

Table A-6   Critical Values of the Pearson Correlation Coefficient r

Table A-7   Critical Values of T for the Wilcoxon Signed-Ranks Test

Table A-8   Critical Values of Spearman's Rank Correlation Coefficient r_s

Table A-9   Critical Values for Number of Runs G

TABLE A-1 | Binomial Probabilities

								p							
n	x	.01	.05	.10	.20	.30	.40	.50	.60	.70	.80	.90	.95	.99	x
2	0	.980	.902	.810	.640	.490	.360	.250	.160	.090	.040	.010	.002	0+	0
	1	.020	.095	.180	.320	.420	.480	.500	.480	.420	.320	.180	.095	.020	1
	2	0+	.002	.010	.040	.090	.160	.250	.360	.490	.640	.810	.902	.980	2
3	0	.970	.857	.729	.512	.343	.216	.125	.064	.027	.008	.001	0+	0+	0
	1	.029	.135	.243	.384	.441	.432	.375	.288	.189	.096	.027	.007	0+	1
	2	0+	.007	.027	.096	.189	.288	.375	.432	.441	.384	.243	.135	.029	2
	3	0+	0+	.001	.008	.027	.064	.125	.216	.343	.512	.729	.857	.970	3
4	0	.961	.815	.656	.410	.240	.130	.062	.026	.008	.002	0+	0+	0+	0
	1	.039	.171	.292	.410	.412	.346	.250	.154	.076	.026	.004	0+	0+	1
	2	.001	.014	.049	.154	.265	.346	.375	.346	.265	.154	.049	.014	.001	2
	3	0+	0+	.004	.026	.076	.154	.250	.346	.412	.410	.292	.171	.039	3
	4	0+	0+	0+	.002	.008	.026	.062	.130	.240	.410	.656	.815	.961	4
5	0	.951	.774	.590	.328	.168	.078	.031	.010	.002	0+	0+	0+	0+	0
	1	.048	.204	.328	.410	.360	.259	.156	.077	.028	.006	0+	0+	0+	1
	2	.001	.021	.073	.205	.309	.346	.312	.230	.132	.051	.008	.001	0+	2
	3	0+	.001	.008	.051	.132	.230	.312	.346	.309	.205	.073	.021	.001	3
	4	0+	0+	0+	.006	.028	.077	.156	.259	.360	.410	.328	.204	.048	4
	5	0+	0+	0+	0+	.002	.010	.031	.078	.168	.328	.590	.774	.951	5
6	0	.941	.735	.531	.262	.118	.047	.016	.004	.001	0+	0+	0+	0+	0
	1	.057	.232	.354	.393	.303	.187	.094	.037	.010	.002	0+	0+	0+	1
	2	.001	.031	.098	.246	.324	.311	.234	.138	.060	.015	.001	0+	0+	2
	3	0+	.002	.015	.082	.185	.276	.312	.276	.185	.082	.015	.002	0+	3
	4	0+	0+	.001	.015	.060	.138	.234	.311	.324	.246	.098	.031	.001	4
	5	0+	0+	0+	.002	.010	.037	.094	.187	.303	.393	.354	.232	.057	5
	6	0+	0+	0+	0+	.001	.004	.016	.047	.118	.262	.531	.735	.941	6
7	0	.932	.698	.478	.210	.082	.028	.008	.002	0+	0+	0+	0+	0+	0
	1	.066	.257	.372	.367	.247	.131	.055	.017	.004	0+	0+	0+	0+	1
	2	.002	.041	.124	.275	.318	.261	.164	.077	.025	.004	0+	0+	0+	2
	3	0+	.004	.023	.115	.227	.290	.273	.194	.097	.029	.003	0+	0+	3
	4	0+	0+	.003	.029	.097	.194	.273	.290	.227	.115	.023	.004	0+	4
	5	0+	0+	0+	.004	.025	.077	.164	.261	.318	.275	.124	.041	.002	5
	6	0+	0+	0+	0+	.004	.017	.055	.131	.247	.367	.372	.257	.066	6
	7	0+	0+	0+	0+	0+	.002	.008	.028	.082	.210	.478	.698	.932	7
8	0	.923	.663	.430	.168	.058	.017	.004	.001	0+	0+	0+	0+	0+	0
	1	.075	.279	.383	.336	.198	.090	.031	.008	.001	0+	0+	0+	0+	1
	2	.003	.051	.149	.294	.296	.209	.109	.041	.010	.001	0+	0+	0+	2
	3	0+	.005	.033	.147	.254	.279	.219	.124	.047	.009	0+	0+	0+	3
	4	0+	0+	.005	.046	.136	.232	.273	.232	.136	.046	.005	0+	0+	4
	5	0+	0+	0+	.009	.047	.124	.219	.279	.254	.147	.033	.005	0+	5
	6	0+	0+	0+	.001	.010	.041	.109	.209	.296	.294	.149	.051	.003	6
	7	0+	0+	0+	0+	.001	.008	.031	.090	.198	.336	.383	.279	.075	7
	8	0+	0+	0+	0+	0+	.001	.004	.017	.058	.168	.430	.663	.923	8

NOTE: 0+ represents a positive probability less than 0.0005.   *(continued)*

| TABLE A-1 | Binomial Probabilities (*continued*) |

							p								
n	*x*	.01	.05	.10	.20	.30	.40	.50	.60	.70	.80	.90	.95	.99	*x*
9	0	.914	.630	.387	.134	.040	.010	.002	0+	0+	0+	0+	0+	0+	0
	1	.083	.299	.387	.302	.156	.060	.018	.004	0+	0+	0+	0+	0+	1
	2	.003	.063	.172	.302	.267	.161	.070	.021	.004	0+	0+	0+	0+	2
	3	0+	.008	.045	.176	.267	.251	.164	.074	.021	.003	0+	0+	0+	3
	4	0+	.001	.007	.066	.172	.251	.246	.167	.074	.017	.001	0+	0+	4
	5	0+	0+	.001	.017	.074	.167	.246	.251	.172	.066	.007	.001	0+	5
	6	0+	0+	0+	.003	.021	.074	.164	.251	.267	.176	.045	.008	0+	6
	7	0+	0+	0+	0+	.004	.021	.070	.161	.267	.302	.172	.063	.003	7
	8	0+	0+	0+	0+	0+	.004	.018	.060	.156	.302	.387	.299	.083	8
	9	0+	0+	0+	0+	0+	0+	.002	.010	.040	.134	.387	.630	.914	9
10	0	.904	.599	.349	.107	.028	.006	.001	0+	0+	0+	0+	0+	0+	0
	1	.091	.315	.387	.268	.121	.040	.010	.002	0+	0+	0+	0+	0+	1
	2	.004	.075	.194	.302	.233	.121	.044	.011	.001	0+	0+	0+	0+	2
	3	0+	.010	.057	.201	.267	.215	.117	.042	.009	.001	0+	0+	0+	3
	4	0+	.001	.011	.088	.200	.251	.205	.111	.037	.006	0+	0+	0+	4
	5	0+	0+	.001	.026	.103	.201	.246	.201	.103	.026	.001	0+	0+	5
	6	0+	0+	0+	.006	.037	.111	.205	.251	.200	.088	.011	.001	0+	6
	7	0+	0+	0+	.001	.009	.042	.117	.215	.267	.201	.057	.010	0+	7
	8	0+	0+	0+	0+	.001	.011	.044	.121	.233	.302	.194	.075	.004	8
	9	0+	0+	0+	0+	0+	.002	.010	.040	.121	.268	.387	.315	.091	9
	10	0+	0+	0+	0+	0+	0+	.001	.006	.028	.107	.349	.599	.904	10
11	0	.895	.569	.314	.086	.020	.004	0+	0+	0+	0+	0+	0+	0+	0
	1	.099	.329	.384	.236	.093	.027	.005	.001	0+	0+	0+	0+	0+	1
	2	.005	.087	.213	.295	.200	.089	.027	.005	.001	0+	0+	0+	0+	2
	3	0+	.014	.071	.221	.257	.177	.081	.023	.004	0+	0+	0+	0+	3
	4	0+	.001	.016	.111	.220	.236	.161	.070	.017	.002	0+	0+	0+	4
	5	0+	0+	.002	.039	.132	.221	.226	.147	.057	.010	0+	0+	0+	5
	6	0+	0+	0+	.010	.057	.147	.226	.221	.132	.039	.002	0+	0+	6
	7	0+	0+	0+	.002	.017	.070	.161	.236	.220	.111	.016	.001	0+	7
	8	0+	0+	0+	0+	.004	.023	.081	.177	.257	.221	.071	.014	0+	8
	9	0+	0+	0+	0+	.001	.005	.027	.089	.200	.295	.213	.087	.005	9
	10	0+	0+	0+	0+	0+	.001	.005	.027	.093	.236	.384	.329	.099	10
	11	0+	0+	0+	0+	0+	0+	0+	.004	.020	.086	.314	.569	.895	11
12	0	.886	.540	.282	.069	.014	.002	0+	0+	0+	0+	0+	0+	0+	0
	1	.107	.341	.377	.206	.071	.017	.003	0+	0+	0+	0+	0+	0+	1
	2	.006	.099	.230	.283	.168	.064	.016	.002	0+	0+	0+	0+	0+	2
	3	0+	.017	.085	.236	.240	.142	.054	.012	.001	0+	0+	0+	0+	3
	4	0+	.002	.021	.133	.231	.213	.121	.042	.008	.001	0+	0+	0+	4
	5	0+	0+	.004	.053	.158	.227	.193	.101	.029	.003	0+	0+	0+	5
	6	0+	0+	0+	.016	.079	.177	.226	.177	.079	.016	0+	0+	0+	6
	7	0+	0+	0+	.003	.029	.101	.193	.227	.158	.053	.004	0+	0+	7
	8	0+	0+	0+	.001	.008	.042	.121	.213	.231	.133	.021	.002	0+	8
	9	0+	0+	0+	0+	.001	.012	.054	.142	.240	.236	.085	.017	0+	9
	10	0+	0+	0+	0+	0+	.002	.016	.064	.168	.283	.230	.099	.006	10
	11	0+	0+	0+	0+	0+	0+	.003	.017	.071	.206	.377	.341	.107	11
	12	0+	0+	0+	0+	0+	0+	0+	.002	.014	.069	.282	.540	.886	12

NOTE: 0+ represents a positive probability less than 0.0005.   (*continued*)

TABLE A-1 Binomial Probabilities (*continued*)

n	x	.01	.05	.10	.20	.30	.40	.50	.60	.70	.80	.90	.95	.99	x
13	0	.878	.513	.254	.055	.010	.001	0+	0+	0+	0+	0+	0+	0+	0
	1	.115	.351	.367	.179	.054	.011	.002	0+	0+	0+	0+	0+	0+	1
	2	.007	.111	.245	.268	.139	.045	.010	.001	0+	0+	0+	0+	0+	2
	3	0+	.021	.100	.246	.218	.111	.035	.006	.001	0+	0+	0+	0+	3
	4	0+	.003	.028	.154	.234	.184	.087	.024	.003	0+	0+	0+	0+	4
	5	0+	0+	.006	.069	.180	.221	.157	.066	.014	.001	0+	0+	0+	5
	6	0+	0+	.001	.023	.103	.197	.209	.131	.044	.006	0+	0+	0+	6
	7	0+	0+	0+	.006	.044	.131	.209	.197	.103	.023	.001	0+	0+	7
	8	0+	0+	0+	.001	.014	.066	.157	.221	.180	.069	.006	0+	0+	8
	9	0+	0+	0+	0+	.003	.024	.087	.184	.234	.154	.028	.003	0+	9
	10	0+	0+	0+	0+	.001	.006	.035	.111	.218	.246	.100	.021	0+	10
	11	0+	0+	0+	0+	0+	.001	.010	.045	.139	.268	.245	.111	.007	11
	12	0+	0+	0+	0+	0+	0+	.002	.011	.054	.179	.367	.351	.115	12
	13	0+	0+	0+	0+	0+	0+	0+	.001	.010	.055	.254	.513	.878	13
14	0	.869	.488	.229	.044	.007	.001	0+	0+	0+	0+	0+	0+	0+	0
	1	.123	.359	.356	.154	.041	.007	.001	0+	0+	0+	0+	0+	0+	1
	2	.008	.123	.257	.250	.113	.032	.006	.001	0+	0+	0+	0+	0+	2
	3	0+	.026	.114	.250	.194	.085	.022	.003	0+	0+	0+	0+	0+	3
	4	0+	.004	.035	.172	.229	.155	.061	.014	.001	0+	0+	0+	0+	4
	5	0+	0+	.008	.086	.196	.207	.122	.041	.007	0+	0+	0+	0+	5
	6	0+	0+	.001	.032	.126	.207	.183	.092	.023	.002	0+	0+	0+	6
	7	0+	0+	0+	.009	.062	.157	.209	.157	.062	.009	0+	0+	0+	7
	8	0+	0+	0+	.002	.023	.092	.183	.207	.126	.032	.001	0+	0+	8
	9	0+	0+	0+	0+	.007	.041	.122	.207	.196	.086	.008	0+	0+	9
	10	0+	0+	0+	0+	.001	.014	.061	.155	.229	.172	.035	.004	0+	10
	11	0+	0+	0+	0+	0+	.003	.022	.085	.194	.250	.114	.026	0+	11
	12	0+	0+	0+	0+	0+	.001	.006	.032	.113	.250	.257	.123	.008	12
	13	0+	0+	0+	0+	0+	0+	.001	.007	.041	.154	.356	.359	.123	13
	14	0+	0+	0+	0+	0+	0+	0+	.001	.007	.044	.229	.488	.869	14
15	0	.860	.463	.206	.035	.005	0+	0+	0+	0+	0+	0+	0+	0+	0
	1	.130	.366	.343	.132	.031	.005	0+	0+	0+	0+	0+	0+	0+	1
	2	.009	.135	.267	.231	.092	.022	.003	0+	0+	0+	0+	0+	0+	2
	3	0+	.031	.129	.250	.170	.063	.014	.002	0+	0+	0+	0+	0+	3
	4	0+	.005	.043	.188	.219	.127	.042	.007	.001	0+	0+	0+	0+	4
	5	0+	.001	.010	.103	.206	.186	.092	.024	.003	0+	0+	0+	0+	5
	6	0+	0+	.002	.043	.147	.207	.153	.061	.012	.001	0+	0+	0+	6
	7	0+	0+	0+	.014	.081	.177	.196	.118	.035	.003	0+	0+	0+	7
	8	0+	0+	0+	.003	.035	.118	.196	.177	.081	.014	0+	0+	0+	8
	9	0+	0+	0+	.001	.012	.061	.153	.207	.147	.043	.002	0+	0+	9
	10	0+	0+	0+	0+	.003	.024	.092	.186	.206	.103	.010	.001	0+	10
	11	0+	0+	0+	0+	.001	.007	.042	.127	.219	.188	.043	.005	0+	11
	12	0+	0+	0+	0+	0+	.002	.014	.063	.170	.250	.129	.031	0+	12
	13	0+	0+	0+	0+	0+	0+	.003	.022	.092	.231	.267	.135	.009	13
	14	0+	0+	0+	0+	0+	0+	0+	.005	.031	.132	.343	.366	.130	14
	15	0+	0+	0+	0+	0+	0+	0+	0+	.005	.035	.206	.463	.860	15

NOTE: 0+ represents a positive probability less than 0.0005.

From Frederick C. Mosteller, Robert E. K. Rourke, and George B. Thomas, Jr., *Probability with Statistical Applications,* 2nd ed., © 1970 Addison-Wesley Publishing Co., Reading, MA. Reprinted with permission.

NEGATIVE z Scores

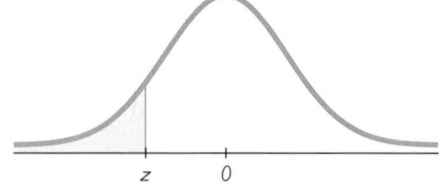

| TABLE A-2 | Standard Normal (z) Distribution: Cumulative Area from the LEFT |

z	.00	.01	.02	.03	.04	.05	.06	.07	.08	.09
−3.50 and lower	.0001									
−3.4	.0003	.0003	.0003	.0003	.0003	.0003	.0003	.0003	.0003	.0002
−3.3	.0005	.0005	.0005	.0004	.0004	.0004	.0004	.0004	.0004	.0003
−3.2	.0007	.0007	.0006	.0006	.0006	.0006	.0006	.0005	.0005	.0005
−3.1	.0010	.0009	.0009	.0009	.0008	.0008	.0008	.0008	.0007	.0007
−3.0	.0013	.0013	.0013	.0012	.0012	.0011	.0011	.0011	.0010	.0010
−2.9	.0019	.0018	.0018	.0017	.0016	.0016	.0015	.0015	.0014	.0014
−2.8	.0026	.0025	.0024	.0023	.0023	.0022	.0021	.0021	.0020	.0019
−2.7	.0035	.0034	.0033	.0032	.0031	.0030	.0029	.0028	.0027	.0026
−2.6	.0047	.0045	.0044	.0043	.0041	.0040	.0039	.0038	.0037	.0036
−2.5	.0062	.0060	.0059	.0057	.0055	.0054	.0052	.0051	* .0049	.0048
−2.4	.0082	.0080	.0078	.0075	.0073	.0071	.0069	.0068	.0066	.0064
−2.3	.0107	.0104	.0102	.0099	.0096	.0094	.0091	.0089	.0087	.0084
−2.2	.0139	.0136	.0132	.0129	.0125	.0122	.0119	.0116	.0113	.0110
−2.1	.0179	.0174	.0170	.0166	.0162	.0158	.0154	.0150	.0146	.0143
−2.0	.0228	.0222	.0217	.0212	.0207	.0202	.0197	.0192	.0188	.0183
−1.9	.0287	.0281	.0274	.0268	.0262	.0256	.0250	.0244	.0239	.0233
−1.8	.0359	.0351	.0344	.0336	.0329	.0322	.0314	.0307	.0301	.0294
−1.7	.0446	.0436	.0427	.0418	.0409	.0401	.0392	.0384	.0375	.0367
−1.6	.0548	.0537	.0526	.0516	.0505	* .0495	.0485	.0475	.0465	.0455
−1.5	.0668	.0655	.0643	.0630	.0618	.0606	.0594	.0582	.0571	.0559
−1.4	.0808	.0793	.0778	.0764	.0749	.0735	.0721	.0708	.0694	.0681
−1.3	.0968	.0951	.0934	.0918	.0901	.0885	.0869	.0853	.0838	.0823
−1.2	.1151	.1131	.1112	.1093	.1075	.1056	.1038	.1020	.1003	.0985
−1.1	.1357	.1335	.1314	.1292	.1271	.1251	.1230	.1210	.1190	.1170
−1.0	.1587	.1562	.1539	.1515	.1492	.1469	.1446	.1423	.1401	.1379
−0.9	.1841	.1814	.1788	.1762	.1736	.1711	.1685	.1660	.1635	.1611
−0.8	.2119	.2090	.2061	.2033	.2005	.1977	.1949	.1922	.1894	.1867
−0.7	.2420	.2389	.2358	.2327	.2296	.2266	.2236	.2206	.2177	.2148
−0.6	.2743	.2709	.2676	.2643	.2611	.2578	.2546	.2514	.2483	.2451
−0.5	.3085	.3050	.3015	.2981	.2946	.2912	.2877	.2843	.2810	.2776
−0.4	.3446	.3409	.3372	.3336	.3300	.3264	.3228	.3192	.3156	.3121
−0.3	.3821	.3783	.3745	.3707	.3669	.3632	.3594	.3557	.3520	.3483
−0.2	.4207	.4168	.4129	.4090	.4052	.4013	.3974	.3936	.3897	.3859
−0.1	.4602	.4562	.4522	.4483	.4443	.4404	.4364	.4325	.4286	.4247
−0.0	.5000	.4960	.4920	.4880	.4840	.4801	.4761	.4721	.4681	.4641

NOTE: For values of z below −3.49, use 0.0001 for the area.
*Use these common values that result from interpolation:

z score	Area
−1.645	0.0500
−2.575	0.0050

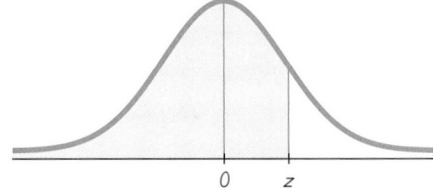

POSITIVE *z* Scores

| TABLE A-2 | (*continued*) Cumulative Area from the LEFT |

z	.00	.01	.02	.03	.04	.05	.06	.07	.08	.09
0.0	.5000	.5040	.5080	.5120	.5160	.5199	.5239	.5279	.5319	.5359
0.1	.5398	.5438	.5478	.5517	.5557	.5596	.5636	.5675	.5714	.5753
0.2	.5793	.5832	.5871	.5910	.5948	.5987	.6026	.6064	.6103	.6141
0.3	.6179	.6217	.6255	.6293	.6331	.6368	.6406	.6443	.6480	.6517
0.4	.6554	.6591	.6628	.6664	.6700	.6736	.6772	.6808	.6844	.6879
0.5	.6915	.6950	.6985	.7019	.7054	.7088	.7123	.7157	.7190	.7224
0.6	.7257	.7291	.7324	.7357	.7389	.7422	.7454	.7486	.7517	.7549
0.7	.7580	.7611	.7642	.7673	.7704	.7734	.7764	.7794	.7823	.7852
0.8	.7881	.7910	.7939	.7967	.7995	.8023	.8051	.8078	.8106	.8133
0.9	.8159	.8186	.8212	.8238	.8264	.8289	.8315	.8340	.8365	.8389
1.0	.8413	.8438	.8461	.8485	.8508	.8531	.8554	.8577	.8599	.8621
1.1	.8643	.8665	.8686	.8708	.8729	.8749	.8770	.8790	.8810	.8830
1.2	.8849	.8869	.8888	.8907	.8925	.8944	.8962	.8980	.8997	.9015
1.3	.9032	.9049	.9066	.9082	.9099	.9115	.9131	.9147	.9162	.9177
1.4	.9192	.9207	.9222	.9236	.9251	.9265	.9279	.9292	.9306	.9319
1.5	.9332	.9345	.9357	.9370	.9382	.9394	.9406	.9418	.9429	.9441
1.6	.9452	.9463	.9474	.9484	.9495 *	.9505	.9515	.9525	.9535	.9545
1.7	.9554	.9564	.9573	.9582	.9591	.9599	.9608	.9616	.9625	.9633
1.8	.9641	.9649	.9656	.9664	.9671	.9678	.9686	.9693	.9699	.9706
1.9	.9713	.9719	.9726	.9732	.9738	.9744	.9750	.9756	.9761	.9767
2.0	.9772	.9778	.9783	.9788	.9793	.9798	.9803	.9808	.9812	.9817
2.1	.9821	.9826	.9830	.9834	.9838	.9842	.9846	.9850	.9854	.9857
2.2	.9861	.9864	.9868	.9871	.9875	.9878	.9881	.9884	.9887	.9890
2.3	.9893	.9896	.9898	.9901	.9904	.9906	.9909	.9911	.9913	.9916
2.4	.9918	.9920	.9922	.9925	.9927	.9929	.9931	.9932	.9934	.9936
2.5	.9938	.9940	.9941	.9943	.9945	.9946	.9948	.9949 *	.9951	.9952
2.6	.9953	.9955	.9956	.9957	.9959	.9960	.9961	.9962	.9963	.9964
2.7	.9965	.9966	.9967	.9968	.9969	.9970	.9971	.9972	.9973	.9974
2.8	.9974	.9975	.9976	.9977	.9977	.9978	.9979	.9979	.9980	.9981
2.9	.9981	.9982	.9982	.9983	.9984	.9984	.9985	.9985	.9986	.9986
3.0	.9987	.9987	.9987	.9988	.9988	.9989	.9989	.9989	.9990	.9990
3.1	.9990	.9991	.9991	.9991	.9992	.9992	.9992	.9992	.9993	.9993
3.2	.9993	.9993	.9994	.9994	.9994	.9994	.9994	.9995	.9995	.9995
3.3	.9995	.9995	.9995	.9996	.9996	.9996	.9996	.9996	.9996	.9997
3.4	.9997	.9997	.9997	.9997	.9997	.9997	.9997	.9997	.9997	.9998
3.50 and up	.9999									

NOTE: For values of *z* above 3.49, use 0.9999 for the area.

*Use these common values that result from interpolation:

z score	Area
1.645	0.9500 ←
2.575	0.9950 ←

Common Critical Values

Confidence Level	Critical Value
0.90	1.645
0.95	1.96
0.99	2.575

TABLE A-3	t Distribution: Critical t Values				
			Area in One Tail		
	0.005	0.01	0.025	0.05	0.10
Degrees of Freedom			Area in Two Tails		
	0.01	0.02	0.05	0.10	0.20
1	63.657	31.821	12.706	6.314	3.078
2	9.925	6.965	4.303	2.920	1.886
3	5.841	4.541	3.182	2.353	1.638
4	4.604	3.747	2.776	2.132	1.533
5	4.032	3.365	2.571	2.015	1.476
6	3.707	3.143	2.447	1.943	1.440
7	3.499	2.998	2.365	1.895	1.415
8	3.355	2.896	2.306	1.860	1.397
9	3.250	2.821	2.262	1.833	1.383
10	3.169	2.764	2.228	1.812	1.372
11	3.106	2.718	2.201	1.796	1.363
12	3.055	2.681	2.179	1.782	1.356
13	3.012	2.650	2.160	1.771	1.350
14	2.977	2.624	2.145	1.761	1.345
15	2.947	2.602	2.131	1.753	1.341
16	2.921	2.583	2.120	1.746	1.337
17	2.898	2.567	2.110	1.740	1.333
18	2.878	2.552	2.101	1.734	1.330
19	2.861	2.539	2.093	1.729	1.328
20	2.845	2.528	2.086	1.725	1.325
21	2.831	2.518	2.080	1.721	1.323
22	2.819	2.508	2.074	1.717	1.321
23	2.807	2.500	2.069	1.714	1.319
24	2.797	2.492	2.064	1.711	1.318
25	2.787	2.485	2.060	1.708	1.316
26	2.779	2.479	2.056	1.706	1.315
27	2.771	2.473	2.052	1.703	1.314
28	2.763	2.467	2.048	1.701	1.313
29	2.756	2.462	2.045	1.699	1.311
30	2.750	2.457	2.042	1.697	1.310
31	2.744	2.453	2.040	1.696	1.309
32	2.738	2.449	2.037	1.694	1.309
33	2.733	2.445	2.035	1.692	1.308
34	2.728	2.441	2.032	1.691	1.307
35	2.724	2.438	2.030	1.690	1.306
36	2.719	2.434	2.028	1.688	1.306
37	2.715	2.431	2.026	1.687	1.305
38	2.712	2.429	2.024	1.686	1.304
39	2.708	2.426	2.023	1.685	1.304
40	2.704	2.423	2.021	1.684	1.303
45	2.690	2.412	2.014	1.679	1.301
50	2.678	2.403	2.009	1.676	1.299
60	2.660	2.390	2.000	1.671	1.296
70	2.648	2.381	1.994	1.667	1.294
80	2.639	2.374	1.990	1.664	1.292
90	2.632	2.368	1.987	1.662	1.291
100	2.626	2.364	1.984	1.660	1.290
200	2.601	2.345	1.972	1.653	1.286
300	2.592	2.339	1.968	1.650	1.284
400	2.588	2.336	1.966	1.649	1.284
500	2.586	2.334	1.965	1.648	1.283
1000	2.581	2.330	1.962	1.646	1.282
2000	2.578	2.328	1.961	1.646	1.282
Large	2.576	2.326	1.960	1.645	1.282

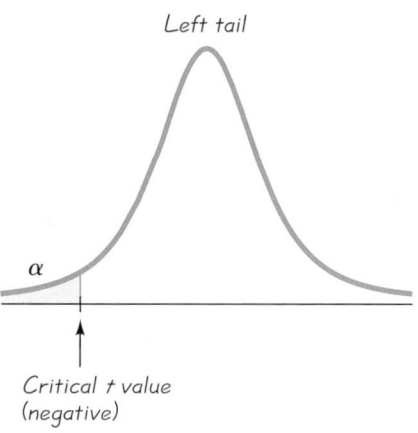

Left tail

α

Critical t value
(negative)

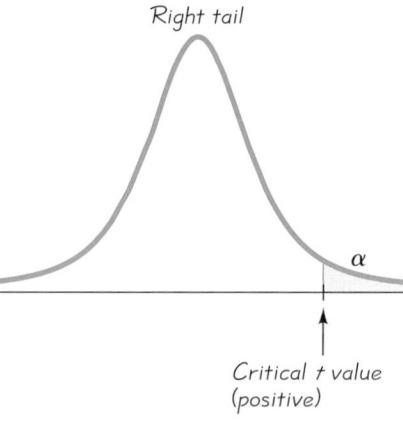

Right tail

α

Critical t value
(positive)

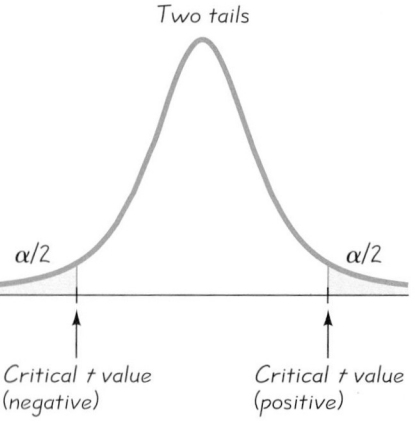

Two tails

$\alpha/2$ $\alpha/2$

Critical t value   Critical t value
(negative)   (positive)

TABLE A-4 Chi-Square (χ^2) Distribution

Degrees of Freedom	Area to the Right of the Critical Value									
	0.995	0.99	0.975	0.95	0.90	0.10	0.05	0.025	0.01	0.005
1	—	—	0.001	0.004	0.016	2.706	3.841	5.024	6.635	7.879
2	0.010	0.020	0.051	0.103	0.211	4.605	5.991	7.378	9.210	10.597
3	0.072	0.115	0.216	0.352	0.584	6.251	7.815	9.348	11.345	12.838
4	0.207	0.297	0.484	0.711	1.064	7.779	9.488	11.143	13.277	14.860
5	0.412	0.554	0.831	1.145	1.610	9.236	11.071	12.833	15.086	16.750
6	0.676	0.872	1.237	1.635	2.204	10.645	12.592	14.449	16.812	18.548
7	0.989	1.239	1.690	2.167	2.833	12.017	14.067	16.013	18.475	20.278
8	1.344	1.646	2.180	2.733	3.490	13.362	15.507	17.535	20.090	21.955
9	1.735	2.088	2.700	3.325	4.168	14.684	16.919	19.023	21.666	23.589
10	2.156	2.558	3.247	3.940	4.865	15.987	18.307	20.483	23.209	25.188
11	2.603	3.053	3.816	4.575	5.578	17.275	19.675	21.920	24.725	26.757
12	3.074	3.571	4.404	5.226	6.304	18.549	21.026	23.337	26.217	28.299
13	3.565	4.107	5.009	5.892	7.042	19.812	22.362	24.736	27.688	29.819
14	4.075	4.660	5.629	6.571	7.790	21.064	23.685	26.119	29.141	31.319
15	4.601	5.229	6.262	7.261	8.547	22.307	24.996	27.488	30.578	32.801
16	5.142	5.812	6.908	7.962	9.312	23.542	26.296	28.845	32.000	34.267
17	5.697	6.408	7.564	8.672	10.085	24.769	27.587	30.191	33.409	35.718
18	6.265	7.015	8.231	9.390	10.865	25.989	28.869	31.526	34.805	37.156
19	6.844	7.633	8.907	10.117	11.651	27.204	30.144	32.852	36.191	38.582
20	7.434	8.260	9.591	10.851	12.443	28.412	31.410	34.170	37.566	39.997
21	8.034	8.897	10.283	11.591	13.240	29.615	32.671	35.479	38.932	41.401
22	8.643	9.542	10.982	12.338	14.042	30.813	33.924	36.781	40.289	42.796
23	9.260	10.196	11.689	13.091	14.848	32.007	35.172	38.076	41.638	44.181
24	9.886	10.856	12.401	13.848	15.659	33.196	36.415	39.364	42.980	45.559
25	10.520	11.524	13.120	14.611	16.473	34.382	37.652	40.646	44.314	46.928
26	11.160	12.198	13.844	15.379	17.292	35.563	38.885	41.923	45.642	48.290
27	11.808	12.879	14.573	16.151	18.114	36.741	40.113	43.194	46.963	49.645
28	12.461	13.565	15.308	16.928	18.939	37.916	41.337	44.461	48.278	50.993
29	13.121	14.257	16.047	17.708	19.768	39.087	42.557	45.722	49.588	52.336
30	13.787	14.954	16.791	18.493	20.599	40.256	43.773	46.979	50.892	53.672
40	20.707	22.164	24.433	26.509	29.051	51.805	55.758	59.342	63.691	66.766
50	27.991	29.707	32.357	34.764	37.689	63.167	67.505	71.420	76.154	79.490
60	35.534	37.485	40.482	43.188	46.459	74.397	79.082	83.298	88.379	91.952
70	43.275	45.442	48.758	51.739	55.329	85.527	90.531	95.023	100.425	104.215
80	51.172	53.540	57.153	60.391	64.278	96.578	101.879	106.629	112.329	116.321
90	59.196	61.754	65.647	69.126	73.291	107.565	113.145	118.136	124.116	128.299
100	67.328	70.065	74.222	77.929	82.358	118.498	124.342	129.561	135.807	140.169

Degrees of Freedom

$n - 1$	for confidence intervals or hypothesis tests with a standard deviation or variance
$k - 1$	for goodness-of-fit with k categories
$(r - 1)(c - 1)$	for contingency tables with r rows and c columns
$k - 1$	for Kruskal-Wallis test with k samples

TABLE A-5 — F Distribution ($\alpha = 0.025$ in the right tail)

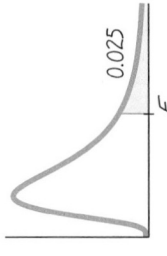

Denominator degrees of freedom (df$_2$)	Numerator degrees of freedom (df$_1$)								
	1	2	3	4	5	6	7	8	9
1	647.79	799.50	864.16	899.58	921.85	937.11	948.22	956.66	963.28
2	38.506	39.000	39.165	39.248	39.298	39.331	39.335	39.373	39.387
3	17.443	16.044	15.439	15.101	14.885	14.735	14.624	14.540	14.473
4	12.218	10.649	9.9792	9.6045	9.3645	9.1973	9.0741	8.9796	8.9047
5	10.007	8.4336	7.7636	7.3879	7.1464	6.9777	6.8531	6.7572	6.6811
6	8.8131	7.2599	6.5988	6.2272	5.9876	5.8198	5.6955	5.5996	5.5234
7	8.0727	6.5415	5.8898	5.5226	5.2852	5.1186	4.9949	4.8993	4.8232
8	7.5709	6.0595	5.4160	5.0526	4.8173	4.6517	4.5286	4.4333	4.3572
9	7.2093	5.7147	5.0781	4.7181	4.4844	4.3197	4.1970	4.1020	4.0260
10	6.9367	5.4564	4.8256	4.4683	4.2361	4.0721	3.9498	3.8549	3.7790
11	6.7241	5.2559	4.6300	4.2751	4.0440	3.8807	3.7586	3.6638	3.5879
12	6.5538	5.0959	4.4742	4.1212	3.8911	3.7283	3.6065	3.5118	3.4358
13	6.4143	4.9653	4.3472	3.9959	3.7667	3.6043	3.4827	3.3880	3.3120
14	6.2979	4.8567	4.2417	3.8919	3.6634	3.5014	3.3799	3.2853	3.2093
15	6.1995	4.7650	4.1528	3.8043	3.5764	3.4147	3.2934	3.1987	3.1227
16	6.1151	4.6867	4.0768	3.7294	3.5021	3.3406	3.2194	3.1248	3.0488
17	6.0420	4.6189	4.0112	3.6648	3.4379	3.2767	3.1556	3.0610	2.9849
18	5.9781	4.5597	3.9539	3.6083	3.3820	3.2209	3.0999	3.0053	2.9291
19	5.9216	4.5075	3.9034	3.5587	3.3327	3.1718	3.0509	2.9563	2.8801
20	5.8715	4.4613	3.8587	3.5147	3.2891	3.1283	3.0074	2.9128	2.8365
21	5.8266	4.4199	3.8188	3.4754	3.2501	3.0895	2.9686	2.8740	2.7977
22	5.7863	4.3828	3.7829	3.4401	3.2151	3.0546	2.9338	2.8392	2.7628
23	5.7498	4.3492	3.7505	3.4083	3.1835	3.0232	2.9023	2.8077	2.7313
24	5.7166	4.3187	3.7211	3.3794	3.1548	2.9946	2.8738	2.7791	2.7027
25	5.6864	4.2909	3.6943	3.3530	3.1287	2.9685	2.8478	2.7531	2.6766
26	5.6586	4.2655	3.6697	3.3289	3.1048	2.9447	2.8240	2.7293	2.6528
27	5.6331	4.2421	3.6472	3.3067	3.0828	2.9228	2.8021	2.7074	2.6309
28	5.6096	4.2205	3.6264	3.2863	3.0626	2.9027	2.7820	2.6872	2.6106
29	5.5878	4.2006	3.6072	3.2674	3.0438	2.8840	2.7633	2.6686	2.5919
30	5.5675	4.1821	3.5894	3.2499	3.0265	2.8667	2.7460	2.6513	2.5746
40	5.4239	4.0510	3.4633	3.1261	2.9037	2.7444	2.6238	2.5289	2.4519
60	5.2856	3.9253	3.3425	3.0077	2.7863	2.6274	2.5068	2.4117	2.3344
120	5.1523	3.8046	3.2269	2.8943	2.6740	2.5154	2.3948	2.2994	2.2217
∞	5.0239	3.6889	3.1161	2.7858	2.5665	2.4082	2.2875	2.1918	2.1136

(continued)

TABLE A-5 F Distribution (α = 0.025 in the right tail) (continued)

Numerator degrees of freedom (df$_1$)

df$_2$	10	12	15	20	24	30	40	60	120	∞
1	968.63	976.71	984.87	993.10	997.25	1001.4	1005.6	1009.8	1014.0	1018.3
2	39.398	39.415	39.431	39.448	39.456	39.465	39.473	39.481	39.490	39.498
3	14.419	14.337	14.253	14.167	14.124	14.081	14.037	13.992	13.947	13.902
4	8.8439	8.7512	8.6565	8.5599	8.5109	8.4613	8.4111	8.3604	8.3092	8.2573
5	6.6192	6.5245	6.4277	6.3286	6.2780	6.2269	6.1750	6.1225	6.0693	6.0153
6	5.4613	5.3662	5.2687	5.1684	5.1172	5.0652	5.0125	4.9589	4.9044	4.8491
7	4.7611	4.6658	4.5678	4.4667	4.4150	4.3624	4.3089	4.2544	4.1989	4.1423
8	4.2951	4.1997	4.1012	3.9995	3.9472	3.8940	3.8398	3.7844	3.7279	3.6702
9	3.9639	3.8682	3.7694	3.6669	3.6142	3.5604	3.5055	3.4493	3.3918	3.3329
10	3.7168	3.6209	3.5217	3.4185	3.3654	3.3110	3.2554	3.1984	3.1399	3.0798
11	3.5257	3.4296	3.3299	3.2261	3.1725	3.1176	3.0613	3.0035	2.9441	2.8828
12	3.3736	3.2773	3.1772	3.0728	3.0187	2.9633	2.9063	2.8478	2.7874	2.7249
13	3.2497	3.1532	3.0527	2.9477	2.8932	2.8372	2.7797	2.7204	2.6590	2.5955
14	3.1469	3.0502	2.9493	2.8437	2.7888	2.7324	2.6742	2.6142	2.5519	2.4872
15	3.0602	2.9633	2.8621	2.7559	2.7006	2.6437	2.5850	2.5242	2.4611	2.3953
16	2.9862	2.8890	2.7875	2.6808	2.6252	2.5678	2.5085	2.4471	2.3831	2.3163
17	2.9222	2.8249	2.7230	2.6158	2.5598	2.5020	2.4422	2.3801	2.3153	2.2474
18	2.8664	2.7689	2.6667	2.5590	2.5027	2.4445	2.3842	2.3214	2.2558	2.1869
19	2.8172	2.7196	2.6171	2.5089	2.4523	2.3937	2.3329	2.2696	2.2032	2.1333
20	2.7737	2.6758	2.5731	2.4645	2.4076	2.3486	2.2873	2.2234	2.1562	2.0853
21	2.7348	2.6368	2.5338	2.4247	2.3675	2.3082	2.2465	2.1819	2.1141	2.0422
22	2.6998	2.6017	2.4984	2.3890	2.3315	2.2718	2.2097	2.1446	2.0760	2.0032
23	2.6682	2.5699	2.4665	2.3567	2.2989	2.2389	2.1763	2.1107	2.0415	1.9677
24	2.6396	2.5411	2.4374	2.3273	2.2693	2.2090	2.1460	2.0799	2.0099	1.9353
25	2.6135	2.5149	2.4110	2.3005	2.2422	2.1816	2.1183	2.0516	1.9811	1.9055
26	2.5896	2.4908	2.3867	2.2759	2.2174	2.1565	2.0928	2.0257	1.9545	1.8781
27	2.5676	2.4688	2.3644	2.2533	2.1946	2.1334	2.0693	2.0018	1.9299	1.8527
28	2.5473	2.4484	2.3438	2.2324	2.1735	2.1121	2.0477	1.9797	1.9072	1.8291
29	2.5286	2.4295	2.3248	2.2131	2.1540	2.0923	2.0276	1.9591	1.8861	1.8072
30	2.5112	2.4120	2.3072	2.1952	2.1359	2.0739	2.0089	1.9400	1.8664	1.7867
40	2.3882	2.2882	2.1819	2.0677	2.0069	1.9429	1.8752	1.8028	1.7242	1.6371
60	2.2702	2.1692	2.0613	1.9445	1.8817	1.8152	1.7440	1.6668	1.5810	1.4821
120	2.1570	2.0548	1.9450	1.8249	1.7597	1.6899	1.6141	1.5299	1.4327	1.3104
∞	2.0483	1.9447	1.8326	1.7085	1.6402	1.5660	1.4835	1.3883	1.2684	1.0000

Denominator degrees of freedom (df$_2$)

From Maxine Merrington and Catherine M. Thompson, "Tables of Percentage Points of the Inverted Beta (F) Distribution," *Biometrika 33* (1943): 80–84. Reproduced with permission of the Biometrika Trustees.

TABLE A-5 F Distribution ($\alpha = 0.05$ in the right tail) (continued)

Numerator degrees of freedom (df_1)

Denominator degrees of freedom (df_2)	1	2	3	4	5	6	7	8	9
1	161.45	199.50	215.71	224.58	230.16	233.99	236.77	238.88	240.54
2	18.513	19.000	19.164	19.247	19.296	19.330	19.353	19.371	19.385
3	10.128	9.5521	9.2766	9.1172	9.0135	8.9406	8.8867	8.8452	8.8123
4	7.7086	6.9443	6.5914	6.3882	6.2561	6.1631	6.0942	6.0410	5.9988
5	6.6079	5.7861	5.4095	5.1922	5.0503	4.9503	4.8759	4.8183	4.7725
6	5.9874	5.1433	4.7571	4.5337	4.3874	4.2839	4.2067	4.1468	4.0990
7	5.5914	4.7374	4.3468	4.1203	3.9715	3.8660	3.7870	3.7257	3.6767
8	5.3177	4.4590	4.0662	3.8379	3.6875	3.5806	3.5005	3.4381	3.3881
9	5.1174	4.2565	3.8625	3.6331	3.4817	3.3738	3.2927	3.2296	3.1789
10	4.9646	4.1028	3.7083	3.4780	3.3258	3.2172	3.1355	3.0717	3.0204
11	4.8443	3.9823	3.5874	3.3567	3.2039	3.0946	3.0123	2.9480	2.8962
12	4.7472	3.8853	3.4903	3.2592	3.1059	2.9961	2.9134	2.8486	2.7964
13	4.6672	3.8056	3.4105	3.1791	3.0254	2.9153	2.8321	2.7669	2.7144
14	4.6001	3.7389	3.3439	3.1122	2.9582	2.8477	2.7642	2.6987	2.6458
15	4.5431	3.6823	3.2874	3.0556	2.9013	2.7905	2.7066	2.6408	2.5876
16	4.4940	3.6337	3.2389	3.0069	2.8524	2.7413	2.6572	2.5911	2.5377
17	4.4513	3.5915	3.1968	2.9647	2.8100	2.6987	2.6143	2.5480	2.4943
18	4.4139	3.5546	3.1599	2.9277	2.7729	2.6613	2.5767	2.5102	2.4563
19	4.3807	3.5219	3.1274	2.8951	2.7401	2.6283	2.5435	2.4768	2.4227
20	4.3512	3.4928	3.0984	2.8661	2.7109	2.5990	2.5140	2.4471	2.3928
21	4.3248	3.4668	3.0725	2.8401	2.6848	2.5727	2.4876	2.4205	2.3660
22	4.3009	3.4434	3.0491	2.8167	2.6613	2.5491	2.4638	2.3965	2.3419
23	4.2793	3.4221	3.0280	2.7955	2.6400	2.5277	2.4422	2.3748	2.3201
24	4.2597	3.4028	3.0088	2.7763	2.6207	2.5082	2.4226	2.3551	2.3002
25	4.2417	3.3852	2.9912	2.7587	2.6030	2.4904	2.4047	2.3371	2.2821
26	4.2252	3.3690	2.9752	2.7426	2.5868	2.4741	2.3883	2.3205	2.2655
27	4.2100	3.3541	2.9604	2.7278	2.5719	2.4591	2.3732	2.3053	2.2501
28	4.1960	3.3404	2.9467	2.7141	2.5581	2.4453	2.3593	2.2913	2.2360
29	4.1830	3.3277	2.9340	2.7014	2.5454	2.4324	2.3463	2.2783	2.2229
30	4.1709	3.3158	2.9223	2.6896	2.5336	2.4205	2.3343	2.2662	2.2107
40	4.0847	3.2317	2.8387	2.6060	2.4495	2.3359	2.2490	2.1802	2.1240
60	4.0012	3.1504	2.7581	2.5252	2.3683	2.2541	2.1665	2.0970	2.0401
120	3.9201	3.0718	2.6802	2.4472	2.2899	2.1750	2.0868	2.0164	1.9588
∞	3.8415	2.9957	2.6049	2.3719	2.2141	2.0986	2.0096	1.9384	1.8799

(continued)

TABLE A-5 F Distribution ($\alpha = 0.05$ in the right tail) (continued)

Numerator degrees of freedom (df$_1$)

df$_2$	10	12	15	20	24	30	40	60	120	∞
1	241.88	243.91	245.95	248.01	249.05	250.10	251.14	252.20	253.25	254.31
2	19.396	19.413	19.429	19.446	19.454	19.462	19.471	19.479	19.487	19.496
3	8.7855	8.7446	8.7029	8.6602	8.6385	8.6166	8.5944	8.5720	8.5494	8.5264
4	5.9644	5.9117	5.8578	5.8025	5.7744	5.7459	5.7170	5.6877	5.6581	5.6281
5	4.7351	4.6777	4.6188	4.5581	4.5272	4.4957	4.4638	4.4314	4.3985	4.3650
6	4.0600	3.9999	3.9381	3.8742	3.8415	3.8082	3.7743	3.7398	3.7047	3.6689
7	3.6365	3.5747	3.5107	3.4445	3.4105	3.3758	3.3404	3.3043	3.2674	3.2298
8	3.3472	3.2839	3.2184	3.1503	3.1152	3.0794	3.0428	3.0053	2.9669	2.9276
9	3.1373	3.0729	3.0061	2.9365	2.9005	2.8637	2.8259	2.7872	2.7475	2.7067
10	2.9782	2.9130	2.8450	2.7740	2.7372	2.6996	2.6609	2.6211	2.5801	2.5379
11	2.8536	2.7876	2.7186	2.6464	2.6090	2.5705	2.5309	2.4901	2.4480	2.4045
12	2.7534	2.6866	2.6169	2.5436	2.5055	2.4663	2.4259	2.3842	2.3410	2.2962
13	2.6710	2.6037	2.5331	2.4589	2.4202	2.3803	2.3392	2.2966	2.2524	2.2064
14	2.6022	2.5342	2.4630	2.3879	2.3487	2.3082	2.2664	2.2229	2.1778	2.1307
15	2.5437	2.4753	2.4034	2.3275	2.2878	2.2468	2.2043	2.1601	2.1141	2.0658
16	2.4935	2.4247	2.3522	2.2756	2.2354	2.1938	2.1507	2.1058	2.0589	2.0096
17	2.4499	2.3807	2.3077	2.2304	2.1898	2.1477	2.1040	2.0584	2.0107	1.9604
18	2.4117	2.3421	2.2686	2.1906	2.1497	2.1071	2.0629	2.0166	1.9681	1.9168
19	2.3779	2.3080	2.2341	2.1555	2.1141	2.0712	2.0264	1.9795	1.9302	1.8780
20	2.3479	2.2776	2.2033	2.1242	2.0825	2.0391	1.9938	1.9464	1.8963	1.8432
21	2.3210	2.2504	2.1757	2.0960	2.0540	2.0102	1.9645	1.9165	1.8657	1.8117
22	2.2967	2.2258	2.1508	2.0707	2.0283	1.9842	1.9380	1.8894	1.8380	1.7831
23	2.2747	2.2036	2.1282	2.0476	2.0050	1.9605	1.9139	1.8648	1.8128	1.7570
24	2.2547	2.1834	2.1077	2.0267	1.9838	1.9390	1.8920	1.8424	1.7896	1.7330
25	2.2365	2.1649	2.0889	2.0075	1.9643	1.9192	1.8718	1.8217	1.7684	1.7110
26	2.2197	2.1479	2.0716	1.9898	1.9464	1.9010	1.8533	1.8027	1.7488	1.6906
27	2.2043	2.1323	2.0558	1.9736	1.9299	1.8842	1.8361	1.7851	1.7306	1.6717
28	2.1900	2.1179	2.0411	1.9586	1.9147	1.8687	1.8203	1.7689	1.7138	1.6541
29	2.1768	2.1045	2.0275	1.9446	1.9005	1.8543	1.8055	1.7537	1.6981	1.6376
30	2.1646	2.0921	2.0148	1.9317	1.8874	1.8409	1.7918	1.7396	1.6835	1.6223
40	2.0772	2.0035	1.9245	1.8389	1.7929	1.7444	1.6928	1.6373	1.5766	1.5089
60	1.9926	1.9174	1.8364	1.7480	1.7001	1.6491	1.5943	1.5343	1.4673	1.3893
120	1.9105	1.8337	1.7505	1.6587	1.6084	1.5543	1.4952	1.4290	1.3519	1.2539
∞	1.8307	1.7522	1.6664	1.5705	1.5173	1.4591	1.3940	1.3180	1.2214	1.0000

Denominator degrees of freedom (df$_2$)

From Maxine Merrington and Catherine M. Thompson, "Tables of Percentage Points of the Inverted Beta (F) Distribution," *Biometrika* 33 (1943): 80–84. Reproduced with permission of the Biometrika Trustees.

TABLE A-6	Critical Values of the Pearson Correlation Coefficient r	
n	$\alpha = .05$	$\alpha = .01$
4	.950	.990
5	.878	.959
6	.811	.917
7	.754	.875
8	.707	.834
9	.666	.798
10	.632	.765
11	.602	.735
12	.576	.708
13	.553	.684
14	.532	.661
15	.514	.641
16	.497	.623
17	.482	.606
18	.468	.590
19	.456	.575
20	.444	.561
25	.396	.505
30	.361	.463
35	.335	.430
40	.312	.402
45	.294	.378
50	.279	.361
60	.254	.330
70	.236	.305
80	.220	.286
90	.207	.269
100	.196	.256

NOTE: To test H_0: $\rho = 0$ against H_1: $\rho \neq 0$, reject H_0 if the absolute value of r is greater than the critical value in the table.

TABLE A-7 Critical Values of *T* for the Wilcoxon Signed-Ranks Test

	α			
n	.005 (one tail) .01 (two tails)	.01 (one tail) .02 (two tails)	.025 (one tail) .05 (two tails)	.05 (one tail) .10 (two tails)
5	*	*	*	1
6	*	*	1	2
7	*	0	2	4
8	0	2	4	6
9	2	3	6	8
10	3	5	8	11
11	5	7	11	14
12	7	10	14	17
13	10	13	17	21
14	13	16	21	26
15	16	20	25	30
16	19	24	30	36
17	23	28	35	41
18	28	33	40	47
19	32	38	46	54
20	37	43	52	60
21	43	49	59	68
22	49	56	66	75
23	55	62	73	83
24	61	69	81	92
25	68	77	90	101
26	76	85	98	110
27	84	93	107	120
28	92	102	117	130
29	100	111	127	141
30	109	120	137	152

NOTES:

1. *indicates that it is not possible to get a value in the critical region.
2. Reject the null hypothesis if the test statistic *T* is less than or equal to the critical value found in this table. Fail to reject the null hypothesis if the test statistic *T* is greater than the critical value found in the table.

From *Some Rapid Approximate Statistical Procedures,* Copyright © 1949, 1964 Lederle Laboratories Division of American Cyanamid Company. Reprinted with the permission of the American Cyanamid Company.

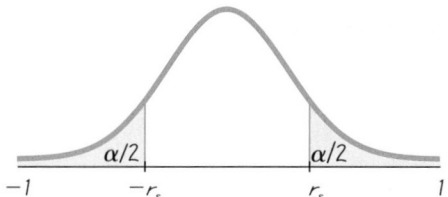

TABLE A-8	Critical Values of Spearman's Rank Correlation Coefficient r_s			
n	$\alpha = 0.10$	$\alpha = 0.05$	$\alpha = 0.02$	$\alpha = 0.01$
5	.900	—	—	—
6	.829	.886	.943	—
7	.714	.786	.893	.929
8	.643	.738	.833	.881
9	.600	.700	.783	.833
10	.564	.648	.745	.794
11	.536	.618	.709	.755
12	.503	.587	.678	.727
13	.484	.560	.648	.703
14	.464	.538	.626	.679
15	.446	.521	.604	.654
16	.429	.503	.582	.635
17	.414	.485	.566	.615
18	.401	.472	.550	.600
19	.391	.460	.535	.584
20	.380	.447	.520	.570
21	.370	.435	.508	.556
22	.361	.425	.496	.544
23	.353	.415	.486	.532
24	.344	.406	.476	.521
25	.337	.398	.466	.511
26	.331	.390	.457	.501
27	.324	.382	.448	.491
28	.317	.375	.440	.483
29	.312	.368	.433	.475
30	.306	.362	.425	.467

NOTES:
1. For $n > 30$, use $r_s = \pm z/\sqrt{n-1}$ where z corresponds to the level of significance. For example, if $\alpha = 0.05$, then $z = 1.96$.
2. If the absolute value of the test statistic r_s exceeds the positive critical value, then reject $H_0: \rho_s = 0$ and conclude that there is a correlation.

Based on data from "*Biostatistical Analysis,* 4th edition," © 1999, by Jerrold Zar, Prentice Hall, Inc., Upper Saddle River, New Jersey, and "Distribution of Sums of Squares of Rank Differences to Small Numbers with Individuals," *The Annals of Mathematical Statistics,* Vol. 9, No. 2.

TABLE A-9	Critical Values for Number of Runs G

	Value of n_2																		
	2	3	4	5	6	7	8	9	10	11	12	13	14	15	16	17	18	19	20
2	1	1	1	1	1	1	1	1	1	1	2	2	2	2	2	2	2	2	2
	6	6	6	6	6	6	6	6	6	6	6	6	6	6	6	6	6	6	6
3	1	1	1	1	2	2	2	2	2	2	2	2	2	3	3	3	3	3	3
	6	8	8	8	8	8	8	8	8	8	8	8	8	8	8	8	8	8	8
4	1	1	1	2	2	2	3	3	3	3	3	3	3	3	4	4	4	4	4
	6	8	9	9	9	10	10	10	10	10	10	10	10	10	10	10	10	10	10
5	1	1	2	2	3	3	3	3	3	4	4	4	4	4	4	4	5	5	5
	6	8	9	10	10	11	11	12	12	12	12	12	12	12	12	12	12	12	12
6	1	2	2	3	3	3	3	4	4	4	4	5	5	5	5	5	5	6	6
	6	8	9	10	11	12	12	13	13	13	13	14	14	14	14	14	14	14	14
7	1	2	2	3	3	3	4	4	5	5	5	5	5	6	6	6	6	6	6
	6	8	10	11	12	13	13	14	14	14	14	15	15	15	16	16	16	16	16
8	1	2	3	3	3	4	4	5	5	5	6	6	6	6	6	7	7	7	7
	6	8	10	11	12	13	14	14	15	15	16	16	16	16	17	17	17	17	17
9	1	2	3	3	4	4	5	5	5	6	6	6	7	7	7	7	8	8	8
	6	8	10	12	13	14	14	15	16	16	16	17	17	18	18	18	18	18	18
10	1	2	3	3	4	5	5	5	6	6	7	7	7	7	8	8	8	8	9
	6	8	10	12	13	14	15	16	16	17	17	18	18	18	19	19	19	20	20
11	1	2	3	4	4	5	5	6	6	7	7	7	8	8	8	9	9	9	9
	6	8	10	12	13	14	15	16	17	17	18	19	19	19	20	20	20	21	21
12	2	2	3	4	4	5	6	6	7	7	7	8	8	8	9	9	9	10	10
	6	8	10	12	13	14	16	16	17	18	19	19	20	20	21	21	21	22	22
13	2	2	3	4	5	5	6	6	7	7	8	8	9	9	9	10	10	10	10
	6	8	10	12	14	15	16	17	18	19	19	20	20	21	21	22	22	23	23
14	2	2	3	4	5	5	6	7	7	8	8	9	9	9	10	10	10	11	11
	6	8	10	12	14	15	16	17	18	19	20	20	21	22	22	23	23	23	24
15	2	3	3	4	5	6	6	7	7	8	8	9	9	10	10	11	11	11	12
	6	8	10	12	14	15	16	18	18	19	20	21	22	22	23	23	24	24	25
16	2	3	4	4	5	6	6	7	8	8	9	9	10	10	11	11	11	12	12
	6	8	10	12	14	16	17	18	19	20	21	21	22	23	23	24	25	25	25
17	2	3	4	4	5	6	7	7	8	9	9	10	10	11	11	11	12	12	13
	6	8	10	12	14	16	17	18	19	20	21	22	23	23	24	25	25	26	26
18	2	3	4	5	5	6	7	8	8	9	9	10	10	11	11	12	12	13	13
	6	8	10	12	14	16	17	18	19	20	21	22	23	24	25	25	26	26	27
19	2	3	4	5	6	6	7	8	8	9	10	10	11	11	12	12	13	13	13
	6	8	10	12	14	16	17	18	20	21	22	23	23	24	25	26	26	27	27
20	2	3	4	5	6	6	7	8	9	9	10	10	11	12	12	13	13	13	14
	6	8	10	12	14	16	17	18	20	21	22	23	24	25	25	26	27	27	28

Value of n_1 (row labels)

NOTES:
1. The entries in this table are the critical G values, assuming a two-tailed test with a significance level of $\alpha = 0.05$.
2. The null hypothesis of randomness is rejected if the total number of runs G is less than or equal to the smaller entry or greater than or equal to the larger entry.

From "Tables for Testing Randomness of Groupings in a Sequence of Alternatives," *The Annals of Mathematical Statistics,* Vol. 14, No. 1. Reprinted with permission of the Institute of Mathematical Statistics.

Appendix B: Data Sets

Data Set 1:   Health Exam Results

Data Set 2:   Body Temperatures (in degrees Fahrenheit) of Healthy Adults

Data Set 3:   Freshman 15 Data

Data Set 4:   Cigarette Tar, Nicotine, and Carbon Monoxide

Data Set 5:   Passive and Active Smoke

Data Set 6:   Bears (wild bears anesthetized)

Data Set 7:   Alcohol and Tobacco Use in Animated Children's Movies

Data Set 8:   Word Counts by Males and Females

Data Set 9:   Movies

Data Set 10:   NASA Space Transport System Data

Data Set 11:   Forecast and Actual Temperatures

Data Set 12:   Electricity Consumption of a Home

Data Set 13:   Voltage Measurements from a Home

Data Set 14:   Rainfall (in inches) in Boston for One Year

Data Set 15:   Old Faithful Geyser

Data Set 16:   Cars

Data Set 17:   Weights and Volumes of Cola

Data Set 18:   M&M Plain Candy Weights (grams)

Data Set 19:   Screw Lengths (inches)

Data Set 20:   Coin Weights (grams)

Data Set 21:   Axial Loads of Aluminum Cans

Data Set 22:   Weights of Discarded Garbage for One Week

Data Set 23:   Home Sales

Data Set 24:   FICO Credit Rating Scores

Additional data sets are available at the Web site aw.com/Triola.

Data Set 1: Health Exam Results

AGE is in years, HT is height (inches), WT is weight (pounds), WAIST is circumference (cm), Pulse is pulse rate (beats per minute), SYS is systolic blood pressure (mm Hg), DIAS is diastolic blood pressure (mm Hg), CHOL is cholesterol (mg), BMI is body mass index, Leg is upper leg length (cm), Elbow is elbow breadth (cm), Wrist is wrist breadth (cm), and Arm is arm circumference (cm). Data are from the U.S. Department of Health and Human Services, National Center for Health Statistics, Third National Health and Nutrition Examination Survey.

STATDISK: Data set name for males is Mhealth.

Minitab: Worksheet name for males is MHEALTH.MTW.

Excel: Workbook name for males is MHEALTH.XLS.

TI-83/84 Plus: App name for male data is MHEALTH and the file names are the same as for text files.

Text file names for males: MAGE, MHT, MWT, MWAST, MPULS, MSYS, MDIAS, MCHOL, MBMI, MLEG, MELBW, MWRST, MARM.

Male	Age	HT	WT	Waist	Pulse	SYS	DIAS	CHOL	BMI	Leg	Elbow	Wrist	Arm
	58	70.8	169.1	90.6	68	125	78	522	23.8	42.5	7.7	6.4	31.9
	22	66.2	144.2	78.1	64	107	54	127	23.2	40.2	7.6	6.2	31.0
	32	71.7	179.3	96.5	88	126	81	740	24.6	44.4	7.3	5.8	32.7
	31	68.7	175.8	87.7	72	110	68	49	26.2	42.8	7.5	5.9	33.4
	28	67.6	152.6	87.1	64	110	66	230	23.5	40.0	7.1	6.0	30.1
	46	69.2	166.8	92.4	72	107	83	316	24.5	47.3	7.1	5.8	30.5
	41	66.5	135.0	78.8	60	113	71	590	21.5	43.4	6.5	5.2	27.6
	56	67.2	201.5	103.3	88	126	72	466	31.4	40.1	7.5	5.6	38.0
	20	68.3	175.2	89.1	76	137	85	121	26.4	42.1	7.5	5.5	32.0
	54	65.6	139.0	82.5	60	110	71	578	22.7	36.0	6.9	5.5	29.3
	17	63.0	156.3	86.7	96	109	65	78	27.8	44.2	7.1	5.3	31.7
	73	68.3	186.6	103.3	72	153	87	265	28.1	36.7	8.1	6.7	30.7
	52	73.1	191.1	91.8	56	112	77	250	25.2	48.4	8.0	5.2	34.7
	25	67.6	151.3	75.6	64	119	81	265	23.3	41.0	7.0	5.7	30.6
	29	68.0	209.4	105.5	60	113	82	273	31.9	39.8	6.9	6.0	34.2
	17	71.0	237.1	108.7	64	125	76	272	33.1	45.2	8.3	6.6	41.1
	41	61.3	176.7	104.0	84	131	80	972	33.2	40.2	6.7	5.7	33.1
	52	76.2	220.6	103.0	76	121	75	75	26.7	46.2	7.9	6.0	32.2
	32	66.3	166.1	91.3	84	132	81	138	26.6	39.0	7.5	5.7	31.2
	20	69.7	137.4	75.2	88	112	44	139	19.9	44.8	6.9	5.6	25.9
	20	65.4	164.2	87.7	72	121	65	638	27.1	40.9	7.0	5.6	33.7
	29	70.0	162.4	77.0	56	116	64	613	23.4	43.1	7.5	5.2	30.3
	18	62.9	151.8	85.0	68	95	58	762	27.0	38.0	7.4	5.8	32.8
	26	68.5	144.1	79.6	64	110	70	303	21.6	41.0	6.8	5.7	31.0
	33	68.3	204.6	103.8	60	110	66	690	30.9	46.0	7.4	6.1	36.2
	55	69.4	193.8	103.0	68	125	82	31	28.3	41.4	7.2	6.0	33.6
	53	69.2	172.9	97.1	60	124	79	189	25.5	42.7	6.6	5.9	31.9
	28	68.0	161.9	86.9	60	131	69	957	24.6	40.5	7.3	5.7	32.9
	28	71.9	174.8	88.0	56	109	64	339	23.8	44.2	7.8	6.0	30.9
	37	66.1	169.8	91.5	84	112	79	416	27.4	41.8	7.0	6.1	34.0
	40	72.4	213.3	102.9	72	127	72	120	28.7	47.2	7.5	5.9	34.8
	33	73.0	198.0	93.1	84	132	74	702	26.2	48.2	7.8	6.0	33.6
	26	68.0	173.3	98.9	88	116	81	1252	26.4	42.9	6.7	5.8	31.3
	53	68.7	214.5	107.5	56	125	84	288	32.1	42.8	8.2	5.9	37.6
	36	70.3	137.1	81.6	64	112	77	176	19.6	40.8	7.1	5.3	27.9
	34	63.7	119.5	75.7	56	125	77	277	20.7	42.6	6.6	5.3	26.9
	42	71.1	189.1	95.0	56	120	83	649	26.3	44.9	7.4	6.0	36.9
	18	65.6	164.7	91.1	60	118	68	113	26.9	41.1	7.0	6.1	34.5
	44	68.3	170.1	94.9	64	115	75	656	25.6	44.5	7.3	5.8	32.1
	20	66.3	151.0	79.9	72	115	65	172	24.2	44.0	7.1	5.4	30.7

(continued)

Data Set 1: Health Exam Results (*continued*)

STATDISK:	Data set name for females is Fhealth.
Minitab:	Worksheet name for females is FHEALTH.MTW.
Excel:	Workbook name for females is FHEALTH.XLS.
TI-83/84 Plus:	App name for female data is FHEALTH and the file names are the same as for text files.
Text file names for females:	FAGE, FHT, FWT, FWAST, FPULS, FSYS, FDIAS, FCHOL, FBMI, FLEG, FELBW, FWRST, FARM.

Female Age	HT	WT	Waist	Pulse	SYS	DIAS	CHOL	BMI	Leg	Elbow	Wrist	Arm
17	64.3	114.8	67.2	76	104	61	264	19.6	41.6	6.0	4.6	23.6
32	66.4	149.3	82.5	72	99	64	181	23.8	42.8	6.7	5.5	26.3
25	62.3	107.8	66.7	88	102	65	267	19.6	39.0	5.7	4.6	26.3
55	62.3	160.1	93.0	60	114	76	384	29.1	40.2	6.2	5.0	32.6
27	59.6	127.1	82.6	72	94	58	98	25.2	36.2	5.5	4.8	29.2
29	63.6	123.1	75.4	68	101	66	62	21.4	43.2	6.0	4.9	26.4
25	59.8	111.7	73.6	80	108	61	126	22.0	38.7	5.7	5.1	27.9
12	63.3	156.3	81.4	64	104	41	89	27.5	41.0	6.8	5.5	33.0
41	67.9	218.8	99.4	68	123	72	531	33.5	43.8	7.8	5.8	38.6
32	61.4	110.2	67.7	68	93	61	130	20.6	37.3	6.3	5.0	26.5
31	66.7	188.3	100.7	80	89	56	175	29.9	42.3	6.6	5.2	34.4
19	64.8	105.4	72.9	76	112	62	44	17.7	39.1	5.7	4.8	23.7
19	63.1	136.1	85.0	68	107	48	8	24.0	40.3	6.6	5.1	28.4
23	66.7	182.4	85.7	72	116	62	112	28.9	48.6	7.2	5.6	34.0
40	66.8	238.4	126.0	96	181	102	462	37.7	33.2	7.0	5.4	35.2
23	64.7	108.8	74.5	72	98	61	62	18.3	43.4	6.2	5.2	24.7
27	65.1	119.0	74.5	68	100	53	98	19.8	41.5	6.3	5.3	27.0
45	61.9	161.9	94.0	72	127	74	447	29.8	40.0	6.8	5.0	35.0
41	64.3	174.1	92.8	64	107	67	125	29.7	38.2	6.8	4.7	33.1
56	63.4	181.2	105.5	80	116	71	318	31.7	38.2	6.9	5.4	39.6
22	60.7	124.3	75.5	64	97	64	325	23.8	38.2	5.9	5.0	27.0
57	63.4	255.9	126.5	80	155	85	600	44.9	41.0	8.0	5.6	43.8
24	62.6	106.7	70.0	76	106	59	237	19.2	38.1	6.1	5.0	23.6
37	60.6	149.9	98.0	76	110	70	173	28.7	38.0	7.0	5.1	34.3
59	63.5	163.1	104.7	76	105	69	309	28.5	36.0	6.7	5.1	34.4
40	58.6	94.3	67.8	80	118	82	94	19.3	32.1	5.4	4.2	23.3
45	60.2	159.7	99.3	104	133	83	280	31.0	31.1	6.4	5.2	35.6
52	67.6	162.8	91.1	88	113	75	254	25.1	39.4	7.1	5.3	31.8
31	63.4	130.0	74.5	60	113	66	123	22.8	40.2	5.9	5.1	27.0
32	64.1	179.9	95.5	76	107	67	596	30.9	39.2	6.2	5.0	32.8
23	62.7	147.8	79.5	72	95	59	301	26.5	39.0	6.3	4.9	31.0
23	61.3	112.9	69.1	72	108	72	223	21.2	36.6	5.9	4.7	27.0
47	58.2	195.6	105.5	88	114	79	293	40.6	27.0	7.5	5.5	41.2
36	63.2	124.2	78.8	80	104	73	146	21.9	38.5	5.6	4.7	25.5
34	60.5	135.0	85.7	60	125	73	149	26.0	39.9	6.4	5.2	30.9
37	65.0	141.4	92.8	72	124	85	149	23.5	37.5	6.1	4.8	27.9
18	61.8	123.9	72.7	88	92	46	920	22.8	39.7	5.8	5.0	26.5
29	68.0	135.5	75.9	88	119	81	271	20.7	39.0	6.3	4.9	27.8
48	67.0	130.4	68.6	124	93	64	207	20.5	41.6	6.0	5.3	23.0
16	57.0	100.7	68.7	64	106	64	2	21.9	33.8	5.6	4.6	26.4

Data Set 2: Body Temperatures (in degrees Fahrenheit) of Healthy Adults

Data provided by Dr. Steven Wasserman, Dr. Philip Mackowiak, and Dr. Myron Levine of the University of Maryland.

STATDISK:	Data set name for the 12 A.M. temperatures on Day 2 is Bodytemp.
Minitab:	Worksheet name for the 12 A.M. temperatures on Day 2 is BODYTEMP.MTW.
Excel:	Workbook name for the 12 A.M. temperatures on Day 2 is BODYTEMP.XLS.
TI-83/84 Plus:	App name for 12 A.M. temperatures on Day 2 is BTEMP and the file name is BTEMP.
Text files:	Text file name is BTEMP.

Subject	Age	Sex	Smoke	Temperature Day 1		Temperature Day 2	
				8 AM	12 AM	8 AM	12 AM
1	22	M	Y	98.0	98.0	98.0	98.6
2	23	M	Y	97.0	97.6	97.4	—
3	22	M	Y	98.6	98.8	97.8	98.6
4	19	M	N	97.4	98.0	97.0	98.0
5	18	M	N	98.2	98.8	97.0	98.0
6	20	M	Y	98.2	98.8	96.6	99.0
7	27	M	Y	98.2	97.6	97.0	98.4
8	19	M	Y	96.6	98.6	96.8	98.4
9	19	M	Y	97.4	98.6	96.6	98.4
10	24	M	N	97.4	98.8	96.6	98.4
11	35	M	Y	98.2	98.0	96.2	98.6
12	25	M	Y	97.4	98.2	97.6	98.6
13	25	M	N	97.8	98.0	98.6	98.8
14	35	M	Y	98.4	98.0	97.0	98.6
15	21	M	N	97.6	97.0	97.4	97.0
16	33	M	N	96.2	97.2	98.0	97.0
17	19	M	Y	98.0	98.2	97.6	98.8
18	24	M	Y	—	—	97.2	97.6
19	18	F	N	—	—	97.0	97.7
20	22	F	Y	—	—	98.0	98.8
21	20	M	Y	—	—	97.0	98.0
22	30	F	Y	—	—	96.4	98.0
23	29	M	N	—	—	96.1	98.3
24	18	M	Y	—	—	98.0	98.5
25	31	M	Y	—	98.1	96.8	97.3
26	28	F	Y	—	98.2	98.2	98.7
27	27	M	Y	—	98.5	97.8	97.4
28	21	M	Y	—	98.5	98.2	98.9
29	30	M	Y	—	99.0	97.8	98.6
30	27	M	N	—	98.0	99.0	99.5
31	32	M	Y	—	97.0	97.4	97.5
32	33	M	Y	—	97.3	97.4	97.3
33	23	M	Y	—	97.3	97.5	97.6
34	29	M	Y	—	98.1	97.8	98.2
35	25	M	Y	—	—	97.9	99.6
36	31	M	N	—	97.8	97.8	98.7
37	25	M	Y	—	99.0	98.3	99.4
38	28	M	N	—	97.6	98.0	98.2
39	30	M	Y	—	97.4	—	98.0
40	33	M	Y	—	98.0	—	98.6
41	28	M	Y	98.0	97.4	—	98.6
42	22	M	Y	98.8	98.0	—	97.2
43	21	F	Y	99.0	—	—	98.4
44	30	M	N	—	98.6	—	98.6

(continued)

Data Set 2: Body Temperatures (*continued*)

Subject	Age	Sex	Smoke	Temperature Day 1		Temperature Day 2	
				8 AM	12 AM	8 AM	12 AM
45	22	M	Y	—	98.6	—	98.2
46	22	F	N	98.0	98.4	—	98.0
47	20	M	Y	—	97.0	—	97.8
48	19	M	Y	—	—	—	98.0
49	33	M	N	—	98.4	—	98.4
50	31	M	Y	99.0	99.0	—	98.6
51	26	M	N	—	98.0	—	98.6
52	18	M	N	—	—	—	97.8
53	23	M	N	—	99.4	—	99.0
54	28	M	Y	—	—	—	96.5
55	19	M	Y	—	97.8	—	97.6
56	21	M	N	—	—	—	98.0
57	27	M	Y	—	98.2	—	96.9
58	29	M	Y	—	99.2	—	97.6
59	38	M	N	—	99.0	—	97.1
60	29	F	Y	—	97.7	—	97.9
61	22	M	Y	—	98.2	—	98.4
62	22	M	Y	—	98.2	—	97.3
63	26	M	Y	—	98.8	—	98.0
64	32	M	N	—	98.1	—	97.5
65	25	M	Y	—	98.5	—	97.6
66	21	F	N	—	97.2	—	98.2
67	25	M	Y	—	98.5	—	98.5
68	24	M	Y	—	99.2	97.0	98.8
69	25	M	Y	—	98.3	97.6	98.7
70	35	M	Y	—	98.7	97.5	97.8
71	23	F	Y	—	98.8	98.8	98.0
72	31	M	Y	—	98.6	98.4	97.1
73	28	M	Y	—	98.0	98.2	97.4
74	29	M	Y	—	99.1	97.7	99.4
75	26	M	Y	—	97.2	97.3	98.4
76	32	M	N	—	97.6	97.5	98.6
77	32	M	Y	—	97.9	97.1	98.4
78	21	F	Y	—	98.8	98.6	98.5
79	20	M	Y	—	98.6	98.6	98.6
80	24	F	Y	—	98.6	97.8	98.3
81	21	F	Y	—	99.3	98.7	98.7
82	28	M	Y	—	97.8	97.9	98.8
83	27	F	N	98.8	98.7	97.8	99.1
84	28	M	N	99.4	99.3	97.8	98.6
85	29	M	Y	98.8	97.8	97.6	97.9
86	19	M	N	97.7	98.4	96.8	98.8
87	24	M	Y	99.0	97.7	96.0	98.0
88	29	M	N	98.1	98.3	98.0	98.7
89	25	M	Y	98.7	97.7	97.0	98.5
90	27	M	N	97.5	97.1	97.4	98.9
91	25	M	Y	98.9	98.4	97.6	98.4
92	21	M	Y	98.4	98.6	97.6	98.6
93	19	M	Y	97.2	97.4	96.2	97.1
94	27	M	Y	—	—	96.2	97.9
95	32	M	N	98.8	96.7	98.1	98.8
96	24	M	Y	97.3	96.9	97.1	98.7
97	32	M	Y	98.7	98.4	98.2	97.6
98	19	F	Y	98.9	98.2	96.4	98.2
99	18	F	Y	99.2	98.6	96.9	99.2
100	27	M	N	—	97.0	—	97.8
101	34	M	Y	—	97.4	—	98.0
102	25	M	N	—	98.4	—	98.4
103	18	M	N	—	97.4	—	97.8
104	32	M	Y	—	96.8	—	98.4
105	31	M	Y	—	98.2	—	97.4
106	26	M	N	—	97.4	—	98.0
107	23	M	N	—	98.0	—	97.0

Data Set 3: Freshman 15 Data

Weights are in kilograms, and BMI denotes measured body mass index. Measurements were made in September of freshman year and then later in April of freshman year. Results are published in "Changes in Body Weight and Fat Mass of Men and Women in the First Year of College: A Study of the 'Freshman 15'" by Hoffman, Policastro, Quick, and Lee, *Journal of American College Health,* Vol. 55, No. 1. Reprinted with permission of the Helen Dwight Reid Educational Foundation. Published by Heldref Publications, 1319 Eighteenth St., NW, Washington, DC 20036-1802. Copyright © (2006).

	STATDISK:	Data set name is Freshman15.
	Minitab:	Worksheet name is FRESH15.MTW.
	Excel:	Workbook name is FRESH15.XLS.
	TI-83/84 Plus:	App name is FRESH and the file names are the same as for text files.
	Text file names:	WTSEP, WTAPR, BMISP, BMIAP.

SEX	Weight in September	Weight in April	BMI in September	BMI in April
M	72	59	22.02	18.14
M	97	86	19.70	17.44
M	74	69	24.09	22.43
M	93	88	26.97	25.57
F	68	64	21.51	20.10
M	59	55	18.69	17.40
F	64	60	24.24	22.88
F	56	53	21.23	20.23
F	70	68	30.26	29.24
F	58	56	21.88	21.02
F	50	47	17.63	16.89
M	71	69	24.57	23.85
M	67	66	20.68	20.15
F	56	55	20.97	20.36
F	70	68	27.30	26.73
F	61	60	23.30	22.88
F	53	52	19.48	19.24
M	92	92	24.74	24.69
F	57	58	20.69	20.79
M	67	67	20.49	20.60
F	58	58	21.09	21.24
F	49	50	18.37	18.53
M	68	68	22.40	22.61
F	69	69	28.17	28.43
M	87	88	23.60	23.81
M	81	82	26.52	26.78
M	60	61	18.89	19.27
F	52	53	19.31	19.75
M	70	71	20.96	21.32
F	63	64	21.78	22.22
F	56	57	19.78	20.23
M	68	69	22.40	22.82
M	68	69	22.76	23.19
F	54	56	20.15	20.69
M	80	82	22.14	22.57
M	64	66	20.27	20.76
F	57	59	22.15	22.93
F	63	65	23.87	24.67
F	54	56	18.61	19.34
F	56	58	21.73	22.58
M	54	56	18.93	19.72
M	73	75	25.88	26.72
M	77	79	28.59	29.53
F	63	66	21.89	22.79
F	51	54	18.31	19.28
F	59	62	19.64	20.63
F	65	68	23.02	24.10
F	53	56	20.63	21.91
F	62	65	22.61	23.81
F	55	58	22.03	23.42
M	74	77	20.31	21.34
M	74	78	20.31	21.36
M	64	68	19.59	20.77
M	64	68	21.05	22.31
F	57	61	23.47	25.11
F	64	68	22.84	24.29
F	60	64	19.50	20.90
M	64	68	18.51	19.83
M	66	71	21.40	22.97
F	52	57	17.72	19.42
M	71	77	22.26	23.87
F	55	60	21.64	23.81
M	65	71	22.51	24.45
M	75	82	23.69	25.80
F	42	49	15.08	17.74
M	74	82	22.64	25.33
M	94	105	36.57	40.86

Data Set 4: Cigarette Tar, Nicotine, and Carbon Monoxide

All measurements are in milligrams per cigarette. CO denotes carbon monoxide. The king size ciga-rettes are nonfiltered, nonmenthol, and non-light. The menthol cigarettes are 100 mm long, filtered, and non-light. The cigarettes in the third group are 100 mm long, filtered, nonmenthol, and non-light. Data are from the Federal Trade Commission.

STATDISK:	Data set name is Cigaret.
Minitab:	Worksheet name is CIGARET.MTW.
Excel:	Workbook name is CIGARET.XLS.
TI-83/84 Plus:	App name is CIGARET and the file names are the same as for text files.
Text file names:	KGTAR, KGNIC, KGCO, MNTAR, MNNIC, MNCO, FLTAR, FLNIC, FLCO (where KG denotes the king size cigarettes, MN denotes the menthol cigarettes, and FL denotes the filtered cigarettes that are not menthol types).

King Size				Menthol				Filtered 100 mm Non-menthol			
Brand	Tar	Nicotine	CO	Brand	Tar	Nicotine	CO	Brand	Tar	Nicotine	CO
Austin	20	1.1	16	Alpine	16	1.1	15	Barclay	5	0.4	4
Basic	27	1.7	16	Austin	13	0.8	17	Basic	16	1.0	19
Bristol	27	1.7	16	Basic	16	1.0	19	Camel	17	1.2	17
Cardinal	20	1.1	16	Belair	9	0.9	9	Highway	13	0.8	18
Cavalier	20	1.1	16	Best Value	14	0.8	17	Jacks	13	0.8	18
Chesterfield	24	1.4	17	Cavalier	13	0.8	17	Kent	14	1.0	13
Cimarron	20	1.1	16	Doral	12	0.8	15	Lark	15	1.1	17
Class A	23	1.4	15	Focus	14	0.8	17	Marlboro	15	1.1	15
Doral	20	1.0	16	GPC	14	0.9	15	Maverick	15	1.1	15
GPC	22	1.2	14	Highway	13	0.8	17	Merit	9	0.8	12
Highway	20	1.1	16	Jacks	13	0.8	17	Monaco	13	0.8	18
Jacks	20	1.1	16	Kool	16	1.2	15	Monarch	13	0.8	17
Marker	20	1.1	16	Legend	13	0.8	17	Mustang	13	0.8	18
Monaco	20	1.1	16	Marker	13	0.8	17	Newport	15	1.0	16
Monarch	20	1.1	16	Maverick	18	1.3	18	Now	2	0.2	3
Old Gold	10	1.8	14	Merit	9	0.7	11	Old Gold	15	1.1	18
Pall Mall	24	1.6	16	Newport	19	1.4	18	Pall Mall	15	1.0	15
Pilot	20	1.1	16	Now	2	0.2	3	Pilot	13	0.8	18
Prime	21	1.2	14	Pilot	13	0.8	17	Players	14	1.0	15
Pyramid	25	1.5	18	Players	14	1.0	14	Prime	15	0.9	17
Raleigh Extra	23	1.3	15	Prime	14	0.8	15	Raleigh	16	1.1	15
Sebring	20	1.1	16	Pyramid	15	0.8	22	Tareyton	15	1.1	15
Summit	22	1.3	14	Salem	16	1.2	16	True	7	0.6	7
Sundance	20	1.1	16	True	6	0.6	7	Viceroy	17	1.3	16
Worth	20	1.1	16	Vantage	8	0.7	9	Winston	15	1.1	14

Data Set 5: Passive and Active Smoke

All values are measured levels of serum cotinine (in ng/mL), a metabolite of nicotine. (When nicotine is absorbed by the body, cotinine is produced.) Data are from the U.S. Department of Health and Human Services, National Center for Health Statistics, Third National Health and Nutrition Examination Survey.

	STATDISK:	Data set name is Cotinine.
	Minitab:	Worksheet name is COTININE.MTW.
	Excel:	Workbook name is COTININE.XLS.
	TI-83/84 Plus:	App name is COTININE and the file names are the same as for text files.
	Text file names:	SMKR, ETS, NOETS.

SMKR (smokers, or subjects reported tobacco use)

1	0	131	173	265	210	44	277	32	3
35	112	477	289	227	103	222	149	313	491
130	234	164	198	17	253	87	121	266	290
123	167	250	245	48	86	284	1	208	173

ETS (nonsmokers exposed to environmental tobacco smoke at home or work)

384	0	69	19	1	0	178	2	13	1
4	0	543	17	1	0	51	0	197	3
0	3	1	45	13	3	1	1	1	0
0	551	2	1	1	1	0	74	1	241

NOETS (nonsmokers with no exposure to environmental tobacco smoke at home or work)

0	0	0	0	0	0	0	0	0	0
0	9	0	0	0	0	0	0	244	0
1	0	0	0	90	1	0	309	0	0
0	0	0	0	0	0	0	0	0	0

Data Set 6: Bears (wild bears anesthetized)

Age is in months, Month is the month of measurement (1 = January), Sex is coded with 1 = male and 2 = female, Headlen is head length (inches), Headwth is width of head (inches), Neck is distance around neck (in inches), Length is length of body (inches), Chest is distance around chest (inches), and Weight is measured in pounds. Data are from Gary Alt and Minitab, Inc.

	STATDISK:	Data set name is Bears.
	Minitab:	Worksheet name is BEARS.MTW.
	Excel:	Workbook name is BEARS.XLS.
	TI-83/84 Plus:	App name is BEARS and the file names are the same as for text files.
	Text file names:	BAGE, BMNTH, BSEX, BHDLN, BHDWD, BNECK, BLEN, BCHST, BWGHT.

Age	Month	Sex	Headlen	Headwth	Neck	Length	Chest	Weight
19	7	1	11.0	5.5	16.0	53.0	26.0	80
55	7	1	16.5	9.0	28.0	67.5	45.0	344
81	9	1	15.5	8.0	31.0	72.0	54.0	416
115	7	1	17.0	10.0	31.5	72.0	49.0	348
104	8	2	15.5	6.5	22.0	62.0	35.0	166
100	4	2	13.0	7.0	21.0	70.0	41.0	220
56	7	1	15.0	7.5	26.5	73.5	41.0	262
51	4	1	13.5	8.0	27.0	68.5	49.0	360
57	9	2	13.5	7.0	20.0	64.0	38.0	204
53	5	2	12.5	6.0	18.0	58.0	31.0	144
68	8	1	16.0	9.0	29.0	73.0	44.0	332
8	8	1	9.0	4.5	13.0	37.0	19.0	34
44	8	2	12.5	4.5	10.5	63.0	32.0	140
32	8	1	14.0	5.0	21.5	67.0	37.0	180
20	8	2	11.5	5.0	17.5	52.0	29.0	105
32	8	1	13.0	8.0	21.5	59.0	33.0	166
45	9	1	13.5	7.0	24.0	64.0	39.0	204
9	9	2	9.0	4.5	12.0	36.0	19.0	26
21	9	1	13.0	6.0	19.0	59.0	30.0	120
177	9	1	16.0	9.5	30.0	72.0	48.0	436
57	9	2	12.5	5.0	19.0	57.5	32.0	125
81	9	2	13.0	5.0	20.0	61.0	33.0	132
21	9	1	13.0	5.0	17.0	54.0	28.0	90
9	9	1	10.0	4.0	13.0	40.0	23.0	40
45	9	1	16.0	6.0	24.0	63.0	42.0	220
9	9	1	10.0	4.0	13.5	43.0	23.0	46
33	9	1	13.5	6.0	22.0	66.5	34.0	154
57	9	2	13.0	5.5	17.5	60.5	31.0	116
45	9	2	13.0	6.5	21.0	60.0	34.5	182
21	9	1	14.5	5.5	20.0	61.0	34.0	150
10	10	1	9.5	4.5	16.0	40.0	26.0	65
82	10	2	13.5	6.5	28.0	64.0	48.0	356
70	10	2	14.5	6.5	26.0	65.0	48.0	316
10	10	1	11.0	5.0	17.0	49.0	29.0	94
10	10	1	11.5	5.0	17.0	47.0	29.5	86
34	10	1	13.0	7.0	21.0	59.0	35.0	150
34	10	1	16.5	6.5	27.0	72.0	44.5	270
34	10	1	14.0	5.5	24.0	65.0	39.0	202
58	10	2	13.5	6.5	21.5	63.0	40.0	202
58	10	1	15.5	7.0	28.0	70.5	50.0	365
11	11	1	11.5	6.0	16.5	48.0	31.0	79
23	11	1	12.0	6.5	19.0	50.0	38.0	148
70	10	1	15.5	7.0	28.0	76.5	55.0	446
11	11	2	9.0	5.0	15.0	46.0	27.0	62
83	11	2	14.5	7.0	23.0	61.5	44.0	236
35	11	1	13.5	8.5	23.0	63.5	44.0	212
16	4	1	10.0	4.0	15.5	48.0	26.0	60
16	4	1	10.0	5.0	15.0	41.0	26.0	64
17	5	1	11.5	5.0	17.0	53.0	30.5	114
17	5	2	11.5	5.0	15.0	52.5	28.0	76
17	5	2	11.0	4.5	13.0	46.0	23.0	48
8	8	2	10.0	4.5	10.0	43.5	24.0	29
83	11	1	15.5	8.0	30.5	75.0	54.0	514
18	6	1	12.5	8.5	18.0	57.3	32.8	140

Data Set 7: Alcohol and Tobacco Use in Animated Children's Movies

The data are based on "Tobacco and Alcohol Use in G-Rated Children's Animated Films," by Goldstein, Sobel, and Newman, *Journal of the American Medical Association,* Vol. 281, No. 12.

STATDISK: Data set name is Chmovie.

Minitab: Worksheet name is CHMOVIE.MTW.

Excel: Workbook name is CHMOVIE.XLS.

TI-83/84 Plus: App name is CHMOVIE and the file names are the same as for text files.

Text file names: CHLEN, CHTOB, CHALC.

Movie	Company	Length (min)	Tobacco Use (sec)	Alcohol Use (sec)
Snow White	Disney	83	0	0
Pinocchio	Disney	88	223	80
Fantasia	Disney	120	0	0
Dumbo	Disney	64	176	88
Bambi	Disney	69	0	0
Three Caballeros	Disney	71	548	8
Fun and Fancy Free	Disney	76	0	4
Cinderella	Disney	74	37	0
Alice in Wonderland	Disney	75	158	0
Peter Pan	Disney	76	51	33
Lady and the Tramp	Disney	75	0	0
Sleeping Beauty	Disney	75	0	113
101 Dalmatians	Disney	79	299	51
Sword and the Stone	Disney	80	37	20
Jungle Book	Disney	78	0	0
Aristocats	Disney	78	11	142
Robin Hood	Disney	83	0	39
Rescuers	Disney	77	0	0
Winnie the Pooh	Disney	71	0	0
Fox and the Hound	Disney	83	0	0
Black Cauldron	Disney	80	0	34
Great Mouse Detective	Disney	73	165	414
Oliver and Company	Disney	72	74	0
Little Mermaid	Disney	82	9	0
Rescuers Down Under	Disney	74	0	76
Beauty and the Beast	Disney	84	0	123
Aladdin	Disney	90	2	3
Lion King	Disney	89	0	0
Pocahontas	Disney	81	6	7
Toy Story	Disney	81	0	0
Hunchback of Notre Dame	Disney	90	23	46
James and the Giant Peach	Disney	79	206	38
Hercules	Disney	92	9	13
Secret of NIMH	MGM	82	0	0
All Dogs Go to Heaven	MGM	89	205	73
All Dogs Go to Heaven 2	MGM	82	162	72
Babes in Toyland	MGM	74	0	0
Thumbelina	Warner Bros	86	6	5
Troll in Central Park	Warner Bros	76	1	0
Space Jam	Warner Bros	81	117	0
Pippi Longstocking	Warner Bros	75	5	0
Cats Don't Dance	Warner Bros	75	91	0
An American Tail	Universal	77	155	74
Land Before Time	Universal	70	0	0
Fievel Goes West	Universal	75	24	28
We're Back: Dinosaur Story	Universal	64	55	0
Land Before Time 2	Universal	73	0	0
Balto	Universal	74	0	0
Once Upon a Forest	20th Century Fox	71	0	0
Anastasia	20th Century Fox	94	17	39

Data Set 8: Word Counts by Males and Females

The columns are counts of the numbers of words spoken in a day by male (M) and female (F) subjects in six different sample groups. Column 1M denotes the word counts for males in Sample 1, 1F is the count for females in Sample 1, and so on.

Sample 1: Recruited couples ranging in age from 18 to 29
Sample 2: Students recruited in introductory psychology classes, aged 17 to 23
Sample 3: Students recruited in introductory psychology classes in Mexico, aged 17 to 25
Sample 4: Students recruited in introductory psychology classes, aged 17 to 22
Sample 5: Students recruited in introductory psychology classes, aged 18 to 26
Sample 6: Students recruited in introductory psychology classes, aged 17 to 23

Results were published in "Are Women Really More Talkative Than Men?" by Matthias R. Mehl, Simine Vazire, Nairan Ramirez-Esparza, Richard B. Slatcher, James W. Pennebaker, *Science,* Vol. 317, No. 5834.

	STATDISK:	Data set name is WORDS.
	Minitab:	Worksheet name is WORDS.MTW.
	Excel:	Workbook name is WORDS.XLS.
	TI-83/84 Plus:	App name is WORDS, and the file names are M1, F1, M2, F2, M3, F3, M4, F4, M5, F5, M6, F6.
	Text file names:	Text file names correspond to the columns below: 1M, 1F, 2M, 2F, 3M, 3F, 4M, 4F, 5M, 5F, 6M, 6F.

1M	1F	2M	2F	3M	3F	4M	4F	5M	5F	6M	6F
27531	20737	23871	16109	21143	6705	47016	11849	39207	15962	28408	15357
15684	24625	5180	10592	17791	21613	27308	25317	20868	16610	10084	13618
5638	5198	9951	24608	36571	11935	42709	40055	18857	22497	15931	9783
27997	18712	12460	13739	6724	15790	20565	18797	17271	5004	21688	26451
25433	12002	17155	22376	15430	17865	21034	20104		10171	37786	12151
8077	15702	10344	9351	11552	13035	24150	17225		31327	10575	8391
21319	11661	9811	7694	11748	24834	24547	14356		8758	12880	19763
17572	19624	12387	16812	12169	7747	22712	20571			11071	25246
26429	13397	29920	21066	15581	3852	20858	12240			17799	8427
21966	18776	21791	32291	23858	11648	3189	10031			13182	6998
11680	15863	9789	12320	5269	25862	10379	13260			8918	24876
10818	12549	31127	19839	12384	17183	15750	22871			6495	6272
12650	17014	8572	22018	11576	11010	4288	26533			8153	10047
21683	23511	6942	16624	17707	11156	12398	26139			7015	15569
19153	6017	2539	5139	15229	11351	25120	15204			4429	39681
1411	18338	36345	17384	18160	25693	7415	18393			10054	23079
20242	23020	6858	17740	22482	13383	7642	16363			3998	24814
10117	18602	24024	7309	18626	19992	16459	21293			12639	19287
20206	16518	5488	14699	1118	14926	19928	12562			10974	10351
16874	13770	9960	21450	5319	14128	26615	15422			5255	8866
16135	29940	11118	14602		10345	21885	29011				10827
20734	8419	4970	18360		13516	10009	17085				12584
7771	17791	10710	12345		12831	35142	13932				12764
6792	5596	15011	17379		9671	3593	2220				19086
26194	11467	1569	14394		17011	15728	5909				26852
10671	18372	23794	11907		28575	19230	10623				17639
13462	13657	23689	8319		23557	17108	20388				16616
12474	21420	11769	16046		13656	23852	13052				
13560	21261	26846	5112		8231	11276	12098				
18876	12964	17386	8034		10601	14456	19643				

(continued)

Data Set 8: Word Counts by Males and Females (*continued*)

1M	1F	2M	2F	3M	3F	4M	4F	5M	5F	6M	6F
13825	33789	7987	7845		8124	11067	21747				
9274	8709	25638	7796			18527	26601				
20547	10508	695	20910			11478	17835				
17190	11909	2366	8756			6025	14030				
10578	29730	16075	5683			12975	7990				
14821	20981	16789	8372			14124	16667				
15477	16937	9308	17191			22942	5342				
10483	19049		8380			12985	12729				
19377	20224		16741			18360	18920				
11767	15872		16417			9643	24261				
13793	18717		2363			8788	8741				
5908	12685		24349			7755	14981				
18821	17646					9725	1674				
14069	16255					11033	20635				
16072	28838					10372	16154				
16414	38154					16869	4148				
19017	25510					16248	5322				
37649	34869					9202					
17427	24480					11395					
46978	31553										
25835	18667										
10302	7059										
15686	25168										
10072	16143										
6885	14730										
20848	28117										

Data Set 9: Movies

STATDISK:	Data set name is Movies.
Minitab:	Worksheet name is MOVIES.MTW.
Excel:	Workbook name is MOVIES.XLS.
TI-83/84 Plus:	App name is MOVIES and the file names are the same as for text files.
Text file names:	MVBUD, MVGRS, MVLEN, MVRAT.

Title	MPAA Rating	Budget ($) in Millions	Gross ($) in Millions	Length (min)	Viewer Rating
8 Mile	R	41.0	117	110	6.7
Alone in the Dark	R	20.0	5	96	2.2
Aviator	PG-13	116.0	103	170	7.6
Big Fish	PG-13	70.0	66	125	8.0
Bourne Identity	PG-13	75.0	121	119	7.4
Break-Up	PG-13	52.0	116	105	5.8
Charlie's Angels: Full Throttle	PG-13	120.0	101	106	4.8
Collateral	R	65.0	100	120	7.7
Crash	R	6.5	55	113	8.3
Daddy Day Care	PG	60.0	104	92	5.7
DaVinci Code	PG-13	125.0	213	149	6.5
Eternal Sunshine	R	20.0	34	108	8.6
From Justin to Kelly	PG	5.0	12	81	1.9
Harry Potter Goblet of Fire	PG-13	150.0	290	157	7.8
Hostel	R	4.5	47	94	5.8
House of the Dead	R	7.0	10	90	2.0
Last Samurai	R	100.0	111	154	7.8
Million Dollar Baby	PG-13	30.0	100	132	8.4
Pirates of the Caribbean (II)	PG-13	225.0	322	150	7.5
Rollerball	PG-13	70.0	19	97	2.7
S.W.A.T.	PG-13	80.0	117	117	6.0
Secret Window	PG-13	40.0	48	96	6.3
Signs	PG-13	70.0	228	106	7.0
Silent Hill	R	50.0	47	127	6.6
Son of the Mask	PG	74.0	17	94	2.0
Spider-Man 2	PG-13	200.0	373	127	7.8
Star Wars III	PG-13	113.0	380	140	8.0
Sum of All Fears	PG-13	68.0	118	124	6.4
The Pianist	R	35.0	33	150	8.5
The Village	PG-13	72.0	114	108	6.6
Van Helsing	PG-13	160.0	120	132	5.3
Vanilla Sky	R	68.0	101	136	6.8
Walk the Line	PG-13	29.0	120	136	8.1
War of the Worlds	PG-13	132.0	234	116	6.7
Wedding Crashers	R	40.0	209	119	7.3

Data Set 10: NASA Space Transport System Data

STATDISK: Data set name is NASA.
Minitab: Worksheet name is NASA.MTW.
Excel: Workbook name is NASA.XLS.
TI-83/84 Plus: App name is NASA and the file names are
                     the same as for text files.
Text file names: NASA1, NASA2.

Shuttle Flight Durations (hours)

54	54	192	169	122	120	146	145	247	191	167	144	197	191	73	167	168	169	190	170
97	168	165	146	0	97	105	119	96	121	119	120	261	106	121	98	117	215	143	199
218	213	128	166	193	214	213	331	191	190	236	175	143	222	239	239	236	336	259	174
335	269	353	262	269	262	198	399	235	214	260	381	196	214	377	221	240	405	243	423
244	239	95	221	376	284	259	376	211	381	235	213	283	235	118	191	269	237	283	309
259	309	307	285	306	285	283	262	259	332	259	330	382	333	306					

Numbers of Shuttle Flights by Astronauts

2	4	2	3	2	3	1	0	2	0	4	0	2	3	2	3	4	0	3	2
2	0	5	3	4	3	0	4	2	0	5	1	4	4	1	6	1	2	4	2
1	4	4	0	0	0	0	3	1	1	1	1	4	3	0	0	7	0	2	1
3	1	2	3	3	5	2	3	2	4	2	4	0	3	4	3	1	2	4	3
0	4	1	5	1	1	1	2	1	2	0	0	1	1	1	6	0	0	1	0
0	1	2	0	2	2	2	3	4	1	5	0	2	4	0	2	3	4	0	0
1	1	2	4	2	1	5	0	2	4	2	1	3	3	5	5	1	4	1	1
0	4	1	1	1	5	0	4	0	2	1	5	1	5	3	0	0	4	3	0
1	1	1	2	0	4	3	4	3	1	1	3	2	3	0	2	3	0	3	1
0	4	3	3	5	1	1	1	3	0	3	1	2	1	1	2	3	1	3	0
2	0	0	1	0	1	3	6	4	3	4	2	0	0	4	2	0	0	0	2
3	2	4	2	0	1	1	1	1	0	1	4	3	2	2	0	1	1	4	3
0	2	5	1	7	3	3	1	1	1	3	2	3	3	0	2	1	3	2	0
3	1	0	4	1	2	4	0	2	0	2	3	0	1	1	3	5	3	4	3
4	2	3	2	2	2	0	5	5	1	4	0	4	2	2	6	0	1	1	4
0	2	1	0	0	0	4	3	0	1	6	0								

Data Set 11: Forecast and Actual Temperatures

Temperatures are in degrees Fahrenheit. All measurements were recorded near the author's home.

STATDISK:	Data set name is Weather.
Minitab:	Worksheet name is WEATHER.MTW.
Excel:	Workbook name is WEATHER.XLS.
TI-83/84 Plus:	App name is WEATHER and the file names are the same as for text files.
Text file names:	ACTHI, ACTLO, PHI1, PLO1, PHI3, PLO3, PHI5, PLO5, PREC.

Date	Actual High	Actual Low	1 Day Predicted High	1 Day Predicted Low	3 Day Predicted High	3 Day Predicted Low	5 Day Predicted High	5 Day Predicted Low	Precip. (in.)
Sept. 1	80	54	78	52	79	52	80	56	0.00
Sept. 2	77	54	75	53	86	63	80	57	0.00
Sept. 3	81	55	81	55	79	59	79	59	0.00
Sept. 4	85	60	85	62	83	59	80	56	0.00
Sept. 5	73	64	76	53	80	63	79	64	0.00
Sept. 6	73	51	75	58	76	61	82	57	0.00
Sept. 7	80	59	79	66	80	63	76	61	0.00
Sept. 8	72	61	74	66	79	67	73	63	0.47
Sept. 9	83	68	75	66	76	66	77	59	1.59
Sept. 10	81	62	80	53	79	58	83	61	0.07
Sept. 11	75	53	75	51	78	58	77	58	0.01
Sept. 12	78	52	79	55	75	54	79	56	0.00
Sept. 13	80	56	80	53	74	48	74	50	0.01
Sept. 14	71	56	70	53	73	55	75	52	0.00
Sept. 15	73	54	72	60	73	59	76	54	0.00
Sept. 16	78	64	79	63	76	60	78	62	0.06
Sept. 17	75	62	75	60	76	56	76	58	0.01
Sept. 18	63	55	67	43	73	52	75	60	2.85
Sept. 19	63	48	64	43	75	53	77	55	0.00
Sept. 20	70	40	69	46	68	53	71	50	0.00
Sept. 21	77	47	77	50	77	51	74	54	0.00
Sept. 22	82	49	81	55	83	54	73	56	0.00
Sept. 23	81	53	81	51	78	57	75	53	0.01
Sept. 24	76	51	80	53	75	54	79	56	0.00
Sept. 25	77	54	78	54	77	51	74	53	0.00
Sept. 26	76	58	76	50	72	46	71	44	0.00
Sept. 27	74	48	76	60	74	56	70	45	0.01
Sept. 28	66	61	70	49	74	49	73	52	1.99
Sept. 29	66	57	69	41	68	41	72	48	0.67
Sept. 30	62	53	68	45	72	50	69	47	0.21
Oct. 1	71	51	75	49	72	49	70	47	0.02
Oct. 2	68	46	71	47	73	52	68	46	0.01
Oct. 3	66	44	68	42	66	43	67	46	0.05
Oct. 4	71	39	69	44	68	44	61	40	0.00
Oct. 5	58	43	56	29	62	38	64	42	0.00

Data Set 12: Electricity Consumption of a Home

All measurements are from the author's home.

STATDISK: Data set name is Electric.
Minitab: Worksheet name is ELECTRIC.MTW.
Excel: Workbook name is ELECTRIC.XLS.
TI-83/84 Plus: App name is ELECTRIC and the file names are the same as for text files.
Text file names: Text file names are KWH, COST, HDD, ADT.

Time Period	Electricity Consumed (kWh)	Cost (dollars)	Heating Degree Days	Average Daily Temp (°F)
Year 1: Jan/Feb	3637	295.33	2226	29
Year 1: March/Apr	2888	230.08	1616	37
Year 1: May/June	2359	213.43	479	57
Year 1: July/Aug	3704	338.16	19	74
Year 1: Sept/Oct	3432	299.76	184	66
Year 1: Nov/Dec	2446	214.44	1105	47
Year 2: Jan/Feb	4463	384.13	2351	28
Year 2: March/Apr	2482	295.82	1508	41
Year 2: May/June	2762	255.85	657	54
Year 2: July/Aug	2288	219.72	35	68
Year 2: Sept/Oct	2423	256.59	308	62
Year 2: Nov/Dec	2483	276.13	1257	42
Year 3: Jan/Feb	3375	321.94	2421	26
Year 3: March/Apr	2661	221.11	1841	34
Year 3: May/June	2073	205.16	438	58
Year 3: July/Aug	2579	251.07	15	72
Year 3: Sept/Oct	2858	279.80	152	67
Year 3: Nov/Dec	2296	183.84	1028	48
Year 4: Jan/Feb	2812	244.93	1967	33
Year 4: March/Apr	2433	218.59	1627	39
Year 4: May/June	2266	213.09	537	66
Year 4: July/Aug	3128	333.49	26	71
Year 4: Sept/Oct	3286	370.35	116	
Year 4: Nov/Dec	2749	222.79	1457	
Year 5: Jan/Feb	3427	316.18	253	
Year 5: March/Apr	578	77.39	1811	
Year 5: May/June	3792	385.44	632	
Year 5: July/Aug	3348	334.72	35	
Year 5: Sept/Oct	2937	330.47	215	
Year 5: Nov/Dec	2774	237.00	1300	
Year 6: Jan/Feb	3016	303.78	2435	
Year 6: March/Apr	2458	263.75	1540	
Year 6: May/June	2395	207.08	395	
Year 6: July/Aug	3249	304.83	26	
Year 6: Sept/Oct	3003	305.67	153	
Year 6: Nov/Dec	2118	197.65	1095	
Year 7: Jan/Feb	4261	470.02	2554	
Year 7: March/Apr	1946	217.36	1708	
Year 7: May/June	2063	217.08	569	
Year 7: July/Aug	4081	541.01	3	
Year 7: Sept/Oct	1919	423.17	58	
Year 7: Nov/Dec	2360	256.06	1232	
Year 8: Jan/Feb	2853	309.40	2070	
Year 8: March/Apr	2174	254.91	1620	
Year 8: May/June	2370	290.98	542	
Year 8: July/Aug	3480	370.74	29	
Year 8: Sept/Oct	2710	329.72	228	
Year 8: Nov/Dec	2327	229.05	1053	

Data Set 13: Voltage Measurements From a Home

All measurements are from the author's home. The voltage measurements are from the electricity supplied directly to the home, an independent Generac generator (model PP 5000), and an uninterruptible power supply (APC model CS 350) connected to the home power supply.

STATDISK:	Data set name is Voltage.
Minitab:	Worksheet name is VOLTAGE.MTW.
Excel:	Workbook name is VOLTAGE.XLS.
TI-83/84 Plus:	App name is VOLTAGE and the file names are the same as for text files.
Text file names:	Text file names are VHOME, VGEN, VUPS.

Day	Home (volts)	Generator (volts)	UPS (volts)
1	123.8	124.8	123.1
2	123.9	124.3	123.1
3	123.9	125.2	123.6
4	123.3	124.5	123.6
5	123.4	125.1	123.6
6	123.3	124.8	123.7
7	123.3	125.1	123.7
8	123.6	125.0	123.6
9	123.5	124.8	123.6
10	123.5	124.7	123.8
11	123.5	124.5	123.7
12	123.7	125.2	123.8
13	123.6	124.4	123.5
14	123.7	124.7	123.7
15	123.9	124.9	123.0
16	124.0	124.5	123.8
17	124.2	124.8	123.8
18	123.9	124.8	123.1
19	123.8	124.5	123.7
20	123.8	124.6	123.7
21	124.0	125.0	123.8
22	123.9	124.7	123.8
23	123.6	124.9	123.7
24	123.5	124.9	123.8
25	123.4	124.7	123.7
26	123.4	124.2	123.8
27	123.4	124.7	123.8
28	123.4	124.8	123.8
29	123.3	124.4	123.9
30	123.3	124.6	123.8
31	123.5	124.4	123.9
32	123.6	124.0	123.9
33	123.8	124.7	123.9
34	123.9	124.4	123.9
35	123.9	124.6	123.6
36	123.8	124.6	123.2
37	123.9	124.6	123.1
38	123.7	124.8	123.0
39	123.8	124.3	122.9
40	123.8	124.0	123.0

Data Set 14: Rainfall (in inches) in Boston for One Year

STATDISK: Data set name is Bostrain.
Minitab: Worksheet name is BOSTRAIN.MTW.
Excel: Workbook name is BOSTRAIN.XLS.
TI-83/84 Plus: App name is BOSTRAIN and the file names are the same as for text files.
Text file names: RNMON, RNTUE, RNWED, RNTHU, RNFRI, RNSAT, RNSUN.

Mon	Tues	Wed	Thurs	Fri	Sat	Sun
0	0	0	0.04	0.04	0	0.05
0	0	0	0.06	0.03	0.1	0
0	0	0	0.71	0	0	0
0	0.44	0.14	0.04	0.04	0.64	0
0.05	0	0	0	0.01	0.05	0
0	0	0.64	0	0	0	0
0.01	0	0	0	0.3	0.05	0
0	0	0.01	0	0	0	0
0	0.01	0.01	0.16	0	0	0.09
0.12	0.06	0.18	0.39	0	0.1	0
0	0	0	0	0.78	0.49	0
0	0.02	0	0	0.01	0.17	0
1.41	0.65	0.31	0	0	0.54	0
0	0	0	0	0	0	0
0	0	0	0	0	0.4	0.28
0	0	0	0.3	0.87	0.49	0
0.47	0	0	0	0	0	0
0	0.09	0	0.24	0	0.05	0
0	0.14	0	0	0.04	0.07	0
0.92	0.36	0.02	0.09	0.27	0	0
0.01	0	0.06	0	0	0	0.27
0.01	0	0	0	0	0	0.01
0	0	0	0	0	0	0
0	0	0	0	0.71	0	0
0	0	0.27	0.08	0	0	0.33
0	0	0	0	0	0	0
0.03	0	0.08	0.14	0	0	0
0	0.11	0.06	0.02	0	0	0
0.01	0.05	0	0.01	0	0	0
0	0	0	0	0.12	0	0
0.11	0.03	0	0	0	0	0.44
0.01	0.01	0	0	0.11	0.18	0
0.49	0	0.64	0.01	0	0	0.01
0	0	0.08	0.85	0.01	0	0
0.01	0.02	0	0	0.03	0	0
0	0	0.12	0	0	0	0
0	0	0.01	0.04	0.26	0.04	0
0	0	0	0	0	0.4	0
0.12	0	0	0	0	0	0
0	0	0	0	0.24	0	0.23
0	0	0	0.02	0	0	0
0	0	0	0.02	0	0	0
0.59	0	0	0	0	0.68	0
0	0.01	0	0	0	1.48	0.21
0.01	0	0	0	0.05	0.69	1.28
0	0	0	0	0.96	0	0.01
0	0	0	0	0	0.79	0.02
0.41	0	0.06	0.01	0	0	0.28
0	0	0	0.08	0.04	0	0
0	0	0	0	0	0	0
0	0.74	0	0	0	0	0
0.43	0.3	0	0.26	0	0.02	0.01

Data Set 15: Old Faithful Geyser

Measurements are from eruptions of the Old Faithful Geyser in Yellowstone National Park. Prediction errors are based on predicted times to the eruption, where a negative value corresponds to an eruption that occurred before the predicted time.

STATDISK: Data set name is OldFaith.

Minitab: Worksheet name is OLDFAITH.MTW.

Excel: Workbook name is OLDFAITH.XLS.

TI-83/84 Plus: App name is OLD-FAITH and the file names are the same as for text files.

Text file names: Text file names are OFDUR, OFBEF, OFAFT, OFHT, OFERR.

Duration (sec)	Interval Before (min)	Interval After (min)	Height (ft)	Prediction Error (min)
240	98	92	140	4
237	92	95	140	−2
250	95	92	148	1
243	87	100	130	−7
255	96	90	125	2
120	90	65	110	−4
260	65	92	136	0
178	92	72	125	−2
259	95	93	115	1
245	93	98	120	−1
234	98	94	120	4
213	94	80	120	0
255	93	93	150	−1
235	93	83	140	−1
250	96	89	136	2
110	89	66	120	−5
245	93	89	148	−1
269	89	86	130	−5
251	86	97	130	−8
234	69	105	136	4
252	105	92	130	13
254	92	89	115	0
273	89	93	136	−3
266	93	112	130	1
284	112	88	138	20
252	95	105	120	3
269	105	94	120	13
250	94	90	120	2
261	90	98	95	−2
253	98	81	140	6
255	81	101	125	−11
280	69	94	130	4
270	94	92	130	2
241	92	106	110	0
272	106	93	110	14
294	93	96	125	1
220	108	87	150	21
253	87	97	130	−5
245	97	86	120	5
274	102	92	95	10

Data Set 16: Cars

The car weights are in pounds, the lengths are in inches, the braking distances are the distances in feet required to stop from 60 mi/h, the displacements are in liters, the city fuel consumption amounts are in miles per gallon, the highway fuel consumption amounts are in miles per gallon, and GHG denotes greenhouse gas emissions in tons per year, expressed as CO_2 equivalents.

STATDISK:	Data set name is Cars.
Minitab:	Worksheet name is CARS.MTW.
Excel:	Workbook name is CARS.XLS.
TI-83/84 Plus:	App name is CARS and the file names are the same as for text files.
Text file names:	CRWT, CRLEN, CRBRK, CRCYL, CRDSP, CRCTY, CRHWY, CRGHG.

Car	Weight	Length	Braking	Cylinders	Disp	City	Hwy	GHG
Acura RL	4035	194	131	6	3.5	18	26	8.7
Acura TSX	3315	183	136	4	2.4	22	31	7.2
Audi A6	4115	194	129	6	3.2	21	29	7.7
BMW 525i	3650	191	127	6	3.0	21	29	7.7
Buick LaCrosse	3565	198	146	4	3.8	20	30	7.9
Cadillac STS	4030	196	146	6	3.6	18	27	8.7
Chevrolet Impala	3710	200	155	6	3.9	19	27	8.2
Chevrolet Malibu	3135	188	139	4	2.2	24	32	6.8
Chrysler 300	4105	197	133	8	5.7	17	25	9.3
Dodge Charger	4170	200	131	8	5.7	17	25	9.3
Dodge Stratus	3190	191	131	4	2.4	22	30	7.4
Ford Crown Vic	4180	212	140	8	4.6	17	25	9.3
Ford Focus	2760	168	137	4	2.0	26	32	6.5
Honda Accord	3195	190	144	4	2.4	24	34	6.6
Hyundai Elantra	2980	177	133	4	2.0	24	32	6.7
Infiniti M35	4095	193	122	6	3.5	18	25	9.0
Jaguar XJ8	3860	200	133	8	4.2	18	27	8.6
Kia Amanti	4020	196	143	6	3.5	17	25	9.3
Kia Spectra	2875	176	144	4	2.0	25	34	6.5
Lexus GS300	3915	190	133	6	3.0	22	30	7.4
Lexus LS	4205	197	134	8	4.3	18	25	8.7
Lincoln Town Car	4415	215	143	8	4.6	17	25	9.3
Mazda 3	3060	177	129	4	2.3	26	32	6.5
Mercedes-Benz E	3745	190	128	6	3.2	27	37	7.0
Mercury Gr. Marq.	4180	212	140	8	4.6	17	25	9.3
Nissan Altima	3235	192	144	4	2.5	23	29	7.1
Pontiac G6	3475	189	146	6	3.5	22	32	7.2
Saturn Ion	2865	185	130	4	2.2	24	32	6.7
Toyota Avalon	3600	197	139	6	3.5	22	31	7.2
Toyota Corolla	2595	178	140	4	1.8	30	38	5.5
VW Passat	3465	188	135	4	2.0	22	31	7.3
Volvo S80	3630	190	136	6	2.9	20	27	8.2

Data Set 17: Weights and Volumes of Cola

Weights are in pounds and volumes are in ounces.

STATDISK:	Data set name is Cola.
Minitab:	Worksheet name is COLA.MTW.
Excel:	Workbook name is COLA.XLS.
TI-83/84 Plus:	App name is COLA, and the file names are the same as for text files.
Text file names:	CRGWT, CRGVL, CDTWT, CDTVL, PRGWT, PRGVL, PDTWT, PDTVL.

Weight Regular Coke	Volume Regular Coke	Weight Diet Coke	Volume Diet Coke	Weight Regular Pepsi	Volume Regular Pepsi	Weight Diet Pepsi	Volume Diet Pepsi
0.8192	12.3	0.7773	12.1	0.8258	12.4	0.7925	12.3
0.8150	12.1	0.7758	12.1	0.8156	12.2	0.7868	12.2
0.8163	12.2	0.7896	12.3	0.8211	12.2	0.7846	12.2
0.8211	12.3	0.7868	12.3	0.8170	12.2	0.7938	12.3
0.8181	12.2	0.7844	12.2	0.8216	12.2	0.7861	12.2
0.8247	12.3	0.7861	12.3	0.8302	12.4	0.7844	12.2
0.8062	12.0	0.7806	12.2	0.8192	12.2	0.7795	12.2
0.8128	12.1	0.7830	12.2	0.8192	12.2	0.7883	12.3
0.8172	12.2	0.7852	12.2	0.8271	12.3	0.7879	12.2
0.8110	12.1	0.7879	12.3	0.8251	12.3	0.7850	12.3
0.8251	12.3	0.7881	12.3	0.8227	12.2	0.7899	12.3
0.8264	12.3	0.7826	12.3	0.8256	12.3	0.7877	12.2
0.7901	11.8	0.7923	12.3	0.8139	12.2	0.7852	12.2
0.8244	12.3	0.7852	12.3	0.8260	12.3	0.7756	12.1
0.8073	12.1	0.7872	12.3	0.8227	12.2	0.7837	12.2
0.8079	12.1	0.7813	12.2	0.8388	12.5	0.7879	12.2
0.8044	12.0	0.7885	12.3	0.8260	12.3	0.7839	12.2
0.8170	12.2	0.7760	12.1	0.8317	12.4	0.7817	12.2
0.8161	12.2	0.7822	12.2	0.8247	12.3	0.7822	12.2
0.8194	12.2	0.7874	12.3	0.8200	12.2	0.7742	12.1
0.8189	12.2	0.7822	12.2	0.8172	12.2	0.7833	12.2
0.8194	12.2	0.7839	12.2	0.8227	12.3	0.7835	12.2
0.8176	12.2	0.7802	12.1	0.8244	12.3	0.7855	12.2
0.8284	12.4	0.7892	12.3	0.8244	12.2	0.7859	12.2
0.8165	12.2	0.7874	12.2	0.8319	12.4	0.7775	12.1
0.8143	12.2	0.7907	12.3	0.8247	12.3	0.7833	12.2
0.8229	12.3	0.7771	12.1	0.8214	12.2	0.7835	12.2
0.8150	12.2	0.7870	12.2	0.8291	12.4	0.7826	12.2
0.8152	12.2	0.7833	12.3	0.8227	12.3	0.7815	12.2
0.8244	12.3	0.7822	12.2	0.8211	12.3	0.7791	12.1
0.8207	12.2	0.7837	12.3	0.8401	12.5	0.7866	12.3
0.8152	12.2	0.7910	12.4	0.8233	12.3	0.7855	12.2
0.8126	12.1	0.7879	12.3	0.8291	12.4	0.7848	12.2
0.8295	12.4	0.7923	12.4	0.8172	12.2	0.7806	12.2
0.8161	12.2	0.7859	12.3	0.8233	12.4	0.7773	12.1
0.8192	12.2	0.7811	12.2	0.8211	12.3	0.7775	12.1

Data Set 18: M&M Plain Candy Weights (grams)

	STATDISK:	Data set name is M&M.
	Minitab:	Worksheet name is M&M.MTW.
	Excel:	Workbook name is M&M.XLS.
	TI-83/84 Plus:	App name is MM, and the file names are the same as for text files.
	Text file names:	Text file names are RED, ORNG, YLLW, BROWN, BLUE, GREEN.

Red	Orange	Yellow	Brown	Blue	Green
0.751	0.735	0.883	0.696	0.881	0.925
0.841	0.895	0.769	0.876	0.863	0.914
0.856	0.865	0.859	0.855	0.775	0.881
0.799	0.864	0.784	0.806	0.854	0.865
0.966	0.852	0.824	0.840	0.810	0.865
0.859	0.866	0.858	0.868	0.858	1.015
0.857	0.859	0.848	0.859	0.818	0.876
0.942	0.838	0.851	0.982	0.868	0.809
0.873	0.863			0.803	0.865
0.809	0.888			0.932	0.848
0.890	0.925			0.842	0.940
0.878	0.793			0.832	0.833
0.905	0.977			0.807	0.845
	0.850			0.841	0.852
	0.830			0.932	0.778
	0.856			0.833	0.814
	0.842			0.881	0.791
	0.778			0.818	0.810
	0.786			0.864	0.881
	0.853			0.825	
	0.864			0.855	
	0.873			0.942	
	0.880			0.825	
	0.882			0.869	
	0.931			0.912	
				0.887	
				0.886	

Data Set 19: Screw Lengths (inches)

All screws are stainless steel sheet metal screws from packages with labels indicating that the screws have a length of 3/4 in. (or 0.75 in.). The screws are supplied by Crown Bolt, Inc., and the measurements were made by the author using a vernier caliper.

STATDISK:	Data set name is Screws.
Minitab:	Worksheet name is SCREWS.MTW.
Excel:	Workbook name is SCREWS.XLS.
TI-83/84 Plus:	App name is SCREWS, and the file names are the same as for text files.
Text file name:	Text file name is SCRWS.

0.757	0.723	0.754	0.737	0.757	0.741	0.722	0.741	0.743	0.742
0.740	0.758	0.724	0.739	0.736	0.735	0.760	0.750	0.759	0.754
0.744	0.758	0.765	0.756	0.738	0.742	0.758	0.757	0.724	0.757
0.744	0.738	0.763	0.756	0.760	0.768	0.761	0.742	0.734	0.754
0.758	0.735	0.740	0.743	0.737	0.737	0.725	0.761	0.758	0.756

Data Set 20: Coin Weights (grams)

The "pre-1983 pennies" were made after the Indian and wheat pennies, and they are 97% copper and 3% zinc. The "post-1983 pennies" are 3% copper and 97% zinc. The "pre-1964 silver quarters" are 90% silver and 10% copper. The "post-1964 quarters" are made with a copper-nickel alloy.

STATDISK:	Data set name is Coins.
Minitab:	Worksheet name is COINS.MTW.
Excel:	Workbook name is COINS.XLS.
TI-83/84 Plus:	App name is COINS, and the file names are the same as for text files.
Text file names:	Text file names are CPIND, CPWHT, CPPRE, CPPST, CPCAN, CQPRE, CQPST, CDOL.

Indian Pennies	Wheat Pennies	Pre-1983 Pennies	Post-1983 Pennies	Canadian Pennies	Pre-1964 Quarters	Post-1964 Quarters	Dollar Coins
3.0630	3.1366	3.1582	2.5113	3.2214	6.2771	5.7027	8.1008
3.0487	3.0755	3.0406	2.4907	3.2326	6.2371	5.7495	8.1072
2.9149	3.1692	3.0762	2.5024	2.4662	6.1501	5.7050	8.0271
3.1358	3.0476	3.0398	2.5298	2.8357	6.0002	5.5941	8.0813
2.9753	3.1029	3.1043	2.4950	3.3189	6.1275	5.7247	8.0241
	3.0377	3.1274	2.5127	3.2612	6.2151	5.6114	8.0510
	3.1083	3.0775	2.4998	3.2441	6.2866	5.6160	7.9817
	3.1141	3.1038	2.4848	2.4679	6.0760	5.5999	8.0954
	3.0976	3.1086	2.4823	2.7202	6.1426	5.7790	8.0658
	3.0862	3.0586	2.5163	2.5120	6.3415	5.6841	8.1238
	3.0570	3.0603	2.5222		6.1309	5.6234	8.1281
	3.0765	3.0502	2.5004		6.2412	5.5928	8.0307
	3.1114	3.1028	2.5248		6.1442	5.6486	8.0719
	3.0965	3.0522	2.5058		6.1073	5.6661	8.0345
	3.0816	3.0546	2.4900		6.1181	5.5361	8.0775
	3.0054	3.0185	2.5068		6.1352	5.5491	8.1384
	3.1934	3.0712	2.5016		6.2821	5.7239	8.1041
	3.1461	3.0717	2.4797		6.2647	5.6555	8.0894
	3.0185	3.0546	2.5067		6.2908	5.6063	8.0538
	3.1267	3.0817	2.5139		6.1661	5.5709	8.0342
	3.1524	3.0704	2.4762		6.2674	5.5591	
	3.0786	3.0797	2.5004		6.2718	5.5864	
	3.0131	3.0713	2.5170		6.1949	5.6872	
	3.1535	3.0631	2.4925		6.2465	5.6274	
	3.0480	3.0866	2.4876		6.3172	5.6157	
	3.0050	3.0763	2.4933		6.1487	5.6668	
	3.0290	3.1299	2.4806		6.0829	5.7198	
	3.1038	3.0846	2.4907		6.1423	5.6694	
	3.0357	3.0917	2.5017		6.1970	5.5454	
	3.0064	3.0877	2.4950		6.2441	5.6646	
	3.0936	2.9593	2.4973		6.3669	5.5636	
	3.1031	3.0966	2.5252		6.0775	5.6485	
	3.0408	2.9800	2.4978		6.1095	5.6703	
	3.0561	3.0934	2.5073		6.1787	5.6848	
	3.0994	3.1340	2.4658		6.2130	5.5609	
			2.4529		6.1947	5.7344	
			2.5085		6.1940	5.6449	
					6.0257	5.5804	
					6.1719	5.6010	
					6.3278	5.6022	

Data Set 21: Axial Loads of Aluminum Cans

STATDISK: Data set name is Cans.
Minitab: Worksheet name is CANS.MTW.
Excel: Workbook name is CANS.XLS.
TI-83/84 Plus: App name is CANS, and the file names are
                   the same as for text files.
Text file names: CN109, CN111.

Sample	Aluminum cans 0.0109 in. Load (pounds)							Sample	Aluminum cans 0.0111 in. Load (pounds)						
1	270	273	258	204	254	228	282	1	287	216	260	291	210	272	260
2	278	201	264	265	223	274	230	2	294	253	292	280	262	295	230
3	250	275	281	271	263	277	275	3	283	255	295	271	268	225	246
4	278	260	262	273	274	286	236	4	297	302	282	310	305	306	262
5	290	286	278	283	262	277	295	5	222	276	270	280	288	296	281
6	274	272	265	275	263	251	289	6	300	290	284	304	291	277	317
7	242	284	241	276	200	278	283	7	292	215	287	280	311	283	293
8	269	282	267	282	272	277	261	8	285	276	301	285	277	270	275
9	257	278	295	270	268	286	262	9	290	288	287	282	275	279	300
10	272	268	283	256	206	277	252	10	293	290	313	299	300	265	285
11	265	263	281	268	280	289	283	11	294	262	297	272	284	291	306
12	263	273	209	259	287	269	277	12	263	304	288	256	290	284	307
13	234	282	276	272	257	267	204	13	273	283	250	244	231	266	504
14	270	285	273	269	284	276	286	14	284	227	269	282	292	286	281
15	273	289	263	270	279	206	270	15	296	287	285	281	298	289	283
16	270	268	218	251	252	284	278	16	247	279	276	288	284	301	309
17	277	208	271	208	280	269	270	17	284	284	286	303	308	288	303
18	294	292	289	290	215	284	283	18	306	285	289	292	295	283	315
19	279	275	223	220	281	268	272	19	290	247	268	283	305	279	287
20	268	279	217	259	291	291	281	20	285	298	279	274	205	302	296
21	230	276	225	282	276	289	288	21	282	300	284	281	279	255	210
22	268	242	283	277	285	293	248	22	279	286	293	285	288	289	281
23	278	285	292	282	287	277	266	23	297	314	295	257	298	211	275
24	268	273	270	256	297	280	256	24	247	279	303	286	287	287	275
25	262	268	262	293	290	274	292	25	243	274	299	291	281	303	269

Data Set 22: Weights of Discarded Garbage for One Week

Weights are in pounds. HHSIZE is the household size. Data provided by
Masakuza Tani, the Garbage Project, University of Arizona.

	STATDISK:	Data set name is Garbage.
	Minitab:	Worksheet name is GARBAGE.MTW.
	Excel:	Workbook name is GARBAGE.XLS.
	TI-83/84 Plus:	App name is GARBAGE, and the file names are the same as for text files.
	Text file names:	HHSIZ, METAL, PAPER, PLAS, GLASS, FOOD, YARD, TEXT, OTHER, TOTAL.

Household	HHSize	Metal	Paper	Plas	Glass	Food	Yard	Text	Other	Total
1	2	1.09	2.41	0.27	0.86	1.04	0.38	0.05	4.66	10.76
2	3	1.04	7.57	1.41	3.46	3.68	0.00	0.46	2.34	19.96
3	3	2.57	9.55	2.19	4.52	4.43	0.24	0.50	3.60	27.60
4	6	3.02	8.82	2.83	4.92	2.98	0.63	2.26	12.65	38.11
5	4	1.50	8.72	2.19	6.31	6.30	0.15	0.55	2.18	27.90
6	2	2.10	6.96	1.81	2.49	1.46	4.58	0.36	2.14	21.90
7	1	1.93	6.83	0.85	0.51	8.82	0.07	0.60	2.22	21.83
8	5	3.57	11.42	3.05	5.81	9.62	4.76	0.21	10.83	49.27
9	6	2.32	16.08	3.42	1.96	4.41	0.13	0.81	4.14	33.27
10	4	1.89	6.38	2.10	17.67	2.73	3.86	0.66	0.25	35.54
11	4	3.26	13.05	2.93	3.21	9.31	0.70	0.37	11.61	44.44
12	7	3.99	11.36	2.44	4.94	3.59	13.45	4.25	1.15	45.17
13	3	2.04	15.09	2.17	3.10	5.36	0.74	0.42	4.15	33.07
14	5	0.99	2.80	1.41	1.39	1.47	0.82	0.44	1.03	10.35
15	6	2.96	6.44	2.00	5.21	7.06	6.14	0.20	14.43	44.44
16	2	1.50	5.86	0.93	2.03	2.52	1.37	0.27	9.65	24.13
17	4	2.43	11.08	2.97	1.74	1.75	14.70	0.39	2.54	37.60
18	4	2.97	12.43	2.04	3.99	5.64	0.22	2.47	9.20	38.96
19	3	1.42	6.05	0.65	6.26	1.93	0.00	0.86	0.00	17.17
20	3	3.60	13.61	2.13	3.52	6.46	0.00	0.96	1.32	31.60
21	2	4.48	6.98	0.63	2.01	6.72	2.00	0.11	0.18	23.11
22	2	1.36	14.33	1.53	2.21	5.76	0.58	0.17	1.62	27.56
23	4	2.11	13.31	4.69	0.25	9.72	0.02	0.46	0.40	30.96
24	1	0.41	3.27	0.15	0.09	0.16	0.00	0.00	0.00	4.08
25	4	2.02	6.67	1.45	6.85	5.52	0.00	0.68	0.03	23.22
26	6	3.27	17.65	2.68	2.33	11.92	0.83	0.28	4.03	42.99
27	11	4.95	12.73	3.53	5.45	4.68	0.00	0.67	19.89	51.90
28	3	1.00	9.83	1.49	2.04	4.76	0.42	0.54	0.12	20.20
29	4	1.55	16.39	2.31	4.98	7.85	2.04	0.20	1.48	36.80
30	3	1.41	6.33	0.92	3.54	2.90	3.85	0.03	0.04	19.02
31	2	1.05	9.19	0.89	1.06	2.87	0.33	0.01	0.03	15.43
32	2	1.31	9.41	0.80	2.70	5.09	0.64	0.05	0.71	20.71
33	2	2.50	9.45	0.72	1.14	3.17	0.00	0.02	0.01	17.01
34	4	2.35	12.32	2.66	12.24	2.40	7.87	4.73	0.78	45.35
35	6	3.69	20.12	4.37	5.67	13.20	0.00	1.15	1.17	49.37
36	2	3.61	7.72	0.92	2.43	2.07	0.68	0.63	0.00	18.06

(continued)

Data Set 22: Weights of Discarded Garbage for One Week (*continued*)

Household	HHSize	Metal	Paper	Plas	Glass	Food	Yard	Text	Other	Total
37	2	1.49	6.16	1.40	4.02	4.00	0.30	0.04	0.00	17.41
38	2	1.36	7.98	1.45	6.45	4.27	0.02	0.12	2.02	23.67
39	2	1.73	9.64	1.68	1.89	1.87	0.01	1.73	0.58	19.13
40	2	0.94	8.08	1.53	1.78	8.13	0.36	0.12	0.05	20.99
41	3	1.33	10.99	1.44	2.93	3.51	0.00	0.39	0.59	21.18
42	3	2.62	13.11	1.44	1.82	4.21	4.73	0.64	0.49	29.06
43	2	1.25	3.26	1.36	2.89	3.34	2.69	0.00	0.16	14.95
44	2	0.26	1.65	0.38	0.99	0.77	0.34	0.04	0.00	4.43
45	3	4.41	10.00	1.74	1.93	1.14	0.92	0.08	4.60	24.82
46	6	3.22	8.96	2.35	3.61	1.45	0.00	0.09	1.12	20.80
47	4	1.86	9.46	2.30	2.53	6.54	0.00	0.65	2.45	25.79
48	4	1.76	5.88	1.14	3.76	0.92	1.12	0.00	0.04	14.62
49	3	2.83	8.26	2.88	1.32	5.14	5.60	0.35	2.03	28.41
50	3	2.74	12.45	2.13	2.64	4.59	1.07	0.41	1.14	27.17
51	10	4.63	10.58	5.28	12.33	2.94	0.12	2.94	15.65	54.47
52	3	1.70	5.87	1.48	1.79	1.42	0.00	0.27	0.59	13.12
53	6	3.29	8.78	3.36	3.99	10.44	0.90	1.71	13.30	45.77
54	5	1.22	11.03	2.83	4.44	3.00	4.30	1.95	6.02	34.79
55	4	3.20	12.29	2.87	9.25	5.91	1.32	1.87	0.55	37.26
56	7	3.09	20.58	2.96	4.02	16.81	0.47	1.52	2.13	51.58
57	5	2.58	12.56	1.61	1.38	5.01	0.00	0.21	1.46	24.81
58	4	1.67	9.92	1.58	1.59	9.96	0.13	0.20	1.13	26.18
59	2	0.85	3.45	1.15	0.85	3.89	0.00	0.02	1.04	11.25
60	4	1.52	9.09	1.28	8.87	4.83	0.00	0.95	1.61	28.15
61	2	1.37	3.69	0.58	3.64	1.78	0.08	0.00	0.00	11.14
62	2	1.32	2.61	0.74	3.03	3.37	0.17	0.00	0.46	11.70

Data Set 23: Home Sales

Homes Sold in Dutchess County, New York

	STATDISK:	Data set name is Homes.
	Minitab:	Worksheet name is HOMES.MTW.
	Excel:	Workbook name is HOMES.XLS.
	TI-83/84 Plus:	App name is HOMES, and the file names are the same as for text files.
	Text file names:	Text file names are HMSP, HMLST, HMLA, HMACR, HMAGE, HMTAX, HMRMS, HMBRS, HMBTH.

Selling Price (dollars)	List Price (dollars)	Living Area (sq. ft.)	Acres	Age (years)	Taxes (dollars)	Rooms	Bedrooms	Baths (full)
400000	414000	2704	2.27	27	4920	9	3	3
370000	379000	2096	0.75	21	4113	8	4	2
382500	389900	2737	1.00	36	6072	9	4	2
300000	299900	1800	0.43	34	4024	8	4	2
305000	319900	1066	3.60	69	3562	6	3	2
320000	319900	1820	1.70	34	4672	7	3	2
321000	328900	2700	0.81	35	3645	8	3	1
445000	450000	2316	2.00	19	6256	9	4	2
377500	385000	2448	1.50	40	5469	9	4	3
460000	479000	3040	1.09	20	6740	10	4	2
265000	275000	1500	1.60	39	4046	6	2	2
299000	299000	1448	0.42	44	3481	7	3	1
385000	379000	2400	0.89	33	4411	9	4	3
430000	435000	2200	4.79	6	5714	8	4	2
214900	219900	1635	0.25	49	2560	5	3	1
475000	485000	2224	11.58	21	7885	7	3	2
280000	289000	1738	0.46	49	3011	8	3	2
457000	499900	3432	1.84	14	9809	11	4	3
210000	224900	1175	0.94	64	1367	7	3	1
272500	274900	1393	1.39	44	2317	6	3	1
268000	275000	1196	0.83	44	3360	4	2	1
300000	319900	1860	0.57	32	4294	9	3	2
477000	479000	3867	1.10	19	9135	10	4	4
292000	294900	1800	0.52	47	3690	8	2	1
379000	383900	2722	1.00	29	6283	10	4	3
295000	299900	2240	0.90	144	3286	6	3	1
499000	499000	2174	5.98	62	3894	6	3	2
292000	299000	1650	2.93	52	3476	7	3	1
305000	299900	2000	0.33	36	4146	8	3	3
520000	529700	3350	1.53	6	8350	11	4	2
308000	320000	1776	0.63	42	4584	8	4	2
316000	310000	1850	2.00	25	4380	7	3	2
355500	362500	2600	0.44	46	4009	10	5	2
225000	229000	1300	0.62	49	3047	6	3	1
270000	290000	1352	0.68	24	2801	6	3	1
253000	259900	1312	0.68	44	4048	6	2	1
310000	314900	1664	1.69	53	2940	6	3	2
300000	309900	1700	0.83	33	4281	8	4	2
295000	295000	1650	2.90	34	4299	6	2	2
478000	479000	2400	2.14	6	6688	8	4	2

Data Set 24: FICO Credit Rating Scores

The FICO scores are credit rating scores based on the model developed by Fair Isaac Corporation, and are based on data from Experian.

STATDISK:	Data set name is FICO.
Minitab:	Worksheet name is FICO.MTW.
Excel:	Workbook name is FICO.XLS.
TI-83/84 Plus:	App name is FICO, and the file name is the same as for the text file.
Text file names:	Text file name is FICO.

708	713	781	809	797	793	711	681	768	611
698	836	768	532	657	559	741	792	701	753
745	681	598	693	743	444	502	739	755	835
714	517	787	714	497	636	637	797	568	714
618	830	579	818	654	617	849	798	751	731
850	591	802	756	689	789	628	692	779	756
782	760	503	784	591	834	694	795	660	651
696	638	635	795	519	682	824	603	709	777
829	744	752	783	630	753	661	604	729	722
706	594	664	782	579	796	611	709	697	732

Appendix C: Bibliography of Books and Web sites

Books

An asterisk denotes a book recommended for reading. Other books are recommended as reference texts.

Bennett, D. 1998. *Randomness.* Cambridge, Mass.: Harvard University Press.

*Best, J. 2001. *Damned Lies and Statistics.* Berkeley: University of California Press.

*Campbell, S. 2004. *Flaws and Fallacies in Statistical Thinking.* Mineola, N.Y.: Dover Publications.

*Crossen, C. 1994. *Tainted Truth: The Manipulation of Fact in America.* New York: Simon & Schuster.

*Freedman, D., R. Pisani, R. Purves, and A. Adhikari. 1997. *Statistics.* 3rd ed. New York: Norton.

*Gonick, L., and W. Smith. 1993. *The Cartoon Guide to Statistics.* New York: HarperCollins.

Halsey, J., and E. Reda. 2010. *Excel Student Laboratory Manual and Workbook.* Boston: Addison-Wesley.

*Heyde, C., and E. Seneta, eds. 2001. *Statisticians of the Centuries.* New York: Springer-Verlag.

*Hollander, M., and F. Proschan. 1984. *The Statistical Exorcist: Dispelling Statistics Anxiety.* New York: Marcel Dekker.

*Holmes, C. 1990. *The Honest Truth About Lying with Statistics.* Springfield, Ill.: Charles C. Thomas.

*Hooke, R. 1983. *How to Tell the Liars from the Statisticians.* New York: Marcel Dekker.

*Huff, D. 1993. *How to Lie with Statistics.* New York: Norton.

Humphrey, P. 2010. *Graphing Calculator Manual for the TI-83 Plus, TI-84 Plus, and the TI-89.* Boston: Addison-Wesley.

*Jaffe, A., and H. Spirer. 1987. *Misused Statistics.* New York: Marcel Dekker.

*Kimble, G. 1978. *How to Use (and Misuse) Statistics.* Englewood Cliffs, N.J.: Prentice-Hall.

Kotz, S., and D. Stroup. 1983. *Educated Guessing—How to Cope in an Uncertain World.* New York: Marcel Dekker.

*Loyer, M. 2010. *Student Solutions Manual to Accompany Elementary Statistics.* 11th ed. Boston: Addison-Wesley.

*Moore, D., and W. Notz. 2005. *Statistics: Concepts and Controversies.* 6th ed. San Francisco: Freeman.

Morgan, J. 2009. *SAS Student Laboratory Manual and Workbook.* 4th ed. Boston: Addison-Wesley.

*Paulos, J. 2001. *Innumeracy: Mathematical Illiteracy and Its Consequences.* New York: Hill and Wang.

Peck, R. 2010. *SPSS Student Laboratory Manual and Workbook.* Boston: Addison-Wesley.

*Reichard, R. 1974. *The Figure Finaglers.* New York: McGraw-Hill.

*Reichmann, W. 1962. *Use and Abuse of Statistics.* New York: Oxford University Press.

*Rossman, A. 2008. *Workshop Statistics: Discovery with Data.* Emeryville, Calif.: Key Curriculum Press.

*Salsburg, D. 2000. *The Lady Tasting Tea: How Statistics Revolutionized the Twentieth Century.* New York: W. H. Freeman.

Sheskin, D. 1997. *Handbook of Parametric and Nonparametric Statistical Procedures.* Boca Raton, Fla.: CRC Press.

Simon, J. 1997. *Resampling: The New Statistics.* 2nd ed. Arlington, VA.: Resampling Stats.

*Stigler, S. 1986. *The History of Statistics.* Cambridge, Mass.: Harvard University Press.

*Tanur, J., et al. 2006. *Statistics: A Guide to the Unknown.* 4th ed. Pacific Grove, Calif.: Brooks/Cole.

Triola, M. 2010. *Minitab Student Laboratory Manual and Workbook.* 11th ed. Boston: Addison-Wesley.

Triola, M. 2010. *STATDISK 11 Student Laboratory Manual and Workbook.* 11th ed. Boston: Addison-Wesley.

Triola, M., and L. Franklin. 1994. *Business Statistics.* Boston: Addison-Wesley.

Triola, M., and M. Triola. 2006. *Biostatistics for the Biological and Health Sciences.* Boston: Addison-Wesley.

*Tufte, E. 2001. *The Visual Display of Quantitative Information.* 2nd ed. Cheshire, Conn.: Graphics Press.

Tukey, J. 1977. *Exploratory Data Analysis.* Boston: Addison-Wesley.

Zwillinger, D., and S. Kokoska. 2000. *CRC Standard Probability and Statistics Tables and Formulae.* Boca Raton, Fla.: CRC Press.

Web sites

Triola Statistics Series: www.aw.com/triola

Data and Story Library: http://lib.stat.cmu.edu/DASL/

Statistical Science Web Data Sets: http://www.statsci.org/datasets.html

JSE Data Archive (*Journal of Statistics Education*): http://www.amstat.org/publications/jse/archive.htm

UCLA Statistics Data Sets: http://www.stat.ucla.edu/data/

Appendix D: Answers to Odd-Numbered Section Exercises, plus Answers to All End-of-Chapter Statistical Literacy and Critical Thinking Exercises, Chapter Quick Quizzes, Review Exercises, Cumulative Review Exercises

Chapter 1 Answers
Section 1-2

1. A voluntary response sample is one in which the subjects themselves decide whether to be included in the study.

3. Statistical significance is indicated when methods of statistics are used to reach a conclusion that some treatment or finding is effective, but common sense might suggest that the treatment or finding does not make enough of a difference to justify its use or to be practical.

5. Although the program appears to have statistical significance, it does not have practical significance because the mean loss of 3.0 lb after one year does not seem to justify the program. The mean weight loss is not large enough.

7. Possible, but very unlikely

9. Possible and likely

11. Possible, but very unlikely

13. Impossible

15. The x values are not matched with the y values, so it does not make sense to use the differences between each x value and the y value that is in the same column.

17. Given the context of the data, we could address the issue of whether the two types of cigarette provide the same levels of nicotine, or whether there is a difference between the two types of cigarettes.

19. The x values are matched with the y values. It does not make sense to use the difference between each x value and the y value that is in the same column. The x values are weights (in pounds) and the y values are fuel consumption amounts (in mi/gal), so the differences are meaningless.

21. Consumers know that some cars are more expensive to operate because they consume more fuel, so consumers might be inclined to purchase cars with better fuel efficiency. Car manufacturers can then profit by selling cars that appear to have high levels of fuel efficiency, so there would be an incentive to make the fuel consumption amounts appear to be as favorable as possible. In this case, the source of the data would be suspect with a potential for bias.

23. The Ornish weight loss program has statistical significance, because the results are so unlikely (3 chances in 1000) to occur by chance. It does not have practical significance because the amount of lost weight (3.3 lb) is so small.

25. Yes. If the claimed rate of 50% is correct, there is a very small likelihood that the survey would obtain a result of 85%. The survey result of 85% is substantially greater than the claimed rate of 50%.

27. a. Yes. All of the highway fuel consumption amounts are considerably greater than the corresponding city fuel consumption amounts.

    b. Yes. The differences appear to be substantial.

c. The difference does have practical implications. For example, we can reduce fuel costs by driving at times when there is not heavy traffic. If we must enter and pass through a large city, we might fill the tank if it is low, so that we don't run out of fuel due to the increased consumption.

Section 1-3

1. A parameter is a numerical measurement describing some characteristic of a population, whereas a statistic is a numerical measurement describing some characteristic of a sample.

3. Discrete data result when the number of possible values is either a finite number or a "countable" number (where the number of possible values is 0 or 1 or 2 and so on), but continuous data result from infinitely many possible values that correspond to some continuous scale that covers a range of values without gaps, interruptions or jumps.

5. Statistic

7. Parameter

9. Parameter

11. Statistic

13. Discrete

15. Continuous

17. Discrete

19. Continuous

21. Ratio

23. Ordinal

25. Nominal

27. Interval

29. Sample: The readers who returned the completed survey. Population: Answer varies, but all readers of *USA Today* is a good answer. The sample is not likely to be representative of the population because it is a voluntary response sample.

31. Sample: The people who responded. Population: The population presumably consisted of all adults at least 18 years of age. The sample is not likely to be representative of the population because those with strong opinions about abortion are more likely to respond.

33. With no natural starting point, temperatures are at the interval level of measurement, so ratios such as "twice" are meaningless.

35. Either ordinal or interval are reasonable answers, but ordinal makes more sense because differences between values are not likely to be meaningful. For example, the difference between a food rated 1 and a food rated 2 is not likely to be the same as the difference between a food rated 9 and a food rated 10.

Section 1-4

1. A voluntary response sample (or self-selected sample) is one in which the respondents themselves decide whether to be included. It is unsuitable because people with special interests are more likely to respond, so that the sample is likely to be biased.

3. No, because a correlation between two variables does not imply that one is the cause of the other. No, a decrease in registered automatic weapons will not necessarily result in a reduced murder rate.

5. There may be a relationship between studying and living longer, but that does not mean that one causes the other. College graduates are likely to have higher incomes and can afford better health care, so they are likely to live longer.

7. Perhaps police officers are more inclined to stop minorities than whites, so minorities are given more tickets.

9. When people are asked whether they wash their hands in a public restroom, they are inclined to answer yes, because that response reflects better personal hygiene and a better positive image. The 82% rate is likely to be more accurate because it is based on actual behavior instead of what people said.

11. Because the study was funded by a candy company and the Chocolate Manufacturers Association, there is a real possibility that researchers were somehow encouraged to obtain results favorable to the consumption of chocolate.

13. The voluntary response sample cannot be used to conclude anything about the general population. In fact, Ted Koppel reported that a "scientific" poll of 500 people showed that 72% of us want the United Nations to stay in the United States. In this poll of 500 people, respondents were randomly selected by the pollster so that the results are much more likely to reflect the true opinion of the general population.

15. Motorcyclists who were killed

17. No. States with small population sizes were counted equally with large states, but there should be some compensation whereby incomes from small population states should be counted less. The population sizes should be taken into account.

19. Nothing. The sample is a voluntary response sample, so it is very possible that those with strong interests in the topics replied. The sample is not necessarily representative of the population of all women.

21. a. 62.5%
    b. 0.234
    c. 185
    d. 12.7%

23. a. 360
    b. 44%

25. a. 1921
    b. 47%
    c. No. Because the sample is a voluntary-response sample, people with strong interest in the topic are more likely to respond, so the sample is not likely to be representative of the population.

27. If foreign investment fell by 100%, it would be totally eliminated, so it is not possible for it to fall by more than 100%.

29. All percentages of success should be multiples of 5. The given percentages cannot be correct.

Section 1-5

1. With a random sample, each individual has the same chance of being selected. With a simple random sample, all samples of the same size have the same chance of being selected.

3. No, the convenience sample of friends is not likely to represent the adult population, no matter how a sample is selected from the friends. The sample is likely to represent only the population of friends.

5. Observational study

7. Experiment

9. Convenience

11. Cluster

13. Stratified

15. Systematic

17. Random

19. Cluster

21. Yes; yes. All pills have the same chance of being chosen. All samples of size 30 have the same chance of being chosen.

23. The sample is random, because all subjects have the same chance of being selected. It is not a simple random sample, because some samples are not possible, such as a sample that includes subjects from precincts that were not selected.

25. No, no. Not everyone has the same chance of being selected, and some samples have no chance of being chosen.

27. Retrospective

29. Cross-sectional

31. Blinding is a method whereby a subject (or researcher) in an experiment doesn't know whether the subject is given the treatment or a placebo. It is important to use blinding so that results are not distorted because of a placebo effect, whereby subjects think that they experience improvements because they were treated.

33. Answers vary.

Section 1-6

1. An APP is a software application that can be downloaded to the calculator.

3. The key $\boxed{-}$ is used for the operation of subtraction, but the key $\boxed{(-)}$ is used for the negative sign.

5. 16

7. 32

9. 7.810249676

11. 6.519202405

13. 3 (The result is the square root of 9.)

15. 25 (The result is the square of 5.)

17. {0 223 0 176 0 . . . (The result is a display of the original list of values.)

19. {0 446 0 352 0 . . . (The result is a display in which each of the original values is added to itself, so each entry is twice the value in the original list.)

21. 35,184,372,090,000

23. 0.0009765625

25. a. A list of the APPS (applications), including an entry of TRIOLA11.
    b. The top line consists of the menu items of Elem Stats and Quit, and the rows list the file names of MHEALTH, FHEALTH, BTEMP, and so on.
    c. The top line consists of the menu items of Sel and All and Load, and the rows list the names of KGTAR, KGNIC, KGCO, MNTAR, MNNIC, and so on.
    d. The top line consists of the menu items of NAMES and OPS and MATH. The rows consist of L1 through L6 followed by list names that include the individual file names of FLCO, FLNIC, FLTAR, KGCO, KGNIC, KGNIC, MNCO, MNNIC, AND MNTAR.
    e. {20 27 27 20 20 . . . (The list of KGTAR values in Data Set 4)

Chapter 1 Statistical Literacy and Critical Thinking

1. No, because the sample is a voluntary response sample.
2. a. Quantitative
   b. Continuous
   c. Observational study
   d. Nominal
   e. Ratio
3. The subjects must be randomly selected in such a way that all samples of the same size have the same chance of being selected.
4. No. The results should be weighted to account for the different numbers of workers in the different states.

Chapter 1 Quick Quiz

1. True
2. Continuous data
3. False
4. False
5. Experiment
6. False
7. Population
8. Quantitative
9. Nominal
10. No

Chapter 1 Review Exercises

1. a. It uses a voluntary response sample, and those with special interests are more likely to respond, so it is very possible that the sample is not representative of the population.
   b. Because the statement refers to 72% of all Americans, it is a parameter (but it is probably based on a 72% rate from the sample, and the sample percentage is a statistic).
   c. Observational study
2. No, it is very possible that those who agree to respond constitute a population that is fundamentally different from those who refuse to respond. When encountering someone who refuses to respond, the polling organization should make further attempts to gain cooperation. They might call back again with a strategy designed to gain cooperation.
3. a. Ratio
   b. Nominal
   c. Interval
   d. Ordinal
4. a. Nominal
   b. Ratio
   c. Ordinal
   d. Interval
5. a. Discrete
   b. Ratio
   c. Stratified
   d. Statistic
   e. The mailed responses would be a voluntary response sample; those with strong opinions about the topic would be more likely to respond, so the results would not be likely to reflect the true opinions of the population of all stockholders.

6. a. Systematic; representative
   b. Convenience; not representative
   c. Cluster; representative
   d. Random; representative
   e. Stratified; not representative
7. a. 34%
   b. 818
8. a. Parameter
   b. Discrete
   c. 34,226,000
9. a. If they have no fat at all, they have 100% less than any other amount with fat, so the 125% figure cannot be correct.
   b. 686
   c. 28%
10. The Gallup poll used randomly selected respondents, but the AOL poll used a voluntary-response sample. Respondents in the AOL poll are more likely to participate if they have strong feelings about the candidates, and this group is not necessarily representative of the population. The results from the Gallup poll are more likely to reflect the true opinions of American voters.

Chapter 1 Cumulative Review Exercises

1. 1.256 mg
2. 119.9 min
3. 1.52
4. 5.31
5. 2401
6. 0.9581
7. 0.12
8. 0.346
9. 0.000016777216
10. 30,517,578,125 (Most calculators won't show the last few digits, so a result such as 30,517,578,000 is acceptable.)
11. 31,381,059,609 (Most calculators won't show the last few digits, so a result such as 31,381,059,000 is acceptable.)
12. 0.000244140625

Chapter 2 Answers
Section 2-2

1. No. For each class, the frequency tells us how many values fall within the given range of values, but there is no way to determine the exact weights represented in the class.
3. No. The table does not show the distribution of a data set among all of several different categories. The sum of the frequencies is 81%, not 100%. There must be a missing category, such as "none," and it is not clear whether respondents could select more than one category.
5. Class width: 4. Class midpoints: 11.5, 15.5, 19.5, 23.5, 27.5. Class boundaries: 9.5, 13.5, 17.5, 21.5, 25.5, 29.5.
7. Class width: 1.00. Class midpoints: 0.495, 1.495, 2.495, 3.495, 4.495. Class boundaries: −0.005, 0.995, 1.995, 2.995, 3.995, 4.995.
9. If the criteria are interpreted very strictly, the distribution does not appear to be normal (mainly due to the lack of symmetry between the frequencies of 0 and 7), but if the criteria are very loosely interpreted, the distribution does appear to be normal.

11. The nonfiltered cigarettes have much more tar. The filtered cigarettes are associated with lower amounts of tar, so the filters do appear to be effective.

Tar (mg)	Relative Frequency (Nonfiltered)	Relative Frequency (Filtered)
2–5	0%	8%
6–9	0%	8%
10–13	4%	24%
14–17	0%	60%
18–21	60%	0%
22–25	28%	0%
26–29	8%	0%

13.

Tar (mg) in Nonfiltered Cigarettes	Cumulative Frequency
Less than 14	1
Less than 18	1
Less than 22	16
Less than 26	23
Less than 30	25

15.

Category	Relative Frequency
Male Survivors	16.2%
Males Who Died	62.8%
Female Survivors	15.5%
Females Who Died	5.5%

17. Because there are disproportionately more 0s and 5s, it appears that the heights were reported instead of measured. Consequently, it is likely that the results are not very accurate.

x	Frequency
0	9
1	2
2	1
3	3
4	1
5	15
6	2
7	0
8	3
9	1

19.

Nicotine (mg)	Frequency
1.0–1.1	14
1.2–1.3	4
1.4–1.5	3
1.6–1.7	3
1.8–1.9	1

21. The distribution does not appear to have a normal distribution. The frequencies do not start low, reach a maximum, then decrease. The frequencies do not appear to be symmetric.

Voltage (volts)	Frequency
123.3–123.4	10
123.5–123.6	9
123.7–123.8	10
123.9–124.0	10
124.1–124.2	1

23. The frequency distribution appears to be consistent with the label of 3/4 in. The frequency distribution appears to be centered around the length of 0.75 in.

Length (in.)	Frequency
0.720–0.729	5
0.730–0.739	10
0.740–0.749	11
0.750–0.759	17
0.760–0.769	7

25. The distribution does not appear to be normal because the frequencies are not symmetric about the maximum frequency.

FICO Score	Frequency
400–449	1
450–499	1
500–549	5
550–599	8
600–649	12
650–699	16
700–749	19
750–799	27
800–849	10
850–899	1

27.

Weight (g)	Frequency
6.0000–6.0499	2
6.0500–6.0999	3
6.1000–6.1499	10
6.1500–6.1999	8
6.2000–6.2499	6
6.2500–6.2999	7
6.3000–6.3499	3
6.3500–6.3999	1

29.

Blood Group	Frequency
O	22
A	20
B	5
AB	3

31. An outlier can dramatically affect the frequency table.

Weight (lb)	With Outlier	Without Outlier
200–219	6	6
220–239	5	5
240–259	12	12
260–279	36	36
280–299	87	87
300–319	28	28
320–339	0	
340–359	0	
360–379	0	
380–399	0	
400–419	0	
420–439	0	
440–459	0	
460–479	0	
480–499	0	
500–519	1	

Section 2-3

1. It is easier to see the shape of the distribution of the data by examining the graph of the histogram than by examining the numbers in a frequency distribution.

3. With a data set that is so small, the true nature of the distribution cannot be seen with a histogram.

5. 60; 20

7. 2500 miles; 42,500 miles

9. The digits of 0 and 5 occur disproportionately more often than the others, so it appears that the heights were reported, not measured.

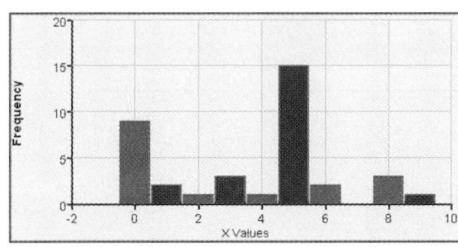

11.

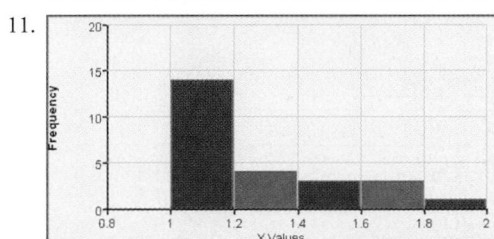

13. The distribution does not appear to have a normal distribution. The frequencies do not start low, reach a maximum, then decrease. The frequencies do not appear to be symmetric.

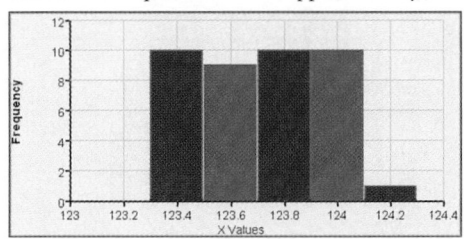

15. The lengths appear to be centered around the length of 0.75 in. The frequency distribution appears to be consistent with the label of 3/4 in.

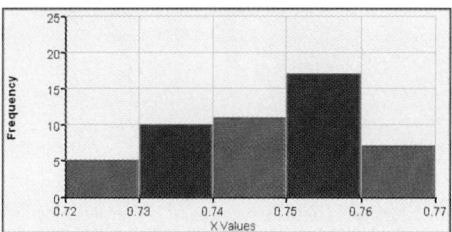

17. The distribution does not appear to be normal because the graph is not symmetric.

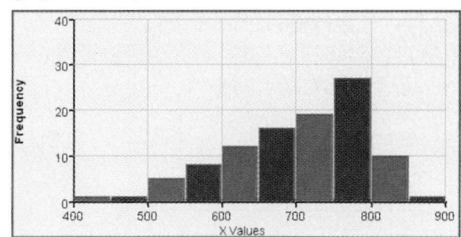

19.

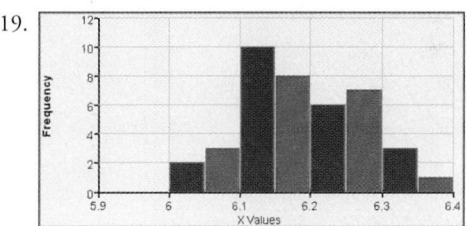

21.

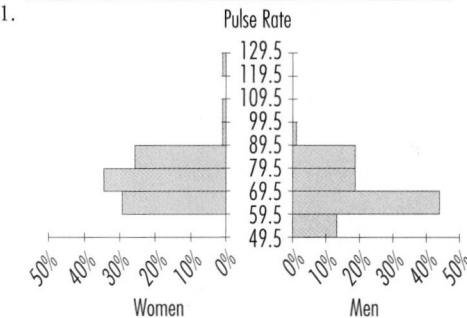

Section 2-4

1. The dotplot allows you to identify all of the original data values. The dotplot is simpler, and easier to construct.

3. By using relative frequencies consisting of proportions or percentages, both data sets use comparable measures. If frequency polygons are constructed with two samples having very different numbers of elements, the comparison is made difficult because the frequencies will be very different.

5. The dotplot suggests that the amounts have a distribution that is approximately normal, and it is centered around 150 millibecquerels.

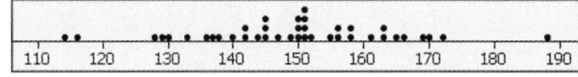

7.

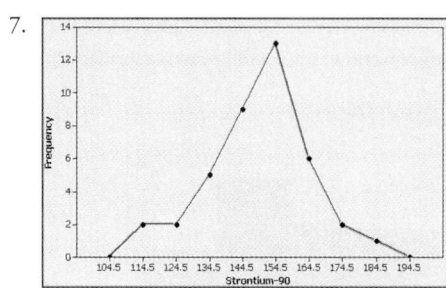

9. The stemplot suggests that the weights of discarded plastic have a distribution that is skewed to the right, but it is not too far from being a normal distribution.

0.	12356677888999
1.	11234444444445556678
2.	00111111333466888899
3.	0345
4.	36
5.	2

11. 59 of the weights are below 4 lb.

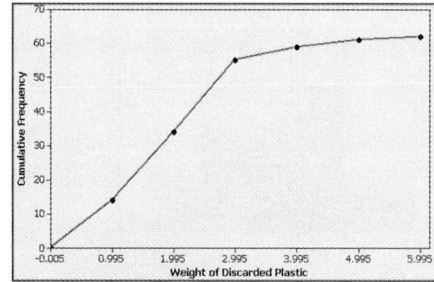

13.

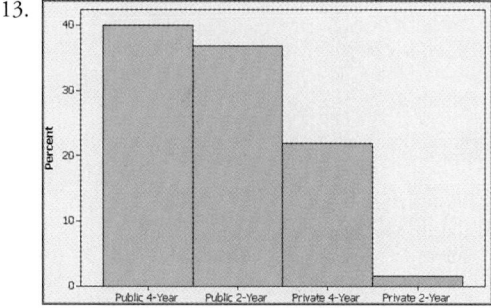

15.

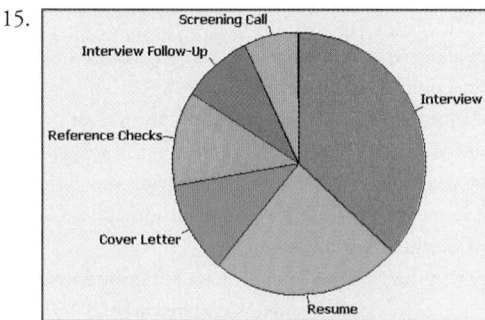

17.

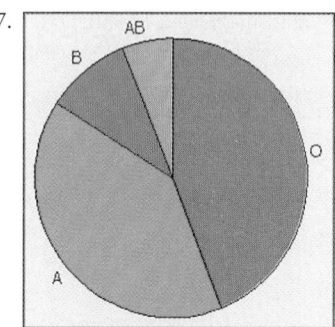

19.

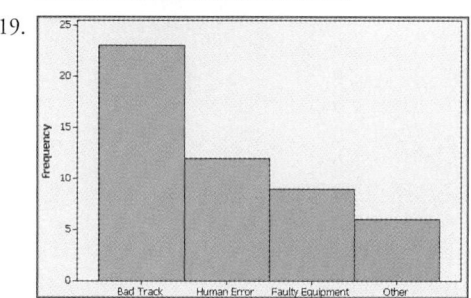

21. In king-size cigarettes, there does not appear to be a relationship between tar and carbon monoxide.

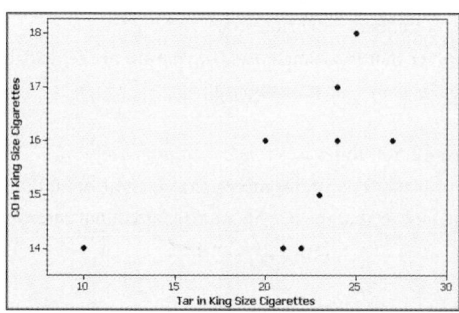

23.

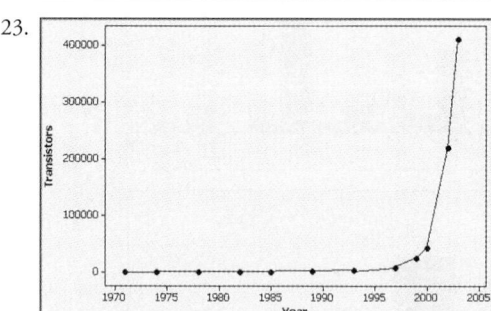

25. Total numbers in both categories have increased due to increasing population, so the rates give better information about trends. The marriage rate has remained somewhat steady, but the divorce rate grew more during the period from 1900 to 1980, and it has declined slightly since then.

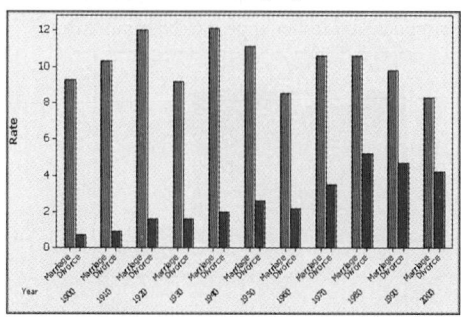

27. The pulse rates of men appear to be generally lower than the pulse rates of women.

Women	Stem (tens)	Men
	5	666666
888884444000	6	00000004444444888
66666622222222	7	22222266
88888000000	8	44448888
6	9	6
4	10	
	11	
4	12	

Section 2-5

1. The illustration uses images of dollar bills, which are objects of area, but the amounts of purchasing power are one-dimensional, so the graph is misleading.

3. No. The graph should be created in a way that is fair and objective. The readers should be allowed to judge the information on their own, instead of being manipulated by a misleading graph.

5. No. The weights are one-dimensional measurements, but the graph uses areas, so the average weight of men appears to be much larger than it actually is. (Based on the values given, men weigh about 25% more than women, but by using areas, the graph makes it appear that men weigh about 58% more than women.)

7. No. By using areas, the graph creates the impression that the average income for men is roughly twice that of women, but the actual values show that the average income for men is about 43% more than that of women.

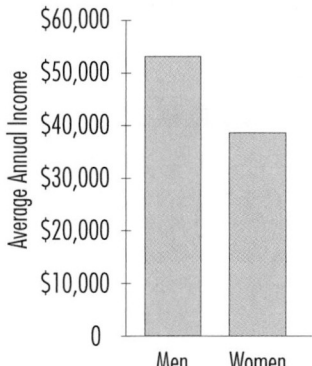

9. By starting the horizontal scale at 100 ft, the graph cuts off the left portions of the bars so that perceptions are distorted. The graph creates the false impression that the Acura RL has a braking distance that is more than twice that of the Volvo S80, but a comparison of the actual numbers shows that the braking distance of the Acura RL is roughly 40% more than that of the Volvo S80, not 100% more.

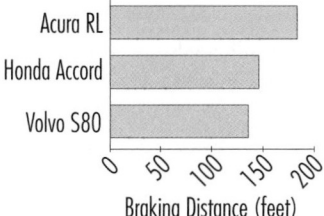

11. The graph is misleading because it is essentially a pie chart that is distorted by overlaying it atop the head of a stereotypical CEO. The graph could be better represented by a standard pie chart, and an even better graph would be a bar graph with a horizontal axis representing the successive categories of the age groups listed in order from youngest to oldest. The vertical axis should represent the percentages, and it should start at 0%.

Chapter 2 Statistical Literacy and Critical Thinking

1. The histogram is more effective because it provides a picture, which is much easier to understand than the table of numbers.

2. The time-series graph would be better because it would show changes over time, whereas the histogram would hide the changes over time.

3. By using images with heights and widths, the graph tends to create a distorted perception. The number of men is roughly twice the number of women, but the graph exaggerates that difference by using objects with areas.

4. The bars of the histogram start relatively low, then they increase to some maximum height, then they decrease. The histogram should be symmetric with the left half being roughly a mirror image of the right half.

Chapter 2 Quick Quiz

1. 10
2. −0.5 and 9.5
3. No
4. False
5. Variation
6. 52, 52, 59
7. Scatterplot
8. True
9. Distribution of the data
10. Pareto chart

Chapter 2 Review Exercises

1. The pulse rates of men are generally lower than those of women.

Pulse Rate	Frequency
50–59	6
60–69	17
70–79	8
80–89	8
90–99	1

2. The distribution shape is roughly the same, but the pulse rates of males appear to be generally lower than those of females.

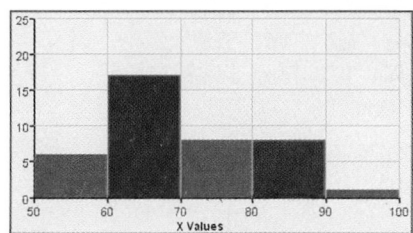

3. The dotplot shows pulse rates that are generally lower than the pulse rates of females.

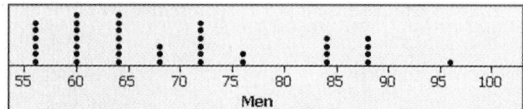

4. The pulse rates of males are generally lower than the pulse rates of females.

5	666666
6	00000004444444888
7	22222266
8	44448888
9	6

5. There does not appear to be a relationship between weight and braking distance.

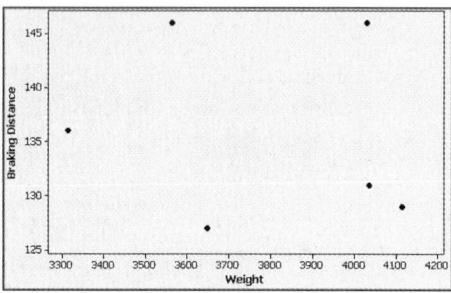

6. There is a trend that consists of a repeating cycle.

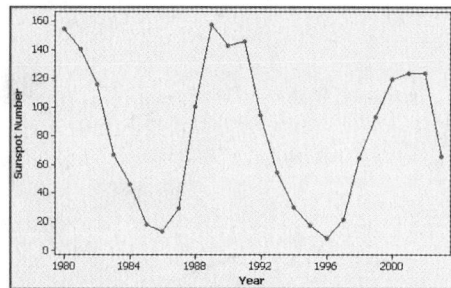

7. The graph is misleading. The differences are exaggerated because the vertical scale does not start at zero.

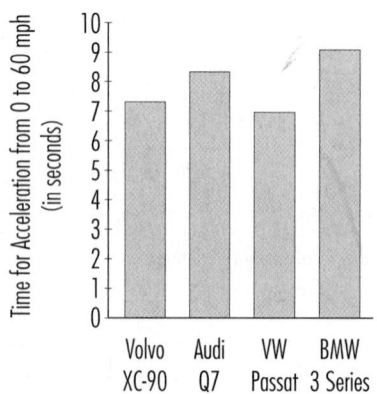

8. a. 25
   b. 100 and 124
   c. 99.5 and 124.5
   d. No

Chapter 2 Cumulative Review Exercises

1. Yes. All of the responses are summarized in the table, and the sum of the percentages is 100%.
2. Nominal, because the responses consist of "yes," "no," or "maybe." Those responses do not measure or count anything, and they cannot be arranged in order. The responses consist of names only.
3. Yes: 884; no: 433; maybe: 416
4. Voluntary response sample (or self-selected sample). The voluntary response sample is not likely to be representative of the population, because those with special interests or strong feelings about the topic are more likely to respond, and they might be very different from the general population.
5. a. In a random sample, every subject has the same chance of being selected.
   b. In a simple random sample of subjects, every sample with the same number of subjects has the same chance of being selected.
   c. Yes; no
6. a. 100
   b. −0.5 and 99.5
   c. 0.275 or 27.5%
   d. Ratio
   e. Quantitative
7. The data have a normal distribution if the histogram is approximately bell-shaped. Using a fairly strict interpretation of normality, the histogram does not represent data with a normal distribution.

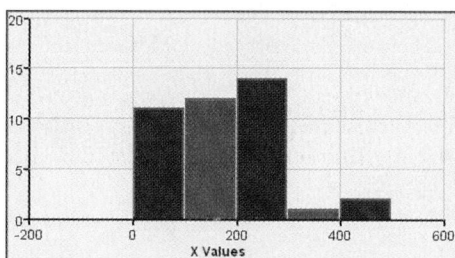

8. Statistic. In general, a statistic is a measurement describing some characteristic of a sample, whereas a parameter is a numerical measurement describing some characteristic of a population.

Chapter 3 Answers
Section 3-2

1. They use different approaches for providing a value (or values) of the center or middle of a set of data values.
3. No. The price exactly in between the highest and lowest is the midrange, not the median.
5. $\bar{x} = 53.3$; median = 52.0; mode: none; midrange = 56.5. Using the mean of 53.3 words per page, the estimate of the total number of words is 77,765. Because the mean is based on a small sample and because the numbers of defined words on pages appear to vary by a large amount, the estimate is not likely to be very accurate.
7. $\bar{x} = \$6412.2$; median = $6374.0; mode: none; midrange = $6664.0. The different measures of center do not differ by very large amounts.

9. $\bar{x}$ = \$20.4 million; median = \$14.0 million; mode: none; midrange = \$23.2 million. Because the sample values are the 10 highest, they give us almost no information about the salaries of TV personalities in general. Such *top 10* lists are not valuable for gaining insight into the larger population.

11. $\bar{x}$ = 217.3 hours; median = 235.0 hours; mode = 235.0 hours; midrange = 190.5 hours. The duration time of 0 hours is very unusual. It could represent a flight that was aborted. (It actually represents the duration time of the *Challenger* on the flight that resulted in a catastrophic explosion shortly after takeoff.)

13. $\bar{x}$ = 2.4 mi/gal; median = 2.0 mi/gal; mode = 2 mi/gal; midrange = 2.5 mi/gal. Because the amounts of decrease do not appear to vary much, there would not be a large error.

15. $\bar{x}$ = 259.6 sec; median = 246.0 sec; mode: 213 sec and 246 sec; midrange = 330.5 sec. The time of 448 sec is very different from the others.

17. $\bar{x}$ = 6.5 years; median = 4.5 years; mode = 4 and 4.5; midrange = 9.5 years. It is common to earn a bachelor's degree in four years, but the typical college student requires more than four years.

19. $\bar{x}$ = 110.0; median = 97.5; mode: none; midrange = 192.5. Stricter federal bankruptcy laws went into effect after October, so the bankruptcy filings increased up to the time of the new laws, then decreased substantially after that.

21. 30 Days in Advance: $\bar{x}$ = \$272.6; median = \$264.0. 1 Day in Advance: $\bar{x}$ = \$690.3; median = \$614.0. There is a substantial savings by purchasing the tickets 30 days in advance instead of 1 day in advance.

23. Nonfiltered: $\bar{x}$ = 1.26 mg; median = 1.10 mg. Filtered: $\bar{x}$ = 0.92 mg; median = 1.00 mg. The filtered cigarettes do appear to have less nicotine. It appears that cigarettes with filters are associated with lower amounts of nicotine.

25. $\bar{x}$ = 98.20°F; median = 98.40°F. These results suggest that the mean is less than 98.6°F.

27. Home: $\bar{x}$ = 123.66 volts; median = 123.70 volts. Generator: $\bar{x}$ = 124.66 volts; median = 124.70 volts. UPS: $\bar{x}$ = 123.59 volts; median = 123.70 volts. The three different groups appear to have about the same voltage.

29. $\bar{x}$ = 20.9 mg. The mean from the frequency table is close to the mean of 21.1 mg for the original list of data values.

31. 46.8 mi/h, which is very close to the value found from the original list of values. The speeds are well above the speed limit of 30 mi/h (probably because the police ticketed only those who were traveling well above the posted speed limit).

33. 2.93; no

35. a. 642.51 micrograms
    b. $n - 1$

37. $\bar{x}$ = 703.1; 10% trimmed mean: 709.7; 20% trimmed mean: 713.7. The results are not dramatically different. The results appear to show a trend of increasing values as the percentage of trim increases, so the distribution of the data appears to be skewed to the left.

39. 1.056; 5.6%. The result is not the same as the mean of 10%, 5%, and 2%, which is 5.7%.

41. 20.9 mg; the value of 20.0 mg is better because it is based on the original data and does not involve interpolation.

Section 3-3

1. Variation is a general descriptive term that refers to the amount of dispersion or spread among the data values, but the variance refers specifically to the square of the standard deviation.

3. The incomes of adults from the general population should have more variation, because the statistics teachers are a much more homogeneous group with incomes that are likely to be closer together.

5. Range = 45.0 words; s^2 = 245.1 words^2; s = 15.7 words. The numbers of words appear to have much variation, so we do not have too much confidence in the accuracy of the estimate.

7. Range = \$4774.0; s^2 = 3,712,571.7 dollars^2; s = \$1926.8. Damage of \$10,000 is not unusual because it is within two standard deviations of the mean.

9. Range = \$29.6 million; s^2 = 148.1 (million dollars)^2; s = \$12.2 million. Because the sample values are the 10 highest, they give us almost no information about the variation of salaries of TV personalities in general.

11. Range = 381.0 hours; s^2 = 11,219.1 hours^2; s = 105.9 hours. The duration time of 0 hours is unusual because it is more than two standard deviations below the mean.

13. Range = 3.0 mi/gal; s^2 = 0.6 (mi/gal)^2; s = 0.7 mi/gal. The decrease of 4 mi/gal is unusual because it is more than two standard deviations above the mean.

15. Range = 235.0 sec; s^2 = 2975.6 sec^2; s = 54.5 sec. Yes, it changes from 54.5 sec to 22.0 sec.

17. Range = 11.0 years; s^2 = 12.3 year^2; s = 3.5 years. No, because 12 years is within two standard deviations of the mean.

19. Range = 357.0; s^2 = 8307.3; s = 91.1. The only unusual value is 371.

21. 30 Days in Advance: 8.5% (TI-83/84 Plus: 8.6%). 1 Day in Advance: 33.8% (TI-83/84 Plus: 33.7%). The costs for the tickets purchased 30 days in advance have much less variation than the costs for the tickets purchased 1 day in advance. There is a substantial savings when purchasing the tickets 30 days in advance instead of 1 day in advance.

23. Nonfiltered: 18.3% (TI-83/84 Plus: 18.5%). Filtered: 27.2% (TI-83/84 Plus: 27.1%). The nonfiltered cigarettes have less variation.

25. Range = 3.10°F; s^2 = 0.39 (degrees Fahrenheit)^2; s = 0.62°F.

27. Home: Range = 0.90 volt; s^2 = 0.06 volt^2; s = 0.24 volt. Generator: Range = 1.20 volt; s^2 = 0.08 volt^2; s = 0.29 volt. UPS: Range = 1.00 volt; s^2 = 0.09 volt^2; s = 0.31 volt. The three different groups appear to have about the same amount of variation, which is very small in each case. (The coefficients of variation are 0.19%, 0.23%, and 0.25%.)

29. s = 3.2 mg. The standard deviation is the same as for the original list of data values.

31. $s \approx 1.5$ years

33. a. 68%
    b. 95%

35. At least 75% of the heights are within 2 standard deviations of the mean. The minimum is 147 cm and the maximum is 175 cm.

37. a. 32.7
    b. 32.7
    c. 16.3

   d. Part (b), because repeated samples result in variances that target the same value (32.7) as the population variance. Use $n-1$.

   e. No. The mean of the sample variances (32.7) equals the population variance (32.7), but the mean of the sample standard deviations (4.1) does not equal the mean of the population standard deviation (5.7).

Section 3-4

1. Below the mean. Her age is 0.61 standard deviations below the mean.

3. 0 hours is the length of the shortest flight, the first quartile Q_1 is 166 hours, the second quartile Q_2 (or median) is 215 hours, the third quartile Q_3 is 269 hours, and the maximum is 423 hours.

5. a. 25.2 years
   b. 2.23
   c. 2.23
   d. Unusual

7. a. 135 sec
   b. 3.71
   c. −3.71
   d. Unusual

9. a. 4.52; unusual
   b. −2.10; unusual
   c. −1.97; usual

11. 2.67; unusual

13. The score of 1840 converts to $z = 0.99$, and the score of 26.0 converts to $z = 1.02$, so the score of 26.0 is relatively better because it has the larger z score.

15. 38th percentile

17. 50th percentile

19. 39 (TI-83/84 Plus: 40.2)

21. 60 (TI-83/84 Plus: 59.5)

23. 53.5

25. 42 (TI-83/84 Plus: 42.5)

27.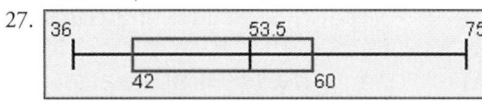

29.

31. The weights of diet Coke appear to be generally less than those of regular Coke, probably due to the sugar in cans of regular Coke.

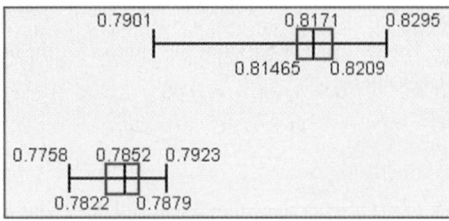

33. The weights of the pre-1964 quarters appear to be considerably greater than those of the post-1964 quarters.

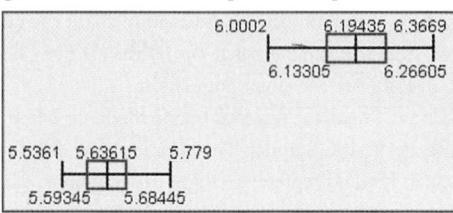

35. Outliers: 27.0 cm, 31.1 cm, 32.1 cm, 48.6 cm.

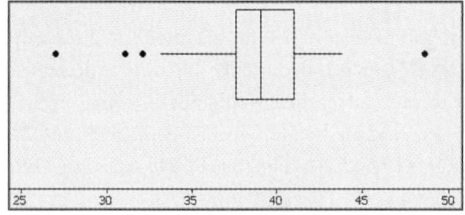

37. The result is 33.75 with interpolation and 35 without interpolation. These two results are not too far apart.

Chapter 3 Statistical Literacy and Critical Thinking

1. No. In addition to having a mean of 12 oz, the amounts of cola must not vary too much. If they have a mean of 12 oz but they vary too much, some of the cans will be underfilled by unacceptable amounts, while others will be overfilled, so the production process would need to be fixed.

2. Because the ZIP codes are not measurements or counts of anything and there isn't a consistent ordering, the ZIP codes are at the nominal level of measurement, so the mean and standard deviation are meaningless in this context.

3. The mean, standard deviation, and range will change by a fairly large amount, but the median will not change very much.

4. Trend or pattern over time

Chapter 3 Quick Quiz

1. 4.0 cm
2. 3.0 cm
3. 2 cm
4. 25.0 ft²
5. −3
6. Variation
7. Sample: s; population: σ
8. Sample: $\bar{x}$; population: μ
9. 75
10. True

Chapter 3 Review Exercises

1. a. 19.3 oz
   b. 19.5 oz
   c. 20 oz
   d. 19.0 oz
   e. 4.0 oz
   f. 1.3 oz

g. 1.8 oz²

h. 18 oz

i. 20 oz

2.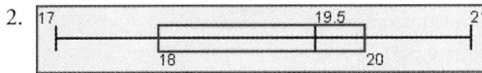

3. a. 918.8 mm

b. 928.0 mm

c. 936 mm

d. 907.0 mm

e. 58.0 mm

f. 23.0 mm

g. 530.9 mm²

h. 923 mm

i. 934 mm

4. The z score is -1.77. Because the z score is between -2 and 2, the sitting height of 878 mm is not unusual.

5. Because the median and quartiles are located to the far right portion of the boxplot, the boxplot suggests that the data are from a population with a skewed distribution, not a distribution that is normal (or bell-shaped).

6. The score of 1030 converts to $z = -1.50$, and the score of 14.0 converts to $z = -1.48$, so the score of 14.0 is relatively better because it has the larger z score.

7. a. Answer varies, but an estimate around 4 years is reasonable.

b. Answer varies, but using a minimum of 0 year and a maximum of 20 years results in $s = 5.0$ years.

8. a. Answer varies, but assuming that the times are between 15 sec and 90 sec, a mean of about 50 sec or 60 sec is reasonable.

b. Answer varies, but assuming that the times are between 15 sec and 90 sec, a standard deviation around 18 sec or 20 sec is reasonable.

9. Minimum: 1110 mm; maximum: 1314 mm. Because the design should accommodate most women pilots, the grip reach of 1110 mm is more relevant. No instruments or controls should be located more than 1110 mm from a sitting adult woman.

10. Minimum: 83.7 cm; maximum: 111.3 cm. The height of 87.8 cm is not unusual, so the physician should not be concerned.

Chapter 3 Cumulative Review Exercises

1. a. Continuous

b. Ratio

2.

Sitting Height (mm)	Frequency
870–879	1
880–889	1
890–899	0
900–909	0
910–919	0
920–929	3
930–939	4

3. The distribution appears to be skewed.

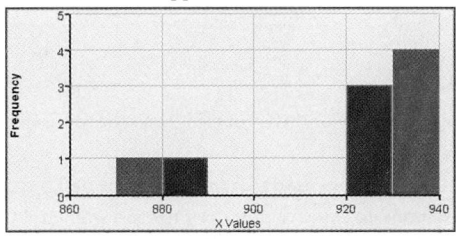

4.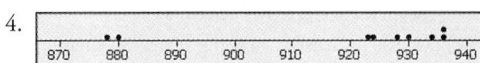

5.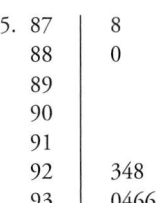

87	8
88	0
89	
90	
91	
92	348
93	0466

6. a. The mode is most appropriate because it identifies the most common choice. The other measures of center do not apply to data at the nominal level of measurement.

b. Convenience

c. Cluster

d. The standard deviation should be lowered.

7. No. Even though the sample is large, it is a voluntary response sample, so the responses cannot be considered to be representative of the population of the United States.

8. A simple random sample of n subjects is selected in such a way that every possible *sample of the same size n* has the same chance of being chosen. A voluntary response sample is one in which the subjects decide themselves whether to be included in the study. A simple random sample is generally better than a voluntary response sample.

9. In an observational study, we observe and measure specific characteristics, but we don't attempt to modify the subjects being studied. In an experiment, we apply some *treatment* and then proceed to observe its effects on the subjects.

10. By not starting the vertical scale at 0, the graph exaggerates the differences among the frequencies.

Chapter 4 Answers
Section 4-2

1. The chance of an injury from sports or recreation equipment is very small. Such injuries occur roughly one time in 500. Because the probability is so small, such an injury is unusual.

3. $\overline{A}$ denotes the complement of event A, meaning that event A does not occur. If $P(A) = 0.995$, $P(\overline{A}) = 0.005$. If $P(A) = 0.995$, $\overline{A}$ is unusual because it is very unlikely to occur.

5. 4/21 or 0.190

7. 1/2 or 0.5

9. 6/36 or 1/6 or 0.167

11. 0

13. 3:1, 5/2, -0.5, 321/123

15. a. 3/8

b. 3/8

c. 1/8

17. a. 98
    b. 41
    c. 41/98
    d. 0.418
19. 0.153. The probability of this error is high, so the test is not very accurate.
21. 0.16. No, the probability is too far from the value of 0.5 that would correspond to equal chances for men and women. It appears that more men are elected as Senators.
23. 0.000000966. No, the probability of being struck is much greater in an open golf course during a thunderstorm. He or she should seek shelter.
25. a. 1/365
    b. Yes
    c. He already knew.
    d. 0
27. 0.0571. No, a firmware malfunction is not unusual among malfunctions.
29. 0.0124; yes, it is unusual. There might be more reluctance to sentence women to death, but very serious crimes are committed by many more men than women.
31. 0.0811; not unusual
33. a. bb, bg, gb, gg
    b. 1/4
    c. 1/2
35. a. brown/brown, brown/blue, blue/brown, blue/blue
    b. 1/4
    c. 3/4
37. 423:77 or roughly 5.5:1 or 11:2
39. a. $12.20
    b. 6.1:1 or roughly 6:1
    c. 443:57 or roughly 8:1
    d. $17.54 or roughly $18
41. Relative risk: 0.939; odds ratio: 0.938; the probability of a headache with Nasonex (0.0124) is slightly less than the probability of a headache with the placebo (0.0132), so Nasonex does not appear to pose a risk of headaches.
43. 1/4

Section 4-3

1. The two events cannot happen at the same time.
3. $P(A \text{ and } B)$ denotes the probability that on a single trial, events A and B both occur at the same time.
5. Not disjoint
7. Disjoint
9. Not disjoint
11. Not disjoint
13. $P(\overline{M})$ denotes the probability of getting a STATDISK user who is not using a Macintosh computer, and $P(\overline{M}) = 0.95$.
15. 0.21
17. 89/98 or 0.908
19. 32/98 or 16/49 or 0.327
21. 512/839 or 0.610
23. 615/839 or 0.733
25. 713/839 or 0.850

27. 156/1205 = 0.129. Yes. A high refusal rate results in a sample that is not necessarily representative of the population, because those who refuse may well constitute a particular group with opinions different from others.
29. 1060/1205 = 0.880
31. 1102/1205 = 0.915
33. a. 300
    b. 178
    c. 178/300 = 0.593
35. 0.603
37. 27/300 = 0.090. With an error rate of 0.090 (or 9%), the test does not appear to be highly accurate.
39. 3/4 = 0.75
41. $P(A \text{ or } B) = P(A) + P(B) - 2P(A \text{ and } B)$
43. a. $1 - P(A) - P(B) + P(A \text{ and } B)$
    b. $1 - P(A \text{ and } B)$
    c. Different

Section 4-4

1. Answers vary.
3. The events are dependent. Because different numbers are used, the selections are made without replacement, so the probability of an adult being selected changes as the selections are made.
5. Independent
7. Dependent
9. Dependent (same power supply)
11. Independent
13. 0.0221. Yes, it is unusual because the probability is very low, such as less than 0.05.
15. 0.109. The event is not unusual because its probability is not very small, such as less than 0.05.
17. a. 0.152
    b. 0.150
19. a. 0.0000130
    b. 0.00000383
21. a. Yes
    b. 1/8 = 0.125
    c. No, it would be better to be prepared.
23. $0.95^9 = 0.630$. With nine independent polls, there is a reasonably good chance (0.370) that at least one of them is not within the claimed margins of error.
25. 1/8 = 0.125. The likelihood of getting 3 girls in 3 births by chance is high, so the results do not indicate that the method is effective.
27. a. 0.1
    b. 0.01
    c. 0.99
    d. Yes, the probability of not being awakened drops from 1 chance in 10 to one chance in 100.
29. a. 0.849
    b. 0.0169. Yes, because there is a very small chance that all 100 tires are good.
31. a. 0.9999
    b. 0.9801
    c. The series arrangement provides better protection.
33. 0.0192

Section 4-5

1. The exact number of defects is 1 or 2 or 3 or 4 or 5 or 6 or 7 or 8 or 9 or 10.

3. No, the two outcomes are not equally likely. The actual survival rate was not considered, but it should have been.

5. All 15 players test negative.

7. At least one of the four males has the X-linked recessive gene.

9. 0.984. Yes.

11. 0.590. There is a good chance of passing with guesses, but it isn't reasonable to expect to pass with guesses.

13. 0.5. No.

15. 0.179. The four cars are not randomly selected. They are driven by people in the same family, and they are more likely to have similar driving habits, so the probability might not be correct.

17. 0.944. Yes, the probability is quite high, so they can be reasonably confident of getting at least one offspring with vestigial wings.

19. $15/47 = 0.319$. This is the probability of unfairly indicating that the subject lied when the subject did not lie.

21. $9/41 = 0.220$. The results are different.

23. a. 1/3 or 0.333
    b. 0.5

25. 0.5

27. 0.999. Yes. With only one alarm clock, there is a 0.10 chance of the alarm clock not working, but with three alarm clocks there is only a 0.001 chance that they all fail. The likelihood of a functioning alarm clock increases dramatically with three alarm clocks. The results change if the alarm clocks run on electricity, because they are no longer independent. The student should use battery-powered alarm clocks to maintain independence.

29. 0.271

31. a. 0.431
    b. 0.569

33. 1/12; 35

Section 4-6

1. No. The generated numbers between 2 and 12 would be equally likely, but they are not equally likely with real dice.

3. Yes, it does. Each of the 365 birthdays has the same chance of being selected, and the cards are replaced, so it is possible to select the same birthday more than once.

5. Randomly generate 50 random numbers, with each number between 1 and 100. Consider the numbers 1 through 95 to be adults who recognize the brand name of McDonald's, while the numbers 96 through 100 represent adults who do not recognize McDonald's.

7. Randomly generate a number between 1 and 1000. Consider an outcome of 1 through 528 to be a free throw that is made, and consider an outcome of 529 through 1000 to be a free throw that is missed.

9. a. Answer varies.
   b. It would be unusual to randomly select 50 consumers and find that about half of them recognize McDonald's.

11. a. Answer varies.
    b. It would be unusual for O'Neal to make all of five free throws in a game.

13. The exact answer is 0.5, so an answer from a simulation should be around 0.5. Such runs are not unusual.

15. For the simulation, generate 152 random numbers between 0 and 1 inclusive. Sum the results to find the number of boys. Repeating the simulation should show that a result of 127 boys is unusual, suggesting that the YSORT method is effective.

17. With switching, $P(\text{win}) = 2/3$. With sticking, $P(\text{win}) = 1/3$.

19. No; no

Section 4-7

1. With permutations, different orderings of the same items are counted separately, but they are not counted separately with combinations.

3. Because the order of the first three finishers does make a difference, the trifecta involves permutations.

5. 120

7. 1326

9. 15,120

11. 5,245,786

13. 1/25,827,165

15. 1/376,992

17. a. 1/1,000,000,000
    b. 1/100,000. No, but if they could keep trying with more attempts, they could eventually get your social security number.

19. 27,907,200

21. 10

23. 256; yes, because the typical keyboard has about 94 characters.

25. 120 ways; JUMBO; 1/120

27. a. 7920
    b. 330

29. a. 16,384
    b. 14
    c. 14/16,384 or 0.000854
    d. Yes, because the probability is so small and is so far away from the 7 girls/7 boys that is typical.

31. 2/252 or 1/128. Yes, if everyone treated is of one sex while everyone in the placebo group is of the opposite sex, you would not know if different reactions are due to the treatment or sex.

33. 1/146,107,962

35. 144

37. 2,095,681,645,538 (about 2 trillion)

39. a. Calculator: 3.0414093×10^{64}; approximation: 3.0363452×10^{64}
    b. 615

41. 293

Chapter 4 Statistical Literacy and Critical Thinking

1. The results are very unlikely to occur by chance. The results would occur an average of only 4 times in 1000.

2. No, because they continue to have a common power source. If the home loses all of its power, all of the smoke detectors would be inoperable.

3. No. The calculation assumes that the events are independent, but the same detective is handling both cases, and both cases are likely to be in the same general location, so they are not independent. The probability of clearing both cases depends largely on her commitment, ability, resources, and support.

4. No. The principle of independence applies here. Based on the way that lottery numbers are drawn, past results have no effect on the probability of future outcomes, so there is no way to predict which numbers are more likely or less likely to occur.

Chapter 4 Quick Quiz

1. 0.30
2. 0.49
3. Answer varies, but an answer such as 0.01 or lower is reasonable.
4. No, because the results could have easily occurred by chance.
5. 0.6
6. $427/586 = 0.729$
7. $572/586 = 0.976$
8. 0.00161
9. $10/586 = 0.0171$
10. $10/24 = 0.417$

Chapter 4 Review Exercises

1. $576/3562 = 0.162$
2. $752/3562 = 0.211$
3. $1232/3562 = 0.346$
4. $3466/3562 = 0.973$
5. $96/3562 = 0.0270$
6. $2330/3562 = 0.654$
7. 0.0445 (not 0.0446)
8. 0.0261
9. $480/576 = 0.833$
10. $656/752 = 0.872$
11. Based on data from J.D. Power and Associates, 16.7% of car colors are black, so any estimate between 0.05 and 0.25 is reasonable.
12. a. 0.65
    b. 0.0150
    c. Yes. Because the probability is so small (0.0150), it is not likely to happen by chance.
13. a. 1/365
    b. 31/365
    c. Answer varies, but it is probably small, such as 0.02.
    d. Yes
14. a. 0.000152
    b. 0.0000000231
    c. 0.999696
15. a. 0.60
    b. No. Because the poll uses a voluntary response sample, its results apply only to those who chose to respond. The results do not necessarily apply to the greater population of all Americans.
16. 0.0315. It is not likely that such combined samples test positive.
17. $0.85^{30} = 0.00763$. Because the probability of getting 30 Democrats by chance is so small, chance is not a reasonable explanation, so it appears that the pollster is lying.

18. a. 0.998856
    b. 0.998
    c. 0.999
    d. Males are more likely to die as a result of military engagement, car crashes, violent crimes, and other such high-risk activities.
19. 501,942; 1/501,942. It is unusual for a particular person to buy one ticket and win, but it is not unusual for *anyone* to win, because wins often do occur.
20. 10,000,000,000,000

Chapter 4 Cumulative Review Exercises

1. a. 19.3 oz
   b. 19.5 oz
   c. 1.1 oz
   d. 1.2 oz^2
   e. No. The mean is well below the desired value of 21 oz, and all of the steaks weigh less than 21 oz before cooking.
2. a. 58.2%
   b. 0.582
   c. It is a voluntary response sample (or self-selected sample), and it is not suitable for making conclusions about any larger population; it is useful for describing the opinions of only those who choose to respond.
   d. The poll had 5226 respondents. A simple random sample of 5226 subjects would be selected in such a way that all samples of 5226 subjects have the same chance of being selected. A simple random sample is definitely better than a voluntary response sample.
3. a. Regular: 371.5 g; diet: 355.5 g. Diet Coke appears to weigh substantially less.
   b. Regular: 371.5 g; diet: 356.5 g. Diet Coke appears to weigh substantially less.
   c. Regular: 1.5 g; diet: 2.4 g. The weights of diet Coke appear to vary more.
   d. Regular: 2.3 g^2; diet: 5.9 g^2
   e. No, diet Coke appears to weigh considerably less than regular Coke.
4. a. Yes, because 38 is more than two standard deviations below the mean.
   b. 1/1024 or 0.000977. Because the probability is so small, the outcome of a perfect score is unusual.
5. a. Convenience sample
   b. If the students at the college are mostly from a surrounding region that includes a large proportion of one ethnic group, the results will not reflect the general population of the United States.
   c. 0.75
   d. 0.64
6. 14,348,907

Chapter 5 Answers
Section 5-2

1. A random variable is a variable that has a single numerical value, determined by chance, for each outcome of a procedure. The random variable x is the number of winning tickets in 52 weeks, and its possible values are 0, 1, 2, . . . , 52.

3. The random variable x represents all possible events in the entire sample space, so we are certain (with probability 1) that one of the events will occur.

5. a. Discrete
   b. Continuous
   c. Continuous
   d. Discrete
   e. Continuous

7. Probability distribution with $\mu = 1.5$ and $\sigma = 0.9$.

9. Not a probability distribution because $\Sigma P(x) = 0.984 \neq 1$.

11. Probability distribution with $\mu = 2.8$ and $\sigma = 1.3$.

13. $\mu = 6.0$ peas; $\sigma = 1.2$ peas

15. a. 0.267
    b. 0.367
    c. Part (b)
    d. No, because the probability of 7 or more peas with green pods is high. (0.367 is greater than 0.05).

17. a. Yes
    b. $\mu = 5.8$ games and $\sigma = 1.1$ games.
    c. No, because the probability of 4 games is not small (such as less than 0.05), the event of a four-game sweep is not unusual.

19. a. Yes
    b. $\mu = 0.4$ bumper sticker and $\sigma = 1.3$ bumper stickers.
    c. -2.2 to 3.0
    d. No, because the probability of more than 1 bumper sticker is 0.093, which is not small.

21. $\mu = 1.5$; $\sigma = 0.9$. It is not unusual to get 3 girls, because the probability of 3 girls is high (1/8), indicating that 3 girls could easily occur by chance.

23. $\mu = 4.5$; $\sigma = 2.9$. The probability histogram is flat.

25. a. 1000
    b. 1/1000 or 0.001
    c. $249.50
    d. -25¢
    e. Because both games have the same expected value, neither bet is better than the other.

27. a. -39¢
    b. The bet on the number 13 is better because its expected value of -26¢ is greater than the expected value of -39¢ for the other bet.

29. a. $-$161 and $99,839
    b. $-$21
    c. Yes. The expected value for the insurance company is $21, which indicates that the company can expect to make an average of $21 for each such policy.

31. The A bonds are better because the expected value is $49.40, which is higher than the expected value of $26 for the B bonds. She should select the A bond because the expected value is positive, which indicates a likely gain.

33. Put 1, 2, 3, 4, 5, 6 on one die and put 0, 0, 0, 6, 6, 6 on the other die.

Section 5-3

1. The given event describes only one way in which 2 people have blue eyes among 5 randomly selected people, but there are other ways of getting 2 people with blue eyes among 5 people and their corresponding probabilities must also be included in the result.

3. Yes. Although the selections are not independent, they can be treated as being independent by applying the 5% guideline. Only 30 people are randomly selected from the group of 1236, which is less than 5%.

5. Binomial

7. Not binomial; there are more than two possible outcomes.

9. Not binomial. Because the Senators are selected without replacement, the events are not independent, and they cannot be treated as being independent.

11. Binomial. Although the events are not independent, they can be treated as being independent by applying the 5% guideline.

13. a. 0.128
    b. WWC, WCW, CWW; 0.128 for each
    c. 0.384

15. 0.420

17. 0+

19. 0.075

21. 0.232

23. 0.182

25. 0.950; yes

27. 0.0185. Yes, with a probability that is small (such as less than 0.05), it is unusual.

29. 0.387; 0.613

31. 0.001 or 0.000609. It is unusual because the probability is very small, such as less than 0.05.

33. 0.244. It is not unusual because the probability of 0.244 is not very small, such as less than 0.05.

35. a. 0.882
    b. Yes, because the probability of 7 or fewer graduates is small (less than 0.05).

37. a. 0.00695
    b. 0.993
    c. 0.0461
    d. Yes. With a 22 share, there is a very small chance of getting at most 1 household tuned to *NBC Sunday Night Football*, so it appears that the share is not 22.

39. a. 0.209
    b. Unlike part (a) the 20 selected subjects are more than 5% of the 320 subjects available, so independence cannot be assumed by the 5% guideline. The independence requirement for the binomial probability formula is not satisfied.

41. 0.662. The probability shows that about 2/3 of all shipments will be accepted.

43. 0.000139. The very low probability value suggests that the low number of hired women is not the result of chance. The low probability value does appear to support the charge of discrimination based on gender.

45. 0.030. The result is unusual, but with 580 peas, any individual number is unusual. The result does not suggest that 0.75 is wrong. The probability value of 0.75 would appear to be wrong

if 428 peas with green pods is an unusually low number, which it is not.

47. a. 0.0000000222
    b. 0.00000706
    c. 0.0104
    d. 0.510

Section 5-4

1. Yes. Because $q = 1 - p$, the two expressions are equivalent in the sense that they will always give the same results.
3. 59.3 people^2 (or 58.7 people^2 if unrounded results are used)
5. $\mu = 10.0$; $\sigma = 2.8$; minimum = 4.4; maximum = 15.6
7. $\mu = 144.0$; $\sigma = 8.7$; minimum = 126.6; maximum = 161.4
9. a. $\mu = 37.5$ and $\sigma = 4.3$
   b. It would not be unusual to pass by getting at least 45 correct answers, because the range of usual numbers of correct answers is from 28.9 to 46.1, and 45 is within that range of usual values.
11. a. $\mu = 16.0$ and $\sigma = 3.7$.
    b. The result of 19 green M&Ms is not unusual because it is within the range of usual values, which is from 8.6 to 23.4. The claimed rate of 16% does not appear to be wrong.
13. a. $\mu = 287.0$ and $\sigma = 12.0$
    b. The result of 525 girls is unusual because it is outside of the range of usual values, which is from 263 to 311. The results suggest that the XSORT method is effective.
15. a. $\mu = 160.0$ and $\sigma = 8.9$
    b. 142.2 to 177.8
    c. 250, which is unusual because it is outside of the range of usual values given in part (b). The results suggest that the headline is justified.
    d. Because the sample of 320 respondents is a voluntary response sample, all of the results are very questionable.
17. a. $\mu = 611.2$, $\sigma = 15.4$
    b. No, because the range of usual values is 580.4 to 642.0. The result of 701 is unlikely to occur because it is outside of the range of usual values.
    c. No, it appears that substantially more people say that they voted than the number of people who actually did vote.
19. a. $\mu = 10.2$ and $\sigma = 3.2$
    b. 30 is unusual because it is outside of the range of usual values, which is from 3.8 to 16.6.
    c. Chantrix does appear to be the cause of some nausea, but the nausea rate with Chantrix is still quite low, so nausea appears to be an adverse reaction that does not occur very often.
21. $\mu = 3.0$ and $\sigma = 1.3$ (not 1.5)

Section 5-5

1. The random variable x is the number of occurrences of an event over some interval, the occurrences are random, the occurrences are independent of each other, and the occurrences are uniformly distributed over the interval.
3. The probability is so small that, for all practical purposes, we can consider it to be zero.

5. 0.180
7. 0.0332
9. a. 0.0970
   b. 0.000141
   c. Yes, because the probability is very small, such as less than 0.05.
11. a. 62.2
    b. 0.0155 (0.0156 using rounded mean)
13. a. 0.728
    b. 0.231
    c. 0.0368
    d. 0.00389
    e. 0.000309
    Using the computed probabilities, the expected frequencies are 266, 84, 13, 1.4, and 0.1, and they agree quite well with the actual frequencies.
15. a. 2.24
    b. 0.811 (or 0.812 if binomial is used). No, the probability of a profit is high, but not high enough to be "almost sure."
17. The Poisson probabilities are not close enough to the binomial probabilities. For example, $P(6) = 0.205$ with the binomial distribution, but $P(6) = 0.146$ with the Poisson distribution.

Chapter 5 Statistical Literacy and Critical Thinking

1. A random variable is a variable that has a single numerical outcome, determined by chance, for each outcome of a procedure. It is possible for a discrete random variable to have an infinite number of values.
2. A discrete random variable has either a finite number of values or a countable number of values, but a continuous random variable has infinitely many values, and those values can be associated with measurements on a continuous scale without gaps or interruptions.
3. The sum of p and q is 1, or $p + q = 1$, or $p = 1 - q$, or $q = 1 - p$.
4. No. There are many discrete probability distributions that don't satisfy the requirements of a binomial distribution or a Poisson distribution.

Chapter 5 Quick Quiz

1. No
2. 0.7
3. 200
4. 10
5. Yes
6. Yes
7. 0.5904 or 0.590
8. 0.0016
9. 0.1792
10. Yes

Chapter 5 Review Exercises

1.

x	p
0	0.004
1	0.031
2	0.109
3	0.219
4	0.273
5	0.219
6	0.109
7	0.031
8	0.004

2. $\mu = 4.0$ and $\sigma = 1.4$. Usual values are from 1.2 to 6.8. The occurrence of 8 deaths in the week before Thanksgiving is unusual because it is outside of the range of usual values (or because its probability of 0.004 is so low).

3. a. 0.0370
   b. Yes, because the probability of 0.0370 is very small.
   c. No. Determining whether 14 is an unusually high number of deaths should be based on the probability of 14 or more deaths, not the probability of exactly 14 deaths (or the range of usual values could be used).

4. $315,075. Because the offer is well below her expected value, she should continue the game (although the guaranteed prize of $193,000 had considerable appeal). (She accepted the offer of $193,000, but she would have won $500,000 if she continued the game and refused all further offers.)

5. a. 1.2¢
   b. 1.2¢ minus cost of stamp.

6. No, because $\Sigma P(x) = 0.9686$, but the sum should be 1. (There is a little leeway allowed for rounding errors, but the sum of 0.9686 is too far below 1.)

7. a. 1/10,000 or 0.0001
   b.

x	P(x)
−$1	0.9999
$4999	0.0001

   c. 0.0365
   d. 0.0352
   e. −50¢

8. a. 0.00361
   b. This company appears to be very different because the event of at least four firings is so unlikely with a probability of only 0.00361.

9. a. 236.0
   b. $\mu = 236.0$ and $\sigma = 12.8$
   c. 210.4 to 261.6
   d. Yes, because 0 is well outside of the range of usual values.

10. a. 535/576 = 0.929
    b. 0.395
    c. 227.5 regions
    d. The actual result of 229 is very close to the computed result of 227.5.

Chapter 5 Cumulative Review Exercises

1. a. $159.674
   b. $142.94
   c. $102.60
   d. $41.985
   e. 1762.700 dollars^2
   f. $75.704 to $243.644
   g. No, because all of the sample values are within the range of usual values.
   h. Ratio
   i. Discrete
   j. Convenience
   k. $21,396.32

2. a. 0.026
   b. 0.992 (or 0.994)
   c. $\mu = 8.0$ and $\sigma = 1.3$.
   d. 5.4 to 10.6

3. a. $\mu = 15.0$, $\sigma = 3.7$
   b. Twelve positive cases falls within the range of usual values (7.6 to 22.4), so 12 is not unusually low. Because 12 positive cases could easily occur with an ineffective program, we do not have enough justification to say that the program is effective.

4. a. 706/2223 = 0.318
   b. 0.101
   c. 0.466

5. No. The mean should take the state populations into account. The 50 values should be weighted, with the state populations used as weights.

Chapter 6 Answers
Section 6-2

1. The word "normal" has a special meaning in statistics. It refers to a specific bell-shaped distribution that can be described by Formula 6-1.
3. The mean and standard deviation have the values of $\mu = 0$ and $\sigma = 1$.
5. 0.5
7. 0.75
9. 0.7734
11. 0.6106 (TI-83/84 Plus: 0.6107)
13. 2.05
15. 1.24
17. 0.0668
19. 0.8907
21. 0.0132
23. 0.9599
25. 0.1498 (TI-83/84 Plus: 0.1499)
27. 0.1574 (TI-83/84 Plus: 0.1573)
29. 0.8593
31. 0.9937 (TI-83/84 Plus: 0.9938)
33. 0.9999 (TI-83/84 Plus: 0.9998)
35. 0.5000
37. 68.26% (TI-83/84 Plus: 68.27%)
39. 99.74% (TI-83/84 Plus: 99.73%)

41. 1.645
43. 1.28
45. 0.9500
47. 0.0100
49. 1.645
51. −1.96, 1.96
53. a. 95.44% (TI-83/84 Plus: 95.45%)
    b. 31.74% (TI-83/84 Plus: 31.73%)
    c. 5.00%
    d. 99.74% (TI-83/84 Plus: 99.73%)
    e. 0.26% (TI-83/84 Plus: 0.27%)
55. a. 1.75
    b. −2.00
    c. 1.50
    d. 0.82
    e. 0.12

Section 6-3

1. The standard normal distribution has a mean of 0 and a standard deviation of 1, but a nonstandard distribution has a different value for one or both of those parameters.
3. The mean is 0 and the standard deviation is 1.
5. 0.9082 (TI-83/84 Plus: 0.9088)
7. 0.5899 (TI-83/84 Plus: 0.5889)
9. 103.8
11. 75.3
13. 0.8413
15. 0.4972 (TI-83/84 Plus: 0.4950)
17. 92.2 (TI-83/84 Plus: 92.1)
19. 110.0 (TI-83/84 Plus: 110.1)
21. a. 85.77% (TI-83/84 Plus: 85.80%)
    b. 99.96%
    c. The height is not adequate because 14% of adult men will need to bend, so it would be better to have doorways with a higher opening, but other design considerations probably make that impractical.
    d. 74.7 in. (TI-83/84 Plus: 74.8 in.)
23. a. 3.67% (TI-83/84 Plus: 3.71%)
    b. 0.52%
    c. No. The percentage of eligible men is greater than the percentage of eligible women.
25. a. 98.74% (TI-83/84 Plus: 98.75%). No, only about 1% of women are not eligible.
    b. Minimum: 57.8 in.; maximum: 68.7 in.
27. a. 4.09%
    b. 2630 g
    c. Without a specific cutoff birth weight, there would be no way to know if a baby is in the bottom 3% until all babies have been born, but waiting would deny treatment for those babies who need special treatment.
29. a. 0.01%; yes
    b. 99.22°
31. a. 0.0038; either a very rare event has occurred or the husband is not the father.
    b. 242 days

33. a. $\bar{x} = 118.9$, $s = 10.5$, histogram is roughly bell shaped.
    b. 101.6, 136.2
35. a. The z scores are real numbers that have no units of measurement.
    b. $\mu = 0$; $\sigma = 1$; distribution is normal.
37. a. 75; 5
    b. No, the conversion should also account for variation.
    c. 31.4, 27.6, 22.4, 18.6
    d. Part (c), because variation is included in the conversion.
39. 0.0074 (TI-83/84 Plus: 0.0070)

Section 6-4

1. A sampling distribution of a statistic is the distribution of all values of that statistic when all possible samples of the same size are taken from the same population.
3. The mean of all of the sample means is equal to the population mean.
5. The sample is not a simple random sample from the population of all U.S. college students. It is likely that the students at New York University do not accurately reflect the behavior of all U.S. college students.
7. Normal (approximately)
9. a.

Sample Median	Probability
2	1/9
2.5	2/9
3	1/9
6	2/9
6.5	2/9
10	1/9

   b. The population median is 3, but the mean of the sample medians is 5. The values are not equal.
   c. The sample medians do not target the population median of 3, so sample medians do not make good estimators of population medians.

11. a.

s^2	Probability
0	3/9
0.5	2/9
24.5	2/9
32	2/9

   b. The population variance is 12.667 and the mean of the sample variances is also 12.667. The values are equal.
   c. The sample variances do target the population variance of 12.667, so the sample variances do make good estimators of the population variance.

13. a. (56, 56), (56, 49), (56, 58), (56, 46), (49, 56), (49, 49), (49, 58), (49, 46), (58, 56), (58, 49), (58, 58), (58, 46), (46, 56), (46, 49), (46, 58), (46, 46)

b.

$\bar{x}$	Probability
46	1/16
47.5	2/16
49	1/16
51	2/16
52	2/16
52.5	2/16
53.5	2/16
56	1/16
57	2/16
58	1/16

c. The mean of the population is 52.25 and the mean of the sample means is also 52.25.

d. The sample means target the population mean. Sample means make good estimators of population means because they target the value of the population mean instead of systematically underestimating or overestimating it.

15. a. Same as Exercise 13 part (a).

b.

Range	Probability
0	4/16
2	2/16
3	2/16
7	2/16
9	2/16
10	2/16
12	2/16

c. The range of the population is 12, but the mean of the sample ranges is 5.375. Those values are not equal.

d. The sample ranges do not target the population range of 12, so sample ranges do not make good estimators of population ranges.

17.

Proportion of Odd Numbers	Probability
0	4/9
0.5	4/9
1	1/9

Yes. The proportion of odd numbers in the population is 1/3, and the sample proportions also have a mean of 1/3, so the sample proportions do target the value of the population proportion. The sample proportion does make a good estimator of the population proportion.

19. a. The proportions of 0, 0.5, 1 have the corresponding probabilities of 1/16, 6/16, 9/16.

b. 0.75

c. Yes; yes

21. The formula yields $P(0) = 0.25$, $P(0.5) = 0.5$, and $P(1) = 0.25$, which does describe the sampling distribution of the proportions. The formula is just a different way of presenting the same information in the table that describes the sampling distribution.

Section 6-5

1. It is the standard deviation of the sample means, which is denoted as $\sigma_{\bar{x}}$ or $\sigma/\sqrt{n}$.

3. The notation $\mu_{\bar{x}}$ represents the mean of the population consisting of all sample means. The notation $\sigma_{\bar{x}}$ represents the standard deviation of the population consisting of all sample means.

5. a. 0.4761 (TI-83/84 Plus: 0.4779)
   b. 0.2912 (TI-83/84 Plus: 0.2898)

7. a. 0.0316 (TI-83/84 Plus: 0.0304)
   b. 0.1227 (TI-83/84 Plus: 0.1210)
   c. If the original population is normally distributed, then the distribution of sample means is normally distributed for *any* sample size.

9. a. 0.3897 (TI-83/84 Plus: 0.3913)
   b. 0.1093 (TI-83/84 Plus: 0.1087)
   c. Yes. The probability of exceeding the 3500 lb safe capacity is 0.1093, which is far too high.

11. a. 0.5675 (TI-83/84 Plus: 0.5684)
    b. 0.7257 (TI-83/84 Plus: 0.7248)
    c. The gondola is probably designed to safely carry a load somewhat greater than 2004 lb, but the operators would be wise to avoid a load of 12 men, especially if they appear to have high weights.

13. a. 0.0274 (TI-83/84 Plus: 0.0272)
    b. 0.0001
    c. Because the original population is normally distributed, the sampling distribution of sample means will be normally distributed for any sample size.
    d. No, the mean can be less than 140 while individual values are above 140.

15. a. 0.8577 (TI-83/84 Plus: 0.8580)
    b. 0.9999 (TI-83/84 Plus: Most technologies provide a result of 1, but the actual probability is very slightly less than 1.)
    c. The probability from part (a) is more relevant because it shows that 85.77% of male passengers will not need to bend. The result from part (b) gives us information about the mean for a group of 100 men, but it doesn't give us useful information about the comfort and safety of individual male passengers.
    d. Because men are generally taller than women, a design that accommodates a suitable proportion of men will necessarily accommodate a greater proportion of women.

17. a. 0.5302 (TI-83/84 Plus: 0.5317)
    b. 0.7323 (TI-83/84 Plus: 0.7326)
    c. Part (a), because the seats will be occupied by individual women, not groups of women.

19. 0.0001 (TI-83/84 Plus: 0.0000). The results do suggest that the Pepsi cans are filled with an amount greater than 12.00 oz.

21. a. 73.6 in.
    b. 69.5 in.
    c. The result from part (a) is more relevant because it indicates that the doorway height of 73.6 in. is adequate for 95% of individual men. When designing the doorway height, we are concerned with the distribution of heights of men, not the sampling distribution of mean heights of men.

23. a. Yes. Sampling is without replacement (because the 12 passengers are different people) and the sample size of 12 is greater than 5% of the finite population of size 210.

   b. 175 lb

   c. 0.0918 (TI-83/84 Plus: 0.0910). This probability is too high; the elevator could be overloaded about 9% of the time.

   d. 10 passengers

Section 6-6

1. The histogram should be approximately normal or bell shaped, because sample proportions tend to approximate a normal distribution.

3. No. With $n = 4$ and $p = 0.5$, the requirements of $np \geq 5$ and $nq \geq 5$ are not satisfied, so the normal distribution is not a good approximation.

5. The area to the right of 8.5

7. The area to the left of 4.5

9. The area to the left of 15.5

11. The area between 4.5 and 9.5

13. 0.117; normal approximation: 0.1140 (TI-83/84 Plus: 0.1145)

15. 0.962; normal approximation is not suitable

17. 0.0018 (TI-83/84 Plus using binomial: 0.0020). Because that probability is very low, it is not likely that the goal of at least 1300 will be reached.

19. 0.0001 (TI-83/84 Plus: 0.0000). The method appears to be effective, because the probability of getting at least 525 girls by chance is so small.

21. a. 0.0318 (TI-83/84 Plus using normal approximation: 0.0305; TI-83/84 Plus using binomial: 0.0301)

   b. 0.2676 (TI-83/84 Plus using normal approximation: 0.2665; TI-83/84 Plus using binomial: 0.2650)

   c. Part (b)

   d. No, the results could easily occur by chance with a 25% rate.

23. 0.2709 (TI-83/84 Plus using normal approximation: 0.2697; TI-83/84 Plus using binomial: 0.2726). Media reports appear to be wrong.

25. 0.7704 (TI-83/84 Plus using normal approximation: 0.7717; TI-83/84 Plus using binomial: 0.7657). There is a 0.7704 probability that the pool of 200 volunteers is sufficient. Because blood banks are so important, it would probably be much better to get more donors so that the probability of reaching the goal of at least 10 universal donors becomes more likely.

27. 0.2776 (TI-83/84 Plus using binomial: 0.2748). No, 27 blue M&Ms is not unusually high because the probability of 27 or more blue M&Ms is 0.2776, which is not small.

29. 0.3015 (TI-83/84 Plus using normal approximation: 0.3000; TI-83/84 Plus using binomial: 0.2900). Lipitor users do not appear to be more likely to have flu symptoms.

31. 0.4168 (TI-83/84 Plus using normal approximation: 0.4153; TI-83/84 Plus using binomial: 0.4264). No, the probability is too high, so it is not wise to accept 236 reservations.

33. a. 6; 0.4602 (TI-83/84 Plus using normal approximation: 0.4583; TI-83/84 Plus using binomial: 0.4307)

   b. 101; 0.3936 (TI-83/84 Plus using normal approximation: 0.3933; TI-83/84 Plus using binomial: 0.3932)

   c. The roulette game provides a better likelihood of making a profit.

35. a. 0.821

   b. 0.9993 (TI-83/84 Plus using binomial: 0.9995)

   c. 0.0000165

   d. 0.552

Section 6-7

1. A normal quantile plot can be used to determine whether sample data are from a population having a normal distribution.

3. Because the weights are likely to have a normal distribution, the pattern of the points should be approximately a straight-line pattern.

5. Not normal. The points are not reasonably close to a straight-line pattern.

7. Normal. The points are reasonably close to a straight-line pattern and there is no other pattern that is not a straight-line pattern.

9. Normal

11. Not normal

13. Normal

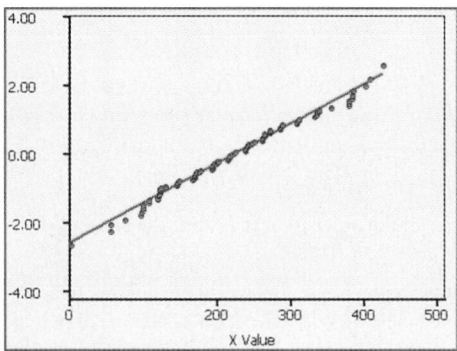

15. Not normal

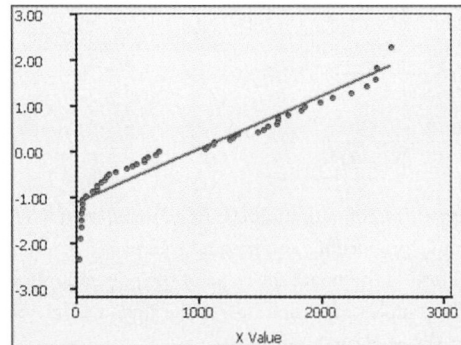

17. Heights appear to be normal, but cholesterol levels do not appear to be normal. Cholesterol levels are strongly affected by diet, and diets might vary in dramatically different ways that do not yield normally distributed results.

19. Normal; -1.28, -0.52, 0, 0.52, 1.28.

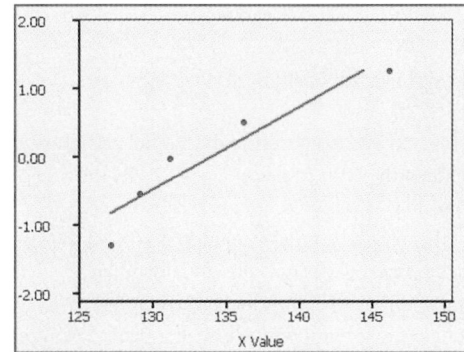

21. a. Yes
    b. Yes
    c. No

Chapter 6 Statistical Literacy and Critical Thinking

1. A normal distribution is the probability distribution of a continuous random variable with the properties that a graph of the distribution is symmetric and bell shaped, and it can be described by Formula 6-1. A standard normal distribution is a normal distribution that has mean of 0 and a standard deviation of 1.

2. No. The term "normal" distribution has a special meaning in statistics that is different from the ordinary usage of the word "normal." One property of a normal distribution is that it is bell shaped and symmetric, so the skewed distribution of incomes is not a normal distribution.

3. The sample means will tend to have a normal distribution.

4. No. The sample might well be biased. It is a convenience sample, not a simple random sample. Those with strong feelings about the issues involved might be more likely to respond, and that group might be very different from the general population.

Chapter 6 Quick Quiz

1. 1.88
2. Normal
3. $\mu = 0$ and $\sigma = 1$.
4. 0.1587
5. 0.9270
6. 0.8413
7. 0.1151
8. 0.5762 (TI-83/84 Plus: 0.5763)
9. 0.8413
10. 0.0228

Chapter 6 Review Exercises

1. a. 1.62% (TI-83/84 Plus: 1.61%)
    b. 0.01% (TI-83/84 Plus: 0.00%)
    c. The day bed length appears to be adequate, because very few men or women have heights that exceed it.

2. 73.6 in.

3. a. Men: 0.07%; women: 0.01% (TI-83/84 Plus: 0.00%)
    b. 75.5 in.

4. 12.22%. Yes, because the minimum acceptable height is greater than the mean height of all women. All of the Rockettes are taller than the mean height of all women.

5. 0.2296 (TI-83/84 Plus using normal approximation: 0.2286; TI-83/84 Plus using binomial: 0.2278). The occurrence of 787 offspring plants with long stems is not unusually low, because its probability is not small. The results are consistent with Mendel's claimed proportion of 3/4.

6. Parts (a), (b), (c) are true.

7. a. 0.0222 (TI-83/84 Plus: 0.0221)
    b. 0.2847 (TI-83/84 Plus: 0.2836)
    c. 0.6720 (TI-83/84 Plus: 0.6715)
    d. 254.6 mg/100 mL

8. 0.0778 (TI-83/84 Plus using normal approximation: 0.0774; TI-83/84 Plus using binomial: 0.0769). Because the probability is not very small, such as less than 0.05, the results could easily occur by chance. There is not strong evidence to charge that the Newport Temp Agency is discriminating against women.

9. a. 0.44
    b. -2.65
    c. 1.96

10. a. Normal
    b. 3420 g
    c. 53.7 g

11. a. 0.4602 (TI-83/84 Plus: 0.4588)
    b. 0.0655 (TI-83/84 Plus: 0.0656). Yes, because the weight limit will be exceeded about 7% of the time, or about one flight out of every 15 flights.

12. Yes. A histogram is approximately bell shaped, and a normal quantile plot results in points that are reasonably close to a straight-line pattern without any other systematic pattern.

Cumulative Review Exercises

1. a. $327,900
    b. $223,500
    c. $250,307
    d. 62,653,600,000 square dollars (TI-83/84 Plus: 62,653,655,570 square dollars)
    e. -0.37
    f. Ratio
    g. Discrete

2. a. A simple random sample is a sample selected in such a way that every possible sample of the same size has the same chance of being selected.
    b. A voluntary response is a sample obtained in such a way that the respondents themselves decide whether to be included. Because the respondents decide whether to be included, it is common that those with strong or special interests are more likely to respond, so the sample is biased and not representative of the population.

3. a. 0.0000412
    b. 0.1401 (TI-83/84 Plus using normal approximation: 0.1399; TI-83/84 Plus using binomial: 0.1406)
    c. No. The probability of 40 or more viral infections is not small, such as less than 0.05, so getting as many as 40 or more viral infections could easily occur, and it is not unusually high.

d. No. We would need evidence to show that the rate of viral infections among Nasonex users is significantly greater than the rate of viral infections among those not treated with Nasonex, but that information was not given here.

4. By using a vertical scale that does not begin at 0, the graph distorts the data by exaggerating the differences.

5. a. 0.001

   b. 0.271

   c. The requirement that $np \geq 5$ is not satisfied, indicating that the normal approximation would result in errors that are too large.

   d. 5.0

   e. 2.1

   f. No, 8 is within two standard deviations of the mean and is within the range of values that could easily occur by chance.

Chapter 7 Answers
Section 7-2

1. The level of confidence (such as 95%) was not provided.

3. The point estimate of 43% or 0.43 does not reveal any information about the accuracy of the estimate. By providing a range of values associated with a probability (confidence level), a confidence interval does reveal important information about accuracy.

5. 2.575 (TI-83/84 Plus: 2.5758293)

7. 1.645 (TI-83/84 Plus: 1.6448536)

9. 0.350 ± 0.150

11. 0.483 ± 0.046

13. 0.370; 0.050

15. 0.480; 0.047

17. 0.0304

19. 0.0325

21. $0.145 < p < 0.255$

23. $0.0674 < p < 0.109$

25. 475

27. 1996 (TI-83/84 Plus: 1998)

29. a. 0.915

   b. $0.892 < p < 0.937$

   c. Yes. The true proportion of girls with the XSORT method is substantially above the proportion of (about) 0.5 that is expected with no method of gender selection.

31. a. 0.505

   b. $0.496 < p < 0.514$

   c. No, because the proportion could easily equal 0.5. The proportion is not substantially less than 0.5 the week before Thanksgiving.

33. a. $0.226 < p < 0.298$

   b. No, the confidence interval includes 0.25, so the true percentage could easily equal 25%.

35. a. $0.0267\% < p < 0.0376\%$

   b. No, because 0.0340% is included in the confidence interval.

37. $0.714 < p < 0.746$. Although the confidence interval limits do not include the value of 3/4 or 0.75, the statement gives the value of 3/4, which is not likely to be wrong by very much.

39. $0.615 < p < 0.625$. No. The sample is a voluntary response sample that might not be representative of the population.

41. a. 4145 (TI-83/84 Plus: 4147)

   b. 3268 (TI-83/84 Plus: 3270)

43. 1068

45. $11.3\% < p < 26.7\%$. Yes. Because the confidence interval limits do contain the claimed percentage of 16%, the true percentage of green M&Ms could be equal to 16%.

47. Wednesday: $0.178 < p < 0.425$; Sunday: $0.165 < p < 0.412$. The confidence intervals are not substantially different. Precipitation does not appear to occur more on either day.

49. 985; the sample size is not too much smaller than the sample of 1068 that would be required for a very large population.

51. $0.0395 < p < 0.710$; no

53. a. The requirement of at least 5 successes and at least 5 failures is not satisfied, so the normal distribution cannot be used.

   b. 0.15

Section 7-3

1. A point estimate is a single value used to estimate the population parameter. The best point estimate of the population mean is found by computing the value of the sample mean $\bar{x}$.

3. Assuming that the standard deviation of the heights of all women is 2.5 in., the mean height of all women is estimated to be 63.20 in., with a margin of error of 0.78 in. In theory, in 95% of such studies, the mean should differ by no more than 0.78 in. in either direction from the mean that would be found by using the heights of all women.

5. 1.645 (TI-83/84 Plus: 1.6448536)

7. 1.28 (TI-83/84 Plus: 1.2815516)

9. $E = 18.8$; $658.2 < \mu < 695.8$

11. The margin of error and confidence interval cannot be calculated by the methods of this section.

13. 1974

15. 2981 (TI-83/84 Plus: 2982)

17. 21.120 mg

19. 21.120 mg ± 1.267 mg

21. a. 146.22 lb

   b. 136.66 lb $< \mu < 155.78$ lb

23. a. 58.3 sec

   b. 55.4 sec $< \mu < 61.2$ sec

   c. Yes, because the confidence interval limits do contain 60 sec.

25. a. $1464 < \mu < 1580$

   b. $1445 < \mu < 1599$

   c. The 99% confidence interval is wider. We need a wider range of values to be more confident that the interval contains the population mean.

27. $128.7 < \mu < 139.2$; ideally, all of the measurements would be the same, so there would not be an interval estimate.

29. $\$89.9$ million $< \mu < \$156.1$ million

31. 35

33. 6907. The sample size is too large to be practical.

35. 38,416

37. $1486 < \mu < 1558$. The confidence interval becomes much narrower because the sample is such a large segment of the population.

Section 7-4

1. The amount refers to an average, which is likely to be the mean or median, but the margin of error is appropriate for a proportion, not a mean or median. The margin of error should be an amount in dollars, not percentage points.

3. No. The simple random sample is obtained from Californians, so it would not necessarily yield a good estimate for the United States population of adults. It is very possible that Californians are not representative of the United States. A simple random sample of the population is needed.

5. $t_{\alpha/2} = 2.074$

7. Neither normal nor t distribution applies.

9. $z_{\alpha/2} = 1.645$

11. $t_{\alpha/2} = 3.106$

13. $E = \$266; \$8738 < \mu < \$9270$

15. $8.0518 \text{ g} < \mu < 8.0903 \text{ g}$. There is 95% confidence that the limits of 8.0518 g and 8.0903 g contain the mean weight of the population of all dollar coins.

17. a. 3.2
    b. $-2.3 < \mu < 8.7$. Because the confidence interval limits contain 0, it is very possible that the mean of the changes in LDL cholesterol is equal to 0, suggesting that the Garlicin treatment did not affect the LDL cholesterol levels. It does not appear that the Garlicin treatment is effective in lowering LDL cholesterol.

19. a. 98.20°F
    b. $98.04°F < \mu < 98.36°F$; because the confidence interval limits do not contain 98.6°F, the results suggest that the mean is less than 98.6°F.

21. a. $5.8 \text{ days} < \mu < 6.2 \text{ days}$
    b. $5.9 \text{ days} < \mu < 6.3 \text{ days}$
    c. The two confidence intervals are very similar. The echinacea treatment group does not appear to fare any better than the placebo group, so the echinacea treatment does not appear to be effective.

23. a. $3.9 < \mu < 6.1$
    b. $3.3 < \mu < 6.1$
    c. The confidence intervals do not differ by very much, so it appears that the magnet treatment and the sham treatment produce similar results. The magnet treatment does not appear to be effective.

25. $-0.471 < \mu < 3.547$; the confidence interval is likely to be a poor estimate because the value of 5.40 appears to be an outlier, suggesting that the assumption of a normally distributed population is not correct.

27. a. $\$11.7 \text{ million} < \mu < \$29.1 \text{ million}$
    b. No.
    c. Reasonable answers for the population are the salaries of all TV personalities, or the top 10 TV salaries for different years, but the given sample is not representative of either of these populations.
    d. No. The nature of the sample makes the confidence interval meaningless as an estimate of the mean of some larger population.

29. A normal quantile plot shows that the data appear to be from a normally distributed population. $4092.2 \text{ sec} < \mu < 4594.1 \text{ sec}$ (TI-83/84 Plus: $4092.3 \text{ sec} < \mu < 4594.1 \text{ sec}$)

31. a. $1.16 < \mu < 1.35$
    b. $0.81 < \mu < 1.02$
    c. The filtered cigarettes result in a confidence interval that is completely less than the confidence interval for the nonfiltered cigarettes. Assuming that the king-size cigarettes and the 100 mm cigarettes contain about the same amount of tobacco, it appears that the filters are effective in reducing the amount of nicotine.

33. $40.5 \text{ years} < \mu < 86.9 \text{ years}$ (TI-83/84 Plus: $40.5 < \mu < 86.8 \text{ years}$). This result is substantially different from the confidence interval of $52.3 \text{ years} < \mu < 57.4 \text{ years}$ that was found in Exercise 30. Confidence intervals are very sensitive to outliers. Outliers should be carefully examined and discarded if they are found to be errors. If an outlier is a correct value, it might be very helpful to see its effects by constructing the confidence interval with and without the outlier included.

35. $0.8462 \text{ g} < \mu < 0.8668 \text{ g}; 0.8474 \text{ g} < \mu < 0.8656 \text{ g}$ the second confidence interval is narrower, indicating that we have a more accurate estimate when the relatively large sample is from a relatively small finite population.

Section 7-5

1. There is 95% confidence that the limits of 0.0455 g and 0.0602 g contain the true value of the standard deviation of the population of all M&Ms.

3. No. The outcomes are equally likely and they have a uniform distribution, not a normal distribution as required. Because the requirement of normality is not satisfied, the resulting confidence interval cannot be expected to provide a good estimate.

5. 2.180, 17.535

7. 51.172, 116.321

9. $265 < \sigma < 448$

11. $1.148 < \sigma < 6.015$ (TI-83/84 Plus: $1.148 < \sigma < 6.016$)

13. 19,205; no, the sample size is excessively large for most applications.

15. 101; yes, the sample size is small enough to be practical.

17. $586 \text{ g} < \sigma < 717 \text{ g}$; no, because the confidence interval limits contain 696 g.

19. a. $17.7 \text{ min} < \sigma < 32.4 \text{ min}$
    b. $14.7 \text{ min} < \sigma < 35.3 \text{ min}$
    c. The two confidence intervals are not substantially different. There does not appear to be a difference between the standard deviation of lengths of movies rated PG or PG-13 and lengths of movies rated R.

21. $253.2 \text{ sec} < \sigma < 811.8 \text{ sec}$

23. $1.195 < \sigma < 4.695$; yes, the confidence interval is likely to be a poor estimate because the value of 5.40 appears to be an outlier, suggesting that the assumption of a normally distributed population is not correct.

25. $80.6 < \sigma < 106.5$ (TI-83/84 Plus: $81.0 < \sigma < 107.1$)

27. 152.3644 and 228.4771 are close to the STATDISK values.

Chapter 7 Statistical Literacy and Critical Thinking

1. A point estimate is the single value of the sample proportion, and it is the best estimate of the population proportion. A confidence interval is a range of values associated with a confidence level, and it consists of the range of values that is likely to contain the value of the population proportion. By including a range of values with an associated confidence level, the confidence interval has the advantage of giving meaningful information about how accurate the estimate of the population proportion is likely to be.

2. There is 95% confidence that the limits of 0.0262 and 0.0499 contain the true value of the population proportion.

3. The confidence interval has a confidence level of 95% or 0.95. In general, the confidence level is the proportion of times that the confidence interval actually does contain the value of the population parameter, assuming that the process of sampling and creating a confidence interval is repeated a large number of times.

4. Because the poll uses a voluntary response sample, it is very possible that the respondents are not representative of the general population, so the confidence interval should not be used to form conclusions about the general population.

Chapter 7 Quick Quiz

1. There is 95% confidence that the limits of 10.0 and 20.0 contain the true value of the population mean μ. This means that if the sampling procedure is repeated many times, in the long run 95% of the resulting confidence interval limits will contain the true value of the population mean μ.

2. (2) The election is too close to call.

3. 2.093

4. 1.645

5. 2401

6. 0.4

7. $0.361 < p < 0.439$

8. $36.6 \text{ years} < \mu < 43.4 \text{ years}$

9. $36.7 \text{ years} < \mu < 43.3 \text{ years}$

10. 2213

Chapter 7 Review Exercises

1. $76.1\% < p < 82.0\%$. There is 95% confidence that the limits of 76.1% and 82.0% contain the true value of the percentage of all adults who believe that it is morally wrong to not report all income on tax returns.

2. 4145 (TI-83/84 Plus: 4147)

3. 21,976 (TI-83/84 Plus: 21,991). The sample size is too large to be practical, unless the data can be processed automatically.

4. $2.492 \text{ g} < \mu < 2.506 \text{ g}$. The confidence interval does contain the required mean value of 2.5 g, so the manufacturing process appears to be OK.

5. $368.4 < \mu < 969.2$

6. $145.0 < \sigma < 695.6$ (TI-83/84 Plus: $145.0 < \sigma < 695.3$)

7. a. 0.890
   b. $0.871 < p < 0.910$
   c. Yes. Because there is 95% confidence that the limits of 0.871 and 0.910 contain the true value of the population proportion, it appears that there is strong evidence that the population proportion is greater than 0.5.

8. a. 601
   b. 1383
   c. 1383

9. a. 3.840 lb
   b. $1.786 \text{ lb} < \mu < 5.894 \text{ lb}$
   c. $1.686 \text{ lb} < \mu < 5.994 \text{ lb}$

10. a. $1.624 \text{ lb} < \sigma < 5.000 \text{ lb}$
    b. $2.639 \text{ lb}^2 < \sigma^2 < 25.002 \text{ lb}^2$ (TI-83/84 Plus: $2.639 \text{ lb}^2 < \sigma^2 < 25.004 \text{ lb}^2$)

Chapter 7 Cumulative Review Exercises

1. 118.4 lb, 121.0 lb, 9.2 lb

2. Ratio

3. $111.8 \text{ lb} < \mu < 125.0 \text{ lb}$

4. 55

5. a. 0.962
   b. 0.00144
   c. 0.4522 (TI-83/84 Plus using normal approximation: 0.4534; TI-83/84 Plus using binomial probability distribution: 0.4394)

6. a. 0.5910 (TI-83/84 Plus: 0.5906)
   b. 0.8749 (TI-83/84 Plus: 0.8741)
   c. 27.2 (TI-83/84 Plus: 27.3)

7. A sample of n items is a simple random sample if every sample of the same size has the same chance of being selected. A sample is a voluntary response sample if the subjects decide themselves whether to be included.

8. 1

9. With random guesses, the probability of 12 correct answers to 12 true–false questions is 1/4096 or 0.000244. It is possible that random guesses were made, but it is highly unlikely.

10. It is a convenience sample, and it is not likely to be representative of the population.

Chapter 8 Answers
Section 8-2

1. Given the large sample size and the sample percentage of 20%, the sample data appear to support the claim that less than 50% of people believe that bosses are good communicators. Given that the survey subjects constitute a voluntary response sample instead of a simple random sample, the sample results should not be used to form any conclusions about the general population.

3. The claim that the mean is equal to 325 mg becomes the null hypothesis. The hypothesis testing procedure allows us to reject that null hypothesis or we can fail to reject it, but we can never support a null hypothesis. Hypothesis testing can never be used to support a claim that a parameter is equal to some particular value.

5. There is sufficient evidence to support the claim that the coin favors heads (because 90 heads in 100 tosses is very unlikely to occur by chance with a fair coin).

7. There is not sufficient evidence to support the claim that the mean pulse rate is less than 75.

9. H_0: μ = \$60,000. H_1: μ > \$60,000.

11. H_0: σ = 0.62°F. H_1: σ ≠ 0.62°F.

13. H_0: σ = 40 sec. H_1: σ < 40 sec.

15. H_0: p = 0.80. H_1: p ≠ 0.80.

17. z = ±2.575 (TI-83/84 Plus: ±2.5758)

19. z = 2.05

21. z = ±1.96

23. z = −2.575 (TI-83/84 Plus: −2.5758)

25. z = 0.67

27. z = 2.31

29. 0.1056; fail to reject the null hypothesis.

31. 0.0802 (TI-83/84 Plus: 0.0801); fail to reject the null hypothesis.

33. 0.0060; reject the null hypothesis.

35. 0.0107; reject the null hypothesis.

37. There is not sufficient evidence to support the claim that the percentage of blue M&Ms is greater than 5%.

39. There is not sufficient evidence to warrant rejection of the claim that the percentage of Americans who know their credit score is equal to 20%.

41. Type I error: Reject the claim that the percentage of nonsmokers exposed to secondhand smoke is equal to 41%, when that percentage is actually 41%. Type II error: Fail to reject the claim that the percentage of nonsmokers exposed to secondhand smoke is equal to 41% when that percentage is actually different from 41%.

43. Type I error: Reject the claim that the percentage of college students who use alcohol is equal to 70% when that percentage is actually 70%. Type II error: Fail to reject the claim that the percentage of college students who use alcohol is equal to 70% when that percentage is actually greater than 70%.

45. a. No, because rejection at the 0.01 significance level requires a sample statistic that differs from the claimed value of the parameter by an amount that is more extreme than the difference required for rejection at the 0.05 significance level. If H_0 is rejected at the 0.05 level, it may or may not be rejected at the 0.01 level.

    b. Yes, because rejection at the 0.01 significance level requires a sample statistic that differs from the claimed value of the parameter by an amount that is more extreme than the difference required for rejection at the 0.05 significance level, so rejection of H_0 at the 0.01 level requires that H_0 must also be rejected at the 0.05 level.

47. a. 0.7852 (TI-83/84 Plus: 0.7857)

    b. Assuming that p = 0.5, as in the null hypothesis, the critical value of z = 1.645 corresponds to $\hat{p}$ = 0.6028125, so any sample proportion greater than 0.6028125 causes us to reject the null hypothesis, as shown in the shaded critical region of the top graph. If p is actually 0.65, then the null hypothesis of p = 0.5 is false, and the actual probability of rejecting the null hypothesis is found by finding the area greater than $\hat{p}$ = 0.6028125 in the bottom graph, which is the shaded area in the bottom graph. That is, the shaded area in the bottom graph represents the probability of rejecting the false null hypothesis.

Section 8-3

1. $\hat{p}$ = 0.230. The symbol $\hat{p}$ is used to represent a sample proportion.

3. The sample proportion is $\hat{p}$ = 0.439. Because the P-value of 0.9789 is high, we should not reject the null hypothesis of p = 0.5. We should conclude that there is not sufficient evidence to support the claim that p > 0.5.

5. a. z = 1.90

   b. z = ±2.575 (TI-83/84 Plus: ±2.576)

   c. 0.0574 (TI-83/84 Plus: 0.0578)

   d. There is not sufficient evidence to reject the null hypothesis that the percentage of college applications submitted online is equal to 50%.

   e. No, a hypothesis test cannot be used to prove that a population proportion is equal to some claimed value.

7. H_0: p = 0.75. H_1: p ≠ 0.75. Test statistic: z = −2.17. P-value = 0.0301. Reject H_0. There is sufficient evidence to warrant rejection of the claim that the proportion of adults who use cell phones while driving is equal to 75%.

9. H_0: p = 0.75. H_1: p ≠ 0.75. Test statistic: z = 2.56. Critical values: z = ±2.575. P-value: 0.0104 (TI-83/84 Plus: 0.0105). Fail to reject H_0. There is not sufficient evidence to warrant rejection of the claim that 75% of adults say that it is morally wrong to not report all income on tax returns.

11. H_0: p = 1/3. H_1: p > 1/3. Test statistic: z = 3.47. Critical value: z = 2.33. P-value: 0.0003. Reject H_0. There is sufficient evidence to support the claim that the proportion of challenges that are successful is greater than 1/3. When evaluating the quality of referee calls, it should be noted that only close calls are likely to be challenged, and 327 successful challenges among thousands of calls is a very small proportion of wrong calls, so the quality of the referees is quite good, but they are not perfect.

13. H_0: p = 0.06. H_1: p > 0.06. Test statistic: z = 4.47. Critical value: z = 1.645. P-value: 0.0001 (TI-83/84 Plus: 0.000004). Reject H_0. There is sufficient evidence to support the claim that the rate of nausea is greater than the 6% rate experienced by flu patients given a placebo. The rate of nausea with the Tamiflu treatment appears to be greater than the rate for those given a placebo, and the sample proportion of 0.0994 suggests that about 10% of those treated experience nausea, so nausea is a concern for those given the treatment.

15. H_0: p = 0.000340. H_1: p ≠ 0.000340. Test statistic: z = −0.66. Critical values: z = ±2.81. P-value: 0.5092 (TI-83/84 Plus: 0.5122). Fail to reject H_0. There is not sufficient evidence to support the claim that the rate is different from 0.0340%. Cell phone users should not be concerned about cancer of the brain or nervous system.

17. H_0: p = 0.20. H_1: p < 0.20. Test statistic: z = −2.93. Critical value: z = −2.33. P-value: 0.0017. Reject H_0. There is sufficient evidence to support the claim that less than 20% of Michigan gas pumps are inaccurate. The percentage of inaccurate pumps should be very low, but this hypothesis test shows only that it appears to be less than 20%. The sample proportion of 0.186 suggests that too many pumps are not accurate.

19. H_0: $p = 0.80$. H_1: $p < 0.80$. Test statistic: $z = -1.11$. Critical value: $z = -1.645$. *P*-value: 0.1335 (TI-83/84 Plus: 0.1332). Fail to reject H_0. There is not sufficient evidence to support the claim that the polygraph results are correct less than 80% of the time. Based on the sample proportion of correct results in 75.5% of the 98 cases, polygraph results do not appear to have the high degree of reliability that would justify the use of polygraph results in court, so polygraph test results should be prohibited as evidence in trials.

21. H_0: $p = 0.20$. H_1: $p < 0.20$. Test statistic: $z = -8.84$. Critical value: $z = -2.33$. *P*-value: 0.0001 (TI-83/84 Plus: 0.0000). Reject H_0. There is sufficient evidence to support the claim that less than 20% of the TV sets in use were tuned to *60 Minutes*.

23. H_0: $p = 0.5$. H_1: $p \neq 0.5$. Test statistic: $z = -0.66$ (using $\hat{p} = 0.473$) or $z = -0.65$ (using $x = 71$). Critical values: $z = \pm 1.96$ (assuming a 0.05 significance level). *P*-value: 0.5092 or 0.5156 (TI-83/84 Plus: 0.5136). Fail to reject H_0. There is not sufficient evidence to warrant rejection of the claim that among senior executives, 50% say that the most common job interview mistake is to have little or no knowledge of the company. The important lesson is to always prepare for a job interview by learning about the company.

25. H_0: $p = 3/4$. H_1: $p \neq 3/4$. Test statistic: $z = -2.53$ (using $\hat{p} = 0.73$) or $z = -2.54$ (using $x = 2198$). Critical values: $z = \pm 1.96$ (assuming a 0.05 significance level). *P*-value: 0.0114 or 0.0110 (TI-83/84 Plus: 0.0112). Reject H_0. There is sufficient evidence to warrant rejection of the claim that 3/4 of all adults use the Internet. The proportion appears to be different from 3/4 or 0.75. The reporter should not write that 3/4 of all adults use the Internet.

27. H_0: $p = 0.5$. H_1: $p \neq 0.5$. Test statistic: $z = -1.08$ (using $\hat{p} = 0.43$) or $z = -1.17$ (using $x = 25$). Critical values: $z = \pm 1.96$. *P*-value: 0.2802 or 0.2420 (TI-83/84 Plus: 0.2413). Fail to reject H_0. There is not sufficient evidence to warrant rejection of the claim that these women have no ability to predict the sex of their baby. The results for these women with 12 years of education or less suggest that their percentage of correct predictions is not very different from results expected with random guesses.

29. H_0: $p = 1/4$. H_1: $p > 1/4$. Test statistic: $z = 5.73$. Critical value: $z = 1.645$. *P*-value: 0.0001 (TI-83/84 Plus: 0.0000). Reject H_0. There is sufficient evidence to support the claim that more than 1/4 of people say that bosses scream at employees. If the sample is a voluntary response sample, the conclusion about the population might not be valid.

31. H_0: $p = 0.5$. H_1: $p > 0.5$. Test statistic: $z = 5.83$ (using $\hat{p} = 0.61$) or $z = 5.85$ (using $x = 429$). Critical value: $z = 1.645$. *P*-value: 0.0001 (TI-83/84 Plus: 0.0000). Reject H_0. There is sufficient evidence to support the claim that most workers get their jobs through networking.

33. H_0: $p = 0.20$. H_1: $p \neq 0.20$. Test statistic: $z = -1.75$. Critical values: $z = \pm 1.96$ (assuming a 0.05 significance level). *P*-value: 0.0802 (TI-83/84 Plus: 0.0801). Fail to reject H_0. There is not sufficient evidence to warrant rejection of the claim that 20% of plain M&M candies are red.

35. H_0: $p = 0.5$. H_1: $p \neq 0.5$. Test statistic: $z = 2.18$. Critical values: $z = \pm 1.96$. *P*-value: 0.0292 (TI-83/84 Plus: 0.0295). Reject H_0. There is sufficient evidence to warrant rejection of the claim that the percentage of males in the population is equal to 50%.

37. H_0: $p = 0.55$. H_1: $p \neq 0.55$. Using the binomial probability distribution with an assumed value of $p = 0.55$ and with $n = 35$, the probability of 12 or fewer movies with R ratings is 0.0109367, so the *P*-value is 0.02187. Because that *P*-value is greater than the significance level of 0.01, fail to reject H_0. There is not sufficient evidence to warrant rejection of the claim that the movies in Data Set 9 are from a population in which 55% of the movies have R ratings.

39. H_0: $p = 0.10$. H_1: $p \neq 0.10$. Test statistic: $z = -2.36$. Critical values: $z = \pm 2.575$. *P*-value: 0.0182 (TI-83/84 Plus: 0.0184). Fail to reject H_0. Even though no blue candies are obtained, there is not sufficient evidence to warrant rejection of the claim that 10% of the candies are blue.

Section 8-4

1. The sample must be a simple random sample, the population standard deviation must be known, and the population must be normally distributed (because the sample size is not greater than 30).

3. 98% or 0.98

5. H_0: $\mu = 5$ cm. H_1: $\mu \neq 5$ cm. Test statistic: $z = 1.34$. Critical values: $z = \pm 1.96$ (assuming a 0.05 significance level). *P*-value: 0.1797. Fail to reject H_0. There is not sufficient evidence to warrant rejection of the claim that women have a mean wrist breadth equal to 5 cm.

7. H_0: $\mu = 210$ sec. H_1: $\mu > 210$ sec. Test statistic: $z = 4.93$. Critical value: $z = 1.645$. *P*-value: 0.0001 (TI-83/84 Plus: 0.0000). Reject H_0. There is sufficient evidence to support the claim that the sample is from a population of songs with a mean length greater than 210 sec. These results suggest that the advice of writing a song that must be no longer than 210 seconds is not sound advice.

9. H_0: $\mu = 0.8535$ g. H_1: $\mu \neq 0.8535$ g. Test statistic: $z = 0.77$. Critical values: $z = \pm 1.96$. *P*-value: 0.4412 (TI-83/84 Plus: 0.4404). Fail to reject H_0. There is not sufficient evidence to warrant rejection of the claim that green M&Ms have a mean weight of 0.8535 g. The green M&Ms appear to have weights consistent with the package label.

11. H_0: $\mu = 0$ lb. H_1: $\mu > 0$ lb. Test statistic: $z = 3.87$. Critical value: $z = 2.33$. *P*-value: 0.0001. Reject H_0. There is sufficient evidence to support the claim that the mean weight loss is greater than 0. Although the diet appears to have statistical significance, it does not appear to have practical significance, because the mean weight loss of only 3.0 lb does not seem to be worth the effort and cost.

13. H_0: $\mu = 91.4$ cm. H_1: $\mu \neq 91.4$ cm. Test statistic: $z = 2.33$. Critical values: $z = \pm 1.96$. *P*-value: 0.0198 (TI-83/84 Plus: 0.0196). Reject H_0. There is sufficient evidence to support the claim that males at her college have a mean sitting height different from 91.4 cm. Because the student selected male friends, she used a convenience sample that might have different characteris-

tics than the population of males at her college. Because the requirement of a simple random sample is not satisfied, the method of hypothesis testing given in this section does not apply, so the results are not necessarily valid.

15. H_0: $\mu = \$500,000$. H_1: $\mu < \$500,000$. Test statistic: $z = -1.15$. Critical value: $z = -1.645$. P-value: 0.1251 (TI-83/84 Plus: 0.1257). Fail to reject H_0. There is not sufficient evidence to support the claim that the mean salary of a football coach in the NCAA is less than $500,000.

17. H_0: $\mu = 235.8$ cm. H_1: $\mu \neq 235.8$ cm. Test statistic: $z = -0.56$. Critical values: $z = \pm 1.96$. P-value: 0.5754 (TI-83/84 Plus: 0.5740). Fail to reject H_0. There is not sufficient evidence to warrant rejection of the claim that the new baseballs have a mean bounce height of 235.8 cm. The new baseballs do not appear to be different.

19. H_0: $\mu = 678$. H_1: $\mu \neq 678$. Test statistic: $z = 4.93$. Critical values: $z = \pm 1.96$. P-value: 0.0002 (TI-83/84 Plus: 0.0000). Reject H_0. There is sufficient evidence to warrant rejection of the claim that these sample FICO scores are from a population with a mean FICO score equal to 678.

21. H_0: $\mu = 3/4$ in. H_1: $\mu \neq 3/4$ in. Test statistic: $z = -1.87$. Critical values: $z = \pm 1.96$. P-value: 0.0614 (TI-83/84 Plus: 0.0610). Fail to reject H_0. There is not sufficient evidence to warrant rejection of the claim that the mean length is equal to 0.75 in. or 3/4 in. The lengths appear to be consistent with the package label.

23. a. The power of 0.2296 shows that there is a 22.96% chance of supporting the claim that $\mu > 166.3$ lb when the true mean is actually 170 lb. This value of the power is not very high, and it shows that the hypothesis test is not very effective in recognizing that the mean is greater than 166.3 lb when the actual mean is 170 lb.

   b. $\beta = 0.7704$ (TI-83/84 Plus: 0.7719). The probability of a type II error is 0.7704. That is, there is a 0.7704 probability of making the mistake of not supporting the claim that $\mu > 166.3$ lb when in reality the population mean is 170 lb.

Section 8-5

1. Because the sample size is not greater than 30, the sample data must come from a population with a normal distribution. To determine whether the requirement of a normal distribution is satisfied, examine a histogram to determine whether it is approximately normal, verify that there are no outliers (or at most one outlier), or examine a normal quantile plot, or use some formal test of normality, such as the Ryan-Joiner test described in Section 6-7.

3. A t test is a hypothesis test that uses the Student t distribution. It is called a t test because it involves use of the Student t distribution.

5. Student t

7. Neither normal nor Student t

9. H_0: $\mu = 210$ sec. H_1: $\mu > 210$ sec. Test statistic: $t = 4.93$. P-value: 0.000. Reject H_0. There is sufficient evidence to support the claim that the sample is from a population of songs with a mean length greater than 210 sec. These results suggest that the advice of writing a song that must be no longer than 210 seconds is not sound advice.

11. H_0: $\mu = 21.1$ mg. H_1: $\mu < 21.1$ mg. Test statistic: $t = -10.676$. Critical value: $t = -1.711$. P-value < 0.005 (TI-83/84 Plus: 0.0000). Reject H_0. There is sufficient evidence to support the claim that filtered 100 mm cigarettes have a mean tar amount less than 21.1 mg. The results suggest that the filters are effective in reducing the amount of tar.

13. H_0: $\mu = 2.5$ g. H_1: $\mu \neq 2.5$ g. Test statistic: $t = -0.332$. Critical value: $t = \pm 2.028$. P-value > 0.20 (TI-83/84 Plus: 0.7417). Fail to reject H_0. There is not sufficient evidence to warrant rejection of the claim that pennies have a mean weight equal to 2.5 g. The pennies appear to conform to the specifications of the U.S. Mint.

15. H_0: $\mu = 4.5$ years. H_1: $\mu > 4.5$ years. Test statistic: $t = 1.227$. Critical value: $t = 1.664$. P-value > 0.10 (TI-83/84 Plus: 0.1117). Fail to reject H_0. There is not sufficient evidence to support the claim that the mean time is greater than 4.5 years.

17. H_0: $\mu = 49.5$ cents. H_1: $\mu < 49.5$ cents. Test statistic: $t = -8.031$. Critical value: $t = -2.364$ (approximately). P-value < 0.01 (TI-83/84 Plus: 0.0000). Reject H_0. There is sufficient evidence to support the claim that the mean is less than 49.5 cents. The results suggest that the cents portions of check amounts are such that the values from 0 cents to 99 cents are not equally likely.

19. H_0: $\mu = 8.00$ tons. H_1: $\mu \neq 8.00$ tons. Test statistic: $t = -1.152$. Critical values: $t = \pm 2.040$. P-value > 0.20 (TI-83/84 Plus: 0.2580). Fail to reject H_0. There is not sufficient evidence to warrant rejection of the claim that cars have a mean greenhouse gas emission of 8.00 tons.

21. H_0: $\mu = 1000$ hic. H_1: $\mu < 1000$ hic. Test statistic: $t = -2.661$. Critical value: $t = -3.365$. P-value is between 0.01 and 0.025 (TI-83/84 Plus: 0.0224). Fail to reject H_0. There is not sufficient evidence to support the claim that the population mean is less than 1000 hic. There is not strong evidence that the mean is less than 1000 hic, and one of the booster seats has a measurement of 1210 hic, which does not satisfy the specified requirement of being less than 1000 hic.

23. H_0: $\mu = \$5000$. H_1: $\mu \neq \$5000$. Test statistic: $t = 1.639$. Critical values: $t = \pm 2.776$. P-value is between 0.10 and 0.20 (TI-83/84 Plus: 0.1766). Fail to reject H_0. There is not sufficient evidence to reject the claim that the mean damage cost is $5000.

25. H_0: $\mu = 3/4$ in. H_1: $\mu \neq 3/4$ in. Test statistic: $t = -1.825$. Critical values: $t = \pm 2.009$ (approximately). P-value is between 0.05 and 0.10 (TI-83/84 Plus: 0.0741). Fail to reject H_0. There is not sufficient evidence to warrant rejection of the claim that the mean length is equal to 0.75 in. or 3/4 in. The lengths appear to be consistent with the package label.

27. H_0: $\mu = 98.6°F$. H_1: $\mu \neq 98.6°F$. Test statistic: $t = -6.611$. Critical values: $t = \pm 1.984$ (approximately, assuming a 0.05 significance level). P-value < 0.01 (TI-83/84 Plus: 0.0077). Reject H_0. There is sufficient evidence to warrant rejection of the claim that the sample is from a population with a mean equal to 98.6°F. The common belief appears to be wrong.

29. $H_0: \mu = 100.$ $H_1: \mu \neq 100.$ Test statistic: $t = 1.999.$ Critical values: $t = \pm 2.040.$ $0.05 < P\text{-value} < 0.10$ (TI-83/84 Plus: 0.0545). Fail to reject H_0. There is not sufficient evidence to warrant rejection of the claim that the mean equals 100. Using the alternative method with 15.0 used for σ, we get test statistic $z = 2.00$, critical values $z = \pm 1.96$, $P\text{-value} = 0.0456$, so we reject H_0 and conclude that there is sufficient evidence to warrant rejection of the claim that the mean equals 100. The conclusions are different. The alternative method does not always yield the same conclusion as the t test.

31. The same value of 1.666 is obtained.

33. Using Table A-3, the power is between 0.90 and 0.95 (TI-83/84 Plus: 0.9440). β is between 0.05 and 0.10 (TI-83/84 Plus: 0.0560). Because the power is high, the test is very effective in supporting the claim that $\mu > 166.3$ lb when the true population mean is 180 lb.

Section 8-6

1. The normality requirement for a hypothesis test of a claim about a standard deviation is much more strict, meaning that the distribution of the population must be much closer to a normal distribution.

3. No. Unlike the situation with sample means, the use of large samples does not compensate for the lack of normality. The methods of this section cannot be used with sample data from a population with a distribution that is far from normal.

5. $\chi^2 = 20.612.$ Critical values: $\chi^2 = 12.401$ and $39.364.$ $P\text{-value} > 0.20$ (and also less than 0.80). There is not sufficient evidence to support the alternative hypothesis.

7. $\chi^2 = 26.331.$ Critical value: $\chi^2 = 29.141.$ $0.01 < P\text{-value} < 0.025.$ There is not sufficient evidence to support the alternative hypothesis.

9. $H_0: \sigma = 0.0230$ g. $H_1: \sigma > 0.0230$ g. Test statistic: $\chi^2 = 98.260.$ Critical value of χ^2 is between 43.773 and 55.758. $P\text{-value}: 0.0000.$ Reject H_0. There is sufficient evidence to support the claim that pre-1983 pennies have a standard deviation greater than 0.0230 g. Weights of pre-1983 pennies appear to vary more than those of post-1983 pennies.

11. $H_0: \sigma = 3.2$ mg. $H_1: \sigma \neq 3.2$ mg. Test statistic: $\chi^2 = 32.086.$ Critical values: $\chi^2 = 12.401$ and $\chi^2 = 39.364.$ $P\text{-value}: 0.2498.$ Fail to reject H_0. There is not sufficient evidence to support the claim that the tar content of filtered 100 mm cigarettes has a standard deviation different from 3.2 mg.

13. $H_0: \sigma = 2.5$ in. $H_1: \sigma < 2.5$ in. Test statistic: $\chi^2 = 2.880.$ Critical value: $\chi^2 = 2.733.$ $P\text{-value}: 0.0583.$ Fail to reject H_0. There is not sufficient evidence to support the claim that supermodels have heights that vary less than women in the general population. The supermodels might well have heights that vary less than women in general, but the sample evidence is not strong enough to justify that conclusion.

15. $H_0: \sigma = 10.$ $H_1: \sigma \neq 10.$ Test statistic: $\chi^2 = 60.938.$ Critical values: $\chi^2 = 24.433$ (approximately) and 59.342 (approximately). $P\text{-value}: 0.0277.$ Reject H_0. There is sufficient evidence to warrant rejection of the claim that pulse rates of women have a standard deviation equal to 10.

17. $H_0: \sigma = 1.34.$ $H_1: \sigma \neq 1.34.$ Test statistic: $\chi^2 = 7.053.$ Critical values: $\chi^2 = 1.735$ and $23.589.$ $P\text{-value}: 0.7368.$ Fail to reject H_0. There is not sufficient evidence to support the claim that the recent winners have BMI values with variation different from that of the 1920s and 1930s.

19. $H_0: \sigma = 32.2$ ft. $H_1: \sigma > 32.2$ ft. Test statistic: $\chi^2 = 29.176.$ Critical value: $\chi^2 = 19.675.$ $P\text{-value}: 0.0021.$ Reject H_0. There is sufficient evidence to support the claim that the new production method has errors with a standard deviation greater than 32.2 ft. The variation appears to be greater than in the past, so the new method appears to be worse, because there will be more altimeters that have larger errors. The company should take immediate action to reduce the variation.

21. 65.673 and 137.957

Chapter 8 Statistical Literacy and Critical Thinking

1. There is sufficient evidence to support the claim that the mean is greater than 0 in. In general, a small $P\text{-value}$ indicates that the null hypothesis should be rejected. If the $P\text{-value}$ is large (such as greater than 0.05), there is not sufficient evidence to reject the null hypothesis.

2. The low $P\text{-value}$ indicates that there is sufficient evidence to support the given claim, so there is statistical significance. There is not practical significance, because the mean difference is only 0.019 in., which is roughly 1/50th in. That tiny amount does not have practical significance.

3. A voluntary response sample (or self-selected sample) is one in which the respondents themselves decide whether to be included. In general, a voluntary response sample cannot be used with a hypothesis test to make a valid conclusion about the larger population.

4. A test is robust against departures from normality if it works quite well even when the sample data are from a population with a distribution that is somewhat different from a normal distribution. The t test is robust against departures from normality, but the χ^2 test is not.

Chapter 8 Quick Quiz

1. $H_0: p = 0.5.$ $H_1: p > 0.5.$
2. t distribution
3. Chi-square distribution
4. True
5. 0.1336
6. $z = -2.04$
7. $t = -2.093$ and $t = 2.093$
8. 0.2302 (TI-83/84 Plus: 0.2301)
9. There is not sufficient evidence to support the claim that the population proportion is greater than 0.25.
10. False

Chapter 8 Review Exercises

1. H_0: $p = 1/4$. H_1: $p < 1/4$. Test statistic: $z = -0.77$. Critical value: $z = -1.645$. P-value: 0.2206. Fail to reject H_0. There is not sufficient evidence to support the claim that less than $1/4$ of adults between the ages of 18 and 44 smoke.

2. H_0: $p = 0.5$. H_1: $p > 0.5$. Test statistic: $z = 3.06$. Critical value: $z = 2.33$. P-value: 0.0011. Reject H_0. There is sufficient evidence to support the claim that among college students seeking bachelor's degrees, most earn that degree within five years.

3. H_0: $\mu = 3700$ lb. H_1: $\mu < 3700$ lb. Test statistic: $t = -1.068$. Critical value: $t = -2.453$. P-value > 0.10 (TI-83/84 Plus: 0.1469). Fail to reject H_0. There is not sufficient evidence to support the claim that the mean weight of cars is less than 3700 lb. When considering weights of cars for the purpose of constructing a road that is strong enough, the weight of the heaviest car is more relevant than the mean weight of all cars.

4. H_0: $\mu = 3700$ lb. H_1: $\mu < 3700$ lb. Test statistic: $z = -1.03$. Critical value: $z = -2.33$. P-value $= 0.1515$. Fail to reject H_0. There is not sufficient evidence to support the claim that the mean weight of cars is less than 3700 lb.

5. H_0: $p = 0.20$. H_1: $p < 0.20$. Test statistic: $z = -4.81$. Critical value: $z = -2.33$. P-value: 0.0001 (TI-83/84 Plus: 0.0000). Reject H_0. There is sufficient evidence to support the claim that fewer than 20% of adults consumed herbs within the past 12 months.

6. H_0: $\mu = 281.8$ lb. H_1: $\mu < 281.8$ lb. Test statistic: $t = -8.799$. Critical value: $t = -2.345$ (approximately). P-value < 0.005 (TI-83/84 Plus: 0.0000). Reject H_0. There is sufficient evidence to support the claim that the thinner cans have a mean axial load less than 281.8 lb. Given the values of the sample mean and standard deviation, the thinner cans appear to have axial loads that can easily withstand the top pressure of 158 lb to 165 lb.

7. H_0: $\mu = 74$. H_1: $\mu \neq 74$. Test statistic: $z = 0.32$. Critical values: $z = \pm 1.96$. P-value: 0.7490. Fail to reject H_0. There is not sufficient evidence to warrant rejection of the claim that the mean is equal to 74. The calculator appears to be working correctly.

8. H_0: $\mu = 74$. H_1: $\mu \neq 74$. Test statistic: $t = 0.342$. Critical values: $t = \pm 1.984$ (approximately). P-value > 0.20 (TI-83/84 Plus: 0.7332). Fail to reject H_0. There is not sufficient evidence to warrant rejection of the claim that the mean is equal to 74. The calculator appears to be working correctly.

9. H_0: $\sigma = 12.5$. H_1: $\sigma \neq 12.5$. Test statistics: $\chi^2 = 86.734$. Critical values: $\chi^2 = 74.222$ and 129.561 (approximately). P-value: 0.3883. Fail to reject H_0. There is not sufficient evidence to warrant rejection of the claim that the standard deviation of such generated values is equal to 12.5.

10. H_0: $\sigma = 520$ lb. H_1: $\sigma < 520$ lb. Test statistic: $\chi^2 = 28.856$. Critical value: $\chi^2 = 14.954$ (approximately). P-value: 0.4233. Fail to reject H_0. There is not sufficient evidence to support the claim that the standard deviation of the weights of cars is less than 520 lb.

Chapter 8 Cumulative Review Exercises

1. a. 10.907 sec
   b. 10.940 sec
   c. 0.178 sec
   d. 0.032 sec^2
   e. 0.540 sec

2. a. Ratio
   b. Continuous
   c. No
   d. Changing pattern over time
   e. Time series plot

3. 10.770 sec $< \mu <$ 11.043 sec. The confidence interval should not be used to estimate winning times, because the listed values show a pattern of decreasing times, so there isn't a fixed population mean.

4. H_0: $\mu = 11$ sec. H_1: $\mu < 11$ sec. Test statistic: $t = -1.576$. Critical value: $t = -1.860$. $0.05 < P$-value < 0.10 (TI-83/84 Plus: 0.0768). Fail to reject H_0. There is not sufficient evidence to support the claim that the mean winning time is less than 11 sec. Because of the changing pattern over time, which shows a trend of decreasing times, future winning times are likely to be less than 11 sec.

5. a. Yes
   b. 100
   c. 0.02 gram
   d. 0.86 gram
   e. No

6. By using a vertical scale that does not begin with 0, differences among the outcomes are exaggerated.

7. The mean is 3.47.

Outcome	Frequency
1	16
2	13
3	22
4	21
5	13
6	15

8. 0.0025

9. a. 0.1190 (TI-83/84 Plus: 0.1186)
   b. 37.07% (TI-83/84 Plus: 36.94%)
   c. 0.0268 (TI-83/84 Plus: 0.0269)
   d. 781.2 mm (TI-83/84 Plus: 781.3 mm)

10. The sample appears to be from a normally distributed population. A histogram is roughly bell-shaped. A normal quantile plot has 20 points that are close to a straight-line pattern, and there is no other pattern. There are no outliers.

Chapter 9 Answers
Section 9-2

1. The sample is a convenience sample, not a simple random sample. For the sample of males, the requirement of at least 5 successes and at least 5 failures is not satisfied.

3. $\hat{p}_1 = 15/1583$ or 0.00948; $\hat{p}_2 = 8/157$ or 0.0510; $\bar{p} = 23/1740$ or 0.0132. The symbol p_1 denotes the proportion of all Zocor users who experience headaches. The symbol p_2 represents the proportion of headaches among all people treated with a placebo.

5. 1396

7. a. $\bar{p} = 27/63 = 0.429$
   b. $z = -1.25$
   c. ± 1.96
   d. 0.2112 (TI-83/84 Plus: 0.2115)

9. $E = 0.245$; $-0.403 < p_1 - p_2 < 0.0878$

11. H_0: $p_1 = p_2$. H_1: $p_1 \neq p_2$. Test statistic: $z = -0.73$. Critical values: $z = \pm 1.96$. P-value: 0.4638. Fail to reject H_0. There is not sufficient evidence to warrant rejection of the claim that the rate of infections is the same for those treated with Lipitor and those given a placebo.

13. H_0: $p_1 = p_2$. H_1: $p_1 < p_2$. Test statistic: $z = -2.25$. Critical value: $z = -1.645$. P-value: 0.0122 (TI-83/84 Plus: 0.0123). Reject H_0. There is sufficient evidence to support the claim that the proportion of college students using illegal drugs in 1993 was less than it is now.

15. $0.00558 < p_1 - p_2 < 0.0123$. Because the confidence interval limits do not include 0, it appears that the two fatality rates are not equal. Because the confidence interval limits include only positive values, it appears that the fatality rate is higher for those not wearing seat belts. The use of seat belts appears to be effective in saving lives.

17. H_0: $p_1 = p_2$. H_1: $p_1 \neq p_2$. Test statistic: $z = 2.04$. Critical values: $z = \pm 1.96$. P-value: 0.0414 (TI-83/84 Plus: 0.0416). Reject H_0. There is sufficient evidence to support the claim that the percentage of women who agree is different from the percentage of men who agree. There does appear to be a difference in the way that women and men feel about the issue.

19. H_0: $p_1 = p_2$. H_1: $p_1 > p_2$. Test statistic: $z = 0.89$. Critical value: $z = 1.645$. P-value: 0.1867 (TI-83/84 Plus: 0.1858). Fail to reject H_0. There is not sufficient evidence to support the claim that the proportion of wins at home is higher with a closed roof than with an open roof. The closed roof does not appear to be a significant advantage. (But Houston lost this World Series.)

21. $-0.0798 < p_1 - p_2 < 0.149$. Because the confidence interval limits do contain 0, there is not a significant difference between the two proportions. Echinacea does not appear to have a significant effect.

23. H_0: $p_1 = p_2$. H_1: $p_1 < p_2$. Test statistic: $z = -8.47$. Critical value: $z = -2.33$. P-value: 0.0001 (TI-83/84 Plus: 0.0000). Reject H_0. There is sufficient evidence to support the claim that the rate of Norovirus sickness on the *Freedom of the Seas* is less than the rate on the *Queen Elizabeth II*. When a Norovirus outbreak occurs on cruise ships, the proportions of infected passengers can vary considerably.

25. $-0.0364 < p_1 - p_2 < 0.138$ (TI-83/84 Plus: $-0.0364 < p_1 - p_2 < 0.139$). Because the confidence interval limits do contain 0, there is not a significant difference between the two proportions. There does not appear to be a significant difference between the success rates of men and women.

27. H_0: $p_1 = p_2$. H_1: $p_1 \neq p_2$. Test statistic: $z = 11.35$. Critical values: $z = \pm 2.575$. P-value: 0.0002 (TI-83/84 Plus: 0.0000). Reject H_0. There is sufficient evidence to support the claim that the female survivors and male survivors have different rates of thyroid diseases.

29. Using $x_1 = 504$ and $x_2 = 539$: $-0.0497 < p_1 - p_2 < 0.0286$. Using $\hat{p}_1 = 0.69$ and $\hat{p}_2 = 0.70$: $-0.0491 < p_1 - p_2 < 0.0291$. Because the confidence interval limits do contain 0, there is not a significant difference between the two proportions. It appears that males and females agree.

31. Using $x_1 = 69$ and $x_2 = 22$: H_0: $p_1 = p_2$. H_1: $p_1 > p_2$. Test statistic: $z = 6.37$. Critical value: $z = 2.33$. P-value: 0.0001 (TI-83/84 Plus: 0.0000). Reject H_0. There is sufficient evidence to support the claim that the percentage of returns designating the \$3 for the campaign was greater in 1976 than it is now.

33. Using $x_1 = 117$ and $x_2 = 29$: H_0: $p_1 = p_2$. H_1: $p_1 > p_2$. Test statistic: $z = 7.60$. Critical value: $z = 2.33$. P-value: 0.0001 (TI-83/84 Plus: 0.0000). Reject H_0. There is sufficient evidence to support the claim that the proportion of headaches is greater for those treated with Viagra. Headaches do appear to be a concern for those who take Viagra.

35. $0.0131 < p_1 - p_2 < 0.0269$. Because the confidence interval limits do not contain 0, there is a significant difference between the two proportions. Because the confidence interval includes only positive values, it appears that the proportion of women is greater than the proportion of men. Because the sample is a voluntary response sample, the confidence interval is not necessarily a good estimate of the general population.

37. a. $0.0227 < p_1 - p_2 < 0.217$; because the confidence interval limits do not contain 0, it appears that $p_1 = p_2$ can be rejected.
   b. $0.491 < p_1 < 0.629$; $0.371 < p_2 < 0.509$; because the confidence intervals do overlap, it appears that $p_1 = p_2$ cannot be rejected.
   c. H_0: $p_1 = p_2$. H_1: $p_1 \neq p_2$. Test statistic: $z = 2.40$. P-value: 0.0164. Critical values: $z = \pm 1.96$. Reject H_0. There is sufficient evidence to reject $p_1 = p_2$.
   d. Reject $p_1 = p_2$. Least effective: Using the overlap between the individual confidence intervals.

39. H_0: $p_1 - p_2 = 0.15$. H_1: $p_1 - p_2 \neq 0.15$. Test statistic: $z = 2.36$. Critical values: $z = \pm 2.575$. P-value: 0.0182. Fail to reject H_0. There is not sufficient evidence to warrant rejection of the claim that the rate of thyroid disease among female atom bomb survivors is equal to 15 percentage points more than that for male atom bomb survivors.

Section 9-3

1. $1.6 < \mu_1 - \mu_2 < 12.2$
3. 98%
5. Independent
7. Dependent

9. H_0: $\mu_1 = \mu_2$. H_1: $\mu_1 \neq \mu_2$. Test statistic: $t = -0.452$. Critical values: $t = \pm 2.014$ (TI-83/84 Plus: $t = \pm 1.987$). P-value > 0.20 (TI-83/84 Plus: 0.6521). Fail to reject H_0. There is not sufficient evidence to warrant rejection of the claim that the two groups are from populations with the same mean. This result suggests that the increased humidity does not help in the treatment of croup.

11. 6.2 mg $< \mu_1 - \mu_2 <$ 9.6 mg (TI-83/84 Plus: 6.3 mg $< \mu_1 - \mu_2 <$ 9.5 mg). The confidence interval limits do not include 0, which suggests that the mean tar content of unfiltered king size cigarettes is greater than the mean for filtered 100 mm cigarettes.

13. H_0: $\mu_1 = \mu_2$. H_1: $\mu_1 < \mu_2$. Test statistic: $t = -5.137$. Critical value: $t = -1.660$ approximately (TI-83/84 Plus: $t = -1.653$). P-value < 0.005 (TI-83/84 Plus: 0.0000). Reject H_0. There is sufficient evidence to support the claim that the cents portions of the check amounts have a mean that is less than the mean of the cents portions of the credit card charges. One reason for the difference might be that many check amounts are for payments rounded to the nearest dollar, so there are disproportionately more cents portions consisting of 0 cents.

15. H_0: $\mu_1 = \mu_2$. H_1: $\mu_1 > \mu_2$. Test statistic: $t = 10.343$. Critical value: $t = 2.896$ (TI-83/84 Plus: $t = 2.512$). P-value < 0.005 (TI-83/84 Plus: 0.0000). Reject H_0. There is sufficient evidence to support the claim that supermodels have heights with a mean that is greater than the mean of women who are not supermodels.

17. -4.6 ft $< \mu_1 - \mu_2 <$ 7.0 ft (TI-83/84 Plus: -4.4 ft $< \mu_1 - \mu_2 <$ 6.8 ft). The confidence interval includes 0 ft, which suggests that the two population means could be equal. There does not appear to be a significant difference between the two means.

19. H_0: $\mu_1 = \mu_2$. H_1: $\mu_1 \neq \mu_2$. Test statistic: $t = -0.721$. Critical values: $t = \pm 2.064$ (TI-83/84 Plus: $t = \pm 2.011$). P-value > 0.20 (TI-83/84 Plus: 0.4742). Fail to reject H_0. There is not sufficient evidence to support the claim that menthol cigarettes and nonmenthol cigarettes have different amounts of nicotine. The menthol content does not appear to have an effect on the nicotine content.

21. H_0: $\mu_1 = \mu_2$. H_1: $\mu_1 < \mu_2$. Test statistic: $t = -59.145$. Critical value: $t = -2.426$ (TI-83/84 Plus: $t = -2.379$). P-value < 0.005 (TI-83/84 Plus: 0.0000). Reject H_0. There is sufficient evidence to support the claim that the mean FICO score of borrowers with high-interest mortgages is lower than the mean FICO score of borrowers with low-interest mortgages. It appears that the higher FICO credit rating scores are associated with lower mortgage interest rates.

23. H_0: $\mu_1 = \mu_2$. H_1: $\mu_1 > \mu_2$. Test statistic: $t = 1.678$. Critical value: $t = 1.717$ (TI-83/84 Plus: $t = 1.682$). P-value > 0.05 but less than 0.10 (TI-83/84 Plus: 0.0504). Fail to reject H_0. There is not sufficient evidence to support the claim that the unsuccessful applicants are from a population with a greater mean age than the mean age of successful applicants. Based on the result, there does not appear to be discrimination based on age.

25. H_0: $\mu_1 = \mu_2$. H_1: $\mu_1 > \mu_2$. Test statistic: $t = 2.790$. Critical value: $t = 2.390$ (approximately). P-value is less than 0.005 (TI-83/84 Plus: 0.0031). Reject H_0. There is sufficient evidence to support the claim that heavy users have a lower mean than light users. Because marijuana appears to adversely affect mental abilities, it should be a serious concern.

27. H_0: $\mu_1 = \mu_2$. H_1: $\mu_1 > \mu_2$. Test statistic: $t = 0.132$. Critical value: $t = 1.729$. P-value > 0.10 (TI-83/84 Plus: 0.4480). Fail to reject H_0. There is not sufficient evidence to support the claim that the magnets are effective in reducing pain. It is valid to argue that the magnets might appear to be effective if the sample sizes are larger.

29. a. H_0: $\mu_1 = \mu_2$. H_1: $\mu_1 < \mu_2$. Test statistic: $t = -2.335$. Critical value: $t = -1.833$ (TI-83/84 Plus: $t = -1.738$). P-value is between 0.01 and 0.025 (TI-83/84 Plus: 0.0160). Reject H_0. There is sufficient evidence to support the claim that recent winners have a lower mean BMI than the mean BMI from the 1920s and 1930s.

   b. $-2.50 < \mu_1 - \mu_2 < -0.30$ (TI-83/84 Plus: $-2.44 < \mu_1 - \mu_2 < -0.36$). The confidence interval includes only negative values, which suggests that recent winners have a lower mean BMI than the mean BMI from the 1920s and 1930s.

31. a. H_0: $\mu_1 = \mu_2$. H_1: $\mu_1 < \mu_2$. Test statistic: $t = -1.812$. Critical value: $t = -2.650$ (TI-83/84 Plus: $t = -2.574$). P-value > 0.025 but less than 0.05 (TI-83/84 Plus: 0.0441). Fail to reject H_0. There is not sufficient evidence to support the claim that popes have a mean longevity after election that is less than the mean for British monarchs after coronation.

   b. -23.6 years $< \mu_1 - \mu_2 <$ 4.4 years (TI-83/84 Plus: -23.2 years $< \mu_1 - \mu_2 <$ 4.0 years). The confidence interval includes 0 years, which suggests that the two population means could be equal.

33. a. H_0: $\mu_1 = \mu_2$. H_1: $\mu_1 > \mu_2$. Test statistic: $t = 2.746$. Critical value: $t = 2.718$ (TI-83/84 Plus: $t = 2.445$). P-value < 0.01 (TI-83/84 Plus: 0.0049). Reject H_0. There is sufficient evidence to support the claim that movies with ratings of PG or PG-13 have a higher mean gross amount than movies with R ratings.

   b. 0.8 million dollars $< \mu_1 - \mu_2 <$ 153.2 million dollars (TI-83/84 Plus: 8.4 million dollars $< \mu_1 - \mu_2 <$ 145.5 million dollars). The confidence interval includes only positive values, which suggests that movies with ratings of PG or PG-13 have a higher mean gross amount than movies with ratings of R. The confidence interval suggests that PG and PG-13 movies gross much more than R movies.

35. H_0: $\mu_1 = \mu_2$. H_1: $\mu_1 \neq \mu_2$. Test statistic: $t = -16.830$. Critical values: $t = \pm 2.023$ (TI-83/84 Plus: $t = \pm 1.992$). P-value < 0.01 (TI-83/84 Plus: 0.0000). Reject H_0. There is sufficient evidence to warrant rejection of the claim that the population of home voltages has the same mean as the population of generator voltages. Although there is a statistically significant difference, the sample means of 123.66 V and 124.66 V suggest that the difference does not have practical significance. The generator could be used as a substitute when needed.

37. H_0: $\mu_1 = \mu_2$. H_1: $\mu_1 \neq \mu_2$. Test statistic: $t = -0.452$. Critical values: $t = \pm 1.987$. P-value > 0.20 (TI-83/84 Plus: 0.6521). Fail to reject H_0. There is not sufficient evidence to warrant rejection of the claim that the two groups are from populations with the same mean. This result suggests that the increased humidity does not help in the treatment of croup. In this case, the results are basically the same as in Exercise 9.

39. 6.3 mg $< \mu_1 - \mu_2 <$ 9.5 mg. The confidence interval limits are closer together here than in Exercise 11, so the confidence interval provides an estimate with a smaller range of values.

41. In the hypothesis test, the test statistic changes from -1.812 to -1.083, which is a dramatic change. The P-value changes from a value between 0.025 and 0.05 to a value greater than 0.10, which is a dramatic change. (Using a TI-83/84 Plus Calculator, the P-value changes from 0.0441 to 0.1493, which is a dramatic change). The conclusion remains the same in this case, but introduction of an outlier could cause the conclusion to change. The confidence interval limits change to -447.5 years and 187.9 years, and they are dramatically different from the original confidence interval limits. The outlier causes a large increase in variation, and that is reflected in the confidence interval that becomes much wider.

43. H_0: $\mu_1 = \mu_2$. H_1: $\mu_1 \neq \mu_2$. Test statistic: $t = 15.322$. Critical values: $t = \pm 2.080$. P-value < 0.01 (TI-83/84 Plus: 0.0000). Reject H_0. There is sufficient evidence to warrant rejection of the claim that the two populations have the same mean.

Section 9-4

1. $\bar{d} = -0.2$ min; $s_d = 8.0$ min. μ_d represents the mean of the differences from the paired data in the population.

3. The methods of this section will produce numerical results, but they will make no sense. The differences between pulse rates and cholesterol levels themselves make no sense, because they are different types of quantities. The methods of this section should not be used.

5. a. -8.3
   b. 0.5
   c. -33.000
   d. ± 3.182

7. $-9.0 < \mu_d < -7.5$

9. H_0: $\mu_d = 0$. H_1: $\mu_d \neq 0$. Test statistic: $t = 0.029$. Critical values: $t = \pm 2.776$. P-value > 0.20 (TI-83/84 Plus: 0.9783). Fail to reject H_0. There is not sufficient evidence to warrant rejection of the claim that for students in their freshman year, the mean change in BMI is equal to 0. The BMI does not appear to change during freshman year.

11. H_0: $\mu_d = 0$. H_1: $\mu_d < 0$. Test statistic: $t = -4.712$. Critical value: $t = -1.761$. P-value < 0.005 (TI-83/84 Plus: 0.0002). Reject H_0. There is sufficient evidence to support the claim that best actresses are younger than best actors. Yes.

13. $-2.32°F < \mu_d < 0.52°F$. The confidence interval includes $0°F$, which suggests that the mean of the differences could be 0. Based on the available data, body temperatures appear to be about the same at both times.

15. H_0: $\mu_d = 0$. H_1: $\mu_d \neq 0$. Test statistic: $t = -2.712$. Critical values: $t = \pm 2.571$. P-value is between 0.02 and 0.05 (TI-83/84 Plus: 0.0422). Reject H_0. There is sufficient evidence to warrant rejection of the claim of no effect. Hospital admissions appear to be affected.

17. $-\$647.9 < \mu_d < \$2327.0$. The confidence interval includes $0, which suggests that the mean of the differences could be 0, so there does not appear to be a significant difference between the costs of repairing the front end and the costs of repairing the rear end.

19. H_0: $\mu_d = 0$. H_1: $\mu_d > 0$. Test statistic: $t = 14.104$. Critical value: $t = 2.539$. P-value < 0.005 (TI-83/84 Plus: 0.0000). Reject H_0. There is sufficient evidence to support the claim that the old ratings are higher than the new ratings.

21. a. H_0: $\mu_d = 0$. H_1: $\mu_d \neq 0$. Test statistic: $t = 1.054$. Critical values: $t = \pm 2.023$. P-value > 0.20 (TI-83/84 Plus: 0.2983). Fail to reject H_0. There is not sufficient evidence to support the claim that these paired sample values have differences that are from a population with a mean of 0 volts. The home voltage and the UPS voltage appear to be about the same.
   b. The gasoline-powered generator is completely independent of the voltage supplied by the power company to the home, so the home voltage amounts and the generator voltage amounts are two independent samples.

23. 6.610 lb $< \mu_d <$ 8.424 lb (TI-83/84 Plus: 6.611 lb $< \mu_d <$ 8.424 lb). Discarded paper appears to weigh more.

25. The five pairs of data are (97, 171), (116, 196), (116, 191), (165, 207), (191, 224). The answers may differ somewhat due to the accuracy that the graph provides. H_0: $\mu_d = 0$. H_1: $\mu_d \neq 0$. Test statistic: $t = -6.286$. Critical values: $t = \pm 2.776$. P-value < 0.01 (TI-83/84 Plus: 0.0033). Reject H_0. There is sufficient evidence to warrant rejection of the claim that there is no difference between the right hand and left hand reaction times. There does appear to be a difference.

Section 9-5

1. Because F is so close to 1, the two sample standard deviations are very close in value.

3. The F test is sensitive to departures from normal distributions, which means that it performs poorly by giving bad results when either or both of the populations does not have a normal distribution. Alternatives are the count five test and the Levene-Brown-Forsyth test.

5. H_0: $\sigma_1 = \sigma_2$. H_1: $\sigma_1 > \sigma_2$. Test statistic: $F = 1.1842$. Critical F value: 2.4034. P-value: 0.3738. Fail to reject H_0. There is not sufficient evidence to support the claim that weights of babies born to mothers given placebos vary more than weights of babies born to mothers given zinc supplements.

7. H_0: $\sigma_1 = \sigma_2$. H_1: $\sigma_1 \neq \sigma_2$. Test statistic: $F = 1.5777$. P-value $= 0.0028$, which is less than the significance level of 0.01, so reject H_0. There is sufficient evidence to warrant rejection of the claim that the two samples are from populations with the same standard deviation.

9. H_0: $\sigma_1 = \sigma_2$. H_1: $\sigma_1 \neq \sigma_2$. Test statistic: $F = 1.7778$. Upper critical F value is between 1.8752 and 2.0739 (TI-83/84 Plus:

1.8907).P-value: 0.0762. Fail to reject H_0. There is not sufficient evidence to warrant rejection of the claim that subjects from both treatment groups have ages with the same amount of variation. If treatment groups have different characteristics, comparisons of treatments become unclear, because differences might be due to the treatments or they might be due to the different group characteristics.

11. H_0: $\sigma_1 = \sigma_2$. H_1: $\sigma_1 > \sigma_2$. Test statistic: $F = 9.3364$. Critical F value is between 2.0540 and 2.0960 (TI-83/84 Plus: 2.0842). P-value: 0.0000. Reject H_0. There is sufficient evidence to support the claim that the treatment group has scores that vary more than the scores of the placebo group.

13. H_0: $\sigma_1^2 = \sigma_2^2$. H_1: $\sigma_1^2 > \sigma_2^2$. Test statistic: $F = 2.1267$. The critical F value is between 2.1555 and 2.2341. (TI-83/84 Plus: P-value $= 0.0543$.) Fail to reject H_0. There is not sufficient evidence to support the claim that the pain reductions for the sham treatment group vary more than the pain reductions for the magnet treatment group.

15. H_0: $\sigma_1 = \sigma_2$. H_1: $\sigma_1 > \sigma_2$. Test statistic: $F = 4.1648$. Critical F value is between 2.7876 and 2.8536 (TI-83/84 Plus: 2.8179).P-value: 0.0130. Reject H_0. There is sufficient evidence to support the claim that amounts of strontium-90 from Pennsylvania vary more than amounts from New York.

17. H_0: $\sigma_1 = \sigma_2$. H_1: $\sigma_1 \neq \sigma_2$. Test statistic: $F = 1.5061$. Upper critical F value is between 2.1540 and 2.2131 (TI-83/84 Plus: 2.1812). P-value: 0.2990. Fail to reject H_0. There is not sufficient evidence to warrant rejection of the claim that both samples are from populations having the same standard deviation.

19. H_0: $\sigma_1 = \sigma_2$. H_1: $\sigma_1 > \sigma_2$. Test statistic: $F = 3.4381$. Critical value of F is between 1.6928 and 1.8409 (TI-83/84 Plus: 1.7890). P-value: 0.0003. Reject H_0. There is sufficient evidence to support the claim that near the end of the freshman year, weights of male college students vary more than weights of female college students.

21. $c_1 = 4$, $c_2 = 0$, critical value is 5. Fail to reject $\sigma_1^2 = \sigma_2^2$.

23. 0.2727, 2.8365

Chapter 9 Statistical Literacy and Critical Thinking

1. A method is robust if the requirement of normally distributed populations is a somewhat loose requirement in the sense that the method works well even with populations that do not have normal distributions, provided that the distribution is not radically far from being normal. The F test for comparing two population standard deviations or variances is not robust against departures from normal distributions.

2. Nothing. The results are from a voluntary response (or self-selected) sample, so the results apply only to those who chose to respond. The results may or may not be representative of the general population, so methods of statistics should not be used to make inferences about the general population. To do so would be a ginormous mistake.

3. The samples are independent. The pairing of the cans is arbitrary, so a can of cola within a particular pair is not actually related to the other can in any way.

4. The different states have different population sizes, so weighted means should be used.

Chapter 9 Quick Quiz

1. H_0: $p_1 = p_2$. H_1: $p_1 > p_2$.
2. 0.5
3. $z = -1.73$
4. 0.0404
5. There is sufficient evidence to support the claim that $\mu_1 > \mu_2$.
6. H_0: $\mu_d = 0$. H_1: $\mu_d \neq 0$.
7. H_0: $\mu_1 = \mu_2$. H_1: $\mu_1 < \mu_2$.
8. F distribution
9. There is sufficient evidence to support the claim that the two populations have different means.
10. True

Chapter 9 Review Exercises

1. H_0: $p_1 = p_2$. H_1: $p_1 > p_2$. Test statistic: $z = 3.12$. Critical value: $z = 2.33$. P-value: 0.0009. Reject H_0. There is sufficient evidence to support a claim that the proportion of successes with surgery is greater than the proportion of successes with splinting. When treating carpal tunnel syndrome, surgery should generally be recommended instead of splinting.

2. H_0: $\mu_1 = \mu_2$. H_1: $\mu_1 < \mu_2$. Test statistic: $t = -2.908$. Critical value: $t = -1.653$ (approximately). P-value < 0.005 (TI-83/84 Plus: 0.0019). Reject H_0. There is sufficient evidence to support the claim that the children exposed to cocaine have a lower mean score.

3. a. H_0: $\mu_d = 0$. H_1: $\mu_d \neq 0$. Test statistic: $t = -1.532$. Critical values: $t = \pm 2.228$. P-value is between 0.10 and 0.20 (TI-83/84 Plus: 0.1565). Fail to reject H_0. There is not sufficient evidence to warrant rejection of the claim that there is no difference. There does not appear to be a difference.

   b. $-2.7 < \mu_d < 0.5$

   c. No, there is not a significant difference.

4. H_0: $p_1 = p_2$. H_1: $p_1 \neq p_2$. Test statistic: $z = -4.20$. Critical values: $z = \pm 2.575$. P-value: 0.0002 (TI-83/84 Plus: 0.000). Reject H_0. There is sufficient evidence to warrant rejection of the claim that the acceptance rate is the same with or without blinding. Without blinding, reviewers know the names and institutions of the abstract authors, and they might be influenced by that knowledge.

5. H_0: $\mu_1 = \mu_2$. H_1: $\mu_1 > \mu_2$. Test statistic: $t = 5.529$. Critical value: $t = 1.796$. P-value < 0.005 (TI-83/84 Plus: 0.0000). Reject H_0. There is sufficient evidence to support the claim that *Harry Potter* is easier to read than *War and Peace*. The result is as expected, because *Harry Potter* was written for children, but *War and Peace* was written for adults.

6. a. $9.5 < \mu_d < 27.6$

   b. H_0: $\mu_d = 0$. H_1: $\mu_d > 0$. Test statistic: $t = 6.371$. Critical value: $t = 2.718$ (assuming a 0.01 significance level). P-value < 0.005 (TI-83/84 Plus: 0.0000). Reject H_0. There is sufficient evidence to support the claim that the blood pressure levels are lower after taking captopril.

7. H_0: $p_1 = p_2$. H_1: $p_1 > p_2$. Test statistic: $z = 1.59$. Critical value: $z = 1.645$. P-value: 0.0559 (TI-83/84 Plus: 0.0557). Fail to reject H_0. There is not sufficient evidence to support the claim that the proportion of men who smoke is greater than the proportion of women who smoke.

8. H_0: $\mu_1 = \mu_2$. H_1: $\mu_1 < \mu_2$. Test statistic: $t = -9.567$. Critical value: $t = -2.429$. P-value < 0.005 (TI-83/84 Plus: 0.0000). Reject H_0. There is sufficient evidence to support the claim that workers with a high school diploma have a lower mean annual income than workers with a bachelor's degree. Solving this exercise contributes to a bachelor's degree and to a higher income.

9. Using the sample of incomes from workers with bachelor's degrees as sample 1: H_0: $\sigma_1 = \sigma_2$. H_1: $\sigma_1 \neq \sigma_2$. Test statistic: $F = 2.9460$. Upper critical F value is between 1.6141 and 1.8152 (TI-83/84 Plus: 1.6935). P-value: 0.0001. Reject H_0. There is sufficient evidence to warrant rejection of the claim that the two samples are from populations with the same standard deviation.

10. H_0: $\sigma_1 = \sigma_2$. H_1: $\sigma_1 \neq \sigma_2$. Test statistic: $F = 1.2874$. Upper critical F value is between 1.4327 and 1.6668 (TI-83/84 Plus: 1.5874). P-value: 0.2825. Fail to reject H_0. There is not sufficient evidence to warrant rejection of the claim that the two samples are from populations with the same amount of variation.

Chapter 9 Cumulative Review Exercises

1. a. The samples are dependent, because the data are from couples that are paired.
   b. $\bar{x} = 14.9$ thousand words, median: 14.5 thousand words, mode: 8 thousand words and 14 thousand words, range: 17.0 thousand words, $s = 5.7$ thousand words.
   c. ratio

2. H_0: $\mu_d = 0$. H_1: $\mu_d < 0$. Test statistic: $t = -2.163$. Critical value: $t = -1.833$. P-value is between 0.025 and 0.05, but less than 0.05 (TI-83/84 Plus: 0.0294). Reject H_0. There is sufficient evidence to support the claim that among couples, females are more talkative than males.

3. H_0: $\mu_1 = \mu_2$. H_1: $\mu_1 \neq \mu_2$. Test statistic: $t = -1.990$. Critical values: $t = \pm 2.262$ (TI-83/84 Plus: $t = \pm 2.132$). P-value > 0.05 but less than 0.10 (TI-83/84 Plus: 0.0652). Fail to reject H_0. There is not sufficient evidence to reject the claim that the two samples are from populations with the same mean.

4. 10.8 thousand words $< \mu < 19.0$ thousand words or 10,800 words $< \mu < 19,000$ words

5.

Word Count	Frequency
6 – 9	3
10 – 13	0
14 – 17	4
18 – 21	2
22 – 25	1

6. a. 0.3707 (TI-83/84 Plus: 0.3694)
   b. 0.1587
   c. 22,680 (TI-83/84 Plus: 22,689)

7. 1083

8. 32.8% $< p <$ 43.3%

9. a. 25.0
   b. 3.5

c. 0.9406 (TI-83/84 Plus using normal approximation: 0.9401; TI-83/84 Plus using binomial: 0.9405).

10. $0.984^{20} = 0.724$. Such an event is not unusual, because the probability is not low (such as less than 0.05).

Chapter 10 Answers
Section 10-2

1. r represents the value of the linear correlation coefficient computed by using the paired sample data. ρ represents the value of the linear correlation coefficient that would be computed by using all of the paired data in the population. The value of r is estimated to be 0 (because there is no correlation between points scored in a football game and the numbers of new cars sold).

3. No. The presence of a linear correlation between two variables does not imply that one of the variables is the cause of the other variable.

5. Yes. The value of $|0.758|$ is greater than the critical value of 0.254 (approximately).

7. No. The value of $|0.202|$ is less than the critical value 0.312.

9. a.

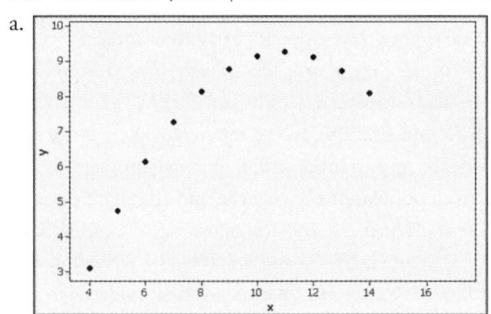

   b. $r = 0.816$. Critical values: $r = \pm 0.602$. P-value $= 0.002$. There is sufficient evidence to support the claim of a linear correlation between the two variables.
   c. The scatterplot reveals a distinct pattern that is not a straight-line pattern.

11. a. There appears to be a linear correlation.
    b. $r = 0.906$. Critical values: $r = \pm 0.632$ (for a 0.05 significance level). There is a linear correlation.
    c. $r = 0$. Critical values: $r = \pm 0.666$ (for a 0.05 significance level). There does not appear to be a linear correlation.
    d. The effect from a single pair of values can be very substantial, and it can change the conclusion.

13. $r = 0.985$. Critical values: $r = \pm 0.811$. P-value $= 0.000$. There is sufficient evidence to support the claim of a linear correlation between CPI and cost of a slice of pizza.

15. $r = 0.867$. Critical values: $r = \pm 0.878$. P-value $= 0.057$. There is not sufficient evidence to support the claim of a linear correlation between the systolic blood pressure measurements of the right and left arm.

17. $r = 0.948$. Critical values: $r = \pm 0.811$. P-value $= 0.004$. There is sufficient evidence to support the claim of a linear correlation between the overhead width of a seal in a photograph and the weight of a seal.

19. $r = 0.709$. Critical values: $r = \pm 0.754$. P-value $= 0.075$. There is not sufficient evidence to support the claim of a linear correlation between the costs of tickets purchased 30 days in advance and those purchased one day in advance.

21. $r = -0.283$. Critical values: $r = \pm 0.754$. P-value $= 0.539$. There is not sufficient evidence to support the claim of a linear correlation between the repair costs from full-front crashes and full-rear crashes.

23. $r = 0.892$. Critical values: $r = \pm 0.632$. P-value $= 0.001$. There is sufficient evidence to support the claim of a linear correlation between global temperature and the concentration of CO_2.

25. $r = 0.872$. Critical values: $r = \pm 0.754$. P-value $= 0.011$. There is sufficient evidence to support the claim of a linear correlation between the given differences (between runs scored and runs allowed) and the proportions of wins.

27. $r = 0.179$. Critical values: $r = \pm 0.632$. P-value $= 0.620$. There is not sufficient evidence to support the claim of a linear correlation between brain size and intelligence. It does not appear that people with larger brains are more intelligent.

29. $r = 0.744$. Critical values: $r = \pm 0.335$. P-value $= 0.000$. There is sufficient evidence to support the claim of a linear correlation between movie budget amounts and the amounts that the movies grossed.

31. $r = 0.319$. Critical values: $r = \pm 0.254$ (approximately). P-value $= 0.017$. There is sufficient evidence to support the claim of a linear correlation between the numbers of words spoken by men and women who are in couple relationships.

33. A linear correlation between two variables does not necessarily mean that one of the variables is the *cause* of (or directly affects) the other variable.

35. Averages tend to suppress variation among individuals, so a linear correlation among *averages* does not necessarily mean that there is a linear correlation among *individuals*.

37. a. 0.942
    b. 0.839
    c. 0.9995 (largest)
    d. 0.983
    e. -0.958

Section 10-3

1. $\hat{y}$ represents the predicted value of cholesterol. The predictor variable represents weight. The response variable represents cholesterol level.

3. If r is positive, the regression line has a positive slope and it rises from left to right. If r is negative, the slope of the regression line is negative and it falls from left to right.

5. 6.40 people

7. 76.3 beats per minute

9. $\hat{y} = 3.00 + 0.500x$. The data have a pattern that is not a straight line.

11. a. $\hat{y} = 0.264 + 0.906x$
    b. $\hat{y} = 2 + 0x$ (or $\hat{y} = 2$)
    c. The results are very different, indicating that one point can dramatically affect the regression equation.

13. $\hat{y} = -0.162 + 0.0101x$; $1.68 (which might be rounded up to the more convenient value of $1.75)

15. $\hat{y} = 43.6 + 1.31x$; 163.2 mm Hg

17. $\hat{y} = -157 + 40.2x$; 205 kg

19. $\hat{y} = -1240 + 7.07x$; $690

21. $\hat{y} = 2060 - 0.186x$; $1615. The predicted cost of $1615 is very different from the actual cost of $982.

23. $\hat{y} = 10.5 + 0.0109x$; 14.5°. Yes, the predicted temperature is the same as the actual temperature.

25. $\hat{y} = 0.494 + 0.000490x$; 0.519. The predicted proportion of wins is reasonably close to the actual proportion of 0.543.

27. $\hat{y} = 71.8 + 0.0286x$; 103 (or the known mean IQ score of 100)

29. $\hat{y} = 20.6 + 1.38x$; $186 million

31. $\hat{y} = 13,400 + 0.302x$; 15,200 words in a day

33. With $\beta_1 = 0$, the regression line is horizontal so that different values of x result in the same y value, and there is no correlation between x and y.

35. The residual plot does not suggest that the regression equation is a bad model, because there is no obvious pattern and the residual plot does not become thicker (or thinner). The scatterplot suggests that the regression equation is a bad model because the points do not fit the pattern of a straight line very well.

RESIDUAL PLOT

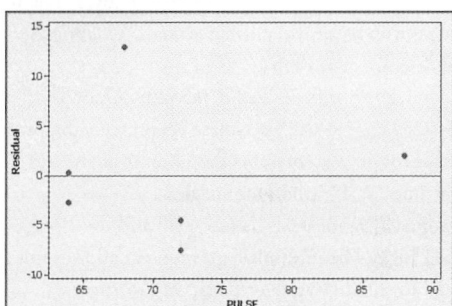

Section 10-4

1. It is the standard error of estimate, which is a measure of the differences between the observed weights and the weights predicted from the regression equation.

3. Unlike the predicted value of 180 lb, the prediction interval provides us with a range of likely weights, so that we have a sense of how accurate the prediction is likely to be. The terminology of *prediction interval* is used for an interval estimate of a variable, whereas the terminology of *confidence interval* is used for an interval estimate of a parameter.

5. 0.762; 76.2%

7. 0.748; 74.8%

9. $r = -0.806$ (r is negative because the highway fuel consumption amounts decrease as the weight of the cars increase, as shown by the fact that the slope of the regression line is negative). The critical value of r is between 0.361 and 0.335 (assuming a 0.05 significance level). P-value $= 0.000$. There is sufficient evidence to support a claim of a linear correlation between the weights of cars and their highway fuel consumption amounts.

11. 27.028 mi/gal

13. a. 2.64829

   b. 0.0800433

   c. 2.728333

   d. 0.9706622

   e. 0.1414596

15. a. 8880.182

   b. 991.1515

   c. 9871.333

   d. 0.8995929

   e. 15.74128

17. a. $1.72

   b. $1.27 < y < $2.17

19. a. 205 kg

   b. 157 kg < y < 253 kg

21. $1.32 < y < $2.72

23. $0.12 < y < $0.90

25. $-0.229 < \beta_0 < 0.298$; $0.738 < \beta_1 < 1.15$

Section 10-5

1. b_1 is the coefficient of x_1 and β_1 is the coefficient of x_1, but β_1 is a population parameter obtained by using all of the data in the population, whereas b_1 is a sample statistic that is an estimate of β_1. The value of b_1 is found from the available sample data.

3. No, because the data are categorical (or qualitative). The methods of this section require quantitative data. (Exceptions can occur with logistic regression.)

5. Nicotine = 1.59 + 0.0231 Tar − 0.0525 CO, or $\hat{y}$ = 1.59 + 0.0231x_1 − 0.0525x_2, where $\hat{y}$ represents the predicted amount of nicotine, x_1 represents the amount of tar, and x_2 represents the amount of carbon monoxide.

7. No. The adjusted R^2 of 0.017 is too small and the P-value of 0.317 is too high. The multiple regression equation is not a good model for predicting the amount of nicotine.

9. LP (list price), because it has the best combination of lowest P-value (0.000) and highest adjusted R^2 (0.990).

11. $\hat{y}$ = 99.2 + 0.979 LP, because it has the lowest P-value of 0.000 and the highest adjusted R^2 of 0.990.

13. The best regression equation is $\hat{y}$ = 0.127 + 0.0878x_1 − 0.0250x_2, where x_1 represents tar and x_2 represents carbon monoxide. It is best because it has the highest adjusted R^2 value of 0.927 and the lowest P-value of 0.000. It is a good regression equation for predicting nicotine content because it has a high value of adjusted R^2 and a low P-value.

15. The regression equation $\hat{y}$ = 47.9 − 0.00440x_1 − 0.823x_2 (where x_1 = weight and x_2 = engine displacement) has the highest adjusted R^2 of 0.644 and the lowest P-value of 0.000. It would be better to use the regression equation $\hat{y}$ = 50.5 − 0.00587x (where x represents weight), because it also has a P-value of 0.000 and its adjusted R^2 value of 0.639 is only a little less than 0.644, but it is better to use a regression equation with fewer predictor variables.

17. For H_0: β_1 = 0, the test statistic is t = 5.49, the P-value is 0.000, and the critical values are t = ±2.110, so reject H_0 and conclude that the regression coefficient of b_1 = 0.707 should be kept. For H_0: β_2 = 0, the test statistic is t = 1.29, the P-value

is 0.213, and the critical values are t = ±2.110, so fail to reject H_0 and conclude that the regression coefficient of b_2 = 0.164 should be omitted. It appears that the regression equation should include the height of the mother as a predictor variable, but the height of the father should be omitted.

19. $\hat{y}$ = 3.06 + 82.4x_1 + 2.91x_2, where x_1 represents sex and x_2 represents age. Female: 61 lb; male: 144 lb. The sex of the bear does appear to have an effect on its weight. The regression equation indicates that the predicted weight of a male bear is about 82 lb more than the predicted weight of a female bear with other characteristics being the same.

Section 10-6

1. The value of R^2 = 1 indicates that the model fits the data perfectly. Given the inherent variation and changing economic conditions, it is very unlikely that the model will fit the data perfectly, so the claim is likely to be wrong.

3. 8.2% of the variation in Super Bowl points can be explained by the quadratic model that relates the variable of year and the variable of points scored. Because such a small percentage of such variation is explained by the model, the model is not very useful.

5. Linear: y = 8 + 2x

7. Quadratic: d = −16t^2 + 500

9. Exponential: y = 0.158270(1.05935^x), where 1960 is coded as 1. With the high R^2 value of 0.970, the model appears to be good. The projected value for 2020 is $5.33.

11. Quadratic: y = 0.00802808x^2 + 2.45538x + 15.3223, where 1980 is coded as 1. The quadratic model is not much better than the linear, exponential, and power models. The projected number of 87 isn't too far from the actual number of 92.

13. Quadratic: y = 4.90x^2 − 0.0286x + 0.00476. The model yields a distance of 705 m, but the building is only 50 m tall, so the distance cannot exceed 50 m.

15. Quadratic: y = 0.00447053x^2 + 0.0198551x + 13.8205, where 1950 is coded as 1, 1955 is coded as 2, and so on. The predicted temperature for 2010 is 14.8°C.

17. a. Exponential: $y = 2^{\frac{2}{3}(x-1)}$ (or y = (0.629961)(1.587401)^x for an initial value of 1 that doubles every 1.5 years)

   b. Exponential: y = (1.33197)(1.43035)^x, where 1971 is coded as 1.

   c. Moore's law does appear to be working reasonably well. With R^2 = 0.988, the model appears to be very good.

19. a. 6641.8

   b. 73.2

   c. The quadratic sum of squares of residuals (73.2) is less than the sum of squares of residuals from the linear model (6641.8).

Chapter 10 Statistical Literacy and Critical Thinking

1. With the method of correlation discussed in Section 10-2, the objective is to determine whether there is an association between two variables. With the methods of Section 9-4, the objective is to make inferences about the mean of the differences between the values in the population of matched pairs.

2. There is sufficient evidence to support the claim of a linear correlation between chest size and weight. Although there is a linear correlation, we cannot conclude that a larger chest size is the *cause* of a larger weight.

3. She is not correct. The value of $r = 1$ indicates that there is sufficient evidence to support a claim of a linear correlation between the two sets of prices, but that does not necessarily mean that both sets of prices are the same. For example, if each ring at the discount company is priced at 50% of the Tiffany price, the value of r will be 1, but the corresponding prices are very different.

4. The conclusion is not necessarily correct. The value of $r = 0$ suggests that there is no linear correlation between the two variables, but there might be some other nonlinear correlation.

Chapter 10 Quick Quiz

1. An error was made in the calculations, because r must always be between -1 and 1.
2. There is sufficient evidence to support the claim of a linear correlation between the two variables.
3. True
4. There is not sufficient evidence to support the claim of a linear correlation between the two variables.
5. False. There might be a nonlinear relationship between the two variables.
6. ± 0.514
7. $r = -1$
8. 15
9. 0.160
10. False

Chapter 10 Review Exercises

1. a. The scatterplot suggests that there is not a linear correlation between the two variables.

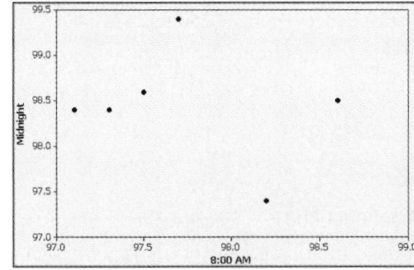

   b. $r = -0.254$. Critical values: $r = \pm 0.811$ (assuming a 0.05 significance level). P-value $= 0.627$. There is not sufficient evidence to support a claim of a linear correlation between 8:00 AM temperatures and midnight temperatures.
   c. $\hat{y} = 126 - 0.285x$
   d. 98.45°F (the value of $\bar{y}$)

2. a. $r = 0.522$, the critical values are $r = \pm 0.312$ (assuming a 0.05 significance level), and the P-value is 0.001, so there is sufficient evidence to support a claim of a linear correlation between heights and weights of males.
   b. 27.2%

   c. $\hat{y} = -139 + 4.55x$
   d. 189 lb

3. a. The scatterplot shows that as length increases, weight also tends to increase, so there does appear to be a correlation between length and weight.

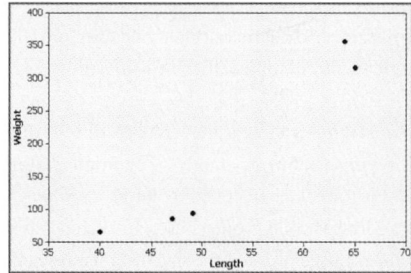

   b. $r = 0.964$. Critical values: $r = \pm 0.878$ (assuming a 0.05 significance level). P-value $= 0.008$. There is sufficient evidence to support a claim of a linear correlation between lengths of bears and their weights.
   c. $\hat{y} = -468 + 12.3x$
   d. 418 lb

4. a. The scatterplot suggests that there is not a linear correlation between upper leg lengths and heights of males.

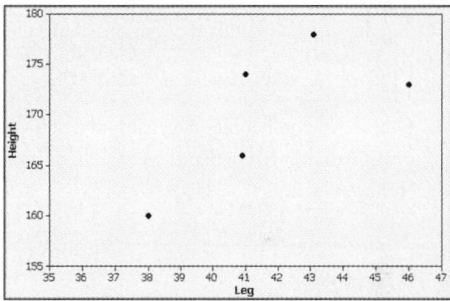

   b. $r = 0.723$. Critical values: $r = \pm 0.878$ (assuming a 0.05 significance level). P-value $= 0.168$. There is not sufficient evidence to support a claim of a linear correlation between upper leg lengths and heights of males.
   c. $\hat{y} = 97.3 + 1.74x$
   d. 170.2 cm (the value of $\bar{y}$)

5. $\hat{y} = 140 + 2.50x_1 - 2.27x_2$; $R^2 = 0.977$; adjusted $R^2 = 0.954$; P-value $= 0.023$. Using a 0.05 significance level, the multiple regression equation is a good model and can be used to predict the height of a male when given his upper leg length and arm circumference.

Chapter 10 Cumulative Review Exercises

1. From 1877: $\bar{x} = 66.2$ in.; median: 66 in.; $s = 2.5$ in. From recent results: $\bar{x} = 68.1$ in.; median: 68.5 in.; $s = 3.6$ in.

2. H_0: $\mu_1 = \mu_2$. H_1: $\mu_1 < \mu_2$. Test statistic: $t = -1.372$. Critical value: $t = -1.833$. P-value > 0.05 (TI-83/84 Plus: 0.094). Fail to reject H_0. There is not sufficient evidence to support the claim that males in 1877 had a mean height that is less than the mean height of males today.

3. H_0: $\mu = 69.1$. H_1: $\mu < 69.1$. Test statistic: $t = -3.690$. Critical value: $t = -1.833$. P-value < 0.005 (TI-83/84 Plus: 0.0025). There is sufficient evidence to support the claim that heights of men from 1877 have a mean less than 69.1 in.

4. 64.4 in. $< \mu <$ 68.0 in.

5. -1.2 in. $< \mu_1 - \mu_2 <$ 5.0 in. (TI-83/84 Plus: -1.0 in. $< \mu_1 - \mu_2 <$ 4.8 in.). The confidence interval limits do include 0, which indicates that the two population means could be the same, so there is not a significant difference between the two population means.

6. The two sets of sample data are not matched as required, so the value of the linear correlation coefficient is meaningless in this situation.

7. a. A statistic is a numerical measurement describing some characteristic of a *sample*, but a parameter is a numerical measurement describing some characteristic of a *population*.

   b. A simple random sample is one chosen in such a way that every possible sample of the same size has the same chance of being chosen.

   c. A voluntary response sample is one in which the respondents themselves decide whether to be included. Results from such samples are generally unsuitable because people with strong interests in the topic are more likely to respond, and the result is a sample that is not representative of the population.

8. Yes. Different explanations are possible, including these: 40 is more than two standard deviations away from the mean; only 0.26% of BMI values are 40 or above; 40 converts to a z score of 2.8, indicating that it is 2.8 standard deviations above the mean.

9. a. 0.3446

   b. 0.0548

10. 0.000207. Because the probability of getting four subjects with green eyes is so small, it is likely that the researcher did not randomly select the subjects.

Chapter 11 Answers

Section 11-2

1. The random digits should all be equally likely, so the distribution is uniform. The test for goodness-of-fit is a hypothesis test that the sample data agree with or fit the uniform distribution with all of the digits being equally likely.

3. O represents the observed frequencies, and they are 5, 8, 7, 9, 13, 17, 11, 10, 10, 12, 8, 10. E represents the expected frequencies, and the twelve expected frequencies are each 10.

5. Critical value: $\chi^2 = 16.919$. P-value > 0.10 (TI-83/84 Plus: 0.516). There is not sufficient evidence to warrant rejection of the claim that the observed outcomes agree with the expected frequencies. The slot machine appears to be functioning as expected.

7. Test statistic: $\chi^2 = 70.160$. Critical value: $\chi^2 = 7.815$. P-value < 0.005 (TI-83/84 Plus: 0.000). There is sufficient evidence to warrant rejection of the claim that the four categories are equally likely. The results appear to support the expectation that the frequency for the first category is disproportionately high.

9. Test statistic: $\chi^2 = 6.320$. Critical value: $\chi^2 = 7.815$. P-value > 0.05 (TI-83/84 Plus: 0.097). There is not sufficient evidence to warrant rejection of the claim that the four categories are equally likely. The results do not support the expectation that the frequency for the first category is disproportionately high.

11. Test statistic: $\chi^2 = 5.860$. Critical value: $\chi^2 = 11.071$. (TI-83/84 Plus: P-value = 0.3201.) There is not sufficient evidence to support the claim that the outcomes are not equally likely. The outcomes appear to be equally likely, so the loaded die does not appear to behave differently from a fair die.

13. Test statistic: $\chi^2 = 13.193$. Critical value: $\chi^2 = 16.919$. P-value > 0.10 (TI-83/84 Plus: 0.154). There is not sufficient evidence to warrant rejection of the claim that the likelihood of winning is the same for the different post positions. Based on these results, post position should not be considered when betting on the Kentucky Derby race.

15. Test statistic: $\chi^2 = 1159.820$. Critical value: $\chi^2 = 19.675$. P-value < 0.005 (TI-83/84 Plus: 0.000). There is sufficient evidence to warrant rejection of the claim that UFO sightings occur in the different months with equal frequency. July and August have the highest frequencies, and those summer months are times when people are outdoors more than other months.

17. Test statistic: $\chi^2 = 15.822$. Critical value: $\chi^2 = 7.815$. P-value < 0.005 (TI-83/84 Plus: 0.001). There is sufficient evidence to warrant rejection of the claim that the observed frequencies agree with the proportions that were expected according to principles of genetics.

19. Test statistic: $\chi^2 = 6.682$. Critical value: $\chi^2 = 11.071$ (assuming a 0.05 significance level). P-value > 0.10 (TI-83/84 Plus: 0.245). There is not sufficient evidence to warrant rejection of the claim that the color distribution is as claimed.

21. Test statistic: $\chi^2 = 3650.251$. Critical value: $\chi^2 = 20.090$. P-value < 0.005 (TI-83/84 Plus: 0.000). There is sufficient evidence to warrant rejection of the claim that the leading digits are from a population with a distribution that conforms to Benford's law. It does appear that the checks are the result of fraud.

23. Test statistic: $\chi^2 = 49.689$. Critical value: $\chi^2 = 20.090$. P-value < 0.005 (TI-83/84 Plus: 0.000). There is sufficient evidence to warrant rejection of the claim that the leading digits are from a population with a distribution that conforms to Benford's law. The contribution amounts do not appear to be legitimate.

25. The test statistic changes from 11.250 to 247.280. (TI-83/84 Plus: The P-value changes from 0.259 to 0.000.) The effect of the outlier is dramatic.

Section 11-3

1. P-value = 0.0000000000173. Because the P-value is so low, we should reject the claim that getting paralytic polio is independent of whether the child was treated with the Salk vaccine or was given a placebo. The Salk vaccine appears to be effective.

3. The P-value of 0.0000000000173 is the probability of getting sample results at least as extreme as those given in the contingency table, assuming that getting paralytic polio is independent of whether a child was treated with the Salk vaccine or was given a placebo.

5. Fail to reject the null hypothesis that home/visitor wins are independent of the sport. It appears that the home-field advantage does not depend on the sport.

7. Test statistic: $\chi^2 = 2.235$. Critical value: $\chi^2 = 3.841$. P-value > 0.10 (TI-83/84 Plus: 0.135). There is not sufficient

evidence to warrant rejection of the claim of independence between success in challenges and sex of the player. Neither sex appears to be more successful.

9. Test statistic: $\chi^2 = 25.571$. Critical value: $\chi^2 = 3.841$. P-value < 0.005 (TI-83/84 Plus: 0.000). There is sufficient evidence to warrant rejection of the claim that whether a subjects lies is independent of the polygraph test indication. The results suggest that polygraphs are effective in distinguishing between truths and lies, but there are many false positives and false negatives, so they are not highly reliable.

11. Test statistic: $\chi^2 = 0.751$. Critical value: $\chi^2 = 3.841$. P-value > 0.10 (TI-83/84 Plus: 0.386). There is not sufficient evidence to warrant rejection of the claim of independence between the type of restoration and adverse health conditions. Amalgam restorations do not appear to affect health conditions.

13. Test statistic: $\chi^2 = 42.557$. Critical value: $\chi^2 = 3.841$. P-value < 0.005 (TI-83/84 Plus: 0.000). There is sufficient evidence to warrant rejection of the claim that the sentence is independent of the plea. The results encourage pleas for guilty defendants.

15. Test statistic: $\chi^2 = 9.750$. Critical value: $\chi^2 = 6.635$. P-value < 0.005 (TI-83/84 Plus: 0.002). There is sufficient evidence to warrant rejection of the claim that success is independent of the type of treatment. The results suggest that the surgery treatment is better.

17. Test statistic: $\chi^2 = 0.792$. Critical value: $\chi^2 = 5.991$. P-value > 0.10 (TI-83/84 Plus: 0.673). There is not sufficient evidence to warrant rejection of the claim that the sex of the respondent is independent of the choice for the cause of global warming. Men and women appear to generally agree.

19. Test statistic: $\chi^2 = 42.568$. Critical value: $\chi^2 = 9.210$. P-value < 0.005 (TI-83/84 Plus: 0.000). There is sufficient evidence to warrant rejection of the claim that experiencing an adverse reaction in the digestive system is independent of the treatment group. Treatments with Campral appear to be associated with a decrease in adverse effects on the digestive system.

21. Test statistic: $\chi^2 = 0.773$. Critical value: $\chi^2 = 7.815$. P-value > 0.10 (TI-83/84 Plus: 0.856). There is not sufficient evidence to warrant rejection of the claim that getting an infection is independent of the treatment. The atorvastatin treatment does not appear to have an effect on infections.

23. Test statistic: $\chi^2 = 51.458$. Critical value: $\chi^2 = 6.635$. P-value < 0.005 (TI-83/84 Plus: 0.000). There is sufficient evidence to warrant rejection of the claim that the proportions of agree/disagree responses are the same for the subjects interviewed by men and the subjects interviewed by women. It appears that the gender of the interviewer affected the responses of women.

25. Test statistics: $\chi^2 = 2.234562954$ and $z = 1.494845462$, so that $z^2 = \chi^2$. Critical values: $\chi^2 = 3.841$ and $z = \pm 1.96$, so that $z^2 = \chi^2$ (approximately).

Section 11-4

1. Neither format appears to be better. Considering the results from the discordant cases, we see that the frequencies of 31 and 33 are quite close, indicating that students performed about the same with both formats.

3. When trying to determine whether there is a significant *difference* between the two different formats, we are not helped by subjects whose results from the two different formats are the same, as in the mastery/mastery results and the nonmastery/nonmastery results. We can detect a difference by analyzing those cases in which one of the formats resulted in mastery while the other format did not, as in the discordant pairs.

5. 1017

7. 652

9. *b, c*

11. 6.635

13. Test statistic: $\chi^2 = 0.237$. Critical value: $\chi^2 = 3.841$. P-value > 0.10 (TI-83/84 Plus: 0.626). There is not sufficient evidence to warrant rejection of the claim that the following two proportions are the same: (1) the proportion of subjects with no hip fracture on the protected hip and a hip fracture on the unprotected hip; (2) the proportion of subjects with a hip fracture on the protected hip and no hip fracture on the unprotected hip. The hip protectors do not appear to be effective.

15. Test statistic: $\chi^2 = 2.382$. Critical value: $\chi^2 = 3.841$. (TI-83/84 Plus: P-value $= 0.123$.) There is not sufficient evidence to warrant rejection of the null hypothesis that the following two proportions are the same: (1) The proportion of subjects with no cure on the fungicide-treated foot and a cure on the foot treated with a placebo; (2) the proportion of subjects with a cure on the fungicide-treated foot and no cure on the foot treated with a placebo. The fungicide treatment does not appear to be effective.

17. Test statistic: $\chi^2 = 6.750$. Critical value: $\chi^2 = 3.841$ (assuming a 0.05 significance level). (TI-83/84 Plus: P-value $= 0.009$.) Reject the null hypothesis that the following two proportions are the same: (1) The proportion of tumors with incorrect staging from MRI and correct staging from PET/CT; (2) the proportion of tumors with correct staging from MRI and incorrect staging from PET/CT. The PET/CT technology appears to be more accurate.

19. The uncorrected test statistic is 11.267. The uncorrected value is greater than the corrected value of 9.600. The conclusion is the same in this case. There could be cases in which the uncorrected test statistic leads to rejection of the null hypothesis while the corrected test statistic does not.

21. When rounded to three decimal places, the same P-value of 0.289 is obtained. With a P-value of 0.289, fail to reject the null hypothesis that the following two proportions are the same: (1) The proportion of subjects with one foot cured with the Pedacream treatment while the other foot is not cured with the Fungacream treatment; (2) the proportion of subjects with one foot not cured with the Pedacream treatment while the other foot is cured with the Fungacream treatment. There does not appear to be a significant difference between the two treatments.

Chapter 11 Statistical Literacy and Critical Thinking

1. The numbers are frequency counts for the six different categories corresponding to the six different cells in the table. The categories are nausea with Celebrex, nausea with Ibuprofen, and so on.

2. The frequency counts are categorized according to two different variables: (1) whether the subject experienced nausea; (2) whether the subject was treated with Celebrex, Ibuprofen, or a placebo.

3. No, we can only conclude that nausea is associated with the treatment, but we cannot attribute the *cause* of the nausea to the treatment.

4. If we assume that experiencing or not experiencing nausea is independent of the treatment (Celebrex, Ibuprofen, placebo), the expected number of subjects who experience nausea with the Celebrex treatment is 160.490.

Chapter 11 Quick Quiz

1. H_0: $p_1 = p_2 = p_3 = p_4 = p_5 = p_6 = p_7$. H_1: At least one of the proportions is different from the others.

2. Observed: 40; expected: 30.857

3. $\chi^2 = 12.592$

4. There is not sufficient evidence to warrant rejection of the claim that fatal DWI crashes occur equally on the different days of the week.

5. Right-tailed

6. Chi-square

7. H_0: Response is independent of whether the person responding is a worker or senior-level boss.
   H_1: Response and whether the person is a worker or senior-level boss are dependent.

8. $\chi^2 = 3.841$

9. There is sufficient evidence to warrant rejection of the claim that response is independent of whether the person is a worker or senior-level boss. Response appears to be somehow associated with whether the person is a worker or senior-level boss.

10. There is not sufficient evidence to warrant rejection of the claim that response is independent of whether the person is a worker or senior-level boss. Response and whether the person is a worker or senior-level boss appear to be independent of each other.

Chapter 11 Review Exercises

1. Test statistic: $\chi^2 = 9.294$. Critical value: $\chi^2 = 5.991$. *P*-value < 0.01 (TI-83/84 Plus: 0.0096). There is sufficient evidence to warrant rejection of the claim of independence between experiencing nausea and the type of treatment. The adverse reaction of nausea does not appear to be about the same for the different treatments.

2. Test statistic: $\chi^2 = 36.366$. Critical value: $\chi^2 = 16.812$. *P*-value < 0.005 (TI-83/84 Plus: 0.000). There is sufficient evidence to warrant rejection of the claim that deaths from lightning occur on the days of the week with equal frequency. Sunday appears to have disproportionately more deaths from lightning, and that might be explained by the fact that many people engage in outdoor recreational activities on Sundays.

3. Test statistic: $\chi^2 = 51.270$. Critical value: $\chi^2 = 9.488$. *P*-value < 0.005 (TI-83/84 Plus: 0.000). There is sufficient evidence to warrant rejection of the claim that that the participants fit the same distribution as the U.S. population. If study participants are not representative of the population, the results might be misleading because some groups might have cancer rates different from others, and they might skew the outcomes.

4. Test statistic: $\chi^2 = 10.732$. Critical value: $\chi^2 = 3.841$. *P*-value < 0.005 (TI-83/84 Plus: 0.001). There is sufficient evidence to warrant rejection of the claim that whether a subject smokes is independent of whether the subject was treated with bupropion hydrochloride or a placebo. It appears that the bupropion hydrochloride treatment is effective in the sense that it is better than a placebo, but that the treatment is not highly effective because many of those in the treatment group continued to smoke.

5. Test statistic: $\chi^2 = 16.264$. Critical value: $\chi^2 = 3.841$. *P*-value < 0.005 (TI-83/84 Plus: 0.000). There is sufficient evidence to warrant rejection of the claim that the two given proportions are the same. The results indicate that when reporting about coughing of children, there is substantial disagreement between the children and their parents.

Chapter 11 Cumulative Review Exercises

1. a. Observational study
   b. Discrete
   c. Statistics
   d. The organizations releasing the data have a special interest in the topic and those who conducted the research may have been influenced by them.

2. H_0: $p_1 = p_2$. H_1: $p_1 \neq p_2$. Test statistic: $z = -20.35$. Critical values: $z = \pm1.96$. *P*-value: 0.0002 (TI-83/84 Plus: 0.000). Reject H_0. There is sufficient evidence to warrant rejection of the claim that the proportion of men who wash their hands is equal to the proportion of women who wash their hands. There is a significant difference between men and women.

3. Test statistic: $\chi^2 = 414.230$. Critical value: $\chi^2 = 3.841$. *P*-value < 0.005 (TI-83/84 Plus: 0.000). There is sufficient evidence to warrant rejection of the claim that hand washing is independent of gender. It appears that whether a person washes their hands is related to the gender of the person.

4. First round: $\bar{x} = 71.2$, median: 71.5, range $= 8.0$, $s = 3.2$. Fourth round: $\bar{x} = 70.3$, median: 69.5, range $= 4.0$, $s = 1.8$. The fourth round scores are slightly lower and closer together.

5. $r = -0.406$. Critical values: $r = \pm0.811$. *P*-value $= 0.425$. There is not sufficient evidence to support the claim of a linear correlation between the first round scores and the fourth round scores.

6. $67.8 < \mu < 74.5$. We have 95% confidence that the limits of 67.8 and 74.5 contain the true value of the mean first round golf score. This means that if we were to repeat the first round under the same conditions and construct the 95% confidence interval from the six scores, 95% of the confidence intervals would contain the true population mean score.

7. $0.828 < p < 0.932$. Given that such a high percentage of senior executives believe that a thank-you note is helpful, it would be wise to send a thank-you note after every job interview.

8. H_0: $p = 0.75$. H_1: $p > 0.75$. Test statistic: $z = 3.68$. Critical value: $z = 2.33$. *P*-value: 0.0001. Reject H_0. There is sufficient evidence to support the claim that more than 75% of all senior executives believe that a thank-you note after a job interview increases the chances of the applicant.

9. a. 0.2033 (TI-83/84 Plus: 0.2040)
   b. 0.0005
   c. The result from part (a) is more relevant. The designers need to consider the upper leg lengths of individual men who will occupy the cockpit; a cockpit will never be occupied by a group of 16 men.
10. 0.677. It is not unusual, because the probability is not low, such as 0.05 or less.

Chapter 12 Answers
Section 12-2

1. a. The data are categorized according to the one characteristic or factor of epoch.
   b. The terminology of *analysis of variance* refers to the method used to test for equality of the three means. That method is based on two different estimates of the population variance.
3. We should reject the null hypothesis that the three epochs have the same mean skull breadth. There is sufficient evidence to conclude that at least one of the means is different from the others.
5. Test statistic: $F = 9.4695$. Critical value of F is approximately 3.3158 (TI-83/84 Plus: 3.2849). P-value: 0.000562. Reject H_0: $\mu_1 = \mu_2 = \mu_3$. There is sufficient evidence to warrant rejection of the claim that the three books have the same mean Flesch Reading Ease score.
7. Test statistic: $F = 0.3521$. Critical value: $F = 2.6626$. P-value: 0.7877. Fail to reject H_0: $\mu_1 = \mu_2 = \mu_3 = \mu_4$. There is not sufficient evidence to warrant rejection of the claim that the mean weight loss is the same for the diets. The diets do not appear to be very effective.
9. Test statistic: $F = 5.313$. Critical value of F is approximately 3.0718 (TI-83/84 Plus: 3.0738). P-value: 0.010. Reject H_0: $\mu_1 = \mu_2 = \mu_3$. There is sufficient evidence to warrant rejection of the claim that PG movies, PG-13 movies, and R movies have the same mean gross amount.
11. Test statistic: $F = 0.3974$. Critical value: $F = 3.3541$. P-value: 0.6759. Fail to reject H_0: $\mu_1 = \mu_2 = \mu_3$. There is not sufficient evidence to warrant rejection of the claim that the different car categories have the same mean. These data do not suggest that larger cars are safer.
13. Test statistic: $F = 27.2488$. Critical value: $F = 3.8853$. P-value: 0.0000. Reject H_0: $\mu_1 = \mu_2 = \mu_3$. There is sufficient evidence to warrant rejection of the claim that the three different miles have the same mean time. These data suggest that the third mile appears to take longer, and a reasonable explanation is that the third lap has a hill.
15. Test statistic: $F = 18.9931$. Critical value of F is approximately 3.1504 (TI-83/84 Plus: 3.1239). P-value: 0.0000. Reject H_0: $\mu_1 = \mu_2 = \mu_3$. There is sufficient evidence to warrant rejection of the claim that the three different types of cigarettes have the same mean amount of nicotine. Given that the king size cigarettes have the largest mean of 1.26 mg per cigarette, compared to the other means of 0.87 mg per cigarette and 0.92 mg per cigarette, it appears that the filters do make a difference (although this conclusion is not justified by the results from analysis of variance).

17. The Tukey test results show that the mean from small cars is significantly different from the mean from large cars. The conclusion is the same as that obtained with the Bonferroni test results.

Section 12-3

1. The data are categorized using the *two* different variables of sex and age bracket.
3. Because each cell contains the same number of sample values (five), this is a balanced design.
5. Test statistic: $F = 0.24$. P-value: 0.625. Fail to reject the null hypothesis of no interaction effect. There is not sufficient evidence to warrant rejection of the claim that heights are not affected by an interaction between sex and age bracket. There does not appear to be an effect from an interaction between sex and age bracket.
7. Test statistic: $F = 2.95$. P-value: 0.091. Fail to reject the null hypothesis of no effect from age bracket. There is not sufficient evidence to support the claim that age bracket (under 30, over 30) has an effect on height.
9. Test statistic: $F = 2.25$. P-value: 0.159. Fail to reject the null hypothesis of no effect from the type of car. There is not sufficient evidence to support the claim that whether the car is foreign or domestic has an effect on head injury measurements.
11. Test statistic: $F = 3.73$. P-value: 0.029. Reject the null hypothesis of no interaction effect. There is sufficient evidence to warrant rejection of the claim that measurements of self-esteem are not affected by an interaction between the subject's self-esteem and the target's self-esteem. There does appear to be an effect from an interaction between the self-esteem of the subject and their perceived self-esteem of the target.
13. Test statistic: $F = 1.58$. P-value: 0.213. Fail to reject the null hypothesis of no effect from self-esteem of the subject. There is not sufficient evidence to support the claim that the self-esteem of the subject (low, medium, high) has an effect on the measurements of self-esteem.
15. For interaction, the test statistic is $F = 41.38$ and the P-value is 0.000, so there is a significant interaction effect. The rating appears to be affected by an interaction between the use of the supplement and the amount of whey.
17. Row factor: Test statistic is $F = 0.10$ and P-value $= 0.771$, so fail to reject the null hypothesis of no effect from the row factor of supplement; there is not sufficient evidence to support the claim that ratings are affected by the use of the supplement. Column factor: Test statistic is $F = 1.23$ and P-value $= 0.434$, so fail to reject the null hypothesis of no effect from the column factor of the amount of whey; there is not sufficient evidence to support the claim that ratings are affected by the amount of whey.

Chapter 12 Statistical Literacy and Critical Thinking

1. Because two of the samples are from the same source at the same times, they are not independent. One-way analysis of variance should not be used because it has a strict requirement of independent data sets.
2. One-way analysis of variance is used with sample data categorized according to one factor, but two-way analysis of variance is used with sample data categorized according to two different factors.

3. No, because the MPAA ratings of PG, PG-13, and R are qualitative data, but methods of one-way analysis of variance or two-way analysis of variance require quantitative data.

4. Using the least expensive cars could easily result in biased samples consisting of cars that include fewer safety features and are not as well built, so such biased samples should not be used for making inferences about all cars.

Chapter 12 Quick Quiz

1. Test a null hypothesis that three or more samples are from populations with equal means.
2. Right-tailed
3. Smaller P-values
4. $F = 8.98$
5. There is sufficient evidence to reject the null hypothesis that the three populations of grade level reading scores have the same mean.
6. It appears that at least one of the books has a population mean grade level reading score that is different from the other population means.
7. With one-way analysis of variance, the different samples are categorized using only one factor, but with two-way analysis of variance, the sample data are categorized into different cells determined by two different factors.
8. Fail to reject the null hypothesis of no interaction. There does not appear to be an effect due to an interaction between sex and major.
9. There is not sufficient evidence to support a claim that the length estimates are affected by the sex of the subject.
10. There is not sufficient evidence to support a claim that the length estimates are affected by the subject's major.

Chapter 12 Review Exercises

1. a. One (type of diet)
   b. One-way analysis of variance
   c. Because the P-value is high, it appears that the four samples have means that do not differ by significant amounts. It appears that the mean ages of the four treatment groups are about the same.
   d. A small P-value would indicate that at least one of the treatment groups has a mean age that is significantly different from the others, so we would not know if differences from the diet treatments are due to the diets or to differences in age. A small P-value would undermine the effectiveness of the experiment.
2. Test statistic: $F = 54.70$. P-value: 0.000. Reject $H_0: \mu_1 = \mu_2 = \mu_3$. There is sufficient evidence to warrant rejection of the claim that the different car categories have the same mean weight. Because the sample means are 3095.0 lb (4 cylinders), 3835.0 lb (six cylinders), and 4159.3 (eight cylinders), it does appear that cars with more cylinders weigh more.
3. Test statistic: $F = 0.37$. P-value: 0.698. Fail to reject the null hypothesis of no interaction effect. There is not sufficient evidence to warrant rejection of the claim that left femur loads are not affected by an interaction between the type of car (foreign, domestic) and size of the car (small, medium, large). There does not appear to be an effect from an interaction.

4. Test statistic: $F = 1.82$. P-value: 0.202. Fail to reject the null hypothesis of no effect from the type of car. There is not sufficient evidence to support the claim that whether the car is foreign or domestic has an effect on left femur load measurements.
5. Test statistic: $F = 0.48$. P-value: 0.632. Fail to reject the null hypothesis of no effect from the size of the car. There is not sufficient evidence to support the claim that whether the car is small, medium, or large has an effect on left femur load measurements.
6. Test statistic: $F = 0.5010$. Critical value of F is approximately 3.1504 (TI-83/84 Plus: 3.1239). P-value: 0.6080. Fail to reject $H_0: \mu_1 = \mu_2 = \mu_3$. There is not sufficient evidence to warrant rejection of the claim that the three different types of cigarettes have the same mean amount of carbon monoxide. It appears that the filters do not make a difference in the amount of carbon monoxide.
7. For interaction, the test statistic is $F = 0.8733$ and the P-value is 0.3685, so there is no significant interaction effect. For gender, the test statistic is $F = 0.0178$ and the P-value is 0.8960, so there is no significant effect from gender. For smoking, the test statistic is $F = 3.0119$ and the P-value is 0.1082, so there is no significant effect from smoking.
8. Test statistic: $F = 3.1095$. P-value $= 0.0506$. Using a 0.05 significance level, fail to reject $H_0: \mu_1 = \mu_2 = \mu_3$. There is not sufficient evidence to warrant rejection of the claim that the three groups have the same mean longevity. The survival times appear to have means that are the same.

Chapter 12 Cumulative Review Exercises

1. a. 15.5, 13.1, 22.7
   b. 9.7, 9.0, 18.6
2. Test statistic: $t = -1.383$. Critical values: $t = \pm 2.160$ (assuming an 0.05 significance level). (TI-83/84 Plus: P-value $= 0.1860$.) Fail to reject $H_0: \mu_1 = \mu_2$. There is not sufficient evidence to support the claim that there is a difference between the means for the two groups.
3. Normal, because the histogram is approximately bell-shaped.
4. 12.3 years $< \mu < 18.7$ years
5. a. Interval
   b. Continuous
   c. 97.87°F
   d. 98.00°F
   e. 2.40°F
   f. 0.71°F
   g. 0.50 degree^2
6.

Rating	Frequency
1.0–1.9	1
2.0–2.9	4
3.0–3.9	0
4.0–4.9	1
5.0–5.9	4
6.0–6.9	9
7.0–7.9	9
8.0–8.9	7

7. No. The shape of the histogram is not reasonably close to the bell shape of a normal distribution.

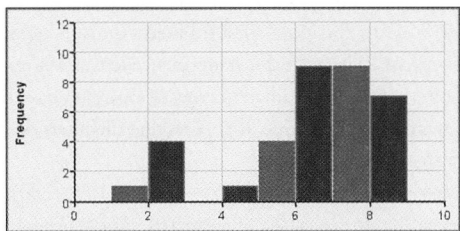

8. The sample means of 6.63, 6.43, and 6.18 do not appear to be very different. Analysis of variance test: Test statistic is $F = 0.0886$ and the P-value is 0.915, so fail to reject $H_0: \mu_1 = \mu_2 = \mu_3$ and conclude that there is not sufficient evidence to warrant rejection of the claim that PG movies, PG-13 movies, and R movies have the same mean viewer rating.

9. a. $r = 0.603$. Critical values: $r = \pm 0.878$ (assuming a 0.05 significance level). P-value = 0.282. There is not sufficient evidence to support the claim of a linear correlation between amounts of discarded paper and plastic.
   b. $\hat{y} = -2.17 + 0.405x$
   c. No. Because there is not a linear correlation between the amounts of discarded paper and plastic, the regression line does not fit the data well enough.

10. Test statistic: $\chi^2 = 2.909$. Critical value: $\chi^2 = 5.991$. P-value > 0.10 (TI-83/84 Plus: 0.2335). There is not sufficient evidence to warrant rejection of the claim of independence between injury category and whether the firearm was a handgun or a rifle or shotgun.

Chapter 13 Answers
Section 13-2

1. The sign test does not have a requirement that the sample data come from a population with a particular distribution.

3. The alternative hypothesis is that the proportion of girls is greater than 0.5, but the sample proportion of girls is 39/211, which is less than 0.5. The test is not necessary because there is no way that a sample proportion of 39/211 (or 0.185) could ever support a claim that the population proportion is greater than 0.5.

5. Test statistic is $x = 1$. TI-83/84 Plus: P-value = 0.0018. There is sufficient evidence to warrant rejection of the claim of no difference. There does appear to be a difference.

7. Test statistic is $x = 360$. TI-83/84 Plus: P-value = 0.6314. There is not sufficient evidence to warrant rejection of the claim of no difference. There does not appear to be a difference.

9. Test statistic is $x = 2$. TI-83/84 Plus: P-value = 0.0074. There is sufficient evidence to reject the claim of no difference. Based on the sample data, it appears that when awards are presented, there is a difference between the age of the best actress and the best actor.

11. Test statistic is $x = 3$. TI-83/84 Plus: P-value = 0.1719. There is not sufficient evidence to reject the claim of no difference. There is not sufficient evidence to support the claim that the time spent depicting the use of alcohol is less than the time spent depicting the use of tobacco.

13. Test statistic is $x = 39$. TI-83/84 Plus: P-value = 0.0000. There is sufficient evidence to warrant rejection of the claim of no difference. The YSORT method appears to be effective in increasing the likelihood of a boy.

15. Test statistic is $x = 47$. TI-83/84 Plus: P-value = 0.1888. There is not sufficient evidence to warrant rejection of the claim of no difference. It appears that women do not have the ability to predict the sex of their babies.

17. Test statistic is $x = 12$. TI-83/84 Plus: P-value = 0.0166. There is not sufficient evidence to warrant rejection of the claim that the median is equal to 5.670 g. The quarters appear to be minted according to specifications.

19. Test statistic is $x = 1$. TI-83/84 Plus: P-value = 0.0000. There is sufficient evidence to warrant rejection of the claim that the median amount of Coke is equal to 12 oz. Consumers are not being cheated because they are generally getting more than 12 oz of Coke, not less.

21. Second approach: The test statistic of $z = -4.29$ is in the critical region bounded by $z = -1.645$, so the conclusions are the same as in Example 4. Third approach: The test statistic of $z = -2.82$ is in the critical region bounded by $z = -1.645$, so the conclusions are the same as in Example 4. The different approaches can lead to very different results; see the test statistics of -4.61, -4.29, and -2.82. The conclusions are the same in this case, but they could be different in other cases.

Section 13-3

1. The sign test converts the sample data to positive and negative signs, whereas the Wilcoxon signed-ranks test uses ranks, so conclusions will sometimes be different. Because the Wilcoxon signed-ranks test uses more information about the data, it is likely to yield better results.

3. $n = 10$. The critical value is $T = 8$.

5. Test statistic: $T = 1.5$. Critical value: $T = 1$. Fail to reject the null hypothesis that the population of differences has a median of 0. Based on the sample data, it appears that when Friday falls on the 13th, the numbers of hospital admissions are not affected.

7. Test statistic: $T = 3$. Critical value: $T = 25$. Reject the null hypothesis that the population of differences has a median of 0. Based on the sample data, it appears that when awards are presented, there is a difference between the age of the best actress and the best actor. Reject the null hypothesis that the population of differences has a median of 0. Based on the sample data, it appears that when awards are presented, there is a difference between the age of the best actress and the best actor.

9. Test statistic: $T = 13.5$. Critical value: $T = 8$. Fail to reject the null hypothesis that the population of differences has a median of 0. There is not sufficient evidence to support the claim that the time spent depicting the use of alcohol is less than the time spent depicting the use of tobacco.

11. Convert $T = 196$ to the test statistic $z = -2.88$. Critical values: $z = \pm 2.575$. (TI-83/84 Plus: P-value = 0.004.) There is sufficient evidence to warrant rejection of the claim that the median is equal to 5.670 g. The quarters do not appear to be minted according to specifications.

13. Convert $T = 15.5$ to the test statistic $z = -4.82$. Critical values: $z = \pm1.96$. (TI-83/84 Plus: P-value $= 0.000$.) There is sufficient evidence to warrant rejection of the claim that the median amount of Coke is equal to 12 oz. Consumers are not being cheated because they are generally getting more than 12 oz of Coke, not less.

15. a. 975

  b. $\dfrac{n(n + 1)}{2} - k$

Section 13-4

1. Because the samples are small and they are from populations not having normal distributions, the parametric t test should not be used.

3. a. 500.5

  b. No, we are testing the null hypothesis that the two populations have the same median, which is not specified.

5. $R_1 = 692$, $R_2 = 739$, $\mu_R = 621$, $\sigma_R = 55.723$, test statistic: $z = 1.27$. Critical values: $z = \pm1.96$. (TI-83/84 Plus: P-value $= 0.2026$.) Fail to reject the null hypothesis that the populations have the same median.

7. $R_1 = 416$, $R_2 = 325$, $\mu_R = 468$, $\sigma_R = 33.045$, test statistic: $z = -1.57$. Critical values: $z = \pm1.96$. (TI-83/84 Plus: P-value $= 0.1156$.) Fail to reject the null hypothesis that the populations have the same median.

9. $R_1 = 863$, $R_2 = 412$, $\mu_R = 637.5$, $\sigma_R = 51.539$, test statistic: $z = 4.38$. Critical value: $z = 2.33$. (TI-83/84 Plus: P-value $= 0.0000$.) Reject the null hypothesis that the populations have the same median. There is sufficient evidence to support the claim that the nonfiltered king size cigarettes have a median amount of nicotine that is greater than the median amount of nicotine in the 100 mm filtered cigarettes.

11. $R_1 = 487.5$, $R_2 = 142.5$, $\mu_R = 414$, $\sigma_R = 28.775$, test statistic: $z = 2.55$. Critical value: $z = 1.645$. (TI-83/84 Plus: P-value $= 0.0053$.) Reject the null hypothesis that the populations have the same median. There is sufficient evidence to support the claim that movies with ratings of PG or PG-13 have a higher median gross amount than movies with R ratings.

13. $z = -0.63$. The test statistic is the same value with opposite sign.

Section 13-5

1. Because the values from each sample are not matched in any way with values from the other samples, the samples are independent.

3. $1 + 4 + 6 + 7.5 + 7.5 + 10 + 12 + 21.5 + 21.5 = 91$

5. Test statistic: $H = 6.6305$. Critical value: $\chi^2 = 5.991$. (TI-83/84 Plus: P-value $= 0.0363$.) Reject the null hypothesis of equal medians. Interbreeding is suggested by the data.

7. Test statistic: $H = 1.2239$. Critical value: $\chi^2 = 5.991$. (TI-83/84 Plus: P-value $= 0.5423$.) Fail to reject the null hypothesis of equal medians. The available data do not provide sufficient evidence to suggest that larger cars are safer.

9. Test statistic: $H = 20.9247$. Critical value: $\chi^2 = 5.991$. (TI-83/84 Plus: P-value $= 0.000$.) Reject the null hypothesis of equal medians. It appears that cars with more cylinders produce larger amounts of greenhouse gases.

11. Test statistic: $H = 27.9098$. Critical value: $\chi^2 = 5.991$. (TI-83/84 Plus: P-value $= 0.000$.) Reject the null hypothesis of equal medians. There is sufficient evidence to warrant rejection of the claim that the three different types of cigarettes have the same median amount of nicotine. It appears that the filters do make a difference.

13. Corrected H is 5.835 (using $\Sigma T = 282$ and $N = 30$), which is not very different from the uncorrected H of 5.774.

Section 13-6

1. The methods of Section 10-3 should not be used for predictions. The regression equation is based on a *linear* correlation between the two variables, but the methods of this section do not require a linear relationship. The methods of this section could suggest that there is a correlation with paired data associated by some nonlinear relationship, so the regression equation would not be a suitable model for making predictions.

3. The symbol r_s is used to represent the rank correlation coefficient computed from the given sample data, and the symbol ρ_s represents the rank correlation coefficient of the paired data for the entire population. The subscript s is used so that the rank correlation coefficient can be distinguished from the linear correlation coefficient r. The subscript does not represent the standard deviation s. It is used to honor Charles Spearman, who introduced the rank correlation method.

5. $r_s = 1$. Critical values are -0.886 and 0.886. Reject the null hypothesis of $\rho_s = 0$. There is sufficient evidence to support a claim of a correlation between distance and time.

7. a. ±0.521

  b. ±0.521

  c. ±0.197

  d. ±0.322

9. $r_s = -0.929$. Critical values: -0.786, 0.786. Reject the null hypothesis of $\rho_s = 0$. There is sufficient evidence to support the claim of a correlation between the two judges. Examination of the results shows that the first and third judges appear to have opposite rankings.

11. $r_s = -0.007$. Critical values: -0.587, 0.587. Fail to reject the null hypothesis of $\rho_s = 0$. There is not sufficient evidence to support the claim of a correlation between the conviction rates and recidivism rates. Conviction rates do not appear to be related to recidivism rates.

13. $r_s = 0.664$. Critical values: -0.618, 0.618. Reject the null hypothesis of $\rho_s = 0$. There is sufficient evidence to support the claim of a correlation between quality score and cost. It appears that higher quality is associated with higher cost, so you can expect to get higher quality by purchasing a more expensive LCD TV.

15. $r_s = 1$. Critical values: -0.886, 0.886. Reject the null hypothesis of $\rho_s = 0$. There is sufficient evidence to conclude that there is a relationship between overhead widths of seals from photographs and the weights of the seals.

17. $r_s = 0.231$. Critical values: -0.264, 0.264. Fail to reject the null hypothesis of $\rho_s = 0$. There is not sufficient evidence to conclude that there is a correlation between the numbers of words spoken by men and women who are in couple relationships.

19. Using Formula 10-1 for the case of ties among ranks, $r_s = 0.109$. Using the formula for the case of no ties among ranks (even though there are ties), $r_s = 0.385$. There is a substantial difference between the two results. The first result is better because it is exact (except for rounding), whereas 0.385 is not exact. Using a 0.05 significance level, the critical values are -0.274 and 0.274, so the two results lead to different conclusions.

Section 13-7

1. No. The runs test can be used to determine whether the sequence of World Series wins by American League teams and National League teams is random, but the runs test does not show whether the proportion of wins by the American League is significantly greater than 0.5.

3. No. There might be other problems with the data selection process. For example, a voluntary response sample might appear to be random, but it would not be suitable for most statistical purposes.

5. $n_1 = 9$, $n_2 = 7$, $G = 4$, critical values: 4, 14. Reject randomness. There is sufficient evidence to reject the claim that the genders of the younger winners occur randomly.

7. $n_1 = 19$, $n_2 = 14$, $G = 15$, critical values: 11, 23. Fail to reject randomness. There is not sufficient evidence to warrant rejection of the claim that we elect Democrats and Republicans in a random sequence.

9. $n_1 = 20$, $n_2 = 10$, $G = 16$, critical values: 9, 20. Fail to reject randomness. There is not sufficient evidence to reject the claim that the dates before and after July 1 are randomly selected.

11. $n_1 = 23$, $n_2 = 18$, $G = 16$, $\mu_G = 21.1951$, $\sigma_G = 3.11307$. Test statistic: $z = -1.67$. Critical values: $z = \pm 1.96$. (TI-83/84 Plus: P-value $= 0.0952$.) Fail to reject randomness. The sequence appears to be random. The runs test suggests that the sequence appears to be random, but it does not test for disproportionately more occurrences of one of the two categories, so the runs test does not suggest that either conference is superior.

13. The median is 1956, and the value of 1956 is deleted to yield $n_1 = 21$, $n_2 = 21$, $G = 2$, $\mu_G = 22$, $\sigma_G = 3.20061$. Test statistic: $z = -6.25$. Critical values: $z = \pm 1.96$. (TI-83/84 Plus: P-value $= 0.0000$.) Reject randomness. The sequence does not appear to be random when considering values above and below the median. There appears to be an upward trend, so the stock market appears to be a profitable investment.

15. b. The 84 sequences yield 2 runs of 2, 7 runs of 3, 20 runs of 4, 25 runs of 5, 20 runs of 6, and 10 runs of 7.

    c. With $P(2 \text{ runs}) = 2/84$, $P(3 \text{ runs}) = 7/84$, $P(4 \text{ runs}) = 20/84$, $P(5 \text{ runs}) = 25/84$, $P(6 \text{ runs}) = 20/84$, and $P(7 \text{ runs}) = 10/84$, each of the G-values of 3, 4, 5, 6, 7 can easily occur by chance, whereas $G = 2$ is unlikely because $P(2 \text{ runs})$ is less than 0.025. The lower critical value of G is therefore 2,

and there is no upper critical value that can be equaled or exceeded.

d. Critical value of $G = 2$ agrees with Table A-9. The table lists 8 as the upper critical value, but it is impossible to get 8 runs using the given elements.

Chapter 13 Statistical Literacy and Critical Thinking

1. A nonparametric test is a hypothesis test that does not require that sample data come from a population having a normal distribution or any other specific distribution. A parametric test is a hypothesis test with a requirement that the sample data are from a population having a particular distribution, and the normal distribution is most common.

2. There is no difference. They are different names for the same category of hypothesis tests that do not require that populations have normal distributions or any other particular distribution.

3. A rank is a number assigned to an individual sample item according to its order in the sorted list. The first item is assigned a rank of 1, the second item is assigned a rank of 2, and so on. The given values have ranks of 1.5, 1.5, 5, 4, and 3, respectively.

4. Efficiency is a measure of how much stronger sample evidence must be for the nonparametric test to produce the same results as a corresponding parametric test. For example, the sign test has an efficiency of 0.63, meaning that under the same conditions, the sign test requires 100 sample observations to achieve the same results as 63 sample observations analyzed with a corresponding parametric test. We should use a nonparametric test when requirements of the corresponding parametric test (such as the requirement of a normally distributed population) are not satisfied.

Chapter 13 Quick Quiz

1. 2, 1, 3.5, 3.5, 5
2. The runs test for randomness is used to test the null hypothesis that sample data in a sequence are in a random order. This test is based on sample data that have two characteristics.
3. Rank correlation can be used in a wider variety of circumstances than the parametric method of linear correlation. Rank correlation does *not* require a normal distribution for any population. Rank correlation can be used to detect some (not all) relationships that are not linear.
4. When compared to the parametric test, the nonparametric test requires stronger evidence, such as a larger sample or greater differences, in order to reject a null hypothesis.
5. Distribution-free test
6. False
7. Four independent samples
8. Two independent samples
9. Matched pairs
10. Four independent samples

Chapter 13 Review Exercises

1. The test statistic is $x = 42$. TI-83/84 Plus: P-value $= 0.0756$. Fail to reject the null hypothesis of $p = 0.5$. There is not sufficient evidence to warrant rejection of the claim that in each World Series, the American League team has a 0.5 probability of winning.

2. The test statistic is $x = 1$. TI-83/84 Plus: P-value $= 0.0391$. Reject the null hypothesis that the median equals 98.6°F. There is sufficient evidence to warrant rejection of the claim that body temperatures have a median equal to 98.6°F.

3. The test statistic of $T = 1.5$ is less than or equal to the critical value of 6. Reject the null hypothesis that the median equals 98.6°F. There is sufficient evidence to warrant rejection of the claim that body temperatures have a median equal to 98.6°F.

4. $n_1 = 8$, $n_2 = 2$, and the number of runs is $G = 5$. The critical values are 1 and 6. Fail to reject the null hypothesis of randomness. There is not sufficient evidence to warrant rejection of the claim that the sequence of odd and even digits is random.

5. $r_s = 0.714$. Critical values: -0.738, 0.738. Fail to reject the null hypothesis of $\rho_s = 0$. There is not sufficient evidence to support the claim that there is a correlation between the student ranks and the magazine ranks. When ranking colleges, students and the magazine do not appear to agree.

6. Test statistic: $x = 2$. TI-83/84 Plus: P-value $= 0.1797$. Fail to reject the null hypothesis that the population of differences has a median of 0. Based on the sample data, it appears that the predictions are reasonably accurate, because there does not appear to be a difference between the actual high temperatures and the predicted high temperatures.

7. The test statistic of $T = 11.5$ is not less than or equal to the critical value of 6. Fail to reject the null hypothesis that the population of differences has a median of 0. Based on the sample data, it appears that the predictions are reasonably accurate, because there does not appear to be a difference between the actual high temperatures and the predicted high temperatures.

8. Test statistic: $H = 0.559$. Critical value: $\chi^2 = 5.991$. (TI-83/84 Plus: P-value $= 0.756$.) Fail to reject the null hypothesis of equal medians. The available data do not provide sufficient evidence to suggest that larger cars are safer.

9. $R_1 = 273$, $R_2 = 192$, $\mu_R = 232.5$, $\sigma_R = 24.1091$, test statistic: $z = 1.68$. Critical values: $z = \pm 1.96$. (TI-83/84 Plus: P-value $= 0.0930$.) Fail to reject the null hypothesis that the populations have the same median. There is not sufficient evidence to support the claim that the region becomes cooler as the fall season progresses.

10. $n_1 = 17$, $n_2 = 13$, and the number of runs is $G = 12$. The critical values are 10 and 22. Fail to reject the null hypothesis of randomness. There is not sufficient evidence to warrant rejection of the claim that the sequence of odd and even temperatures is random.

Chapter 13 Cumulative Review Exercises

1. Mean: 213.2 mg; median: 155.5 mg; range: 469.0 mg; standard deviation: 150.1 mg; variance: 22,529.5 mg^2.

2.

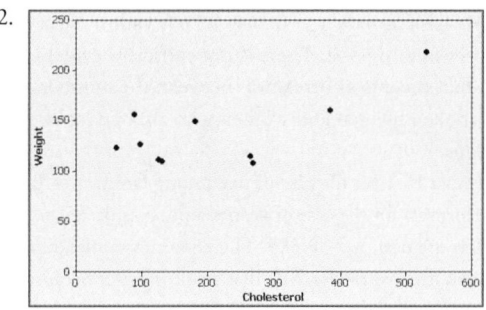

3. $r = 0.689$. Critical values: $r = \pm 0.632$. P-value $= 0.0277$. There is sufficient evidence to support the claim of a linear correlation between cholesterol and weight.

4. $\hat{y} = 104 + 0.158x$; 120 lb

5. $r_s = 0.212$. Critical values: -0.648, 0.648. Fail to reject the null hypothesis of $\rho_s = 0$. There is not sufficient evidence to support the claim that there is a correlation between cholesterol and weight.

6. Test statistic: $F = 8.9785$. Critical value of F is between 3.2317 and 3.3158 (TI-83/84 Plus: 3.2849). P-value: 0.0008. Reject the null hypothesis of equal population means. There is sufficient evidence to warrant rejection of the claim that the three books have the same mean Flesch-Kincaid Grade Level. The three books do not appear to be written with the same reading level.

7. Test statistic: $H = 13.218$. Critical value: $\chi^2 = 5.991$. (TI-83/84 Plus: P-value $= 0.0013$.) Reject the null hypothesis of equal medians. The available data provide sufficient evidence to warrant rejection of the claim that the three books have the same median Flesch-Kincaid Grade Level. The three books do not appear to be written with the same reading level.

8. $H_0: \mu_1 = \mu_2$. $H_1: \mu_1 \neq \mu_2$. Test statistic: $t = 1.820$. Critical values: $t = \pm 2.201$ (TI-83/84 Plus: $t = \pm 2.123$). P-value > 0.05 (TI-83/84 Plus: 0.0878). Fail to reject H_0. There is not sufficient evidence to reject the claim that that the samples of Flesch-Kincaid Grade Level scores from Clancy and Rowling have the same mean.

9. $R_1 = 175.5$, $R_2 = 124.5$, $\mu_R = 150$, $\sigma_R = 17.32051$, test statistic: $z = 1.47$. Critical values: $z = \pm 1.96$. (TI-83/84 Plus: P-value $= 0.1410$.) Fail to reject the null hypothesis that the samples of Flesch-Kincaid Grade Level scores from Clancy and Rowling are from populations having the same median.

10. There must be an error, because the rates of 13.7% and 10.6% are not possible with samples of size 100.

Chapter 14 Answers
Section 14-2

1. No. If the process is within statistical control, we know that the cans are being filled in such a way that the mean and range do not vary too much, but being within statistical control indicates nothing about whether the product specifications are being met. It is possible to be within statistical control while underfilling all of the cans by a substantial amount.

3. Random variation is the type of variation inherent in the process that is not capable of filling every can of Coke with exactly 12 ounces. There are different examples of assignable variation, so answers will vary, but one example is the changing amounts of Coke that result from machinery that becomes worn through extensive use.

5. a. Within statistical control
   b. Not applicable
   c. The variation is too large, so that some cans are overfilled too much while others are under-filled too much.

7. a. Out of statistical control.
   b. There are 8 consecutive points all below the centerline. There are also 8 consecutive points all above the centerline.
   c. The process does not appear to be behaving as it should. There appears to be a shift in the process values, with all of the low values occurring at the beginning.

9. $\bar{\bar{x}} = 319.26$, $\bar{R} = 6.69$. R chart: LCL = 1.49, UCL = 11.89. $\bar{x}$ chart: LCL = 317.20, UCL = 321.32.

11. The process variation is out of statistical control. All three criteria for statistical stability are violated. There is a pattern of increasing values that is not random, there are points lying beyond the control limits, and there are 8 points lying below the centerline.

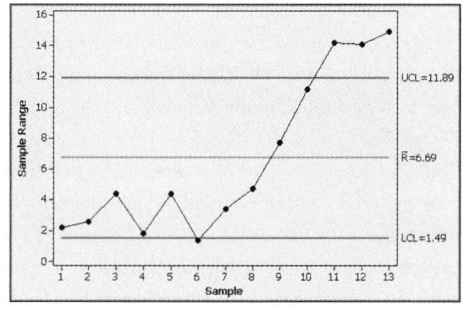

13. $\bar{\bar{x}} = 2781.71$, $\bar{R} = 1729.38$. R chart: LCL = 0, UCL = 3465.67. $\bar{x}$ chart: LCL = 1946.42, UCL = 3617.00.

15. The process mean is within statistical control.

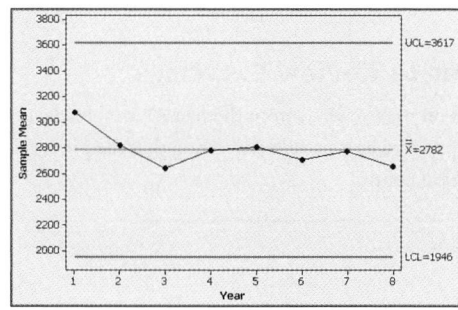

17. $\bar{\bar{x}} = 123.66$, $\bar{R} = 0.375$. R chart: LCL = 0, UCL = 0.79. $\bar{x}$ chart: LCL = 123.45, UCL = 123.88.

19. There is no pattern suggesting that the process is not within statistical control. Based on the runs chart, the process appears to be within statistical control.

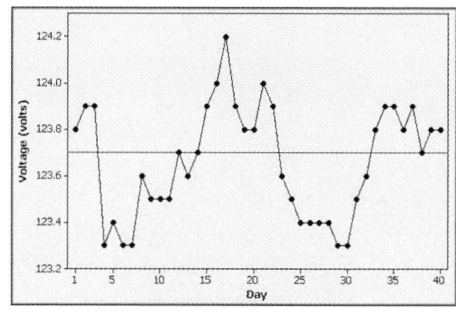

21. The s chart is very similar to the R chart. Because the s chart uses sample standard deviations instead of ranges, it is much more difficult to construct than the R chart.

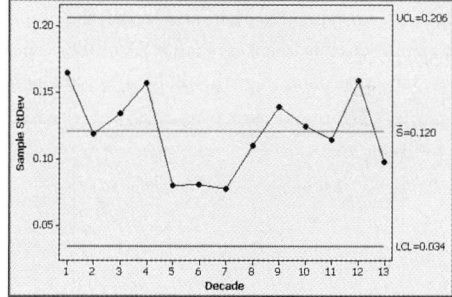

Section 14-3

1. No. It is possible that the proportions of defects in repeated samplings behave in a way that makes the process within statistical control, but the actual proportions of defects could be very high, such as 90%, so that almost all of the tablets fail to meet the manufacturing specifications.

3. The LCL of -0.022 is replaced by 0, because it is impossible to have negative values of proportions. UCL = 0.062.

5. Within statistical control

7. The process appears to be out of control because there is a pattern of an upward trend, there is a point lying beyond the upper control limit, and there are 8 consecutive points that are all above the centerline.

9. The process appears to be within statistical control.

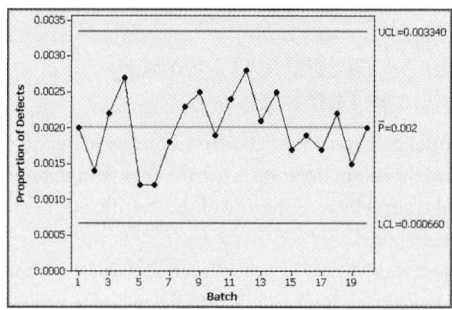

11. The process is within statistical control. College enrollments are fluctuating between 60% and 68%; ideally, the rate of college enrollments would be much higher.

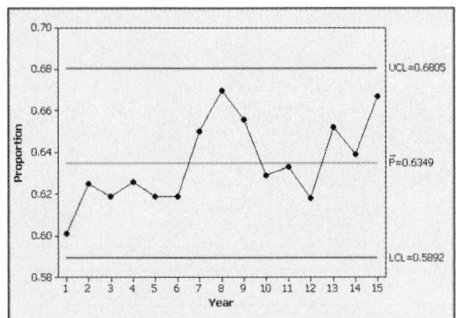

13. The process is not within statistical control, because there are points lying beyond the upper and lower control limits, and there is a nonrandom pattern of proportions that consistently fall, then rise. It appears that higher proportions of people vote in years with a presidential election. Overall, the proportions of voters are inexcusably low. Less than 60% of those eligible actually vote, and that percentage should be substantially higher.

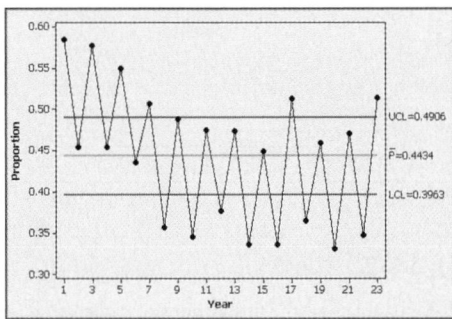

15. Except for the vertical scale, the control chart is identical to the one included with Example 1.

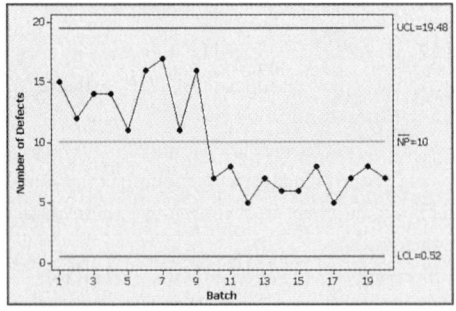

Chapter 14 Statistical Literacy and Critical Thinking

1. Statistical process control consists of methods used to monitor data over time and to ensure that the data do not go out of statistical control by meeting any of the specific out-of-control criteria being used.
2. Due to a wide variety of causes, such as changing employees and worn machinery, it is possible that the manufacturing process could result in far too many defective tablets with too much or too little of the drug atorvastatin. If the process is not monitored, the manufacturer Pfizer could be forced to halt production of Lipitor, with the result that Pfizer would lose some or all of the annual $13 billion income from that drug.
3. The process could go out of statistical control because of a changing mean, or increased variation, or both.
4. No. The limits of 9.5 mg and 10.5 mg are specifications that are independent of upper and lower control limits used for $\bar{x}$ charts and R charts. It is possible for the manufacturing process to be within control with almost all tablets containing an unacceptable amount of atorvastatin, such as 2 mg instead of some value near the desired level of 10 mg.

Chapter 14 Quick Quiz

1. Process data are data arranged according to some time sequence. They are measurements of a characteristic of goods or services that result from some combination of equipment, people, materials, methods, and conditions.
2. Random variation is due to chance, but assignable variation results from causes that can be identified, such as defective machinery or untrained employees.
3. There is a pattern, trend, or cycle that is obviously not random. There is a point lying outside of the region between the upper and lower control limits. There are eight consecutive points all above or all below the centerline.
4. An R chart uses ranges to monitor variation, but an $\bar{x}$ chart uses sample means to monitor the center (mean) of a process.
5. No. The R chart shows a point lying beyond the upper control limit.
6. $\bar{R} = 21.2$. In general, a value of $\bar{R}$ is found by first finding the range for the values within each individual subgroup; the mean of those ranges is the value of $\bar{R}$.
7. The process mean is within statistical control.
8. $\bar{\bar{x}} = 6.45$. In general, a value of $\bar{\bar{x}}$ is found by first finding the mean of the values within each individual subgroup; the mean of those subgroup means is the value of $\bar{\bar{x}}$.
9. A p chart is a control chart of the proportions of some attribute, such as defective items.
10. False

Chapter 14 Review Exercises

1. Based on the run chart from the first 21 axial loads, there does not appear to be a pattern suggesting that the process is out of statistical control.

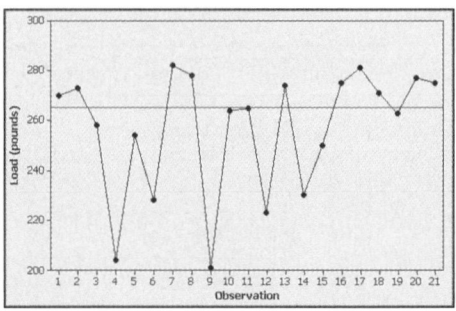

2. $\bar{R} = 62$; LCL $= 4.7$; UCL $= 119.3$

3. The variation of the process is within statitistical control because none of the three out-of-control conditions are met.

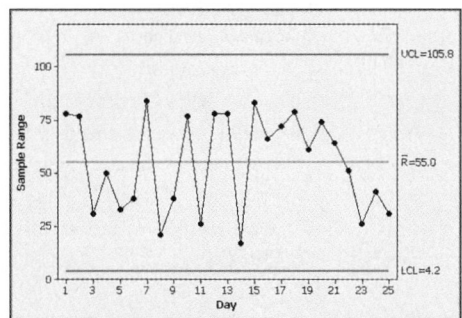

4. $\bar{\bar{x}} = 256.95$; LCL $= 230.97$; UCL $= 282.93$

5. The mean of the process is within statitistical control because none of the three out-of-control conditions are met.

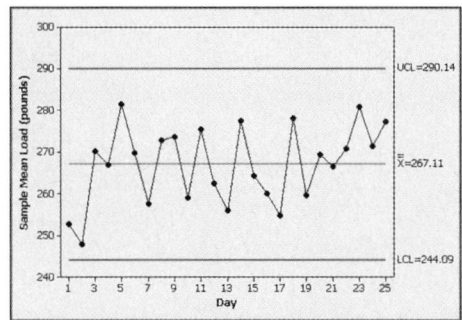

6. The proportions do not appear to be within statistical control because there is a pattern of declining proportions. The chart suggests that in recent years, the homicide rate has been declining, which is good, but it has leveled off in the last four years.

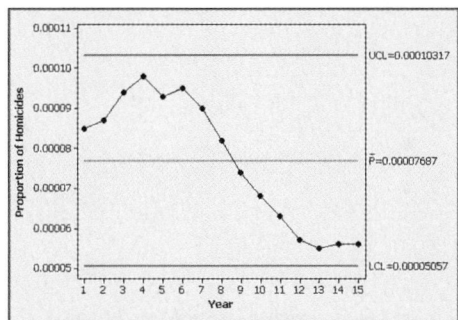

7. The process is out of statistical control, because there is a pattern of increasing variation. If left uncorrected, the process will eventually result in production runs with very high proportions of defective tablets.

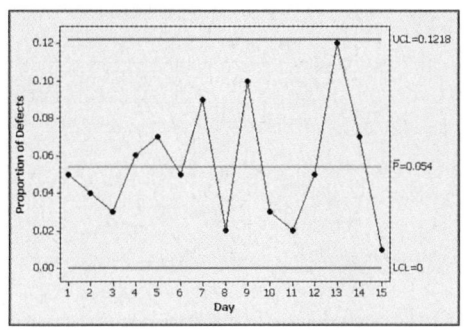

Chapter 14 Cumulative Review Exercises

1. a. $r = 0.691$. Critical values: $r = \pm 0.632$. P-value $= 0.0270$. There is sufficient evidence to support the claim of a linear correlation between carbon dioxide concentration and temperature.
   b. No. The presence of a correlation between two variables does not imply that one of the variables is the cause of the other.
   c. $\hat{y} = 9.54 + 0.0135x$
   d. The predicted value of 13.47°C is somewhat close to the actual temperature of 13.88°C. Because $r = 0.691$ is not extremely high, predicted values are not likely to be extremely accurate.

2. The process is out of statistical control because there are points lying beyond the upper control limit. Also, there appears to be a pattern of increasing variation.

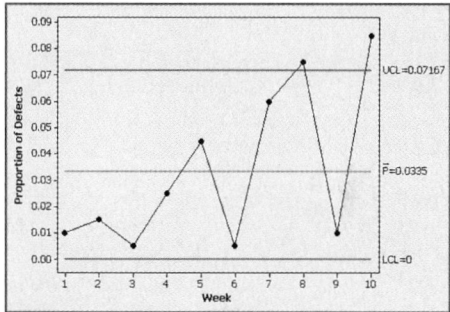

3. $0.0256 < p < 0.0414$

4. H_0: $p = 0.03$. H_1: $p > 0.03$. Test statistic: $z = 0.92$. Critical value: $z = 1.645$. P-value $= 0.1794$. Fail to reject H_0. There is not sufficient evidence to support the claim that the rate of defects is greater than 3%.

5. a. $1/256$
   b. $1/256$
   c. $1/128$

6. a. 15.87%
   b. 4.4 in. and 7.6 in.

7. No. The online poll resulted in a voluntary-response sample that represents only those who chose to respond. It is likely that those who chose to respond are not representative of the larger population.

8. The values appear to be from a population with a distribution that is not normal. The presence of 12 zeros among the 30 values causes a histogram to be skewed to the right instead of being bell-shaped. A normal quantile plot shows a pattern of points that is not reasonably close to a straight-line pattern.

9. $\bar{x} = 66.0$ cents; median $= 21.5$ cents; $s = 108.1$ cents

10. H_0: $p = 0.5$. H_1: $p > 0.5$. Test statistic: $z = 1.10$ (using $\hat{p} = 18/30 = 0.6$). Critical value: $z = 1.645$. P-value $= 0.1357$ (TI-83/84 Plus: 0.1367). Fail to reject H_0. There is not sufficient evidence to support the claim that most students have some change in their possession.

Credits

PHOTOGRAPHS

Chapter 1
Page 2, Time & Life Pictures/Getty Images.
Page 5, Jeff Greenberg/PhotoEdit.
Page 6, Shutterstock.
Page 8, Shutterstock.
Page 12, PhotoDisc.
Page 13, Shutterstock.
Page 13 (margin), Shutterstock.
Page 18, Shutterstock.
Page 19, Shutterstock.
Page 20, PhotoDisc (PP).
Page 27, Getty RF.
Page 28, Shutterstock.
Page 30, PhotoDisc Red.
Page 49, OJO Images/Getty Royalty Free.

Chapter 2
Page 50, AP Wideworld Photos.
Page 52, Shutterstock.
Page 54, U.S. Mint.
Page 55, Image Source/Getty Royalty Free.
Page 56, Kelly/Mooney Photography/Corbis.
Page 67, Stockbyte Gold RF.
Page 68, public domain.
Page 71, Brand X Pictures.
Page 87, Beth A. Keiser/Corbis.

Chapter 3
Page 88, Blend Images/Getty Royalty Free.
Page 91, Shutterstock.
Page 92, Digital Vision.
Page 93, PhotoDisc.
Page 95, Shutterstock.
Page 96, Shutterstock.
Page 105, Shutterstock.
Page 106, AP Wideworld Photo.
Page 108, PhotoDisc.
Page 120, Shutterstock.
Page 121, Shutterstock.
Page 125, Shutterstock.
Page 141, Shutterstock.

Chapter 4
Page 142, Everett Collection.
Page 145, Shutterstock.
Page 146, (margin), PhotoDisc.
Page 146 (Figure 4.1a), Shutterstock.
Page 146 (Figure 4.1b), Shutterstock.
Page 146 (Figure 4.1c), NASA/John F. Kennedy Space Center.
Page 147 and 165, PhotoDisc.
Page 148, PhotoDisc.
Page 150, Shutterstock.
Page 158, PhotoDisc.
Page 160, public domain.
Page 162, Shutterstock.
Page 166, Stone/Getty Images.

Page 169, Shutterstock.
Page 172, Shutterstock.
Page 173, Shutterstock.
Page 177, Dorling Kindersley.
Page 178 (margin), The Everett Collection.
Page 178, Shutterstock.
Page 180, Corbis.
Page 185 (margin), PhotoDisc.
Page 185, Shutterstock.
Page 186, Shutterstock.
Page 189, Shutterstock.
Page 190, PhotoDisc.
Page 191, PhotoDisc.
Page 192, Shutterstock.
Page 193, PhotoDisc.
Page 194, PhotoDisc.
Page 207, Shutterstock.

Chapter 5
Page 208, Shutterstock.
Page 213, Shutterstock.
Page 216, PhotoDisc.
Page 219, Shutterstock.
Page 226, PhotoDisc.
Page 227, Shutterstock.
Page 237, PhotoDisc.
Page 240, Shutterstock.
Page 241, Shutterstock.
Page 253, National Geographic/Getty Royalty Free.

Chapter 6
Page 254, Digital Vision.
Page 257, Shutterstock.
Page 258, Shutterstock.
Page 272, Blend Images/Getty Royalty Free.
Page 283, Anthony Redpath/Corbis.
Page 295, Image Source/Getty Royalty Free.
Page 299, iStockphoto.
Page 307 (margin), public domain.
Page 307, Shutterstock.
Page 317, PhotoDisc.
Page 319, PhotoDisc.
Page 331, PhotoEdit.

Chapter 7
Page 332, Photographer's Choice/Getty RF.
Page 336, PhotoDisc Red.
Page 338, Shutterstock.
Page 342, Shutterstock.
Page 353, PhotoDisc.
Page 354, Shutterstock.
Page 362, Shutterstock.
Page 364, Shutterstock.
Page 367, AP Wideworld Photo.
Page 368, Shutterstock.
Page 377, Shutterstock.

Chapter 8
Page 396, Shutterstock.
Page 399, Shutterstock.
Page 403, Shutterstock.
Page 404, Getty/Image Bank.
Page 407, PhotoDisc.
Page 410, PhotoDisc.
Page 422, Dorling Kindersley.
Page 423, Digital Vision.
Page 424, PhotoDisc.
Page 432, Photodisc/Getty RF.
Page 440, Beth Anderson.
Page 442, PhotoDisc Blue.
Page 449, Shutterstock.
Page 451, Shutterstock.
Page 463, Shutterstock.

Chapter 9
Page 464, Blend Images/Getty Royalty Free.
Page 468, PhotoDisc.
Page 470, USPS.
Page 471, Shutterstock.
Page 473, Bill Wisser/Getty Images.
Page 477, AP Wideworld Photos.
Page 477 (margin), PhotoDisc.
Page 480, Shutterstock.
Page 481, Lawrence Manning/Corbis.
Page 482, Comstock RF.
Page 483, Shutterstock.
Page 491, Shutterstock.
Page 493, Shutterstock.
Page 494, PhotoDisc.
Page 503, PhotoDisc.
Page 519, Shutterstock.

Chapter 10
Page 520, Digital Vision.
Page 523, Shutterstock.
Page 525, Shutterstock.
Page 527, PhotoDisc.
Page 529, Shutterstock.
Page 540, Digital Vision.
Page 545, Getty RF.
Page 546, PhotoDisc.
Page 548, Comstock RF.
Page 556, PhotoDisc Red.
Page 564, PhotoDisc.
Page 567, PhotoDisc.
Page 568, Shutterstock.
Page 568 (margin), Mark Asnin/Corbis Saba.
Page 569, PhotoDisc Red.
Page 575, PhotoDisc.
Page 587, Getty Editorial.

Chapter 11
Page 588, Image Source/Getty RF.
Page 593, Shutterstock.
Page 594, Shutterstock.
Page 596, Digital Vision.
Page 602, PhotoDisc.
Page 605, Shutterstock.
Page 608, Shutterstock.
Page 617, Shutterstock.
Page 629, Beth Anderson.

Chapter 12
Page 630, Shutterstock.
Page 635, Shutterstock.
Page 647, Digital Vision.
Page 663, Shutterstock.

Chapter 13
Page 664, MIXA/Getty Royalty Free.
Page 669, Shutterstock.
Page 670, PhotoDisc.
Page 672, Shutterstock.
Page 679, Shutterstock.
Page 682, AP Wideworld Photo.
Page 688, PhotoDisc.
Page 695, PhotoDisc.
Page 696, Shutterstock.
Page 702, Rubberball/Getty RF.
Page 704, Noren Trotman/NBAE/Getty Images.
Page 705, Shutterstock.
Page 715, Shutterstock.

Chapter 14
Page 716, Getty Royalty Free.
Page 719, US Army.
Page 720, NASA.
Page 721, David Lees/Corbis.
Page 723, Owen Franken/Corbis.
Page 724, Getty Editorial.
Page 732, Shutterstock.
Page 734 (margin), Shutterstock.
Page 734, Stockbyte/Getty RF.
Page 745, Shutterstock.

ILLUSTRATIONS
Chapter 2
Page 72, Figure 2.8 "Car Reliability Data": Copyright © 2003 by Consumers Union of U.S., Inc. Yonkers NY 10703-1057, a nonprofit organization. Reprinted with permission from the April 2003 issue of Consumer Reports® for educational purposes only. No commercial use or reproduction permitted. www.ConsumerReports.org

Index

A

Academy Awards problem, 86, 138, 516
Accuracy, 22–23
Achievement tests, 151, 564
Addition rule, 158–162, 173
Adjusted coefficient of determination, 566–567
Adjusted Wald confidence interval, 344–345
Advertising claims, 432
Age discrimination problem, 86, 138, 516
Airbag study, 468–471
Aircraft
    engine reliability, 172–173
    equipment accuracy, 362
    seat design, 329
    seat safety, 596
Analysis of variance (ANOVA).
        See also Variance
    definition, 633
    degrees of freedom, 637
    experimental design, 640
    F distribution, 632, 636–638
    interaction, 647–652
    means, 640–642
    MS(total), 640
    MS(treatment), 639
    one-way, 632–643
    sample size, 636–642
    SS(error), 639
    SS(total), 638–639
    SS(treatment), 639
    TI-83/84 Plus calculator, one-way ANOVA, 643
    TI-83/84 Plus calculator, two-way ANOVA,
        652–653
    treatment (factor), 633
    two-way, 632, 646–653
AOL survey problem, 149
Applet project, 47
Appliance voltage level problem, 380–381
Area, 258, 259, 273
    TI-83/84 Plus calculator, 271, 277
    z score, 263–267
    vs. *z* score, 273
Arithmetic mean. *See* Mean
Aspirin, 399
Assignable variation, 720
Athletic streaks study, 704
"at least one," 177–178
Attributes, control charts, 732–735
Average, 94
Average American, 91

B

b_1 slope, 540–543
Bacterial growth problem, 697–698
Bad graph, 76–78
Bar codes, 190
Bar graph, 68
Baseball problems, 106, 424, 594–595
Basketball study, 424
Bayes' theorem, 199
Bell-shaped distribution, 112–113
    empirical rule, 112–113
    normal distribution, 56, 63, 256
Bias, publication, 20
Bimodal data, 93
Binomial probability
    formula, 227–228, 229–230
    table, 753–755
Binomial probability distribution, 225–231
    binomial probability formula, 227–228,
        229–230
    definition, 225
    exact method, 423–424
    mean, variance, standard deviation, 235–237
    normal distribution approximation, 305–311
    notation, 226
    Poisson distribution, 241, 242–243
    TI-83/84 Plus calculator, 228, 230–231, 311
Birthday problems, 171, 185
Birth weight problem, 272–273, 275
Bivariate data, 522
Bivariate normal distribution, 524
Blinding, 32
Blocks, experimental, 33
Blood pressure problem, 695–696
Boat overloading problem, 433–434, 440–441
Body mass index problem, 370
Body temperature problem, 15, 672–673, 679
Bonferroni multiple comparison test, 641–642
Bootstrap resampling, 390–391
Boundary, class, 53, 54
Boxplot (box-and-whisker diagram),
        127–129
    construction, 127–128
    modified, 130–131, 132
    skeletal, 130
    TI-83/84 Plus calculator, 130, 132
Boystov, Artem, 395
Braking distance problem, 684–685
Bribery, 723
Burger, Joanna, 715

C
Calculator. *See* TI-83/84 Plus calculator
Cancer, 695
Car braking distance problem, 684–685
Car crash problems, 147, 593, 631, 634–635, 642,
    650–652, 689–690
Card shuffles, 191
Car reliability data, 72
Carvalho, Barbara, 331
Case-control study, 30, 31
Causality, vs. correlation, 19–20
CD disc study, 567
Cell phone study, 214
Censored data, 104
Census, 4, 336
Center, 90–100
    definition, 52, 90
    mean, 90–92, 94, 96–97, 98. *See also* Mean
    median, 92–93, 94, 98. *See also* Median
    midrange, 93–94
    mode, 93, 98
Centerline, 721
Central limit theorem, 293–301
    applications, 296–298
    continuity correction, 309–311
    finite population correction, 301
Chang, Andrew, 745
Chebyshev's theorem, 113
Children Out of School in America, 20
Children's Defense Fund, 20
Children's movies, 319–320
Chi-square distribution, 376–379
    critical value, 377–379
    degrees of freedom, 376–377, 448–449
    degrees of freedom (table), 759
    hypothesis testing, 447–452, 455, 591–592
    properties, 377, 449–450
Chronological order problem, 190
Cigarette smoking problems, 58, 695
Citrus sugar study, 368
Class
    boundary, 53, 54
    limit, 53, 54
    midpoint, 53, 62–63
    width, 53
Class attendance, 669
Class size, 93, 483
Clinical trial problems, 193–194, 423, 575
Cluster, scatterplots, 70
Cluster sampling, 28, 29
Coefficient of determination, 557–558
    adjusted, 566–567
Coefficient of variation, 114–115
Cohort study, 30, 31
Coincidence, 178
Coin flip problem, 186, 217

Coin quality problem, 449–451
Coke can filling problem, 298–299
College admission problem, 665, 695
College enrollment, 58
College rankings, 14, 665, 695
College success study, 564
Combinations rule, 193–194
    TI-83/84 Plus calculator, 194–195
Commercials, 432
Compact disc manufacturer, 567
Complement, 177–178
Complementary events, 150–153
    rule of, 161–162
Completely randomized experiment, 32–33, 640
Composite sampling, 194
Compound event, 158–161
Conditional probability, 167–168, 179–180
Condominium prices, 548
Confidence interval (interval estimate)
    adjusted Wald, 344–345
    bootstrap method, 390–391
    data comparison, 337, 370
    definition, 335
    dependent samples, 492, 495–496
    interpretation, 336–337, 353
    margin of error, 343–344, 363–364, 367–369
    mean, 352–355, 363–366, 434, 479, 481–482, 483
    mean (standard deviation not known), 353, 363–364,
        370–371, 441–442
    overlapping, 370, 380
    point estimate, 343–344, 369
    prediction intervals, 558–560
    proportion, 335–337, 339–341, 421, 469–471
    rationale, 354–355, 381–382
    round-off rule, 340, 353
    standard deviation, 379–383, 451
    TI-83/84 Plus calculator, 345, 353, 357, 370–371, 383
    two proportions, 467
    variance, 379–383
    Wilson score, 345
Confidence level (confidence coefficient, degree
        of confidence), 335–337, 352
Confounding, 32
Confusion of the inverse, 180–181
Consumer's Report, 72
Contingency, *P*-values, 607, 608
Contingency tables, 602–610
    definition, 602
    expected frequency, 604–607
    test of homogeneity, 607–609
    test of independence, 602–607
    TI-83/84 Plus calculator, 610
Continuity correction, 309–311
Continuous (numerical) data, 13
Continuous random variables, 212–213, 256
Control charts, 718–729

attributes, 732–735
centerline, 721
interpretation, 724–725
mean, 725–729
p chart, 732–735
process data, 718
R chart, 721–724
run charts, 718–721
TI-83/84 Plus calculator, 727–729, 735
upper/lower control limits, 721
variation, 718–725
$\bar{x}$ chart, 725–729
Convenience sample, 28, 29
Coronary heart disease study, 414, 449
Correlation, 522–534
vs. causality, 19–20, 530
definition, 522
errors, 530
explained variation, 529
formal hypothesis test, 530–534
linear correlation coefficient, 523–529
vs. linearity, 530
rank, 665, 693–698
scatterplot, 522–523
TI-83/84 Plus calculator, 534
Cost of Laughing Index, 121
Count five method, 505
Counting, 189–195
combinations rule, 193–194
factorial rule, 191
fundamental counting rule, 189–190
permutations rule, 191–192
Course grades, 14
Crest toothpaste study, 491
Cricket chirp problem, 70
Crimean War, 68
Critical region (rejection region), 405, 407
Critical thinking, 17–23, 90
age discrimination, 86, 138, 516
aircraft seating design, 329
axial load control, 743
bad graphs, 76–78
beyond formulas, 52
"Do Not Call" registry, 392
draft lottery, 713
Duragesic, 585
frequency distribution, 56–58
histogram, 63
jury selection, 251
Lipitor study, 662
measures of center, 95–96
misuse of statistics, 17–23
poll results, 460
pregnancy test, 205
six degrees of separation, 48
Titanic sinking, 627

Critical value, 266, 337–338, 405
chi-square distribution, 377–379
definition, 337
dependent samples, 492
goodness-of-fit test, 591
hypothesis testing, 405
Kruskal-Wallis test, 689
linear correlation, 530
margin of error, 338–339
McNemar's test, 616
mean, 431, 439
normal distribution, 266
rank correlation test, 694
runs test for randomness, 703
standard deviation/variance, 502
Student *t* distribution, 362–363
test of independence, 603
TI-83/84 Plus calculator, 338
two variances, 502
variance, 377–379
Wilcoxon tests, 676, 683
Cross-sectional study, 30, 31
Croup treatment study, 478
Crowd size, 369
Cumulative frequency distribution, 55–56
Curbstoning, 336
CVDOT, 52, 748

D
Data, 11–15
bad, 76–78
bell-shaped, 112–113
bimodal, 93
bivariate, 522, 524
categorical (qualitative, attribute), 12
censored, 104
center, 52, 90–100. *See also* Center
characteristics, 52
collection, 26–34
continuous, 13
definition, 4
discrete, 13
distribution, 52
interval, 14–15
missing, 22, 71
multimodal, 93
nominal, 13–14, 15, 671–672
ordinal, 14, 15
outlier, 52, 121–122, 129, 130, 546–547
paired, 522, 524
phony, 19
procedures, 748–749
process, 718
qualitative (categorical, attribute), 12
quantitative (numerical), 12–13
range, 106

Data (*continued*)
  ratio, 15
  reliability, 237
  runs test, 701–707
  skewed, 98
  sources, 6
  symmetric, 98
  time, 52
  transformed, 320
  types, 11–15
  validity, 237
  variation, 52
  visualization, 65–73. *See also* Graph(s)
Data sets
  aluminum can axial loads, 793
  bears, 777
  body temperatures, 772–773
  cars, 788
  children's movies, 778
  cigarette components, 775
  coin weights, 792
  Coke/Pepsi, 789
  electricity consumption, 784
  FICO score, 797
  forecast temperatures, 783
  freshman 15, 774
  garbage weights, 794–795
  health exam results, 770–771
  home sales, 796
  M&M candy weights, 790
  movies, 781
  NASA space transport system, 782
  Old Faithful Geyser, 787
  passive and active smoke, 776
  rainfall, 786
  screw lengths, 791
  voltage measurements, 785
  word counts, 779–780
Death penalty, 471
Death postponement study, 546
Decimal, percentage conversion, 21
Decision theory, 218
Degrees of freedom, 479, 637
  chi-square distribution, 376–377, 448
  definition, 362
  Student *t* distribution, 362–363
Density curve (probability density function), 257–258
Dependent events, 168, 170–171
De Toro, Karen, 207
Deviation, 113–114. *See also* Standard deviation; Variation
Dice rolling problems, 218–219, 283, 285, 286
Diet pills, 481
Discordant pairs, 618
Discrete data, 13
Discrete random variables, 212, 218–219
Disjoint events, 160–161

Disobedience study, 13
Dispersion. *See* Variation
Distance, 15
Distortions, 23
Distribution, 52. *See also* Frequency distribution
Distribution-free tests. *See* Nonparametric methods
"Do Not Call" registry, 392
Doorway height problem, 271–272, 274
Dotplot, 67
Double-blind experiment, 32
Dow Jones Industrial Average, 71
Draft lottery problem, 713
Drug approval process, 423
Duragesic, 585

E
Earthquake problem, 241–242
Echinacea experiment problem, 602–603, 605–606
Efficiency rating, 666–667
Elevator failure problem, 148
Empirical rule, 112–113
Envelope stuffing study, 193
Ergonomics, 255
  aircraft seats, 329
  doorway height problem, 271–272, 274
  water taxi, 296–298
Errors, 410–413
  controlling, 411
  nonsampling, 34
  notation, 410
  sampling, 34
  type I, 410–411
  type II, 410–411
ESP study, 177
Estimating, 332–394
  chi-square distribution, 376–379
  confidence intervals, 335–337. *See also* Confidence interval (interval estimate)
  critical value, 337–338. *See also* Critical value
  degrees of freedom, 362–363, 376–377
  margin of error, 338–339. *See also* Margin of error
  mean, 351–357, 361–371
  point estimate, 334–335, 343–344, 351, 369, 379
  population mean, 351–357
  population mean (standard deviation not known), 361–371
  population variance, 376–383
  probability, 146
  proportion, 334–335
  sample mean, 351, 361
  sample proportion, 335, 339–341
  sample size, 342–344, 351–352, 355–357, 382–383
  sample variance, 379–383
  sampling distribution, 286–289
  standard deviation, 110–112, 379–383
  Student *t* distribution, 362–368
  variance, 376–383

Estimator, 109–110, 286–289, 379
Ethics, 6, 682
Event(s)
    certain, 150
    complementary, 150–153, 161–162
    compound, 158–161
    definition, 145
    dependent, 168, 170–171
    disjoint (mutually exclusive), 160–161
    impossible, 150
    independent, 168
    simple, 145
    unusual, 152
Exacta bet problem, 192
Exact method, 423–424
Expected frequency, 591–597, 604–607
Expected value, 218–219
Experiment, 26–27
    blinding, 32
    blocks, 33
    completely randomized, 32–33, 640
    confounding, 32
    definition, 26
    design, 30–34, 496
    double-blind, 32
    ethics, 6
    matched pairs, 33
    power, 414
    randomization, 31
    randomized block design, 33
    replication, 32
    rigorously controlled, 33, 640
    sample size, 32
    sampling errors, 34
    sampling strategies, 26–34
    types, 26–27
    variables, 32–34
Experimental units (subjects), 26
Experimenter effect, 28
Explained/unexplained deviation, 555–558
Explained/unexplained variation, 557–558
Extrasensory perception, 422
Extremely improbable event, 147

F
Factorial rule, 191
    TI-83/84 Plus calculator, 194–195
Factorial symbol (!), 190
False negative, 143
False positive, 143, 602
FBI arrest problem, 162
F distribution, 501–505, 632, 636–638
    degrees of freedom (table), 760–763
Federalist Papers, 54
Feeding tube removal survey, 51
Fertilizer problem, 496

Fewer, vs. *less,* 13
Finite population correction, 301
Fisher exact test, 609
Five-number summary, 127–129
    TI-83/84 Plus calculator, 130
Fluoride, 491
Flynn effect, 719
Football kicker study, 569
Forensics study, 180
Formal addition rule, 160
Formal multiplication rule, 168
Foy, Jeffrey, 663
Fraction, percentage conversion, 21
Frequency
    definition, 53
    expected, 591–597, 604–607
    relative, 55
Frequency distribution, 52–59
    classes, 53–54
    construction, 54–55
    cumulative, 55–56
    definition, 53
    interpretation, 56–58
    mean, 96–97
    relative, 55
    TI-83/84 Plus calculator, 58–59, 100, 115
Frequency histogram, relative, 63
Frequency polygon, 66
    TI-83/84 Plus calculator, 73
Freshmen 15 weight problem, 5–7, 465, 492–496,
        670–671, 677–678, 704
Fundamental counting rule, 189–190
Fuzzy central limit theorem, 295

G
Gallup poll, 3, 19
Gambling problems, 146, 153, 172, 216,
        218–219, 243
Gaps, 57, 70
Garbage problems, 369, 568
Garlic study, 364–365
Gender problems. *See also* Gender selection problems
    "at least one," 177–178
    birth probabilities, 8, 145, 149, 158, 177–178, 283
    discrimination, 82, 138, 516
    income, 68
    medical tests, 670
    survey, 608–609
    words spoken, 89, 91–93, 97, 108–109, 128–129,
        479–482
Gender selection problems
    effectiveness study, 172
    Gender Choice, 399–400
    MicroSort method, 8, 184–185, 192–193, 397
    sampling distribution, 289–290
    sex prediction guesses, 422–423

Gender selection problems (*continued*)
    XSORT method, 397, 400–401, 402, 403–404, 407,
        410, 413–414, 419–420, 670, 672
    YSORT method, 397
Genetics problems, 209, 211, 215, 217–218, 226–227,
    228, 236–237
Genotype problem, 148
Geometric distribution, 235
Geometric mean, 105
Global warming problem, 705–706, 717, 718,
    720, 723–724, 725, 726–727
    poll, 333, 335, 340–341
Goodness-of-fit test, 590–602
    TI-83/84 Plus calculator, 597
Grade point average problem, 97
Graph(s), 65–73
    bad, 76–78
    bar, 68
    boxplot, 127–128
    commercial use, 67
    design, 71
    dotplot, 67
    frequency polygon, 66, 73
    histogram, 62–64, 213–214
    misleading, 18, 76–78
    modified boxplot, 130–131
    nonzero axis, 76–77
    normal quantile plot, 315–317
    ogive, 67
    Pareto chart, 69
    pictograph, 76–78
    pie chart, 69–70
    probability histogram, 213–214
    purposes, 72
    relative frequency histogram, 63
    relative frequency polygon, 66
    residual plots, 548–550
    scatterplot, 70, 73
    stemplot, 67–68
    time-series, 70–71
    tree diagram, 166
    Venn diagram, 161, 162
Growth chart study, 56

H
Handedness, 442
Harmonic mean, 104
Haskell, Mark D., 587
Hawthorne effect, 28
Heart attack study, 449
Heart defibrillator problem, 734–735
Height problems, 115, 565–566, 570
Hip protector study, 617
Histogram, 62–64, 213–214
    definition, 62
    interpretation, 63

relative frequency, 63
    TI-83/84 Plus calculator, 63–64
HIV test study, 602
Holzman, Robert S., 141
Home field advantage, 424
Horizontal scale, histogram construction, 62–63
Horse racing problem, 192
Household size study, 568
H test, 688–691
Hydroxyurea study, 575
Hypergeometric distribution, 235
Hypothesis. *See also* Hypothesis testing
    alternative, 401–403
    definition, 398
    null, 401–403, 408–410
    research, 402
Hypothesis testing, 298–301, 396–463. *See also* Test statistic
    chi-square distribution, 448–450, 591–592
    components, 411–413
    conclusions, 408–410
    confidence interval method, 412–413, 421, 434, 441–442,
        451, 469–470
    critical region, 405, 407
    critical value, 405
    definition, 398
    dependent samples, 491–497
    errors, 410–413
    inferences from two samples, 467–469, 491–497,
        501–506
    left-tailed, 408, 449, 532
    linear correlation, 530–534
    mean, 431–435, 455, 482–485
    mean (standard deviation not known), 438–443, 455
    multiple-negative conclusions, 410
    one-tailed, 408, 532
    power, 413–414
    proportion, 418–426, 455, 467–471
    P-value method, 406–407, 408, 412, 419, 420, 422–424,
        432–433, 440–441, 450–451, 468–469, 479–481,
        492–495, 504–505, 531–533, 606
    right-tailed, 408, 449, 532
    significance level, 405
    simulations, 459–460
    standard deviation/variance, 447–452, 455, 501–505
    summary, 455
    test statistic, 403–404. *See also* Test statistic
    TI-83/84 Plus calculator, mean, 435
    TI-83/84 Plus calculator, mean (standard deviation not
        known), 440–441, 443
    TI-83/84 Plus calculator, proportion, 425–426
    TI-83/84 Plus calculator, two variances/standard
        deviations, 451–452, 506
    traditional method, 412, 421, 433–434, 440,
        449–450, 469, 503–505, 606
    two means, 479–482
    two-tailed, 408, 449

I

Identity theft, 189–190
Improbable event, 147
Income problem
    gender, 68
    per capita, 95–96
    pictograph, 76–78
Independent events, 168
Inferences, 748
Inferences from two samples, 464–519
    confidence interval, 467, 469–471, 479, 483, 492
    count five method, 505
    dependent samples, 491–497
    F distribution, 501–505
    hypothesis testing, 467–471, 479–482, 483, 492–496
    Levene-Brown-Forsythe test, 505–506
    means, 477–485
    proportions, 466–472
    TI-83/84 Plus calculator, 471–472, 484–485, 497
    variance/standard deviation, 501–505
Influential points, 546–547
Insulation thickness study, 732
Interaction, 647–652
Internet penetration survey, 309–311, 342–343
Internet project, 47
    analysis of variance, 661
    central limit theorem, 328
    comparing populations, 515
    computing probabilities, 205
    confidence intervals, 392
    contingency tables, 626
    control charts, 742
    data source, 85
    data summary, 138
    graphs, 85
    hypothesis testing, 460, 515
    linear regression, 584
    nonparametric tests, 712
    probabilities, 205
    probability distribution, 250
    Web site, 47
Internet use problem, 342–343
Interquartile range, 126
Interval data, 14–15
Interval estimate. *See* Confidence interval (interval estimate)
Intuitive addition rule, 160
Intuitive multiplication rule, 168
Inverse, confusion of, 180–181
iPod random shuffle study, 702
IQ score problems, 56, 109, 112–113, 356–357

J

Jet engine problem, 171
Job search problem, 69
Journalism, 18
Jury selection problem, 251

K

Katsingris, Peter, 49
Kruskal-Wallis test, 688–691
    TI-83/84 Plus calculator, 691

L

Laughing index, 121
Law of large numbers, 147
Lead monitoring problem, 442–443
Least-squares property, 547–548
Left-handedness study, 442
Left-tailed test, 408, 449, 532
Less, vs. *fewer,* 13
Levene-Brown-Forsythe test, 505–506
Lie detection, 404, 407. *See also* Polygraph problem
Life data analysis, 213
Limit, class, 53, 54
Linear correlation, formal hypothesis test, 530–531
Linear correlation coefficient, 523–525
    calculation, 525–527
    definition, 523
    interpretation, 524–525, 527–529
    properties, 525
    rounding, 525
Linearity, vs. correlation, 530
Lipitor study, 662
Literary Digest poll, 3, 4, 19, 26
Loaded questions, 21
Logistic regression, 570
Lognormal distribution, 320
Longitudinal study, 30, 31
Lottery problems, 148, 194, 216, 218–219, 243, 272, 307
Lower class limits, 53
Lower control limits, 721–724
Lung cancer, 695
Lurking variable, 530
Lycett, Mark T., 519

M

Macmullan, Jackie, 629
Madison, James, 54
Mail survey problem, 307–309
Mann-Whitney *U* test, 682
Marginal change, 546
Margin of error
    definition, 339
    mean, 352–353, 363–364, 369
    polling, 468
    proportion, 338–339, 343–344
Mars Climate Orbiter, 720
Matched pairs, 477
    confidence interval, 492
    experimental design, 510
    hypothesis testing, 492–496
    inferences, 491–497
    McNemar's test, 615–619

Matched pairs (*continued*)
 sign test, 670–671
 Wilcoxon signed-ranks test, 676–679
Mathematical expectation, 218
Mathematical modeling, 574–576
 TI-83/84 Plus calculator, 576
McDonald's brand recognition, 228–229
McNemar's test, 615–619
 TI-83/84 Plus calculator, 618–619
Mean, 90–92, 98, 214–216. *See also* Estimating; Standard
  deviation
 analysis of variance, 640–642
 binomial distribution, 235–237
 confidence interval, 352–355, 363–366
 control chart, 725–729
 definition, 90–92
 expected value, 218–219
 frequency distribution, 96–97
 geometric, 105
 harmonic, 104
 hypothesis testing, 431–435, 455, 482–485
 inference about two means, 477–485
 margin of error, 369
 multiple comparison test, 641–642
 point estimate, 369
 probability distribution, 214–215
 quadratic, 105
 randomness above and below, 704–706
 range test, 641
 round-off rule, 94
 sample mean, 282–283, 351
 sample size, 351–352, 355–357, 370
 sampling distribution, 282–283
 standard deviation known, 351–357, 455
 standard deviation not known, 361–371, 438–443, 455,
  479–482, 484–485
 standard error, 295
 testing a claim, 431–435, 438–443, 455
 test statistic, 403–404, 455
 TI-83/84 Plus calculator, 99, 215, 435, 443, 484–485
 trimmed, 104
 usage for, 93
 weighted, 97
Mean absolute deviation, 113–114
Mean square (MS), 639
Measurement
 interval, 14–15
 nominal, 13–14, 15
 ordinal, 14, 15
 ratio, 15
Measures of center, 90–100. *See also* Center
Median, 92–93, 98
 median class, 105
 randomness above and below, 704–706
 round-off rule, 94
 sign test, 672–673

TI-83/84 Plus calculator, 99
 usage for, 93
 Wilcoxon signed-ranks test, 679–680
Mendel, Gregor, 209, 594
Mesnick, Sarah, 253
Meta-analysis, 440
MicroSort problem, 8, 184–185, 192–193, 397
Midpoint, class, 53
Midrange, 93–94
Migraine headache study, 440
Milgram, Stanley, 13, 48
Minitab, Pareto chart, 74
Mode, 93, 98
 TI-83/84 Plus calculator, 99
Modeling, 574–576
 TI-83/84 Plus calculator, 576
Morrison, Kathleen, 519
Motor vehicle deaths problem, 705
Movie budget problem, 122–123, 126
Movie length problem, 318–319
MS(total), 640
MS(treatment), 639
Multimodal data, 93
Multinomial distribution, 235
Multiple choice test problem, 151
Multiple coefficient of determination, 566–567
Multiple comparison test, 641–642
Multiple negatives, 410
Multiple regression, 564–571
 TI-83/84 Plus calculator, 570–571
Multiplication rule, 165–173, 177–181
Multistage sampling design, 28, 30

N
Napster web site survey problem, 18–19
National Children's Study, 30
National park visiting problem, 191
NBA salaries, 568
NFL kicker, 569
Nicotine patch therapy, 344
Nielsen ratings, 49, 338
Nightingale, Florence, 68
Nominal data, 13–14, 15, 671–672
 sign test, 671–672
No mode, 93
Nonparametric methods, 664–715
 advantages, 666
 definition, 666
 disadvantages, 666
 efficiency rating, 666–667
 Kruskal-Wallis test, 688–691
 matched pairs, 670–671, 676–679
 median of single population, 672–673, 679–680
 nominal data, 671–672
 rank correlation, 666, 693–698
 ranks, 667

runs test for randomness, 701–707
sign test, 667–673
summary, 709
ties in ranks, 667
Wilcoxon rank-sum test, 682–685
Wilcoxon signed-ranks test, 676–679
Nonresponse, 22
Nonsampling errors, 34
Normal distribution, 56, 63, 254–331. *See also* Standard
    normal distribution
    applications, 270–277
    assessment, 315–320
    bell-shaped, 56, 63, 256
    binomial distribution approximation, 305–311
    bivariate, 524
    central limit theorem, 293–301
    data transformation, 320
    definition, 256
    determination, 315–317
    finding values, 273–276
    histogram, 63
    nonstandard, 270
    normal quantile plot, 315–318
    sampling distribution, 282–283
    TI-83/84 Plus calculator, 260
    z score, 270–271, 276
Normality, 351, 439
Normal quantile plot, 315–317
    manual construction, 317–318
    TI-83/84 Plus calculator, 320
Null hypothesis, 401–403, 406–407
    accept/fail to reject, 409
Numbers, precise, 22–23
Numerical data, 12–13

O
Observational study, 26–27, 30, 31
Odds, 152–153
Odds ratio, 158
Ogive, 67
One-tailed test, 532
Online classes, 746–747
Oral contraceptive survey, 635
Oral report, 746
Oranges, 368
Ordinal data, 14, 15
Outcomes, possible, 148–149
Outdoor darkness problem, 181
Outliers, 52, 129, 130, 546–547

P
Pacemaker quality control problem, 169–171
Pain medicine study, 585
Paired data, 522, 524
Palmistry, 529
Parachuting safety, 319

Parameter, 12
Pareto chart, 69
Passwords, 189–190
Payoff odds, 153
p chart, 732–735
Pearson product moment correlation coefficient, 523–525
    critical values (table), 764
Penny weight problem, 57, 70
Per capita income problem, 95–96
Percentages
    decimal conversion, 21
    misleading, 20–21
    principles, 21
    proportion conversion, 334
Percentiles, 122–125
    TI-83/84 Plus calculator, 127
Permutations rule, 191–192
    TI-83/84 Plus calculator, 194–195
Personal security codes, 189–190
Perspectives, 749–750
Pictograph, 76–78
Pie chart, 69–70
Pizza/subway fare price problem, 521, 525–527, 531,
    532–533, 542–543, 544–547, 549–550,
    557–558, 559, 560
Placebo effect, 28, 32, 257
Plant hybridization experiment, 209, 211, 215, 217–218,
    226–227, 228, 236–237
Point estimate, 334–335, 343–344, 351, 369
Poisson distribution, 240–243
    TI-83/84 Plus calculator, 243
Polio study, 470
Polls. *See* Surveys
Polygraph problem, 143, 147–148, 159–161, 167, 179–180
Pooled estimate, 484–485
Pooled sample proportion, 467
Population
    definition, 4
    finite, 301
    size, 342
    standard deviation, 109
    variance, 109–110
Population mean. *See* Mean
Population model, 575–576
Population proportion. *See* Proportion
Population size problem, 575–576
Population variance. *See* Variance
Power, 413–414
Power supply problem, 257–258
Practical significance, 7–8
Precise numbers, 22–23
Prediction, regression equation, 543–546
Prediction intervals, 558–560
    TI-83/84 Plus calculator, 561
Predictor variable, 540
Pregnancy test problem, 205

Presidential election problem, 148
Prices, 15
Primary sampling units, 28
Prisoner experimentation study, 451
Probability, 142–205
    addition rule, 158–162
    "at least one," 177–178
    basic concepts, 144–153
    classical approach, 145–146, 149
    combinations rule, 193–194
    complementary events, 150–153, 161–162
    complements, 177–178
    conditional, 167–168, 179–180
    counting, 189–195
    FAA interpretation, 147
    factorial rule, 191
    intuitive approach, 179–180
    law of large numbers, 147
    multiplication rule, 165–173, 177–181
    notation, 145
    odds, 152–153
    permutations rule, 191–192
    rare event rule, 144
    relative frequency approach, 145, 146, 149
    rounding off, 151–152
    simulations, 150, 184–186, 204
    subjective, 146, 148
    z score, 257–264, 267
Probability distribution, 208–253. *See also* Normal
        distribution; Standard normal distribution
    binomial, 225–231
    definition, 211
    expected value, 218–219
    mean, 214–216, 220
    multinomial, 235
    Poisson distribution, 240–243
    probability histogram, 213–214, 220
    random variables, 211–220
    requirements, 213–214
    simulations, 250
    standard deviation, 214–216, 220
    TI-83/84 Plus calculator, 215, 220
    unusual results, 216–218
    variance, 214–216
Probability histogram, 213–214
Probability sample, 27
Probability value. *See* *P*-value
Probable event, 147
Procedures, 748–749
Process data, 718
Projects, 746–747
Proportion, 334–335
    confidence interval, 335–337, 339–341, 467, 469–471
    critical values, 337–338
    definition, 335
    hypothesis testing, 418–426, 467–471

    inferences about two proportions, 466–472
    margin of error, 338–339, 343–344
    notation, 466
    point estimate, 334–335, 343–344
    sample size, 342–344
    sampling distribution, 285–291
    testing a claim, 418–426, 455
    test statistic, 419, 467
    TI-83/84 Plus calculator, 425–426
Prosecutor's fallacy, 180
Prospective (longitudinal, cohort) study, 30, 31
Publication bias, 20
Pulse rate problem, 53, 54–55, 56–57, 66, 67–68, 129, 132
Push polling, 377
P-value method
    chi-square distribution, 591–592, 597
    goodness-of-fit test, 591–592, 597
    hypothesis testing, 406–407, 408, 419, 420, 422–424,
        432–433, 440–441, 450–451, 468–469, 479–481,
        492–495, 504–505, 531–533, 606
    multiple regression, 567
    Student *t* distribution, 440–441

Q
Quadratic mean, 105
Qualitative (categorical, attribute) data, 12
Quality control. *See* Control charts
Quality control problem, 169–171, 449–451
Quantitative data, 12–13
Quarter weight problem, 503–505
Quartiles, 125–127
    TI-83/-84 Plus calculator, 127
Questions
    loaded, 21
    nonresponse, 22
    order of, 21–22
    wording of, 21
Queuing theory, 241

R
Randomization, 31
Randomized block design, 33
Randomness, runs test, 701–707
Random numbers, 185–186
Random sample, 28, 29, 333
Random selection, 4
Random variables, 211–220
    binomial probability distribution, 225–231
    continuous, 212–213, 256
    definition, 211
    discrete, 212, 218–219
    expected value, 218–219
    probability histogram, 213–214
    uniform distribution, 257–258
Range, 106
Range chart, 721–724

Range rule of thumb, 110–112, 216
Range tests, 641
Rank, 14, 667
Rank correlation, TI-83/84 Plus calculator, 698
Rank correlation test, 666, 693–698
Rare event rule, 144, 152, 216–217, 298, 399
Ratio data, 15
R chart, 721–724
   TI-83/84 Plus calculator, 728–729
Real estate sales study, 480
Redundancy problem, 166, 172–173
Refusals, 647
Regression, 540–551
   coefficient of determination, 557–558
   guidelines, 550, 567–568
   influential points, 546–547
   interpretation, 546
   least-squares property, 547–548
   logistic, 570
   marginal change, 546
   multiple, 564–571
   outliers, 546–547
   predictions, 543–546
   regression equation/line, 540–546
   residuals, 547–548
   standard error of estimate, 558–559
   TI-83/84 Plus calculator, 550–551
Regression equation/line, 540–543
   interpretation, 546
   TI-83/84 Plus calculator, 550–551
   predictions, 543–546
Relative frequency distribution, 55
Relative frequency histogram, 63
Relative frequency polygon, 66
Relative risk, 158
Relative standing, 120–132
   percentiles, 122–125, 127
   quartiles, 125–127
   z scores, 120–122
Reliability, 237
Replication, 32
Residual, 547–548
Residual plots, 548–550
Retrospective (case-control) study, 30, 31
Right-tailed test, 408, 449, 532
Rigorously controlled experiment, 33, 640
Roosevelt, Franklin D., 3
Root mean square, 105
Rosenthall effect, 28
Roulette problem, 153, 218–219
Rounding, error, 96
Round-off rule
   confidence interval, 340, 353
   linear correlation coefficient, 525
   mean, 94, 216
   measures of variation, 106
   median, 94
   midrange, 94
   probabilities, 151–152
   proportion, 340
   regression, 541
   sample size, 342, 356
   slope, 541
   standard deviation, 216
   variance, 216
   y-intercept, 541
   z score, 120
Rule of complementary events, 162
Run charts, 718–721
   TI-83/84 Plus calculator, 727–728
Runs test for randomness, 701–707
   STATDISK, 706–707
Ryan-Joiner test, 319–320

S
Saccucci, Michael, 463
Salk vaccine study, 26, 31, 152
Sample(s)
   bad, 18–19
   biased, 18, 410
   definition, 4
   dependent, 477, 491–497
   independent, 467, 477
   mean, 282–283, 292–295, 351
   multistage design, 28, 30
   probability, 27
   random, 28, 29, 333
   self-selected (voluntary response), 6, 18–19, 149
   simple random, 27, 28, 29
   small, 20, 317
   space, 145–146
   standard deviation, 379. *See also* Standard deviation
   variance, 109–110, 379
   variance between, 636–640
   variance within, 636–640
   voluntary (self-selected), 6, 18–19, 149
Sample proportion, 335, 343–344
   pooled, 467
Sample size *(n)*, 91
   equal, 636–638
   mean, 351–352, 355–356, 370
   proportion, 342–344
   round-off rule, 342, 356
   standard deviation, 356–357
   unequal, 638–640
   variance, 382–383
Sampling
   bootstrap, 390–391
   cluster, 28, 29
   composite, 194
   convenience, 28, 29
   distribution, 282–291

Sampling (*continued*)
  errors, 34
  multistage, 28
  random, 27, 28, 29, 333
  with replacement, 290–291
  without replacement, 301
  simple random, 27
  stratified, 28, 29
  systematic, 28, 29
Sampling distribution, 282–291
  central limit theorem, 293–301
  definition, 282
  estimators, 286–289
  mean, 282–283
  proportion, 285–291
  range, 288–289
  of statistic, 282
  variance, 283–285
  $\bar{x}$, 282–283, 294–295
SAT test, 151, 564
Scatterplot, 70, 522–523
  TI-83/84 Plus calculator, 73
Schiavo, Terri, 51
Screw length problem, 299–300
Seat belt study, 92
Second (time), definition, 105
Sehlinger, Bob, 87
Self-interest study, 22
Self-selected (voluntary) sample, 6
Serial killer problem, 589, 606–607
Shakespeare, William, 160
Significance, 7–8
Significance level, 405
Sign test, 667–673
  TI-83/84 Plus calculator, 668–669, 673
Simple event, 145
Simple random sample, 27, 28, 29
Simulations, 150, 184–186, 204
  TI-83/84 Plus calculator, 185–186, 204, 328, 459–460, 514–515
Six degrees of separation, 48
Six Sigma programs, 734
Skewness, 98
Slope, regression equation, 541–543
Soda can filling problem, 298–299
Spearman's rank correlation coefficient, 694
  critical values (table), 766
Spearman's rank correlation test, 666, 693–698
Speeding ticket study, 523
Speed limit problem, 125
Spread. *See* Variation
SS(error), 639
SS(total), 639
SS(treatment), 639
Standard deviation, 214–216
  binomial distribution, 235–237

  calculation, 107–108
  Chebyshev's theorem, 113
  comparing two samples, 501–505
  confidence interval, 451
  definition, 107
  empirical rule, 112–113
  estimation, 110–112, 379–383
  finding, 107–108
  interpretation, 110–112
  mean absolute deviation, 113–114
  pooled estimate, 484–485
  population, 109
  probability distribution, 214–216
  properties, 110–112
  range rule of thumb, 110–112
  rationale, 113–114
  sample size, 356–357
  testing a claim, 447–452, 455
  TI-83/84 Plus calculator, 215, 451–452
Standard error of estimate, 558–559
Standard error of the mean, 295
Standard normal distribution, 257–267
  critical value, 266, 267
  definition, 259
  density curve, 259–264
  finding probabilities, 259–264
  finding z scores, 264–266
Standard score. *See* z score
STATDISK
  boxplots, 128–129
  exact probabilities, 424
  runs test for randomness, 706–707
  sample size, 357, 383
Statistic
  definition, 12
  sampling distribution of, 282
Statistically stable process, 720–721
Statistical significance, 7–8
Statistical study, evaluation of, 5
Statistical thinking, 4–9
Statistics, 4
  descriptive, 90
  in education, 750
  ethics, 6
  inferential, 90
  misleading, 17–23
  origin of, 12
Stemplot, 67–68
Stock investment study, 108
Stratified sampling, 28, 29
Student evaluation study, 527
Student t distribution, 361–368, 439
  critical t values (table), 758
  properties, 365–366
  P-value, 440–441
  vs. z score, 366–368

Subjective probability, 146, 148
Successes *(x)*, 422–423
Sugging, 22
Super Bowl game, 482, 556
Surveys
   AOL, 460
   callbacks, 226
   distorted, 23
   ethics, 682
   feeding tube removal, 51
   gender effects, 608–609
   global warming, 333, 335, 340–341
   Internet penetration, 309–311, 342–343
   mail, 307–309
   margin of error, 468
   nonresponse, 22
   personal vs. mail/phone, 635
   project, 747
   question order, 21–22
   question wording, 21, 688
   refusal rates, 647
   self-interest, 22
   self-selected, 18–19
   voluntary response, 18–19, 149
Symmetric data, 98
Syphilis testing study, 194
Systematic sampling, 28, 29

T
Tank production study, 354
t distribution. *See* Student *t* distribution
Telekinesis study, 177
Television viewing study, 338
Temperature
   body, 15, 672–673, 679–680
   environmental, 719, 723–724, 725, 726–727
   global, 705–706, 717, 718
Test of homogeneity, 607–609
Test of independence, 602–607
Test of significance. *See* Hypothesis testing
Test statistic, 403–404
   ANOVA, 640
   assessment, 404–407
   critical region, 405, 407
   critical value, 405
   dependent sample, 492
   goodness-of-fit test, 591, 597
   Kruskal-Wallis test, 689
   linear correlation, 530
   McNemar's test, 616
   mean, 403–404, 439, 478–479, 482, 483, 484
   one-way ANOVA, 636–638, 640
   proportion, 419, 420, 421
   P-value, 406–407
   rank correlation, 694
   runs test for randomness, 703

   significance level, 405
   sign test, 668
   standard deviation/variance, 448
   test of homogeneity, 607–609
   test of independence, 603
   two means, 478–479, 483
   two proportions, 467, 471
   two variances, 502
   Wilcoxon rank-sum test, 683–684
   Wilcoxon signed-ranks test, 676
Thanksgiving Day, 150
Theft identification study, 477
Thermometer testing problem, 261–266
Time, 52
   data change, 52
Time-series graph, 70–71
TI-83/84 Plus calculator, 37–41. *See also* STATDISK
   airline overbooking, 250
   APPS, 41
   areas, 271, 277
   baseball statistics, 582–583
   batteries, 38
   binomial distribution, 311
   binomial probabilities, 228, 230–231
   bootstrap resampling, 390–391
   boxplot, 130
   calculator mode, 38
   calculator project, 741–742
   clearing a list, 39–40
   combinations rule, 194–195
   confidence interval, 357
   confidence interval (mean), 353, 370–371
   confidence interval (standard deviation), 383
   confidence interval estimate of *p,* 345
   contingency tables, 610
   control charts, 727–729, 735
   correlation, 534
   critical values, 338
   data entering, 40
   data generation, 514–515, 626, 712
   data sets, 138
   descriptive statistics, 99–100
   exact results, 311
   factorial rule, 194–195
   five-number summary, 130
   frequency distribution, 58–59, 100, 115
   frequency polygon, 73
   goodness-of-fit, 597
   graphs, 84
   histogram, 63–64
   hypothesis testing, 425
   hypothesis testing, with two variances/
      standard deviations, 506
   inferences about mean of two samples, 484–485
   inferences about two proportions, 471–472
   inferences from dependent samples, 497

TI-83/84 Plus calculator (*continued*)
   IQ scores, 514–515
   keys, 38–39
   Kruskal-Wallis test, 691
   list naming, 40–41
   McNemar's test, 618–619
   mean, 215
   measures of variation, 115–116
   median, 99
   mode, 99
   modeling, 576
   modified boxplot, 132
   multiple regression, 570–571
   normal distribution area, 260
   normal quantile plot, 320
   one-way ANOVA, 643
   p charts, 735
   percentiles, 127
   permutations rule, 194–195
   Poisson distribution, 243
   prediction interval, 561
   probabilities, 204
   probability distributions, 220
   pulse rates, 84
   quartiles, 127
   random digit generation, 626
   random numbers, 185–186
   rank correlation, 698
   R chart, 728–729
   regression, 550–551
   run chart, 727–728
   sample size, 345, 357
   scatterplot, 73
   scientific notation, 39
   screen darkness, 38
   sign test, 668–669, 673
   simulations, 185–186, 204, 328, 459–460, 514–515
   special keys, 38
   standard deviation, 215
   Student t distribution, 362
   testing a claim about a mean, 435, 443
   testing a claim about a proportion, 425–426
   testing a claim about a standard deviation, 451–452
   TRIOLA11, 41
   two-way ANOVA, 652–653
   values, 277
   variation, measures of, 115–116
   voltage variation, 138
   Wilcoxon rank-sum test, 685
   Wilcoxon signed-ranks test, 680
   words spoken problem, 47, 661
   $\bar{x}$ chart, 729
   z score, 266–267
Tire defect problem, 178
Tobacco studies, children's movies, 319–320
Toothpaste study, 491

Topics, project, 747
Total deviation, 556–557
Transformed data, 320
Transmission gearbox quality, 503
Traveling salesman problem, 191
Treatment (factor), 633
Tree diagram, 166
Trimmed mean, 104
Twin research, 491, 494
Two-tailed test, 408, 449
Type I errors, 410–411
Type II errors, 410–411

U
Unbiased estimator, 109–110, 286–288, 379
Unemployment statistics, 28, 30
Unexplained deviation, 555–558
Uniform distribution, 257–258
Unusual results/values, 121–122, 216–218, 246, 311
Upper class limits, 53
Upper control limits, 721–724

V
Validity, 237
Variable(s)
   controlling, 32–34
   dichotomous, 568–570
   dummy, 568–570
   explanatory (predictor, independent), 540
   lurking, 530
   predictor, 540
   random. *See* Random variables
   response (dependent), 540
Variance, 214–216. *See also* Analysis of variance (ANOVA)
   between/within samples, 636–640
   binomial distribution, 235–237
   chi-square distribution, 376–379
   comparing two samples, 501–505
   confidence interval, 379–383
   count five method, 505
   definition, 109
   estimating, 379–383
   F test, 501–505
   Levene-Brown-Forsythe test, 505–506
   mean absolute deviation, 113–114
   population variance, 109–110
   probability distribution, 214–216
   sample size, 382–383
   sample variance, 109–110, 379
   sampling distribution, 283–285
   testing a claim, 447–452, 455
   TI-83/84 Plus calculator, 383
Variation, 52, 105–116, 555–561
   assignable, 721
   coefficient of determination, 557–558
   coefficient of variation, 114–115

control chart, 718–725
   definition, 52
   explained/unexplained deviation, 529, 555–558
   measures, 105–116
   prediction intervals, 558–560
   random, 721
   range, 106
   standard deviation, 107–109, 110–112
   standard error of estimate, 558–559
   TI-83/84 Plus calculator, 115–116
Venn diagram, 161, 162
Vertical scale, histogram construction, 62–63
Video game study, 368
Vocabulary size study, 160
Voltage level problem, 257, 258, 380–381
Voltaire, 307
Voluntary response (self-selected) sample, 6, 18–19, 149
Voting, 20, 185

W
Water taxi safety problem, 296–298
Weather forecasting, 540
Weighted mean, 97
Weight problems
   freshman fifteen, 5–7, 465, 492–496, 670–671,
      677–678, 704
   height-to-weight, 115
   last digits, 592–593
   mean weight, 354–355
Width, class, 53
Wilcoxon rank-sum test, 682–685
   TI-83/84 Plus calculator, 685

Wilcoxon signed-ranks test, 676–679
   critical values of T (table), 765
   TI-83/84 Plus calculator, 680
Wildlife population size, 353
Wilson score confidence interval, 345
Words spoken problem, 89, 91–93, 97, 108–109,
      128–129, 479–482
World Series problem, 594–595
Written report, 746

X
$\bar{x}$, sampling distribution, 282–283, 294–295
$\bar{x}$ chart, 725–729
   TI-83/84 Plus calculator, 729

Y
y-intercept, 540–543

Z
z score, 120–122, 266
   vs. area, 273
   area between, 263–264
   critical value, 266, 337–338
   definition, 120
   finding, 264–266, 273–275
   negative (table), 756
   notation, 337
   positive (table), 757
   standard normal distribution, 259–264
   vs. Student t distribution, 366–368
   TI-83/84 Plus calculator, 264, 266–267
   unusual values, 121–122

TABLE A-3	_t_ Distribution: Critical _t_ Values				

Degrees of Freedom	Area in One Tail				
	0.005	0.01	0.025	0.05	0.10
	Area in Two Tails				
	0.01	0.02	0.05	0.10	0.20
1	63.657	31.821	12.706	6.314	3.078
2	9.925	6.965	4.303	2.920	1.886
3	5.841	4.541	3.182	2.353	1.638
4	4.604	3.747	2.776	2.132	1.533
5	4.032	3.365	2.571	2.015	1.476
6	3.707	3.143	2.447	1.943	1.440
7	3.499	2.998	2.365	1.895	1.415
8	3.355	2.896	2.306	1.860	1.397
9	3.250	2.821	2.262	1.833	1.383
10	3.169	2.764	2.228	1.812	1.372
11	3.106	2.718	2.201	1.796	1.363
12	3.055	2.681	2.179	1.782	1.356
13	3.012	2.650	2.160	1.771	1.350
14	2.977	2.624	2.145	1.761	1.345
15	2.947	2.602	2.131	1.753	1.341
16	2.921	2.583	2.120	1.746	1.337
17	2.898	2.567	2.110	1.740	1.333
18	2.878	2.552	2.101	1.734	1.330
19	2.861	2.539	2.093	1.729	1.328
20	2.845	2.528	2.086	1.725	1.325
21	2.831	2.518	2.080	1.721	1.323
22	2.819	2.508	2.074	1.717	1.321
23	2.807	2.500	2.069	1.714	1.319
24	2.797	2.492	2.064	1.711	1.318
25	2.787	2.485	2.060	1.708	1.316
26	2.779	2.479	2.056	1.706	1.315
27	2.771	2.473	2.052	1.703	1.314
28	2.763	2.467	2.048	1.701	1.313
29	2.756	2.462	2.045	1.699	1.311
30	2.750	2.457	2.042	1.697	1.310
31	2.744	2.453	2.040	1.696	1.309
32	2.738	2.449	2.037	1.694	1.309
33	2.733	2.445	2.035	1.692	1.308
34	2.728	2.441	2.032	1.691	1.307
35	2.724	2.438	2.030	1.690	1.306
36	2.719	2.434	2.028	1.688	1.306
37	2.715	2.431	2.026	1.687	1.305
38	2.712	2.429	2.024	1.686	1.304
39	2.708	2.426	2.023	1.685	1.304
40	2.704	2.423	2.021	1.684	1.303
45	2.690	2.412	2.014	1.679	1.301
50	2.678	2.403	2.009	1.676	1.299
60	2.660	2.390	2.000	1.671	1.296
70	2.648	2.381	1.994	1.667	1.294
80	2.639	2.374	1.990	1.664	1.292
90	2.632	2.368	1.987	1.662	1.291
100	2.626	2.364	1.984	1.660	1.290
200	2.601	2.345	1.972	1.653	1.286
300	2.592	2.339	1.968	1.650	1.284
400	2.588	2.336	1.966	1.649	1.284
500	2.586	2.334	1.965	1.648	1.283
1000	2.581	2.330	1.962	1.646	1.282
2000	2.578	2.328	1.961	1.646	1.282
Large	2.576	2.326	1.960	1.645	1.282

NEGATIVE z Scores

TABLE A-2	Standard Normal (z) Distribution: Cumulative Area from the LEFT									
z	.00	.01	.02	.03	.04	.05	.06	.07	.08	.09
−3.50 and lower	.0001									
−3.4	.0003	.0003	.0003	.0003	.0003	.0003	.0003	.0003	.0003	.0002
−3.3	.0005	.0005	.0005	.0004	.0004	.0004	.0004	.0004	.0004	.0003
−3.2	.0007	.0007	.0006	.0006	.0006	.0006	.0006	.0005	.0005	.0005
−3.1	.0010	.0009	.0009	.0009	.0008	.0008	.0008	.0008	.0007	.0007
−3.0	.0013	.0013	.0013	.0012	.0012	.0011	.0011	.0011	.0010	.0010
−2.9	.0019	.0018	.0018	.0017	.0016	.0016	.0015	.0015	.0014	.0014
−2.8	.0026	.0025	.0024	.0023	.0023	.0022	.0021	.0021	.0020	.0019
−2.7	.0035	.0034	.0033	.0032	.0031	.0030	.0029	.0028	.0027	.0026
−2.6	.0047	.0045	.0044	.0043	.0041	.0040	.0039	.0038	.0037	.0036
−2.5	.0062	.0060	.0059	.0057	.0055	.0054	.0052	.0051 *	.0049	.0048
−2.4	.0082	.0080	.0078	.0075	.0073	.0071	.0069	.0068	.0066	.0064
−2.3	.0107	.0104	.0102	.0099	.0096	.0094	.0091	.0089	.0087	.0084
−2.2	.0139	.0136	.0132	.0129	.0125	.0122	.0119	.0116	.0113	.0110
−2.1	.0179	.0174	.0170	.0166	.0162	.0158	.0154	.0150	.0146	.0143
−2.0	.0228	.0222	.0217	.0212	.0207	.0202	.0197	.0192	.0188	.0183
−1.9	.0287	.0281	.0274	.0268	.0262	.0256	.0250	.0244	.0239	.0233
−1.8	.0359	.0351	.0344	.0336	.0329	.0322	.0314	.0307	.0301	.0294
−1.7	.0446	.0436	.0427	.0418	.0409	.0401	.0392	.0384	.0375	.0367
−1.6	.0548	.0537	.0526	.0516	.0505 *	.0495	.0485	.0475	.0465	.0455
−1.5	.0668	.0655	.0643	.0630	.0618	.0606	.0594	.0582	.0571	.0559
−1.4	.0808	.0793	.0778	.0764	.0749	.0735	.0721	.0708	.0694	.0681
−1.3	.0968	.0951	.0934	.0918	.0901	.0885	.0869	.0853	.0838	.0823
−1.2	.1151	.1131	.1112	.1093	.1075	.1056	.1038	.1020	.1003	.0985
−1.1	.1357	.1335	.1314	.1292	.1271	.1251	.1230	.1210	.1190	.1170
−1.0	.1587	.1562	.1539	.1515	.1492	.1469	.1446	.1423	.1401	.1379
−0.9	.1841	.1814	.1788	.1762	.1736	.1711	.1685	.1660	.1635	.1611
−0.8	.2119	.2090	.2061	.2033	.2005	.1977	.1949	.1922	.1894	.1867
−0.7	.2420	.2389	.2358	.2327	.2296	.2266	.2236	.2206	.2177	.2148
−0.6	.2743	.2709	.2676	.2643	.2611	.2578	.2546	.2514	.2483	.2451
−0.5	.3085	.3050	.3015	.2981	.2946	.2912	.2877	.2843	.2810	.2776
−0.4	.3446	.3409	.3372	.3336	.3300	.3264	.3228	.3192	.3156	.3121
−0.3	.3821	.3783	.3745	.3707	.3669	.3632	.3594	.3557	.3520	.3483
−0.2	.4207	.4168	.4129	.4090	.4052	.4013	.3974	.3936	.3897	.3859
−0.1	.4602	.4562	.4522	.4483	.4443	.4404	.4364	.4325	.4286	.4247
−0.0	.5000	.4960	.4920	.4880	.4840	.4801	.4761	.4721	.4681	.4641

NOTE: For values of z below −3.49, use 0.0001 for the area.
*Use these common values that result from interpolation:

z score	Area
−1.645	0.0500
−2.575	0.0050